Methods in Enzymology

Volume 119

INTERFERONS

Part C

METHODS IN ENZYMOLOGY

EDITORS-IN-CHIEF

Sidney P. Colowick   Nathan O. Kaplan

Methods in Enzymology

Volume 119

Interferons

Part C

EDITED BY

Sidney Pestka

DEPARTMENT OF MOLECULAR GENETICS AND MICROBIOLOGY
UNIVERSITY OF MEDICINE AND DENTISTRY OF NEW JERSEY
ROBERT WOOD JOHNSON MEDICAL SCHOOL
PISCATAWAY, NEW JERSEY

1986

ACADEMIC PRESS, INC.

Harcourt Brace Jovanovich, Publishers

Orlando San Diego New York Austin Boston
London Montreal Sydney Tokyo Toronto

ACADEMIC PRESS, INC.
Orlando, Florida 32887

United Kingdom Edition published by
ACADEMIC PRESS INC. (LONDON) LTD.
24–28 Oval Road, London NW1 7DX

LIBRARY OF CONGRESS CATALOG CARD NUMBER: 54-9110

ISBN 0–12–182019–X

PRINTED IN THE UNITED STATES OF AMERICA

86 87 88 89 9 8 7 6 5 4 3 2 1

To Harry Pestka and Bernice Pestka,
who have provided a wealth of aspirations from birth's light

To John D. Sparacin and Lillian Sparacin,
who have brought imagination and perfection to great height

Table of Contents

Contributors to Volume 119 xvii

Preface xxv

Volumes in Series xxvii

Section I. Introduction

1. Interferon from 1981 to 1986	Sidney Pestka	3
2. Interferon Standards and General Abbreviations	Sidney Pestka	14

Section II. Induction of Interferons

A. Human Interferons

3. Effect of Purification Procedures on the Composition of Human Leukocyte Interferon Preparations	Hanna-Leena Kauppinen, Sinikka Hirvonen, and Kari Cantell	27
4. Large-Scale Production of Human Interferon from Lymphoblastoid Cells	A. W. Phillips, N. B. Finter, C. J. Burman, and G. D. Ball	35
5. Large-Scale Production and Recovery of Human Leukocyte Interferon from Peripheral Blood Leukocytes	Bernard Horowitz	39
6. Induction of Human Interferon Gamma with Phorbol Esters and Phytohemagglutinin	Jan Vilček, Junming Le, and Y. K. Yip	48
7. Production and Partial Purification of Human Immune Interferon	Kari Cantell, Sinikka Hirvonen, and Hanna-Leena Kauppinen	54
8. Production of Human Immune Interferon from Leukocytes Cocultured with Exogenous Cells	Philip C. Familletti and Donna Stremlo	63

9. Induction of Human Immune Interferon with Ionophores — FERDINANDO DIANZANI, GUIDO ANTONELLI, AND MARIA R. CAPOBIANCHI — 69

10. Production of Human Immune Interferon with Normal Human Leukocytes by the Inducers A-23187 and Mezerein — IRWIN A. BRAUDE AND CRISS TARR — 72

11. Large-Scale Production and Recovery of Human Immune Interferon — ROBERT VAN REIS, STUART E. BUILDER, AND ANTHONY S. LUBINIECKI — 77

12. Large-Scale Production, Concentration, and Partial Purification of Human Immune Interferon — JERZY A. GEORGIADES — 83

13. Production and Partial Purification of Human Immune Interferon Induced with Concanavalin A, Staphylococcal Enterotoxin A, and OK-432 — MARC DE LEY, JOZEF VAN DAMME, AND ALFONS BILLIAU — 88

14. Induction of Human Immune Interferon from Peripheral Blood Mononuclear Cells Induced with Combinations of T Cell Mitogens — MASAFUMI TSUJIMOTO AND NAOKI HIGASHI — 93

15. Preparation and Partial Purification of Human Interferon δ — MILES F. WILKINSON AND ALAN G. MORRIS — 96

16. Preparation and Assay of Poly ICL-CM Dextran, An Interferon Inducer of Reduced Toxicity — JAKE BELLO, EDWARD N. GRANADOS, MICHAEL MCGARRY, AND JUDITH A. O'MALLEY — 103

B. Non-Human Interferons

17. Interferon Induction: Dose–Response Curves — PHILIP I. MARCUS — 106

18. Induction of High Titer Chicken Interferon — MARGARET J. SEKELLICK AND PHILIP I. MARCUS — 115

19. *In Vitro* Induction of Interferon from Guinea Pig Fibroblasts — TIMOTHY R. WINSHIP — 125

20. Induction of Equine Interferons — TILAHUN YILMA — 130

21. Induction and Characterization of Bovine Leukocyte Interferon — SARA COHEN, BARUCH VELAN, TAMAR BINO, HAGAI ROSENBERG, AND AVIGDOR SHAFFERMAN — 136

22. Production and Purification of Mouse Immune Interferon — MARC DE LEY, HUBERTINE HEREMANS, AND ALFONS BILLIAU — 145

Section III. Purification of Interferons

23. Large-Scale Purification of Recombinant Human Leukocyte Interferons — S. Joseph Tarnowski, Swapan K. Roy, Raymond A. Liptak, David K. Lee, and Robert Y. Ning 153

24. Purification of Recombinant Human IFN-α2 — David R. Thatcher and Nikos Panayotatos 166

25. Purification of Recombinant Human Fibroblast Interferon Produced in *Escherichia coli* — John A. Moschera, Diana Woehle, Kelly Peng Tsai, Chien-Hwa Chen, and S. Joseph Tarnowski 177

26. Purification of Recombinant Human Interferon β Expressed in *Escherichia coli* — Leo S. Lin, Ralph Yamamoto, and Robert J. Drummond 183

27. Purification of Natural Human Immune Interferon Induced by A-23187 and Mezerein — Irwin A. Braude 193

28. Nickel Chelate Chromatography of Human Immune Interferon — Dorian H. Coppenhaver 199

29. Purification of Recombinant Human Immune Interferon — Hsiang-fu Kung, Yu-ching E. Pan, John Moschera, Kelly Tsai, Eva Bekesi, May Chang, Hiromu Sugino, and Susumu Honda 204

30. Production and Purification of Bovine Interferon β — Reinhard Ahl and M. Gottschalk 211

31. Production, Purification, and Characterization of Rat Interferon from Transformed Rat Cells — Peter H. van der Meide, Jacqueline Wubben, Kitty Vijverberg, and Huub Schellekens 220

Section IV. Modification and Analysis of Interferons

32. Chemical Modifications of Human and Murine Interferons — Karl H. Fantes 233

33. Analysis of Different Forms of Recombinant Human Interferons by High-Performance Liquid Chromatography — Arthur M. Felix, Edgar P. Heimer, and S. Joseph Tarnowski 242

34. Procedure for Reduction and Reoxidation of Human Leukocyte Interferon — JEROME A. LANGER AND SIDNEY PESTKA — 248

35. Radiation Inactivation and Target Size Analysis of Interferons — ELLIS S. KEMPNER AND SIDNEY PESTKA — 255

Section V. Radiolabeling of Interferons

36. Radiolabeling of Human Leukocyte and Immune Interferons with ^{125}I and Lactoperoxidase — FAZLUL H. SARKAR AND SOHAN L. GUPTA — 263

37. Radioiodination of Human Alpha Interferons by the Chloramine T Method — KNUD ERIK MOGENSEN AND GILLES UZÉ — 267

38. Radiolabeling of Human Interferon Alphas with ^{125}I-Labeled Bolton–Hunter Reagent — DOROTHY L. ZUR NEDDEN AND KATHRYN C. ZOON — 276

39. Radiolabeling of Human Immune Interferon with ^{125}I-Labeled Bolton–Hunter Reagent — PAUL ANDERSON — 281

40. *In Vivo* Radiolabeling of Human IFN-β — ERNEST KNIGHT, JR. AND DIANA FAHEY — 284

41. Labeling of Recombinant Interferon with [^{35}S]Methionine *in Vivo* — ABBAS RASHIDBAIGI AND SIDNEY PESTKA — 286

42. Labeling of Interferons with [^{35}S]Methionine in a Cell-Free DNA-Dependent System — HSIANG-FU KUNG — 292

43. Phosphorylation of Human Immune Interferon (IFN-γ) — HSIANG-FU KUNG AND EVA BEKESI — 296

Section VI. Procedures for Studying the Interferon Receptor and Uptake of Interferon

44. Procedures for Studying Binding of Interferon to Human Cells in Suspension Cultures — JEROME A. LANGER AND SIDNEY PESTKA — 305

45. Procedures for Studying the Binding of Interferon to Human and Bovine Cells in Monolayer Culture — KATHRYN C. ZOON, DOROTHY ZUR NEDDEN, AND HEINZ ARNHEITER — 312

46. Binding of ^{32}P-Labeled Human Recombinant Immune Interferon to U937 Cells — ABBAS RASHIDBAIGI, HSIANG-FU KUNG, AND SIDNEY PESTKA — 315

47. Procedures for Studying Binding of Interferon to Mouse Cells — MICHEL AGUET — 321

48. Procedures for Binding an Antibody to Receptor-Bound Interferon — HEINZ ARNHEITER AND KATHRYN C. ZOON — 326

49. Procedures for Measuring Receptor-Mediated Binding and Internalization of Human Interferon — KATHRYN C. ZOON, HEINZ ARNHEITER, AND DAVID FITZGERALD — 332

50. Identification of Interferon Receptors by Chemical Cross-Linking — SOHAN L. GUPTA AND ARATI RAZIUDDIN — 340

51. Extraction of Alpha Interferon–Receptor Complexes with Digitonin — PIERRE EID AND KNUD ERIK MOGENSEN — 347

52. Measurement of a Receptor for (2′-5′)-Oligoadenylate(trimer) on Macrophages — XIN-YUAN LIU, BO-LIANG LI, AND SHI WU LI — 351

Section VII. Procedures for Isolation of Genes and Expression of Interferons in Bacterial and Heterologous Cells

53. A Procedure for Isolation of Alpha Interferon Genes with Short Oligonucleotide Probes — ARTHUR P. BOLLON, MOTOHIRO FUKE, AND RICHARD M. TORCZYNSKI — 359

54. Use of the Phage Lambda P_L Promoter for High-Level Expression of Human Interferons in *Escherichia coli* — ERIK REMAUT, PATRICK STANSSENS, GUUS SIMONS, AND WALTER FIERS — 366

55. Expression of Human Interferon Genes in *E. coli* with the Lambda P_L Promoter — ROBERT CROWL — 376

56. Use of Rous Sarcoma Viral Genome to Express Human Fibroblast Interferon — LINDA MULCAHY, MARIA KAHN, BRUCE KELDER, EDWARD REHBERG, SIDNEY PESTKA, AND DENNIS W. STACEY — 383

57. Procedures for Expression, Modification, and Analysis of Human Fibroblast Interferon (IFN-β) Genes in Heterologous Cells — MICHAEL A. INNIS AND FRANK MCCORMICK — 397

58. Procedures for *in Vitro* DNA Mutagenesis of Human Leukocyte Interferon Sequences — THOMAS M. DECHIARA, FRAN ERLITZ, AND S. JOSEPH TARNOWSKI — 403

59. Yeast Vectors for Production of Interferon — MICHAEL D. SCHABER, THOMAS M. DECHIARA, AND RICHARD A. KRAMER — 416

60. Construction of Expression Vectors for Secretion of Human Interferons by Yeast — Ronald A. Hitzeman, Chung Nan Chang, Mark Matteucci, L. Jeanne Perry, William J. Kohr, John J. Wulf, James R. Swartz, Christina Y. Chen, and Arjun Singh 424

61. Procedures for Isolation of Murine Alpha Interferon Genes and Expression of a Murine Leukocyte Interferon in *E. coli* — Bruce L. Daugherty and Sidney Pestka 434

62. Cloning, Expression, and Purification of Rat IFN-α1 — Peter H. van der Meide, Rein Dijkema, Martin Caspers, Kitty Vijverberg, and Huub Schellekens 441

63. Cloning, Expression, and Purification of Rat IFN-γ — Rein Dijkema, Peter H. van der Meide, Martin Dubbeld, Martin Caspers, Jacqueline Wubben, and Huub Schellekens 453

64. Isolation of Bovine IFN-α Genes and Their Expression in Bacteria — Baruch Velan, Sara Cohen, Haim Grosfeld, and Avigdor Shafferman 464

65. The Detection of Individual Cells Containing Interferon mRNA by *in Situ* Hybridization with Specific Interferon DNA Probes — Rainer Zawatzky, Jaqueline De Maeyer-Guignard, and Edward De Maeyer 474

66. Detection of a Single Base Substitution between Human Leukocyte IFN-αA and -α2 Genes with Octadecyl Deoxyoligonucleotide Probes — Kunimoto Hotta, Kenneth J. Collier, and Sidney Pestka 481

Section VIII. Enzymes in Interferon Action

67. RNase L, a (2′-5′)-Oligoadenylate-Dependent Endoribonuclease: Assays and Purification of the Enzyme; Cross-Linking to a (2′-5′)-Oligoadenylate Derivative — Georgia Floyd-Smith and Peter Lengyel 489

68. Purification of Double-Stranded RNA-Dependent Protein Kinase from Mouse Fibroblasts — CHARLES E. SAMUEL, GRACE S. KNUTSON, MARLA J. BERRY, JONATHAN A. ATWATER, AND STEPHEN R. LASKY 499

69. Enzymatic Synthesis of (2′-5′)-Linked Oligoadenylates on a Preparative Scale — HIMADRI SAMANTA AND PETER LENGYEL 516

70. Methods for the Synthesis of Analogs of (2′-5′)-Oligoadenylic Acid — PAUL F. TORRENCE, JIRO IMAI, JEAN-CLAUDE JAMOULLE, AND KRYSTYNA LESIAK 522

Section IX. Assay of Interferons

A. Antiviral Assays

71. Objective Antiviral Assay of the Interferons by Computer Assisted Data Collection and Analysis — ROBERT L. FORTI, SHIRLEY S. SCHUFFMAN, HUGH A. DAVIES, AND WILLIAM M. MITCHELL 533

72. Measurement of Interferon in Human Amniotic Fluid and Placental Blood Extract — PAULETTE DUC-GOIRAN, PIERRE LEBON, AND CHARLES CHANY 541

73. Use of Ovine and Caprine Cells to Measure Antiviral Effects of Interferon — TILAHUN YILMA 551

74. Quantitation of Neutralization of Interferon by Antibody — YOSHIMI KAWADE 558

B. Antiproliferative Assays

75. A Convenient Microassay for Cytolysis and Cytostasis — STEPHEN TYRING, W. ROBERT FLEISCHMANN, JR., AND SAMUEL BARON 574

76. A Simplified Antigrowth Assay Based on Color Change of the Medium — PAUL AEBERSOLD AND SALLY SAMPLE 579

C. Immunoassays

77. A Sandwich Radioimmunoassay for Human IFN-γ — BRUCE KELDER, ABBAS RASHIDBAIGI, AND SIDNEY PESTKA 582

78. Procedures for Measurement of Interferon Dimers and Higher Oligomers by Radioimmunoassay — SIDNEY PESTKA, BRUCE KELDER, AND S. JOSEPH TARNOWSKI — 588

Section X. Biology of Interferon Action

79. Selection and Screening of Transformed NIH3T3 Cells for Enhanced Sensitivity to Human Interferons α and β — VINCENT JUNG AND SIDNEY PESTKA — 597

80. Measurement of the Effect of Interferons on Cellular Differentiation in Murine and Human Melanoma Cells — PAUL B. FISHER, ARMAND F. MIRANDA, AND LEE E. BABISS — 611

81. Measurement of the Effect of Interferons on Cellular Differentiation of Human Skeletal Muscle Cells — ARMAND F. MIRANDA, LEE E. BABISS, AND PAUL B. FISHER — 619

82. Measurement of the Effect of Interferons on the Proliferative Capacity and Cloning Efficiency of Normal and Leukemic Human Myeloid Progenitor Cells in Culture — PAUL B. FISHER, STEVEN GRANT, JOHN W. GREINER, AND JEFFREY SCHLOM — 629

83. Measurement of Effect of Interferons on Cloning Efficiency of Primary Tumor Cells in Culture — SYDNEY E. SALMON — 635

84. Measurement of the Antiproliferative Effect of Interferon: Influence of Growth Factors — NICOLETTE EBSWORTH, ENRIQUE ROZENGURT, AND JOYCE TAYLOR-PAPADIMITRIOU — 643

85. Animal Models for Investigating Antitumor Effects of Interferon — FRANCES R. BALKWILL — 649

86. Measurement of Antagonistic Effects of Growth Factors and Interferons — ANNA D. INGLOT AND ELŻBIETA PAJTASZ — 657

87. Measurement of Effect of (2′-5′)-Oligoadenylates and Analogs on Protein Synthesis and Growth of Cells — ROBERT J. SUHADOLNIK, CHOONGEUN LEE, AND DAVID H. WILLIS, JR. — 667

88. Assay of Effect of (2′-5′)-Oligoadenylate on Macrophages — XIN-YUAN LIU, HONG DA ZHENG, NING WANG, WEN HUA REN, AND T. P. WANG — 676

89. Radioimmunoassay for Detection of Changes in Cell Surface Tumor Antigen Expression Induced by Interferon — J. W. GREINER, P. HORAN HAND, D. WUNDERLICH, AND D. COLCHER — 682

90. Measurements of Changes in Histocompatibility Antigens Induced by Interferons — MARIANNE HOKLAND, IVER HERON, PETER HOKLAND, PER BASSE, AND KURT BERG 688

91. Purification, Assay, and Characterization of the Interferon Antagonist: Sarcolectin — FRANÇOISE CHANY-FOURNIER, PAN HONG JIANG, AND CHARLES CHANY 694

92. Assay of Effect of Interferon on Virus-Induced Cell Fusion — YOSHIMI TOMITA AND TSUGUO KUWATA 702

93. Measurement of Hyporesponsiveness to Interferon and Interferon Induction with Prostaglandins — DALE A. STRINGFELLOW 707

Section XI. Measurement of Effect of Interferons on Drug Metabolism

94. Measurement of Effect of Interferon on Metabolism of Diphenylhydantoin — GERALD SONNENFELD AND DONALD E. NERLAND 715

95. Measurement of Effect of Interferon on Drug Metabolism — GILBERT J. MANNERING 718

Section XII. Interferon and Plant Cells

96. Production, Preparation, and Assay of an Antiviral Substance from Plant Cells — ABDULLAH GERA, SARA SPIEGEL, AND GAD LOEBENSTEIN 729

97. Preparation and Measurement of an Antiviral Protein Found in Tobacco Cells after Infection with Tobacco Mosaic Virus — ILAN SELA 734

98. Assay of Effect of Human Interferons on Tobacco Protoplasts — ILAN SELA 744

99. Enzymatic Synthesis of Plant Oligoadenylates *in Vitro* — YAIR DEVASH, ILAN SELA, AND ROBERT J. SUHADOLNIK 752

100. Measurement of Effect of (2′-5′)-Oligoadenylates and Analogs on Tobacco Mosaic Virus Replication — YAIR DEVASH, ROBERT J. SUHADOLNIK, AND ILAN SELA 759

AUTHOR INDEX 763

SUBJECT INDEX 791

Contributors to Volume 119

Article numbers are in parentheses following the names of contributors.
Affiliations listed are current.

PAUL AEBERSOLD (76), *Immunopharmacology Laboratory, Bionetics Research, Inc., Rockville, Maryland 20850*

MICHEL AGUET (47), *Institute of Immunology and Virology, University of Zurich, CH-8028 Zurich, Switzerland*

REINHARD AHL (30), *Federal Research Center for Virus Diseases of Animals, D-7400 Tübingen, Federal Republic of Germany*

PAUL ANDERSON (39), *Department of Internal Medicine, Brigham and Women's Hospital, Boston, Massachusetts 02115*

GUIDO ANTONELLI (9), *The Institute of Virology, University "La Sapienza," Rome 00185, Italy*

HEINZ ARNHEITER (45, 48, 49), *Institute for Immunology and Virology, University of Zurich, CH-8028 Zurich, Switzerland*

JONATHAN A. ATWATER (68), *Section of Biochemistry and Molecular Biology, Department of Biological Sciences, University of California, Santa Barbara, California 93106*

LEE E. BABISS (80, 81), *Department of Microbiology, College of Physicians and Surgeons, Columbia University, New York, New York 10032*

FRANCES R. BALKWILL (85), *Interferon Laboratory, Imperial Cancer Research Fund, London WC2A 3PX, England*

G. D. BALL (4), *Virology Research and Development Department, Wellcome Biotechnology Ltd, Beckenham, Kent BR3 3BS, England*

SAMUEL BARON (75), *Department of Microbiology, University of Texas Medical Branch, Galveston, Texas 77550*

PER BASSE (90), *Institute of Medical Microbiology, University of Aarhus, DK-8000 Aarhus C, Denmark*

EVA BEKESI (29, 43), *Department of Molecular Genetics, Hoffmann-La Roche, Inc., Nutley, New Jersey 07110*

JAKE BELLO (16), *Department of Biophysics, Roswell Park Memorial Institute, Buffalo, New York 14263*

KURT BERG (90), *Institute of Medical Microbiology, University of Copenhagen, DK-2100 Copenhagen, Denmark*

MARLA J. BERRY (68), *Section of Biochemistry and Molecular Biology, Department of Biological Sciences, University of California, Santa Barbara, California 93106*

ALFONS BILLIAU (13, 22), *Rega Institute for Medical Research, University of Leuven, B-3000 Leuven, Belgium*

TAMAR BINO (21), *Department of Biochemistry, Israel Institute for Biological Research, Ness Ziona 70450, Israel*

ARTHUR P. BOLLON (53), *Department of Molecular Genetics, Wadley Institutes of Molecular Medicine, Dallas, Texas 75235*

IRWIN A. BRAUDE (10, 27), *Department of Immunology, Bristol-Meyers Company, Syracuse, New York 13221*

STUART E. BUILDER (11), *Genentech, Inc., South San Francisco, California 94080*

C. J. BURMAN (4), *Virology Research and Development Department, Wellcome Biotechnology Ltd, Beckenham, Kent BR3 3BS, England*

KARI CANTELL (3, 7), *National Public Health Institute, SF-00280 Helsinki, Finland*

MARIA R. CAPOBIANCHI (9), *The Institute of Virology, University "La Sapienza," Rome 00185, Italy*

MARTIN CASPERS (62, 63), *Medical Biological Laboratory TNO, 2280 AA Rijswijk, The Netherlands*

Chung Nan Chang (60), *Genentech, Inc., South San Francisco, California 94080*

May Chang (29), *Biopolymer Research Department, Hoffmann-La Roche, Inc., Nutley, New Jersey 07110*

Charles Chany (72, 91), *Institut National de la Santé et de la Recherche Médicale, Hôpital Saint Vincent-de-Paul, Paris 75674, Cedex 14, France*

Françoise Chany-Fournier (91), *Institut National de la Santé et de la Recherche Médicale, Hôpital Saint Vincent-de-Paul, Paris 75674, Cedex 14, France*

Chien-Hwa Chen (25), *The Biosciences Department, Hoffmann-La Roche, Inc., Nutley, New Jersey 07110*

Christina Y. Chen (60), *Genentech, Inc., South San Francisco, California 94080*

Sara Cohen (21, 64), *Department of Biochemistry, Israel Institute for Biological Research, Ness Ziona 70450, Israel*

D. Colcher (89), *Laboratory of Tumor Immunology and Biology, National Cancer Institute, National Institutes of Health, Bethesda, Maryland 20205*

Kenneth J. Collier (66), *Department of Molecular Genetics, Hoffmann-La Roche, Inc., Nutley, New Jersey 07110*

Dorian H. Coppenhaver (28), *Department of Microbiology, University of Texas Medical Branch, Galveston, Texas 77550*

Robert Crowl (55), *Department of Molecular Genetics, Hoffmann-La Roche, Inc., Nutley, New Jersey 07110*

Bruce L. Daugherty (61), *Department of Biochemistry, Roche Institute of Molecular Biology, Nutley, New Jersey 07110*

Hugh A. Davies (71), *Department of Pathology, School of Medicine, Vanderbilt University, Nashville, Tennessee 37235*

Thomas M. DeChiara (58, 59), *Department of Human Genetics and Development, College of Physicians and Surgeons, Columbia University, New York, New York 10032*

Marc De Ley (13, 22), *Laboratory of Biochemistry, University of Leuven, B-3000 Leuven, Belgium*

Edward De Maeyer (65), *Section de Biologie, Institut Curie, Université de Paris–Sud, Orsay 91405, France*

Jaqueline De Maeyer-Guignard (65), *Section de Biologie, Institut Curie, Universitè de Paris–Sud, Orsay 91405, France*

Yair Devash (99, 100), *Repligen Corporation, Cambridge, Massachusetts 02142*

Ferdinando Dianzani (9), *The Institute of Virology, University "La Sapienza," Rome 00185, Italy*

Rein Dijkema (62, 63), *Organon Int., 5340 BH Oss, The Netherlands*

Robert J. Drummond (26), *Department of Protein Chemistry, Cetus Corporation, Emeryville, California 94608*

Martin Dubbeld (63), *Department of Virology, Primate Center TNO, 2288 GJ Rijswijk, The Netherlands*

Paulette Duc-Goiran (72), *Institut National de la Santé et de la Recherche Médicale, Hôpital Saint Vincent-de-Paul, Paris 75674, Cedex 14, France*

Nicolette Ebsworth (84), *Imperial Cancer Research Fund, London WC2A 3PX, England*

Pierre Eid (51), *Laboratoire d'Oncologie Virale, Institut de Recherches Scientifiques sur le Cancer, 94802 Villejuif, France*

Fran Erlitz (58), *Department of Molecular Genetics, Hoffmann-La Roche, Inc., Nutley, New Jersey 07110*

Diana Fahey (40), *Central Research and Development Department, E.I. duPont deNemours and Company, Wilmington, Delaware 19898*

Philip C. Familletti (8), *Department of Molecular Genetics, Hoffmann-La Roche, Inc., Nutley, New Jersey 07110*

Karl H. Fantes (32), *8 Orchard Way, Shirley, Croydon, Surrey, England*

Arthur M. Felix (33), *Biological Re-*

search Department, Hoffmann-La Roche, Inc., Nutley, New Jersey 07110

WALTER FIERS (54), *Laboratory of Molecular Biology, State University of Ghent, B-9000 Ghent, Belgium*

N. B. FINTER (4), *Virology Research and Development Department, Wellcome Biotechnology Ltd, Beckenham, Kent BR3 3BS, England*

PAUL B. FISHER (80, 81, 82), *Department of Microbiology, Cancer Center/Institute of Cancer Research, College of Physicians and Surgeons, Columbia University, New York, New York 10032*

DAVID FITZGERALD (49), *Laboratory of Molecular Biology, National Cancer Institute, National Institutes of Health, Bethesda, Maryland 20205*

W. ROBERT FLEISCHMANN, JR. (75), *Department of Microbiology, University of Texas Medical Branch, Galveston, Texas 77550*

GEORGIA FLOYD-SMITH (67), *Department of Zoology, Arizona State University, Tempe, Arizona 85282*

ROBERT L. FORTI (71), *Department of Medicine, School of Medicine, Vanderbilt University, Nashville, Tennessee 37235*

MOTOHIRO FUKE (53), *Department of Molecular Genetics, Wadley Institutes of Molecular Medicine, Dallas, Texas 75235*

JERZY A. GEORGIADES (12), *Immuno Modulators Laboratories, Inc., Stafford, Texas 77477*

ABDULLAH GERA (96), *Virus Laboratory, Agricultural Research Organization, The Volcani Center, Bet Dagan 50250, Israel*

M. GOTTSCHALK (30), *Botanisches Institut, Universität Tübingen, D-7400 Tübingen, Federal Republic of Germany*

EDWARD N. GRANADOS (16), *Abbott Laboratories, North Chicago, Illinois 60064*

STEVEN GRANT (82), *Department of Medicine, Cancer Center/Institute of Cancer Research, College of Physicians and Surgeons, Columbia University, New York, New York 10032*

JOHN W. GREINER (82, 89), *Laboratory of Tumor Immunology and Biology, National Cancer Institute, National Institutes of Health, Bethesda, Maryland 20205*

HAIM GROSFELD (64), *Department of Biochemistry, Israel Institute for Biological Research, Ness Ziona 70450, Israel*

SOHAN L. GUPTA (36, 50), *Molecular Biology and Virology Program, Memorial Sloan-Kettering Cancer Center, New York, New York 10021*

P. HORAN HAND (89), *Laboratory of Tumor Immunology and Biology, National Cancer Institute, National Institutes of Health, Bethesda, Maryland 20205*

EDGAR P. HEIMER (33), *Biological Research Department, Hoffmann-La Roche, Inc., Nutley, New Jersey 07110*

HUBERTINE HEREMANS (22), *Rega Institute for Medical Research, University of Leuven, B-3000 Leuven, Belgium*

IVER HERON (90), *Institute of Medical Microbiology, University of Aarhus, DK-8000 Aarhus, Denmark*

NAOKI HIGASHI (14), *Pharmaceutical Development Department, Pharmaceutical Division, Suntory Ltd., Tokyo 107, Japan*

SINIKKA HIRVONEN (3, 7), *National Public Health Institute, SF-00280 Helsinki, Finland*

RONALD A. HITZEMAN (60), *Genentech, Inc., South San Francisco, California 94080*

MARIANNE HOKLAND (90), *Institute of Medical Microbiology, University of Aarhus, DK-8000 Aarhus C, Denmark*

PETER HOKLAND (90), *University Department of Medicine and Haematology, Aarhus Amtssygehus, DK-8000 Aarhus C, Denmark*

SUSUMU HONDA (29), *Central Research Division, Takeda Chemical Industries, Ltd., Yodogawa-ku, Osaka 532, Japan*

BERNARD HOROWITZ (5), *Blood Derivatives Program, New York Blood Center, New York, New York 10021*

Kunimoto Hotta (66), *Department of Antibiotics, National Institute of Health, Shinagawa-ku, Tokyo 141, Japan*

Jiro Imai (70), *Biomedical Research Laboratory, Morinaga Milk Industry Co., Ltd., Meguro-ku, Tokyo, Japan*

Anna D. Inglot (86), *Laboratory of Tumor Virology, Ludwik Hirszfeld Institute of Immunology and Experimental Therapy, Polish Academy of Sciences, 53-114 Wroclaw, Poland*

Michael A. Innis (57), *Microbial Genetics Department, Cetus Corporation, Emeryville, California 94608*

Jean-Claude Jamoulle (70), *Institut de Pharmacie, University of Liege, 4000 Leige, Belgium*

Pan Hong Jiang (91), *National Institute of Vaccines and Serum, Beijing, China*

Vincent Jung (79), *Department of Biochemistry, Roche Institute of Molecular Biology, Nutley, New Jersey 07110*

Maria Kahn (56), *Roche Institute of Molecular Biology, Nutley, New Jersey 07110*

Hanna-Leena Kauppinen (3, 7), *Finnish Red Cross, Blood Transfusion Service, SF-00310 Helsinki, Finland*

Yoshimi Kawade (74), *Institute for Virus Research, Kyoto University, Sakyo-ku, Kyoto 606, Japan*

Bruce Kelder (56, 77, 78), *Department of Animal Drug Discovery, Merck, Sharpe & Dohme Research Laboratories, Rahway, New Jersey 07065*

Ellis S. Kempner (35), *Laboratory of Biochemical Pharmacology, National Institute of Arthritis, Diabetes, and Digestive and Kidney Diseases, National Institutes of Health, Bethesda, Maryland 20205*

Ernest Knight, Jr. (40), *Central Research and Development Department, E.I. duPont deNemours and Company, Wilmington, Delaware 19898*

Grace S. Knutson (68), *Section of Biochemistry and Molecular Biology, Department of Biological Sciences, University of California, Santa Barbara, California 93106*

William J. Kohr (60), *Genentech, Inc., South San Francisco, California 94080*

Richard A. Kramer (59), *Department of Molecular Genetics, Hoffmann-La Roche, Inc., Nutley, New Jersey 07110*

Hsiang-fu Kung (29, 42, 43, 46), *Department of Molecular Genetics, Hoffmann-La Roche, Inc., Nutley, New Jersey 07110*

Tsuguo Kuwata (92), *Department of Microbiology, School of Medicine, Chiba University, Chiba 280, Japan*

Jerome A. Langer (34, 44), *Department of Biochemistry, Roche Institute of Molecular Biology, Nutley, New Jersey 07110*

Stephen R. Lasky (68), *Immunex Corporation, Seattle, Washington 98101*

Junming Le (6), *Department of Microbiology, New York University School of Medicine, New York, New York 10016*

Pierre Lebon (72), *Institut National de la Santé et de la Recherche Médicale, Hôpital Saint Vincent-de-Paul, Paris 75674, Cedex 14, France*

Choongeun Lee (87), *Department of Biochemistry, Temple University School of Medicine, Philadelphia, Pennsylvania 19140*

David K. Lee (23), *Bioprocess Development, Technical Development Department, Hoffmann-La Roche, Inc., Nutley, New Jersey 07110*

Peter Lengyel (67, 69), *Department of Molecular Biophysics and Biochemistry, Yale University, New Haven, Connecticut 06511*

Krystyna Lesiak (70), *Radiation Biophysics, Nuclear Reactor Center, The University of Kansas, Lawrence, Kansas 66045*

Bo-liang Li (52), *Shanghai Institute of Biochemistry, Academia Sinica, Shanghai 200031, China*

Shi Wu Li (52), *Beijing University of Medicine, Beijing, China*

Leo S. Lin (26), *Department of Protein*

Chemistry, Cetus Corporation, Emeryville, California 94608

RAYMOND A. LIPTAK (23), *Bioprocess Development, Technical Development Department, Hoffmann-La Roche, Inc., Nutley, New Jersey 07110*

XIN-YUAN LIU (52, 88), *Shanghai Institute of Biochemistry, Academia Sinica, Shanghai 200031, China*

GAD LOEBENSTEIN (96), *Virus Laboratory, Agricultural Research Organization, The Volcani Center, Bet Dagan 50250, Israel*

ANTHONY S. LUBINIECKI (11), *Genentech, Inc., South San Francisco, California 94080*

GILBERT J. MANNERING (95), *Department of Pharmacology, University of Minnesota, Minneapolis, Minnesota 55455*

PHILIP I. MARCUS (17, 18), *Department of Molecular and Cell Biology, The University of Connecticut, Storrs, Connecticut 06268*

MARK MATTEUCCI (60), *Genentech, Inc., South San Francisco, California 94080*

FRANK MCCORMICK (57), *Human Genetics Department, Cetus Corporation, Emeryville, California 94608*

MICHAEL MCGARRY (16), *Department of Animal Services, Roswell Park Memorial Institute, Buffalo, New York 14263*

ARMAND F. MIRANDA (80, 81), *Department of Pathology, College of Physicians and Surgeons, Columbia University, New York, New York 10032*

WILLIAM M. MITCHELL (71), *Department of Pathology, School of Medicine, Vanderbilt University, Nashville, Tennessee 37235*

KNUD ERIK MOGENSEN (37, 51), *Laboratoire d'Oncologie Virale, Institut de Recherches Scientifiques sur le Cancer, 94802 Villejuif, France*

ALAN G. MORRIS (15), *Department of Biological Sciences, University of Warwick, Coventry CV4 7AL, England*

JOHN A. MOSCHERA (25, 29), *Biopolymer Research Department, Hoffmann-La Roche, Inc., Nutley, New Jersey 07110*

LINDA MULCAHY (56), *Roche Institute of Molecular Biology, Nutley, New Jersey 07110*

DONALD E. NERLAND (94), *Department of Pharmacology and Toxicology, School of Medicine, University of Louisville, Louisville, Kentucky 40292*

ROBERT Y. NING (23), *Hybritech, Inc., 11095 Torreyana Road, San Diego, California 92126*

JUDITH A. O'MALLEY (16), *Department of Biological Resources, Roswell Park Memorial Institute, Buffalo, New York 14263*

ELŻBIETA PAJTASZ (86), *Laboratory of Cellular Immunology, Ludwik Hirszfeld Institute of Immunology and Experimental Therapy, Polish Academy of Sciences, 53-114 Wroclaw, Poland*

YU-CHING E. PAN (29), *Biopolymer Research Department, Hoffmann-La Roche, Inc., Nutley, New Jersey 07110*

NIKOS PANAYOTATOS (24), *Biogen S.A., Geneva 1211, Switzerland*

L. JEANNE PERRY (60), *Genentech, Inc., South San Francisco, California 94080*

SIDNEY PESTKA (1, 2, 34, 35, 41, 44, 46, 56, 61, 66, 77, 78, 79), *Department of Molecular Genetics and Microbiology, University of Medicine and Dentistry of New Jersey, Robert Wood Johnson Medical School, Piscataway, New Jersey 08854*

A. W. PHILLIPS (4), *Virology Research and Development Department, Wellcome Biotechnology Ltd, Beckenham, Kent BR3 3BS, England*

ABBAS RASHIDBAIGI (41, 46, 77), *Department of Biochemistry, Roche Institute of Molecular Biology, Nutley, New Jersey 07110*

ARATI RAZIUDDIN (50), *Memorial Sloan-Kettering Cancer Center, New York, New York 10021*

EDWARD REHBERG (56), *Roche Institute of Molecular Biology, Nutley, New Jersey 07110*

ERIK REMAUT (54), *Laboratory of Molecular Biology, State University of Ghent, B-9000 Ghent, Belgium*

WEN HUA REN (88), *Department of Pathology, Shanghai Second Medical University, Shanghai, China*

HAGAI ROSENBERG (21), *Department of Biochemistry, Israel Institute for Biological Research, Ness Ziona 70450, Israel*

SWAPAN K. ROY (23), *Bioprocess Development, Technical Development Department, Hoffmann-La Roche, Inc., Nutley, New Jersey 07110*

ENRIQUE ROZENGURT (84), *Imperial Cancer Research Fund, London WC2A 3PX, England*

SYDNEY E. SALMON (83), *Department of Internal Medicine and the Arizona Cancer Center, University of Arizona College of Medicine, Tucson, Arizona 85724*

HIMADRI SAMANTA (69), *Eugene Tech International, Allendale, New Jersey 07401*

SALLY SAMPLE (76), *Department of Pediatrics, University of California, San Francisco, California 94143*

CHARLES E. SAMUEL (68), *Section of Biochemistry and Molecular Biology, Department of Biological Sciences, University of California, Santa Barbara, California 93106*

FAZLUL H. SARKAR (36), *Department of Biological Sciences, Oakland University, Rochester, Michigan 48063*

MICHAEL D. SCHABER (59), *Department of Molecular Genetics, Hoffmann-La Roche, Inc., Nutley, New Jersey 07110*

HUUB SCHELLEKENS (31, 62, 63), *Department of Virology, Primate Center TNO, 2280 GJ Rijswijk, The Netherlands*

JEFFREY SCHLOM (82), *Laboratory of Tumor Immunology and Biology, National Cancer Institute, National Institutes of Health, Bethesda, Maryland 20205*

SHIRLEY S. SCHUFFMAN (71), *Department of Pathology, School of Medicine, Vanderbilt University, Nashville, Tennessee 37235*

MARGARET J. SEKELLICK (18), *Department of Molecular and Cell Biology, The University of Connecticut, Storrs, Connecticut 06268*

ILAN SELA (97, 98, 99, 100), *Virus Laboratory, The Hebrew University of Jerusalem, Faculty of Agriculture, Rehovot 76100, Israel*

AVIGDOR SHAFFERMAN (21, 64), *Department of Biochemistry, Israel Institute for Biological Research, Ness Ziona 70450, Israel*

GUUS SIMONS (54), *Dutch Institute for Dairy Research, 6710 BA Ede, The Netherlands*

ARJUN SINGH (60), *Genentech, Inc., South San Francisco, California 94080*

GERALD SONNENFELD (94), *Department of Microbiology and Immunology and Department of Oral Biology, Schools of Medicine and Dentistry, University of Louisville, Louisville, Kentucky 40292*

SARA SPIEGEL (96), *Virus Laboratory, Agricultural Research Organization, The Volcani Center, Bet Dagan 50250, Israel*

DENNIS W. STACEY (56), *Roche Institute of Molecular Biology, Nutley, New Jersey 07110*

PATRICK STANSSENS (54), *Plant Genetics Systems, State University of Ghent, B-9000 Ghent, Belgium*

DONNA STREMLO (8), *Department of Molecular Genetics, Hoffmann-La Roche, Inc., Nutley, New Jersey 07110*

DALE A. STRINGFELLOW (93), *Bristol-Myers Company, Pharmaceutical Research and Development Division, Syracuse, New York 13221*

HIROMU SUGINO (29), *Central Research Division, Takeda Chemical Industries, Ltd., Yodogawa-ku, Osaka 532, Japan*

ROBERT J. SUHADOLNIK (87, 99, 100), *Department of Biochemistry, Temple University School of Medicine, Philadelphia, Pennsylvania 19140*

JAMES R. SWARTZ (60), *Genentech, Inc., South San Francisco, California 94080*

S. JOSEPH TARNOWSKI (23, 25, 33, 58, 78), *Process Development Department, Interferon Sciences, Inc., New Brunswick, New Jersey 08901*

CRISS TARR (10), *Revlon Health Care Group, Meloy Laboratories, Inc., Springfield, Virginia 22151*

JOYCE TAYLOR-PAPADIMITRIOU (84), *Imperial Cancer Research Fund, London WC2A 3PX, England*

DAVID R. THATCHER (24), *Biogen Research Corporation, 14 Cambridge Center, Cambridge, Massachusetts 02142*

YOSHIMI TOMITA (92), *Department of Microbiology, School of Medicine, Chiba University, Chiba 280, Japan*

RICHARD M. TORCZYNSKI (53), *Department of Molecular Genetics, Wadley Institutes of Molecular Medicine, Dallas, Texas, 75235*

PAUL F. TORRENCE (70), *Laboratory of Chemistry, National Institute of Arthritis, Diabetes, and Digestive and Kidney Diseases, National Institutes of Health, Bethesda, Maryland 20205*

KELLY PENG TSAI (25, 29), *Biopolymer Research Department, Hoffmann-La Roche, Inc., Nutley, New Jersey 07110*

MASAFUMI TSUJIMOTO (14), *Laboratory of Molecular Biology, Suntory Institute for Biomedical Research, Shimamoto-cho, Osaka 618, Japan*

STEPHEN TYRING (75), *Department of Microbiology, University of Texas Medical Branch, Galveston, Texas 77550, and Department of Dermatology, University of Alabama at Birmingham, Birmingham, Alabama 35294*

GILLES UZÉ (37), *Laboratoire d'Oncologie Virale, Institut de Recherches Scientifiques sur le Cancer, 94802 Villejuif, France*

JOZEF VAN DAMME (13), *Rega Institute for Medical Research, University of Leuven, B-3000 Leuven, Belgium*

PETER H. VAN DER MEIDE (31, 62, 63), *Department of Virology, Primate Center TNO, 2280 GJ Rijswijk, The Netherlands*

ROBERT VAN REIS (11), *Genentech, Inc., South San Francisco, California 94080*

BARUCH VELAN (21, 64), *Department of Biochemistry, Israel Institute for Biological Research, Ness Ziona 70450, Israel*

KITTY VIJVERBERG (31, 62), *Department of Virology, Primate Center TNO, 2280 GJ Rijswijk, The Netherlands*

JAN VILČEK (6), *Department of Microbiology, New York University School of Medicine, New York, New York 10016*

NING WANG (88), *Department of Biochemistry, Kunming Medical College, Kunming, China*

T. P. WANG (88), *Shanghai Institute of Biochemistry, Academia Sinica, Shanghai 200031, China*

MILES F. WILKINSON (15), *University of California, San Diego, La Jolla, California 92093*

DAVID H. WILLIS, JR. (87), *Department of Biochemistry, Temple University School of Medicine, Philadelphia, Pennsylvania 19140*

TIMOTHY R. WINSHIP (19), *Daryl Laboratories, Inc., Santa Clara, California 95050*

DIANA WOEHLE (25), *Biopolymer Research Department, Hoffmann-La Roche, Inc., Nutley, New Jersey 07110*

JACQUELINE WUBBEN (31, 63), *Department of Virology, Primate Center TNO, 2280 GJ Rijswijk, The Netherlands*

JOHN J. WULF (60), *Genentech, Inc., South San Francisco, California 94080*

D. WUNDERLICH (89), *Laboratory of Tumor Immunology and Biology, National Cancer Institute, National Institutes of Health, Bethesda, Maryland 20205*

RALPH YAMAMOTO (26), *Department of Protein Chemistry, Cetus Corporation, Emeryville, California 94608*

TILAHUN YILMA (20, 73), *Department of Veterinary Microbiology and Pathology, Washington State University, Pullman, Washington 99164*

Y. K. YIP (6), *Department of Microbiology,*

New York University School of Medicine, New York, New York 10016

RAINER ZAWATZKY (65), *German Cancer Research Center, D-6900 Heidelberg, Federal Republic of Germany*

HONG DA ZHENG (88), *Shanghai Institute of Biochemistry, Academia Sinica, Shanghai 200031, China*

KATHRYN C. ZOON (38, 45, 48, 49), *Division of Virology, Office of Biologics Research and Review, Center for Drugs and Biologics, Food and Drug Administration, Bethesda, Maryland 20205*

DOROTHY L. ZUR NEDDEN (38, 45), *Division of Virology, Office of Biologics Research and Review, Center for Drugs and Biologics, Food and Drug Administration, Bethesda, Maryland 20205*

Preface

In an area of research that is evolving dynamically, a volume on the methods used in the field is extremely useful, yet has the potential to be outdated quickly. Since the chapters represent basic methodology on a variety of topics, it is not likely they will be supplanted by entirely new methodology, but by modifications. Thus, these volumes on interferons should serve as a resource for years to come. The most likely way in which these volumes will become dated is not by the aging of the contributions, but by the omission of methods that should have been included and the inability to keep pace with newly published procedures relevant to interferon research.

I thank the numerous contributors who have made this volume possible by their detailed texts and insights into subtleties not obviously evident. It is these insights that often determine whether a procedure "flies" or "fails." The inclusion of such critical yet almost trivial detail is a feature that distinguishes contributions in *Methods in Enzymology* from the methods sections of reports published in most journals. With pressure on limiting publishing space, experimental procedures have often been relegated to miniprint and curtailed. Thus, texts devoted to making the methodology of research flow smoothly will find great use. One additional advantage of a methods book devoted to one subject is the presence of a variety of procedures in a single volume. Thus, it is easier to compare methods before adopting a specific one.

This volume is published after the death of the founders of this series, Drs. Sidney P. Colowick and Nathan O. Kaplan. They contributed to each of these volumes by their critical appraisal of good scientific research. Both of them made many contributions to the organization of these volumes. I am grateful to have been the recipient of Sidney Colowick's generous advice; and I am grateful to have had Nathan Kaplan's wealthy scientific repetoire and experience at my disposal. Their efforts have improved these volumes immeasurably.

During the preparation of this volume, as during the preparation of the preceding two, my family has continually provided the sustenance to foster its birth. Special thanks are given to them, to the staff of Academic Press, and to my many colleagues who have contributed to it. I am particularly grateful to Robert K. Pestka, Joseph V. Campellone, and Sharon D. Pestka who were instrumental in the preparation of the subject index for this volume.

SIDNEY PESTKA

METHODS IN ENZYMOLOGY

EDITED BY

Sidney P. Colowick and Nathan O. Kaplan

VANDERBILT UNIVERSITY
SCHOOL OF MEDICINE
NASHVILLE, TENNESSEE

DEPARTMENT OF CHEMISTRY
UNIVERSITY OF CALIFORNIA
AT SAN DIEGO
LA JOLLA, CALIFORNIA

I. Preparation and Assay of Enzymes
II. Preparation and Assay of Enzymes
III. Preparation and Assay of Substrates
IV. Special Techniques for the Enzymologist
V. Preparation and Assay of Enzymes
VI. Preparation and Assay of Enzymes (*Continued*)
Preparation and Assay of Substrates
Special Techniques
VII. Cumulative Subject Index

METHODS IN ENZYMOLOGY

EDITORS-IN-CHIEF

Sidney P. Colowick and Nathan O. Kaplan

Volume VIII. Complex Carbohydrates
Edited by Elizabeth F. Neufeld and Victor Ginsburg

Volume IX. Carbohydrate Metabolism
Edited by Willis A. Wood

Volume X. Oxidation and Phosphorylation
Edited by Ronald W. Estabrook and Maynard E. Pullman

Volume XI. Enzyme Structure
Edited by C. H. W. Hirs

Volume XII. Nucleic Acids (Parts A and B)
Edited by Lawrence Grossman and Kivie Moldave

Volume XIII. Citric Acid Cycle
Edited by J. M. Lowenstein

Volume XIV. Lipids
Edited by J. M. Lowenstein

Volume XV. Steroids and Terpenoids
Edited by Raymond B. Clayton

Volume XVI. Fast Reactions
Edited by Kenneth Kustin

Volume XVII. Metabolism of Amino Acids and Amines (Parts A and B)
Edited by Herbert Tabor and Celia White Tabor

VOLUME XVIII. Vitamins and Coenzymes (Parts A, B, and C)
Edited by DONALD B. MCCORMICK AND LEMUEL D. WRIGHT

VOLUME XIX. Proteolytic Enzymes
Edited by GERTRUDE E. PERLMANN AND LASZLO LORAND

VOLUME XX. Nucleic Acids and Protein Synthesis (Part C)
Edited by KIVIE MOLDAVE AND LAWRENCE GROSSMAN

VOLUME XXI. Nucleic Acids (Part D)
Edited by LAWRENCE GROSSMAN AND KIVIE MOLDAVE

VOLUME XXII. Enzyme Purification and Related Techniques
Edited by WILLIAM B. JAKOBY

VOLUME XXIII. Photosynthesis (Part A)
Edited by ANTHONY SAN PIETRO

VOLUME XXIV. Photosynthesis and Nitrogen Fixation (Part B)
Edited by ANTHONY SAN PIETRO

VOLUME XXV. Enzyme Structure (Part B)
Edited by C. H. W. HIRS AND SERGE N. TIMASHEFF

VOLUME XXVI. Enzyme Structure (Part C)
Edited by C. H. W. HIRS AND SERGE N. TIMASHEFF

VOLUME XXVII. Enzyme Structure (Part D)
Edited by C. H. W. HIRS AND SERGE N. TIMASHEFF

VOLUME XXVIII. Complex Carbohydrates (Part B)
Edited by VICTOR GINSBURG

VOLUME XXIX. Nucleic Acids and Protein Synthesis (Part E)
Edited by LAWRENCE GROSSMAN AND KIVIE MOLDAVE

VOLUME XXX. Nucleic Acids and Protein Synthesis (Part F)
Edited by KIVIE MOLDAVE AND LAWRENCE GROSSMAN

VOLUME XXXI. Biomembranes (Part A)
Edited by SIDNEY FLEISCHER AND LESTER PACKER

VOLUME XXXII. Biomembranes (Part B)
Edited by SIDNEY FLEISCHER AND LESTER PACKER

VOLUME XXXIII. Cumulative Subject Index Volumes I–XXX
Edited by MARTHA G. DENNIS AND EDWARD A. DENNIS

VOLUME XXXIV. Affinity Techniques (Enzyme Purification: Part B)
Edited by WILLIAM B. JAKOBY AND MEIR WILCHEK

VOLUME XXXV. Lipids (Part B)
Edited by JOHN M. LOWENSTEIN

VOLUME XXXVI. Hormone Action (Part A: Steroid Hormones)
Edited by BERT W. O'MALLEY AND JOEL G. HARDMAN

VOLUME XXXVII. Hormone Action (Part B: Peptide Hormones)
Edited by BERT W. O'MALLEY AND JOEL G. HARDMAN

VOLUME XXXVIII. Hormone Action (Part C: Cyclic Nucleotides)
Edited by JOEL G. HARDMAN AND BERT W. O'MALLEY

VOLUME XXXIX. Hormone Action (Part D: Isolated Cells, Tissues, and Organ Systems)
Edited by JOEL G. HARDMAN AND BERT W. O'MALLEY

VOLUME XL. Hormone Action (Part E: Nuclear Structure and Function)
Edited by BERT W. O'MALLEY AND JOEL G. HARDMAN

VOLUME XLI. Carbohydrate Metabolism (Part B)
Edited by W. A. WOOD

VOLUME XLII. Carbohydrate Metabolism (Part C)
Edited by W. A. WOOD

VOLUME XLIII. Antibiotics
Edited by JOHN H. HASH

VOLUME XLIV. Immobilized Enzymes
Edited by KLAUS MOSBACH

VOLUME XLV. Proteolytic Enzymes (Part B)
Edited by LASZLO LORAND

VOLUME XLVI. Affinity Labeling
Edited by WILLIAM B. JAKOBY AND MEIR WILCHEK

VOLUME XLVII. Enzyme Structure (Part E)
Edited by C. H. W. HIRS AND SERGE N. TIMASHEFF

VOLUME XLVIII. Enzyme Structure (Part F)
Edited by C. H. W. HIRS AND SERGE N. TIMASHEFF

VOLUME XLIX. Enzyme Structure (Part G)
Edited by C. H. W. HIRS AND SERGE N. TIMASHEFF

VOLUME L. Complex Carbohydrates (Part C)
Edited by VICTOR GINSBURG

VOLUME LI. Purine and Pyrimidine Nucleotide Metabolism
Edited by PATRICIA A. HOFFEE AND MARY ELLEN JONES

VOLUME LII. Biomembranes (Part C: Biological Oxidations)
Edited by SIDNEY FLEISCHER AND LESTER PACKER

VOLUME LIII. Biomembranes (Part D: Biological Oxidations)
Edited by SIDNEY FLEISCHER AND LESTER PACKER

VOLUME LIV. Biomembranes (Part E: Biological Oxidations)
Edited by SIDNEY FLEISCHER AND LESTER PACKER

VOLUME LV. Biomembranes (Part F: Bioenergetics)
Edited by SIDNEY FLEISCHER AND LESTER PACKER

VOLUME LVI. Biomembranes (Part G: Bioenergetics)
Edited by SIDNEY FLEISCHER AND LESTER PACKER

VOLUME LVII. Bioluminescence and Chemiluminescence
Edited by MARLENE A. DELUCA

VOLUME LVIII. Cell Culture
Edited by WILLIAM B. JAKOBY AND IRA PASTAN

VOLUME LIX. Nucleic Acids and Protein Synthesis (Part G)
Edited by KIVIE MOLDAVE AND LAWRENCE GROSSMAN

VOLUME 73. Immunochemical Techniques (Part B)
Edited by JOHN J. LANGONE AND HELEN VAN VUNAKIS

VOLUME 74. Immunochemical Techniques (Part C)
Edited by JOHN J. LANGONE AND HELEN VAN VUNAKIS

VOLUME 75. Cumulative Subject Index Volumes XXXI, XXXII, and XXXIV–LX
Edited by EDWARD A. DENNIS AND MARTHA G. DENNIS

VOLUME 76. Hemoglobins
Edited by ERALDO ANTONINI, LUIGI ROSSI-BERNARDI, AND EMILIA CHIANCONE

VOLUME 77. Detoxication and Drug Metabolism
Edited by WILLIAM B. JAKOBY

VOLUME 78. Interferons (Part A)
Edited by SIDNEY PESTKA

VOLUME 79. Interferons (Part B)
Edited by SIDNEY PESTKA

VOLUME 80. Proteolytic Enzymes (Part C)
Edited by LASZLO LORAND

VOLUME 81. Biomembranes (Part H: Visual Pigments and Purple Membranes, I)
Edited by LESTER PACKER

VOLUME 82. Structural and Contractile Proteins (Part A: Extracellular Matrix)
Edited by LEON W. CUNNINGHAM AND DIXIE W. FREDERIKSEN

VOLUME 83. Complex Carbohydrates (Part D)
Edited by VICTOR GINSBURG

VOLUME 84. Immunochemical Techniques (Part D: Selected Immunoassays)
Edited by JOHN J. LANGONE AND HELEN VAN VUNAKIS

VOLUME 85. Structural and Contractile Proteins (Part B: The Contractile Apparatus and the Cytoskeleton)
Edited by DIXIE W. FREDERIKSEN AND LEON W. CUNNINGHAM

VOLUME 86. Prostaglandins and Arachidonate Metabolites
Edited by WILLIAM E. M. LANDS AND WILLIAM L. SMITH

VOLUME 87. Enzyme Kinetics and Mechanism (Part C: Intermediates, Stereochemistry, and Rate Studies)
Edited by DANIEL L. PURICH

VOLUME 88. Biomembranes (Part I: Visual Pigments and Purple Membranes, II)
Edited by LESTER PACKER

VOLUME 89. Carbohydrate Metabolism (Part D)
Edited by WILLIS A. WOOD

VOLUME 90. Carbohydrate Metabolism (Part E)
Edited by WILLIS A. WOOD

VOLUME 91. Enzyme Structure (Part I)
Edited by C. H. W. HIRS AND SERGE N. TIMASHEFF

VOLUME 92. Immunochemical Techniques (Part E: Monoclonal Antibodies and General Immunoassay Methods)
Edited by JOHN J. LANGONE AND HELEN VAN VUNAKIS

VOLUME 93. Immunochemical Techniques (Part F: Conventional Antibodies, Fc Receptors, and Cytotoxicity)
Edited by JOHN J. LANGONE AND HELEN VAN VUNAKIS

VOLUME 94. Polyamines
Edited by HERBERT TABOR AND CELIA WHITE TABOR

VOLUME 95. Cumulative Subject Index Volumes 61–74 and 76–80
Edited by EDWARD A. DENNIS AND MARTHA G. DENNIS

VOLUME 96. Biomembranes [Part J: Membrane Biogenesis: Assembly and Targeting (General Methods; Eukaryotes)]
Edited by SIDNEY FLEISCHER AND BECCA FLEISCHER

VOLUME 97. Biomembranes [Part K: Membrane Biogenesis: Assembly and Targeting (Prokaryotes, Mitochondria, and Chloroplasts)]
Edited by SIDNEY FLEISCHER AND BECCA FLEISCHER

VOLUME 98. Biomembranes [Part L: Membrane Biogenesis (Processing and Recycling)]
Edited by SIDNEY FLEISCHER AND BECCA FLEISCHER

VOLUME 99. Hormone Action (Part F: Protein Kinases)
Edited by JACKIE D. CORBIN AND JOEL G. HARDMAN

VOLUME 100. Recombinant DNA (Part B)
Edited by RAY WU, LAWRENCE GROSSMAN, AND KIVIE MOLDAVE

VOLUME 101. Recombinant DNA (Part C)
Edited by RAY WU, LAWRENCE GROSSMAN, AND KIVIE MOLDAVE

VOLUME 102. Hormone Action (Part G: Calmodulin and Calcium-Binding Proteins)
Edited by ANTHONY R. MEANS AND BERT W. O'MALLEY

VOLUME 103. Hormone Action (Part H: Neuroendocrine Peptides)
Edited by P. MICHAEL CONN

VOLUME 104. Enzyme Purification and Related Techniques (Part C)
Edited by WILLIAM B. JAKOBY

VOLUME 105. Oxygen Radicals in Biological Systems
Edited by LESTER PACKER

VOLUME 106. Posttranslational Modifications (Part A)
Edited by FINN WOLD AND KIVIE MOLDAVE

VOLUME 107. Posttranslational Modifications (Part B)
Edited by FINN WOLD AND KIVIE MOLDAVE

VOLUME 108. Immunochemical Techniques (Part G: Separation and Characterization of Lymphoid Cells)
Edited by GIOVANNI DI SABATO, JOHN J. LANGONE, AND HELEN VAN VUNAKIS

VOLUME 109. Hormone Action (Part I: Peptide Hormones)
Edited by LUTZ BIRNBAUMER AND BERT W. O'MALLEY

VOLUME 110. Steroids and Isoprenoids (Part A)
Edited by JOHN H. LAW AND HANS C. RILLING

VOLUME 111. Steroids and Isoprenoids (Part B)
Edited by JOHN H. LAW AND HANS C. RILLING

VOLUME 112. Drug and Enzyme Targeting (Part A)
Edited by KENNETH J. WIDDER AND RALPH GREEN

VOLUME 113. Glutamate, Glutamine, Glutathione, and Related Compounds
Edited by ALTON MEISTER

VOLUME 114. Diffraction Methods for Biological Macromolecules (Part A)
Edited by HAROLD W. WYCKOFF, C. H. W. HIRS, AND SERGE N. TIMASHEFF

VOLUME 115. Diffraction Methods for Biological Macromolecules (Part B)
Edited by HAROLD W. WYCKOFF, C. H. W. HIRS, AND SERGE N. TIMASHEFF

VOLUME 116. Immunochemical Techniques (Part H: Effectors and Mediators of Lymphoid Cell Functions)
Edited by GIOVANNI DI SABATO, JOHN J. LANGONE, AND HELEN VAN VUNAKIS

VOLUME 117. Enzyme Structure (Part J)
Edited by C. H. W. HIRS AND SERGE N. TIMASHEFF

VOLUME 118. Plant Molecular Biology
Edited by ARTHUR WEISSBACH AND HERBERT WEISSBACH

VOLUME 119. Interferons (Part C)
Edited by SIDNEY PESTKA

VOLUME 120. Cumulative Subject Index Volumes 81–94, 96–101

VOLUME 121. Immunochemical Techniques (Part I: Hybridoma Technology and Monoclonal Antibodies)
Edited by JOHN J. LANGONE AND HELEN VAN VUNAKIS

VOLUME 122. Vitamins and Coenzymes (Part G)
Edited by FRANK CHYTIL AND DONALD B. MCCORMICK

VOLUME 123. Vitamins and Coenzymes (Part H)
Edited by FRANK CHYTIL AND DONALD B. MCCORMICK

VOLUME 124. Hormone Action (Part J: Neuroendocrine Peptides)
Edited by P. MICHAEL CONN

VOLUME 125. Biomembranes (Part M: Transport in Bacteria, Mitochondria, and Chloroplasts: General Approaches and Transport Systems)
Edited by SIDNEY FLEISCHER AND BECCA FLEISCHER

VOLUME 126. Biomembranes (Part N: Transport in Bacteria, Mitochondria, and Chloroplasts: Protonmotive Force)
Edited by SIDNEY FLEISCHER AND BECCA FLEISCHER

VOLUME 127. Biomembranes (Part O: Protons and Water: Structure and Translocation)
Edited by LESTER PACKER

VOLUME 128. Plasma Lipoproteins (Part A: Preparation, Structure, and Molecular Biology)
Edited by JERE P. SEGREST AND JOHN J. ALBERS

VOLUME 129. Plasma Lipoproteins (Part B: Characterization, Cell Biology, and Metabolism)
Edited by JOHN J. ALBERS AND JERE P. SEGREST

VOLUME 130. Enzyme Structure (Part K) (in preparation)
Edited by C. H. W. HIRS AND SERGE N. TIMASHEFF

VOLUME 131. Enzyme Structure (Part L) (in preparation)
Edited by C. H. W. HIRS AND SERGE N. TIMASHEFF

VOLUME 132. Immunochemical Techniques (Part J: Phagocytosis and Cell-Mediated Cytotoxicity) (in preparation)
Edited by GIOVANNI DI SABATO AND JOHANNES EVERSE

VOLUME 133. Bioluminescence and Chemiluminescence (Part B) (in preparation)
Edited by MARLENE A. DELUCA AND WILLIAM D. MCELROY

VOLUME 134. Structural and Contractile Proteins (Part C: The Contractile Apparatus and the Cytoskeleton) (in preparation)
Edited by RICHARD B. VALLEE

Section I

Introduction

[1] Interferon from 1981 to 1986

By SIDNEY PESTKA

Since the first two volumes, Interferons, Part A[1] and Part B,[2] were published, progress in this field has developed rapidly and many of the human interferons are in clinical trials. This volume has incorporated new methodologies that have sprung from the achievements in purification, cloning, and bacterial expression of the interferons as well as in other areas. The section on induction of interferons provides some new procedures and novel insights into the production and synthesis of the interferons by animal cells. Despite the fact that some general principles can be applied to the induction of the interferons from all animal cells,[3] the specific details vary remarkably from cell to cell and from species to species. Thus, in this volume, the induction of interferons from guinea pig, equine, bovine, chicken, and mouse cells is described. Although purification of the interferons has been covered in previous volumes, new and scaled up procedures for purification of IFN-α, IFN-β, and IFN-γ from human and other animal species are described in detail. The production and purification of the rat interferons should be of particular interest and provide material for use in an important experimental animal and open up new model systems for evaluation. Chemical modification of the interferons has not been performed extensively. The little that has been done has provided some insight into its secondary structure, in particular disulfide linkages, and estimates of the size of the functional unit of the interferons.

Radiolabeling of the interferons is a prerequisite for studies of their binding and internalization. With the purification[4–6] and cloning[4–38] of the interferons, various strategies for obtaining radiolabeled biologically active interferons were employed (see Section V of this volume). Ra-

[1] S. Pestka, ed., this series, Vol. 78.
[2] S. Pestka, ed., this series, Vol. 79.
[3] S. Pestka and S. Baron, this series, Vol. 78, p. 3.
[4] S. Pestka, *Arch. Biochem. Biophys.* **221,** 1 (1983).
[5] S. Pestka, *Sci. Am.* **249,** 36 (1983).
[6] J. A. Langer and S. Pestka, *Pharmacol. Ther.* **27,** 371 (1985).
[7] C. Weissmann, S. Nagata, W. Boll, M. Fountoulakis, A. Fujisawa, J.-I. Fujisawa, J. Haynes, K. Henco, N. Mantei, H. Ragg, C. Schein, J. Schmid, G. Shaw, M. Streuli, H. Taira, K. Todokoro, and U. Weidle, *Philos. Trans. R. Soc. London, Ser. B* **299,** 7 (1982).
[8] S. Maeda, R. McCandliss, M. Gross, A. Sloma, P. C. Familletti, J. M. Tabor, M. Evinger, W. P. Levy, and S. Pestka, *Proc. Natl. Acad. Sci. U.S.A.* **77,** 7010 (1980); **78,** 4648 (1981).

[9] D. V. Goeddel, E. Yelverton, A. Ullrich, H. L. Heyneker, G. Miozzari, W. Holmes, P. H. Seeburg, T. Dull, L. May, N. Stebbing, R. Crea, S. Maeda, R. McCandliss, A. Sloma, J. M. Tabor, M. Gross, P. C. Familletti, and S. Pestka, *Nature* (*London*) **287,** 411 (1980).

[10] S. Maeda, R. McCandliss, T.-R. Chiang, L. Costello, W. P. Levy, N. T. Chang, and S. Pestka, *in* "Developmental Biology Using Purified Genes" (D. D. Brown, ed.), p. 85. Academic Press, New York, 1981.

[11] D. V. Goeddel, D. W. Leung, T. J. Dull, M. J. Gross, R. M. Lawn, R. McCandliss, P. H. Seeburg, A. Ullrich, E. Yelverton, and P. W. Gray, *Nature* (*London*) **290,** 20 (1981).

[12] R. M. Lawn, J. Adelman, T. J. Dull, M. Gross, and D. V. Goeddel, *Science* **212,** 1159 (1981).

[13] R. M. Lawn, J. Adelman, A. E. Franke, C. M. Houck, M. Gross, R. Najarian, and D. V. Goeddel, *Nucleic Acids Res.* **9,** 1045 (1981).

[14] R. M. Lawn, M. Gross, C. M. Houck, A. E. Franke, P. W. Gray, and D. V. Goeddel, *Proc. Natl. Acad. Sci. U.S.A.* **78,** 5435 (1981).

[15] S. Pestka, S. Maeda, D. S. Hobbs, T.-R.C. Chiang, L. L. Costello, E. Rehberg, W. P. Levy, N. T. Chang, N. R. Wainwright, J. B. Hiscott, R. McCandliss, S. Stein, J. A. Moschera, and T. Staehelin, *in* "Recombinant DNA" (A. G. Walton, ed.), p. 51. Elsevier/North-Holland Biomedical Press, Amsterdam, 1981.

[16] N. Mantei, M. Schwartzstein, M. Streuli, S. Panem, S. Nagata, and C. Weissmann, *Gene* **10,** 1 (1980).

[17] M. Streuli, S. Nagata, and C. Weissmann, *Science* **209,** 1343 (1980).

[18] X.-Y. Liu, A. Rashidbaigi, V. Jung, and S. Pestka, in preparation.

[19] D. W. Bowden, J. Mao, T. Gill, K. Hsiao, J. S. Lilquist, D. Testa, and G. F. Vovis, *Gene* **27,** 87 (1984).

[20] R. M. Torczynski, M. Fuke, and A. P. Bollon, *Proc. Natl. Acad. Sci. U.S.A.* **81,** 6451 (1984).

[21] A. P. Bollon, M. Fuke, and R. M. Torczynski, this volume [53].

[22] A. Sloma, European Patent Application 83102893.1 (1983).

[23] M. A. Innis, International Patent Application PCT/US83/00032 (1983).

[24] F. Meyer, European Patent Application 82109094.1 (1983).

[25] E. Dworkin-Rastl, M. B. Dworkin, and P. Swetly, *J. Interferon Res.* **2,** 575 (1982).

[26] P. Slocombe, A. Easton, P. Boseley, and D. C. Burke, *Proc. Natl. Acad. Sci. U.S.A.* **79,** 5455 (1982).

[27] Y. Chernajovsky, Y. Mory, B. Vaks, S. I. Feinstein, D. Segev, and M. Revel, *Ann. N.Y. Acad. Sci.* **413,** 88 (1983).

[28] T. Taniguchi, M. Sakai, Y. Fujii-Kuriyama, M. Muramatsu, S. Kobayashi, and T. Sudo, *Proc. Jpn. Acad., Ser. B* **55,** 464 (1979).

[29] T. Taniguchi, S. Ohno, Y. Fujii-Kuriyama, and M. Muramatsu, *Gene* **10,** 11 (1980).

[30] R. Derynck, J. Content, E. De Clercq, G. Volckaert, J. Tavernier, R. Devos, and W. Fiers, *Nature* (*London*) **285,** 542 (1980).

[31] M. Houghton, A. G. Stewart, S. M. Doel, J. S. Emtage, M. A. W. Eaton, J. C. Smith, T. P. Patel, H. M. Lewis, A. G. Porter, J. R. Birch, T. Cartwright, and N. H. Carey, *Nucleic Acids Res.* **8,** 1913 (1980).

[32] D. V. Goeddel, H. M. Shepard, E. Yelverton, D. Leung, R. Crea, A. Sloma, and S. Pestka, *Nucleic Acids Res.* **8,** 4057 (1980).

[33] G. Gross, U. Mayr, W. Bruns, F. Grosveld, H.-H.M. Dahl, and J. Collins, *Nucleic Acids Res.* **9,** 2495 (1981).

[34] K. Siggens, P. Slocombe, A. Easton, P. Boseley, A. Meager, J. Tinsley, and D. Burke, *Biochim. Biophys. Acta* **741,** 65 (1983).

dioiodination by several different methods has been the most versatile and common method used. Very recently direct labeling of Hu-IFN-γ with ^{32}P has proved to be extremely useful and convenient.[39–42] These labeled interferons have provided the initial tools to begin dissecting the interaction of the interferons with their surface receptors and studying their uptake (see Section VI of this volume).

The cloning of the sequences coding for the interferons[4–38] and their expression in bacteria and other heterologous hosts has provided great impetus to interferon research. A large number of DNA sequences corresponding to the interferons from various species have been isolated and identified. The human leukocyte interferons (IFN-αs) are the most extensive family so far characterized. Members of this family so far identified are shown in Fig. 1. As can be seen, they represent a large number of substantially homologous sequences isolated by many different laboratories. Presenting this vast array of data in a useful form is a challenge (Fig. 1). Data bases available on computer are extremely useful for storing, comparing and manipulating these sequences. Some new sequences related to the leukocyte interferons have been described and isolated.[43,44] In addition, the sequences of human fibroblast interferon (IFN-β)[4,10,28–32] and immune interferon (IFN-γ)[35,36] are shown in Figs. 2 and 3, respectively. It is worth noting that the original sequence predicted from the DNA sequence for the mature form of the Hu-IFN-γ monomer was incorrect.[35,36] More recent work describing the amino acid sequence of the Hu-IFN-γ isolated from induced human cells[45] showed that the amino terminus of IFN-γ was Glx-Asp-Pro rather than Cys-Tyr-Cys, which are very likely to

[35] R. Devos, H. Cheroutre, Y. Taya, W. Degrave, H. van Heuverswyn, and W. Fiers, *Nucleic Acids Res.* **10,** 2487 (1982).

[36] P. W. Gray, D. W. Leung, D. Pennica, E. Yelverton, R. Najarian, C. C. Simonsen, R. Derynck, P. J. Sherwood, D. M. Wallace, S. L. Berger, A. D. Levinson, and D. V. Goeddel, *Nature (London)* **295,** 503 (1982).

[37] X.-Y. Liu, S. Maeda, K. Collier, and S. Pestka, *Gene Anal. Techn.* **2,** 83 (1985).

[38] M. Fuke, L. C. Hendrix, and A. P. Bollon, *Gene* **32,** 135 (1984).

[39] H.-F. Kung and E. Bekesi, this volume [43].

[40] A. Rashidbaigi, H.-F. Kung, and S. Pestka, this volume [46].

[41] B. Robert-Galliot, M. J. Commoy-Chevalier, P. Georges, and C. Chany, *J. Gen. Virol.* **66,** 1439–1448 (1985).

[42] A. Rashidbaigi, H.-F. Kung, and S. Pestka, *J. Biol. Chem.* **260,** 8514 (1985).

[43] S. I. Feinstein, Y. Mory, Y. Chernajovsky, L. Maroteaux, U. Nir, V. Lavie, and M. Revel, *Mol. Cell. Biol.* **5,** 510 (1985).

[44] D. J. Capon, H. M. Shepard, and D. V. Goeddel, *Mol. Cell. Biol.* **5,** 768 (1985).

[45] R. van Reis, S. E. Builder, and A. S. Lubiniecki, this volume [11].

	S1 S10	S20	S23	1 10	20	30	40	50	60	70
IFN-α consensus	MALSFSLLMA	VLVLSYKSIC	SLG	CDLPQTHSLG	NRRALILLAQ	MGRISPFSCL	KDRHDFGFPQ	EEFDGNQFQK	AQAISVLHEM	IQQTFNLFST
IFN-αK (α6)	...P.A....	LV...C..S.	..D	....	H..TMM....	.R...L....	......R...	....	.E.......V	....
IFN-α5 (G)	...P.V....	LV..NC....	...	.........S	...T.MIM..	....	....	....	....	....
IFN-αA (α2)	...T.A..V.	L....C..S.	.V.	....	S..T.M....	.RK..L....	....	...-......	.ET.P.....	...I......
IFN-αD (α1)	..SP.A...V	LV...C..S.	...	....E....D	...T.M....	.S....S...	M.........	....	.P.......L	...I....T.
IFN-αH1 (αH2)	...P...M..	LV...C..S.	...	.N.S.....N	...T.M.M..	.R........	......E...	....	....	M.........
IFN-αB2 (α8)	...T.Y..V.	LV......FS	...	....	....	.R........	......E...	....DK....	....	....
IFN-αB	...T.Y.MV.	LV......FS	...	....	....	.R........	......E...	....DK....	....	....
IFN-α4b	....	....	...	....	....	.....H....	.........E	.....H....	T.........	....
IFN-αC	....	....	...	....	........G.	....	......RI..	....	....	....
IFN-αL (Ψα10)	....	.........*	...	........T.R	........G.	....	......RI..	....	....	....
IFN-αJ1 (α7)	..R......V	....	...	.........R	....	....	....E.R..E	.....H....	T.........	....
IFN-αJ2	..R......V	....	...	.........R	....	....	....E.R..E	.....H....	T.........	....
IFN-αI	....	....	...	....	....	....	...P...L..	....	T.........	....
IFN-αF	....	....	...	....	....	....	....	....	....	....
IFN-αWA	....	....	...	....	....	.....H....	...Y......	.V........	.....AF...	....
IFN-αGx-1	...P...M..	LV...C..S.	...	.N.S.....N	...T.MIM..	....	....	....	....	....
IFN-α76	....	....	...	....	....	.....H....	.........E	.....H....	....	....

	80	90	100	110	120	130	140	150	160	166
IFN-α consensus	KDSSAAWDES	LLEKFSTELY	QQLNDLEACV	IQEVGVEETP	LMNEDSILAV	RKYFQRITLY	LTEKKYSPCA	WEVVRAEIMR	SFSFSTNLQK	RLRRKD
IFN-αK (α6)	....V....R	..D.LY....	....	M...W.GG..	....	....	....	....	...S.R...E	.....E
IFN-αG (α5)	.....T...T	..D..Y....	.........M	M......D..	...V....T.	....	....	....	...L.A...E	.....E
IFN-αA (α2)	.........T	..D..Y....	....	..G...T...	..K.......	....	.K........	....	...L.....E	S..S.E
IFN-αD (α1)	.........D	..D..C....	....	M..ER.G...	...V......	K...R.....	....	....	.L.L.....E	.....E
IFN-αH1 (αH2)	.N.......T	.....YI..F	..M.......	....	....	K.........	.M........	....	....	......
IFN-αB2 (α8)	......L..T	..DE.YI..D	.......S..	M.....I.S.	..Y.......	....	.......S..	....	...L.I....	..KS.E
IFN-αB	......L..T	..DE.YI..D	VLC	D.....I.S.	..Y.......	....	.......S..	....	...L.I....	..KS.E
IFN-α4b	E......EQ.	....	....	....	...V......	....	....	....	.L........	......
IFN-αC	E......EQ.	....	....	....	....	....	.I.R......	....	.L........	......
IFN-αL (Ψα10)	E......EQ.	........I.	....	....	....	....	.I.R......	....	.L........	......
IFN-αJ1 (α7)	E......EQ.	....	....	....	.....F....	....	.M........	....	........K.	G.....
IFN-αJ2	E......EQ.	....	....	....	.....F....	....	.M........	....	....	......
IFN-αI	E......EQ.	....	....N.....	.....M....	....	....	....	....	.L........	I.....
IFN-αF	.....T.EQ.	.........N	.....M....	....	...V......	K.........	....	....	...L.KIF.E	.....E
IFN-αWA	.........T	..D..YI..F	....	T.......IA	....	....	.MG.......	....	....	G.....
IFN-αGx-1	.....T...T	..D..Y....	.........M	M......D..	...V....T.	....	....	....	...L.A...E	.....E
IFN-α76	E......EQ.	....	....	....	....	....	....	....	.L........	......

be the last three amino acids of the signal peptide sequence.[46] Because of these original uncertainties, both forms were expressed in bacteria and have been used in basic and clinical research. It is important, therefore, for researchers to designate which form was used in their studies. Reports have appeared without this information. The methods for cloning and expression of these interferons are described in Section VII of this volume. A summary of the molecular weights and amino acid compositions of the human interferons is presented in the table.

After DNA recombinants coding for the human interferons were obtained, these were used as probes to obtain the corresponding animal

[46] E. Rinderknecht, B. H. O'Connor, and H. Rodriguez, *J. Biol. Chem.* **259**, 6790 (1984).

FIG. 1. The amino acid sequences of human leukocyte interferon (IFN-α) species derived from cDNA or genomic DNA sequences. Sequences, including the signal peptide (S1–S23), are presented in comparison with a consensus sequence. Corresponding residues are presented only when they differ from the consensus sequence. Residues which are common to all listed sequences are underlined in the consensus sequence. Sequences A–L are from the laboratories of Pestka, Goeddel, and colleagues,[4–6,8–15] whereas sequences with numeric designations are from the laboratory of Weissmann and colleagues.[7,16,17] Closely related sequences are listed together (see Langer and Pestka[6]). As suggested,[7] sequences are listed as two classes (Class I: αK, α5, αA, and αD; Class II: α4b, αC, αL, αJ1, αJ2, αI, and αF) and an intermediate class (αH1, αH2, αB2, and αB). Within each class, they are ordered according to increasing difference from the consensus sequence. The α5 sequence is complete, whereas the G sequence begins at residue 34, since only a partial length clone was isolated; identity of G and α5 is proposed on the basis of the extant sequences. The cloned L ($\Psi\alpha$10) sequence has a termination codon within the signal sequence (indicated by an asterisk) and may therefore represent a pseudogene; however, since the coding region for the mature protein is otherwise normal, it is possible that *in vivo* suppression might lead to the production of a normal protein with this sequence. Indeed, it has been shown IFN-αL exhibits antiviral activity when expressed in *E. coli* without the signal peptide[18] or with a modified signal peptide.[38] The clone IFN-αWA is from Torczynski *et al.*[20,21] The interferons IFN-αGx-1[22] and IFN-α76[23] appear to represent recombinants between, respectively, IFN-αH1(1–16) and IFN-αG(17–166) on the one hand and between IFN-α4b(1–50)/IFN-αC(51–131)/IFN-α4b(132–166) on the other. The signal peptide of IFN-αGx-1 may be as shown in this figure rather than as reported: the codon at position S1 was reported as AAT (Asn or N) rather than ATG (Met or M) as in all other IFN-α genes. Interferons reported by Meyer[24] appear to be similar to those listed in the figure although there are some minor differences of one amino acid in some cases. The following interferons are of identical amino acid sequence: αK = α6; αJ1 = α7; αB2 = α8; αL = $\Psi\alpha$10. The following interferons are identical except for the residues shown in parentheses: IFN-αA (Lys 23), IFN-α2 (Arg 23); IFN-αD (Val 114), IFN-α1 (Ala 114); IFN-αJ1 (Lys 159, Gly 161), IFN-αJ2 (Gln 159, Arg 161), IFN-α7 (Lys 159, Gly 161); IFN-αH1 (Phe 152), IFN-αH2 (Leu 152). A gap was introduced at position #44 in IFN-αA to provide for maximum alignment with the other species. Gaps are indicated by dashes. In several positions, alternative amino acids could be chosen for the consensus sequence: #S11, V or L; #78, D or E; #79, E or Q; #80, S or T; #83, D or E; #86, S or Y; #154, F or L; #166, D or E.

```
                                     S10                    S20  S21
IFN-β consensus    M T x R C L L Q x A   L L L C F S T T A L   S
Hu-IFN-β           . . N K . . . . I .   . . . . . . . . . .   .
Mo-IFN-β           . . N K . . . . I .   . . . . . . . . . .   .
Bo-IFN-β1          . . Y . . . . . M V   . . . . . . . . . .   .
Bo-IFN-β2          . . H . . . . . M V   . . . . . . . . . .   .
Bo-IFN-β3          . . Y . . . . P M V   . . . . . . . . . .   .
Mu-IFN-β           . N N . W I . H A .   F . . . . . . . . .   .

                                      10                    20                    30
IFN-β consensus    x S Y x L L x F Q Q   R x S x x x C Q K L   L x Q L x x x x x x
Hu-IFN-β           M . . N . . G . L .   . S . N F Q . . . .   . W . . N G R L E Y
Mo-IFN-β           M . . N . . G . L .   . S . S F Q . . . .   . W . . N G S L E Y
Bo-IFN-β1          R . . S . . R . . .   . Q . L K E . . . .   . G . . P S T S Q H
Bo-IFN-β2          R . . S . . R . . .   . R . L A L . . . .   . R . . P S T P Q H
Bo-IFN-β3          R . . S . . R . . .   . R . A E V . . . .   . G . . H S T P Q H
Mu-IFN-β           I N . K Q . Q L . E   . T N I R K . . E .   . E . . N G K I - -

                                      40                    50                    60
IFN-β consensus    C L x x R M D F x x   P E E M K Q x Q Q F   Q K E D A A L x I Y
Hu-IFN-β           . . K D . . N . D I   . . . I . . L . . .   . . . . . . . T . .
Mo-IFN-β           . . K D . . N . D I   . . . I . . P . . .   . . . . . . . T . .
Bo-IFN-β1          . . E A . . . . Q M   . . . . . . E . . .   . . . . . I . V M .
Bo-IFN-β2          . . E A . . . . Q M   . . . . . . A . . .   . . . . . I . V . .
Bo-IFN-β3          . . E A K . . . Q V   . . . . N . A . . .   R . . . . I . V . .
Mu-IFN-β           N . T Y . A . . K I   . . . . T E - K M -   . . S Y T . F A . Q

                                      70                    80                    90
IFN-β consensus    E M L Q N I F x I F   R x D F S S T G W N   E T I V E x L L x E
Hu-IFN-β           . . . . . . . A . .   . Q . S . . . . . .   . . . . . N . . A N
Mo-IFN-β           . . . . . . . A . .   . Q . L . . . . . .   . . . . . K . . A N
Bo-IFN-β1          . V . . H . . G . L   T R . . . . . . . S   . . . I . D . . K .
Bo-IFN-β2          . . . . Q . . N . L   T R . . . . . . . S   . . . I . D . . E .
Bo-IFN-β3          . . . . Q . . N . L   T R . . . . . . . S   . . . I . D . . V .
Mu-IFN-β           . . . . . V . L V .   . N N . . . . . . .   . . . . V R . . D .

                                     100                   110                        120
IFN-β consensus    L Y x Q x N x L K T   V L E E K x E K E N   x T x G x x M S S - - L
Hu-IFN-β           V . H . I . H . . .   . . . . . L . . . D   F . R . K L . . . - - .
Mo-IFN-β           V . H . I D H . . .   I . . . . L . . . D   F . R . K F . . . - - .
Bo-IFN-β1          . . W . M . R . Q P   I Q K . I M Q . Q .   S . T E D T I V - - - P
Bo-IFN-β2          . . E . M . H . E P   I Q K . I M Q . Q .   S . M . D T T V - - - .
Bo-IFN-β3          . . G . M . R . Q P   I Q K . I M Q E Q .   F . M . D T T V - - - .
Mu-IFN-β           . H Q . T V F . . .   . . . . . Q . - . R   L . W E - - . . . T A .

                                     130                   140                   150
IFN-β consensus    H L K x Y Y x R x x   x Y L K x K E Y x x   C A W T V V R V E I
Hu-IFN-β           . . . R . . G . I L   H . . . A . . . S H   . . . . I . . . . .
Mo-IFN-β           . . . R . . G . I L   H . . . A . . . S H   . . . . I . . . . .
Bo-IFN-β1          . . G K . . F N L M   Q . . E S . . . D R   . . . . . . Q . Q .
Bo-IFN-β2          . . R K . . F N L V   Q . . . S . . . N R   . . . . . . . . Q .
Bo-IFN-β3          . . . K . . F N L V   Q . . E S . . . N R   . . . . . . . . Q .
Mu-IFN-β           . . . S . . W . V Q   R . . . L M K . N S   Y . . M . . . A . .

                                     160        166
IFN-β consensus    L R N F x F I x R L   T G Y L R N
Hu-IFN-β           . . . . Y . . N . .   . . . . . .
Mo-IFN-β           . . . . F . . N K .   . . . . . .
Bo-IFN-β1          . T . V S . L M . .   . . . V . D
Bo-IFN-β2          . . . . S . L T . .   . . . . . E
Bo-IFN-β3          . T . . S . L M . .   . A S . . D
Mu-IFN-β           F . . . L I . R . .   . R N F Q .
```

interferons in a variety of species. Nucleic acid hybridization of genomic DNA from various animal species with DNA probes corresponding to Hu-IFN-α and Hu-IFN-β showed that there was extensive cross-hybridization across the mammalian species and was also observed in avian species as well.[47] No cross-hybridization was observed with DNA from reptiles or insects, however. With this in mind, many groups used the Hu-IFN-α, Hu-IFN-β, and Hu-IFN-γ probes to isolate corresponding DNA sequences from a number of animal species. Accordingly, recombinants for murine,[48–54] rat,[55–58] bovine,[59–61] and monkey[62–64] interferons α, β, and

[47] V. Wilson, A. J. Jeffreys, P. A. Barrie, P. G. Boseley, P. M. Slocombe, A. Easton, and D. C. Burke, *J. Mol. Biol.* **166,** 457 (1983).

[48] G. D. Shaw, W. Boll, H. Taira, N. Mantei, P. Lengyel, and C. Weissmann, *Nucleic Acids Res.* **11,** 555 (1983).

[49] B. Daugherty, D. Martin-Zanca, B. Kelder, K. Collier, T. C. Seamans, K. Hotta, and S. Pestka, *J. Interferon Res.* **4,** 635 (1984).

[50] B. L. Daugherty and S. Pestka, this volume [61].

[51] E. C. Zwarthoff, A. T. A. Mooren, and J. Trapman, *Nucleic Acids Res.* **13,** 805 (1985).

[52] K. A. Kelley and P. M. Pitha, *Nucleic Acids Res.* **13,** 805 (1985).

[53] Y. Higashi, Y. Sokawa, Y. Watanabe, Y. Kawade, S. Ohno, C. Takaoka, and T. Taniguchi, *J. Biol. Chem.* **258,** 9522 (1983).

[54] P. W. Gray and D. V. Goeddel, *Proc. Natl. Acad. Sci. U.S.A.* **80,** 5842 (1983).

[55] R. Dijkema, P. Pouwels, A. de Reus, and H. Schellekens, *Nucleic Acids Res.* **12,** 1227 (1984).

[56] P. H. van der Meide, R. Dijkema, M. Caspers, K. Vijverberg, and H. Schellekens, this volume [62].

[57] R. Dijkema, P. H. van der Meide, M. Dubbeld, M. Caspers, J. Wubben, and H. Schellekens, this volume [63].

[58] R. Dijkema, P. H. van der Meide, P. H. Pouwels, M. Caspers, M. Dubbeld, and H. Schellekens, *EMBO J.* **4,** 761 (1985).

[59] B. Velan, S. Cohen, H. Grosfeld, and A. Schafferman, this volume [64].

[60] D. W. Leung, D. J. Capon, and D. V. Goeddel, *Biotechnology* **2,** 458 (1984).

[61] R. Derynck, *in* "Interferon" (I. Gresser, ed.), Vol. 5, p. 183. Academic Press, New York, 1984.

[62] P. Gold and S. Pestka, *Fed. Proc., Fed. Am. Soc. Exp. Biol.* **44,** 1613 (1985).

[63] P. Gold and S. Pestka, in preparation.

[64] P. Gold, B. L. Daugherty, and S. Pestka, in preparation.

FIG. 2. Comparison of amino acid sequences of IFN-β from several sources. The amino acid sequences predicted from DNA sequences for IFN-β from human,[10,29–32] bovine,[60] murine,[53] and monkey[62–64] genes are presented as they differ from a consensus sequence. Numbering is according to the human sequence. Amino acids of the consensus sequence which are common to all the sequences are underlined. Positions where no clear consensus exists are indicated in the consensus sequence by "x." To accommodate the murine sequence, a gap of 2 residues (indicated by hyphens) was introduced in the consensus (and other) sequences between residues 119 and 120; other gaps have been introduced into the murine sequence to facilitate alignment.

	S1 S10	S20	S23
Hu-IFN-γ	M K Y T S Y I L A F	Q L C I V L G S L G	C Y C
Bo-IFN-γ	. . . . . . F . . L	L . . G L . . F S .	S . G
Mu-IFN-γ	. N A . H C . . . L	. . F L M A V . - .	. . .
Ra-IFN-γ	. S A . R R V . V L	. . . L M A L . - .	. . .

	1 10	20	30
Hu-IFN-γ	Q D P Y V K E A E N	L K K Y F N A G H S	D V A D N G T L F L
Hu-IFN-γ (protein)	e	. . . .	. . . .
Bo-IFN-γ	. G Q F F R . I . .	. . E . . . . S S P	. . . K G . P . . S
Mu-IFN-γ	H G T V I E S L . S	. N N . . . S S G I	. . - E E K S . . .
Ra-IFN-γ	. G T L I E S L . S	. . N . . . S S S M	. A M E G K S . L .

	40	50	60
Hu-IFN-γ	G I L K N W K E E S	D R K I M Q S Q I V	S F Y F K L F K N F
Hu-IFN-γ (protein)	. . . .	. . . .	. . . .
Bo-IFN-γ	D . . . . . . D . .	. K . . I	. . . . E . L
Mu-IFN-γ	D . W R . . Q K D G	. M . . L I	. . . L R . . E V L
Ra-IFN-γ	D . W R . . Q K D G	N T . . L E . . . I	. . . L R . . E V L

	70	80	90
Hu-IFN-γ	K D D Q S I Q K S V	E T I K E D M N V K	F F N S N K K K R D
Hu-IFN-γ (protein)	. . . .	. . . .	. . . .
Bo-IFN-γ	. . N . V . . R . M	D I . . Q . . F Q .	. L . G S S E . L E
Mu-IFN-γ	. . N . A . S N N I	S V . E S H L I T T	. . S N S . A . K .
Ra-IFN-γ	. . N . A . S N N I	S V . E S H L I T N	. . S N S . A . K .

	100	110	120
Hu-IFN-γ	D F E K L T N Y S V	T D L N V Q R K A I	H E L I Q V M A E L
Hu-IFN-γ (protein)	. . . .	. . . .	. . . .
Bo-IFN-γ	. . K . . I Q I P .	D . . Q I	N . . . K . . N D .
Mu-IFN-γ	A . M S I A K F E .	N N P Q . . . Q . F	N . . . R . V H Q .
Ra-IFN-α1	A . M S I A K F E .	N N P Q I . H . . V	N . . . R . I H Q .

	130	140	143
Hu-IFN-γ	S P A A K T G K R K	R S Q M L F R G R R	A S Q
Hu-IFN-γ (protein)	. . . . . . † † † †	† † † †	
Bo-IFN-γ	. . K S N L R . . .	. . . N	. . M
Mu-IFN-γ	L . E S S L R . . .	. . R C*	
Ra-IFN-γ	. . E S S L R . . .	. . R C*	

FIG. 3. Comparison of amino acid sequences of IFN-γ from several sources. The amino acid sequences predicted from the DNA sequences of human,[35,36] murine,[54] rat,[57,58] and bovine[61] IFN-γ have been aligned. The structure derived by direct amino acid sequencing of natural IFN-γ is denoted as "Hu-IFN-γ (protein)" (Rinderknecht *et al.*[46]). The first residue of this protein is pyroglutamate, indicated as "e." Various residues which were found as the COOH terminus of the isolated protein are indicated by "†." Only differences from the human sequence are indicated for the bovine, murine, and rat molecules. Gaps (indicated by hyphens) in the mouse sequence were introduced to allow alignment. The end of the murine and rat sequences are indicated by an asterisk.

AMINO ACID COMPOSITION OF HUMAN RECOMBINANT INTERFERONS[a]

IFN	Asn	Asp	Thr	Ser	Gln	Glu	Pro	Gly	Ala	Cys	Val	Met	Ile	Leu	Tyr	Phe	His	Lys	Arg	Trp	Total	MW
IFN-αA	4	8	10	14	12	14	5	5	8	4	7	5	8	21	5	10	3	11	9	2	165	19,241
IFN-α2	4	8	10	14	12	14	5	5	8	4	7	5	8	21	5	10	3	10	10	2	165	19,269
IFN-αB	4	12	6	15	13	15	4	2	9	4	7	4	10	22	5	10	3	10	10	1	166	19,494
IFN-α8 (αB2)	4	11	6	16	13	15	4	2	9	4	7	5	10	21	5	10	3	10	10	1	166	19,484
IFN-αC	6	8	7	14	14	15	5	5	9	4	7	4	10	20	4	9	3	7	13	2	166	19,406
IFN-αD	6	11	9	13	10	15	6	3	9	5	7	6	7	22	4	8	3	8	12	2	166	19,414
IFN-α1	6	11	9	13	10	15	6	3	10	5	6	6	7	22	4	8	3	8	12	2	166	19,386
IFN-αF	6	7	8	14	14	15	5	5	9	4	8	5	9	18	3	11	3	10	10	2	166	19,330
IFN-α5	6	10	11	13	13	13	5	4	8	4	7	8	7	18	5	10	3	8	11	2	166	19,524
IFN-αH	9	7	8	13	13	15	4	2	9	4	7	9	7	17	4	12	3	10	11	2	166	19,741
IFN-αH2	9	7	8	13	13	15	4	2	9	4	7	9	7	18	4	11	3	10	11	2	166	19,707
IFN-αI	7	7	9	14	14	15	6	5	9	4	6	5	9	21	4	9	2	8	10	2	166	19,259
IFN-αJ	5	7	8	13	12	17	5	4	9	4	7	5	8	19	4	12	4	9	12	2	166	19,606
IFN-αJ2	5	7	8	13	13	17	5	3	9	4	7	5	8	19	4	12	4	8	13	2	166	19,705
IFN-αK (α6)	5	9	8	13	12	14	4	4	8	4	9	6	6	20	5	9	4	8	15	3	166	19,745
IFN-αL (Ψα10)	6	8	8	13	14	15	5	4	9	4	7	4	11	19	4	9	3	7	14	2	166	19,519
IFN-α4B	5	8	9	14	13	15	4	5	9	4	8	4	8	20	4	10	5	8	11	2	166	19,379
IFN-αWA	6	10	7	12	13	11	3	7	12	4	7	5	9	18	5	13	3	9	10	2	166	19,282
IFN-β	12	5	7	9	11	13	1	6	6	3	5	4	11	24	10	9	5	11	11	3	166	20,027
IFN-γ	10	10	5	11	9	9	2	5	8	0	8	4	7	10	4	10	2	20	8	1	143	16,775

[a] The molecular weight of IFN-γ may be 16,757 if the NH_2-terminal residue is 2-pyrrolidone-5-carboxylic acid (L-pyroglutamic acid; 5-oxo-2-pyrrolidinecarboxylic acid) instead of glutamine.[32]

γ have been isolated and identified. As can be seen in Section VII, some of these have been expressed in bacteria and other heterologous hosts. Figure 4 presents a summary of the murine, rat, monkey, and human IFN-α sequences, Fig. 2 a comparison of IFN-β from several sources, and Fig. 3 a comparison of the IFN-γ sequences from several species.

Additional purification and understanding of the enzymes involved in interferon action has developed. Yet there is no clear insight as to what enzymatic mechanisms define the antiviral, antiproliferative, natural killer cell responses, and other biological activities. It is, indeed, very likely that the enzymatic mechanisms involved vary with the cell type as well as with the virus. Despite these uncertainties, isolation of the enzymes (and in some cases their corresponding genes) involved in interferon action are slowly providing a picture of the biochemical architecture involved.

Known effects of interferons on cells have increased immensely with the availability of pure interferons of various types for research. Improved assays for measuring interferons through their antiviral or antiproliferative activity and with monoclonal antibodies have been developed (Section IX). Effects on cellular differentiation, the expression of cell surface antigens, and cytotoxicity of cells have been reported in many different systems and species (Section X). With the need to develop extensive preclinical data prior to therapeutic use in humans, antitumor effects of the human interferons on xenografts in athymic mice have provided useful and valuable data. It is important to note, however, that the antitumor efficacy of an interferon on nude mouse xenografts frequently does not correlate with the antiproliferative effects of that same interferon on the identical cultured cells.

Initial clinical trials with pure interferons have required an understanding of the effects of interferons on drug metabolism. Various assays to evaluate these effects of interferon are given in detail in Section XI.

The finding that human leukocyte and fibroblast interferons are active on tobacco plants and protoplasts has, indeed, been intriguing.[65–67] Such results suggest that the antiviral substances produced by plant cells may indeed be analogous to the interferon system of animal cells. For these reasons, a section devoted to the methodology relating to these effects on plant cells was included in this volume (Section XI).

New developments in interferon research are progressing rapidly, and as our horizons expand, so does our awareness of the unknown. The

[65] A Gera, S. Spiegel, and G. Loebenstein, this volume [96].

[66] I. Sela, this volume [97].

[67] I. Sela, this volume [98].

```
                     S1       S10           S20   S24   1        10         20         30         40         50         60         70
Mu-IFN-α consensus   MARLCAFLMV  LAVMSY-WST  CSLG  CDLPQTHNLR  NKRALTLLVQ  MRRLSPLSCL  KDRKDFGFPQ  EKVDAQQIQK  AQAIPVLSEL  TQQILNIFTS
Mu-IFN-αA            .........T  .L....-...  ....  ..........  ..........  ..........  ......R...  .........N  .......Q..  ...V......
Mu-IFN-α1            ..........  ......-.P.  ....  ..........  ..........  ..........  ..........  ........K.  ..........  ..........
Mu-IFN-α2            ........VM  .I....-..I  ....  ....H.Y...  .....KV.A.  ....PF....  ...Q.....L  ....N.....  ........RD.  ...T..L...
Mu-IFN-α4            .........I  .VM...Y..A  ....  ....H.Y..G  ......V.EE  ....P.....  .........L  ....N.....  ....L..RD.  ......L...
Mu-IFN-α5            ..........  .P.L..-.P.  ....  ..........  .........K  ..........  ..........  ...G.....E  ..........  ...V......
Mu-IFN-α6            ..........  ......-...  ....  .......K..  ........I.  ..........  ..........  ....TLK...  EK.......V  ..........
Mu-IFN-α6k           ..........  ...L..-.P.  ....  ..........  ...L.....K  ..........  ..........  ...G.....E  ........T..  .....TL...
Ra-IFN-α1            .........S  .V.V..-..A  .C..  ....H.....  ...VF...A.  .......V...  ....Y....L  ....G.....  ........H..  .....SL...
Mo-IFN-αD            ..LPF.L..A  .V.L.C-K.G  ....  ....E..S.D  .R.TMM..K.  .S.I..S...  M..H......  QEF.GN.F..  .P..S..H..  I..TF.L..T
Hu-IFN-α consensus   MALSFSLLMA  VLVLSY-KSI  CSLG  CDLPQTHSLG  NRRALILLAQ  MGRISPFSCL  KDRHDFGFPQ  EEFDGNQFQK  AQAISVLHEM  IQQTFNLFST

                            80         90        100        110        120        130        140        150        160      167 169
Mu-IFN-α consensus   KDSSAAWNAT  LLDSFCNDLH  QQLNDLQACL  MQEVGVQEPP  LTQEDSLLAV  RKYFHRITVY  LREKKHSPCA  WEVVRAEVWR  ALSSSANLLA  RLSEEKE
Mu-IFN-αA            .......D.S  ..........  ......K..V  ..........  .....Y....  .T........  ..........  ..........  .M....K...  ......E
Mu-IFN-α1            ..........  ..........  .......G..  ..Q.....F.  .....A....  ..........  ..........  ..........  .......V.G  ..R...
Mu-IFN-α2            .A........  ..........  .......T..  ..Q.......  .....A....  ..........  ..........  ..........  .....V...P  ......E
Mu-IFN-α4            ..L..T....  ..........  ......K..V  ..-----...  ..........  .T........  ..K....L..  ...I......  .....T....  .......
Mu-IFN-α5            ..........  ........EV.  ......K..V  ........S.  ..........  ..........  .......L..  ..........  .....V....  ...K.E.
Mu-IFN-α6            .......D..  ...T.....Y  ..........  V...RL....  ....V.....  ..........  ..........  ..........  .......V.G  ..R...
Mu-IFN-α6k           ..........  .........H  .L.....G..  ..Q.EI.AL.  ..........  .T.......F  ..........  ..........  ......K...  ..N.DE
Ra-IFN-α1            .E..T..D..  .........Q  ...SG.....  ..Q.....S.  ..........  .E........  ...N......  ....K.....  ........MG  ..R..RNES
Mo-IFN-αD            .......DED  ...K..TE.Y  ......E..V  ..QER.G.T.  .MNA..T...  K...R...L.  .T...Y....  ........IM.  SF.L.T..QE  ..RRKE
Hu-IFN-α consenus    KDSSAAWDES  LLEKFSTELY  QQLNDLEACV  IQEVGVEETP  LMNEDSILAV  RKYFQRITLY  LTEKKYSPCA  WEVVRAEIMR  SFSFSTNLQK  RLRRKD
```

FIG. 4. Amino acid sequences of murine, rat, and monkey IFN-α. Amino acid sequences of Mu-IFN-αA (Daugherty *et al.*[49,50]), Mu-IFN-α1, and Mu-IFN-α2 (Shaw *et al.*[48]), and Mu-IFN-α4, Mu-IFN-α5, Mu-IFN-α6 (Zwarthoff *et al.*[51] and Kelley and Pitha[52]), Ra-IFN-α1 (Dijkema *et al.*[55] and van der Meide *et al.*[56]), and monkey Mo-IFN-αD (Gold and Pestka[62–64]) derived from cloned DNA sequences are compared. Only differences from a murine consensus sequence are listed. Amino acids which are found in all Mu-IFN-α sequences *and* in all IFN-α sequences are underlined in the murine consensus sequence. The human consensus sequence is presented for comparison, with underlined residues indicating amino acids which are common to all human DNA-derived sequences (see Fig. 1). Dashes indicate no amino acid present in position. The mature proteins vary in size from 162 amino acids (Mu-IFN-α4), 166 amino acids for Mu-IFN-α1, Mu-IFN-α6, and Mu-IFN-α6k, and 167 amino acids for Mu-IFN-αA, Mu-IFN-α2, and Mu-IFN-α5. The Ra-IFN-α1 consists of 169 amino acids. The Mu-IFN-α4 and Mu-IFN-α6 sequences are from Zwarthoff *et al.*[51] The Mu-IFN-α5 and Mu-IFN-α6k sequences are from Kelley and Pitha.[52] Although the Mu-IFN-α4 from Zwarthoff *et al.*[51] and Kelley and Pitha[52] are identical, the Mu-IFN-α6 sequence from these reports differ significantly so that the latter is shown as Mu-IFN-α6k. The monkey interferon sequence, Mo-IFN-αD, is from the data of Gold and Pestka.[63] Gaps in the sequences (dashes) were introduced to provide for maximum alignment.

interval from the previous two volumes[1,2] to this one has expanded our knowledge of the interferons and their actions immensely. Despite that it has been a tantalizing and humbling experience to realize we still do not know the physiological roles of these molecules with reasonable certainty. Nevertheless, there are many avenues to pursue. Those that hold the greatest surprises are most likely to be the most enlightening.

Acknowledgment

I thank Jerry Langer for reviewing this chapter and the sequences in detail, Paul Gold for the monkey interferon sequences, and Sophie Cuber for typing the manuscript.

[2] Interferon Standards and General Abbreviations

By SIDNEY PESTKA

Standardization of interferons and interferon assays have been described in the first volume of Interferons.[1–13] Virtually all assays for standardization and titration of interferons are based on antiviral activity.[1,14–16] Because some radioimmunoassays and enzyme immunoassays

[1] N. B. Finter, this series, Vol. 78, p. 14.
[2] M. P. Langford, D. A. Weigent, G. J. Stanton, and S. Baron, this series, Vol. 78, p. 339.
[3] D. A. Weigent, G. J. Stanton, M. P. Langford, R. E. Lloyd, and S. Baron, this series, Vol. 78, p. 346.
[4] G. J. Stanton, M. P. Langford, and F. Dianzani, this series, Vol. 78, p. 351.
[5] P. Jameson and S. E. Grossberg, this series, Vol. 78, p. 357.
[6] J. J. Sedmak and S. E. Grossberg, this series, Vol. 78, p. 369.
[7] N. Hahon, this series, Vol. 78, p. 373.
[8] J. A. Armstrong, this series, Vol. 78, p. 381.
[9] P. C. Familletti, S. Rubinstein, and S. Pestka, this series, Vol. 78, p. 387.
[10] M. D. Johnston, N. B. Finter, and P. A. Young, this series, Vol. 78, p. 394.
[11] D. J. Giron, this series, Vol. 78, p. 399.
[12] J. Suzuki, M. Iisuka, and S. Kobayashi, this series, Vol. 78, p. 403.
[13] F. Dianzani and S. Baron, this series, Vol. 78, p. 409.
[14] Standardization of Interferons. Report of a World Health Organization Informal Consultation, *W.H.O. Tech. Rep. Ser.* **687,** 35–60 (1983).
[15] *Bull. W.H.O.* **62,** 696 (1984).
[16] "NIAID Catalog of Research Reagents." Research Resources Branch, National Institute of Allergy and Infectious Diseases, National Institutes of Health, Bethesda, Maryland, 963–966, 1978–1980.

have correlated closely to antiviral activity,[17–25] they are useful for monitoring interferon during purification and pharmacokinetic studies, where the number of samples is large. Indeed, many of the monoclonal antibodies generated against Hu-IFN-α seem to react preferentially with the biologically active interferon molecule.[20] Some of these antibodies can even be used to discriminate between interferon monomers and oligomers in a sandwich radioimmunoassay.[26,27]

It must be emphasized that, whenever possible, interferon assays must be reported with respect to a recognized international standard reference preparation. Since the quantity of such standard preparations is limited, a laboratory standard that has been well characterized and is as close as possible in purity and form to the reference preparation should be used daily for adjustment of all titers. This laboratory preparation should be stored in small portions (such as 10–100 μl) in a liquid nitrogen freezer and not refrozen after thawing.

Interferon Reference Standards

A summary of the international reference standards is given in Table I. This table is an updated version of the data in Volume 78 of this series.[1] Although these standard preparations are relatively stable at 4° for several years, it is generally recommended that the original vials be stored at −20° or below. Once thawed and reconstituted, the reference preparations should be stored in small volumes in a liquid nitrogen freezer.

Reference Antisera to Interferons

A number of antisera have been prepared to various interferons. Several reference antisera and their corresponding controls are available as

[17] T. Staehelin, C. Stähli, D. S. Hobbs, and S. Pestka, this series, Vol. 79, p. 589.

[18] H. Gallati, *J. Clin. Chem. Clin. Biochem.* **20,** 907 (1982).

[19] S. Pestka, H.-F. Kung, B. Durrer, J. Schmidt, and T. Staehelin, *in* "Interferon" (J. Vilcek and E. De Maeyer, eds.), Vol. 2, p. 249. Elsevier, Amsterdam, 1984.

[20] S. Pestka, B. Kelder, J. A. Langer, and T. Staehelin, *Arch. Biochem. Biophys.* **224,** 111 (1983).

[21] D. S. Secher, *Nature (London)* **290,** 501 (1981).

[22] M. Inoue and Y. H. Tan, *Infect. Immun.* **33,** 763 (1981).

[23] A. R. Neurath, N. Strick, N. B. K. Raj, and P. M. Pitha, *J. Interferon Res.* **2,** 51 (1982).

[24] P. Daubas and K. E. Mogensen, *J. Immunol. Methods* **48,** 1 (1982).

[25] B. Kelder, A. Rashidbaigi, and S. Pestka, this volume [77].

[26] S. Pestka, B. Kelder, D. K. Tarnowski, and S. J. Tarnowski, *Anal. Biochem.* **132,** 328 (1983).

[27] S. Pestka, B. Kelder, and S. J. Tarnowski, this volume [78].

TABLE I

INTERNATIONAL REFERENCE PREPARATIONS OF INTERFERONS[a]

Interferon	Current designation	Source	Preparation No.	Defined activity (units/container)
Human leukocyte (Le)	Human alpha [Hu-IFN-α(leukocyte/Sendai)]	NIBSC	69/19 (MRC Research Standard B)	5,000
Human leukocyte (Le)	Human alpha [Hu-IFN-α(leukocyte/Sendai)]	NIH	G-023-901-527	20,000
Human leukocyte (Le)	Human alpha [Hu-IFN-α(leukocyte/Sendai)]	NIH	Ga23-902-530	12,000
Human lymphoblastoid (Namalva) (Ly)	Human alpha [Hu-IFN-α(Namalwa/Sendai)]	NIH	Ga23-901-532	25,000
Human leukocyte αA	Human recombinant alpha A [rHu-IFN-αA]	NIH	Gxa01-901-535	9,000
Human leukocyte α2	Human recombinant alpha 2 [rHu-IFN-α2]	NIBSC	82/576	In progress
Human leukocyte αD	Human recombinant alpha 1 [rHu-IFN-αD]	NIBSC	83/514	In progress
Human fibroblast	Human beta [Hu-IFN-β]	NIH	G-023-902-527	10,000
Human fibroblast	Human beta [Hu-IFN-β]	NIH	Gb23-902-531	In progress
Human fibroblast	Human beta [Hu-IFN-β]	NIBSC	79/520	In progress
Human recombinant fibroblast	Human recombinant beta [rHu-IFN-β]	NIH	Gxb01-901-535	In progress
Human recombinant fibroblast Ser	Human recombinant beta-Ser ([Ser^{17}]rHu-IFN-β)	NIH	Gxb02-901-535	In progress
Human immune	Human gamma [Hu-IFN-γ]	NIH	Gg23-901-530	4,000
Human immune	Human gamma [Hu-IFN-γ]	NIBSC	82/587	In progress
Human recombinant immune	Human recombinant gamma [rHu-IFN-γ]	NIH	Gxg01-901-535	In progress
Mouse	Murine alpha/beta [Mu-IFN-α/β (L cell/NDV)]	NIH	G-002-904-511	12,000
Murine alpha	Murine alpha [Mu-IFN-α]	NIH	Ga02-901-511	In progress
Murine beta	Murine beta [Mu-IFN-β]	NIH	Gb02-902-511	In progress
Murine alpha/beta	Murine alpha/beta [Mu-IFN-α/β]	NIH	Gu02-901-511	In progress
Murine immune	Murine gamma [Mu-IFN-γ]	NIH	Gg02-901-533	In progress
Rabbit		NIH	G-019-902-528	10,000
Chick		NIBSC	67/18	80

[a] NIBSC, The International Laboratory for Biological Standards, National Institute for Biological Standards and Control, Holly Hill, Hampstead, London NW3 6RB. NIH, Research Resources Branch, National Institute of Allergy and Infectious Diseases, National Institutes of Health, Bethesda, Maryland 20205. The NIH human leukocyte interferon reference preparations G-023-901-527 (unitage 20,000) and Ga23-902-530 (unitage 12,000) were calibrated against the MRC 69/19 leukocyte standard. These two reference interferon reagents represent useful resources and have been designated international working reference preparations, but are not formally considered international reference standards by the World Health Organization (W.H.O.).[14–16] All the other completed preparations listed in the table have been designated international reference standards by the W.H.O. Qualified investigators can obtain an ampoule of an international reference preparation by writing the director of the appropriate organization as shown in the table.[28] The murine G-002-904-511 L cell interferon standard is no longer available.

TABLE II

ANTISERA TO INTERFERONS AVAILABLE AS REFERENCE REAGENTS[a]

Antiserum or control	Interferon preparation used as antigen	Catalog No.	Neutralizing titer[b]
Sheep anti-human fibroblast interferon	From human diploid cells induced with poly(I) · poly(C) (1×10^4 units/mg)	G028-501-568	12,000 against 8–10 units IFN
Control sheep anti-human fibroblast interferon		G029-501-568	<50 against 8–10 units IFN[1]
Sheep anti-human leukocyte interferon	From buffy coat with Sendai virus (1×10^4 units/mg)	G026-502-568	750,000 against 8–10 units IFN[2]
Control sheep anti-human leukocyte interferon		G027-501-568	<50 against 8–10 units human IFN
Bovine anti-human lymphoblastoid (Namalwa) interferon	From Namalwa cells induced with Sendai virus	G030-501-553	40,000 units Namalwa IFN Neutralized by contents of 1 vial[3]
Control bovine anti-human lymphoblastoid (Namalwa) interferon		G031-501-553	<25 against Namalwa IFN[4]
Rabbit anti-human gamma interferon	Hu-IFN-γ prepared from peripheral blood leukocytes with staphylococcal enterotoxin A	G034-501-565	1300 units/ml against 10 units of Hu-IFN-γ
Control rabbit anti-human gamma interferon		G035-501-565	None
Sheep anti-mouse L cell interferon	Murine L_{929} cells induced with NDV (1×10^4 units/mg)	G024-501-568	300,000 against 8–10 units IFN[5]
Control sheep anti-mouse L cell interferon		G025-501-568	<50 against 8–10 units mouse IFN
Rabbit anti-murine gamma interferon	Murine spleen cells stimulated with staphylococcal enterotoxin A	G032-501-565	800 units/ml against 10 units of Mu-IFN-γ
Control rabbit anti-murine gamma interferon		G033-501-565	Negative

[a] These titers are those stated in the NIH reference reagent notes describing each preparation.[28] The anti-human lymphoblastoid interferon and control sera were absorbed with Namalwa cell protein and Sendai virus. The data of this table were modified from Refs. 14 and 28.

[b] Heterologous antibody activity: [1] May be toxic at $\leq 1:100$ dilutions. [2] 2000 titer against 8–10 units human fibroblast IFN. [3] The contents of one vial neutralizes 40,000 units of leukocyte IFN and 10,000 units of fibroblast IFN.[28] [4] <25 titer against leukocyte IFN; <25 titer against fibroblast IFN. [5] Low levels of antibody to human leukocyte IFN were observed.

reagents from the National Institutes of Health (Table II).[28] These preparations are extremely useful in characterizing interferon preparations and in standardizing neutralization assays for the interferons.

The neutralization titer of an antiserum is "defined as the reciprocal of the antiserum dilution that reduces the potency from 10 interferon units/ml to 1 interferon unit/ml."[14] However, for practical purposes most laboratories use a microtiter assay where the interferon is kept at a constant level and the antiserum is diluted serially with the endpoint taken as that antiserum dilution where the interferon exhibits a level of 50% protection as in most cytopathic effect inhibition assays.[9] Often, but not necessarily, the 50% protection endpoint is approximately at 1 unit/ml so the assays in these instances are nearly comparable. To determine neutralization titers, a series of interferon–antiserum mixtures are prepared in microtiter plates with a constant quantity of interferon (usually 5 or 10 units/well containing 0.2 ml) and a series of antibody dilutions. These are incubated together for 1 hr at 37° and then the contents of the wells assayed for antiviral activity. This serves as a useful operational definition for neutralization units. For theoretical and experimental reasons, however, there are many difficulties in defining standard neutralizing units as discussed in detail by Kawade[29] in this volume.

Relative Activities of Human IFN-αA

Although the interferons generally show species specificity, they also exhibit activity in cells of heterologous species to one extent or another. A summary of the antiviral activity of Hu-IFN-αA on various cells is shown in Table III. At high concentrations, activity on murine and rat cells is significant: for antiviral effects equivalent to those observed on human cells a concentration 14,000-fold and 540,000-fold that required for human cells is necessary for murine and rat cells, respectively.[30]

Units Versus Molar Concentration. With the availability of purified interferon preparations, it is essential to use molar concentrations (or mass) wherever possible as well as units. Indeed, it is often deceiving to use antiviral units when comparing activities across species or in nonanti-

[28] "NIAID Interferon Reagents," A Program of the National Institute of Allergy and Infectious Diseases, 461–320:4903, p. 29. U.S. Department of Health and Human Services, Public Health Service, National Institutes of Health, U.S. Govt. Printing Office, Washington, D.C., 1985.

[29] Y. Kawade, this volume [74].

[30] E. Rehberg, B. Kelder, E. G. Hoal, and S. Pestka, *J. Biol. Chem.* **257,** 11497 (1982).

TABLE III
RELATIVE ANTIVIRAL ACTIVITY OF IFN-αA ON VARIOUS CELL LINES[a]

Cell	Species	Relative antiviral activity
WISH	Human	100
VERO	Monkey	9
MDBK	Bovine	150
Felung	Feline	130
Transformed guinea pig cells	Guinea pig	30

[a] Assays on murine, rat, rabbit, hamster, and horse cells yielded relative antiviral titers less than 1% that observed with WISH cells. Information from Research Reference Reagents Note No. 31 for Freeze-dried Reference Human Recombinant Alpha 2 (Alpha A) Interferon Catalog Number Gxa01-901-535 from the Research Resources Section, National Institute of Allergy and Infectious Diseases, National Institutes of Health, Bethesda, MD 20205 (U.S.A.), November 1983.

viral assays.[30-33] With the use of units alone, it is not possible to distinguish which interferons are most potent on the basis of mass. Thus, it is recommended that in the future molar concentration (or mass) be used when the materials employed permit it. When the pure interferon is not available, however, results can only be expressed in terms of antiviral units.

Designation of Interferons

The recommended abbreviations for the interferons will be followed in this volume where possible.[34] A summary of these and other abbreviations used in the literature are given in Table IV. It must be emphasized

[31] S. Pestka, *Arch. Biochem. Biophys.* **221,** 1 (1983).

[32] S. Pestka, B. Kelder, E. Rehberg, J. R. Ortaldo, R. B. Herberman, E. S. Kempner, J. A. Moschera, and S. J. Tarnowski, *in* "The Biology of the Interferon System" (E. De Maeyer and H. Schellekens, eds.), p. 535. Elsevier, Amsterdam, 1983.

[33] S. Pestka, B. Kelder, E. Rehberg, J. R. Ortaldo, R. B. Herberman, E. S. Kempner, J. A. Moschera, and S. J. Tarnowski, *in* "Gene Expression: The Translational Step and Its Control" (B. F. C. Clark and H. U. Petersen, eds.), p. 459. Munksgaard, Copenhagen, 1984.

[34] Interferon Nomenclature, *Cell. Immunol.* **80,** 432 (1983).

TABLE IV
ABBREVIATIONS USED AND DESIGNATIONS OF HUMAN INTERFERONS

Interferon class	Recommended abbreviations	Individual molecular components	Other past abbreviations
Leukocyte	IFN-α	IFN-αA, IFN-αB, IFN-αC, ... IFN-α1, IFN-α2, IFN-α3, ...	IFL, LeIF, LIF
Fibroblast	IFN-β	IFN-β	IFF, FIF
Immune	IFN-γ	IFN-γ	IFI, ImIF

that IFN-α should not be used as the abbreviation for crude leukocyte interferon preparations in which up to 10% (or more) of its activity may be associated with IFN-β. Thus, the specific designations should be used for the pure species uncontaminated by other interferons. To designate the animal species of origin, a prefix should be employed such as Hu-, Bo-, Mu-, and Ra- for human, bovine, murine, and rat species, respectively. Since multiple IFN-β interferons have been described in bovine cells, these individual species have been designated IFN-β1, IFN-β2, etc.

Because the genes for many of the interferons have been isolated and expressed in bacteria, it has been possible to generate mutated and hybrid interferon species. These should be designated by the standard nomenclature recommended by the "IUPAC-IUB Joint Commission on Biochemical Nomenclature" on *Nomenclature and Symbolism for Amino Acids and Peptides*.[35] Single letter and three letter abbreviations for the amino acids may be used. Table V summarizes the abbreviations for the amino acids. Substitutions, should provide the amino acid and number from the NH_2-terminal end. Thus, a substitution of serine for phenylalanine at position #116 of IFN-αJ would be abbreviated [Ser^{116}]IFN-αJ. Hybrid interferons should be designated from the NH_2-terminal end as IFN-$\alpha A_{1-91}/D_{93-166}$ to indicate the NH_2-terminal end consists of residues 1–91 of IFN-αA and the COOH-terminal end, residues 93–166 of IFN-αD with residue 91 of IFN-αA covalently linked to residue 93 of IFN-αD. Extensions at the NH_2 and COOH terminus would be designated by [Cys-Tyr-Cys]IFN-γ and IFN-γ[Cys-Tyr-Cys], respectively. Insertion of a residue would be indicated by the prefix "endo" as endo-Tyr^{4a}-IFN-αA designates an insertion of Tyr after amino acid residue #4. Removal of residues is indicated by the prefix "des" so that des-Pro^4-IFN-αA designates IFN-αA with Pro^4 deleted.

[35] IUPAC-IUB Joint Commission on Biochemical Nomenclature (JCBN), *Eur. J. Biochem.* **138,** 9 (1984).

TABLE V
ABBREVIATIONS FOR THE AMINO ACIDS[a]

Amino acid	Three-letter symbol	One-letter symbol
Alanine	Ala	A
Aspartic acid or asparagine	Asx	B
Cysteine	Cys	C
Aspartic acid	Asp	D
Glutamic acid	Glu	E
Phenylalanine	Phe	F
Glycine	Gly	G
Histidine	His	H
Isoleucine	Ile	I
Lysine	Lys	K
Leucine	Leu	L
Methionine	Met	M
Asparagine	Asn	N
Proline	Pro	P
Glutamine	Gln	Q
Arginine	Arg	R
Serine	Ser	S
Threonine	Thr	T
Valine	Val	V
Tryptophan	Trp	W
Unknown or "other" amino acid	Xaa	X
Tyrosine	Tyr	Y
Glutamic acid or glutamine (or substance such as 4-carboxyglutamic acid and 5-oxoproline that yield glutamic acid on acid hydrolysis of peptides)	Glx	Z

[a] The abbreviations and table are taken from Ref. 20. They are presented here for a ready reference for figures and articles throughout this volume.

No differences should be used for interferons produced by recombinant DNA technology or the equivalent interferon obtained from induction of human cells if they are identical molecules. The source, however, should be designated.

Glycosylation or other substitution of side chain residues of an interferon should be noted by adding the substitution to the name as $CHO^{25,97}$-IFN-γ should indicate glycosylation of Asn^{25} and Asn^{97} of IFN-γ (see this volume, Fig. 3, Chapter [1]).

TABLE VI
GENERAL ABBREVIATION LIST

2,5-A_n: (2′-5′)-oligoadenylate; (2′-5′)-oligo(adenylic) acid; oligoadenylic acid with 2′,5′-phosphodiester linkages; also abbreviated as (2′-5′)-oligo(A), pppA(2′p5′A)$_n$, pppA(2′-5′)A_n, or (2′-5′)p_3A_3; occasionally in the text and figures 2-5A has been used as well as 2,5-p_3A_n.

A_{260} unit: the quantity of material that yields an absorbance of 1.0 when measured at 260 nm in a cuvette with a path length of 1.0 cm. Analogously, A_{600} represents measurements at 600 nm, A_{550} at 550 nm; etc. This is also sometimes designated as OD_{260}, OD_{550}, OD_{600}, etc.

Act D: actinomycin D

BES: *N,N*-bis[2-hydroxyl]-2-aminoethanesulfonic acid; 2-[bis(2-hydroxyethyl)amino]-ethanesulfonic acid

BP: base pairs; usually used as lower case bp

Bq: becquerel; 1 Bq = 1 dps or 60 dpm; 1 Ci = 3.7×10^{10} Bq; TBq = 10^{12} Bq

BSA: bovine serum albumin

Con A: concanavalin A

CPD: citrate phosphate dextrose solution

CPE: cytopathic effect

DEAE cellulose: diethylaminoethylcellulose

DHFR: dihydrofolate reductase

DNA: deoxyribonucleic acid
  cDNA: complementary DNA
  dsDNA: double-stranded DNA

DRB: 5,6-dichloro-1-β-D-ribofuranosylbenzimidazole

dsRNA: double-stranded RNA

DTT: dithiothreitol

EDTA: ethylenediaminetetraacetic acid

EGTA: ethylene glycol bis(β-aminoethyl ether)-*N,N′*-tetraacetic acid

EID_{50}: egg infectious dose; concentration at which half the eggs are infected

EMCV: encephalomyocarditis virus

EMEM: Eagle's minimal essential medium

ESS: Earle's salt solution

FCS: fetal calf serum

GPT: guanosine phosphoribosyltransferase (recommended name is guanosine phosphorylase; systematic name is guanosine : orthophosphate ribosyltransferase)

HBSS: Hanks' balanced salt solution

HEPES: *N*-2-hydroxyethylpiperazine-*N′*-2-ethanesulfonic acid

IU: units of interferon with respect to the appropriate international reference standard

KBP: kilobase pairs; usually used as lower case kbp

MDMP: 2-(4-methyl-2,6-dinitroanilino)-*N*-methylpropionamide

MEM: minimal essential medium (EMEM, unless otherwise noted)

$Mg(OAc)_2$: magnesium acetate

MOI: multiplicity of infection

NAD: nicotinamide adenine dinucleotide

NaOAc: sodium acetate

Natural interferon: this refers to interferon produced by animal cells after induction in contrast to interferon produced by recombinant DNA technology. Indeed, in some cases the two (i.e., the interferon produced by animal cells and the same species by recombinant DNA technology) may be chemically identical

TABLE VI (*continued*)

NDV: Newcastle disease virus
PBL: peripheral blood leukocytes
PBMC: peripheral blood mononuclear cells
PBS: phosphate-buffered saline
PFC: plaque-forming cell
PFU: plaque-forming unit
PHA: phytohemagglutinin
Poly(I) · poly(C): polyinosinic acid · polycytidylic acid, double-stranded synthetic homopolymers [alternatively poly(rI) · poly(rC)]
RNA: ribonucleic acid
  dsRNA: double-stranded RNA
SDS: sodium dodecyl sulfate
SDS–PAGE: SDS–polyacrylamide gel electrophoresis
SEA: staphylococcal enterotoxin A
SEB: staphylococcal enterotoxin B
SSC: standard saline citrate solution, 0.15 *M* NaCl/0.015 *M* sodium citrate, adjusted to pH 7.0 with NaOH
TCA: trichloroacetic acid
$TCID_{50}$: tissue culture infectious dose; concentration at which half the cultures are infected
TEMED: *N,N,N′,N′*-tetramethylethylenediamine
TES: *N*-tris(hydroxymethyl)methyl-2-aminoethanesulfonic acid
Tricine: *N*-tris(hydroxymethyl)methylglycine
TPA: the phorbol ester 12-*O*-tetradecanoylphorbol-13-acetate
Tris: tris(hydroxymethyl)aminomethane
VSV: vesicular stomatitis virus

Other Abbreviations

A large number of standard media have been used in the methods described in this volume. For reference, the most commonly used media and formulations were summarized in Chapter [3] of Volume 78.[36] In addition, many nonstandard abbreviations were used throughout this volume. Because the chapters of this volume span several different fields, the reader should find the list of these nonstandard abbreviations given in Table VI convenient as a ready source of reference.

Acknowledgment

I am grateful to Dr. Maureen Myers of the National Institute of Allergy and Infectious Diseases, Dr. Anthony Meager of the National Institute for Biological Standards and Control, and Dr. Jan Vilcek for reviewing this chapter and for their useful suggestions.

[36] S. Pestka, this series, Vol. 78, p. 21.

Section II

Induction of Interferons

A. Human Interferons
Articles 3 through 16

B. Non-Human Interferons
Articles 17 through 22

[3] Effect of Purification Procedures on the Composition of Human Leukocyte Interferon Preparations

By HANNA-LEENA KAUPPINEN, SINIKKA HIRVONEN, and KARI CANTELL

The preparations of human leukocyte IFN are mixtures of many IFN-α subtypes.[1-3] More than 20 genes for different human IFN-α subtypes have been identified so far[4] and it is not impossible that most of them are activated after the induction of human leukocyte suspensions with Sendai virus.[5-7] The major IFN mRNA species in such cell suspensions are α1(D), α2(A) and α4.[7] This is consistent with the identification of IFN-α2(A) and IFN-α1(D) as major species of natural interferons produced by leukocytes after induction with Newcastle disease virus.[3,8,9] The IFN-α subtypes show marked differences in their antiviral, antiproliferative and NK-cell stimulatory activities.[3,8,10,11] The antiviral activity of α1(D) is much lower in human cells than that of the α2(A). The different subtypes may have interactions. It has been reported that the mixture of α1 and α2 has a much higher antiviral activity in human cells than either IFN alone.[12] Thus it would be interesting to know the subtype composition of the IFN preparations used for clinical trials. This chapter gives some data about the effect of purification procedures on the composition of human leukocyte IFN preparations and compares IFN preparations from different sources.

[1] S. Nagata, N. Mantei, and C. Weissman, *Nature (London)* **287,** 401 (1980).

[2] D. V. Goeddel, D. W. Leung, T. J. Dull, M. Gross, R. M. Lawn, R. McCandliss, P. H. Seeburg, A. Ullrich, E. Yelverton, and P. W. Gray, *Nature (London)* **290,** 20 (1981).

[3] S. Pestka, *Arch. Biochem. Biophys.* **221,** 1 (1983).

[4] M. Revel, *in* "Interferon" (I. Gresser, ed.), Vol. 5, p. 205. Academic Press, New York, 1984.

[5] K. Cantell, S. Hirvonen, H.-L. Kauppinen, and G. Myllylä, this series, Vol. 78, p. 29.

[6] K. Cantell, S. Hirvonen, and V. Koistinen, this series, Vol. 78, p. 499.

[7] J. Hiscott, K. Cantell, and C. Weissman, *Nucleic Acids Res.* **12,** 3727 (1984).

[8] M. Rubinstein, W. P. Levy, J. A. Moschera, C.-Y. Lai, R. D. Hershberg, R. T. Bartlett, and S. Pestka, *Arch. Biochem. Biophys.* **210,** 307 (1981).

[9] W. P. Levy, M. Rubinstein, J. Shively, U. Del Valle, C.-Y. Lai, J. Moschera, L. Brink, L. Gerber, S. Stein, and S. Pestka, *Proc. Natl. Acad. Sci. U.S.A.* **78,** 6186 (1981).

[10] P. K. Weck, S. Apperson, L. May, and N. Stebbing, *J. Gen. Virol.* **57,** 233 (1981).

[11] S. Pestka, B. Kelder, E. Rehberg, J. R. Ortaldo, R. B. Herberman, E. S. Kempner, J. A. Moshera, and S. J. Tarnowski, *in* "The Biology of the Interferon System" (E. DeMeyer and H. Schellekens, eds.), p. 535. Elsevier, Amsterdam, 1983.

[12] P. Orchansky, T. Goren, and M. Rubinstein, *in* "The Biology of the Interferon System" (E. DeMeyer and H. Schellekens, eds.), p. 183. Elsevier, Amsterdam, 1983.

Purification by Ethanol Fractionation

The leukocyte IFN preparations used in clinical trials have been mostly purified by fractionating precipitation in acidic ethanol.[6,13] The preparations are called by a common name P-IF. The original method has been more or less modified in various laboratories. The procedure includes the final concentration of partially purified IFN by means of KSCN at pH values ranging from 5.1 to 2.8. The proteins precipitating first when lowering the pH, i.e. at pH around 5 are poorly soluble. In our laboratories this fraction, P-IFB is dissolved separately from the other fraction P-IFA, which is precipitated at pH 2.8, to gain the maximum recovery of IFN activity. We routinely pool the fractions for clinical use, but in some other laboratories the P-IFB fraction is discarded or the final precipitation is done in one step.

Our earlier studies suggested that P-IFB and P-IFA differ in their composition of IFN-α subtypes.[6] To study this point further we performed the final KSCN precipitation in 5 successive steps at pH 5.1, 4.8, 4.5, 4.2, and 2.8 and assayed the concentrates for antiviral activity both in human HEp2 and bovine NBL-cells. The results (Table I) show that the activity of the fractions in bovine cells increases with the decrease of the pH. Some IFN-α subtypes such as α2(A) are characterized by a low relative antiviral activity in bovine cells and others such as α1(D) by a high bovine activity.[3,8,10] Thus it appears the P-IFB has a high content of α2(A) and P-IFA contains much α1(D). In any case, discarding of any fractions of P-IF is likely to affect the subtype composition of the final product.

Purification by Anti-interferon Immunoadsorbents

In addition to the partially purified P-IF, highly purified preparations of human leukocyte IFN have also been used in some clinical trials.[14,15] The further purification has been achieved by immunoadsorption chromatography on a Sepharose column containing mouse monoclonal antibody NK2[16,17] anti-human IFN-α. We have used the following procedure.

[13] H.-L. Kauppinen, *in* "Interferons" (N. B. Finter, ed.), Vol. 4, p. 73. Elsevier/North-Holland Biomedical Press, Amsterdam, 1985.

[14] K. Mattsson, A. Niiranen, M. Iivanainen, M. Fárkkilá, L. Bergstróm, L. R. Holsti, H.-L. Kauppinen, and K. Cantell, *Cancer Treat. Rep.* **67,** 958 (1983).

[15] G. M. Scott, D. Secher, D. Flowers, J. Bate, K. Cantell, and D. Tyrrell, *Br. Med. J.* **282,** 1345 (1981).

[16] D. S. Secher and D. C. Burke, *Nature (London)* **285,** 446 (1980).

[17] D. S. Secher, *Nature (London)* **290,** 501 (1981).

TABLE I

ANTIVIRAL ACTIVITY OF DIFFERENT FRACTIONS OF PARTIALLY PURIFIED HUMAN LEUKOCYTE IFN IN HUMAN AND BOVINE CELLS

Step	Volume (ml)	IFN assayed with: Human HEp2 (units × 10^6/ml)	IFN assayed with: Bovine NBL-1 (units × 10^6/ml)	Ratio of activity bovine/human	Spec. act. in HEp2 (units × 10^6/mg)	Yield (%)
Crude	70,000	0.049	0.264	5.4	0.027	100
KSCN precipitates[a]						
1. pH 5.10	71	6.5	21.7	3.3	1.0	13.1
2. pH 4.80	72	5.7	28.6	5.0	2.5	11.5
3. pH 4.50	72	6.5	31.6	4.9	3.0	13.1
4. pH 4.20	72	5.3	43.3	8.2	2.4	10.7
5. pH 2.80	72	1.8	23.6	13.1	0.5	3.6

[a] Seventy liters of crude IFN was purified in 3 batches by KSCN and ethanol fractionation as described earlier.[6] The precipitate containing IFN was dissolved in phosphate buffer, pH 8 containing 0.5 *M* KSCN. The solution was concentrated in 5 successive steps by means of KSCN precipitation at different pHs between 5.1 and 2.8.

Twelve milliliters of NK2 Sepharose 4B (Celltech, England) was packed in a column of 26 mm diameter (K 26/40 column with two adaptors from Pharmacia Fine Chemicals) with phosphate buffered saline (PBS), pH 7.3. The column was rinsed well with PBS. The pyrogenicity of the column was checked by testing the rinsing buffer by Limulus assay (Millipore). Two hundred milliliters of P-IF diluted in PBS to 5 million units/ml, in total 1×10^9 units, was loaded onto the column at a flow rate of 50 to 250 ml/hr. The flow through, NK2-effluent, was saved for IFN analysis and further purification. The column was rinsed with about 50 ml of PBS containing 0.5% human albumin. The bound interferon was eluted with 36 ml (three column volumes) of 0.1 *M* glycine·HCl buffer, pH 2, containing 0.2 *M* NaCl and 0.5% human albumin. The NK2-eluate was collected into a plastic cylinder containing 3.6 ml of 1 *M* phosphate buffer, pH 8, for immediate neutralization of the eluate. The pH of the eluate was about 7. The eluate was stored at −70°. The eluates from multiple runs were pooled for clinical use. The pH was adjusted to 7.3 and the salt concentration to isotonicity. Before vialing the preparation, NK2-IFN, was filtered for sterility (Durapore 0.22-μm filter, Millipore).

All the buffers used in the procedure were sterile and prepared in pyrogen-free water. The procedure was carried out at room temperature on a laminar flow bench.

The recovery of IFN after the NK2-immunoadsorption (Table II) was 67% in the NK2-eluate and 23% in the effluent when assayed by the plaque reduction method with human cells. The recovery of IFN in the eluate was about 10% lower when albumin was omitted from the buffer. This is apparently due to the instability of the highly purified NK2-IFN preparations at low protein concentrations. According to our experience the minimum protein concentration required to stabilize the NK2-IFN is 0.1%. The recovery of IFN estimated by radioimmunoassays with the ^{125}I-labeled NK2-antibody[17] was 72 to 78% in the NK2-eluate, but no IFN was detected in the NK2-effluent. The antiviral activity found in the effluent probably represents the IFN-α subtypes D and F not bound to NK2-Sepharose.[18,19]

Recently the rat monoclonal YOK anti-human IFN-α immunoadsorbent[20] has become available (Celltech, England). This antibody is reported to have a high affinity to IFN-α1(D).[20] The effluent from one NK2-

[18] G. Allen, K. H. Fantes, D. C. Burke, and J. Morser, *J. Gen. Virol.* **63,** 207 (1982).

[19] K. Alton, Y. Stabinsky, R. Richards, B. Ferguson, L. Goldstein, B. Altrock, L. Miller, and N. Stebbing, *in* "The Biology of the International System" (E. DeMeyer and H. Schellekens, eds.), p. 119. Elsevier, Amsterdam, 1983.

[20] R. E. Hawkins, J. S. Spragg, and D. S. Secher, *Int. T.N.O. Meet., 2nd, 1983* p. 9 (1983).

TABLE II
RECOVERY OF IFN IN SUCCESSIVE STEPS OF PURIFICATION BY MONOCLONAL NK2 AND YOK ANTI-IFN-α IMMUNOADSORBENTS

	IFN antiviral activity (%)[a]	IFN (%) assayed by FRC RIAs[b]		IFN (%) assayed by Celltech IRMAs[c]	
		NK2-RIA	YOK-RIA	NK2-IRMA	YOK-IRMA
P-IF	100	100	100	100	100
NK2-eluate	67	72	27	78	8
NK2-effluent	23 100	0	60 100	0	79 100
YoK-eluate	6 26	0	50 83	0	64 81
YoK-effluent	15 65	0	0 0	0	0 0

[a] Assayed by VSV plaque reduction in human HEp2 cells.

[b] The FRC (Finnish Red Cross) RIA is a sandwich radioimmunoassay. The ^{125}I-labeled monoclonal NK2 and YOK anti-IFN were obtained from Celltech and the polyclonal sheep anti-leukocyte IFN (Iivar antiserum) was prepared as described previously (K. E. Mogensen, L. Pyhälä, and K. Cantell, *Acta Path. Microbiol. Scand.* **83B,** 443 (1975). A crude leukocyte IFN calibrated to NIH reference 69/19 was used as standard. "Iivar" is the name of the sheep in which the anti-leukocyte serum was prepared.

[c] Radioimmunoassay kits from Celltech, England, contained the same monoclonal antibodies as in the FRC RIA and another polyclonal sheep anti-IFN-α from Celltech.

chromatography representing about 250 million units in total was loaded onto a YOK-Sepharose column of 5 ml and the bound IFN was eluted (YOK-eluate). The buffers as well as the other experimental conditions were the same as for NK2-purification. The IFN activities of the YOK-effluent and -eluate were determined as described above. The results (Table II) show that only 26% of the biological activity was recovered in the YOK-eluate (YOK-IFN) whereas 65% passed through. By immunoassays with ^{125}I-labeled YOK-antibody the recovery was 81-83%, but no IFN was found in the effluent. Thus the YOK-effluent reprepresenting 15% of the antiviral activity of the P-IF preparation must contain IFN subtypes such as F which are not recognized by either NK2- or YOK-immunoadsorbents. However, the IFN-α subtypes can have different relative activities against various viruses.[10] Thus the proportion of the antiviral activity in the eluates and the effluents could have been different if other cell lines or viruses had been used for the biological assay.

The immunochromatography purifications were also made without albumin in the buffers in order to analyze the purity of the IFN preparations recovered. The NK2-IFN had a specific activity of 2.0×10^8 units/mg protein and the YOK-IFN 2.5×10^7 units/mg when assayed with HEp2 cells and VSV. The SDS–polyacrylamide gel electrophoresis (Fig. 1) of

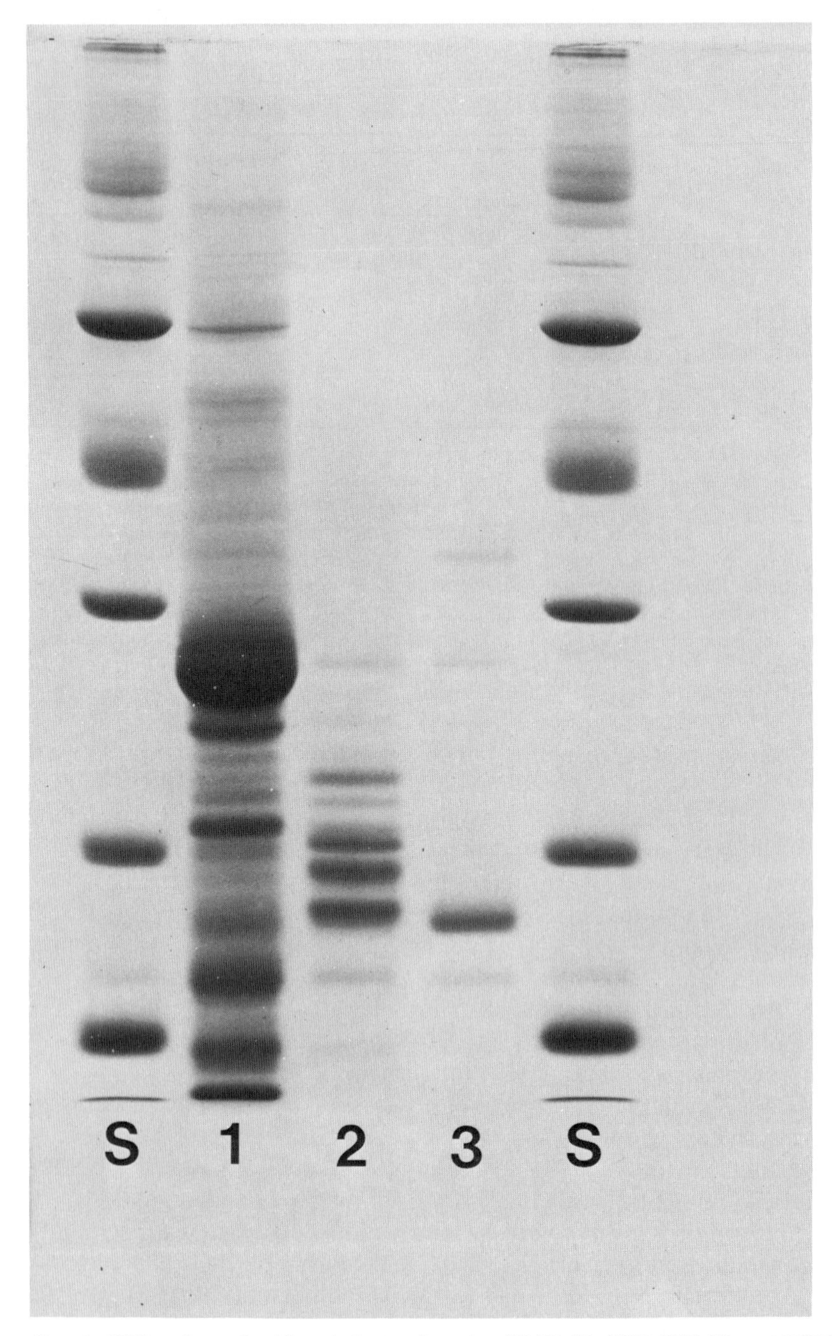

FIG. 1. SDS–polyacrylamide gel electrophoresis of P-IF (1), NK2-IFN (2), and YOK-IFN (3). The molecular weight standards (S) from the top are phosphorylase b (94,000), bovine serum albumin (67,000), ovalbumin (43,000), carbonic anhydrase (30,000), soybean trypsin inhibitor (20,100), and lactalbumin (14,000).

NK2-IFN shows 7 to 9 bands in the MW range of 16,400 to 25,000 and the electrophoresis of YOK-IFN shows one major band of MW 18,000 and another weak band of MW 16,400.

For comparison, 2500 ml of crude leukocyte IFN, a total of 175 million units, was purified on NK2- and YOK-columns and assayed as described. The protein bands in the SDS–PAGE were similar to those described above for P-IF.

Comparison of P-IF Preparations from Different Sources

Four P-IF preparations from different laboratories and different countries were compared in the assays described above. The results (Table III) show that the ratio of the YOK- and NK2-RIA titers of the different preparations can vary considerably. This is mainly due to differences in the YOK-titers. The result confirms the earlier finding by Hawkins *et al.*[20] that the content of IFN-α1(D) in P-IF preparations from different sources can vary markedly as revealed by YOK-RIA.

There are many possible reasons for the differences among P-IF preparations. The titer and the subtype composition of the crude IFN preparations may vary from one laboratory to another. This could be due to differences among the blood donors or to conditions of the IFN production. The recovery of interferon after purification varies considerably in different laboratories. In our hands the total recovery is 60 to 70%.[13]; the corresponding figures are generally lower in other laboratories. In the laboratory where the P-IF I (Table III) was prepared the recovery is 30 to 40%. The volumes of the purification batches vary from a few liters to several hundreds of liters in different laboratories and the duration of the routine purification process can also vary considerably. It has been reported that the subtype D has a somewhat lower pH 2 stability than the subtypes A, B, C, and F.[10] Thus a prolonged storage of leukocyte IFN in acid ethanol may cause a selective inactivation of the subtype α2(D).

Concluding Remarks

At present, no good methods are available for the quantitative analysis of the IFN-α subtypes in human leukocyte IFN preparations. Specific monoclonal antibodies against all the subtypes would be needed for such analysis by radioimmunoassay. Specific probes are available for the detection of different IFN-α mRNA molecules, but their amounts in the induced leukocytes do not reliably reflect the quantities of the corresponding interferon proteins.[7]

TABLE III

COMPARISON OF P-IF PREPARATIONS FROM FOUR LABORATORIES IN DIFFERENT ASSAYS[a]

P-IF preparation	Antiviral titer (units × 10^6 ml)	FRC RIA titers (units × 10^6/ml)		Ratio YOK-RIA/NK2-RIA	Celltech IRMA-titers (units × 10^6/ml)		Ratio YOK-IRMA/NK2-IRMA
		YOK-RIA	NK2-RIA		YOK-IRMA	NK2-IRMA	
I	2.9	3.3	3.4	0.97	0.6	4.6	0.13
II	2.5	4.2	3.6	1.17	1.6	5.3	0.30
III	2.3	2.9	2.2	1.32	1.4	3.3	0.42
Finnish Red Cross	6.7	9.3	6.6	1.41	3.4	9.3	0.37

[a] The assay methods are described in Table II.

Our results give only very rough and preliminary information about the composition of the preparations of human leukocyte interferon used for clinical trials. The main purpose of this chapter is to draw attention to the fact that such preparations do differ depending on how and where they are produced and purified. The clinical significance of such differences will have to be unraveled during the coming years.

[4] Large-Scale Production of Human Interferon from Lymphoblastoid Cells

By A. W. PHILLIPS, N. B. FINTER, C. J. BURMAN, and G. D. BALL

Earlier in this series, Mizrahi[1] and Bodo[2] gave accounts of the development of their production systems for interferon based upon the lymphoblastoid cell, Namalwa. In the larger of the two systems outlined, a culture vessel with a capacity of 100 liters was used for production of the cells. At about the same time we were making Namalwa cell interferon for clinical trials in a 1000 liter production vessel. The scale of this work has since been expanded in two stages and our largest vessels now routinely yield gram quantities of crude interferon alpha in a single cycle. Although it is not the aim of this communication to provide details of a commercial manufacturing process, an outline of the system used is given below.

Maintenance of the Cell Line

Following the establishment of an actively growing line of Namalwa cells, a quantity of culture together with a cryoprotective agent was dispensed into vials and frozen in liquid nitrogen. Samples were tested to demonstrate freedom from moulds, bacteria, mycoplasmas, and other infective agents. The frozen stock has subsequently been maintained as a Master Cell Bank from which fresh cells are withdrawn as needed for revival and establishment in culture. The cells have been grown in RPMI 1640 medium[3] supplemented with γ-irradiated bovine serum at concentra-

[1] A. Mizrahi, this series, Vol. 78, p. 54.

[2] G. Bodo, this series, Vol. 78, p. 69.

[3] G. E. Moore, R. E. Gerner, and H. A. Franklin, *JAMA, J. Am. Med. Assoc.* **199,** 519 (1968).

tions up to 7.5% (v/v). Many investigators have used either fetal serum or calf serum, while Mizrahi *et al.*[4] and Zoon *et al.*[5] used fractions of bovine serum from which most γ-globulin had been removed. We initially used only the serum of young animals but more recently have found that of adults to be equally satisfactory.

For production purposes the cells are grown in tanks and stirred with an impeller which is coupled magnetically to the drive motor, thus avoiding the need for a complicated gland and allowing the tank to be operated while completely sealed. Cultures are subjected to aeration in order to control the redox potential: temperature and pH are also controlled. Under our conditions, doubling times from 1 to 2 days are observed and densities of 3–4 million viable cells/ml are regularly obtained.

A major technical goal has been to establish and maintain on a long-term basis large "breeder" cultures from which cells can be withdrawn at intervals and used to produce batches of interferon. Maintenance of sterility of such cultures was at first difficult to achieve but the problem has now been solved by meticulous attention to detail in the design and operation of the plant. As a result, it is now normal for the sterility of the largest vessels to be maintained between successive cleaning operations carried out at intervals of 4–6 months to remove the proteinaceous deposits which form.

Induction

Although a number of authors[6–8] have reported the spontaneous formation of interferon by lymphoblastoid cells, the amounts produced in this way are very small and it is necessary in practice to induce the cells with virus. Newcastle disease virus has been used for this purpose but we have employed Sendai virus exclusively for our production. Although this virus is easy to prepare in embryonated hens' eggs, Johnston[9] has pointed out that in order to prepare an effective inducer, it is necessary to employ conditions of manufacture which lead to the production of a proper mixture of infectious virus and defective interfering particles.

[4] A. Mizrahi, S. Renveny, A. Traub, and M. Minai, *Biotechnol. Lett.* **2,** 267 (1980).

[5] K. C. Zoon, P. J. Bridgen, and M. E. Smith, *J. Gen. Virol.* **44,** 227 (1980).

[6] G. Henle and W. Henle, *J. Bacteriol.* **89,** 252 (1965).

[7] B. A. Zajac, W. Henle, and G. Henle, *Cancer Res.* **29,** 1467 (1969).

[8] A. T. Haase, J. S. Johnson, J. A. Kasal, S. Margolis, and H. B. Levy, *Proc. Soc. Exp. Biol. Med.* **133,** 1076 (1970).

[9] M. D. Johnston, *J. Gen. Virol.* **56,** 175 (1981).

Stimulators

Treatment of Namalwa cells with even an optimum mixture of the two types of virus particle was found not of itself to be adequate to guarantee the consistent production of large amounts of interferon. A solution to the problem of variable yields was provided by observations made by Johnston[10,11] and independently by Adolf and Swetley[12] that the addition of short-chain fatty acids to the culture prior to induction resulted in consistently high yields. We now use *n*-butyric acid at a concentration of about 1 m*M* and add the inducing virus 48 hr later.

Enhancers

In the course of their study on stimulators of production of interferon Adolf and Swetley[13] observed that dimethyl sulfoxide has a similar effect to that of butyrate. More recently, Johston[14] has shown that these two compounds can act additively. A number of other substances when added at about the time of induction, were found to extend the duration of production of interferon with a consequent increase in yield of up to 3-fold.[14]

Harvesting

At the end of the period of induction, purification begins with the removal of the cellular components of the culture to yield a crude solution of interferon. This step can be carried out either by the use of a sterilizable continuous centrifuge or by filtration. The former approach is convenient but the equipment is complex, difficult to clean effectively without dismantling, and expensive to repair and replace. Filtration systems can be time consuming to develop and involve a significant operating cost in terms of preformed filters. In one of our plants we have installed both systems for removing cells in order to evaluate their cost efficiency and reliability in the long term.

[10] M. D. Johnston (The Wellcome Foundation Limited), European Patent application 0,000,520 (1979).

[11] M. D. Johnston, *J. Gen. Virol.* **50,** 191 (1980).

[12] G. R. Adolf and P. Swetly, *Virology* **99,** 158 (1979).

[13] G. R. Adolf and P. Swetly, *Nature (London)* **282,** 736 (1979).

[14] M. D. Johnston (The Wellcome Foundation Limited), European Patent 0,097,353 (1984).

Yield of Crude Interferon

The concentration of interferon in the crude harvest lies in the range 7,000–50,000 units/ml. The quantity produced on any particular occasion depends in part upon the numbers of cells used and more significantly upon their condition. This is reflected in the degree of vigor of their growth and is influenced both by nutrition and by physical conditions.

Purification

The culture filtrate obtained as the product of the primary production process contains interferon in admixture with about 40,000 times its weight of contaminating protein, mostly of bovine origin. The procedure devised for the purification of the interferon uses general stages including precipitation, extraction into an organic solvent, and chromatography on columns: it yields a product in which the content of bovine proteins has been reduced at least 80,000-fold compared with the crude harvest. It has been demonstrated[15] that the process also eliminates each of a wide range of infectious agents and nucleic acids when deliberately added to crude interferon. The product obtained after purification contains a mixture of alpha interferons (at least eight[16] and probably at least twice as many subtypes) with a mean specific activity of at least 10^8 units/mg total protein. Namalwa cells when stimulated with Sendai virus produce significant quantities of beta interferon[17]: This is separated during the purification procedure to such a degree as to be barely detectable in the final product.[18]

Scale of Production

The Namalwa cell system has been used for production on scales which range from a few liters to many thousands of liters. In these laboratories the capacity of our production vessels has been increased from 1000 liters to 8000 liters and production in still larger vessels seems technically possible.

[15] N. B. Finter and K. H. Fantes, *in* "Interferon" (I. Gresser, ed.), Vol. 2, p. 65. Academic Press, New York, 1981.

[16] G. Allen and K. H. Fantes, *Nature (London)* **287,** 408 (1980).

[17] E. A. Havell, Y. K. Yip, and J. Vilček, *J. Gen. Virol.* **38,** 51 (1978).

[18] A. Billiau, personal communication.

[5] Large-Scale Production and Recovery of Human Leukocyte Interferon from Peripheral Blood Leukocytes

By BERNARD HOROWITZ

Advances in the assessment of the biological and clinical efficacy of human leukocyte interferons (IFN-α) have depended on the establishment of satisfactory protocols for their induction and purification from peripheral blood leukocytes or lymphoblastoid cell lines capable of continuous growth in culture. More recently IFN-α has been produced in prokaryotic organisms through recombinant DNA technology. Together, these approaches have assured an ample supply of IFN-α for characterization. It is now evident that IFN-α derived from leukocytes consists of multiple subtypes, each with subtle differences in structure and biologic activity whereas that derived following gene cloning consists of a single subtype with a restricted expression of biologic activity.[1]

The New York Blood Center has been engaged in the pilot plant scale production of IFN-α for therapeutic evaluation since 1979, relying on the production scheme outlined by Cantell and coworkers,[2-4] adapted in part to large-scale process equipment. More recently a two-step chromatographic process has been used for the purification of IFN-α. It is the purpose of this chapter to describe the processes used for IFN-α production in our facility and to describe some of the characteristics of the isolated interferon.

Materials and Solutions

Phosphate buffered saline (PBS): 0.01 *M* sodium dihydrogen phosphate containing 0.14 *M* NaCl and adjusted to pH 7.4 with NaOH.

Agamma plasma: agamma plasma was routinely prepared by adding 209 g ammonium sulfate to each liter of plasma at 4°. After 30–60 min, the precipitate was removed by centrifugation (in a Sharples centrifuge) and the supernatant dialyzed against 3 changes of 10 volumes of 0.9% sodium chloride at 4°.

[1] S. Pestka, *Arch. Biochem. Biophys.* **221,** 1 (1983).
[2] K. Cantell and S. Hirvonen, *Tex. Rep. Biol. Med.* **35,** 138 (1977).
[3] K. Cantell and S. Hirvonen, *J. Gen. Virol.* **39,** 541 (1978).
[4] K. Cantell, S. Hirvonen, H.-L. Kauppinen, and G. Myllylä, this series, Vol. 78, p. 29.

Sendai virus, Cantell strain: obtained from either Flow Laboratories (McLean, Va.) or A. S. Benzon (Denmark; U.S. distributor: TMC Inc., New York).

Culture medium: the culture medium contained, per liter, 9.53 g Eagle's MEM (Gibco 410-1100), 700 mg $NaHCO_3$, 3 g tricine, 25 mg neomycin, and 0.92 g agamma plasma. The pH was adjusted to 7.4 with 2 *N* NaOH.

Acidified ethanol: 94% ethanol containing 2.25 m*M* HCl.

Assays

Assay of interferon titer was by microscopic estimation of cytopathology following incubation of WISH cells with interferon and subsequent exposure to vesicular stomatitis virus.[5] The W.H.O. human leukocyte interferon reference preparation (Lot #260379) was used as standard. Protein concentration was determined by Coomassie Blue dye binding[6] or by Lowry[7] with human serum albumin as standard.

Collection of Leukocytes and Interferon Induction

Whole blood was collected from voluntary donors into CPD or CPDA-1 anticoagulant by standard phlebotomy techniques. Within 4 hr of collection, the units of whole blood were separated into components by centrifugation for 4.75 min at 1230 *g* at 22 ± 2° in Lourdes model 30-R centrifuges. The layer containing the buffy coat (35 ± 5 g) was expressed into plastic satellite bags and stored at 4° overnight. The leukocytes were removed from the plastic bags and pooled under laminar flow hoods by Masterflex peristaltic pumps (model 7015) driven continuously at 1 liter/min by Cole Palmer drives (model 7520-00). To penetrate the seal on the satellite bags, a Fenwal sampling site coupler (with rubber septum removed) was attached to the pump tubing. Pooled leukocytes were transferred into glass 2-liter graduate cylinders to which was added one-half volume of a sterile solution of 4.5% (w/v) hydroxyethyl starch (American Critical Care, McGaw Park, Ill.) in 0.9% sodium chloride. The cell suspension was mixed by inversion, covering the cylinder with nonsterile Parafilm. After 1.5 hr the upper leukocyte phase was harvested by pumping with the assembly described above through a 25-ml disposable glass pipet into three volumes of 0.83% (w/v) ammonium chloride in 0.017 *M*

[5] M. Ho and J. F. Enders, *Proc. Natl. Acad. Sci. U.S.A.* **45,** 385 (1959).

[6] BioRad Protein Determination Kit.

[7] O. H. Lowry, N. J. Rosebrough, A. L. Farr, and R. J. Randall, *J. Biol. Chem.* **193,** 265 (1951).

Tris·HCl, pH 7.2, at 4° cold room. After 10 to 15 min the cell suspension from a maximum of 150 leukocyte concentrates was pumped at 150 ml/min through an IEC model CH continuous flow centrifuge set at 4000-4500 rpm. This was immediately followed by one-third to one-half liter of cold PBS. The supernatant fluid remaining in the bowl following centrifugation was removed by aspiration and discarded. The leukocyte pellet was resuspended in 2 ml culture medium per leukocyte concentrate. Resuspension was achieved by aspiration through the tip of a 25-ml pipet with a Clay Adams pipetaid. The resuspended leukocytes were placed on ice immediately and stirred by magnetic stirrer for at least 15 min.

Induction and biosynthesis proceeded essentially as described by Cantell and colleagues.[2–4] Six liters of culture medium to which was added 125 units of IFN-α/ml was prewarmed at 37.5° in 12-liter round bottom flasks (Corning 4260). This temperature was maintained throughout the incubation period with heaters (Haake model E3) and polypropylene water baths [United Utensils Co., Inc., Port Washington, N.Y.; Series R, 15 × 48 × 12 in. (W × L × D), optional flange]. Purified leukocytes were then added to a final concentration of 10^7 leukocytes/ml and kept in suspension with egg-shaped magnetic stirring bars (Cole Palmer C4768-60) and Wheaton Biostir (model II) drives. The stir rate was the maximum consistent with avoidance of foaming. Following a 2 hr preincubation, Sendai virus was added to bring its concentration to 125 HA units/ml. The incubation proceeded overnight (usually 18–20 hr) after which the cells and cell debris were removed by filtration through AMF Cuno Zeta Plus cartridges (Cat. No. 45116-12-10CP) with 3.7 sq. ft for up to 100 liters of crude interferon (CIF) and a pump rate of 3 liters/min generated by 2 Masterflex 7018 pumpheads. Prior to filtration the filters were washed with 50 liters of 0.9% sodium chloride. The filtrate was either processed immediately or stored frozen at −30° in stainless steel milk cans. Recent batches have been concentrated approximately 10-fold with a Pellicon ultrafiltration cell (Millipore).

Purification

All process steps were carried out at 2 ± 2° unless otherwise indicated.

Alternative 1: Preparation of Partially Purified Interferon (P-IFA)

Step 1: CIF to EIF. Sufficient 5 *M* KSCN was added to bring its concentration to 0.5 *M* over a 5- to 15-min period with an overhead propeller mixer (Lightnin model NS-1-SCR or ND-1-SCR). The pH was lowered to 3.5 over a 30-min interval by pumping 2 *N* HCl (approximately 5.5 ml/liter) into a point of high turbulence beneath the propeller blade.

Here and elsewhere the pH electrode used was obtained from Ingold, model 465-TT. Calibration was with aqueous pH standards. The suspension was mixed for an additional 30 min and clarified by centrifugation with Sharples AS16 and/or Sharples AS26 continuous flow centrifuges. The pump rate was 750 ml/min per centrifuge. The supernatant was discarded. The pellet was homogenized in a cold room (+4°) with Waring blender model CB-2-10 and 1 liter of acidified ethanol (initial temperature $\leq -10°$) per 10 liters of CIF. Homogenization was accomplished with 5 pulses of 5–6 sec duration such that the temperature did not exceed 4°. The homogenate was further diluted with acidified ethanol such that the total volume of acidified ethanol used was 1/5 the volume of CIF. The suspension was adjusted if necessary to pH 3.5 ± 0.2 with 0.5 *N* HCl or NaOH and mixed at 4° overnight. The suspension was clarified by pumping through a Sharples centrifuge at 500 ml/min. This clarified supernatant is designated EIF. The precipitate was discarded.

Step 2: EIF to P-IF. Additional 5 *M* KSCN was added to EIF to bring the KSCN concentration to 0.1 *M*. The initial KSCN concentration was estimated by conductivity with a standard curve by adding known concentrations of KSCN to EIF. The pH of EIF was raised to approximately 5.2 over a 60-min period by pumping 0.5 *N* NaOH (approximately 6.7 ml/liter) into an area of high turbulence beneath the propeller blade as described above. Due to variability in pH electrode performance and the subtlety of this adjustment, the titration was terminated when the turbidity of the sample, first diluted with 2 volumes of cold 94% ethanol, reached 700 as measured with a turbidometer (HF Instruments, model DRT1000). The suspension was mixed for an additional 30 min and clarified by centrifugation at 0.5 liter/min with either Sharples centrifuge. The supernatant, labeled EIF-5A was brought to pH 5.5 with 0.1 *N* NaOH (approximately 5 ml/liter) over a 90-min period, mixed for an additional 30 min, and clarified as for EIF-5A. The supernatant was labeled EIF-5C. The precipitate from this and the preceding centrifugation was extracted into acidified ethanol to determine the losses of IFN-α before discarding these precipitates. EIF-5C was adjusted to pH 8.0 ± 0.2 with 0.1 *N* NaOH (approximately 5.9 ml/liter) over a 60-min period, mixed for an additional 30 min, and clarified as for EIF-5A. The supernatant was discarded. The pellet was extracted for 48 hr by stirring with a magnetic stirrer into 0.1 *M* sodium phosphate buffer, pH 8.0, containing 0.5 *M* KSCN and 0.25% neomycin, with one-tenth the volume of EIF. The suspension was clarified by centrifugation for 30 min at 2000 rpm in a bucket centrifuge (Beckman J6 or equivalent); the supernatant was labeled P-IF; the precipitate was discarded.

Step 3: P-IF to P-IF-A. P-IF was adjusted to pH 5.2 over a 30-min period with 2 *N* HCl (approximately 35 ml/liter) added beneath the solution surface, mixed for an additional 30 min, and clarified as for P-IF. The supernatant (P-IFA), containing virtually all of the IFN-α, was adjusted to pH 3.0 by addition of 0.5 *N* HCl (approximately 27 ml/liter), added below the surface of the liquid over a 30-min period. The suspension was incubated for an additional 30 min and clarified by centrifugation for 30 min at 2000 rpm in a bucket centrifuge (Beckman J6 or equivalent). The supernatant was discarded. The pellet was extracted for 48 hr by stirring with a magnetic stirrer into 0.1 *M* sodium phosphate buffer, pH 8.0, containing 0.25% neomycin, with a volume equal to 1/30 of the volume of P-IF for this extraction.The pH at the outset of this extraction was raised to 7.0, if necessary, with 0.1 *N* NaOH. The extract was clarified in a Sorval RC5 centrifuge with a GSA rotor at 8000 rpm for 30 min, and the supernatant was labeled P-IFA. P-IFA was dialyzed, first against PBS buffer containing 0.25% neomycin and then twice against PBS, each time with a 100-fold volume excess of dialysate. A small amount of hemoglobin, which tended to precipitate if not removed, remained as a contaminant. Much of this hemoglobin was eliminated by several freeze/thaw cycles. P-IFA was stored at $-80°$, and sterilized by filtration with a Pall NR filter, 47 mm diameter.

Alternative 2: Affinity-Purified IFN-α

Adsorption to and Elution from Controlled Pore Glass. Controlled pore glass (CPG-10-75, Electronucleonics, Fairfield, N.J.) was added to CIF in a ratio of 5 g (dry weight) per liter and mixed overnight by axial rotation of the vessel with a Wheaton tissue culture roller apparatus. After allowing the CPG to settle for 30 min, 80–90% of the supernatant fluid was removed by pumping, and the CPG was transferred to a chromatographic column, whose height to width ratio was approximately 10. The glass was washed with 2 column volumes of PBS buffer and 2 column volumes of PBS buffer containing an additional 1.5 *M* NaCl, both at a linear flow rate of 50 cm/hr. Bound IFN-α was eluted with 3 column volumes of 0.02 *M* sodium phosphate, pH 7.4, containing 1.5 *M* sodium chloride and 50% (v/v) ethylene glycol at a linear flow rate of 10 cm/hr. The CPG eluate was diluted 5-fold or dialyzed prior to further processing.

Immune Affinity Purification. The CPG eluate was applied to a column (height to width ratio ≥ 5) of NK2-Sepharose (Celltech, Ltd.; Slough, United Kingdom) equilibrated in PBS buffer at a linear flow rate of 20 cm/hr. The column was washed with 10 column volumes of PBS at 20 cm/hr and IFN-α eluted at 10 cm/hr with 0.1 *M* sodium citrate, pH 3.5, containing 0.3 *M* NaCl. The interferon was neutralized by addition of 1 and 0.1 *N* NaOH, and sufficient human serum albumin was added to achieve a final concentration of 3 mg/ml. The solution was sterilized by filtration with a Pall NR filter, 47 mm diameter.

Comments

IFN-α yields obtained for a period of 9 consecutive months were as follows. The processing of 51,688 leukocyte concentrates resulted in the synthesis of 629 billion units of crude IFN-α, or 12.0 million units per leukocyte concentrate. Interferon synthesis was quite variable from batch-to-batch. Titers ranged from a low of 25,100 to a high of 219,000, and a frequency histogram for all available daily titers is given in Fig. 1. The IFN-α titer was, on average, 83,100 $\pm$ 38,200, a value close to the observed median of 78,000. The cause of the observed variation is unknown.

Clarified crude IFN-α solutions were stored either at 4 or at $-30°$ prior to further processing, and studies designed to indicate the effect of duration of storage through 6 months indicated no effect. Nonetheless, we observed an apparent decline when the interferon titer immediately following induction was compared to the titer of CIF pooled at the initiation of purification (on average: 83,000 vs 57,000). Perhaps CIF contains an especially labile subspecies of interferon.

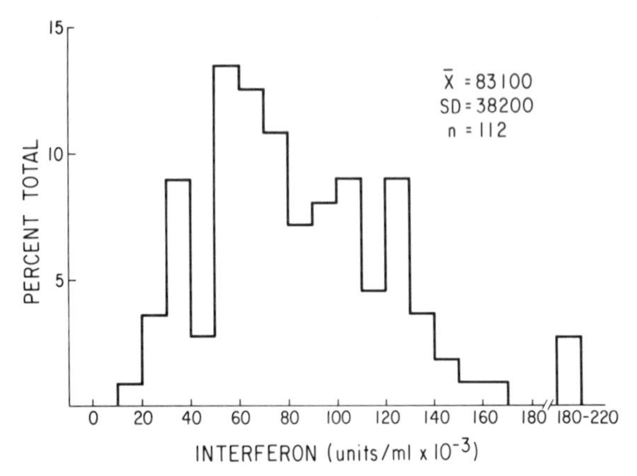

FIG. 1. Frequency histogram of daily titers of IFN-α. The data where available from each batch of CIF produced for human use during 9 months were tabulated with regard to the titer of IFN-α and presented as a frequency histogram.

TABLE I
PURIFICATION OF PIF-A (PILOT PLANT SCALE)[a]

Step	Volume (liters)	Protein g/liter	Protein Total (g)	Interferon units/ml	Interferon Total ($\times 10^{-9}$)	Yield (%)	Spec. act. (units/mg $\times 10^{-6}$)
CIF	306	1.09	334	54,500	16.7	100	0.050
EIF	60	4.00	240	260,000	15.6	93	0.065
EIF-5A	58	1.69	98.0	266,000	15.4	92	0.157
PPT					1.4	8.5	
EIF-5C	57	0.20	11.4	266,000	15.1	90	1.32
PPT					1.1	6.7	
P-IF	5.9	0.64	3.78	1.67×10^6	9.85	59	2.61
P-IF-5.2	6.1	0.88	5.37	1.42×10^6	8.66	52	1.61
P-IFA	0.28	11.0	3.08	21.5×10^6	6.02	36	1.95

[a] The results in the table represent the average of 16 consecutive batches.

The results from 16 consecutive purification batches of P-IFA are summarized in Table I. On average, 306 liters of CIF containing 16.7×10^9 units of IFN-α was pooled for processing as a single batch. Most of the interferon initially present could be recovered in the ethanol extract (EIF) following precipitation at pH 3.5 in the presence of KSCN, and no detectable IFN-α remained in the aqueous supernatant from this precipitation. The subsequent two precipitation steps, EIF-5A and EIF-5C, removed the bulk of the remaining protein. The specificity of these steps depended greatly on the presence of sufficient KSCN, without which the loss of IFN-α into these precipitates was quite large. Variability and difficulty with these fractionation steps reported elsewhere may also have resulted from insufficient KSCN. With the KSCN concentration adjusted to 0.1 *M*, 15% of the IFN-α initially present was detected in these precipitates (see below for additional detail). The principal loss of IFN-α occurred in processing after EIF-5C; however little or no IFN-α could be detected in the ethanolic supernatant from P-IF, the precipitate of P-IF-5.2, or the aqueous supernatant or residual pellet from P-IFA. Application of the process resulted in an overall yield of IFN-α of 36% or 2.9×10^6 units per leukocyte concentrate processed. The final specific activity of IFN-α averaged 1.95×10^6 units/mg, and the specific activity of all batches exceeded 1.0×10^6 units/mg.

The results from several, recent, consecutive purification batches of affinity-purified IFN-α are summarized in Table II. Relative to alterna-

TABLE II

AFFINITY PURIFICATION OF IFN-α[a]

Step	Volume (liter)	Protein		Interferon		Yield (%)	Spec. act. (units/mg $\times 10^{-6}$)
		mg/ml	Total (mg)	units/ml ($\times 10^{-6}$)	Total		
CIF	5.61	9.06	50,800	0.426	2,390	100	0.05
CPG							
Unadsorbed	5.14	8.81	45,300	0.007	34.9	1.5	0.001
Eluate	1.15	0.89	1,023	1.56	1,790	79	1.75
NK2-Sepharose							
Start	1.29	0.42	541	0.33	428	100	0.79
Unadsorbed	1.16	0.43	499	0.11	128	30	0.26
Eluate	0.0455	0.017	0.77	6.15	280	65	362

[a] The CPG step is the averge of the 6 latest batches. The NK2-Sepharose step is the average of the 4 latest batches; however, earlier batches of CPG purified IFN were used. The CIF preparation used was concentrated 10-fold.

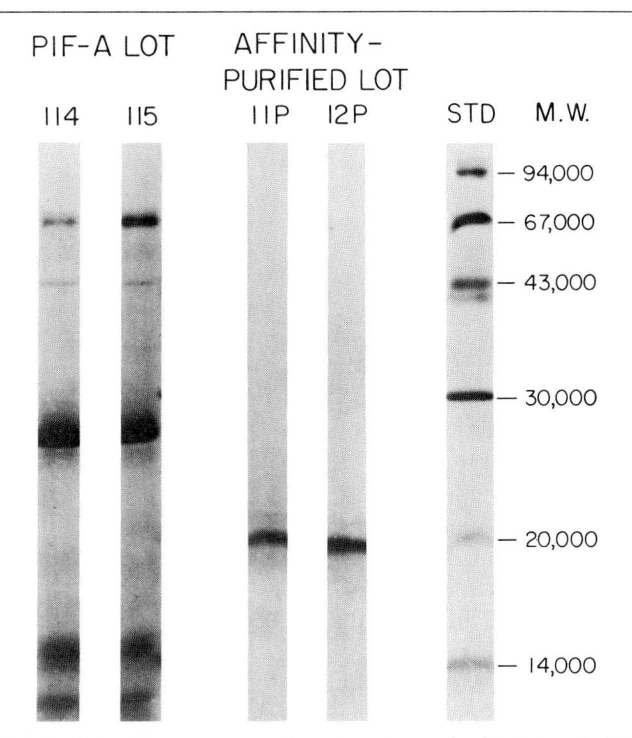

FIG. 2. SDS–PAGE and IFN-α preparations. Two lots each of P-IFA and affinity purified IFN-α were analyzed by SDS–PAGE and protein was stained with Coomassie Blue.

tive 1, the affinity process is simple, consistent, and results in essentially pure IFN-α. SDS–PAGE analysis of P-IFA and affinity purified IFN-α is given in Fig. 2. The overall yield of affinity purified IFN-α was approximately 65%. Approximately 30% of the IFN-α applied to NK2-Sepharose did not bind and was lost from the final product. This nonbinding fraction contained at least one IFN-α subtype characterized as having an approximate molecular weight of 21,500 and p*I* of 6.5.[8]

Acknowledgments

This work was supported in part by an award, #NOI-HB-0-2919, received from the National Heart, Lung, and Blood Institute. Special mention should be given to Mr. Richard J. Bonomo, Mr. William Swiggard, and Dr. Richard W. Shulman for their innovative roles in defining process requirements.

[8] B. Horowitz and M. Horowitz, *in* "Interferon: Research, Clinical Application, and Regulatory Consideration" (K. C. Zoon, P. D. Noguchi, and T.-Y. Liu, eds.), p. 41. Am. Elsevier, New York, 1984.

[6] Induction of Human Interferon Gamma with Phorbol Esters and Phytohemagglutinin

By JAN VILČEK, JUNMING LE, and Y. K. YIP

It is well known that immune interferon (IFN-γ) is produced by lymphocytes in response to specific antigens or nonspecific mitogens. The T cell mitogen, phytohemagglutnin (PHA) has been widely employed as an inducer of IFN-γ in cultures of human mononuclear cells from peripheral blood.[1] However, IFN-γ yields obtained from cultures stimulated with PHA are variable. The method of IFN-γ production based on the combined stimulation of peripheral blood leukocyte cultures with PHA and the phorbol ester, 12-*O*-tetradecanoylphorbol-13-acetate (TPA), generally results in the production of higher yields of IFN-γ than stimulation with PHA or other T cell mitogens alone.[2,3] In contrast, in the human cutaneous T cell lymphoma line Hut 102-B2 and several T cell hybridoma lines stimulation with TPA alone results in maximal IFN-γ production.[4,5]

TPA (also termed phorbol myristate acetate or PMA) is a potent tumor-promoting agent exerting hormone-like pleiotropic effects on growth, differentiation, and many other cell functions. Recently it was shown that the cellular receptor for TPA and related phorbol esters is likely to be protein kinase C.[6] Activation of the Ca^{2+}-dependent protein kinase C by TPA and related phorbol esters is probably responsible for the wide repertoire of biological activities associated with TPA. TPA was found to stimulate the production of IFN-α in human lymphoblastoid cell lines.[7,8] This finding, along with the earlier demonstrated mitogenic and comitogenic

[1] E. F. Wheelock, *Science* **149,** 310 (1965).
[2] J. Vilček, I. T. Sulea, F. Volvovitz, and Y. K. Yip, *in* "Biochemical Characterization of Lymphokines" (A. L. de Weck, F. Kristensen, and M. Landy, eds.), p. 323. Academic Press, New York, 1980.
[3] Y. K. Yip, R. H. L. Pang, J. D. Oppenheim, M. S. Nachbar, D. Henriksen, I. Zerebeckyj-Eckhardt, and J. Vilček, *Infect. Immun.* **34,** 131 (1981).
[4] J. Le, W. Prensky, D. Henriksen, and J. Vilček, *Cell. Immunol.* **72,** 157 (1982).
[5] J. Le, J. Vilček, C. Saxinger, and W. Prensky, *Proc. Natl. Acad. Sci. U.S.A.* **79,** 7857 (1982).
[6] Y. Nishizuka, *Nature (London)* **308,** 693 (1984).
[7] G. Klein and J. Vilček, *J. Gen. Virol.* **46,** 111 (1980).
[8] G. R. Adolph and P. Swetly, *J. Gen. Virol.* **51,** 61 (1980).

activity of TPA on T lymphocytes,[9,10] prompted us to examine the effect of TPA and other phorbol esters on IFN-γ production.[2-5]

Materials

Plateletpheresis residues: These lymphocyte-rich cell concentrates are a by-product of plateletpheresis and can be obtained from major blood banks. They should be used within 24 hr of collection.

T cell lines and hybridomas: The cloned human T lymphoblast line Hut 102-B2 was established from lymph node cells of a male patient with mycosis fungoides.[11] This cell line can be obtained from Dr. Robert C. Gallo (National Cancer Institute, National Institutes of Health, Bethesda, MD). Cloned T cell hybridoma cell lines L 265-K and L 265-O were derived by Le *et al.*[5] from hybridoma cultures obtained by fusion of human peripheral blood lymphocytes with the 6-thioguanine-resistant mutant cell line SH9. The SH9 line was derived from the Hut 102-B2 cell line after gamma irradiation.[5] Hybridoma lines L 265-K and L 265-O can be obtained from Dr. Junming Le (Department of Microbiology, New York University Medical Center, New York, NY).

Culture medium: RPMI 1640 (Gibco, Grand Island, NY), supplemented with gentamycin (50 μg/ml), *N*-2-hydroxyethyl-piperazine-*N*′-2-ethanesulfonic acid (HEPES, 6 m*M*) and tricine (3 m*M*), is employed for primary cultures of peripheral blood leukocytes. The composition of the medium used for T cell lines and T cell hybridomas is described below.

The diterpene esters, 12-*O*-tetradecanoylphorbol-13-acetate (TPA) or mezerein were purchased from LC Services Corp. (Woburn, MA). They are dissolved either in ethanol or dimethyl sulfoxide at 1 mg/ml and stored at −70°.

Phytohemagglutinin (PHA) was prepared by Dr. J. D. Oppenheim of New York University Medical Center. PHA preparations with similar activity can be purchased from Burroughs Wellcome Co. (Research Triange Park, NC) or Pharmacia P-L Biochemicals (Milwaukee, WI). PHA is dissolved in phosphate-buffered saline at 10 mg/ml and stored at −20°.

[9] J. L. Wang, D. A. McClain, and G. M. Edelman, *Proc. Natl. Acad. Sci. U.S.A.* **72,** 1917 (1975).

[10] A. M. Mastro and G. C. Mueller, *Exp. Cell Res.* **88,** 40 (1974).

[11] B. J. Poiesz, F. W. Ruscetti, A. F. Gazdar, P. A. Bunn, J. D. Minna, and R. C. Gallo, *Proc. Natl. Acad. Sci. U.S.A.* **77,** 7415 (1980).

IFN-γ Production in Peripheral Blood Leukocyte Cultures Derived from Plateletpheresis Residues

Background

Initially, we compared interferon yields in cultures of mononuclear cells obtained from fresh blood. Cells were stimulated with PHA alone, with TPA alone, or by combined treatment with PHA and TPA. Although PHA is an effective inducer of IFN-γ production in T lymphocytes, the yields of interferon obtained after stimulation with PHA are rather variable, probably depending to a large measure on the reactivity of lymphocytes from individual donors. TPA treatment alone also induced some interferon, probably due to its known mitogenic effect in T lymphocytes.[9] Combined stimulation with TPA and PHA produced high interferon yields that were far in excess of the sum of yields obtained when either stimulus was applied separately.[2,3]

Subsequently we established the optimal conditions for IFN-γ production in cultures of mononuclear cells isolated from plateletpheresis residues by Ficoll–Hypaque centrifugation and treated with a combination of TPA and PHA.[3] TPA added to cultures 2–3 hr before PHA resulted in a somewhat higher interferon yield than simultaneous addition of the two mitogens. The optimal concentration for PHA was found to be near 5 μg/ml, while TPA showed a concentration optimum between 5 and 50 ng/ml. The magnitude of the enhancing effect of TPA was dependent on cell concentration; we obtained optimal stimulation at a density of 6×10^6 cells/ml. Maximum yields of interferon were attained by about 48 hr after stimulation. These preliminary experiments led us to design the procedure for IFN-γ production described in detail below.

Procedure for IFN-γ Production

1. Plateletpheresis residues obtained from blood banks usually require the addition of anticoagulants to prevent clotting on subsequent dilution of the cells in culture medium. In our work, heparin is added to a final concentration of 40 USP units per ml of blood concentrate before processing.

2. Each residue supplied in a sealed plastic bag is first diluted to 200 ml with serum-free PRMI 1640 medium. The cells are sedimented by centrifugation for 5 min at 2000 rpm and resuspended in 200 ml of the same medium. (If the presence of a small amount of residual human plasma is not a problem, the sedimentation and washing of cells can be

omitted.) Cell counts are performed after diluting 0.1 ml of the diluted cell suspension in 3.9 ml (40-fold dilution) of a 3% (v/v) acetic acid/water solution. Each residue usually contains 6 to 12 × 10^9 white blood cells, of which 80–90% are mononuclear cells. There is also a large number of contaminating red blood cells and platelets.

Alternatively, mononuclear cells are separated from the red blood cells, platelets, and polymorphonuclear leukocytes before seeding the cultures. For this purpose, the heparinized blood cell concentrate is diluted with an equal volume of serum-free RPMI 1640 medium and then overlaid on an equal volume of Ficoll-Hypaque (Pharmacia Fine Chemicals, Piscataway, NJ). After centrifugation at 400 *g* for 30 min, cells at the interface are collected and washed three times with RPMI 1640 medium. Cell counts are taken as described above.

3. For seeding of the cultures, the cell suspension prepared in step 2 is diluted to 6 × 10^6 white blood cells per ml of serum-free RPMI 1640 medium. Therafter, 20 μl of the TPA or mezerein stock solution (1 mg/ml) is added to 1 liter of cell suspension (final concentration of 20 ng/ml) in a plastic tissue culture roller bottle. After thorough mixing by rotating the capped bottle, the cell suspension with added TPA or mezerein is dispensed into 150 × 25-mm plastic Petri dishes in a volume of approximately 120 ml per dish.

4. After 2 to 3 hr of incubation at 37° in a humidified CO_2 incubator, 60 μl of the PHA stock solution (10 mg/ml) is added to each 120 ml culture (final concentration at 5 μg/ml) and mixed well by gently swirling the dishes.

5. After further stationary incubation for 48 to 72 hr, the culture medium is collected by pipetting after tilting the culture dishes by resting them on their covers. By careful pipetting, a clear supernatant can be obtained and it may not be necessary to remove cells and cell debris by centrifugation. A yield of approximately 5000 IFN-γ reference units/ml can be expected with a usual range of 2000 to 10,000 units/ml. (These titers are based on the value of IFN-γ reference standard Gg 23-901-530.) The conditioned medium can be stored at 4° for several months without a significant loss of IFN-γ activity.

Comments

Buffy coats can be employed as a source of cells instead of plateletpheresis residues. However, the average IFN-γ yields obtained from buffy coat-derived cultures were only about half of the yields obtained with cells from plateletpheresis residues.[3] The reason for this difference is not known. Concanavalin A, pokeweed mitogen, numerous other mitogenic plant lectins,[3] as well as monoclonal antibody OKT-3[12] can be employed, instead of PHA, in combination with TPA.

The induction procedure described above results in the production of several cytokines in addition to IFN-γ, including T cell growth factor (IL-2),[13] lymphotoxin, and a monocyte cytotoxin.[14] This fact should be kept in mind when interpreting various biological activities of IFN-γ preparations produced in this way. Another potential problem is the presence of the inducing agents, TPA and PHA, in IFN-γ preparations. In view of its varied biological effects, the presence of TPA (or mezerein) can pose a problem in various assay systems. Fortunately, it is relatively easy to separate TPA from IFN-γ by simple purification steps.[13]

The fact that TPA is a potent tumor promoting agent should be kept in mind and any material containing TPA should be handled with caution. Mezerein is less potent as a tumor-promoting agent in mouse skin and this feature makes it preferable to TPA.[15] It is interesting that various structurally unrelated toxins isolated from marine algae exert biological effects similar to TPA and other tumor promoting diterpene esters, including the stimulation of IFN-γ production.[13]

IFN-γ Production in the Human T Cell Lymphoma Line Hut 102-B2 or T Cell Hybridoma Lines

Background

The production of many types of interferons was greatly aided by the availability of cell lines producing high yields of interferon on suitable stimulation. Availability of continuous cell lines consistently producing high yields of IFN-γ also would facilitate large-scale production. The T cell line Mo derived from a patient with hairy cell leukemia was shown to produce IFN-γ in response to PHA, and the addition of TPA together with PHA increased IFN-γ yields.[16] In contrast, T cell line Hut 102-B2 employed in our experiments did not respond to PHA alone (see the

[12] R. H. L. Pang, Y. K. Yip, and Vilček, *Cell. Immunol.* **64,** 304 (1981).

[13] Y. K. Yip, H. C. Kelker, D. S. Stone-Wolff, K. Pearlstein, C. Urban, and J. Vilček, *Cell. Immunol.* **79,** 389 (1983).

[14] D. S. Stone-Wolff, Y. K. Yip, H. C. Kelker, J. Le, D. Henriksen-DeStefano, B. Rubin, E. Rinderknecht, B. B. Aggarwal, and J. Vilček, *J. Exp. Med.* **159,** 828 (1984).

[15] J. Vilček, Y. K. Yip, D. S. Stone-Wolff, and R. H. L. Pang, *Tex. Rep. Biol. Med.* **41,** 108 (1982).

[16] I. Nathan, J. E. Groopman, S. G. Quan, N. Bersch, and D. W. Golde, *Nature (London)* **292,** 842 (1981).

INTERFERON INDUCTION IN T CELL LINES AND HYBRIDOMAS[a]

Cell line	IFN units/ml after stimulation with: No stimulation	TPA (20 ng/ml)	PHA (5 μg/ml)	TPA/PHA[b]	Type of interferon produced
Hut 102-B2	<10	2000	<10	330	$\alpha + \gamma$
SH9	<20	<20	<20	<20	—
L 265-K	330	5330	330	4000	γ
L 265-0	500	4000	170	2000	γ

[a] The cell lines were cultured at 8×10^5 cells/ml with or without inducers as indicated. Culture supernatants were collected for interferon assay after 48 hr.

[b] Where TPA and PHA were both used (TPA/PHA), PHA (5 μg/ml) was added 3 hr after the seeding of cultures in the presence of TPA (20 ng/ml).

table). Interferon was produced in response to TPA, but the addition of PHA or Con A together with TPA did not increase interferon yields above those obtained with TPA alone.[4] Interferon produced by the Hut 102-B2 line in response to TPA induction is a mixture of IFN-γ and IFN-α made in approximately equal amounts in terms of antiviral activity.[4]

T cell hybridomas were established by fusion of Con A stimulated human peripheral blood lymphocytes with a 6-thioguanine-resistant mutant cell line SH9, derived from the Hut 102-B2 cell line.[5] Some cloned T cell hybridoma lines, e.g., L 265-K and L 265-O, produced IFN-γ spontaneously and production was enhanced by exposure to TPA, but not by exposure to PHA (see the table). All interferon activity produced by lines L 265-K or L 265-O, either spontaneously or in response to TPA, was neutralized by an antibody specific for human IFN-γ, indicating that no other type of interferon was produced.[5]

Procedure for IFN-γ Production

The Hut 102-B2 cell line and the cloned T cell hybridoma lines L 265-K and L 265-O are cultured at 37° in RPMI 1640 medium supplemented with 10% heat-inactivated fetal bovine serum, 0.1 m*M* MEM with nonessential amino acids, 1 m*M* sodium pyruvate, 2 m*M* L-glutamine, 100 U/ml penicillin, and 100 μg/ml streptomycin (growth medium). The cells are routinely maintained in suspension in 25- or 75-cm² tissue culture flasks, and passaged twice a week by 1:5 dilution in growth medium.

For interferon induction, Hut 102-B2, L 265-K, and L 265-O cells growing in exponential phase are washed once with growth medium, and then resuspended and adjusted to 8×10^5 cells/ml in the same medium. The cell suspensions thus obtained are dispensed in 24-well tissue culture plates at 1.5 to 2 ml per well or at 10 ml into 25-ml flasks. TPA is added to the cultures for a final concentration of 20 ng/ml. The cells are then incubated at 37° in 5% CO_2 for 2 days before harvesting the interferon-containing culture supernatants. The table shows interferon yields obtained from these cells.

T cell hybridomas have not yet been employed for large-scale production of IFN-γ. One reason is that the Hut 102-B2 cell line as well as various mutant and hybrid lines derived from it produce human T cell leukemia virus (HTLV-I).[11] Consequently these cells and their products must be handled with care.

[7] Production and Partial Purification of Human Immune Interferon

By KARI CANTELL, SINIKKA HIRVONEN, and HANNA-LEENA KAUPPINEN

Human leukocytes from normal donors are used in many laboratories around the world for the production of alpha interferons for clinical use. Either of two paramyxoviruses, Sendai[1,2] or NDV,[3] is an effective inducer of many subtypes of alpha interferons and monocytes are the main producer cells.[4] Human leukocytes are also used for the production of immune or gamma interferon for clinical studies.[5,6] Mitogens are commonly used as inducers and the immune interferon is produced by T lymphocytes.[7] This volume contains many articles dealing with the production and purification of human immune interferon. The method developed in our laboratory during the past 5 years is described in the present chapter.

[1] K. Cantell, S. Hirvonen, H.-L. Kauppinen, and G. Myllylä, this series, Vol. 78, p. 29.

[2] K. Cantell and S. Hirvonen, this series, Vol. 78, p. 299.

[3] A. A. Waldman, R. S. Miller, P. C. Familletti, S. Rubinstein, and S. Pestka, this series, Vol. 78, p. 39.

[4] E. Saksela, I. Virtanen, T. Hovi, D. S. Secher, and K. Cantell, *Prog. Med. Virol.* **30,** 78 (1984).

[5] J. A. Georgiades, *Tex. Rep. Biol. Med.* **41,** 179 (1982).

[6] I. A. Braude, *Prep. Biochem.* **13,** 177 (1983).

[7] I. Schober, R. Braun, H. Reiser, K. Munk, M. Leroux, and H. Kirchner, *Exp. Cell Res.* **152,** 348 (1984).

Our aim has been (1) to mimic the routine methodology for the production of alpha interferons so that both natural leukocyte interferon and natural immune interferon could be conveniently prepared in the same laboratory, and (2) to avoid the use of reagents which would make the product unsuitable for clinical use.

Biologicals, Chemicals, and Media

IMDM, Iscove's modified Dulbecco's medium[8] is prepared in batches of 40 liters in our institute. All containers are sterilized at 200° for 1.5 hr. Pyrogen-free distilled water is used. L-tyrosine and L-cystine are dissolved in hot water containing 1 *N* HCl. Folic acid and riboflavin are dissolved in hot water containing 1 *N* NaOH. The HEPES buffer is dissolved in water and the pH is raised to 7.4 with 30% NaOH. The concentration of $NaHCO_3$ is 6.048 g/liter, i.e., twice as high as in the original formula. Phenol red was omitted from the medium in early experiments so that protein determinations could be made without dialysis of the samples. Routinely the concentration of phenol red is 0.00375 g/liter, i.e., one-fourth of that in the original formula. The medium is stored at 4° and used within 1 month. One percent of concentrated tricine (Sigma, St. Louis, Mo) solution is added to give a final concentration of 3 mg/ml and 0.25% of concentrated neomycin (Sigma) solution is added to give 25 μg/ml. The pH of the complete medium is 7.4.

Human agamma serum is prepared as described earlier.[1] It is stored at 4°. The IMDM is supplemented with 4 mg of human agamma serum per ml.

LCL, *lens culinaris* lectin or lentil lectin is obtained from Dr. B. Ersson, Uppsala University Biomedical Center, Uppsala, Sweden. It is dissolved in sterile PBS to give a stock solution of 6.67 mg/ml and stored in aliquots of 12.5 ml at −70°. It is thawed just before the use, filtered through a disposable Millex-GV 0.22-μm filter (Millipore, Molsheim, France), and mixed well by vortexing.

PEG 4000, polyethylene glycol MW 4000, Fluka Buchs, Switzerland.

Ammonium sulfate, Merck, Darmstadt, BRD.

CM-Sephadex C-50, Pharmacia Fine Chemicals, Uppsala, Sweden.

PBS, phosphate-buffered saline, pH 7.3 (NaCl 8.0 g, KCl, 0.2 g, $Na_2HPO_4 \cdot 2H_2O$ 1.4 g, and KH_2PO_4 0.2 g/liter).

PB, 0.1 *M* phosphate buffer, pH 8.0 ($Na_2HPO_4 \cdot 12H_2O$, 34 g, $NaH_2PO_4 \cdot 2H_2O$, 0.78 g/liter).

pH 6.5 buffer, 0.02 *M* phosphate buffer (about 3 parts of 0.02 *M*

[8] N. Iscove and F. Melchers, *J. Exp. Med.* **147,** 923 (1978).

KH_2PO_4 and about 1 part of 0.02 *M* $Na_2HPO_4 \cdot 2H_2O$ to give pH 6.5).
pH 9.5 buffer, 0.05 *M* glycine · NaOH, 0.2 *M* NaCl, pH 9.5.

Glassware, Plastics, Membranes

Glass beads, diameter 5 mm, Dragon Werk Wild, Bayreuth, BRD.
Falcon 3027 bottles, Becton Dickinson, Oxnard, Calif.
Polycarbonate 1000 ml bottles, Mekalasi, Helsinki, Finland.
Dialysis bag, Kalle 50, Kalle, Wiesbaden-Biebrich, BRD.
Nylon network, P-41 nybolt, Schweizerische Seidengatzefabrik, Zürich, Switzerland.
PM 10 Diaflo membrane, Amicon, Danvers, Mass.
Pall Ultipor NR 0.2-μm membrane, Pall, Portsmouth, England.

Laboratory Equipment

Vortex-Genie mixer, Scientific Industries, Bohemia, N.Y.
Roller Apparatus 7730-00509, Bellco, Vineland, N.J.
MSE Mistral 6L and Coolspin centrifuges, MSE Scientific Instruments, Sussex, England.
Sorvall RC-5B, RC-2 and SS-3 centrifuges, Du Pont, Newtown, Conn.
Sartorius filtration equipment SM 16201, Sartorius-Instruments, Surrey, England.
Amicon 2000 and 8400 cells, Amicon, Danvers, Mass.

Assay of Interferon

The interferons are assayed on vesicular stomatitis virus plaque reduction in HEp2 cells. A laboratory standard is included in every assay. One laboratory unit corresponds to 1.1 units of the human IFN-γ international standard Gg 23-901-530. All titers are expressed as international units per ml.

Collection and Purification of Leukocytes

Procedure. The leukocyte buffy coats are collected as for the preparation of the leukocyte (alpha) interferon. The volume of the buffy coat has been reduced from 40 to 30 ml. The recovery of the leukocytes is 0.8×10^9 per buffy coat. The pooled leukocytes are stored overnight at 4° and the red cells are lysed by two cycles of ammonium chloride treatment. The purified leukocytes are suspended in cold incubation medium (see below) to give 2×10^8 cells per ml. They are kept in a glass cylinder in ice water at 0 to 4°.

Comments. It is convenient that the early stages of the methods for the production of leukocyte and immune interferons are identical. Thus a proportion of the purified leukocyte suspension can be used for the production of leukocyte interferon and a proportion for the production of immune interferon. Braude[6,9] reported that better yields of immune IFN are obtained from diluted, nonpurified buffy coats than from purified leukocyte suspension. In our hands leukocytes purified by only one cycle of NH_4Cl treatment gave lower yields of immune IFN than twice purified cells. Very little IFN was obtained from nonpurified diluted buffy coat leukocytes.

Induction and Incubation

Procedure. The following ingredients are added into sterile Falcon 3027 bottles to a final volume of 1600 ml: IMDM at room temperature, human agamma serum 4 mg of protein/ml, LCL 10 μg/ml, and the concentrated purified leukocyte suspension to give 3.75×10^6 cells per ml. The cell suspension is gently mixed by rotating movements. Two hundred-milliliter aliquots of the suspension are poured into five 1000-ml polycarbonate bottles. They are closed tightly with screw caps. Two rubber bands are put around each bottle to increase friction. The bottles including the Falcon bottle with 600 ml of the cell suspension are incubated for 3 days in the roller at 1 rpm at 36.5°. The cell suspensions in the polycarbonate and Falcon bottles are pooled, respectively, and centrifuged at 800 *g* for 45 min at 4°. The supernatant is the crude immune interferon. It is stored for 2 to 4 weeks at 4° or for longer periods at −20°. The titer of the crude interferon is usually 6000 units/ml (range 3500 to 11000 units/ml).

Comments. Better yields of immune IFN are obtained with IMDM than with Eagle's minimum essential medium (MEM). We have not systematically compared IMDM with other media. We observed by chance that an increase of the sodium bicarbonate concentration improved the production of immune interferon. Twice the usual concentration in IMDM appears to give optimum yields. Human agamma serum gives at least as high titers as whole serum. The minimum agamma serum concentration for optimum yields is 4 mg protein per ml. The optimum cell concentration is approximately 3.0×10^6 per ml. Lower titers are ob-

[9] I. A. Braude, *J. Immunol. Methods* **63,** 237 (1983).

tained at lower cell concentrations. Higher cell concentrations do not clearly improve the yields.

Rönnblom *et al.*[10] reported that LCL was the best inducer of immune IFN among a number of lectins tested. Likewise, in our experiments LCL is a more potent inducer than staphylococcal enterotoxin A, concanavalin A, or phytohemagglutinin. The minimum LCL concentration for optimum titers is 10 μg/ml. In our hands, clearly better titers are obtained from roller cultures than from suspension cultures such as those used for leukocyte IFN. Good titers are obtained both in the disposable Falcon bottles and in the polycarbonate bottles. The latter are much cheaper and they can be used for years. The titers in various glass bottles have not been as good. The volume of the cell suspension affects the titers. Volumes exceeding 600 ml in the Falcon 3027 bottles and 200 ml in the 1000-ml polycarbonate bottles tend to decrease the yields. The titers of the immune IFN rise up to 3 days of incubation and remain at the same level up to 4 days at least.

PEG-$(NH_4)_2SO_4$ Concentration

Procedure. Three batches of 4000 ml are concentrated at the same time; 1620 g of PEG 4000 is added to each batch to give 30% w/v. They are kept in 5-liter flasks on magnetic stirrers for 2 hr at 4°. The suspension is then distributed into 1000-ml polypropylene centrifuge bottles and stored overnight at 4°. After centrifugation in MSE 6L or Coolspin centrifuge for 1 hr at 1500 *g* at 4° the sediment is dissolved with the aid of glass beads to a part of the supernatant to give a 6-fold concentration. The suspension is recentrifuged for 1 hr at 6800 *g* in a Sorvall RC-58 centrifuge at 4°. The precipitate is dissolved in 2000 ml of PBS containing ammonium sulfate. The final concentration of ammonium sulfate is 18%. After overnight storage at 4° the solution is placed in an ice bath on a magnetic stirrer while PBS containing 44% ammonium sulfate is added slowly dropwise to give a final concentration of 31% $(NH_4)_2SO_4$. The solution is transferred to a 5-liter separatory funnel in which it is kept at 4° for 2 to 3 days. Two phases are separated. The cloudy upper phase contains most of the PEG. The lower phase, containing the immune IFN, is collected. Enough solid ammonium sulfate is then added to raise the $(NH_4)_2SO_4$ concentration to 45%. After stirring for 2 hr at 4° the suspension is centrifuged for 1 hr at 8300 *g* at 4°. The sediment is dissolved in PB with the aid of glass beads. At this stage the crude IFN has been concentrated 31-fold,

[10] L. Rönnblom, K. Funa, B. Ersson, and G. V. Alm, *Scand. J. Immunol.* **16**, 327 (1982).

i.e., 387 ml of concentrated IFN is obtained from 12 liters of crude IFN. The concentrates are stored at −70°.

Comments. All immune IFN and most other proteins of the crude preparation are precipitated in the presence of 30% PEG 4000. At lower PEG concentrations the precipitation of immune IFN is not quantitative. Thus the PEG precipitation does not give any essential purification. Most of the PEG can be separated from the protein solution by an increase of the salt concentration.[11] The separation of the phases occurs at 31% ammonium sulfate. It is convenient to use ammonium sulfate for this purpose, because immune IFN can be precipitated later by a further increase of the ammonium sulfate concentration to 45%. Again no essential purification is achieved in relation to proteins, because most other proteins are also precipitated and the precipitation of immune IFN is not quantitative at lower ammonium sulphate concentrations. However, the PEG–$(NH_4)_2SO_4$ concentration aids the next step. The recovery of IFN in the PEG–$(NH_4)_2SO_4$ concentration is about 90% (see the table).

Ion Exchange Purification

Procedure. Six immune IFN concentrates representing 72 liters of crude IFN are thawed overnight in a coldroom and centrifuged for 2 hr at 6800 *g* at 4°. Most of the remaining PEG is pressed on the walls of the centrifuge bottles.

The supernatants are dialyzed in 6 sterile Kalle dialysis bags against 20 volumes of the pH 6.5 buffer with three changes of the dialysate overnight. Increase in volume is prevented by twisting about 7 cm of the upper part of the bag which is then covered with cotton and tied with thread. Knots are made above this tied part of the bag. After dialysis the interferon is harvested by cutting the thread and unwinding the bag. The interferon solution can then be poured off.

Twenty-two grams of CM-Sephadex C-50 is swollen in about 1.5 liters of sterile pH 6.5 buffer for at least 1 day. The gel is poured into an Amicon 2000 cell which has a nylon network (P-41 nybolt) above the membrane support. The filtrate outlet of the cell is connected to a vacuum flask under water suction. The gel is equilibrated in the pH 6.5 buffer by washing it twice with about 1000 ml of the buffer. The volume of the gel is about 900 ml. The dialyzed IFN concentrates are added to the gel in three steps about one liter at a time. The gel is stirred gently for 40 min at 4°. The

[11] Y. L. Hao, K. C. Ingham, and M. Wickerhauser, *in* "Methods of Plasma Protein Fractionation" (J. M. Curling, ed.), p. 57. Academic Press, London, 1980.

PURIFICATION OF HUMAN IMMUNE INTERFERON

Step	Volume (ml)	IFN		Protein		Specific activity (units/mg)	Degree of purification	Yield (%)
		units/ml[a]	Total × 10^6	mg/ml	Total g			
Crude	72,000	4,500	324	4.0	288	1,100	0	100
PEG–$(NH_4)_2SO_4$ concentrate	2,272	127,000	289	79.3	180	1,600	1.5	89
pH 6.5 dialysis	2,940	76,000	223	58.5	172	1,300	1.2	69
CM-Sephadex								
Effluent	3,035	3,600	11	48.4	147	74		3
Washes I + II + III	2,280	480	1	1.4	3.2	343		0.3
Wash IV	610	4,800	3	2.4	1.5	2,000		0.9
Eluates I + II	1,160	220,000	255	1.7	2.0	129,000	117	79
Amicon 2000 concentrate	200	1,350,000	270	9.8	2.0	138,000	125	83
Amicon 8400 concentrate	20	11,300,000	226	70.0	1.4	161,000	146	70

[a] Units of interferon are expressed with reference to the international standard for human immune interferon (see text) and represent the mean of 12 assays.

effluent is collected into the suction flask. The gel is washed 3 times with 1000 ml of sterile pH 6.5 buffer. The washing fluids are collected into the suction flask. The fourth washing is performed with 400 ml of a mixture which contains equal volumes of the pH 6.5 and 9.5 buffers. The pH of this mixture is 8.2. The pH of the effluent after the fourth wash is 6.7.

The immune IFN is eluted from the gel with the pH 9.5 buffer containing 1 *M* NaCl. The elution is done in two steps. Both times 400 ml of the cold pH 9.5 buffer is added and the gel is stirred for 10 min. The eluates are pooled. Their total volume is about 1000 ml.

Comments. We routinely load the gel with PEG–$(NH_4)_2SO_4$ concentrates derived from 72 liters of crude IFN. The gel can bind the interferon from at least 96 liters of crude preparation. We do not know the maximum number of immune IFN concentrates the 20 g of the CM-Sephadex gel can bind.

It is simple and convenient to do the ion-exchange purification in the Amicon 2000 cell. The gel can be mixed and washed gently but efficiently. The fourth wash releases some proteins but little IFN from the gel. A further increase in the pH and salt concentration of the washing solution causes a substantial release of IFN from the gel.

About 80% of the recovered IFN is in the first eluate, the rest in the second eluate. The two eluates contain 70 to 80% of the total IFN and about 1% of the protein of the starting material, the crude immune IFN preparation (see the table).

Amicon Concentration

Procedure. An Amicon 2000 cell with a PM 10 membrane (MW cutoff 10,000) is used for the concentration of the above CM-Sephadex eluates. A pressure of 3.8 kg/cm^2 is obtained from a nitrogen tank. First a batch of about 1000 ml is concentrated about 5-fold. Further 10-fold concentration is achieved in an Amicon 8400 cell with a PM 10 membrane. The concentrate is diluted 10-fold with sterile PBS and centrifuged for 40 minutes at 31,500 *g*. The supernatant is filtered through a Pall 0.2-μm membrane in Sartorius filtration equipment with the aid of water suction. The sterile filtrate is reconcentrated 10-fold in an autoclaved Amicon 8400 cell. During this step the nitrogen gas is sterilized by filtration through a Millipore FG 0.2-μm filter. The sterile, partially purified and concentrated immune IFN is collected from the cell.

Comments. The concentration in the Amicon 2000 cell takes about 1 hr and the concentration in the Amicon 8400 cell 2 to 2.5 hr. The recovery of IFN after these concentration steps is close to 100% (see the table).

Concluding Comments

Human leukocytes can be collected and purified by the same procedure for the production of both the leukocyte and the immune interferons. However, the subsequent production steps for the interferons differ substantially as far as the inducer, cell concentration, medium, conditions, and time of the incubation are concerned. This is not surprising because the two interferons are produced by different cells.

About 1×10^6 units of human immune IFN are obtained from each buffy coat by the method described above.

Routinely 72 liters of crude immune IFN is processed through the different steps described above to obtain about 20 ml of sterile, physiologic, partially purified immune IFN. The recovery is about 70% and the purification over 100-fold. The final product contains approximately 1×10^7 units/ml and its specific activity is 1 to 2×10^5 units/mg of protein. The two main impurities are transferrin and hemopexin as revealed by immunoelectrophoresis. The content of the inducer, LCL, is reduced almost 200-fold in relation to IFN during the purification. The final product contains about 70 μg of LCL/ml as measured by a RIA.

We routinely store the purified immune IFN at $-70°$. Repeated freezing and thawing does not destroy its activity. The stability of the product at different temperatures is under study.

We have attempted further purification by controlled pore glass (CPG).[5,6] Preliminary results suggest that about 5-fold additional purification can be achieved whereby the content of both transferrin and hemopexin can be greatly reduced, but LCL follows immune IFN through the CPG purification step.

Several chapters in this volume describe methods for the production and purification of human immune IFN. In most of these the purity of the final product exceeds that described here. However, our method may have some advantages for clinical application. No highly toxic or carcinogenic substances are used. We have produced and purified about 500 liters of human immune IFN during the past year. The results are reproducible. The specific activity of the product is not itself very important for clinical safety, however, the quality and quantity of the impurities are important. The two chief impurities of our partially purified preparations are physiologic serum proteins. LCL is also present, but its clinical significance is not known. The preparations of human leukocyte IFN used for clinical studies also contain impurities derived from the inducer used.[12] However,

[12] S. Ingimarsson, K. Cantell, G. Carlström, B. Dalton, K. Paucker, and H. Strander, *Acta Med. Scand.* **209,** 17 (1981).

these Sendai and chick proteins have not caused serious adverse effects in clinical studies.

We have used most of our partially purified human immune IFN for pharmacokinetic animal experiments so far,[13,14] but our aim is to start clinical studies with the combination of natural human leukocyte and immune interferons.

Addendum

The purification procedure has been modified as follows: The interferon preparation is centrifuged in Sorvall SS-3 at 31,000 *g* for 30 min after the Amicon 2000 concentration. Thereafter, it is sterile filtered with pressure through Millipore GVWP 09050, diameter 90 mm, dialyzed against PBS, and further concentrated in an autoclaved Amicon 8400 cell. We now have produced and purified about 2000 liters of human immune interferon by the above methods and the results have been very reproducible.

Tumor necrosis factor is present in the final products (D. Wallach, personal communication).

Further purification has been achieved by affinity chromatography with a monoclonal antibody.

[13] K. Cantell, S. Hirvonen, L. Pyhälä, A. DeReus, and H. Schellekens, *J. Gen. Virol.* **64,** 1823 (1983).

[14] K. Cantell, W. Fiers, S. Hirvonen, and L. Pyhälä, *J. Interferon Res.* **4,** 291 (1984).

[8] Production of Human Immune Interferon from Leukocytes Cocultured with Exogenous Cells

By PHILIP C. FAMILLETTI and DONNA STREMLO

Human gamma interferon (IFN-γ) is produced *in vitro* from peripheral blood leukocytes stimulated with mitogens[1–6] or with the calcium ionophore A-23187.[7,8] In this section a detailed procedure for the isolation

[1] J. Vilček, J. Le, and Y. K. Yip, this volume [6].

[2] K. Cantell, S. Hirvonen, and H.-L. Kauppinen, this volume [7].

[3] R. Van Reis, S. E. Builder, and A. S. Lubiniecki, this volume [11].

[4] J. A. Georgiades, this volume [12].

[5] M. de Ley, J. van Damme, and A. Billiau, this volume [13].

[6] M. Tsujimoto and N. Higashi, this volume [14].

[7] F. Dianzani, G. Antonelli, and M. R. Capobianchi, this volume [9].

[8] I. A. Braude and C. Tarr, this volume [10].

of human leukocytes, the initiation of the leukocytes into 10-liter suspension cell cultures, and induction of the cells with a lectin, phytohemagglutinin-A (PHA), to produce IFN-γ is described. Yields of interferon are augmented by the addition of a phorbol ester to the culture medium and by the cocultivation of the leukocytes with an established human promyloblastic cell line.

Materials and Reagents

Induction Vessel. Human IFN-γ is produced in 10-liter volumes with 15-liter spinner culture flasks (Bellco Biotechnology, Vineland, NJ, No. 1967-1500) modified with an overhead-driven paddle stirrer set at 25 rpm for gentle agitation of the cell culture. The vessel is incubated at 37° in a water bath.

Media. RPMI medium 1640 (RPMI 1640) supplemented with 25 m*M* HEPES buffer (Gibco Laboratories, Grand Island, NY, No. 430-1800), 100 units/ml of penicillin, 100 μg/ml of streptomycin (Gibco, No. 200-5140), and 10% (v/v) fetal calf serum (FCS, Gibco, No. 200-6140) is used as the induction medium and as the growth medium for the supplemental cell cultures. Eagle's minimum essential medium (Gibco No. 410-1500), supplemented with 10% (v/v) of FCS (EMEM), is the growth medium for the cells used in the interferon antiviral assays.

Interferon Inducers. Human leukocytes are induced to produce IFN-γ by the addition of PHA obtained from California Medicinal Chemistry Corp., San Francisco, CA (PHA—L, No. YL1801), or Difco Laboratories, Detroit, MI (PHA-P, No. 3110-57), and a phorbol ester, 12-*O*-tetradecanoyl-phorbol-13-acetate (TPA), obtained from Consolidated Midland Corp., Brewster, NY.

Cell Cultures

Isolation of Human Leukocytes. Human leukocyte concentrates, collected and prepared from normal donors as previously described,[9] are obtained from the American Red Cross, Lansing, MI. The contents from 50 leukocyte concentrates are aseptically removed from the collection bags and pooled. The leukocyte pool is mixed with a half volume of a 6% solution of hetastarch (Hespan, American Hospital Supply Corp., Irvine, CA) in a separatory funnel and allowed to stand for 3–3.5 hr at room temperature. This procedure differentially sediments the contaminating

[9] A. A. Waldman, R. S. Miller, P. C. Familletti, S. Rubinstein, and S. Pestka, this series, Vol. 78, p. 39.

red blood cells from the leukocytes. The volume of the leukocyte–Hespan mixture should not exceed two-thirds of the capacity of the separatory funnel to ensure proper sedimentation. The red blood cells sediment to the bottom of the separatory funnel and cell separation is complete when a sharp band of white cells is apparent just above the red cell layer at the interface. The "low density" white blood cells are used for the production of IFN-γ. These cells are removed from the separatory funnel by carefully aspirating only the uppermost layer of cells located above the white cells at the interface. The low density leukocytes are removed from the Hespan by sedimentation in 250-ml conical centrifuge tubes (Corning Glass Works, Corning, NY, No. 25350-250) at 500 *g* for 20 min. The pellet is resuspended in 9 volumes of a 0.83% solution of ammonium chloride for 5 min to lyse any remaining red blood cells. The leukocytes are removed from the ammonium chloride solution by sedimentation at 500 *g* for 10 min and the pellet is resuspended to a concentration not exceeding 3.5×10^7 cells/ml in RPMI-1640 medium prewarmed to 37°. No FCS is added to the medium to avoid clumping of the concentrated cells.

Cell Lines. HL-60, a human promyelocytic cell line and RPMI-1788, a hematopoietic cell line, were obtained from the American Type Culture Collection (ATCC), Rockville, MD (Nos. CCL 240 and CCL 156, respectively). Each of these cell lines is grown in RPMI-1640 medium in conventional 3-liter spinner culture flasks (Bellco, No. 1969-03000) at cell densities ranging from 0.1 to 2.0×10^6 cells/ml. The HL-60 and RPMI-1788 cells have a doubling time of approximately 20 hr in log phase. WISH, a human amnion cell line, MDBK, a bovine kidney cell line, and L cells, a mouse fibroblast cell line (ATCC No. CCL 25, CCL 22, and CCL 1, respectively) are propagated in EMEM in monolayer culture in T-75 flasks (Corning, No. 25110 or equivalent) and are used for the interferon assays.

Interferon Assay

The assay for human IFN-γ quantitates the reduction of the cytopathic effect (CPE) of vesicular stomatitis virus (VSV) on the assay cell as previously described for human IFN-α and IFN-β with some modifications.[10] WISH cells are incubated with the IFN-γ sample for 6–20 hr at 37° prior to the addition of the challenge virus. The assay is stained and the endpoints of the IFN-γ standard and samples are determined approximately 30 hr after the addition of the VSV when 90–100% CPE is noted in the virus control wells. All IFN-γ titers are adjusted to a laboratory standard preparation which has been calibrated according to the NIH

[10] P. C. Familletti, S. Rubinstein, and S. Pestka, this series, Vol. 78, p. 387.

human IFN-γ standard (No. Gg23-901-530) obtained from the National Institute of Allergy and Infectious Diseases, Research Resources Branch, National Institutes of Health, Bethesda MD, and expressed in reference units/ml.

Production of Human Gamma Interferon

The freshly isolated leukocytes, free of red blood cell contamination, are diluted in the induction vessel with the appropriate volume of RPMI-1640 medium supplemented with 10% FCS to make a final concentration of 2×10^6 cells/ml. A "priming" concentration of PHA, which does not induce a detectable level of IFN-γ, is added to the medium [0.5 μg/ml of PHA-L or 0.05% (v/v) of PHA-P] and the culture is incubated at 37° for 20 hr. A volume of medium containing enough HL-60 cells in log growth phase is added to the leukocyte culture to make the HL-60 cells 4% of the total cell concentration. Incubation of the cocultivated leukocyte and HL-60 suspension is continued for 20 hr. TPA (10 ng/ml) is added to the culture followed 3 hr later by the first of two induction level doses of PHA [2 μg/ml of PHA-L or 0.1% (v/v) of PHA-P]. Incubation of the culture continues for 4 hr at which time a second induction dose of PHA, identical to the first, is added. Incubation of the culture continues for 48 hr. The cells are removed from the culture medium by sedimentation at 1000 *g* and discarded. The interferon containing culture fluid is concentrated and partially purified by adsorption to silica by a procedure described elsewhere in this volume.[5]

Concluding Remarks

Interferon Yield and Characterization. An average titer of 10,000 reference units/ml of IFN-γ was obtained from human leukocytes cocultivated with HL-60 cells in spinner culture, however, due to variation of the leukocyte pools, titers ranged from 2500 to 40,000 units/ml. The interferon produced from the leukocyte/HL-60 cultures was not active on MDBK cells, a cell line sensitive to human leukocyte interferon or on mouse L-cells, but was active on human WISH cells (Table I).[11,12] Antisera to human IFN-α or IFN-β, obtained from the National Institute of Allergy and Infectious Disease (glob. 24, 5174 and 29-33 glob), did not neutralize the antiviral activity of the interferon produced by the cocul-

[11] P. C. Familletti, *Adv. Biotechnol. Processes* **2,** 169 (1983).

[12] R. Dennin, C. Kramer, S. Evans, and M. J. Kramer, *Adv. Biotechnol. Processes* **2,** 160 (1983).

TABLE I
ASSAY OF INTERFERON ON VARIOUS CELLS[a]

	Interferon (units/ml)			
Interferon	Bovine MDBK cells	Human WISH cells	Mouse L-929 cells	Ratio MDBK/WISH
Immune	<40	2560	<40	<0.02
Leukocyte	640	640	<40	1.00
Fibroblast	240	20,480	<40	0.01

[a] The IFN-γ was obtained from primary leukocytes induced with PHA and TPA and cocultured with HL-60 cells. IFN-α was obtained from primary leukocytes induced with Sendai virus.[11] IFN-β was obtained from human fibroblasts induced with poly(I) · poly(C).[12]

ture, however, the antiviral activity was neutralized by antisera containing polyclonal antibody to human IFN-γ obtained from Interferon Sciences, Inc., New Brunswick, NJ (Table II).

Addition of Exogenous Cells. HL-60 cells proliferate in suspension culture; therefore, the time of addition of the cells to the leukocyte induction medium and the final concentration of the HL-60 cells is critical for optimum IFN-γ yields (Fig. 1). Low concentrations of HL-60 cells in the induction medium (<10%) increased the yield of human IFN-γ from the leukocytes, while high concentrations of HL-60 cells (>10%) decreased

TABLE II
NEUTRALIZATION OF INTERFERON WITH ANTI-INTERFERON SERA[a]

	Neutralization Titer		
Interferon	Anti-gamma	Anti-alpha	Anti-beta
Immune	1280	<40	<16
Leukocyte	<40	1280	<16
Fibroblast	<40	<40	256

[a] Ten units of each interferon was incubated in wells of a microtiter plate containing serial 2-fold dilutions of antiserum for 1 hr at 37° in a volume of 0.1 ml. WISH cells were used in the assay. Neutralization titers are expressed as the reciprocal of the dilution of antiserum in the well in which 50% of the cells were destroyed by vesicular stomatitus virus.

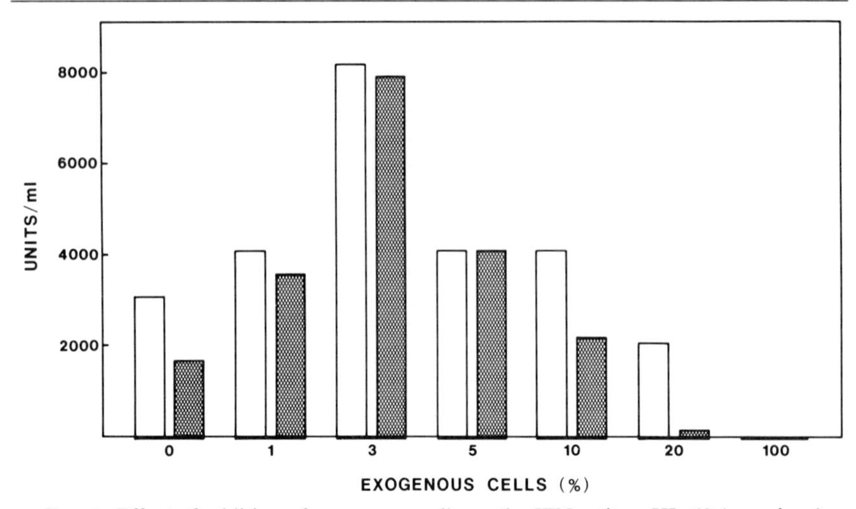

FIG. 1. Effect of addition of exogenous cells on the IFN-γ titer. HL-60 (open bars) or RPMI-1788 (shaded bars) cells were added to primary leukocyte cultures at the percentages indicated as described in the text. The medium was assayed for interferon activity 48 hr after the addition of PHA to induction medium.

the interferon titers. A similar augmentation of the IFN-γ yield was obtained by the addition of RPMI-1788 cells to the leukocyte induction medium. HL-60 or RPMI-1788 cell cultures alone, induced with PHA and TPA, did not produce detectable levels of IFN-γ.

Cells Obtained by Leukapheresis. Human leukocytes, obtained by leukapheresis of normal donors, may be used for the production of human IFN-γ in place of the normal buffy coat leukocytes. Blood preparations obtained by leukapheresis have less red blood cell contamination and separate more quickly in the Hespan (1.5 hr) than the buffy coat preparations. The following modifications to the IFN-γ induction procedure are necessary when using cells obtained by leukapheresis. The contaminating red blood cells are removed by sedimentation in a 2% solution of Hespan and by lysis in ammonium chloride as described above. The leukocytes free of red blood cells are placed into culture at a concentration of 1×10^6 cells/ml with the same "priming" concentration of PHA used for the buffy coat leukocytes. The culture is incubated at 37° for 20 hr. HL-60 or RPMI-1788 cells are added to the leukocytes as described and incubation of the coculture continues for 20 hr. TPA (10 ng/ml) is added and the coculture is induced for IFN-γ production 3 hr later with a single dose of PHA [2.0 μg/ml of PHA-L or 0.1% (v/v) of PHA-P]. The IFN-γ-containing medium is harvested 48–72 hr after the addition of the PHA as described.

Ultrafiltration. Tangential-flow ultrafiltration with a Pellicon Cassette System (Millipore Corp., Bedford, MA) is used as an alternative procedure to concentrate the IFN-γ containing culture medium. The cells are removed from the culture medium by filtration through a 0.45-μm membrane cassette (HVLP 000 C5). The interferon-containing medium, devoid of cells, is concentrated to 1/20 of its original volume by ultrafiltration with a 10,000 molecular-weight cutoff membrane cassette (PTGC 00001) without loss of biological activity.

[9] Induction of Human Immune Interferon with Ionophores

By FERDINANDO DIANZANI, GUIDO ANTONELLI, and MARIA R. CAPOBIANCHI

It has been shown that activation of a calcium flux through the lymphocytic membrane following oxidation of galactose residues of surface glycoproteins appears to be critical for induction of human immune interferon by mitogens and specific antigens. This view is substantiated by the finding that calcium depletion by the chelating agent EGTA or blockade of calcium channels by calcium entry blockers completely prevent interferon production by mitogen or antigen stimulated peripheral blood mononuclear cells (PBMC).[1,2]

The finding that immune interferon production may be induced by the calcium ionophore A-23187, and not by the potassium ionophore Nonactin, suggests that calcium influx is a sufficient inducing stimulus and therefore capable of bypassing the specific interaction between conventional inducers and the recognition sites at the plasma membrane.[3] By this point of view immune interferon induction by calcium ionophores may provide an efficient tool to study the intracellular events leading to lymphokine production and lymphocyte activation. Specificity of this type of induction must be, therefore, carefully verified. Namely, ionophore induction must be prevented by calcium depletion; conversely, since ionophores are capable of causing calcium intake through the membrane by bypassing regular calcium channels, their action should not be prevented by specific blockers of calcium channels.

[1] F. Dianzani, T. M. Monahan, and M. Santiano, *Infect. Immun.* **36,** 915 (1982).
[2] F. Dianzani, M. R. Capobianchi, and J. Facchini, *J. Virol.* **50,** 964 (1984).
[3] F. Dianzani, T. M. Monahan, J. Georgiades, and I. Alperin, *Infect. Immun.* **29,** 561 (1980).

Reagents

Peripheral blood mononuclear cell, PBMC, cultures (10^6 cells per ml) are established by Ficoll–Hypaque gradient sedimentation of peripheral blood from healthy adult donors. RPMI or McCoy's media supplemented with 5% fetal bovine serum may be used as culture media. The cultures may be used immediately or after an overnight incubation at 37°.

Ionophore A-23187 (Calbiochem) is dissolved at 10 m*M* in DMSO as stock solution, aliquoted, and stored at −20°. Ionomycin, a polyether antibiotic identified as a highly specific ionophore for calcium,[4] may be obtained from Squibb, Princeton, NJ. It may be dissolved in ethanol at 5 mg/ml and kept at −20° as a stock solution. EGTA (Sigma Chemical Co.) is used at a final concentration of 1.5 m*M* in calcium-free medium. Calcium entry blocker Verapamil (Sigma Chemical Co.) may be dissolved in ethanol at 10^{-2} *M;* further dilutions are performed in culture medium. Staphylococcal enterotoxin B (SEB, Sigma Chemical Co.) is dissolved in culture medium at 100 μg/ml and stored at −20° until use.

Immune interferon titration and characterization are carried out by standard techniques.

Procedure for Induction of IFN-γ

Induction procedures are simple and straightforward. The stock solution of A-23187 is diluted 200-fold in culture medium to yield a 50 μ*M* working solution that is added to the cell cultures (0.1 ml/ml of culture) for interferon induction. Usually a final concentration of 5 μ*M* provides the highest interferon yield and a cell mortality at 24 hr lower than 10%. However, since different lots of ionophore may have slightly different inducing capabilities, it is advisable to test other concentrations (e.g., 1, 5, and 10 μ*M*) with each new batch. The cultures are then incubated at 37° without further treatment. Agitation of the cultures is not necessary. The kinetics of interferon production is shown in Fig. 1. It may be seen that interferon is detectable as early as 6 hr after treatment but reaches a maximum titer between 12 and 24 hr. As with any other inducer, interferon yield may vary in different experiments but an average titer of 300 units/ml may be expected regularly. Similar results are obtained with ionomycin as the inducer, although lower concentrations of this ionophore are required (0.5 μg/ml).

Specificity of induction may be tested on the assumption that activation of calcium flux through specific channels on the cell membrane is required for immune interferon induction by any inducer; however, calcium ionophores are capable of opening new, nonspecific channels.

[4] C. M. Liu and T. E. Hermann, *J. Biol. Chem.* **253,** 5892 (1978).

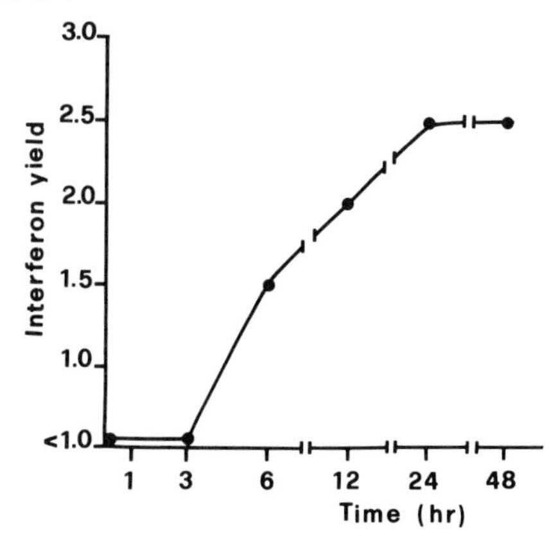

FIG. 1. Kinetics of IFN production by PBMC after treatment with the calcium ionophore A-23187 (5 μM). Interferon yield is given as the $\log_{10}$ units/ml.

Therefore, while calcium depletion in the culture medium prevents interferon induction by both conventional inducers and ionophores, treatment with specific blockers of calcium entry abolishes interferon induction only by conventional inducers.

The experiment may be carried out by comparing interferon induction by a mitogen (e.g., SEB, 2 μg/ml) and by the ionophore under study in regular cultures and in cultures treated with EGTA (1.5 mM in calcium-free medium) or with a calcium channel blocker (e.g., verapamil, 10^{-4} M), both added to the cultures at least 5 min before the ionophore. Some experimental results are presented in the table.

Comments

Induction of immune interferon by calcium ionophores is a simple and reproducible procedure which offers several advantages of both practical

INHIBITION OF INTERFERON INDUCTION BY CALCIUM DEPLETION OR BY CALCIUM CHANNEL BLOCKADE

Interferon inducer	Interferon production (units/ml) by PBMC treated with		
	None	EGTA	Verapamil
SEB	300	10	10
A-23187	300	10	300

and theoretical value. In fact, the interferon production is rapid, as compared to mitogen and antigen induction, and within wide limits the response is not strictly dose dependent, since the optimal inducing dose may be increased by severalfold without loss of activity, in spite of the fact that under these conditions the cell viability at 24 hr is sharply decreased. Another interesting feature is that the presence of macrophages does not seem to be required. Finally, some pathologic conditions have been identified where PBMC cultures from patients do not respond to mitogens but can be regularly stimulated by A-23187, suggesting that in these cases the defect resides in the membrane recognition site(s) rather than at the genomic level. If this is the case, calcium ionophores could be used as a diagnostic tool to distinguish among different clinical situations.

Acknowledgment

We wish to express our gratitude to Dr. Domenico Delia for kindly providing Ionomycin.

[10] Production of Human Immune Interferon with Normal Human Leukocytes by the Inducers A-23187 and Mezerein

By IRWIN A. BRAUDE and CRISS TARR

Human leukocytes are an abundant and available source for the production of natural immune interferon. This chapter describes the induction of these leukocytes with the combination of mezerein and the calcium ionophore, A-23187. This novel procedure offers the use of small organic compounds which, by virtue of their physicochemical properties, can be readily removed from the final product.

The method, summarized in Fig. 1, has been previously reported[1] and will be described, in detail, in the following sections.

Cell Preparation

Buffy coats from individual donors are pooled into sterile Erlenmeyer flasks and held at room temperature until use. An aliquot from each pool is removed and the leukocyte concentration quantitated by diluting the buffy coats 1 : 200 with aqueous acetic acid containing 0.01% crystal violet. Generally 4 squares in a hemocytometer are counted.

[1] I. A. Braude, *J. Immunol. Methods* **63,** 237 (1983).

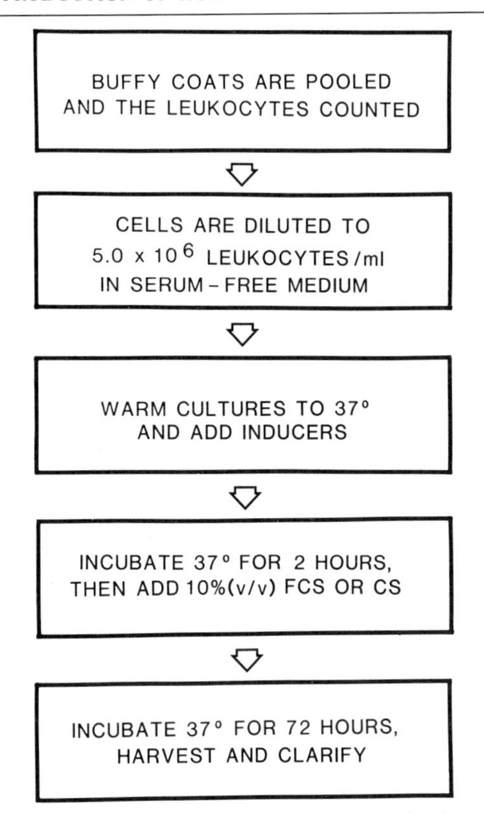

FIG. 1. A diagram outlining the sequential steps involved in the production of natural human immune interferon.

Unlike the production of human leukocyte interferon,[2] lysis of erythrocytes with NH_4Cl is undesirable. As previously reported,[1] NH_4Cl treatment of buffy coat cells, prior to the addition of A-23187 and mezerein, leads to a 5-fold reduction in immune interferon production. A second NH_4Cl treatment decreases the yields still further. Consequently, after counting, the unprocessed buffy coat material is diluted to 5.0×10^6 leukocytes/ml in Dulbecco's minimal essential medium (DMEM; Meloy Laboratories). If open spiner culture conditions are employed (see below), cells are diluted in DMEM containing 3 mg/ml tricine (Calbiochem), 25 μg/ml neomycin, and 1.3 mg/ml sodium bicarbonate (Meloy Laboratories). More comprehensive media such as RPMI-1640 medium did not enhance the yields.

[2] K. Cantell, S. Hirvonen, H.-L. Kauppinen, and G. Myllylä, this series, Vol. 78, p. 29.

Culture Conditions

The cell suspension is then placed either in stationary or spinner culture vessels. A variety of stationary culture vessels have been successfully employed. Optimally the surface to volume ratio should be greater than 2 cm^2 : 1 ml. For the sake of convenience, 1750-cm^2 disposable roller bottles (Falcon, catalog number 3029) are routinely utilized. The loosely capped bottles, held in a fixed horizontal position with tape, are stored in a CO_2 incubator. The incubator is maintained at 37° and 5% CO_2.

Two spinner vessel devices are employed. The first is a 36 liter capacity closed Bellco μ spinner system. Its design employs a wide flexible paddle and a magnetic stirrer geared for slow speeds. Batches as large as 36 liters have been processed in this device. The unit is stored in an enclosed hot box set at 37° and a positive flow (through a Millipore Twin-90 filter) of 5% CO_2 is passed over the culture. The magnetic stirrer is set for 15–20 rpm.

The second spinner vessel is an open system as essentially described by Cantell *et al.*[2] Ten liter cultures are placed into 22-liter round bottom flasks (Corning, catalog number 4260) and the flasks are submerged in 37° water baths. Sterile 2.5 in. long egg-shaped stirring bars (Fisher Scientific) are added and the magnetic stirrer (Bellco, Bell-stirrer) speed increased till a vortex is created. Evaporation of water from the bath is minimized by the addition of ping-pong balls. After the cultures are brought to 37° (this usually requires at least 1 hr), the inducers are added.

Inducers

Stock solutions of the inducers A-23187 (Calbiochem) and mezerein (Chemicals for Cancer Research) are prepared in dimethyl sufoxide (Fisher Scientific, Spectra Analyzed) at concentrations of 50 and 1 mg/ml, respectively. Working stock solutions are then prepared in the appropriate culture media.

As shown in Fig. 2 A and B, A-23187 and mezerein either mixed with suboptimal concentrations of coinducer or alone (data not shown) are poor human immune interferon inducers, whereas, when used in the appropriate mixture, the combination is very potent. The data in Fig. 2 also demonstrate that regardless of the culture system employed, a limited optimal concentration for A-23187 either in stationary culture (Fig. 2A) or spinner culture (Fig. 2C) exists.

Generally it has been our experience that 0.5 and 2.0 μg/ml of A-23187 is optimal for stationary and spinner cultures, respectively. However, some lot-to-lot variability of A-23187 has been observed and therefore it is advisable to determine the optimal concentrations of new batches prior to

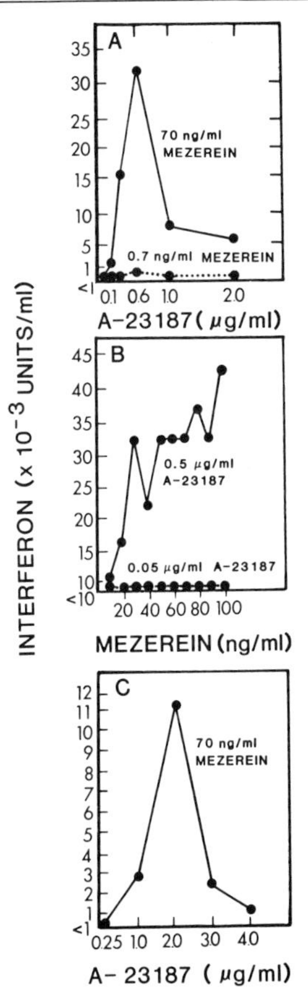

FIG. 2. Dose–response of A-23187 and mezerein. A represents increasing concentrations of A-23187 in the presence of 70 ng/ml mezerein (solid line) or 0.7 ng/ml mezerein (broken line) under stationary culture conditions. B depicts increasing concentrations of mezerein in the presence of 0.5 μg/ml A-23187 (solid line) or 0.05 μg/ml A-23187 (broken line) under stationary culture conditions. C shows increasing concentrations of A-23187 in the presence of 70 ng/ml mezerein under spinner culture conditions.

use on a large scale. A wide range of concentrations can be used with mezerein (Fig. 2C). Typically concentrations of 70 ng/ml have been employed.

Postinduction Procedures

After inducing the cells, the cultures are incubated at 37° for 2 hr. Finally, 10% (v/v) fetal calf serum (FCS) or calf serum (CS), brought to

37°, is added (the source of the serum does not appear critical). The cultures are then incubated at 37°, an additional 72 hr. Incubation periods less than 72 hr resulted in lower titers, while longer times did not appear to increase yields significantly.

The material is then harvested by pouring it into 1-liter centrifuge bottles (IEC, catalog number 2939) and clarifying at 2600 *g* for 1 hr at 4°. The supernatants are decanted into sterile roller bottles and stored until use at 4°. Typically the crude immune interferon can be held for as long as 2 months without appreciable loss of activity. However, upon long storage a precipitate forms which requires an additional clarification (2600 *g* for 1 hr, at 4°).

Concluding Remarks

Over the past 2 years, both spinner culture systems and stationary culture systems have been utilized. Generally, the average titer for both systems has been approximately 12,000 units/ml (the bioassay is described in a subsequent chapter[3]). The advantage in employing the spinner system is its yield per volume, whereas the stationary system requires considerably less A-23187.

The method described in this chapter for the production of human immune interferon has several advantages. First, the reagents employed (e.g., A-23187, mezerein, CS, and DMEM) are readily available and inexpensive. Second, unlike proteinaceous inducers such as PHA,[4] concanavalin A,[5] or staphylococcal enterotoxin A,[6] these low-molecular-weight organic compounds can be readily removed from the final product. Last, because this procedure does not require lysis of erythrocytes and is suitable in spinner culture, this method is quite adaptable to large-scale practice.

Acknowledgment

We wish to thank Ms. Pat Gregory for her skillful secretarial assistance.

[3] I. A. Braude, this volume [27].

[4] J. Vilček, I. T. Sulea, F. Volvovitz, and Y. K. Yip, *in* "Biochemical Characterization of Lymphokines" (A. L. de Weck, I. Kristensen, and M. Landy, eds.), p. 323. Academic Press, New York, 1980.

[5] M. de Ley, J. van Damme, H. Claeys, H. Weening, J. W. Heine, A. Billiau, C. Vermylen, and P. DeSomer, *Eur. J. Immunol.* **10,** 877 (1980).

[6] H. M. Johnson, G. J. Stanton, and S. Baron, *Proc. Soc. Exp. Biol. Med.* **154,** 138 (1977).

[11] Large-Scale Production and Recovery of Human Immune Interferon[1,2]

By Robert van Reis, Stuart E. Builder, and Anthony S. Lubiniecki

Large amounts of lymphocytes are collected each year in the form of buffy coats as a by-product of erythrocyte/plasma separation. These cells provide a source for production of natural human immune interferon. Yip *et al.*[3] have demonstrated high levels of immune interferon induction when using cultures derived from lymphocytes costimulated with mitogens such as concanavalin A (Con A)[4] or phytohemagglutinin (PHA) and phorbol myristate acetate (PMA) or the antileukemic agent mezerein (MZN). This chapter reports the development of a method for large-scale production of human immune interferon based on this induction scheme with the addition of a cell fractionation process and a novel method for product recovery. A 10- to 15-fold enhancement in interferon yield and a 50-fold increase in purity of crude product was accomplished by nylon wool column fractionation of lymphocytes combined with simultaneous reduction of autologous plasma protein levels prior to induction.

Preparation of Fractionated Human Lymphocytes

Buffy coats were obtained from the Tidewater Regional Red Cross Blood Services (Norfolk, VA). Ten to 20 buffy coats were pooled within 24 hr of collection. Each bag was emptied into a sterile container and rinsed with 20 ml of RPMI 1640 medium (Flow Laboratories, McLean, VA) containing 50 units/ml of heparin (Flow Laboratories). A total viable white blood cell (WBC) count was performed by the standard trypan blue exclusion method. In addition, a sample was microscopically examined after acetic acid treatment for determination of total WBC nuclei count and differential WBC counts were determined by standard procedures

[1] Research described in this chapter was carried out at the American Red Cross Blood Services Laboratories, Bethesda, MD and at Flow Laboratories Inc., McLean, VA.

[2] Parts of this material have been adapted from R. van Reis, A. S. Lubiniecki, R. A. Olson, R. R. Stromberg, J. A. Madsen, and L. I. Friedman, *J. Immunol.* **133,** 758 (1984).

[3] Y. K. Yip, R. H. L. Pang, C. Urban, and J. Vilĉek, *Proc. Natl. Acad. Sci. U.S.A.* **78,** 1601 (1981).

[4] Abbreviations used: Con A, concanavalin A; PHA, phytohemagglutinin; PMA, phorbol myristate acetate; MZN, mezerein; WBC, white blood cell; PBS, phosphate buffered saline.

with Wright stain. The pooled cell suspension was diluted with RPMI 1640 medium containing 3 m*M* Tricine (Sigma, St. Louis, MO) and 4 m*M* L-glutamine (Flow Laboratories, McLean, VA) to a final cell concentration of 12×10^6 cells/ml. The diluted cells were then immediately pumped through a nylon wool column (Leuko-Pak, Fenwal Laboratories, Deerfield, IL) with a Masterflex pump (Cole-Parmer, Chicago, IL) at a rate of 40 ml/min. The column was subsequently rinsed with an equal volume of medium. Total and differential WBC counts performed pre- and postfractionation demonstrated 85% recovery of lymphocytes and a 90% reduction of neutrophils. Monocytes were reduced from 4.0 to 0.5%, eosinophils decreased from 3.0 to less than 0.2% and basophils were depleted from 1% to undetectable levels. Total viable cell counts were in excess of 95%. The cell suspension was centrifuged for 15 min at 200 *g* and the supernatant was removed and replaced with an equal amount of medium. This resulted in a reduction of average plasma protein content from 3.85 to 1.07 mg/ml. The final WBC concentration was 2×10^6 cells/ml. A total of 6 liters of cell suspension was obtained from 20 buffy coats. Initial experiments demonstrated that ammonium chloride lysis of erythrocytes prior to induction adversely affected immune interferon yields. These cells were therefore not removed from the cell suspension.

Induction Procedures and Incubation Conditions

The cell suspensions were incubated asceptically in either a warm room or a water bath at 37° in siliconized glass spinner flasks with working volumes of one liter. Agitation was provided by means of magnetic stirrers at rates sufficient to avoid sedimentation. The cultures were allowed to reach 37° prior to addition of 5 ng/ml of PMA (LC Services, Woburn, MA). After 3 hr of incubation the cultures were induced with 5 μg/ml of PHA (P-L Biochemicals, Milwaukee, WI). Fractionated cell cultures had greater than 95% cell viability after 48 hr of incubation in contrast to unfractionated cell cultures which possessed a large percentage of nonviable cells as determined by trypan blue staining. Optimal yields of interferon were obtained after 48–72 hr of incubation in fractionated cell cultures.

Recovery of Crude Interferon Preparation

Two different methods of recovery were employed. In a simple procedure for small-scale processing the cells were centrifuged in one liter bottles for 15 min at 7000 *g* in a Sorvall RC-3 centrifuge. The supernatant was removed and stored at −20° for subsequent purification. Another method aimed at large-scale production and recovery involved the use of a cascade filtration system initially developed for natural human alpha interferon production.[5] A flow diagram for the filtration system is shown in Fig. 1. The spinner flask used for interferon production was connected to an Asahi Plasmaflo® (Parker-Hannifin, Irvine, CA) hollow fiber cartridge with a sterile tubing harness (TS-110, Extracorporeal, King of Prussia, PA) and a peristaltic pump (Model 5500, Sarns, Ann Arbor, MI). The Asahi hollow fiber cartridge provided 7 square feet of filtration area with a nominal pore size of 0.2 μm. A siliconized glass spinner flask used for collection of interferon solution was connected in the same way to a parallel set of Amicon Diafilter 20 (Amicon Corporation, Lexington, MA) hollow fiber cartridges. These cartridges provided a total of 4.3 square feet of membrane area with a nominal molecular weight cut-off of 10,000. Two glass reservoirs equipped with sterile air filters were connected to the system via a three-way valve and a peristaltic pump. The reservoirs contained 4 liters of PBS (Flow Laboratories, McLean, VA) and 4 liters of media, respectively. A total of 1500 ml of cell suspension was added to the production vessel and subsequently diafiltered with media to achieve the desired concentration of autologous plasma protein. The volume of media used was determined by

$$V_M = V_S \ln(C_0/C_t)$$

where V_M = media volume, V_S = cell suspension volume, C_0 = initial protein concentration, and C_t = final protein concentration. The cell suspension was recirculated through the Asahi cartridge at a flow rate of 300 ml/min and an inlet pressure of 30 mm Hg. No cellular damage was detected as determined by total and viable cell counts. The filtration rate was 100 ml/min. Part of the protein-containing filtrate was used to pretreat the Amicon hollow fiber cartridges in order to avoid binding of interferon to the polysulfone membrane. This was accomplished by recirculating the protein solution through the cartridges for 15 min. The remaining protein solution was discarded. Induction was carried out as described above. During incubation the cell suspension was recirculated through the hollow fiber at a flow rate of approximately 50 ml/min with the filtrate port closed. After 72 hr the interferon was recovered as follows. The recirculation rate of the cell suspension was increased to 300 ml/min and the filtrate taken off was collected in a separate spinner vessel. When 1000 ml of interferon solution had been collected in the receiving vessel, the second recirculation system was started. A filtration rate of 100 ml/min was obtained at a recirculation rate of 250 ml/min and an inlet pressure of 80 mm Hg. Media was simultaneously pumped into the pro-

[5] R. van Reis, R. R. Stromberg, L. I. Friedman, J. Kern, and J. Franke, *J. Interferon Res.* **2**, 533 (1982).

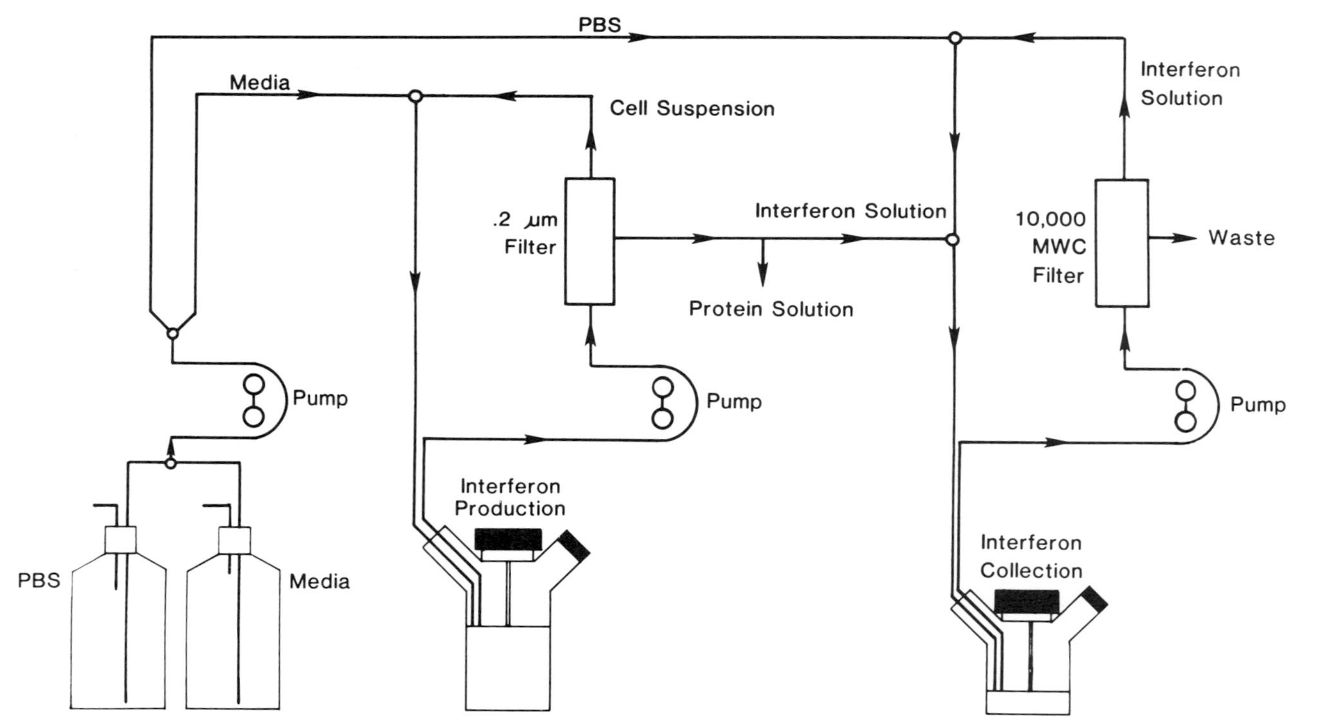

FIG. 1. Flow diagram of the cascade filtration system used for interferon production and recovery. Prior to induction, the cell suspension was recirculated through a microporous hollow fiber cartridge allowing a cell-free protein solution to be taken off by continuous replacement with media. After production of interferon was completed, the cell suspension was again recirculated through the same cartridge which permitted a particle-free interferon solution to be removed and subsequently concentrated in a second hollow fiber system with an ultrafiltration membrane. The interferon solution in the cell culture was replaced with an equal volume of PBS. After recovery, the interferon solution was further concentrated and diafiltered with PBS in order to remove low molecular weight contaminants.

duction vessel at the same flow rate as the filtrate. After perfusion of 4 liters of media, the first filtration system was shut off and 4 liters of PBS was pumped into the collection vessel at a rate of 100 ml/min. The filtrate containing low molecular weight contaminants was discarded. Interferon was quantitatively recovered and stored at −20° for subsequent purification.

Attempts to increase immune interferon production by continuous removal during incubation were unsuccessful, in contrast to the demonstrated utility for leukocyte interferon production.[5] Cell viability was maintained, however, and initial experiments demonstrated the feasibility of extended interferon production by use of a multiple harvest/reinduction scheme.

Interferon and Protein Assays

Interferon activities were determined on cell-free culture solutions by a cytopathic effect inhibition assay with human HT1080 tumor cells (Flow Laboratories, McLean, VA) challenged by the Indiana strain of vesicular stomatitis virus. An internal gamma interferon standard (Flow Laboratories) was used. Details of the procedure can be found in Ref. 2.

Total protein determinations were performed on cell-free culture solutions according to the method of Lowry with a bovine serum albumin standard (Bio-Rad, Richmond CA).

Concluding Remarks

The production method described here is suitable for large-scale preparation of natural human immune interferon. Its principal advantages include greater interferon yield and purity without substantial increase in process costs. The process provides a method for production of both higher and more consistent yields of interferon (Table I). A major factor in

TABLE I
IMMUNE INTERFERON YIELD IN FRACTIONATED AND IN UNTREATED CELL CULTURES[a]

	Experiment number				
	1	2	3	4	5
Fractionated (units/ml)	24,000	23,000	51,000	58,000	16,000
Untreated (units/ml)	420	4,300	4,000	8,700	370

[a] Adapted from van Reis *et al.*[2]

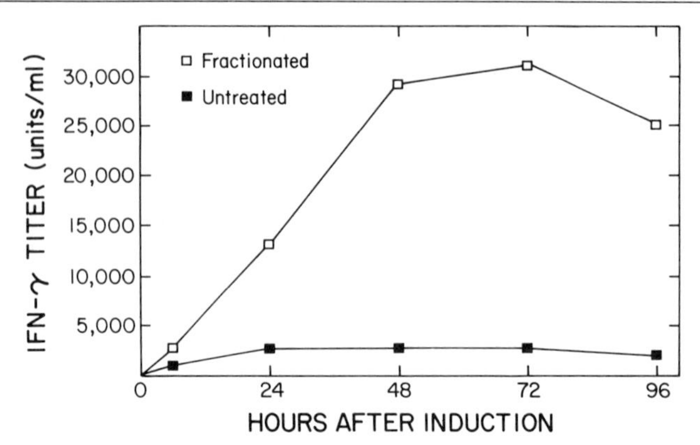

FIG. 2. Kinetics of immune interferon production in fractionated and in untreated cell cultures. Each data point represents an average titer from five experiments. (Adapted from van Reis *et al.*[2])

the higher yield from nylon wool fractionated buffy coat cells is prolongation of the active production period compared to unfractionated cell preparations (Fig. 2). Higher purity without concomitant yield loss results from removal of about 70% of buffy coat plasma, a step which results in reduced yields when using unfractionated cells. This sequential application of nylon wool fractionation and protein removal steps combine to provide 50-fold improvement in initial purity (Table II), which should lead to higher purification yields and a reduction in purification costs. The nylon absorbent material is prepared for blood transfusion applications; therefore, it is not expected to compromise the safety of biologicals resulting from this process. Enhanced production of other essential lymphokines such as interleukin-2 and lymphotoxin might also result from employing this procedure since these proteins are produced by the same induction schemes.

TABLE II
SPECIFIC ACTIVITY OF CRUDE IMMUNE INTERFERON IN FRACTIONATED/REDUCED PROTEIN CELL CULTURES AND IN UNTREATED CELL CULTURES[a]

	Experiment number				
	1	2	3	4	5
Fract./red. (units/mg)	21,000	27,000	39,000	18,000	10,000
Untreated (units/mg)	70	980	620	350	45

[a] Adapted from van Reis *et al.*[2]

Finally, the method presented consists of procedures which represent no significant barriers to large-scale operations. This method is recommended for general applications in both small and large-scale cases where minimal cost and maximal yield and purity of crude immune interferon are important considerations.

[12] Large-Scale Production, Concentration, and Partial Purification of Human Immune Interferon

By JERZY A. GEORGIADES

Induction of the synthesis of gamma interferon by lymphocytes is connected with the activation of a variety of peripheral white blood cells (PWBC) with nonspecific mitogens (SEA, SEB, PHA, or Con A). Normally PWBC from a large number of donors are employed for large-scale production. In such situations, a mixed lymphocyte reaction occurs in addition to mitogen activation. As a result of the combined activation, different types of PWBC release into the media gamma interferon plus a variety of enzymes (proteases, phosphatases, oxygenases, phosphorylases, lipases, etc.) and other lymphokines (interleukins 1, 2, and 3, inhibitors, other molecules, and growth factors). In media from stimulated PWBC, more than 100 different types of activities have been described.[1] It is not clear how many distinct molecules account for all these different activities.

Isolation of gamma interferon from the products of activated PWBC and the proteins and mitogens introduced in the incubation medium creates formidable purification problems. In addition, released proteases (J. A. Georgiades, unpublished observation) present in this mixture cause constant degradation of gamma interferon. At the time of harvest, the medium contains intact gamma interferon molecules as well as partially degraded interferon products which may have biological activity but differ physicochemically and in stability from the native molecules[2] (J. A. Georgiades, unpublished observation). In earlier studies, we have observed such heterogeneity in purified gamma interferon preparations.[3,4]

[1] J. Georgiades, S. Baron, W. Fleichmann, Jr., M. Langford, D. Weigent, and G. Stanton, *in* "Interferons" (P. E. Came and W. A. Carter, eds.), p. 305. Springer-Verlag, Berlin and New York.

[2] J. Sims, *Proc. 1984 TNO-ISIR Meet. Interferon Syst., 1984* (1984).

[3] J. Georgiades, *in* "Texas Reports on Biology and Medicine" (S. Baron, F. Dianziani, and G. Stanton, eds.), p. 260. University of Texas Medical Branch, Galveston, 1982.

[4] J. Friedlander, D. Fischer, and M. Rubinstein, *Anal. Biochem.* **137,** 115 (1984).

The properties of gamma interferon, especially its pH and heat sensitivity, its destruction by organic solvents and ionic detergents are additional factors which make the purification procedure complex.[1]

The observation of Pestka *et al.*[5] that the functional unit of gamma interferon is trimeric or tetrameric adds another dimension of difficulty for purification. Use of any drastic separation methodology leads to disassociation and appearance of monomeric forms of gamma interferon.[6] Thus, it is desirable to purify gamma interferon in a manner designed to produce the native form.

Existing purification methods utilize adsorption to controlled pore glass beads.[7-10] Eluted with ethylene glycol or TMAC, gamma interferon is further purified by partition chromatography, by affinity chromatography columns[6-8] or by ion-exchange matrices.[8] Recently, partially purified material was also purified on an immunoaffinity column prepared from monoclonal antibodies and successfully eluted at high pH.[11]

Though these methods are successful on a laboratory or analytical scale, they are laborious and frequently require intermediate steps like dialysis. Frequently end products, although very highly purified, show lack of stability under conditions of clinical use and appear to be in the monomeric form.[8] Furthermore, use of such methods (ethylene glycol or high pH) for prolonged periods of time during purification may cause irreversible losses of other biologically active molecules.[1] These could also be of clinical interest and could be recovered during gamma interferon purification should the process be gentle.

Taking all of the above into consideration, we developed a purification procedure for gamma interferon which would satisfy the following criteria:

1. The method should be applicable for large-scale production.
2. It should be possible to purify 100–500 liters of crude material without difficulty.

[5] S. Pestka, B. Kelder, P. Familletti, J. Moschera, R. Crowl, and E. Kempner, *J. Biol. Chem.* **258,** 9706 (1983).

[6] Y. Yip, B. Barrowclough, C. Urban, and J. Vilček, *Science* **215,** 411 (1982).

[7] J. Georgiades, M. Langford, G. Stanton, and H. Johnson, *IRCS Med. Sci.: Libr. Compend.* **7,** 559 (1979).

[8] I. A. Braude, U.S. Patent 4,440,675 (1984).

[9] M. Wiranowska-Stewart, L. Lin, I. A. Braude, and W. Stewart, II. *Mol. Immunol.* **17,** 625 (1980).

[10] V. Papermaster and S. Baron, *in* "Texas Reports on Biology and Medicine" (S. Baron, F. Dianzani, and G. Stanton, eds.), p. 672. University of Texas Medical Branch, Galveston, 1982.

[11] D. Novick, Z. Eshhar, D. Fischer, J. Friedlander, and M. Rubinstein, *EMBO J.* **2,** 1527 (1983).

3. The procedure must be fast enough to accomplish the process within 36–48 hr.
4. Each step must be carried out under mild chromatographic conditions so that fractions which do not contain gamma interferon may be used for further purification of other biologically active molecules.
5. The purified interferon must be free from other commonly known lymphokines.
6. Purified interferon must be stable for at least 12 weeks at 2–8°.
7. The resulting product should be in its native form.

In developing the purification procedure described below,[12] we tried to take into consideration all the above mentioned factors. This method of purification has now been in routine use for more than two years. The procedure gives a reproducible, consistent product.

Interferon Gamma Induction

To achieve success in purification, it was necessary to achieve adequate induction of gamma interferon by the PWBC. Gamma interferon batches which gave us an initial titer below 300 units/ml usually gave poor results during purification.[13] For that reason, we prepared crude gamma interferon from cultures of PWBC stimulated for 4 days with SEB (early induction), followed by 2 days stimulation with PHA (late induction). Purification procedures described herein are applicable to both types of induction of crude gamma interferon. For our purposes, the induction media from which the PWBC have been removed is called "crude material."

Purification Procedure

Step I

One hundred (100) liters of crude material with an average titer of 10^3 units/ml and containing 1 mM EDTA tetrasodium salt was clarified through a 0.5-μm Millipore membrane at room temperature by means of a Procon pump. The starting material typically contained a total of 1×10^8 units of gamma interferon and 180 g of protein (specific activity 600 units/mg protein).

[12] J. Georgiades, Gumulka, and Sulkowski, in preparation.

[13] J. Georgiades, *in* "Texas Reports on Biology and Medicine" (S. Baron, F. Dianzani, and G. Stanton, eds.), p. 179. University of Texas Medical Branch, Galveston, 1982.

Retained material, approximately 2000 ml, contained about 1.4 mg/ml of protein and was not used. The ultrafiltrate from the 0.5-μm cassette filter (containing 1×10^8 units of gamma interferon) was directly concentrated five times (to 20 liters) on a 10,000 molecular weight cut-off membrane. Five separate times, 20 liters of deionized water was added and the mixture concentrated to the original volume of 20 liters. The ultrafiltrate did not contain a detectable level of gamma interferon activity. With this step, specific activity of gamma interferon in the retained material increased to 900 units/mg of protein. Decrease of the ionic strength on the cassette system following addition of deionized water did not affect the interferon titer. This step gave approximately 1.5-fold purification.

Step II

The concentrated material was acidified to pH 6.5 to 6.7 with 50% acetic acid while being cooled quickly to 10 to 12°, then mixed with two liters of activated CM-Sepharose Cl-6B. After mixing for one hr at 4°, the supernatant was separated from the beads by means of filtration. CM-Sepharose with adsorbed interferon was loaded into a column. The matrix was washed with 0.02 *M* sodium acetate buffer, pH 6.3, until all unadsorbed proteins were removed, then washed with 0.02 *M* sodium phosphate buffer, pH 7.0 to 7.05, to remove proteins of noninterest. Proteins released at this pH were discarded.

Gamma interferon was eluted with phosphate buffer containing 1.0 *M* sodium chloride, pH 7.2 to 7.4, at a linear flow rate of 10 cm/hr. Approximately 90% of the applied interferon was found in the major fraction. The collected fraction of 1100 ml contained 605 mg (i.e., 0.55 mg/ml) of protein and 10^6 units/ml of interferon (specific activity 1.8×10^6 units/mg protein). Purification was 300-fold; recovery was 200%.

Step III

The interferon fraction from the CM column was directly applied to a zinc-chelate column[14,15] equilibrated with 0.02 *m* sodium phosphate/1 *M* NaCl, pH 7.4, at a temperature of 4°. Material not adsorbed to the column contained 90% of the applied interferon activity. The phosphate buffer with 1.0 *M* NaCl wash removed the remaining 10% of activity. This step gave total recovery of applied interferon and threefold decrease in protein

[14] K. J. W. Heine and A. Billiau, this series, Vol. 78, p. 448.

[15] V. G. Edy, A. Billiau, and P. DeSomer, *J. Biol. Chem.* **252,** 5934 (1977).

concentration. Specific activity increased from 1.8×10^6 to 5.4×10^6 units/mg.

Step IV

The fraction which was unadsorbed to the zinc chelate column and wash fraction were pooled and applied directly onto a Con A-agarose column at 4°. This breakthrough fraction contained 1% of the interferon activity and 75% of applied proteins. After washing with PBS, interferon was eluted with 0.1 *M* methyl-D-α-mannopyranoside containing 1.0 *M* NaCl. The major interferon fraction contains 80% of applied activity with specific activity of 1×10^7 units/mg of protein.

The final gamma interferon product was free of inducers, IL-1, IL-2, IL-3, and had an undetectable level of lymphotoxins, MIF, and chemotactic factors. Levels of other types of interferons were undetectable as shown by absence of interferon activity on bovine embryo kidney cells (BEKC) and its sensitivity to heating, and pH 2. In addition, it was not neutralized by specific antibodies to Hu-IFN-α, Hu-IFN-β, or a mixture of antibodies to IFN-α and IFN-β, but was neutralized by a specific antibody against Hu-IFN-γ.

Conclusions

The above purification method is simple, effective, and relatively fast (purification time does not exceed 48 hr). In addition, this method can be applied to large quantities and can be totally automated without need for intermediate steps which would otherwise require adjustment of the material between steps of purification. Moreover, this method minimizes the possibility of bacterial and fungal contamination, is reproducible, and IFN-γ appears to be in tetrameric form. Because of the very mild conditions of the purification, protein denaturation or alteration is reduced. Intermediate fractions obtained during IFN-γ purification can be utilized for purification of other lymphokines (IL-2 for example).

[13] Production and Partial Purification of Human Immune Interferon Induced with Concanavalin A, Staphylococcal Enterotoxin A, and OK-432

By MARC DE LEY, JOZEF VAN DAMME, and ALFONS BILLIAU

Introduction

The availability of large amounts of Hu-IFN-γ is a prerequisite for the study of its biochemical and biological properties as well as for its clinical application. The methods described in this chapter were routinely carried out on an average of 30 pooled buffy coats, but they can in principle be scaled up at will.

Preparation of the White Blood Cells

Depending on the number of red blood cells that is allowed as a contamination of the white cell population two different methods are available for further processing of the pooled buffy coats.

1. Erythrocytes are removed by ammonium chloride treatment as described in detail by Cantell *et al.*[1] Briefly, buffy coats are collected and pooled in 0.4% EDTA and treated twice with two volumes of 0.83% NH_4Cl for 10 min at 4°. After a final centrifugation they are resuspended in RPMI 1640 medium supplemented with 0.2 m*M* glutamine, 100 units/ml penicillin, 100 μg/ml streptomycin, 1% (v/v) fetal bovine serum, 6 m*M* HEPES, 13 m*M* tricine, pH 7.4 (induction medium). This method results in a population of white blood cells virtually devoid of erythrocytes.

2. Alternatively the erythrocytes may be removed by an agglutination and an accelerated sedimentation at unit gravity.[2,3] Briefly, pooled buffy coats are mixed with Dulbecco's phosphate-buffered saline (PBS) and 6% hydroxyethyl starch (Plasmasteril, Fresenius, Bad Hamburg, FRG; v/v/v, 1 : 1 : 1) and allowed to sediment for 30 min at 37°. The upper phase is then collected and centrifuged at 350 *g* for 10 min. The cells are washed twice with PBS, and resuspended in induction medium. The leukocytes obtained by this procedure are more contaminated by erythrocytes than

[1] K. Cantell, S. Hiervonen, K. E. Mogensen, and L. Pyhälä, *In Vitro* **3,** 35 (1974).

[2] H.-L. Kauppinen and G. Myllylä, *in* "Interferon: Properties and Clinical Uses" (A. Khan, N. O. Hill, and G. L. Dora, eds.), p. 1. Leland Fikes Foundation Press of Wadley Institutes of Molecular Medicine, Dallas, Texas, 1979.

[3] M. De Ley and H. Claeys, *Int. Arch. Allergy Appl. Immunol.* **74,** 21 (1984).

those obtained by the previous method. This method however is less complicated and contact of the leukocytes with ammonium chloride is avoided.

Preparation of the Inducers

Concanavalin A (Con A)

Commercially available Con A (Calbiochem, San Diego, CA) is dissolved in PBS at 1 mg/ml and sterilized by filtration through a 0.22-μm Millipore filter.

Staphylococcal Enterotoxin A (SEA)

The culture broth of *Staphylococcus aureus*, strain 13N2909, is used as the source of SEA.[4] *S. aureus* 13N2909, known to be a high producer of SEA, was kindly provided by Dr. J. F. Metzger (U.S. Army Medical Research Institute of Infectious Diseases, Fort Detrick, MD). After growing in trypticase soy broth for 48 hr at 27° the cells are removed by centrifugation (10,000 rpm, 30 min). The culture broth is sterilized by ultrafiltration through a 0.22-μm filter (Type GS, Millipore, Molsheim, France) and stored at −20°.

Streptococcal Preparation OK-432

The streptococcal preparation OK-432 (Picibanil; Chugai Pharmaceutical Co., Tokyo, Japan) is an inactivated and lyophilized preparation of a low virulence strain of *Streptococcus pyogenes* A3. Immediately before use it is reconstituted by the addition of sterile water. The coccus content of OK-432 is expressed in Klinische Einheit (KE), 1 KE containing 0.1 mg of dried cocci (10^7–10^8 cocci).

Preparation and Induction of the White Blood Cell Cultures

The white blood cells isolated as mentioned above (either by the NH_4Cl method or the unit gravity sedimentation procedure depending on the inducer to be used) are resuspended in induction medium at a cell density of 5×10^6 cells/ml. The cells are then induced to produce IFN-γ by addition of either Con A (10 μg/ml), SEA (culture broth diluted about 1/1000, the optimal dilution being determined experimentally) or OK-432 (0.02 KE/ml). Due to the greater contamination of white blood cells with erythrocytes in the unity gravity sedimentation method, only NH_4Cl-

[4] L. Spero and J. F. Metzger, this series, Vol. 78, p. 331.

treated buffy coats are suitable for induction with Con A. With SEA or OK-432 as inducers the best results are obtained with hydroxyethyl starch-treated buffy coats. The induced cultures are then transferred in 250 ml portions to 1/2 gallon roller bottles (2 rpm) and incubated at 37°.

Harvesting of the Cultures and Initial Purification on Silicic Acid

After 2 days of incubation at 37° in the case of Con A or 3 days in the case of SEA or OK-432 the crude IFN is harvested, pooled, and clarified by centrifugation (3000 rpm, 10 min, 4°). The crude IFN-γ is then partially purified and concentrated by adsorption to silicic acid and subsequent elution with 50% ethylene glycol or 0.4 *M* tetraethylammonium chloride in PBS.[3,5,6]

All purification steps are carried out at 4°. Crude IFN-γ is mixed with silicic acid (60 Å, LC 50–100 MY; Grace GmbH, Worms, FRG) at a ratio of 7.5 g/liter crude IFN-γ and stirred for 2 hr. The silicic acid is then removed by centrifugation (3000 rpm, 10 min, 4°) and washed twice with PBS at 1/10 of the starting volume of crude IFN. The bound IFN-γ is finally eluted by stirring the silicic acid for 15 min with 0.4 *M* tetraethylammonium chloride in PBS at 1/20 of the starting volume twice. The two eluted fractions are then either pooled or separately dialyzed against 15% (w/v) polyethylene glycol 20,000 in 20 m*M* Tris · HCl buffer, pH 8.0.

Treatment with DEAE-Sephacel

In order to remove the majority of proteins with an isoionic point below 8, the silicic acid-purified IFN-γ is treated with DEAE-Sephacel at pH 8.[7] The silicic acid-purified interferon is extensively equilibrated with 20 m*M* Tris · HCl buffer, pH 8. The interferon is then transferred to a spinner flask and stirred with DEAE-Sephacel (1/10 of the IFN-γ sample volume) for 15 min at 4°. The DEAE-Sephacel is then removed by centrifugation at 3000 rpm for 10 min or by filtration through a sinterred glass filter.

Affinity Chromatography on Con A-Sepharose

Con A-Sepharose columns (0.9 × 15 cm; Pharmacia Fine Chemicals, Uppsala, Sweden) are equilibrated with at least 10 column volumes of 0.5

[5] M. De Ley, J. Van Damme, H. Claeys, H. Weening, J. W. Heine, A. Billiau, C. Vermylen, and P. De Somer, *Eur. J. Immunol.* **10,** 877 (1980).

[6] J. Van Damme, M. De Ley, H. Claeys, A. Billiau, C. Vermylen, and P. De Somer, *Eur. J. Immunol.* **11,** 937 (1981).

[7] Y. K. Yip, B. S. Barrowclough, C. Urban, and J. Vilček, *Proc. Natl. Acad. Sci. U.S.A.* **79,** 1820 (1982).

M NaCl in 20 m*M* Tris · HCl, pH 8.0. The DEAE-Sephacel-treated interferon in 20 m*M* Tris · HCl, pH 8.0, is adjusted to 0.5 *M* NaCl by the addition of 4 *M* NaCl in distilled water and then applied to Con A-Sepharose at a flow rate of 10 ml/hr. Then the column is washed with 0.5 *M* NaCl/20 m*M* Tris · HCl buffer, pH 8.0, until the absorbance at 280 nm (monitored with a LKB 2238 Uvicord SII, Bromma, Sweden) again reaches the baseline. Then the bound IFN-γ is eluted with 0.5 *M* α-methyl-D-mannopyranoside (α-MM), 0.5 *M* NaCl in 20 m*M* Tris · HCl buffer, pH 8.0.

Affinity Chromatography on Poly(U)- and Blue Sepharose

The Hu-IFN-γ containing fractions from the Con A-Sepharose chromatography are extensively dialyzed against 10 m*M* Tris · HCl buffer, pH 7.4. After equilibration of a poly(U)-Sepharose column (0.9 × 15 cm; Pharmacia Fine Chemicals, Uppsala, Sweden) against the same buffer, the preparation is applied at a flow rate of 10 ml/hr, after which the column is washed with starting buffer until the absorbance reaches the baseline again. The bound IFN-γ is then eluted with 1 *M* NaCl in 10 m*M* Tris · HCl, pH 7.4. The fractions containing maximal antiviral activity are then pooled and again dialyzed against 10 m*M* Tris · HCl, pH 7.4.

The subsequent affinity chromatography on blue Sepharose (0.9 × 15 cm; Pharmacia Fine Chemicals, Uppsala, Sweden) is performed in exactly the same way as in the case for poly(U)-Sepharose.[8]

Concluding Comments

The results of the production of Hu-IFN-γ after stimulation of human leukocytes (isolated either by the sedimentation or the NH_4Cl method) with different inducers are summarized in Table I. Two conclusions can be drawn from this table. First, both SEA and OK-432 are far superior to Con A with respect to IFN-γ production. Second, if Con A is used as an inducer, leukocytes isolated by the NH_4Cl method are far superior to those isolated by the sedimentation method. This could be explained by the greater contamination with erythrocytes in the latter method as compared to the former one. Erythrocytes, by also binding Con A, make it less available for induction of leukocytes.

The adsorption/elution step on silicic acid and the treatment in batch with DEAE-Sephacel are both efficient and straightforward purification steps. They result in a marked purification with high recovery (Table II).

[8] A. Mizrahi, J. A. O'Malley, W. A. Carter, A. Takatsuki, G. Tamura, and E. Sulkowski, *J. Biol. Chem.* **253**, 7612 (1979).

TABLE I
PRODUCTION OF Hu-IFN-γ WITH DIFFERENT INDUCERS AND DIFFERENT LEUKOCYTE ISOLATION METHODS

Method of leukocyte isolation	Inducer	Titer crude IFN ($\log_{10}$ units/ml)[a]
Sedimentation	Con A	2.2
	SEA	2.9
	OK-432	3.6
NH_4Cl	Con A	2.8
	SEA	3.2
	OK-432	4.1

[a] As no internationally accepted reference preparation for Hu-IFN-γ was available at the time of these experiments, our internal laboratory standard was assigned an arbitrary value (units/ml) which corresponded to the reciprocal of the average titration end point on human diploid cells.[5] This allowed the expression of all results in arbitrary units.

TABLE II
PURIFICATION OF Hu-IFN-γ

Purification step	Purification factor	Stepwise recovery (%)
Silicic acid	17×	70
DEAE-Sephacel	8×	100
Con A-Sepharose	10×	50
Poly(U)-Sepharose	4×	80
Blue-Sepharose	2×	66

The DEAE-Sephacel treatment is particularly efficient in that it only takes 30 min, proceeds with essentially 100% recovery, and avoids overloading of the subsequent Con A-Sepharose affinity column. On elution with 0.5 *M* α-MM/0.5 *M* NaCl in 20 m*M* Tris · HCl, pH 8.0, a 10-fold purification with only 50% recovery was achieved. Introduction of 20% (w/v) ethylene glycol in the elution buffer did not increase the yield of Hu-IFN-γ obtained in this step.

[14] Induction of Human Immune Interferon from Peripheral Blood Mononuclear Cells Induced with Combinations of T Cell Mitogens

By MASAFUMI TSUJIMOTO and NAOKI HIGASHI

We describe our procedure for the induction of immune interferon from human mononuclear cells by the combined treatment with OK-432[1,2] and staphylococcal enterotoxin B (SEB)[3] in serum-free medium.[4] We can obtain a high titer of immune interferon because OK-432 and SEB act synergistically on the induction of immune interferon.

Materials

OK-432. Penicillin- and H_2O_2-killed lyophilized preparation of Su strain of *Streptococcus pyrogenes* (Chugai Pharmaceutical Co., Ltd., Tokyo, Japan). One KE (Klinische Einheit) of OK-432 corresponds to 0.1 mg of lyophilized preparation.

SEB. Staphylococcal enterotoxin B used without further purification (Maker Chemicals, Ltd., Jerusalem, Israel). These materials are stored at 4° until use.

Serum-Free Induction Medium. We found the following medium to be excellent for the induction of human immune interferon. Medium (100 ml) contains RPMI-1640 medium (96 ml, Nissui Pharmaceutical Co., Ltd., Tokyo, Japan), essential amino acid solution (×50) (2 ml, Flow Laboratories Inc., McLean, VA), nonessential amino acid solution (×100) (1 ml, Flow Laboratories), human serum albumin (fraction V) (350 mg, Seikagaku Kogyo Co., Ltd., Tokyo, Japan), *N*-2-hydroxyethylpiperazine-*N*′-2-ethanesulfonic acid (10 m*M*, Sigma Chemical Co., St. Louis, MO), and glutamine (20 m*M*, Nissui Pharmaceutical Co., Ltd.). Medium also contains penicillin G (100 units/ml) and streptomycin (100 μg/ml).

[1] S. Matsubara, F. Suzuki, and N. Ishida, *Cancer Immunol. Immunother.* **6,** 41 (1979).
[2] M. Saito, T. Ebina, M. Koi, T. Yamaguchi, Y. Kawade, and N. Ishida, *Cell. Immunol.* **68,** 187 (1982).
[3] P. von Wussow, Y. Chen, M. Wiranowska-Stewart, and W. E. Stewart, II, *J. Interferon Res.* **2,** 11 (1982).
[4] M. Tsujimoto, T. Tarutani, K. Ogawa, K. Akashi, Y. Okubo, and N. Higashi, *Infect. Immun.* **41,** 181 (1983).

Interferon Assay

Antiviral assay of immune interferon is performed by a cytopathic effect inhibition assay with sindbis virus and FL amnion cells. Units were determined with respect to a laboratory reference preparation (crude immune interferon) which was titrated against the human leukocyte interferon reference standard from the National Institutes of Health (G-023-901-527).

Procedure

For induction of immune interferon, human mononuclear cells are cultured in the serum-free induction medium. Blood is collected in heparin by venipuncture from healthy donors. Buffy coats are collected by centrifugation at 2000 rpm for 10 min at 15°. The cells are resuspended in conical plastic centrifuge tubes (15 or 50 ml, Corning) and an equal volume of Ficoll–Paque (Pharmacia) is layered at the bottom of the suspension with the use of a 14-cm long needle. Mononuclear cells are isolated by sequential centrifugation at 800 rpm for 5 min, 1000 rpm for 5 min, and

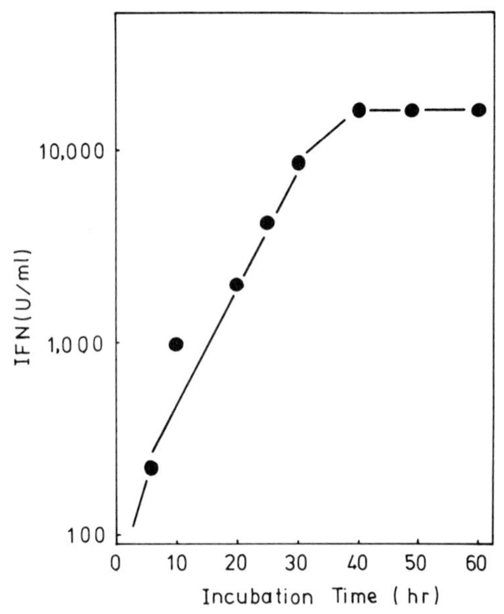

Fig. 1. Kinetics of interferon production from human mononuclear cells treated with OK-432 (0.025 KE/ml) and SEB (50 ng/ml).

SYNERGISTIC EFFECT OF OK-432 AND SEB ON IFN PRODUCTION

Amount of OK-432 (KE/ml)	IFN (U/ml) at SEB concentration (ng/ml)			
	0	6	20	60
0	0	4	16	64
0.0075	0	32	32	250
0.025	16	259	2000	2000
0.075	64	1000	2000	4000

then 2000 rpm for 10 min at 19°. Cells at the interface are collected and washed twice with Dulbecco's PBS (minus calcium and magnesium), then with the induction medium, and finally are suspended in a 24-well plate or a tissue culture flask (Corning) or desired volume at 5×10^6 cells per ml. Usually, the inducers, OK-432 and SEB, are added simultaneously to final concentrations of 0.025 KE/ml and 50 ng/ml, respectively. Incubation is carried out in stationary culture at 37° in a humidified atmosphere of 95% air and 5% CO_2 without further treatment. After incubation, the culture fluid is collected by centrifugation at 1500 rpm for 10 min and is stored at −80° until use. Interferon activity is detectable as early as 6 hr after treatment, and reaches a maximum titer between 30 and 40 hr (Fig. 1). We usually collect the culture fluid 42 hr after treatment.

In this combination system, we could obtain human immune interferon with a relatively high titer especially from low responder cells (see the table). Almost all the interferon activity is neutralized by anti-IFN-γ serum, whereas anti-IFN-α serum or anti-IFN-β serum has no effect on interferon activity.

It seems that stimulation of mononuclear cells by the combined treatment with OK-432 and SEB eliminated variability of interferon yield among different cell preparations because of their synergistic effects on the induction of immune interferon. Furthermore, the purification procedure for immune interferon is simplified because the induction medium contains human serum albumin instead of serum.

[15] Preparation and Partial Purification of Human Interferon δ

By MILES F. WILKINSON and ALAN G. MORRIS

The present recognition of three antigenic types of interferon[1] does not exclude the existence of further types. We have recently studied the interferon activity of supernatants from fresh human peripheral blood mononuclear leukocytes (PBML) induced with mitogen together with a tumor promoter. The main activity in such supernatants is clearly IFN-γ: equally clearly there are components other than IFN-γ as has been reported by others.[2,3] The question arises whether these additional components belong to known interferon types or represent novel interferons. As we have argued elsewhere,[4,5] there is good evidence for such a novel interferon type, distinct from IFN-α, -β, and -γ by antigenic and physicochemical criteria. By definition, we refer to this species as IFN-δ. The most surprising characteristic of this interferon is that it appears to induce the antiviral state only in cells trisomic for chromosome 21, which are more sensitive than diploid cells to the classical interferons. This suggests that the major role of this interferon may be other than protection of cells against virus infection, and that its antiviral effect is indeed incidental to its main (unknown) function. In this respect, it is worth noting that macrophage-activating factor (MAF) has been shown to be identical with IFN-γ, and that indeed this protein is more potent in activating macrophages than as an antiviral agent.[6]

Our study of IFN-δ has not aimed at total purification, but instead the objective was to separate IFN-δ from IFN-γ in order to characterize it. This chapter describes simple techniques to achieve this separation.

[1] Committee on Nomenclature, *Nature (London)* **286,** 110 (1980).

[2] M. Wiranowska-Stewart, *J. Interferon Res.* **1,** 315 (1981).

[3] J. van Damme, M. de Ley, H. Claeys, A. Billiau, C. Vermylen, and P. DeSomer, *Dev. Biol. Stand.* **50,** 369 (1982).

[4] M. Wilkinson and A. Morris, *Biochem. Biophys. Res. Commun.* **111,** 498 (1983).

[5] M. Wilkinson and A. Morris, *in* "The Biology of the Interferon System 1983" (E. De-Maeyer and H. Schellekens, eds.), p. 149. Elsevier, Amsterdam, 1983.

[6] G. Nathan, H. W. Murray, M. E. Wieke, and B. Y. Rubin, *J. Exp. Med.* **158,** 670 (1983).

Materials, Cells, and General Procedures

Materials

Phytohemagglutinin (PHA) reagent grade from Sigma or pure from Wellcome.

Teleocidin[7] from Dr H. Fujiki, National Cancer Institute, Tokyo.

Con A-Sepharose from Pharmacia.

Matrex Blue from Dr. P. Dean, University of Liverpool, U.K.

Antiserum to IFN-γ was prepared in this laboratory by immunization of a rabbit with Hu-IFN-γ partially purified by absorption to silicic acid; the IgG from this antiserum was prepared and coupled to Sepharose by standard techniques to give anti-IFN-γ-Sepharose.

Cell Lines

GM2504, GM2767 cells (human fibroblasts trisomic for chromosome 21) were obtained from the NIGMS Genetic Mutant Human Cell Repository, Camden, New Jersey.

WISH cells [human amnion CCL25] were from Flow Laboratories (American Type Culture Collection catalog numbers are given in brackets.)

HFF (human diploid fibroblasts) were the kind gift of Dr. J. Vilcek, New York University Medical Center.

Hep2c cells [human embryo lung fibroblasts, CCL23], HeLa cells [human epithelial, CCL2], and L929 cells [mouse embryo fibroblasts, CCL1] were all laboratory strains.

EBTr cells [bovine turbinate, CCL44] and MDBK cells [bovine kidney, CCL22] were also from Dr. J. Vilcek.

All cell lines were cultured in GMEM (Glasgow Modification[8]) (Flow) supplemented with 10% fetal calf serum (Gibco).

Human PBML

Human buffy coat cells (from the Blood Transfusion Centre, Vincent Drive, Birmingham, U.K.) were overlaid on Ficoll–Hypaque (Pharmacia) and centrifuged at 800 *g* for 30 min. The interface cells (mononuclear leukocytes) were harvested and washed 3 times with PBS, then resuspended in RPMI-1640 medium (Flow) supplemented with 10% fetal calf

[7] H. Fujiki, M. Mori, M. Nakayasu, M. Teruda, T. Sugimura, and R. Moore, *Proc. Natl. Acad. Sci. U.S.A.* **78,** 38 (1981).

[8] I. A. McPherson and M. G. P. Stoker, *Virology* **16,** 147 (1962).

TABLE I
pH 2 STABLE ANTIVIRAL ACTIVITY PRODUCED BY PBML IN RESPONSE TO PHA/TELEOCIDIN[a]

Donor	PHA	Concentration (μg/ml)	Antiviral activity (units/ml)	
			pH 7	pH 2
1	Wellcome (purified)	5	40,000	10,000
1	Wellcome (purified)	20	13,000	4,000
1	Sigma (reagent grade)	10	200,000	8,000
1	Sigma (reagent grade)	20	100,000	6,300
1	Sigma (reagent grade)	40	80,000	6,300
2	Sigma (reagent grade)	10	6,300	1,300
3	Sigma (reagent grade)	10	130,000	16,000

[a] PBML at 2×10^6 cells/ml were induced with the type and concentration of PHA shown along with 10 ng/ml teleocidin. Media from the cells at 2 days were split into two portions: one was dialyzed against PBS for 2 days, the other was dialyzed against 87 m*M* KCl, adjusted with HCl to pH 2, for 24 hr at 4°, and then dialyzed against PBS (pH 7) for a further 24 hr. The antiviral activity in these samples was then assayed on GM2767 fibroblasts. The types of PHA used were purified PHA from Wellcome and reagent grade PHA from Sigma. In all other experiments, 10 μg/ml of reagent grade PHA from Sigma was used for induction.

serum (Gibco) at $1–2 \times 10^6$ cells/ml. PBML were cultured at 37° in 5% CO_2/air.

Production of IFN-δ and IFN-γ

Cells were treated with PHA (5–40 μg/ml) and teleocidin (10 ng/ml). Supernatants were harvested 2 days later and stored at −70°. Table I shows titers on GM2767 cells on antiviral activity of medium from PBML from three different donors after dialysis against pH 7 or pH 2 buffers. Since IFN-γ is pH 2 labile, the pH 2-resistant fraction is (as a first approximation) a measure of the non-IFN-γ component. As can be seen, this may be as much as 25% of the total activity.

Antiviral Assay

The $INAS_{50}$ method[9] (50% inhibition of nucleic acid synthesis) was used. Cells were treated with dilutions of interferon, left overnight, and

[9] G. J. Atkins, M. D. Johnston, L. M. Westmacott, and D. C. Burke, *J. Gen. Virol.* **25,** 381 (1974).

then challenged with Semliki Forest virus. GM2767 cells were used as the standard cell for assay of interferon; the other cell types listed above were used in comparative assays. A laboratory IFN-γ standard was used in all assays. At the time this work was done, no IFN-γ international standard was available. In our assay, 1 international unit of IFN-α titrates to about 10 units on GM2767 cells: that is, at a concentration of 0.1 unit/ml of the IFN-α reference standard, 50% inhibition of virus replication is observed. All titers quoted in the text are laboratory units based on assay with GM2767 cells.

Purification Procedures

Ammonium Sulfate Precipitation

Crude preparations of interferon (cell supernatants) were concentrated by ammonium sulfate precipitation. Solid ammonium sulfate was added to a concentration of 30% (w/v) at 4° for 2 hr. The precipitate at this stage was sedimented and discarded. Additional ammonium sulfate was added to a final concentration of 50% (w/v), left overnight at 4°, and the precipitate sedimented. This precipitate contained about 20–30% of the interferon activity. It was dissolved in a small volume of distilled water and dialyzed against PBS.

Chromatographic Techniques

The Con A-Sepharose column (1 ml) equilibrated in PBS was loaded in PBS and the void volume collected. The column was then washed with PBS and eluted with 0.5 *M* α-methyl mannoside (Sigma) in 0.5 *M* NaCl buffered with 10 m*M* sodium phosphate, pH 7.2. The Matrex blue column (1 ml) was loaded as described in the footnotes to Tables II and III, the void volume collected, washed with Tris · HCl, pH 7.5, or PBS for Tables II and III, respectively, and eluted with 40% ethylene glycol/1 *M* NaCl in the same buffer. The anti-IFN-γ-Sepharose column (1 ml) equilibrated in PBS was loaded in PBS and the void volume collected. No attempt was made to elute material from the column since standard elution techniques (e.g., pH 2) destroyed IFN-γ activity. As shown in Tables II and III, the void volumes from these three columns contain the bulk of the IFN-δ activity.

Separation of IFN-δ from IFN-γ

Table II shows that chromatography on either Con A-Sepharose or Matrex blue separates the pH 2-stable and -unstable components: the pH 2-

TABLE II
SEPARATION OF pH 2 STABLE ANTIVIRAL ACTIVITY FROM IFN-γ[a]

		Antiviral activity			
		pH 7		pH 2	
Chromatography	Step	Total units ($\times 10^4$)	Yield (%)	Total units ($\times 10^4$)	Yield (%)
Con A-Sepharose	Load	6.4	100	2.6	100
	Void	2.5	40	1.6	60
	Eluate	5.0	80	<0.1	<4
Matrex blue	Load	6.4	100	2.6	100
	Void	0.9	14	2.2	85
	Eluate	3.8	60	<0.2	<8

[a] Two day supernatants from PHA/teleocidin-induced PBML were loaded onto 1 ml Con A-Sepharose and Matrex blue columns after dialysis against PBS and 50 m*M* Tris · HCl, pH 7.5, respectively. The total antiviral units that were loaded on the columns is shown, as well as the antiviral activity in the void volume and the eluate. Load volume was 2 ml, void volume was 7 ml, and elution volumes for the Con A and Matrex blue columns were 4 and 6 ml, respectively. The proportion of the antiviral activity remaining after pH 2 treatment was determined as in Table I. Antiviral activity was assessed on GM2767 cells. The yields indicate recovery as a percentage of the activity applied to the column.

TABLE III
SEPARATION OF IFN-γ FROM IFN-δ[a]

		Total units		
		GM2767 cells		
Step	Void volume	Total units ($\times 10^5$)	Yield (%)	WISH cells
Con A-Sepharose	4 ml	5.2	65	1.28×10^4
Matrex Blue	6 ml	4.8	60	$<3.0 \times 10^3$
Anti-IFN-γ-Sepharose	8 ml	6.4	80	$<3.0 \times 10^3$

[a] A crude preparation of interferon from PBML as described in Table I was precipitated by ammonium sulfate, dialyzed against PBS, and chromatographed as described in the text. Two milliliters with a total activity of 8.0×10^5 units was loaded onto the Con A-Sepahrose column. All the columns were equilibrated with PBS prior to use. The void volumes are as indicated. The yields indicate recovery on GM2767 cells as a percentage of the starting activity in the ammonium sulfate precipitate.

TABLE IV
TITRATION OF IFN-δ AND IFN-α ON DIFFERENT CELL LINES[a]

Cell line	IFN (units/ml)	
	IFN-δ	IFN-α
GM2767	16,000	16,000
WISH	<3	1,000
HEp/2C	<3	1,600
HFF	<3	—
HeLa	<3	500
Murine L-929	<3	13
Bovine MDBK	4	32,000
Bovine EBTr	<3	6,300

[a] The partially purified IFN-δ preparation was compared with an IFN-α preparation on a panel of cell lines. The IFN-α preparation was a laboratory standard prepared from human lymphoblastoid cells (Namalva) by induction with Sendai virus.

stable component passes directly through and appears in the void volume, while the eluates are essentially completely pH 2 labile.

These chromatographic techniques form the basis of a separation of IFN-γ and IFN-δ. A crude supernatant containing a total of 2.5×10^6 units was concentrated by ammonium sulfate precipitation then dialyzed against PBS. It was then successively chromatographed on Con A-Sepharose, Matrex blue, and then anti-IFN-γ-Sepharose. Aliquots from the void volumes were retained and titrated for interferon activity; the remainder was loaded directly onto the next column. The total yield at each stage is shown in Table III. Interferon activity was assessed on both GM2767 and WISH cells, both of which are sensitive to Hu-IFN-α, -β, and -γ. The void volume of these columns displayed antiviral activity on GM2767 cells, but not WISH cells indicating the presence of a novel interferon species. The final preparation, which was about 20% of the total starting activity in the culture medium, was stable at pH 2, and was not neutralized by antisera to IFN-α (lymphoblastoid or leukocyte), IFN-β, or IFN-γ.[4,5] IFN-δ was titrated on a panel of different cell lines (Table IV). As can be seen, it had high activity on GM2767 cells. In a separate experiment, the same preparation had a similar titer on GM2504 cells.

[10] D. J. Giron, this series, Vol. 78, p. 399.

[11] J. Suzuki, M. Iizuka, and S. Kobayashi, this series, Vol. 78, p. 403.

TABLE V
SEROLOGICAL, PHYSICOCHEMICAL, AND BIOLOGICAL PROPERTIES OF IFN-δ

Property	IFN-α	IFN-β	IFN-γ	IFN-δ
Antigenic specificity	α	β	γ	Non-α, -β, -γ
Stability				
pH 2 (overnight)	Stable	Stable	Labile	Stable
0.1% SDS (1 hr)	Stable	Stable	Labile	Stable
56° (1 hr)	Stable	Stable	Labile	Stable
Apparent molecular weight under nondenaturing conditions	20–25K	20–22K	40–45K	30–35K
Chromatographic behavior				
Con A-Sepharose	Binds	Binds	Binds	No binding
Matrex blue	Binds	Binds	Binds	No binding
Antiviral activity on[a]				
Human trisomic 21 fibroblasts GM2767	100	100	100	100
Human foreskin fibroblasts	4	3	2	<0.01
Human WISH cells	21	15	21	<0.01
Bovine EBTr cells	40	0.2	<0.01	0.01
Murine L-929 cells	0.2	0.03	<0.01	<0.01

[a] Relative to activity on GM2767 cells taken as 100. Data taken from Wilkinson and Morris.[4]

However, it had little or no activity on other cell lines including WISH cells which are sensitive to all other known interferon types or on two bovine cell lines which are very sensitive to Hu-IFN-α. For comparison, the titers of an IFN-α (lymphoblastoid) preparation is shown on the same cell types.

Concluding Remarks

A novel interferon type, present in supernatants from human PBML induced with phytohemagglutinin and teleocidin, may be easily separated from the main component (IFN-γ) by simple chromatographic procedures. The activity is easily detected on human cells trisomic for chromosome 21, but not on other cell lines commonly used for assay of interferon. A summary of its characteristics is outlined in Table V.

Acknowledgment

We wish to thank the Cancer Research Campaign for financial support.

[16] Preparation and Assay of Poly ICL-CM Dextran, An Interferon Inducer of Reduced Toxicity

By JAKE BELLO, EDWARD N. GRANADOS, MICHAEL MCGARRY, and JUDITH A. O'MALLEY

A complex of poly(I) · poly(C) with polylysine and carboxymethylcellulose, abbreviated as poly ICLC, was developed by Levy *et al.*[1] as an interferon (IFN) inducer. This complex also has activity as an immunological adjuvant, radioprotective agent, antitumor, and antiviral agent.[2–4] Poly ICLC has shown significant toxicity in phase I trials on patients with advanced malignancies.[5] CMC is poorly degradable, may leave long-term irritating residues, and has been included in the NIOSH list of suspected carcinogens.[6] It would be desirable to replace the CMC with a less toxic component. We have prepared a complex, poly ICL-CMD, which has carboxymethyldextran (CMD) as a replacement for CMC, and which is as effective an IFN inducer in mice and monkeys as is poly ICLC, but is less toxic.[7]

Reagents and Procedures

Preparation of CMD. Carboxymethylation of dextran was done by the method of Chang *et al.*[8] The degree of carboxymethylation was estimated by a colorimetric procedure,[9] and was proportional to the ratio of chloroacetic acid to dextran.[7]

Dextrans of M_r 10,000, 40,000 and 70,000 were carboxymethylated at about 0.22 and 0.43 carboxymethyl groups per glucose residue. Com-

[1] H. B. Levy, G. Baer, S. Baron, C. E. Buckler, C. J. Gibbs, M. J. Iadarola, W. T. London, and J. A. Rice, *J. Infect. Dis.* **132,** 434 (1975).

[2] H. B. Levy, *in* "Interferon and Interferon Inducers" (D. Stringfellow, ed.), p. 167. Dekker, New York, 1980.

[3] H. B. Levy, E. S. Stephen, D. Harrington, K. Engel, F. Riley, and E. Lvovsky, *in* "Augmenting Agents in Cancer Therapy" (E. M. Hersch *et al.*, eds.), p. 135. Raven Press, New York, 1980.

[4] E. Lvovsky, W. B. Baze, D. E. Hilmas, and H. B. Levy, *Tex. Rep. Biol. Med.* **35,** 388 (1977).

[5] B. G. Leventhal, H. Kashima, A. S. Levine, and H. B. Levy, *J. Pediatr.* **99,** 614 (1981).

[6] U.S. Environmental Protection Agency, Office of Toxic Substances, March 1976.

[7] E. N. Granados, J. Dawidzik, J. O'Malley, M. McGarry, and J. Bello, *J. Interferon Res.* **4,** 155 (1984).

[8] R. L. S. Chang, M. P. Crawford, and M. D. West, *J. Biomed. Eng.* **2,** 41 (1980).

[9] Hercules Chemical Co., Bulletin VC-472A. Wilmington, Delaware.

plexes soluble in 1/200 or 1/20 PBS were obtained with all CMDs except M_r 10,000 at 0.24 degree of substitution.

Preparation of Poly ICL-CMD. A solution of polylysine · HBr, 1.5 mg/ml in water was filtered through a sterile 0.45-μm filter. A solution of CMD, 5 mg/ml in water, was autoclaved. All mixing was done under sterile conditions. The solution of polylysine was added, with stirring, to an equal volume of CMD. This solution was added to an equal volume of poly(I) · poly(C), 1 mg/ml in 1/100 PBS, to give the complex with a final poly(I) · poly(C) concentration of 0.5 mg/ml.

Poly ICLC. Batches of this material were obtained from Dr. H. B. Levy at NIH, and had been prepared by Flow Laboratories. A batch was also prepared in this laboratory by the method of Levy.[1]

Biological Testing. BALB/c mice and adult rhesus monkeys were used for testing interferon induction. Mice were injected (iv) with complexes, at a dosage of 10 μg of poly(I) · poly(C) per mouse. Blood samples were obtained after 3 hr. Monkeys were injected, also iv, with complexes at a dosage of 500 μg poly(I) · poly(C) per kg. Blood samples were obtained 0 and 6 hr after a single injection.

Interferon assays of the serum samples were done by a colorimetric procedure.[10] Two-fold dilutions of the serum were made in minimal essential medium (MEM) containing 2% fetal calf serum (FCS). The diluted samples were applied in 0.5 ml volumes to monolayers of cells which were established in 16-mm wells in MEM containing 10% FCS 4–5 days prior to the assay. For mouse serum, mouse L cells were used, whereas for monkey serum GM 2504 human cells were used since monkey interferon is active on human cells. The cultures were incubated for 20 hr, the medium aspirated, and vesicular stomatitis virus, at a multiplicity of infection of 0.15 plaque-forming units per cell, in MEM containing 2% FCS, was added to the cells. The cultures were further incubated until virus controls showed marked cytopathic effects (24–48 hr). The medium was aspirated and the antiviral activity determined by a standard colorimetric procedure which measures uptake of a vital dye, neutral red.[10] Interferon titers are expressed in international reference units, with mouse interferon obtained from the Research Resources Branch, National Institute of Allergy and Infectious Diseases, National Institutes of Health, Bethesda, Maryland, used as a reference.

Examples

The IFN induction in mice and the survival of the mice are shown in Table I. IFN induction is similar for poly ICLC and poly ICL-CMD made

[10] N. B. Finter, *J. Gen. Virol.* **5,** 419 (1969).

TABLE I
INTERFERON INDUCTION AND SURVIVAL OF BALB/c MICE TREATED WITH POLY ICL-CMD[a]

	IFN[b] (units/ml)	Survival
ICLC (Flow lots 3 and 4 from NIH)	1464	7/36
ICLC (our preparation)	1950	4/36
Poly ICL-CMD (M_r 70,000, 0.43)[c]	1482	31/36
Poly ICL-CMD (M_r 10,000, 0.43)[c]	790	14/16

[a] Data from Granados *et al.*[7]

[b] Average IFN titer (units/ml) is the mean value for 35 mice. For IFN induction mice were injected with poly ICL-CMD containing 10 μg of poly(I) · poly(C). For survival studies half of the mice received three once-daily injections containing 150 μg poly(I) · poly(C) and half received four such successive injections.

[c] The M_r value given is for the starting dextran. The M_r of the CMD is not known. The second number in the parenthesis is the degree of carboxymethylation per glucose residue.

TABLE II
INTERFERON INDUCTION IN ADULT RHESUS MONKEYS AFTER ADMINISTRATION OF POLY ICLC AND POLY ICL-CM DEXTRAN[a]

Inducer[b]	Number of monkeys	IFN (units/ml)[c] 0 hr	6 hr
Poly ICLC (Flow lot #4)	7	<3	98
Poly ICLC (our preparation)	4	<2	26
Poly ICL-CMD (T-70, 0.43)	7	<3	69
Poly ICL-CMD (T-70, 0.21)	3	<5	121

[a] Data from Granados *et al.*[7]

[b] Each animal was administered complex containing 0.5 mg of poly(I) · poly(C) per kg.

[c] The values for units/ml of IFN in monkey serum refer to the mean value for the numbers of monkeys tested.

with M_r 70,000 dextran, but is less for that made with M_r 10,000 dextran. Survival was tested at higher doses than for IFN induction, and was found to be better with poly ICL-CMD than with poly ICLC. Overall survival with poly ICLC was 15% (average of 20 and 11%, respectively, for the Flow lots and our preparation), and with poly ICL-CMD was 87%. The difference in toxicity may indeed be greater, since our preparation of poly ICLC gave 11% survival and the poly ICL-CMD gave 87%, and both of these were made with the same batches of poly(I) · poly(C) and polylysine, while the Flow lots of poly ICLC were made with different batches in a different facility. We do not know if CMD is intrinsically less toxic than CMC or if the toxicity of the latter arises from impurities, or if the toxicity difference results from a difference in the nature of the complex.

The results in Table II show that poly ICL-CMD is as effective an interferon inducer in the rhesus monkey as poly ICLC, although the number of monkeys is small. Physiological parameters (pulse, temperature, WBC, etc.) of the monkeys have been measured and no significant differences were observed between animals given poly ICLC and those given poly ICL-CMD.

Acknowledgments

This work was supported in part by Dept. of Health and Human Services NIH Contract AI-02657. We are grateful to Steven Lewinski for technical assistance, and to Jean Dawidzik for biological testing.

[17] Interferon Induction: Dose–Response Curves

By PHILIP I. MARCUS

The IFN-inducing capacity intrinsic to a virus preparation can be assessed quantitatively only by generating a complete dose (multiplicity)–response (IFN yield) curve, and only under conditions that preserve the initial multiplicity of infection. Yet tests for IFN induction are often carried out at a single, high multiplicity, and often under conditions that permit cycling infection by inducer virus. Such tests can lead to erroneous conclusions regarding the IFN-inducing capacity of a virus stock, and have led us to develop a uniform approach to the study of IFN induction by viruses through the use of dose–response curve analysis.[1-4] This analysis has been developed to the extent that virus populations can be described both by the numbers of particles they contain that are capable of inducing IFN (termed IFN-inducing particles, IFP), and the amount of IFN these particles can induce (termed the quantum yield).[1-3,5]

This chapter describes the generation and analysis of the two types of IFN induction dose–response curves most commonly observed in virus–host cell systems, and how such analysis can be used to detect and enumerate IFN-inducing particles and quantum IFN yields. Implicit in the analysis of dose–response curves are five basic assumptions regarding IFN induction by viruses: (1) the induction process is initiated by the action of a single particle—the IFN-inducing particle, (2) IFN-inducing particles need not have the capacity to reproduce, (3) viral dsRNA is the proximal inducer molecule, (4) the threshold for IFN induction is a single molecule of dsRNA per cell, and (5) induction results in the production of a quantum yield of IFN.[1-3,5]

Generation of IFN Induction Dose–Response Curves: General Aspects

Typically, IFN induction dose–response curves are generated by infecting monolayers of susceptible cells with increasing multiplicities of virus and 24 hr later harvesting the medium bathing the cells and assaying it for IFN.[1-3,5] Following aspiration of the growth medium, and without washing, virus infection is carried out in a small volume, typically 0.3 ml of medium (NCI or MEM with 6% calf serum)[6] on a confluent monolayer of cells in a 50-mm plate.[7] The monolayers are incubated at 37° for 30–60 min to ensure ≥90% attachment of the virus. Following virus attachment, the 0.3 ml inoculum is removed by aspiration and the monolayers are washed twice with serum-free basal medium. Each 50-mm plate then receives 3.0 ml of serum-free medium to bathe the cells during IFN production—usually for 24 hr at a temperature appropriate for each cell type. For example, "aged" primary chick embryo cells[6,7] are incubated in NCI medium[6] at 40.5° whereas most other cell types receive MEM and incubation at 37.5°. We have observed that serum-free medium invariably results in higher yields of IFN than when serum is present.[7]

[1] P. I. Marcus, *Tex. Rep. Biol. Med.* **41,** 70 (1981–1982).

[2] P. I. Marcus, *J. Interferon Res.* **2,** 511 (1982).

[3] P. I. Marcus, *in* "Interferon" (I. Gresser, ed.), Vol. 5, p. 115. Academic Press, 1983.

[4] P. I. Marcus and M. J. Sekellick, *Nature (London)* **266,** 815 (1977).

[5] P. I. Marcus, P. T. Guidon, Jr., and M. J. Sekellick, *J. Interferon Res.* **1,** 601 (1981).

[6] D. H. Carver and P. I. Marcus, *Virology* **32,** 247 (1967).

[7] M. J. Sekellick and P. I. Marcus, this volume [18].

To generate accurate IFN induction dose–response curves it is critical that the initial (input) multiplicity of infection be preserved. If cycling infection takes place then the effective multiplicity of infection will remain an unknown variable and a complex dose–response curve will result along with attendant difficulties of interpretation.[3,8]

When infectious virus is the inducer in a permissive cell, it is possible to preserve the initial (input) multiplicity by adding specific virus antiserum to the medium once virus attachment and entry are complete (usually within the 30–60 min adsorption period)—but before the end of the virus latent period. Occasionally, some viral antisera will have an adverse effect on the yield of IFN, hence each new batch of antiserum must be screened for this undesirable property.[7]

Virus antibody need not be added (and indeed, has no effect) when nonmultiplying viruses are used as inducers. For examples, infectious virus in nonpermissive cells (human reovirus in chick embryo cells[9]), *ts* mutants at nonpermissive temperatures (vesicular stomatitis virus mutant *ts*G41 at 40.5°[10]), UV (254 nm) or heat-inactivated virus (Newcastle disease virus[8]), or defective virus particles (VSV [±]DI-011[4]).

In order to ensure that sufficient time elapses following infection to accommodate a full (quantum) yield of IFN it is prudent to establish the time-course of IFN production for each new virus–cell system. This procedure also establishes whether or not the IFN which accumulates during induction is stable. (Occasionally, the level of IFN in the medium will decline significantly with time—possible because of cellular proteases released during the cell destruction that sometimes accompanies induction by infectious viruses.)

Type $r \geq 1$ Dose–Response Curves

In the most commonly observed type of IFN induction dose–response curve, termed $r \geq 1$, cells subjected to increasing multiplicities of virus respond by producing, within a finite period (usually 12–24 hr), increasing amounts of IFN until a plateau is reached. Data points from a typical $r \geq 1$ curve are illustrated in Fig. 1. Once a plateau level is reached increasing the multiplicity of the inducing virus has no effect on the absolute level of the plateau, i.e., the IFN yield remains constant. This type of dose–response curve shows a good fit to a theoretical curve generated by assuming that all cells infected with 1, or more than 1 (≥ 1), particle with the

[8] P. I. Marcus, C. Svitlik, and M. J. Sekellick, *J. Gen. Virol.* **62,** 2419 (1983).

[9] T. R. Winship and P. I. Marcus, *J. Interferon Res.* **1,** 155 (1980).

[10] M. J. Sekellick and P. I. Marcus, *Virology* **95,** 36 (1979).

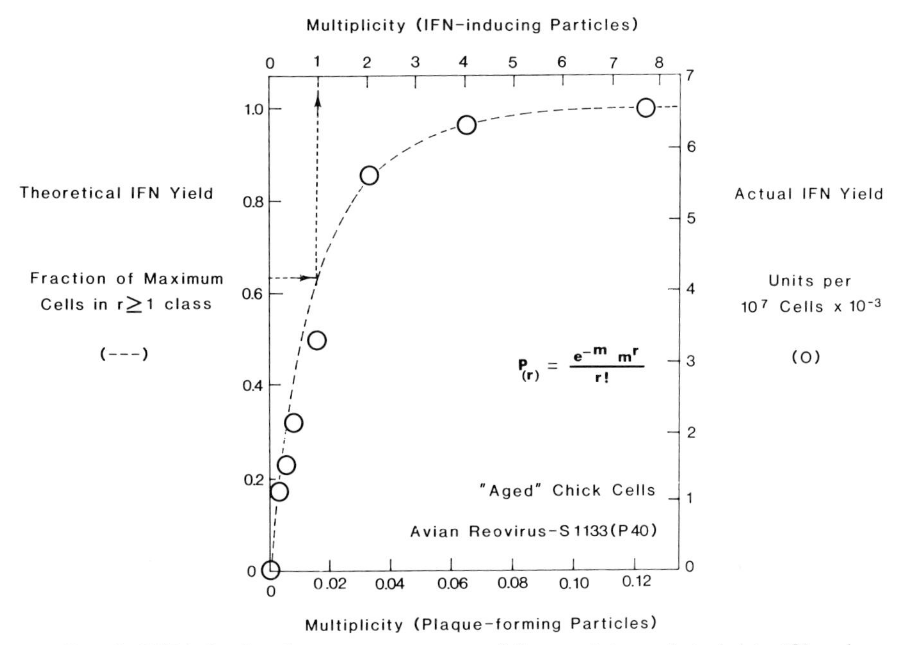

FIG. 1. IFN induction dose–response curve of the $r \geq 1$ type. Actual data (○) and a theoretical curve (---) are compared. The experimental points represent acid-stable IFN produced by a monolayer of "aged" primary chick embryo cells[6,7] infected with various multiplicities of plaque-forming particles of avian reovirus-S1133(P40) (the lower abscissa). The multiplicity of IFN-inducing particles is displayed in the upper abscissa and was calculated as described in the text. The theoretical curve represents the fraction of the cell population infected with 1, or more than 1, IFN-inducing particle as a function of multiplicity, and was derived from the Poisson distribution formula shown within the figure. Data points from Winship and Marcus.[9] In our assays 1 PR_{50} (VSV) unit of chicken interferon is equivalent to about 10–25 units of the MRC Research Standard A 62/4 provided by the National Institute of Allergy and Infectious Diseases, Research Resources Branch.

capacity to induce IFN (IFN-inducing particles, IFP), produce the same quantum yield of IFN. The plateau of an $r \geq 1$ dose–response curve represents the yield of IFN achieved when *all* cells in the population have been infected, i.e., receive ≥ 1 IFP. A theoretical curve of the $r \geq 1$ type can be generated by plotting as a function of multiplicity (m), the probability (fraction) of cells infected with 1 or more IFP, $P_{(r \geq 1)}$, calculated from the Poisson distribution:

$$P_{(r)} = \frac{e^{-m} m^r}{r!}$$

where $P_{(r)}$ equals the fraction of cells that receives r particles when the multiplicity is m. This curve is plotted as a dashed line in Fig. 1. According to the Poisson distribution the value of m which induces 63% (0.63) of the maximum yield (plateau value) of IFN represents an average of 1 IFP per cell ($m_{ifp} = 1$) in the population. The upper abscissa in Fig. 1 was determined in this way. At $m_{ifp} = 1$, 37% ($0.37 = e^{-1}$) of the cells escape infection, by chance, and do not contribute to the yield of IFN. In a practical sense, at values of $m_{ifp} \geq 4$ virtually all of the cells have received at least 1 IFP and are producing full quantum yields of IFN. The ratio IFP : PFP can be determined by comparing the lower (m_{pfp}) and upper (m_{ifp}) abscissas thereby providing an assessment of the number of IFP in a virus stock. In the example shown in Fig. 1 there were about 60 times more IFP than PFP in this stock of avian reovirus, i.e., IFP : PFP = 60.

Type $r = 1$ Dose–Response Curves

In the less commonly observed type of IFN induction curve, termed $r = 1$, cells subjected to increasing multiplicities of inducer virus respond by producing amounts of IFN which increase linearly until a peak is reached. Any further increase in multiplicity results in a marked decline in the amount of IFN produced. Data points from a typical $r = 1$ dose–response curve are plotted in Fig. 2. As in an $r \geq 1$ type curve, the maximum amount of IFN produced was reached usually within 12 to 24 hr after infection, independent of multiplicity.

The $r = 1$ type of IFN induction dose–response fits best a curve generated by assuming that cells infected with 1, and only 1, inducing particle are capable of producing a full (quantum) yield of IFN, and that cells infected with ≥ 2 particles produce little or no IFN. According to this interpretation, the peak yield of IFN occurs when the fraction of cells infected with only 1 IFN-inducing particle is maximal. Assuming a Poisson distribution of IFP among the population of cells, this maximum is reached when $m_{ifp} = 1$. In Fig. 2 the upper abscissa was derived in this way. A theoretical curve which represents the fraction of maximum cells infected with only 1 IFP ($r = 1$) as a function of m_{ifp} in the Poisson distribution is presented as a dashed line in Fig. 2. At $m_{ifp} = 1$, 37% of the cells are, by chance, infected with only 1 IFP. This represents the maximum fraction of cells infected with 1, and only 1, IFP. Presumably, these are the cells responsible for the peak yield of IFN in an $r = 1$ type curve. At this peak, 63% of the cells are not producing interferon, 37% because by chance they escaped infection, and 26% because they were infected with 2 or more particles. Since only 37% of the cell's population is pro-

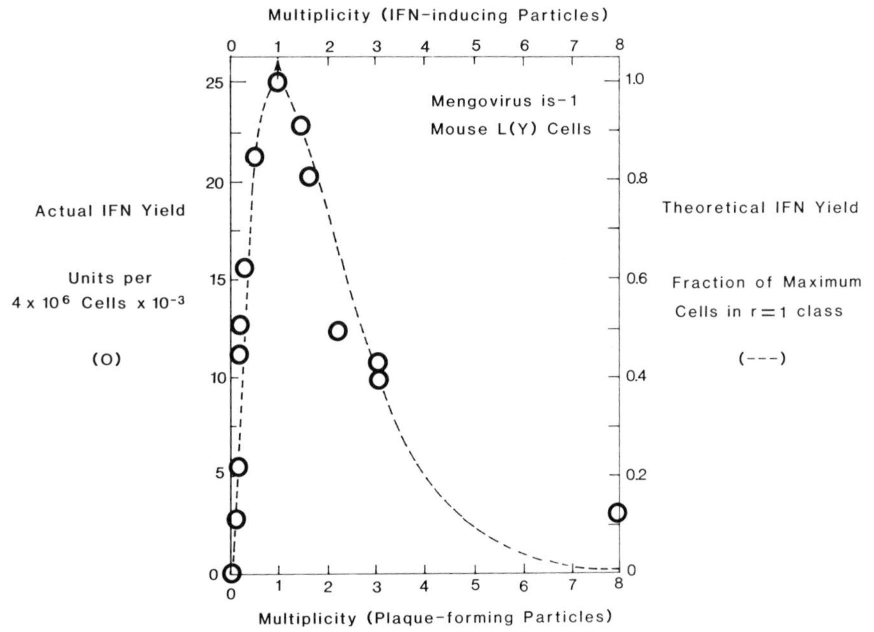

FIG. 2. IFN induction dose–response curve of the $r = 1$ type. Actual data (○) and a theoretical curve (---) are compared. The experimental points represent IFN produced by a monolayer of mouse L(Y) cells infected with various multiplicities of plaque-forming particles of Mengovirus is-1(*ifp*$^+$) (the lower abscissa). The multiplicity of IFN-inducing particles is displayed in the upper abscissa and was calculated as described in the text. The theoretical curve represents (normalized to 1.0) the fraction of cells in the population infected with 1, and only 1, IFN-inducing particle, as a function of multiplicity, and was derived from the Poisson distribution formula. Data points from Marcus *et al.*[5] In our assays 1 PR_{50} (VSV) unit of mouse interferon is equivalent to about 2 units of the standard G-002-904-511, provided by the National Institute of Allergy and Infectious Diseases, Research Resources Branch.[24]

ducing IFN at the peak of an $r = 1$ dose–response curve, the actual quantum yield of IFN on a *per cell* basis is 2.7 times greater than calculated when assuming that *all* cells of the monolayer are contributing to the yield of IFN (1.0/0.37 = 2.7).

A comparison of the lower abscissa (m_{pfp}) with the upper (m_{ifp}) in Fig. 2 reveals that the stock of Mengovirus is-1(*ifp*$^+$)[5] used in this induction experiment contained essentially equal numbers of PFP and IFP, i.e., IFP : PFP = 1.

It is not understood why simultaneous infection of a cell with two or more particles induces little or no IFN. In effect, when these IFN-inducing particles are present at 2 or more per cell they function as IFN induction-suppressing particles and can be assayed as such.[2,3,11]

Quantum Yields of Interferon

Implicit in the good fit of the data points to the theoretical curves for $r = 1$ and $r \geq 1$ dose–responses is that each cell induced to produce IFN responds by producing the same quantity of IFN—termed a quantum yield. The absolute value of an IFN quantum may vary in different virus–cell systems, and even within a system depending upon the physiology of the cell, and the strain of virus used as inducer. For example, primary chick embryo cells "aged" *in vitro* for several days, produce 50 to 100 times more IFN than "unaged" cells.[6,8,9] Furthermore, as illustrated in Fig. 3, three different strains of Newcastle disease virus, all of which induce $r \geq 1$ type dose–response curves in mouse L(Y) cells, produce different quantum yields of IFN. Thus, the N.J.-LaSota strain consistently induces 5 times more IFN than the AV strain, with the B_1-Hitchner strain consistently inducing an intermediate level of IFN.[12]

For $r \geq 1$ type dose–response curves the quantum yield of IFN can be calculated from the yield of IFN at the plateau, i.e., where $m_{ifp} \geq 4$. At these multiplicities essentially all of the cells are producing a quantum yield of IFN, hence, the total amount of IFN produced by a cell population (as measured from the plateau yield), divided by the total number of cells, is a measure of the IFN quantum yield. For example, in Fig. 1,[9] the IFN quantum yield is 6.6×10^{-3} units/cell. However, 1 of our laboratory units is equivalent to about 10–25 units of the MRC Research Standard A 62/4 provided by the NIAID Research Resources Branch. This means that it only takes about 6 to 15 "aged" chick embryo cells to produce 1 standard unit of IFN when infected with ≥ 1 IFP of avian reovirus (S1133/P40).

In an $r = 1$ dose–response curve the peak yield of IFN (observed at $m_{ifp} = 1$) represents the production of an IFN quantum from only 37% of the cell population. Therefore, the quantum yield actually represents the total yield of IFN measured at the peak ($m_{ifp} = 1$) divided by 0.37 of the total number of cells in the population. For example, in Fig. 2 the IFN quantum yield is 1.7×10^{-2} units/cell [$(2.5 \times 10^4$ units$)/(0.37)(4 \times 10^6$ cells$)$]. Since 1 of our laboratory units is equivalent to 2 units of the mouse reference standard, it takes about 30 mouse L(Y) cells, each infected with 1, and only 1, particle of Mengovirus is-1(ifp^+) to produce 1 unit of IFN.

[11] P. I. Marcus and M. J. Sekellick, *Virology* **142,** 411 (1985).

[12] C. Svitlik and P. I. Marcus, unpublished observations.

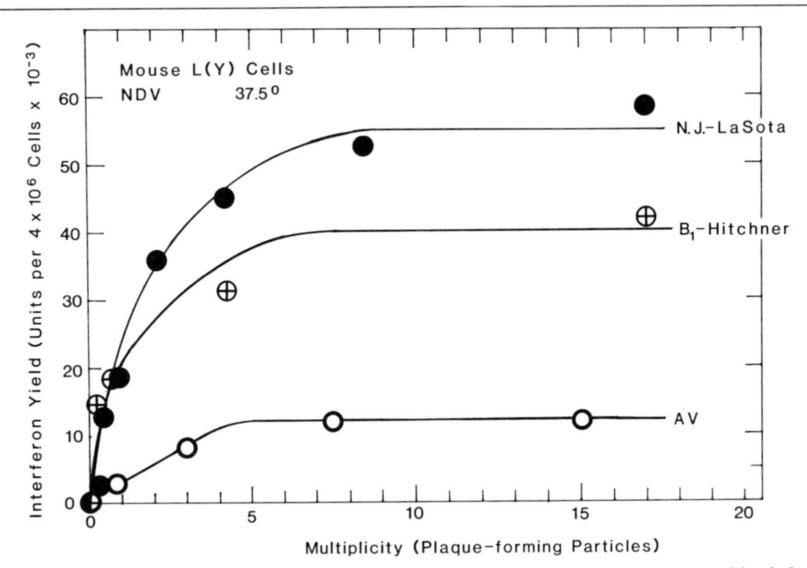

FIG. 3. A family of $r \geq 1$ type IFN induction dose–response curves generated by infecting mouse L(Y) cells with three different strains of Newcastle disease virus, NDV. The relative quantum yields (plateau levels) of IFN for these strains of NDV are approximately 1 : 3 : 5 for AV, B_1-Hitchner and N.J.-LaSota, respectively. Unpublished data from Svitlik and Marcus.[12]

The molecular events that affect the absolute value of an IFN quantum are not understood.

Interferon-Inducing Particles (IFP)

Once the nature of the IFN induction dose–response curve has been established, the virus dilution which contains, on average, one IFP ($m_{ifp} = 1$) can be determined. As noted, for $r \geq 1$ curves, $m_{ifp} = 1$ occurs at 0.63 of the maximum (plateau) value of the IFN yield (Fig. 1). For $r = 1$ curves, $m_{ifp} = 1$ occurs at the peak of the IFN yield (Fig. 2).

The simplest way to calculate the IFP titer is from the ratio of IFP : PFP, however, assays other than infectivity can be used to establish this ratio if they can be quantified: for example, defective-interfering particles,[13,14] or a nonbiological assay, namely physical particles determined by electron microscopy.[4]

[13] A. J. D. Bellett and P. D. Cooper, *J. Gen. Microbiol.* **21,** 498 (1959).

[14] M. J. Sekellick and P. I. Marcus, *Virology* **104,** 247 (1980).

Concluding Comments

Data from IFN induction dose–response curves often are plotted as linear by logarithmic or logarithmic by logarithmic plots of inducer multiplicity and IFN yield.[15-19] However, the exact nature of the dose–response curve is difficult to intuit from such graphs and the use of linear by linear plots is recommended.

Complete characterization of the IFN-inducing capacity of a virus population can be accomplished only if full dose–response curves are established for each virus–cell system. For example, a virus–host cell combination that produces an $r = 1$ type of dose–response curve would score as a "poor" inducer when tested at $m_{ifp} \geq 5$ and yet be an excellent inducer under conditions of natural infection where each cell may be infected initially with a single particle (Fig. 2). Recent reports confirm the $r = 1$ type of dose–response.[20-22] It is also important to recognize that virus populations may contain a large excess of IFN-inducing particles relative to infectious particles, and that extremely high dilutions of virus may be required to reveal the portion of a dose–response curve that is responsive to changes in m_{ifp} and hence allow calculation of IFP titers (Fig. 1).

Through analysis of IFN induction dose–response curves it has been possible to define the genes of a virus that must be expressed for it to function as an IFN-inducing particle.[5,9,23,24] In addition, the $r = 1$ dose–response, first observed in "aged" primary chick embryo cells[7] infected with [±]DI-011 particles of VSV, led to the establishment of the threshold for IFN induction as a single molecule of dsRNA.[3,4]

Acknowledgments

This work was supported by NIAID Grant AI-18381 and benefited from the use of a Cell Culture Facility supported in part by NCI Grant CA-14733 and by The University of Connecticut Research Foundation. I thank Dr. Margaret J. Sekellick for helpful discussions and a critical review of this contribution.

[15] W. F. Long and D. C. Burke, *J. Gen. Virol.* **12,** (1971).

[16] B. Lomniczi, *J. Gen. Virol.* **21,** 305 (1973).

[17] W. R. Fleischmann, Jr. and E. H. Simon, *J. Gen. Virol.* **25,** 337 (1974).

[18] M. Kohase and J. Vilček, *Jpn. J. Med. Sci. Biol.* **32,** 281 (1979).

[19] T. K. Frey, E. V. Jones, J. J. Cardamone, Jr., and J. S. Youngner, *Virology* **99,** 95 (1979).

[20] S. H. Cohen and G. W. Jordan, *Infect. Immun.* **42,** 605 (1983).

[21] E. Slattery, H. Taira, R. Broeze, and P. Lengyel, *J. Gen. Virol.* **49,** 91 (1980).

[22] T. R. Winship, C. K. Y. Fong, and G. D. Hsiung, *J. Interferon Res.* **3,** 71 (1983).

[23] P. I. Marcus and F. J. Fuller, *J. Gen. Virol.* **44,** 169 (1979).

[24] P. I. Marcus and M. J. Sekellick, *J. Gen. Virol.* **47,** 89 (1980).

[18] Induction of High Titer Chicken Interferon

By MARGARET J. SEKELLICK and PHILIP I. MARCUS

Primary cultures of chicken (chick) embryo cells can be used to produce high titers of chick interferon(s). However, in order to obtain high titers routinely care must be exercised during (1) the initial preparation of the primary cells from the embryos, (2) incubation and treatment of the cultures, and (3) the induction process per se. When all three of these conditions are optimized, a properly prepared, "aged," and induced culture of primary chick embryo cells (CEC) can produce tens of thousands of units of interferon (IFN). This chapter provides a guide to produce high titers of chicken IFN.

Preparation of Primary Chick Embryo Cells

Primary chick embryo cell (CEC) cultures are prepared from 10-day-old embryonated chicken eggs (utility grade, SPAFAS, Inc., Norwich, CT). Embryonated eggs from other sources (Department of Animal Genetics, University of Connecticut, Storrs, CT) and with different genetic backgrounds (e.g., scaleless mutant) have all produced primary CEC cultures capable of producing high titers of chicken IFN. Ten-day-old embryonated eggs are used because, in general, younger embryos yield significantly fewer cells; those that are much younger than 9 days old produce less IFN. Embryos older than 10 days will produce primary cell cultures that yield high titers of IFN, but embryos older than 13 days are more difficult to mince and process for cell plating.

Routinely, primary CEC cultures are prepared from about 24 10-day-old embryos at a time by the following procedure: (1) sterilize the egg shell by wiping it with a Kimwipe moistened with 70% aqueous ethanol. From this point on asceptic technique is used with sterile materials. (2) Remove embryos from the eggs and transfer them to a 100-mm Petri dish containing approximately 10 ml of Ca^{2+}- and Mg^{2+}-free phosphate-buffered saline (PBS), pH 7. Remove the heads by pinching the neck with forceps and discard them. (3) Transfer decapitated embryos to a clean, dry Petri dish and mince with a fine-pointed scissors until a slurry of tissue mass is obtained. (4) Transfer the minced tissue to a 250-ml-wide mouth Erlenmeyer flask which contains approximately 50 ml of cold PBS. Allow this mixture to stand for 3 min and decant the supernatant fluid. Repeat this step with warm 0.2% trypsin (Trypsin 1-300 ICN, Nutritional Biochemicals, Cleveland, Ohio) prepared in PBS with penicillin and streptomycin at 200 units or μg/ml, respectively. Discard the supernatant fluid after allowing the bulk of tissue mass to settle for 3 min. (5) Add 100 ml of warm 0.2% trypsin–PBS solution to the flask now containing the washed fragments of tissue and a 2-in. sterile Teflon-coated magnetic spin bar. Cap the flask with foil and stir the mixture on a magnetic stirrer at 35° for 30 min. Stirring should create a deep vortex but no foaming. (6) Remove the flask from the magnetic stirrer and allow any larger fragments of tissue to settle for about 3 min. Decant the cell suspension into a wide-mouth Erlenmeyer flask covered with 8 layers of gauze as a filter (Curity cheesecloth, grade #40, The Kendel Company, Wellesley Hills, MA) and containing 15 ml of cold calf serum. (All calf serum is screened, and used only if shown to be nontoxic to CEC, and to support their growth at relatively low cell densities.) The serum–cell suspension is mixed well and stored on crushed ice. The bulk of the tissue mass remains settled in the original flask. (7) Repeat the trypsinization procedure described above (steps 5 and 6) with the remaining tissue fragments. This time add 80 ml of fresh warm trypsin–PBS solution and stir for 20 min. Add an additional 15 ml of calf serum to the filter flask, collect the cell suspension as before, and pool it with the material from the first trypsinization procedure. Trypsinize the remaining tissue fragments a third time by adding 60 ml of warm trypsin–PBS and stirring for 15 min. Collect this third and final suspension of cells and pool it with the others along with an additional 15 ml of calf serum. (8) After pouring the third batch of dissociated cells through the cheesecloth filter, allow it to drain, then remove and discard it. Pour the pooled suspension of cells/serum/trypsin into two 200-ml screw-cap glass conical centrifuge tubes, and centrifuge at 900 rpm (IEC Model PR-6 or PR-6000) for 15 min. Remove the supernatant fluid by aspiration and resuspend the cell pellets in 50–75 ml of NCI medium[1] supplemented with 6% calf serum. This mixture of NCI medium containing 6% calf serum, termed AS medium, is our standard growth medium for chick embryo cells. (9) Prepare a 1 : 200 dilution of the cell suspension for counting in a hemacytometer (the eliptical red blood cells should be excluded from the count). Keep the cell suspension on ice until the volume can be adjusted according to the cell count and the cells are dispensed into plastic tissue culture dishes for incubation and "aging." Each embryo should yield at least 2×10^8 cells.

Preparation of Primary Chick Embryo Cell Monolayers

Prepare monolayers of cells by inoculating 1×10^7 cells into 5 ml of AS medium per 50-mm plastic tissue culture dish. This inoculum (5.2×10^3

[1] P. I. Marcus and D. H. Carver, *Science* **149**, 983 (1965).

cells per cm^2) should produce a monolayer of cells that is at least 95% confluent in 24 hr. Monolayers prepared from lower or higher inocula will display an altered rate of "aging" (see below) and hence affect the optimal time for enhanced IFN production from the monolayer. We have always used NCI as the basal medium for plating and "aging" cells, however, from limited experience it seems likely that any rich basal growth medium may be substituted, for example, MEM or DME. Fetal bovine serum (also screened for its growth properties) can be substituted for calf serum, but, for economy, calf serum at 6% is routinely used.

It is essential that monolayers of primary chick embryo cells be established from freshly prepared cells. Overnight refrigeration of newly made CEC results in a cell population that produces excellent monolayers, however, they produce very low yields of IFN when induced. Likewise, secondary CEC prepared from primary CEC capable of producing high titers of IFN, produce less IFN, even when "aged."

"Aging": Incubation without a Medium Change

Perhaps the most critical step in producing primary chick embryo cell monolayers that will yield high levels of IFN upon appropriate induction is to "age" the cultures *in vitro,* i.e., incubate the monolayers undisturbed for 7 to 10 days.[2] Cell cultures incubated at 37.5 or 38.5° in a CO_2-gassed water-jacketed incubator adjusted to produce pH 7 give equivalent results with respect to the "aging" phenomenon.[2] Primary CEC tolerate a slightly acidic environment, but grow poorly if the pH becomes too alkaline. Any medium changes during the period of "aging" severely compromises the IFN producing capacity of the cells, hence, the cultures are not disturbed until the time of induction![2]

A period of 7 to 10 days is considered ideal for aging the primary cultures (Fig. 1). Aging the cells for shorter periods of time results in a considerably reduced yield of IFN from the monolayers. Aging cells for longer periods of time does not improve the yield of IFN, and indeed it may actually decline. Also, prolonged "aging" is often accompanied by extensive development of muscle straps and a tendency for the monolayer to contract into a ball of cells.

"Aged" monolayers that are capable of producing high titers of IFN are characterized by numerous muscle straps (fused myoblasts) crisscrossing the culture, islands of specialized cells such as hepatocytes, and a profusion of spherical cells with phagocytic activity that accumulate on top of the dense confluent monolayer.

[2] D. H. Carver and P. I. Marcus, *Virology* **32,** 247 (1967).

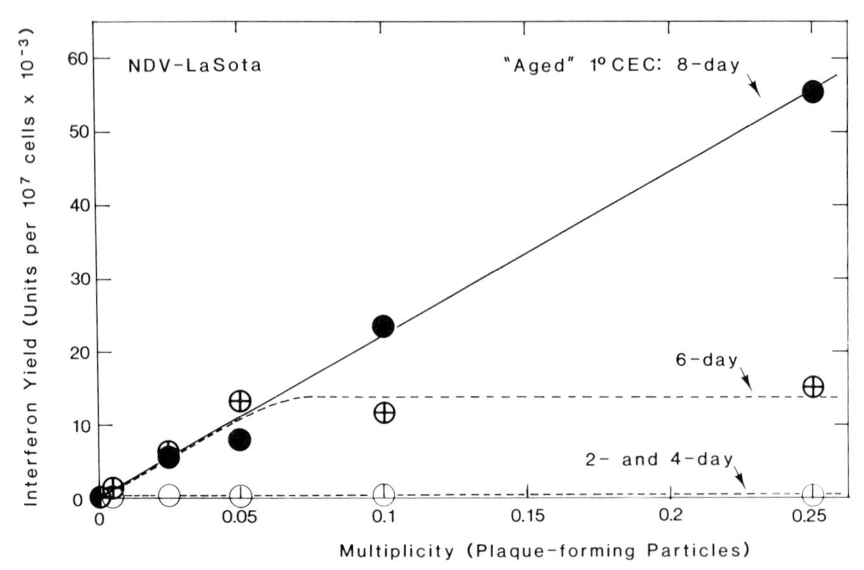

FIG. 1. A family of IFN induction dose–response curves generated from monolayers of primary chick embryo cells "aged" for 2, 4, 6, or 8 days before infection with the N.J.-LaSota strain of NDV. In this experiment the highest multiplicity was not sufficient to induce a peak (plateau) yield of IFN in the 8-day-old cells in what would otherwise be an $r \geq 1$ type dose–response.[11,16] The virtual absence of IFN induction in the 2- or 4-day-old cells is in keeping with the numerous reports of active NDV as a poor inducer in "young" CEC.[11] A more complete dose–response curve has been reported.[23]

Other Cell Types

Primary mouse embryo cell cultures also appear to respond to "aging" *in vitro* by enhancing their capacity to produce IFN upon appropriate induction. One example of an established line of cells that appears to "age" *in vitro* is the L(Y) strain of mouse cells. Higher titers of IFN are produced if these cells are "aged" for about 4 days prior to induction.[3,4]

Choice of Inducer: Viral

Certain viral inducers, both replicating and nonreplicating, induce high titers of IFN from "aged" monolayers of CEC. Presumably, efficient inducers reflect the ability of the virus particle to present double-stranded (ds)RNA to the cell in such a way that the IFN induction/production

[3] T. R. Winship and P. I. Marcus, cited on p. 158 in P. I. Marcus, *in* "Interferon" (I. Gresser, ed.), Vol. 5, p. 115. Academic Press, New York, 1983.

[4] P. T. Guidon, Jr., Masters Thesis, University of Connecticut, Storrs (1982).

system is stimulated maximally before the virus shuts off host–cell directed macromolecular synthesis.[5,6] In our experience, among the best viral inducers of chicken IFN are (1) a defective-interfering particle of vesicular stomatitis virus (VSV), [±]DI-011,[7,8] (2) certain *ts* mutants of VSV at nonpermissive temperatures,[9] (3) Sindbis wildtype virus,[10] (4) active Newcastle disease virus, strains LaSota, AV, or California,[11] (5) avian reovirus strain S1133(P40) at a nonpermissive temperature,[12] and (6) many *ts* mutants of human reovirus, especially *ts*G(107) and *ts*C(447), at nonpermissive temperatures.[13]

Poly(rI) · Poly(rC)

The synthetic dsRNA molecule poly(rI) · poly(rC) is a relatively poor inducer of IFN in "aged" primary CEC even though it induces a very efficient antiviral state.[6] Under optimal conditions of induction, which include the presence of DEAE-dextran and thoroughly washed cells to rid the monolayer of dsRNases adsorbed from the serum in the growth medium,[14] about 300 units of IFN are induced per 10^7 aged CEC. This compares with about 100,000 units obtained with various viral inducers.

Multiplicity of Inducing Virus

Choosing the optimal multiplicity of infection for the inducing virus ranks of paramount importance next to proper "aging" of the cells. Infecting cells with high multiplicities of virus does not necessarily result in high yields of IFN, indeed, just the opposite may take place.[15,16] Consequently, it is important first to determine the exact relationship between virus dose (multiplicity) and cell response (IFN yield). This is especially critical since information concerning the type of IFN induction dose–response curve, coupled with information on the ratio of IFN-inducing

[5] P. I. Marcus, *J. Interferon Res.* **2,** 511 (1982).

[6] P. I. Marcus, *in* "Interferon" (I. Gresser, ed.), Vol. 5, p. 115. Academic Press, New York, 1983.

[7] P. I. Marcus and M. J. Sekellick, *Nature (London)* **266,** 815 (1977).

[8] M. J. Sekellick and P. I. Marcus, *Virology* **85,** 175 (1978).

[9] M. J. Sekellick and P. I. Marcus, *Virology* **95,** 36 (1979).

[10] P. I. Marcus and F. J. Fuller, *J. Gen. Virol.* **44,** 169 (1979).

[11] P. I. Marcus, C. Svitlik, and M. J. Sekellick, *J. Gen. Virol.* **64,** 2419 (1983).

[12] T. R. Winship and P. I. Marcus, *J. Interferon Res.* **1,** 155 (1980).

[13] T. R. Winship and P. I. Marcus, unpublished observation (1981).

[14] F. J. Fuller and P. I. Marcus, *J. Cell. Physiol.* **98,** 1 (1979).

[15] P. I. Marcus, P. T. Guidon, Jr., and M. J. Sekellick, *J. Interferon Res.* **1,** 601 (1981).

particles (IFP) to plaque-forming particles permits the proper choice of an optimal multiplicity for induction.[16]

To achieve and maintain multiplicities for optimal induction of IFN when using an infectious virus in a permissive cell it is necessary to restrict the spread of infectious virus.[11,15] This may be accomplished by including virus-specific antiserum in the otherwise serum-free medium during the production of IFN added after virus attachment and entry, but before the end of the latent period. Viral antiserum prepared in rabbits and not otherwise purified must be preadsorbed onto monolayers of uninfected CEC in order to clear it of any anti-chick cell activity that might compromise the induction process. Preadsorption of crude antiserum involves three successive 30 min adsorptions of 1 ml volumes of serum onto 50-mm dishes containing "aged" CEC. Antiserum treated in this way has no deleterious effect on the yield of IFN from monolayers of "aged" CEC.

Virus Infection/Interferon Induction

To induce IFN in monolayers of "aged" primary chick embryo cells, the medium from the 7- to 10-day-old cultures is removed and, without washing the monolayer, the inducer virus is added in a 0.3 ml volume of AS medium per 50-mm dish. The virus is allowed to attach for 60 min at 37.5°, and unattached virus removed by aspiration. The infected monolayers are then washed two times with 1.5 ml of warm (37°) serum-free NCI medium before the addition of 3 ml of warm serum-free NCI medium used to bathe the cells during the production and release of IFN. (Serum is omitted from the medium because most batches have an adverse effect on the yield of IFN as shown by the data in the table.) The infected monolayers are incubated for about 24 hr at 40.5° before harvesting the medium for IFN. Monolayers serving as mock controls are processed similarly except that the inducer virus is omitted. Spontaneous production of IFN was never observed in mock-infected cultures when tested under conditions that would detect 1 unit of IFN.

The time-course experiment illustrated in Fig. 2 demonstrates that for VSV-[±]DI-011 particles or VSV-*ts*G11(I), the maximum yield of IFN is obtained after about 15 hr. Most of the other inducers mentioned above induce full yields of IFN well within a 24-hr period. Longer incubation may lead to a loss of IFN activity; however, the time to maximum production of IFN should be determined for each new inducer. The choice of 40.5° as an induction temperature pertains only to "aged" primary chick

[16] P. I. Marcus, this volume [17].

Assay of Test Lots of Calf and Fetal Bovine Serum for Their Effect on IFN Production by "Aged" Chick Embryo Cells[a]

Serum and lot #	Interferon yield (units/10^7 cells)	Percentage of control IFN yield
Control—no serum	25,500	100
Calf serum Lot #1[b]	1,110	4
Calf serum Lot #2	5,400	21
Calf serum Lot #3	5,100	20
Calf serum Lot #4	7,800	31
Calf serum Lot #5	9,600	38
Calf serum Lot #6	4,050	16
Calf serum Lot #7	5,400	21
Calf serum Lot #8	<1,500	<6
Calf serum Lot #9	4,800	19
Calf serum Lot #10	18,000	71
Calf serum Lot #11	3,000	12
Calf serum Lot #12	9,600	38
Calf serum Lot #13	9,600	38
Fetal bovine serum Lot #14	30,000	118

[a] "Aged" primary chick embryo cells (8 days) were infected with [±]DI-011 (m_{ifp} = 1). DI particles were adsorbed to cells in a 0.3 ml volume for 30 in at 37.5° in the presence of 10 μg/ml DEAE-dextran. After removal of unattached DI particles by aspiration, monolayers were washed 2 times with warm NCI medium (serum was omitted). Fourteen batches of AS medium were prepared each with a different test lot of serum. Three milliliters from each batch of AS medium was added to separate plates of DI-011-infected cells. The control represents a DI-011-infected cell monolayer with 3 ml of serum-free NCI medium added. Plates were incubated at 40.5° for 24 hr. At the end of this time the medium was processed as described in the text to determine the interferon titers. In our assays 1 PR_{50} (VSV) unit of chicken interferon is equivalent to about 10–25 units of the MRC Research Standard A 62/4 provided by the National Institute of Allergy and Infectious Diseases Research Resources Branch.

[b] In one experiment, yields of mouse interferon from mouse L(Y) cells were found to be unaffected by the presence of calf serum Lot #1 in the medium during the induction period.

embryo cells, and was determined from the data shown in Fig. 3. IFN production by primary mouse embryo cells is maximal at 37.5°.

Processing of Interferon to Remove Residual Viral Inducer

It is important to remove residual IFN-inducing particles from the IFN samples before bioassay, otherwise they may induce IFN endogenously

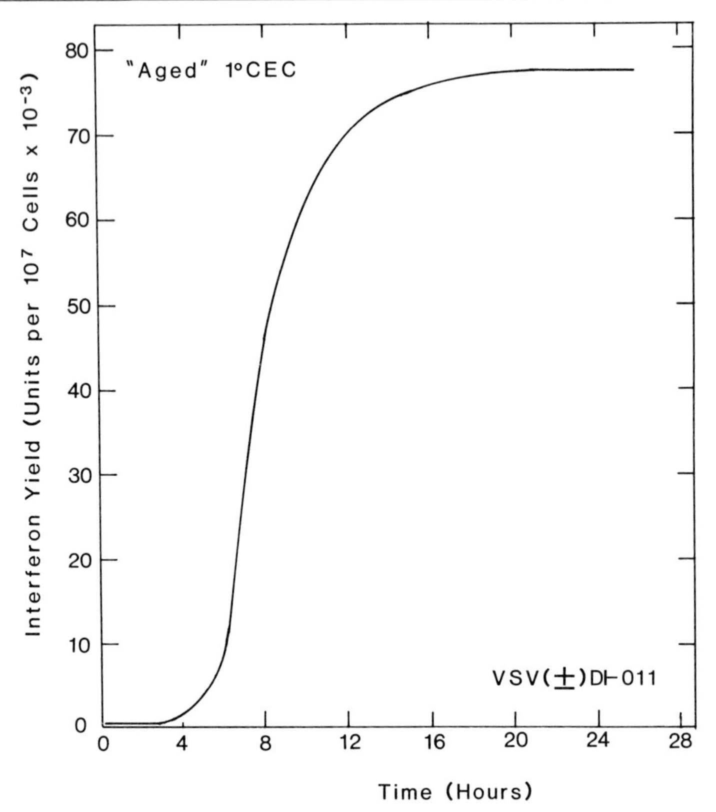

FIG. 2. Time course for the production of IFN. Monolayers of "aged" (7-day-old) primary chick embryo cells were induced with the [±]DI-011 particle of VSV, or VSV *ts*G11(I), at m_{ifp} = 1. The solid line represents the composite of 6 experiments.

during the assay process and contribute spuriously to the titer.[15] This is accomplished by using fetal bovine serum as a carrier to precipitate inducer virus (and non-IFN macromolecules) in the presence of perchloric acid (PCA). The IFN-containing medium is chilled on ice, fetal bovine serum is added to a final concentration of 6%, and cold PCA is added to a final concentration of 0.15 *M* (a 1.5 *M* stock solution of PCA is diluted 1 : 10 in the IFN medium) for 20 to 24 hr at 4°. Perchloric acid acidification of the IFN preparation in the presence of a large amount of carrier protein is considered preferable to dialysis against HCl to pH 2 because it physically removes any potential inducer virus from the IFN medium. (Hydrochloric acid treatment was found to be much less effective than PCA in

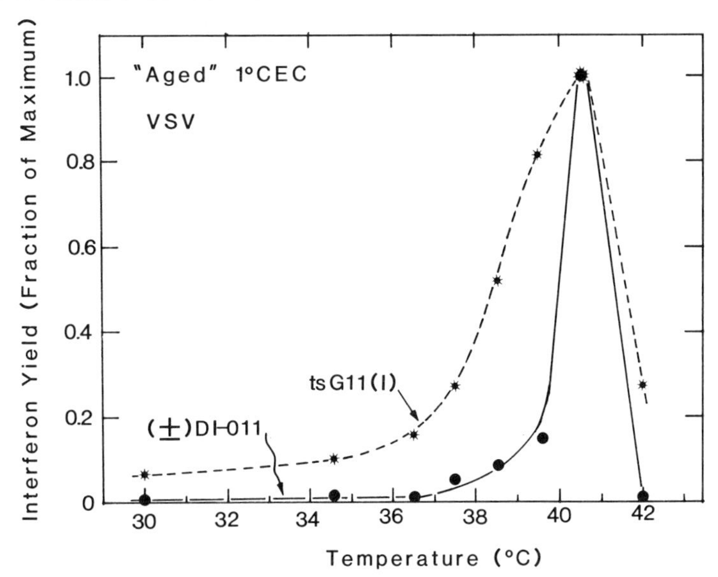

FIG. 3. The IFN-inducing capacity of VSV-[±]DI-011 and VSV-*ts*G11(I) in "aged" (6-day-old) primary CEC as a function of temperature during IFN production. Infection with [±]DI-011 particles was at $m_{ifp} = 1$. This multiplicity produces a peak yield of IFN in a type $r = 1$ dose–response[16]; in this case, over 1,000,000 reference standard units per 10^7 cells were produced. Maximal IFN production under these conditions occurs upon infection of a cell with a single physical particle of [±]DI-011.[7,16] The peak yield for *ts*G11(I) was over 290,000 reference standard units per 10^7 cells, and was achieved at $m_{ifp} = 1$, also under the conditions of an $r = 1$ type of dose–response.[16]

inactivating residual viral inducers in the IFN stocks; also see Gifford.[17]) For example, residual IFN-inducing activity was demonstrable in reovirus preparations inactivated by HCl at pH 2 by the procedure of Lei and Joklik,[18] but not by the PCA procedure.[13] The precipitate that results from PCA treatment is removed by centrifugation at 2000 rpm for 10 min. The IFN-containing supernatant fluid is transferred by decantation to a sterile vessel where the PCA is neutralized by the dropwise addition of 4 *N* KOH, with phenol red in the NCI medium as a visual indicator of pH. Dilute PCA (0.15 *M*) is used to back-titrate the IFN preparation so that the final pH of the sample is slightly acidic, about pH 6.5–6.8. The $KClO_4$ precipitate that forms in copious amounts upon neutralization of the PCA

[17] G. E. Gifford, this series, vol. 78 [27].

[18] M. T. Lei and W. K. Joklik, *Virology* **51,** 191 (1973).

with KOH is removed by centrifugation: 2000 rpm for 5 min. The supernatant fluid contains the crude acid-stable chicken IFN and can be stored at −20° until assayed. Under these conditions of handling the chicken IFN is stable indefinitely and can be frozen and thawed many times without any loss of antiviral activity.

Bioassay of Chicken Interferon

Chick IFN is assayed by a modification of the cytopathic effect inhibition assay of Rubinstein *et al.*[19] Microtiter wells in 96-well trays are inoculated with 2.5×10^5 primary CEC per well in 100 μl of AS medium. Cell monolayers that develop in the microwells are "aged" (incubated without a medium change) for 4 days at 37.5° under the same conditions used to age cells for IFN induction. In order to maximize gas circulation and exchange and improve the uniformity of cell growth, the microwell tray covers are raised by about 5 mm from their snug position. This can be accomplished by inserting 5-mm lengths of sterilized 3/16 in. o.d. plastic Tygon tubing held in place with a small amount of sterile Dow Corning high vacuum grease at the four corners of the bottom portion of the microwell tray. If the lids of the microwell trays are not raised in this manner, then as the cells "age," there is uneven growth of the cells in the center portion of the microtiter tray.

After 4 days of "aging" 100 μl of the IFN samples to be assayed is added to the aged medium in duplicate wells. Twofold serial dilutions are accomplished with a 100 μl Co-Star Octapette. After 24 hr incubation at 37.5° the IFN-containing medium is removed by aspiration and the cells are challenged with a non-IFN-inducing strain of wild-type (Ind.-HR) vesicular stomatitis virus.[9] A 250 μl volume of AS medium containing 1.0×10^3 plaque-forming particles is added to each well. Incubation for the development of virus cytopathic effects is carried out at 37.5° for 48 hr. The medium from each well is then aspirated and the cells stained vitally by adding 250 μl of neutral red (0.05 g/liter) in slightly acidified AS medium for 1 hr at 37.5°. The neutral red is removed by aspiration and the monolayers washed twice with PBS at room temperature. Following this, the monolayers are allowed to dry. Neutral red is then released into 100 μl of 3 *M* guanidine hydrochloride and the cytopathic effect inhibition is quantified by determining the percentage dye uptake with an Artek Model 210 Elisa reader containing a 540-nm filter. Interferon titers are determined as the reciprocal dilution of interferon resulting in a 50% cytopathic effect inhibition. In our hands, this procedure results essentially in equiv-

[19] S. Rubinstein, P. C. Familletti, and S. Pestka, *J. Virol.* **37,** 755 (1981).

alence between the 50% cytopathic effect inhibition and the 50% plaque reduction (PR_{50} VSV) endpoints. In our assay 1 PR_{50} (VSV) unit is equivalent to about 10–25 units of the chick interferon MRC Research Standard A 62/4 provided by the National Institute of Allergy and Infectious Diseases Research Resources Branch.[20-22] This means that the peak yield of 75,000 units of IFN per 10^7 "aged" CEC shown in Fig. 2 would convert to an apparent yield of between 750,000 and 1,875,000 standard units, i.e., about 0.13 unit per cell. However, since IFN induction under these conditions results in a type $r = 1$ dose–response curve,[6,16] the actual yield per cell is 2.7 times higher (only 37% of the cells are producing IFN[16]), namely, 0.35 unit per cell. As an additional control, in each interferon assay we include our own chick interferon standard [a single batch of chick interferon containing 20,000 PR_{50} (VSV) units/ml stored at $-20°$] in order to ensure that the different batches of chick embryo cell monolayers used for interferon assays throughout the course of our work respond uniformly to interferon treatment.

Acknowledgments

This work was supported by NIAID Grant AI-18381 and benefited from the use of a Cell Culture Facility supported in part by NCI Grant CA-14733 and by The University of Connecticut Research Foundation.

[20] N. B. Finter, *in* "Interferons and Interferon Inducers" (N. B. Finter, ed.), p. 161. Elsevier, Amsterdam, 1973.
[21] N. B. Finter, this series, Vol. 78 [2].
[22] S. Pestka, this volume [2].
[23] M. J. Sekellick and P. I. Marcus, *J. Interferon Res.* **5,** 651 (1985).

[19] *In Vitro* Induction of Interferon from Guinea Pig Fibroblasts

By TIMOTHY R. WINSHIP

Introduction

The guinea pig (*Cavia cobaya,* family Caviidae) is an attractive model for studies of infectious disease since cavids are larger than most rodents but relatively easy to care for and manipulate. The production and use of biological response modifiers in guinea pigs, however, have been frustrating due in part to the difficulties associated with the *in vitro* culture of guinea pig cells.

Several investigators have reported attempts to produce guinea pig interferons, both *in vitro* and *in vivo*, with various levels of success. These efforts have been reviewed by Sonnenfeld in a previous volume of this series.[1] Interferon yields were usually low and difficult to assay.

Our own efforts to produce guinea pig interferon in quantities sufficient for *in vivo* purposes centered around the use of fibroblasts originating from whole guinea pig embryos (GPE cells) for interferon induction and assay. We found that crude guinea pig fibroblast interferons do not share all the typical properties of classical fibroblast-derived interferons and require careful attention to induction and storage conditions.

Cell Culture

Suspensions of guinea pig embryo cells were prepared from minced and gently trypsinized whole guinea pig embryos of about 30 days gestation by a previously published technique.[2] Primary cells are seeded at a density of 3×10^4 cells per cm^2 of growth area in 0.25 ml of minimal essential medium with Hanks' salts (HMEM) plus 10% newborn calf serum (Flow Laboratories or Kansas City Biological, Inc.). This medium also included extra glutamine (to 494 mg/liter) and nonessential amino acids. Primary cells seeded in this way generally achieve confluence after 48–72 hr incubation at 37°. The seeding density used is important since the interferon-producing capacity of GPE cells decreases with cell passage number.

Confluent GPE monolayers can be maintained in MEM with Earle's salts (EMEM), extra glutamine (494 mg/ml), nonessential amino acids, and 5% newborn calf serum for several days. Optimal interferon induction conditions, which are somewhat different than those suitable for cell maintenance, are described in detail below.

Monolayers of primary guinea pig embryo fibroblasts are useful for the production of relatively small volumes (1–2 liters) of interferon. Such monolayers often give the highest yields of interferons per milliliter of any guinea pig cell (up to 6000 PR_{50} units/ml). Primary GPE cells do not, however, readily lend themselves to larger scale techniques. One can produce larger volumes of crude guinea pig fibroblast interferon at useful concentrations through the use of passaged guinea pig embryo fibroblasts. While GPE cells can be successfully passaged more than 10 times, the cells lose their ability to produce significant quantities of interferon after the fourth passage (counting the initial growth of cells from trypsinized

[1] G. Sonnenfeld, this series, Vol. 78, p. 162.

[2] G. D. Hsiung, R. B. Tenser, and C. K. Y. Fong, *Infect. Immun.* **13,** 926 (1976).

tissue as passage 1) when seeded at a density of 3×10^4 cells per cm^2 growth area.

Passaged GPE cells can be adapted for growth on roller vessels. Particular attention must be paid to the degree of cleanliness of the vessels if glass roller bottles are used since the cells are intolerant of even minute quantities of detergent residue. Several brands of plastic disposable roller bottles were tried. We found both Corning and Falcon bottles equally suitable for GPE cell culture.

Adaptation of GPE cells to roller culture is best achieved by the following conditions. The primary cell contents of two 120-cm^2 culture flasks (32-oz glass prescription bottles) are transferred to a single 690-cm^2 glass roller bottle (Wheaton) with 175 ml of HMEM containing 10% newborn calf serum prepared as described above. The cells should achieve confluence in 48–72 hr at 37°.

When these initial roller cultures become confluent they can be expanded to other roller flasks as long as a split ratio of 1 : 3 is not exceeded. The volume of medium added to any size roller flask should be 0.25 ml/cm^2 of growth area to achieve reproducible cell growth and maintenance. It is important to initially seed these cells in medium whose pH will not change beyond the range of 6.8–7.2, particularly during the first 24 hr of culture. GPE cells used in interferon induction experiments showed extreme sensitivity to changes in pH.

Interferon Induction

Flasks or roller bottles of GPE cells are usually not suitable for interferon induction if the cells do not become confluent within 72 hr of seeding. When the cells reach confluence, the medium in each culture vessel is discarded and fresh EMEM plus 10% newborn bovine serum added. This medium is allowed to remain unchanged on the cells for an "aging" period of 7 days at 37°. Careful pH control of the medium during this period is essential for cell survival and optimal interferon production. We maintained medium pH between 6.8 and 7.2 by the periodic addition of small quantities of 7.5% sodium bicarbonate solution, but other buffering systems could be used for equal effect. We found an aging period to be essential for production of optimal interferon quantities. This requirement has also been observed in some other cell–virus systems, and was first noted by Carver and Marcus[3] (see also Sekellick and Marcus[4]).

At the end of the aging period the medium is removed from each

[3] D. H. Carver and P. I. Marcus, *Virology* **32,** 247 (1967).

[4] M. J. Sekellick and P. I. Marcus, this volume [18].

culture vessel and the monolayers washed with warmed Hanks' buffered salt solution (HBSS). The cells are then infected with an inducing virus. We found several viruses that would induce interferon from guinea pig embryo fibroblasts, including Newcastle disease virus (NDV), Sindbis virus, and vesicular stomatitis virus (VSV) T1026R1. Maximum interferon production with each inducing virus was very dependent on multiplicity of infection. Of the viruses we tested, VSV T1026R1 induced the greatest quantities of interferon.[5] VSV T1026R1 is a non-*ts* revertant of a mutant originated by C. P. Stanners.[6] It is best prepared in Vero cells since yields of this virus tend to be low when cell lines that will produce interferon are used. Wild type VSV did not induce measurable quantities of interferon from infected GPE cells.

After an appropriate adsorption period at 37° (1 hr for VSV T1026R1 or Sindbis virus, 2 hr for NDV) the viral inoculum is removed from the infected culture and the monolayers rinsed three times with warmed HBSS. The infected cells are then fed with EMEM containing 0.5% newborn bovine serum. We were not able to induce significant amounts of guinea pig interferon in totally serum-free media. The infected cultures are then incubated 24 hr at 37°. An incubation temperature below 36 or above 38° is detrimental to the yield of interferon.

Harvest and Storage of Interferon

Interferon-containing supernatant medium from infected monolayers can be harvested 18–24 hr after inoculation. Time of harvest is important since the period of maximum interferon production is brief and the crude product is unstable at 37°.[7] The crude material can be stored at −70° for at least 6 months. It loses some activity after 2 weeks storage at −20° or if held overnight at 4°.

Interferon Assay

A standard plaque reduction assay was our method of choice for the quantitation of guinea pig interferon levels. Passaged guinea pig embryo cells were used for all such assays. The sensitivity of such cells to the antiviral effects of interferon was not affected by density of seeding, cell "age," or passage number. We did note, however, that lots of passaged

[5] T. R. Winship, C. K. Y. Fong, and G. D. Hsiung, *J. Interferon Res.* **3,** 71 (1983).
[6] C. P. Stanners, A. M. Francoeur, and T. Lam, *Cell* **11,** 273 (1977).
[7] T. R. Winship, C. K. Y. Fong, and G. D. Hsiung, *J. Gen. Virol.* **65,** 843–847 (1984).

guinea pig fibroblasts originating from different groups of embryos could vary significantly (up to 3-fold) in their sensitivity to a "standard" interferon preparation. This may have been due to the use of outbred Hartley strain guinea pigs as a source of cells. Therefore experiments dealing with guinea pig interferon need to be assayed on a single lot of GPE fibroblasts to avoid excessive variation of titers. Wild-type VSV was used as the challenge virus in all such assays. Results were expressed in 50% plaque reduction (PR_{50}) units per ml.

The standard method of inactivating the inducing virus by treatment at pH 2 for 24 hr was inappropriate for guinea pig interferon since it resulted in the loss of more than half the antiviral activity of crude preparations.[5,7] Of the other methods available for the removal of inducing virus, irradiation of samples with shortwave ultraviolet light was rapid and reliable. One milliliter quantities of crude interferon were irradiated in 35-mm Petri dishes with a single 15-W germicidal lamp at a distance of 25 cm. An 8 min exposure under these conditions was used routinely to inactivate the virus in crude preparations. This exposure was approximately four times that needed to completely inactivate the inducing virus. VSV T1026R1, NDV, or Sindbis virus inactivated in this way could not induce interferon in guinea pig cells and did not interfere with wild-type VSV in interferon assays.

Remarks

Our experiments with the production of crude guinea pig interferon showed the critical role cell culture and type of viral inducers can have on interferon yields. It was clear from this work that not all antiviral substances produced by cell cultures behave as "classical" Type I interferons.[8] We strongly recommend that investigators attempting to induce interferon from uncharacterized animals or cell cultures initially avoid subjecting their preparation to heat, low pH, or other chemical and physical regimens to which interferons are "known" to be resistant.

[8] S. Pestka and S. Baron, this series, Vol. 78, p. 3.

[20] Induction of Equine Interferons

By TILAHUN YILMA

Introduction

Horses are widely used for the sport of racing, as pleasure animals for riding, and as working animals. They also serve as an animal model for a number of human diseases such as severe combined immunodeficiency disease (SCID).[1] Among the greatest health problems in horses are viral infections of the respiratory system, such as influenza and equine rhinopneumonitis (Herpes), and there is hope that recombinant DNA-derived interferon may be used for therapy in the near future.

In addition to its wide range of antimicrobial and anticellular activity, interferon has been shown to profoundly regulate both humoral and cellular immunity.[2] In this regard, we have attempted to study the role of interferon in Arabian foals with SCID, a fatal genetic disease inherited as an autosomal recessive trait and similar to some forms of SCID of children. In our initial study we demonstrated that cells obtained from foals with SCID produce normal levels of fibroblast and leukocyte interferons but were unable to produce immune interferon.[3] In an earlier study a preliminary characterization of equine interferons was described.[4] Here a brief narrative on the induction of the 3 major types of equine interferons is provided.

Preparation of Cell Cultures

Equine Dermal Cell Cultures. Equine dermal cells can be obtained from the American Type Culture Collection (12301 Parklawn Drive, Rockville, Maryland 20852). These cells are available at the tenth passage and are usually useful up to 40 passages. Alternatively, primary equine dermal cell cultures can be easily initiated from the skin of a fetus or a foal. Briefly, the hair from the skin is completely removed by shaving. Several pieces of the skin are removed from a fetus or by biopsy from a foal and then transferred to a beaker containing adequate amounts of modified Dulbecco's minimum essential medium (DMEM) supplemented

[1] T. C. McGuire and M. J. Poppie, *Infect. Immun.* **8,** 272 (1973).
[2] B. R. Brodeur and T. C. Merigan, *J. Immunol.* **114,** 1323 (1975).
[3] T. Yilma, L. E. Perryman, and T. C. McGuire, *J. Immunol.* **129,** 931 (1982).
[4] T. Yilma, T. C. McGuire, and L. E. Perryman, *J. Interferon Res.* **2,** 363 (1982).

with 10% fetal calf serum and antibiotics (growth medium). The underlying subcutaneous layer, mainly containing fatty tissue, is removed by dissection. The glistening, white dermal tissue is minced with a pair of scalpel blades to a size of 1 mm^3 or less in a glass Petri dish containing growth medium. Primary cultures are then established as described under the section on ovine choroid plexus cell cultures elsewhere in this volume.[5] Equine dermal cells established in this manner have a useful life span for interferon assay up to 40 passages.

Isolation of Mononuclear Cells from Equine Blood. Blood is withdrawn aseptically from the jugular vein of the horse with a 16-gauge needle and a 50-ml syringe. After removing the needle, blood is poured gently into a glass bottle containing heparin (10 units heparin/ml of blood). Due to the unique characteristics of equine whole blood, the red blood cells separate from the plasma without centrifugation when allowed to settle for about 30 min at room temperature. After the red blood cells have settled, the plasma is transferred to 50-ml glass centrifuge tubes (Corning Glass Works, Palo Alto, California) in 15-ml aliquots per tube. To each tube an equal volume of Ca^{2+}- and Mg^{2+}-free Hanks' balanced salt solution (HBSS) is added.

With a 6-in. cannula and syringe, 12 ml of Ficoll–Hypaque solution (*d* 1.079 g/ml) (Winthrop Laboratories, New York, New York) is gently layered below the plasma/HBSS mixture. Tubes are then centrifuged at 1360 rpm (400 *g*) for 40 min in a Beckman TJ-6 centrifuge. The mononuclear cells will band at the interface of the Ficoll–Hypaque and plasma/HBSS mixture.

The cell band is removed and then transferred to another 50-ml centrifuge tube. The remaining volume is filled with HBSS and then centrifuged at 1100 rpm (350 *g*) for 12 min. After the supernatant is aspirated, the pellet is first resuspended in a small amount of HBSS, and then the total volume is brought up to about 30 ml with HBSS. Cells are then centrifuged at 900 rpm (150 *g*) for 8 min, resuspended, and centrifuged again under the same conditions. After the last wash, the supernatant is aspirated and the pellet is resuspended in RPMI-1640 medium containing 10 m*M* HEPES and antibiotics. The total number of mononuclear cells and their viability are determined by the dye exclusion test with trypan blue. About 10^8 cells per 50 ml of blood are routinely obtained. The percentages of lymphocytes, monocytes, and neutrophils are determined by a differential count with Wright's stain. In general, greater than 90% lymphocytes, 5–10% monocytes, and less than 1% neutrophils can be obtained.[4]

[5] T. Yilma, this volume [73].

Interferon Induction

Poly(I)·Poly(C)-Induced Fibroblast Interferon. Stock solution (1 mg/ml) of the sodium or potassium salt of poly(I)·poly(C) (Sigma Chemical Co., St. Louis, Missouri) is prepared in double-distilled water and then sterilized by filtration. Confluent equine fibroblast cultures in 25-cm^2 plastic flasks are rinsed twice with DMEM and then induced with 30–90 μg of poly(I)·poly(C) and 60 μg DEAE-dextran per ml of medium. After all cultures are incubated at 37° for 1 hr, monolayers are rinsed twice, and 4 ml of growth medium is added to each flask. After overnight incubation at 37°, supernatant fluid is harvested and stored at 4° until assayed for interferon. The antiviral activity of all 3 types of interferons are tested by the semimicrotitration method.[6] Briefly, interferon samples are tested for the prevention of cytopathic effect (CPE) in 2×10^4 equine dermal cells inoculated with 10^4 plaque-forming units of vesicular stomatitis virus in 96-well microtiter plates (Costar, Cambridge, Massachusetts). Interferon units are expressed as the reciprocal of the highest dilution which gives 50% protection. Since no international standard interferons are available for equine interferons, the use of internal laboratory standards is recommended.

In our laboratory, the use of poly(I)·poly(C) alone failed to induce detectable levels of interferon in equine dermal cells. Furthermore, we have not observed significant differences among dermal cultures of different passages in the amount of interferon produced after induction with varying concentration (30–90 μg/ml) of poly(I)·poly(C) plus a constant level (60 μg/ml) of DEAE-dextran. Also, the sensitivity of the test was not significantly different among samples assayed for interferon on dermal cultures of different passages. By and large, unprimed or nonsuperinduced equine dermal cultures did not produce more than 100 units of interferon.[4]

Priming and Superinduction of Fibroblast Interferon. As previously reported,[7] older cultures of equine dermal cells produce higher levels of interferon than fresh cultures. Optimum levels of interferon production are attained when 2-week-old dermal cultures in 25-cm^2 flasks are first primed with 100 units of interferon (total of 2 ml per flask) for 16 hr at 37°. After several rinses with DMEM, cells are superinduced with 75 μg/ml of poly(I)·poly(C) and 120 μg/ml of DEAE-dextran in the presence of 15–25 μg/μg of cyclohexamide (total volume of 2 ml per flask) for the first 3 hr of

[6] P. C. Familletti, S. Rubinstein, and S. Pestka, this series, Vol. 78, p. 387.

[7] W. E. Stewart, II, "The Interferon System," p. 23. Springer-Verlag, Berlin and New York, 1979.

incubation and then with 2 μg/ml of actinomycin D for an additional 2 hr of incubation. All cultures are then rinsed with DMEM 3 consecutive times with a 10-min incubation at 37° between rinses. Finally 4 ml of growth medium is added, and after overnight incubation at 37° the supernatant is harvested. On the average, priming in combination with superinduction increases the amount of interferon induced by about 100-fold. In this laboratory up to 10,000 units/ml of interferon has been obtained through the use of the above procedure. Equine dermal cells are quite sensitive to actinomycin D and display severe CPE when greater than 4 μg/ml are used.

Newcastle Disease Virus-Induced Fibroblast Interferon. The lentogenic strain of Newcastle disease virus (NDV) is a nonpathogenic, vaccine strain. It is produced to a very high titer in the allantoic cavity of 10-day-old embryonated chicken eggs.[8] The virus can be titered by several means, including hemagglutination of 0.4% chicken erythrocytes in PBS or plaque assay on primary chick fibroblast cells. Since the virus induces CPE in equine dermal cells, it is necessary to treat the virus with either UV or heat before using it for the induction of interferon. In our laboratory, we have found that 0.5 ml of 10^8 PFU/ml of NDV in a 2-ml glass vial when heat inactivated for 6 min in a 56° water bath induces high levels of interferon without causing CPE in cells. Equine dermal cells induced with NDV at a multiplicity of 10 give the best results. Equine fibroblast interferon produced by induction with either poly(I) · poly(C) or NDV is acid resistant at pH 2.

Induction of Leukocyte Interferon. For the induction of leukocyte interferon, 6-well Linbro plates (Flow Laboratories, Inglewood, California) or other types of plastic flasks can be used. Each well (10 cm^2) is seeded with 2×10^6 cells in 2 ml of RPMI-1640 medium supplemented with 10% FCS, 10 m*M* HEPES, and antibiotics. When larger flasks are used for the production of large quantities of interferon, it is essential that the above cell concentration be maintained for optimum induction. Cultures are induced with 15 μg/ml of poly(I) · poly(C) and 30 μg/ml of DEAE-dextran or with partially heat-inactivated NDV at an MOI of 10. Maximum levels of interferon are induced when cells are cultured for 24 to 48 hr at 37° in a humidified CO_2 incubator. At the end of the induction period, culture fluid is harvested and then stored at −70° until assayed for interferon. Using the above procedure, one can obtain 3,000 to 10,000 units of leukocyte interferon/ml. While NDV-induced equine leukocyte interferon is acid labile at pH 2, poly(I) · poly(C)-induced equine leukocyte interferon is acid resistant at that pH.

[8] K. C. Zoon and F. E. Campbell, this series, Vol. 78, p. 301.

Induction of Immune Interferon. The culture conditions for the induction of immune interferon are virtually identical with those employed for induction of leukocyte interferon, except for the type of inducer used and the length of the incubation period. Blood mononuclear cells seeded at 2 $\times 10^6$ cells in 2 ml of RPMI-1640 medium supplemented with 10% FCS, 10 m*M* HEPES, and antibiotics are induced with 15 μg/ml of phytohemagglutinin (PHA) (Difco, PHA-P3110-57, Detroit, Michigan) for 3 or 4 days. Culture fluid is collected and then stored at $-70°$ until assayed for interferon. Similar to the induction of leukocyte interferon, the cell concentration is critical and must be maintained when large volumes of immune interferon are produced. The use of greater than 15 μg/ml of PHA is not recommended and may even be toxic to mononuclear cells. Using the above protocol, one can obtain 2000–6000 units of immune interferon/ml. In our laboratory, concanavalin A (Con A) was found to be inferior to PHA for the induction of equine immune interferon. Induction of interferon with Con A was highly variable from batch to batch and often too unpredictable for routine production of immune interferon. Since equine immune interferon is highly labile, storage beyond 6 months at $-70°$ is not recommended, particularly for use on equine cells. However, its antiviral activity is retained on ovine and bovine cells (see concluding remarks). Similar to human immune interferon, equine immune interferon is acid labile at pH 2.

Cross-Species Antiviral Activities of Equine Interferons

Although the antiviral activities of interferons are often described as host or species specific, there are numerous examples in which their antiviral activity has been demonstrated in closely related or unrelated species. For example, Gresser *et al.*[9] demonstrated greater activity of human leukocyte interferon on bovine cells than on human cells. Recently the greater sensitivity of the ovine choroid plexus[10] and caprine synovial membrane cells[11] to human and other animal interferons has been demonstrated in our laboratory. A summary of the cross-species antiviral activity of equine interferons is presented in the table.

Briefly, all 3 types of equine interferons have high levels of antiviral activity on bovine and ovine cells. In particular, equine leukocyte interferon has greater antiviral activity on bovine and ovine cells than on equine cells and demonstrable activity on human cells. Both equine fibro-

[9] I. Gresser, T. M. Bandu, D. Brouty-Boye, and M. Tovey, *Nature (London)* **251,** 543 (1974).

[10] T. Yilma, *J. Gen. Virol.* **64,** 2013 (1983).

[11] T. Yilma, R. G. Breeze, and S. R. Leib, *Am. J. Vet. Res.* **45,** 2094 (1984).

CROSS-SPECIES ANTIVIRAL ACTIVITY OF EQUINE INTERFERONS

Equine interferon type[a]	Original titer on equine cells	Interferon titer (units/ml) on indicated cells			
		Equine dermal cells	Bovine turbinate cells	Ovine choroid plexus cells	Human dermal cells (FS-7)
Fibroblast	NA[b]	4570	3510	3510	0
Leukocyte	NA	700	2890	17660	550
Immune	1030[c]	0	890	1970	0
Immune	New	3160	1000	1000	ND[d]

[a] Fibroblast interferon was obtained by superinducing equine dermal cells. Leukocyte interferon was induced in blood mononuclear cells with NDV. Immune interferon was induced in blood mononuclear cells with PHA (see text).
[b] NA, not applicable.
[c] This refers to immune interferon samples which have been in storage at −70° for over 6 months and have lost their antiviral activity upon retesting on equine dermal cells but not on bovine and ovine cells.
[d] ND, not done.

blast and immune interferons have approximately equivalent antiviral activity on equine, bovine, and ovine cells. Interestingly, equine immune interferon lost its antiviral activity on equine cells but not on bovine or ovine cells after 6 months storage at −70°. No detectable levels of antiviral activity of equine interferons were observed on either feline (feline Crandell kidney) or murine (L-929) cell lines.

Concluding Comments

Equine interferons are very similar to the well-studied human interferons in terms of their antiviral activity and their physiochemical properties, with few exceptions. For example, it is interesting to note that NDV-induced human leukocyte interferon is acid stable at pH 2, similar to poly(I) · poly(C)-induced leukocyte interferon; however, NDV-induced equine leukocyte interferon is acid labile at that pH. There are at least two possible explanations for this observation. First, there have already been demonstrated a number of non-allelic members within the human interferon alpha (α)[12,13] and the bovine interferon alpha gene families (Daniel Capon, Genentech, Inc., personal communication). Similarly, it is very

[12] S. Nagata, C. Brack, K. Henco, A. Schambrock, and C. Weissman, *J. Interferon Res.* **1,** 333 (1981).
[13] S. Pestka, *Arch. Biochem. Biophys.* **221,** 1 (1983).

likely that several subtypes within the equine interferon alpha gene family could exist, since interferon genes are highly conserved in vertebrates.[14] It is possible that the products of these genes could have variable sensitivity to acid. If so, NDV might preferentially induce those subtypes of equine interferon α-genes whose products are acid labile compared to those induced with poly(I)·poly(C). An alternative explanation could be that poly(I)·poly(C) is primarily inducing the acid-resistant interferon beta in equine leukocytes instead of the acid-sensitive equine interferon alpha, since fibroblast and leukocyte interferons are often a mixture of both the alpha and beta interferon types.[15] Unfortunately, specific antisera are presently unavailable for definitive classification of equine interferons.

Noteworthy is the loss of the antiviral activity of equine immune interferon on homologous cells after several months of storage at $-70°$, in contrast with the retention of its full antiviral activity on both bovine and ovine cells under the same conditions. At this point it would be too speculative to attempt to explain this interesting observation.

Acknowledgments

The author wishes to acknowledge Mr. Steven Leib and Ms. Barbara Nakata for editing and Ms. Rose Cheff for secretarial work on the manuscript. The work was supported in part by Public Health Service Grant RR05465-19 and by Grants 2 SO7 RR 05465 and 83-CRSR-2-2189 from the National Institute of Health and the United States Department of Agriculture, respectively.

[14] M. Streuli, S. Nagata, and C. Weissman, *Science* **209,** 1343 (1980).
[15] E. A. Havell, T. G. Hayes, and J. Vilček, *Virology* **89,** 330 (1978).

[21] Induction and Characterization of Bovine Leukocyte Interferon

By SARA COHEN, BARUCH VELAN, TAMAR BINO, HAGAI ROSENBERG, and AVIGDOR SHAFFERMAN

The bovine interferons (Bo-IFN), unlike their counterparts from human or murine sources, have not been extensively studied. Yet the Bo-IFN may prove to be a potent agent for the prevention of virus mediated cattle diseases such as foot-and-mouth disease, infectious bovine rhinotracheitis, pseudorabies, bluetongue, and neonatal bovine diarrhea. Indeed when Bo-IFN preparations were tested *in vitro* they exerted antiviral activity against some of the viruses involved in these diseases.[1–3] The methods used for production and characterization of the Bo-IFN from leukocytes are outlined here.

Purification of Leukocytes, Incubation Medium, and Interferon Induction

Isolation of leukocytes and viral induction of interferon synthesis is essentially according to the well-established procedure for human leukocyte interferon. The reader should refer to the detailed description by Cantell *et al.*[4] The procedure was slightly modified and adapted for bovine leukocytes.

For large-scale prepartions, 20–30 liters of blood is collected at the point of slaughter into stainless-steel buckets containing citrate dextrose solution (trisodium citrate, 66 g/liter, citric acid monohydrate, 0.15 g/liter, dextrose, 100 g/liter). Sixty milliliters of citrate dextrose is added per liter blood. Leukocytes are separated in a Sorvall H-6000 A swinging bucket rotor at 5000 g for 7 min at 4°. The buffy coats are pooled and purified by ammonium chloride treatment.[4] The leukocyte suspension recovered from a liter of blood contains $2–2.4 \times 10^9$ cells.

The freshly purified leukocytes are used immediately for IFN induction. The leukocytes are diluted to 10^7 cells per ml in Eagle's minimal essential medium containing 2.4 mg/ml of human agamma plasma. The human agamma plasma can be replaced by bovine agamma plasma or by bovine serum albumin without any noticeable effect on induction. Sendai virus at a final concentration of 100–200 hemagglutination units/ml is used to induce interferon production. Priming with bovine or human leukocyte interferon does not improve the yield significantly. Incubation of the culture with Sendai virus is continued for about 20 hr with gentle stirring with a magnetic stirrer. Cells and debris are removed by centrifugation and the supernatant is stored at −70°. This supernatant is the crude IFN preparation. The same procedure for leukocyte IFN induction can be applied for evaluation of IFN production from an individual animal. In this case it is sufficient to collect 0.5 liter of blood from the jugular vein.

[1] R. Ahl and A. Rump, *Infect. Immun.* **14,** 603 (1976).
[2] L. A. Babiuk and B. T. Rouse, *Intervirology* **8,** 250 (1977).
[3] R. P. Basnal, R. C. Joshi, and H. K. L. Sumar, *Acta Virol. Engl. Ed.* **25,** 61 (1981).
[4] K. Cantell, S. Hirvonen, H. L. Kauppinen, and G. Myllylá, this series, Vol. 78, [4].

Interferon Assay

Antiviral activity of bovine IFN was determined by the cytopathic effect (CPE) inhibition assay which was performed in microassay plates.[5] The bovine kidney cell line, MDBK was used as cell substrate and vesicular stomatitis virus (VSV) as a challenge virus. NIH human leukocytes interferon standard G-023-901-527 was used as a reference.

Inactivation or removal of the inducing Sendai virus is done in one of the following ways.

1. Adjustment to pH 2.0 with 1 *N* HCl and incubation at 4° for 24 hr. The pH of the IFN preparation is then readjusted to 7.0 and antiviral activity determined.

2. Neutralization of the Sendai virus by anti-Sendai antiserum. The IFN preparation is incubated for 1 hr at 37° with goat anti-Sendai serum (titer 1:320) prior to the assay.

3. Chromatography of crude IFN preparations on Affi-Gel Blue (Bio-Rad Laboratories) column (see below).

It is a common practice to treat human IFN preparations with acid to inactivate the inducing virus. Such a procedure, however, may result in loss of activity of some IFN species which are acid labile. Indeed, the antiviral activities of acid-treated crude bovine interferon preparations are consistently lower (30–50%) than that of anti-Sendai-treated preparations. This is demonstrated in Fig. 1 where production kinetics are monitored.

Kinetics of Bo-IFN Production by Leukocytes

Kinetics of interferon production by bovine leukocytes were compared to that of human leukocytes (Fig. 1). The production rate of the Bo-IFN is somewhat slower than that of the Hu-IFN. Yields of Bo-IFN are about one-tenth (1000 units/ml) that of leukocyte Hu-IFN induced and tested in the same way.

Chromatography of Bo-IFN on Affi-Gel Blue Columns

Affi-Gel Blue (Bio-Rad Laboratories) columns allow the separation and partial purification of two subclasses of Bo-IFNs from the crude leukocyte Bo-IFN preparation. It is also useful for removal of Sendai virus particles by means other than denaturing conditions (low pH) or the neutralization of Sendai virus by antibodies.

[5] J. G. Tilles and M. Finland, *Appl. Microbiol.* **16,** 1706 (1968).

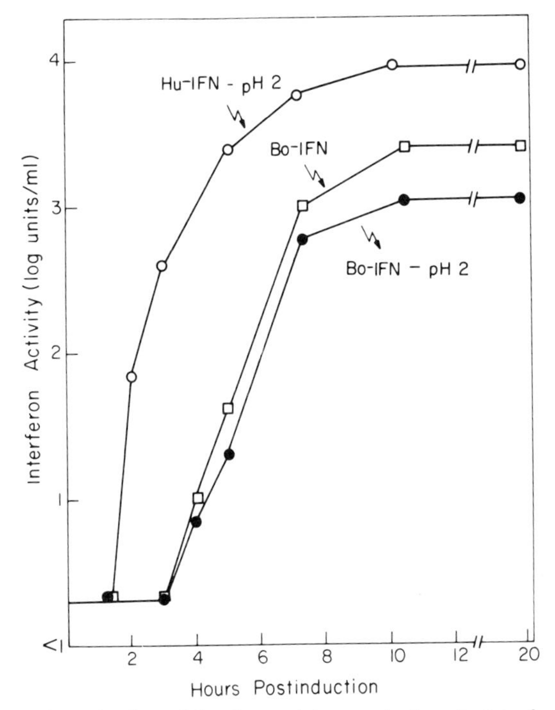

FIG. 1. Rates of production of bovine and human leukocyte interferon. Bovine and human buffy coats were induced with Sendai virus. Samples were assayed for antiviral activity on MDBK cells after treatment either with anti-Sendai antiserum (Bo-IFN), or with acid (Bo-IFN pH 2; Hu-IFN pH 2).

Crude leukocyte IFN preparation (160 ml) is brought to 0.15 *M* NaCl and 0.02 *M* sodium phosphate buffer, pH 7.2 (PBS) and applied on a 3.5 × 1.6 cm column equilibrated with PBS. The column is then washed with 50 ml PBS. Sendai virus particles pass through the column and the antiviral activity is retained. For the elution of the antiviral activity, the column is washed with 60 ml of 1 *M* NaCl, 0.02 *M* phosphate buffer, pH 7.2 (Fraction I) followed by 60 ml of a solution containing 1 *M* NaCl and 50% ethylene glycol in PBS (Fraction II). Results presented in Fig. 2 indicate that the antiviral activity expressed by bovine leukocytes can be attributed to two subpopulations of molecules: molecules that interact electrostatically with the chromophore (Fraction I) and molecules that interact hydrophobically (Fraction II).

Rechromatography of each of these Bo-IFN fractions on the same column indicates that the separation is efficient without any detectable cross contamination between the two subclasses. Acid treatment does not

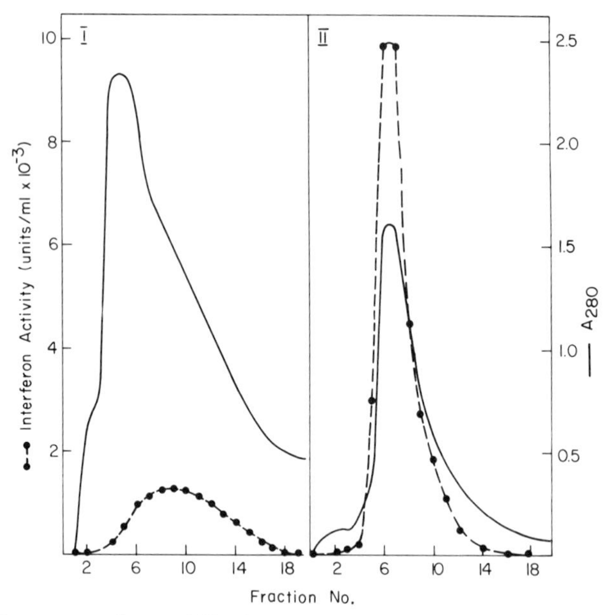

FIG. 2. Chromatography on Affi-gel Blue column. Crude IFN preparation (10^5 units) (specific activity of 5×10^2 units/mg protein) in PBS was applied on the column and washed with PBS. The flow-through contains the virus (anti-Sendai antiserum neutralized its activity on MDBK cells) and 80% of the protein. The column was then washed as described in text. One milliliter fractions were collected and assayed for IFN activity and for absorption at 280 nm.

affect the antiviral activity of Fraction I but results in a 50–70% loss in the activity of Fraction II.

Cross-Species Antiviral Activity of Bo-IFN

Bo-IFN preparations were assayed for their ability to induce antiviral activity in various mammalian cell cultures (see the table). The Bo-IFN antiviral activity is compared to that promoted by leukocyte Hu-IFN and lymphoblastoid Hu-IFN prepared from Sendai-induced Namalva cells. All bovine and human preparations were assayed simultaneously on the cell species indicated in the table by the CPE inhibition method with VSV as challenge virus. Acid stable leukocyte Bo-IFN appears to be quite specific to cells from bovine origin. Hu-IFNs have marked activity on bovine cells, while Bo-IFN fails to protect human cells (less than 4% of the activity on MDBK cells).

CROSS-SPECIES ANTIVIRAL ACTIVITY OF BOVINE LEUKOCYTE INTERFERON[a]

	Cell origin[c]							
	Bovine		Human			Simian	Rodent	
IFN type[b]	MDBK	EBTr	FSII	WISH	HeLa	Vero	BHK	L929
Leu Bo-IFN	240	70	10	<2	<2	<2	<2	<2
Leu Hu-IFN	320	160	320	40	<2	<2	<2	<2
Nam Hu-IFN	260	130	130	130	<2	4	<2	<2

[a] The values in the table represent the reciprocal of the dilution at which 50% inhibition of viral cytopathic effect was observed and thus represent relative activity of various interferon preparations on the diverse cells.

[b] Leu Bo-IFN is an acid-treated crude bovine leukocyte IFN preparation. Leu Hu-IFN is the NIH standard human leukocyte IFN. Nam Hu-IFN represents human IFN prepared from induced lymphoblastoid Namalva cell line, kindly provided by A. Mizrahi.

[c] Cells were obtained from the American type culture collection. FSIIA cells were kindly provided by D. Gurari-Rotman and M. Revel.

Molecular Weights of Bovine Leukocyte IFN

Crude IFN preparation was partially purified to a specific activity of 10^5 units/mg protein (with 30% recovery) by the first 5 steps in Cantell's procedure for human leukocyte IFN purification.[6] This preparation was analyzed by SDS–polyacrylamide gel electrophoresis and compared with a human leukocyte IFN preparation. Electrophoresis was performed in 15% polyacrylamide slab gels containing 0.1 *M* sodium phosphate buffer, pH 7.0, and 0.1% SDS.

The major interferon activity recovered from the gel appears at a molecular weight of 16,000. A minor interferon activity is also detected at a molecular weight of 25,000 (Fig. 3A). The human leukocyte products analyzed by the same procedure (Fig. 3B) reveal the two well-characterized interferon species at molecular weights of 16,000 and 19,000.[7] From this comparative analysis it appears that leukocyte Bo-IFN preparation contains IFN molecules which comigrate with the 16,000-Da human leukocyte IFN fraction. However, it lacks the 20,000-Da class and on the other hand contains polypeptides migrating at 25,000 Da.

[6] K. Cantell, S. Hirvonen, and V. Koistinen, this series, Vol. 78 [71].

[7] J. Vilček, E. A. Havell, and S. Yamazaki, *Ann. N.Y. Acad. Sci.* **284,** 703 (1977).

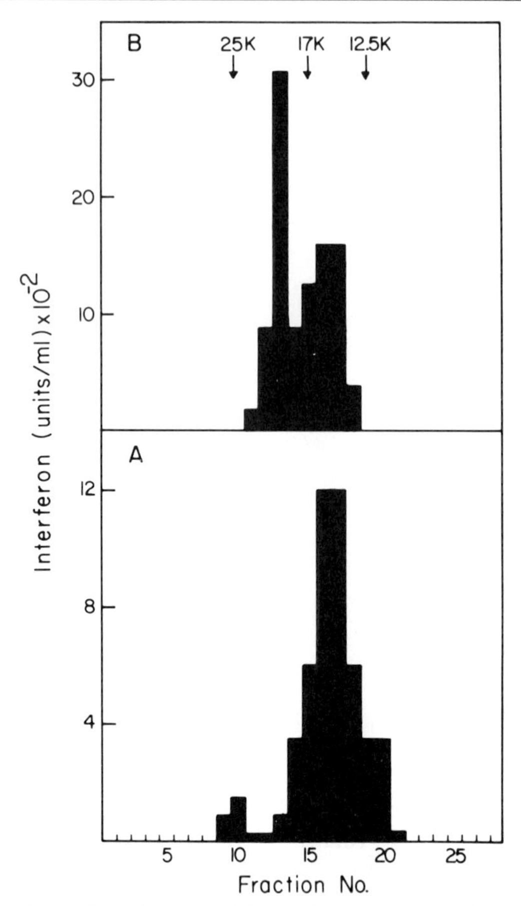

FIG. 3. Comparative SDS–polyacrylamide gel electrophoresis of interferon derived from bovine leukocytes (A) and human leukocytes (B). Electrophoresis was performed in a 15% acrylamide slab gel (0.12 cm thick, 16 cm long containing 0.1% SDS and phosphate buffer). The buffer system was that described by Weber and Osborn.[8] Bo-IFN (2×10^4 units) (specific activity of about 10^5 units/mg protein) and 1×10^5 units of Hu-IFN (specific activity of about 5×10^6 units/mg protein) were applied to the slots of the slab gel. The samples were heated for 2 min at 100° in 0.1 ml of a solution containing 0.01 *M* Na phosphate pH 7, 2% SDS, 10% glycerol, and 0.001% bromophenol blue. The gel was stained for 15 min with 0.125% Coomassie Brilliant blue in methanol : acetic acid : water (5 : 1 : 4) and destained in 7% acetic acid 10% methanol. Slices (2 mm) were cut and IFN was eluted by incubation of the slices in 0.5 ml 0.05 *M* Tris·HCl buffer, pH 8, containing 0.01% SDS for 4 hr at 4°. Eagle's minimal essential medium (MEM) (1.5 ml) containing 10% calf serum was added and interferon activity was assayed after overnight incubation at 4°. The migration position of protein markers, run in a parallel lane, are indicated by arrows: chymotrypsinogen (25K), myoglobin (17K), and cytochrome *c* (12.5). To check possible distortion of the migration pattern by the presence of large amounts of contaminating proteins, myoglobin was included in both interferon preparations (no distortion was observed).

Antigenic Cross-Reactivity of Bo-IFN and Hu-IFN

Neutralization. Neutralization tests were done by preincubating the Bo-IFN for 1 hr at 37° with either anti-α or anti-β human interferon antibodies (research reference G-026-502-568 NIH). Serial dilutions of antibodies starting with a 200-fold excess of antibody units were used. Bo-IFN activity on MDBK cells is not neutralized by antiserum to Hu-IFN-α or to Hu-IFN-β. It seems therefore that leukocyte Bo-IFN and Hu-IFN (α or β) do not carry common antigenic determinants involved in induction of the antiviral state.

Binding. Binding of Bo-IFN to anti-Hu-IFN was determined by using immunoadsorbent Sepharose columns. γ-Globulin fractions were bound to CNBr-activated Sepharose 4B (Pharmacia) as recommended by the manufacturer. Antibodies used were either standard anti-human IFNs (see above) or antibodies prepared against Namalva IFN (kindly supplied by A. Traub). Partially purified Bo-IFN preparation (see previous section) resuspended in 0.15 *M* NaCl and 0.02 *M* sodium phosphate pH 8.0 binds to the anti-Namalva-IFN column (the Namalva preparation contains predominantly Hu-IFN-α). The Bo-IFN activity is recovered quantitatively from the column by elution with 0.1 *M* sodium acetate pH 2.4 (Fig. 4). Binding of the leukocyte Bo-IFN can be attributed to the presence of anti-IFN-α antibodies in the anti-Namalva-IFN rather than anti-human IFN-β antibodies, since similar results were obtained with immobilized standard research anti-Hu-IFN-α. Immunoadsorbent columns prepared with anti-human leukocyte IFN antibodies are useful for purification of Bo-IFN.

The binding and the neutralization experiments suggest that leukocyte Bo-IFN preparation shares antigenic determinants with Hu-IFN-α, although these were not involved in the induction of the antiviral state on human cells.

Concluding Remarks

Bovine leukocytes induced by Sendai virus produce acid stable and acid labile interferons.

The antigenic data and the similarities in molecular weight of the acid stable Bo-IFN and Hu-IFN-α suggest that these molecules are structurally related. Furthermore, translation experiments in *Xenopus laevis* oocytes injected with mRNA from induced bovine leukocytes indicate that Bo-IFN like the Hu-IFN-α[9,10] are translated from 13 S mRNA (unpublished data).

[8] K. Weber and M. Osborn, *J. Biol. Chem.* **244,** 4406 (1969).

[9] R. McCandliss, A. Sloma and S. Pestka, this series, Vol. 79, p. 51.

[10] S. L. Berger, M. J. M. Hitchcock, K. C. Zoon, C. S. Birkenmeier, R. M. Friedman, and E. H. Chany, *J. Biol. Chem.* **255,** 2955 (1980).

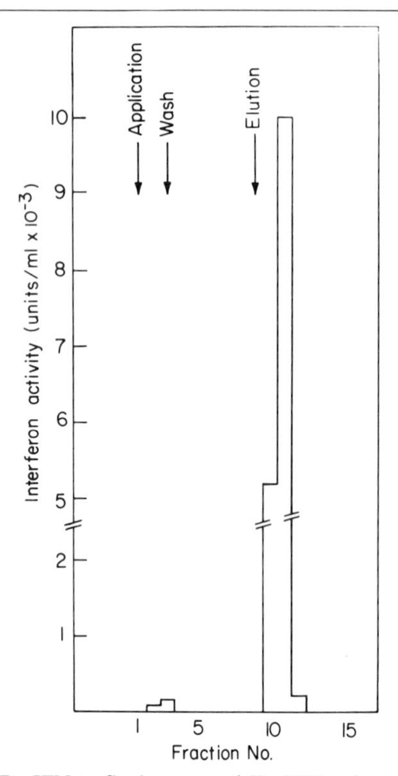

FIG. 4. Binding of Bo-IFN to Sepharose anti-Hu-IFN column. A partially purified Bo-IFN preparation (2×10^4 units) was loaded on a 5 ml column of anti-Namalva-IFN bound to Sepharose 4B (the binding capacity of the column was 10^7 units of Namalva IFN). Binding, washing and elution were performed as described in the text. One milliliter fractions were collected and assayed for IFN activity. Eighty percent of activity was recovered from these columns.

The nucleotide sequence of three of the Bo-IFN-α genes[11] sustains these observations. The putative polypeptides coded by these genes share a 65% amino acid homology with the human IFN-αs. In this context it is worth noting that the antiviral activity of Bo-IFN is specific to cells of bovine origin while Hu-IFN is about equally effective on human and bovine cells.

The characteristics of the acid labile product remain to be elucidated. This activity binds strongly to Affi-Gel Blue and can be recovered in the 1 *M* NaCl, 50% ethylene glycol eluate. However, this eluate probably contains acid stable molecules as well. This is indicated by the fact that the genetically engineered acid stable Bo-IFN-α is recovered from the column in the same fraction.[11]

[11] B. Velan, S. Cohen, H. Grosfeld, and A. Shafferman, this volume [64].

[22] Production and Purification of Mouse Immune Interferon

By MARC DE LEY, HUBERTINE HEREMANS, and ALFONS BILLIAU

Introduction

Crude murine immune IFN preparations contain a variety of lymphokines and biologically active molecules other than Mu-IFN-γ. The execution of *in vivo* experiments demands large quantities of Mu-IFN-γ which should be largely devoid of these substances. Also, the investigations on the biochemical properties of the Mu-IFN-γ molecules, and the generation of both polyclonal and monoclonal antibodies (e.g., in rat), require highly purified antigen preparations. Methods for obtaining such preparations are described in this chapter.

Preparation of Murine Spleen Cell Cultures

Murine spleen cell cultures are prepared as described previously.[1] Briefly, spleens are removed aseptically from NMRI mice (more than 8 weeks old) and teased with the plunger of a syringe through a fine stainless-steel wire screen. After washing the cells with RPMI 1640 medium they are resuspended in RPMI 1640 medium supplemented with 2% heat-inactivated (0.5 hr, 56°) fetal calf serum, 2 m*M* glutamine, 100 units/ml penicillin, 50 μg/ml streptomycin, 10^{-5} *M* 2-mercaptoethanol, 6 m*M* HEPES, 13 m*M* tricine, pH 7.4. Cell suspensions of 2 ml are seeded at a concentration of 3×10^6 cells/ml in 24-well culture plates.

Preparation of the Inducer

When staphylococcal enterotoxin A (SEA) is used as the inducer, the culture broth of *S. aureus* 13N2909, known to be a high producer of SEA, is prepared as described elsewhere.[2,3] Alternatively the killed bacteria may also be used as an inducer of Mu-IFN-γ. The bacteria grown as described above are therefore isolated by centrifugation (10,000 rpm, 30 min) and washed in phosphate-buffered saline (PBS). Then they are treated with 1.5% formalin for 1.5 hr at 27°, washed with and resuspended

[1] H. Heremans, M. De Ley, A. Billiau, and P. De Somer, *Cell. Immunol.* **71,** 353 (1982).
[2] M. De Ley, J. Van Damme, and A. Billiau, this volume [13].
[3] L. Spero and J. F. Metzger, this series, Vol. 78, p. 331.

in PBS containing 0.02% sodium azide to 1/10 of the original volume. Finally, they are heated for 10 min at 50° and lysed in a French press.

Induction and Harvesting of the Cultures

After seeding the spleen cells in 24-well plates, either culture broth or cell lysates of *S. aureus* 13N2909 are added to the suspensions, the optimal dilutions of these inducers being determined in a preliminary experiment. Generally they are in the range of 2 to 20 μl/ml cell suspension for culture broth and in the range of 0.05 to 0.5 μl/ml cell suspension for bacterial lysates. The induced cultures are then incubated in a humidified atmosphere (5% CO_2, 95% air) at 37° for 48 to 72 hr. The supernatants are collected, clarified by centrifugation (3000 rpm, 20 min), and stored at −70° until further processing.

Initial Purification on Silicic Acid

The crude Mu-IFN-γ is stirred for 3 hr at 4° with autoclaved silicic acid (1 g/100 ml of crude IFN). The silicic acid is then sedimented by centrifugation (2000 rpm, 10 min) and washed with PBS (one-third the original volume). The adsorbed Mu-IFN-γ is eluted by stirring the silicic acid for 15 min with 50% (v/v) ethylene glycol, 1.4 *M* NaCl in PBS (1/10 of the starting volume) twice. The eluted Mu-IFN-γ is further concentrated by dialysis against 20% (w/v) PEG-20,000 in 1 *M* NaCl, PBS.

Affinity Chromatography on Con A-Sepharose

The silicic acid-purified and concentrated Mu-IFN-γ is then extensively dialyzed against 20 m*M* Tris·HCl, pH 8.0, and applied to a Con A-Sepharose column (0.9 × 15 cm; 10 ml/hr) equilibrated with the same buffer. The column is rinsed with the same buffer until the absorbance monitored at 280 nm (LKB 2238 Uvicord SII, Bromma, Sweden) reaches the base line level. Elution of the bound Mu-IFN-γ is then performed with 0.5 *M* α-methylmannoside, 0.5 *M* NaCl in 20 m*M* Tris·HCl, pH 8.0. A typical elution pattern obtained by titrating the fractions on L929 cells, as shown in Fig. 1, reveals that 50% of the applied antiviral activity passes unadsorbed through the column. The other 50% is bound and can be specifically eluted with 0.5 *M* α-methylmannoside. Similar results were obtained by Wietzerbin and coworkers.[4] The latter Mu-IFN-γ-containing

[4] J. Wietzerbin, S. Stefanos, M. Lucero, E. Falcoff, J. A. O'Malley, and E. Sulkowski, *J. Gen. Virol.* **44,** 773 (1979).

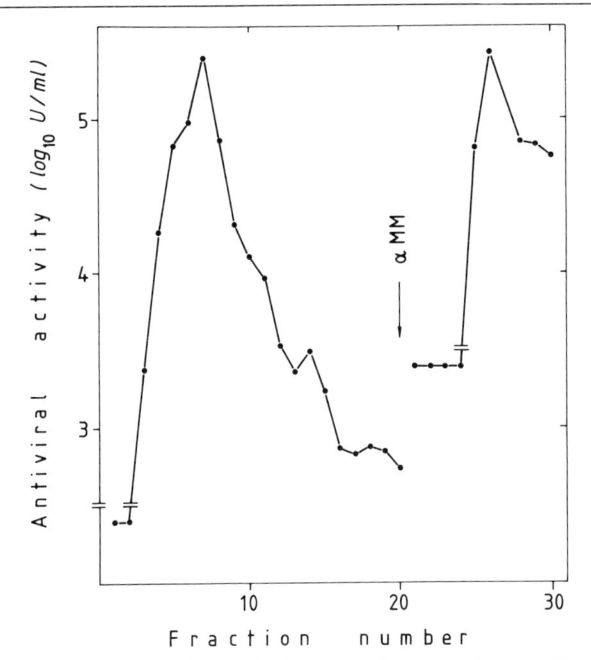

FIG. 1. Affinity chromatography of Mu-IFN-γ on Con A-Sepharose. Three million units of MuIFN-γ (in 0.15 *M* NaCl, 20 m*M* Tris·HCl, pH 8.0) was applied to Con A-Sepharose (0.9 × 12 cm; 10 ml/hr; 4°). The combined fractions of the unbound Mu-IFN-γ contained 1.3 × 10^6 units of antiviral activity, the α-methylmannoside (αMM)-eluted fractions 1.35 × 10^6 units in total.

fractions are then pooled and subjected to a subsequent purification by fast protein liquid chromatography (FPLC).

Ion-Exchange Chromatography by FPLC Mono Q

The pooled Mu-IFN-γ containing fractions from Con A-Sepharose chromatography are extensively dialyzed against 20 m*M* triethanolamine, pH 7.3 and sterilized through a Millipore 0.22-μm filter. They are then applied to a mono Q column (0.5 × 5 cm, 1 ml/min, room temperature, Pharmacia Fine Chemicals, Uppsala, Sweden). Following a rinsing step with starting buffer, the bound proteins are eluted by applying a linear gradient from 0 to 0.5 *M* NaCl in the same buffer. As shown by the results of the titration of the fractions on L929 cells and the absorbance pattern at 280 nm of a typical purification run on a mono Q column (Fig. 2) the bulk of the proteins is bound to the ion-exchange resin, while 16% of the applied Mu-IFN-γ flows through. Upon application of the NaCl gradient

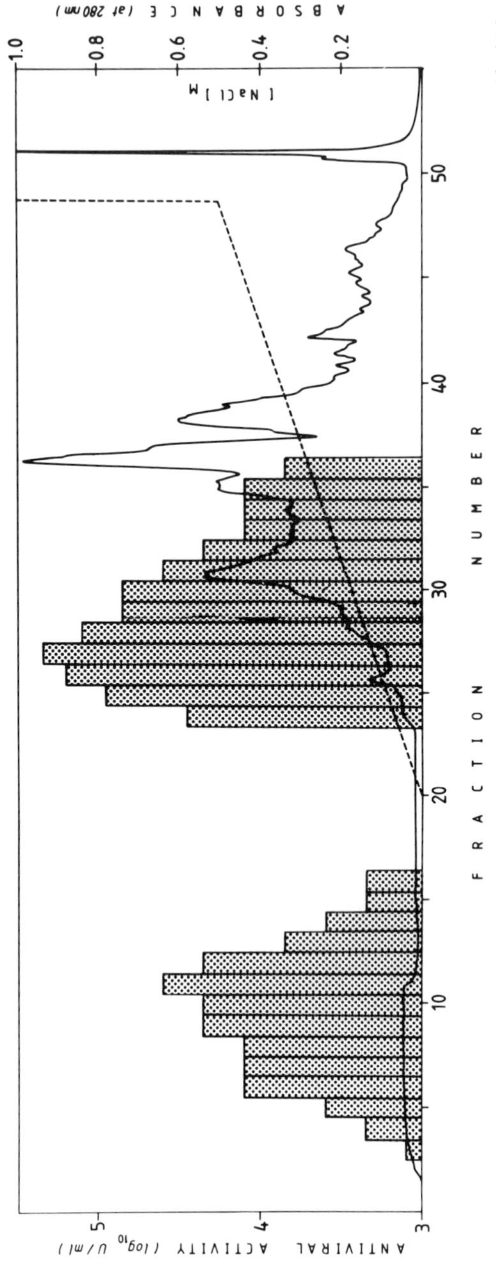

FIG. 2. Fast protein liquid chromatography of Mu-IFN-γ on a mono Q column. One million units of Con A-Sepharose-purified Mu-IFN-γ (in 20 m*M* triethanolamine, pH 7.3) was applied to a mono Q column (5×50 mm; 1 ml/min; 22°). After washing with starting buffer, a linear gradient from 0 to 0.5 *M* NaCl (---) was applied. The total antiviral activity (bar graph) in the void fractions amounted to 1.7×10^5 units, in the NaCl-eluted fractions 8.3×10^5 units.

PURIFICATION OF Mu-IFN-γ

Purification step	Purification factor	Recovery (%)	Specific activity ($\log_{10}$ units/mg protein)
Silicic acid	25–80	77	4.1
Con A Sepharose	7	42	5.25
FPLC mono Q	45	58	6.7

the bound Mu-IFN-γ (84% of the total activity) is eluted as a single peak at 0.11 *M* NaCl.

Concluding Comments

The highest levels of Mu-IFN-γ production were obtained with the lysate of *S. aureus* 13N2909. On average a level of $10^{3.1}$–$10^{4.3}$ units/ml was reached after 72 hr of incubation. This crude Mu-IFN-γ was purified by a sequence of silicic acid adsorption/elution, affinity chromatography on Con A-Sepharose, and anion-exchange chromatography on FPLC mono Q. The average purification factor, recovery, and specific activity obtained after these different steps are summarized in the table.

Section III

Purification of Interferons

[23] Large-Scale Purification of Recombinant Human Leukocyte Interferons

By S. JOSEPH TARNOWSKI, SWAPAN K. ROY, RAYMOND A. LIPTAK, DAVID K. LEE, and ROBERT Y. NING

The large-scale purification of recombinant human leukocyte alpha A interferon (rIFN-αA) has been facilitated by two events: first, the cloning and expression of the rIFN-αA gene in *Escherichia coli*[1] and second, the purification of rIFN-αA from crude, cell-free *E. coli* extracts with the use of columns of immobilized monoclonal antibodies raised against human leukocyte interferon.[2] Analysis of the isolated molecules by nonreducing SDS–PAGE[3] revealed the presence of oligomers formed by intermolecular disulfide bonds.[4] The recovery and separation of these monomers of rIFN-αA from oligomers is described here.

Materials and Solutions

Filter aids: Hyflo and Standard Super Cel (Johns-Manville Products, Englewood Cliffs, NJ).

E. coli W3110 *Trp* R$^-$/pLIF A *Trp* 55 grown as described.[1]

Cell extraction buffer: 2 *M* guanidine hydrochloride (GuHCl); 2% (v/v) Triton X-100; 0.1 *M* Tris · HCl, pH 7.5.

AB Buffer 1: 0.3 *M* GuHCl; 0.3% Triton X-100; 0.1 *M* Tris · HCl, pH 7.5.

AB Buffer 2: 0.5 *M* NaCl; 0.2% Triton X-100; 0.025 *M* Tris · HCl, pH 7.5.

AB Buffer 3: 1.0 *M* NaSCN; 0.1% Triton X-100; 0.025 *M* Tris · HCl, pH 7.5.

AB Buffer 4: 0.15 *M* NaCl; 0.1% Triton X-100.

AB Buffer 5: 0.2 *M* acetic acid (HoAc); 0.1% Triton X-100; 0.15 *M* NaCl.

[1] D. V. Goeddel, E. Yelverton, A. Ullrich, H. L. Heyneker, G. Miozzari, W. Holmes, P. H. Seeburg, T. Dull, L. May, N. Stebbing, R. Crea, S. Maeda, R. McCandliss, A. Sloma, J. M. Tabor, M. Gross, P. C. Familletti, and S. Pestka, *Nature (London)* **287,** 311 (1980).

[2] T. Staehlin, D. S. Hobbs, H. F. Kung, C.-Y. Lai, and S. Pestka, *J. Biol. Chem.* **256,** 9750 (1981).

[3] S. Pestka, B. Kelder, D. K. Tarnowski, and S. J. Tarnowski, *Anal. Biochem.* **132,** 328 (1983).

[4] T. DeChiara, F. Erlitz, and S. J. Tarnowski, this volume [58].

CM Buffer 1: 0.025 *M* NH_4OAc, pH 4.5; 0.024 *M* NaCl
CM Buffer 2: 0.025 *M* NH_4OAc, pH 5.0; 0.12 *M* NaCl.

All solutions were prepared and all chromatography steps were performed at 4°.

Extraction and Purification

Cell Extract Preparation

Frozen, acid-treated *E. coli* cell paste (6 kg) was ground in a meat grinder and added to cell extraction buffer[5] (24 liters) at 4°. The slurry was stirred with an overhead propeller attached to an air-driven mixing motor until homogeneous, and the pH was adjusted to 7.5 by gradual addition of 5 *N* NaOH. The mixture was stirred for an additional 2 hr and then, diluted in a 50-gallon refrigerated tank to 150 liters with cold distilled water.

The dilution step enhanced floculation of debris and lowered the GuHCl concentration. Dry Hyflo filter aid (40 g/liter) was added to the diluted extraction mixture. The slurry was continuously stirred and pumped to the reservoir of a Komline-Sanderson rotary drum vacuum minifilter (Komline-Sanderson, Peapack, NJ).

The vacuum drum wheel was previously precoated with 6–10 kg of Standard Super Cel slurried into water and washed with several hundred liters of water to remove fine particles. The solid shavings containing cell debris and precipitated proteins that resulted from the advance of the knife blade were incinerated. The cell-free, debris-free filtrate was collected into a receiver under vacuum and 2-octanol was added, as needed, to eliminate foam. Each time the receiver filled, the filtrate was pumped through a 0.3-μm filter cartridge (Pall Corp., Cortland, NY) to remove remaining Super Cel fines, then through a stainless-steel coil immersed in an ice-water bath to a holding tank at 4°.

The filtrate was concentrated about 3-fold to 53 liters by tangential flow ultrafiltration on a Millipore Pellicon unit equipped with 50 ft^2 PTGC (10,000 MWCO) membranes (Millipore Corp., Bedford, MA). The concentrated drum filtrate was subsequently filtered through a 1.2-μm filter cartridge (Pall Corp, Cortland, NY) and stored at 4° until loading onto the immunosorbent column.

Comments. The filtration rate through the precoated bed on the rotary drum minifilter was 60–70 liters/hr. This allowed a batch to be clarified in about 2.5 hr. Continuous centrifugation of the diluted extract at low *g*-

[5] S. J. Tarnowski, U.S. Patent 4,432,895 (1984).

TABLE I
COMPONENT LIST FOR AUTOMATION OF THE IMMUNOSORBENT COLUMN

Item	Quantity	Manufacturer
Eldex Chromat-A-Trol Model II	1	Eldex Laboratories, Menlo Park, CA
Eldex Selector Valve (6-position)	2	Eldex Laboratories
Eldex Power Module	1	Eldex Laboratories
Fluorocarbon Solenoid Valve (2-way)	1	Fluorocarbon, Anaheim, CA
Fluorocarbon Solenoid Valve (3-way)	6	Fluorocarbon
LKB Uvicord S Monitor	1	LKB Instruments, Rockville, MD
LKB Dual Channel Recorder	1	LKB Instruments
Masterflex Peristaltic Pump (7535-00)	1	Cole-Parmer, Inc., Chicago, IL
Masterflex Pump Head (7016)	1	Cole-Parmer, Inc.
Bubble detector and relay	1	Skan O Matic, Elbridge, NY
Pressure switch and relay	1	Built in-house
Pall DPA 4001 11012 P Disposable Filters	2	Pall Trincor, Cortland, NY
pH controller and probe	1	Fisher Scientific, Springfield, NJ
K 100 × 45 column	1	Pharmacia Inc., Piscataway, NJ

force (i.e., 10,000 to 14,000 *g*) was attempted but never clarified the extract satisfactorily. The combination of binding and depth filtration on the precoated bed contributed to obtaining a cleared filtrate.

Immunosorbent Column Purification

Monoclonal antibody from hybridoma LI-8 was coupled to *N*-hydroxysuccinimide ester activated Sepharose 4B[6] (Pharmacia Fine Chemicals, Piscataway, NJ) to a density of 12.7 mg/ml of settled gel. A column of immunosorbent gel (10.0 × 7.5 cm) totaling 720 ml was packed and automated[6] for recycling with a microprocessor interfaced with the components listed in Table I and according to the schematic in Fig. 1.

The microprocessor was programmed to deliver concentrated drum filtrate 15 times in 3.5 liter portions from the holding tank to the column. This quantity of extract exceeded the binding capacity of the column for

[6] S. J. Tarnowski and R. A. Liptak, Jr., *Adv. Biotechnol. Processes* **2,** 271 (1983).

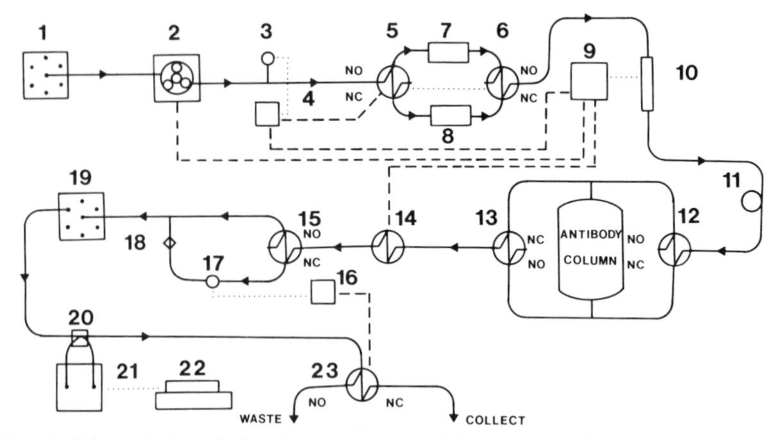

FIG. 1. Schematic interfacing the components of the automated immunosorbent column. NO, normally open: NC, normally closed. (1) Input six-way valve; (2) peristaltic pump; (3) pressure switch; (4) pressure relay; (5,6) filter solenoid; (7,8) filters (1.2 μm); (9) detector power supply; (10) bubble detector; (11) delay loop; (12,13) loading solenoids; (14) antisyphon solenoid; (15) stream diverter; (16) pH controller; (17) pH electrode: (18) check valve; (19) output six-way valve; (20) stream splitter; (21) UV detector; (22) recorder; (23) collection solenoid. →, fluid path and direction; – – –, control circuit ···, associated units.

rIFN-αA by 25%. The packed column was first equilibrated with 2.5 column volumes of AB Buffer 1, and then a 3.5 liter portion of the drum filtrate was pumped onto the column at 3 liters/hr (linear velocity: 38.2 cm/hr). The column was washed sequentially with five column volumes each of AB Buffer 1, AB Buffer 2, AB Buffer 3, and AB Buffer 4. The column breakthrough and washes were discarded to waste. Following the wash steps, AB Buffer 5 (elution) was pumped onto the column and the effluent stream was diverted through a pH electrode and subsequently through a flow cell spectrophotometer measuring absorbance at 280 nm. The column effluent was collected as waste until the pH decreased to 5.0, at which time the pH controller switched the solenoid valve to collect a separate pool. The protein peak that elutes was collected for about 1.5 column volumes. The column was then reequilibrated with AB Buffer 1 and another cycle was begun.

Each load/elution cycle required 7.25 hr, or 4.5 days to process the total drum filtrate volume. The eluate pools were accumulated as one volume of 16.5 liters. Between each cycle the pool was adjusted to pH 3.0 with HCl to prevent proteolytic degradation of rIFN-αA during storage before the next step.

TABLE II
SUMMARY OF THE PURIFICATION OF rIFN-αA[a]

Step	Volume (1)	Total protein (g)	Total activity ($\times 10^{-12}$ units)	Specific activity (units/mg)	Recovery (%)	Purification factor
1. Crude extract	24.0	394.0	2.6	7.0×10^6	100	1.0
2. Drum filtrate (concentrate)	52.5	192.0	3.1	1.6×10^7	119	2.3
3. Immunosorbent eluate (Ab-Pool)	16.5	8.7	1.9	2.2×10^8	73	31
4. CM-52 Pool (concentrate)	0.98	4.9	1.2	2.5×10^8	46	35
5. Sephadex G-50 chromatography (monomer pool concentrate)	0.48	2.4	0.6	2.5×10^8	23	36

[a] Protein concentrations were determined by the Bradford[7] dye binding assay with the Bio-Rad reagent and purified rIFN-αA monomer as standard. Interferon titers were determined by a colormetric solid-phase enzyme immunoassay (EIA) with a pair of monoclonal antibodies.[8,9] One of the monoclonal antibody pair was covalently linked to horseradish peroxidase.

Table II[7-9] shows the recovery of rIFN-αA in the combined pool is 73% based on total activity. Only 3% of the loaded activity was found in the column breakthrough even though the column was overloaded by 25%. Thirty percent of the activity was, however, found in the combined washes suggesting that the overloaded activity was not tightly bound to antibody.

The SDS–PAGE profiles in Fig. 2 show that the immunosorbent column provides a one-step purification.[10] Lane 3 on the gel shows a single protein band migrating at approximately 20,000 Da free of contaminating *E. coli* proteins. When a duplicate sample of column eluate is electrophoresed under nonreducing conditions (lane 4), two monomers (slow and fast moving monomers, SMM and FMM, respectively), dimers, trimers, and higher oligomers are present. Extraction of *E. coli* cells in the presence of *N*-ethylmaleimide showed that the predominant molecular

[7] M. M. Bradford, *Anal. Biochem.* **72,** 248 (1976).
[8] T. Staehelin, C. Stähli, D. S. Hobbs, and S. Pestka, this series, Vol. 79, p. 589.
[9] H. Gallati, *J. Clin. Chem. Clin. Biochem.* **20,** 907 (1982).
[10] U. K. Laemmli, *Nature (London)* **227,** 680 (1970).

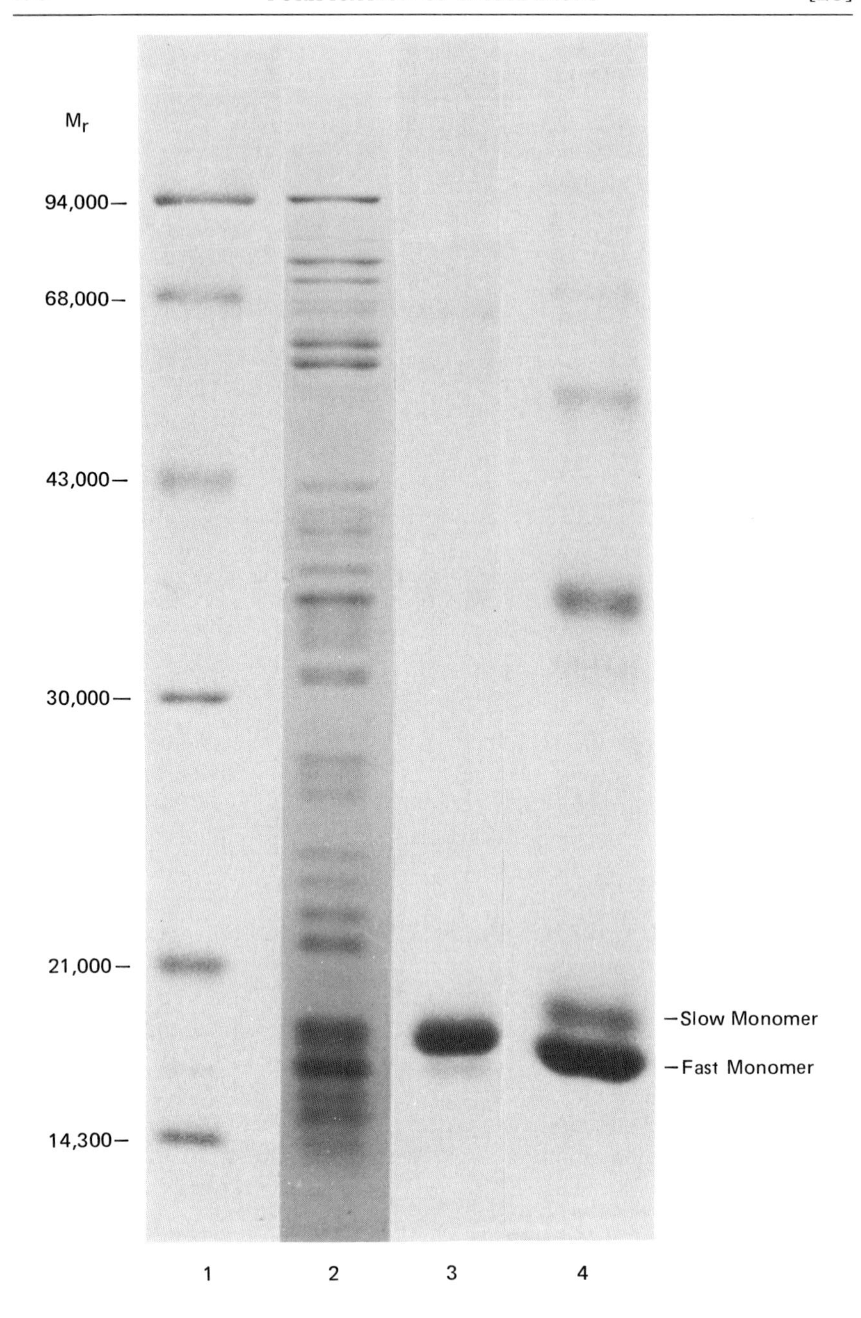
Mr
94,000—
68,000—
43,000—
30,000—
21,000—
14,300—
—Slow Monomer
—Fast Monomer
1
2
3
4

form of rIFN-αA in the cell is a partially or completely reduced molecule.[11] This alkylated derivative was isolated from cell extract and comigrated with SMM on nonreducing SDS–PAGE. We believe that these molecular forms are not isolated from the cell extract, but form from a reduced monomer as the result of intermolecular disulfide bond "scrambling" after isolation. The cysteine residues at positions 1 and 98 of the primary sequence have been implicated in the formation of intermolecular disulfide bonded oligomers.[4]

Comments. Conceptually, monoclonal antibody columns can be developed into one-step purification processes because they are highly specific and they recover antigens in high yield and purity. However, if monoclonal antibody columns are to be competitive on a manufacturing scale operation in the purification of protein pharmaceuticals, several events must occur. First, the supply of antibody must become virtually unlimited. For example, an 11-liter column containing 140 g of immobilized monoclonal antibody LI-8 would have been required to process the extract from 6 kg of *E. coli* cell paste in a single cycle. Manufacturers are responding to these demands by moving away from mouse ascitic fluid to revolutionary *in vitro* and *in vivo* cell culture techniques.

Second, the physical supports to which monoclonal antibodies are coupled must be more rigid to maximize the yield per unit gel per unit time. The Sepharose 4B used here is a soft agarose that begins to compress under the flow rates used. The net effect is decreased flow and

[11] S. K. Roy, unpublished data.

FIG. 2. Sodium dodecyl sulfate–polyacrylamide gel electrophoresis (SDS–PAGE) of immunosorbent purified rIFN-αA. Samples were dried under vacuum in a Savant spinvac and subjected to electrophoresis according to Laemmli[10] on a 12.5% polyacrylamide slab gel (10 × 15 × 0.15 cm) with a 6.0% (3.5 × 15 × 0.15 cm) stacking gel containing 0.1% SDS. The dried samples were dissolved in 10 μl of sample buffer containing 2-mercaptoethanol for reducing conditions and sample buffer without 2-mercaptoethanol for nonreducing conditions. The electrophoresis was performed first at room temperature for 2 hr at 40 V and then for 3 hr at 100 V with cooling. The gel was stained for 1 hr at room temperature with 0.2% Coomassie brilliant blue R-250 dissolved in methanol/acetic acid/water (25:10:65) and destained by diffusion overnight at room temperature with 5% (v/v) methanol in 10% (v/v) acetic acid. Lane 1, standard protein markers (reducing) from Bio-Rad Laboratories (Richmond, CA): M_r = 94,000 phosphorylase *b*; M_r = 68,000, bovine serum albumin; M_r = 43,000, ovalbumin; M_r = 30,000, carbonic anhydrase; M_r = 21,000, soybean trypsin inhibitor; M_r = 14,300, lysozyme; Lane 2, concentrated drum filtrate, 30 μg (reducing); Lane 3, immunosorbent column purified rIFN-αA (reducing), 10 μg; Lane 4, immunosorbent column purified rIFN-αA (nonreducing), 10 μg.

increased back pressure. Attempts were made to minimize bed compression and increase flux by adding coaxial, cylindrical baffles[12] to the gel beds. While this was successful initially, channeling of the column bed occurred after a few cycles, and accordingly the concept was abandoned. Improved agarose support media, such as Sepharose Fast-Flow, have higher cross-linking and are reported to attain a flux 10 times that of Sepharose 4B. Recently, monoclonal antibody LI-8 was coupled to polyhydroxyphase silica.[13] The rigidity of this support allowed flow rates to be achieved which were limited only by the ability of the antibody to capture the antigen.

Third, the longevity of immobilized antibodies packed in columns must be demonstrated. The column used here was recycled hundreds of times and has been packed for over 1 year. The quality (clarity) of the batch loaded onto the column is important, as well as the prevention of microbial growth during long-term storage. Agents such as sodium azide have been used successfully.

Cation-Exchange Chromatography

The cumulative antibody eluate pool (16.5 liters; 0.53 mg/ml protein; 5.6 mS conductivity) was concentrated by ultrafiltration to 1 liter at a rate of 150 ml/min with a Millipore Pellicon cassette system equipped with 5 ft^2 PTGC membranes (10,000 MWCO). The concentrated retentate was diafiltered by diluting with 9 liter of CM Buffer 1, concentrated again to 0.5 liter, and then transferred to a suitable vessel. The cassette membranes were washed with about 1 liter of DM Buffer 1 and this wash was combined with the protein concentrate. The final solution (CM-Load) that was loaded onto the cation exchange column was about 5 mg/ml protein, pH 4.4, and the conductivity was 2.2 mS.

Carboxymethyl (CM)-52 cellulose (Whatman, Inc., Clifton, NJ) was washed and equilibrated prior to use. First, the cellulose was stirred into an excess volume of 0.5 *N* NaOH, filtered, washed with distilled water to neutrality, slurried into an excess volume of 0.5 *N* HCl (two times), filtered, and washed with distilled water to neutrality.

Swollen cellulose was packed into a Pharmacia K 50/30 column (5.0 × 20 cm; 400 ml bed volume) and equilibrated with greater than 2 liters of CM Buffer 1. The CM-Load was pumped onto the column at 20 ml/min and the effluent was monitored at 280 nm with a flow cell spectrophotometer (LKB Uvicord S Monitor, LKB Instruments, Rockville, MD) and

[12] E. Sada, J. Ketoh, and M. Shizawa, *Biotechnol. Bioeng.* **224,** 2279 (1982).

[13] S. K. Roy, D. V. Weber, and W. C. McGregor, *J. Chromatogr.* **303,** 225 (1984).

chart recorder. Following loading, the column was washed with CM Buffer 1 until the absorbance returned to zero baseline. This indicated that Triton X-100 was removed from the sample. The loaded protein was eluted from the column with CM Buffer 2 and the peak was collected as one fraction (CM-Pool; 2.1 liters, 2.35 mg/ml protein). The CM-Pool was concentrated on the Millipore Pellicon cassette system to a protein concentration of 5.0 mg/ml in a total volume of 982 ml. The concentrate was filtered through a 0.45-μm filter just prior to loading onto the gel filtration column.

Comments. There is a significant protein loss (44%) on the CM-52 column (see Table II) due to precipitation of polymeric forms of rIFN-αA on the resin bed. These insoluble aggregates are eluted only with 0.1 *N* NaOH. SDS–PAGE analysis of these aggregates (data not shown) demonstrated that they contained only low levels of monomer.

In addition to removing Triton X-100, the CM-52 column also separates the trace amount of cleavage enzyme from rIFN-αA. ^{14}C-Methylated rIFN-αA[14] is used as a substrate to monitor "protease" that specifically cleaves rIFN-αA at the NH_2-terminus to yield a 15,000 Da fragment. This potentially novel "protelytic enzyme" is found in the breakthrough fraction from the CM-52 column.

Sephadex G-50 Chromatography

Dry Sephadex G-50 Superfine gel (Pharmacia Fine Chemicals, Inc., Piscataway, NJ) was boiled with pyrogen-free distilled water for 1 hr. The slurry was cooled at 4° overnight and poured into a Pharmacia K 100/100 column with a packing reservoir. When the gel settled, the reservoir was replaced by an adapter to form a column of gel 10 × 95 cm with a 7.5 liter bed volume. The column was equilibrated with CM Buffer 2 at a flow rate of 4 ml/min.

The CM-Pool concentrate was divided into five 196-ml aliquots (2.6% of total bed volume) and each was pumped separately onto the column. The flow direction was from bottom to top.

Column effluent was monitored at 280 nm absorbance with a flow cell spectrophotometer (LKB Uvicord S Monitor, LKB Instruments, Rockville, MD) and chart recorder. An LKB fraction collector was used to collect 24 ml fractions at 6 min per fraction. A typical elution profile is shown in Fig. 3. The peak eluting first contained oligomers of rIFN-αA and the second peak contains rIFN-αA monomers (Fig. 4).

After each of the five 196 ml portions were consecutively chromato-

[14] S. J. Tarnowski and R. T. Bartlett, unpublished data.

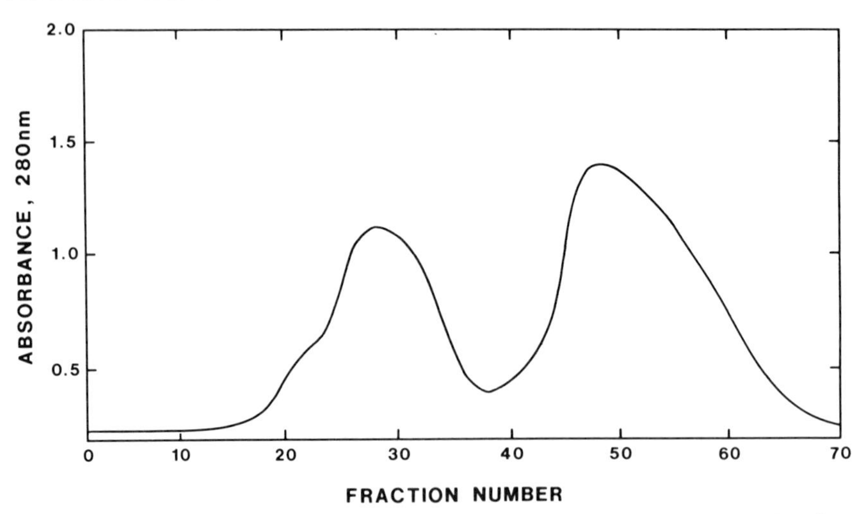

FIG. 3. Sephadex G-50 Superfine gel filtration chromatography of rIFN-αA. Fractions were collected and pooled according to peak areas.

graphed, the fractions for each peak were combined into pools. The monomer pool (1.0 mg/ml protein; 2.9 liters) contained approximately 49% of the protein loaded on the column. The monomer pool was concentrated to 5 mg/ml protein in 480 ml with the Millipore Pellicon cassette system. The concentrate was subdivided into polypropylene bottles and stored at −20°.

Interferon activity was determined on the pooled oligomers and monomers with a cytopathic effect inhibition assay with vesicular stomatitis virus and bovine MDBK cells.[15] The oligomer pool yielded a specific activity of 0.7×10^8 units/mg protein compared to 2.2×10^8 units/mg protein for the monomer pool.

Comments. Chromatographic resolution of the oligomers and monomers of rIFN-αA requires that the column be equilibrated at pH 5.0. When pH values greater than 5.0 are used, resolution disappears and all the protein is eluted in the void volume of the G-50 column. Avoiding this pH-dependent polymerization[16] is critical for the separation process.

Furthermore, resolution of the rIFN-αA oligomers has been accomplished with Sephadex G-75 Superfine gel.[3] Sephadex G-75 was not chosen for the gel filtration described here because it is a soft gel that compresses easily. G-50, a more rigid gel, resists compression and, therefore,

[15] S. Rubinstein, P. C. Familletti, and S. Pestka, *J. Virol.* **37,** 755 (1981).

[16] S. J. Shire, *Biochemistry* **22,** 2664 (1983).

has superior flow properties for scale-up. Resolution of SMM and FMM (see Fig. 4) cannot be achieved with either gel because globular proteins require a difference of about 5000 Da to be separated on Sephadex gel filtration columns. The apparent molecular weight difference between SMM and FMM (~ 500) shown on nonreducing SDS–PAGE may represent improperly folded molecules, molecules with mismatched disulfide bonds, or molecules that have incomplete disulfide bond formation (partially reduced).

The CM-52 column used prior to this step cannot be eliminated from the purification scheme because the equilibrium between Triton X-100 micelles and monomers prevents its complete removal from rIFN-αA during any of the other steps.

Concluding Remarks

Immunosorbents made by covalently coupling monoclonal antibodies to agarose gels have high specificities, and, therefore, are powerful tools for the purification of proteins. Specificity is demonstrated in the case of rIFN-αA, where essentially crude extracts of *E. coli* are passed over a column of immunosorbent gel and rIFN-αA is purified therefrom in essentially one step. Recovery yields are high after this step, but the formation of rIFN-αA oligomers via intermolecular disulfide bonds requires additional steps in the purification scheme to obtain purified monomers. The lower specific activity of oligomers and the possibility they might cause adverse reactions in human clinical trials made it desirable to obtain a monomer preparation.

The monomer preparation shown here contains at least two molecular forms, slow-moving monomer (SMM) and fast-moving monomer (FMM). Ten monomer forms are predicted if all the possible combinations of disulfide bonds, and free sulfhydryl forms, are considered. Radiolabeing the preparation with the alkylating agent, *N*-ethylmaleimide, showed that SMM contained free sulfhydryl groups, but FMM did not. The free —SH in SMM potentially could give rise to oligomers on storage. It is assumed that FMM contains two disulfide bonds. Separation of these two species is desirable to obtain a stable product form. Recently, Bodo and Fogy[17] isolated at least 7 monomer forms of rIFN-α2C(Arg) (identical to [Arg^{23},Arg^{34}]IFN-αA) by affinity purification and reversed-phase chromatography. Subsequent characterization of these component monomers showed that disulfide bonds were partially and completely reduced, "scrambled" and correctly formed between cysteine residues 1–98 and

[17] G. Bodo and I. Fogy, *in* "The Interferon System" (F. Dianzani and G. B. Rossi, eds.), Vol. 24. Raven, New York.

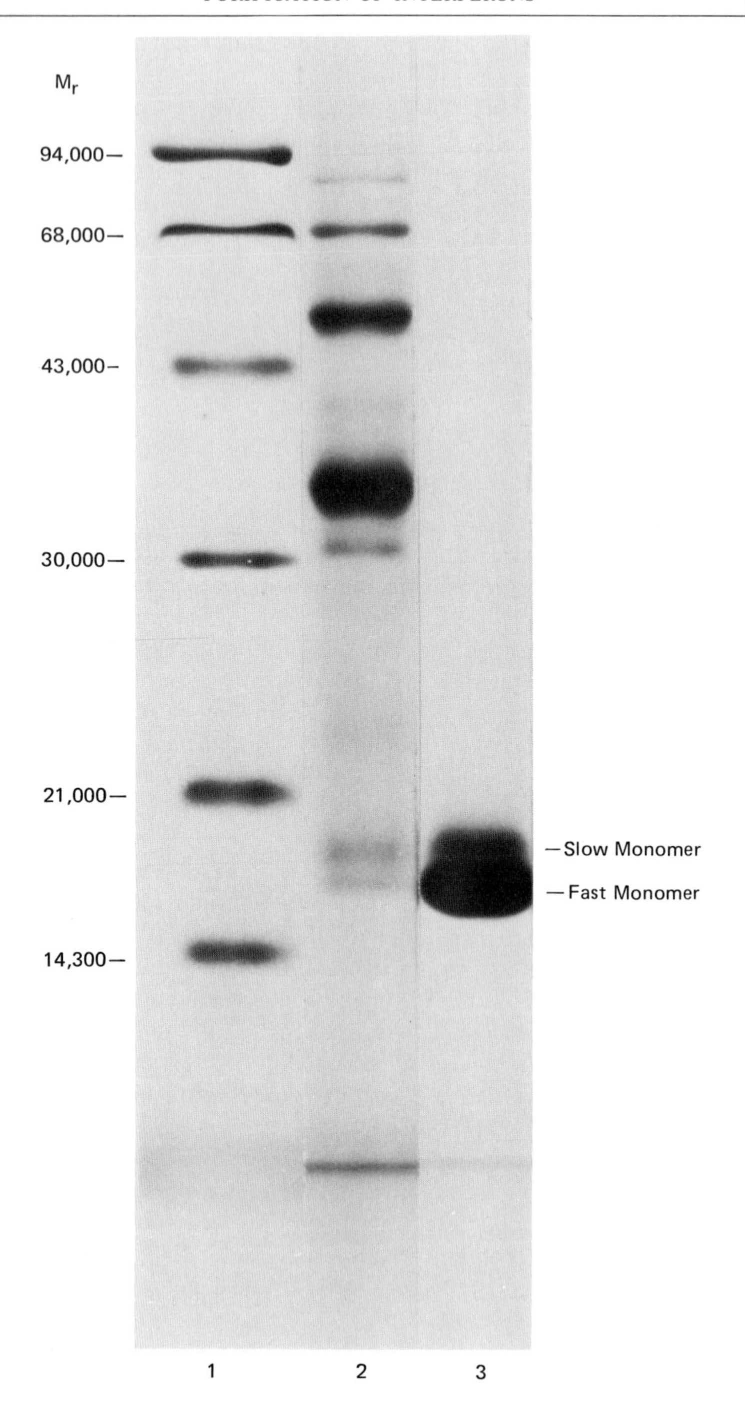
M_r
94,000—
68,000—
43,000—
30,000—
21,000—
14,300—
—Slow Monomer
—Fast Monomer
1
2
3

29–138. They concluded, as we have, that these various species may arise from the reduced monomer.

The formation and removal of rIFN-αA oligomers adversely impacts on the overall recovery for the purification process. Since most oligomers are disulfide linked, and pH-dependent polymerizations can be controlled, conversion to fully active monomer has been demonstrated with reduction–oxidation systems.[5,18] Incorporation of a redox system into the purification scheme could make the process one of extraordinary yield.

The high homology among the leukocyte interferon alpha subtypes engenders an epitope on many that is recognized by monoclonal antibody LI-8. Consequently, other recombinant interferons, rIFN-αA/D (*Bgl*II), rIFN-αC, rIFN-αD, rIFN-αI, rIFN-αJ, rIFN-αK, and rIFN-α2 have been purified by procedures similar to the process described here.

Despite initial skepticism, monoclonal antibody immunosorbent chromatography has proven to be a viable procedure for the purification of proteins for pharmaceutical applications. The ability to reuse the same column hundreds of times makes these columns practical because the quantity of monoclonal antibody required is reasonable and readily obtainable without heroic efforts. Accordingly, the commercial manufacture of protein therapeutics by immobilized monoclonal antibodies for purification should be considered when planning the research and development of these pharmaceuticals.

Acknowledgments

The authors wish to thank Mr. Edward Jenkins for *E. coli* fermentations, Mr. James Feuker for enzyme immunoassays, Ms. Linda Petervary for antiviral assays, Mr. Pascal Bailon for the preparation of immunosorbent gels, and Miss Barbara Thode for her patience in the preparation of this manuscript.

[18] J. Langer and S. Pestka, this volume [34].

FIG. 4. SDS-PAGE of rIFN-αA purified by Sephadex G-50 Superfine chromatography. See legend to Fig. 2 for experimental details. Lane 1, standard protein markers (reducing); Lane 2, rIFN-αA oligomers (nonreducing), 10 μg; Lane 3, rIFN-αA monomers (nonreducing), 10 μg.

[24] Purification of Recombinant Human IFN-α2

By DAVID R. THATCHER and NIKOS PANAYOTATOS

Alpha or leukocyte interferon represents a homologous family of proteins with antiviral activity secreted by nearly all virus-infected human cells.[1] Because *in vivo* levels of synthesis are extremely low, interferon purification from natural sources has been a challenging technical problem.[2] Large volumes of either peripheral blood leukocytes[1-3] (buffy coats) or immortalized cell lines[4-10] are required in conjunction with sophisticated separation techniques such as reversed phase HPLC[2,11] or monoclonal immunoaffinity chromatography.[12,13] The advent of gene technology has afforded a more efficient and cost effective approach to production of interferon. Microorganisms, expressing individual interferon genes at high level,[2] now provide an abundant source of protein for purification. As described in this chapter, purification of recombinant interferon can be achieved by standard methodology and can easily be scaled up to meet the precise and reproducible criteria required for manufacturing. In fact, human IFN-α2 has been purified and crystallized on a commercial scale by Nagabushan and co-workers.[14]

The Hu-IFN-α2 and the closely related Hu-IFN-αA genes were isolated, respectively, by Nagata *et al.*[15] and Streuli *et al.*[16] on the one hand and by Maeda *et al.*[17] and Goeddel *et al.*[18] on the other. The cDNA clones

[1] S. Pestka and S. Baron, this series, Vol. 78, p. 3.

[2] S. Pestka, *Arch. Biochem. Biophys.* **221,** 1 (1983).

[3] K. Berg, C. A. Ogburn, K. Paucker, K. E. Mogensen, and K. Cantell, *J. Immunol.* **114,** 640 (1975).

[4] H. Strander, K. E. Mogensen, and K. Cantell, *J. Clin. Microbiol.* **1,** 116 (1975).

[5] A. Mizrahi, this series, Vol. 78, p. 54.

[6] G. Bodo, this series, Vol. 78, p. 69.

[7] F. Klein and R. T. Ricketts, this series, Vol. 78, p. 75.

[8] P. C. Familietti, L. Costello, C. A. Rose, and S. Pestka, this series, Vol. 78, p. 83.

[9] T. M. Powledge, *Bio/Technology* 214 (1984).

[10] A. W. Phillips, N. B. Finter, C. J. Burman, and G. D. Ball, this volume [4].

[11] M. Rubinstein, S. Rubinstein, P. C. Familletti, R. S. Miller, A. A. Waldman, and S. Pestka, *Proc. Natl. Acad. Sci. U.S.A.* **76,** 640 (1979).

[12] D. Secher and D. C. Burke, *Nature (London)* **285,** 446 (1980).

[13] T. Staehelin, D. S. Hobbs, H.-F. Kung, and S. Pestka, this series, Vol. 78, p. 505.

[14] T. L. Nagabushan and P. Leibowitz, *in* "Interferon alpha 2: preclinical and clinical evaluation," pp. 1–12. Nijhoff, The Hague, 1985.

[15] S. Nagata, H. Taira, A. Hall, L. Johnsrud, M. Streuli, J. Escodi, W. Boll, K. Cantell, and C. Weissmann, *Nature (London)* **284,** 316 (1980).

METHODS IN ENZYMOLOGY, VOL. 119

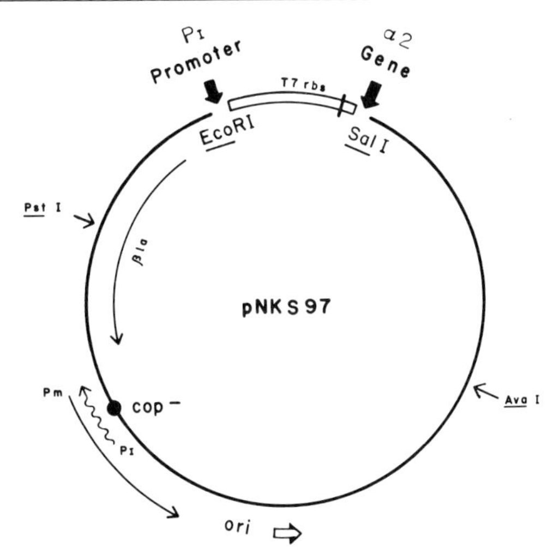

FIG. 1. Structure of the pI.T7.α2.cop1 expression vector. pI.T7.α2.cop1 consists of the pBR322 sequence between the unique *Sal*I (position 650) and *Eco*RI (position 4359) restriction sites and 112 bp of bacteriophage T7 DNA containing the ribosome binding site for gene 1.1.[19] The promoter pI described in this work introduced at the *Eco*RI site and the human leukocyte IFN-α2 gene at the *Sal*I site. The positions of the unique *Pst*I and *Ava*I restriction sites are shown. The origin and direction of replication, the positions of the primer RNA (Pm) and the RNA I (pI) promoters, as well as the cop1 point mutation at position 3029 of the pBR322 map are also indicated.

were prepared from mRNA derived from human leukocytes induced with Sendai virus. The region coding for the natural interferon signal peptide (which cannot be processed correctly by *Escherichia coli*) was removed, and the gene sequence corresponding to the extracellular protein was expressed in *E. coli*.[18]

The Plasmid Vector

A plasmid containing the IFN-α2 gene was kindly provided by C. Weissmann. A 720 bp *Hin*dIII–*Pst*I fragment was digested with S1 nuclease to provide blunt ends and was inserted into plasmid pNKS97 at a unique *Sal*I site rendered blunt with S1 nuclease (Fig. 1). Plasmid

---

[16] M. Streuli, S. Nagata, and C. Weissmann, *Science* **209,** 1343 (1980).

[17] S. Maeda, R. McCandliss, M. Gross, A. Sloma, P. C. Familletti, J. M. Tabor, M. Evinger, W. P. Levy, and S. Pestka, *Proc. Natl. Acad. Sci. U.S.A.* **77,** 7010 (1980); **78,** 4648 (1981).

[18] D. V. Goeddel, E. Yelverton, A. Ullrich, H. L. Heyneker, G. Miozzari, W. Holmes, P. H. Seeburg, T. Dull, L. May, N. Stebbing, R. Crea, S. Maeda, R. McCandliss, A. Sloma, J. M. Tabor, M. Gross, P. C. Familletti, and S. Pestka, *Nature (London)* **287,** 411 (1980).

pNKS97 is a versatile vector for gene expression in *E. coli*.[19] This plasmid consists of 3709 bp of the pBR322 sequence between the SalI and *Eco*RI sites containing the β-lactamase (amp^R) and replication regions as well as a 112 bp fragment from bacteriophage T7 which contains the ribosome binding site for gene 1.1 (Fig. 1).

A unique *Eco*RI site allows the insertion of promoter fragments upstream from the ribosome binding site whereas the unique *Sal*I site permits the insertion of the gene to be expressed (in this case the IFN-α2 gene) in phase with an initiating methionine codon. The promoter used was the PI promoter from ColE1 which promotes the transcription of a small RNA molecule (RNAI) which is involved in the control of plasmid replication. A copy number mutation of this plasmid [pI.T7.α2.cop1] was isolated by selection on L-broth agar plates containing 20 mg/ml methicillin.[20]

Fermentation

The bacteria harboring the pI.T7.α2.cop1 plasmid were maintained on L-agar plates containing 10 mg/ml ampicillin and were subcultured at weekly intervals. Cultures were allowed to grow at 37° for 20 hr into stationary phase in overfilled 3l shake flasks containing 1.6 liter of standard L-broth at low agitation. IFN-α2 expression was of the order of 5–15 $\times$ 10^9 units/liter/OD_{590} (16–47 mg/liter), or 10–30% of total cellular protein.

Purification of IFN-α2

Increased levels of interferon expression in this system lead to a proportionately increasing fraction of interferon protein laid down as an intracellular insoluble aggregate (Fig. 2). At high levels of expression, over 95% of the interferon synthesized by the pI.T7.α2.cop1 system is in an insoluble physical state. This aggregation of foreign or mutant proteins expressed at high levels in *E. coli* is a well documented phenomenon[21,22] and can be exploited to advantage in protein purification strategies. Upon cell breakage, the soluble cell constituents, which include the majority of the contaminating *E. coli* proteins, can be washed away by successive

[19] N. Panayotatos and K. Truong, *Nucleic Acids Res.* **9,** 5679 (1981).

[20] N. Panayotatos, A. Fontaine, and K. Truong, *J. Cell. Biochem., Suppl.* **7B,** Abstr. No. 765 (1983).

[21] A. L. Goldberg and A. C. St. John, *Annu. Rev. Biochem.* **45,** 747 (1976).

[22] D. C. Williams, R. M. van Frank, W. L. Muth, and J. P. Burnett, *Science* **215,** 687 (1982).

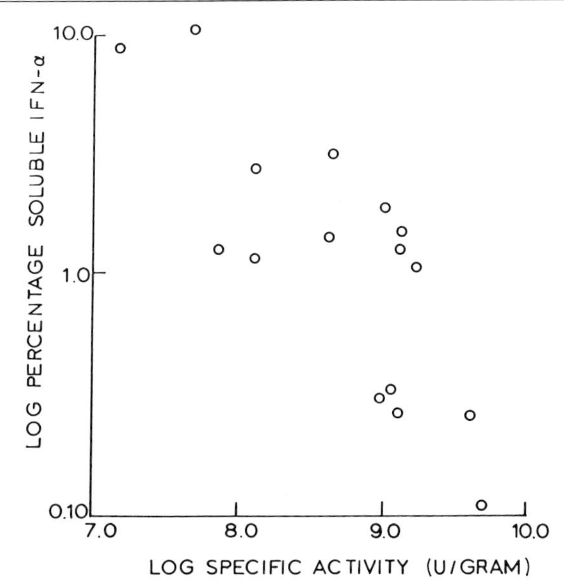

FIG. 2. The fraction of soluble intracellular IFN-α2 at different levels of expression. Total interferon expression (abscissa) was determined by bioassay of cells lysed in 0.1% SDS and is expressed in terms of units per gram fresh weight of *E. coli* cells. To determine the fraction of soluble interferon, cells were first sheared with a french press and then centrifuged at 10000 *g* for 1 hr. The supernatant was then bioassayed and the resulting value expressed as a percentage of the total interferon produced.

resuspension and centrifugation steps. Subsequently, the insoluble contaminating membrane proteins of this washed pellet may be selectively solubilized to leave an insoluble pellet whose major proteinaceous constituent is IFN-α2.

The interferon aggregate is then solubilized by a chaotropic solvent and renatured to give a near homogeneous solution of IFN-α2 which can be purified by ion-exchange, metal chelate, and gel permeation chromatography, followed by recrystallization.

Buffers

Buffer A: 0.1 *M* Tris · HCl, 0.05 *M* EDTA, pH 7.5
Buffer B: 2 *M* guanidine · HCl treated with activated charcoal and filtered through a 0.22-μm filter
Buffer C: 8 *M* guanidine · HCl, purified as above
Buffer D: 0.04 *M* sodium dihydrogen phosphate
Buffers E and F as D but adjusted to pH 5.5 and 6.0, respectively, with 1 *M* sodium hydroxide.

Cell Breakage

The inherent advantage offered by the observed partitioning of IFN-α2 into inclusion bodies can only be exploited fully if near total cell breakage is achieved. As the cells are fermented into late stationary phase and are consequently difficult to rupture by mechanical shear, pretreatment with lysozyme has been adopted in order to destroy the integrity of the peptidoglycan layer of the cell wall. This treatment, as well as increasing the efficiency of cell breakage, also reduces the level of carbohydrate in the final washed pellet. The method described below is designed for 100 g to 1 kg of cell pellet.

Cells were harvested from fermentation broth by centrifugation at 5000 *g*. The cells were weighed and placed in a chilled Waring Blendor. A 2-fold (w/v) excess of Buffer A was added, with blending at low speed for 30 sec. Solid sucrose was then dissolved in the cell suspension to give a final concentration of 30% (w/v) followed by the addition of 1 mg of lysozyme hydrochloride for each gram of original cell pellet. After incubation for 1 hr at 30°, the cell suspension was diluted with an equal volume of chilled deionized water. The viscous suspension was thinned by blending with a Kinematica Polytron PT45 and passed once through a french press at a pressure of 800 bar (about 12,000 lb/in.^{2}). The solid components of the suspension, interferon aggregate and cell membrane, were then collected by centrifugation at 10000 *g* for 30 min. The pellet was resuspended in Buffer B and recentrifuged. This washing procedure was then repeated two more times.

Extraction

The final washed pellet was solubilized in a 10-fold (w/v) excess of Buffer C and immediately diluted 4-fold with Buffer A. The suspension was then centrifuged at 10000 *g* for 30 min and the supernatant dialyzed four times against 20 volumes of distilled water at 4° over a 24-hr period. The cloudy brown precipitate that formed was sedimented at 10000 *g* for 30 min, the precipitate discarded, and the clear colorless supernatant collected.

Column Chromatography

The IFN-α2 at this stage comprised >90% of the protein in solution as judged by SDS–polyacrylamide gel electrophoresis (SDS–PAGE) on samples reduced with 2-mercaptoethanol. Nonreduced samples, however, run on the same SDS–PAGE system, showed a doublet band; the upper band comigrated with reduced interferon and the lower band with native oxidized interferon.[23]

The pH of the solution was reduced to 4.0 by the addition of concentrated orthophosphoric acid and the solution applied to a 5 × 10 cm column of SP Sephadex A-50 equilibrated with Buffer D. The column was then washed with 1 liter of Buffer D followed by 500 ml each of Buffers E and F.

Fractions (50 ml) were collected and those containing interferon, as judged by SDS–PAGE, were pooled and passed through a sintered glass funnel containing a bed of DEAE-Sephadex A20 (5 × 10 cm diameter) equilibrated with Buffer F. The flow through from this column was then adjusted to 1 *M* with respect to NaCl concentration and applied to a copper chelate column. A column (2.5 × 5 cm) of chelating Sepharose 6B (Pharmacia, Uppsala) was prepared as described by the manufacturer. After washing with 9 bed volumes of 1 *M* NaCl the column was charged with Cu^{2+} ions[24] by applying a solution of 1 mg/ml copper sulfate. The column was then repacked with 1 cm of unchelated matrix at the bottom of the column to mop up any leached copper ion. The column was then washed successively with Buffers F, E, and D all containing 1 *M* NaCl. Fractions (50 ml) were collected and those containing oxidized interferon as judged by SDS–PAGE run under nonreducing conditions were pooled. The pooled solution was raised to 80% ammonium sulfate saturation by the addition of the solid. The flocculent white precipitate which formed was collected by centrifugation and redissolved in a minimum volume of Buffer F. The concentrate was then applied to a 5 × 100 cm column of Sephacryl S-200 equilibrated with Buffer D containing 1 *M* NaCl. Fractions (5 ml) were collected and the elution profile monitored at 280 nm. The protein peak eluted as an asymmetric peak with a retention volume corresponding to a molecular weight of approximately 40,000 to 60,000. The major portion of this peak was pooled and the ammonium sulfate saturation raised to 80%. The white precipitate which formed was collected by centrifugation and suspended in a minimum volume of Buffer F. The suspension was then dialyzed against 5 liters of buffer F diluted 100-fold. The precipitate disappeared within 30 min and after 48 hr the dialysis bag was filled with microcrystals (Fig. 3). These crystals were collected by low speed centrifugation in a bench centrifuge and stored moist at 4°.

[23] S. Pestka, B. Kelder, D. K. Tarnowski, and S. J. Tarnowski, *Anal. Biochem.* **132,** 328 (1983).

[24] K. C. Chada, this series, Vol. 78, p. 220.

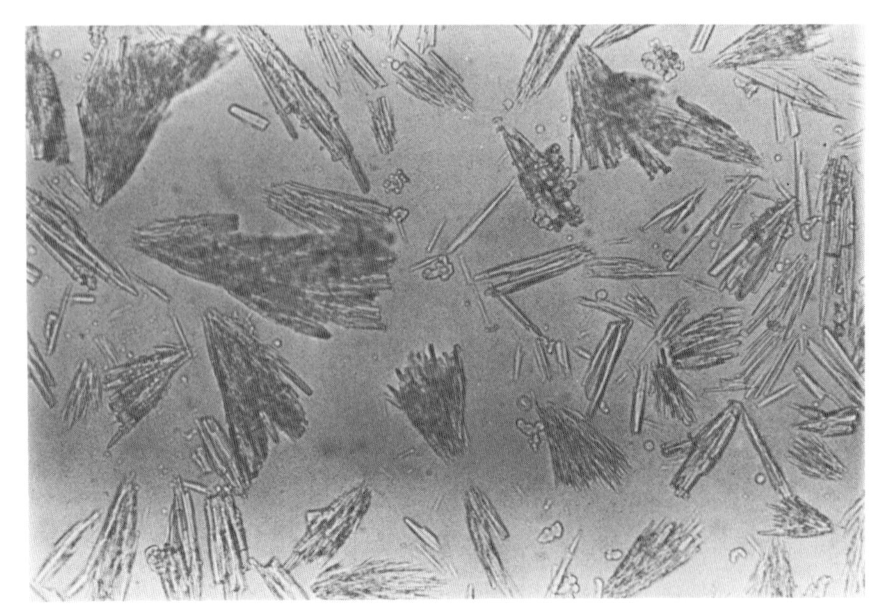

FIG. 3. Phase-contrast microscopy of IFN-α2 microcrystals. Magnification, $\times$100.

PURIFICATION OF IFN-α2[a]

Stage	Volume (ml)	Activity/ml (units/ml)	Total (units)	Yield (%)	Specific activity (units/mg)
Suspended cells	1500	1.1×10^9	1.6×10^{12}	100	2.9×10^6
Cell homogenate	3400	1.1×10^9	3.7×10^{12}	—	9.6×10^7
Supernatant	3140	1.2×10^8	3.7×10^{11}	23	1.5×10^7
Supernatant of Tween wash	1080	1.1×10^7	1.2×10^{10}	0.8	5.9×10^6
Dialysed guanidine extract	3160	1.2×10^8	3.7×10^{11}	23	—
Supernatant	3060	1.2×10^8	3.7×10^{11}	23	—
SP-Sephadex eluate	550	1.1×10^9	6×10^{11}	38	—
Combined Cu-chelate eluate	500	9.3×10^8	1.4×10^{12}	87	—
Combined Sephacryl 5200 fractions	308	8×10^8	2.4×10^{11}	15	—
Crystallized	612	2.7×10^8	1.7×10^{11}	11	3.2×10^8

[a] About 770 g of *E. coli* containing IFN-α2 was used as the starting material.

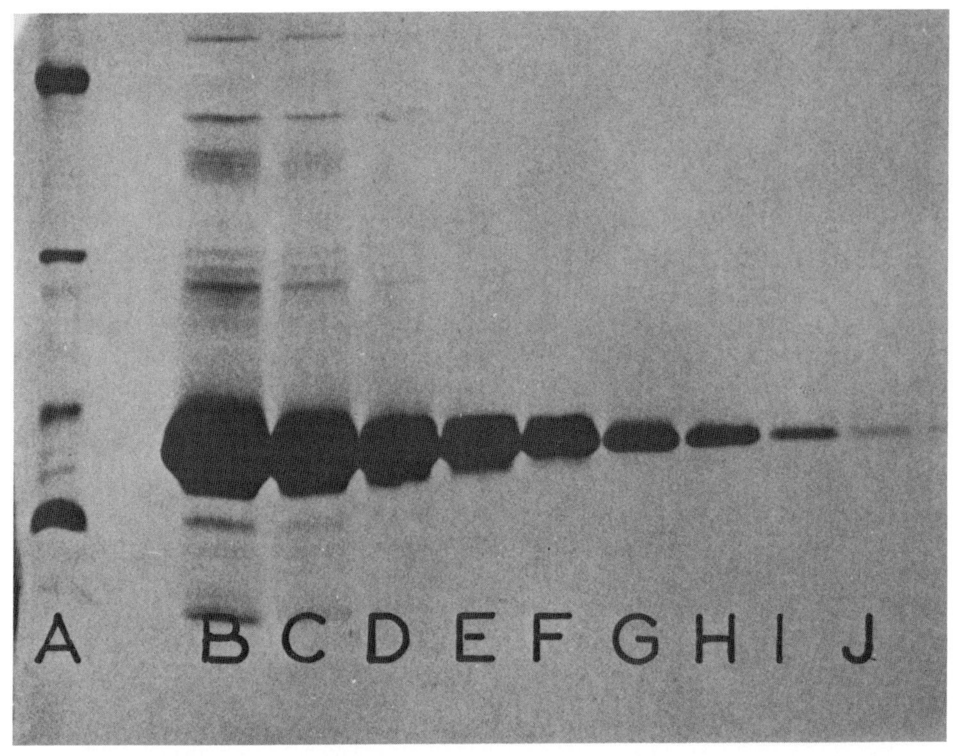

FIG. 4. Purity of IFN-α2 determined by SDS–polyacrylamide gel electrophoresis. The electrophoresis buffer was the standard Laemmli[25] system and the percentage of acrylamide 12%. The gel was stained with silver. The marker proteins (A) were lysozyme (14 kDa), soya bean trypsin inhibitor (21.5 kDa), carbonic anhydrase (31 kDa), ovalbumin (43 kDa), and serum albumin (68 kDa). The loading concentrations were (from left to right) 100 (B), 50 (C), 25 (D), 12.5 (E), 6.25 (F), 3.15 (G), 1.56 (H), 0.78 (I), and 0.39 (J) μg per slot.

The yields and specific activities obtained at each step are shown in the table.

Properties of Purified Hu-IFN-α2

Interferon produced by this method is of high purity as judged by silver stained gel (Fig. 4)[25] and migrates as a single species with an apparent molecular weight of 18,000 on SDS–polyacrylamide gel electrophoresis. The homogeneity of the preparation is suggested by the formation of crystals (Fig. 3). However, gel electrophoresis by the method of Na-

[25] U. K. Laemmli, *Nature (London)* **227,** 680 (1970).

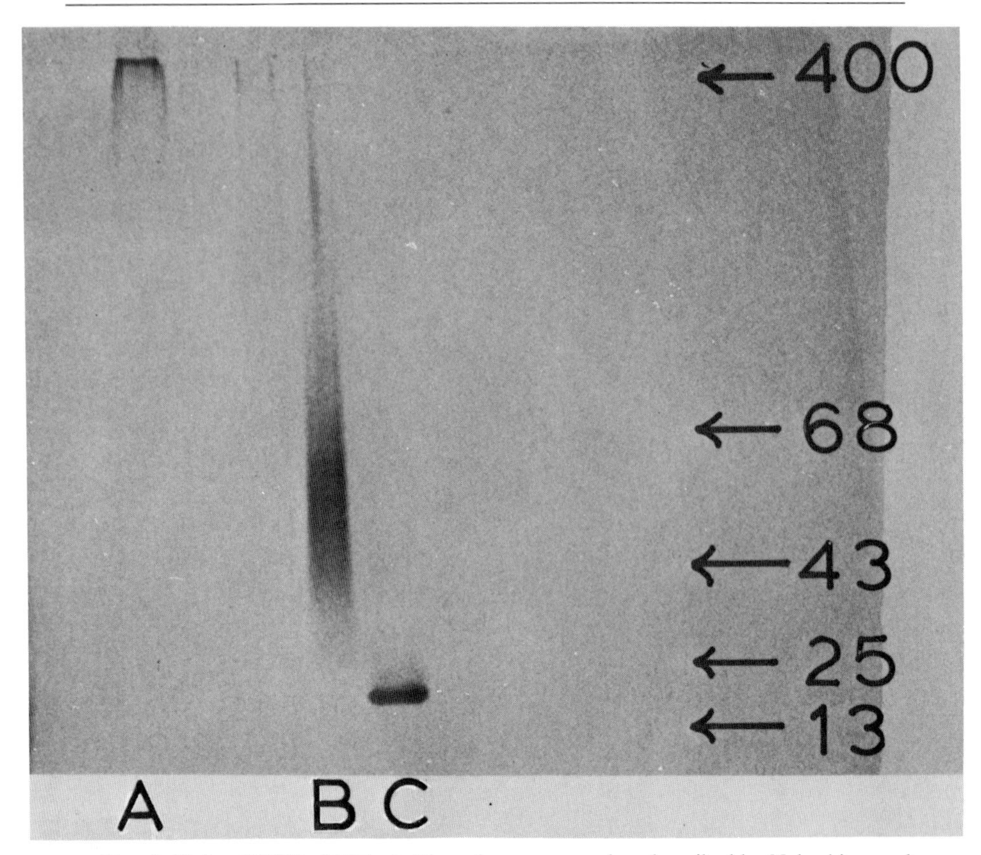

FIG. 5. Native PAGE of IFN-α2. The gel was prepared as described by Nakeshima and Makino.[26] The buffer used was the standard pH 4.0 β-alanine discontinuous system and the polyacrylamide gradient ranged from 5 to 30% (w/v). A is thyroglobulin (did not enter gel), B is IFN-α2, and C is bovine growth hormone. The positions of several marker proteins run on the same gel are shown: 13 kDa is ribonuclease, 25 kDa is chymotrypsinogen, 43 kDa is ovalbumin, 68 kDa is bovine serum albumin, and 400 kDa is ferritin.

kashima and Makino[26] in the absence of ionic detergents shows smearing of the protein with apparent molecular weight distribution of 30,000–60,000 (Fig. 5). This apparent absence of a species of homogeneous molecular radius under native conditions correlates with the anomalous elution behavior observed on gel filtration and with the reported sedimentation properties in the analytical ultracentrifuge.[27]

The purified recombinant IFN-α2 behaved as a relatively hydrophobic molecule on C-18 reverse phase HPLC, eluting as a single major species

[26] H. Nakashima and S. Makino, *J. Biochem.* (*Tokyo*) **88,** 933 (1980).
[27] S. Shire, *Biochemistry* **22,** 2664 (1983).

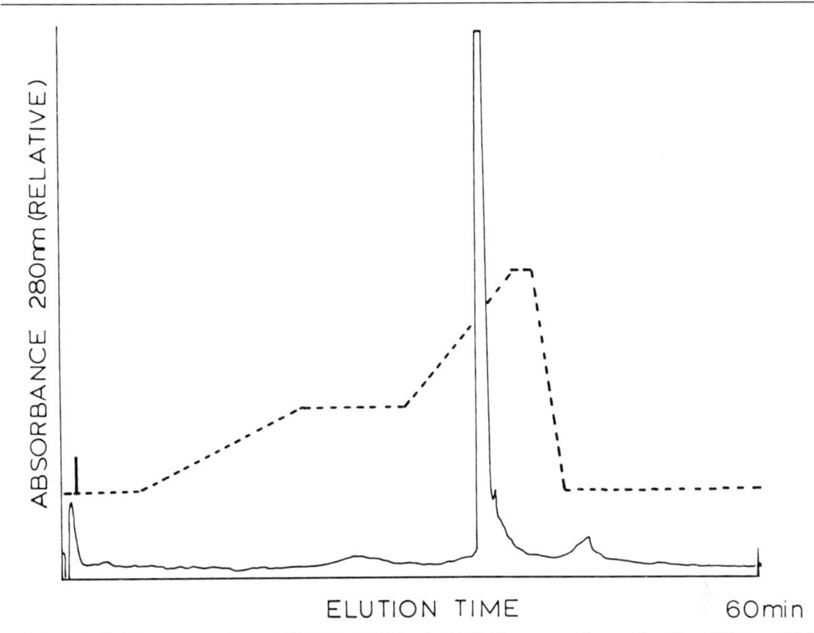

FIG. 6. C-18 reverse phase HPLC of IFN-α2. HPLC was performed with a Waters 6000A system and a 4 × 250 mm Synchro Pak C-18 analytical column (Synchrom Inc, Linden, Indiana). For this chromatogram, 100 μg of protein was applied and the elution conditions were 0 to 35% acetonitrile in 20 min, isocratic at 50% acetonitrile for 10 min, and finally a linear gradient to 75% acetonitrile in 10 min. The flow rate was 1 ml/min and the mobile phase contained 0.1% (v/v) trifluoroacetic acid. The elution profile was monitored by absorbance at 280 nm, where 0.05 A_{280} unit represents maximum deflection.

at an acetonitrile concentration of approximately 55% (Fig. 6). (See also Staehelin *et al.*[28] and Felix *et al.*[29]) Isoelectric focusing in agarose gels gave rise to multiple bands. The major component had a p*I* of 5.9 accompanied by three lesser anodic bands within 0.3 of a p*I* unit (Fig. 7).

IFN-α2 prepared by this method had specific activity of 3.2×10^8 units/mg when assayed by CPE inhibition with HEp-2 cells and EMCV.

Comments

The exploitation of gene cloning and in particular the development of high level expression systems in the bacterium *E. coli* has changed the nature of the problems involved in preparing homogeneous IFN-α.

[28] T. Staehelin, D. S. Hobbs, H.-F. Kung, C.-Y. Lai, and S. Pestka, *J. Biol. Chem.* **256,** 9750 (1981).

[29] A. Felix, E. Heimer, and S. J. Tarnowski, this volume [33].

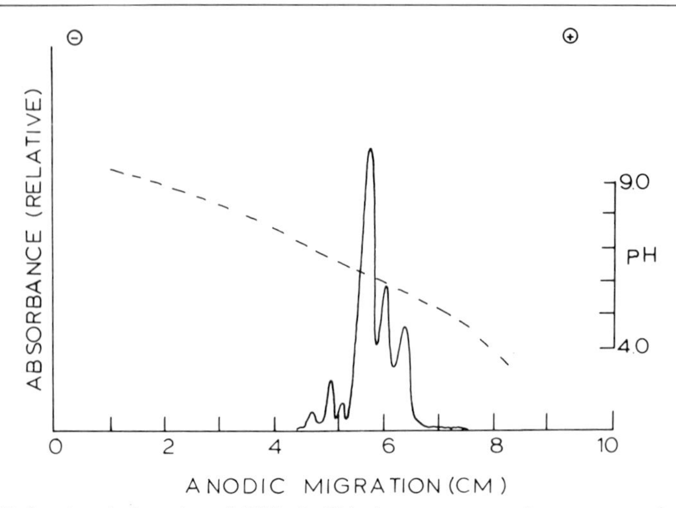

FIG. 7. Isoelectric focusing of IFN-α2. Thin layer agarose gels were poured with 3–10 pharmalytes (Pharmacia, Sweden) as a source of ampholytes. The focused gel was stained with Coomassie Blue, dried, and then scanned with a Joyce Loebel Chromoscan 3 densitometer.

Whereas isolation of IFN-α from naturally occurring sources involves a several thousand fold enrichment of starting material, recombinant techniques require only 3- to 10-fold purification to achieve homogeneity. The genetic engineering strategy must, however, solve a novel problem, namely the isolation, in the correct conformation, of an extracellular protein synthesized in an intracellular environment. The intracellular environment of microorganisms has a net reducing redox potential[30] and the formation of the two disulfide bridges of natural IFN-α2 is consequently not favored. Another complication which may be related to the synthesis of a reduced molecular species is the finding that during synthesis, the IFN-α2 is laid down as insoluble aggregates in a morphologically distinct inclusion body.

The challenge therefore is to solubilize this inclusion body and renature IFN-α2 in the natural conformation with both disulfide bridges correctly formed. As we described above this aggregate was found to be soluble in strongly chaotropic solvents such as guanidine hydrochloride and could be renatured in high yield to an active conformation simply on dilution into aqueous solvents. The problem of formation of disulfide bridges from reduced interferon has been studied in detail elsewhere.[31]

[30] P. Apontoweil and W. Berends, *Biochim. Biophys. Acta* **339,** 1 (1975).
[31] H. Morehead, P. N. Johnston, and R. Wetzel, *Biochemistry* **23,** 2500 (1984).

This oxidation process occurs in the method described here either spontaneously or by copper catalyzed air oxidation during column chromatography on chelating sepharose. Also the fully oxidized form may be resolved from the other forms during purification.

The separation of correctly refolded IFN-α2 from contaminating conformational variants is not a trivial problem. These variants have similar properties to the native molecule and purification strategies are complicated by the tendency of IFN-α2 to form oligomers transiently in concentrated solution (>1 mg/ml).[23] Conformational variants which participate in this association are consequently almost impossible to completely separate from the "native" conformer. The purification approach outlined above, incorporates separation techniques selected for their compatibility with buffer systems favoring the monomeric form of the protein: high ionic strength and low pH. The presence of conformational variants trapped in relatively concentrated preparations may contribute to the observed immunogenicity of the IFN-α preparations used in clinical trials.[32]

Acknowledgments

We wish to thank Dr. A. Pickett of the Fermentation Department and Dr. M. Hirschi of the Cell Biology Department, Biogen S.A. for their continued help on this project. We also wish to acknowledge the provision of the gel electrophoretic and HPLC data by Dr. Denis Bergher, Biogen S.A.

[32] P. W. Trown, M. J. Kramer, R. A. Dennin, E. V. Connell, A. V. Palleroni, J. Quesada, and J. U. Gutterman, *Lancet* 81 (1983).

[25] Purification of Recombinant Human Fibroblast Interferon Produced in *Escherichia coli*

By JOHN A. MOSCHERA, DIANA WOEHLE, KELLY PENG TSAI, CHIEN-HWA CHEN, and S. JOSEPH TARNOWSKI

Many interferons have been expressed in *Escherichia coli*.[1] Their use, however, requires purification to homogeneity so that traces of *E. coli* endotoxin and other contaminants are undetectable or at acceptable lev-

[1] S. Pestka, this volume [1].

els. In this chapter, the purification of Hu-IFN-β expressed in *E. coli*[2] is described. The following procedure has been used to isolate recombinant human fibroblast interferon from up to 3 kg of *Escherichia coli* paste.

Materials, Solutions, and General Procedures

Escherichia coli W3110/pFIF-A-347 was grown in 10-liter fermenters in medium as described[2] and harvested by centrifugation.

Buffer A: 0.1 *M* Tris · HCl, pH 7.9, 0.05 *M* Na_2EDTA, 0.1 m*M* PMSF

Buffer B: 10% PEG-6000, 8% K_2HPO_4, pH 7.9

Buffer C: 0.05 *M* $Na_2HPO_4 \cdot 7H_2O$, pH 7.2, 1.0 *M* NaCl

Buffer D: 15% ethylene glycol, 0.05 *M* $Na_2HPO_4 \cdot 7H_2O$, pH 7.2, 1.0 *M* NaCl

Buffer E: 30% ethylene glycol, 0.05 *M* $Na_2HPO_4 \cdot 7H_2O$, pH 7.2, 1.0 *M* NaCl

Buffer F: 50% ethylene glycol, 0.05 *M* $Na_2HPO_4 \cdot 7H_2O$, pH 7.2, 1.0 *M* NaCl

Buffer G: 0.05 *M* sodium acetate, pH 5.0, 10% propylene glycol

General Procedures. Protein was determined by the method of Lowry *et al.*[3] with crystalline bovine serum albumin as standard. Bioactivity was measured by a cytopathic effect inhibition assay with vesicular stomatitis virus and AG-1732 cells (human fibroblast) as described.[4] Sodium dodecyl sulfate–polyacrylamide gel electrophoresis (SDS–PAGE) was performed according to the procedure of Laemmli[5] with samples run under reducing conditions and bands visualized by silver staining.[6]

Purification of Recombinant Human Fibroblast Interferon (Hu-IFN-β)

Cell Disruption and Precipitation of Interferon. Unless specified otherwise, all operations were carried out at 4°. Frozen live *E. coli* cell paste (3 kg wet weight) was thawed by suspension in 4 volumes of Buffer A for 1 hr, or until smooth. After the addition of lysozyme (1 mg/ml), the suspension was incubated at 4° for 1 hr, while stirring vigorously to eliminate viscosity contributed by DNA. The suspension was filtered through two

[2] D. V. Goeddel, H. M. Shepard, E. Yelverton, D. Leung, R. Crea, A. Sloma, and S. Pestka, *Nucleic Acids Res.* **8,** 4057 (1980).

[3] O. H. Lowry, N. J. Rosebrough, A. L. Farr, and R. J. Randall, *J. Biol. Chem.* **193,** 265 (1951).

[4] P. C. Familletti, S. Rubinstein, and S. Pestka, this series, Vol. 78, p. 387.

[5] U. K. Laemmli, *Nature (London)* **277,** 680 (1970).

[6] R. C. Switzer, III, C. R. Merril, and S. Shifrin, *Anal. Biochem.* **98,** 231 (1979).

layers of cheesecloth and then passed twice through a Manton-Gaulin homogenizer set at 7000 psi. During the process, the exit temperature of the solution was maintained at about 4° by means of a heat exchange coil. One volume of Buffer B was added per two volumes of lysate and, after 15 min of stirring, the mixture was decanted into 1-liter bottles and centrifuged at 3500 rpm for 1–2 hr in a Sorvall RC-3B centrifuge. After removal of the supernatant by decantation, the pellet was scraped into a tared vessel and weighed.

Solubilization and Initial Fractionation of Interferon. The pellet was extracted by stirring with five volumes (per weight) of 7 *M* guanidine hydrochloride (GuHCl) in distilled water for 2–7 hr. The mixture was diluted tenfold with Buffer A and allowed to stand overnight. The diluted extract was centrifuged to remove debris in an R-K High Speed continuous centrifuge (Electro-Nucleonics) equipped with a J-1 rotor. Typically, 70–80 liters of diluted extract was maintained at 4° and pumped at 200 ml/min through the rotor, which was revolving at 55,000 rpm (150,000 *g*). The clarified supernatant was concentrated about 10-fold on a Millipore Pellicon ultrafiltration unit fitted with 50 ft^2 membranes (PTGC, 10,000 MW cut off).

Blue Sepharose Chromatography. The concentrate, which was slightly opalescent, was loaded at 35 ml/min onto a 10 × 10 cm column of Blue Sepharose (Pharmacia),[7–10] previously equilibrated with 3–4 column volumes of Buffer C. The column effluent was monitored at 280 nm. After loading, the column was washed with Buffer C until the A_{280} of the effluent returned to near baseline level and then eluted sequentially with 2–3 column volumes of Buffer D and 2–3 column volumes of Buffer E. The IFN-β was then eluted in the reverse direction of loading with Buffer F and collected in a refrigerated (4°) fraction collector. Fractions with a specific activity of 5×10^7 units/mg, or greater, were pooled and stored at −20° until final processing was carried out. Pooled fractions have been stored for more than a year without loss of biological activity.

The Blue Sepharose column was regenerated *in situ* with 2 column volumes of 7 *M* guanidine hydrochloride at room temperature and reequilibrated to pH 7.2 with Buffer C. Between runs, the column was stored in 0.01% sodium azide solution.

[7] H.-J. Friesen, S. Stein, M. Evinger, P. C. Familletti, J. Moschera, J. Meienhofer, J. Shively, and S. Pestka, *Arch. Biochem. Biophys.* **206,** 432 (1981).

[8] S. Stein, C. Kenny, H.-J. Friesen, J. Shively, U. Del Valle, and S. Pestka, *Proc. Natl. Acad. Sci. U.S.A.* **77,** 5102 (1980).

[9] E. Knight, Jr., M. W. Hunkapiller, B. D. Korant, R. W. F. Hardy, and L. E. Hood, *Science* **207,** 525 (1980).

[10] C. Kenny, J. A. Moschera, and S. Stein, this series, Vol. 78, p. 435.

Dialysis and Ultrafiltration. The Blue Sepharose eluate was placed in dialysis bags which were then suspended in 100 volumes of Buffer F and dialyzed for 1–3 hr at 4°. The bags were transferred to chambers containing 100 volumes of fresh Buffer F and dialysis continued for an additional 4 hr. This step was repeated as necessary to attain a low level of ethylene glycol. The retentates were pooled, centrifuged at 15,000 rpm for 20 min (4–8°), and concentrated to 1–2 mg/ml or greater by ultrafiltration with a stirred cell diafiltration apparatus (Amicon) fitted with a Diaflo YM-5 membrane (Millipore). The concentrate was centrifuged at 15,000 rpm for 20 min (4–8°) and stored at 4°. Concentrates have been stored for more than a year without loss of protein or biological activity. The purification, through the Blue Sepharose chromatography step, is summarized in Table I. The results of a typical concentration of Blue Sepharose Eluate are presented in Table II.

Characterization of Recombinant Hu-IFN-β

Purified IFN-β has a molecular weight of 17,600, as determined by mobility measurements on SDS–polyacrylamide gel electrophoresis (Fig.

TABLE I
PURIFICATION OF RECOMBINANT HUMAN FIBROBLAST INTERFERON THROUGH BLUE SEPHAROSE CHROMATOGRAPHY[a]

Step	Volume (liter)	Total protein (mg)	Total activity (units × 10^{-6})	Specific activity (units/mg)	Recovery (%)
Cell paste	(3 kg)	310,200	230,000	7.4×10^5	100
Manton-Gaulin supernatant	14.5	252,445	1,000	4×10^3	0.4
PEG/phosphate supernatant	20.0	15,940	4,000	2.5×10^5	1.7
7 *M* GuHCl extract	6.5	70,265	4,500	6.4×10^5	19.6
Diluted extract	65.0	52,650	3,900	7.4×10^5	17.0
R-K centrifuge supernatant	65.0	32,955	4,500	1.4×10^6	19.6
Blue-Sepharose load	12.0	24,804	3,400	1.4×10^6	14.8
Flow through	12.0	21,108	460	2.2×10^4	0.2
15% ethylene glycol wash	1.4	54	26	4.8×10^4	0.01
30% ethylene glycol wash	0.9	47	97	2.1×10^6	0.04
50% ethylene glycol eluant	2.7	524	4,900	9.4×10^7	21.3

[a] Protein was measured by the method of Lowry *et al.*[3] with crystalline bovine serum albumin as standard. Bioactivity was determined by a cytopathic effect inhibition assay with vesicular stomatitis virus and AG-1732 cells (human fibroblast) as described.[4] The total activity in the cell paste was determined by extracting 1 gram of cell paste with 20 volumes of 7 *M* guanidine hydrochloride.

TABLE II
PURIFICATION OF RECOMBINANT HUMAN FIBROBLAST INTERFERON: FINAL PROCESSING STEPS[a]

Step	Volume (ml)	Total protein (mg)	Total activity (units $\times 10^{-7}$)	Specific activity (units/mg)	Recovery (%)
50% eluant	310	127	1519	1.2×10^8	100
Dialysis retentate	565	122	1703	1.4×10^8	117
Concentrate	200	142	1880	1.4×10^8	117

[a] The data shown summarize the dialysis and concentration of a typical 50% ethylene glycol eluate. Protein and bioactivity were determined by the methods indicated in Table I. Recovery was calculated on the basis of bioactivity.

1). SDS–PAGE analysis, carried out under reducing conditions, shows that the product is nearly homogeneous, with only trace amounts of impurities appearing upon silver staining of highly overloaded gels. Analysis under nonreducing conditions is complicated by intermolecular disulfide bond formation during sample preparation for electrophoresis leading to varying amounts of dimer.

The average amino acid composition for IFN-β is consistent with that of naturally occurring fibroblast interferon. Amino acid sequence analysis through at least 20 cycles of Edman degradation revealed a sequence that corresponded to that of the natural protein, with approximately 30% of the protein chains showing the NH_2-terminal methionine residue and 70% of the chains showing an NH_2-terminal serine residue. The partial removal of NH_2-terminal methionine residues is a disadvantage of the production in *E. coli* of recombinant IFN-β and other proteins normally containing a terminal methionine.

The specific activity of the recombinant IFN-β isolated by the described procedure usually falls within the range of $1–2 \times 10^8$ units/mg, which is slightly lower than that reported for the naturally derived material. The recombinant molecule also exhibits a considerably greater degree of hydrophobicity than natural fibroblast interferon. The hydrophobicity of recombinant IFN-β, which is most likely exacerbated by the absence of carbohydrate, makes purification by techniques such as reverse-phase HPLC difficult and contributes markedly to the solubility/stability profile of the product. This is an important factor to be considered in the further purification and/or formation of recombinant IFN-β for clinical use.

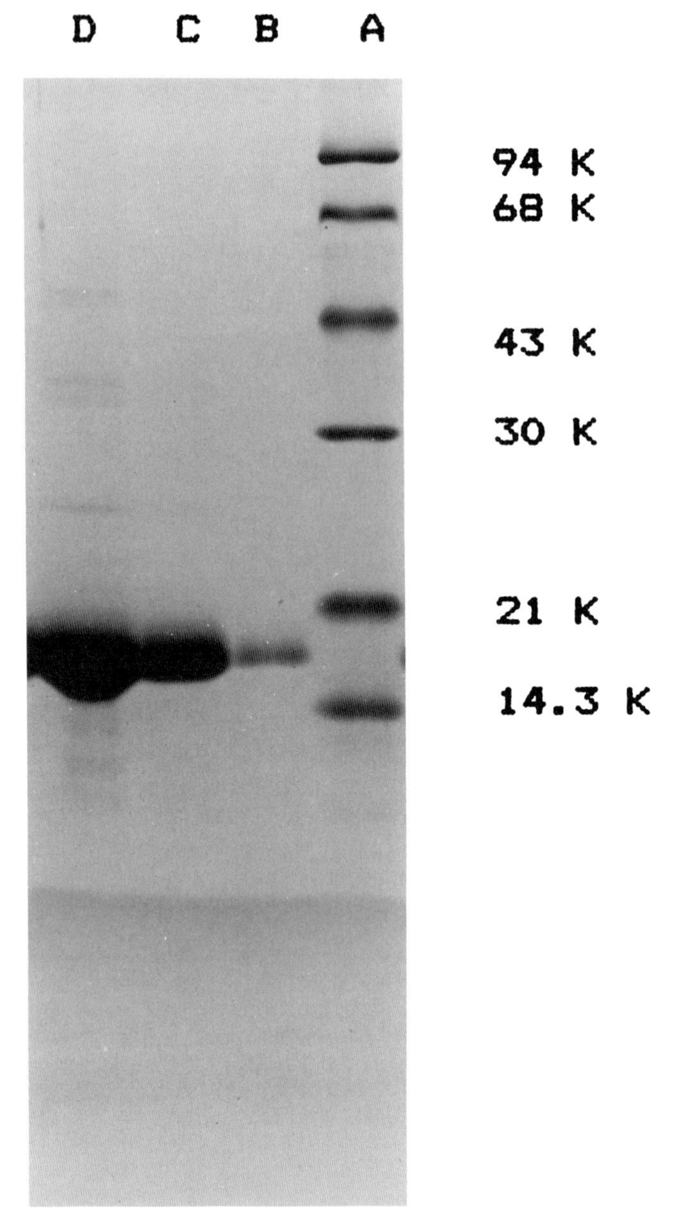

FIG. 1. Sodium dodecyl sulfate–polyacrylamide gel electrophoresis of recombinant human fibroblast interferon. Samples were run under reducing conditions on a 12.5% polyacrylamide gel and visualized with silver stain. (A) MW calibration mixture, (B) 0.11 μg of dialysis retentate, (C) 9 μg of concentrate, (D) 18 μg of concentrate.

Acknowledgments

The authors would like to thank R. Liptak and R. Keeney for their expert technical assistance, R. Harkins for collaboration in procedural development, P. Familletti and L. Petervary for assay support, E. Ernst and E. Russoman for polyacrylamide gel analysis, S. Pestka and A. H. Nishikawa for assistance during the preparation of this manuscript, and S. Cuber for typing the manuscript.

[26] Purification of Recombinant Human Interferon β Expressed in *Escherichia coli*

By LEO S. LIN, RALPH YAMAMOTO, and ROBERT J. DRUMMOND

Many purification methods developed for native (glycosylated) human IFN-β are not applicable to nonglycosylated recombinant IFN-β obtained after expression in *Escherichia coli*. The following method was developed for large-scale purification of recombinant IFN-β and IFN-β (Ser 17). IFN-β (Ser 17) is a highly active and stable form of IFN-β containing a cysteine to serine substitution at position 17 introduced by *in vitro* site-specific mutagenesis.[1] This protocol was tailored for the isolation of IFN-β and IFN-β (Ser 17) expressed at high levels in *E. coli*. These proteins are hydrophobic and, at neutral pH, are readily soluble only in the presence of ionic detergents such as sodium dodecyl sulfate (SDS) or strong chaotropic agents such as guanidine hydrochloride. Unlike other proteins, IFN-β and IFN-β (Ser 17) do not lose biological activity after exposure to SDS. The interaction of SDS and IFN-β has been utilized in the purification scheme presented below. The effectiveness of this method for purification of other interferons from crude extracts of nonmicrobial origin has not been examined.

Expression of Recombinant IFN-β and IFN-β (Ser 17) in *E. coli*

Recombinant human interferons have been expressed in mammalian, insect, yeast, and bacterial cells.[2-6] High-level expression of IFN-β and

[1] D. Mark, S. Lu, A. Creasey, R. Yamamoto, and L. Lin, *Proc. Natl. Acad. Sci. U.S.A.* **81,** 5662 (1984).

[2] N. Mantei and C. Weissmann, *Nature (London)* **297,** 128 (1982).

[3] S. Ohno and T. Taniguchi, *Nucleic Acids Res.* **10,** 967 (1982).

[4] G. Smith, M. Summers, and M. Fraser, *Mol. Cell. Biol.* **3,** 2156 (1983).

[5] R. Hitzeman, F. Hagie, L. Howard, D. Goeddel, G. Ammerer, and B. Hall, *Nature (London)* **293,** 717 (1981).

IFN-β (Ser 17) in *E. coli* cells under control of the tryptophan promoter has been described elsewhere.[1,7,8] When high concentrations of IFN-β accumulate in *E. coli,* the bacterial cells elongate and cytoplasmic inclusion bodies, presumably aggregates of IFN-β, are formed.[7] Such cells provide the starting material for purification of recombinant IFN-β or IFN-β (Ser 17), both of which can be purified by the following procedure.

Purification of IFN-β (Ser 17) in *E. coli*

Interferon Assay. Routine interferon assays were performed with a human cell line trisomic for chromosome 21 (GM2504) and vesicular stomatitis virus in a cytopathic effect inhibition assay.[9] An NIH reference standard for fibroblast interferon, catalog number G-023-902-527, was obtained and used to calibrate an in-house reference standard of poly(I) · poly(C) superinduced IFN-β.

Preparation of Bacterial Extract. At the end of fermentation, culture fluid and cells are chilled to 15°, and cells are concentrated by centrifugation or cross-flow filtration. Concentrated cells are resuspended in 0.1 *M* sodium phosphate, pH 7.4, 0.15 *M* NaCl (PBS) buffer at 100–200 OD at 680 nm/ml and are disrupted by sonication (Heat Systems Model W-375, maximum power, 5 min) or pressure cycling (three passes through a Manton-Gaulin Press, Model M-15). Temperature is maintained at 5° during this stage and at 20–25° during all subsequent stages of purification. Following disruption, particulates containing cell wall, membrane, and IFN-β aggregates are collected by centrifugation (10,000 *g* for 10 min); the pellet is saved and may be stored frozen (−20°) for months without significant loss of biological activity. The particulate material (4 g wet wt.) is resuspended in 100 ml of PBS containing 2% SDS and 10 m*M* DTT and is briefly sonicated (15–30 sec, maximum power setting) to obtain complete dissolution. The turbid suspension rapidly clarifies during sonication. A slight temperature rise (5–10°) at this point is not deleterious to recovery of biological activity and accelerates dissolution. Interferon activity in this crude extract is typically 5×10^7 units/ml with a specific activity of approximately 1.6×10^7 units per mg (Table I).

[6] D. Goeddel, E. Yelverton, A. Ullrich, H. Heyneker, G. Miozzari, W. Holmes, P. Seeburg, T. Dull, L. May, N. Stebbing, R. Crea, S. Maeda, R. McCandliss, A. Sloma, J. Tabor, M. Gross, P. Familletti, and S. Pestka, *Nature* (*London*) **287,** 411 (1980).

[7] B. Khosrovi, *in* "Interferon: Research, Clinical Application, and Regulatory Consideration" (K. C. Zoon, P. D. Noguchi, and T.-Y. Liu, eds.), p. 89. Elsevier, New York, 1984.

[8] M. Konrad and L. Lin, U.S. Patent, PCT Int. Appl. WO 83 03,103 (1983).

[9] J. A. Armstrong, this series, Vol. 78, p. 381.

TABLE I
PURIFICATION OF IFN-β (SER 17)[a]

	Volume (ml)	Total activity (units)	Total protein (mg)	Specific activity (units/mg)	Recovery (%)
Total extract (whole cell extract)	100	1.0×10^{10}	1800	5.5×10^{6}	100
Crude extract (particulate pellet)	200	1.0×10^{10}	600	1.6×10^{7}	100
Organic extract	205	5.3×10^{9}	72	7.3×10^{7}	53
Redissolved acid precipitate	9	4.8×10^{9}	54	7.4×10^{7}	48
S-200 column pool	40	3.7×10^{9}	41	9.0×10^{7}	37
G-75 column pool	15	2.3×10^{9}	19	1.2×10^{8}	23

[a] Interferon antiviral assays were performed as reported.[9] Protein was determined by the method of Lowry with bovine serum albumin as standard except for the G-75 fraction whose protein was determined by absorbance at 280 nm ($\varepsilon_{280}^{0.1\%} = 1.8$ as determined by spectroscopy and quantitative amino acid analysis).

Organic Extraction. An equal volume of 2-butanol (reagent grade) is added to the solution of solubilized particulates and the solutions are vigorously mixed by shaking or stirring. Two liquid phases form when mixing ceases. Complete phase separation is accomplished by centrifugation (10,000 *g* for 15 min in a swinging bucket rotor). After centrifugation the upper organic phase, which is separated from the lower aqueous phase by a leathery layer of insoluble material, is collected by aspiration and contains most of the interferon biological activity (Table I).

Acid Precipitation. Interferon protein is recovered from the organic phase by slow dilution (1–2 min duration per 100 ml extract) of the butanol extract into a 4-fold excess of 0.1 *M* sodium phosphate, pH 7.4, 0.1% SDS. Upon dilution, the mixture becomes slightly turbid as protein begins to precipitate. At this point the suspension is acidified to pH 5.0 by slow addition of glacial acetic acid and stirred for 10 min. Precipitated protein is collected by centrifugation (25°, 10,000 *g* for 10 min). The supernatant solution is discarded and the protein pellet is dissolved in 0.1 *M* sodium phosphate, pH 7.4, 10% SDS, 10 m*M* dithiothreitol, and 0.5 m*M* EDTA.

Sephacryl S-200 Chromatography. The redissolved and reduced acid precipitate is chromatographed on dual 2.6 × 80 cm columns packed with Sephacryl S-200 Superfine. Columns are equilibrated and eluted with 50 m*M* sodium acetate, pH 5.5, 2 m*M* DTT, and 0.5 m*M* EDTA. Sample volumes of 10–20 ml (5–10 mg protein/ml) and flow rates of 1–2 ml/min

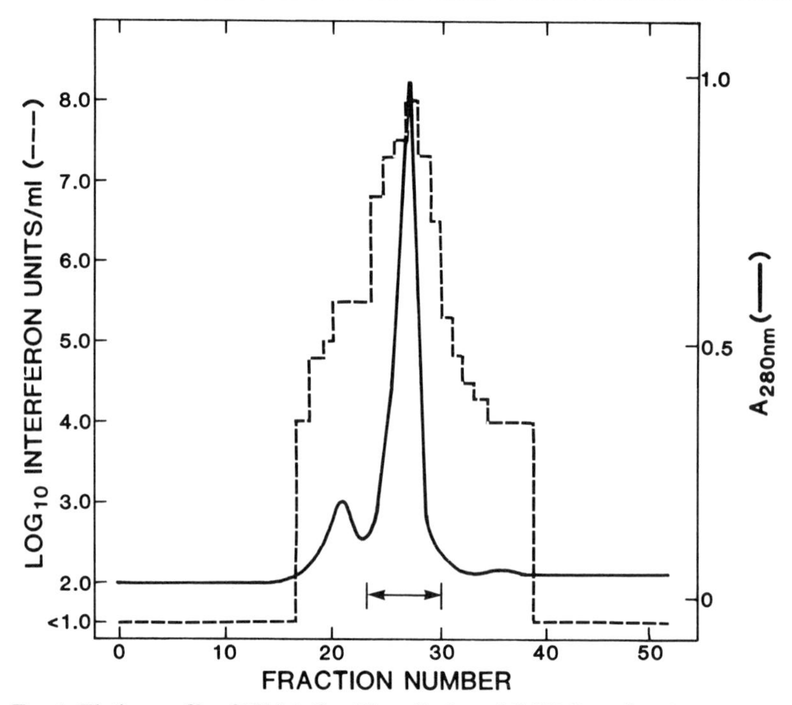

FIG. 1. Elution profile of IFN-β (Ser 17) on Sephacryl S-200 Superfine chromatography. The arrow indicates the fractions which were pooled for further purification or for biological and biochemical characterization.

are optimal. Protein elution is monitored by absorbance at 280 nm and interferon activity is quantitated by interferon assay (CPE). An example of a typical Sephacryl S-200 separation is presented in Fig. 1. Fractions of maximum interferon biological activity and purity (as assessed by SDS–polyacrylamide gel electrophoresis[10]) are pooled. At this stage samples of interferon of greater than 95% purity are obtained. For many biological and chemical analyses, this level of purity is adequate. If material of greater than 99% purity is required, further gel permeation chromatography (Sephadex G-75) is performed.

Sephadex G-75 Chromatography. The major contaminants after Sephacryl S-200 chromatography (low-molecular-weight species which appear to be a mixture of IFN-β (Ser 17) fragments and *E. coli* proteins) are efficiently removed by additional gel filtration on Sephadex G-75 (Superfine). After Sephacryl S-200 chromatography the IFN-β (Ser 17) pool

[10] U. K. Laemmli, *Nature (London)* **227**, 680 (1970).

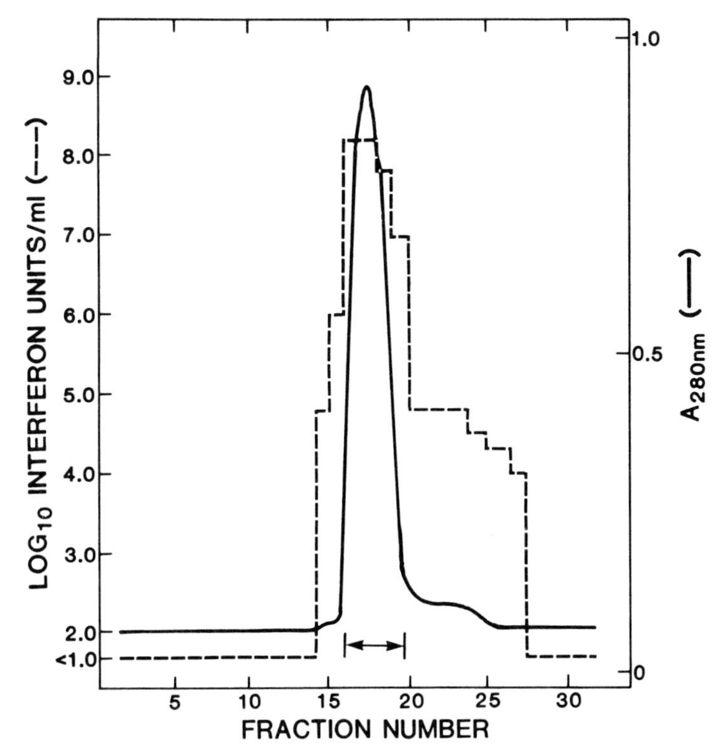

FIG. 2. Elution profile of IFN-β (Ser 17) on Sephadex G-75 Superfine chromatography. The arrow indicates the fractions which were pooled for biological and biochemical characterization.

is concentrated and loaded onto a 2.6 × 80 cm column of Sephadex G-75 previously equilibrated with 50 m*M* sodium acetate, pH 5.5, 2 m*M* DTT, and 0.5 m*M* EDTA. This buffer is also used for protein elution. Sample volumes of 5–15 ml and flow rates of 0.5–1.0 ml/min are optimal. Effluent is again monitored by UV absorbance at 280 nm and fractions are evaluated by SDS-PAGE analysis and CPE assay (Fig. 2). At this point IFN-β (Ser 17) is greater than 99% pure.

Purity of IFN-β (Ser 17). SDS–PAGE analysis of IFN-β (Ser 17) at various stages of the purification is presented in Fig. 3. Quantitative densitometry of Coomassie brilliant blue or fast green-stained[11] polyacrylamide gradient gels yields final purity values of greater than 98–99%. These purity estimates are in accord with quantitative reverse-phase

[11] M. Gorovsky, K. Carlson, and J. Rosenbaum, *Anal. Biochem.* **35,** 359 (1970).

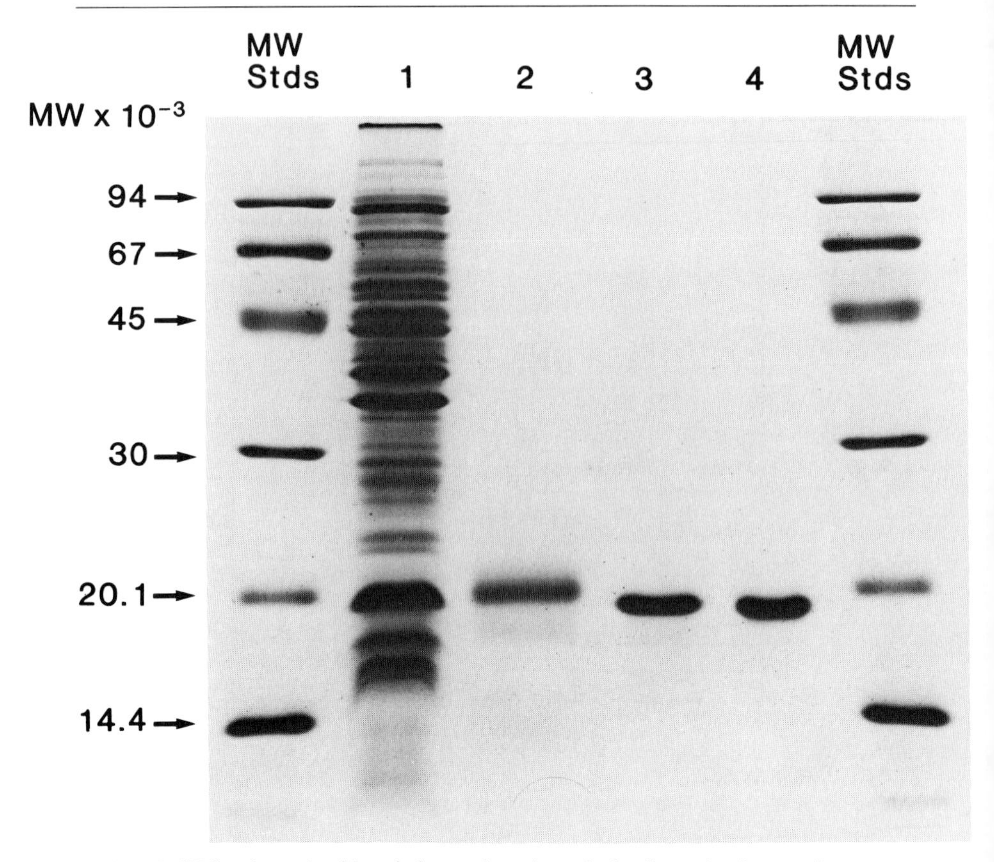

FIG. 3. SDS polyacrylamide gel electrophoretic analysis of proteins from various stages of purification of IFN-β (Ser 17). Lane 1, crude bacterial particulate fraction. Lane 2, organic extract. Lane 3, S-200 chromatography pooled fractions. Lane 4, G-75 chromatography pooled fractions.

HPLC results (Fig. 4). Amino acid analysis (Table II) and amino-terminal amino acid sequence determination (Table III) confirm identity of the purified IFN-β and IFN-β (Ser 17).

Concluding Comments

A method for the purification of recombinant human IFN-β and an active variant, IFN-β (Ser 17), obtained through genetic engineering, has been described. Both proteins are nonglycosylated, lack the amino-terminal methionine present in the native protein, and occur as aggregates

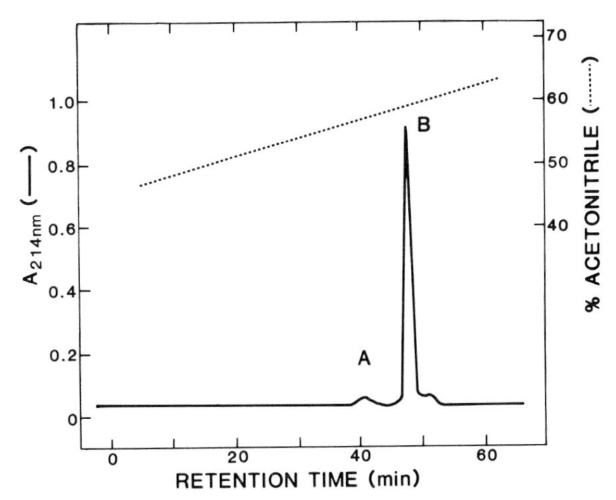

FIG. 4. Reverse-phase high-pressure liquid chromatographic analysis of IFN-β (Ser 17) from G-75 chromatography pooled fractions. The minor peak, A, has SDS–PAGE mobility, amino acid composition, and amino-terminal amino acid sequence identical to that of peak B, IFN-β (Ser 17). Peak A has a specific activity comparable to that for peak B. The basis for different retention times for the two forms of IFN-β (Ser 17) has not been determined. The ratio of the two peaks varies slightly in different preparations, but peak A never exceeds 5% of the total protein.

when expressed at high levels of *E. coli*. Aggregation is presumably a consequence of the intrinsic hydrophobicity of IFN-β.[12] It was found that detergents, such as sodium dodecyl sulfate or sarkosyl, and chaotropic agents such as guanidine hydrochloride, effectively solubilize the biological activity present in *E. coli* cells. This observation, and the fact that IFN-β at neutral pH is poorly soluble in the absence of ionic detergents, led to a purification scheme based on the interaction of IFN-β with dodecyl sulfate.

The explanation for the selective extraction of recombinant IFN-β and IFN-β (Ser 17) into 2-butanol is unclear. A variety of organic solvents have been examined for their ability to extract IFN-β, but only 2-butanol and 2-methyl-2-butanol have been found to be effective. Extraction will not occur in the absence of an ionic detergent. It appears that the preferential formation of an IFN-β–dodecyl sulfate ion-pair may be responsible for the specificity of the effect. The general utility of the method has not been investigated.

[12] E. Sulkowski and D. Goeddel, *in* "Interferon: Properties, Mode of Action, Production, Clinical Application" (K. Munk and H. Kirchner, eds.), p. 106. Karger, Basel, 1982.

TABLE II
AMINO ACID COMPOSITION OF PURIFIED IFN-β (SER 17)

Residue	Hydrolysis time (hr) 24	48	72	Cumulative mean[a]	Predicted value from DNA sequence[b]
Asx	16.9	16.9	16.7	16.8 ± 0.5	17
Thr	7.2	7.1	6.8	7.0 ± 0.3	7
Ser	9.7	8.8	8.4	9.7 ± 0.3[c]	10
Glx	24.4	24.6	24.7	24.5 ± 0.7	24
Gly	6.3	6.4	6.4	6.3 ± 0.2	6
Ala	6.3	6.3	6.3	6.3 ± 0.2	6
Val	4.9	5.3	5.2	5.1 ± 0.3	5
Met	3.0	3.0	3.1	3.1 ± 0.2	3
Ile	10.2	10.7	10.7	10.7 ± 0.3[d]	11
Leu	24.6	24.6	24.6	24.6 ± 0.3	24
Tyr	9.9	9.8	9.9	9.9 ± 0.3	10
Phe	9.0	9.2	9.2	9.2 ± 0.4	9
Lys	10.6	10.7	11.1	10.8 ± 0.5	11
His	4.8	4.8	4.9	4.8 ± 0.2	5
Arg	11.1	10.9	11.1	11.0 ± 0.4	11
Trp	2.5	—	—	2.5 ± 0	3
Cys	2.0	—	—	2.0 ± 0.1[e]	2
Pro	1.0	—	—	1.0 ± 0.1	1

[a] The numbers representing the mean residues/molecule are averages from four separate hydrolysis series, each performed in duplicate. Cumulative mean values represent three hydrolysis times except where indicated. Uncertainties represent 1/2 the range of values averaged for cumulative mean.

[b] NH_2-terminal methionine omitted.

[c] 24 hr values only.

[d] 48 and 72 hr values only.

[e] Analyzed separately from the timed hydrolyses by performic acid oxidation.

Native and recombinant human IFN-β contain three cysteines in positions 17, 31, and 141 (residue numbers based on amino acid sequence containing amino-terminal methionine).[13-16] Based on sequence homology

[13] T. Taniguchi, S. Ohno, Y. Fujii-Kurizama, and M. Muramatsu, *Gene* **10,** 11 (1980).

[14] E. Knight, M. W. Hunkapiller, B. D. Korant, R. W. F. Hardy, and L. E. Hood, *Science* **207,** 525 (1980).

[15] S. Stein, C. Kenny, H.-J. Friesen, J. Shively, U. Del Valle, and S. Pestka, *Proc. Natl. Acad. Sci. U.S.A.* **77,** 5716 (1980).

[16] H.-J. Friesen, S. Stein, M. Evinger, P. C. Familletti, J. Moschera, J. Meienhofer, J. Shively, and S. Pestka, *Arch. Biochem. Biophys.* **206,** 432 (1982).

TABLE III
PARTIAL AMINO ACID SEQUENCE OF PURIFIED IFN-β (SER 17)

Residue number	Minor residue	Yield (nmol)[a]	Minor residue	Yield (nmol)
1	Ser[b]	1.40		
2	Tyr	22.1		
3	Asn	13.9	Asp	0.91
4	Leu	17.0		
5	Leu	20.4		
6	Gly	14.8		
7	Phe	17.8		
8	Leu	15.4		
9	Gln	13.0	Glu	2.54
10	Arg	1.07		
11	Ser			
12	Ser			
13	Asn	6.46	Asp	0.51
14	Phe	9.05		
15	Gln	7.88	Glu	1.43
16	Ser			
17	Gln	6.94	Glu	1.72
18	Lys	2.31		
19	Leu	10.3		
20	Leu	10.2		
21	Trp	3.37		
22	Gln	4.54	Glu	2.11
23	Leu	9.93		
24	Asn	2.10	Asp	1.10
25	Gly	3.69		
26	Arg	1.94		
27	Leu	4.04		
28	Glu	4.37		
29	Tyr	4.17		
30	Cys[c]			

[a] A 35.7 nmol sample of IFN-β (Ser 17) was subjected to automated Edman degradation, and the PTH amino acids were analyzed by reverse-phase HPLC.

[b] Serine was recovered primarily as PTH-dehydroserine which could be detected at 313 nm.

[c] Cysteine was identified as PTH-cystine. Dehydroserine and cystine were not quantitated.

to IFN-α[6,17] and the proven position of disulfide bridges in IFN-α,[18] native IFN-β is assumed to have a disulfide bridge between Cys 31 and Cys 141. The unpaired cysteine at position 17 presents a purification problem because disulfide-linked IFN-β oligomers readily form in the absence of reducing agents. The purification scheme presented above avoids this problem by inclusion of the reducing agent, DTT, at an early step in the purification and at each step thereafter. IFN-β (Ser 17) has had Cys 17 replaced by serine to prevent the formation of incorrect intramolecular and intermolecular disulfide bridges.[1] At the end of the purification, IFN-β (Ser 17) is easily reoxidized to the disulfide-containing species after removal of DTT from dilute protein solutions by dialysis and by allowing oxidation with atmospheric oxygen to occur at pH 8.0. A slight increase in biological activity is observed upon formation of the intramolecular disulfide bridge in IFN-β (Ser 17). Oxidation of reduced IFN-β is believed to occur during bioassay, thereby making measurements of biological activity for the reduced protein difficult to correlate with protein structure. Oxidation of IFN-β containing three sulfhydryl groups is generally accompanied by formation of disulfide-linked oligomers and a mixture of the three possible disulfide containing monomers. Oxidized IFN-β has a lower specific activity than oxidized IFN-β (Ser 17).[1] Recombinant DNA technology has allowed the production of a new protein, IFN-β (Ser 17), which cannot undergo incorrect intramolecular disulfide bridge formation but which is functionally equivalent to native IFN-β and recombinant IFN-β in all bioassays examined to date.[1]

Acknowledgments

The authors wish to thank C. Vitt and the assay group for performing interferon bioassays, K. Watt and Al Boosman for amino acid composition and sequence determinations, M. Kunitani, P. Fernandes, W. Hanisch, M. Konrad, K. Bauer, K. Koths, W. Bloch, T. White, and B. Khosrovi. Without their encouragement and contributions this work could not have been completed. We would also like to acknowledge Shell Oil Company of Houston who are joint partners with Cetus in the development of interferons.

[17] T. Taniguchi, N. Mantei, M. Schwarzstein, S. Nagata, M. Muramatsu, and C. Weismann, *Nature (London)* **285**, 547 (1980).

[18] R. Wetzel, L. Perry, D. Estell, N. Lin, H. L. Levine, B. Slinker, F. Fields, M. Ross, and J. Shively, *J. Interferon Res.* **1**, 381 (1981).

[27] Purification of Natural Human Immune Interferon Induced by A-23187 and Mezerein

By IRWIN A. BRAUDE

The greatest obstacle in purifying natural immune interferon has been due to its relative instability. This is particularly relevant when the material has been subjected to a variety of manipulations[1-3] (e.g., dialysis, concentration, and adjustments in pH). In this chapter a procedure is described for the purification of natural human immune interferon which overcomes this obstacle. The method takes advantage of sequential chromatographic techniques[4] and immune interferon's biochemical characteristics.

Production and Bioassay

The production of natural immune interferon by the coinducers A-23187 and mezerein is described in a previous chapter.[5] The interferon end point is determined by a cytopathic effect inhibition assay,[6] and is described in detail elsewhere.[7] All interferon titers are calibrated against the international human immune interferon standard (G-g23-901-530) and are expressed as international units/ml.

Column Preparation

Controlled Pore Glass Beads (*CPG*). The CPG is purchased from Electro-Nucleonics. The pore size is 350 Å and the mesh 120–200. Fresh beads (100 g) are added to 300 ml of 66% nitric acid, stirred, and allowed to settle. The nitric acid is then decanted and the beads are washed (by stirring and settling) repeatedly in approximately 2 liters of water until the pH of the water is approximately 5.0. Again in 2 liter increments, the beads are then washed once in distilled H_2O followed by PBS. The

[1] M. deLey, J. vanDamme, H. Claeys, H. Weening, J. W. Heine, A. Billiau, C. Vermylen, and P. DeSomer, *Eur. J. Immunol.* **10,** 877 (1980).

[2] M. P. Langford, G. J. Stanton, and H. M. Johnson, *Infect. Immun.* **22,** 62 (1978).

[3] Y. K. Yip, B. S. Barrowclough, C. Urban, and J. Vilček, *Proc. Natl. Acad. Sci. U.S.A.* **79,** 1820 (1982).

[4] I. A. Braude, *Prep. Biochem.* **13,** 177 (1983).

[5] I. A. Braude and C. Tarr, this volume [10].

[6] J. A. Armstrong, *Appl. Microbiol.* **21,** 723 (1971).

[7] I. A. Braude, *Biochemistry* **23,** 5603 (1984).

column is packed, by gravity, with a 50% slurry of CPG and PBS. After packing, the flow adaptor is added and the column is washed with 2 liters of PBS. The CPG is regenerated by removing the beads into a beaker and resuspending in 2 liters of distilled H_2O. The beads are allowed to settle and the supernatant decanted. In 500 ml increments, 2 liters of 1% (v/v) SDS is added while stirring. The beads are allowed to settle and the supernatant is decanted. If the beads are not white in appearance, the procedure should be repeated. Following the same format, the SDS is removed by sequentially washing the beads with 25% ethanol (EtOH) in H_2O, 50% EtOH, 75% EtOH, 50% EtOH, 25% EtOH, distilled H_2O, and finally PBS.

Concanavalin A-Sepharose (*Con A/S*). Con A/S (Pharmacia) available in preswollen form is packed into a column by gravity, and the flow adaptor set in place. Three column volumes of 100 m*M* acetate with 500 m*M* NaCl, pH 5.5 (the pH is adjusted with 100 m*M* acetic acid containing 500 m*M* NaCl). The column is then washed with 3 bed volumes each of 100 m*M* Tris · HCl, pH 8.5, and PBS containing 50% ethylene glycol. Finally the column is equilibrated with 5 bed volumes of PBS. After use, the Con A/S can be regenerated by the same procedure.

Heparin-Sepharose (*H/S*). H/S is also supplied by Pharmacia. Generally 10 g of H/S is swollen in 100 ml of 20 m*M* phosphate buffer, pH 7.2 (PB). The beads are packed by gravity and the flow adaptor set in place. The column is then equilibrated with 5 bed volumes of PB. After use, the beads are regenerated, as described for Con A/S, with 100 m*M* sodium acetate, pH 5.5 and 100 m*M* Tris · HCl, pH 8.5. The resin is reequilibrated with 5 bed volumes of PB.

Carboxymethyl BioGel Agarose (*CM-BGA*). CM-BGA (Bio-Rad), available in preswollen form, is packed into a column by gravity, and the flow adaptor set in place. The resin is then washed with 3 bed volumes of 10 m*M* Tris · HCl, pH 9.5. After use, the beads can be regenerated with 3 bed volumes of 20 m*M* Tris · HCl, pH 9.5 containing 2 *M* NaCl. In practice, however, because small volumes of resin were utilized, they were discarded after use.

Ultrogel AcA 54. This resin (LKB) is also provided in preswollen form. Two columns are packed by gravity and the flow adaptors are set in place. The final packing and column equilibration is performed by washing the columns, at a linear flow rate of 19 cm/hr, with 2 bed volumes of PBS containing 2 *M* NaCl.

Sequential Chromatography

The elution profile for the sequential chromatographic steps is shown in Fig. 1A. The pH of 16 liters of crude immune interferon is adjusted to

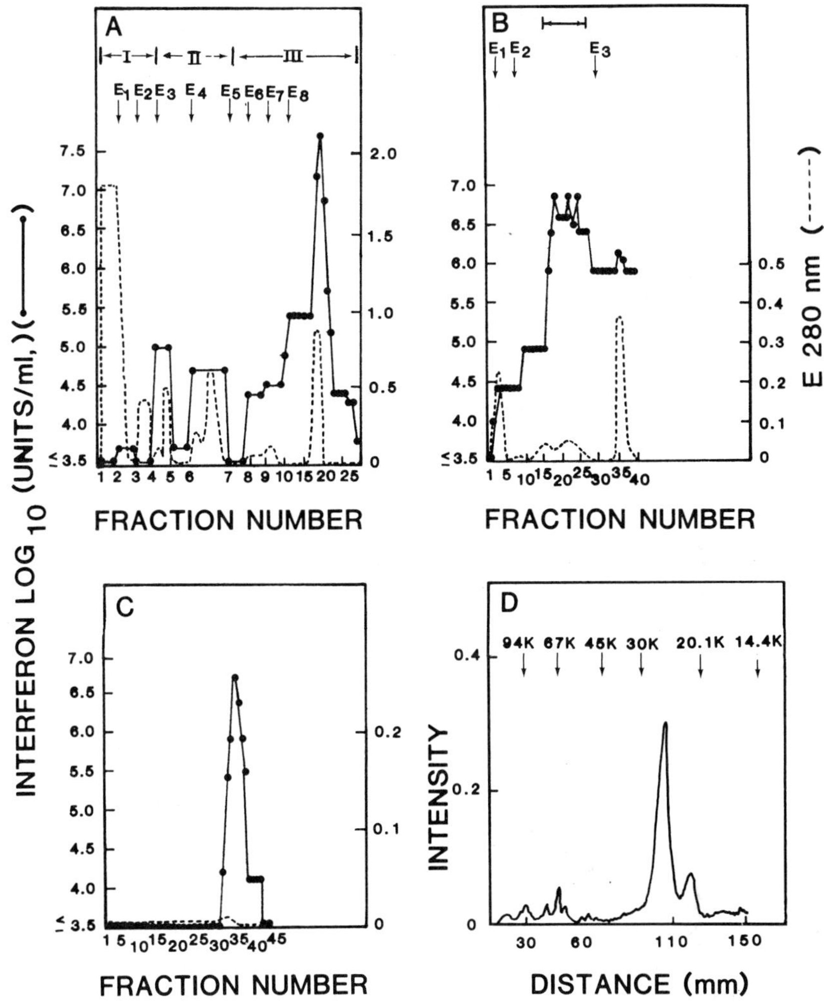

FIG. 1. The chromatographic purification of human immune interferon. A represents the sequential chromatographic procedure. The columns are I, CPG; II, Con A/S; III, H/S. The eluants are E_1, 0.5 *M* Tris · HCl, pH 9.5; E_2, PBS; E_3, 2 *M* $(NH_4)_2SO_4$, pH 9.0; E_4, PBS; E_5, PBS containing 1.0 *M* α-methyl-D-mannopyranoside; E_6, PB; E_7, PBS; E_8, PBS containing 2 *M* NaCl. Fractions were assayed for bioactivity (solid line) and the protein monitored at 280 mm (broken line). B shows the CM-BGA step. The eluants are E_1, 10 m*M* Tris · HCl, pH 9.5; E_2, 20 m*M* Tris · HCl, pH 9.5 containing 50 m*M* NaCl; E_3, 20 m*M* Tris · HCl, pH 9.5 containing 2 *M* NaCl. The bordered arrow represents fractions pooled and loaded on to the gel filtration column. C depicts the gel filtration step as described in the text. The left and right ordinates of C represent interferon activity (solid line) and E280 (dashed line), respectively, as A and B. D represents a spectrophotometric scan of gel filtration fraction 35. Adapted with permission from *Biochemistry* **23,** 5603, 1984, by the American Chemical Society.

9.5 with 0.5 *M* Tris. All steps are performed at 4°. The material is then clarified in 1 liter centrifuge bottles (IEC, catalog number 2939) at 2600 *g* for 1 hr. The supernatants are pooled and loaded onto a 5.0 × 20.4 cm column of CPG (equilibrated in PBS) at a flow rate of 900 ml/hr (linear flow rate 46 cm/hr). Unbound material is removed by washing (at 900 ml/hr) the column first with 0.5 *M* Tris · HCl, pH 9.5, followed by PBS, until the absorbance at 280 nm (A_{280}) returns to baseline. The bound material is eluted, at a flow rate of 200 ml/hr (10 cm/hr), from the CPG column with 2 *M* $(NH_4)_2SO_4$, pH 9.0. The bioactivity resides within the second elution peak and is directly loaded, at a flow rate of 200 ml/hr (38 cm/hr), onto a 2.6 × 18.9 cm column of Con A/S equilibrated in PBS. After the A_{280} has returned to baseline (i.e., loading is completed) the CPG is disconnected and the Con A/S column is washed, at 200 ml/hr (38 cm/hr), with 3 column volumes of 2 *M* $(NH_4)_2SO_4$, pH 9.0. The column is then washed, at a flow rate of 20 ml/hr (3.8 cm/hr), with PBS until the A_{280} returns to baseline. The bound material is then eluted (flow rate 64 ml/hr or 12 cm/hr) from the Con A/S column with PBS containing 1.0 *M* α-methyl-D-mannopyranoside and loaded directly onto a 2.6 × 7.5 cm column of H/S equilibrated in PB. After loading (i.e., A_{280} returns to baseline) the Con A/S column is disconnected and the H/S column is washed with one column volume each of PB and PBS. The bound material is eluted, at a flow rate of 64 ml/hr (12 cm/hr) from the H/S by reversing the flow to an ascending direction. Three milliliter fractions are collected into polypropylene tubes and the peak protein fractions pooled and dialyzed twice against 10 m*M* Tris · HCl, pH 9.5. Normally, a retentate to dialysate ratio of 1 : 100 is suitable.

Cationic-Exchange Chromatography

The chromatographic profile for the cationic-exchange procedure is shown in Fig. 1B. Sequentially purified material is loaded on to a 1.6 × 10 cm column of CM-BGA (equilibrated in 10 m*M* Tris · HCl, pH 9.5) at a flow rate of 20 ml/hr (10 cm/hr). The unbound material is removed by washing the column with 10 m*M* Tris · HCl, pH 9.5 until the A_{280} returns to baseline. The immune interferon is eluted, in an ascending direction, at a flow rate of 20 ml/hr (10 cm/hr), with 20 m*M* Tris · HCl, pH 9.5 containing 50 m*M* NaCl. Two milliliter fractions are collected into polypropylene tubes and the appropriate fractions pooled. Generally the peak bioactivity will elute somewhere between the ascending limb of the first peak and the descending limb of the second peak. It is advisable during the first few trial runs to assay the fractions (although some loss in bioactivity may be experienced) prior to pooling for the last step. Additional protein can be removed from the resin by washing the column with 20 mM Tris · HCl, pH 9.5 containing 2 M NaCl.

Gel-Filtration Chromatography

The last step in the purification procedure is gel filtration and is shown in Fig. 1C. The pooled CM-BGA fractions are loaded by gravity (approximate flow rate 10 ml/hr or 1.9 cm/hr) onto a drained bed of Ultrogel AcA 54 equilibrated in PBS containing 2 M NaCl. Upon loading and washing the sample onto the bed, the column is topped with the same buffer and the sample allowed to flow at 20 ml/hr (3.8 cm/hr) under positive pressure. A pair of 2.6×90 cm columns is employed where the flow of the first column is in the descending direction and the flow of the second column is in the ascending direction. Approximately 10 ml fractions are collected into polypropylene tubes.

Assessment of Purity of Human Immune Interferons

The peak fraction from the gel-filtration column (number 35) was analyzed for purity on 12% SDS–PAGE. The details are described elsewhere.[7] The gel was stained with Coomassie Blue R-250 (Bio-Rad) and spectrophotometrically scanned at 565 nm. As shown in Fig. 1D, one major band with a molecular weight of 26,000, and four minor bands, with molecular weights of 74,000, 67,000, 56,000, and 22,000, are observed. Based on results with Western hybridization techniques,[8] measurements of residual bioactivities from SDS–PAGE[3] and immunoabsorbent chromatography,[7] the 26,000 and 22,000 molecular weight bands have been identified as human immune interferon. Furthermore, based on measuring the area under the peaks shown in Fig. 1D, the final product is calculated to be 92% pure.

A cyanogen bromide digest of the 26,000 molecular weight species is shown to have the amino acid sequence: Ala-Glu-Leu-X-Pro-Ala-Ala-Lys-Thr-X-Lys-Arg. The 22,000 molecular weight form is shown to have the sequence: Ala-Glu-Leu-Ser-Pro-Ala-Ala-Lys-Thr-Gly-Lys. Both these sequences match the predicted sequence deduced from the known human gamma interferon gene.[9,10]

[8] H. K. Hochkeppel and M. de Ley, *Nature (London)* **296,** 258 (1982).

[9] P. W. Gray, D. W. Leung, D. Pennica, E. Yelverton, R. Najarian, C. C. Simonsen, R. Derynck, P. J. Scherwood, D. M. Wallace, S. L. Berger, A. D. Levinson, and D. Goeddel, *Nature (London)* **295,** 503 (1982).

[10] R. Devos, H. Cheroutre, T. Yoichi, W. Degrave, H. van Heuverswyn, and W. Fiers, *Nucleic Acids Res.* **10,** 2487 (1982).

PURIFICATION OF HUMAN IMMUNE INTERFERON[a]

Step		Total units	Total protein (mg)	Specific activity (units/mg)	Degree of purification (fold)	Overall recovery
Crude		3.2×10^8	384,000	8.3×10^2	—	—
Sequential chromatography		3.0×10^8	57.6	5.2×10^6	6,265	94
CM-BGA		9.8×10^7	3.4	2.9×10^7	34,939	31
AcA 54						
overall[b]		1.0×10^8	1.69	5.9×10^7	71,084	33
Fraction[c]	34	9.2×10^6	0.86	1.0×10^7	12,048	3
	35	5.7×10^7	0.23	2.5×10^8	301,204	18
	36	2.9×10^7	0.31	9.3×10^7	112,048	9
	37	9.2×10^6	0.29	3.1×10^7	37,349	3

[a] Reprinted with permission from *Biochemistry* 23,5603, 1984, by the American Chemical Society.

[b] The results of the total activity and protein recovered from the gel filtration step.

[c] These represent the four fractions of the gel filtration step which contained the majority of the recovered bioactivity.

Concluding Remarks

The preceding method is described for the purification of 16 liters of crude immune interferon. However if the linear flow rates for the various chromatographic steps are followed, 50 liter batches can readily be processed. Furthermore with the deletion of the CM-BGA step, the entire process can be semiautomated.[4]

The data in the table demonstrate the efficiency of the procedure. The overall purification of the final product is greater than 70,000-fold. The overall recovery is approximately 33%. The peak gel-filtration fraction (number 35) represented a greater than 300,000-fold purification and a recovery of 18%.

The specific activity of fraction 35 is 2.5×10^8 units/mg protein. As the purity of this fraction is judged to be 92% (see above), the specific activity of pure gamma interferon is about 2.7×10^8 units/mg protein. This value is in close agreement with the specific activities reported for human alpha interferon[11,12] and human beta interferon.[13–15]

[11] M. Rubinstein, S. Rubinstein, P. C. Familletti, R. S. Miller, A. A. Waldman, and S. Pestka, *Proc. Natl. Acad. Sci. U.S.A.* **76,** 640 (1979).

[12] K. C. Zoon, M. E. Smith, P. J. Bridgen, D. Zur Nedden, and C. B. Anfinsen, *Proc. Natl. Acad. Sci. U.S.A.* **76,** 5601 (1979).

[13] E. Knight, Jr., *Proc. Natl. Acad. Sci. U.S.A.* **73,** 520 (1976).

Acknowledgments

I wish to thank Ms. Pat Gregory for her skillful secretarial assistance. I would also like to thank Drs. W. D. Terry and C. G. Smith and Mr. D. Lewis for their support and encouragement.

[14] Y. H. Tan, F. Barakat, W. Berthold, H. Smith-Johannsen, and C. Tan, *J. Biol. Chem.* **254,** 8067 (1979).

[15] H.-J. Friesen, S. Stein, M. Evinger, P. C. Familletti, J. Moschera, J. Meienhofer, J. Shively, and S. Pestka, *Arch. Biochem. Biophys.* **206,** 432 (1981).

[28] Nickel Chelate Chromatography of Human Immune Interferon

By Dorian H. Coppenhaver

Introduction

Since its introduction by Porath and colleagues[1,2] metal chelate chromatography has been successfully applied to the purification of a variety of proteins. The technique was quickly applied to interferon, with human fibroblast interferon being successfully chromatographed on Zn^{2+} chelate columns.[3] Neither human IFN-γ nor IFN-α is adsorbed to Zn^{2+} chelates under standard conditions.[3] Very different results are obtained when a Cu^{2+} charged metal chelate column is used for chromatography of interferons. Cu^{2+} chelates are more retentive than are Zn^{2+} chelates[2]; practically all serum proteins are strongly adsorbed to Cu^{2+} chelate columns.[4] Human α and β interferons are all strongly adsorbed to Cu^{2+} chelates, which are eluted from the columns only with difficulty.[5] Human IFN-γ is also strongly adsorbed to Cu^{2+} chelates,[6] but does not interact with most

[1] L. Lundberg and J. Poráth, *J. Chromatogr.* **90,** 87 (1975).

[2] J. Poráth, J. Carlsson, I. Olsson, and G. Belfrage, *Nature (London)* **258,** 598 (1975).

[3] V. G. Edy, A. Billiau, and P. DeSomer, *J. Biol. Chem.* **252,** 5934 (1977).

[4] D. H. Coppenhaver, N. P. Sollenne, and B. H. Bowman, *Arch. Biochem. Biophys.* **226,** 218 (1983).

[5] K. C. Chada, P. M. Grob, A. J. Mikulski, L. R. Davis, Jr., and E. Sulkowski, *J. Gen. Virol.* **43,** 701 (1979).

[6] E. Sulkowski, K. Vastola, D. Oleszek, and W. von Meunchhausen, *in* "Affinity Chromatography and Related Techniques" (T. C. J. Gribnau, J. Visser, and R. J. F. Nivard, eds.), p. 313. Elsevier, Amsterdam, 1982.

other metal chelates, which Porath[2] has determined to be weaker adsorbents.[7] Human IFN-γ is retained on iminodiacetic acid activated agarose columns which have been charged with Ni^{2+}, however. This moderately weak metal chelate gel shows an unexpected interaction with IFN-γ, the mechanism of which has not been determined, which can be useful in its purification.

Materials

Sepharose 6B, Pharmacia Fine Chemicals
1,4-Butanediol diglycidylether, Aldrich Chem. Co.
Iminodiacetic acid disodium salt, Aldrich Chem. Co.
Sodium Borohydride, Sigma
$NiSO_4 \cdot 7H_2O$ (Gold Label), Aldrich Chem. Co.
Tris(hydroxymethyl) aminomethane
Na_2CO_3
EDTA
NaCl
$NaC_2H_3O_2$
NaH_2PO_4

Solutions

0.02 *M* Tris · HCl, 0.15 *M* NaCl, pH 8.5 (Buffer A)
0.1 *M* NaH_2PO_4, 0.15 *M* NaCl, pH 6.5 (Buffer B)
0.1 *M* $NaC_2H_3O_2$, 1.0 *M* NaCl, pH 4.5 (Buffer C)
0.05 *M* EDTA, 1.0 *M* NaCl

Interferon and Interferon Assay

Human immune interferon was prepared from peripheral blood lymphocytes which had been stimulated with staphylococcal enterotoxin A (SEA) as previously described.[8,9] Interferon was concentrated by chromatography on columns of silicic acid and eluted with 0.3 *M* $(CH_3)_4NCl$ in PBS.[10] In some experiments, silicic acid concentrated interferon was further purified by gel filtration on Ultragel AcA54 eluted with 1.0 *M* NaCl in PBS before Ni^{2+} chelate chromatography was performed.[10] Interferon

[7] D. H. Coppenhaver, unpublished observations.
[8] M. P. Langford, J. A. Georgiades, G. J. Stanton, F. Dianzani, and H. M. Johnson, *Infect. Immun.* **26,** 36 (1979).
[9] H. M. Johnson, F. Dianzani, and J. A. Georgiades, this series, Vol. 78, p. 158.
[10] V. M. Papermaster and S. Baron, *Tex. Rep. Biol. Med.* **41,** 672 (1981–1982).

was assayed on human WISH cells by estimation of reduction of cytopathic effect[11]; interferon activity was expressed in terms of National Institutes of Health reference standard.

Preparation of Iminodiacetic Acid-Activated Sepharose and Ni^{2+} Chelate Column

Sepharose 6B is epoxy activated essentially as described by Porath.[1,2] Swelled resin (150 ml) is dried in air (30 min on a sintered glass filter) before being added to a mixture of 100 ml 1,4-butanediol diglycidyl ether and 200 mg of $NaBH_4$ in 100 ml of 0.5 *M* NaOH. The mixture is sealed and incubated with gentle agitation in a 25° water bath for 24 hr. The resin is then removed and exhaustively washed with 100–200 volumes of deionized water. The washed resin is suction dried on a sintered glass filter before being added to 20 g of iminodiacetic acid, disodium salt, dissolved in 100 ml of 2 *M* Na_2CO_3. The resin/iminodiacetic acid mixture is then placed in a 65° water bath and incubated with constant gentle agitation for 24 hr. After incubation, the iminodiacetic acid activated Sepharose 6B is exhaustively washed (100–200 volumes) with deionized water after which it is ready for chromatographic use.

Columns of epoxy activated Sepharose 6B which have not been previously used are flushed with 0.05 *M* EDTA in 1.0 *M* NaCl to remove any metal ions which have been inadvertently bound to the resin. The EDTA is then flushed from the column with Buffer A before charging the resin with Ni^{2+}. The column is charged by passing a 2–3 mg/ml solution of $NiSO_4$ in water or Buffer A through the column until the top 80% of the resin is completely saturated. The resin which has complexed with Ni^{2+} will be pale blue; the uncomplexed resin serves as a reservoir to scavenge Ni^{2+} ions which are displaced during the course of the chromatography. Finally, the column is reequilibrated with Buffer A before sample application.

Nickel Chelate Chromatography

Samples of partially purified immune interferon that have been dialyzed against Buffer A are loaded onto the column which is then developed at a linear flow rate of 10–15 ml/cm^2/hr at 22°. The volume of the applied sample does not appear to be critical. The column is flushed with 4–5 column volumes of Buffer A, until unbound material is totally eluted. An intermediate pH buffer (Buffer B) may then be applied to the column

[11] P. C. Familletti, S. Rubinstein, and S. Pestka, this series, Vol. 78, p. 387.

Quantitative Data on Nickel Chelate Chromatography of Human IFN-γ

Sample	Volume (ml)	Units/ml	Total units	Specific activity	Yield (%)	Purification factor
Expt. A						
Silicic acid concentrated IFN-γ	4.0	2.0×10^3	8.0×10^3	1.0×10^3	—	—
Eluate from Ni^{2+} column	8.7	0.9×10^3	7.8×10^3	7.8×10^4	98	78
Peak fraction	2.9	2.0×10^3	5.8×10^3	2.0×10^5	72	200
Expt. B						
AcA-54 purified IFN-γ	9.2	3.0×10^3	2.7×10^4	4.3×10^3	—	—
Eluate from Ni^{2+} column	12.0	2.5×10^3	3.0×10^4	1.7×10^6	108	395
Peak fraction	4.0	5.0×10^3	2.0×10^4	5.0×10^6	74	1,163
Expt. C						
AcA-54 purified IFN-γ	29.0	3.0×10^3	2.7×10^4	4.0×10^4	—	—
Eluate from Ni^{2+} column	14.4	3.2×10^3	4.6×10^4	1.6×10^6	53	40
Peak fraction	2.4	8.0×10^3	1.9×10^4	8.9×10^6	41	222

to remove weakly adsorbed material. Interferon activity is recovered by flushing with 4–5 column volumes of Buffer C. The entire chromatographic procedure may be carried out at 4°; however, slightly better resolution and recovery of interferon activity are obtained when the column is maintained at room temperature. The column effluent should be maintained at 4° to preserve interferon activity. The IFN-γ does not seem to be adversely affected by short (24 hr) exposure to the pH 4.5 buffer at 4°. Total recovery of applied interferon activity has ranged from 53 to 125% in various experiments; greater than 90% recovery is typical. From 80 to 96% of the recovered IFN-γ activity is associated with the material eluting at pH 4.5. The remainder is found in the initial wash. Typical results from Ni^{2+} chelate chromatography of IFN-γ preparations are given in the table, and a representative column profile is depicted in Fig. 1.

Comments

Chromatography on columns of epoxy activated agarose charged with Ni^{2+} ions has proven to be a useful step in IFN-γ purification. The tech-

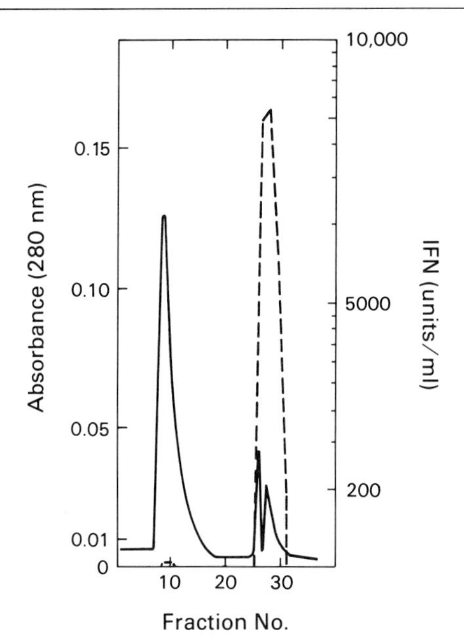

FIG. 1. Nickel chelate chromatography of silicic acid concentrated human IFN-γ. A volume of 4.0 ml of concentrated, dialyzed IFN-γ was applied to a 1.5 × 12 cm column of iminodiacetic acid-activated Sepharose 6B which had been charged with Ni^{2+} and equilibrated in 20 m*M* Tris · HCl, 0.15 *M* NaCl, pH 8.5. After 20 fractions of 2.9 ml had been collected, the elution buffer was changed to 0.1 *M* $NaC_2H_3O_2$, 1.0 *M* NaCl, pH 4.5. Units of IFN activity (dashed line) and total UV absorbance (solid line) are shown. Quantitative data for this experiment are given in the table (Expt. A).

nique in general appears to be forgiving, in that useful results have been obtained under a variety of conditions. Experience has shown that some precautions will enhance the success of the technique. Although chromatography can be carried out under cold room conditions, total recovery of IFN-γ activity is enhanced if the column is maintained at 22°. No deterioration of the separation resin has been observed after dozens of experiments, but results are enhanced if the column is stripped of bound metal ions with 50 m*M* EDTA and recharged with Ni^{2+} after each run. As might be expected theoretically,[1,2] maintenance of a basic pH (8.0–8.5) appears to be important to maximize the initial adsorption of the IFN-γ to the Ni^{2+} chelate column. Adsorption to metal chelates is effective at a wide range of ionic strengths, but appears to undergo a minimum at about 1 *M* NaCl.[12] Hence, an increased appearance of IFN activity in the wash fraction can be expected if the sample is loaded in buffers containing 1 *M*

[12] J. Poráth and B. Olin, *Biochemistry* **22,** 1621 (1983).

NaCl or maintained at neutral pH. Sample volume, however, does not appear to be a critical variable. Although slightly better results have been obtained with more concentrated samples, even dilute preparations have been successfully chromatographed on these columns. In these cases (e.g., Expt. C, the table) a significant concentration of the IFN-γ activity has been obtained, particularly in peak fractions. Thus, Ni^{2+} chelate chromatography may be a useful technique for both purification and concentration of IFN-γ.

[29] Purification of Recombinant Human Immune Interferon

By HSIANG-FU KUNG, YU-CHING E. PAN, JOHN MOSCHERA, KELLY TSAI, EVA BEKESI, MAY CHANG, HIROMU SUGINO, and SUSUMU HONDA

Full length cDNA coding for mature human immune interferon (IFN-γ) has been isolated and the coding sequence for mature IFN-γ has been expressed in *Escherichia coli,*[1-3] yeast,[4] and monkey cells.[1,2] We have purified recombinant immune interferon (IFN-γ) produced in *E. coli.* The purification as well as structural characterization of the bacterial product is described here.

Materials and General Procedures

Bacteria and Bacterial Fermentation

Escherichia coli RR1 (pRK248cIts, pRC231/IFN-γ) was used for fermentations to produce IFN-γ.[3] pRC231/IFN-γ differs from pRC23/IFN-γ[3] in nucleotide spacing between the Shine-Dalgaro sequence and the ATG initiation codon. pRC231[4a] is a derivative of pBR322 containing the phage λ P_L promoter. Expression of the IFN-γ gene is controlled by the

[1] P. W. Gray, D. W. Leung, D. Pennica, E. Yelverton, R. Najarian, C. C. Simonsen, R. Derynck, P. J. Sherwood, D. M. Wallace, S. L. Berger, A. D. Levinson, and D. V. Goeddel, *Nature (London)* **295,** 503 (1982).

[2] R. Devos, H. Cheroutre, Y. Taya, W. Degrave, H. van Heuverswyn, and W. Fiers, *Nucleic Acids Res.* **10,** 2487 (1982).

[3] R. Crowl, this volume [55].

[4] R. Derynck, A. Singh, and D. V. Goeddel, *Nucleic Acids Res.* **11,** 1819 (1983).

[4a] R. Crowl *et al.* In preparation.

temperature-sensitive λ cI represssor produced from the compatible plasmid pRK248cIts.[3,5]

Overnight cultures of *E. coli* RR1 (pRK248cIts, pRC231/IFN-γ) were grown in LB broth at 30°. Five hundred milliliters of the overnight culture was diluted to 10 liters with minimal M-9 medium containing casamino acids. Bacteria were harvested by centrifugation and the bacterial pellets were stored at −20° until used. All fermentations and procedures were performed in accordance with recombinant DNA guidelines of the National Institutes of Health.

Preparation and Purification of Monoclonal Antibodies from Ascitic Fluid

Procedures were as described previously.[6] Monoclonal antibodies were made against a synthetic COOH-terminal peptide of IFN-γ (last 16 amino acid residues). One of the monoclonal antibodies (MOγ2-11.1) was used for the purification of IFN-γ.[6a]

Preparation of Immunoadsorbent Column

Coupling of monoclonal antibody (MOγ2-11.1) to Affi-Gel 10 (Bio-Rad) was described previously.[6]

Chemicals

Iodo[1-^{14}C]acetic acid was obtained from New England Nuclear, cyanogen bromide from Pierce, fluorescamine from Roche, C-8 and C-18 reverse-phase chromatographic columns from Supelco, Inc. All the solvents used for protein characterization were redistilled over ninhydrin.

Peptide Fragmentation and Separation

The IFN-γ was treated with CNBr (100-fold molar excess over methionine) in 70% formic acid as previously described.[7] CNBr peptides were separated by HPLC on a C-18 reverse-phase column. A linear gradient of 0 to 70% of CH_3CN in 0.1% trifluoroacetic acid was used for peptide elution.

[5] H. Bernhard and D. Helinski, this series, Vol. 68, p. 482.

[6] T. Staehelin, D. S. Hobbs, H. F. Kung, C. Y. Lai, and S. Pestka, *J. Biol. Chem.* **256,** 9750 (1981).

[6a] Y. Ichimori, T. Kurokawa, S. Honda, N. Suzuki, M. Wakimasu, and K. Tsukamoto, *J. Immunol. Methods* **80,** 55 (1985).

[7] R. A. Wolfe and S. Stein, "Modern Methods in Pharmacology," p. 55. Alan R. Liss, Inc., New York, 1982.

Amino Acid Analysis

Protein or peptide samples were hydrolyzed for 20–24 hr in sealed, N_2-flushed, evacuated tubes in constant boiling HCl containing 4% thioglycolic acid. Amino acid analyses were performed with a fluorescamine amino acid analyzer as previously described.[8]

NH_2-Terminal Sequencing by Edman Degradation

An ABI (Applied Biosystem, Inc.) gas-phase sequencer 470A was used for sequence analyses of carboxymethylated proteins[9] Samples of PTH-amino acids were identified by reverse-phase HPLC on an ultrasphere ODS column.[10]

Other General Procedures

Sodium dodecyl sulfate–polyacrylamide gel electrophoresis was performed as described by Laemmli.[11] Protein was determined by fluorescamine analysis with crystalline bovine serum albumin as the reference standard.[8] Interferon activity was determined by a cytopathic effect inhibition assay with vesicular stomatitis virus and human WISH cells as reported.[12,13] All interferon titers are expressed in reference units/ml calibrated against the reference standard of partially purified human immune interferon prepared in our laboratory.

Purification of Recombinant Human Immune Interferon (IFN-γ)

Recombinant immune interferon was extracted from *E. coli* cells in two ways. All purification steps were carried out at 4°. (1) Guanidine hydrochloride extraction. Frozen cells (25 g) were suspended in three volumes (75 ml) of 7 *M* guanidine · HCl (pH 7). The mixture was stirred for 1 hr and then centrifuged for 1 hr at 30,000 *g*. The supernatant was diluted 10-fold with Dulbecco's phosphate-buffered saline (PBS) or 0.15 *M* sodium borate buffer (pH 9.5) and then centrifuged for 30 min at 30,000 *g*. (2) Sonication. Alternatively, frozen cells (25 g) were suspended in 1.5 volumes (37.5 ml) of 0.15 *M* sodium borate buffer (pH 9.5) and stirred for

[8] S. Stein and L. Brink, this series, Vol. 79, p. 20.

[9] R. M. Hewick, M. W. Hunkapillar, L. E. Hood, and W. I. Dreyer, *J. Biol. Chem.* **256,** 7990 (1981).

[10] D. Hawke, P. M. Yuan, and J. E. Shively, *Anal. Biochem.* **120,** 302 (1982).

[11] U. K. Laemmli, *Nature (London)* **227,** 680 (1970).

[12] S. Rubinstein, P. C. Familletti, and S. Pestka, *J. Virol.* **37,** 755 (1981).

[13] P. C. Familletti, S. Rubinstein, and S. Pestka, this series, Vol. 78, p. 3817.

PURIFICATION OF IFN-γ[a]

Purification step	Total protein (mg)	Total activity (units)	Specific activity (unit/mg)	Purification (-fold)	Yield (%)
I. Guanidine extract					
Supernatant	2806	2.5×10^8	9×10^4	—	100
Silica	98	1.0×10^8	1×10^6	11	42
Monoclonal antibody	8	0.8×10^8	1×10^7	110	32
II. Sonication					
Supernatant	6136	8.0×10^7	1.3×10^4	—	100
Silica	87	4.5×10^7	5.2×10^6	400	56
Monoclonal antibody	2	2.0×10^7	1.0×10^7	769	25

[a] Details of purification procedure are described in the text.

1 hr. The mixture was sonicated 5 times for a total of 150 sec at 765 W and then centrifuged for 1 hr at 30,000 *g*.

The supernatants from either guanidine extraction or sonication were mixed for 1 hr on a rotating shaker with 25 ml silica (NuGel-952AC, Separation Industries, Metuchin, NJ), previously equilibrated with phosphate-buffered saline. The mixture was poured into a column tube and the column was washed with 20 to 30 bed volumes of 1 *M* NaCl. The column was then eluted with 0.5 *M* tetramethylammonium chloride in 0.01 *M* sodium borate buffer (pH 8.0). Interferon activity was recovered in about 200 ml and separated into 4 pools. Each pool was loaded onto a monoclonal antibody (MOγ2-11.1) affinity column (4 ml bed volume) equilibrated with phosphate-buffered saline. After washing with 10 bed volumes of phosphate-buffered saline, the column was eluted with 50% ethylene glycol containing 1 *M* NaCl and 20 m*M* sodium phosphate buffer (pH 7.0). Interferon activity was recovered in the first 20 ml.

A summary of the purification procedures is presented in the table. The overall recovery was 25–32% and the purification was 110- and 769-fold for the samples prepared by guanidine extraction and sonication, respectively. The product showed an average specific activity of about 1×10^7 units/mg. Of the two extraction methods, guanidine hydrochloride afforded 3 to 4 time more total activity than soniçation (see the table). Sodium dodecyl sulfate–polyacrylamide gel electrophoresis of the final products is shown in Fig. 1.[14] The material purified from guanidine extraction showed a single band at about 18,000 D (18K IFN-γ) and a dimer at

[14] J. H. Morrissey, *Anal. Biochem.* **117,** 307 (1981).

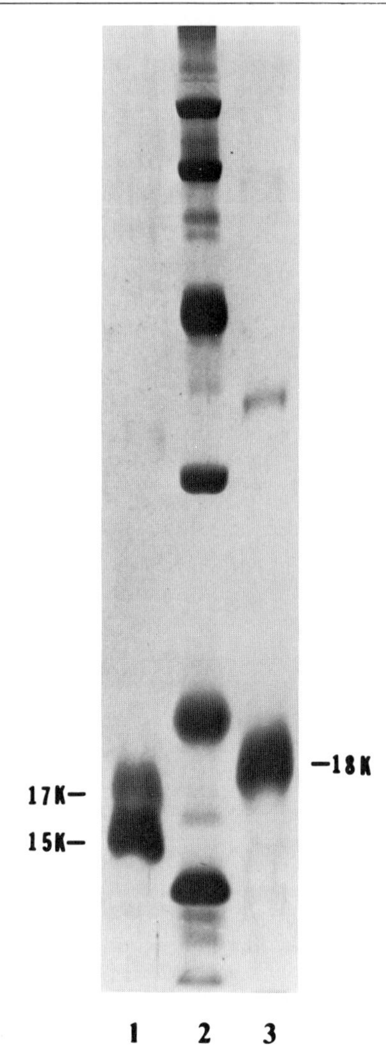

FIG. 1. Sodium dodecyl sulfate–polyacrylamide gel electrophoresis of purified recombinant human immune interferon. Ten microliters (about 2 μg) of purified proteins was subjected to electrophoresis according to the procedure of Laemmli[11] on a 12.5% polyacrylamide gel containing 0.1% sodium dodecyl sulfate. The gels were stained by the silver stain protocol of Morrisey.[14] Protein standard markers were purchased from Bio-Rad: 94,000, phosphorylase *b*; 68,000, bovine serum albumin; 43,000, ovalbumin; 30,000, carbonic anhydrase; 21,000, soybean trypsin inhibitor; lysozyme, 14,300. Lane 1, 15K and 17K IFN-γ; lane 2, standards (STDS); and lane 3, 18K IFN-γ.

about 36,000 D whereas the sonication procedure yielded a major band at about 15,000 D (15K IFN-γ) and a second band at about 17,000 D (17K IFN-γ) (Fig. 1).

The 18K IFN-γ was homogeneous and the amino acid composition was consistent with that predicted from the DNA sequence of IFN-γ. Similarly, the NH_2-terminal sequences of first 32 residues and 25 residues of the 15K and 18K proteins, respectively, were in accord with that predicted by the DNA sequence of IFN-γ.

The COOH-terminal residues predicted from the DNA sequence were determined by analyzing and sequencing the COOH-terminal peptides obtained from CNBr cleavage. COOH-terminal peptides were separated on the HPLC C-18 reverse-phase column. Our results indicated that the 18K species was the intact IFN-γ molecule, whereas the 15K species was a proteolytic product, cleaved between amino acid residues Lys-Arg, with the last 15 COOH-terminal amino acid residues missing.

Although the MOγ2-11.1 antibody had been prepared against the COOH-terminal hexadecapeptide of IFN-γ, "Western Blot" analysis[15] showed that it could bind to the 17K IFN-γ. Our results indicated that the 17K IFN-γ was missing the last four COOH-terminal residues. By contrast the Western Blot analysis of the 15K IFN-γ was negative with MOγ2-11.1, which suggested that this interferon species was more severely truncated at the COOH-terminus. Thus the presence of 15K IFN-γ in preparations of 17K IFN-γ obtained from gels containing immobilized MOγ2-11.1 was perplexing. In light of the tendency for IFN-γ to aggregate, we surmise that the 15K IFN-γ copurifies by association with the 17K species which is selectively bound to the immobilized antibody.

Antiviral activities of both the 18K and 15K IFN-γ were about the same on human cells. Little or no activity was seen on bovine and murine cells. Since the 18K and 15K IFN-γ had the same specific activity, last 15 COOH-terminal amino acid residues are not essential for antiviral activity. The 15K IFN-γ is the smallest active IFN-γ so far identified.

Concluding Comments

The purification of IFN-γ can be carried out simply, efficiently and rapidly, with the use of silica gel and monoclonal antibodies to the COOH-terminal peptide of IFN-γ. IFN-γ appeared to be degraded during the purification process. The degradation appeared to be prevented with the use of guanidine hydrochloride in the initial step of purification and homogeneous intact molecules could be obtained (18K IFN-γ). By contrast, sonication of frozen cells in the absence of guanidine yielded mainly

[15] W. N. Burnette, *Anal. Biochem.* **112,** 195 (1981).

the truncated product (15K IFN-γ, where 15 COOH-terminal amino acid residues were lost). Amino acid compositions and NH_2-terminal sequences were consistent with that expected from the DNA sequence. The specific activities of the 18K and 15K IFN-γ were the same (10^7 units/mg) and comparable to that of the homogeneous natural human immune interferon.[12] Capon *et al.*[16] have previously reported that the last 10 COOH-terminal amino acids of IFN-γ are not essential for antiviral activity. Our present finding of full activity in the 15K IFN-γ suggests that the last 15 COOH-terminal amino acids are also not required for antiviral activity. Further elucidation of nonessential regions as well as active sites of IFN-γ will depend on more protein modification and DNA mutation studies. Since *E. coli* extracts contain protease(s) which degrade foreign proteins,[18] it will be important to characterize the protease(s) responsible for the degradation of COOH-terminal amino acids of IFN-γ and to find *E. coli* mutants deficient in these protease(s). IFN-γ should be an ideal model system for such studies.

Using the same monoclonal antibody column (MOγ2-11.1), we have also purified natural immune interferon from normal human buffy coat lymphocytes induced with phytohemmaglutinin (PHA). On sodium dodecyl sulfate–polyacrylamide gel electrophoresis, the purified material showed a major band at about 25,000 D (25K IFN-γ) and a minor band at about 20,000 D (20K IFN-γ). The two species seen on SDS–polyacrylamide gel electrophoresis are probably due to differences in the glycosylation of natural immune interferon. Yip *et al.*[17] have reported the purification of natural immune interferon with apparent molecular weights of approximately 25,000 and 20,000 which is consistent with our present findings.

Acknowledgments

We are grateful to Robert Crowl for providing *E. coli* clones and to Kyozo Tsukamoto of Takeda Co. (Osaka, Japan) for providing monoclonal antibody. We wish to thank Philip Familletti for performing interferon assays, Richard Chizzonite for advice on performing the immunoblot analysis, S. Edward Lee for fermentation of *E. coli* cells, and Juli Farruggia for typing the manuscript.

[16] D. J. Capon, D. W. Leung, R. A. Hitzeman, L. J. Perry, W. J. Kohr, P. W. Gray, R. Derynck, and D. V. Goeddel, *3rd Annu. Int. Congr. Interferon. Res.* (1982).

[17] Y. K. Yip, B. S. Barrowclough, C. Urban, and J. Vilček, *Proc. Natl. Acad. Sci. U.S.A.* **79,** 1820 (1982).

[18] A. L. Goldberg and A. C. St. John, *Annu. Rev. Biochem.* **45,** 747 (1976).

[30] Production and Purification of Bovine Interferon β

By REINHARD AHL and M. GOTTSCHALK

Although desirable for both theoretical and practical reasons information on the production, purification, and characterization of bovine interferons has been scarce for many years. This seems surprising in view of the possibility of studying the effect of interferons in calves experimentally infected with viruses having properties similar or equivalent to those of corresponding human viruses. The difficulty of producing sufficient amounts of interferon α in bovine leukocytes and of interferon β in bovine cell cultures[1–5a,b,c] has limited their use.

We have developed a method for production of higher titer interferon in calf or adult bovine primary kidney cell cultures which provided a basis for purification and characterization of bovine interferon β.[6,7] This procedure is briefly outlined before describing its purification.

Production of Interferon in Bovine Kidney Cell Cultures

Media and Solutions

PBS, Dulbecco's phosphate-buffered saline

PBS(−), i.e., PBS without Ca^{2+} and Mg^{2+}

Dispase, grade II (Cat. No. 165859, Boehringer-Mannheim, Mannheim, Germany), dissolved in PBS(−) at a concentration of 0.4 units/ml

ESS modified to contain 0.9 g/liter of $NaHCO_3$ and supplemented with 5 g/liter lactalbumin hydrolysate (Serva Feinbiochemica, Heidelberg, Germany)

[1] E. Peterhans, J. Charrey, Y. Richoz, and R. Wyler, *Res. Vet. Sci.* **20,** 99 (1976).

[2] B. D. Rosenquist and R. W. Loan, *Am. J. Vet. Res.* **28,** 619 (1967).

[3] L. A. Babiuk and B. T. Rouse, *Intervirology* **8,** 250 (1977).

[4] M. G. Tovey, M.-T. Bandu, J. Begon-Lours, D. Brouty-Boyé, and I. Gresser, *J. Gen. Virol.* **36,** 341 (1976).

[5] C. R. Rossi and G. K. Kiesel, *Am. J. Vet. Res.* **41,** 557 (1980).

[5a] R. W. Fulton and N. J. Pearson, *Can. J. Comp. Med.* **46,** 100 (1982).

[5b] G. J. Letchworth and L. E. Carmichael, *Vet. Microbiol.* **8,** 69 (1983).

[5c] V. E. Reyes Luna, A. D. H. Luk, S. K. Tyring, J. M. Hellman, and S. S. Lefkowitz, *Experientia* **40,** 1410 (1984).

[6] R. Ahl, *Dev. Biol. Stand.* **50,** 159 (1982).

[7] R. Ahl, *J. Interferon Res.* **1,** 203 (1981).

EMEM in Earle's salts
BSA
Active calf serum

Preparation of Cell Cultures

Primary monolayer cultures are prepared from kidneys received from the local abattoir. Fat tissue and capsule are removed aseptically and the cortex tissue sliced off. The minced pieces are washed with PBS(−) and incubated on a magnetic stirrer overnight at room temperature in Dispase solution (600 ml for one kidney). Dissociated cells are separated from clumped fibrous tissue by sieving through gauze and centrifuged twice at 200 *g* for 8 min. After washing, the cell slurry is suspended in medium at a concentration of about 15 ml/liter medium. Growth medium is modified ESS with lactalbumin hydrolysate and 10% calf serum. After 2 days the medium is replaced by EMEM containing 10% serum. Roller flasks (6 × 35 cm) are filled with 100 ml of medium. Dense monolayers of epithelioid cells are grown after eight days incubation at 37°.

Interferon Induction

The following viruses have been found to be good interferon inducers in bovine kidney cells: a mutant of subtype O_1 Lombardy of foot-and-mouth disease virus (FMDV) named O_1L_{if}; NDV strain CG and strain Italien; Bluetongue virus (BTV) types 8 and 10. FMDV and BTV are grown in BHK-21 cell cultures. NDV is propagated in embryonated eggs. Poly(I) · poly(C) is a relatively poor interferon inducer in roller bottle cultures of bovine kidney cells, even if used in the presence of DEAE dextran.[6]

For interferon induction, medium is removed from cultures and cells are washed with PBS. Virus is added in 5 ml of production medium (EMEM without serum, supplemented with 0.05% BSA) at a multiplicity of infection (MOI) of 10 to 50. After adsorption for 1 hr at 37° 20 ml production medium is added and the bottles are incubated for further 24 hr at 37°. When FMDV strain O_1L_{if} is used as inducer, guanidine hydrochloride (Merck, Darmstadt, Germany) is added to the production medium at a concentration of 0.3 mg/ml to enhance interferon yield. It is an advantage of this method that the cytopathic effect is suppressed in the cultures. Likewise, NDV strain CG induces interferon without a strong cytopathic effect.

The time course of interferon production in kidney cells is shown for FMDV and NDV strain CG in Fig. 1. Titers of about 50,000–100,000

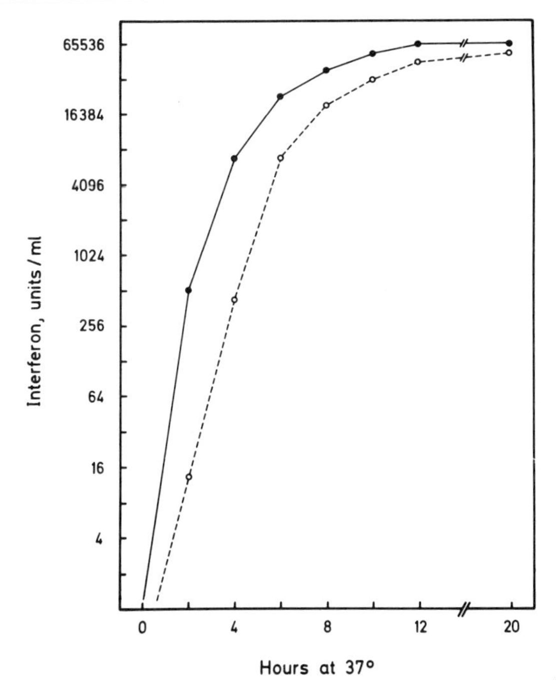

FIG. 1. Time course of interferon production in bovine kidney cells in rolling flasks induced with either FMDV mutant O_1L_{if} (plus 0.3 mg/ml guanidine) (●) or NDV strain CG (○). (From Ahl.[6])

units/ml are regularly obtained, corresponding to 2000–4000 units per cm^2 cell sheet.[6,7]

Medium containing interferon is harvested and clarified by centrifugation (30 min at 2200 *g*). The crude preparation is then acidified with concentrated HCl in order to inactivate the virus. Whereas NDV requires pH 2 to destroy the infectivity, FMDV and BTV are inactivated at pH values below 5.0 within an hour. The preparation is then adjusted to pH 5.5 with NaOH and again centrifuged. Preparations are stored at −20°. Interferon activity is assayed as described in cultures of the porcine cell line IB-RS-2 with FMDV as challenge virus. One unit is equivalent to 0.8 unit of human leukocyte interferon.[7]

Purification of Bovine Interferon β

For concentration and partial purification of bovine interferon β affinity chromatography and hydrophobic chromatography methods as developed for the purification of human interferon β[8–13] were adopted. However, the methods had to be modified to achieve selective binding of interferon and extensive separation from contaminating proteins.[14,15]

Materials and Solutions

Columns: C 16/40 (16 mm diameter) and C 10/20 (10 mm diameter), (Pharmacia, Freiburg, Germany)
Siliconized glassware and polystyrene tubes
Blue-Sepharose CL-6B (Pharmacia, Freiburg, Germany)
Phenyl-Sepharose CL-4B (Pharmacia, Freiburg, Germany)
Sodium azide (Merck, Darmstadt, Germany)
Ethylene glycol (Merck, Darmstadt, Germany)
0.02 *M* HEPES, 0.5 *M* NaCl, pH 5.5; dissolve 4.8 g HEPES and 29.22 g NaCl in 1 liter of distilled water
0.02 *M* sodium phosphate buffer, 0.5 *M* NaCl, pH 7.4; 2.93 g $Na_2HPO_4 \cdot 2H_2O$, 0.49 g $NaH_2PO_4 \cdot 2H_2O$, and 29.22 g NaCl dissolved in 1 liter of water
0.02 *M* sodium phosphate buffer, 1 *M* NaCl, pH 7.4; as above, but with 58.44 g NaCl per liter
0.04 *M* sodium phosphate buffer, 2 *M* NaCl, pH 7.4; 5.86 g $Na_2HPO_4 \cdot 2H_2O$, 0.98 g $NaH_2PO_4 \cdot 2H_2O$, and 116.88 g NaCl dissolved in 1 liter of water. This "double concentrated" buffer is mixed with ethylene glycol and water as required.
Buffers are sterilized by filtration and degassed before use.

Preparation of Columns

Freeze-dried Blue-Sepharose is allowed to swell and washed in distilled water. A C 16/40 column is prepared and filled to contain about 25 ml of settled gel. The column is washed and equilibrated in 0.02 *M* HEPES buffer, 0.5 *M* NaCl, pH 5.5. Phenyl-Sepharose is suspended in water and a prepared C 10/20 column filled with about 4 ml of gel. This column is equilibrated with 0.02 *M* sodium phosphate buffer, 1 *M* NaCl, pH 7.4.

[8] W. J. Jankowski, W. von Muenchhausen, E. Sulkowski, and W. A. Carter, *Biochemistry* **15,** 5182 (1976).
[9] E. Knight, this series, Vol. 78, p. 417.
[10] E. Knight and D. Fahey, *J. Biol. Chem.* **256,** 3609 (1981).
[11] C. Kenny, J. A. Moschera, and S. Stein, this series, Vol. 78, p. 435.
[12] W. Berthold, C. Tan, and Y. H. Tan, *J. Biol. Chem.* **253,** 5206 (1978).
[13] A. J. Mikulski, J. W. Heine, H. V. Le, and E. Sulkowski, *Prep. Biochem.* **10,** 103 (1980).
[14] R. Ahl and M. Gottschalk, *Antiviral Res. Abstr.* **1** (2), 40 (1983).
[15] M. Gottschalk and R. Ahl, *Zentralbl. Bakteriol., Parasitenkd., Infektionskr. Hyg., Abt. 1: Orig., Reihe A* **255,** 170 (1983).

Affinity Chromatography on Blue-Sepharose

One to two liters of crude interferon, adjusted to pH 5.5, is made 0.5 *M* in NaCl. NaN_3 at 0.2 g/liter is added to prevent contamination during loading. The Blue-Sepharose column is loaded at room temperature at a flow rate of 50 ml/hr. Under the conditions applied about 70 to 90% of the interferon activity is bound to the gel. In contrast to human interferon β, bovine interferon is insufficiently bound to Blue-Sepharose in 1 *M* NaCl.

The loaded column is washed with 0.02 *M* sodium phosphate buffer, 0.5 *M* NaCl, pH 7.4, and subsequently eluted at a flow rate of 36 ml/hr in 0.02 *M* sodium phosphate buffer, 1 *M* NaCl, pH 7.4, containing 10% ethylene glycol (EG) (Fig. 2). This step removes most of the contaminating proteins (mainly BSA) and a minor fraction (10 to 15%) of interferon activity. The bulk of interferon and a small protein peak is eluted from the column by increasing the concentration of EG in the buffer to 25%. The broad interferon peak appears somewhat delayed after the protein peak. Residual bound interferon is released by increasing the concentration of EG to 50%. Recovery of interferon in peak fractions is in the range of 50–

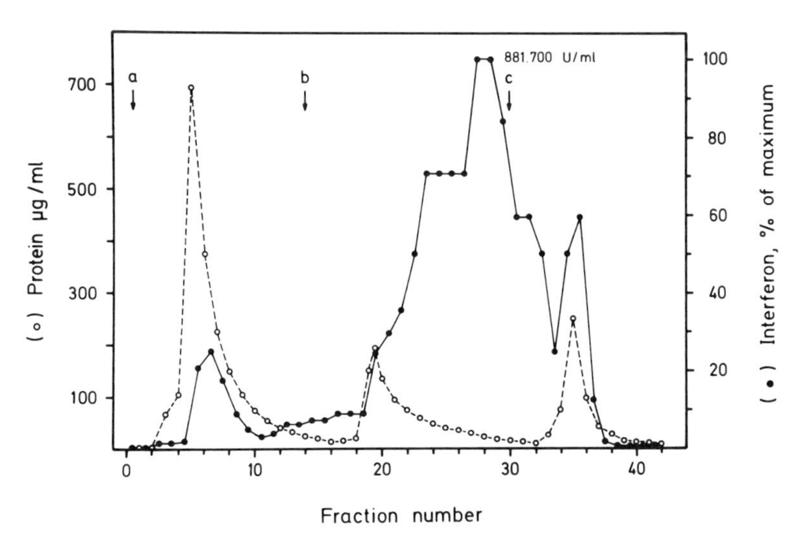

FIG. 2. Chromatography of bovine interferon β on Blue-Sepharose CL-6B. One liter of crude interferon in 0.5 *M* NaCl, pH 5.5, was bound to a Blue-Sepharose column (1.6 × 12 cm). After washing with 0.02 *M* sodium phosphate buffer, 0.5 *M* NaCl, pH 7.4, the column was developed in sodium phosphate buffer, 1 *M* NaCl, pH 7.4, containing (a) 10% ethylene glycol (EG), (b) 25% EG, and (c) 50% EG. Fractions (5.2 ml) contained up to 881,700 units/ml interferon. Recovery was 70%. (From Ahl and Gottschalk[25] by permission of the Commission of the European Communities.)

PURIFICATION OF BOVINE INTERFERON β

Purification step	Volume (ml)	Activity		Protein		Spec. act. (units/mg)	Yield (%)	Purification factor
		units/ml	Total units	mg/ml	Total (mg)			
1. Crude interferon preparation	728	1.3×10^5	9.5×10^7	0.9	656	1.4×10^5	100	1
2. Blue-Sepharose column								
Unbound fraction	—	—	2.7×10^7	—	—	—	28	—
Eluted in 1 *M* NaCl, 10% EG	—	—	1.5×10^7	—	—	—	16	—
Peak fractions in 1 *M* NaCl, 25% EG	70	8.1×10^5	5.7×10^7	0.06	4.2	1.4×10^7	56	100
3. Phenyl-Sepharose column								
Unbound fraction	—	—	0.1×10^7	—	—	—	1.2	—
Eluted in 1 *M* NaCl, 40% EG	—	—	1.1×10^7	—	—	—	11	—
Peak fractions in 1 *M* NaCl, 50% EG	15	27.3×10^5	4.1×10^7	0.044	0.66	6.2×10^7	48	440

70% of input activity (see the table). The regenerated column can be reused.

Hydrophobic Chromatography on Phenyl-Sepharose

As a second step in purification, the peak fractions of interferon activity recovered from Blue-Sepharose in 25% EG arc directly applied onto a small Phenyl-Sepharose column equilibrated in 0.02 *M* sodium phosphate buffer, 1 *M* NaCl, pH 7.4 (Fig. 3). Flow rate is 15 ml/hr. Interferon is almost completely adsorbed. To remove as many contaminating proteins as possible the column is washed with 0.02 *M* sodium phosphate buffer, 1 *M* NaCl, pH 7.4, containing 40% EG, although by this step some activity is lost. Interferon is then eluted by increasing the concentration of EG to 50%. In peak fractions specific activities up to 10^8 units/mg protein can be obtained, and overall concentration of interferon activity is about 20- to 50-fold to 6×10^6 units/ml (see the table), depending on the initial titer of the crude preparation.

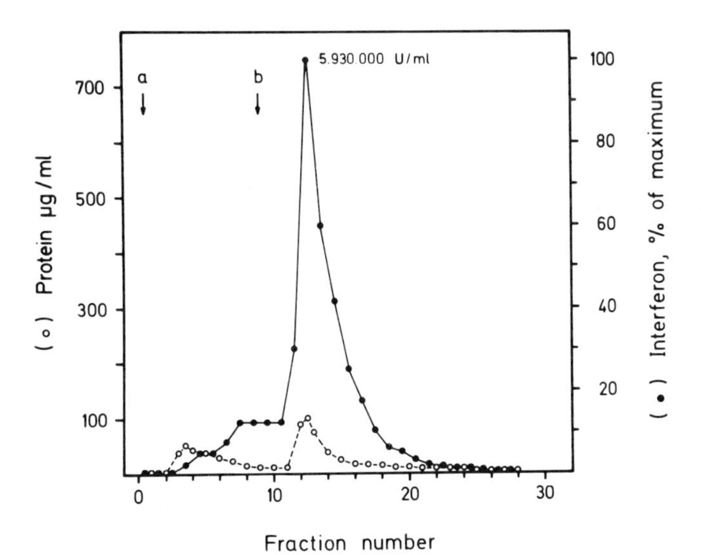

FIG. 3. Chromatography of bovine interferon β on Phenyl-Sepharose CL-4B. Interferon recovered from Blue-Sepharose at 25% EG was directly applied onto a column of Phenyl-Sepharose (1 × 5 cm) equilibrated in 0.02 *M* sodium phosphate buffer, 1 *M* NaCl, pH 7.4. Elution proceeded in the same buffer containing (a) 40% EG and (b) 50% EG. Fractions (2.5 ml) contained up to 5.9×10^6 units/ml interferon. Recovery was nearly 100%. (From Ahl and Gottschalk[25] by permission of the Commission of the European Communities.)

Tandem Chromatography on Blue-Sepharose and Phenyl-Sepharose

Both steps of purification can be combined in a tandem chromatography procedure as recommended by Sulkowski and co-workers.[16,17] Crude interferon is loaded onto a Blue-Sepharose column as described, and proteins are largely separated from interferon by washing with sodium phosphate buffer, 1 *M* NaCl, pH 7.4, containing 10% EG. Interferon is eluted in the same buffer with 25% EG with a delay to the second protein peak (Fig. 2) which is discarded before the eluate is adsorbed on Phenyl-Sepharose by means of a three-way valve. Loading is continued until the extinction is dropped to the baseline of the recorder. Then the Phenyl-Sepharose column is eluted as described with (1) 40% EG and (2) 50% EG in sodium phosphate buffer, 1 *M* NaCl, pH 7.4. The column is regenerated by washing first with 70% EG in the same buffer and then with buffer.

Concluding Comments

Analysis of purified interferon on SDS-polyacrylamide gels after labeling with ^{125}I has revealed that complete purification is not achieved by the method described. The main contaminants are BSA and a 50,000 Da protein. Interferon activity has been recovered from the acrylamide gels. It should be noted that all components of the system (except for the virus) are of bovine origin and therefore should not be harmful to bovine cells and calves.

Additional procedures tested for purification have not been very successful. Remazol Brilliant Blue R coupled to Sepharose[18] strongly binds bovine interferon produced in cell cultures but separation from other proteins also requires additional steps of purification.

In contrast to human interferon β[19,20] bovine interferon β preparations did not bind to zinc chelate Sepharose or to nickel chelate Sepharose.

Further methods such as high-performance liquid chromatography[21] or antibody affinity chromatography[22] may be used to achieve complete purification.

[16] E. Sulkowski, E. Bollin, W. J. Jankowski, and W. A. Carter, *Fed. Proc., Fed. Am. Soc. Exp. Biol.* **36,** 741 (1977).

[17] E. Bollin, this series, Vol. 78, p. 178.

[18] E. Sulkowski, *Tex. Rep. Biol. Med.* **41,** 234 (1982).

[19] V. G. Edy, A. Billiau, and P. DeSomer, *J. Biol. Chem.* **252,** 5934 (1977).

[20] J. W. Heine, J. Van Damme, M. De Ley, A. Billiau, and P. De Somer, *J. Gen. Virol.* **54,** 47 (1981).

[21] H. J. Friesen, S. Stein, and S. Pestka, this series, Vol. 78, p. 430.

[22] Y. H. Tan, this series, Vol. 78, p. 422.

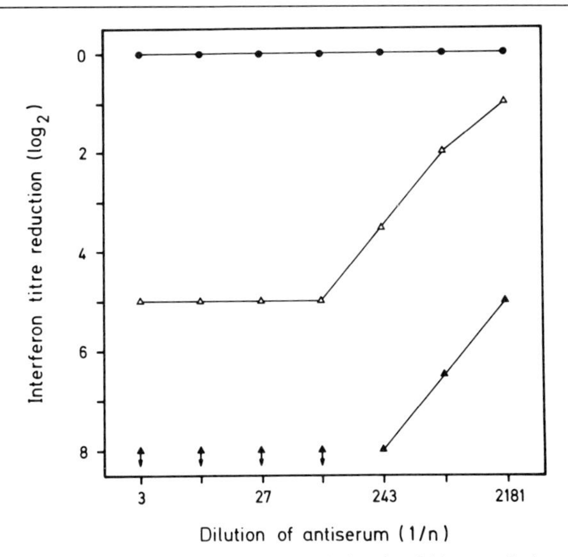

FIG. 4. Neutralization of interferon produced in bovine kidney cells by rabbit antiserum to interferon purified as described. ▲, Purified interferon β; △, interferon not bound to Blue-Sepharose in 0.5 *M* NaCl; ●, bovine leukocyte interferon, produced in peripheral blood leukocytes *in vitro* with NDV-Italien.

Bovine interferon β purified as described is already unstable and for storage stabilization by 50% EG or 0.1% BSA is essential.[23]

The methods required for purification have shown that bovine interferon β differs significantly from human interferon β which is eluted from Blue-Sepharose in 50% EG[9,10] and from Phenyl-Sepharose in 75% EG.[13] Thus, bovine interferon β appears to be less hydrophobic than human interferon β.

Neutralization tests with antiserum to purified bovine interferon β revealed that in crude preparations from cell cultures at least three interferon components can be distinguished (Fig. 4).[14,15] The main component which binds to Blue-Sepharose is strongly neutralized whereas a second fraction which does not bind to Blue-Sepharose is weakly neutralized. A third fraction which likewise is not bound to Blue-Sepharose is not neutralized at all. Tentatively we refer to the strongly neutralized component as bovine interferon β_1 and to the weakly neutralized fraction as bovine interferon β_2. The nonneutralized fraction appears to be a bovine interferon α according to its physicochemical properties and its activity in heterologous cells.

[23] J. W. Heine, A. J. Mikulski, E. Sulkowski, and W. A. Carter, *Arch. Virol.* **57**, 185 (1978).

It is noteworthy that the proportion of the nonneutralized fraction of interferon produced in bovine kidney cell cultures depends on the species of the inducing virus. Bluetongue virus which is a double-stranded RNA virus repeatedly induced relatively more nonneutralizable interferon than FMDV or NDV. On the contrary, only the β_1 component is induced by poly(I) · poly(C).

Application of genetic engineering techniques has recently resulted in the isolation of three distinct bovine interferon β genes.[24] This supports our conclusion that more than one species of bovine interferon β exists.

[24] D. W. Leung, D. J. Capon, and D. V. Goeddel, *Bio/Technology* **2,** 458 (1984).
[25] R. Ahl and M. Gottschalk, Official Publication of the European Communities No. EUR 8675 EN, pp. 31–36 (1984).

[31] Production, Purification, and Characterization of Rat Interferon from Transformed Rat Cells

By PETER H. VAN DER MEIDE, JACQUELINE WUBBEN, KITTY VIJVERBERG, and HUUB SCHELLEKENS

Despite the numerous reports on interferons produced in human and mouse systems, surprisingly little is known about interferon of rat origin. The scarcity of information is most likely attributable to the reasons cited by Poindron[1] and the lack of good methods for its production and purification. In this chapter, we will briefly describe our work on the production, purification, and characterization of rat interferon produced in cultures of a continuous cell line of embryonic origin.

Cell Culture and Interferon Production

A transformed rat embryonic cell line (Ratec) obtained from embryos of WAG/Rij rats was used for the preparation of cell cultures in which the interferon was induced by Sendai virus. Cells were grown to confluency in 150-cm^2 plastic flasks in 50 ml Dulbecco's modified Eagle's medium (DME) supplemented with 300 mg glutamine per liter, 100 units/ml of penicillin, 130 μg/ml of streptomycin, and 10% heat inactivated fetal calf serum. Cells were incubated at 37° in a humidified atmosphere with 5%

[1] P. Poindron, G. Coupin, D. Illinger, and B. Fauconnièr, this series, Vol. 78, p. 165.

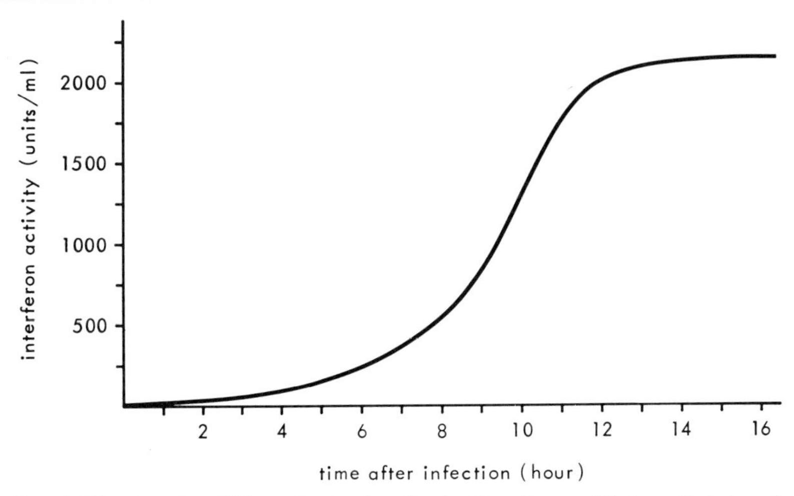

FIG. 1. Kinetics of rat IFN production in roller-bottle cultures of Ratec cells induced with Sendai virus. At specified times after infection, samples of 0.2 ml were removed from the culture fluid and assayed for interferon activity.

CO_2. When cells were kept in culture with a split ratio of 1/10, confluency was routinely reached within 4 days, after which the cells were dispersed by trypsin and suspended in 10 ml DME with the same additives mentioned above. Then 90 ml of fresh DME (plus additives) was added and the cells were transferred to 2-liter roller bottles (Falcon 3027) and incubated at 37° (0.38 rpm). After reaching confluency (usually within 5 days), cells were washed with Hanks' balanced salt solution and reincubated with a 10 ml virus solution (Sendai virus concentration, 10^3 HAU/ml) in serum-free medium for 1 hr at 37° to permit attachment and penetration of the virus. This medium was then discarded and the cells were incubated in 50 ml of serum-free medium for 18 hr at 37°, the time of maximal interferon production (see Fig. 1). These crude interferon preparations contained 2–3 × 10^3 units/ml and about 4 × 10^4 units/mg of protein.

Interferon Assay

For estimation of the interferon activity, the cytopathic reduction assay described by Armstrong[2] was used. Ratec cells were incubated with vesicular stomatitis virus (VSV) to determine the cytopathic effect. Because no internationally accepted reference standard for rat interferon is available, all titers were determined by comparison with a laboratory

[2] J. A. Armstrong, *Appl. Microbiol.* **21,** 723 (1971).

preparation of interferon in which 1 unit is roughly defined as the concentration that results in a 50% cytopathic effect in a microtiter plate seeded with Ratec cells and challenged with VSV. In our test system 1 unit is the equivalent of 6 units of standard rat interferon as produced by Lee Biomolecular Lab., San Diego, CA (lot no. 83010, Cat. no. 40001). The interferon activity in this chapter is expressed in our own laboratory units.

Concentration and Partial Purification of Rat Interferon

The interferon containing culture medium was harvested, adjusted to pH 2.0 by the addition of 4 *N* HCl, and stored for 72 hr at 4° to inactivate residual Sendai virus. Cell debris was removed from the fluid by low-speed centrifugation (10,000 *g*) for 25 min. The interferon preparation was then concentrated 25-fold by use of a 76 mm YM-10 Amicon Diaflo membrane at 4° (5 days). Afterward the concentrated interferon was centrifuged at 150,000 *g* for 3 hr and the supernatant was adjusted to pH 7.0 with 2.0 *M* NaOH. To this solution was added crystalline ammonium sulfate to 40% saturation at 0°. The precipitate was removed (it contained less than 10% of the starting activity) by centrifugation at 20,000 *g* for 20 min at 4°. To the supernatant fraction, an additional amount of ammonium sulfate was added to a final saturation of 85%. After centrifugation the supernatant was discarded and the precipitate was dissolved in phosphate-buffered saline and exhaustively dialyzed. The interferon activity was remarkably stable during these manipulations and about 90% of the initial activity was recovered in the 40 to 85% ammonium sulfate pellet fraction.

Gel Filtration on Ultrogel AcA 44

The 40 to 85% ammonium sulfate concentrated interferon was first purified by gel filtration through a 60-cm-long ultrogel AcA 44 column at 4°. The preparation with an activity of 1.2×10^6 units in a volume of 6 ml was dialyzed against a buffer containing 40 m*M* Tris · HCl, pH 7.2, 50 m*M* NaCl, 10% glycerol, and 1 m*M* dithioerythritol (DTE) and then applied to the column (1.6×60 cm) which was equilibrated with the same buffer. The interferon was pumped through at a flow rate of 10 ml/hr. Fractions of 8 ml were collected and the protein elution profile together with the antiviral activity is shown in Fig. 2. Fractions 14 and 15 showed the highest interferon activity, with a specific activity of 1.1×10^6 units/mg, corresponding to a 15-fold purification; 84% of the initial activity was recovered.

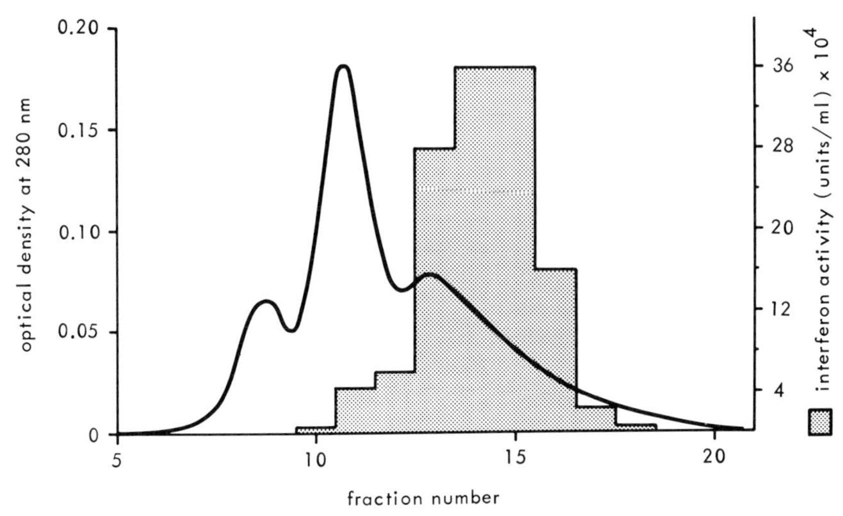

FIG. 2. Purification of rat IFNs on Ultragel AcA 44 size-exclusion column. The column was run as described in the text.

Application of Fast Protein Liquid Chromatography (FPLC) in the Purification of Rat Interferon

The introduction of uniform spherical particles for use in ion-exchange chromatography has dramatically improved the speed and resolution of liquid chromatography. In our purification scheme the binding of rat interferon to a prepared Mono-Q column (anion-exchanger; type HR 5/5 from Pharmacia) connected to a FPLC-system was studied with two preparations of rat interferon of different degrees of purification. Approximately 1.5×10^5 units of interferon (extensively dialyzed against the buffer mentioned below) with a specific activity of 6×10^4 units/mg of protein was applied to the column in a volume of 0.5 ml. The column was previously equilibrated with 40 m*M* Tris · HCl, pH 7.2, 50 m*M* NaCl, 1 m*M* DTE, 0.1 m*M* phenylmethylsulfonylfluoride (PMSF), and 15% (v/v) glycerol. The column was eluted with a gradient from 50 to 1000 m*M* NaCl at room temperature at a flow rate of 1 ml/min. All buffers were filtered through 0.22-μm single use filter units (Millex-GV Millipore). The absorbance (280 nm) elution profile and the position of the interferon activity are indicated in Fig. 3. This typical run (including column regeneration) took about 50 min. The recovery from this system was remarkably high: about 95% of the initial interferon activity was recovered with a roughly 10-fold purification.

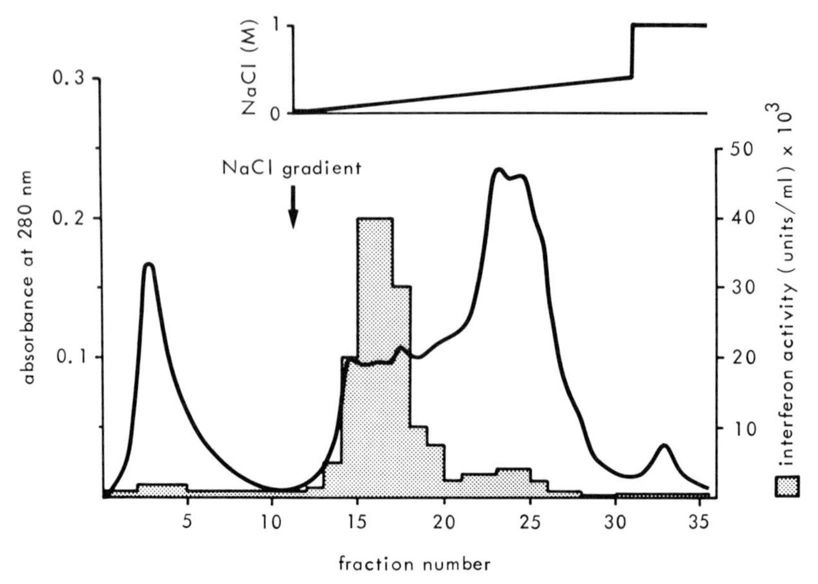

FIG. 3. Elution profile of rat IFN by anion-exchange chromatography. Approximately 1.5×10^5 units was chromatographed on a Mono-Q column (HR 5/5, Pharmacia) connected to a FPLC-system. Fractions of 0.5 ml were collected and assayed for IFN activity. Further details are outlined in the text.

With this system, we attempted to increase the specific activity of partially purified interferon which was obtained by Ultrogel AcA 44 chromatography. The pooled peak fractions from the Ultrogel column were applied to the MonoQ column and eluted with the same buffers as above. Despite a significant increase in the specific activity of the peak fractions

PURIFICATION STEPS FOR RAT INTERFERON PRODUCED IN VIRUS-INDUCED RATEC CELLS

	Activity (units/ml)	Volume (ml)	Specific activity (units/mg)	Percentage of original activity
Crude rat interferon	2000	1000	4×10^4	100
Ultrafiltration	4.8×10^4	40	4.5×10^4	96
40–85% ammonium sulfate pellet fraction	18.6×10^4	10	7×10^4	93
AcA 44 gel filtration peak fractions	5.7×10^4	24	1.1×10^6	68
MonoQ anionic exchange chromatography peak fractions	3.0×10^5	1.8	10^7	27

(10^7 units/mg of protein) a significant loss of activity was observed (see the table). Obviously the interferon becomes less stable in solutions from which contaminating proteins are removed.

Some Characteristics of Rat Interferon

SDS–Polyacrylamide Gel Electrophoresis of Interferon Preparations Derived from Ratec Cells

The 40–85% ammonium sulfate pellet fractions were analyzed by electrophoresis on 15% polyacrylamide gels according to the method described by Laemmli.[3] Two interferon samples were incubated in a solution containing 50 m*M* Tris · HCl, pH 6.8, 0.1% SDS, and 10% glycerol in the absence (Fig. 4A) or presence of 5% 2-mercaptoethanol (Fig. 4B). After electrophoresis, the gel was cut into 5-mm slices. The gel slices were eluted with 20 m*M* Tris · HCl, pH 7.0, 0.5% sarcosine, and 1% 2-mercaptoethanol during 48 hr at room temperature prior to being assayed for interferon activity. Recovery of the initial activity was about 70%. The use of SDS–polyacrylamide gel electrophoresis resulted in the separation of three distinct peaks of interferon with molecular weights of approximately 80,000, 40,000, and 20,000 (Fig. 4A). The 80,000 component was absent when 2-mercaptoethanol was present in the sample buffer (Fig. 4B) suggesting that the interferon molecules tend to aggregate in the absence of a reducing agent. Inactivation of the 80,000 component by this agent is not likely, since the total recovery of interferon activity from the two samples was the same. It is notable that the interferon activity was found to be heterogeneous in size. Interferons from different animal cells are known to be heterogeneous in molecular weight. The activity profile of naturally derived rat interferon suggests that there are at least two distinct classes of molecules each of which consists of a continuous gradient of various size molecules.

Isoelectric Focusing of Rat Interferon

A crude interferon preparation (400 μg protein) was subjected to isoelectric focusing on cylindrical gels of 6 mm diameter in perspex tubes. The gel was run as described by O'Farrell *et al.*[4] After electrophoresis it was cut into 5 mm slices which were eluted in 0.5 ml buffer containing 20 m*M* Tris · HCl, pH 7.2, 0.5% sarcosine, and 1% 2-mercaptoethanol for 2 days at room temperature. Most of the interferon activity was found in a

[3] U. K. Laemmli, *Nature* (*London*) **227,** 680 (1970).

[4] P. Z. O'Farrell, H. M. Goodman, and P. H. O'Farrell, *Cell* **12,** 1133 (1977).

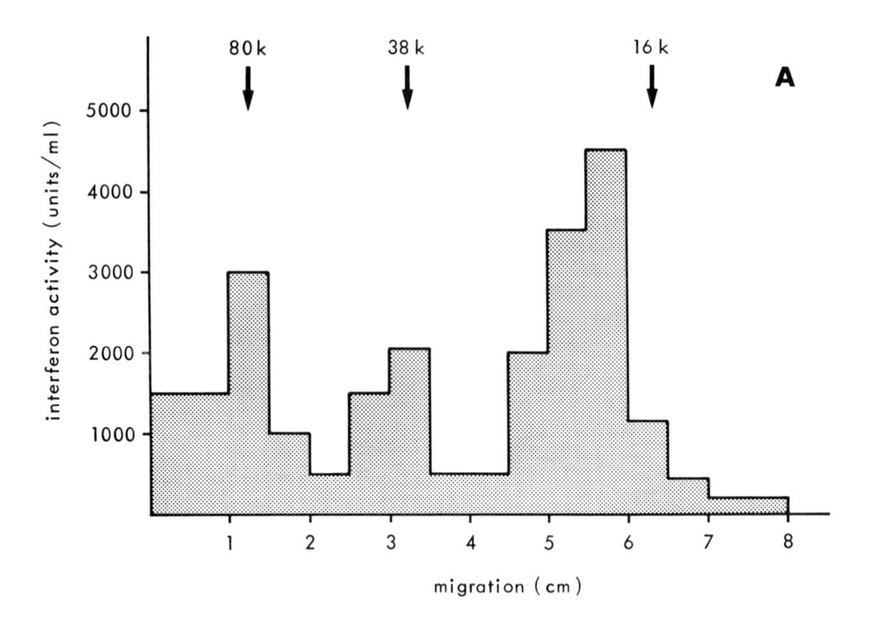

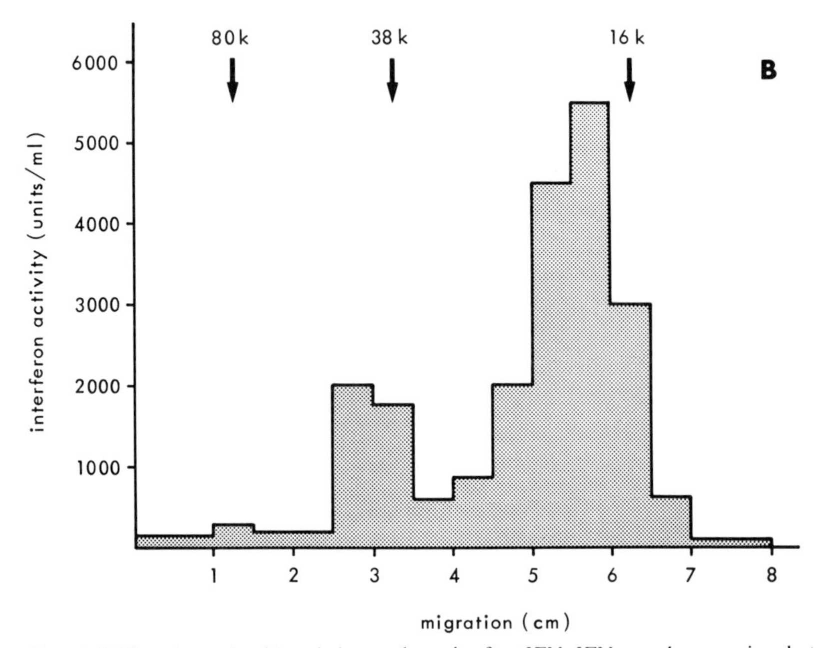

FIG. 4. SDS–polyacrylamide gel electrophoresis of rat IFN. IFN samples were incubated in a sample buffer (see text) with (B) or without (A) 5% 2-mercaptoethanol before electrophoresis. Gel slices (5 mm) were eluted with 0.5 ml of solution containing 20 m*M* Tris · HCl, 0.5% sarcosine, and 1% 2-mercaptoethanol (48 hr at room temperature) prior to being assayed for IFN activity. Molecular weight markers are indicated by arrows.

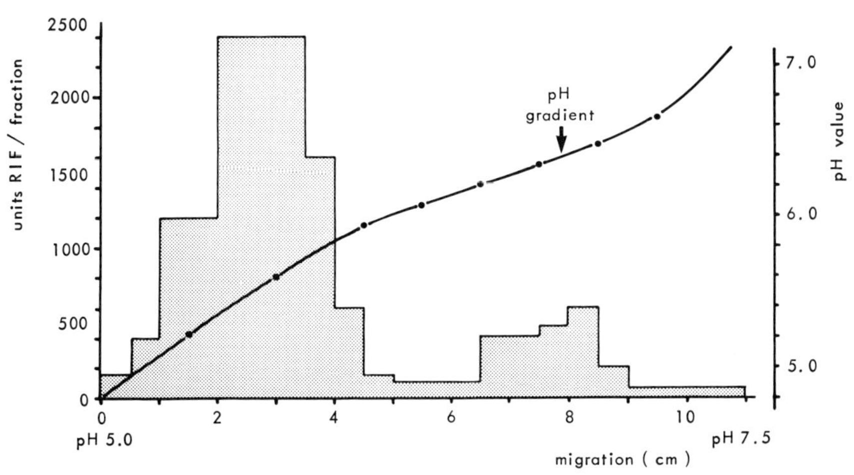

FIG. 5. Isoelectric focusing of rat IFN. About 400 μg of protein (containing 2×10^4 units) was subjected to analytical electrofocusing as described in the text. Gelrods were cut into 5 mm slices and eluted with 0.5 ml of a solution containing 20 m*M* Tris–HCl (pH 7.0), 0.5% sarcosine, and 1% 2-mercaptoethanol (48 hr at room temperature) prior to being assayed for rat IFN (RIF) activity.

rather broad area between pH 5.0 and 5.8. A minor peak of activity was present at pH 6.2 to 6.6 (usually less than 20% of the total activity). The overall recovery from the gel was about 60%. This experiment (Fig. 5) also suggests that there are at least two distinct classes of molecules, varying in both size and charge.

Structural Homology between Human α2 Interferon and Naturally Derived Rat Interferon

There are several reports[5–7] which indicate that certain subtypes of human and mouse interferon are antigenetically related to each other and that antisera raised against these subtypes can cross-react in neutralization reactions. The same cross-reactivity was found between human interferon α2 (Hu-IFN-α2 prepared in *E. coli*) and rat interferons, produced from Ratec cells (nRV-IFN) or an alpha subtype produced in *E. coli* (rR-IFN-α1). While rR-IFN-α1 can be neutralized with an antibody prepara-

[5] L. A. Pickering, L. H. Kronenberg, and W. E. Stewart, *Proc. Natl. Acad. Sci. U.S.A.* **77,** 5938 (1980).

[6] W. E. Stewart, II and E. A. Havell, *Virology* **101,** 315 (1980).

[7] Y. Kawade, Y. Watanabe, Y. Yamamoto, J. Fujisawa, B. J. Dalton, and K. Paucker, *Antiviral Res.* **1,** 167 (1981).

tion directed against Hu-IFN-α2,[8] the nR-IFN activity was only marginally affected (not shown).

In an attempt to isolate the recombinant and naturally derived rat interferons by affinity chromatography on an anti-Hu-IFN-α2 antibody column, the IgG fraction from antiserum directed against Hu-IFN-α2 was purified and coupled to CNBr-Sepharose. For that purpose, the crude serum (20 ml) was fractionated by the addition of one volume saturated ammonium sulfate. The 50% saturated solution was gently stirred at 0° for 24 hr. The solution was then centrifuged for 20 min at 10,000 *g* at 4°. The pellet was well drained and dissolved in 10 ml of potassium phosphate buffer (0.0175 *M*, pH 7.5; 0.45 g/liter KH_2PO_4 and 2.47 g/liter K_2HPO_4) and dialyzed overnight against the same buffer with three changes. The dialyzed solution was chromatographed on a DEAE-cellulose (DE32, Whatman) column. This column (32 × 2.7 cm) was equilibrated with the same potassium phosphate buffer at 4°. The crude IgG fraction was applied to the column and pumped through at a flow rate of 20 ml/hr. The purified IgG proteins passed directly through the column and were concentrated by pressure ultrafiltration with a 62-mm YM-30 Amicon Diaflo membrane. The bound proteins could be eluted from the column with 0.3 *M* phosphate buffer (7.74 g/liter KH_2PO_4 and 42.28 g/liter K_2HPO_4). Swelling, washing, regeneration, and preservation of the cellulose were performed as described by the manufacturer. For the coupling of the IgG fraction, 2 g of freeze-dried CNBr-activated Sepharose 4B (Pharmacia) was swollen and washed on a Whatman 3 MM filter with 1 m*M* HCl for 12 min. A total volume of 500 ml HCl was aspirated off between successive additions. This procedure was followed by a washing step with 200 ml coupling buffer (0.1 *M* $NaHCO_3$, 0.2 *M* NaCl, and 0.2 m*M* dithioerythritol, pH 9.0). Without delay, the gel slurry (7 ml) was transferred to a plastic tube and mixed with an equal volume of purified IgG (25 mg of protein) which had been exhaustively dialyzed against the coupling buffer. The mixture was rotated end-over-end at 4° for 24 hr. After reaction, the residual active groups were blocked by adding ethanolamine (pH 9.0) and dithioerythritol to final concentrations of 0.2 *M* and 5 m*M*, respectively. The mixture was again rotated for 12 hr at 4°. The gel slurry was then washed free of uncoupled antibodies and reactants with phosphate-buffered saline and stored in this buffer in the presence of 0.02% sodium azide at 4°.

Before use, the gel was washed with two volumes of PBS to remove the sodium azide and mixed with an equal volume of a preparation of rat

[8] P. H. van der Meide, R. Dijkema, M. Caspers, K. Vijverberg, and H. Schellekens, this volume [62].

interferon (a 40–85% ammonium sulfate pellet fraction with an activity of 1.2×10^5 units/ml). The mixture was rotated end-over-end during 20 hr at 4°. The gel was poured into a column (13 × 0.9 cm) and eluted with about 6 column volumes of PBS (the flow through fraction). Nonspecifically bound proteins were removed by washing with 1 *M* NaCl in PBS. The retained fraction was subsequently eluted with 0.2 *M* glycine · HCl, pH 2.3. Only 15% of the original nR-IFN activity was found in the retained fraction and 85% in the flow through (Fig. 6). When the concentrated flow through fraction was rerun on the column, less than 1% of the activity was retained, indicating that only a small fraction of nR-IFN shows structural homology with Hu-IFN-α2. The same experiment performed with interferon of mouse origin (from L929 cells induced with Sendai virus) revealed percentages of 30 and 70% respectively (Fig. 6). These results may be interpreted in different ways. As reported for mouse interferon, nR-IFN might be a mixture of the α and β type (as suggested by gel electrophoresis and isoelectric focusing) of which only the α type is reacting and

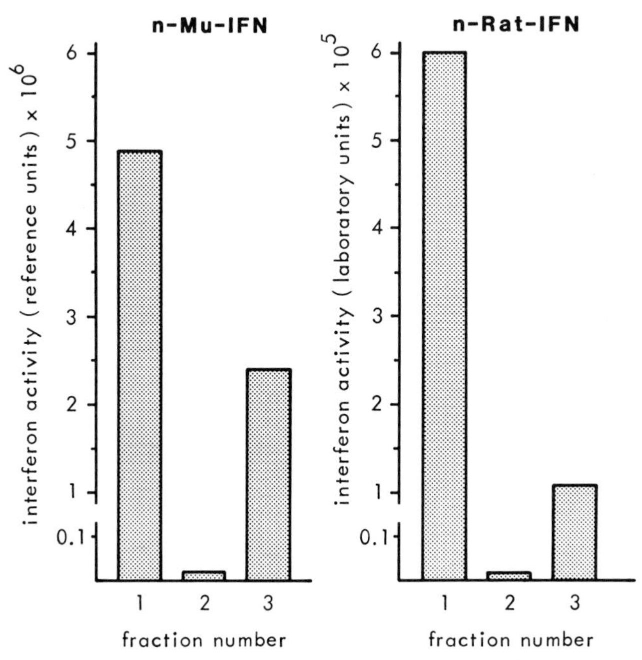

FIG. 6. Fractionation of Sendai induced mouse (n-Mu-IFN) and rat (n-R-IFN) interferon by a column containing immobilized anti-Hu-IFN-α2. The recovery of the initial activity applied to the column was 76 and 58% for n-Mu-IFN and n-R-IFN, respectively. Fraction no. 1: flow through; fraction no. 2: 1 *M* NaCl wash; and fraction no. 3: retained activity.

β is not. Second, nR-IFN preparations contain different alpha subtypes of which only a minor population reacts with anti-Hu-IFN-α2. Third, the sugar moiety of the nR-IFN molecules plays a role in the antigenicity of this type of interferon. Most likely, a combination of these three possibilities is responsible for the poor neutralization and the low binding of nR-IFN by antibodies to Hu-IFN-α2. Our finding that rR-IFN-α1 is neutralizable with and binds effectively to anti-Hu-IFN-α2 is particularly suggestive for the first and third interpretations. In our opinion, the second interpretation is less likely because the cloned rat IFN-α1 gene is primarily expressed after viral infection, as suggested by the finding that all cDNA clones derived from rat IFN-α mRNA (extracted from Sendai induced Ratec cells) were of the same α1 subtype.[8]

Section IV

Modification and Analysis of Interferons

[32] Chemical Modifications of Human and Murine Interferons

By KARL H. FANTES

Modification of interferons could serve several purposes: to increase biological activity or stability, to decrease toxicity, to prolong persistence in the circulation, to assist targetting to specific cells or sites, and to facilitate structural studies. Some such modifications have been achieved in a number of ways: by the addition of various agents during biosynthesis, by enzymatic methods, by the action of chemical reagents, by reactions affecting cysteine or cystine residues, by complexing with other substances, and by manipulation of recombinant DNAs for production by bacteria or other vehicles.

Agents Added during Biosynthesis

After it was found that chick interferon consisted of several components[1] and after it was subsequently shown that the heterogeneity of rabbit interferon was greatly reduced by treatment with neuraminidase,[2] it was assumed by many authors that all heterogeneity of interferons was due to carbohydrate. Many attempts were therefore made to produce carbohydrate-free, homogeneous human and mouse interferons by either adding carbohydrate-synthesis inhibitors during the period of biosynthesis or to remove carbohydrate from already formed interferons (for the latter see the relevant sections). Claims to have produced homogeneous, or at least less heterogeneous molecules, were made by using 2-deoxyglucose or D-glucosamine during the production of Hu-IFN-α,[3,4] Hu-IFN-β,[5,6] and Mu-IFN[7] or by using tunicamycin during the biosynthesis of

[1] K. H. Fantes, *Science* **163,** 1198 (1979).

[2] E. Schonne, A. Billiau, and P. DeSomer, *Symp. Ser. Immunobiol. Stand.* **14,** 61 (1970).

[3] B. Cuncliffe and J. Morser, *J. Gen. Virol.* **43,** 457 (1979).

[4] W. E. Stewart, M. Wiranowska-Stewart, V. Koistinen, and K. Cantell, *Virology* **97,** 473 (1979).

[5] E. A. Havell, J. Vilcek, E. Falcoff, and B. Berman, *Virology* **63,** 475 (1975).

[6] E. A. Havell, S. Yamazaki, and J. Vilček, *J. Biol. Chem.* **252,** 4425 (1977).

[7] W. E. Stewart, T. Chudzio, L. S. Lin, and M. Wiranowska-Stewart, *Proc. Natl. Acad. Sci. U.S.A.* **75,** 4814 (1978).

most types of human and murine interferons (e.g., Hu-IFN-α,[8-10] Hu-IFN-β,[11,12] Hu-IFN-γ,[13] Mu-IFN[14,15]).

However, it is likely that these carbohydrate-synthesis inhibitors also have other functions: e.g., total protein synthesis was inhibited in some instances[3] and the relative abundance of individual Hu-IFN-α components was changed.[10] In any case, any effect on the production of Hu-IFN-α should not involve carbohydrate if the claims that this species is essentially free from saccharide is correct.[16,17]

Apart from additives affecting carbohydrate content of interferons, other additives (butyrate, 5-bromodeoxyuridine, dexamethazone, dimethyl sulfoxide, tetradecanylphorbol acetate, 5-azacytidine) have been found to enhance Hu-IFN-α biosynthesis, but insufficient work has been done to establish whether these alter the distribution of individual subcomponents.[18-21]

Changes Effected by Enzymatic Methods

Enzymes have been used to modify both the polypeptide and the carbohydrate portions of various interferons.

While aminopeptidases and carboxypeptidases have been reported to have no or only slight effects on various parameters of interferons,[22,23] it has been known since the early days of interferon research[24] that endoproteases inactivate all interferon activity. Yet by controlled mild treatment with α-chymotrypsin Braude *et al.*[25] managed to preserve the antivi-

[8] K. C. Chadha, P. M. Grob, R. L. Hamill, and E. Sulkowski, *Arch. Virol.* **64,** 109 (1980).
[9] K. C. Chadha, this series, Vol. 78, p. 220.
[10] J. Morser, *J. Gen. Virol.* **63,** 213 (1982).
[11] Y. K. Yip and J. Vilček, this series, Vol. 78, p. 212.
[12] Y. Watanabe and Y. Kawade, *J. Gen. Virol.* **64,** 1391 (1983).
[13] A. Mizrahi, J. A. O'Malley, W. A. Carter, A. Takatsuki, G. Tamura, and E. Sulkowski, *J. Biol. Chem.* **253,** 7612 (1978).
[14] N. B. K. Raj and P. M. Pitha, *J. Interferon Res.* **1,** 595 (1981).
[15] E. A. Havell and W. A. Carter, *Virology* **108,** 80 (1981).
[16] G. Allen and K. H. Fantes, *Nature (London)* **287,** 408 (1980).
[17] M. Rubinstein, W. P. Levy, J. A. Moschera, C.-Y. Lai, R. D. Hershberg, R. T. Bartlett, and S. Pestka, *Arch. Biochem. Biophys.* **210,** 307 (1981).
[18] M. D. Johnston, *J. Gen. Virol.* **56,** 175 (1981).
[19] G. R. Adolf and P. Swetly, *J. Interferon Res.* **2,** 261 (1982).
[20] J. Shuttleworth, J. Morser, and D. C. Burke, *Biochim. Biophys. Acta* **698,** 1 (1982).
[21] M. G. Tovey, C. Vincent, I. Gresser, and M. Revel, *in* "The Biology of the Interferon System" (E. De Maeyer and H. Schellekens, eds.), p. 45. Elsevier, Amsterdam, 1983.
[22] K. H. Fantes and C. F. O'Neill, *Nature (London)* **203,** 1048 (1964).
[23] M. J. Otto, J. J. Sedmak, and S. E. Grossberg, *J. Virol.* **35,** 390 (1980).
[24] A. Isaacs and J. Lindenmann, *Proc. R. Soc. London, Ser. B* **147,** 258 (1957).
[25] I. A. Braude, L. S. Lin, and W. E. Stewart, *Biochem. Biophys. Res. Commun.* **89,** 612 (1979).

ral activity of Hu-IFN-α (leukocyte) on bovine (though not on human) cells. However, one cannot deduce from the results whether they represent complete inactivation of some of the subspecies or partial proteolysis, e.g., by cleavage at tryptophanyl, phenylalanyl, or tyrosyl residues on the carboxy side of the molecules. Rose *et al.*[26] isolated DNA recombinant-produced biologically active Hu-IFN-γ, that had lost its 15 carboxy-terminal amino acids by the action of *E. coli* proteases. Similarly truncated, active forms of Hu-IFN-α and -β have been described by others,[27] and Wetzel *et al.*[28] removed the 13 carboxy-terminal amino acids from Hu-IFN-αA by limited digestion with thermolysin without impairing its biological activities.

Glycosidases, including neuraminidase, have been used more often than proteolytic enzymes in attempts to modify various interferons. There have been several reports on changes of Hu-IFN-α,[23,29] due to these enzymes, which, however, were not confirmed by others.[30,31] Hu-IFN-β, a known glycoprotein[32] was, as expected, always altered by the action of glycosidases,[23,32,33] as was also Hu-IFN-γ.[34]

As Hu-IFN-α (unlike -β and -γ) is supposed not to contain carbohydrate,[16,17] it is difficult to interpret effects of carbohydrases on such a substance, unless the enzyme preparations were contaminated with proteases, or unless Hu-IFN-α does indeed contain some neutral sugar, as has been suggested by some authors.[35,36] A recent report, however, has described carbohydrate residues on some species of human leukocyte interferon.[37]

[26] K. Rose, M. G. Simona, R. E. Offord, C. P. Prior, B. Otto, and D. R. Thatcher, *Biochem. J.* **215,** 273 (1983).

[27] W. P. Levy, M. Rubinstein, J. Shively, U. del Valle, C. Y. Lai, J. Moschera, L. Brink, L. Gerber, S. Stein, and S. Pestka, *Proc. Natl. Acad. Sci. U.S.A.* **78,** 6186 (1981).

[28] R. Wetzel, H. L. Levine, D. A. Estell, S. Shire, J. Finer-Moore, R. M. Stroud, and T. A. Bewley, *in* "Interferons" (T. C. Merigan and R. M. Friedman, eds.), p. 365. Academic Press, New York, 1982.

[29] S. Bose, D. Gurari-Rotman, U. T. Ruegg, L. Corley, and C. B. Anfinsen, *J. Biol. Chem.* **251,** 1659 (1976).

[30] K. E. Mogensen, L. Pyhälä, E. Torma, and K. Cantell, *Acta Pathol. Microbiol. Scand., Ser. B: Microbiol. Immunol.* **82B,** 305 (1974).

[31] J. Morser, J. P. Kabayo, and D. W. Hutchinson, *J. Gen. Virol.* **41,** 175 (1978).

[32] E. Knight, *Proc. Natl. Acad. Sci. U.S.A.* **73,** 520 (1976).

[33] E. Knight and D. Fahey, *J. Interferon Res.* **2,** 421 (1982).

[34] Y. K. Yip, H. Chroboczek Kelker, J. Le, P. Anderson, B. S. Barrowclough, C. Urban, and J. Vilček, *in* "The Biology of the Interferon System" (E. De Maeyer and H. Schellekens, eds.), p. 129. Elsevier, Amsterdam, 1983.

[35] P. Meindl, G. Bodo, and H. Tuppy, personal communication (1981).

[36] S. Yoshikawa, T. Tanimoto, S. Koyama, Y. Sato, K. Masuda, K. Yokobayashi, and M. Kurimoto, personal communication (1983).

[37] J. E. Labdon, K. D. Gibson, S. Sun, and S. Pestka, *Arch. Biochem. Biophys.* **237,** 422 (1984).

Reactions with Chemicals

The first attempts to show with specific reagents whether certain amino acids in an interferon (chick) could be modified without loss of activity revealed that disulfide bonds and free amino, but not sulfhydryl groups, were essential for activity.[22] These findings were confirmed and extended by others for different interferons.

Labeling with ^{125}I, a reaction that involves tyrosine residues, was achieved, usually with little loss of activity, for Hu-IFN-α,[38,39] Hu-IFN-β,[40] and Mu-IFN.[41] However, Tan *et al.*[42] lost all activity when they iodinated (or nitrated) Hu-IFN-β, although they used, like Knight,[40] the Bolton–Hunter reagent for the iodination. From this they concluded that one or more tyrosyl residues were indispensable for activity.

Hu-IFN-α was also modified in its amino (and hydroxyl?) group(s) by reaction with [^{3}H]dansylchloride[38] and in its tryptophan residue(s) by reaction with 2-hydroxy-5-nitrobenzyl bromide.[43] Both these reactions resulted in some, but by no means complete, loss of biological activity.

The effect of various chemical reagents on Mu-IFN-α and -β was also examined by Rusckowski *et al.*[43] Increasing concentrations of fluorescamine increasingly destroyed various biological activities of both interferons, implicating the essentialness of amino groups (in agreement with Ref. 22, but not with Ref. 38). 5,5′-dithiobis(2-nitrobenzoic acid), a chemical that reacts with —SH groups, enhanced some of the biological activities of the α-type, but decreased those of Mu-IFN-β. Similar respective enhancement and decrease of the activities of the two Mu-IFN species were obtained with the tryptophan affecting reagent 2-methoxy-5-nitrobenzyl bromide.

McCray and Weil[44] found that halomethyl ketone derivatives of phenylalanine (to a lesser extent of tryptophan, but not at all of lysine) inactivated mouse (and rabbit) interferons. It was suggested that the ketone derivative first binds (via its aromatic ring) to a hydrophobic site of the interferon and then reacts irreversibly with a nearby nucleophilic

[38] M. E. Smith, K. Zoon, P. Bridgen, and C. B. Anfinsen, *Fed. Proc., Fed. Am. Soc. Exp. Biol.* **38,** A2897 (1979).

[39] D. S. Secher and D. C. Burke, *Nature (London)* **285,** 446 (1980).

[40] E. Knight, *J. Gen. Virol.* **40,** 681 (1978).

[41] Y. Iwakura, S. Yonehara, and Y. Kawade, *Biochem. Biophys. Res. Commun.* **84,** 557 (1978).

[42] Y. H. Tan, F. Barakat, W. Bertholt, H. Smith-Johannsen, and C. Tan, *J. Biol. Chem.* **254,** 8067 (1979).

[43] M. Rusckowski, M. Paucker, B. Dalton, and C. A. Ogburn, *J. Interferon Res.* **2,** 177 (1982).

[44] J. W. McCray and R. Weil, *Proc. Natl. Acad. Sci. U.S.A.* **79,** 4829 (1982).

amino acid residue, probably histidine. Histidine is therefore considered to be essential for activity by these authors.

In interpreting some of the above results, especially where there was residual activity after a reaction, one cannot be sure whether this activity was due to residual unchanged molecules or to modified molecules with lower intrinsic activity.

Stewart and his colleagues treated Hu-IFN-α (leukocyte)[45] and Mu-IFN[46] with periodate. This treatment reduced the heterogeneity of the interferons, a result that was interpreted as being due to deglycosylation of the "higher molecular weight" species. Although the authors adduce a number of arguments in favor of their interpretation, it need not be correct: native Hu-IFN-α is (1) probably essentially carbohydrate free[16,17] and (2) consists of a mixture of different polypeptides with different mobilities in SDS–PAGE gels.[16,17] It seems more likely that the periodate treatment affected the polypeptide rather than a saccharide portion of the interferons: some amino acids are attacked by periodate more readily than some sugars,[47] and Braude *et al.*[25] have shown that proteolysis need not abolish all biological activity.

Reactions Affecting Cysteine and Cystine Residues

It has been known that one or more intact disulfide bonds were essential for interferon activity.[22,48,49] The linear peptide chains of Hu-IFN-α,[48] of Hu-IFN-β, and of Mu-IFN,[50] in which all noncovalent and all cross-linking disulfide bonds were broken by treatment with 2-mercaptoethanol in the presence of guanidine hydrochloride or urea, were found to be inactive, but fully active after reoxidation in the presence of SDS. Oxidative cleavage with performic acid, of one or more disulfide bonds, or carboxymethylation with iodoacetamide or reduced Hu-IFN-α led to irreversible loss of activity; but activity was not impaired by the sulfhydryl reagent *p*-chloromercuribenzoate.[48] In numerous publications, Stewart and collaborators reported results on Hu-IFN-α, Hu-IFN-β, and Mu-IFN, that were obtained after various treatments with SDS, urea, and periodate under reducing and nonreducing conditions. The results are not

[45] W. E. Stewart, L. S. Lin, M. Wiranowska-Stewart, and K. Cantell, *Proc. Natl. Acad. Sci. U.S.A.* **74,** 4200 (1977).

[46] W. E. Stewart, T. Chudzio, L. S. Lin, and M. Wiranowska-Stewart, *Proc. Natl. Acad. Sci. U.S.A.* **75,** 4814 (1978).

[47] B. Sklarz, *Q. Rev., Chem. Soc.* **21,** 3 (1967).

[48] K. E. Mogensen and K. Cantell, *J. Gen. Virol.* **22,** 95 (1974).

[49] W. E. Stewart, E. De Clercq, and P. DeSomer, *Nature (London)* **249,** 460 (1974).

[50] W. E. Stewart and J. Desmyter, *Virology* **67,** 68 (1975).

always easily interpreted, nor is it often clear what kind of chemical changes have been effected: e.g., the two species (MW 15,000 and 21,000) of Hu-IFN-α (leukocyte) were fully recovered after boiling in SDS and removal of excess detergent, but only the 21,000 component showed "human" activity when the boiling took place in the presence of 2-mercaptoethanol. But the "cat" activity of both species survived under these conditions.[50,51]

In another paper[52] it was found that Hu-IFN-α that was recovered after boiling with SDS under nonreducing conditions consisted of several species, all with sedimentation constants greater than that of the untreated material: association was assumed, but it was not known whether just between interferon molecules or between interferon and contaminating protein. To complicate matters even further, it was later found[53] that the single 15,000 MW component obtained by periodate treatment of Hu-IFN-α (see above), when boiled with SDS under reducing conditions, migrated in SDS–PAGE gels with an apparent MW of 26,000–30,000, but with a MW of 19,000–22,000 after reoxidation. Moreover the apparently single (MW 21,000) species obtained under reducing conditions was now thought to be composed of both the larger and the smaller components.

Hu-IFN-β lost most of its activity when boiled in dilute SDS, unless this was done under reducing conditions. The lost activity could be restored by the joint addition of SDS and 2-mercaptoethanol.[54]

When Mu-IFN was boiled in SDS under reducing conditions,[55] the antiviral activity of the 38,000 MW component increased, but that of the 22,000 MW species decreased. Again, the underlying molecular changes are not clear.

Further extensive investigations into the function of disulfide bonds were carried out by Wetzel *et al.*, who had the advantage of working with a pure interferon species (Hu-IFN-αA). Their findings, described in several earlier and two more recent publications[28,56] can be summarized as follows.

The molecule has two disulfide bonds, linking cysteine residues in positions 1 and 98, and 29 and 138, respectively. Breaking these by oxidative sulfitolysis creates a labile form, in which all cysteine and cystine

[51] J. Desmyter and W. E. Stewart, *Virology* **70,** 451 (1976).

[52] P. T. Allen and W. E. Stewart, *J. Gen. Virol.* **32,** 133 (1976).

[53] I. A. Braude, L. S. Lin, and W. E. Stewart, *Biochem. J.* **193,** 947 (1981).

[54] W. E. Stewart, P. DeSomer, V. G. Edy, K. Paucker, K. Berg, and C. A. Ogburn, *J. Gen. Virol.* **26,** 327 (1975).

[55] W. E. Stewart, S. Le Goff, and M. Wiranowska-Stewart, *J. Gen. Virol.* **37,** 277 (1977).

[56] R. Wetzel, P. D. Johnston, and C. W. Czarniecki, *in* "The Biology of the Interferon System" (E. De Maeyer and H. Schellekens, eds.), p. 101. Elsevier, Amsterdam, 1983.

residues are converted to cysteine-*S*-sulfonates. (The preparation of the *S*-carboxymethyl and *S*-carboxamidomethyl derivatives of disulfide-free interferon and the *S*-carboxamidomethyl derivative containing only the single 29–138 disulfide bond are also described.) This oxidized structure retains about 4% of the activity of the native IFN-αA,[57] but resembles the native molecule antigenically. Mild reduction of it with 2-mercaptoethanol leads to a new, biologically active, but rather labile compound with only one (Cys 29 to Cys 138) disulfide bond, which is readily and irreversibly destroyed by urea or heat. This shows that the other (Cys 1 to Cys 98) disulfide bond is not needed for activity and is thought to play a minor role in structural stabilization. Reduction of the native molecule with 2-mercaptoethanol results in inactive monomeric and oligomeric (up to 15 or more subunits) materials.[57] But the native interferon itself can also form (biologically active) multimers via intermolecular disulfide bonds, which are then unusually resistant toward reduction.[58–60] In addition, neither of the two near-equal halves of the molecule obtained after cleavage with CNBr is biologically active, implying that the whole tertiary structure is essential for maintaining the active site(s); also, approximately half of its high (60–70%) α-helix content is reversibly destroyed by acidification at pH 1.5.[28,56]

Conjugates with Other Substances

Probably the first covalent conjugation of an interferon to another substance was the preparation of an active antiviral Mu-IFN–CNBr-Sepharose complex by Ankel and his colleagues.[61,62] The binding must have involved one or more of the primary amino groups of the interferon.

Traub *et al.*[63] linked Hu-IFN-α (Namalwa), via its amino group(s), to *N*-succinimidyl-3-(2-pyridylthio)propionate and then conjugated the resulting compound via a disulfide link(s) to a sulfhydryl group(s) of serum albumin. The interferon–albumin compound formed in this way did not dissociate and possessed full antiviral activity. Pelham *et al.*[64] used a

[57] J. A. Langer and S. Pestka, this volume [44].

[58] S. Pestka, B. Kelder, D. K. Tarnowski, and S. J. Tarnowski, *Anal. Biochem.* **132,** 328 (1983).

[59] T. DeChiara, F. Erlitz, and S. J. Tarnowski, this volume [58].

[60] S. Pestka, B. Kelder, and S. J. Tarnowski, this volume [78].

[61] H. Ankel, C. Chany, B. Galliot, M. J. Chevalier, and M. Robert, *Proc. Natl. Acad. Sci. U.S.A.* **70,** 2300 (1973).

[62] F. Besançon and H. Ankel, *Nature (London)* **250,** 786 (1974).

[63] A. Traub, B. Payess, S. Reuveny, and A. Mizrahi, *J. Gen. Virol.* **53,** 389 (1981).

[64] J. M. Pelham, J. D. Gray, G. R. Flannery, M. V. Pimm, and R. W. Baldwin, *Cancer Immunol. Immunother.* **15,** 210 (1983).

similar reaction to form a compound from Hu-IFN-α (Namalwa) and a murine monoclonal antibody specific for human osteosarcoma cells. The conjugate retained interferon as well as antibody activity and was shown to target onto human osteosarcoma cell xenografts in immuno-deprived mice. Ng[65] used glutaraldehyde to prepare a fully active conjugate from Hu-IFN-β and immunoglobulin.

Modified Interferons by cDNA Manipulation

Several groups[66–73] isolated DNA recombinants coding for Hu-IFN-α proteins. Hu-IFN-β,[74–76] Hu-IFN-γ[77,78] and Mu-IFN[79,80] were prepared by similar methods.

It is much easier to obtain chemically modified interferons by altering the nucleotide sequence of coding cDNAs than it is to introduce changes into the amino acid chain of already formed interferon molecules. In this way a number of new, unnatural human and murine interferons with often markedly changed biological properties have been constructed.

[65] M. H. Ng, British Patent 1,564,666 (1980).

[66] S. Nagata, H. Taira, A. Hall, L. Johnsrud, M. Streuli, J. Escodi, W. Ball, K. Cantell, and C. Weissmann, *Nature (London)* **284,** 316 (1980).

[67] M. Streuli, A. Hall, W. Boll, W. E. Stewart, S. Nagata, and C. Weissmann, *Proc. Natl. Acad. Sci. U.S.A.* **78,** 4848 (1981).

[68] C. Brack, S. Nagata, N. Mantei, and C. Weissmann, *Gene* **15,** 379 (1981).

[69] D. V. Goeddel, D. W. Leung, T. J. Dull, M. Gross, R. M. Lawn, R. M. McCandliss, P. H. Seeburg, A. Ullrich, E. Yelverton, and P. W. Gray, *Nature (London)* **290,** 20 (1981).

[70] P. K. Weck, E. Rinderknecht, D. A. Estell, and N. Stebbing, *Infect. Immun.* **35,** 660 (1982).

[71] E. Rehberg, B. Kelder, E. G. Hoal, and S. Pestka, *J. Biol. Chem.* **257,** 11497 (1982).

[72] S. Pestka, *Sci. Am.* **249,** 29 (1983).

[73] S. Pestka, *Arch. Biochem. Biophys.* **221,** 1 (1983).

[74] T. Taniguchi, L. Guarente, T. M. Roberts, D. Kimelman, J. Douhan, and M. Ptashne, *Proc. Natl. Acad. Sci. U.S.A.* **77,** 5230 (1980).

[75] D. V. Goeddel, H. M. Shepard, E. Yelverton, D. Leung, R. Crea, A. Sloma, and S. Pestka, *Nucleic Acids Res.* **8,** 4057 (1980).

[76] R. Derynck, J. Content, E. De Clercq, G. Volckaert, J. Tavernier, R. Devos, and W. Fiers, *Nature (London)* **285,** 542 (1980).

[77] K. Alton, Y. Stabinsky, R. Richards, B. Ferguson, L. Goldstein, B. Altrock, L. Miller, and N. Stebbing, *in* "The Biology of the Interferon System" (E. De Mayer and H. Schelleken, eds.), p. 119. Elsevier, Amsterdam, 1983.

[78] S. J. Scahill, R. Devos, J. van der Heyden, and W. Fiers, *Proc. Natl. Acad. Sci. U.S.A.* **80,** 4654 (1983).

[79] Y. Higashi, Y. Sokawa, Y. Watanabe, Y. Kawade, S. Ohno, C. Takaoka, and T. Taniguchi, *J. Biol. Chem.* **258,** 9522 (1983).

[80] B. Daugherty, D. Martin-Zanca, T. C. Seamans, K. Collier, and S. Pestka, *J. Interferon Res.* **4,** 635 (1984).

Streuli *et al.*,[67] Weck *et al.*,[70,81] and Pestka *et al.*[71,82] constructed hybrid interferons, one of them with 100 times greater activity on murine cells than that of either of the parent molecules. Alton *et al.*[77] constructed among other analogs a consensus molecule (based on all the known 13 subtypes) that contained the most frequently observed amino acid residue at each position.

Fewer attempts have been made to construct Hu-IFN-β analogs by recombinant techniques, possibly because there probably is only one gene that codes for this species,[73,83,84] (but see Sehgal *et al.*[85]). However, replacing a cysteine residue with serine increased the stability and activity of Hu-IFN-β.[86] As to Hu-IFN-γ, Alton *et al.*[77] produced four new analogs with only a few amino acid substitutions and somewhat changed (usually lower) biological activity.

Summary and Conclusions

Methods that modify the carbohydrate and peptide portions of human and murine interferons have been described. Such work has helped considerably in our understanding of these fascinating substances. The most powerful of the methods relies on gene manipulation: it has already led, and will lead in the future, to further species with altered biological activities. Some of them may prove clinically advantageous, but one will have to be aware of the fact that some of these "unnatural" interferons could lead to immunological complications.

[81] P. K. Weck, S. Apperson, N. Stebbing, P. W. Gray, D. Leung, H. M. Shepard, and D. V. Goeddel, *Nucleic Acids Res.* **9,** 6153 (1981).

[82] S. Pestka, B. Kelder, E. Rehberg, J. R. Ortaldo, R. B. Herberman, E. S. Kempner, J. A. Moschera, and S. J. Tarnowski, *in* "The Biology of the Interferon System" (E. De Maeyer and H. Schellekens, eds.), p. 535. Elsevier, Amsterdam, 1983.

[83] W. Fiers, R. Devos, H. Cheroutre, W. Degrave, F. Duerinck, D. Gheysen, G. Plaetinck, E. Remaut, S. Scahill, G. Simons, P. Stanssens, J. Tavernier, and J. van der Heyden, *in* "The Biology of the Interferon System" (E. De Maeyer and H. Schellekens, eds.), p. 3. Elsevier, Amsterdam, 1983.

[84] S. Maeda, R. McCandliss, T.-R. Chiang, L. Costello, W. P. Levy, N. T. Chang, and S. Pestka, *in* "Developmental Biology Using Purified Genes" (D. Brown and C. F. Fox, eds.), p. 85. Academic Press, New York, 1981.

[85] P. B. Sehgal, L. T. May, and A. D. Sagar, *in* "The Biology of the Interferon System" (E. De Maeyer and H. Schellekens, eds.), p. 15. Elsevier, Amsterdam, 1983.

[86] L. S. Lin, R. Yamamoto, and R. J. Drummond, this volume [26].

[33] Analysis of Different Forms of Recombinant Human Interferons by High-Performance Liquid Chromatography

By ARTHUR M. FELIX, EDGAR P. HEIMER, and S. JOSEPH TARNOWSKI

The cloning and expression of the recombinant human leukocyte interferon A (IFN-αA) gene in *E. coli*[1] and subsequent purification by the use of immobilized monoclonal antibodies[2] has led to the scaled-up production[3] of IFN-αA for human clinical trials.[4] Analysis of the resultant IFN-αA by nonreducing SDS–polyacrylamide gel electrophoresis (SDS–PAGE) revealed the presence of higher oligomers (dimers, trimers, tetramers, etc.) and at least two monomeric forms.[5] The two major monomeric forms have been termed "slow" and "fast" migrating monomers (SMM and FMM, respectively) to designate their relative mobilities on SDS–PAGE.

Although disulfide linkages[6] in IFN-αA were determined to reside between Cys 1–Cys 98 and Cys 29–Cys 138, it has been suggested that only the Cys 29–Cys 138 disulfide bridge is required for biological activity.[7] Cys 1 and Cys 98 have been replaced[8] by *in vitro* DNA mutagenesis yielding only the SMM which suggested the involvement of Cys 1 and Cys 98 in intermolecular disulfide bonded oligomers. Thus, it appears that the FMM could represent a monomer with both disulfide bonds intact whereas the SMM represents a partially reduced form with only the Cys 29–Cys 138 bond intact. This mixture of monomers has recently been separated by a modification of the metal chelate chromatographic method of Porath.[9,10]

[1] D. V. Goeddel, E. Yelverton, A. Ullrich, H. L. Heyneker, G. Miozzari, W. Holmes, P. H. Seeburg, T. Dull, L. May, N. Stebbing, R. Crea, S. Maeda, R. McCandliss, A. Sloma, J. M. Tabor, M. Gross, P. C. Familletti, and S. Pestka, *Nature (London)* **287,** 311 (1980).

[2] T. Staehlin, D. S. Hobbs, H. F. Kung, C.-Y. Lai, and S. Pestka, *J. Biol. Chem.* **256,** 9750 (1981).

[3] S. J. Tarnowski and R. A. Liptak, *Adv. Biotechnol. Processes* **2,** 271 (1983).

[4] J. U. Gutterman, S. Fine, J. Quesada, S. J. Horning, J. F. Levine, R. Alexanian, L. Bernhardt, M. Kramer, H. Spiegel, W. Colburn, P. Trown, T. Merigan, and Z. Dziewanowska, *Ann. Intern. Med.* **96,** 549 (1982).

[5] S. Pestka, B. Kelder, D. K. Tarnowski, and S. J. Tarnowski, *Anal. Biochem.* **132,** 328 (1983).

[6] R. Wetzel, *Nature (London)* **289,** 606 (1981).

[7] R. Wetzel, H. L. Levine, D. E. Estell, and S. Shire, *J. Cell. Biochem., Suppl.* **6,** 89 (1982).

[8] T. DeChiara, F. Erlitz, and S. J. Tarnowski, this volume [58].

[9] J. Poráth, *Nature (London)* **258,** 598 (1975).

[10] E. Houchuli, personal communication.

Although the components of IFN-αA can be separated by SDS–PAGE, high-performance liquid chromatography (HPLC) is an alternative method which may be advantageous for routine analysis of multiple samples. Preparative high-performance liquid chromatography of human leukocyte interferon had been successfully carried out by reverse phase using LiChrosorb RP-8.[11,12] We have examined a number of analytical HPLC procedures for possible use in the rapid analysis of IFN-αA and the related proteins IFN-αD, IFN-αA/D, IFN-γ, as well as for chemically synthesized peptide fragments.

Apparatus and Materials

The HPLC system was comprised of two Constametric pumps (Models I and IIG), Gradient Master and a Spectromonitor III variable wavelength UV detector linked to a Model 3400 Recorder (all from Laboratory Data Control). Samples were introduced through a Rheodyne injector (#7125). The analytical columns [Synchropak C_{18} RP-P (0.41 × 25 cm) or (1 × 25 cm), SynChrom, Inc.] were used with a precolumn of Copell ODS pellicular packing (Whatman). The unretarded peaks correspond to the presence of ammonium acetate, NaCl, and Tween-20 detergent present in the samples.

Trifluoroacetic acid (TFA) (Chemical Dynamics, Sequalog grade) was freshly distilled and acetonitrile was HPLC grade (Fisher Scientific). Distilled water was further purified with a Hydro Model DC 1-18 system consisting of a 0.5 μm filter cartridge, organic adsorption unit, and mixed bed ion-exchange unit giving water with specific resistance in excess of 18 $M\Omega/cm^3$. Perchloric acid (70%) was of reagent grade purity (Fisher Scientific) and used to prepare a 0.1% aqueous solution which was adjusted to pH 2.5 with 1 *M* NaOH. All solutions were deaerated with helium prior to use.

IFN-αA (and oligomers), IFN-αD, and the IFN-αA/D hybrid were prepared and purified as previously described.[3,5,13] Immune interferon, IFN-γ, was prepared by a procedure similar to that previously reported.[14–16] Synthetic peptide fragments comprising residues (105–125) of IFN-αA and the fragment Gly (140–166) of IFN-αD were prepared by the

[11] M. Rubinstein, S. Rubinstein, P. C. Familletti, R. S. Miller, A. A. Waldman, and S. Pestka, *Proc. Natl. Acad. Sci. U.S.A.* **76,** 640 (1979).

[12] M. Rubinstein and S. Pestka, U.S. Patent 4,289,690 (1981).

[13] E. Rehberg, B. Kelder, E. G. Hoal, and S. Pestka, *J. Biol. Chem.* **257,** 11497 (1982).

[14] P. W. Gray, D. W. Leung, D. Pennica, E. Yelverton, R. Najarian, C. C. Simonsen, R. Derynck, P. J. Sherwood, D. M. Wallace, S. L. Berger, A. D. Levinson, and D. V. Goeddel, *Nature (London)* **295,** 503 (1982).

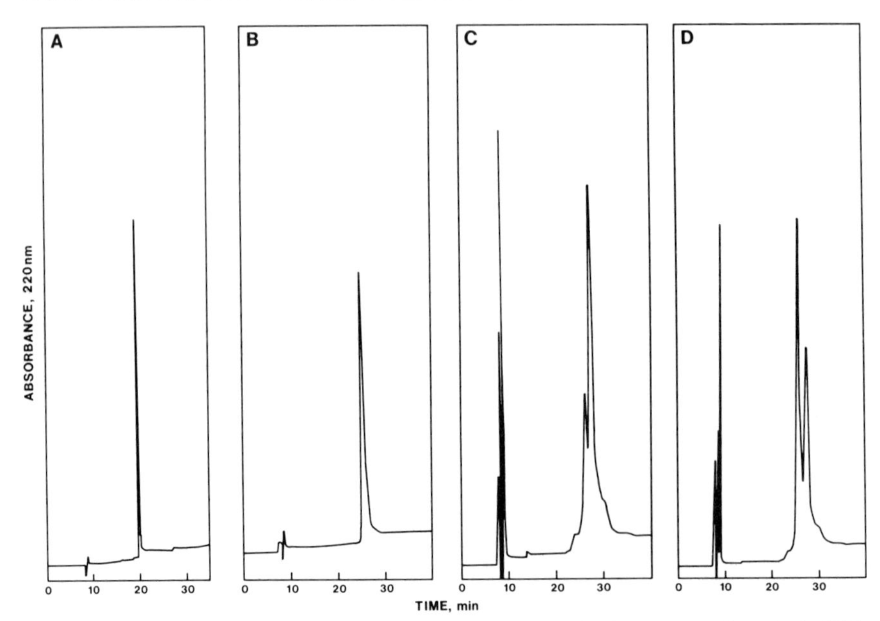

FIG. 1. High-performance liquid chromatographic analysis of (A) 10 μg of synthetic IFN-αA (105–125), (B) 16.4 μg of IFN-αA [purified, "fast moving monomer" (FMM)], (C) 40 μg of IFN-αA [partially reduced "slow moving monomer" (SMM)], and (D) admixture of IFN-αA (16.4 μg of FMM and 26.6 μg of SMM) on a Synchropak C_{18} RP-P (1 × 25 cm) column eluted with a linear gradient of 30–60% solution II for 30 min. Solution I consists of 0.025% trifluoroacetic acid (TFA) in H_2O; solution II consists of 0.025% TFA in CH_3CN. The linear gradient for (A) was from 20 to 50% solution II for 30 min. Detection was measured at 220 nm; flow rate: 2 ml/min; sensitivity: 0.2 Absorption Units Full Scale (AUFS).

solid phase procedure[17] and purified to homogeneity by preparative HPLC.[18]

HPLC Analysis of Human Interferons and Synthetic Peptide Fragments

Purified IFN-αA with both disulfide bonds intact (FMM) was found to be homogeneous by HPLC (Fig. 1B) with retention time of approximately

[15] R. Devos, H. Cheroutre, Y. Taya, W. Degrave, M. van Heuverswyn, and W. Fiers, *Nucleic Acids res.* **10,** 2487 (1982).

[16] R. Crowl, this volume [55].

[17] R. B. Merrifield, *J. Am. Chem. Soc.* **85,** 2149 (1963).

[18] A. M. Felix, E. P. Heimer, C.-T. Wang, T. J. Lambros, J. Swistok, M. Ahmad, M. Roszkowski, D. Confalone, and J. Meienhofer, *Pept: Struct. Funct., Proc. Am. Pept. Symp., 8th, 1983* p. 889 (1983).

26 min. The purity of FMM is in agreement with the detection of one major band for IFN-αA (FMM) by SDS–PAGE.[5] Analytical HPLC of the "slow moving monomer" (SMM) revealed the presence of 2 major components (Fig. 1C) with retention times of approximately 27 and approximately 28 min. Chromatograms of mixtures of IFN-αA (FMM) with IFN-αA (SMM) revealed that the earlier peak in SMM (approximately 27 min) did not correspond to FMM and it was concluded that FMM is not present in SMM (Fig. 1D). The synthetic peptide fragment corresponding to IFN-αA (105–125) emerged much earlier than the intact proteins under the same conditions of analytical HPLC. Modification of the gradient resulted in a retention time of approximately 20 min (Fig. 1A) and demonstrates the homogeneity of the synthetic peptide.

Purified IFN-αD, which was homogeneous by SDS–PAGE,[19] was shown to contain a single peak (retention time 29.5 min) by HPLC (Fig. 2B). A synthetic peptide fragment corresponding to IFN-αD, Gly (140–166), emerged earlier than the corresponding IFN-αD as observed above for the IFN-αA proteins (Fig. 1). Modification of the gradient resulted in retention time of approximately 24.5 min and established the homogeneity of the synthetic peptide.

The hybrid leukocyte interferon, IFN-αA/D, was partially purified by adsorption onto immobilized LI-8 monoclonal antibody,[2] and found to contain 1 major component by HPLC (Fig. 3A) with a retention time of approximately 27.5 min. A similar observation was made with a Synchropak C_{18} RP-P column (0.41 × 25 cm) with a 0.1% $HClO_4$ (pH 2.5)–acetonitrile gradient system. Purified immune interferon, IFN-γ (structure as in review by Pestka,[20] p. 25), known to be pure by SDS–PAGE, was shown to be homogeneous by HPLC and emerged earlier (retention time approximately 18.5 min) than IFN-αA, IFN-αD, or IFN-αA/D (Fig. 3B).

Oligomers of IFN-αA are known to consist of dimers, trimers, etc. containing intermolecular disulfide bonds involving Cys 1 and Cys 98 (e.g., Cys 1–Cys 1, Cys 1–Cys 98, and Cys 98–Cys 98).[8] Analytical HPLC of the oligomers was best achieved on the Synchropak C_{18} RP-P column with the 0.1% $HClO_4$ (pH 2.5)–acetonitrile gradient system. These oligomers (Fig. 4A) were shown to be complex mixtures containing numerous components. The earliest peak (retention time approximately 20.5 min), present in only minor amounts, was shown to correspond to FMM by admixture studies (Fig. 4D). The dimers (Fig. 4B) and trimers (Fig. 4C) can each be resolved into several components corresponding probably to

[19] S. J. Tarnowski, S. K. Roy, R. Y. Ning, D. K. Lee, R. A. Liptak, and D. K. Lee, this volume [23].

[20] S. Pestka, *Arch. Biochem. Biophys.* **221,** 1 (1983).

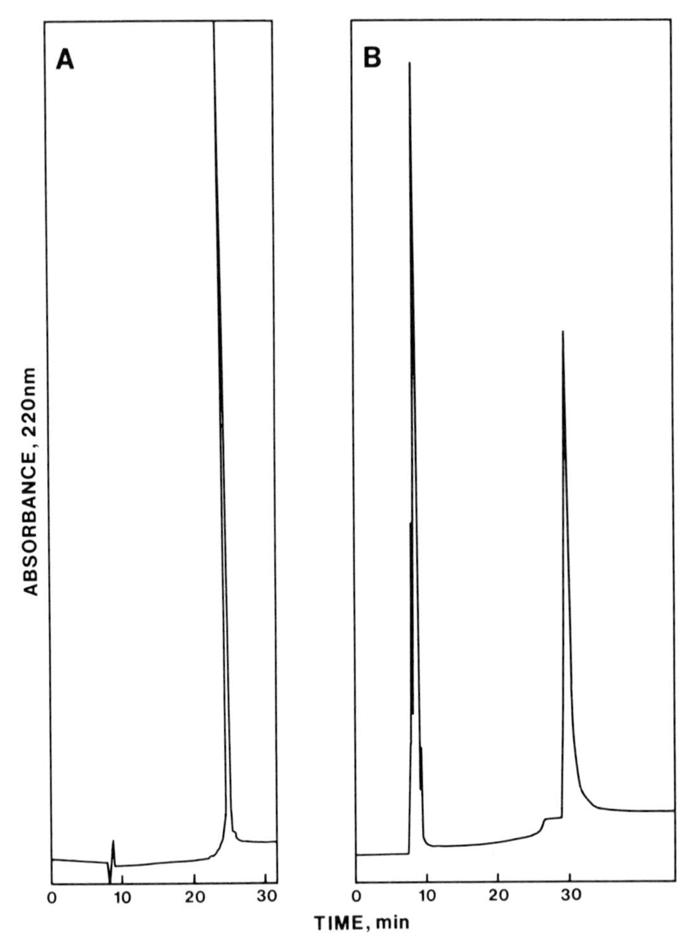

FIG. 2. High performance liquid chromatography of (A) 15 μg of a synthetic IFN-αD peptide: Gly-(140–166) and (B) 21 μg of IFN-αD on a Synchropak C_{18} RP-P (1 × 25 cm) column eluted with a linear gradient of 30–60% solution II for 30 min. Solution I consists of 0.025% TFA in H_2O; solution II consists of 0.025% TFA in CH_3CN. The linear gradient for (A) was from 20 to 50% solution II for 30 min. Detection was measured at 220 nm; flow rate: 2 ml/min; sensitivity: 0.2 AUFS.

the various combinations of disulfide bonds and are readily distinguished from the monomeric forms of IFN-αA.

Concluding Remarks

Analytical HPLC is a powerful tool for the evaluation of human leukocyte interferons (IFN-αA, IFN-αD), hybrids (IFN-αA/D), immune inter-

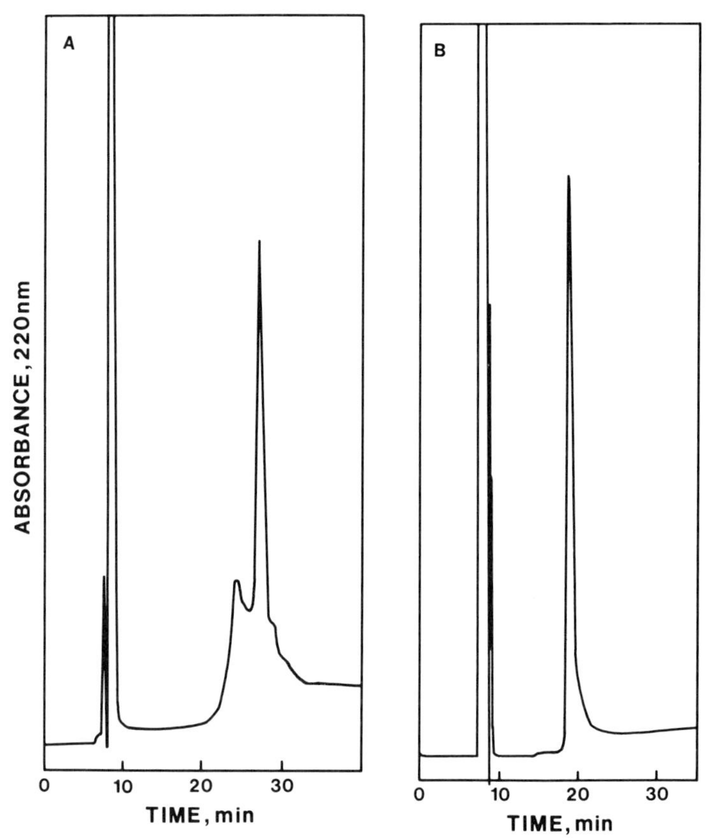

FIG. 3. High-performance liquid chromatography of (A) 45 μg of partially purified IFN-αA/D and (B) 40 μg of purified IFN-γ on a Synchropak C_{18} RP-P (1 × 25 cm) column eluted with a linear gradient of 30–60% solution II for 30 min. Solution I consists of 0.025% TFA in H_2O; solution II consists of 0.025% TFA in CH_3CN. Detection was measured at 220 nm; flow rate: 2 ml/min; sensitivity: 0.2 AUFS.

feron, (IFN-γ), as well as related synthetic peptide fragments. The three forms of human leukocyte interferon can be resolved and chromatograms of admixtures can be readily carried out to make positive identification of known impurities. In addition to determining the homogeneity of a sample, analytical HPLC can be used to measure quantitatively interconversion of forms, degradation products, and to carry out stability studies of human leukocyte interferons. Analytical HPLC can also be carried out at multiple wavelengths (e.g., 206 and 280 nm) to evaluate and distinguish the presence of proteins containing aromatic residues. The potential also exists for the extrapolation of these analytical systems to the preparative scale as an alternative to gel filtration and ion-exchange chromatography.

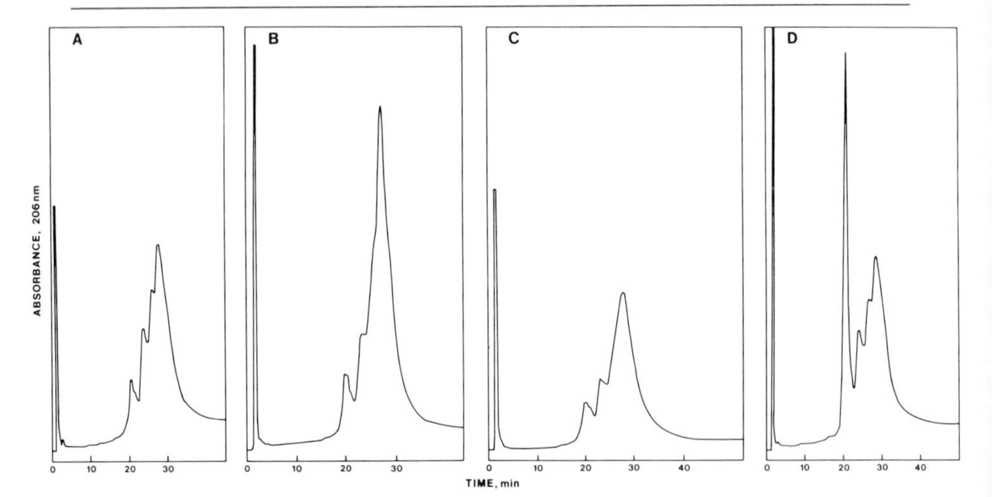

FIG. 4. High-performance liquid chromatographic analysis of (A) IFN-αA (oligomers), (B) IFN-αA (dimers), (C) IFN-αA (trimers), and (D) admixture of IFN-αA (oligomers) with FMM on a Synchropak C_{18} RP-P (0.41 × 25 cm) column eluted with a linear gradient of 35–50% solution II for 30 min. Solution I consists of 0.1% $HClO_4$ (pH 2.5) in H_2O; solution II consists of CH_3CN. Detection was measured at 206 nm; flow rate: 2 ml/min; sensitivity: 0.1 AUFS.

Acknowledgments

The authors are grateful to Dr. R. Y. Ning, Mr. D. K. Lee, and Mr. C.-H. Chen for providing purified samples of IFN-αA/D, IFN-αD, and IFN-γ.

[34] Procedure for Reduction and Reoxidation of Human Leukocyte Interferon

By JEROME A. LANGER and SIDNEY PESTKA

Considerably before the availability of purified interferons, it was observed that interferon activity was lost in the presence of disulfide-reducing agents.[1–5] It was also found that chaotropic agents and denaturants

[1] K. E. Mogensen and K. Cantell, *J. Gen. Virol.* **22,** 95 (1974).
[2] K. H. Fantes and C. F. O'Neill, *Nature (London)* **203,** 1048 (1964).
[3] T. C. Merigan, C. A. Winget, and C. B. Dixon, *J. Mol. Biol.* **13,** 679 (1965).
[4] Y. H. Ke and M. Ho, *Proc. Soc. Exp. Biol. Med.* **129,** 433 (1968).
[5] L. W. Marshall, P. M. Pitha, and W. A. Carter, *Virology* **48,** 607 (1972).

protected interferon activity in the presence of reducing agents and restored activity lost by reduction or heat denaturation.[1,6–10] These observations are consistent with the often utilized procedure of recovering activity from SDS–polyacrylamide gels after the interferon samples have been boiled in SDS containing high concentrations of 2-mercaptoethanol (see, e.g., Ref. 6).

The sequencing of the proteins or cDNAs corresponding to the proteins of various human and murine leukocyte interferons has demonstrated that there are usually 4 or 5 cysteine residues in these interferons (usually 4 in humans; thus far usually 5 in murine species, though there are fewer murine sequences available) and that their positions in the polypeptide are conserved (reviewed in Refs. 11–14). The disulfide linkages in IFN-αA have been determined[15,16] and it has been demonstrated that the disulfide linking cysteines 29 and 138 seems necessary for optimal activity.[17,18]

The reduction and reoxidation of IFN is of interest both in understanding the structure–function relationship of the disulfides in IFN and toward the practical goal of recovering or regenerating active interferon after reductive procedures. Three procedures are described here: (1) the reduction and (presumably) direct reoxidation of IFN-αA; (2) the reduction and reversible covalent modification of the sulfhydryls by sulfitolysis, with subsequent reoxidation (modified from Ref. 17); and (3) the irreversible modification of sulfhydryls by reaction with iodoacetate or iodoacet-

[6] W. E. Stewart, II, E. De Clercq, and P. DeSomer, *Nature (London)* **249,** 460 (1974).

[7] W. E. Stewart, II, P. DeSomer, and E. De Clercq, *Biochim. Biophys. Acta* **359,** 364 (1974).

[8] W. E. Stewart, II, P. De Somer and E. De Clercq, *J. Gen. Virol.* **24,** 567 (1974).

[9] R. J. Jariwalla, J. J. Sedmak, and S. E. Grossberg, *Experientia* **36,** 1390 (1980).

[10] J. J. Sedmak and S. E. Grossberg, *Tex. Rep. Biol. Med.* **41,** 274 (1982).

[11] S. Pestka, *Sci. Am.* **249** (2), 37 (1983).

[12] C. Weissmann, S. Nagata, W. Boll, M. Fountoulakis, A. Fujisawa, J.-I. Fujisawa, J. Haynes, K. Henco, N. Mantei, H. Ragg, C. Schein, J. Schmid, G. Shaw, M. Streuli, H. Taira, K. Todokoro, and U. Weidele, *Philos. Trans. R. Soc. London, Ser. B* **299,** 7 (1982).

[13] G. D. Shaw, W. Boll, H. Taira, N. Mantei, P. Lengyel, and C. Weissmann, *Nucleic Acids Res.* **11,** 555 (1983).

[14] B. Daugherty, D. Martin-Zanca, B. Kelder, K. Collier, T. C. Seamans, K. Hotta, and S. Pestka, *J. Interferon Res.* **4,** 635 (1985).

[15] R. Wetzel, *Nature (London)* **289,** 606 (1981).

[16] R. Wetzel, L. J. Perry, D. A. Estell, N. Lin, H. L. Levine, B. Slinker, F. Fields, M. J. Ross, and J. Shively, *J. Interferon Res.* **1,** 381 (1981).

[17] R. Wetzel, H. L. Levine, D. A. Estell, S. Shire, J. Finer-Moore, R. M. Stroud, and T. A. Bewley, *in* "Interferons" (T. C. Merigan, R. M. Friedman, and C. F. Fox, eds.), p. 365. Academic Press, New York, 1983.

[18] H. Morehead, P. D. Johnston, and R. Wetzel, *Biochemistry* **23,** 2500 (1984).

tamide. Many of the procedures and results presented here are similar to those previously reported.[17,18]

Materials

Recombinant human leukocyte interferon A (IFN-αA), isolated as previously described.[19,20] Phosphate-buffered saline (PBS): Dulbecco's, lacking Ca^{2+} and Mg^{2+}.

Guanidine hydrochloride (GuHCl; Schwarz/Mann Ultrapure): 8 *M* stock solution in PBS or 0.1 *M* sodium phosphate, pH 7.0, filtered through a 0.20-μm filter.

Urea: 10 *M* stock solution (Schwarz/Mann Ultrapure) in PBS or 0.1 *M* sodium phosphate, pH 7.0, filtered through a 0.20-μm filter.

2-Mercaptoethanol: Bio-Rad, Electrophoresis grade.

Redox mixture: 1.0 m*M* reduced glutathione and 0.1 m*M* oxidized glutathione (both from Sigma).

Sulfitolysis solution (final concentrations after addition of protein): sodium phosphate, pH 8.0, 0.1 *M;* disodium ethylenediaminetetraacetate (Na_2EDTA), 1.0 m*M;* sodium sulfite (Na_2SO_3), 20 mg/ml; sodium tetrathionate ($Na_2S_4O_6 \cdot 2H_2O$, Pierce), 10 mg/ml.

Analytical Procedures and Assays

The viral cytopathic effect (CPE) inhibition assay was performed on bovine MDBK cells or human WISH cells, as described previously.[21] The "sandwich" radioimmunoassay with monoclonal antibodies LI-1 and LI-8 or LI-9 was performed as described.[22] Interferon samples were prepared for antiviral assay or radioimmunoassay (RIA) by dilution into a solution of 0.1% bovine serum albumin (BSA) in PBS. SDS–polyacrylamide gel electrophoresis (SDS–PAGE) was performed in 12.5% acrylamide gels as described.[23] In this system, the 2-mercaptoethanol-reduced form of IFN-αA migrates more slowly than the nonreduced form when 2-mercaptoethanol is omitted from the sample buffer (it is prudent to leave at least one empty lane between samples containing 2-mercaptoethanol and those lacking it). This faster moving nonreduced form will be referred to as the "oxidized" form.

[19] T. Staehelin, D. S. Hobbs, H.-F. Kung, and S. Pestka, this series, Vol. 78, p. 505.

[20] T. Staehelin, D. S. Hobbs, H.-F. Kung, C-Y. Lai, and S. Pestka, *J. Biol. Chem.* **256,** 9750 (1981).

[21] P. C. Familletti, S. Rubinstein, and S. Pestka, this series, Vol. 78, p. 387.

[22] T. Staehelin, C. Stähli, D. S. Hobbs, and S. Pestka, this series, Vol. 79, p. 589.

[23] U. K. Laemmli, *Nature* (*London*) **227,** 680 (1970).

TABLE I
CHARACTERIZATION OF NATIVE, REDUCED, AND OXIDIZED IFN-αA[a]

Treatment	Dialysis	Antiviral activity (%)	RIA (%)	Analysis by SDS–PAGE
None	—	100	100	Oxidized
Reduced	—	0.1	0.7	Reduced
Reduced	#1	0.004	0.3	Reduced
Reduced + 6 *M* GuHCl	—	80	84	Oxidized
Reduced + 6 *M* GuHCl	#2	40	25	Oxidized

[a] Dialysis buffers were as follows: #1, 0.1 *M* sodium phosphate, pH 7.0; #2, same as #1 plus redox mixture.

Reduction and Oxidation of Interferon

The incubation of IFN-αA with 5% (v/v) 2-mercaptoethanol at 37° for 1 hr leads to precipitation of the protein. This material is inactive in the CPE assay and is not recognized in the sandwich radioimmunoassay (Table I). Dialysis of this material against PBS or against PBS containing the redox mixture had no effect on restoring activity.

Direct addition of 0.1 volume of 10% SDS (w/v, in water) to the IFN in the presence of 2-mercaptoethanol causes the precipitate to dissolve immediately. This material, suitably diluted into 0.1% BSA/PBS, was active in both the RIA and CPE assay. SDS at a final concentration of 0.05–0.1% could restore most of the activity to reduced IFN-αA.

Similarly, direct addition of GuHCl to a final concentration of 6 *M* was effective in restoring the activity of reduced IFN-αA[24] (Table I, line 4). Concentrations of less than 4–5 *M* GuHCl had reduced effectiveness. Urea, at up to 8 *M*, was not as effective as 6 *M* GuHCl and produced a reactivation of 15–30% of the starting activity.

It was desirable to find conditions whereby the denaturing agents which were added to facilitate reoxidation could be subsequently eliminated. We focused on the use of GuHCl, since it can be removed simply by dialysis, whereas the strong binding of SDS makes it more difficult to eliminate quantitatively. As mentioned above, the direct dialysis of reduced IFN-αA against PBS did not lead to the restoration of activity. However, following addition of GuHCl to the reduced IFN, dialysis

[24] S. Pestka, B. Kelder, J. A. Langer, and T. Staehelin, *Arch. Biochem. Biophys.* **224,** 111 (1983).

against PBS resulted in restoration of 20–40% of the input activity; a precipitate reappeared during this dialysis and may partially account for the incomplete recovery of activity. Most of the activity could be regained by dialyzing the reduced protein in 6 *M* GuHCl against PBS or 0.1 *M* sodium phosphate, pH 7.0, containing the redox mixture (Table I, line 5). Slightly more consistent results were obtained when 6 *M* GuHCl was included in the first stage of dialysis together with the redox mixture. Not only was the activity regained by this procedure, but the IFN migrated as the oxidized form on SDS gels. Finally, this renatured and dialyzed IFN appeared identical to native IFN on high-performance liquid chromatography (HPLC).

Comments

The procedure described here is a simple one for the regeneration of active, presumably native IFN-αA from the inactive reduced form. We have found, for example, that the activity can be restored to heat-inactivated IFN-αA by addition of 2-mercaptoethanol and 1% SDS.[24] A further use might be in obtaining fully active IFN when it is produced in high yield in *E. coli* or other organisms.

Sulfitolysis of IFN-αA and Reoxidation

The formation of the tetrasulfonate form of IFN-αA by Wetzel *et al.*[17,18] (see also Refs. 24, 25) and the controlled reoxidation to the disulfonate form containing the single disulfide Cys 29–Cys 138 was productively exploited by those workers to demonstrate the functional requirement for that disulfide and the lack of such a requirement for the Cys 1–Cys 98 disulfide. We have modified the procedure of Wetzel *et al.*[17] to produce the completely renatured (oxidized) form of IFN-αA.

Sulfitolysis was carried out by adding IFN-αA to the sulfitolysis solution for a final protein concentration of 1 mg/ml. The reaction proceeded at room temperature (21–23°) for 16 hr. Reagents were eliminated by dialysis against 200 volumes of 0.05 *M* sodium phosphate, pH 8.0, at 23° for 10 hr. This material was characterized as shown in Table II (also Ref. 24), and on this basis was taken to be the tetrasulfonate derivative.

Complete reoxidation was accomplished by dialyzing the tetrasulfonate derivative against 200 volumes of 0.05 *M* sodium phosphate, pH 8.0, containing 1 m*M* EDTA and the redox mixture. The reoxidized material was similar to the starting IFN in terms of its antiviral activity and immu-

[25] P. G. Katsoyannis, A. Tometsko, C. Zalut, S. Johnson, and A. C. Trakatellis, *Biochemistry* **6,** 2635 (1967).

TABLE II
CHARACTERIZATION OF TETRASULFONATE AND REGENERATED IFN-αA[a]

Treatment	Dialysis	Antiviral activity (%)	Analysis by SDS–PAGE	Analysis by HPLC (retention time) (min)
None	—	100	Oxidized	24
Tetrasulfonate	#1	2	Reduced	30
"Reoxidized" from tetrasulfonate	#2	75	Oxidized	24

[a] Dialysis buffers were as follows: #1, 0.1 *M* sodium phosphate, pH 8.0; #2, same as #1 plus redox mixture. It should be noted that the low activity of the tetrasulfonate was measured on bovine MDBK cells. When the tetrasulfonate derivative was assayed on human WISH cells, it retained 25–40% of the activity of unmodified IFN-αA.

nological recognition in the sandwich RIA (not shown), and identical to the starting IFN in its migration under nonreducing conditions on an SDS gel and retention time on HPLC.

Irreversible Derivatization of the Disulfides

The tetrasulfonate was found to have low but measurable activity in the antiviral assay on MDBK cells (but see footnote to Table II). Although this derivative was stated by Wetzel *et al.*[17] to be "inactive," his data also demonstrate that it exhibits low activity in the antiviral assay. Therefore, to determine whether the tetrasulfonate has low intrinsic activity or whether an alternative explanation might account for this residual activity (for example, a partial reoxidation of the tetrasulfonate during the assay by glutathione or other agents in the serum) irreversible derivatization of the thiol groups was performed. To eliminate the possibility of reversibility, IFN-αA was irreversibly modified by reaction with iodoacetate and iodoacetamide[26,27] and the derivatized IFN-αA was tested for activity.

IFN-αA was adjusted to 0.2 mg/ml in 0.1 *M* sodium phosphate, pH 8.0. The protein was denatured and reduced by the addition of 3 vol of 8 *M* GuHCl (in 0.1 *M* sodium phosphate, pH 8.0), followed by Na_2EDTA and dithiothreitol (DTT) to a final concentration of 1 and 10 m*M*, respectively. Argon was bubbled gently through the solution and the tube was flushed with argon, sealed, and incubated at 55° for 4 hr.

[26] G. E. Means and R. E. Feeney, "Chemical Modification of Proteins." Holden-Day, San Francisco, California, 1971.

[27] A. N. Glazer, R. J. DeLange, and D. S. Sigman, "Chemical Modification of Proteins." North-Holland Publ., New York, 1975.

Iodoacetate (Sigma) and iodoacetamide (Sigma) were made up as stocks of 150 m*M* in 8 *M* GuHCl. Each reagent was added at a final concentration of 10 m*M* to the IFN-αA solution. After an hour at room temperature, 2-mercaptoethanol was added to a concentration of 5% (v/v) (0.71 *M*). Each reaction was then dialyzed against 250 volumes of 0.1 *M* sodium phosphate, pH 8.0, or 6 *M* GuHCl in 0.1 *M* sodium phosphate, pH 8.0, both buffers containing the glutathione redox mix. This dialysis was done so that any unreacted IFN would be reoxidized and would be observable both in terms of its bioactivity and its position on an SDS gel (see previous section). Analysis of the products is summarized in Table III. On the SDS gel, no sample migrated as native oxidized interferon. Bioactivity was about 0.2–1.3% of untreated IFN and about 0.6–4% of reduced and reoxidized material. Samples which were finally dialyzed in the absence of GuHCl had about 8% activity in the RIA, whereas those which were dialyzed in the presence of GuHCl had about 15–20% activity in the RIA relative to untreated IFN or almost 45% activity relative to reduced and reoxidized IFN. However, the derivatized IFN which had been dialyzed migrated as a reduced form on a nonreducing gel, with no material evident as the faster oxidized form. This eliminates the possibility that the significant material recognized by the antibodies was oxidized. Hence, the antibodies must be capable of recognizing to some degree the reduced and derivatized (but soluble) IFN. It cannot be excluded, however, that the low bioactivity was from a small amount of reoxidized IFN-αA which could have been undetected on the gel. Thus, it is possible that neither disulfide bond is strictly required for activity although maximum activity is seen with an intact Cys 29–Cys 138 disulfide bridge.[17,18] The isolation of

TABLE III
CHARACTERIZATION OF IFN-αA REACTED WITH IODOACETATE OR IODOACETAMIDE[a]

Treatment	Dialysis	Antiviral activity (%)	RIA (%)	Analysis by SDS–PAGE
None	—	100	100	Oxidized
Reduced then reoxidized	#2	38	39	Oxidized
Iodoacetate	#2	0.2	7	Reduced
Iodoacetate	#3	0.4	18	—
Iodoacetamide	#2	0.6	8	Reduced
Iodoacetamide	#3	1.0	17	—

[a] Dialysis buffers: #1, 0.1 *M* sodium phosphate, pH 8.0; #2, as #1 plus redox mixture; #3, 6 *M* GuHCl in Buffer #2.

an active fragment of interferon containing only the first 129 residues of IFN-αA by Ackerman *et al.*[28] is consistent with these results.

Concluding Remarks

The importance of the disulfides for maximal interferon activity has been clear from the early experiments. The procedures presented here give several methods for their manipulation, including the controlled recovery of the native-like active form from the completely reduced insoluble form. Unlike the reduced form, which is insoluble at the concentrations used here, the various derivatized forms remain soluble and thus amenable to standard biochemical manipulations. It is likely that future studies on the role of disulfides will be enhanced by the synthesis of proteins of modified sequence generated by oligonucleotide-directed site-specific mutagenesis of the corresponding DNA coding sequences.[29]

Acknowledgments

We gratefully acknowledge Drs. S. J. Tarnowski (Biopolymer Research Department, Hoffmann-La Roche Inc.), R. Wetzel (Genentech), and A. S. Acharya (Rockefeller University) for useful discussions, Dr. E. Heimer (Chemical Research Division, Hoffmann-La Roche Inc.) for HPLC analysis of samples, Mr. Bruce Kelder, Mrs. Judith Altman, and Mrs. Cynthia Rose for bioassays, and Drs. S. J. Tarnowski, C. McGregor, and R. Ning for samples of IFN-αA.

[28] S. K. Ackerman, D. Zur Nedden, M. Heintzelman, M. Hunkapiller, and K. Zoon, *Proc. Natl. Acad. Sci. U.S.A.* **81,** 1045 (1984).
[29] T. M. DeChiara, F. Erlitz, and S. J. Tarnowski, this volume [58].

[35] Radiation Inactivation and Target Size Analysis of Interferons

By ELLIS S. KEMPNER and SIDNEY PESTKA

The conventional methods of determining the size of macromolecules depend on physical properties of the structures. Measurements of hydrodynamic or electrophoretic mobility, changes in osmotic pressure, or scattering of light are related to molecular weight. None of these measurements has any direct connection with the biological activity, if any, of these molecules. Furthermore, the presence of inactive molecules or contaminating materials confounds analysis by these methods.

Recently, there has been a resurgence of interest in an alternative technique of molecular size determination: radiation inactivation. This is undoubtedly stimulated by the unique principles and properties of this approach: that the loss of biological activity resulting from molecular damage is related to the mass of the structure, and the fact that the results are independent of the presence of other molecules. The size measurement obtained from radiation inactivation must be interpreted as the mass of a biologically active unit.

Principles and Theory

All ionizing radiation, whether X rays, γ rays, or high energy particles, travel unabated through free space; only when they encounter mass is there any interaction. The greater the mass, the greater the chance of an impact. For the γ rays and high energy electrons used in target analysis, most of the interactions involve orbital electrons. These interactions distribute randomly in the mass of the irradiated material and occur in discrete events called primary ionizations. At each occurrence, there is a transfer of radiation energy to the target molecule; the average amount of energy in these events is 60 eV, or 1500 kcal/mol. In frozen and in lyophilized material, the absorption and dissipation of this energy are principally confined to the covalent structure in which the primary ionization occurs.[1] The absorption of such a large amount of energy results in many changes in the molecules; some of these result in irreversible damage which occurs throughout the polymer.[2] If all the absorbed energy were involved in bond breakage, 20–50 covalent bonds would be lost in a typical polypeptide and 4–7 of these would be in the backbone strand. It is because of this extensive damage that macromolecules lose their biological activity (enzyme action, ligand binding, infectivity, etc.). This loss is complete; there are no partially active molecules remaining. This can be seen by kinetic analysis of enzyme activity: irradiated enzymes show a drop in V_{max} because fewer active molecules remain, but there is no change in K_m[3,4] because the intrinsic activity of each surviving molecule is unaltered. In ligand-binding studies, Scatchard analysis shows that irradiated receptor preparations have fewer binding sites (reduced B) but the

[1] E. S. Kempner and J. H. Miller, *Science* **222,** 586 (1983).

[2] D. L. Aronson and J. W. Preiss, *Radiat. Res.* **16,** 138 (1962).

[3] D. J. Fluke, *Radiat. Res.* **51,** 56 (1972).

[4] S. Ferguson-Miller *et al.*, as quoted in reference 16.

binding constant is unaffected.[5,6] Similarly, there are no partial structures with full activity.

In frozen and lyophilized material, the energy is absorbed within the individual molecules. Thus, neighboring compounds are unaffected. Those molecules which have escaped direct interaction with radiation are undamaged; their structure remains intact and their activity remains complete. The only detectable activity in an irradiated sample is from molecules which have not been hit.

The fact that the target molecules are hit at random, depending only on their mass, and that the energy is confined to a single covalent structure leads to an important characteristic of the target analysis. The study of a particular activity is independent of the presence of other molecules in the preparation if they are not functional in the assay conditions. Thus, purification is not required. A sequela of this fact is that a single irradiated sample can be monitored for as many independent biological functions as desired, and target sizes for each activity can be obtained simultaneously. Enzymes of known size which are present in the preparation can be monitored as an internal control.

The randomness of the destruction leads to a Poisson distribution of the number of hits per unit as a function of radiation exposure. The surviving activity is dependent on the probability of zero hits, which is a simple exponential decay. Therefore, the inactivation curve, the fraction of surviving activity as a function of the radiation dose, is a straight line on a semilogarithmic plot. The slope of this line, k, gives the molecular weight from the relationship[7]

$$MW = 6.4 \times 10^{11} S_t k$$

where 6.4×10^{11} derives from conversion of units and the average energy per primary ionization, and S_t is a factor dependent on the temperature at which the irradiation was performed.[8] The temperature correction factor S_t was experimentally determined to be 2.8 for irradiation at $-135°$[8,9] at which temperature irradiation of interferon samples was performed.[10] The

[5] T. L. Innerarity, E. S. Kempner, D. Y. Hui, and R. W. Mahley, *Proc. Natl. Acad. Sci. U.S.A.* **78,** 4378 (1981).

[6] C. J. Steer, E. S. Kempner, and G. Ashwell, *J. Biol. Chem.* **256,** 5851 (1981).

[7] T. B. Nielsen, P. M. Lad, M. S. Preston, E. Kempner, W. Schlegel, and M. Rodbell, *Proc. Natl. Acad. Sci. U.S.A.* **78,** 722 (1981).

[8] E. S. Kempner and H. T. Haigler, *J. Biol. Chem.* **257,** 13297 (1982).

[9] W. Schlegel, E. S. Kempner, and M. Rodbell, *J. Biol. Chem.* **254,** 5168 (1979).

[10] S. Pestka, B. Kelder, P. C. Familletti, J. A. Moschera, R. Crowl, and E. S. Kempner, *J. Biol. Chem.* **258,** 9706 (1983).

slope of the inactivation curve, k, is determined as $[D_{37}]^{-1}$ where D_{37} is the radiation dose, in rads, which reduced measured activity to 37% of that found in the unirradiated controls. Thus, the target molecular weight is calculated by the following equation:

$$M_r = (1.79 \times 10^{12})/D_{37}$$

The generality of the principles of target analysis permits the technique to be used for any sample which retains activity after freezing and thawing. Any function which can be quantitatively determined in irradiated samples can be studied, and a simple exponential loss of activity can be taken as evidence for a single-sized target. The calculated target size is interpreted in terms of the concept of a functional unit.[11] It is the mass of the minimal assembly of structures capable of performing the measured function.

Several examples of complex radiation inactivation curves have appeared.[9,12–14] One type exhibits a multiphasic decrease in activity and may be due to two or more structures, each of which independently can perform the measured activity.[15] The other type involves an initial increase in measurable activity after low doses of radiation, followed by a decrease at higher radiation doses. These have been interpreted as evidence for an inhibitor or of an inactive complex of activatable subunits.

Methods

Details of the experimental procedures in choice and measurement of radiations, preparation and handling of samples, measurements of activity, and analysis of data appear in another volume of this series.[16]

Applications to Interferons

The three classes of human interferons (IFN-α, IFN-β, and IFN-γ) have monomer molecular weights between 17,000 and 20,000 excluding

[11] E. S. Kempner, J. H. Miller, W. Schlegel, and J. Z. Hearon, *J. Biol. Chem.* **255,** 6826 (1980).

[12] J. T. Harmon, C. R. Kahn, E. S. Kempner, and W. Schlegel, *J. Biol. Chem.* **255,** 3412 (1980).

[13] R. L. Kincaid, E. Kempner, V. C. Manganiello, J. C. Osborne, Jr., and M. Vaughn, *J. Biol. Chem.* **256,** 11351 (1981).

[14] M. Nielsen, T. Honore, and C. Braestrup, *Biochem. Pharmacol.* **32,** 177 (1983).

[15] E. S. Kempner and W. Schlegel, *Anal. Biochem.* **92,** 2 (1979).

[16] J. T. Harmon, T. B. Nielsen, and E. S. Kempner, this series, Vol. 117, p. 65.

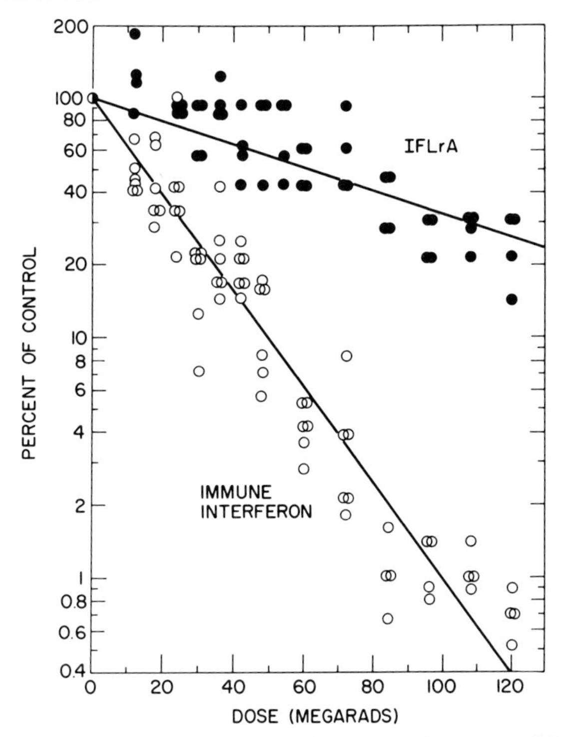

FIG. 1. Irradiation of recombinant human leukocyte (IFN-αA) and immune (IFN-γ) interferons. Data replotted from that of Pestka *et al.*[10]

carbohydrate.[17] Nevertheless, oligomeric forms of all these molecules have been described.[18–24] To determine whether there was any functional significance to the multimeric forms, target analysis was performed with

[17] S. Pestka, *Arch. Biochem. Biophys.* **221,** 1 (1983).

[18] T. Staehelin, D. S. Hobbs, H. Kung, C.-Y. Lai, and S. Pestka, *J. Biol. Chem.* **256,** 9750 (1981).

[19] S. Pestka, B. Kelder, D. K. Tarnowski, and S. J. Tarnowski, *Anal. Biochem.* **132,** 328 (1983).

[20] H.-J. Friesen, S. Stein, M. Evinger, P. C. Familletti, J. Moschera, J. Meienhofer, J. Shively, and S. Pestka, *Arch. Biochem. Biophys.* **206,** 432 (1981).

[21] E. Knight, Jr. and D. Fahey, *J. Biol. Chem.* **256,** 3609 (1981).

[22] P. M. Langford, J. A. Georgiades, G. J. Stanton, F. Dianzani, and H. M. Johnson, *Infect. Immun.* **26,** 36 (1979).

[23] M. De Ley, J. van Damme, H. Claeys, H. Weening, J. W. Heine, A. Billiau, C. Vermylen, and P. De Somer, *Eur. J. Immunol.* **10,** 877 (1980).

[24] Y. K. Yip, R. H. L. Pang, C. Urban, and J. Vilček, *Proc. Natl. Acad. Sci. U.S.A.* **78,** 1601 (1981).

TARGET MOLECULAR WEIGHTS OF INTERFERONS[a]

Interferon	Target M_r ±2 SE	Monomer M_r
Leukocyte		
IFN-α (crude)	20,000 ± 2200	17,000–24,000
IFN-αA (recombinant, pure)	23,000 ± 2200	19,219
IFN-αA (recombinant, pure)	20,000 ± 2300	19,219
Fibroblast		
IFN-β (crude)	31,000 ± 2000	~24,000
IFN-β (crude)	42,000 ± 1900	~24,000
IFN-β (pure, native)	42,000 ± 3200	~24,000
IFN-β (recombinant, pure)	33,000 ± 2300	20,004
Immune		
IFN-γ (crude)	63,000 ± 2800	20,000;25,000
IFN-γ (crude)	66,000 ± 4100	20,000;25,000
IFN-γ (recombinant)	73,000 ± 5800	17,126

[a] The target molecular weight (M_r) is given ±2 SE. The standard error (SE) of the molecular weight was calculated from the standard error of the slope of the inactivation curve. In addition, there are sources of error in the values for the radiation dose and the temperature coefficient, which would not affect the comparison of values in the table, but would influence the accuracy of the calculation. Consideration of these factors leads us to assume 15% is a reasonable estimate of the overall uncertainty in the molecular weight. Data from Pestka *et al.*[10]

natural and recombinant forms of leukocyte, fibroblast, and immune interferons.[10] An example of the data obtained is shown in Fig. 1. It is clear that the target sizes for leukocyte and immune interferons are strikingly different. The target analysis for all three classes of interferons are summarized in the table. These results have led to the substantive conclusion that leukocyte, fibroblast, and immune interferons exhibit target sizes representative of monomer, dimer, and tetramer, respectively.[10] The results seem to be generally consistent with their minimal molecular weights as determined by gel permeation chromatography under nonreducing conditions. Nevertheless, it is not yet clear how these functional sizes relate to the kinetics and thermodynamics of their receptor interactions that lead to generation of the antiviral state.

Section V

Radiolabeling of Interferons

[36] Radiolabeling of Human Leukocyte and Immune Interferons with ^{125}I and Lactoperoxidase

By FAZLUL H. SARKAR and SOHAN L. GUPTA

Introduction

The advent of purified natural and recombinant interferons (IFNs) has made it possible to develop radioimmune assays for them, and to carry out structural and functional studies which were otherwise not feasible. IFNs radiolabeled with ^{125}I are used for radioimmune assays and studies on cell receptors for IFNs. The procedures most commonly employed for radioiodination of proteins make use of chloramine T, lactoperoxidase, or Bolton–Hunter reagent. There are other procedures as well. The choice of the procedure may depend on factors such as the desired specific radioactivity of the labeled product, chemical and biological stability of the sample under the radiolabeling procedure, and the relative cost of the procedure. The chloramine T procedure has been used extensively for radioiodination of polypeptides, such as hormones, to high specific activities. However, during this procedure, the samples are exposed to oxidizing and reducing conditions which may have adverse effects, possibly leading to artifacts. For example, in receptor studies with ^{125}I-labeled epidermal growth factor, it was observed that a fraction of the ^{125}I-labeled hormone bound to cell receptors formed nondissociable, apparently covalent, complexes with the receptor if the hormone was iodinated by the chloramine T procedure. This observation was attributed to the chloramine T procedure used for labeling since iodination with lactoperoxidase did not yield such complexes.[1] Human leukocyte IFN has been successfully labeled by the chloramine T procedure with retention of biological activity.[2,3] Labeling with lactoperoxidase and Bolton–Hunter reagent are generally recognized as mild procedures and have been used for the labeling of IFNs. Chloramine T and lactoperoxidase procedures catalyze the iodination mostly at tyrosine residues in proteins being labeled, whereas with Bolton–Hunter reagent, a radioiodinated phenyl group is transferred to free amino groups of the protein.

[1] P. G. Comens, R. L. Simmer, and J. B. Baker, *J. Biol. Chem.* **257,** 42 (1982).

[2] K. E. Mogensen, M.-T. Bandu, F. Vignaux, M. Aguet, and I. Gresser, *Int. J. Cancer* **28,** 575 (1981).

[3] K. E. Mogensen and G. Uzé, this volume [37].

We have made use of lactoperoxidase to radioiodinate recombinant Hu-IFN-α2,[4] and more recently recombinant Hu-IFN-γ for receptor studies. Lactoperoxidase, in the presence of H_2O_2, catalyzes the oxidation of iodide to iodine which reacts with phenolic groups of proteins.[5,6] The requirement for H_2O_2 can be replaced by glucose and glucose oxidase[7] which generate a slow and steady supply of H_2O_2. Use of lactoperoxidase coupled to a solid support makes it easy to remove the enzyme after radioiodination.[8] Lactoperoxidase and glucose oxidase coupled to solid support are commercially available (Enzymobeads, Bio-Rad). A procedure for the radioiodination of recombinant Hu-IFN-α2 and Hu-IFN-γ is described here.

Materials

Interferon (to be labeled)
^{125}I-labeled sodium iodide (carrier-free, high concentration, high pH, New England Nuclear)
Immobilized lactoperoxidase–glucose oxidase (Enzymobeads, Bio-Rad).
0.03 *N* HCl
0.2 *M* sodium phosphate buffer, pH 7.2
1 *M* sodium azide
β-D-Glucose (Sigma)
Gelatin or bovine serum albumin
L-Tyrosine
Sephadex G-50 (Pharmacia; other grades such as G-25 or G-75 can also be used)

Procedure

1. Enzymobeads are rehydrated in 0.1 *M* sodium phosphate buffer, pH 7.2, at least 1 hr before use.

2. ^{125}I (obtained as NaI in 0.1 *N* NaOH, New England Nuclear) is neutralized (under an exhaust hood) by adding 3 volumes of 0.03 *N* HCl and the volume is brought to 25 μl with 0.2 *M* sodium phosphate buffer, pH 7.2 (e.g., to 2 mCi of ^{125}I in 4 μl, 12 μl of 0.03 *N* HCl, and 9 μl of 0.2 *M*

[4] A. R. Joshi, F. H. Sarkar, and S. L. Gupta, *J. Biol. Chem.* **257,** 13884 (1982).
[5] J. J. Marchalonis, *Biochem. J.* **113,** 299 (1969).
[6] M. Morrison and G. S. Bayse, *Biochemistry* **9,** 2995 (1970).
[7] A. L. Hubbard and Z. A. Cohn, *J. Cell Biol.* **55,** 390 (1972).
[8] G. S. David and R. A. Reisfeld, *Biochemistry* **13,** 1014 (1974).

sodium phosphate buffer are added). The solution is transferred to a 5-ml round bottom polypropylene tube. Enzymobeads (50 μl) are then added, followed by 15 μl of a freshly prepared solution of β-D-glucose (2% in 0.1 *M* sodium phosphate buffer, pH 7.2), and 10 μl of interferon (IFN) to be labeled (about 10 μg protein). The reaction mixture (total volume of 100 μl) is incubated at 4° overnight (about 15 hr) with gentle rocking on a platform rocker.

3. Sodium azide (25 μl of 1 *M* solution) is added and the mixture further incubated for 15 min to stop the reaction. The mixture is transferred to a polypropylene Eppendorf tube. The original incubation tube is rinsed with 125 μl of L-tyrosine (saturated solution in phosphate-buffered saline) and transferred to the same Eppendorf tube. The suspension is mixed and centrifuged for 3 min in an Eppendorf centrifuge (15,000 *g*). The supernatant is applied to a Sephadex G-50 column (0.7 $\times$ 28 cm) equilibrated in phosphate-buffered saline, pH 7.2, containing 0.1% gelatin or bovine serum albumin. Fractions of about 0.5 ml are collected and 10 μl of each is counted in a gamma scintillation spectrometer.

4. The fractions are assayed for IFN activity. We use a microtiter assay based on protection of cells against the cytopathic effect of vesicular stomatitis virus for assaying IFN activity. Human fibroblasts (GM 2767, trisomic for chromosome 21) are used as test cells for human IFN-α and -β and WISH cells for human IFN-γ. International reference standards from the National Institute of Allergy and Infectious Diseases are used as internal standards. IFN assays have been described in a preceding volume of this series.[9]

Figure 1 shows the radioiodination of recombinant Hu-IFN-α2 (panel A) and Hu-IFN-γ (panel B) carried out by the procedure outlined above, except that in panel A, 1 mCi of ^{125}I was used and the reaction mixture was fractionated on a Sephadex G-75 column. As expected, the labeled IFN is eluted in the first peak separated from the free radioactivity. Better separation was obtained with Sephadex G-50 (panel B) than with G-75 (panel A). The peak IFN fractions are pooled and reassayed for ^{125}I radioactivity and IFN activity to determine the specific activity of the preparation. Radiolabeled preparations are stored in accordance with the stability of the IFN. Hu-IFN-α2 is stored in aliquots at $-70°$ and Hu-IFN-γ at 4°.

[9] S. Pestka, ed., this series, Vol. 78.

[10] U. K. Laemmli, *Nature* (*London*) **227,** 680 (1970).

[11] J. Le, W. Prensky, Y. K. Yip, Z. Chang, T. Hoffman, H. C. Stevenson, I. Balazs, J. R. Sadlik, and J. Vilček, *J. Immunol.* **131,** 2821 (1983).

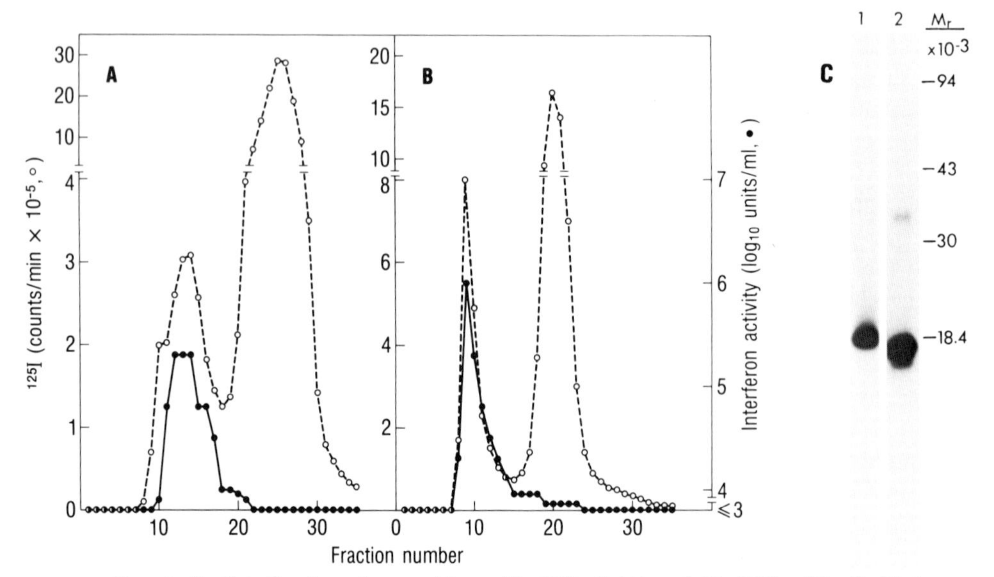

FIG. 1. Radioiodination of recombinant Hu-IFN-α2 (A) and Hu-IFN-γ (B). Purified recombinant Hu-IFN-α2 (provided by Schering-Plough Corporation, Bloomfield, New Jersey; specific activity 1.5×10^8 units/mg of protein) and Hu-IFN-γ (from Genentech, Inc., South San Francisco, California; specific activity 6.8×10^7 units/mg of protein) were radioiodinated as outlined in the text except that in A, 1 mCi of ^{125}I was used and that the reaction mixture was fractionated on a Sephadex G-75 column. (A) Elution profile of radiolabeled Hu-IFN-α2. Aliquots (10 μl) of each fraction were counted for ^{125}I radioactivity (○) and 25 μl of each assayed for IFN activity (●) in a microtiter assay. (B) Elution profile of radiolabeled Hu-IFN-γ. Five microliters of each fraction was taken for determination of radioactivity (○) and 25 μl assayed for IFN activity (●). The peak fractions (in terms of biological activity) were pooled and an aliquot (about 70,000 cpm) of each was used for electrophoresis in a 12% polyacrylamide slab gel by Laemmli procedure.[10] The gel was dried and autoradiographed with a kodak XAR-5 film (C). Lane 1, ^{125}I-labeled Hu-IFN-α2; lane 2, ^{125}I-labeled Hu-IFN-γ. The Hu-IFN-γ consists of two components, a major component (M_r 17,000) and a minor component (M_r 34,000). Both components are biologically active.[11]

Comments

Using the procedure described above, we have radioiodinated recombinant Hu-IFN-α2 and Hu-IFN-γ. Usually, more than 50% of the biological activity (often close to 100%) has been recovered after the entire procedure with 15–30% of the radioisotope incorporated into IFN. Assuming that the minor loss of activity is due to the loss of protein during the procedure, the specific radioactivity of the labeled product is usually $1–3 \times 10^7$ cpm/μg of protein. It can be calculated that less than one atom of ^{125}I is incorporated per 10 IFN molecules. Possibly, higher specific radioactivities may be attained by increasing the ratio of ^{125}I to protein during the incubation. Radioiodination at multiple sites per molecule may be undesirable. Hu-IFN-α2 radioiodinated as outlined above has been stable to storage at −70° for several months. Hu-IFN-γ was labeled by this procedure more recently, and not enough information is available regarding its stability. Hu-IFNs radiolabeled by this procedure have been used successfully for receptor binding studies,[4,12] and for cross-linking studies for the identification of the receptor molecules.[4,13]

The procedure described works at 0–4° as well as at higher temperature (20°). Incubations at 4° for 2 hr have given somewhat lower labeling as compared to overnight incubations. Shorter incubations may be advisable for any labile proteins. The system is inhibited by azide and by reducing agents. Therefore, the presence of these agents in the sample or buffers should be avoided. Due precautions must be taken during the handling of ^{125}I.

Acknowledgments

We are thankful to Schering-Plough Corporation and to Genentech Inc. for kindly providing us purified recombinant Hu-IFN-α2 and Hu-IFN-γ, respectively. This work was supported by Public Health Service Grants CA-29991 and AI-17816 from the National Institutes of Health (to S.L.G.).

[12] A. A. Branca and C. Baglioni, *Nature (London)* **294,** 768 (1981).
[13] F. H. Sarkar and S. L. Gupta, *Proc. Natl. Acad. Sci. U.S.A.* **81,** 5160 (1984).

[37] Radioiodination of Human Alpha Interferons by the Chloramine T Method

By KNUD ERIK MOGENSEN and GILLES UZÉ

The chloramine T method has long been used to incorporate iodine into the tyrosine residues of proteins.[1] The reactants generated are strong oxidants and for sensitive proteins gentler methods may be needed. We were nevertheless encouraged to develop the method for the following reasons.

[1] W. M. Hunter, *in* "Handbook of Experimental Immunology" (D. M. Weir, ed.), 3rd ed., p. 14.3. Blackwell, Oxford, 1978.

1. The biological activity of the alpha interferons is unaffected by low concentrations of oxidants providing the pH is kept above 7.0.

2. Initial studies had suggested that selective iodination at tyrosine residues was feasible and probably easier than selective acylation at amino residues.

3. Calculations showed that complete monoiodination would allow us to work with the subpicomolar quantities of IFN that would be necessary to correlate receptor function with biological activity. In order to avoid confusing the binding of labeled IFN with the activity of unlabeled IFN, we adopted the criterion that, on average, every IFN molecule should be marked, without altering its biological activity.

Materials

Iodine. Our normal source of ^{125}I is New England Nuclear (Boston, MA); code 033A, packed to give 10 mCi/ml on a fixed date. For ^{131}I, comparable material is available from Amersham International (Amersham, Bucks, U.K.) code IBS 30. Both are in the form of sodium iodide in NaOH solution, pH 10.0, and free from reducing agents. The quantities used in the reaction are based on ^{125}I of 99% purity. Reaction is carried out within a day or two of the given assay date. Other sources have proved less satisfactory, with efficiency dependent on the age of the iodine and often falling off rapidly in the days after generation.

IFNs. Successful monoiodination can be achieved with material greater than 50% pure. IFNs of lesser purity can be labeled, but the efficiency falls off with the presence of contaminating protein. The need for purity places limits on the concentration of IFN needed. A stock solution of 100 μg/ml is an absolute minimum for the conditions of reaction described here. Thus for IFN-α2 with a specific activity of 1×10^8 units/mg of protein at 50% purity, 10^7 units/ml represents the lower limit for the stock solution. Because of the different specific biological activities of IFN-αs, and because of the unknown and possibly variable composition of the naturally occurring mixtures,[2,3] biological titer is an inadequate measure of the concentration needed for iodination. Absorbance at 280 nm can be used if the extinction coefficients are known and the contaminating proteins are all minor species. Literature values for 1% extinction coefficients (see the table) lie around 10–11 for IFN-αs so far

[2] C. Weissmann, S. Nagata, W. Boll, M. Fountoulakis, A. Fujisawa, J.-I. Fujisawa, J. Haynes, K. Henco, N. Mantei, H. Ragg, C. Schein, J. Schmid, G. Shaw, M. Streuli, H. Taira, K. Todokoro, and U. Weidle, *Philos. Trans. R. Soc. London, Ser. B* **299,** 7 (1982).

[3] G. Allen, *Biochem. J.* **207,** 397 (1982).

CHARACTERISTICS OF HUMAN IFN-αS LABELED WITH ^{125}I BY THE CHLORAMINE T METHOD

Source of IFN-α	Tyrosines/ molecule	$E_{280}^{1\%}$ [a]	Maximum incorporation tolerated: Molar ratio [b] ^{125}I/IFN-α	Monoiodination [c] (%)
E. coli, IFN-α1[2,8]	4[2,8]	10.7[d]	5.6	85
E. coli, IFN-α2[2,8]	5[2,8]	10.7	8.0	135
Namalwa cells, IFN-α[3]	3–4 mean[3] 3–6 range[3]	~10.0	8.0	85

[a] Extinction coefficient was determined in PBS, pH 7.4, 1% (w/v) of IFN with a cuvette of 1 cm path length at 280 nm against a standard solution of bovine serum albumin. The extinction coefficient was estimated by comparison of ratios with those from the Lowry reaction.

[b] On basis that 1 mCi of ^{125}I (100%) to 9.1 μg protein (MW about 20,000[8]) represents a ratio of one.

[c] Estimated from SDS–polyacrylamide gel electrophoresis of ^{125}I-labeled IFN-α and based on a molecular weight of 20,000.[4,8] The value of 135% means that a significant fraction of IFN-α2 molecules in this preparation contains more than one ^{125}I.

[d] Subsequent batches of α1 have given $E_{280}^{1\%} = 9.5$. The IFN α8 has given $E_{280}^{1\%} = 7.0$. If the published value[5] for α-2 at $E_{280}^{1\%} = 10.6$ is taken as a reference, then extinction coefficients based solely on the relative tryptophan, tyrosine, and phenylalanine contributions to absorbance give $E_{280}^{1\%} = 9.1$ for α1, and $E_{280}^{1\%} = 7.2$ for α8, following the published amino acid compositions.[2]

tested.[4,5] This together with an estimate of the electrophoretic purity gives a suitable basis for calculating the amount of IFN-α needed for the iodination reaction.

The method applies for IFNs in PBS. However, at the time of reaction the total concentration of ^{125}I is about 10^{-5} *M;* that of IFN-α2, 8 times less. If interfering substances are left over from the purification of IFN, then their concentration in PBS should first be reduced to well below that of the reactants. For example, thiocyanate, which is often used in immunoaffinity chromatography in molar concentrations,[4,6] interferes with iodination so that its residual concentration needs to be reduced to below 1 μ*M* by successive dialysis of the stock IFN.

[4] K. E. Mogensen, M.-T. Bandu, F. Vignaux, M. Aguet, and I. Gresser, *Int. J. Cancer* **28,** 575 (1981).

[5] T. A. Bewley, H. L. Levine, and R. Wetzel, *Int. J. Pept. Protein Res.* **20,** 93 (1982).

[6] K. E. Mogensen and K. Cantell, *J. Gen. Virol.* **45,** 171 (1979).

Albumin. The quality of BSA, used as an inert buffer against surface adsorption of protein, depends somewhat on the use intended for the labeled IFN. We use a fraction V (Sigma, St. Louis, MO; code A-7888, RIA grade). The stock solutions for chromatography tend to precipitate on standing and require membrane filtration before use.

Chemicals. Commercial analytical grades suffice. Chloramine T and metabisulfite are best stored in a dry atmosphere. Solutions are made up just before use.

Selective Reaction

The chloramine T reaction involves a slow hydrolysis to generate hypochlorite followed by fast reactions with iodide to form oxidized species that substitute rapidly at positions ortho to the tyrosine hydroxyl, the first substitution facilitating the second.[1] The iodine atom has dimensions similar to those of the phenolate ion so substitution would likely lead to local distortions in protein structure. The method was originally devised to keep hypochlorite and molecular iodine concentrations as low as possible during the reaction, thus reducing damage and the escape of volatile iodine. This, however, makes the velocity of the reaction with tyrosine independent of the concentration of reactive iodine. From initial studies it was clear that such conditions would cause substantial inactivation before complete monoiodination was achieved. The problem was overcome by generating active iodine before allowing it to come into contact with IFN-α then using very short reaction times; in effect, if the method was to work at all, then substitution at the most accessible tyrosine(s) would not affect the biological activity of the molecules. The generation of activated iodine is a complex reaction and acceptable conditions were determined empirically to be the following.

^{125}I (75 μl) at 7.5 mCi and 17.4 Ci/mg iodide is reacted with 75 μl of chloramine T at 0.5 mg/ml, in 0.01 *M* phosphate-buffered saline at pH 7.4 (PBS), for 2 min at room temperature. Of this mixture 140 μl is then added to 14 μg of protein in 170 μl of PBS and rapidly mixed for 5 sec. Sodium metabisulfite (100 μl) at 1.0 mg/ml in PBS is added 10 sec later to reduce the oxidized species and stop the reaction. All reactions are carried out in capped vials to reduce the escape of volatile iodine. For similar reasons dispensing pipettes need to be decontaminated after the reaction.

The quantities given are for human IFN-α2 of greater than 50% purity. The above method was developed with interferon preparations that contained approximately one-third non-IFN protein present as minor contaminants, and 14 μg of protein thus contained 9–10 μg of active IFN

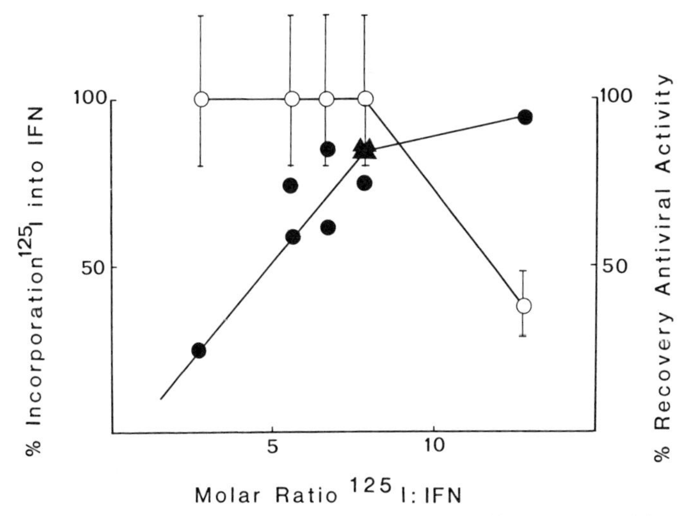

FIG. 1. Relationship between the initial molar ratios of ^{125}I and IFN (origin: Namalwa cells, see the table) and the subsequent incorporation of ^{125}I, with the recovery of biological activity, following reaction by the modified chloramine T method. (●) percentage monoiodination of IFN-α, mean reaction time 15 ± 5 sec; (▲) percentage monoiodination of IFN-α: mean reaction time, 15 ± 2 sec. (○) percentage biological activity recovered, with standard error bars for antiviral assay.

protein.[4,7] The table shows acceptable values for the ratio of ^{125}I to IFN-α for different IFNs. Figure 1 shows that the rationale for the reaction lies in linking the molar ratio of iodide to IFN to the final incorporation. The figure also shows that attention to the time of reaction improves the reproducibility of the procedure. To achieve a proportionally different incorporation, the volume of radioactive iodide is altered; this entails a corresponding change in the volumes of chloramine T and metabisulfite. The final concentration of IFN and the reaction times are maintained.

Separation of Labeled IFN. After the addition of metabisulfite the mixture is allowed to stand for 30 sec to 1 min before being passed onto standardized columns of Sephadex G-25 (PD-10, Pharmacia Fine Chemicals, Uppsala, Sweden) equilibrated with 0.5% bovine serum albumin (BSA) in PBS. For a standard 10 ml column with a void volume of about 2 ml, radioactive protein is collected in good yield between 2 and 4 ml, free iodine being retained beyond that. Inevitably small amounts of albumin

[7] K. E. Mogensen and M.-T. Bandu, *Eur. J. Biochem.* **134,** 355 (1983).

are labeled and these are removed by a second chromatography on a column of G-75 superfine (Pharmacia), bed volume 50 ml, 16 mm diameter, flow 2 ml/hr, in the presence of 0.5% BSA in PBS with 0.02% azide, at room temperature. Peak fractions of IFN-α (K_{av} 0.35–0.4) are pooled, dialyzed free of azide at 4°, and filtered through a 0.2-μm membrane, washed with albumin prior to use. ^{125}I-labeled IFN-α is then stored in convenient aliquots at −80°.

Yield. About one-third of the labeled IFN is recovered (corresponding to a mean recovery of 85% at each manipulation). A second chromatography is not strictly necessary, but the sacrifice in yield pays dividends in binding experiments where backgrounds can be reduced to almost negligible proportions.[4,7] It also offers the chance of removing oligomerized IFN which may be present in the original material. For an input of 2×10^6 units (reference MRC 69/19) of IFN-α2, one obtains a final labeled solution of about 5 ml at $1–1.5 \times 10^5$ units/ml.

Limits. The main determinants of a successful reaction are the time and the molar ratio of iodide to IFN. There is an upper limit on the time of reaction (about 20 sec for interferon from Namalwa cells[4]), and also on the ratio of iodide to IFN (see the table, Fig. 1). There is also a lower limit on the concentration of protein. Adsorption losses of IFNα-s onto glass and plastic surfaces become catastrophic below 25 μg/ml of protein. For the quantities given here, the final concentration in the reaction vial is 35 μg/ml of protein.

Radioactive Contamination

In a properly screened and ventilated hood the reaction poses no great hazard, but particular care is needed to avoid contamination because, in the method described, iodide is allowed to oxidize before it is allowed to react. One of the products is molecular iodine which readily sublimes even from solution. For this reason all material which has been in contact with radioactive iodide is immediately discarded into capped containers. The disposable columns of G-25 used to remove free iodine fit neatly into disposable 50-ml centrifuge tubes. Surfaces which are inevitably exposed can be covered with disposable material. Soluble contamination can be removed by soaking in detergent and washing with water. Surfaces onto which iodine has sublimated can be washed with solutions of sodium or potassium iodide. A particularly refractory contamination can often be removed by repeated applications of an adhesive tape. Disposable gloves are easily contaminated and need to be changed between manipulations.

Integrity of Labeled Interferon

As can be seen from Fig. 1 and the table, the conditions for monoiodination (100% incorporation) are not far from those which cause inactivation. IFN-α2 which appears to contain an additional tyrosine residue compared to some other IFN-αs[4,8] supports the highest incorporation. Small amounts of damage sustained during iodination are difficult to account for and, often, are not immediately apparent. The advantage with the method described is its reproducibility, for when once the product has been well characterized with respect to binding and activity constants, subsequent batches need only a cursory check on their properties. Confidence in the integrity of a labeled IFN (i.e., that it behaves as does the unlabeled form) usually comes only with extensive experimentation. We have found, however, that if a batch does not meet the following criteria, it is unlikely to be of any use.

1. That recovery of biological activity should not be less than 20%. (For the method described, this is equivalent to allowing a mean 80% recovery at each step, i.e., reaction, chromatography, dialysis, etc.).
2. That dpm per unit of biological activity (antiviral titer, reference MRC 69/19) should not exceed 2000 if the starting material is at least 50% pure, i.e., higher ratios are attained only at the expense of activity. The value of 2000 holds for IFNs of specific biological activity $1-2 \times 10^8$ units/mg protein. For IFNs showing a different specific activity on human cells a corresponding value holds well.[9]
3. Biological activity should remain unchanged in the months after iodination.

The behavior of the labeled IFNs, described in the table, has eventually proved to be indistinguishable from that of their unlabeled forms. Figure 2 shows the isoelectric profile of labelled and unlabeled IFN-α2. The curves are superimposable to an extent that suggests little or no change in the physicochemical properties following iodination. We draw attention to the fact that IFN-α2 shows charge heterogeneity despite its homogeneous origin.[2] Both peaks are active and both can be recovered from extracted complexes of IFN and receptor. The ratios of cpm to activity are similar in both peaks. From competitive binding experiments

[8] S. Pestka, *Arch. Biochem. Biophys.* **221,** 1 (1983).

[9] A discussion of the specific biological activities of the human IFN-α and IFN-β is beyond the scope of this chapter. We would note, however, that there is no single biological test system where the relative activities due to all the various IFN-α and -β can be taken, unambiguously, as a faithful reflection of their relative molar concentrations in solution.

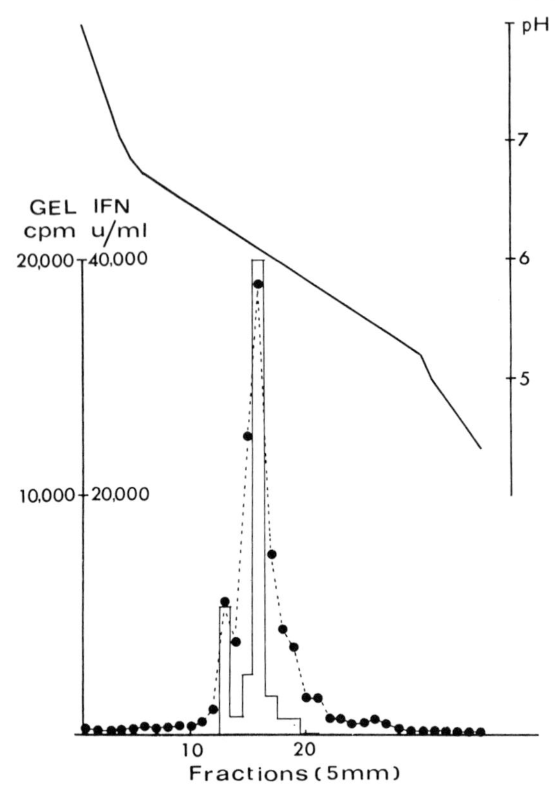

FIG. 2. Isoelectric profile of a mixture of labeled (~100% monoiodination) IFN-α2 (cpm, ●) and unlabeled IFN-α2 (antiviral units/ml, histogram). Input: 83,000 cpm; 150,000 units IFN-α2 in 9 *M* urea and 2% NP 40, 3.78% acrylamide, 0.22% Bis. Dimensions, 17.5 cm, 0.2 mm diameter; run 18 hr, 600 V. Method: P. O'Farell, *J. Biol. Chem.* **250,** 4007 (1975). Sections of the sliced gel were eluted in 0.1% SDS for 24 hr. Eluates were counted for radioactivity (75% efficiency) and titrated for IFN activity.

using either specific antibody[10] or cellular receptors,[4,7] the binding affinity of labeled IFN is the same as that of unlabeled, and in at least one case the equilibrium binding constant has been shown to equal an equilibrium activity constant for both labeled and unlabeled IFN-α2.

The most sensitive tests are physiological. Clearance curves from organ perfusion will often reveal small amounts of damaged protein. Monoexponential clearance rates for acid precipitable radioactivity, super-

[10] P. Daubas and K. E. Mogensen, *J. Immunol. Methods* **48,** 1 (1982).

imposable on rates calculated for biological activity, are a good indication that the molecule has sustained no serious damage during labeling.[11]

Below the limits given in the table for maximum incorporation, we can relate radioactivity to moles of active IFN.[4,7] For IFN-α2, 100% monoiodination results in about 5000 dpm/fmol and about 1250 dpm/unit of biological activity.

Limitations

There are certain important limitations that need to be carefully controlled.

For *in vivo* studies, special treatment is necessary to reduce adsorbed or noncovalently linked ^{125}I. Acid precipitable radioactivity is normally between 5 and 10% of the total when the labeled IFN is stored at $-80°$. Exhaustive dialysis against unlabeled sodium iodide (1.0 m*M* suffices) will reduce backgrounds of acid-soluble radioactivity to 1%.

We have evidence of a temperature dependent decrease in acid-precipitable radioactivity with time. This is the main reason for storing labeled IFN at $-80°$. It may result from a hydrolysis of substituted residues, for it is not accompanied by a loss in biological activity. The loss over 24 hr at 37° is detectable in binding experiments. Fortunately, the rates of dissociation of iodine are slow compared with IFN binding rates, but it does need to be controlled if binding is followed over long periods.

We also have evidence that iodine is not uniformly incorporated into the constituent species of natural mixtures of IFN. This is perhaps not surprising as the tyrosine content varies,[2,3] but it does rather limit the value of binding studies which use labeled mixtures of IFN. We would suggest that modulating effects on the binding of one IFN, in the presence of another are best studied by double labeling techniques. The energies for ^{125}I and ^{131}I are sufficiently different for the two to be separated by appropriate screening on a two channel scintillation spectrometer.

We emphasize that the concept of monoiodination covers a population of molecules that are unlabeled and a population that are substituted twice. This means that with the age of the preparation the more heavily labeled molecules will increase in proportion. If damage is highest among these, then changes in binding affinity may appear with time. Six weeks after iodination with ^{125}I, incorporation will have fallen by one third. However, so far, within this time, we have seen no changes in binding parameters either directly or in competition.

[11] V. Bocci, K. E. Mogensen, M. Muscettola, A. Pacini, L. Paulesu, G. P. Pessina, and S. Skiftas, *J. Lab. Clin. Med.* **101,** 857 (1983).

Acknowledgments

It is a pleasure to acknowledge our indebtedness to Charles Weissmann for IFN-α1 and α2, to Karl Fantes for IFN from Namalwa cells, and to Ion Gresser for his constant encouragement. This work was supported by the following grants: D.G.R.S.T. 82L0074; I.N.S.E.R.M. CRL 811015; C.N.R.S./P.I.R.M.E.D. 100482/60821079; I.N.S.E.R.M. PRC 127012; Fondation de Recherches en Hormonologie 948304; The Richard Lounsbery Foundation, The Simone and Cino del Duca Foundation; and A.R.C., Villejuif.

[38] Radiolabeling of Human Interferon Alphas with ^{125}I-Labeled Bolton–Hunter Reagent

By DOROTHY L. ZUR NEDDEN and KATHRYN C. ZOON

Radioactive derivatives of interferon (IFN) have been prepared by several iodination procedures which chemically add a radioactive iodine atom or iodine-containing group.[1–5] Internal radiolabeling of IFN has also been accomplished with radioactive amino acids.[6,7] This chapter describes procedures for labeling human IFN-α with ^{125}I-Labeled Bolton–Hunter reagent and purification of the radiolabeled ligand.

Preparative Procedures and Methodology

Recombinant DNA-Derived IFNs

Escherichia coli-derived human IFN-α2 purified to a specific activity of 1.7×10^8 units/mg protein (6.7–7.8 mg/ml) was the gift of Schering Corporation, Bloomfield, NJ. *Escherichia coli*-derived human IFN-αA purified to a specific activity of 2×10^8 units/mg protein (4.3 mg/ml) was the gift of Hoffmann-La Roche, Nutley, NJ. These IFNs were diluted 1:10 with 0.1 *M* sodium borate buffer, pH 8.5, 0.05% sodium dodecyl sulfate (borate/SDS) to final protein concentrations of 0.43–0.78 mg/ml and final volumes of 200 μl.

[1] M. Aguet and B. Blanchard, *Virology* **115,** 249 (1981).
[2] A. Branca and C. Baglioni, *Nature* (*London*) **294,** 768 (1981).
[3] K. E. Mogensen, M.-T. Bandu, F. Vignaux, M. Aguet, and I. Gresser, *Int. J. Cancer* **28,** 575 (1981).
[4] K. C. Zoon, D. Zur Nedden, and H. Arnheiter, *J. Biol. Chem.* **257,** 4695 (1982).
[5] P. Anderson, Y. K. Yip, and J. Vilček, *J. Biol. Chem.* **257,** 11301 (1982).
[6] S. Yonehara, M. Yonehara-Takahashi, and A. Ishii, *J. Virol.* **45,** 1168 (1983).
[7] C. W. Czarniecki, C. W. Fennie, D. B. Powers, and D. A. Estell, *J. Virol.* **49,** 490 (1984).

Natural Alpha Interferons

Human (Hu) IFN-α (Ly, MW 18,500), lymphoblastoid IFN derived from Namalwa cells induced with Newcastle disease virus, was purified to a specific activity of 2–2.4 $\times$ 10^8 units/mg protein as previously described.[8]

Prior to iodination 1 to 1.5 ml of Hu-IFN-α (Ly, MW 18,500) (1–4 μg/ml) was either (1) dialyzed against 2 liters of phosphate-buffered saline, pH 7.4, containing 0.05% SDS, lyophilized to dryness in a Rotary Speed-vac system (Savant, Hicksville, NY), and dissolved in 100–200 μl borate/SDS or (2) concentrated to 100–200 μl by Amicon filtration (3 ml cell) with a PM 10 membrane and then washed with 3 ml 0.1 *M* borate/SDS and concentrated again to approximately 150 μl.

^{125}I-Labeled Bolton–Hunter Reagent

N-Succinimidyl 3-(4-hydroxy, 5-[^{125}I]iodophenyl) propionate with a specific activity of approximately 2000 Ci/mmol (74 TBq/mmol) of the monoiodo ester, prepared by a modification of the chloramine T procedure, was obtained in 1 mCi quantities from Amersham, Arlington Heights, IL. The reagent is supplied in dry benzene solution containing 0.2% dimethyl formamide, and should be used within 2 weeks of receipt.

Interferon Assays

IFN assays were performed with human foreskin fibroblasts (Tissue Culture Unit, Office of Biologics Research and Review, Bethesda, MD) or Madin–Darby bovine kidney cells (ATCC CCL22) and vesicular stomatitis virus as the challenge.[9] The human leukocyte standard G-023-901-527 obtained from the Research Resources Branch, National Institute of Allergy and Infectious Diseases, National Institutes of Health, Bethesda, MD 20205, was used to calibrate the assay.

Protein Determination

Protein concentrations of the IFNs were determined by the method of Lowry *et al.*[10] or calculated from the antiviral activity and the specific activity of the interferon alpha species.

[8] K. C. Zoon, M. E. Smith, P. J. Bridgen, D. Zur Nedden, and C. B. Anfinsen, *Proc. Natl. Acad. Sci. U.S.A.* **76,** 5601 (1979).

[9] K. C. Zoon, C. E. Buckler, P. J. Bridgen, and D. Gurari-Rotman, *J. Clin. Microbiol.* **7,** 44 (1978).

[10] O. H. Lowry, N. J. Rosebrough, A. L. Farr, and R. J. Randall, *J. Biol. Chem.* **193,** 265 (1951).

SDS–Polyacrylamide Gel Electrophoresis (PAGE)

SDS–PAGE was performed by a modification of a procedure of Laemmli with a 16% polyacrylamide separating gel and 3.7% polyacrylamide stacking gel.[11,12] Sixteen-centimeter-long and 0.75-mm-thick slab gels were run at constant current (15 mA/slab) at room temperature, with a Hoefer electrophoresis unit (San Francisco, CA).

Radioiodination of Hu-IFN-α (Ly, MW 18,500)

Hu-IFN-αs were radiolabeled with ^{125}I-labeled Bolton–Hunter reagent by a modification of the procedure described by Knight.[13] The IFN in approximately 150 to 200 μl borate/SDS was added to a vial of ^{125}I-labeled Bolton–Hunter reagent (1 mCi) which had been very gently dried immediately before use under a slow stream of N_2 in a well-ventilated fume hood equipped with an activated, impregnated charcoal filter and a lead shield. The sample tube was rinsed with 50 μl borate/SDS which was also added to the reaction vial. The solution was mixed and the reaction was allowed to proceed for 15 min at room temperature with occasional swirling. The reaction mixture was transferred to a second vial containing 1 mCi of ^{125}I-labeled Bolton–Hunter reagent, again very gently dried immediately before use under a slow stream of N_2. The first vial was rinsed with 50 μl of borate/SDS and this was also added to the second vial of 1 mCi ^{125}I-labeled Bolton–Hunter reagent. Again, the reaction was allowed to proceed at room temperature with periodic swirling for 15 min. Subsequently, 0.8 ml of 0.2 *M* glycine in borate/SDS was added to react with any residual reagent. The reaction was allowed to proceed for 30 min at room temperature with periodic swirling. The reaction mixture was applied to a PD-10 column (Pharmacia, Piscaway NJ) equilibrated with phosphate-buffered saline, pH 7.4 with 0.1% SDS (PBS/SDS). This column which separates the ^{125}I-labeled IFN from most of the ^{125}I-labeled Bolton–Hunter reagent was developed with PBS/SDS. Ten drop fractions were collected and the radioactivity was determined in each fraction. The ^{125}I-labeled IFN-enriched fractions were pooled and dialyzed against 1 liter of 0.1% SDS surrounded with lead shielding for 16 hr at room temperature. Approximately 100% of the antiviral activity was recovered at this step (see the table).

To further remove free ^{125}I, ^{125}I-labeled Bolton–Hunter reagent, small molecular derivatives of the reagent and glycine, and oligomers of ^{125}I-

[11] U. K. Laemmli, *Nature (London)* **227,** 680 (1970).

[12] H. Weintraub, K. Palter, and F. van Lente, *Cell* **6,** 85 (1975).

[13] E. Knight, Jr., *J. Gen. Virol.* **40,** 681 (1978).

RECOVERY OF THE ANTIVIRAL ACTIVITY OF Hu-IFN-α2 FOLLOWING IODINATION AND PURIFICATION[a]

Step	Antiviral activity (units × 10^7)	Recovery (%)
Starting material	6.3	—
PD-10/dialysis (0.1% SDS)	8.0	126
SDS–PAGE/dialysis (DPBS/BSA)	2.6	41

[a] Antiviral activity was determined on MDBK cells. Only starting material was assayed prior to iodination. The other two samples were assayed after iodination.

labeled IFN, the ^{125}I-labeled IFN-α preparation was further purified by SDS–PAGE. Prior to electrophoresis, the dialyzed radiolabeled IFN was lyophilized on the Rotary Speed Vac System. Electrophoresis solubilizing solution without reducing agent (80 μl) was added to the dried sample. The ^{125}I-labeled IFN (75 μl) was applied to an SDS gel and electrophoresis was performed as described above.

Following electrophoresis, the gel was covered with clear plastic wrap after removal of the top glass plate. Small squares of filter paper marked with radioactive ink (prepared by rinsing the sample tube with 20 μl ink) were taped onto the plastic wrap at each corner of the gel. Kodak X-OMAT AR film (Rochester, NY) was placed over the gel for 3–5 min in the dark. The film was developed and positioned over the gel. An autoradiogram of a preparative gel of ^{125}I-labeled IFN-α2 is shown in Fig. 1. The area designated by the autoradiogram corresponded to an apparent molecular weight of 18,000–19,000. A sharp instrument such as a hypodermic needle was used to punch holes through the film and gel to delineate this area. The film was removed and the gel inside the ring of holes was cut out with a scalpel and placed in a 75 × 100 mm polypropylene tube containing 3 ml of PBS/SDS. The tube was placed in a lead shielded container overnight at 37° with shaking. The gel pieces were removed and washed one time with 2 ml PBS/SDS and then 1 ml of Dulbecco's phosphate-buffered saline (DPBS) containing 1 mg/ml bovine serum albumin essentially fatty acid and globulin free (BSA, Sigma Chemical Co.) was added. The washes were combined with the eluate and the ^{125}I-labeled IFN was dialyzed for 6 hr at room temperature against 500 ml DPBS containing 1 mg/ml BSA and 0.02% sodium azide, and then twice against 500 ml DPBS containing 1 mg/ml BSA at 4°. The radioactivity of the sample was determined with a Beckman 8500 Gamma Counter with a counting efficiency of approximately 80%. The sample was tested for

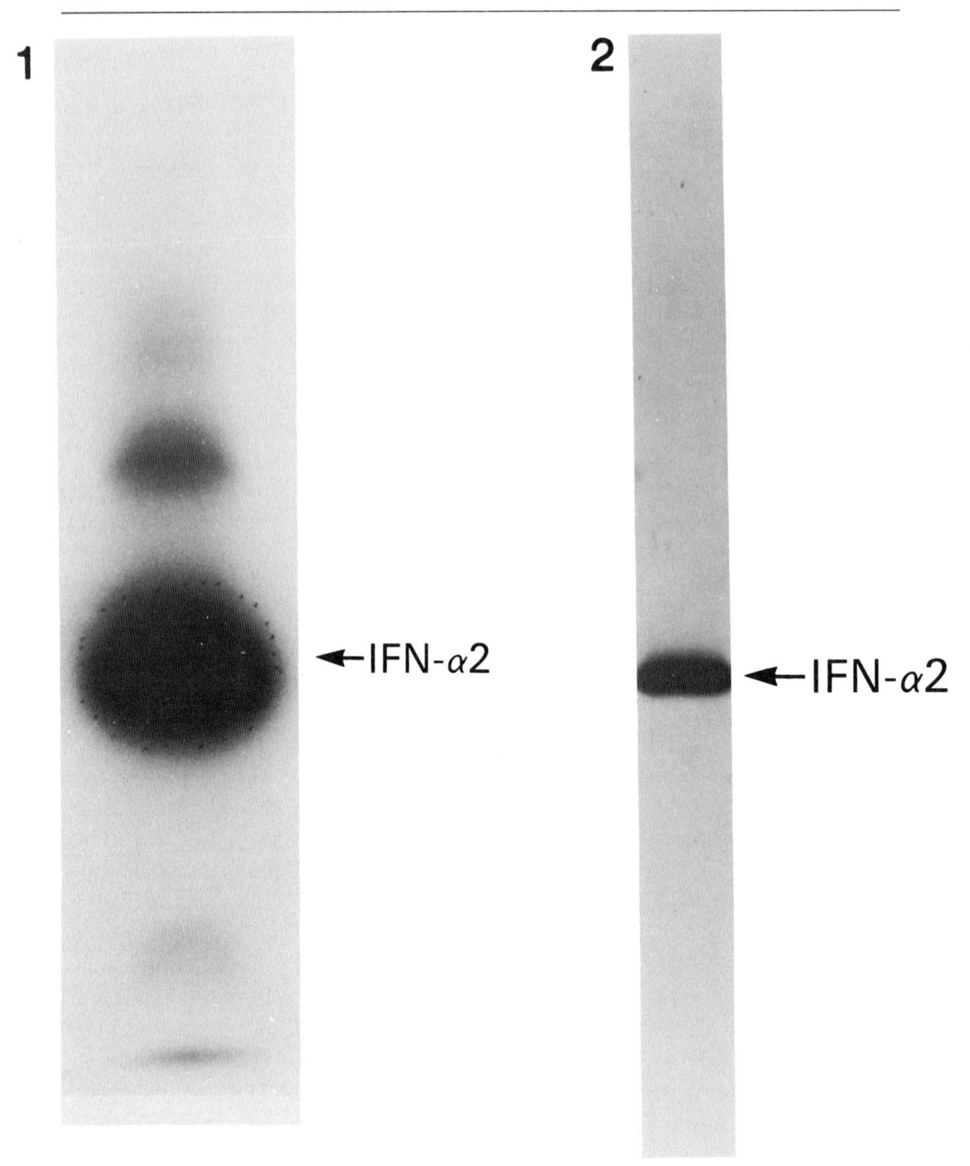

FIG. 1. Autoradiogram of Hu-IFN-α2 radiolabeled with ^{125}I-labeled Bolton–Hunter reagent. SDS–PAGE and autoradiography were performed as described in text.

FIG. 2. Radioactivity profile of SDS–PAGE-purified ^{125}I-labeled Hu-IFN-α2. ^{125}I-labeled Hu-IFN-α2, purified by preparative SDS–PAGE (2×10^6 dpm), was applied to a 16% polyacrylamide separating gel with a 3.7% polyacrylamide stacking gel. SDS–PAGE was performed as described in text and the gel was dried. Kodak X-OMAT AR film was exposed to the gel for approximately 1 hr, and developed.

biological activity, then stored in small aliquots in liquid nitrogen. An autoradiogram of the SDS–PAGE-purified ^{125}I-labeled IFN-α2 after analytical electrophoresis is shown in Fig. 2. A summary of the recovery of antiviral activity during preparation of one batch of ^{125}I-labeled IFN-α2 is shown in the table. The final specific radioactivity of Hu-IFN-α species labeled by this method ranges from 3 to 12 μCi/μg protein (1.1–4.4 $\times$ 10^5 Bq/μg protein).

Comments

^{125}I-labeled Bolton–Hunter reagent is recommended for the iodination of Hu-IFN-α. The reaction of IFN with this reagent takes place under very mild conditions and does not reduce antiviral activity.

Although the radiospecific activity of IFN with this procedure is in general less than that obtained with chloramine T or enzymobeads, it does provide a ^{125}I-labeled IFN preparation which is stable for up to 2 months. In addition to Hu-IFN-α2, Hu-IFN-αA, and Hu-IFN-α (Ly, MW 18,500), Hu-IFN-αD can also successfully be radiolabeled with this protocol.

[39] Radiolabeling of Human Immune Interferon with ^{125}I-Labeled Bolton–Hunter Reagent

By PAUL ANDERSON

The utility of radiolabeled purified interferons as investigational reagents has been amply demonstrated in recent years.[1–11] A number of radioisotopes and radiolabeling protocols have been used. In general, the

[1] M. Aguet, *Nature (London)* **284,** 459 (1980).
[2] A. A. Branca and C. Baglioni, *Nature (London)* **294,** 768 (1981).
[3] K. E. Mogensen, M.-T. Bandu, F. Vignaux, M. Aguet, and I. Gresser, *Int. J. Cancer* **28,** 575 (1981).
[4] K. Zoon, D. Zur Nedden, and H. Arnheiter, *J. Biol. Chem.* **257,** 4695 (1982).
[5] P. Anderson, Y. K. Yip, and J. Vilček, *J. Biol. Chem.* **257,** 11301 (1982).
[6] A. A. Branca, C. R. Faltynek, S. B. D'Alessandro, and C. Baglioni, *J. Biol. Chem.* **257,** 13291 (1982).
[7] P. Anderson, Y. K. Yip, and J. Vilček, *J. Biol. Chem.* **258,** 6497 (1983).
[8] K. E. Mogensen and M.-T. Bandu, *Eur. J. Biochem.* **134,** 355 (1983).
[9] A. R. Joshi, F. H. Sarkar, and S. L. Gupta, *J. Biol. Chem.* **257,** 13884 (1982).
[10] C. R. Faltynek, A. A. Branca, S. McCandliss, and C. Baglioni, *Proc. Natl. Acad. Sci. U.S.A.* **80,** 3269 (1983).
[11] P. Anderson and C. Nagler, *Biochem. Biophys. Res. Commun.* **120,** 828 (1984).

method of choice for any given interferon is one that maximizes specific radioactivity while minimizing loss of biological activity. The Bolton–Hunter reagent (*p*-hydroxyphenylpropionic acid, *N*-hydroxysuccinimide ester) reacts under mild conditions to acylate terminal or lysine amino groups of proteins.[12] We chose this method of iodination because of the mild reaction conditions and because interferon gamma is a basic protein ($pI = 8.6$)[13] with a large number of lysine residues.[14] Using this reagent, we have succeeded in radiolabeling human interferon gamma to a high specific activity, while retaining full biological activity.

Materials

Monoiodinated ^{125}I-labeled Bolton–Hunter reagent was purchased from New England Nuclear Corp. and stored at 4° until ready for use.

Human immune interferon was produced in cultures of lymphocyte-rich plateletpheresis residues stimulated with phytohemagglutinin and 12-*O*-tetradecanoyl-phorbol-13-acetate and purified by sequential processing on silicic acid, concanavalin A Sepharose, DEAE-Sephacel, and BioGel-P200 to within 80–90% of homogeneity (estimated from SDS–PAGE analysis).[5,13,15]

Sephadex G-25 (fine) was loaded into a column made from a 10-ml plastic pipette (Falcon) and equilibrated with PBS, pH 7.4 containing 0.25% gelatin.

Glycine, 2 *M* in PBS, pH 7.4.

Dry nitrogen.

Gelatin, Bloom number approximately 300 (Sigma Chemical Co.).

PBS: NaCl, 8.0 g; KCl, 0.2 g; KH_2PO_4, 0.2 g; Na_2HPO_4, 1.15 g; H_2O to 1000 ml.

Radiolabeling Protocol

^{125}I-labeled Bolton–Hunter reagent (1 mCi in 100 μl benzene) was equilibrated at 4° in the New England Nuclear combi-v-vial in which it

[12] A. E. Bolton and W. M. Hunter, *Biochem. J.* **133,** 529 (1973).

[13] Y. K. Yip, R. H. L. Pang, C. Urban, and J. Vilček, *Proc. Natl. Acad. Sci. U.S.A.* **78,** 1601 (1981).

[14] P. W. Gray, D. W. Leung, D. Pennica, E. Yelverton, R. Najarian, C. Simonsen, R. Derynk, P. J. Sherwood, D. M. Wallace, S. L. Berger, A. D. Levinson, and D. Goeddel, *Nature* (*London*) **295,** 503 (1982).

[15] Y. K. Yip, B. S. Barrowclough, C. Urban, and J. Vilček, *Science* **215,** 411 (1982).

was shipped. Immediately prior to use, the benzene was evaporated under a gentle stream of dry nitrogen in a ventilated fume hood. Purified human interferon gamma (50 μl in 50 mM phosphate buffer, pH 7.4 at a concentration of approximately 100 μg/ml) was added to the glass reaction vial and incubated for 30 min at 4°. Reaction was stopped by the addition of 2 M glycine (10 μl) for an additional 5 min incubation. The solution containing interferon was then removed, pooled with two washes of the reaction vial (0.5 ml PBS containing 0.25% gelatin), and applied to a Sephadex G-25 (fine) column. It is important that buffers containing albumin not be used to equilibrate this column as albumin adsorbs free ^{125}I-labeled Bolton–Hunter reagent. Fractions eluting in the void volume which contained ^{125}I-labeled interferon gamma were pooled and stored at 4°.

Under the reaction conditions described, we routinely obtained specific radioactivities of 20–40 μCi/μg protein. Within the error of our bioassay (inhibition of cytopathic effect of encephalomyocarditis virus on FS-4 fibroblasts), the iodination procedure resulted in no loss of biological activity. The recovery after passage over Sephadex G-25 ranged between 50 and 80%. Similar results have been obtained with recombinant interferon gamma as well.[11]

Comments

^{125}I-labeled human interferon gamma, prepared as described above, has been used to characterize an interferon gamma receptor on the surface of human fibroblasts.[5,7,11] It has also been used to probe the subunit structure of interferon gamma and to investigate the extent of glycosylation of each subunit.[16–18] In all of these studies, its biological activity and immunoreactivity have paralleled those of the native interferon. It is likely that this reagent will prove useful in a wide range of pharmacological, mechanistic and structural studies of interferon gamma.

[16] Y. K. Yip, H. C. Kelker, J. Le, P. Anderson, B. S. Barrowclough, C. Urban, and J. Vilček, *in* "The Biology of the Interferon System 1983" (E. DeMaeyer and H. Schellekens, eds.), p. 129. Elsevier, Amsterdam, 1983.

[17] H. C. Kelker, Y. K. Yip, P. Anderson, and J. Vilček, *J. Biol. Chem.* **258,** 8010 (1983).

[18] H. C. Kelker, J. Le, B. Y. Rubin, Y. K. Yip, C. Nagler, and J. Vilček, *J. Biol. Chem.* **259,** 4301 (1984).

[40] *In Vivo* Radiolabeling of Human IFN-β

By ERNEST KNIGHT, JR. and DIANA FAHEY

Human IFN-β can be labeled with [^{35}S]methionine *in vivo* to a specific radioactivity that is useful for many experiments. The advantage of *in vivo* radiolabeling is that all labeled molecules should have the same specific biological activity as the unlabeled molecules.

Production of [^{35}S]Methionine-Labeled Human IFN-β by Human Diploid Fibroblast Cells

Cells. Human diploid fibroblast cells are grown in 60-mm plastic culture dishes. IFN-β is assayed by a microtiter method[1] with human diploid fibroblast cells and vesicular stomatitis virus as the challenge virus.

Induction and Radiolabeling of IFN-β

Cells are used that have been in culture 8–12 days after passage. Each 60-mm dish contains approximately 4×10^5 cells. The cells are "superinduced" with poly(I) · poly(C), cycloheximide, and actinomycin D to produce IFN-β.[2] Growth medium is removed and 2 ml of Eagle's minimal essential medium (MEM, GIBCO) containing 25 μg/ml poly(I) · poly(C) and 50 μg/ml cycloheximide is added to each dish. Serum is not added to the medium anytime after the growth of the cells. After 4 hr at 37° in a humidified, 5% CO_2–95% air incubator actinomycin D is added to each dish to a concentration of 1 μg/ml. After 2 hr at 37° the medium is removed and the cells are washed 3 times with 2 ml of MEM minus methionine. One milliliter of MEM minus methionine containing 100 μCi/ml of [^{35}S]methionine (NEN, >800 Ci/mmol) is added to each dish. The cells are placed at 37° for 12–20 hr for synthesis and secretion of IFN-β. The ^{35}S-labeled IFN-β is stored at 4° for brief periods (1–7 days) but should be purified as soon as possible. IFN-β activities vary from 3000 to 8000 units/ml and approximately 5×10^5 cpm of [^{35}S]methionine is incorporated per ml of crude IFN-β.

[1] J. A. Armstrong, this series, Vol. 78, p. 381.
[2] E. A. Havell and J. Vilček, *Antimicrob. Agents Chemother.* **2,** 476 (1972).

Purification and Characterization of ^{35}S-Labeled IFN-β

Purification. The ^{35}S-labeled IFN-β is purified to near homogeneity by chromatography on Blue Sepharose.[3] A small column, 0.5×6 cm, of Blue Sepharose (Pharmacia) is equilibrated with 0.02 *M* sodium phosphate, pH 7.2, 1 *M* NaCl (column buffer) at 25°. The crude ^{35}S-labeled IFN-β (10–50 ml) containing 1 *M* NaCl is passed through the column. The column is washed with an equal volume of the column buffer then eluted stepwise with 5 ml of column buffer containing 30% ethylene glycol and 50% ethylene glycol, respectively. Approximately 50% of the original biological activity is recovered with 95% of the recovered activity in the 50% ethylene glycol fraction.

Characterization. ^{35}S-labeled IFN-β is analyzed by electrophoresis on slab gels of polyacrylamide with the buffer system of Laemmli.[4] Samples are prepared for electrophoresis as follows: 1 ml each of crude ^{35}S-labeled IFN-β, ^{35}S-labeled uninduced medium, Blue Sepharose column flow-through, and 0.5 ml of 30 and 50% ethylene glycol fractions are adjusted to 0.1% SDS and dialyzed for 48 hr at 25° against 1 liter of 0.01% SDS with three changes of dialysate. The dialyzed samples are then concentrated to dryness in a Speed-Vac centrifuge (Savant Instruments). Each sample is dissolved in 0.03 ml of 0.05 *M* Tris · HCl, pH 6.8, 1% SDS, 5% glycerol and heated at 100° for 1 min. The 0.03 ml of each is loaded onto a 0.75-mm-thick 10–16% linear gradient polyacrylamide gel. After electrophoresis the gel is fixed in 40% methanol : 10% acetic acid : 50% H_2O for 20 min at 25° then equilibrated in 7% acetic acid. The gel is treated with Enhance (NEN) for 30 min, dried, and the radioactive protein bands are visualized by fluorography. The fluorogram shows that the fraction eluted with 50% ethylene glycol contains primarily only one polypeptide whose molecular weight is 20,000. This radioactive polypeptide is identified as IFN-β by its NH_2-terminal sequence[3] and by its immunoprecipitation with antibody prepared against homogeneous IFN-β.[3]

Purified ^{35}S-labeled IFN-β prepared by the procedure above has a specific radioactivity of 1–3 cpm per unit of IFN-β antiviral activity. Although useful for some studies, this specific radioactivity is not high enough for the IFN-β to be used in receptor binding experiments.

Concluding Comments

The *in vivo* labeling of IFN-β with ^{35}S-labeled methionine and subsequent purification yields ^{35}S-labeled IFN-β that can be used in various

[3] E. Knight, Jr. and M. W. Hunkapiller, *J. Interferon Res.* **1,** 297 (1981).
[4] U. K. Laemmli, *Nature (London)* **227,** 680 (1970).

biochemical studies. ^{35}S-labeled IFN-β has been used in investigating the removal of carbohydrate by glycosidoses from IFN-β[5] and to study the biosynthesis of IFN-β in human diploid fibroblasts.[6]

[5] E. Knight, Jr. and D. Fahey, *J. Interferon Res.* **2,** 421 (1982).
[6] T. P. Chow, W. DeGrado, and E. Knight, Jr., *J. Biol. Chem.* **259,** 12220 (1984).

[41] Labeling of Recombinant Interferon with [^{35}S]Methionine *in Vivo*

By ABBAS RASHIDBAIGI and SIDNEY PESTKA

Study of interferon binding to the plasma membrane receptor is the first step in establishment of interferon function. The number of interferon receptors is relatively low on the cell surface membrane (about 500–5000 receptors per cell). Therefore, in order to characterize the interferon receptor, it is necessary to obtain radiolabeled interferon with high specific radioactivity. Interferon alpha A (IFN-αA) has been iodinated to a reasonably high specific radioactivity with enzymobeads,[1] Bolton–Hunter reagent,[2] and chloramine T.[3] The iodination of interferon does not consistently produce active proteins. This might be due to the oxidation of polypeptide chains or the modification of amino acids which are required for binding to the receptor and for functional activity of interferons. Other interferon species (e.g., fibroblast and immune) are highly sensitive to modification by iodination. Moreover, any chemical modification procedure produces a heterogeneous population of interferon molecules. Here we report a method for uniformally labeling IFN-αA with [^{35}S]methionine by the *in vivo* maxicell labeling technique.[4]

Materials and General Procedures

D-Cycloserine was obtained from Sigma Chemical Co. It was prepared just prior to use as a 20 mg/ml solution in 0.1 *M* sodium phosphate buffer, pH 8.0. L-[^{35}S]Methionine with specific activity of 1110 Ci/mmol was purchased from Amersham. Acrylamide, *N′,N*-methylene bisacrylamide

[1] A. A. Branca and C. Baglioni, *Nature (London)* **294,** 768 (1981).
[2] K. Zoon, D. Zur Nedden, and H. Arnheiter, *J. Biol. Chem.* **257,** 4695 (1982).
[3] K. E. Mogensen and G. Uzé, this volume [37].
[4] A. Sancar, A. M. Hack, and W. D. Rupp, *J. Bacteriol.* **137,** 692 (1979).

(BIS), and sodium dodecyl sulfate were obtained from Bio-Rad. Indolyl-3-acrylic acid was from Aldrich Chemical Co. A stock solution of 25 mg/ml in ethanol was prepared and stored at $-20°$. Other chemicals used were of analytical grade purity.

Media. L-Broth[5] and M-9 medium[6] are prepared as described elsewhere. K-medium is a M-9 medium containing 1% casamino acids (Difco) and 1 μg/ml thiamine.[7] Sulfate-free Hershey salts is a medium containing the following components: NaCl, 5.4 g/liter; KCl, 3.0 g/liter; NH_4Cl, 1.1 g/liter; $FeCl_3 \cdot 6H_2O$, 0.2 g/liter; KH_2PO_4, 87 mg/liter; $CaCl_2 \cdot 2H_2O$, 15 mg/liter; $MgCl_2 \cdot 6H_2O$, 0.2 g/liter; Trisbase, 12.1 g/liter, pH 7.4. Phosphate-buffered saline (PBS) contains 0.01 *M* sodium phosphate, pH 7.3, and 0.15 *M* NaCl. All these media were sterilized by filtration through a 0.20-μm filter unit and stored at 4°.

Analytical Procedures and Assay. Sodium dodecyl sulfate–polyacrylamide gel electrophoresis on a slab gel was performed by the method of Laemmli[8] with a separating gel of 15% (w/v) acrylamide, 0.4% (w/v) BIS, and a stacking gel of 5% (w/v) acrylamide, 0.13% (w/v) BIS.

Antiviral activity was measured by a viral cytopathic inhibition assay on Madin–Darby bovine kidney (MDBK) cells as previously described.[9] The concentration of interferon was calculated from the specific activity of recombinant human IFN-αA of about 2×10^8 units/mg.[10,11] [^{35}S]Methionine incorporation was measured by solubilizing 10 μl of the cell suspension in 0.2 ml of 0.5 *N* KOH containing 0.05 mg/ml bovine serum albumin. The mixture was incubated at 32° for 15 min, and then neutralized with 0.1 ml of 1 *N* HCl. The proteins were precipitated by addition of 2 ml of 10% (w/v) cold trichloroacetic acid followed by incubation at 4° for 10 min. The proteins were filtered on 0.45-μm HA Millipore filters, washed with 3 ml of 2.5% (w/v) cold trichloroacetic acid, and rinsed with 95% ethanol at 4°. The filters were then dried under an infrared lamp for 10 min. Each filter was resuspended in 3 ml of scintillation fluid [0.5% (w/v) PPO and 0.02% (w/v) POPOP in toluene] and the radioactivity was measured in a Beckman model LS 7800 scintillation spectrometer. The efficiency of this counter for ^{35}S was 80%.

[5] S. E. Luria and J. W. Burrous, *J. Bacteriol.* **74,** 461 (1957).

[6] T. Maniatis, E. F. Fritsch, and J. Sambrook, "Molecular Cloning: A Laboratory Manual," p. 440. Cold Spring Harbor Lab., Cold Spring Harbor, New York, 1982.

[7] W. D. Rupp, C. E. Wilde, III, and D. L. Reno, *J. Mol. Biol.* **61,** 25 (1971).

[8] U. K. Laemmli, *Nature (London)* **227,** 680 (1970).

[9] P. C. Familletti, S. Rubinstein, and S. Pestka, this series, Vol. 78, p. 387.

[10] T. Staehelin, D. S. Hobbs, H.-F. Kung, C.-Y. Lai, and S. Pestka, *J. Biol. Chem.* **256,** 9750 (1981).

[11] T. Staehelin, C. Stähli, D. S. Hobbs, and S. Pestka, this series, Vol. 79, p. 589.

[^{35}S]Methionine Labeling of IFN-αA in Maxicells

Escherichia coli CSR 603[4] was transfected with an expression plasmid for IFN-αA, pLIFA-Trp 65.[12] The cells were grown overnight at 37° in L-broth containing 10 μg/ml tetracycline. The overnight culture was diluted, 1 ml to 50 ml of K-medium containing 10 μg/ml tetracycline, and then incubated at 37° for 2.5 hr to reach an OD_{600} of 0.4. The culture (40 ml) was irradiated for 10 sec with a hand held mineral light (model UVSL-58 lamp, Ultraviolet Products, Inc., San Gabriel, CA). The irradiation was performed at a short wavelength (254 nm) in a 150 × 15-mm Petri dish at a distance of 20 cm from the lamp. The cells were then incubated for 1 hr in a 300 ml flask at 37°. Approximately 70% of the cells were killed by this UV treatment. D-Cycloserine was added to the culture to a final concentration of 50 μg/ml. The cells were then incubated for another 18 hr at 37°, followed by centrifugation at 10,000 rpm (12,100 *g*) for 10 min in a Sorvall SS34 rotor. The supernatant was discarded and the cells were washed two times by resuspension and centrifugation in 20 ml of Hershey salts medium without sulfate. The cells were then resuspended in 5 ml of Hershey salts medium containing 0.4% (w/v) glucose, 1 μg/ml thiamine, 10 μg/ml of all unlabeled amino acids except methionine and cysteine, and 10 μg/ml tetracycline. The cell suspension was incubated at 37° for 1 hr while shaking at 300 rpm. [^{35}S]methionine, 32 μl (0.5 mCi), and indolyl-3-acrylic acid, for a final concentration of 50 μg/ml, were added to the culture followed by a 1.5 hr incubation at 37°. The cells were centrifuged at 12,000 rpm (17,300 *g*) in a Sorvall SS34 rotor for 30 min, and the cell pellets were resuspended in 5 ml PBS for purification of [^{35}S]Met-labeled interferon.

Purification of [^{35}S]Met-Labeled IFN-αA. Purification was performed as described,[10] with the modifications described here. All the purification steps were performed at 4°. The cell suspension (5 ml) was passed through a French pressure cell press (American Instrument Inc., Silver Spring, MD) with a 1-cm-diameter piston at 8000 psi. The broken cells were then centrifuged at 12,000 rpm (17,300 *g*) in a Sorvall SS34 rotor for 30 min. The pellet was discarded and the supernatant was diluted to 9 ml with PBS. The diluted sample was applied to a monoclonal antibody affinity column (0.2 ml bed volume, containing 0.26 mg of purified monoclonal antibody LI-8 coupled to Affi-Gel 10, Bio-Rad). Before use, the column was equilibrated with PBS and the flow rate was adjusted to 3.0 ml/hr. After the sample was applied, the column was washed with 1 ml of buffer

[12] D. V. Goeddel, E. R. Yelverton, A. Ullrich, H. L. Heyneker, G. Miozzari, W. Holmes, P. H. Seeburg, T. Dull, L. May, N. Stebbing, R. Crea, S. Maeda, R. McCandliss, A. Sloma, J. M. Tabor, M. Gross, P. C. Familletti, and S. Pestka, *Nature (London)* **287,** 411 (1980).

F [0.5 *M* NaCl, 25 m*M* Tris · HCl, pH 7.5, 0.2% (v/v) Triton X-100], and then rinsed with 1 ml of buffer G [0.15 *M* NaCl, 0.1% (v/v) Triton X-100]. [^{35}S]Met-labeled IFN-αA was eluted with 1 ml of buffer H [0.2 *N* acetic acid, 0.15 *M* NaCl, 0.1% (v/v) Triton X-100, pH 2.5]. This fraction was dried in a Speed Vac concentrator (Savant Instruments, Inc., Hicksville, NY) and its pH was adjusted to 7.3 by resuspending in 0.2 ml of PBS.

Concluding Comments

The maxicell procedure described here is a simple method for obtaining [^{35}S]Met-labeled IFN-αA. In this method chromosomal DNA of *E. coli* is damaged extensively upon UV irradiation. This results in diminished synthesis of *E. coli* proteins. The plasmid DNA which survived UV treatment continues to replicate at a high level.[4]

The IFN-αA gene in pBR322 is under the tryptophan operator/promoter control.[12] The induction of interferon expression was performed with indolyl-3-acrylic acid which is an efficient inducer of transcription from the tryptophan promoter.[13] Total ^{35}S incorporation into trichloroacetic acid precipitable proteins of *E. coli* CSR 603 was measured to be 23% of the total [^{35}S]methionine added after UV irradiation, cycloserine treatment, and indolyl-3-acrylic acid induction. Sodium dodecyl sulfate–polyacrylamide gel electrophoresis of these UV-treated cells revealed the presence of two major labeled protein bands (Fig. 1, lane 2). The presence of cycloserine was required for complete elimination of background labeling (data not shown). The apparent molecular weights of these two proteins were 17,500 and 37,000. The 17,500 Da protein had the same apparent molecular weight as ^{125}I-labeled IFN-αA (Fig. 1, compare lane 2 to lane 1). The 37,000 Da protein is likely to be the tetracycline-resistant polypeptide expressed by the plasmid.[4] When the CSR 603 cells were labeled with [^{35}S]methionine in the absence of UV irradiation, the labeling pattern of Fig. 1, lane 5, was obtained. In the absence of irradiation, all genomic proteins of *E. coli* were expressed.

^{35}S-Labeled interferon was efficiently purified by monoclonal antibody affinity chromatography. The summary of the purification is provided in the table. A homogeneously labeled protein was detected upon sodium dodecyl sulfate–polyacrylamide gel electrophoresis after acid elution from the monoclonal antibody, LI-8, affinity column (Fig. 1, lane 4). The specific radioactivity of the purified IFN-αA was measured to be 320 cpm/unit (477 Ci/mmol). A total of 5 μCi of [^{35}S]Met-labeled IFN-αA was obtained (1% overall incorporation of ^{35}S).

[13] B. P. Nichols and C. Yanofsky, this series, Vol. 101, p. 155.

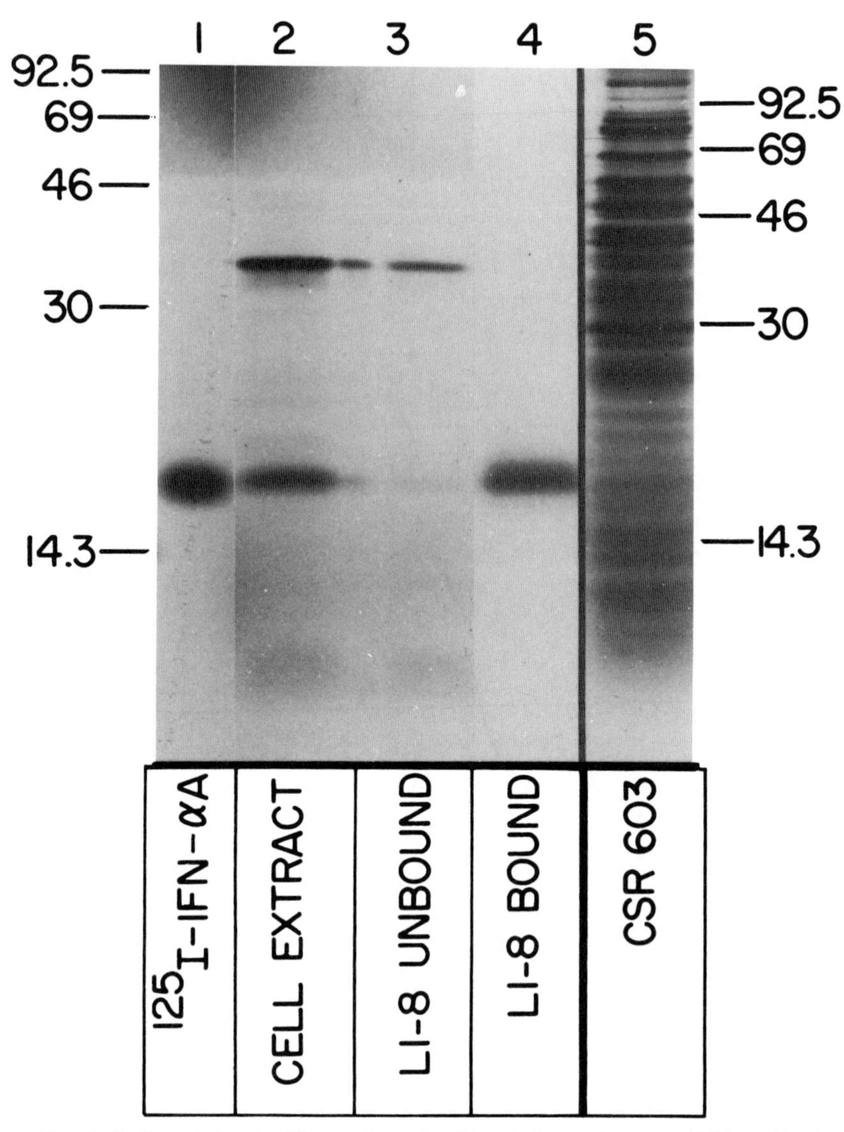

FIG. 1. Sodium dodecyl sulfate–polyacrylamide gel electrophoresis of [^{35}S]methionine-labeled interferon. *E. coli* CSR 603 cells containing an expression plasmid for IFN-αA were irradiated with UV light, incubated with D-cycloserine, induced with indolyl-3-acrylic acid, and labeled with [^{35}S]methionine. The cells were broken, and interferon was purified with a monoclonal antibody, LI-8, affinity column. At different stages of purification samples were analyzed by polyacrylamide gel electrophoresis. Lane 1 contained IFN-αA labeled with ^{125}I-labeled Bolton–Hunter reagent.[2] Lane 2 contained an extract of UV irradiated *E. coli* CSR cells after [^{35}S]methionine labeling. Lane 3 represents the fraction which did not bind to LI-8

PURIFICATION OF [^{35}S]METHIONINE-LABELED RECOMBINANT IFN-αA[a]

Step	Volume (ml)	Radioactivity		Activity	
		Total cpm	Recovered (%)	Total (units)	Recovered (%)
Cell extract	9.0	5.4×10^7	100	49,000	100
LI-8, unbound	9.0	3.9×10^7	72	1,100	2
LI-8, bound	1.0	8.8×10^6	16	32,000	65

[a] Interferon activity was determined by a cytopathic effect inhibition assay with vesicular stomatitis virus and MDBK cells as described.[9]

The labeling of IFN-αA with [^{35}S]methionine was previously reported.[14] Amplification of the plasmid containing the IFN-αA gene with chloramphenicol was used to increase the efficiency of labeling. However, most of the genomic and plasmid coded proteins were labeled by this technique. The resultant labeled proteins were then purified by monoclonal antibody affinity purification.[10,14]

The maxicell method described here is a simple and efficient method for *in vivo* labeling of plasmid coded proteins. It has several advantages over protein iodination and other techniques used for labeling interferons. It provides homogeneously labeled protein. It does not result in diminished activity of the protein of interest due to amino acid oxidation or modification. It is simple and reproducible. Since only the plasmid-coded proteins are specifically labeled, the labeled proteins can be easily visualized upon autoradiography.

Acknowledgments

Human ^{125}I-labeled IFN-αA was supplied by Dr. Jerome A. Langer. We thank Cynthia Rose for bioassays and Drs. S. Joseph Tarnowski, Courtney McGregor, and Robert Ning for samples of IFN-αA.

[14] C. W. Czarniecki, C. W. Fennie, D. B. Powers, and D. A. Estell, *J. Virol.* **49,** 490 (1984).

affinity column. Lane 4 contained the [^{35}S]methionine-labeled IFN-αA eluted from the affinity column with acid. Lane 5 represents unirradiated CSR cells labeled with [^{35}S]methionine. Electrophoresis of this lane was performed on a separate gel. The gels were dried and autoradiographed for 24 hr with Kodak X-Omat AR film with an Ilford tungstate intensifying screen at $-70°$. The standards were ^{14}C-labeled proteins from Amersham; their corresponding molecular weights are indicated ($\times 10^3$).

[42] Labeling of Interferons with [^{35}S]Methionine in a Cell-Free DNA-Dependent System

By HSIANG-FU KUNG

DNA-dependent cell-free systems have been used extensively to study the regulation of gene expression.[1] With the rapid development of recombinant DNA technology, numerous eukaryotic genes (e.g., interferons) have been cloned and expressed in *E. coli*.[2] The recombinant plasmid DNAs were used as templates in the *in vitro* system to label the gene products with [^{35}S]methionine.

In general most investigators have used an unfractionated S-30 extract following the original procedure of Zubay *et al.*[1] In order to remove endogenous templates in S-30 extract, we have used a partially fractionated system dependent on ribosomes, ribosomal wash and a soluble fraction which was partially purified by DEAE-cellulose chromatography.[3]

Materials and Reagents

[^{35}S]Methionine, 5 mCi/ml, 500–1300 Ci/mmol (Amersham)
Nucleotide triphosphates, cyclic adenosine 3′:5′-monophosphate, and phosphenolpyruvate (Sigma)
20 amino acids and dithiothreitol (Calbiochem)
E. coli K-12 tRNA (Schwarz/Mann)
Polyethylene glycol 8000 (formerly polyethylene glycol 6000, Fisher)
Guanosine 5′-diphosphate 2′(3′)-diphosphate (ppGpp) (PL Biochem)

Procedures

Growth of Bacteria. E. coli W4032 HfrC (Δlac, met$^-$, pro$^-$, Sm^S) was grown at 28° to a density of approximately 1×10^9 cells/ml in the medium described by Zubay *et al.*[1] Cells were stored at −80° and used for the preparation of S-30 extract according to the method of Zubay *et al.*[1] with some modifications.

[1] G. Zubay, D. A. Chamers, and L. C. Cheong, *in* "The Lactose Operon" (J. R. Beckwith and D. Zipser, eds.), p. 375. Cold Spring Harbor Lab., Cold Spring Harbor, New York, 1970.
[2] S. Pestka, *Arch. Biochem. Biophys.* **221,** 1 (1983).
[3] H. F. Kung, C. Spears, and H. Weissbach, *J. Biol. Chem.* **250,** 1556 (1975).

Preparation of S-30 Extract. All preparation steps were carried out at 4°. Frozen cells (100 g) were suspended in 150 ml of Buffer A (0.01 *M* Tris · acetate, pH 8.2, 0.014 magnesium acetate, 0.06 *M* potassium acetate, and 0.001 *M* dithiothreitol). The suspension was passed once at 6000 psi through the French Pressure Cell (American Instrument Company, Silver Spring, MD). The solution was then centrifuged at 30,000 *g* for 30 min. Three-fourths of the supernatant was withdrawn carefully with a Pasteur pipette and the S-30 extract was used for further fractionation.

Fractionation of S-30 Extract. The S-30 extract was centrifuged at 200,000 *g* for 1 hr. The supernatant fraction (S-200) was saved and the ribosome pellet was washed once with 20 ml of Buffer A containing 1 *M* NH_4Cl for 4 hr. After centrifugation the supernatant fraction was saved (ribosomal wash) and the ribosomes were washed two more times with Buffer A containing 1 *M* NH_4Cl. Ribosomes (three times washed) were suspended in Buffer A. Ribosomal wash and S-200 were dialyzed overnight against Buffer A. The dialyzed S-200 was loaded on a DE52 column (2.5 × 10 cm) equilibrated with Buffer A. The column was washed with 200 ml of Buffer A and then eluted with 300 ml of 0.25 *M* potassium phosphate buffer (pH 6.5) containing 10 m*M* magnesium acetate and 1 m*M* dithiothreitol. This latter eluate was precipitated by the addition of ammonium sulfate to 85% saturation (0.25 *M* DEAE salt eluate) and the precipitate, dissolved in 20 ml of Buffer A, was dialyzed overnight against the same buffer. The DE52 column was then stripped with 200 ml of 1 *M* potassium phosphate (pH 6.5), and the protein in this eluate was precipitated by the addition of ammonium sulfate to 40% saturation. The precipitate (1 *M* DEAE salt eluate) was dissolved in 2 ml of Buffer A and was dialyzed overnight against the same buffer. The dialyzed fractions were centrifuged to remove precipitates (30,000 *g*, 10 min) and the clear supernatants were used in the cell-free system. The samples were aliquoted and stored in liquid nitrogen.

Plasmid DNA Preparation. Plasmid constructions for the expression of interferons were described previously.[4,5] Plasmid DNA was prepared essentially as described by Clewell and Helinski.[6] A simple procedure for preparation of pure plasmid DNA free from chromosomal DNA has also been used[7] except RNase treatment was omitted.

[4] D. V. Goeddel, E. Yelverton, A. Ullrich, H. L. Heyneker, G. Miozzari, W. Holmes, P. H. Seeburg, T. Dull, L. May, N. Stebbing, R. Crea, S. Maeda, R. McCandliss, A. Sloma, J. M. Tabor, M. Gross, P. C. Familletti, and S. Pestka, *Nature (London)* **287,** 411 (1980).

[5] D. V. Goeddel, H. M. Shepard, E. Yelverton, D. Leung, R. Crea, A. Sloma, and S. Pestka, *Nucleic Acids Res.* **8,** 4057 (1980).

[6] D. B. Clewell and D. R. Helinski, *Biochemistry* **9,** 4428 (1970).

[7] M. Mukhopadhyay and N. C. Mandal, *Anal. Biochem.* **133,** 265 (1983).

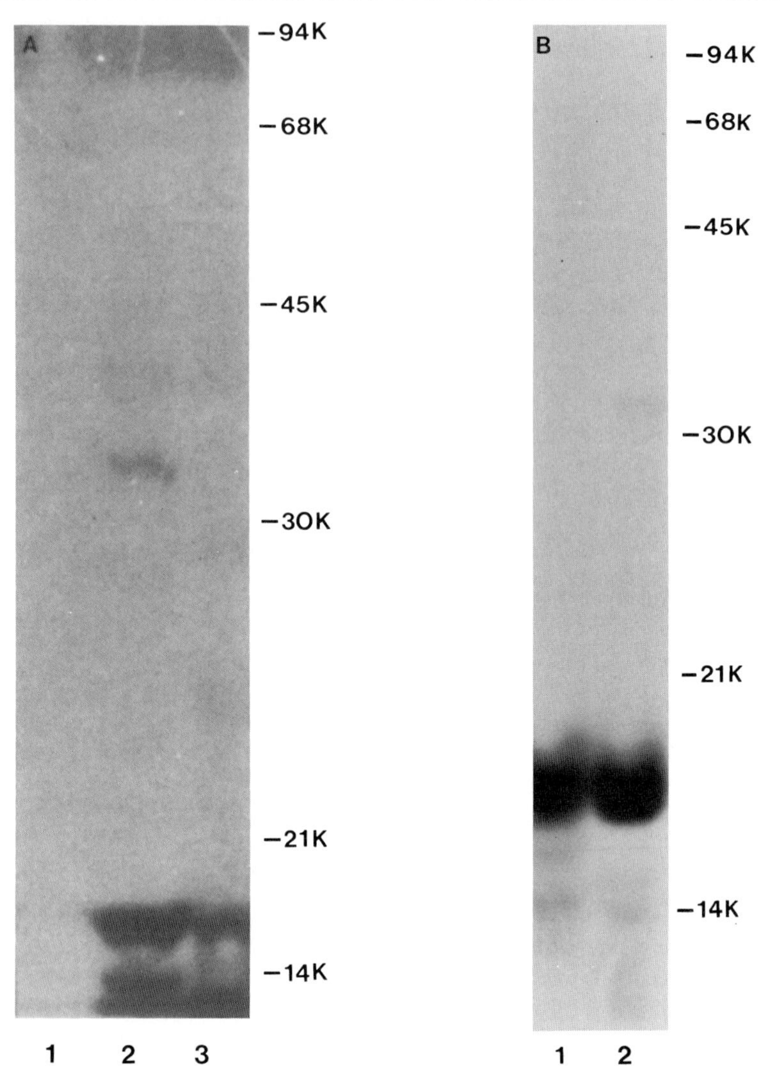

FIG. 1. Sodium dodecyl sulfate–polyacrylamide gel electrophoresis of cell-free products. The [^{35}S]methionine-labeled products were prepared as described in the text. Samples (1 μl) were dissolved in 10 μl of sample buffer and electrophoresis was carried out either overnight at a constant voltage of 30 V or for 3–4 hr at 100 V. Gels were stained with Coomassie brilliant blue, destained, treated with Enhance (NEN), dried, and exposed for autoradiography on X-Omat R film (Kodak). Molecular weight standards used were the following: phosphorylase B, 94 kDa; bovine serum albumin, 68 kDa; ovalbumin, 45 kDa; carbonic anhydrase, 30 kDa; soybean trypsin inhibitor, 21 kDa; and lysozyme, 14 kDa. DNA in each reaction mixture was as follows: (A) Lane 1, no DNA; Lanes 2 and 3, plasmid DNA containing the leukocyte IFN-αA gene. (B) Lane 1, plasmid DNA containing immune interferon gene; Lane 2, plasmid DNA containing fibroblast interferon gene.

DNA-Dependent Cell-Free System. The complete system (17.5 μl) contained 20 m*M* Tris · acetate, pH 8.2, 35 m*M* ammonium acetate, 65 m*M* potassium acetate, 10 m*M* magnesium acetate, 0.8 m*M* spermidine hydrochloride, 2.4 m*M* dithiothreitol, 0.93 m*M* each of UTP, CTP, and GTP, 3 m*M* ATP, 24 m*M* phosphoenolpyruvate, 0.1 μg of pyruvate kinase, 0.05 m*M* ppGpp, 0.7 m*M* 3′,5′-cAMP, 15 m*M* calcium leucovorin, 6 μg of *E. coli* tRNA, 0.3 mg of polyethyleneglycol, 0.112 m*M* each amino acid except methionine, 0.02 m*M* methionine, 20 μCi [^{35}S]methionine, 0.6 A_{260} units of NH_4Cl-washed ribosomes, ribosomal wash (30 μg of protein), a 0.25 *M* DEAE salt eluate (50 μg of protein), a 1 *M* DEAE salt eluate (4 μg of protein), and 2 μg of DNA template. All components except [^{35}S]methionine and DNA template were mixed and stored in small portions in liquid nitrogen. Before incubation, the mixture was thawed and used immediately. [^{35}S]Methionine and DNA template were added to the mixture and the incubations were carried out at 37° for 90 min. Optimal concentration of Mg^{2+} was determined for each batch of *E. coli* extracts.

Analysis and Purification of Gene Products after in Vitro Synthesis. Conditions for *in vitro* synthesis were as described above and the *in vitro* synthesis products were analyzed by sodium dodecyl sulfate–polyacrylamide gel electrophoresis[8] with 12.5% uniform polyacrylamide gels. The stacking gel was 5% polyacrylamide. Figure 1 shows the labeling of leukocyte, fibroblast, and immune interferons with [^{35}S]methionine in the cell-free DNA-dependent system. The system is completely dependent on the addition of DNA templates (Fig. 1A, lane 1) and interferon is the major product synthesized *in vitro*. The labeled interferons can be purified by monoclonal antibody[9] or dye-affinity[10] column.

Concluding Comments

The *in vitro* system is very useful for labeling and identifying gene products encoded by expression vectors in which a foreign DNA sequence has been inserted.[11] It also allows the detection of truncated proteins that may arise from unintended changes in the cloned DNA sequence(s) during the cloning procedure.[12] The *in vitro* system can also be used to examine the relative efficiencies of gene expression. High specific

[8] U. K. Laemmli, *Nature (London)* **227,** 680 (1970).

[9] T. Staehelin, D. S. Hobbs, H. F. Kung, and S. Pestka, this series, Vol. 78, p. 505.

[10] E. Knight, Jr., M. W. Hunkapiller, B. D. Korant, R. W. F. Hardy, and L. E. Hood, *Science* **207,** 525 (1980).

[11] N. Chang, H. F. Kung, and S. Pestka, *Arch. Biochem. Biophys.* **221,** 585 (1983).

[12] H. F. Kung, unpublished results.

radioactive interferon can be synthesized *in vitro* with [^{35}S]methionine (1300 Ci/mmol, Amsersham) in the absence of unlabeled methionine. This material is useful for receptor studies.

Acknowledgments

I wish to thank Ms. Eva Bekesi for her excellent technical assistance and Ms. Sharon Smith for typing the manuscript.

[43] Phosphorylation of Human Immune Interferon (IFN-γ)

By HSIANG-FU KUNG and EVA BEKESI

Posttranslational modifications of proteins (e.g., glycosylations, acetylations, uridylylations, methylations, and phosphorylations) often play an important role in cellular regulatory processes.[1-3] Interest in protein phosphorylation has increased enormously over the past few years.[2,3] The number of proteins known to undergo phosphorylation–dephosphorylation has risen and the scope of this particular type of regulatory mechanism has broadened to include the protein-synthesizing complex[4] and tyrosine phosphorylation in malignant transformation by viruses.[5]

Protein kinases catalyze the transfer of the γ-phosphoryl group of ATP to an acceptor protein substrate. There exist cyclic nucleotide-dependent protein kinases, calcium-dependent protein kinases, tyrosine-specific protein kinases, and cyclic nucleotide- and calcium-independent protein kinases. Lack of absolute substrate specificity seems to be a general property of protein kinases which can be explained, in part at least, by the fact that many protein kinases are related in evolution. In this chapter we describe the phosphorylation of human immune interferon (IFN-γ) with cyclic AMP-dependent protein kinase from bovine heart. The phosphorylation reaction should be useful for the preparation of radioactive IFN-γ for receptor studies.

[1] F. Wold and K. Moldave, this series, Vols. 106 and 107.

[2] E. G. Krebs and J. A. Beavo, *Annu. Rev. Biochem.* **48,** 923 (1979).

[3] J. D. Corbin and J. G. Hardman, this series, Vol. 99.

[4] G. M. Hathaway, T. S. Lundak, S. M. Tahara, and J. A. Traugh, this series, Vol. 60, p. 495.

[5] T. Hunter and J. A. Cooper, *Prog. Nucleic Acid Res.* **29,** 221 (1983).

Materials and Reagents

[γ-^{32}P]ATP, 10 mCi/ml, 5000 Ci/mmol (Amersham)
ATP (Sigma)
Dithiothreitol (Calbiochem)
Cyclic AMP-dependent protein kinase from bovine heart muscle, catalytic subunit, 20,000 units/mg (Sigma)

Recombinant IFN-γ and natural IFN-γ were purified to homogeneity by the procedure described by Kung *et al.*[6]

Phosphorylation Reaction and Product Analysis

The reaction mixture (30 μl) for protein phosphorylation contains 20 m*M* Tris · HCl, pH 7.4, 1 m*M* dithiothreitol, 100 m*M* NaCl, 12 m*M* $MgCl_2$, [γ-^{32}P]ATP (2.5 μCi), 1.4 m*M* ATP, 1 unit of protein kinase, and various amounts of IFN-γ as indicated. Incubations are performed at 37° for 1 hr and the phosphorylation products were analyzed on sodium dodecyl sulfate–polyacrylamide gel electrophoresis[7] with 12.5% uniform polyacrylamide gels. The stacking gels consisted of 5% polyacrylamide. Samples (1 μl) were dissolved in 10 μl of sample buffer and electrophoresis was carried out for 3–4 hr at 100 V. Gels were stained with Coomassie brilliant blue, destained, dried, and exposed 24 hr for autoradiography on X-Omat R film (Kodak). Molecular weight standards used were the following: phosphorylase B, 94,000; bovine serum albumin, 66,000; ovalbumin, 45,000; carbonic anhydrase, 30,000; soybean trypsin inhibitor, 21,000; and lysozyme, 14,000. Figure 1 shows the labeling of IFN-γ with [^{32}P]ATP. All three recombinant IFN-γ species with different NH_2-terminal amino acid sequences were labeled efficiently. The intact recombinant IFN-γ have an apparent molecular weight of about 18,000 on sodium dodecyl sulfate–polyacrylamide gel electrophoresis. The phosphorylated interferons (^{32}P labeled) migrated slightly slower than the standards of nonphosphorylated IFN-γ. The dimer of recombinant IFN-γ also appeared to be phosphorylated. Natural IFN-γ is glycosylated with apparent molecular weights of 25,000 and 20,000 on sodium dodecyl sulfate–polyacrylamide gel electrophoresis.[8] Because of the limited supply of natural IFN-γ, a minimal amount was used in the phosphorylation reaction. Even at limited concentration of natural IFN-γ, we have obtained

[6] H. F. Kung, Y. C. Pan, J. Moschera, K. Tsai, E. Bekesi, M. Chang, H. Sugino, and S. Honda, this volume [29].

[7] U. K. Laemmli, *Nature (London)* **227,** 680 (1970).

[8] Y. K. Yip, B. S. Barrowclough, C. Urban, and J. Vilček, *Proc. Natl. Acad. Sci. U.S.A.* **79,** 1820 (1982).

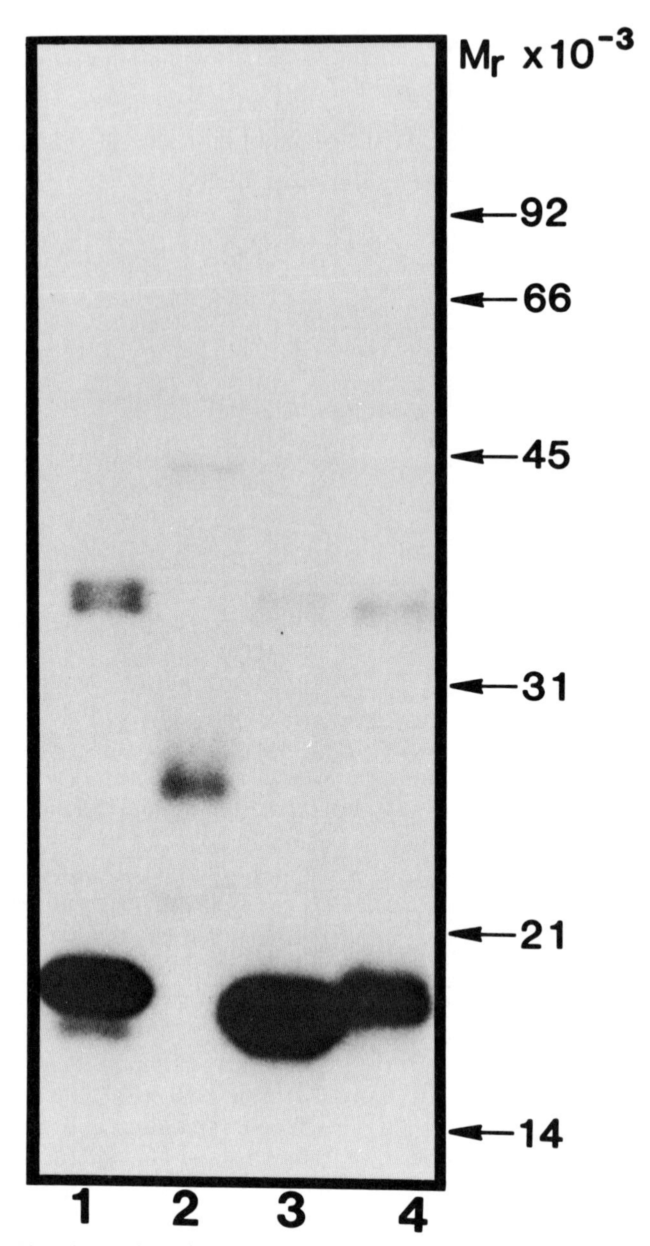

FIG. 1. Phosphorylation of human immune interferons. Experimental details are described in the text. The following immune interferons were used as the substrates: Lane 1, recombinant immune interferon (30 μg) with NH_2-terminal sequence of Ser-Tyr-Ser-Gln-Asp-Pro-; Lane 2, natural immune interferon (3 μg); Lane 3, recombinant immune interferon (50 μg) with NH_2-terminal sequence of Met-Gln-Asp-Pro-; Lane 4, recombinant immune interferon (30 μg) with NH_2-terminal sequence of Cys-Tyr-Cys-Gln-Asp-Pro-; and protein standard markers purchased from Bio-Rad.

EFFECT OF PHOSPHORYLATION ON THE ANTIVIRAL ACTIVITY OF RECOMBINANT Hu-IFN-γ[a]

NH_2-terminal sequence of immune interferon	Protein kinase	Antiviral activity (units/ml) (×10^6) Exp. 1	Exp. 2
1. Cys-Tyr-Cys-Gln-Asp-Pro-	−	7.6	9
	+	15	9
2. Met-Gln-Asp-Pro-	−		13
	+		13

[a] Experimental details are described in the text and in the legend to Fig. 1 (Lanes 3 and 4) except for the omission of [γ-^{32}P]ATP in the reaction mixtures. Protein kinase was added where indicated. Antiviral activity was measured by a cytopathic effect inhibition assay with vesicular stomatitis virus and human WISH cells as reported.[9,10] All titers were adjusted to a laboratory standard preparation of known value which had been adjusted to the NIH human IFN-γ standard (No. Gg23-901-530) and are expressed in reference units/ml.

phosphorylation of the 25,000 and 20,000 species (Fig. 1, lane 2). The 25,000 molecular weight species appeared to be phosphorylated better than the 20,000 molecular weight moiety. In addition, 45,000 and 66,000 molecular weight proteins were phosphorylated (labeling was clearer upon longer exposure of the autoradiograph).

These results indicated that both recombinant and natural IFN-γ can be phosphorylated with the catalytic subunit of protein kinase from bovine heart. In the case of recombinant interferons (Fig. 1, lanes 3 and 4), it has been estimated that about 5–15% of the [γ-^{32}P]ATP was incorporated into an 18,000 MW protein. This suggests that at least 2 molecules of ^{32}P were incorporated per molecule of interferon. The effect of phosphorylation on the antiviral activity of recombinant IFN-γ was determined in a parallel experiment. As shown in the table,[9,10] the phosphorylation does not affect the antiviral activity of the IFN-γ. Since the phosphorylated materials are biologically active, they are suitable for receptor studies. We have obtained ^{32}P-labeled IFN-γ with radioactivity greater than 10,000 cpm/unit of interferon by replacing unlabeled ATP with [γ-^{32}P]ATP and lowering the concentration of IFN-γ in the incubation mixtures.

We have also examined the phosphorylation of other recombinant human interferons under the identical conditions. Figure 2 shows that recombinant human leukocyte interferon (IFN-αA) and fibroblast inter-

[9] S. Rubinstein, P. C. Familletti, and S. Pestka, *J. Virol.* **37,** 755 (1981).
[10] P. C. Familletti, S. Rubinstein, and S. Pestka, this series, Vol. 78, p. 3817.

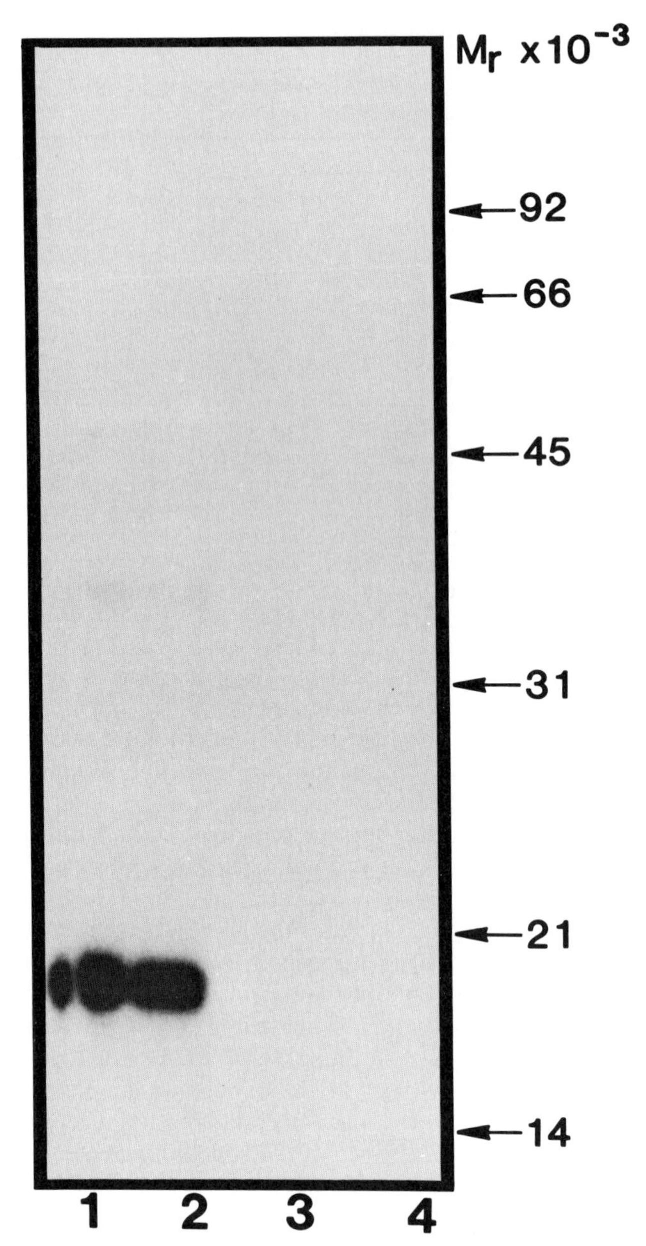

FIG. 2. Phosphorylation of various recombinant human interferons. Experimental details are described in the text. The following recombinant human interferons were used as the substrates: Lanes 1 and 2, recombinant immune interferon (15 and 30 μg, respectively) with NH_2-terminal sequence of Cys-Tyr-Cys-Gln-Asp-Pro-; Lane 3, recombinant IFN-αA; Lane 4, recombinant IFN-β; and protein standard markers purchased from Bio-Rad. Recombinant leukocyte and fibroblast interferons were purified by published procedures.[11,12]

feron (IFN-β) are very poor substrates for bovine heart protein kinase (Fig. 2, lanes 3 and 4).[11,12]

Concluding Remarks

Cyclic AMP (cAMP)-dependent protein kinase catalyzes the transfer of the γ-phosphate of ATP to serine and/or threonine hydroxyl groups in various protein substrates. IFN-γ can be labeled simply and efficiently with the catalytic subunit of cAMP-dependent protein kinase from bovine heart. Beef heart protein kinase catalyzes the phosphorylation of protamine, arginine-rich histone, and lysine-rich histone at relative rates of 10 : 5 : 2. Serum albumin and casein are poor substrates.[13] IFN-γ is a basic protein with high content of arginine and lysine (about 20%).[14,15] Therefore, it is not surprising that IFN-γ can be used as a substrate for bovine heart protein kinase.

Chany *et al.*[16] also recently reported that phosphorylation of recombinant Hu-IFN-γ by protein kinase present in cell culture medium.[16] Although any physiological role of phosphorylation of IFN-γ is unclear, it will be of interest to study the phosphorylation of natural IFN-γ *in vivo* and to compare the biological properties of phosphorylated IFN-γ with nonphosphorylated IFN-γ. The present chapter describes *in vitro* phosphorylation of IFN-γ to high specific radioactivity with [γ-^{32}P]ATP in the absence of unlabeled ATP. This radioactive IFN-γ with full biological activity is suitable for receptor studies. Preliminary data indicate that the ^{32}P-labeled IFN-γ binds specifically to IFN-γ receptors of human cell lines.[17]

Acknowledgments

We are grateful to C. H. Chen, S. J. Tarnowski, K. Tsai, J. Moschera, and R. Chizzonite for supplying us with purified interferons. We wish to thank L. Petervary for performing interferon assays.

[11] T. Staehelin, D. S. Hobbs, H. F. Kung, C. Y. Lai, and S. Pestka, *J. Biol. Chem.* **256,** 9750 (1981).

[12] E. Knight, M. W. Hunkapiller, B. D. Korant, R. W. F. Hardy, and L. E. Hood, *Science* **207,** 525 (1980).

[13] C. S. Rubin, J. Erlichman, and O. M. Rosen, this series, Vol. 99, p. 309.

[14] P. W. Gray, D. W. Leung, D. Pennica, E. Yelverton, R. Majarian, C. C. Simonsen, R. Dergnck, P. J. Sherwood, D. M. Wallace, S. L. Berger, A. D. Levinson, and D. V. Goeddel, *Nature (London)* **295,** 503 (1982).

[15] R. Devos, H. Cheroutre, Y. Taya, W. Degrave, H. van Heuverswyn, and W. Fiers, *Nucleic Acids Res.* **10,** 2487 (1982).

[16] B. Robert-Galliot, M. J. Commoy-Chevalier, P. Georges, and C. Chany, *J. Gen. Virol.* **66,** 1439 (1985).

[17] A. Rashidbaigi, H.-F. Kung, and S. Pestka, *J. Biol. Chem.* **260,** 8514 (1985).

Section VI

Procedures for Studying the Interferon Receptor and Uptake of Interferon

[44] Procedures for Studying Binding of Interferon to Human Cells in Suspension Cultures

By JEROME A. LANGER and SIDNEY PESTKA

It is desirable to have a simple method for measuring the binding of interferon to cells which grow in suspension. Such cells include various naturally occurring circulating cells, both normal and malignant, including many which function in immune responses and for which interferon is a biological effector.[1,2] There are also a large number of established human cell lines which grow in suspension. Many of these lines are derived from leukemic cells (e.g., Daudi, HL60, KG-1) and may prove valuable in studying the various roles of interferon. Indeed, the Daudi lymphoblastoid cell line[3] is probably the best characterized human cell line in terms of its interaction with interferon[1,4–12] and has been useful in correlating the growth inhibition by interferon with its binding.[13] In addition, many suspension lines are relatively easy to grow in large quantities, making such cells attractive for work on the interferon receptor. We describe here a simple and rapid assay for the binding of interferon to cells in suspension culture. Although the method is applicable to cells of any species, human cells are used in these examples.

[1] K. E. Mogensen, M.-T. Bandu, F. Vignaux, M. Aguet, and I. Gresser, *Int. J. Cancer* **28,** 575 (1981).

[2] J. R. Ortaldo, A. Mantovani, D. Hobbs, M. Rubinstein, S. Pestka, and R. B. Herberman, *Int. J. Cancer* **31,** 285 (1983).

[3] E. Klein, G. Klein, J. S. Nadkarni, J. J. Nadkarni, H. Wigzell, and P. Clifford, *Cancer Res.* **28,** 1300 (1968).

[4] A. A. Branca and C. Baglioni, *Nature (London)* **294,** 768 (1981).

[5] A. A. Branca and C. Baglioni, *J. Biol. Chem.* **257,** 13197 (1982).

[6] A. A. Branca, C. R. Faltynek, S. B. D'Alessandro, and C. Baglioni, *J. Biol. Chem.* **257,** 13291 (1982).

[7] A. R. Joshi, F. H. Sarkar, and S. L. Gupta, *J. Biol. Chem.* **257,** 13884 (1982).

[8] P. Eid and K. E. Mogensen, *FEBS Lett.* **156,** 157 (1983).

[9] C. R. Faltynek, A. A. Branca, S. McCandliss, and C. Baglioni, *Proc. Natl. Acad. Sci. U.S.A.* **80,** 3269 (1983).

[10] G. E. Hannigan, D. R. Gewert, E. N. Fish, S. E. Read, and B. R. G. Williams, *Biochem. Biophys. Res. Commun.* **110,** 537 (1983).

[11] S. Yonehara, M. Yonehara-Takahashi, A. Ishii, and S. Nagata, *J. Biol. Chem.* **258,** 9046 (1983).

[12] S. Yonehara, A. Ishii, and M. Yonehara-Takahashi, *J. Gen. Virol.* **64,** 2409 (1983).

[13] K. E. Mogensen and M.-T. Bandu, *Eur. J. Biochem.* **134,** 355 (1983).

Cells

Various suspension cells, including Daudi,[3] HL-60,[14,15] KG-1,[16] and U937[17] were used. These cells, grown in RPMI-1640 supplemented with 10% heat-inactivated fetal calf serum and 50 μg/ml gentamycin, are concentrated by centrifugation (1000 *g;* 10 min) and resuspended in the same or another medium at a concentration of $1-2 \times 10^7$ cells/ml. Virtually any isotonic medium can be used for resuspension and assay. With some cell types, particularly cells from natural sources (e.g., peripheral blood leukocytes) or cultured cells with a significant number of nonviable cells as judged by the uptake of trypan blue, it was found useful to purify the cells on Ficoll (Histopaque-1077; Sigma) according to the supplier's protocol. This often resulted in less nonspecific binding of interferon and higher signal-to-noise ratios in binding experiments.

Preparation of Radioiodinated Interferon (IFN-αA)

IFN-αA was prepared as described[18,19] and was radioiodinated with Bolton–Hunter reagent. Interferon at 0.5 mg/ml was dialyzed extensively at 4° against 0.1 *M* sodium borate, pH 8.5. ^{125}I-labeled Bolton–Hunter reagent (Amersham; about 2000 Ci/mmol; 1 mCi) was dried under a stream of N_2. Twenty microliters of the dialyzed IFN-αA was added to the dry Bolton–Hunter reagent. After 1.5 hr at 0°, the reaction was quenched by the addition of 5 μl of 0.2 *M* glycine, followed by incubation at room temperature for 5 min. Protein was separated from small molecules by chromatography on a 10-ml Sephadex G-25 column (Bio-Rad; PD-10 prepacked column) equilibrated with 1% (w/v) gelatin in Dulbecco's phosphate-buffered saline lacking magnesium and calcium (PBS). Approximately 40–50% of the radioiodine was recovered in the void volume. The ^{125}I-labeled IFN-αA was further purified by immunoaffinity chromatography using monoclonal antibody LI-8,[18,19] which seems to separate active from inactive interferon.[20] The ^{125}I-labeled IFN-αA was stored in solution at 4°. The recovery of antiviral activity was determined

[14] S. J. Collins, R. C. Gallo, and R. E. Gallagher, *Nature (London)* **270,** 347 (1977).

[15] R. Gallagher, S. Collins, J. Trujillo, K. McCredie, M. Ahearn, S. Tsai, R. Metzgar, G. Aulakh, R. Ting, F. Ruscetti, and R. Gallo, *Blood* **54,** 713 (1979).

[16] H. P. Koeffler and D. W. Golde, *Science* **200,** 1153 (1978).

[17] C. Sundström and K. Nilsson, *Int. J. Cancer* **17,** 565 (1976).

[18] T. Staehelin, D. S. Hobbs, H.-F. Kung, C.-Y. Lai, and S. Pestka, *J. Biol. Chem.* **256,** 9750 (1981).

[19] T. Staehelin, D. S. Hobbs, H.-F. Kung, and S. Pestka, this series, Vol. 78, p. 505.

[20] S. Pestka, B. Kelder, J. A. Langer, and T. Staehelin, *Arch. Biochem. Biophys.* **224,** 111 (1983).

on both bovine MDBK and human WISH cells,[21] so that the ratio of activity on the two cell types before and after radioiodination could be determined. This comparison was deemed necessary because of our observation that iodination frequently leads to a large decrease of antiviral activity on human cells while the activity on bovine cells remains largely unchanged (J. A. Langer, unpublished observation). The Bolton–Hunter technique described here consistently yielded ^{125}I-labeled IFN-αA with good activity on both MDBK and WISH cells: The WISH/MDBK antiviral ratio was reduced by no more than a factor of 2–4. The results of labeling interferon with Bolton–Hunter reagent by slightly different protocols[22–24] were not evaluated by us (see also this volume [38]). However, the use of several modifications of the chloramine T method or solid-phase lactoperoxidase procedure (Enzymobeads, Bio-Rad) often produced preparations with drastically reduced (about 5%) activity on human cells, albeit with normal activity on MDBK cells. However, other investigators have claimed success in iodinating interferon with these other techniques[4,7,13] (see this volume [36] and [37]). ^{125}I-labeled IFN-αA had a minimum specific activity of 125 cpm/unit of antiviral activity (MDBK) and frequently exceeded 200 cpm/unit. This is equivalent to 2.5×10^7 cpm/μg (3.6×10^5 Ci/mol), based on a specific activity of 2×10^8 units/mg protein and a molecular weight of 19,200.[18,25]

Binding of ^{125}I-Labeled IFN-αA to Suspension Cultures

The cell suspension was allowed to equilibrate at the desired temperature for at least 30 min. To 100 μl of the cell suspension ^{125}I-labeled IFN-αA was added for a final volume of 125 μl. Interferon was added in the range of 2000 to 1×10^6 cpm (16 to 8000 units) corresponding to 128 to 64,000 units/ml (0.033 to 16 nM) for material at the lowest specific activity. Higher radioactive specific activities allowed experiments to be conducted at correspondingly lower concentrations. A control with at least 100-fold excess nonradioactive IFN-αA was also prepared for each concentration of radioactive interferon. (More precisely, this control should contain IFN-αA at a concentration about 100 times greater than the K_d. In practice, this is generally not necessary.) For the cells which we have tested, maximum binding is reached within 1 hr at 4° or room temperature (20–23°). Longer incubations can be done at 4°. For experiments con-

[21] P. C. Familletti, S. Rubinstein, and S. Pestka, this series, Vol. 78, p. 387.

[22] A. E. Bolton and W. M. Hunter, *Biochem. J.* **133,** 529 (1973).

[23] E. Knight, Jr., *J. Gen. Virol.* **40,** 681 (1978).

[24] K. Zoon, D. Zur Nedden, and H. Arnheiter, *J. Biol. Chem.* **257,** 4695 (1982).

[25] S. Pestka, *Arch. Biochem. Biophys.* **221,** 1 (1983).

ducted at 37°, the incubation is generally done for 30–45 min to avoid internalization. Incubations are performed in polypropylene tubes, usually Eppendorf or Fisher 1.5-ml conical tubes. Following incubation, the entire reaction mixture is layered over a 250 μl cushion of 10% (w/v) sucrose in PBS in a small tube with an elongated tip (Sarstedt sample cup #72.702; Fig. 1). The cells were pelleted by centrifugation (Beckman microfuge B, 1 min; about 11,000 *g*). Tubes were frozen in liquid nitrogen and the tips containing the cell pellets were cut off. Radioactivity of the tip (bound) and the rest of the tube (free) was determined in a gamma counter. "Specific" binding is defined at a given concentration of ^{125}I-labeled IFN-αA as the difference between binding in the absence ("total binding") and the presence ("nonspecific") of excess nonradioactive IFN-αA.

Concluding Comments

The method described here is rapid and convenient. Because the cells are sedimented through a sucrose cushion into an elongated tip, no additional washing of the cells after pelleting is required. This results from the design of the Sarstedt sample cup (Fig. 1), permits rapid handling of a large number of samples, and is a major advantage of the method. A cushion of 10% sucrose has been sufficient for complete separation of bound from unbound interferon. We have found no need to use several sucrose layers of different concentrations or to use mineral oil rather than sucrose.

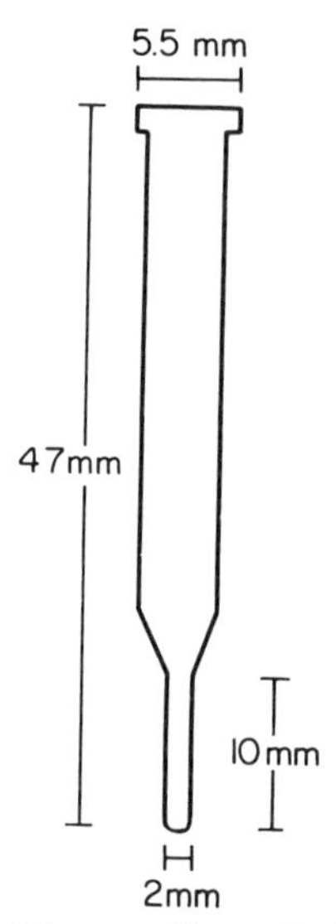

FIG. 1. Schematic diagram of the centrifuge tube (Sarstedt sample cup #72.702).

BINDING OF ^{125}I-LABELED IFN-αA TO SUSPENSION CELLS[a]

Cell type	Origin	Temperature	Molecules IFN per cell	K_{diss}^{app} (M)
Daudi	Human	4°	$7–10 \times 10^3$	$3–5 \times 10^{-9}$
		23°	$2–5 \times 10^3$	$3–5 \times 10^{-10}$
		37°	$2–3 \times 10^3$	$2–3 \times 10^{-10}$
U937	Human	4°	$2–4 \times 10^3$	2×10^{-9}
		23°	$1–2 \times 10^3$	3×10^{-10}
HL60	Human	4°	$1–2 \times 10^3$	$0.5–1 \times 10^{-8}$
		23°	$4–7 \times 10^2$	$4–6 \times 10^{-10}$
HL60 [+DMSO][b]	Human	4°	$2–3 \times 10^3$	$3–5 \times 10^{-9}$
		23°	2×10^3	5×10^{-10}
BL3	Bovine	4°	2×10^3	1×10^{-10}
		23°	2×10^3	7×10^{-11}

[a] The range of values reflects multiple determinations as well as the statistical errors and variation in determination of specific activities and statistical errors in final calculations of these parameters.

[b] HL60 [+DMSO] denotes HL60 cells which were grown in 1.25% DMSO for 4 days to induce differentiation from the promyelocytic to neutrophil-like state.[15,20,32]

An example of some binding parameters obtained with several cell lines is shown in the table (see Ref. 26 for a review of other results). Despite the range of values shown, Daudi cells have more binding sites and/or tighter binding than other human cell types examined thus far.

The optimal temperature and incubation time must be determined experimentally for each cell type; conditions given here are reasonable starting points (Figs. 2 and 3). Although binding at 4° is preferable in terms of limiting the effects of cellular metabolism and possible internalization, we have found that little internalization occurs at 23° during an incubation of 1 hr. A temperature of 23° is more convenient than 4° where the amount of binding, particularly at low concentrations of IFN, is low (Fig. 3). This is equivalent to saying that the K_d is a function of temperature. The measurement of binding at 23 or 37° can thus be an experimental aid, particularly when binding at 4° is difficult to quantitate. However, experiments conducted at the higher temperatures may reflect additional complexities such as internalization and cellular metabolism. Although reliable data for bovine kidney cells have been obtained at 4°,[24] very few reliable data on binding to human cells have been measured at this temperature. Since the binding parameters are a function of temperature, the temperature must always be reported with the measured parameters. The effect of culture age or density on binding should be explored in careful binding studies.

[26] K. C. Zoon and H. Arnheiter, *Pharmacol. Ther.* **24,** 259 (1985).

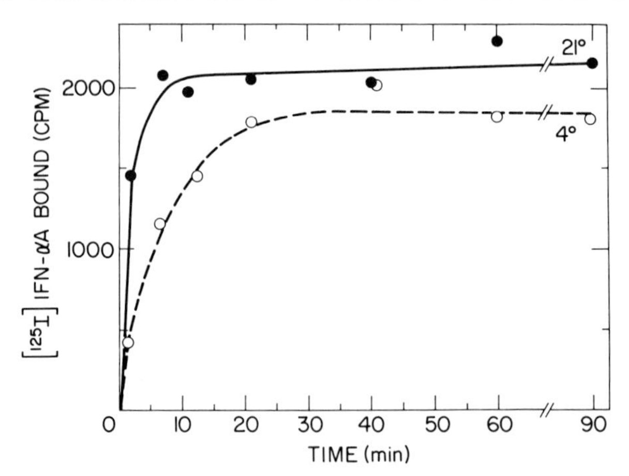

FIG. 2. Time course of specific binding of ^{125}I-labeled IFN-αA to Daudi cells. Daudi cells (1×10^6) were incubated at the indicated temperature with 8×10^4 cpm (6×10^3 units/ml) of ^{125}I-labeled IFN-αA. All manipulations through the final centrifugation were at the same temperatures as the incubations. Nonspecific binding as defined in the text amounted to 2–8% of the total and was subtracted. All points were measured in duplicate.

It should be noted that we have not seen complete saturation of binding sites on Daudi or other human cells (Fig. 3), a situation encountered also by others (see binding curves in Refs. 1,4,5,9,11,13). Analysis of binding curves[27–29] suggests the existence of at least a second class of binding sites with a K_d at 23° of about 10^{-9} to 10^{-8} M. This possible heterogeneity of binding sites must be kept in mind when conditions are chosen for procedures such as chemical crosslinking. Bovine BL-3 cells[30] seem to saturate nicely with human ^{125}I-labeled IFN-αA (J. A. Langer, unpublished results). Because of the variability between experiments, we measure binding to Daudi cells whenever we measure binding to any other suspension cells; the binding to Daudi cells thereby serves as an internal standard.

The only limitation we have found with this procedure is a result of the small sample volume required because of the small tube. The small incubation volume limits our ability to work at very low concentrations of radioiodinated interferon, since a sufficient quantity of ^{125}I-labeled IFN-αA must be maintained to ensure that the number of counts bound is

[27] G. Scatchard, *Ann. N.Y. Acad. Sci.* **51,** 660 (1949).

[28] I. M. Klotz, *Science* **217,** 1247 (1982).

[29] H. A. Feldman, *J. Biol. Chem.* **258,** 12865 (1983).

[30] G. H. Theilen, J. D. Rush, W. A. Nelson-Rees, D. L. Dungworth, R. J. Munn, and J. W. Switzer, *J. Natl. Cancer Inst. (U.S.)* **40,** 737 (1968).

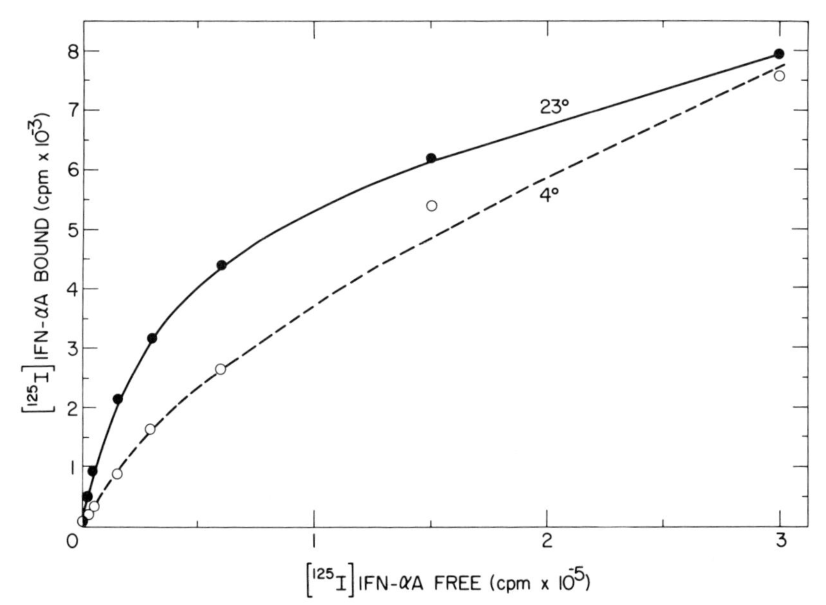

FIG. 3. Concentration dependence of ^{125}I-labeled IFN-αA binding to Daudi cells. Incubations were 1 hr at 23° or 2 hr at 4°. Nonspecific binding was <6% of total binding and has been subtracted. All points were measured in duplicate.

significant. Recent evidence suggests that it may be useful to collect data at concentrations of IFN lower than were used here.[13] The limitation imposed by the small volume can be overcome by using larger incubation volumes and larger tubes for centrifugation, or by using radiolabeled interferon of higher radioactive specific activity. Nevertheless, the method described has been used successfully to evaluate binding of ^{125}I-labeled IFN-αA to various cells (see the table) and was instrumental in establishing that upon differentiation the number of interferon receptors on HL-60 cells increases 3- to 4-fold.[31] Similar experiments can be done on nonhuman suspension cells, as illustrated here with the bovine BL-3 cell line.

Acknowledgments

We greatly appreciate the assistance of Mr. Bruce Kelder, Mrs. Judith Altman, and Mrs. Cynthia Rose in performing the interferon assays, and the cooperation of Drs. S. J. Tarnowski, C. McGregor, and R. Ning for providing samples of IFN-αA.

[31] J. A. Langer and S. Pestka, *J. Interferon Res.* **5,** 637 (1985).

[32] S. J. Collins, F. W. Ruscetti, R. E. Gallagher, and R. C. Gallo, *Proc. Natl. Acad. Sci. U.S.A.* **75,** 2458 (1978).

[45] Procedures for Studying the Binding of Interferon to Human and Bovine Cells in Monolayer Culture

By KATHRYN C. ZOON, DOROTHY ZUR NEDDEN, and HEINZ ARNHEITER

Interferons (IFNs) are a group of proteins which induce a variety of biological responses such as antiviral, antiproliferative, and immunomodulatory activities in sensitive cells. The first event in inducing these responses is the specific binding of these molecules to cell surface receptors.[1-8] The number of IFN receptors per cell in general is very small, less than 5000. The affinity of IFN for its receptor is very high, the apparent dissociation constants are in the order of 10^{-10} to 10^{-11} *M*.[1-8] This chapter describes methods for examining the binding of interferon to bovine and human monolayer cells. The application of Scatchard analysis to determine the apparent dissociation constant and number of binding sites per cell is discussed.

Preparative Procedures and Methodology

Materials

Hu-IFN-αA (2×10^8 units/mg protein) and Hu-IFN-αD (5×10^7 units/mg protein) were the gift of Hoffmann-La Roche, Nutley, NJ. Hu-IFN-α2 (1.7×10^8 units/mg protein) was the gift of Schering Corp., Bloomfield, NJ. Hu-IFN-α (Ly, MW 18,500)[9] (2.0×10^8 units/mg protein) from Namalwa cells was purified to homogeneity as previously described.[10]

[1] M. Aguet, *Nature* (*London*) **284,** 459 (1980).
[2] M. Aguet and B. Blanchard, *Virology* **115,** 249 (1981).
[3] A. Branca and C. Baglioni, *Nature* (*London*) **294,** 768 (1981).
[4] K. E. Mogensen, M.-T. Bandu, F. Vignaux, M. Aguet, and I. Gresser, *Int. J. Cancer* **28,** 575 (1981).
[5] K. C. Zoon, D. Zur Nedden, and H. Arnheiter, *J. Biol. Chem.* **257,** 4695 (1982).
[6] P. Anderson, Y. K. Yip, and J. Vilček, *J. Biol. Chem.* **257,** 11301 (1982).
[7] A. R. Joshi, F. H. Sarkar, and S. L. Gupta, *J. Biol. Chem.* **257,** 13884 (1982).
[8] H. Arnheiter, M. Ohno, M. Smith, B. Gutte, and K. C. Zoon, *Proc. Natl. Acad. Sci. U.S.A.* **80,** 2539 (1983).
[9] Hu-IFN-α (Ly, MW 18,500), a human interferon alpha from the lymphoblastoid (Ly) cell line, Namalwa, with an apparent molecular weight of 18,500.
[10] K. C. Zoon, M. E. Smith, P. J. Bridgen, D. Zur Nedden, and C. B. Anfinsen, *Proc. Natl. Acad. Sci. U.S.A.* **76,** 5601 (1979).

HuIFN-αs were radiolabeled with ^{125}I-labeled Bolton–Hunter reagent to specific activities of $1.1–4.4 \times 10^5$ Bq/μg protein.[11]

Cell Cultures

Madin–Darby bovine kidney (MDBK) cells were obtained from the American Type Culture Collection (ATCC CCL22), Rockville, MD. WISH cells (ATCC CCL 25) were obtained from the Tissue Culture Unit, Office of Biologics Research and Review, Bethesda, MD. Cells were cultured in Eagle's minimal essential medium (EMEM) supplemented with 10% fetal calf serum (Biofluids or S and S Media, Rockville, MD) and 50 μg/ml gentamicin sulfate (complete medium) in 35-mm-diameter 6-well plates (Costar, Cambridge, MA) or 100-mm-diameter tissue culture dishes (Costar, Cambridge, MA) until they reached confluency.

Equilibrium Binding Assay

All steps were performed at 0–4° unless otherwise indicated. Assays were done in triplicate. Ice cold complete medium was added to confluent monolayers of cells on ice (1–2 ml per 35-mm culture plate and 15 ml per 100-mm culture plate). Six cultures were used per assay point; three received radiolabeled IFN alone and three received unlabeled IFN or other test materials followed by the addition of radiolabeled IFN. The cultures were gently swirled on ice and incubated at 4° until they reached equilibrium which was determined by measuring specific binding at different time points. The cells were then placed on ice and the supernatant was removed by suction (or saved if necessary). The cells were washed on ice five times with 2 ml (35-mm well) or 15 ml (100-mm dish) ice cold complete medium. The cells were then solubilized with 2 ml (35-mm well) or 5 ml (100-mm dish) of 0.5% sodium dodecyl sulfate or 0.1 *N* NaOH at room temperature for a minimum of 15 min. The lysates were harvested and the wells or dishes were washed with an additional aliquot of the corresponding solution and pooled with the cell lysate. The radioactivity in the samples was determined with a Beckman Gamma Counter model 8500. Nonsaturable or nonspecific binding is defined as that amount of radiolabeled IFN associated with the cells in the presence of a 100-fold or greater excess of unlabeled IFN of the same type. Nonspecific binding is generally 10–15% of the total binding at equilibrium. Specific binding is defined as the total binding minus the nonspecific binding.

[11] D. Zur Nedden and K. C. Zoon, this volume (38).

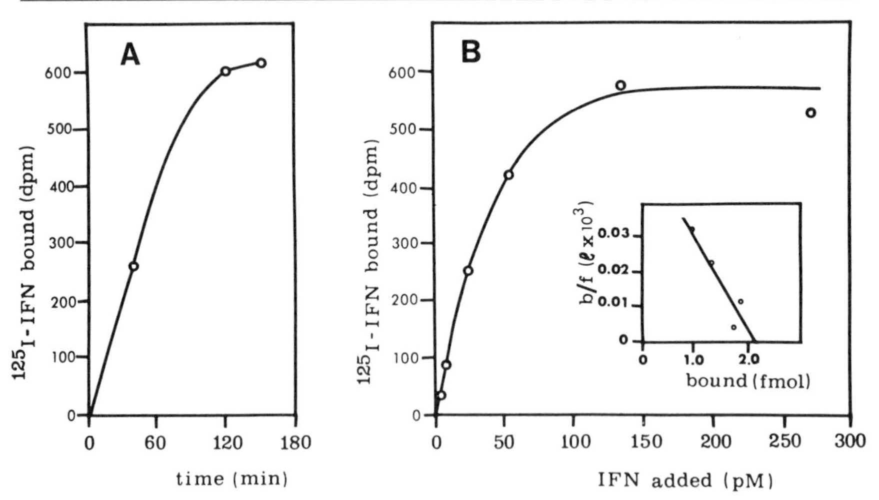

FIG. 1. Binding of radiolabeled human IFN-α2 to MDBK cells. (A) Time course of binding at 4°. (B) Equilibrium binding at 4° as a function of IFN concentration. Inset: Scatchard plot analysis of equilibrium binding data. Data from Refs. 8 and 12.

Comments

The time course of binding of ^{125}I-labeled Hu-IFN-α2 to MDBK cells at 4° is shown in Fig. 1A.[8,12] Maximum binding was observed at 2.5 hr with concentrations of IFN between 5 and 300 pM. Similar binding kinetics were observed with either ^{125}I-labeled Hu-IFN-αA, Hu-IFN-αD, or Hu-IFN-α (Ly, MW 18,500)[5] with MDBK or WISH cells. However, it should be pointed out that the kinetics of binding may vary with both the ligand and cell type. Optimum pH for the binding of radiolabeled Hu-IFN-αs to MDBK cells is approximately pH 7.4.[13]

Scatchard plot analysis was originally proposed to study the binding of small molecules to proteins.[14] The application of this type of analysis to examine protein–protein interactions under equilibrium conditions has been used to assess the complexity of these interactions. Estimates of the apparent dissociation constant and the number of binding sites per cell can be calculated from the slope ($-1/K_d$) and the x intercept, respectively. Scatchard analysis of equilibrium binding data for ^{125}I-labeled Hu-IFN-α2 to MDBK cells is shown in the insert to Fig. 1B.[8] Free ligand was

[12] K. C. Zoon and H. Arnheiter, *Pharmacol. Ther.* **24,** 259 (1984).

[13] K. C. Zoon, H. Arnheiter, and D. Zur Nedden, *in* "Frontiers in Biochemical and Biophysical Studies of Proteins and Membranes" (T.-Y. Liu *et al.*, eds.), p. 475. Am. Elsevier, New York, 1983.

[14] G. Scatchard, *Ann. N.Y. Acad. Sci.* **51,** 660 (1949).

determined by measuring the radioactivity in the cell culture supernatant and the first wash. Approximately 800 binding sites per cell and an apparent K_d of 6×10^{-11} M were calculated from these data. It is important to note that Scatchard analysis of binding data obtained under nonequilibrium conditions is inappropriate. At elevated temperatures such as 37°, human IFNs are internalized and degraded;[15,16] in addition, exposure of cells to Hu-IFNs has been shown in some cases to down regulate the IFN receptors.[12] Thus equilibrium binding experiments are infeasible at temperatures compatible with cell metabolism. Computer systems such as LIGAND[17] are very useful for the analysis of ligand binding data as well as model fitting.

Binding studies can also be carried out in Dulbecco's phosphate-buffered saline, pH 7.4 (S and S Media, Rockville, MD) supplemented with 1 mg/ml bovine serum albumin (Sigma, St. Louis, MO) (DPBS/BSA). However, the nonspecific binding using these conditions is 30–40% of the total binding. Other monolayer cell lines such as WIDR, HEp2 and HFS cells can be used in this binding assay. In addition one can also use these methods to study the binding characteristics of other ligands such as ^{125}I-labeled insulin to monolayer cells. DPBS/BSA is recommended as the binding buffer in studies with radiolabeled insulin and MDBK cells; nonspecific binding is generally 10–15% of total binding.

[15] K. C. Zoon, H. Arnheiter, D. Zur Nedden, D. J. P. FitzGerald, and M. C. Willingham, *Virology* **130,** 195 (1983).

[16] K. C. Zoon, H. Arnheiter, and D. J. P. FitzGerald, this volume [49].

[17] P. J. Munson, this series, Vol. 92, p. 543.

[46] Binding of ^{32}P-Labeled Human Recombinant Immune Interferon to U937 Cells

By ABBAS RASHIDBAIGI, HSIANG-FU KUNG, and SIDNEY PESTKA

For studying the interferon receptors, it is necessary to have radioactively labeled interferons with high affinity and specific radioactivity. Several groups have reported the labeling of human immune interferon (Hu-IFN-γ) with ^{125}I-labeled Bolton–Hunter reagent.[1-3] The specific activity

[1] P. Anderson, Y. K. Yip, and J. Vilček, *J. Biol. Chem.* **257,** 11301 (1982).

[2] P. Orchansky, D. Novick, D. G. Fischer, and M. Rubinstein, *J. Interferon Res.* **4,** 275 (1984).

[3] F. H. Sarkar and S. L. Gupta, *Proc. Natl. Acad. Sci. U.S.A.* **81,** 5160 (1984).

of this ^{125}I-labeled Hu-IFN-γ is barely sufficient for studying the immune interferon receptors in cells with a limited number of binding sites. However, recently it was found that Hu-IFN-γ can be labeled with ^{32}P from [γ-^{32}P]ATP.[4-6] Thus, Hu-IFN-γ with high radiospecific activity can be obtained with [γ-^{32}P]ATP and cyclic-AMP-dependent protein kinase from bovine heart[4,5] to a specific activity (11,000 Ci/mmol) which is 10- to 100-fold higher than ^{125}I-labeled interferons. In this chapter, we describe measurement of the binding of ^{32}P-labeled Hu-IFN-γ to human U937 histiocytic lymphoma cells.

Reagents

All chemicals are of reagent grade unless otherwise specified. Recombinant human immune interferon (Hu-IFN-γ with amino terminus Met-Gln-Asp-Pro-. . .[7]) was generously provided by Drs. Robert Ning and S. Joseph Tarnowski. ^{32}P-labeled Hu-IFN-γ was prepared according to the method of Kung and Bekesi[4] with [γ-^{32}P]ATP and cyclic-AMP-dependent protein kinase from bovine heart to a specific radioactivity of 11,000 Ci/mmol.[5] The ^{32}P-labeled Hu-IFN-γ was stored in liquid nitrogen without significant loss of binding activity when used within 1 month.

Cell Culture

U937 human histiocytic lymphoma cells[8] are routinely grown in RPMI-1640 medium (Gibco) supplemented with 10% heat-inactivated fetal calf serum (FCS), 50 μg/ml gentamycin, and 12.5 mM sodium Hepes in 75-cm^2 flasks (Falcon, 3023). The cells are grown in a humidified incubator supplemented with a 5% CO_2 and 95% air mixture at 37° to a density of 1×10^6 cells/ml as measured with a hemocytometer.

Preparation of Cells for Binding Assay. Cells (50 ml) are harvested by centrifugation at 500 rpm in an International centrifuge (model HN-S) for 10 min at room temperature. The cell pellets are washed by resuspension in 10 ml of fresh medium (RPMI-1640 medium containing 10% FCS), and centrifugation at 500 rpm for 10 min at room temperature. The washed cells are resuspended in fresh medium to a density of about 1×10^7 cells/ml for the binding assay.

[4] H.-F. Kung and E. Bekesi, this volume [43].

[5] A. Rashidbaigi, H.-F. Kung, and S. Pestka, *J. Biol. Chem.* **260,** 8514 (1985).

[6] B. Robert-Galliot, M. J. Commoy-Chevalier, P. Georges, and C. Chany, *J. Gen. Virol.* **66,** 1439 (1985).

[7] S. Pestka, this volume [1].

[8] C. Sundström and K. Nilsson, *Int. J. Cancer* **17,** 565 (1976).

Binding of ^{32}P-Labeled Hu-IFN-γ to U937 Cells

Cell suspensions (0.2 ml of 1×10^7 cells/ml) are incubated with or without 6.5×10^{-8} *M* unlabeled Hu-IFN-γ as a competing ligand for 15 min at 24° and then with ^{32}P-labeled Hu-IFN-γ at 24° for 80 min. The reaction was stopped by layering 50 μl of the incubation mixture in triplicate onto 0.3 ml of 10% sucrose in 0.01 *M* sodium phosphate, pH 7.3, and 0.15 *M* NaCl (PBS) in a 0.4-ml polypropylene tube (Sarstadt No. 72.701) followed by centrifugation for 1 min in a Beckman model 152 microfuge. Tubes were frozen in liquid nitrogen, and the tip containing the cell pellet was cut off. The amount of radioactivity in the tip (bound) and the rest of the tube (free) was then determined by adding 2 ml of a fluor solution (Brays solution, National Diagnostics, Somerville, NJ) and radioactivity determined in a Beckman model LC7800 scintillation spectrometer. Specific binding was defined as the amount of the ^{32}P-labeled Hu-IFN-γ bound in the absence of unlabeled Hu-IFN-γ (total binding) minus the amount bound in the presence of 6.5×10^{-8} *M* unlabeled Hu-IFN-γ (nonspecific binding).

Binding of ^{32}P-labeled Hu-IFN-γ as a function of ^{32}P-labeled Hu-IFN-γ concentration is shown in Fig. 1. ^{32}P-labeled Hu-IFN-γ bound to sites that fulfilled the criteria for a IFN-γ receptor. The binding was saturable, reaching saturation at 5×10^{-10} *M*. The binding was displaceable with unlabeled Hu-IFN-γ but not with Hu-IFN-α or Hu-IFN-β.[5] The nonspecific binding was always less than or equal to 10% of the total binding when the concentration of ^{32}P-labeled Hu-IFN-γ ranged from 1×10^{-12} to 5×10^{-10} *M*. ^{32}P-labeled Hu-IFN-γ should be centrifuged at 170,000 *g* (Beckman Airfuge at 30 psi) for 30 min before use in the binding assay in order to obtain uniform results. The low nonspecific binding is an important property of ^{32}P-labeled Hu-IFN-γ. Nonspecific binding of ^{125}I-labeled Hu-IFN-γ has been reported to be greater than or equal to 30% of the total binding.[1-3]

The equilibrium dissociation constant, K_d, for ^{32}P-labeled Hu-IFN-γ is derived from the simple mathematical assumption that can be made about the interaction of radioactive ligand, L, with a receptor, R, according to the law of mass action to form RL, the receptor–ligand complex:

$$\mathrm{L} + \mathrm{R} \underset{k_{-1}}{\overset{k_1}{\rightleftharpoons}} \mathrm{RL} \tag{1}$$

At equilibrium the K_d is derived from

$$K_d = \frac{[\mathrm{R}][\mathrm{L}]}{[\mathrm{RL}]} \tag{2}$$

where [R], [L], and [RL] represent the concentrations of free receptors, free ligand, and the receptor–ligand complex, respectively. K_d is also

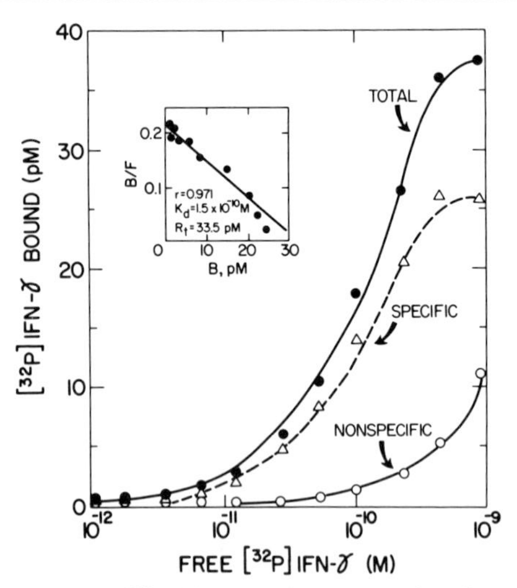

FIG. 1. Equilibrium binding of ^{32}P-labeled Hu-IFN-γ as a function of ^{32}P-labeled Hu-IFN-γ concentration. U937 cells (1×10^7 cells/ml) were incubated with and without 6.5×10^{-8} *M* unlabeled Hu-IFN-γ and then with various concentrations of ^{32}P-labeled Hu-IFN-γ at 24° for 80 min. The amount of radioactivity bound was measured as described in the text. The broken line (open triangles) is the specific binding calculated from the total binding (closed circles) and nonspecific binding (open circles). Each point is the average of triplicate determinations. Inset: Scatchard analysis of the binding data in which B [or [RL] in Eq. (5)] is the concentration of specifically bound ^{32}P-labeled Hu-IFN-γ, and F [or [L] in Eq. (5)] is the concentration of free ^{32}P-labeled Hu-IFN-γ at equilibrium.

equal to the ratio of the absolute reaction rate constants k_{-1} and k_1 for the reverse and forward reactions, respectively. Thus,

$$K_d = \frac{k_{-1}}{k_1} \tag{3}$$

The total concentration of receptor $[R_t]$ is equal to $[R] + [RL]$ and thus

$$K_d = \frac{([R_t] - [RL])[L]}{[RL]} \tag{4}$$

which can be rearranged to

$$\frac{[RL]}{[L]} = \frac{[R_t]}{K_d} - \frac{[RL]}{K_d} \tag{5}$$

This is the equation derived by Scatchard[9] in which the plot of the ratio of bound to free ligand, [RL]/[L], versus the concentration of bound ligand, [RL], yields a line with a slope of $-K_d^{-1}$ and an intercept with the abscissa of $[R_t]$.

[9] G. Scatchard, *Ann. N.Y. Acad. Sci.* **51,** 660 (1949).

Frequently, in the literature [RL], [L], and [R_t] are designated B (bound), F (free), and n (ligand-binding site), respectively, so that Eq. (4) reads

$$\frac{B}{F} = \frac{n}{K_d} - \frac{B}{K_d} \tag{6}$$

or

$$\frac{B}{F} = K_a(n - B) \tag{7}$$

where the association constant K_a is equal to K_d^{-1}.

When the binding data were plotted according to the method of Scatchard[9] [Eq. (6)] a linear plot was obtained (Fig. 1, inset). These data are indicative of a single noncooperative binding site for ^{32}P-labeled Hu-IFN-γ on U937 cells. The equilibrium dissociation constant, K_d, for ^{32}P-labeled Hu-IFN-γ determined from the slope of the binding curve was 1.5×10^{-10} *M*. The concentration of the total binding sites determined by extrapolation of the binding curve to the abscissa was measured to be 33.5 p*M*. This is equivalent to 1800 receptor molecules per cell based on one monomeric IFN-γ molecule per receptor. Because the functional unit of IFN-γ may be a tetramer,[10,11] these values may need to be reevaluated when the tertiary structure of IFN-γ and the receptor are known.

Time Course of Association of ^{32}P-Labeled Hu-IFN-γ Binding. U937 cells (1 ml of 1×10^7 cells/ml) were incubated in polypropylene tubes in the presence and absence of 6.5×10^{-8} *M* unlabeled Hu-IFN-γ at 24° for 20 min, and then with 4×10^{-11} *M* ^{32}P-labeled Hu-IFN-γ. The specific binding of ^{32}P-labeled Hu-IFN-γ was then measured at different times at 24° as described above (Fig. 2). The binding of ^{32}P-labeled Hu-IFN-γ reached equilibrium in 60 min at 24° (Fig. 2). The binding remained at equilibrium for at least 150 min.

Time Course of Dissociation of ^{32}P-Labeled Hu-IFN-γ from Cells. U937 cells (1 ml of 8×10^6 cells/ml) were incubated in the presence and absence of 6.5×10^{-8} *M* Hu-IFN-γ at 24° for 15 min and then with 3.7×10^{-10} *M* ^{32}P-labeled Hu-IFN-γ for 70 min. Specific binding of ^{32}P-labeled Hu-IFN-γ was measured at 24° at various times after addition of 6.5×10^{-8} *M* Hu-IFN-γ to the tube which did not contain unlabeled Hu-IFN-γ (Fig. 3). Dissociation of ^{32}P-labeled Hu-IFN-γ from U937 cells showed a marked deviation from linearity on a semilogarithmic plot (Fig. 3). About 40% of the bound ^{32}P-labeled Hu-IFN-γ dissociates within 40 min. The rest of the binding persisted longer than 120 min. The slow dissociation might be due to the internalization of a receptor–IFN-γ complex.

[10] S. Pestka, B. Kelder, P. C. Familletti, J. A. Moschera, R. Crowl, and E. S. Kempner, *J. Biol. Chem.* **258,** 9706 (1983).

[11] E. S. Kempner and S. Pestka, this volume [35].

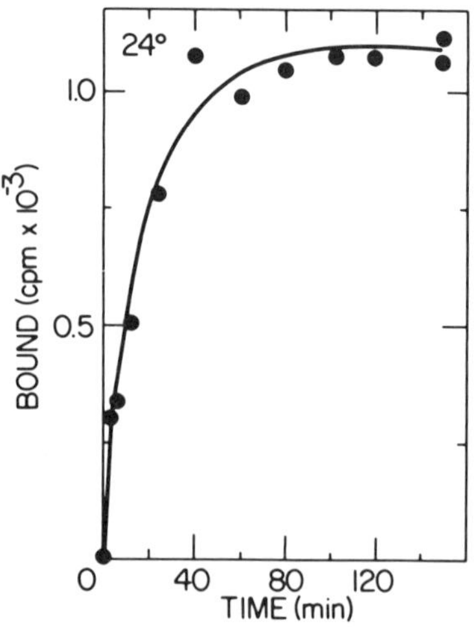

FIG. 2. Binding of ^{32}P-labeled Hu-IFN-γ to cells at 24° as a function of time. U937 cells (1×10^7 cells/ml) were incubated with and without 6.5×10^{-8} *M* unlabeled Hu-IFN-γ and then with 4×10^{-11} *M* ^{32}P-labeled Hu-IFN-γ at 24° as described in the text. Specific binding of ^{32}P-labeled Hu-IFN-γ was then measured at the times shown.

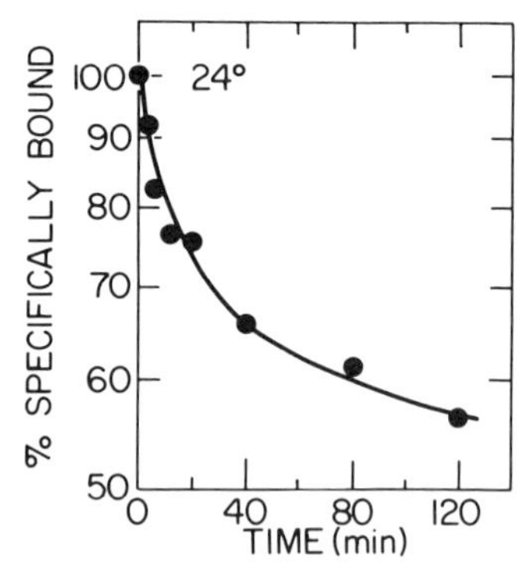

FIG. 3. Dissociation of ^{32}P-labeled Hu-IFN-γ from cells at 24° as a function of time. U937 cells (8×10^6 cells/ml) were incubated with and without 6.5×10^{-8} *M* unlabeled Hu-IFN-γ and then with 3.7×10^{-10} *M* ^{32}P-labeled Hu-IFN-γ at 24° as described in the text. Specific binding of ^{32}P-labeled Hu-IFN-γ was measured at the times shown after addition of unlabeled Hu-IFN-γ to the incubation mixture which did not contain unlabeled Hu-IFN-γ. The solid line is the experimentally obtained curve. The theoretical curve, assuming two first-order dissociation steps, was calculated to have a fast component with a k_{-1} of 0.141 min^{-1} and a $t_{1/2}$ of 5 min and a slow component with a dissociation rate (k'_{-1}) of 0.0027 min^{-1} and a $t_{1/2}$ of 256 min.

Concluding Comments

When recombinant human immune interferon was phosphorylated with [γ-^{32}P]ATP and bovine heart muscle protein kinase, a phosphorylated Hu-IFN-γ was obtained with a specific activity of 11,000 Ci/mmol.[5] This is a specific activity at least 10- to 100-fold higher than reported for other ^{125}I-labeled interferons. This specific activity corresponds to the incorporation of an average of about two molecules of phosphate per molecule of Hu-IFN-γ. The high specific activity of ^{32}P-labeled Hu-IFN-γ and its convenient preparation make it a useful reagent for studying the receptor for human immune interferon.

Acknowledgments

We thank Dr. Jerome A. Langer for careful review of this text and Sophie Cuber for her assistance in the preparation of this manuscript.

NOTE ADDED IN PROOF: The specific radioactivity reported here for ^{32}P-labeled Hu-IFN-γ (11,000 Ci/mmol) is based on the antiviral activity (as determined by the cytopathic effect inhibition assay measured with vesicular stomatitis virus and WISH cells) and the specific activity of Hu-IFN-γ (2×10^7 units/mg); all data in the text were based on this value. The specific radioactivity based on the total input protein was calculated to be 3800 Ci/mmol. This is the minimum value since the protein concentration in the phosphorylated sample was too low to be measured. Based on this minimum value, the parameters were calculated to be as follows: $K_d = 4.3 \times 10^{-10}$ M, $R_t = 97$ pM (5200 receptors/cell), and a total of about one molecule of phosphate incorporated per molecule of Hu-IFN-γ. It is likely that the true values for these parameters are delimited by these values and those given in the text.

[47] Procedures for Studying Binding of Interferon to Mouse Cells

By MICHEL AGUET

The aim of binding experiments is to reveal a specific ligand receptor interaction and thereby provide insight into the early events that initiate a ligand-specific biological response. Therefore, experimental conditions to determine binding site saturation and specific competitive binding inhibition[1] on entire cells have to be compatible with those required to elicit a relevant response. Further experimental restrictions arise from the requirements for interpreting binding data, notably with a view to applying the law of mass action. Isolated purified receptors would provide a more appropriate tool for accurate determination of association and dissociation constants than whole cells. Numerous functional aspects of receptor

[1] P. Cuatrecasas and M. D. Hollenberg, *Adv. Protein Chem.* **30,** 251 (1976).

binding, however, require analysis of binding properties on intact cells. The procedures described below are restricted to the use of living cells.

Media and Reagents

RPMI-1640 tissue culture medium (Gibco), containing 10 m*M* HEPES buffer, supplemented with 10% fetal calf serum (KC Biological)

Dulbecco's modified Eagle's minimal essential medium (Gibco), containing 10 m*M* HEPES buffer, supplemented with 10% fetal calf serum (KC Biological)

Phosphate buffered saline, pH 7.4 (PBS): 0.137 *M* NaCl, 10 m*M* Na_2HPO_4, 3 m*M* KH_2PO_4

$Na^{125}I$ (New England Nuclear), in pH 8–10 aqueous solution (reductant free), carrier free (about 17 Ci/mg), 100–200 mCi/ml

Lactoperoxidase (Sigma), from bovine milk, lyophilized powder, 80–100 units/mg protein

H_2O_2, 30% (8.8 *M*)

Sodium azide, 0.25 *M*

Affi-Gel-10 (Bio-Rad)

Sephadex G-25, prepacked disposable columns PD-10 (Pharmacia)

Labeling of Mouse C243 Cell Interferon with ^{125}I

Highly purified mouse C243 cell interferon induced by Newcastle disease virus comprises three major molecular weight species.[2] As indicated by NH_2-terminal homologies and common antigenic properties with human interferons, the components with molecular weights of 24,000 and 30,000, which represent up to 90% of the total interferon, belong to the β type, whereas the 19,000 molecular weight species behaves as IFN-α.[3] For convenience this heterogeneous material is referred to as interferon in this chapter and the antiviral activity is indicated in terms of reference units/ml. According to our own experience lactoperoxidase catalyzed iodination[4] of this interferon yields the best results with regard to both iodine incorporation and recovery of biological activity.

Preparation of Lactoperoxidase Coupled to Sepharose Beads

About 0.5 ml of wet Affigel-10 freshly washed with ice cold distilled water on a small fritted glass funnel is added to 0.5 ml of 0.1 *M* sodium

[2] M. Aguet and B. Blanchard, *Virology* **115,** 249 (1981).

[3] Y. Kawade, M. Aguet, and M. G. Tovey, *Antiviral Res.* **2,** 155 (1982).

[4] G. S. David and R. A. Reisfeld, *Biochemistry* **13,** 1014 (1974).

phosphate buffer (pH 7.0) containing 400 μg lactoperoxidase per ml as determined by optical density measurement at 412 nm ($E_{412\ nm}^{1\ mM} = 114$). The gel is kept in suspension overnight at 4° and subsequently washed with 20 ml of 1 M NaCl and excess PBS prior to resuspension in 10 ml PBS. Under these conditions lactoperoxidase is entirely coupled to the Sepharose to give a final concentration of 20 μg/ml of gel suspension; unreacted coupling sites of the gel are inert after this incubation period. Provided that the described steps are carried out under sterile conditions this gel suspension is stable at 4° for at least a year. (Addition of sodium azide as preservative would result in inactivation of the enzyme.)

Labeling Procedure

Two millicuries of freshly generated carrier-free ^{125}I (100 mCi/ml) is added to 100 μl of an interferon solution in PBS adjusted to 100 μg protein per ml. Then 20 μl of Affigel-10–lactoperoxidase gel suspension is added prior to initiation of the iodination which is done by addition of 10 μl of diluted H_2O_2 to give a final concentration of 10^{-4} M. After 20 min incubation at room temperature under continuous agitation the reaction is stopped by addition of 20 μl of 0.25 M sodium azide. Separation of free iodine is carried out on a prepacked Sephadex G-25 column equilibrated with PBS containing 1% fetal calf serum. Labeled interferon is eluted in a pool of 2.5 ml and subsequently diluted 8-fold in RPMI-1640 containing 10% fetal calf serum. Aliquots of 1 ml are kept frozen at −70°. Under these conditions the material is stable for approximately 2 months; the loss of specific radioactivity follows the decay of ^{125}I. This procedure yields iodinated interferon with a specific radioactivity of approximately 50 μCi/μg of protein while 25 to 50% of the biological activity is recovered.

Procedure for Studying Binding to Cells Growing in Suspension

The expression of mouse interferon receptors depends on a rather rapid turnover rate[2] and is markedly increased on exponentially growing cells as compared to cells resting at saturation (unpublished observation). To yield reproducible results binding experiments should be carried out on cells which have been cultured for at least 8 hr. Changes in cell culture conditions such as temporary temperature shifts or centrifugation of the cells result in decreased receptor expression with a latency of 1–2 hr upon reincubation at 37°.[2] Therefore, monitoring ^{125}I-labeled interferon binding at 37° requires careful addition of labeled interferon to precultivated samples of cell suspensions while avoiding any perturbation of culture conditions. In contrast receptor expression on cells kept at 4° in tissue culture medium remains virtually unchanged over at least 12 hr.

Cell density is a critical variable in experiments designed to calculate constants defined by mass action law. It is only below a critical limit that the amount of ligand bound per cell at a given ligand concentration becomes independent of cell density.[1,2] This critical density or receptor concentration corresponds to the binding constant K_d and has to be verified experimentally. In the case of mouse L1210 cells which express about 10^3 receptors per cell this critical threshold value corresponds to about 10^7 cells per ml.[2]

In experiments carried out under culture conditions at 37° cells are incubated with ^{125}I-labeled interferon in tissue culture flasks at cell densities compatible with exponential growth (i.e., 0.5–1.5 × 10^6 cells per ml for most lymphoid cells in nonagitated culture). Cell associated radioactivity can be determined simply by washing three times at 200 *g* for 5 min with ice cold culture medium containing 1% fetal calf serum and subsequently transferring the cells into counting vials. Dissociation of specifically bound ligand during this procedure is negligible. Multiple samples can be washed efficiently by transferring the cells into U-bottomed microtiter plates (Greiner) with 1–2 × 10^6 cells per well. The plates are centrifuged at 200 *g* for 5 min in the cold and the supernatants discarded. Prior to addition of fresh medium the cell pellets are disrupted by stirring on a vortex. Before the last centrifugation the cells are transferred by means of a multichannel pipette into U-bottomed minivials placed into the wells of a flat bottomed microtiter plate (Greiner) that serves as a tray. Interferon incubation at 4° can be carried out directly in U-bottomed microtiter plates at a cell density of 0.5–1.0 × 10^7 cells per ml with 200 μl of cell suspension per well. To determine nonspecific binding, specific binding sites have to be saturated with unlabeled interferon. Therefore, an excess of interferon in a final concentration of 5 × 10^4 units/ml is added simultaneously with labeled interferon.

Washing cells by repeated centrifugation to remove unbound ligand is certainly more time consuming than other commonly applied single step procedures such as centrifugation on sucrose gradients or oil. However, it has the advantage of maintaining the cells viable in a physiological medium and therefore of allowing recultivation or consecutive incubations with different ligands (e.g., to reveal specifically bound interferon by means of anti-interferon antibodies[5]).

[5] H. Arnheiter, M. Ohno, M. Smith, B. Gutte, and K. C. Zoon, *Proc. Natl. Acad. Sci. U.S.A.* **80,** 2539 (1983).

Procedure for Studying Binding to Cells Growing as Monolayer Cultures

Binding of ^{125}I-labeled interferon to monolayer cells is best carried out in Costar 3524 24-well plates on exponentially growing cells that reach confluency.[6] To improve cell adherence the plates are pretreated with a 1% solution of gelatine in PBS which is applied for a few minutes at room temperature and subsequently removed prior to drying the plates under sterile conditions. For most monolayer cells beginning confluency corresponds to approximately 10^5 cells per well. To avoid perturbation of culture conditions interferon incubation at 37° is carried out in a waterbath with prewarmed media and interferon samples. After incubation cells are washed three times by addition of 1.0 ml per well of ice cold culture medium which should remain in contact with the cells for about 1 min prior to being removed. Subsequently, cell-associated radioactivity is determined by solubilization of the cells with 0.5 ml 1 *N* NaOH for 1 hr at 37° and transferring the lysates into counting vials. Nondisplaceable binding is determined as described in the previous section.

Interpretation of Binding Data

Binding data are most commonly analyzed by mass action law to estimate binding affinity and number of binding sites involved.[1,7] The limitations have been already emphasized elsewhere,[8,9] yet two aspects should be briefly recalled. As stated above cell density is a critical parameter for interpreting binding data by mass action law. To define a ligand–receptor interaction, mass action law requires that the reaction be reversible, bimolecular, and in an equilibrium state. Obviously these criteria are more convincingly fulfilled at low temperature, whereas at 37° dynamic processes such as receptor turnover, ligand internalization and degradation, or ligand induced modulation of receptor expression represent additional variables difficult to define by mass action law. Therefore, it is important to distinguish between experiments designed to determine or compare physical binding properties and experiments conceived to investigate dynamic events involved or induced by receptor binding under physiological conditions. To our knowledge a generally valid mathematical model to describe these dynamic processes has not yet been proposed.

[6] M. Aguet, I. Gresser, A. G. Hovanessian, M. T. Bandu, B. Blanchard, and D. Blangy, *Virology* **114,** 585 (1981).

[7] G. Scatchard, *Ann. N.Y. Acad. Sci.* **51,** 660 (1949).

[8] I. M. Klotz, *Science* **217,** 1247 (1982).

[9] M. Aguet and K. E. Mogensen, *in* "Interferon" (I. Gresser, ed.), Vol. 5, p. 1. Academic Press, New York, 1984.

[48] Procedures for Binding an Antibody to Receptor-Bound Interferon

By HEINZ ARNHEITER and KATHRYN C. ZOON

Antibodies recognizing receptor-bound IFNs are interesting for various reasons. They may be used to isolate IFN/receptor complexes, to analyze structure/function relationships of IFNs, to analyze the fate of IFNs bound to the surface of living cells, or to identify individual cells with particularly high or particularly low numbers of IFN receptors. Ideally, such antibodies should bind to receptor-bound IFN with high affinity, they should not displace IFN from its receptor, they should not interfere with IFN binding to receptors, and, consequently, not neutralize IFN activity. It appears that antibodies specific for one epitope (or a narrow range of epitopes) of IFNs may satisfy at least some of these criteria.

Here, we briefly describe procedures for binding a mouse monoclonal IgG_1 antibody, III/21, to cell surface-bound IFN. This antibody was made against a synthetic, 56-residue carboxyl-terminal fragment of Hu-IFN-α1.[1] The antibody binds to native recombinant DNA-derived IFN-α1 and -α2[2] at the ultimate 10–16 carboxyl-terminal residues, and it binds both native Hu-IFN-α2 and the synthetic oligopeptide Hu-IFN-α1 (151–166) with an apparent K_d of 6.0×10^{-10} M.[3] It does not neutralize IFN unless used at a 10^6-fold molar excess over IFN, it does not inhibit the equilibrium binding of Hu-IFN-α2 to Madin–Darby bovine kidney (MDBK) cells unless added at a molar excess over IFN of at least 4000-fold, and it does not significantly displace Hu-IFN-α2 previously bound to cells under equilibrium conditions.[3] This suggests that the antibody recognizes IFN at a site which does not play a crucial role for receptor binding.

Preparative Procedures and Methodology

Purification of Antibody III/21

Ascitic fluids collected from female BALB/c mice into which 10^7 cells of hybridoma clone III/21 have been injected intraperitoneally are precip-

[1] H. Arnheiter, R. M. Thomas, T. Leist, M. Fountoulakis, and B. Gutte, *Nature (London)* **294,** 278 (1981).

[2] For some of the studies, Hu-IFN-αA was used instead of Hu-IFN-α2. Hu-IFN-α2 differs from Hu-IFN-αA at position 23 in the amino acid sequence; Hu-IFN-α2 has an arginine at this position, Hu-IFN-αA has a lysine.

[3] H. Arnheiter, M. Ohno, M. Smith, B. Gutte, and K. C. Zoon, *Proc. Natl. Acad. Sci. U.S.A.* **80,** 2539 (1983).

itated at 4° with 50% ammonium sulfate. The precipitate is dissolved in and dialyzed against phosphate-buffered saline (PBS), pH 7.4. Approximately 30 mg of the dialyzed material is applied to an Affi-Gel 10 column (Bio-Rad) to which 20 mg of synthetic Hu-IFN-α1 (111–166) is coupled. Affinity columns to which other corresponding synthetic peptides or Hu-IFN-α1 (1–166) or Hu-IFN-α2 (1–165) are coupled may also be suitable. The column is washed with approximately 30 column volumes of PBS, pH 7.4, subsequently with 10 ml of McIlvaine's citric acid/Na_2HPO_4 buffer at pH 6.0, and then with 10 ml of the same buffer at pH 4.5. Finally the antibody is eluted with 5 ml of the same buffer at pH 2.8. The eluate is dialyzed against PBS, pH 7.4, and stored frozen at −20° or kept at 4° in presence of 0.05% NaN_3. With this procedure we could regularly recover in the pH 2.8 fraction approximately 25% of the protein applied to the column. The recovered antibody was of satisfactory purity (see below).

Radiolabeling of Antibody

A portion of the purified antibody is dialyzed against 0.1 *M* sodium borate buffer, pH 8.5, and 100–200 μg (volume 50–100 μl) of the dialyzed antibody is added to a vial containing 1 mCi of ^{125}I-labeled Bolton–Hunter reagent, for 15 min at room temperature. The reaction is stopped with 0.2 *M* glycine, pH 8.5, for 15 min. The radiolabeled antibody is separated from unreacted bolton–Hunter reagent by passage over a Sephadex G-25 column (Pharmacia) equilibrated with PBS, pH 7.4, and subsequent dialysis against PBS, pH 7.4. Purity of the ^{125}I-labeled antibody is checked by polyacrylamide gel electrophoresis as shown in Fig. 1. Its specific activity is determined by a standard enzyme-linked immunosorbent assay with synthetic Hu-IFN-α1 (111–166) or native Hu-IFN-α coated to plastic plates and a peroxidase-coupled anti-mouse IgG antibody. Unlabeled, purified monoclonal antibody III/21, whose concentration is determined by a Lowry assay, serves to establish the standard curve. We have obtained iodinated antibody preparations with specific activities between 0.7 and 1.5×10^{16} Bq/mol.

Purification and Radiolabeling of Hu-IFN-α2

Purified recombinant DNA-derived Hu-IFN-α2 (from Dr. Charles Weissmann, Zurich, Switzerland) or Hu-IFN-αA (Hoffmann-La Roche, Nutley, New Jersey) is labeled with ^{125}I-labeled Bolton–Hunter reagent as described.[4] The specific activity of both labeled and unlabeled IFN against VSV in MDBK cells is $1.5–2.0 \times 10^8$ units/mg protein.

[4] D. Zur Nedden and K. C. Zoon, this volume [38].

FIG. 1. Autoradiography of a 10% polyacrylamide gel in which purified, ^{125}I-labeled antibody III/21 has been electrophoresed under reducing conditions. HC, Immunoglobulin heavy chain; LC, immunoglobulin light chain.

Cells

Madin–Darby bovine kidney cells (ATCC CCL 22) are cultured in Dulbecco's modified EMEM supplemented with 10% fetal calf serum and 50 μg/ml of gentamicin sulfate (complete medium). Experiments are done with cells grown to confluent monolayers in 50-cm^2 plastic dishes (Falcon) at 37° (5–6 $\times$ 10^6 cells/dish).

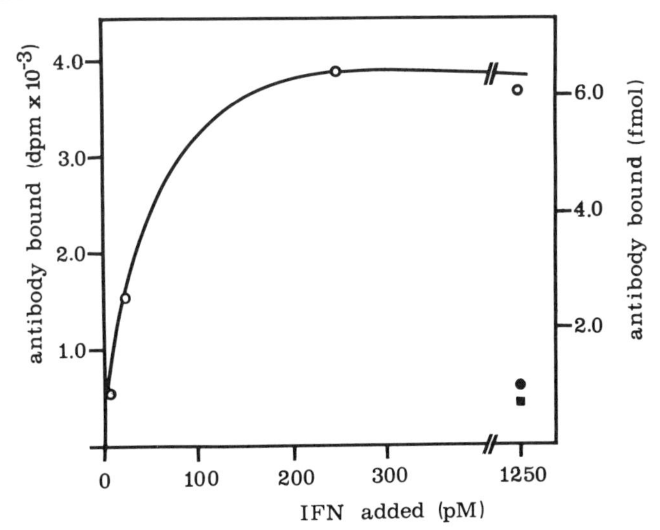

FIG. 2. Binding of ^{125}I-labeled antibody III/21 to Hu-IFN-α2 bound to the surface of monolayer MDBK cells. The cells were treated for 2.5 hr at 4° with unlabeled IFN at the concentrations indicated. The antibody (0.14 nM) was added for another 2 hr (○). Antibody neutralized with Hu-IFN-α1 (151–166), 1.3 μM (●). Antibody bound to plates with no cells (■). Each point represents the mean of the results from six independent culture dishes (see Ref. 3).

Binding of the Radiolabeled Monoclonal Antibody to Cell Surface-Bound IFN

To test the binding of the radiolabeled antibody to cell surface-bound IFN, we add unlabeled IFN dissolved in 10 ml of ice cold complete medium to MDBK cells for 2.5 hr at 4°. The cells are then washed five times with ice cold PBS, pH 7.4, and exposed to radiolabeled antibody (0.14 nM in 10 ml of ice cold PBS, pH 7.4) for 2 hr at 4°. Then, the cells are washed five times with ice cold PBS, pH 7.4, solubilized with 0.5% sodium dodecyl sulfate in PBS, pH 7.4. The samples are then counted in a Beckman 8500 gamma counter.

If antibody III/21 recognized surface-bound Hu-IFN-α2, binding of antibody should reflect the dose dependent binding of IFN to MDBK cells.[5] This appears to be the case. For the representative experiment shown in Fig. 2, unlabeled Hu-IFN-α2 was added to MDBK cells at concentrations between 0 and 1.25 nM and, in a parallel experiment, the amount of IFN bound to the cell surface was controlled by establishing a

[5] K. C. Zoon, D. Zur Nedden, and H. Arnheiter, this volume [45].

binding curve with ^{125}I-labeled Hu-IFN-α2. Each point in Fig. 2 represents the mean value obtained from six individual dishes. Antibody binding to cells reaches half saturation when half saturating IFN concentrations have previously been added to the cells, and reaches saturation when saturating IFN doses have been added (for binding curves of ^{125}I-labeled Hu-IFN-α2 to MDBK cells, see Ref. 5). Over the concentration ranges tested, nonspecific binding of ^{125}I-labeled IFN was less than 10% of the total of the radiolabeled IFN bound to the cell-surface. The molar amounts of cell-bound IFN and IFN-bound antibody are approximately the same. For instance, as shown in Fig. 2, the amount of radiolabeled antibody bound to 7×10^6 MDBK cells to which IFN had been added at a concentration of 250 p*M* is 5.5 fmol. Under the same conditions, the cells bound 5.8 fmol of ^{125}I-labeled IFN specifically.

To test the specificity of antibody binding the synthetic oligopeptide Hu-IFN-α1 (151–166) was used in direct competition assays (closed circle in Fig. 2), and the radiolabeled antibody was added to dishes to which IFN has been added in the absence of cells (close square in Fig. 2). In both controls, antibody binding is as low as its binding to cells not treated with IFN.

Antibody III/21 Allows Determination of Changes in the Amount of Cell Surface-Bound IFN Occurring at 37°

Radiolabeled Hu-IFN-α2 bound to the surface of MDBK cells at 4° is rapidly internalized by these cells when they are shifted to 37°.[6] The corresponding decrease of surface-bound IFN should be detectable with the radiolabeled antibody III/21. To determine the amount of IFN remaining associated with the cell surface after incubation at 37°, the cells are washed five times with ice cold PBS and exposed to the radiolabeled antibody in the cold as described above. A representative experiment is shown in Fig. 3. At time 0, before the cells have been shifted to the warmer temperature, the total of the bound antibody was 3.3 fmol. With increasing time at 37°, the amount of radiolabeled antibody bound to IFN decreased. To determine the amount of ^{125}I-labeled IFN bound to the cell surface before and after the temperature shift, a parallel set of cells is exposed to radiolabeled IFN (125 p*M* in complete medium), and incubated in the cold for 2.5 hr and shifted to 37° as above. Then, the cells are quickly cooled to 4° and washed five times with ice cold PBS, pH 7.4, and exposed to 0.2 *M* acetic acid, 0.5 *M* NaCl, pH 2.5, for 10 min at 4°. This procedure removes surface-bound IFN and leaves internalized IFN asso-

[6] K. C. Zoon, H. Arnheiter, and D. J. P. FitzGerald, this volume [49].

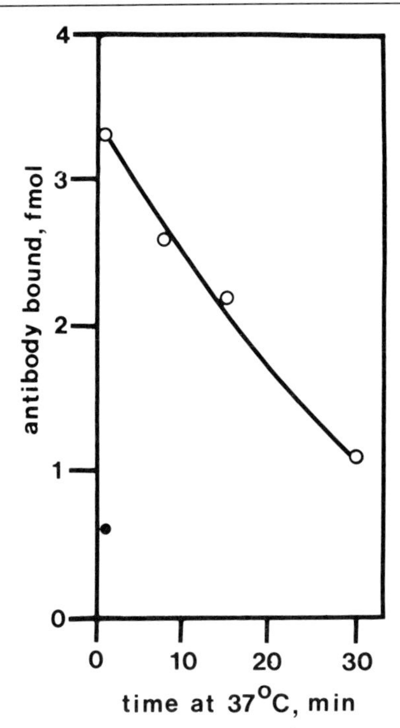

FIG. 3. Decrease of Hu-IFN-$\alpha 2$ on the surface of MDBK cells following incubation at 37°. Unlabeled IFN (125 pM) was added to cells at 4° for 2.5 hr. The cells were washed, incubated at 37° for the time periods indicated, washed again, exposed to ^{125}I-labeled antibody (0.14 nM) for 2 hr at 4°, and cell-associated antibody was assayed (○). Antibody bound to cells not exposed to IFN (●).

ciated with the cells.[6,7] For the experiment shown above, the amount of radiolabeled IFN bound to the cell surface was 3.5 fmol before the shift, and it decreased to approximately 1.5 fmol 30 min after the shift.

Thus, following incubation at 37°, the decrease in acid-releasable material and the decrease in antibody binding are comparable.

Comments

Although the above procedure for binding an antibody to the carboxyl terminus of receptor-bound IFN appears straightforward, there is no guarantee that other monoclonal or polyclonal antibodies recognizing the

[7] K. C. Zoon, H. Arnheiter, D. Zur Nedden, D. J. P. FitzGerald, and M. C. Willingham, *Virology* **130,** 195 (1983).

ultimate carboxyl terminus of IFN are able to recognize receptor-bound IFN. It is conceivable that IFN undergoes several conformational changes after binding to its specific receptor, and that IFN with such conformational changes is recognized only by particular antibodies. Antibody III/21 may be unique in that it binds stoichiometrically to IFN bound to the surface of MDBK cells. It appears that it binds to both that portion of the IFN which is bound specifically to receptors as well as that portion which is bound nonspecifically to the cells.

[49] Procedures for Measuring Receptor-Mediated Binding and Internalization of Human Interferon

By KATHRYN C. ZOON, HEINZ ARNHEITER, and DAVID FITZGERALD

The first step in the action of many polypeptide ligands is the specific binding of the ligand to its receptor on the plasma membrane. The receptor-bound ligand is then either taken up by the cell or processed and/or released at the cell surface. Monoclonal antibodies against the ligands or receptors, radioactive or fluorescent derivatives of ligands, and electron-dense ligand conjugates have been successfully used to study the fate of the receptor-bound ligands and their receptors. Many polypeptide hormones, toxins, and other molecules such as growth factors and low-density lipoprotein are internalized by concentrative adsorptive endocytosis (receptor-mediated endocytosis) following the binding of ligands to their specific cell surface receptors. This pathway consists of clustering of the receptor–ligand complexes and transfer of the complexes to intracellular compartments (receptosomes, the Golgi system, and lysosomes; for reviews, see Refs. 1–3).

Human interferons (IFNs) like many other polypeptide hormones interact with specific cell surface receptors and the complexes appear to be internalized.[4–8] In addition, at least a portion of human leukocyte inter-

[1] J. L. Goldstein, R. G. W. Anderson, and M. S. Brown, *Nature (London)* **279,** 679 (1979).
[2] R. M. Steinman, I. S. Mellman, W. A. Muller, and Z. A. Cohn, *J. Cell Biol.* **96,** 1 (1983).
[3] I. H. Pastan and M. C. Willingham, *Science* **214,** 504 (1981).
[4] K. C. Zoon, H. Arnheiter, D. Zur Nedden, D. J. P. FitzGerald, and M. C. Willingham, *Virology* **130,** 195 (1983).
[5] P. Anderson, Y. K. Yip, and J. Vilček, *J. Biol. Chem.* **258,** 6497 (1983).
[6] A. A. Branca, C. R. Faltynek, S. B. D'Alessandro, and C. Baglioni, *J. Biol. Chem.* **257,** 13291 (1982).
[7] S. Yonehara, A. Ishii, and M. Yonehara-Takahashi, *J. Gen. Virol.* **64,** 2409 (1983).
[8] F. H. Sarkar and S. L. Gupta, *Eur. J. Biochem.* **140,** 461 (1984).

ferons (IFN-α) appears to be taken up by the cell via receptor-mediated endocytosis.[4] This chapter describes the application of several biochemical and electron microscopic techniques to examine the cellular processing of human IFN-α.

Preparative Procedures and Methodology

Interferons. Human IFN-αA (2×10^8 units/mg protein) and human IFN-α2 ($1.5-1.7 \times 10^8$ units/mg protein) were the gifts of Hoffmann-La Roche, Nutley, NJ and Schering Corp., Bloomfield, NJ, respectively. The IFN was radiolabeled as previously described with ^{125}I-labeled Bolton–Hunter reagent.[9] The radiospecific activity of the ligand ranged from 1.1 to 4.4×10^5 Bq/μg protein. No detectable losses of antiviral activity were observed following the radiolabeling procedure.

Cells. Madin–Darby bovine kidney cells (MDBK) (ATCC CCL22) were obtained from the American Type Culture Collection, Rockville, MD and cultured to confluency in 6-well (35-mm-diameter) Costar plates (#3506).

Preparation of Colloidal Gold–Interferon Conjugates

Colloidal gold (5 nm) was prepared essentially as described by Faulk and Taylor[10] and coupled to IFN following methodologies described by DeMey *et al.*[11] and Geoghegan and Ackerman[12]; the details for preparation of IFN–gold and any modifications from the published procedures are given below and in Ref. 4.

Preparation of 5 nm Gold Colloid. Gold chloride (33 mg) was dissolved in 3.3 ml of glass distilled water to make a 1% solution. The gold chloride was handled with a plastic spatula; metal cannot be used. The 3.3 ml gold solution was added to 240 ml of glass distilled water in a 500-ml round-bottom flask. To adjust the pH, 5.4 ml of 0.2 N Na_2CO_3 was added to the 243.3 ml. Finally, 2 ml of phosphorous (yellow) saturated ether was added. The mixture was shaken gently and allowed to stand for 15 min at room temperature. A reflux condenser (uncooled) was fitted to the round-bottom flask and gentle heat applied via a heating cushion. The heat was increased gradually until the solution just began to boil. This level of heating was maintained for a further 5 min and then removed. At this

[9] D. Zur Nedden and K. C. Zoon, this volume [38].

[10] W. P. Faulk and G. M. Taylor, *Immunochemistry* **8,** 1081 (1971).

[11] J. DeMey, M. Moeremans, G. Geuns, R. Nuydens, and M. De Brabander, *Cell Biol. Int. Rep.* **5,** 889 (1981).

[12] W. D. Geoghegan and G. A. Ackerman, *J. Histochem. Cytochem.* **25,** 1187 (1977).

point the reaction was complete, the reflux condenser was removed, and the newly formed gold colloid allowed to cool. The gold solution can be stored in this state at 4° or dialyzed immediately against water. Dialysis against water is recommended before coupling of the gold to a protein. The dialysis bag must be thoroughly washed with glass distilled water to avoid gold floculations forming on the sides of the bag.

Coupling of Proteins to 5 nm Gold. Colloidal gold suspensions by themselves are stable over long periods provided the ionic strength of the storage buffer is kept low. The addition of physiologic saline or salt solutions of similar ionic strength causes the gold to floculate except when it is stabilized with a coating of protein molecules. This property forms the basis of a very simple assay for determining the success or failure of adsorbing a given protein to a gold colloid. Geoghegan and Ackerman outline such an assay with 5 ml of colloidal gold for each test of protein adsorption.[12] We routinely use 0.1 ml of the colloidal gold suspension, then add 5–10 μl of the protein solution and finally 0.02 ml of 10% NaCl. We have not found the need to determine the outcome of the assay spectrophotometrically but look for the presence or absence of gross floculation.

Coupling of IFN to 5 nm Gold. IFN-αA (5 mg/ml) in 0.1 *M* sodium acetate and 0.2 *M* ammonium acetate, pH 5.0, was diluted in distilled water and added to colloidal gold (pH 6.4). To 0.11 ml of IFN (5 mg/ml) was added 1.9 ml of distilled water. Then various concentrations of IFN were tested (as mentioned above) for their ability to stabilize the colloidal gold solution. We determined that 2 ml of IFN (0.5 mg/ml) was sufficient to stabilize 20 ml of the colloidal gold suspension. The IFN was allowed to interact with the colloidal gold for approximately 5 min at which time 3 ml of 8% BSA in 2 m*M* sodium borate (final pH 5.6) was added. IFN–gold conjugate was then centrifuged three times at 100,000 *g* to wash the gold free of any unbound IFN. The washing solution was Tris · HCl, pH 6.0, 1% BSA. The antiviral activity of the conjugated IFN was essentially the same as the unconjugated IFN. Thus, there was no apparent loss of bioactivity following coupling. Calculations indicate there are 40 IFN molecules per gold particle.

Coupling of ^{125}I-Labeled IFN to 5 nm Gold. ^{125}I-labeled IFN-αA was reacted with 5 nm gold essentially by the same procedure and with the same ratio of IFN to gold as described above for unlabeled IFN. ^{125}I-labeled IFN-αA (25 μg) was added to 1.5 ml of colloidal gold suspension. This was allowed to mix together for approximately 5 min after which time 50 μg of unlabeled IFN-αA was added for a further 5 min. Finally, 5 ml of 8% BSA in 2 m*M* borate was added and the IFN-αA–gold washed as described above. After washing the final specific activity of the ^{125}I-labeled IFN-αA–gold was 2×10^6 cpm/μg IFN-αA. The radiolabeled conjugate was used to test conjugate stability. Less than 0.1% of the radioactivity was released following repeated centrifugation at 100,000 *g* (4°, pH 7.4) and no additional radioactivity was released when the conjugate was incubated at pH 4 for 30 min at 37°.

Biochemical Procedures for Determining the Internalization and Processing and of Receptor-Bound IFN-α

Hu-IFN-α2 or Hu-IFN-αA was radiolabeled with ^{125}I-labeled Bolton–Hunter reagent and bound to MDBK cells at 4° as described in this volume.[13] The cells were washed five times with ice cold Eagle's minimal essential medium containing 10% fetal calf serum and 50 μg/ml gentamicin sulfate (complete medium). Each culture then received 2 ml of complete medium at 37° (or 4° for control cultures) and was incubated at the corresponding temperature for various periods of time. The amount of radioactivity in the supernatant (culture medium), associated with the cell surface, and associated with the interior of the cell was determined. Cell surface-associated IFN is defined as the amount of radiolabeled IFN released from the cells by treatment with 1.0 ml 0.2 *M* acetic acid, 0.5 *M* NaCl, pH 2.5 for 10 min at 0–4°. It should be pointed out that the amount of time necessary to remove the surface-bound IFN varies with the cell type. In addition, cells should be greater than 95% viable following exposure to the acidic solution. To determine the IFN associated with the interior of the cell, acid-treated MDBK cells were solubilized with either 0.1 *N* NaOH or 0.5% sodium dodecyl sulfate for at least 15 min at room temperature. Each assay point was done in triplicate. Specific binding (total binding minus nonsaturable binding) was used in the analysis of the internalization and degradation data. Figure 1[14,15] shows the kinetics of internalization and degradation of receptor-bound ^{125}I-labeled Hu-IFN-αA by MDBK cells.

Ligand degradation was analyzed by precipitation with trichloroacetic acid (TCA). An equal volume of ice cold 20% (w/v) TCA was added to each of the above fractions. The samples were incubated on ice for 10 min and centrifuged at 17,000 *g* at 0–4° for 10 min. The radioactivity was determined in the supernatant and pellet which was dissolved in 1 ml 0.1 *N* NaOH.

[13] K. C. Zoon, D. Zur Nedden, and H. Arnheiter, this volume [45].

[14] K. C. Zoon and H. Arnheiter, *Pharmacol. Ther.* **24,** 259 (1984).

[15] K. C. Zoon and H. Arnheiter, *in* "Interferon: Research, Clinical Application, and Regulatory Consideration" (K. C. Zoon *et al.*, eds.), p. 115. Am. Elsevier, New York.

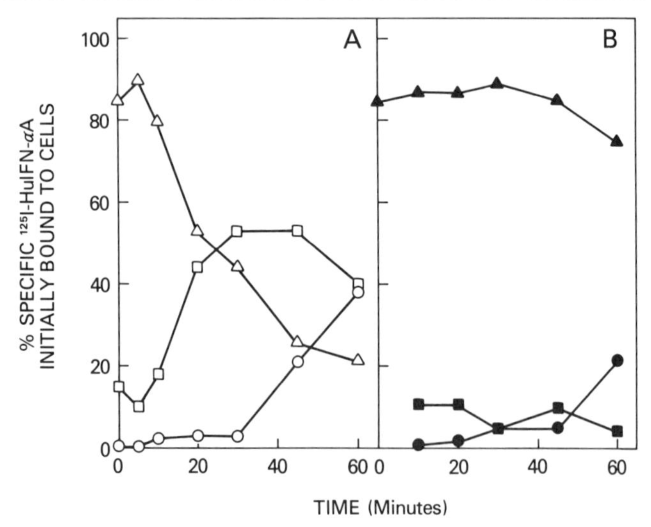

FIG. 1. The uptake and processing of ^{125}I-labeled IFN-αA by MDBK cells. Monolayers of MDBK cells were incubated with 125 p*M* ^{125}I-labeled Hu-IFN-αA in the presence or absence of 15 n*M* unlabeled Hu-IFN-αA at 4° as described. After washing, the cells were incubated at 37° (A) or 4° (B). At the indicated times the radioactivity in the medium and in the acetic acid-releasable and acetic acid-resistant fractions was determined. The amount of radiolabeled interferon specifically bound to the cells was 593 dpm (1.2 fmol). (○) Medium, 37°; (△) acetic acid releasable, 37°; (□) acetic acid resistant, 37°; (●) medium, 4°; (▲) acetic acid releasable, 4°; (■) acetic acid resistant, 4°. Reprinted with permission of *Virology*,[4] *Pharmacology & Therapeutics*,[14] copyright 1984, Pergamon Press, and Elsevier Science Publishing Co. Inc.[15]

Lysosomotropic amines (e.g., chloroquine) and monovalent carboxylic ionophores (e.g., monensin) have been useful in studying the path of internalization and processing of receptor-bound ligands including Hu-IFN-αs.[4-7,14] The former compounds inhibit ligand degradation and in some cases internalization[16]; the latter inhibit receptor recycling and inhibition of intracellular ligand degradation as well as other transport processes.[17] It should be pointed out that these inhibitors have many effects on cellular metabolism, therefore results obtained with these agents are not conclusive but are only supportive in assessing ligand uptake and processing. To examine the effects of these inhibitors on the internalization and degradation of radiolabeled human IFN-α2 on MDBK cells, cultures were incubated at 4° with ^{125}I-labeled IFN until they reached equilibrium, approximately 2.5 hr. The inhibitor (final concentration 1

[16] A. Gonzalez-Noriega, J. H. Grubb, V. Talkad, and W. S. Sly, *J. Cell Biol.* **85,** 839 (1980).
[17] S. K. Basu, J. L. Goldstein, R. G. W. Anderson, and M. S. Brown, *Cell* **24,** 493 (1981).

mM chloroquine or 100 μM monensin) was added and cultures were incubated for an additional 1 hr at 4°. The cultures were washed five times with ice cold complete medium and then incubated at 37° in prewarmed complete medium containing the same respective concentration of inhibitor. A parallel set of cultures without inhibitor was analyzed simultaneously. The amounts of radiolabeled material released into the medium, associated with the cell surface and associated with the cell interior were determined as a function of time at 37°. Ligand degradation was determined with TCA as described above. Figure 2 shows the effect of monensin on the binding, internalization, and degradation of Hu-IFN-α2 by MDBK cells.

Electron Microscopic Visualization of the Internalization of the IFN-αA Conjugated with 5 nm Colloidal Gold. MDBK cells in serum-free Dulbecco's minimal essential medium were incubated at 4° with Hu-IFN-αA conjugated to 5 nm colloidal gold (20 μg IFN/ml, 10^{13} particles/ml) in the presence and absence of a 100-fold excess of unconjugated Hu-IFN-αA (42 μg/ml, 10^{15} molecules/ml) for 3 hr or 24 hr in an atmosphere of 5%

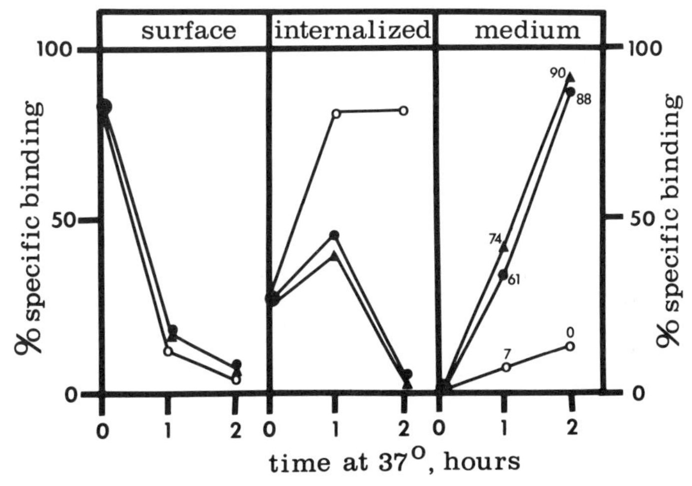

FIG. 2. Effect of monensin on internalization and degradation of Hu-IFN-αA. MDBK cells were incubated with 1 nM ^{125}I-labeled IFN-α2 at 4°. After 2.5 hr, monensin (100 μM) was added for an additional 1 hr at 4°. The cells were washed, and then incubated at 37° in the presence or absence of 100 μM monensin. At the indicated times, the medium was harvested and the cells were treated with acid as described in text. Radioactivity in the supernatant, the acid-releasable fraction, and the acid-resistant fraction was determined. The numbers in the panel "medium" indicate percentage TCA solubility of radiolabeled material. (○) Monensin, 4° and 37°; (●) monensin, 4° only; (▲) control, no monensin. Reprinted with permission of *Pharmacology & Therapeutics*[14] copyright 1984, Pergamon Press.

CO_2. The cells were washed in the same serum-free medium (4°) and then incubated at 37°. At 1 or 8 min the cells were fixed in 3% glutaraldehyde in phosphate-buffered saline, then fixed in OsO_4, dehydrated, and finally embedded in Epon. Thin sections of the samples were prepared with the use of a Sorvall MT 500 ultramicrotome equipped with a diamond knife. Sections were then stained with lead citrate. Transmission electron microscopy was then performed at a magnification 90,000× (Fig. 3). At 1 min IFN–gold conjugates are seen in clathrin-coated pits as well as attached to what appears to be debris on the cell surface. At 8 min particles

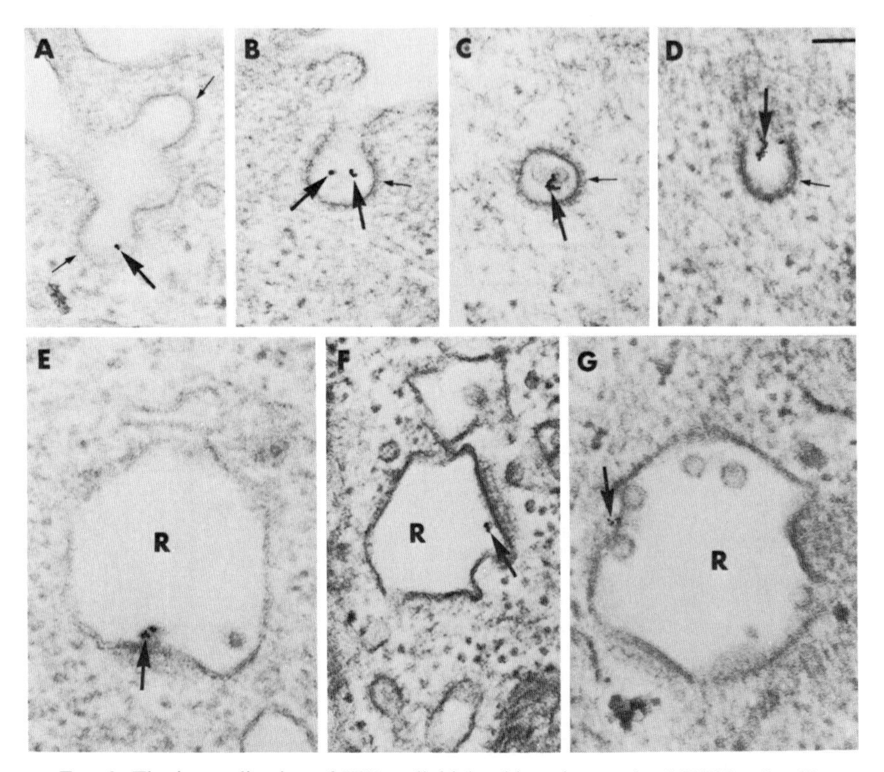

FIG. 3. The internalization of IFN–colloidal gold conjugates by MDBK cells. Human IFN-αA conjugated to 5 nm colloidal gold was incubated with MDBK cells for 3 hr at 4°, washed and warmed to 37° for 1 min (A–D) or 8 min (E–G) prior to fixation. A–D, arrows indicate IFN–gold complex in coated pits on the cell surface; E–G, arrows show complex in receptosomes (R) in the interior of the cell. Reprinted with permission of *Virology*,[4] *Pharmacology & Therapeutics*[14] copyright 1984, Pergamon Press, and Elsevier Science Publishing Co. Inc.[15]

were visible in receptosomes. Approximately 1 gold particle per 20 cell profiles was observed when cells were incubated with conjugate for 3 hr and 1–3 particles per cell profile when incubated for 24 hr (equivalent to 50 particles/cell). For the 24 hr incubation experiment, the labeling index, particles per cell/receptors per cell ($\times 100$), was approximately 6%. Similar results have been observed with other ligands such as α_2-macroglobulin.[18] Specificity of the intracellular localization of the IFN–gold conjugate was examined by morphometric quantitation of the IFN–gold internalized by the MDBK cells. Cells were incubated with IFN–colloidal gold for 24 hr at 4° in the presence and absence of 100-fold excess of unconjugated IFN, shifted to 37° for 8 min and processed for electron microscopy as previously described. Eight hundred sections of each were examined for the presence of the electron-dense conjugate in receptosomes. In our studies, approximately 80% of the conjugated IFN was found specifically associated with the receptosomes.

Comments

The biochemical and ultrastructural techniques described above are useful not only for studying the uptake and intracellular processing of Hu-IFN-α2 but also many other polypeptide ligands. In addition, immunological techniques which employ a monoclonal antibody that recognizes receptor-bound IFN-αs have also been very valuable in studying the internalization of IFN-α2.[19] This method as well is applicable to other ligand systems provided the monoclonal antibody against the ligand of interest can recognize that ligand bound to its receptor.

It should be pointed out that transmission electron microscopy of cells incubated with colloidal gold–ligand conjugates is a powerful technique which permits one to study ligand internalization pathways when the cell surface receptor number is very low (less than 1000 receptors/cell). In the case of Hu-IFN-αA and MDBK cells (800 receptors/cell), it was used to demonstrate that at least a portion of Hu-IFN-αA enters the cell by receptor-mediated endocytosis.[4]

[18] R. B. Dickson, J. C. Nicolas, M. C. Willingham, and I. Pastan, *Exp. Cell Res.* **132,** 488 (1981).

[19] H. Arnheiter and K. C. Zoon, this volume [48].

[50] Identification of Interferon Receptors by Chemical Cross-Linking

By SOHAN L. GUPTA and ARATI RAZIUDDIN

Interferon (IFN) action is apparently mediated through an interaction with receptors on the cell surface. Studies on IFN receptors have been carried out during recent years with the use of purified natural and recombinant IFNs. It has been demonstrated that mouse and human IFNs bind to specific receptors on cells with high affinity.[1–6] The receptors for Hu-IFNs (α, β, and γ) have been identified by chemical cross-linking.[5,7,8] Studies have indicated that after binding to cell receptors, Hu-IFN-αA, Hu-IFN-α2, and Hu-IFN-γ are internalized by the cells.[9–11] The internalization of Hu-IFN-α2 or -αA is associated with a loss of the receptors on the cell surface.[10,12] Hu-IFN-α2 is internalized apparently intact perhaps as a complex with its receptor.[10] However, the manner in which the IFN–receptor interaction leads to an induced expression of specific cellular genes[13] thus resulting in the development of the antiviral and other cellular responses is unclear.

A conventional way to study cellular receptors is by measuring the binding of radiolabeled ligand (IFN in this case) which is displaceable by an excess of the same but unlabeled ligand. This is a measure of specific binding sites. The total number of receptors and the dissociation constant

[1] M. Aguet and B. Blanchard, *Virology* **115,** 249 (1981).
[2] A. A. Branca and C. Baglioni, *Nature* (*London*) **294,** 768 (1981).
[3] K. E. Mogensen, M.-T. Bandu, F. Vignaux, M. Aguet, and I. Gresser, *Int. J. Cancer* **28,** 575 (1981).
[4] K. C. Zoon, D. Zur Nedden, and H. Arnheiter, *J. Biol. Chem.* **257,** 4695 (1982).
[5] A. R. Joshi, F. H. Sarkar, and S. L. Gupta, *J. Biol. Chem.* **257,** 13884 (1982).
[6] P. Anderson, Y. K. Yip, and J. Vilček, *J. Biol. Chem.* **257,** 11301 (1982).
[7] C. R. Faltynek, A. A. Branca, S. McCandliss, and C. Baglioni, *Proc. Natl. Acad. Sci. U.S.A.* **80,** 3269 (1983).
[8] F. H. Sarkar and S. L. Gupta, *Proc. Natl. Acad. Sci. U.S.A.* **81,** 5160 (1984).
[9] A. A. Branca, C. R. Faltynek, S. B. D'Alessandro, and C. Baglioni, *J. Biol. Chem.* **257,** 13291 (1982).
[10] F. H. Sarkar and S. L. Gupta, *Eur. J. Biochem.* **140,** 461 (1984).
[11] P. Anderson, Y. K. Yip, ad J. Vilček, *J. Biol. Chem.* **258,** 6497 (1983).
[12] A. A. Branca and C. Baglioni, *J. Biol. Chem.* **257,** 13197 (1982).
[13] P. Lengyel, *Annu. Rev. Biochem.* **51,** 251 (1982).

of this binding can be estimated by measuring specific displaceable binding at increasing concentrations of the radiolabeled IFN and analysis of the data by Scatchard plot.[1–6,14] In order to determine whether the IFN binds to a specific receptor and to identify the receptor molecule, bifunctional cross-linking reagents have been used to cross-link radiolabeled IFNs bound to cells followed by analysis of their crude membrane fractions by sodium dodecyl sulfate–polyacrylamide gel electrophoresis (SDS–PAGE) to determine whether a specific complex is obtained.[5,7,8] The cross-linking approach has been used successfully to identify and characterize the receptors for insulin and other polypeptide hormones.[15–17] Procedures used to identify the receptors for various human IFNs[5,8] are described here.

Materials

- Radiolabeled and unlabeled Hu-IFNs (α, β, γ; see chapters 36–43, this volume, for radiolabeling of IFNs)
- Human cells (e.g., Daudi cells, grown in suspension in RPMI-1640 containing 20% fetal calf serum,[18] or WISH cells grown in monolayers in Eagle's minimum essential medium containing 10% fetal calf serum)
- Cross-linking reagents, e.g., disuccinimidyl suberate, dithiobis(succinimidyl propionate) (DSS and DTSP, respectively; Pierce Chemical Company, Rockford, Illinois 61105)
- Phosphate-buffered saline (PBS, without Ca^{2+} and Mg^{2+})
- Sucrose solutions (5 and 10% in PBS without Ca^{2+} and Mg^{2+})
- Buffer containing 20 m*M* Tris·HCl, 1 m*M* magnesium acetate, pH 8.0
- Apparatus for slab gel electrophoresis, power supply, etc.
- Dimethyl sulfoxide (DMSO), sodium dodecyl sulfate (SDS), Triton X-100, phenylmethylsulfonyl fluoride (PMSF)

[14] G. Scatchard, *Ann. N.Y. Acad. Sci.* **51,** 660 (1949).

[15] P. F. Pilch and M. P. Czech, *J. Biol. Chem.* **255,** 1722 (1980).

[16] R. Victor Rebois, F. Omedeo-Sale, R. O. Brady, and P. H. Fishman, *Proc. Natl. Acad. Sci. U.S.A.* **78,** 2086 (1981).

[17] S. Paglin and J. D. Jamieson, *Proc. Natl. Acad. Sci. U.S.A.* **79,** 3739 (1982).

[18] Twenty percent serum was used in accordance with the instructions from the supplier (American type Culture Collection), however, Daudi cells quickly adapt to grow in 10% serum.

Procedures

Binding of ^{125}I-Labeled Hu-IFNs to Receptors on Human Cells and Competition by Different IFNs

^{125}I-labeled natural and recombinant IFNs have been used to assay cell receptors.[1–6] The general approach used is outlined here and illustrated with ^{125}I-labeled Hu-IFN-α2 binding to human Daudi lymphoblastoid cells (Fig. 1).

1. Daudi cells grown in suspension are harvested by centrifugation (600 *g*, 10 min) and suspended in growth medium containing fetal calf serum (approximately 5×10^6 cells/ml). The cell suspension is divided into 2-ml aliquots in 35-mm wells of 6-well plates.

2. ^{125}I-labeled Hu-IFN-α2 (50–100 units/ml final concentration) is added to each well either alone or with increasing concentrations of unlabeled Hu-IFN-α2 or other unlabeled IFNs to be tested for their capacity to compete for the receptors recognized by the ^{125}I-labeled IFN. The cultures are incubated either at 4° for 2 hr or at 37° for 30–60 min to allow IFN binding to cells.

3. Following the incubation, the free ^{125}I-labeled IFN is removed from the cell-bound IFN by pelleting cells through two layers of sucrose (5% sucrose over 10% sucrose, 4 ml each) by centrifugation (10,000 *g*, 10 min) at 4°. The supernatants are aspirated carefully (the top first when using sucrose pads), the tubes are inverted, and wiped inside with a tissue paper wrapped around a glass rod to remove any residual supernatant. The cell pellets are dissolved in 0.5 ml of 1% SDS and ^{125}I-radioactivity bound to cells is measured in a γ-counter.

In the case of cells grown in monolayers, cell cultures grown in 35-mm wells are incubated in fresh medium (2 ml) with ^{125}I-labeled Hu-IFN-α2 (50–100 units/ml) either in the absence or presence of increasing concentrations of various unlabeled IFNs as outlined above. The cultures are then washed 2–3 times with growth medium containing serum, the cells are dissolved in 1% SDS, and counted in a γ-counter. This reveals the extent of binding of the ^{125}I-labeled IFN to cells and its displacement by various unlabeled IFNs when added in excess (see Fig. 1).

Identification of Interferon Receptors by Chemical Cross-Linking

The ^{125}I-labeled IFNs bound to receptors on intact cells (or isolated cell membranes) can be cross-linked to the receptors with the use of bifunctional cross linking reagents. Analysis of the membrane fraction by SDS–PAGE allows an identification of the receptor molecule, and whether it may consist of subunits, etc. This approach is illustrated here

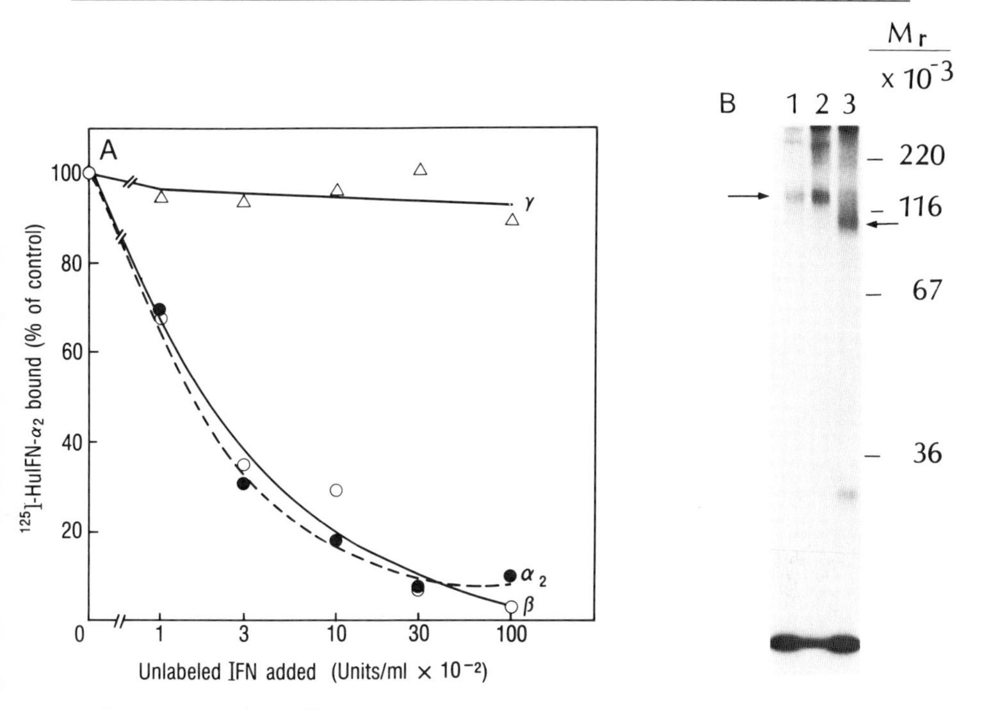

FIG. 1. (A) Binding of ^{125}I-labeled Hu-IFN-α2 to receptors on Daudi cells and its displacement by unlabeled Hu-IFN-α2 (●), Hu-IFN-β (Ser 17) (○), and Hu-IFN-γ (△) added at increasing concentrations. ^{125}I-labeled Hu-IFN-α2 added was 100 units/ml (2.2×10^4 cpm/ml). The 100% binding represents ~2000 cpm (see Procedures for details). (B) Cross-linking of ^{125}I-labeled Hu-IFN-α2 (lane 1), Hu-IFN-β (Ser 17) (lane 2), and Hu-IFN-γ (lane 3) to receptors on human lymphoblastoid cells Daudi: 20×10^6 cells were incubated in growth medium (4 ml) with 2000 units/ml of either ^{125}I-labeled Hu-IFN-α2 (1.1×10^6 cpm), Hu-IFN-β (Ser 17) (8×10^6 cpm), or Hu-IFN-γ (2.6×10^6 cpm). The cells were pelleted and the cell-bound ^{125}I-labeled IFNs were cross-linked with DSS. Their crude membrane fractions were obtained, extracted with Triton X-100, and analyzed by SDS–PAGE in an 8% polyacrylamide slab gel. The cross-linked complexes obtained are indicated by arrows. The recombinant Hu-IFN-γ used consists of two components with M_r of 17,000 and 34,000. Apparently, both components are biologically active.[20] The purified recombinant Hu-IFN-α2, Hu-IFN-β (Ser 17), and Hu-IFN-γ (specific activities 1.5×10^8, 2.6×10^8, and 6.8×10^7 reference units/mg of protein, respectively) used in these experiments were kindly provided by Schering-Plough Corporation, Cetus Corporation, and Genentech, Inc., respectively. Hu-IFN-α2 and Hu-IFN-γ were labeled with ^{125}I by solid phase lactoperoxidase procedure as described by Sarkar and Gupta.[21] Hu-IFN-β (Ser 17) was labeled with ^{125}I by a modified chloramine T procedure.[22]

with cells in suspension, however, the same conditions can be used with cells in monolayers.

1. Human lymphoblastoid cells (Daudi) grown in RPMI-1640 with 10% (or 20%) fetal calf serum are harvested by centrifugation (600 *g*, 10 min) and suspended in fresh growth medium at $\sim 5 \times 10^6$ cells/ml.

2. The cell suspensions (4 ml) are incubated in petri dishes with ^{125}I-labeled Hu-IFN (α, β, or γ) at 2000 international units/ml (which is saturating; lower levels can be used if the ^{125}I-labeled IFN is hot enough) either at 4° for 2 hr or at 37° for 30 min with gentle rocking on a platform rocker. Cells are then pelleted and washed once with PBS by centrifugation (600 *g*, 10 min) at 4°.

3. Each cell pellet is suspended in PBS at room temperature ($\sim 5 \times 10^6$ cells/ml) and a cross-linking reagent (e.g., DSS or DTSP) is added (20 μl of a 30 m*M* solution in DMSO per ml of cell suspension; final concentrations: 0.6 m*M* of cross-linker, 2% DMSO). The suspensions are incubated at room temperature (about 20°) for 20 min to allow cross-linking. DSS and DTSP react with free NH_2 groups in proteins, therefore, use of Tris buffer or presence of substances with free NH_2 groups should be avoided.

4. The cells are then pelleted (600 *g*, 10 min at 4°), and the cell pellets suspended in 2 ml of buffer containing 20 m*M* Tris · HCl, pH 8.0, 1 m*M* magnesium acetate, and 1 m*M* phenylmethylsulfonyl fluoride (PMSF). The cells are disrupted by homogenization in a Dounce grinder (40–50 strokes), an equal volume of 0.5 *M* sucrose (in above buffer) is added and mixed with a few more strokes. The lysates are centrifuged at 500 *g* for 10 min at 4° to remove the nuclei, and the supernatants are then centrifuged at 20,000 *g* for 20 min at 4° to pellet the membranes.

5. The 20,000 *g* pellet fractions (P20) are extracted in 200 μl of PBS containing 2% Triton X-100 and 1 m*M* PMSF by stirring at 4° for 45 min. The suspensions are centrifuged again at 20,000 *g* for 20 min at 4° and the supernatants are precipitated by adding 4 volumes of cold acetone.

6. After storage at −70° for 2 hr (or longer), the samples are centrifuged at 10,000 *g* for 15 min at 4° and the supernatants are discarded. The precipitates are dried under vacuum and dissolved in 100 μl of SDS sample buffer for gel electrophoresis (with or without a reducing agent, e.g., 100 m*M* DTT) by heating in a boiling water bath for 2–3 min. The samples are centrifuged briefly in an Eppendorf centrifuge (12,000 *g*, 5 min) and the clear supernatants are fractionated by SDS–PAGE (in 8 or 10% polyacrylamide slab gels) by the procedure of Laemmli.[19]

7. After drying on a gel drier, the gel is exposed to X-ray film (e.g., Kodak XAR-5) with DuPont Cronex Lightening Plus intensifying screen

[19] U. K. Laemmli, *Nature* (*London*) **227,** 680 (1970).

(3–7 days depending on the amount of radioactivity in the samples). This reveals a band of free ^{125}I-labeled IFN and a band of ^{125}I-labeled IFN cross-linked with its receptor (see Fig. 1B). Molecular weight markers are analyzed in parallel to determine the size of the cross-linked complex.

Figure 1A shows that ^{125}I-labeled Hu-IFN-α2 binds to Daudi cells and this binding is displaced by unlabeled Hu-IFn-α2 and Hu-IFN-β (Ser 17)[23,24] but not by Hu-IFN-γ, indicating that Hu-IFN-α2 and Hu-IFN-β (Ser 17) bind to the same receptors whereas Hu-IFN-γ may interact with different receptors. Similar results were obtained when partially purified natural Hu-IFNs were used for competition.[5] Figure 1B shows an experiment in which ^{125}I-labeled Hu-IFN-α2, Hu-IFN-β (Ser 17), and Hu-IFN-γ bound to Daudi cells were cross-linked with DSS and their crude membrane fractions (P20) analyzed by SDS–PAGE. The cross-linked complexes obtained with ^{125}I-labeled Hu-IFN-α2 and Hu-IFN-β (Ser 17) migrated together (lanes 1 and 2) with an apparent M_r of 150,000 whereas the cross-linked complex with ^{125}I-labeled Hu-IFN-γ (lane 3) migrated faster (M_r 105,000) indicating that the receptor cross-linked with Hu-IFN-γ is different from the receptor cross-linked with Hu-IFN-α2 and Hu-IFN-β (Ser 17). Evidence has been presented that Hu-IFN-α2 and Hu-IFN-β (Ser 17) compete with each other for the formation of the cross-linked complex, indicating that these IFNs recognize the same receptor molecules. However, Hu-IFN-γ does not compete with ^{125}I-labeled Hu-IFN-α2 or Hu-IFN-β (Ser 17). Conversely, the formation of the complex between ^{125}I-labeled Hu-IFN-γ and its receptor is displaced by unlabeled Hu-IFN-γ but not by unlabeled Hu-IFN-α2 or Hu-IFN-β, indicating that the receptors recognized by Hu-IFN-γ are specific for Hu-IFN-γ.[5,8]

Comments

The receptors recognized by various IFNs can be identified by chemical cross-linking of ^{125}I-labeled IFNs bound to intact cells followed by analysis of crude membrane fractions by SDS–PAGE. This approach has provided convincing evidence that Hu-IFN-α and -β interact with the same receptors whereas Hu-IFN-γ binds to different receptors.[5,8] The

[20] J. Le, W. Prensky, Y. K. Yip, Z. Chang, T. Hoffman, H. C. Stevenson, I. Balazs, J. R. Sadlik, and J. Vilček, *J. Immunol.* **131,** 2821 (1983).

[21] F. H. Sarkar and S. L. Gupta, this volume [36].

[22] E. C. O'Rourke, R. J. Drummond, and A. A. Creasey, *Mol. Cell Biol.* **4,** 2745 (1984).

[23] Hu-IFN-β (Ser 17) represents recombinant Hu-IFN-β with serine substituted for cysteine at position 17 (D. F. Mark, S. D. Lu, A. A. Creasey, R. Yamamoto, and L. S. Lin, *Proc. Natl. Acad. Sci. U.S.A.* **81,** 5662 (1984).

[24] L. S. Lin, R. Yamamoto, and R. J. Drummond, this volume [26].

cross-linked complex obtained as described above can be cleaved if a cleavable cross-linking reagent (for example, DTSP, cleaved by reduction with DTT) is used.[5] This excludes the possibility of any putative complex being an artifact (for example due to a minor contaminant in the radiolabeled IFN which may have been enriched after binding to cells), and allows various other experiments where a reversible cross-linking may be desirable.

The cross-linking approach for the identification of the IFN receptors can be used successfully with cells in suspension as well as cells grown in monolayers. In the latter case, the binding of radiolabeled IFN and cross-linking is done in monolayers. The cells are then washed with PBS, scraped, suspended in buffer (20 m*M* Tris · HCl, pH 8.0, 1 m*M* magnesium acetate, 1 m*M* PMSF), and processed as outlined above. The cross-linking approach allows detection of the receptors even when the background is high which makes it difficult to test for receptors simply by binding experiments. For example, ^{125}I-labeled Hu-IFN-β (Ser 17) gave a high background with WISH cells and therefore binding to receptors could not be tested. Nonetheless, receptors were easily identified by cross-linking.[8] However, the quantitation of receptors may be difficult since the efficiency of cross-linking is low and variable. Cross-linking works more efficiently at room temperature than in the cold, especially with cells grown in monolayers. However, it is desirable to do the binding of IFN in the cold (or if at 37°, for a short time, 30 min) to avoid the internalization of the receptor bound IFN and possible loss of the IFN–receptor complex as shown to occur with Hu-IFN-α2 at 37°.[10] Whereas it is necessary to use purified IFN for labeling to use as the binding ligand, partially purified IFNs have been used to test for competition for the receptors.[2,4–6]

Cross-linking experiments may also allow a partial characterization of the receptor molecule. After cross-linking, an analysis of the membrane fraction by SDS–PAGE under reducing and nonreducing conditions reveals whether or not the receptor may consist of subunits linked by disulfide bonds, as in the case of insulin receptors.[15] However, the cross-linked complexes obtained with ^{125}I-labeled Hu-IFN-α2, Hu-IFN-β (Ser 17), and Hu-IFN-γ did not change in size when analyzed with or without reduction with DTT (100 m*M*), indicating that the IFN receptors do not consist of subunits linked by disulfide bonds[5,8,25] (although the possibility of subunits associated through noncovalent bonds is not excluded). Cross-linking may also allow studies on the early events occurring at the receptor level. For example, it could be demonstrated that the IFN–receptor complex

[25] F. H. Sarkar and S. L. Gupta, unpublished results.

formed with ^{125}I-labeled Hu-IFN-α2 at 4° persists on the cell surface at 4°, however, upon incubation of cells at 37°, the complex is rapidly lost.[10] Studies indicated that this is associated with an internalization of the IFN as indicated by its resistance to digestion with trypsin or dissociation at low pH. A 150,000 M_r complex similar to the IFN-α2–receptor complex formed on the cell surface could be demonstrated in the trypsin-resistant fraction, suggesting that the ^{125}I-labeled Hu-IFN-α2 is perhaps internalized as a complex with its receptor.[10] However, the manner in which the IFN–receptor interaction leads to an induced expression of specific cellular genes and the development of the cellular responses remains an open question.

Acknowledgments

We are thankful to Schering-Plough Corporation, Genentech, Inc., and Cetus Corporation for generous gifts of purified recombinant Hu-IFN-α2, Hu-IFN-γ, and Hu-IFN-β (Ser 17), respectively. This work was supported by U.S.P.H.S. Grants CA-29991 and AI-17816 from the National Institutes of Health and a Grant (PCM 8309140) from the National Science Foundation (to S.L.G.).

[51] Extraction of Alpha Interferon–Receptor Complexes with Digitonin

By PIERRE EID and KNUD ERIK MOGENSEN

Our aim was to obtain an extract of the cellular receptor, stable enough to retain bound ligand and in a form suitable for investigating intrinsic enzyme activities. Certain aspects of the binding of IFN-αs to cellular receptors suggested an interaction related to the growth of the cells.[1] For this reason, exponentially growing cells were taken as starting material. Initial experiments suggested that of the commonly used nonionic detergents, only digitonin would serve our purpose. The method described has been developed for ^{125}I-labeled IFN-α2 bound to cells of the Burkitt line, Daudi; but the method works equally well with other IFN-αs and with other human cells of lymphoid origin.

[1] K. E. Mogensen and M.-T. Bandu, *Eur. J. Biochem.* **134,** 355 (1983).

Materials

Digitonin. Of the commercial preparations we have used, those from BDH (Analar grade; British Drug Houses, Ltd., Poole, Dorset, England) and Calbiochem-Behring (La Jolla, CA) have proved the most reliable. Solutions are prepared immediately prior to use, as stock solutions tend to precipitate. Dissolving directly into aqueous buffer is not practicable. An initial solution in a water–miscible organic solvent such as dimethyl sulfoxide is possible, but we have obtained better results with the following method which renders digitonin water soluble.

Commercial digitonin is dissolved in a minimum quantity of boiling absolute ethanol. The solution is then frozen (liquid nitrogen) and lyophilized. It can now be dissolved directly in aqueous buffer at the required concentration. In our hands such material remains in aqueous solution longer than that dissolved directly in dimethyl sulfoxide. The digitonin may be stored in lyophilized form.

Interferons. A certain amount of labeled IFN is required as a marker for the complex. Radioiodination of IFN-αs is described in detail earlier in this volume [37].

Binding of ^{125}I-Labeled IFN-α2 to Daudi Cells

The antiproliferative effect of IFN-α2 on Daudi cells is saturated at 100 units/ml (MRC 69/19). To cover this range and to have enough radioactivity to follow during purification we use ^{125}I-labeled IFN-α2 at a concentration of 300 units/ml. At this concentration maximum binding is attained within 30 min at 37°, on cells in exponential phase.[1,2] IFN-α2 is monoiodinated with ^{125}I as described earlier.[1] Daudi cells are grown in medium RPMI-1640 supplemented with 15% fetal calf serum. Cells are passaged daily by dilution at 100,000 cells/ml (exceptionally at 200,000 cells/ml), for at least 2 days before the experiment. The culture doubling time is 21–22 hr. At these cell concentrations binding varies linearly with the cell concentration. At higher concentrations, as the cells approach saturation density, binding at 37° falls off. This is more striking with some cell lines than with others. A culture of 50 ml provides 10^7 cells, which gives a convenient yield of bound radioactivity: about 40 pmol of bound IFN, yielding 100,000–200,000 dpm, depending upon the age of the radiolabel.

After 30 min incubation with ^{125}I-labeled IFN-α2 at 37°, the culture is placed on ice and centrifuged at 4°. Dissociation of cell bound IFN-α2 has

[2] P. Eid and K. E. Mogensen, *FEBS Lett.* **156,** 157 (1983).

a half-life greater than 10 hr at 4°; about 30 min at 37°. Cells are washed twice with 50 ml of medium containing 1% fetal calf serum, and then once with 1.5 ml 50 m*M* Tris · HCl, 0.15 *M* NaCl, pH 7.4, all at 4°. The cell pellet is resuspended in 0.5 ml of 0.7% digitonin in 50 m*M* Tris · HCl, pH 7.4, containing 1 m*M* phenylmethylsulfonyl fluoride. The extraction is allowed to proceed at room temperature. After 25 min the suspension is centrifuged for 45 min at 35,000 rpm (rotor SW 50.1, Beckman), and the supernatant removed. The distribution of radioactivity between the pellet and the supernatant is about 1 : 2 (a second extraction of the pellet residue yields a similar proportion). Extractions carried out at lower temperatures require a longer incubation with digitonin to obtain the same supernatant yield.

Separation of Complexed IFN

The complex is best separated from low- and high-molecular-weight contaminants immediately after its extraction. The major complex separates with K_{av} 0.4 when sieved on a column of Sephacryl S-400 (Pharmacia Fine Chemicals AB; Uppsala, Sweden), in 0.02% digitonin, 50 m*M* Tris · HCl, pH 7.4 at 4°. This gel sieves effectively in the range M_r: 10^5–10^6, for globular proteins in digitonin, giving an M_r value equivalent to 230,000 for the major complex.[2] With a tightly packed column of Sephacryl S-400, a bed height of 20 cm (diameter 1.6 cm) suffices to separate complexes from free IFN at a flow rate of 13 ml/hr.

Complexes obtained with different IFN-αs, and from different lymphoid cell lines separate identically. About half the IFN radioactivity is found in the complex and about half among the retained proteins of low molecular weight. Separate analysis of low-molecular-weight radioactivity shows that it has similar sieving characteristics to uncomplexed IFN. These proportions depend upon the time of incubation with IFN; and the type of IFN-α used.[2] Once separated, the major complex is stable at 4° with a slow dissociation of IFN, and can be used for further purification for up to 2 weeks afterward. Obviously, when enzyme activities are to be related to the amount of IFN complexed, the experiments are best performed as soon after separation as possible. Details of the extraction and separation are given in the table.

Comments

The method described provides an *in vitro* preparation of a human IFN-α–receptor complex suitable for further biochemical work. The orig-

Extraction of IFN-α2 Complexed to Cellular Receptor[a]

Step	Total bound IFN (pmol)	Absorbance (1 cm, 280 nm)	Volume (ml)
Cell-bound IFN	40	—	50.0
Digitonin extract	25–30	8.5	0.5
Separated complex (Sephacryl S-400)	10–15	0.06	5.0

[a] Following incubation of IFN-α2 at 300 units/ml with 10^7 Daudi cells at 200,000 cells/ml for 30 min at 37°.

inal purpose was to correlate receptor function with biological activity. This inevitably places certain limits on the method.

The complexes obtained are not necessarily those of the initial interaction of IFN with its receptor, as binding of IFN to cells was carried out at 37°. In fact there is reasonable evidence that we have extracted cell-bound IFN in two forms: one of which is stable (M_r 230,000) and one which rapidly dissociates to give free IFN. The relative proportions may vary with the time of incubation of the cells with IFN.[2] Larger aggregates that run close to the exclusion volume of S-400 ($K_{av} \leq 0.1$) may also appear, depending upon the IFN used, the binding temperature, and the time of incubation. In some ways, these aggregates behave similarly to the main complex,[3] but it is not yet known whether aggregation occurs before or after the extraction.

Chromatographic separation of the complex depends upon a well-prepared column. If the gel is too loosely packed, separation is inadequate on a 20 cm column. Columns are washed well with a solution of bovine serum albumin in digitonin prior to use. Initial runs may vary in the values obtained for K_{av}. They generally settle down precisely after 3 or 4 runs so that the content of any given fraction is predictable. The ionic strength is kept low during chromatography, in fact, the lowest compatible with discrete separation. Initial runs on a column may benefit from a slightly higher ionic strength.

Further purification is possible,[4] but techniques which require adsorption–desorption are usually associated with a loss of IFN label from the

[3] P. Eid and K. E. Mogensen, *Pathol. Biol.* **32,** 853 (1984).

[4] Recently we have employed a high-performance liquid chromatography system (FPLC, Pharmacia Fine Chemical AB; Uppsala, Sweden); with gel permeation columns SR6 and SR12 (Superose, Pharmacia Fine Chemicals AB; Uppsala, Sweden). In digitonin, K_{av} is linear on log M_r between 100,000 and 1,000,000. Resolution is improved and two stable complexes at 650,000 and 185,000 are obtained. Excluded radioactivity due to labeled IFN is reduced to less than 2% of the total on the column SR6.

complex. Proteins found in the fractions containing the complex can be dissociated with SDS and separated on SDS–PAGE. If they are first radiolabeled with, for example, ^{125}I, their migration can be marked by autoradiography.

It has not, as yet, been possible to reconstitute the specific IFN–receptor complex *in vitro,* and it seems to us that the isolated complex is one equilibrium form of cell-bound IFN that is stable in solution when isolated from other cellular components.[3]

Acknowledgments

It is a pleasure to acknowledge our indebtedness to Charles Weissmann for IFN-α2, and to Ion Gresser for his constant encouragement. This work was supported by Grants D.G.R.S.T. 82 L0074; I.N.S.E.R.M. CRL 81 10 15; C.N.R.S./PIRMED 100482/60821079; I.N.S.E.R.M. PRC 12 70 12; Fondation de Recherche en Hormonologie 948304; the Richard Lounsbery Foundation, the Simone et Cino del Duca Foundation, and A.R.C. Villejuif.

[52] Measurement of a Receptor for (2′-5′)-Oligoadenylate(trimer) on Macrophages

By XIN-YUAN LIU, BO-LIANG LI, and SHI WU LI

The main component of the (2′-5′)-oligoadenylate family[1] is pppA-2′p5′A2′p5′A, briefly (2′-5′)P_3A_3. It plays an important role in the mechanism of interferon action.[1-3] These may be natural antiviral substances that can regulate the activity of the immune system.[2-5] However, studies on the biological effects of (2′-5′)-oligo(A) usually used $CaCl_2$ or other agents to introduce the compound into cells.[6] In studying the effect of (2′-

[1] A. G. Hovanessian, R. E. Brown, E. M. Martin, W. K. Roberts, M. Knight, and I. M. Kerr, this series, Vol. 79, p. 184.

[2] X.-Y. Liu, Y.-M. Wen, Y. T. Hou, K. Yao, Y.-C. Lou, Z.-Q. Chen, H.-D. Zheng, W.-H. Ren, T.-Z. Lin, Z.-R. Huang, and D. B. Wang, *in* "The Biology of the Interferon System" (De Maeyer *et al.,* eds.), p. 115. Elsevier/North-Holland Biomedical Press, Amsterdam, 1981.

[3] X.-Y. Liu, *J. Exp. Pathol.* (in press).

[4] X.-Y. Liu, H. D. Zheng, N. Wang, W. H. Ren, and T. P. Wang, this volume [88].

[5] X.-Y. Liu, H.-D. Zheng, N. Wang, B.-L. Li, W.-H. Ren, R.-L. Kong, and T. P. Wang, *Sci. Sin.* (*Engl. Ed.*) **26,** 1057 (1983).

[6] G. E. M. Martin, D. M. Reisinger, A. G. Hovanessian, and B. R. G. Williams, this series, Vol. 79, p. 273.

5′)P_3A_3 on the activity of natural killer cells and macrophages, we found that (2′-5′)P_3A_3 can directly activate these cells without the addition of $CaCl_2$ or other substances and that [^{3}H](2′-5′)P_3A_3 can be bound to macrophages.[5] This binding is (1) saturable with high concentration of ligand and also with high concentration of macrophages, (2) reversible, (3) ligand specific and also somewhat cell specific, (4) relevant to its biological effect since there is a correlation of biological effect with binding for various derivatives, and (5) not due to pinocytosis of bulk liquid. Therefore, we concluded that there exists a receptor which mediates the direct action of (2′-5′)P_3A_3 on macrophages.[7,8] These observations seem to represent the first report of an oligonucleotide generating a biological effect through a receptor. Here we describe a method for measurement of this receptor.

Materials and Reagents

Wistar rats were supplied by the Animal Center, Sanghai Branch, Academia Sinica. DE81 chromatography paper was purchased from Whatman Inc.; RPMI-1640 medium, from Serva; [^{3}H]ATP, from Shanghai Institute of Nuclear Research, Academia Sinica; calf serum, from Shanghai Institute of Biological Products, Ministry of Health; heparin, from Shanghai East Wind Biochemical Reagent Factory; dibutylphthalate, from Shanghai Solvent Factory; dinonylphthalate, sodium mercaptoacetate, and sodium dodecyl sulfate (SDS) from Shanghai First Chemical Reagent Factory; bovine serum albumin (BSA) from P-L Biochemicals.

Cell culture medium: RPMI-1640 medium, 10 units heparin/ml, 10% calf serum (inactivated at 56° for 30 min).

Phosphate-buffered saline with potassium, PBS (+K): 136.8 m*M* NaCl; 2.7 m*M* KCl; 8.1 m*M* Na_2HPO_4; 1.5 m*M* KH_2HPO_4, pH 7.4.

Phosphate-buffered saline without potassium, PBS (−K): 140 m*M* NaCl; 8.1 m*M* Na_2HPO_4; 1.5 m*M* NaH_2PO_4, pH 7.4.

PBS–EDTA: 0.2 g disodium ethylenediaminetetraacetic acid and 0.2 g glucose in 1 liter PBS (+K).

BSA–1640 medium: RPMI-1640 medium with 0.2% bovine serum albumin and 10 units heparin/ml, pH 7.4.

BSA–PBS (+K): 0.2% (w/v) BSA in PBS (+K).

BSA–PBS (−K): 0.2% (w/v) BSA in PBS (−K).

BSA–TBS (−K): 0.2% (w/v) BSA in 20 m*M* Tris · HCl, pH 7.4, 140 m*M* NaCl.

[7] B. L. Li and X.-Y. Liu, *Sci. Sin.* (*Engl. Ed.*) **28,** 697 (1985).

[8] B. L. Li and X.-Y. Liu, *Sci. Sin.* (*Engl. Ed.*) **28,** 844 (1985).

Cell separation oil: dibutylphthalate/dinonylphthalate at a ratio of 2/1.5 (v/v).

Aqueous scintillation fluid: 8 g 2,5-diphenyloxazole; 0.5 g 1,4-bis-2-(4-methyl-5-phenyloxazolyl)benzene; 300 ml diethyoxyglycol brought to 1000 ml with dimethylbenzene.

Preparation of [^{3}H]pppA2′p5′A2′p5′A and Related Nonradioactive Compounds

[^{3}H]pppA2′p5′S2′p5′A, briefly [^{3}H](2′-5′)P_3A_3, was synthesized by a method similar to that described by Liu *et al.*[9] with [^{3}H]ATP as substrate, but on a small scale. Products were separated by DE81 paper chromatography with 0.4 *M* NH_4HCO_3 as developing solvent and nonradioactive (2′-5′)P_3A_3 as marker. [^{3}H](2′-5′)P_3A_3 was eluted with 1 *M* NH_4HCO_3, dried by lyophilization, dissolved in distilled H_2O, and analyzed again by DE81 paper chromatography. The final solution with specific activity about 10 Ci/mmol of trimer was adjusted to 50% (v/v) ethanol and stored in small aliquots at −20°.

(2′-5′)P_3A_3 was synthesized and separated by a method[9] which yielded about 1 g total (2′-5′)-oligo(A) and 400–500 mg of (2′-5′)P_3A_3 from reticulocytes from four rabbits.

pppI2′p5′I2′p5′I, abbreviated (2′-5′)P_3I_3, was obtained by deamination of (2′-5′)P_3A_3 with HNO_2 and characterized by paper chromatography,[9] spectrophotometry, circular dichroism, and base analysis.

$A_2'p_5'A_2'p_5'A$, abbreviated (2′-5′)A_3, was prepared by dephosphorylation of (2′-5′)P_3A_3 with alkaline phosphatase.

Preparation of Macrophages

Male Wistar rats (60–80 g) were injected intraperitoneally with 0.8–1.0 ml 3% (w/v) mercaptoacetate in normal saline containing 0.2% agar and killed about 3 days later. Peritoneal macrophages were taken out with a Pasteur pipette, suspended in cell culture medium, and cultured on a glass vessel or large glass Petri dish at 37° for 30–45 min under 3% CO_2, then washed with PBS to remove nonadherent cells.[4,5] Adherent macrophages were detached by repeated pipetting after incubation with PBS–EDTA at 37° for 10–15 min, collected by centrifugation at low speed (less than 1000 *g*, 1 min), washed once with PBS (+K), sedimented at low speed as above, resuspended to 2 $\times 10^7$ cells/ml in cell culture medium. The isolated macrophages were incubated at 37° for 10–15 min to recover their

[9] X.-Y. Liu, Y.-M. Wen, Y.-C. Lou, H. J. Lou, and T. P. Wang, *Kexue tongbao* **26,** 850 (1981) (in Chinese).

TABLE I
BINDING OF $[^3H](2'\text{-}5')P_3A_3$ TO MACROPHAGES OF DIFFERENT ORIGIN[a]

Origin of macrophages	Bound cpm ± SE	Bound cpm/ total cpm	Relative binding (%)
Wistar rats	4341 ± 218	0.15	100
Wild type rats	2092 ± 104	0.07	48
ICR mice	1274 ± 23	0.04	30

[a] Cells (1×10^5) were used in all cases. Peritoneal macrophages were induced by mercaptoacetate and purified by adhesion as described in the text. SE, standard error.

activity then placed in an ice bath. About 95% of the macrophages should still be viable after the above treatment. Mouse macrophages were prepared by a similar procedure.

The ability of macrophages of different origins and those produced by different methods to bind $[^3H](2'\text{-}5')P_3A_3$ varies. Rat macrophages induced by mercaptoacetate and purified by adhesion exhibit greater activity than purified macrophages from noninduced rats which in turn exhibit greater activity than crude macrophages from induced rats. When peritoneal macrophages are induced by mercaptoacetate and purified by adhesion, the ability of macrophages from different animals to bind $[^3H](2'\text{-}5')P_3A_3$ varied in the following order: Wistar rats > wild type rats > mice (Table I). Therefore, for routine studies mercaptoacetate-induced and adhesion-purified macrophages from Wistar rats are best, in which case about 15% of the total isotope added can be bound (Tables I and II).

TABLE II
EFFECT OF REACTION CONDITIONS ON $[^3H](2'\text{-}5')P_3A_3$ BINDING TO MACROPHAGES[a]

Medium	Bound cpm ± SE	Bound cpm/total cpm	Relative binding (%)
10% calf serum–1640	4852 ± 534	0.16	100
BSA–1640	3918 ± 343	0.13	81
BSA–PBS ($+K^+$)	3348 ± 509	0.11	69
BSA–PBS ($-K^+$)	1166 ± 133	0.04	24
BSA–TBS ($-K^+$)	1582 ± 112	0.05	33

[a] Macrophages were from Wistar rats. The 10% calf serum–1640 represents the complete cell culture medium as described in the text.

Measurement of [^{3}H](2′-5′)P_3A_3 Binding to Macrophages

To a siliconized glass tube in an ice bath, 1×10^5 macrophages (5 μl) are added to one side of the tube wall, 5 pmol [^{3}H](2′-5′)P_3A_3 (10 Ci/mmol trimer; in 1–5 μl) to the other side of the tube wall, and they are mixed by the addition of cell culture medium to a final volume of 100 μl. The mixture is incubated at 30° for 45 min, then put in an ice bath. Separation oil (0.4–0.5 ml) was immediately added. After centrifugation (at about 1000 *g*, 1–2 min), the macrophages with bound [^{3}H](2′-5′)P_3A_3 were sedimented to the bottom of the tube and the aqueous layer which contained the free [^{3}H](2′-5′)P_3A_3 floated upon the separation oil which thus divided bound and free ligands. The aqueous layer was aspirated and the tubes were washed 2–3 times with H_2O to remove any isotope remaining in the aqueous phase. Subsequently most of separation oil was aspirated. About 10–20 μl of separation oil was left on the bottom to avoid disturbing the macrophage pellet. Finally, 50 μl of 5% SDS was added to lyse the cell pellet. The entire lysed mixture was transferred into 5 ml of scintillation fluid and counted in a liquid scintillation spectrometer. This represents the total binding. Nonspecific binding was determined in reaction mixtures containing 10^{-4} *M* unlabeled (2′-5′)P_3A_3 in addition to [^{3}H](2′-5′)P_3A_3. The specific binding is defined as total binding minus nonspecific binding. In the case of rat peritoneal macrophages, the nonspecific binding is usually very low, approximately 6% of the specific binding.

Use of an oil to separate bound from free [^{3}H](2′-5′)P_3A_3 is rapid. Because the ligand–receptor complex is easily dissociated, the oil prevented ligand dissociated from the macrophage pellet from entering the aqueous layer. The oil also permits removal of the free ligands without disturbing the bound ligands.

Properties of Binding of [^{3}H](2′-5′)P_3A_3 to Receptor

The binding is pH dependent[7,8] with an optimum about pH 7.5. The time course of binding is shown in Fig. 1. The binding velocity is different at various temperatures. Fixing the incubation time at 45 min and taking the bound cpm at 37° as 100%, the binding at 30° is 97%; 15°, 35%; 0° (ice bath), 11%. At low temperature, the binding reaction is very slow. At 30° for 45 min, the ligands and receptors in the reaction medium are essentially stable so that it is convenient to use 30° in routine assays. Binding of (2′-5′)P_3A_3 to macrophages requires K^+ and changes in different media. In complete cell culture medium containing 10% calf serum, 16% of the added isotope can be bound (Table II). The binding is decreased when the calf serum is replaced by 0.2% bovine serum albumin or the RPMI-1640

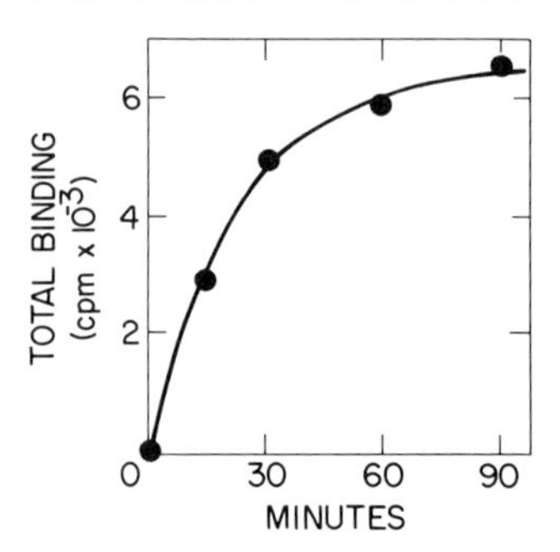

FIG. 1. The time course of (2′-5′)P_3A_3 binding to macrophages. Macrophages were from Wistar rats and binding of $[^3H]$(2′-5′)P_3A_3 was determined at 30°.

medium replaced by PBS (+K). Binding is decreased further when K^+ is omitted from the medium (Table II).

Remarks

There has been no previous report of the action of (2′-5′)-oligo(A) through a membrane receptor. The biological effect of (2′-5′)P_3A_3 is usually studied by adding $CaCl_2$ or hypertonic NaCl to introduce it into cells. The method described here for measuring a receptor for (2′-5′)P_3A_3 provides a way to test the possible direct action of (2′-5′)P_3A_3 or its derivatives on different cells. The binding of $[^3H]$(2′-5′)P_3A_3 to the receptor is ligand specific. A2′p5′A2′p5′A, the dephosphorylated product of (2′-5′)P_3A_3 and pppI2′p5′I2′p5′I, the deamination product of (2′-5′)P_3A_3 are inactive.[7,8] Therefore, the present method may also be useful for studying the biological effects and the structure–function relationship between this receptor and (2′-5′)P_3A_3 or its derivatives. It is also useful for studying the mechanism of action of (2′-5′)P_3A_3 on the immune system.

Acknowledgments

This work was supported by Chinese Academy of Sciences. The authors wish to thank Dr. Sidney Pestka for his support and critical reviewing of the manuscript, Dr. Jerome Langer for reviewing the manuscript, and Ms. Wendy Ewald and Ms. Sophie Cuber for typing the manuscript.

Section VII

Procedures for Isolation of Genes and Expression of Interferons in Bacterial and Heterologous Cells

[53] A Procedure for Isolation of Alpha Interferon Genes with Short Oligonucleotide Probes

By ARTHUR P. BOLLON, MOTOHIRO FUKE, and RICHARD M. TORCZYNSKI

Introduction

A method has been developed which permitted the isolation of several human alpha interferon (IFN-α) genes from a human genomic library in Charon 4A bacteriophages with 17-base synthetic oligonucleotide probes. Previously, human IFN-α genes were isolated from cDNA libraries or genomic libraries with cDNA probes.[1-3] The use of select, short, synthetic probes permits the isolation of subsets of genes within a gene family represented in a genomic library.[4]

Preparative Procedures and Methodology

Probe Selection

Construction of synthetic probes can be based on protein or DNA sequence information.[4] Two IFN-α synthetic probes were constructed based on selected conserved DNA sequences of IFN-α genes A–H.[1] Probe A was based on IFN-α genes A–H, position 412 to 429,[1] and has the sequence CCTGAGGTAGGACCGAC whereas Probe B was based on position 481 to 497[1] and has the sequence GGGAACACGGACCCTCC. Comparison of Probe A, which was synthesized by Bio Logicals Inc., and B, which was synthesized in our laboratory with a Bachem DNA Synthesizer, with the IFN-α genes A–L[3] at the designated gene positions indicated the homologies and mismatches shown in the table.

[1] D. V. Goeddel, D. W. Leung, T. J. Dull, M. Gross, R. M. Lawn, R. McCandliss, P. H. Seeburg, A. Ullrich, E. Yelverton, and P. W. Gray, *Nature* (*London*) **290,** 20 (1981).

[2] S. Nagata, H. Taira, A. Hall, L. Johnsrud, M. Streuli, J. Ecsödi, W. Boll, K. Cantell, and C. Weissmann, *Nature* (*London*) **284,** 316 (1980).

[3] S. Pestka, *Arch. Biochem. Biophys.* **221,** 1 (1983).

[4] A. P. Bollon, E. A. Barron, S. L. Berent, P. W. Bragg, D. Dixon, M. Fuke, C. Hendrix, M. Mahmoudi, R. S. Sidhu, and R. M. Torczynski, *in* "Recombinant DNA Products: Insulin-Interferon-Growth Hormone" (A. P. Bollon, ed.). CRC Press, Boca Raton, Florida, 1984.

PROPERTIES OF SYNTHETIC PROBES

	IFN-α genes[a]									
Probes	A	B	C	D	E	F	G	H	K	L
A	1	0	0	1	0	1	2	0	1	0
B	0	1	0	0	1	1	1	0	0	0

[a] Indicated are the number of mismatches between Probes A and B and IFN-α genes A–L. Perfect homology is represented by 0, one mismatch is 1, and two mismatches are 2.

Human Genomic Library Screening

Probes A and B were used to screen 180,000 PFU (plaque-forming units) of a Charon 4A human genomic library. From 180,000 PFU, five clones appeared to hybridize with both probes but upon further analysis only three individual clones, λ-77, λ-85, and λ-105 hybridized with both probes. The original signals obtained for clones λ-77, λ-85, and λ-105 with the probes A and B are shown in Fig. 1.

5′ Labeling and Filtration of Oligonucleotide Probes. The 17-base probes are labeled at the 5′ ends with T4 polynucleotide kinase and [γ-^{32}P]ATP. Since the oligonucleotides contain free 5′-hydroxyl termini, phosphatase treatment is not required prior to labeling. The procedures for end-labeling and the filtering of DNA samples are given below.

1. Terminal labeling is performed at 37° for 40 min in a 50 μl reaction mixture containing deoxyribooligonucleotide, 0.70 μg/ml; Tris · HCl, 100 mM, pH 7.5; Mg_2Cl_2, 10 mM; dithiothreitol, 10 mM; spermidine, 0.2 mM; [γ-^{32}P]ATP (5000 Ci/mmol, New England Nuclear), 0.4 μM; T4 polynucleotide kinase (P-L Biochemicals), 10 units.

2. The reaction is terminated by one phenol extraction, then one ether extraction.

3. End-labeled DNA is removed from unincorporated [γ-^{32}P]ATP by gel filtration on a Sephadex G-50 column equilibrated in TE buffer (TE: 10 mM Tris · HCl, pH 7.5, 1 mM Na_2EDTA). Sample volume is minimized by using a small (2-ml pipette) siliconized column and single-drop collection. Sephadex G-50 has been found to perform better than G-25 in resolving the labeled DNA from the unincorporated isotope. Peak fractions are pooled and the DNA solution is adjusted to 6× NET by the addition of 20× NET (1× NET: 0.15 M NaCl, 1.0 mM Na_2EDTA, and 3.3 mM Tris · HCl, pH 8.0). Specific activities, assuming 100% recovery following gel filtration, are greater than 2×10^8 cpm/μg.

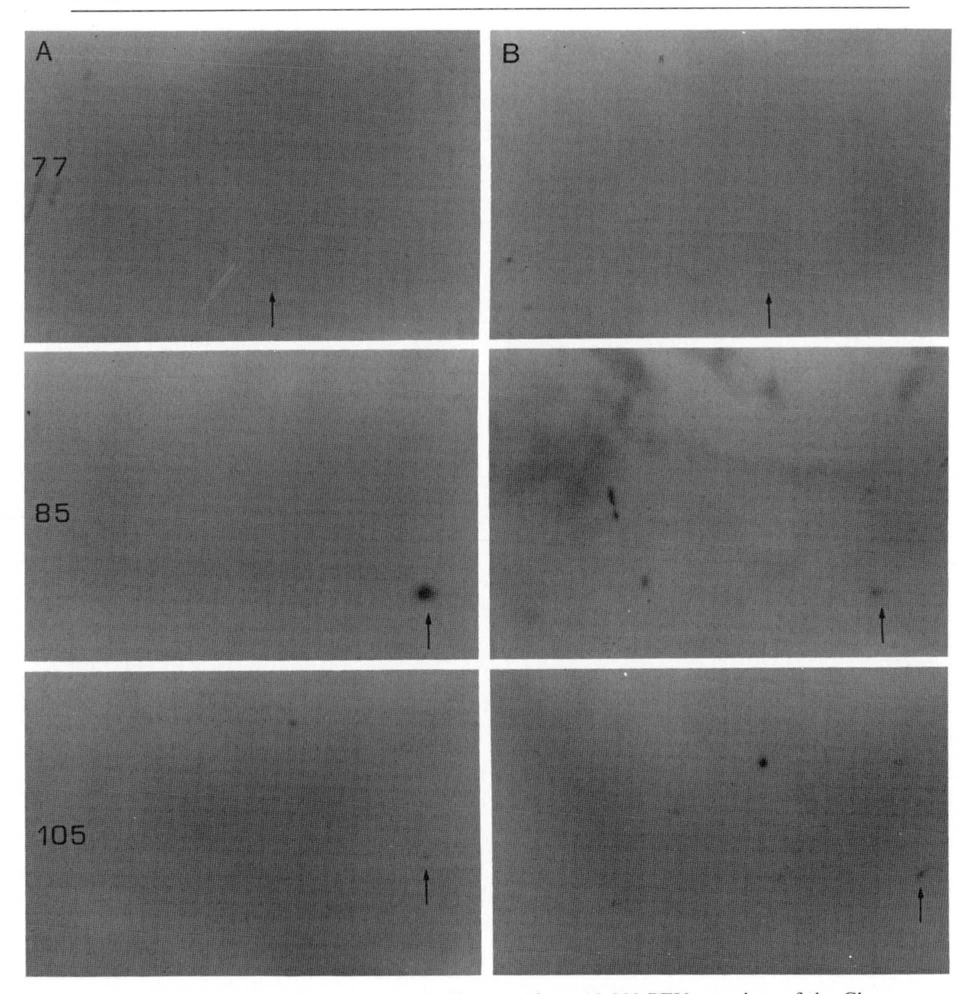

FIG. 1. Screening of human genomic library. About 10,000 PFU per plate of the Charon 4A human genomic library were screened with Probes A and B. Represented are the signals, indicated by arrows, for clone λ-77, λ-85, and λ-105 indicated as the 77, 85, 105 replicas on the first, second, and third row, respectively, with Probes A and B indicated by A and B, respectively. Positive signals were defined as signals recognized by both Probes A and B in the replica screening. About one-half of the original autoradiogram of each of the hybridized filters containing the respective signals is shown.

4. A 0.2-μm Acrodisc filter (Gelman Sciences, Inc.) is pretreated with 200 μg nondenatured tRNA in 6× NET, then washed with 6× NET. Labeled DNA is passed through the filter, collected, and is ready for use as a hybridization probe. Recoveries of labeled DNAs after filtering are

greater than 85%. DNAs are stored, prior to hybridizations, at 0.02–0.025 μg/ml.

Media. NZYDT broth: 10 g NZY amine type A (Humko Sheffield), 5 g yeast extract, 5 g NaCl, and 2 g $MgCl_2 \cdot 6H_2O$ are dissolved in 1 liter H_2O and autoclaved. Diaminopimelic acid and thymidine sterilized by filtration are added to the cooled broth to final concentrations of 0.1 and 0.04 g/liter, respectively.

NZYDT bottom agar: NZYDT broth with 15 g per liter of Bacto-agar.

NZYDT top agarose: NZYDT broth with 7.0 g/liter of agarose (Seakem).

λ *Phage Transfer. Escherichia coli* DP-50, *supF* [F^-, *supF58, supE44, dapD8, lacY1, glnV44, Δ(gal-uvrB)47*, λ^-, *tonA53, tyrT58, gyrA29 (nalA), Δ(thyA57), hsdS3*] is used as the host for phage infections. The human genomic library contained in Charon 4A[5] was obtained from Tom Maniatis. Plaque hybridizations are performed by a modification of the method described by Benton and Davis.[6]

1. A single colony of *E. coli* DP-50 is inoculated into 50 ml of NZYDT broth and incubated at 37° overnight with shaking.

2. For one 80-cm^2 petri dish, 10,000 Charon 4A PFU are mixed with 0.1 ml of an overnight *E. coli* DP-50 culture. After a 15 min incubation at 37°, the infected cells are mixed with 3.0 ml of molten NZYDT top agarose and plated onto NZYDT bottom agar plates.

3. Plates are incubated at 37° for 5–6 hr, then chilled at 4° for 1 hr. Phage growth and incubation time should be carefully monitored. If phage are overgrown and discrete plaques are no longer observed, sharp hybridization signals on the transfer nitrocellulose filter will not be obtained. However, a short incubation time will result in the loss of the hybridization signal due to an insufficient number of phage particles per plaque.

4. Two nitrocellulose filter copies are made from each plate. Precut filters (Schleicher and Schuell, BA85-SD) are used directly from the box. Filters are carefully laid directly on the agarose plate, 5 min for the first copy and 10 min for the second copy. Orientation marks are made on the filter and agarose plate with India ink containing ^{32}P. Filters are then removed and dried, phage-side up, on Whatman 3MM paper for 5 min.

5. Filters are laid, phage-side up, on 3MM paper saturated with 0.5 *N* NaOH and 1.5 *M* NaCl for 5 min, neutralized on paper saturated with 1 *M* Tris · HCl, pH 7.5 for 5 min, and again neutralized on paper saturated with 0.5 *M* Tris · HCl, pH 7.5 and 1.5 *M* NaCl for 5 min. Oversaturation of 3MM paper with alkali or neutralization solutions can result in migration of phage on nitrocellulose filters.

[5] R. M. Lawn, E. F. Fritsch, R. C. Parker, G. Blake, and T. Maniatis, *Cell* **15,** 1157 (1978).
[6] W. D. Benton and R. W. Davis, *Science* **196,** 180 (1977).

6. Filters are dried at room temperature then baked at 80° in a vacuum oven for 2 hr.

Hybridization. Solutions. Prehybridization and hybridization buffer: 6× NET (see above), 5× Denhardt's solution (1× Denhardt's solution is 0.02% each of bovine serum albumin, polyvinylpyrrolidone, and Ficoll 400,000), 0.1% sodium dodecyl sulfate (SDS), and 10% dextran sulfate.

Washing buffer: 6× SSC (1× SSC is 0.15 *M* NaCl and 0.015 *M* sodium citrate, pH adjusted to 7.5 with 10 *N* NaOH).

1. Two filters are placed in a polyethylene bag containing 5 ml of prehybridization buffer. The bag is heat sealed and prehybridized overnight at 37° with rotation.
2. After the prehybridization solution was removed, 2 ml of hybridization solution is added followed by the addition of heated (5 min at 100°) ^{32}P-labeled oligonucleotide at 2 ng/ml. One filter copy of each plate is hybridized with Probe A and the second copy with Probe B. The addition of *E. coli* DNA or yeast tRNA to the hybridization mixture did not reduce background noise and therefore is not used in the final procedure.
3. Filters are hybridized at 37°, with rotation, for 20–24 hr.
4. After hybridization, the filters are washed four times in 250 ml of 6× SSC at room temperature, 15 min per wash, then two times in 250 ml of 6× SSC plus 0.1% SDS at 45° for 1 hr.
5. Filters are dried on 3MM paper, covered with plastic wrap, and exposed to Kodak XAR X-ray with two DuPont Hi-Plus intensifying screens at −80° for 64 hr.

Characterization of IFN-α Genes

The three clones, λ-77, λ-85, and λ-105, which hybridized to Probes A and B, were characterized by cleavage with *Bam*HI, *Dde*I, or *Eco*RI and the digested DNA was electrophoresed on a 1% agarose gel.[4] The DNA was transferred from the gel to a nitrocellulose filter for Southern hybridization[7] to Probes A and B. *Eco*RI fragments from λ-77 and λ-85 (1.8 and 2.1 kb, respectively) and a 3.5 kb *Xba*I at fragment from λ-105 hybridized to both Probes A and B. The *Eco*RI fragments were subcloned into M13mp8[8] and the *Xba*I fragment was subcloned into M13mp11.[8] The subclones were designated mp8-77, mp8-85, and mp11-105.

DNA sequence analysis of the three subclones was performed by the dideoxy sequencing method[9] and the Maxam and Gilbert method[10] for

[7] E. M. Southern, *J. Mol. Biol.* **98,** 503 (1975).
[8] J. Messing and J. Vieira, *Gene* **19,** 269 (1982).
[9] F. Sanger, S. Nicklen, and A. R. Coulson, *Proc. Natl. Acad. Sci. U.S.A.* **74,** 5463 (1977).
[10] A. M. Maxam and W. Gilbert, this series, Vol. 65, p. 499.

select regions. The DNA sequence of the IFN-α gene in the λ-77 *Eco*RI fragment from position 1 to 466 is completely homologous with IFN-αL. The IFN-α gene (IFN-αWA) sequence contained within the λ-85 *Eco*RI fragment was determined and compared to IFN-α genes A–L sequences.[11] IFN-αWA was shown to be novel.[11] It is approximately 88% homologous with IFN-αA, αB, αD, αE, approximately 92.5% homologous with IFN-αF, αG, αH, αK, αL, and 94% homologous with IFN-αC.[11] Sequence analysis of the IFN-αWA gene also indicated that it does not contain any stop codons in the signal or the mature interferon coding region. DNA sequence analysis of the IFN-α gene contained within the λ-105 indicates that it is IFN-αK.

IFN-αWA Expression in *E. coli*

The IFN-αWA gene was engineered for expression in *E. coli* with the M13mp11 lac-fusion expression system as previously described.[12] The *Hin*cII site of the M13mp11 and the *Alu*I site of the IFN-αWA gene was ligated resulting in a fusion product, M13mp11-IFN-αWA, containing coding sequences for the Lac Z gene fragment (Met-Thr-Met-Ile-Thr-Pro-Ser-Leu-Gly-Cys-Arg-Ser) followed by part of the IFN-αWA signal sequence (Tyr-Lys-Ser-Ile-Cys-Ser-Leu-Gly) followed by the mature IFN-αW.[11]

E. coli JM103 was infected with the M13mp11–IFN-αWA and induced with isopropyl β-D-thiogalactoside as previously described.[12] Approximately 5×10^6 units of IFN-αWA per liter of bacterial culture[11] was obtained as determined by the cytopathic effect-reduction assay[13] with WISH cells and vesicular stomatitis virus.

Concluding Comments

The screening procedure with 17-base probes described above should be useful for the isolation of most genes from a human genomic library although some conditions will vary according to the sequence of the probes. It is noteworthy that our procedure yielded three IFN-α genes which had perfect homology with the two probes utilized. Even genes containing introns could be isolated from a genomic library with short and

[11] R. Torczynski, M. Fuke, and A. P. Bollon, *Proc. Natl. Acad. Sci. U.S.A.* **81,** 6451 (1984).

[12] P. Slocombe, A. Easton, P. Boseley, and D. C. Burke, *Proc. Natl. Acad. Sci. U.S.A.* **79,** 5455 (1982).

[13] W. E. Stewart, "The Interferon System." Springer-Verlag, New York, 1979.

S_1									S_{10}										S_{20}
MET	ALA	LEU	SER	PHE	SER	LEU	LEU	MET	ALA	VAL	LEU	VAL	LEU	SER	TYR	LYS	SER	ILE	CYS
			1									10							
SER	LEU	GLY	CYS	ASP	LEU	PRO	GLN	THR	HIS	SER	LEU	GLY	ASN	ARG	ARG	ALA	LEU	ILE	LEU
		20										30							
LEU	ALA	GLN	MET	GLY	ARG	ILE	SER	HIS	PHE	SER	CYS	LEU	LYS	ASP	ARG	TYR	ASP	PHE	GLY
		40										50							
PHE	PRO	GLN	GLU	VAL	PHE	ASP	GLY	ASN	GLN	PHE	GLN	LYS	ALA	GLN	ALA	ILE	SER	ALA	PHE
		60										70							
HIS	GLU	MET	ILE	GLN	GLN	THR	PHE	ASN	LEU	PHE	SER	THR	LYS	ASP	SER	SER	ALA	ALA	TRP
		80										90							
ASP	GLU	THR	LEU	LEU	ASP	LYS	PHE	TYR	ILE	GLU	LEU	PHE	GLN	GLN	LEU	ASN	ASP	LEU	GLU
		100										110							
ALA	CYS	VAL	THR	GLN	GLU	VAL	GLY	VAL	GLU	GLU	ILE	ALA	LEU	MET	ASN	GLU	ASP	SER	ILE
		120										130							
LEU	ALA	VAL	ARG	LYS	TYR	PHE	GLN	ARG	ILE	THR	LEU	TYR	LEU	MET	GLY	LYS	LYS	TYR	SER
		140										150							
PRO	CYS	ALA	TRP	GLU	VAL	VAL	ARG	ALA	GLU	ILE	MET	ARG	SER	PHE	SER	PHE	SER	THR	ASN
		160						166											
LEU	GLN	LYS	GLY	LEU	ARG	ARG	LYS	ASP											

FIG. 2. Amino acid sequence of IFN-αWA. The amino acid sequence is based on the IFN-αWA gene sequence.[11] The signal sequence is represented by amino acids S1 through S23 and the mature interferon is represented by amino acids 1 through 166.

mixed probes provided the intron does not split the sequences of the probes to a size unable to maintain a stable hybrid. Clearly, genomic libraries of appropriate sizes offer the advantage of obtaining genes independent of gene expression such as IFN-αWA which has not previously been found in cDNA libraries. The IFN-αWA amino acid sequence, which is shown in Fig. 2, differs at five amino acid positions (42, 56, 57, 110, and 133) conserved in IFN-αA through L. Changes at positions 42 and 133 result in the replacement of acidic residues with hydrophobic residues.

Acknowledgment

The authors thank Lynda Cheryl Hendrix for assistance in the expression studies, Dr. Paul Bragg for synthesis of one of the 17-base probes, Dr. Kurt Berg and Pia Jensen at Wadley Institutes for interferon assays, and Carol Crumley for preparation of the manuscript. This work was performed in the Oree Meadows Perryman Laboratory for Cancer Research at Wadley Institutes and was funded by the Meadows Foundation and in part by an NIH grant to APB.

[54] Use of the Phage Lambda P_L Promoter for High-Level Expression of Human Interferons in *Escherichia coli*

By ERIK REMAUT, PATRICK STANSSENS, GUUS SIMONS, and WALTER FIERS

Numerous biologically interesting compounds are difficult to obtain from their natural sources in sufficient quantities to permit a detailed study of their biological and biochemical properties and their modes of action. The advent of molecular cloning techniques has offered new approaches to this problem. Today, the extensively studied bacterium, *Escherichia coli,* is still the most widely used host in cloning and expression studies. The mere cloning of a gene into a plasmid and its introduction into *E. coli* is, in general, not a sufficient condition to ensure efficient expression of the cloned gene. In particular, eukaryotic genes, introduced into *E. coli,* are not recognized by the host's transcription–translation machinery. This has been overcome by the development of expression vectors: plasmids which incorporate the essential control elements to ensure efficient transcription and translation of essentially any coding region. Furthermore, as high-level expression of a foreign gene in *E. coli* may be toxic (as it turned out for IFN-β) or at least exert a negative selection pressure, we have developed a temperature-controlled expression system.

This chapter describes the application of expression vectors, based on the inducible leftward promoter of coliphage λ, to obtain efficient synthesis of biologically active human interferons in *E. coli.*

Preparative Procedures and Methodology

Materials and Solutions

1. Restriction enzymes: New England Biolabs, Beverly, Massachusetts, or Boehringer, Mannheim, West Germany
2. T_4-DNA ligase: see below
3. DNA polymerase I (Klenow fragment): Boehringer, Mannheim, West Germany
4. T_4-DNA polymerase: New England Biolabs, Beverly, Massachusetts
5. S1 nuclease: Boehringer, Mannheim, West Germany

6. [U-^{14}C]protein hydrolysate: The Radiochemical Centre, Amersham, England
7. Triton X-100: Sigma Chemical Co., St. Louis, Missouri
8. SDS: sodium dodecyl sulfate (BDH Chemicals, Poole, England)
9. EN^{3}HANCE: New England Nuclear Corp., Boston, Massachusetts
10. Serva Blue R: Serva, Heidelberg, West Germany
11. LB medium: 1% Bacto-tryptone, 0.5% Bacto-yeast extract, 0.5% NaCl (Difco Corp., Detroit, Michigan)
12. TES buffer: 10 m*M* Tris · HCl, pH 8.0; 5 m*M* EDTA; 30 m*M* NaCl

Bacterial Transformation

Bacterial cells were made competent for plasmid DNA uptake by treatment with $CaCl_2$.[1] Competent cells, at a density of 1×10^9/ml were stored in 100-μl aliquots at $-70°$ in the presence of 15% glycerol. Before transformation, an aliquot was thawed on ice. Ligated plasmid DNA was added and the mixture was left on ice for 20 min. The cells were heat-shocked at 34° for 5 min and returned to ice for 15 min. Ten volumes of LB medium were added and the cells incubated at 28° for 1 hr prior to plating on selective media at 28°. This procedure differs from standard transformation procedures in that the heat-shock is performed at 34° rather than at 41°. This modification was adopted because the bacterial hosts used, contain a temperature-sensitive *c*I repressor.[2]

Plasmid DNA Procedures

Plasmid DNA, intended to be used as a source of DNA fragments in ligation and transformation experiments, was usually prepared from 20 ml LB cultures according to the Triton-cleared lysate procedure.[3] Larger amounts of plasmid DNA were isolated by cesium chloride–ethidium bromide density gradient centrifugation. For analytical purposes, a rapid alkaline–SDS lysis procedure was used.[4]

Restriction enzymes were used according to the supplier's specifications. Ligation reactions were carried out for 2–16 hr at 18° in a buffer

[1] E. M. Lederberg and S. N. Cohen, *J. Bacteriol.* **119,** 1072 (1979).

[2] We had earlier observed that plasmids carrying the P_L promoter frequently suffered deletions following transformation when the heat-shock was performed at 41°. This was ascribed to transient expression from the P_L promoter due to heat inactivation of the *c*I repressor of the host.

[3] M. Kahn, R. Kolter, C. Thomas, D. Figurski, R. Meyer, E. Remaut, and D. R. Helinski, this series, Vol. 68, p. 268.

[4] H. C. Birnboim and J. Doly, *Nucleic Acids Res.* **7,** 1513 (1979).

consisting of 50 m*M* Tris · HCl, pH 7.4, 10 m*M* $MgCl_2$, 10 m*M* dithiothreitol, 0.5 m*M* ATP, and T_4-DNA ligase purified from the overproducing strain C600 (p*c*I857) (pPLc28lig8).[5] DNA polymerase I, Klenow fragment, was used to fill in 5′-protruding single-stranded DNA ends. A typical reaction mixture contained 1 pmol of DNA termini and 1 unit of enzyme in 30 μl of a buffer consisting of 25 m*M* Tris · HCl, pH 7.4, 10 m*M* $MgCl_2$, 10 m*M* dithiothreitol, and 0.1 m*M* of all four deoxyribonucleoside triphosphates. The reaction was performed at 18° for 45 min. Resection of 3′-ends was carried out with T_4-DNA polymerase. The reaction was performed at 15° for 3 hr in a 30 μl reaction mixture containing 3 pmol of DNA termini in 65 m*M* Tris · HCl, pH 7.9, 20 m*M* KCl, 10 m*M* $MgCl_2$, 5 m*M* dithiothreitol, and 0.1 m*M* of any one of the four deoxyribonucleoside triphosphates. S1 nuclease was used to blunt DNA fragments with 5′-sticky ends. The reaction buffer consisted of 25 m*M* sodium acetate, pH 4.4, 250 m*M* NaCl, 4.5 m*M* $ZnSO_4$. About 5 μg of plasmid DNA was incubated in 100 μl of buffer with 100 units of enzyme at 18° for 30 min.

Enzymatic reactions were usually stopped by phenol extraction. The aqueous layer was extracted two times with 5 volumes of ethyl ether and the DNA was recovered by precipitation with two volumes of ethanol. Alternatively, restriction enzymes were inactivated by heating at 65° for 10 min.

Specific DNA fragments were recovered by the squeeze-freeze method[6] after electrophoresis in 0.8–2% agarose gels. Although not optimal in recovery of DNA, this procedure allows the extraction of functional DNA with respect to susceptibility to ligation and secondary restriction. To avoid damage to the DNA, the fragment of interest was not exposed to either ethidium bromide or UV light but was localized by means of a stained side track.

Construction of IFN Expression Plasmids

The description of the construction of IFN expression plasmids given below provides an outline of the procedures (see Fig. 1). Intermediate plasmid constructions are not discussed but have been described elsewhere.[7–10]

[5] E. Remaut, H. Tsao, and W. Fiers, *Gene* **22,** 103 (1983).

[6] R. W. J. Thuring, J. P. H. Sanders, and P. Borst, *Anal. Biochem.* **66,** 213 (1975).

[7] R. Derynck, E. Remaut, E. Saman, P. Stanssens, E. De Clercq, J. Content, and W. Fiers, *Nature (London)* **287,** 193 (1980).

[8] E. Remaut, P. Stanssens, and W. Fiers, *Nucleic Acids Res.* **11,** 4677 (1983).

[9] P. Stanssens, Ph.D. Thesis, State University of Ghent, Belgium (1983).

[10] G. Simons, E. Remaut, B. Allet, R. Devos, and W. Fiers, *Gene* **28,** 55 (1984).

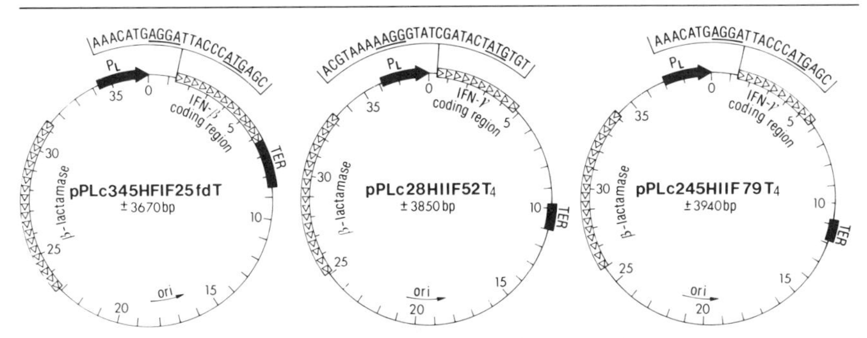

FIG. 1. Functional maps of representative expression plasmids. More detailed maps of the plasmids, including restriction sites, are given in refs. 8 and 10. P_L, position and direction of transcription of the P_L promoter; TER, terminator of transcription; ori, origin of replication.

IFN-β. cDNA clones containing coding sequences of human IFN-β were originally obtained starting from an enriched mRNA fraction prepared from human VGS fibroblasts induced with poly(I)·poly(C) and cycloheximide.[11] Mature IFN-β has an amino-terminal methionine residue coded for by an ATG codon. An *Alu*I site is present eight nucleotides preceding this codon. This site was used as a starting point to trim back the superfluous DNA stretch up to the ATG codon. Following partial *Alu*I cleavage and isolation of the desired DNA fragment, the DNA was treated with T_4-DNA polymerase in the presence of dTTP only. This allowed the progressive shortening of the 3′-end of the DNA fragment up to the first T-residue opposite the A-residue of the ATG codon. The single-stranded 5′-protruding end was removed with S1 nuclease, thus exposing the ATG codon as a blunt end. The blunted end was ligated to a filled-in *Xba*I site. This was done to create a "storage-plasmid" that would allow easy recovery of the coding region of mature IFN-β.[12] In this plasmid, the ATG codon of mature IFN-β was exposed as a blunted end after *Xba*I cleavage and removal of the 5′-protruding end with S1 nuclease. The blunted end was then linked to the S1-blunted *Sal*I site of the expression plasmid, pPLc245.[8] This plasmid contains the P_L promoter and the ribosome binding site of the replicase gene of the RNA phage MS2. The resulting plasmid was designated pPLc245HFIF25. A further

[11] R. Derynck, J. Content, E. De Clercq, G. Volckaert, J. Tavernier, R. Devos, and W. Fiers, *Nature* (*London*) **285,** 542 (1980).

[12] This property has been used to construct expression plasmids containing a variety of ribosome binding sites in which the ATG codon forms part of an *Xba*I site. These plasmids allow easy insertion of IFN-β mature sequences through ligation at the *Xba*I site (P. Stanssens, unpublished work).

derivative, designated pPLc345HFIF25 fdT, was obtained after insertion, downstream from the IFN-β coding sequence, of the central transcription terminator of phage fd.[13,14]

IFN-γ. cDNA clones containing the coding sequences for human IFN-γ were obtained starting from an enriched mRNA fraction prepared from splenocyte cultures stimulated with staphylococcal enterotoxin A.[15] As a starting point for plasmid constructions, we used an *Ava*II site located at the position corresponding to the fifth amino acid residue of mature IFN-γ. Synthetic DNA fragments were used to extend the coding sequence up to a TGT codon corresponding to the amino-terminal cysteine of mature IFN-γ. The linker fragments[16] 5′-TGTTACTGCCAG-3′ and 3′-ACGGTC-CTGGG-5′ were annealed to each other, filled in with DNA polymerase (Klenow fragment) and cleaved with *Ava*II. The fragment was then ligated to the *Ava*II site of IFN-γ. The TGT codon was subsequently linked to the ATG codon of pPLc245 in the same way as described for IFN-β. The resulting plasmid was designated pPLc245HIIF79. In an analogous manner, the TGT codon was linked to a blunted ATG codon of a modified ribosome binding site derived from the attenuator of the *E. coli* tryptophan operon.[10,17] This plasmid was designated pPLc28HIIF52. Further derivatives were constructed carrying a transcription terminator derived from phage T_4 (called pPLc245HIIF79T_4 and pPLc28HIIF52T_4, respectively).[10,18]

Bacterial Synthesis of Interferons

Principle of the Method. The expression vectors used in this work incorporate a 247 bp DNA fragment carrying the leftward operators and leftward promoter (P_L) of phage λ.[19] This promoter is known to be very strong and its activity is tightly regulated by a repressor protein, product of the phage gene *c*I. Mutants are available which synthesize a thermolabile repressor, thus allowing for the experimental control of the activity of the P_L promoter. The PL expression plasmids are propagated at 28° in *E. coli* strains which synthesize a temperature-sensitive repressor, either from a defective lysogen (e.g., *E. coli* strain K12ΔH1Δtrp)[19,20] or from a

[13] Unpublished work of this laboratory (E. Van Mechelen).

[14] B. Reiss, Ph.D. Thesis, University of Heidelberg, F.R.G. (1982).

[15] R. Devos, H. Cheroutre, Y. Taya, W. Degrave, H. Van Heuverswyn, and W. Fiers, *Nucleic Acids Res.* **10,** 2487 (1982).

[16] Kindly supplied by Dr. E. Kawashima (Biogen, S. A., Geneva, Switzerland).

[17] B. Allet (Biogen, S. A., Geneva, Switzerland), unpublished work.

[18] H. M. Krisch and B. Allet, *Proc. Natl. Acad. Sci. U.S.A.* **79,** 4937 (1982).

[19] E. Remaut, P. Stanssens, and W. Fiers, *Gene* **15,** 81 (1981).

[20] H.-U. Bernard, E. Remaut, M. V. Hershfield, H. K. Das, D. R. Helinski, C. Yanofsky, and N. Franklin, *Gene* **5,** 59 (1979).

mutant *c*I gene cloned on a compatible plasmid (pRK248*c*Its or p*c*I857).[5,20] The dual plasmid system allows the use of essentially any *E. coli* strain as a host for PL plasmids. The induction procedure is very simple. While repression of the P_L promoter is virtually complete at 28°, full activity is obtained by raising the temperature to 42°. This procedure does not require the use of special media nor the addition of a chemical compound to the medium.

To ensure translation of foreign (eukaryotic) genes, an efficient ribosome binding site of *E. coli* origin needs to be present in the vector. In this study, we have used ribosome binding sites derived from the replicase gene of the RNA phage MS2 and from the attenuator region of the *E. coli* tryptophan operon. Both systems are known to operate very efficiently in their natural context. The nucleotide sequence at the initiator ATG was manipulated in such a way as to allow easy coupling of the ATG codon to any coding region.[8,17]

Detection of the IFN Protein. Bacteria harboring the various IFN expression plasmids were inoculated at a density of 2×10^6/ml in LB medium containing 5 μCi/ml [U-^{14}C]protein hydrolysate. The cultures were incubated at 28° with vigorous agitation to a density of 2×10^8/ml. Half of the culture was then shifted to 42° and incubation was continued for up to 6 hr. At various time points after induction, aliquots were collected by centrifugation. The pellet was dissolved in sample buffer and boiled for 5 min before electrophoresis in SDS–polyacrylamide gels (15%). The gels were fixed in 10% TCA and stained with 0.05% Serva Blue R in 30% methanol and 7% acetic acid. Autoradiographs were obtained after treatment with EN^{3}HANCE and exposure of the dried gel to X-ray film at −70°. To determine the percentage of IFN protein, the relevant band was excised from the dried gel and its radioactivity compared to the total radioactivity present in the same track. Under the conditions used, the cellular proteins have been uniformly labeled so that the incorporated radioactivity is an accurate measure of the amount of IFN protein synthesized.

Extraction of Active IFN Molecules. Method I: a volume of induced cells was pelleted and washed in one-half volume of TES buffer. The cells were resuspended in one-tenth volume of TES and opened by sonication on ice. As an alternative to sonication, lysis was brought about by the addition of lysozyme (to 1 mg/ml) and either by two freeze-thawing cycles or by the addition of Triton X-100 to 0.1% final concentration. Larger cell masses were opened by means of a French press. Cell debris was removed by centrifugation at 12,000 *g* for 10 min. Method II: induced cells were collected and resuspended in an equal volume of a solution consisting of 50 m*M* Tris · HCl, pH 7.4, 30 m*M* NaCl, 1% 2-mercaptoethanol, 1% SDS, 5 *M* urea, and lysed by immersion in a boiling water bath for 3 min.

The extract was cleared at 12,000 *g* for 10 min. This method is only applicable to IFN-β.

Interferon Assays

IFN-β antiviral activity was measured by a cytopathic effect (CPE) inhibition assay on human FS4 cells. The cells were challenged with EMC virus and CPE was recorded at 24 hr. All assays included an IFN-β reference preparation which was calibrated against the NIH IFN-β reference standard (Cat. Nr. G023-902-527). Similarly, IFN-γ was assayed on T21 cells (human fibroblasts trisomic for chromosome 21). At the time the experiments reported here were performed, no international IFN-γ standard was available. IFN-γ titers were then calibrated against the NIH IFN-α reference standard (Cat. Nr. G-023-901-527). However, the titers given in the table have been recalculated with reference to the recently available NIH IFN-γ standard (Cat. Nr. Gg23-901-530).

Titration was performed in microtiter trays with serial 3-fold dilutions of the bacterial extracts made up in 3% calf serum. Certain samples (especially those containing detergents) were toxic to the cells at the lower dilutions.

Concluding Comments

Figure 2 shows the stained protein profiles obtained after SDS–PAGE of induced bacterial cells synthesizing IFN-β or IFN-γ. In all cases, a new protein band with the expected molecular weight of mature IFN protein can be easily identified. The amount of IFN protein synthesized depended on the type of expression vector used. The values are listed in the table.

Two strong ribosome binding sites, derived from either the replicase gene of the RNA phage MS2 or the *E. coli* tryptophan attenuator region, differ from each other by a factor of 3–4 in their capacity to direct efficient synthesis of IFN-γ.[10] We have as yet no clear insight into the factors which determine the strength of a ribosome binding site. It is thought that local secondary structure of the mRNA plays an important role in the process of initiation.[21,22]

The presence of a transcription terminator downstream from the cloned gene has a favorable effect on the accumulation of IFN-β and IFN-γ. There is roughly a 2- to 3-fold increase in yield.[10,13] This is not caused by an increased efficiency of synthesis but rather is due to a prolonged maintenance of the level of synthesis.[23]

[21] D. Iserentant and W. Fiers, *Gene* **9,** 1 (1980).

[22] D. Gheysen, D. Iserentant, C. Derom, and W. Fiers, *Gene* **17,** 55 (1982).

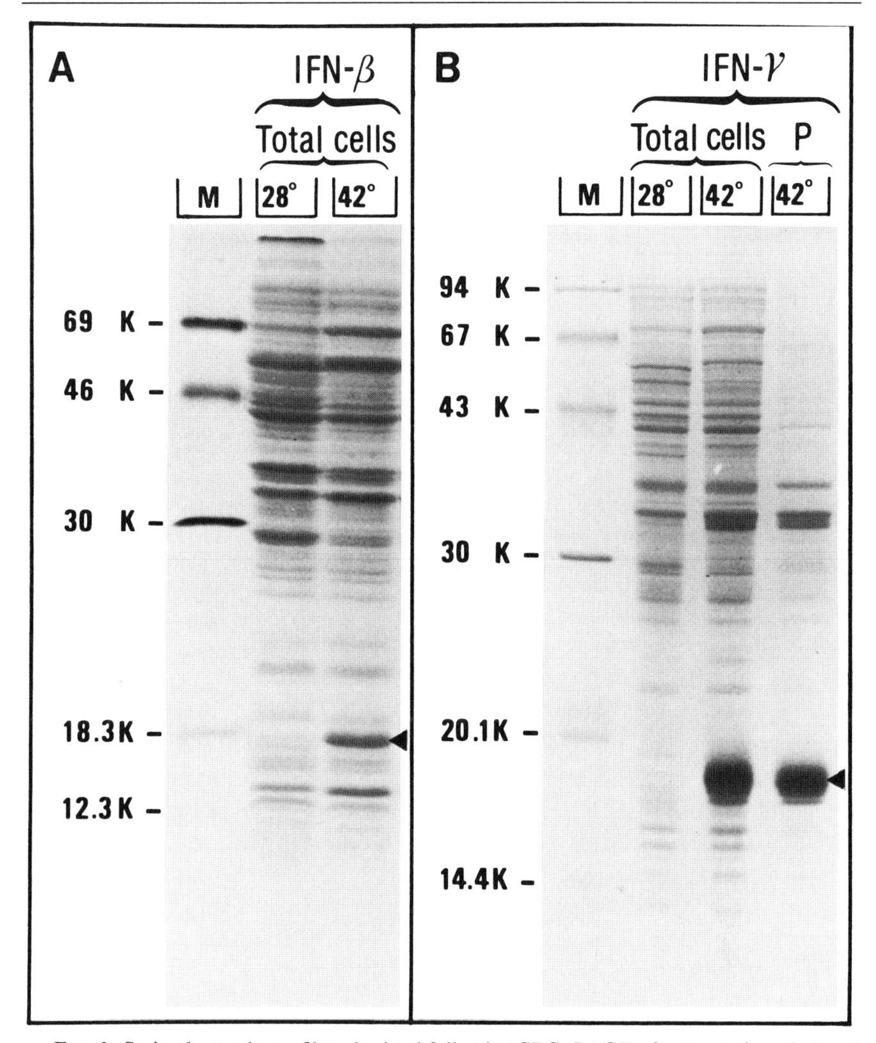

FIG. 2. Stained protein profiles obtained following SDS–PAGE of extracts from induced and uninduced cells (*E. coli* strain K12ΔH1Δtrp[20]). (A) IFN-β synthesis directed by pPLc345HFIF25fdT; (B) IFN-γ synthesis directed by pPLc28HIIF52T_4. M, molecular weight markers; P, pellet fraction of a sonicated culture. The position of the IFN protein is indicated by an arrowhead.

[23] It is our unpublished observation that the initial level of IFN-γ synthesis was independent of the presence of a terminator in the plasmid. However, in the absence of a terminator, the plasmid copy number was found to decrease after prolonged induction. Cloning of an efficient terminator between the cloned gene and the origin of replication stabilized the plasmid copy number, thereby allowing an increased yield of the cloned gene product.

LEVELS OF IFN SYNTHESIS[a]

Expression plasmid	Host (*E. coli* strain)	Percentage of total protein	IFN activity (units/10^8 bacteria) Extraction method I	Extraction method II
IFN-β constructions				
pPLc245HFIF25	K12ΔH1Δtrp	2.0	5×10^2	5×10^5
pPLc345HFIF25fdT	K12ΔH1Δtrp	4.0	ND	ND
IFN-γ constructions				
pPLc245HIIF79	K12ΔH1Δtrp	3.6	8×10^3	NA
pPLc28HIIF52	K12ΔH1Δtrp	14.	8×10^3	NA
pPLc245HIIF79T_4	K12ΔH1Δtrp	9.8	8×10^3	NA
pPLc28HIIF52T_4	K12ΔH1Δtrp	24.1	8×10^3	NA
pPLc28HIIF52T_4	C600(p*c*I857)	24.0	8×10^3	NA

[a] The percentage of total protein was based on long-term, uniform ^{14}C labeling of all proteins, followed by PAGE, excision of appropriate bands, and counting of the radioactivity. IFN-β and IFN-γ units are defined under "Interferon Assays." The extraction procedures are detailed in the text; method I, lysozyme–Triton-X-100–lysis; method II, SDS–urea lysis. ND, not determined; NA, not applicable.

The IFN-β expression plasmids illustrate strikingly the importance of being able to control the expression of a cloned gene. The synthesis of IFN-β was found to drastically inhibit the growth rate of the bacteria.[8] While bacterial cultures harboring IFN-β expression plasmids had growth rates at 28° similar to the parental strain, they stopped further growth very soon after induction at 42°. Regardless of whether this is a direct or indirect effect of the synthesis of IFN-β, this observation indicates that a host–vector system, which does not allow full experimental control over the expression of a cloned gene, is likely to have a negative selection value. Earlier studies with the P_L-regulated expression of the *E. coli* tryptophan synthetase A subunit showed that the level of derepression was at least 300-fold.[19,20] IFN-β titers in uninduced cultures were five orders of magnitude lower than in induced cultures.[8,24]

[24] The apparent repression of IFN-β synthesis at 28° is most likely an overestimate. One possibility is that low amounts of this protein are efficiently degraded by the host's proteases. Pulse-chase experiments, under conditions of partial induction, have indeed shown that low amounts of IFN-β are unstable inside the bacterial cell.

Most of the IFN-β or IFN-γ synthesized is found in the cell pellet fraction (Fig. 2). Gentle lysis of the bacteria does not release an appreciable amount of this material. This is most strikingly exemplified in the case of IFN-β. Soluble titers after sonication or lysozyme–Triton X-100–lysis remain exceedingly low, while after boiling in SDS–urea, very high antiviral activity is obtained (IFN-β is known to be resistant to heat treatment and to detergent action[25]). Since IFN-γ is not resistant to either heating or SDS, the latter extraction procedure is not applicable. The biological activity recovered under mild conditions from induced cells was largely independent of the actual level of accumulated protein (see the table). It would appear that only a minor fraction of the IFN molecules is present in the cell in a soluble form. Lysis of the cells from within, by coinduction of the lysis genes of either phage MS2 or phage λ or both, did not improve the yield of soluble material.[10] We have recently been able to solubilize both IFN-β and IFN-γ and to purify them to apparent homogeneity (unpublished work). The specific biological activity of IFN-β, as determined by the antiviral assay, was approximately 1×10^8 units/mg, a value which agrees well with published values for natural human IFN-β.[26,27] The specific activity of IFN-γ was approximately 5×10^7 units/mg, in good agreement with the value found for glycosylated human IFN-γ purified from Chinese hamster ovary cells.[28,29]

In summary, the PL-expression vectors have been used to direct fully regulated, high-level synthesis of IFN-β and IFN-γ in *E. coli*. Although the bulk of the protein was found to be insoluble inside the cell, appropriate extraction procedures allow its complete recovery in a soluble form. In both instances, the bacterially synthesized interferons were shown to have approximately the same specific antiviral activity as their natural glycosylated counterparts.

Acknowledgments

This research was supported by BIOGEN, S. A. and by a grant from the Gekoncerteerde Onderzoeksakties of the Belgian Ministry of Science.

[25] W. E. Stewart, II, E. De Clercq and P. DeSomer, *Nature (London)* **249,** 460 (1974).

[26] E. Knight, Jr. and D. Fahey, *J. Biol. Chem.* **256,** 3609 (1981).

[27] H.-J. Friesen, S. Stein, M. Evinger, P. C. Familletti, J. Moschera, J. Meienhofer, J. Shively, and S. Pestka, *Arch. Biochem. Biophys.* **206,** 432 (1981).

[28] S. J. Scahill, R. Devos, J. Van der Heyden, and W. Fiers, *Proc. Natl. Acad. Sci. U.S.A.* **80,** 4654 (1983).

[29] R. Devos, C. Opsomer, S. J. Scahill, J. Van der Heyden, and W. Fiers, *J. Interferon Res.* **4,** 461 (1984).

[55] Expression of Human Interferon Genes in *E. coli* with the Lambda P_L Promoter

By ROBERT CROWL

Introduction

The capability of producing mammalian proteins in bacteria has become a valuable tool for studying biologically interesting polypeptides that otherwise cannot be isolated in sufficient amounts. Expression of cloned human interferon genes in *E. coli* is now a classical case in point. Leukocyte interferon is currently being produced and purified in gram quantities, allowing the application of standard biochemical approaches for studying the protein at the molecular level and expanding the number and kinds of experiments designed to understand the mechanism(s) of its biological activities.[1] All three classes of human interferon have been expressed in bacteria with a variety of expression vectors.[2-6] In this chapter I describe the utilization of the expression vector pRC23 containing the phage lambda P_L promoter for the high-level production of human immune interferon (IFN-γ) in *E. coli.*

Rationale for the Design of the Expression Vector

The first requirement for obtaining high-level expression of a heterologous gene in *E. coli* is that the coding sequences be placed under the transcriptional control of a strong promoter, preferably one that can be efficiently and conveniently regulated. Tight transcriptional regulation is important for at least three reasons: (1) the synthesis of some heterolo-

[1] S. Pestka, *Arch. Biochem. Biophys.* **221,** 1 (1983).

[2] D. V. Goeddel, E. Yelverton, A. Ullrich, H. L. Heyneker, G. Miozzari, W. Holmes, P. H. Seeburg, T. Dull, L. May, N. Stebbing, R. Crea, S. Maeda, R. McCandliss, A. Sloma, J. M. Tabor, M. Gross, P. C. Familletti, and S. Pestka, *Nature (London)* **287,** 411 (1980).

[3] D. V. Goeddel, H. M. Shepard, E. Yelverton, D. Leung, R. Crea, A. Sloma, and S. Pestka, *Nucleic Acids Res.* **8,** 4057 (1980).

[4] T. Taniguchi, L. Guarente, T. M. Roberts, D. Kimelman, J. Douhan, and M. Ptashne, *Proc. Natl. Acad. Sci. U.S.A.* **77,** 5230 (1980).

[5] R. Derynck, E. Remaut, E. Saman, P. Stanssens, E. Clereq, J. Content, and W. Fiers, *Nature (London)* **287,** 193 (1980).

[6] P. W. Gray, D. W. Leung, D. Pennica, E. Yelverton, R. Najarian, C. C. Simonsen, R. Derynck, P. J. Sherwood, D. M. Wallace, S. L. Berger, A. D. Levinson, and D. V. Goeddel, *Nature (London)* **295,** 503 (1982).

gous proteins may be toxic to the bacterial cell; (2) transcription from a strong promoter can interfere with plasmid replication, leading to plasmid instability[7,8]; and (3) a high rate of synthesis during a short time period can be crucial for avoiding proteolytic degradation of the foreign gene product.[9] The desirable feature of tight transcriptional regulation makes a promoter derived from a temperate bacteriophage, such as the lambda P_L promoter, an obvious choice since the objective for both the phage and the molecular biologist is the exploitation of the bacterial cell's protein synthetic machinery at the appropriate time.

Transcription initiated from P_L is negatively regulated by the lambda *c*I repressor which binds cooperatively to sites overlapping with those recognized by RNA polymerase.[10] A convenient mechanism for turning transcription on and off is provided by mutations in the *c*I gene that render the repressor temperature sensitive[11]; thus, at 30°, the mutant repressor is functional and transcription is turned off, and, at 42°, the repressor is inactivated allowing gene expression to proceed. Other means of derepression are also available; namely, RecA-mediated cleavage of the repressor triggered by the SOS response.[12] The ability to derepress the promoter at lower temperatures could be advantageous if the desired gene product is thermolabile and/or more susceptible to proteolysis at 42°.

A high rate of transcription is not sufficient to achieve overproduction of a gene product; in addition, the resulting mRNA must be reasonably stable and efficiently translated. Very little is understood about the factors influencing mRNA stability in *E. coli,* however, existing evidence suggests that efficient translation of a mRNA enhances its stability.[13] Although more is known about the mechanism of translation initiation, the subtleties influencing the efficiency of translation have only recently been explored. The nucleotide sequences thought to be necessary for the ribosomes to bind to the mRNA include the AUG (or GUG) initiation codon preceded by a Shine–Delgarno (SD) sequence (3 to 9 bases) that is complimentary to the sequence at the 3′ end of the 16 S ribosomal RNA.[14]

[7] B. P. Nichols and C. Yanofsky, this series, Vol. 101, p. 155.

[8] H.-U. Bernard and D. R. Helinski, this series, Vol. 68, p. 482.

[9] Y.-S.E. Cheng, D. Y. Kwoh, T. J. Kwoh, B. C. Soltvedt, and D. Zipser, *Gene* **14,** 121 (1981).

[10] M. Ptashne, A. Jeffrey, A. D. Johnson, R. Mauer, B. J. Meyer, C. O. Pabo, T. M. Roberts, and R. T. Sauer, *Cell* **19,** 1 (1980).

[11] M. Lieb, *J. Mol. Biol.* **16,** 149 (1966).

[12] J. W. Little and D. W. Mount, *Cell* **29,** 11 (1982).

[13] H. M. Shepard, E. Yelverton, and D. Goeddel, *DNA* **1,** 125 (1982).

[14] J. Shine and L. Delgarno, *Proc. Natl. Acad. Sci. U.S.A.* **71,** 1342 (1974).

It is now clear that recognition of a SD sequence followed by an AUG codon by the ribosome is only part of what is necessary for high-level translation. For instance, the distance between the SD sequence and the AUG, as well as the base composition within this region, can dramatically affect translatability.[13] Moreover, sequences upstream from the SD sequence and within the coding region can also have effects on translational efficiency.[15]

The expression vector pRC23 was constructed by joining the phage lambda P_L promoter to synthetic sequences which were chosen on the basis of a computer-generated model ribosomal binding site[16] (Fig. 1). This sequence contains an 8 base SD sequence which is preceded by an AT-rich region and followed by an *Eco*RI recognition site. *Eco*RI was chosen because it is also AT rich (especially after "filling in" the termini and religating). Having the SD sequence embedded within AT-rich regions seems to be a preferred feature for *E. coli* ribosomal binding sites (RBS).[16] Inserting a heterologous gene at the *Eco*RI site creates a new hybrid RBS, comprised of the synthetic sequences preceding the ATG and the sequence at the 5′ end of the foreign gene. Because different foreign genes result in different hybrid RBS any subtle effects on translation efficiency due to sequences within the coding region are difficult (if not impossible) to predict. The rationale of using part of the extended model RBS is to minimize any negative influence that the foreign gene sequences might have on translatability.

Because the cDNA clones of human interferon genes contain unwanted sequences coding for the signal peptide, it is necessary to reengineer the 5′ end of the gene with synthetic oligonucleotides to create an *Eco*RI terminus and place an ATG initiation codon preceding the coding sequence for the mature form of the gene product. For situations where fusion proteins are acceptable, derivatives of pRC23 have been constructed which contain their own ATG followed by convenient unique restriction sites.[17]

The P_L promoter on pRC23 can be regulated by the *c*I repressor encoded on the low copy-number, compatible plasmid pRK248*c*Its.[8] Alternatively, the *c*Its gene can be isolated from phage lambda DNA on a 2400 bp *Bgl*II fragment and inserted at the *Bgl*II site in pRC23, or the repressor gene can reside on the host chromosome as part of a defective prophage.

[15] M. N. Hall, J. Gabay, M. Debarbouille, and M. Schwartz, *Nature* (*London*) **295,** 616 (1982).

[16] G. F. E. Scherer, M. D. Walkinshaw, S. Arnott, and D. J. More, *Nucleic Acids Res.* **8,** 3895 (1980).

[17] R. Crowl, T. C. Seamans, P. Lomedico, and S. McAndrew, *Gene* **38,** 31 (1985).

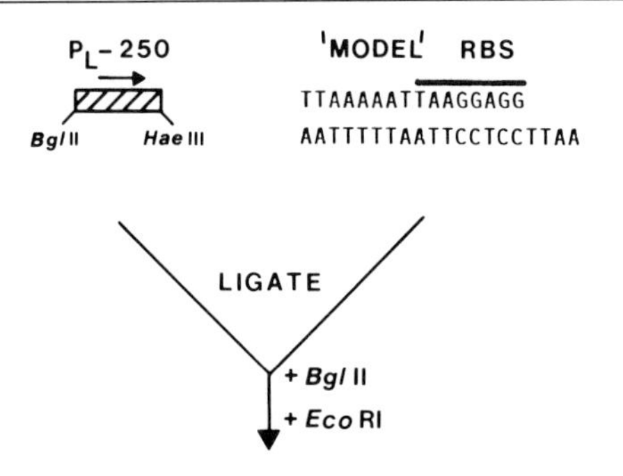

ISOLATE 265 bp FRAGMENT

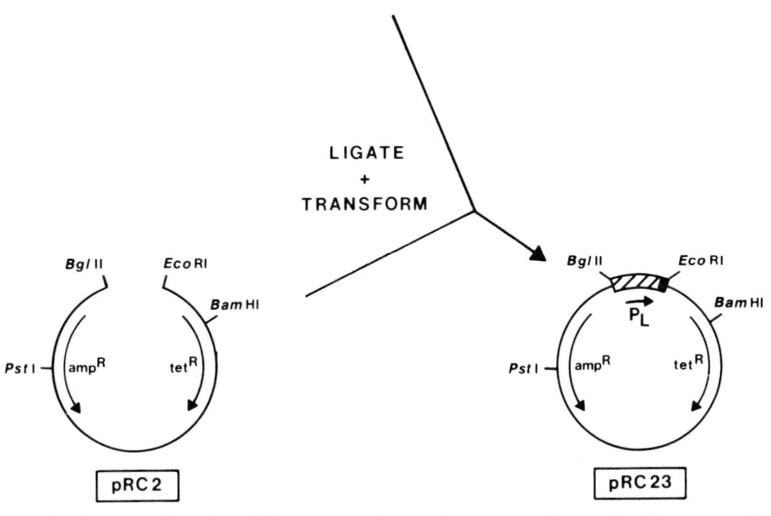

FIG. 1. Construction of pRC23. A pair of complementary oligonucleotides comprising a computer-generated ribosome binding site (RBS) were blunt-end ligated to a 250 bp *Bgl*II–*Hae*III fragment containing the P_L promoter which was isolated from phage lambda DNA. Following digestion with *Bgl*II and *Eco*RI, the ligation product was inserted into pRC2 between the *Bgl*II and *Eco*RI sites. pRC2 is a derivative of pBR322 which contains a *Bgl*II site adjacent to the *Eco*RI site.[17]

Methods for Monitoring Interferon Gene Expression

Cultures of *E. coli* containing the expression vector are grown at 30° in M9 medium[18] containing 0.5% glucose and 0.5% casamino acids. When the cell density reaches an $OD_{600} = 0.5$, the culture is divided and either maintained at 30° or transferred to 42°. At various times, 1 ml samples are

[18] F. Bolivar and K. Backman, this series, Vol. 68, p. 245.

taken and the cells collected by centrifugation (20 sec at 15,000 rpm in an Eppendorf Microfuge). The cell pellets can be frozen and stored at −20°, and then processed as described below.

To analyze total cellular protein, the cell pellet is resuspended in a buffer containing 10 m*M* Tris · HCl (pH 7.4) and 10% glycerol at a concentration of about 10^9 cells/100 μl of buffer. An equal volume of 2× Laemmli sample buffer[19] is added and the mixture is heated at 95° for 5 min. A volume of 20 μl for each sample is subjected to SDS–polyacrylamide gel electrophoresis (SDS–PAGE) followed by staining with Coomassie brilliant blue.[19]

To measure interferon bioactivity, the cell pellets are resuspended and lysed in 50 μl of 7 *M* guanidine · HCl and appropriately diluted in tissue culture medium. Immune interferon antiviral activity is determined on human WISH cells as described.[20]

Expression of the IFN-γ Gene in pRC23

The IFN-γ gene was obtained as a cDNA clone following procedures similar to those described by Gray *et al.* A convenient *Bst*NI site is located within the codon corresponding to the fourth amino acid residue of mature IFN-γ. To engineer the gene for expression in pRC23, a pair of complementary oligonucleotides were synthesized which restore the missing codons, include an ATG initiation codon, and create an *Eco*RI cohesive end (Fig. 2). These molecules were ligated to the 5′-terminal *Bst*NI terminus and to a *Bst*NI terminus generated by cleavage at a second site located in the 3′ noncoding region. The resulting *Eco*RI fragment was inserted at the *Eco*RI site in pRC23 in the appropriate orientation (Fig. 2). The resulting plasmid was transformed into *E. coli* strain RR1 containing the compatible plasmid pRK248*c*Its.

As shown in Fig. 3, IFN-γ is produced in *E. coli* containing pRC23/IFN-γ, accumulating to at least 10% of total cell protein after 2 hr of incubation at 42°. When these cells are lysed by lysozyme/freeze-thaw treatment, at least 90% of the IFN-γ protein sediments with the membrane fraction due to the aggregation of the abundant IFN-γ protein. The protein can be solubilized in 7 *M* guanidine · HCl and recovered in a biologically active form (see the table).

The medium used for cell growth can be crucial for achieving high-level production of interferon in *E. coli*. For example, when a rich medium is used instead of M9 for growth and induction, the IFN-γ protein is

[19] U. K. Laemmli, *Nature* (*London*) **227,** 680 (1970).
[20] P. C. Familletti, S. Rubinstein, and S. Pestka, this series, Vol. 78, p. 387.

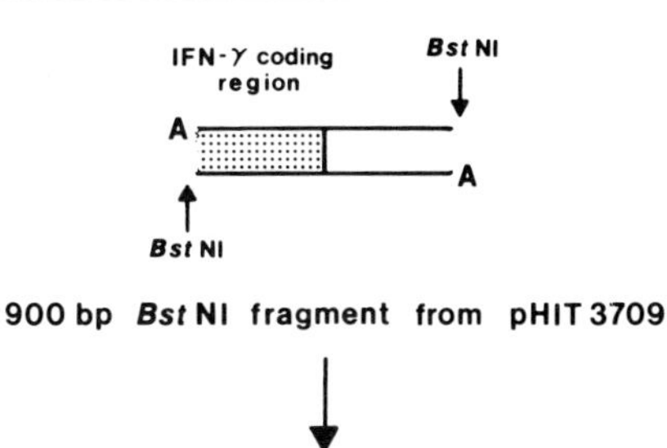

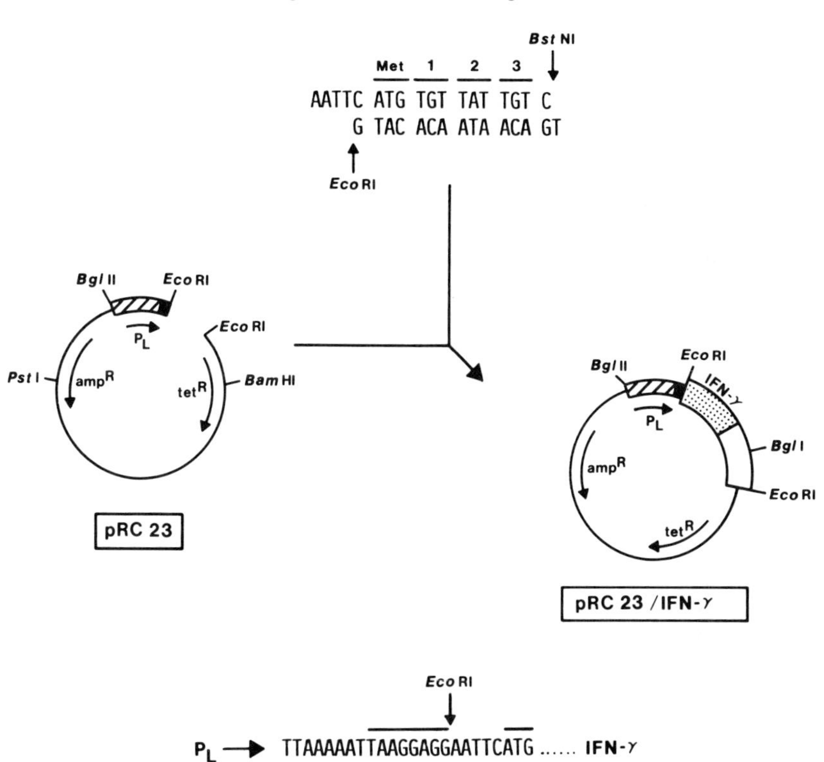

FIG. 2. Construction of an expression vector that directs the synthesis of IFN-γ. The synthetic RBS sequence (shown at the bottom of the figure) is represented by the solid black area in the diagrams. The SD sequence and initiation codon within the RBS are overlined. pHIT3709 is the designation for the pBR322 derivative containing the IFN-γ cDNA sequence. The *Bgl*I site in the 3′ noncoding region was used to confirm the orientation of the *Eco*RI insert.

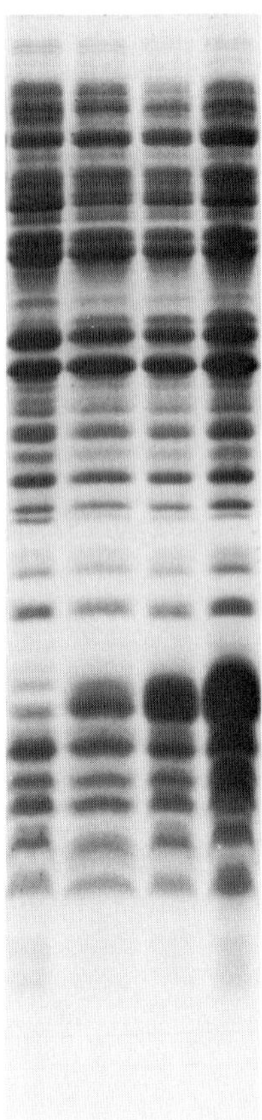

FIG. 3. Analysis of IFN-γ protein synthesized in *E. coli*. Total proteins from uninduced cells (U) and from cells incubated at 42° for 1, 2, and 3 hr were subjected to SDS–polyacrylamide (15%) gel electrophoresis followed by staining with Coomassie brilliant blue. The predominant band that appears in increasing amounts with time at 42° is the 17,000 Da IFN-γ protein.

not detectable by SDS–PAGE. The probable explanation for this result is increased proteolytic activity in cells grown in complex media.

In addition to IFN-γ, IFN-β, and IFN-α[21] have also been expressed in *E. coli* with the vector pRC23 by procedures described here. However, the level of protein that accumulates in the cell, in the case of IFN-β and

[21] T. M. DeChiara, F. Erlitz, and S. J. Tarnowski, this volume [58].

BIOLOGICAL ACTIVITY OF IFN-γ PRODUCED IN *E. Coli*

Plasmid	Induction[a]	IFN-γ antiviral activity (units/ml)[b]
pRC23/IFN-γ	−	8×10^2
pRC23/IFN-γ	+	4×10^5
pRC23	+	$<4 \times 10^2$

[a] Cultures of RR1 (pRK248*c*Its) containing the indicated plasmids were induced by incubating at 42° for 2 hr.

[b] Activity indicated is per milliliter of bacterial culture (at OD_{600} = 1.0).

IFN-α, is approximately one-tenth that observed for IFN-γ. The reasons for this difference are under investigation.

Acknowledgments

I gratefully acknowledge Dr. M. Kikuchi for providing the IFN-γ cDNA clone. I also thank Stephen McAndrew and Daru Young for expert technical assistance, Dr. Mohindar Poonian for providing synthetic oligonucleotides, Sharon DeChiara for performing the antiviral assays, and Dr. Peter Lomedico for helpful advice.

[56] Use of Rous Sarcoma Viral Genome to Express Human Fibroblast Interferon

By LINDA MULCAHY, MARIA KAHN, BRUCE KELDER, EDWARD REHBERG, SIDNEY PESTKA, and DENNIS W. STACEY

The attractiveness of the retroviral genome as a vehicle for the expression of exogenous genes can be attributed primarily to two features. First, the retroviral genome possesses a highly efficient promoter contained within the long terminal repeat (LTR) sequence. The LTR provides a number of functions that are critical to the replication, integration, and expression of retrovirus genomes, including sequences involved in the regulation of transcription. In addition to regulating the expression of the viral genes, the LTR can activate the expression of heterologous DNA sequences.[1–3] The exogenous gene may provide its own translation initiation and termination signals or use those contained within the viral sequences.

The second feature which makes the retroviral genome a potentially useful expression vector concerns the integration of viral DNA into host cell DNA. During the viral infection process, a DNA copy of the viral genome with its LTRs is generated from the viral RNA. This DNA intermediate is integrated into the host chromosomes. It is possible that an exogenous gene, once incorporated into the genome of the retrovirus, would behave as an integral part of the viral sequences, become stably integrated into the host genome, and be constitutively expressed.[4,5]

It has been demonstrated that interferon gene sequences can be expressed when placed under the control of a heterologous eukaryotic promoter.[6–9] In order to determine how the retroviral LTR may be best employed as a promoter for the expression of the human fibroblast interferon (IFN-β) gene, the interferon coding sequence was placed at various locations within the retroviral genome. The efficiency of interferon synthesis was determined as a function of the position of the human IFN-β gene within the retroviral genome.

Reagents and General Procedures

Restriction endonucleases, T_4-DNA ligase, polynucleotide kinase, and nuclease *Bal*31 were obtained from New England Biolabs (Beverly, Mass.). *Sst*II was obtained from Bethesda Research Labs (Gaithersburg, Md.). Enzyme reactions were carried out according to the suppliers' recommendations. DNA fragments were isolated after electrophoresis in 1% low-melting-point agarose gels (Sea Plaque; FMC Corp., Marine Colloids, Rockland, Me.) as described previously.[10] The viral DNA clones used

[1] A. Joyner, Y. Yamamoto, and A. Bernstein, *Proc. Natl. Acad. Sci. U.S.A.* **79,** 1573 (1982).

[2] C. J. Tabin, J. W. Hoffmann, S. P. Goff, and R. A. Weinberg, *Mol. Cell. Biol.* **2,** 426 (1982).

[3] F. Lee, R. Mulligan, P. Berg, and G. Ringold, *Nature (London)* **294,** 228 (1981).

[4] K. Shimotohno and H. Temin, *Cell* **26,** 67 (1981).

[5] C.-M. Wei, M. Gibson, P. G. Spear, and E. M. Scolnick, *J. Virol.* **39,** 935 (1981).

[6] K. Zinn, P. Mellon, M. Ptashne, and T. Maniatis, *Proc. Natl. Acad. Sci. U.S.A.* **79,** 4897 (1982).

[7] S. Ohno and T. Tanaguchi, *Nucleic Acids Res.* **10,** 967 (1982).

[8] D. Canaani and P. Berg, *Proc. Natl. Acad. Sci. U.S.A.* **79,** 5166 (1982).

[9] R. Devos, H. Cheroutre, Y. Taya, W. Degrave, H. van Heuverswyn, and W. Friers, *Nucleic Acids Res.* **10,** 2487 (1982).

[10] J. J. Kopchick and D. W. Stacey, *Mol. Cell. Biol.* **4,** 240 (1984).

were clone pL39td2.4 and clone pLD12. Clone pL39td2.4 contains a transformation-defective sarcoma viral genome.[11] Clone pLD12 contains a leukosis viral recombinant between Rous-associated virus-2 *env* mRNA and RSV.[12] Clone IFF-GI contains the human IFN-β specific sequences located between *Eco*RI sites.[13] *Escherichia coli* (strain RR-1) was transformed with the DNA constructs and chimeric clones were identified initially by colony hybridization with nick-translated DNA probes to IFN-β sequences. Positive colonies were analyzed further by restriction endonuclease and electrophoretic analysis on small agarose gels.[14]

Preparation of chick embryo fibroblast (CEF) cultures, their infection with Bryan RSV to generate RSV(−) cells, and cell culture conditions have been described.[11] QT35 cells originated from a chemically induced quail tumor[15] and were cultured as described previously.[10]

Recombinant chimeric clones were introduced into cultured cells by either microinjection[16] or transfection.[10] Interferon activity released by treated cells was measured in the medium by a cytopathic effect inhibition assay.[17,18] All titers were determined on WISH cells relative to human fibroblast interferon standard G-023-902-527 obtained from the Research Resources Branch, National Institute of Allergy and Infectious Diseases, National Institutes of Health, Bethesda, Maryland.

Molecular Construction and Analysis

The Hu-IFN-β gene coding sequence was positioned within the retroviral genome at multiple locations: (1) near the viral ATG used for the initiation of translation of viral proteins, (2) within the long nontranslated leader sequence closer to the LTR, and (3) downstream from the splice acceptor site used for the generation of env mRNA (Fig. 1).

The sequence of the complete human fibroblast interferon gene has been reported.[13,19,20] For the construction of the chimeric clones it was

[11] J. J. Kopchick, G. Ju, A. M. Skalka, and D. W. Stacey, *Proc. Natl. Acad. Sci. U.S.A.* **78,** 43833 (1981).

[12] L.-H. Wang and D. W. Stacey, *J. Virol.* **41,** 919 (1982).

[13] S. Maeda, R. McCandliss, T. R. Chiang, L. Costello, W. P. Levy, N. T. Chang, and S. Pestka, *in* "Developmental Biology Using Purified Genes" (D. D. Brown and C. F. Fox, eds.), p. 85. Academic Press, New York, 1981.

[14] J. J. Kopchick, B. R. Cullen, and D. W. Stacey, *Anal. Biochem.* **115,** 419 (1981).

[15] C. Moscovici, M. Giovannella Moscovici, and H. Jimenez, *Cell* **11,** 95 (1977).

[16] D. W. Stacey, this series, Vol. 79, p. 76.

[17] P. C. Familletti, S. Rubinstein, and S. Pestka, this series, Vol. 78, p. 387.

[18] S. Rubinstein, P. C. Familletti, and S. Pestka, *J. Virol.* **37,** 755 (1981).

[19] S. Pestka, S. Maeda, D. S. Hobbs, W. P. Levy, R. McCandliss, S. Stein, J. A. Moschera, and T. Staehelin, *Miami Winter Symp.* **18,** 455 (1981).

[20] S. Pestka, *Arch. Biochem. Biophys.* **221,** 1 (1983).

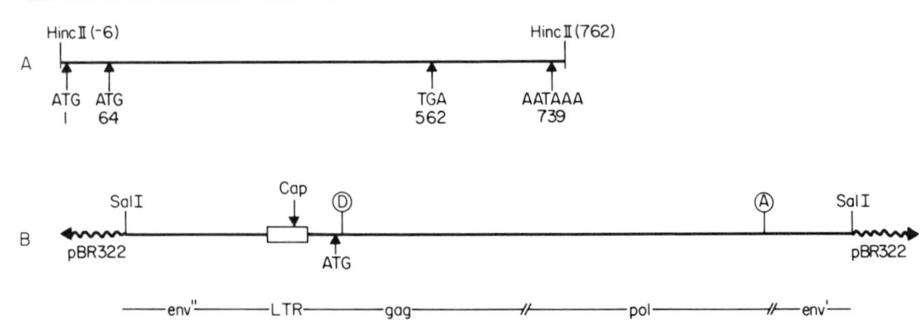

FIG. 1. Restriction endonuclease maps of genes and the viral genomic region employed. The location of important restriction endonuclease sites are indicated along with the locations of transcription initiation (cap site) and termination (AATAAA) sites. In addition, the ATG and TGA codons used for initiation and termination of translation are indicated. (A) The *Hin*cII fragment of the genomic Hu-IFN-β clone contains the entire coding region. The first 21 amino acids comprise the signal peptide and are removed during interferon maturation. The ATG at position 1 is the first translated codon and the ATG at position 64 codes for the first methionine present as amino acid number 1 of the mature protein. (B) The retroviral genome resembles a typical eukaryotic mRNA, possessing a cap structure at its 5′ end, a 5′ noncoding region extending from the cap structure to the initiator AUG, and polyadenylation at the 3′ end. Retrovirus genes are expressed from both the full-length viral RNA (*gag* and *pol* genes) and from a spliced mRNA (*env* gene). The splice joins the long leader sequence from the 5′ end of the genome to the *env* gene sequences, generating a subgenomic *env* mRNA.

first necessary to remove the nucleotide sequences involved in the regulation of transcription from the genomic interferon clone.[20] This was accomplished by isolation of a 762 bp fragment resulting from digestion with the restriction endonuclease *Hin*cII. *Hin*cII conveniently recognizes the first 6 bp preceding the ATG codon used to initiate interferon synthesis (Fig. 1) and another sequence located downstream from the presumptive poly(A) addition site. This fragment, devoid of sequences promoting the expression of the Hu-IFN-β gene, was used in the construction of the chimeric clones.

Positioning of the Interferon Gene Near the Viral ATG

For construction of chimeric clones in which the interferon gene is positioned near the viral ATG, the *Hin*cII fragment was inserted into a plasmid clone of the viral genome, pL39td-2.4. Insertion of the *Hin*cII fragment into the unique *Hpa*I restriction site (located in the viral *pol* gene near position 2730)[21] regenerates a *Hpa*I site at the 5′ junction of the inserted interferon gene but not at the 3′ junction (Fig. 2). Cleavage of the

[21] D. E. Schwartz, R. Tizard, and W. Gilbert, *Cell* **32,** 853 (1983).

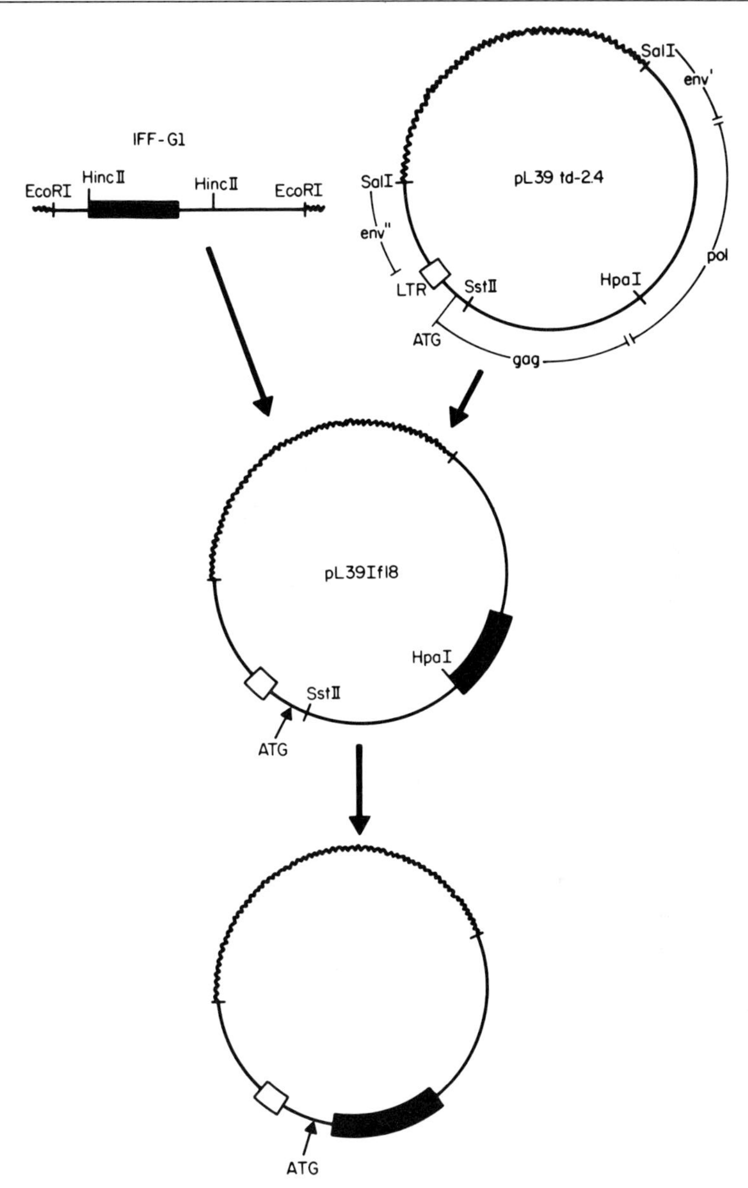

FIG. 2. Construction of viral–interferon chimeric clones positioning the Hu-IFN-β gene near the viral ATG. The interferon-specific sequences of clone IFF-GI are located between *Eco*RI restriction sites. The *Hin*cII fragment of the IFN-β gene contains the coding sequences (heavy line) and was introduced into viral clone pL39td2.4 which contains the entire genome of a transformation defective sarcoma virus inserted into pBR322 (wavy line) at the *Sal*I site. The approximate locations of the viral genes and the ATG used to initiate viral protein translation are indicated. See text for details of construction.

resulting chimeric plasmid IF18 (pL39IF18) at the unique *Hpa*I site generates a terminus with 3 bp preceding the initiation codon for interferon translation.

To position the interferon sequences near the viral ATG at position 380, the plasmid IF18 was cleaved with restriction endonuclease *Sst*II at position 543 (Fig. 2). The DNA was then digested for varying lengths of time with nuclease *Bal*31, a double-strand exonuclease, to remove increasing lengths of DNA from the *Sst*II cleavage site. Then, the DNA was

TABLE I
INTERFERON PRODUCTION FOLLOWING TRANSFECTION OF CHIMERIC DNAs WITH IFN-β GENE NEAR THE VIRAL ATG INTO QT35 CELLS[a]

Clone	Viral position of interferon gene	Interferon activity (units/ml)		
		Expt. 1	Expt. 2	Expt. 3
130	326	512	96	128
66	367	450	256	384
70	367	384	128	256
75	368	272	128	256
76	368	272	128	64
74	408	24	16	8
63	412	46	16	Neg
37	421	72	24	32
69	423	46	32	48
71	441	22	Neg	8
33	463	12	8	8
68	467	60	64	64
31	481	2	8	4
21	523	Neg	Neg	8
73	523	Neg	Neg	Neg
34	524	Neg	ND	Neg

[a] Three separate transfections were performed on RAV-1-infected QT35 cells by the DEAE-dextran method[10] with the addition of 25 μM chloroquine diphosphate (Sigma Chemical Co.) to the culture medium. Monolayer cultures contained approximately 1×10^6 cells per well. The culture medium was replaced at 48 hr and the entire 2 ml of medium was collected at 72 hr. The numbers presented for experiment 1 are the average of four separate determinations performed upon the same sample. In this and subsequent tables clones are indicated by the clone number alone (for example, pL39IF21 is termed 21). ND, Not determined; Neg refers to titers <2 units/ml.

TABLE II
INTERFERON PRODUCTION FOLLOWING MICROINJECTION OF SELECTED CHIMERIC DNAs INTO RSV (−) CELLS[a]

Clone	Viral position of interferon gene	Interferon activity (units/ml)	
		18 hr	24 hr
IF 130	326	384	192
IF 66	367	224	ND
IF 76	368	448	160
IF 21	523	6	6

[a] Two hundred RSV(−) cells were injected and placed in 0.4 ml of culture medium.[16] The medium was replaced at 18 hr and collected at 24 hr. ND, Not determined.

cleaved with *Hpa*I to remove the viral DNA between the *Sst*II site and the interferon sequences. The blunt ends were then ligated to circularize the vector. Sixteen clones were obtained which were shown to have the interferon sequence located in various positions near the viral ATG.

Five of the 16 chimeric clones directed the synthesis of high levels of human fibroblast interferon. Whether the DNA was introduced by microinjection or by transfection, these 5 plasmid DNAs directed the production of similar amounts of interferon, which were at least 5-fold greater than that released by any other clones within this set (Table I). Microinjection of 200 CEFs with DNA of clone IF76 (pL39IF76) resulted in the release of 448 units/ml of interferon activity by 18 hr after injection (Table II). This level compares favorably with that of superinduced human cells, which release approximately one-tenth this level.[22–24] One unit of human fibroblast interferon represents about 2 pg of protein,[25,26] meaning each injected cell would have released approximately 5×10^7 molecules of IFN-β in 18 hr.

[22] E. A. Havell and J. Vilček, *Antimicrob. Agents Chemother.* **2,** 476 (1972).

[23] A. Billeau, V. G. Edy, H. Heremans, J. van Damme, J. Desmyter, J. A. Georgiades, and P. DeSomer, *Antimicrob. Agents Chemother.* **12,** 11 (1977).

[24] J. S. Horoszewicz, S. S. Leong, M. Ito, L. DiBerardino, and W. A. Carter, *Infect. Immun.* **19,** 720 (1978).

[25] S. Stein, C. Kenny, H.-J. Friesen, J. Shively, U. Del Valle, and S. Pestka, *Proc. Natl. Acad. Sci. U.S.A.* **77,** 5716 (1980).

[26] H.-J. Friesen, S. Stein, M. Evinger, P. C. Familletti, J. Moschera, J. Meienhofer, J. Shively, and S. Pestka, *Arch. Biochem. Biophys.* **206,** 432 (1981).

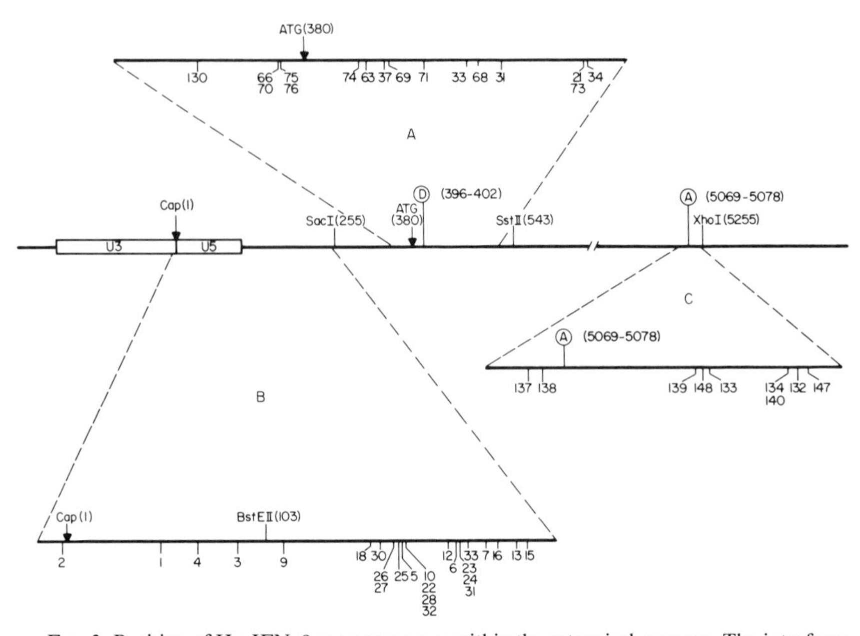

FIG. 3. Position of Hu-IFN-β gene sequence within the retroviral genome. The interferon structural gene was positioned at a variety of locations as indicated. Individual clones are designated by clone number alone (for example, pL39IF21 is termed 21). The position of the beginning of the interferon sequence is shown below the line for constructs in which the IFN-β gene has been placed (A) near the viral ATG, (B) near the viral LTR, and (C) near the splice acceptor site. Biological activity of these clones is presented in Tables I–IV.

In order to relate the biological activities of the various plasmid DNAs to their physical structures, the exact position of the interferon gene within the viral genome was determined by DNA sequencing.[27] The interferon coding sequence was found to be positioned at a variety of locations relative to the ATG used for the initiation of viral protein translation (Fig. 3). The clones most active in directing interferon production, however, all had large deletions of viral sequences such that the interferon gene was located upstream of the viral ATG, position 380 (Table I).

Clones with the interferon gene located downstream from the viral ATG exhibited reduced or no ability to direct the synthesis of interferon. Within this set of clones, those in which the interferon gene is positioned nearer the viral ATG appeared to be more active in directing interferon production. All chimeric DNAs containing the interferon gene within 100 bp downstream from the viral ATG caused the expression of some interferon activity. Clones IF37, IF63, and IF69, with the interferon gene

[27] A. M. Maxam and W. Gilbert, this series, Vol. 65, p. 499.

located from 32 to 43 bp downstream from the viral ATG, expressed nearly 20% the activity of the most active clones. When the interferon gene was positioned over 50 bp downstream, interferon production declined dramatically, except the case of clone IF68. In this case the gene is positioned in the correct reading frame 87 bp downstream and produced nearly 20% of the highest activities detected. Clones in which the gene is located more than 140 bp downstream produced no detectable interferon activity even when positioned in the correct reading frame (clone IF34).

Positioning of the Interferon Gene Near the Viral LTR

For construction of clones in which the interferon gene is positioned near the viral LTR, the interferon sequence was inserted into the plasmid clone pLD12 (Fig. 4). Clone pLD12 contains a unique *Sac*I restriction site

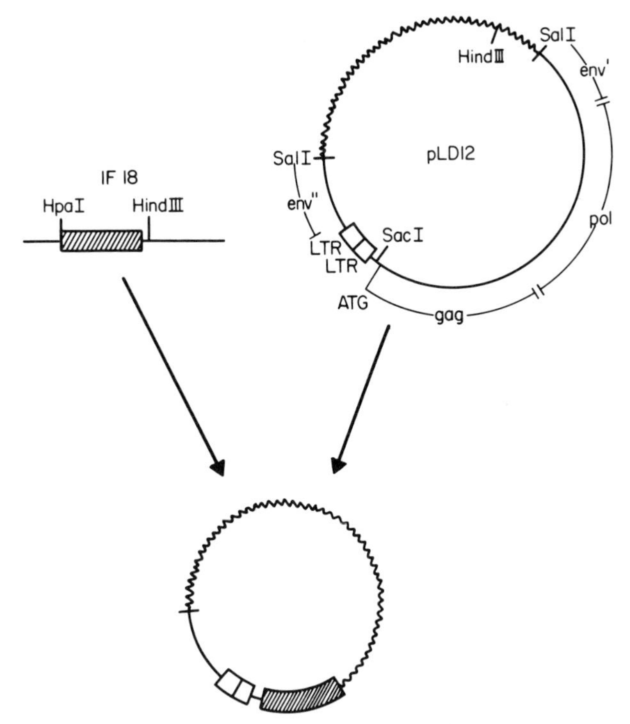

FIG. 4. Construction of viral–interferon chimeric clones positioning the Hu-IFN-β gene within the nontranslated leader sequence. The interferon coding sequence (hatched box) was obtained as the *Hpa*I–*Hin*dIII fragment of clone IF18 (clone pL391F18) (Fig. 2) and was introduced into viral clone pLD12. Clone pLD12 was obtained from a leukosis viral recombinant between Rous-associated virus-2 *env* mRNA and RSV and is inserted into pBR322 (wavy line) at the *Sal*I site. The approximate locations of the viral genes and the ATG used to initiate viral protein translation are indicated. See text for details of construction.

at position 255. Cleavage of clone pLD12 with *Sac*I was followed by digestion with nuclease *Bal*31 to remove varying lengths of DNA from the termini. The *Bal*31-treated DNA was then cleaved with *Hin*dIII to release the DNA sequences between the *Sac*I restriction site and *Hin*dIII restriction site within the pBR322 sequence. This was followed by the insertion of the 773 bp *Hpa*I–*Hin*d III fragment of clone IF18 which contains the IFN-β gene (Fig. 4). Twenty-four clones were obtained in which the interferon sequence had been moved to various positions near the viral LTR.

All 24 chimeric clones directed the synthesis of interferon (Table III). There were no significant differences in the amounts of interferon activity produced among the 24 individual plasmid DNAs. However, these clones were somewhat less efficient in directing the synthesis of interferon than were the 5 best clones with the interferon sequence positioned slightly upstream from the viral ATG. The amounts of interferon produced were approximately 25–50% of the level seen with clone IF66 in the same transfection experiment (Table III).

The exact position of the interferon sequence within the retroviral

TABLE III
INTERFERON PRODUCTION FOLLOWING TRANSFECTION OF CHIMERIC DNAs WITH IFN-β GENE WITHIN NONTRANSLATED LEADER INTO QT35 CELLS[a]

Clone	Viral position of interferon gene	Interferon activity (units/ml)	Clone	Viral position of interferon gene	Interferon activity (units/ml)
2	−1	96	10	(174)	64
1	(49)	64	32	(174)	128
4	(69)	64	28	(174)	32
3	(89)	32	12	(196)	96
9	112	128	6	200	64
18	(156)	64	31	(202)	48
30	(161)	96	23	(202)	128
26	(166)	64	24	(202)	64
27	(166)	128	7	215	32
25	(171)	64	16	(221)	32
5	172	96	13	(231)	32
22	(174)	64	15	(236)	32
			66	367	256

[a] Transfections were performed as described in the footnote in Table I. Clone IF66 was included for comparison with the results listed in Table I. Numbers given for viral position of interferon gene in parentheses were estimated by polyacrylamide gel electrophoresis.

genome was determined for some of the clones by DNA sequencing. The remaining clones were analyzed by appropriate restriction enzyme digestion and polyacrylamide gel electrophoresis to approximate the position of the interferon gene.

This set of plasmid DNAs was found to include chimeric clones with the interferon gene located at a variety of positions, spanning from 1 bp upstream from the cap site to approximately 230 bp downstream (Fig. 3 and Table III). However, all 24 plasmid DNAs directed the synthesis of similar levels of interferon with no obvious effect contributed by the position of the inserted gene.

Positioning of the Interferon Gene Near the Splice Acceptor Site

In these constructions, the interferon sequence was placed at various positions behind the splice acceptor site used in the production of *env* mRNA. This was accomplished by the ligation of 3 fragments: (1) the 6.0 kb *Bam*HI–*Hin*dIII fragment containing the viral LTR isolated from clone pL39td-2.4 and pBR322 sequences, (2) the 1.4 kb *Bam*HI–*Xho*I fragment, containing the viral splice acceptor site, also isolated from clone pL39td-2.4, and (3) the 2.8 kb *Xho*I–*Hin*dIII fragment, containing the interferon sequence, isolated from clone IF18 (Fig. 5). The resulting clone, pLIFenv-1, contained the interferon sequence positioned downstream from the splice acceptor site. Clone pLIFenv-1 was then cleaved at its unique *Xho*I restriction site and digested with nuclease *Bal*31 to remove increasing lengths of DNA from the termini. This was followed by cleavage of the *Bal*31-treated DNA with *Hpa*I to remove viral DNA between the *Xho*I and *Hpa*I sites and ligation to circularize the plasmid. Nine clones were isolated which contained the interferon sequence at various positions near the splice acceptor site.

Of these nine chimeric clones only three were active in directing the synthesis of interferon, and relatively low amounts of interferon activity were detected (Table IV). Transfection of cells with the remaining 6 clones resulted in virtually no detectable interferon activity.

The position of the interferon gene within the retroviral sequences was approximated by polyacrylamide gel electrophoresis. In these nine clones the interferon sequence was found to be positioned over a span from approximately 50 bp upstream to about 110 bp downstream from the splice junction site located at position 5069–5078 (Fig. 3 and Table IV). Two clones contained the interferon gene positioned upstream from the splice junction site and did not direct the synthesis of interferon upon transfection. Of the other 7 plasmid DNAs, 3 clones expressed a low but detectable level of interferon activity, whereas the remaining 4 were virtually inactive in this respect. The levels of interferon activity produced

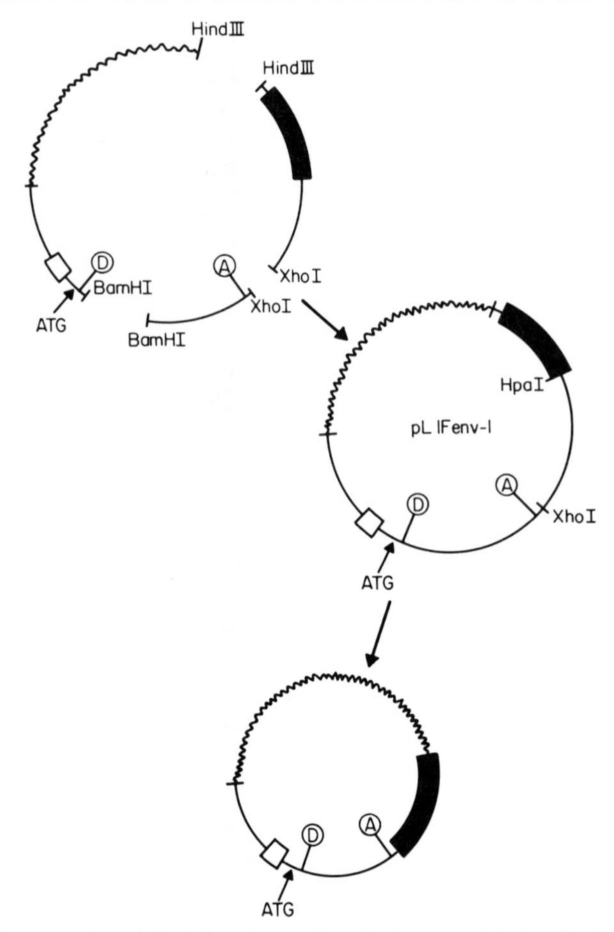

FIG. 5. Construction of viral–interferon chimeric clones positioning the Hu-IFN-β gene near the splice acceptor site. The interferon coding sequence (heavy line) was obtained as the *Xho*I–*Hin*dIII fragment isolated from clone IF18 and was ligated to restriction fragments isolated from clone pL39td2.4 as described in the text. The approximate locations of the splice donor and acceptor sites, as well as the viral ATG used to initiate protein translation, are indicated.

upon transfection of clones IF132, IF133, and IF147 were only 10–25% of that obtained with clone IF66. The synthesis of interferon does not appear to be simply a function of the distance between the splice acceptor site and the interferon gene. Two of the virtually inactive plasmid DNAs (IF134 and IF140) contain the interferon gene at positions intermediate to those of clones IF132, IF133, and IF147 (Table IV). More likely the

TABLE IV
INTERFERON PRODUCTION FOLLOWING TRANSFECTION OF CHIMERIC DNAs WITH IFN-β GENE NEAR SPLICE ACCEPTOR SITE INTO QT35 CELLS[a]

Clone	Viral position of interferon gene	Interferon activity (units/ml)
137	5051	Neg
138	5058	Neg
139	5136	8
148	5139	3
133	5143	24
140	5183	3
134	5183	2
132	5188	64
147	5193	32
66	367	256

[a] Transfections were performed as described in the footnote to Table I. Clone IF66 was included for comparison with the results listed in Table I. Numbers given for viral position of interferon gene have been estimated by polyacrylamide gel electrophoresis. Neg refers to interferon titers <2 units/ml.

differences observed in activity result from whether or not the interferon gene has been positioned in the proper reading frame with respect to the splice junction.

Concluding Comments

The position of the human fibroblast interferon gene inserted into the avian retroviral genome determines to a large extent the level of interferon expression by the chimeric DNA. The 5 clones which were most efficient in directing the synthesis of interferon contained the gene positioned 12 to 54 nucleotides upstream from the viral ATG used for the initiation of translation (Table I). All 5 plasmid DNAs brought about similar levels of interferon activity, which were approximately 10-fold greater than that reported for superinduced human cells.[22-24]

Interferon production by the 11 clones with the interferon genes located downstream from the viral ATG at position 380 was less than 20%

of that observed with the 5 most active clones (Table I). Interferon production following transfection of these plasmid DNAs would require either that translation initiate directly at the downstream interferon codon, or that a viral–interferon fusion protein, possessing an altered signal peptide, be synthesized by initiating translation at the viral AUG. There is some evidence from other systems to support the former hypothesis.[28] The distance between the viral and interferon initiation codons appears to be critical. Clones in which the interferon gene was positioned within 40 bp of viral position 380 were relatively active, while those with the gene located over 100 bp downstream produced no detectable interferon activity.

All 24 clones in which the interferon gene was positioned near the viral LTR directed the synthesis of human fibroblast interferon, regardless of the precise location of the gene (Table III). At best, the amount of interferon activity obtained was 50% of that resulting from transfection with the 5 most active clones. Insertion of an exogenous gene anywhere within the long nontranslated retroviral RNA leader sequence seems to permit reasonably efficient expression of that gene.

Positioning the interferon gene downstream from the splice acceptor site used for generation of env mRNA is moderately effective in producing active interferon. Three of the chimeric clones within this set were able to direct the synthesis of significant levels of interferon, but the amounts obtained were only 10–25% of that produced upon transfection with the 5 most active clones (Table IV). This might be simply a reflection of the proportion of RNA molecules that undergo splicing events. Although the precise location of the interferon gene within the retroviral sequence was not determined, it seems likely that an important requirement for synthesis of active interferon is that the inserted gene be positioned in the correct reading frame relative to the translation of spliced viral *env* mRNA.

[28] P. T. Lomedico and S. J. McAndrew, *Nature* (*London*) **299,** 221 (1982).

[57] Procedures for Expression, Modification, and Analysis of Human Fibroblast Interferon (IFN-β) Genes in Heterologous Cells

By MICHAEL A. INNIS and FRANK MCCORMICK

Human IFN-β, in contrast to multiple subspecies of IFN-α, is a glycoprotein with a single potential N-linked carbohydrate attachment signal located at residues 80–82, Asn*-Glu-Thr. The report that IFN-β synthesized in a cell-free extract exhibited antiviral activity indicated that the carbohydrate was not required.[1] Since then it has become clear that the carbohydrate moiety is dispensable for other biological activities as well since unglycosylated human IFN-β produced in *E. coli* retains all of the antiviral, cell growth-inhibiting and immunomodulatory activities displayed by the natural IFN.[2,3] However, the physical properties and specific activity of *E. coli* IFN-β differ significantly from those of its natural counterpart.[4] We therefore developed procedures to express the IFN-β gene in a mammalian host cell that would provide substantial quantities of glycosylated IFN-β for biological and biochemical characterization. We have previously reported expression levels of glycosylated human IFN-β in Chinese hamster ovary (CHO) cells that are 300 times higher than those from superinduction of primary human fibroblasts, the conventional source of human IFN-β.[5] These procedures can be combined with the techniques for *in vitro* mutagenesis to modify and reintroduce the cloned gene into appropriate host cells. In this chapter, we describe procedures we have used to (1) express human IFN-β in CHO cells, (2) modify the gene sequence encoding the carbohydrate attachment site of IFN-β, and (3) analyze the glycosylation state of the IFN-β produced.

[1] S. Pestka, J. McInnes, E. Havell, and J. Vilček, *Proc. Natl. Acad. Sci. U.S.A.* **72,** 3898 (1975).

[2] D. V. Goeddel, H. M. Shepard, E. Yelverton, D. Leung, R. Crea, A. Sloma, and S. Pestka, *Nucleic Acids Res.* **18,** 4057 (1980).

[3] R. Derynck, E. Remaut, E. Saman, P. Stanssens, E. De Clercq, J. Contents, and W. Fiers, *Nature (London)* **287,** 193 (1980).

[4] D. Mark, S. D. Lu, R. Yamamoto, A. Creasey, and L. Lin, *Proc. Natl. Acad. Sci. U.S.A.* **81,** 5662 (1984).

[5] F. McCormick, M. Trahey, M. Innis, B. Dieckmann, and G. Ringold, *Mol. Cell. Biol.* **4,** 166 (1984).

Expression of Human IFN-β in CHO Cells

There are a number of reports describing expression of the human IFN-β gene in heterologous mammalian cells.[5–12] There are two main advantages of the particular vector/host system described below. First $DHFR^-$ CHO cells are largely resistant to the antiproliferative effects of human IFN-β, up to 10,000 units/ml; in addition, they produce no detectable hamster interferon upon induction. Second, the $DHFR^-$ phenotype allows the use of methotrexate to amplify the unselected (IFN-β) gene along with the selected DHFR gene that is present on the expression vector. Details of the construction of plasmid pM17 containing human DNA encoding the IFN-β gene linked to DNA encoding mouse dihydrofolate reductase (DHFR) have been presented.[5] This plasmid was introduced into $DHFR^-$ CHO cells, and cell lines containing multiple copies of the mouse DHFR gene were selected following exposure to increasing concentrations of the DHFR inhibitor methotrexate. Multiple copies of the unselected IFN-β gene were also present in these cell lines and could be induced to secrete high levels of glycosylated human IFN-β. The protocols for transfecting and amplifying plasmids containing the mouse DHFR gene in $DHFR^-$ CHO cells are similar to those described by Ringold *et al.*[13] and are summarized below.

Transfection and Amplification of Plasmid DNA in CHO Cells

Plasmid DNA (25 μg of ethidium bromide/CsCl band purified) is added to 0.5 ml of a $CaCl_2$ solution to give final concentrations of 40 μg/ml DNA, 250 m*M* $CaCl_2$. This DNA solution is added dropwise to 0.5 ml HeBS (Hepes 21 m*M*, dextrose 6 m*M*, NaCl 137 m*M*, KCl 5 m*M*, $Na_2HPO_4 \cdot 2H_2O$ 0.7 m*M*, adjusted to pH 7.05–7.10 before use) and incubated at room temperature for 30 min. This turbid suspension is layered

[6] P. M. Pitha, D. M. Ciufo, M. Kellum, N. B. K. Raj, G. R. Reyes, and G. S. Hayward, *Proc. Natl. Acad. Sci. U.S.A.* **79,** 4337 (1982).

[7] S. Ohno and T. Taniguchi, *Nucleic Acids Res.* **10,** 967 (1982).

[8] H. Hauser, G. Gross, W. Bruns, H. Hochkeppel, U. Mayr, and J. Collins, *Nature (London)* **297,** 650 (1982).

[9] K. P. Zinn, P. Mellon, M. Ptashne, and T. Maniatis, *Proc. Natl. Acad. Sci. U.S.A.* **79,** 4897 (1982).

[10] D. Canaani and P. Berg, *Proc. Natl. Acad. Sci. U.S.A.* **79,** 5166 (1982).

[11] S. Mitrani-Rosenbaum, L. Maroteaux, Y. Mory, M. Revel, and P. M. Howley, *Mol. Cell. Biol.* **3,** 233 (1983).

[12] J. Tavernier, D. Gheyson, F. Duerinck, J. van der Heyden, and W. Fiers, *Nature (London)* **301,** 634 (1983).

[13] G. Ringold, B. Dieckmann, and F. Lee, *Mol. Appl. Genet.* **1,** 165 (1981).

onto subconfluent cells from which medium has been removed, and left for 30 min at room temperature. Cells are grown in 100-mm dishes and must be in the logarithmic phase of growth. Next, 5 ml of growth medium is added and cells are incubated at 37° for 4 hr. Medium is then removed and 15% glycerol in HeBS is added for about 1 min at room temperature until cells appear to crinkle; cells are then rinsed with PBS nonselective medium added. After 3 days, cells are trypsinized and seeded to low density in selective growth medium (Dulbecco's MEM, proline 35 μg/ml, serum 8%) which lacks thymidine required for growth of the parental cell line. Transformants are picked 10–15 days later.

Clones are now exposed to 10^{-9} *M* methotrexate as described in detail by Ringold *et al.*,[13] and resistant clones are selected. During the course of coamplification of IFN-β genes, a crisis point was observed at which time cells producing relatively high levels of IFN appeared to grow poorly. Cells that grew out of this crisis were resistant to the antigrowth and antiviral effects of IFN-β. This may have been due to phenotypic adaption of the cells (for example, by down-regulation of IFN receptors) or due to a genetic mutation that somehow confers resistance to IFN.[5]

Since this work was started, Simonsen and co-workers have reported the use of a mutant DHFR gene that confers resistance to higher levels of methotrexate than does the wild-type gene.[14] This makes it possible to introduce the gene into wild-type, DHFR$^+$ cells and select directly for those cells containing the gene.

Oligonucleotide Directed *in Vitro* Mutagenesis of the IFN-β Gene Sequence Encoding the Carbohydrate Attachment Site

Indirect methods with tunicamycin or other metabolic inhibitors of glycosylation suggest that some biologically active, unglycosylated IFN-β is produced and secreted into the media in the presence of such inhibitors.[15,16] These experiments yielded very low levels of interferon, and some residual carbohydrate incorporation was always present. In an attempt to circumvent these problems, we used *in vitro* mutagenesis to change the codon for Asn 80 (AAT) to Gln (GAC). Asn 80 is an acceptor for N-linked addition of core saccharides at the Asn*-Glu-Thr glycosylation signal of IFN-β. Gln is not an acceptor for carbohydrate addition, but otherwise should be a conservative amino acid replacement for Asn 80.

[14] C. C. Simonsen and A. D. Levinson, *Proc. Natl. Acad. Sci. U.S.A.* **80,** 2495 (1983).
[15] Y. K. Yip and J. Vilček, this series, Vol. 78, p. 212.
[16] E. A. Havell, J. Vilček, E. Falcoff, and B. Berman, *Virology* **63,** 475 (1975).

The modified gene can be transfected back into DHFR$^-$ CHO cells for analysis.

The protocol for *in vitro* mutagenesis is simplified if the gene is first cloned into a vector such as one of the M13 phage described by Messing[17] or the EMBL plasmid,[18] either of which can yield single-stranded DNA template. One picomole of single-stranded DNA template and 10 pmol of deoxyoligonucleotide (15 long with the mismatch in the center) are incubated 5 min at 70°; then 30 min at 42° in 15 μl of buffer containing 10 m*M* Tris · HCl, pH 7.4, 10 m*M* $MgCl_2$, and 90 m*M* NaCl. The reaction mixture is placed on ice and adjusted to 20 μl by the addition of all four dNTPs to 500 μM final concentration and 3–5 units of the Klenow fragment of *E. coli* DNA polymerase I. Following 5 min on ice, the reaction is incubated at 37° for 30 min. After extension, which can be monitored by gel electrophoresis, the reaction is inactivated by heat at 75° for 15 min. Transformation of an appropriate host, transfer of resulting plaques to filters, and prehybridization are then performed by standard techniques.[19] The deoxyoligonucleotide, labeled to a specific activity of $2–5 \times 10^6$ cpm/pmol with [α^{32}P]ATP and T_4 polynucleotide kinase, is then used as a hybridization probe to select the correctly modified clone. Prehybridization (1 hr), hybridization (1–3 hr), and washing of the filters (5 min) are all done in 6× SSC at the same temperature, a few degrees below the calculated T_d; $T_d = [(GC \times 4°) + (AT \times 2°)]$.[20] In contrast to other procedures, we have found it unnecessary to either phosphorylate the oligonucleotide or to purify the *in vitro* synthesized RF prior to transformation. We routinely observe a frequency of 1–3% of the desired clone.

Metabolic Labeling of IFN

1. *Methionine Labeling*. Cells are seeded at high density and allowed to attach overnight. Typically, 3×10^6 cells are seeded into 6-cm tissue culture wells. High cell densities are required for inducible expression of human IFN-β from CHO cells, and the cells must be maintained at these densities for at least 12 hr prior to superinduction. The superinduction protocol is based on that of Tan and co-workers[21]; poly(rI : rC) at 20 μg/ml and cycloheximide at 2 μg/ml are added to the growth medium, and cells are incubated in this medium for 3 hr at 37°. Actinomycin D is added (2

[17] J. Norrander, T. Kempe, and J. Messing, *Gene* **26,** 101 (1983).

[18] L. Dente, G. Cesareni, and R. Cortese, *Nucleic Acids Res.* **11,** 1645 (1983).

[19] T. Maniatis, E. F. Fritsch, and J. Sambrook, "Molecular Cloning: A Laboratory Manual." Cold Spring Harbor Lab., Cold Spring Harbor, New York, 1982.

[20] G. Dalbadie-McFarland, L. W. Cohen, A. D. Riggs, C. Morin, K. Itakura, and J. H. Richards, *Proc. Natl. Acad. Sci. U.S.A.* **79,** 6409 (1982).

[21] M. Ho, J. A. Armstrong, and Y. H. Tan, *Proc. Soc. Exp. Biol. Med.* **139,** 259 (1972).

μg/ml), and the cells are incubated 1 hr more at 37°. The cells are then washed 3 times in PBS. For production of unlabeled IFN-β, normal growth medium (i.e., 8% fetal calf serum, DME-high glucose, L-proline 35 μg/ml) is added. For production of [^{35}S]methionine-labeled IFN-β, methionine-free, serum-free medium (DME-high glucose, L-proline 35 μg/ml) is added, together with L-[^{35}S]methionine (approximately 1100 Ci/mmol) 100 μCi in 1 ml for 10 hr.

^{35}S-labeled interferon can be visualized by either loading growth medium directly onto an SDS–polyacrylamide gel or by immunoprecipitation followed by SDS–PAGE. In the first case, medium can be concentrated by evaporation to dryness in a rotary vacuum apparatus and resuspended in gel sample buffer (2% SDS, 0.1 *M* DTT, 10 m*M* Tris · glycine, pH 6.8, 0.01% bromophenol blue). This method has the advantage that no assumptions are made as to the ability of an anti-IFN-β antiserum to react with mutant derivatives of the IFN-β molecule. The disadvantage is that a small amount of cell death occurs during the labeling period releasing radiolabeled proteins into the medium that contribute to the background on the gel. In the second case, medium is reacted with appropriate concentrations of anti-IFN-β antiserum for 1 hr at 22°. Immune complexes are collected on protein-A sepharose beads (Sigma), washed with Tris-buffered saline (50 m*M* Tris · HCl, pH 7.4; 150 m*M* NaCl) containing Nonidet P40, 0.5% (w/v). Beads are extracted with sample buffer and loaded onto SDS–PAGE as above.

2. *Labeling with Glucosamine*. The procedure is similar to that described above, except that labeling is performed in complete medium to which DL-[^{14}C]glucosamine has been added (500 μCi/ml, 325 mCi/mmol, New England Nuclear).

3. *Labeling with Mannose*. Cells are labeled for 30 min in glucose-free medium 4 hr after superinduction with 250 μCi/ml of DL-[^{3}H]mannose, 24.3 Ci/mmol, New England Nuclear. ^{3}H-labeled IFN-β can be visualized on autoradiograms by treatment of the SDS gels with Enhance (New England Nuclear) solution before autoradiography.

Chromatography of IFN-β on Concanavalin A-Sepharose

Glycoproteins are known to bind to concanavalin A (con A) and can therefore be separated from nonglycosylated proteins by chromatography on Con A-sepharose. Medium containing IFN-β is dialyzed against loading buffer (0.02 *M* Tris · HCl, pH 7.4, 0.5 *M* NaCl, 1 m*M* $MnCl_2$, 1 m*M* $CaCl_2$). This material is loaded onto a Con A-sepharose column that has been equilibrated with loading buffer. The column is washed with several volumes of loading buffer; portions of the flow-through and wash fractions are taken for IFN-β titration. Elution buffer (0.5 *M* α-methyl manno-

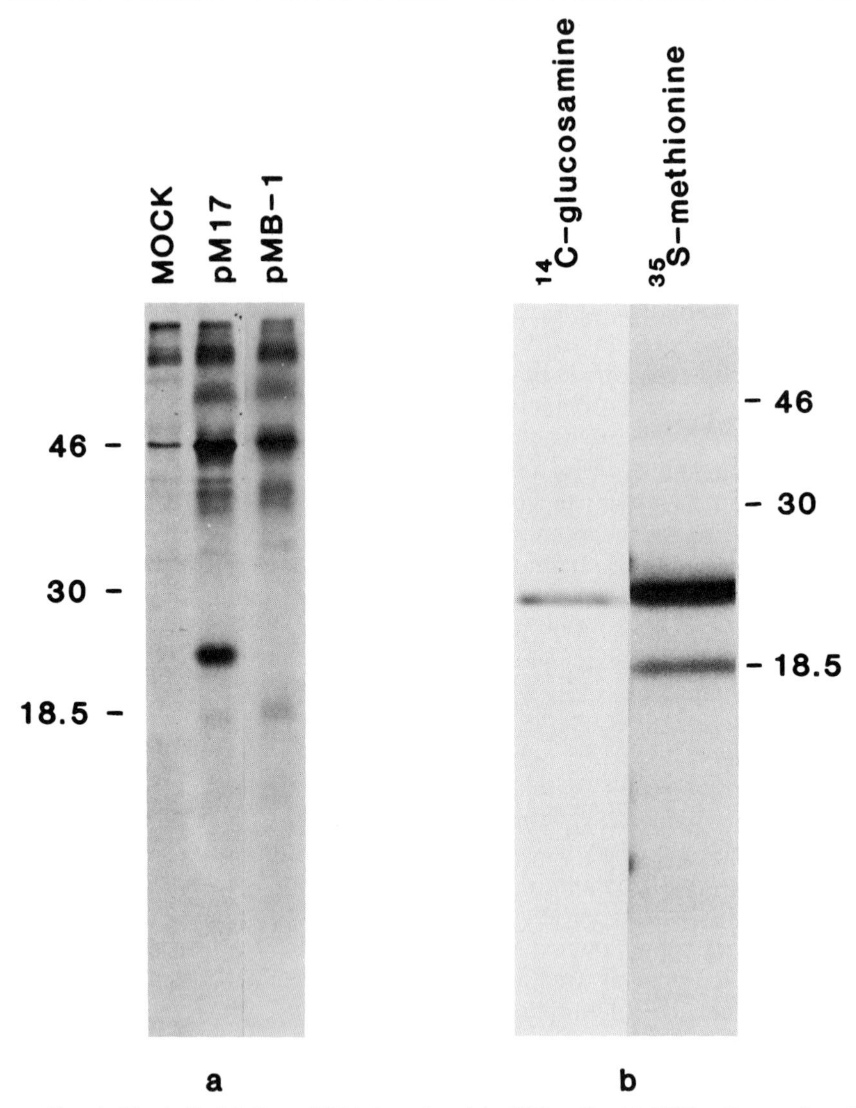

FIG. 1. Metabolic labeling of IFN-β produced in CHO cells. (a) CHO cells transfected with either pMI7 or pMB1 were labeled with [^{35}S]methionine, 100 μCi/ml, 1164 Ci/mmol as described in the text. Supernatants were reacted with anti-IFN-β antibodies, and immunoprecipitates analyzed on SDS–PAGE. The mock culture represents data from wild-type CHO cells. (b) CHO cells containing amplified copies of pMI7 were superinduced and labeled with either [^{14}C]glucosamine or [^{35}S]methionine as indicated; the supernatants were immunoprecipitated as described in the text and analyzed on SDS–PAGE.

side, 10% ethylene glycol in loading buffer) is added to the column; after 15 min on the column, material is collected, dialyzed against PBS, and titrated for IFN activity.

Concluding Comments

We have applied the techniques described above to establish CHO cell lines that can be induced to secrete glycosylated and unglycosylated forms of human IFN-β. While initial levels can be low, about 100 units/ml, we have been able to coamplify the IFN-β gene along with the DHFR gene by methotrexate selection and to obtain very high levels of IFN expression in selected cell lines.[5] Immunoprecipitates of [^{35}S]methionine-labeled supernatants from initial transfectants of CHO cells expressing wild-type IFN-β (pMI7) or a mutant IFN-β (pMB-1) gene where glutamine was substituted for asparagine at position 80 are shown in Fig. 1a. Cells expressing the modified IFN-β (pMB-1) secrete a form of IFN-β that migrates at 18,500 D on SDS–PAGE, and do not produce the major 23,000 form seen in supernatants from cells expressing unmodified IFN-β (pMI7). An 18,500 D form of IFN-β is also produced from the wild-type gene and appears to be unglycosylated, as shown in Fig. 1b. The 18,500 D form of IFN-β produced from the pMB-1 gene appears to be unglycosylated confirming that the asparagine residue at position 80 is the site for glycosylation in native IFN-β. Preliminary results indicate that the unglycosylated IFN-β produced and secreted from the mutant gene has a much lower specific biological activity than glycosylated IFN-β. The availability of these cell lines will permit us to evaluate further the role of glycosylation in the activity and physical properties of IFN-β.

[58] Procedures for *in Vitro* DNA Mutagenesis of Human Leukocyte Interferon Sequences

By THOMAS M. DECHIARA, FRAN ERLITZ, and S. JOSEPH TARNOWSKI

Introduction

The mutagenesis of cloned genes has become a powerful research tool in the analysis of protein function (for a review, see Dalbadie-McFarland and Richards[1]). This chapter describes the production of recombinant

[1] G. Dalbadie-McFarland and J. H. Richards, *Annu. Rep. Med. Chem.* **18,** 237 (1983).

human leukocyte interferon A (IFN-αA) analogs to understand better how the structure of IFN-αA is related to its biological activity. Two approaches were taken to specifically mutate the IFN-αA gene. In one approach, synthetic deoxyoligonucleotides were employed to replace a desired region of IFN-αA DNA located between convenient restriction enzyme recognition sites. These deoxyoligonucleotides were identical in sequence to the segment they replaced, except for a single codon change. The mutated genes encoded the substitution of cysteine residues 1 and 98 in the protein with glycine and serine, respectively. In the second approach, site-specific mutations were directed by single deoxyoligonucleotides in a procedure which utilized a heteroduplex of plasmid DNA. These mutations were in the IFN-αA gene region which encoded the carboxy terminal 27 amino acids.

Solution and Materials

TAE: 40 m*M* Tris-acetate, pH 7.8, 5 m*M* NaOAc, 2 m*M* EDTA
TE: 10 m*M* Tris · HCl, pH 7.4, 1 m*M* EDTA
Ligation solution: 60 m*M* Tris · HCl, pH 7.5, 10 m*M* $MgCl_2$, 10 m*M* dithiothreitol (DTT), 400 μM ATP. All ligation reactions were performed at 15° for 12 hr in a volume of 10 μl except where indicated.
Restriction endonucleases: *Bst*EII, *Eco*RI, *Hin*fI, *Pvu*II, and *Sau*3AI were from New England Biolabs. The buffer conditions used were described by the manufacturer.
Plasmids: (1) pRC23 is described in detail in this volume.[2] (2) pRC234 is a derivative of pRC23 in which the *Pvu*II restriction endonuclease site was deleted by a brief *Bal*31 exonuclease digestion followed by ligation of the blunt ends to recircularize the plasmid. The IFN-αA gene was inserted into the unique *Eco*RI site of these vectors.
Klenow fragment of *E. coli* DNA Polymerase I and T_4-DNA ligase from Boehringer Manheim.

Substitution of Cys 1 and Cys 98 Residues

The nucleotide sequence of the IFN-αA DNA[3] predicts cysteine residues at positions 1, 29, 98, and 138 in the protein (Fig. 1a). Disulfide bond assignments,[4] together with selective reduction studies,[5] have suggested

[2] R. Crowl, this volume [55].
[3] S. Pestka, *Arch. Biochem. Biophys.* **221,** 1 (1983).
[4] R. Wetzel, *Nature (London)* **289,** 606 (1981).
[5] R. Wetzel, H. L. Levine, D. A. Estell, and S. Shire, *J. Cell. Biochem., Suppl.* **6,** 89 (1982).

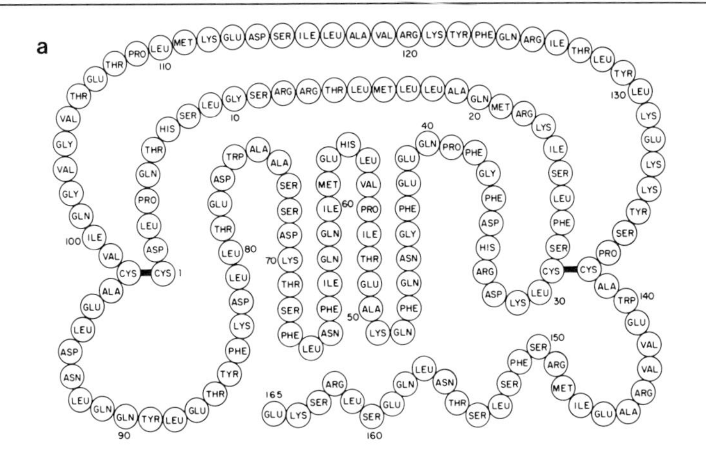

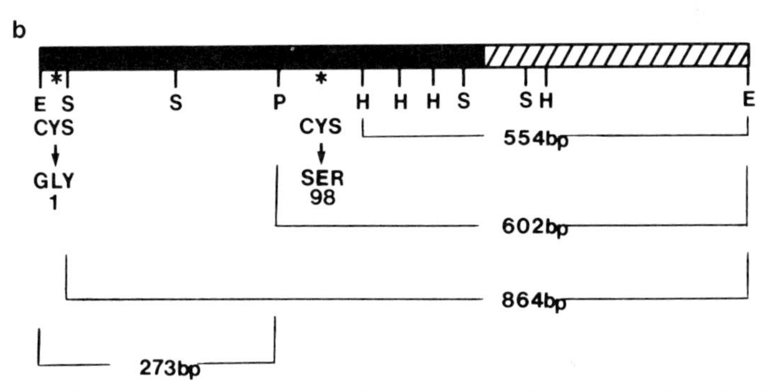

FIG. 1. (a) Schematic representation of IFN-αA (taken with permission from Pestka[3]). (b) The gene encoding IFN-αA is carried on an *Eco*RI fragment.[6,7] The solid region represents about 500 bp of coding sequences and the hatched region, about 350 bp of 3′ untranslated DNA. E, *Eco*RI; S, *Sau*3AI; P, *Pvu*II; H, *Hin*fI.

that, in contrast to the Cys 29/Cys 138 disulfide bond, the Cys 1/Cys 98 bond is less stable and is not a requisite for the antiviral activity of IFN-αA as measured *in vitro*. Therefore, substitution of either the Cys 1 or Cys 98 residues, or both, by closely related amino acids should have no effect on the biological activity of the protein. Synthetic deoxyoligonucleotides (Table I) were used to replace the amino terminal coding region of the gene between the *Eco*RI and *Sau*3AI restriction sites for the substitution of Cys 1, or between the *Pvu*II and *Hin*fI sites to substitute for Cys 98

TABLE I
SYNTHETIC DNA TO REPLACE PARENTAL IFN-αA GENE FRAGMENTS[a]

Synthetic deoxyoligonucleotide	Sequence
	1 Gly
1A (11-mer)	AATTCATGGGC *Sau*3AI
1B (11-mer)	*Eco*RI GTACCCGCTAG
	*Pvu*II
98A (18-mer)	CTGAATGACCTGGAAGCC
98B (27-mer)	GACTTACTGGACCTTCGGTCGCACTAT
	98 Ser
98C (28-mer)	AGCGTGATACAGGGGGTGGGGTGACAG *Hin*fI
98D (22-mer)	GTCCCCCACCCCCACTGTCTGA

[a] Dashed line, restriction site overhang or blunt end; solid line, 9 base complementary sequence for the ligation of 98A, B, C, and D.

(Fig. 1b).[6,7] Each set of synthetic DNA molecules encoded one codon change. The TGT codon for Cys in the parental gene was replaced by GGC (Gly) or by AGC (Ser) where indicated (Table I).

Three mutated IFN-αA genes were constructed: one encoding the Gly 1 substitution, one encoding Ser 98, and one encoding both. The incorporation of glycine in the first amino acid position was achieved by the ligation of a 50-fold molar excess of synthetic deoxyoligonucleotides 1A and 1B (Table I) to the 864 bp *Sau*3AI/*Eco*RI gene fragment. This fragment was generated by *Sau*3AI partial digestion of IFN-αA DNA, necessitated by multiple *Sau*3AI restriction sites (Fig. 1b). Mutagenesis for the incorporation of Ser 98 resulted from the ligation of a 50-fold molar excess of deoxyoligonucleotides 98A, 98B, 98C, and 98D (Table I) to both the 273 bp *Eco*RI/*Pvu*II fragment and the 554 bp *Hin*fI/*Eco*RI fragment. The larger fragment was generated by *Hin*fI partial digestion of the 602 bp *Pvu*II/*Eco*RI gene fragment (Fig. 1b). The double substitution of Gly 1

[6] S. Maeda, R. McCandliss, M. Gross, A. Sloma, P. C. Familletti, J. M. Tabor, M. Evinger, W. P. Levy, and S. Pestka, *Proc. Natl. Acad. Sci. U.S.A.* **77,** 7010 (1980).

[7] D. V. Goeddel, E. Yelverton, A. Ullrich, H. L. Heyneker, G. Miozzari, W. Holmes, P. H. Seeburg, T. Dull, L. May, N. Stebbing, R. Crea, S. Maeda, R. McCandliss, A. Sloma, J. M. Tabor, M. Gross, P. C. Familletti, and S. Pestka, *Nature (London)* **287,** 411 (1980).

and Ser 98 was encoded by ligation of the 273 bp *Eco*RI/*Pvu*II fragment (Gly 1) and the 602 bp *Pvu*II/*Eco*RI fragment (Ser 98).

Following ligation, the DNAs were digested with *Eco*RI and the mutated genes were inserted into the *Eco*RI site of pRC23 for expression in *E. coli* as described in this volume.[2] *E. coli* cells were grown in 10-liter fermenters, harvested, and the cell pastes were lysed in a Manton-Gaulin homogenizer.[8] Each crude extract was passed over a column of immobilized anti-leukocyte interferon monoclonal antibody.[9] The columns were washed to remove extraneous *E. coli* proteins and the analog proteins were desorbed with a dilute acetic acid solution. Samples of the purified analogs were analyzed by SDS–PAGE[10] in the presence or absence of 2-mercaptoethanol, and assayed for their ability to protect MDBK cells from a challenge by vesicular stomatitis virus *in vitro*.[11]

Each of the three analogs possessed a specific activity of 2×10^8 units/mg protein, identical to that of parental IFN-αA. This confirms that formation of the Cys 1/Cys 98 disulfide bond is not necessary for the antiviral activity of the protein. In addition, the protein analogs were purified with monoclonal antibody LI-8 indicating that Cys 1 or Cys 98 residues were not involved in the domain recognized by this immunoglobulin.

Analysis of the purified analogs by nonreducing SDS–PAGE implicates the sulfhydryl groups at positions 1 and 98 to be involved in intermolecular disulfide bond formation. As illustrated in Fig. 2, the Gly 1-substituted analog migrated almost entirely as a "slow" monomer form with some dimer form observed (lane B), compared with the migration pattern of parental IFN-αA (lane A). In contrast, the Ser 98-substituted analog possessed a much higher dimer content (lane C). This indicates that the Cys 1 sulfhydryl group is a more active participant in intermolecular disulfide bond formation than the Cys 98 sulfhydryl. That the dimer forms were disulfide bonded was demonstrated by conversion of the Cys 1/Cys 1 dimer to "slow" monomer form in the presence of 2-mercaptoethanol (lane D). The Gly 1/Ser 98 double-substituted analog migrated as a "slow" monomer form only (lane E). The "slow" monomer form (lane E) appeared to be susceptible to proteolysis as evidenced by the appearance of a discrete fragment which migrated at about 14,000 Da. The appearance of this fragment was dependent on subtle changes in the extraction procedure.

[8] S. J. Tarnowski and R. A. Liptak, *Adv. Biotechnol. Processes* **2,** 271 (1983).

[9] T. Staehelin, D. S. Hobbs, H. Kung, C. Y. Lai, and S. Pestka, *J. Biol. Chem.* **256,** 9750 (1981).

[10] U. K. Laemmli, *Nature (London)* **227,** 680 (1970).

[11] P. C. Familletti, S. Rubinstein, and S. Pestka, this series, Vol. 78, p. 387.

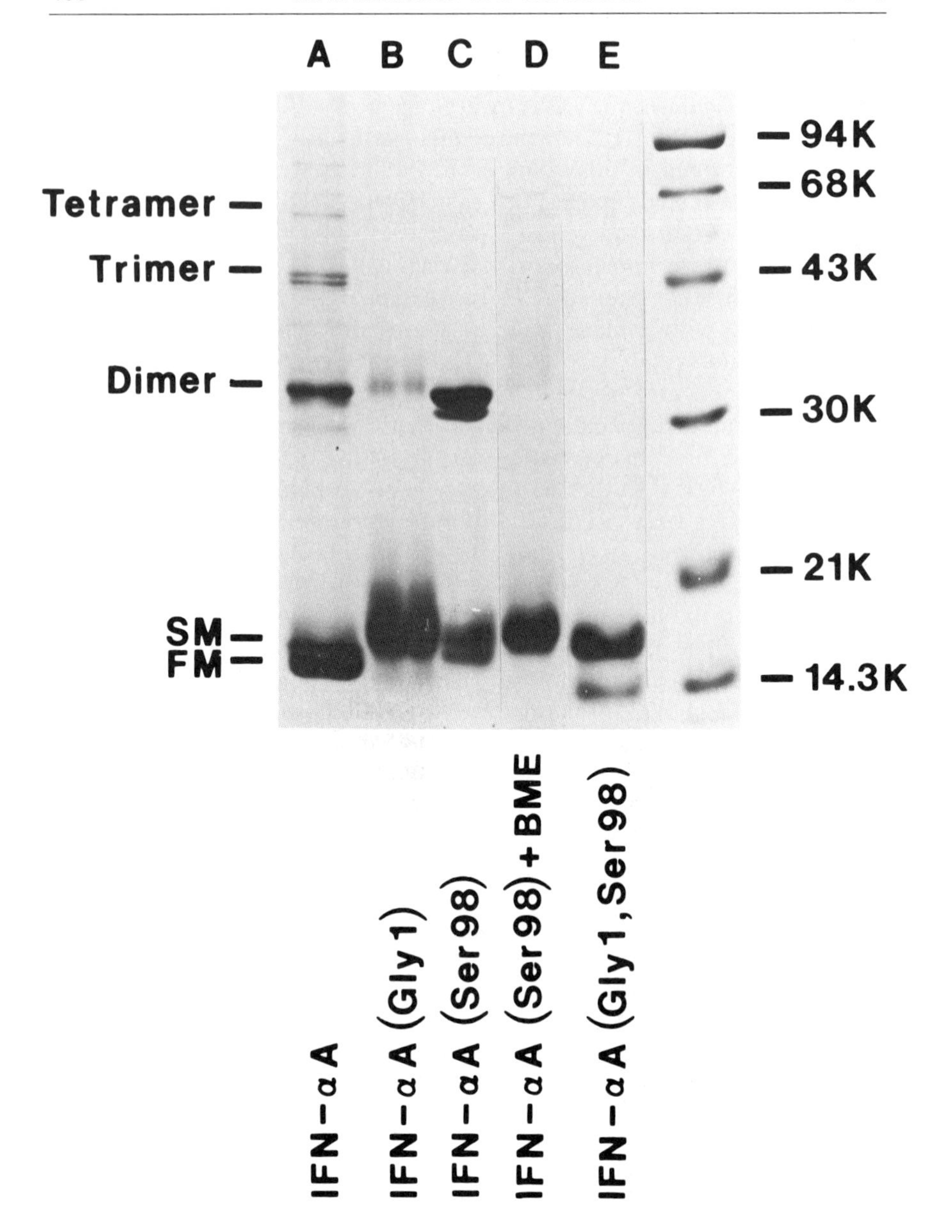

FIG. 2. Coomassie Brilliant blue-stained SDS–PAGE analysis of Cys 1/Cys 98 substituted IFN-αA analogs. Cells produced parental IFN-αA (*E. coli* W 3110 *trp* R-lac[2]/pLiFA *trp* 55), under *trp* promoter–operator control mechanisms.[7] IFN-αA (Gly 1), IFN-αA (Ser 98), or IFN-αA (Gly 1, Ser 98) (*E. coli*. RR1/pRK248*c*Its) were produced at 42° as described by R. Crowl.[2] Frozen cell pastes were homogenized to crude cell extracts[8] and interferon

Modification of the IFN-αA Carboxy-Terminus

Mutagenesis of the IFN-αA gene region encoding the carboxy terminal 27 amino acids (139–165) was performed by generating a plasmid DNA heteroduplex[13] (Fig. 3). The advantages of this method are that (1) the mutagenesis can be performed directly on the expression plasmid so that mutations can be screened rapidly without subcloning as is necessary with the M13 system[14]; (2) the formation of heteroduplexes eliminates the requirement of exonuclease III digestion to create a single-stranded gap from a nick introduced by a restriction endonuclease and a calibrated amount of ethidium bromide[15]; and (3) since heteroduplexes can easily be purified from the linearized and gapped homoduplexes, between 5 and 15% of the resulting transformants contain mutant plasmids. If the mutation is a deletion (as was the case for IFN-αA gene mutations described below), or a base change to introduce a restriction site, then the transformants can easily be screened by a rapid plasmid isolation protocol.[16]

Formation and Isolation of Heteroduplex

Use of the three restriction enzymes (Fig. 3) generated two populations of opened plasmids: a linearized plasmid, and a linearized plasmid with a 550 bp *Pvu*II/*Bst*EII fragment removed. It is desirable that the gap be no larger than about 12% of the plasmid's size. Excessively larger gaps make it more difficult to resolve the heteroduplexes from homoduplex linearized molecules. To obtain sufficient quantities of heteroduplex, at

[12] M. M. Bradford, *Anal. Biochem.* **72,** 248 (1976).

[13] A. Oka, K. Sugimoto, H. Sasaki, and M. Takanami, *Gene* **19,** 59 (1982).

[14] G. Winter, A. K. Fersht, A. J. Wilkinson, M. Zoller, and M. Smith, *Nature* (*London*) **299,** 756 (1982).

[15] G. Dalbadie-McFarland, L. W. Cohen, A. D. Riggs, C. Morin, K. Itakura, and J. H. Richards, *Proc. Natl. Acad. Sci. U.S.A.* **79,** 6409 (1982).

[16] H. C. Birnboim and J. Doly, *Nucleic Acids Res.* **7,** 1513 (1979).

---

was purified on an immunosorbent column of immobilized monoclonal antibody LI-8.[9] Following protein determination[12] aliquots of each antibody pool were dried by evaporation and analyzed by electrophoresis[10] on a 12.5% polyacrylamide slab gel containing 0.1% SDS. Electrophoresis of the samples under nonreducing conditions was performed with sample buffer without 2-mercaptoethanol. Lane A, 15 μg parental IFN-αA, nonreduced; lane B, 10 μg IFN-αA (Gly 1), nonreduced; lane C, 10 μg IFN-αA (Ser 98), nonreduced; lane D, 10 μg IFN-αA (Ser 98), reduced; lane E, 10 μg IFN-αA (Gly 1, Ser 98), nonreduced. Standard protein makers from Bio-Rad Laboratories (Rockville, NY): 94,000, phosphorylase *b;* 68,000, bovine serum albumin; 43,000, ovalbumin; 30,000, carbonic anhydrase; 21,000, soybean trypsin inhibitor; 14,300, lysozyme, all reduced. SM, "slow" monomer; FM, "fast" monomer with both disulfide bonds intact.

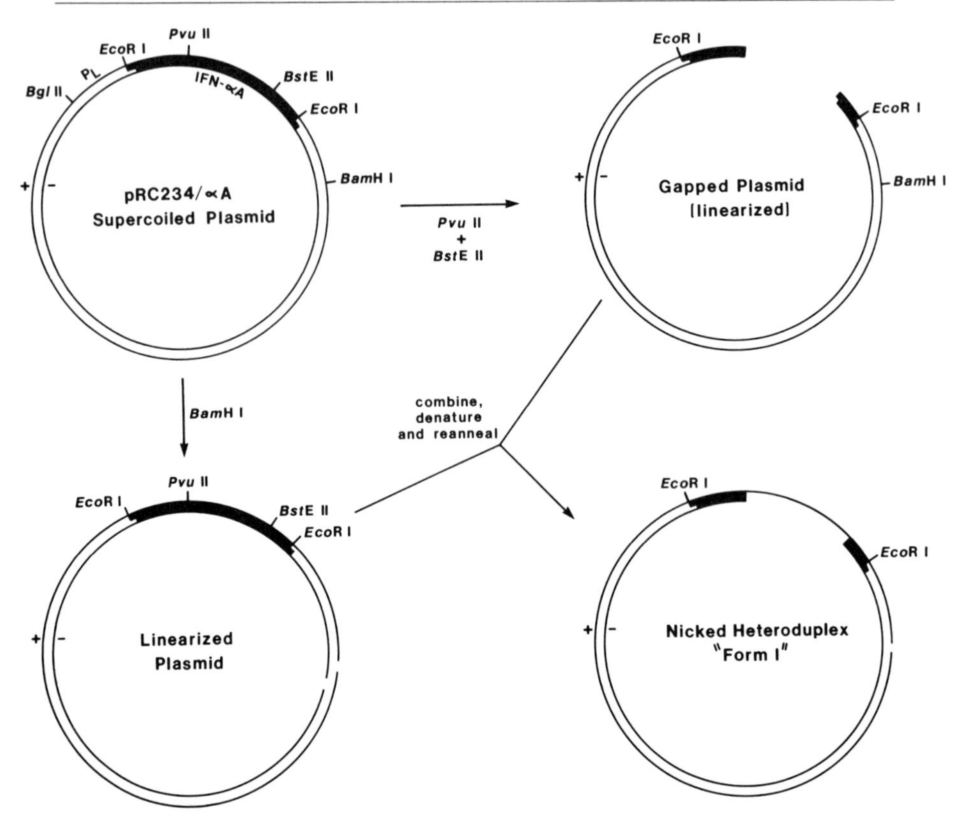

FIG. 3. Formation of plasmid DNA heteroduplex as a base for site-directed mutagenesis. See text for details. pRC 234 is a derivative of pRC 23 with the *Pvu*II site in the plasmid deleted.

least 1 μg of supercoiled plasmid was used for each of the two restriction enzyme digests. After digestion, the linearized plasmids were purified by electrophoresis through agarose, recovered, and precipitated with ethanol.

Purified linearized plasmids were resuspended in 25 μl of H_2O and combined in an Eppendorf tube. After addition of 50 μl of a solution containing 0.2 *N* NaOH and 40 m*M* EDTA, the plasmids were denatured at 23°. After 10 min, 10 μl of 2 *M* Tris (1.8 *M* Tris · HCl, 0.2 *M* Tris base) and 110 μl of deionized formamide were added. Annealling of single-stranded DNA proceeded for about 3 hr at 23°. Two heteroduplex forms resulted during the annealling process, Form I from the annealing of the "+" strand of the linearized plasmid with the "−" strand of the gapped plasmid, and Form II from annealling of the other strands. In Form I, a single-stranded gap of 550 nucleotides was created in the region of the desired mutations on the "+" strand in the IFN-αA gene. It was not necessary to separate the two heteroduplex forms since the deoxyoligonucleotides were complementary to the "+" strand only.

After annealling, 300 μl of 0.3 NaAc, pH 7.0 and 20 μg of carrier tRNA were added and the heteroduplex molecules were precipitated two times with ethanol. After the second centrifugation, the pellet was rinsed with cold ethanol and dried under vacuum. The heteroduplex forms were purified by electrophoresis through a 0.7% agarose gel. The heteroduplex forms migrated with nicked parental plasmid, but slower than both linearized forms.

To recover the heteroduplex forms, it was necessary to use electroelution. The gel strip containing the heteroduplex band was placed in a dialysis bag with 1 ml of TAE, pH 7.8, running buffer with ethidium bromide (50 μg/ml). The heteroduplex molecules were visualized under long wavelength UV as they migrated out of the gel strip into the buffer. After electrophoresis, the elution buffer was recovered and combined with four 1 ml rinses of the dialysis bag with a solution consisting of 10 m*M* Tris · HCl, pH 7.4, 1 *M* EDTA, and 300 m*M* NaCl. 2-Butanol was used to reduce the volume of 0.5 ml. Ten micrograms of carrier tRNA was added and the heteroduplexes were extracted once with 500 μl of phenol/chloroform/isoamyl alcohol (50 : 50 : 1), precipitated two times with ethanol, dried, and resuspended in 3 μl of H_2O. About 50 ng of heteroduplexes (as measured by visual comparison with standardized plasmid DNA on an agarose gel) was recovered from the gel. Electroelution from agarose was necessary due to the strong affinity between the single-stranded region of the heteroduplex and agarose. It was also important not to filter the eluted heteroduplex through nitrocellulose because single-stranded DNA has high affinity for the filter.

Introduction of Deletions in IFN-αA with Synthetic DNA

Deletions in plasmid DNA were used to produce truncated IFN-αA proteins which were terminated after Cys 138, Val 143, or Arg 149 (Fig. 1a). The three deoxyoligonucleotides shown in Table II each directed a loop-out of sequences in the "+" strand to position the TGA stop codon immediately downstream of the sequences encoding Cys 138, Val 143, or Arg 149. By designing the synthetic DNAs to complement 12 nucleotides in the "+" strand on the 5′ side of the splice point and 12 nucleotides on the 3′ side, the intervening sequences not complementary to the deoxyoligonucleotide will loop-out as illustrated in Fig. 4.

TABLE II

SYNTHETIC DNA FOR INTRODUCING DELETIONS INTO IFN-αA[a]

					138					143						149			166				
Parental	Lys	Tyr	Ser	Pro	Cys	Ala	Trp	Glu	Val	Val	Arg	Ala	Glu	Ile	Met	Arg	Ser		Glu	END			
"+" 5'. . .	.AAA	TAC	AGC	CCT	TGT	GCC	TGG	GAG	GTT	GTC	AGA	GCA	GAA	ATC	ATG	AGA	TCT	. . .	.GAA	TGA	AAA	CTG	GTT
					Cys	END																	
Δ138 3'		ATG	TCG	GGA	ACA	ACT	TTT	GAC	CAA														
										Val	END												
Δ143 3'							ACC	CTC	CAA	CAG	ACT	TTT	GAC	CAA									
																Arg	END						
Δ149 3'													CTT	TAG	TAC	TCT	ACT	TTT	GAC	CAA			

[a] Δ138 directs a loop-out of 81 bases; Δ143 directs a loop-out of 66 bases; Δ149 directs a loop-out of 48 bases.

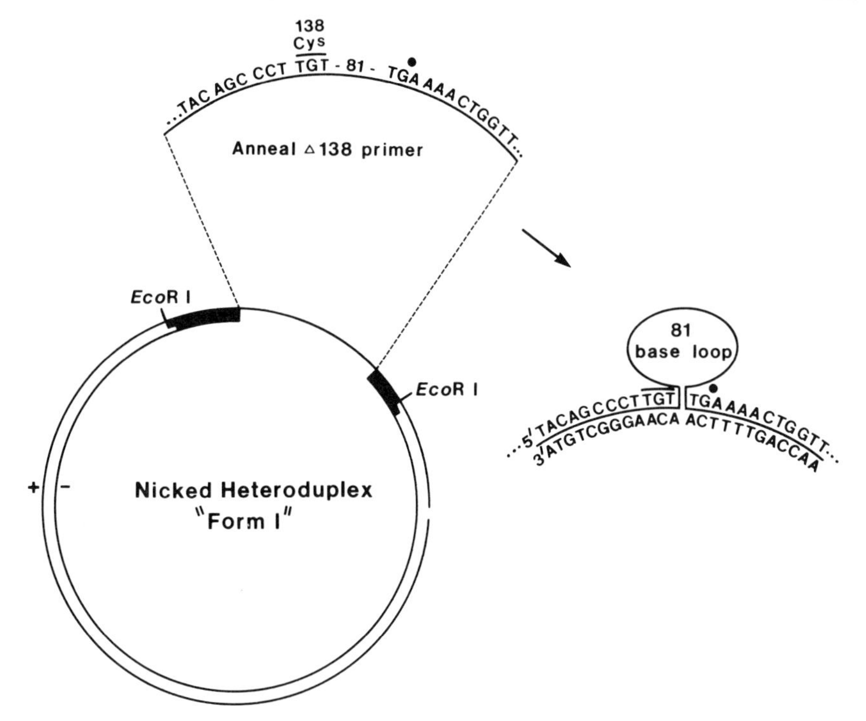

FIG. 4. A synthetic DNA (Δ138) directs the precise splicing of the IFN-αA carboxy-terminal coding sequences. In this example, a deletion of 81 bp resulted from a loop-out of IFN-αA sequences not complementary to the deoxyoligonucleotide. The mutant plasmid results from the replication of the "−" strand.

To effect mutagenesis, the deoxyoligonucleotide was annealled to the "+" strand in the Form I heteroduplex as has been described.[17] Briefly, a 50-fold molar excess of phosphorylated deoxyoligonucleotide was combined with the purified heteroduplex in a volume of 5 μl of 0.1 × TE, pH 7.4. The mixture was heated at 68° for 3 min and quenched at 0°. The rapid cooling prevented the complementary single-stranded regions of both heteroduplex forms from self-annealling to displace the primer. After several minutes at 0°, the volume was adjusted to 20 μl by the addition of 10 m*M* Tris · HCl, pH 7.4, 50 m*M* NaCl, 100 m*M* $MgCl_2$, 1 *M* DTT, 400 μ*M* ATP, 400 μ*M* of each of the four deoxynucleotide triphosphates, 3 units of Klenow fragment of *E. coli* DNA Polymerase I, and 120 units of T_4-DNA Ligase. Filling-in of the single-stranded region and subsequent ligation was performed at 15° for 6 hr. The reaction volume was then adjusted

[17] R. B. Wallace, M. Schold, M. J. Johnson, P. Dembek, and K. Itakura, *Nucleic Acids Res.* **9,** 3647 (1981).

to 250 μl by the addition of 0.3 *M* sodium acetate, pH 7.0, and extracted once with an equal volume of phenol/chloroform/isoamyl alcohol (50 : 50 : 1). The DNA was precipitated twice with ethanol, dried, and resuspended in 5 μl of H_2O for transformation of *E. coli* MC1061.[18] The efficiency of obtaining a mutation was about 5–15% of the colonies screened. Each positive colony detected by mini-prep plasmid preparation had two plasmids present, one plasmid with the deletion and the parental plasmid. Thus, it was necessary to retransform with plasmid DNA to select those colonies which contain the plasmid with the deletion. Similarly, analogs of IFN-αA prematurely terminated at Val 143 and Arg 149 were prepared.

Analysis of Prematurely Terminated IFN-αA Proteins

The results obtained above for the Cys 1 and Cys 98 substitutions were indirect evidence for the importance of Cys 29/Cys 138 intramolecular disulfide bond formation since fully reduced IFN-αA retains only 5% of its activity in the MDBK cell assay.[19] Therefore, it was interesting to determine if amino acids distal to Cys 138 were necessary for biological activity.

E. coli cells which contained plasmids with the specific deletions were induced at 42° and extracts were prepared by 7 *M* guanidine hydrochloride lysis as described.[2] Each extract demonstrated no biological activity in the MDBK cell assay. To determine whether these analogs were inactive as the direct result of the carboxy-terminal deletions or because of *in vivo* instability, plasmids encoding the Cys 138, Val 143, Arg 149, and parental IFN-αA proteins were used in a cell-free, prokaryotic DNA-dependent transcription/translation system.[20] The IFN-αA analogs were synthesized to a level at least equal to that of parental IFN-αA, but were biologically inactive. This result suggests that amino acids distal to the Arg residue at position 149 in the primary structure are required for maximal antiviral activity of IFN-αA.

Concluding Comments

This chapter illustrates two approaches for mutagenesis. The approach used for generating the Cys 1 and Cys 98 substitutions requires extensive synthesis of deoxyoligonucleotide duplexes. The heteroduplex

[18] M. J. Casadaban and S. N. Cohen, *J. Mol. Biol.* **138,** 179 (1980).

[19] J. Langer and S. Pestka, this volume [34].

[20] H.-F. Kung, C. Spears, and H. Weissbach, *J. Biol. Chem.* **250,** 1556 (1975).

method requires only a single synthetic deoxyoligonucleotide and mutagenesis can be performed directly on the expression plasmid. The distance between restriction sites used for making the gap is limited to about 12% of the plasmid length. While this work was in progress, an improvement of the heteroduplex method was reported.[21]

Deletions made in the IFN-αA gene carboxy-terminal coding region provided evidence that amino acid residues distal to Arg^{149} are required for full IFN-αA activity when measured on MDBK cells. When purified preparations (data not shown) of IFN-αA molecules terminated at Cys^{138} and Arg^{149} were assayed, they showed approximately 1/100th of the activity of a molecule terminated at Phe^{151} and full length IFN-αA on MDBK cells. Zoon *et al.*[22] reported that a fragment of IFN-αA produced by thermolysin digestion containing the amino-terminal 110 amino acids also exhibited about 1/100th of the activity of the intact molecule on MDBK cells. Recently, Nisbet *et al.*[23] made a glycine substitution for tyrosine 136 in IFN-α1 that showed significant loss of antiviral activity on bovine cells. It was reported by Levy *et al.*[3,24] that two natural species of leukocyte interferon closely corresponding to IFN-αA in sequence terminated at position 155, 10 amino acids shorter than IFN-αA. These species were fully active interferon molecules. In addition, it has been reported that the carboxy-terminal 10–13 amino acids can be eliminated without loss of IFN-αA activity.[25,26] This information, taken together with the results presented here, suggests that structural features contributed by the carboxy-terminal region (i.e., amino acids 151–154) of IFN-αA may be required for maximal antiviral activity on MDBK cells.

Acknowledgments

We would like to thank Drs. Richard Kramer, Robert Crowl, Kenneth Collier, and Pennina Langer-Safer for helpful discussions and Mr. Dale Mueller for large-scale fermentations of *E. coli* cells. Many thanks to Ms. Sharon Smith for excellence in preparing the manuscript.

[21] Y. Morinaga, T. Franceschini, S. Inouye, and M. Inouye, *Bio/Technology* **July,** 636 (1984).

[22] S. A. Ackerman, D. Z. Nedden, M. Heintzelman, M. Hunkapiller, and K. Zoon, *Proc. Natl. Acad. Sci. U.S.A.* **81,** 1045 (1984).

[23] I. T. Nisbet, M. W. Beilharz, P. J. Hertzog, M. J. Tynus, and A. W. Linnane, *Biochem. Int.* **11,** 301 (1985).

[24] W. P. Levy, M. Rubinstein, J. Shively, U. Del Valle, C.-Y. Lai, J. Moschera, L. Brink, L. Gerber, S. Stein, and S. Pestka, *Proc. Natl. Acad. Sci. U.S.A.* **78,** 6186 (1981).

[25] A. E. Franke, H. M. Shepard, C. M. Houck, D. W. Leung, D. V. Goeddel, and R. M. Lawn, *DNA* **1,** 223 (1982).

[26] N. Chang, H.-F. Kung, and S. Pestka, *Arch. Biochem. Biophys.* **221,** 585 (1983).

[59] Yeast Vectors for Production of Interferon

By MICHAEL D. SCHABER, THOMAS M. DECHIARA, and RICHARD A. KRAMER

The yeast *Saccharomyces cerevisae* has been used for hundreds of years as an industrial microorganism, having proven its utility in both brewing and baking. In addition, yeast has been utilized in the laboratory for many decades as a model system for the study of many cellular processes. Since the development of molecular cloning techniques in microorganisms, yeast has become increasingly popular for the study of heterologous gene expression. As a eukaryote, yeast cells potentially possess much of the cellular machinery necessary for the expression of foreign eukaryotic genes and for the proper modification of the gene product.

Yeast vectors for foreign gene expression contain certain requisite features. The elements of the vector are generally derived from yeast and bacteria to permit propagation of the plasmid in both. The bacterial elements include a selectable marker (antibiotic resistance) and an origin of replication. These features enable the plasmid to be grown and amplified in bacteria (usually *Escherichia coli*) so that DNA manipulations can be made prior to introduction into yeast. The yeast elements include an origin of replication, a selectable marker (usually a gene which complements an auxotrophic mutation in the host), a yeast promoter for the initiation of transcription, and a transcription terminator. In this chapter, we describe the use of two different yeast promoters to direct the synthesis of IFN-α and IFN-γ.

Yeast Strains, Media, and General Procedures

Yeast Strains and Media. W301-18A (αade2-1 leu2-3,112 trp1-1 can1-100 ura3-1 his 3-11,15) was obtained from Rothstein.[1] Yeast growth medium was UMD[2] with adenine and uracil added as described below.

UMD Medium. Each liter of phosphate-free medium contained 20 g dextrose, 2 g asparagine, 0.5 g $MgSO_4 \cdot 7H_2O$, 0.32 g $CaCl_2 \cdot 2H_2O$, 2 g $(NH_4)_2SO_4$, 0.1 mg KI, 200 mg adenine, 20 mg uracil, 35 mg histidine, 35 mg arginine, 52.5 mg tyrosine, 52.5 mg lysine, 105 mg leucine, 87.5 mg phenylalanine, 350 mg threonine, 1.5 g KCl, 1 ml of mineral mix, and 2 ml

[1] R. Rothstein, this series, Vol. 101, p. 202.

[2] K. A. Bostian, J. M. Lemire, and H. O. Halvorson, *Proc. Natl. Acad. Sci. U.S.A.* **77**, 4504 (1980).

of vitamin mix. For high phosphate growth medium, 1.5 g/liter KH_2PO_4 was substituted for the KCl.

Mineral Mix. Per 100 ml: 5.7 mg boric acid, 3.1 mg $MnSO_4 \cdot H_2O$, 30.8 mg $ZnSO_4 \cdot 7H_2O$, 2.5 mg $CuSO_4$, 1.8 mg $(NH_4)_6Mo_7O_{24} \cdot 4H_2O$, 17.8 mg $FeCl_2 \cdot 4H_2O$.

Vitamin Mix. Per 100 ml: 12.5 mg thiamine HCl, 12 mg pyridoxine HCl, 1 mg biotin, 10 mg niacin, 11 mg panthothenic acid, 5 mg riboflavin, 50 mg inositine.

Growth and Induction of Transformants. Yeast transformations were performed as described by Hinnen, Hicks, and Fink.[3] Transformed cells were grown in high phosphate selective medium to a cell density of approximately 1.2×10^7 cells/ml ($OD_{600} \simeq 2$). The cells were then pelleted, washed with sterile water, and resuspended in 10 ml of either phosphate-free medium or high phosphate medium at an OD_{600} of 0.5. During incubation at 30°, aliquots were taken to measure acid phosphatase activity[4] and cell density.

Cell Lysis and Interferon Assays. Extracts were prepared from either whole cells or spheroplasts. To prepare spheroplasts, the cell pellets were resuspended in 1 ml of 1.2 *M* sorbitol, 10 m*M* sodium phosphate, pH 7.5, 1 m*M* DTT, at an OD_{600} of approximately 10. Zymolyase-60,000 (Kirin) was then added to a concentration of 25 μg/ml and the cells incubated at 30° with gentle agitation. The spheroplasts were then pelleted by centrifugation at 1000 *g* for 5 min at room temperature.

For preparation of cell extracts,[5] either whole cells or spheroplasts were harvested by centrifugation and resuspended in 1 ml of 7 *M* guanidine · HCl/1 m*M* dithiothreitol/1 m*M* phenylmethylsulfonyl fluoride. An equal volume of acid-washed glass beads was added, and the samples were mixed on a Vortex mixer for four 30-sec bursts. After 30 min on ice, cell debris and glass beads were removed by centrifugation, and the samples were stored at −20°. For interferon assay, the samples were then diluted 1 : 100 into 0.15 *M* NaCl/20 m*M* $NaPO_4$, pH 7.9. Interferon activity was assayed by the reduction of cytopathic effect of vesicular stomatitis virus on MDBK cells.[6]

In Vivo Labeling of Yeast Proteins. Total cell protein was labeled by adding 15 μCi [^{35}S]methionine (Amersham, >600 Ci/mmol) to a 10 ml culture of cells grown for 4 hr in either high phosphate or phosphate free-medium as described above. Following a 2 min pulse, the cells were

[3] A. Hinnen, J. B. Hicks, and G. R. Fink, *Proc. Natl. Acad. Sci. U.S.A.* **75,** 1929 (1978).

[4] N. Anderson, G. P. Thill, and R. A. Kramer, *Mol. Cell. Biol.* **3,** 562 (1983).

[5] R. A. Kramer, T. M. DeChiara, M. D. Schaber, and S. Hilliker, *Proc. Natl. Acad. Sci. U.S.A.* **81,** 367 (1984).

[6] P. C. Familletti, S. Rubinstein, and S. Pestka, this series, Vol. 78, p. 387.

immediately transferred to an ice bath, then pelleted and washed with ice cold water. At this point, the pellets were either frozen and stored at −20° for later analysis or processed for protein extraction.

Polyacrylamide Gel Electrophoresis/Immunoblot Analysis. Spheroplasts or whole cells were lysed by resuspending the pellets in an equal volume of 2× sample buffer[7] and incubating at 95° for 10 min. Debris was pelleted by centrifugation for 5 min in an Eppendorf microfuge, and the cleared lysates were applied to duplicate 10–20% gradient polyacrylamide gels.[8] One gel was stained, dried and exposed to X-ray film, while the other was used for western blot analysis.[9,10] Proteins were electroblotted onto a 0.1-μm nitrocellulose membrane (Schleicher and Schuell) for 16 hr at 50 V, 0.2 A, in 12.5 m*M* Tris, 96 m*M* glycine, 20% methanol, 0.01% SDS, pH 7.5. Processing of the blot was carried out by the standard procedure.[9,10] The primary reaction was with a mouse monoclonal antibody against either IFN-αA or IFN-γ, followed by goat anti-mouse IgG, and then swine anti-goat IgG antibody–peroxidase conjugate (Boehringer/Mannheim). The immunoreactive bands were visualized by reaction with 4-chloro-1-naphthol (Polysciences, Inc.).

Construction of Yeast Expression Vectors

For the constitutive expression of IFN, we used the promoter from the glyceraldehyde-3-phosphate dehydrogenase gene (GAPDH) from the plasmid pp6y described by Musti *et al.*[11] For regulated expression, we used the promoter and regulatory region from the repressible acid phosphatase (APase) gene, *PH05*. Details on the engineering of the *PH05* promoter are as described.[5] The GAPDH promoter was engineered in a similar fashion to place an *Eco*RI restriction site just downstream of the transcription initiation site. Figure 1A shows the general vector in which these promoters were used. Figure 1B and C show the promoter/IFN junctions of the three plasmids. pYE7 is the yeast expression vector with the *PHO5* promoter, and pYE8 is the vector with the GAPDH promoter. These are called direct expression vectors since the normal translation initiation codon has been deleted and must be provided by the IFN gene.

In addition to direct expression of IFN genes in pYE7 and pYE8, the expression of the two types of *PH05*/IFN fusions illustrated in Fig. 1B

[7] U. K. Laemmli, *Nature (London)* **227,** 680 (1970).

[8] F. W. Studier, *J. Mol. Biol.* 237 (1973).

[9] H. Towbin, T. Staehelin, and J. Gordon, *Proc. Natl. Acad. Sci. U.S.A.* **76,** 4350 (1979).

[10] W. N. Burnette, *Anal. Biochem.* **112,** 195 (1981).

[11] A. M. Musti, Z. Zehner, K. A. Bostian, B. M. Paterson, and R. A. Kramer, *Gene* **25,** 133 (1983).

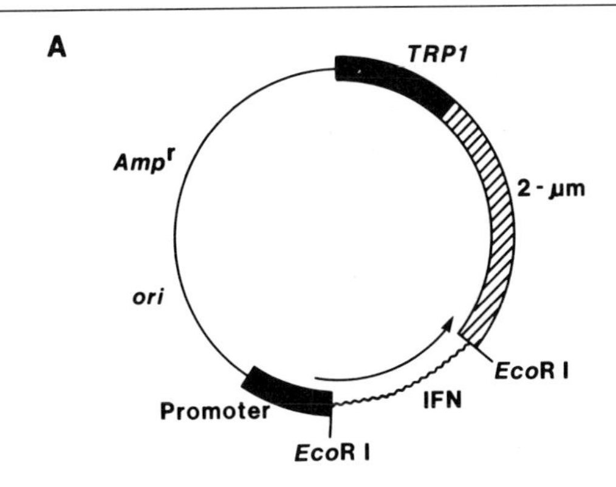

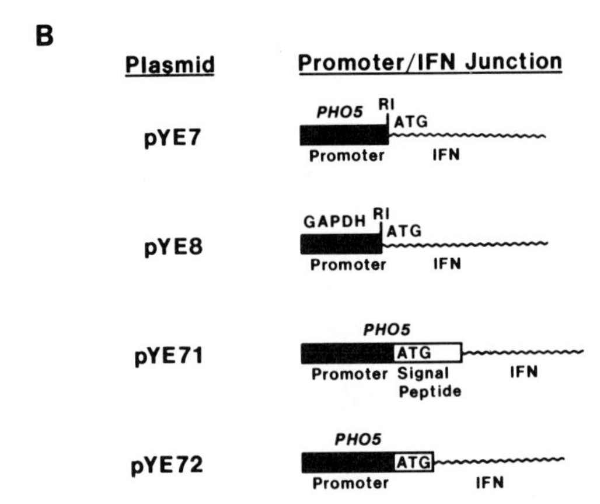

FIG. 1. Vectors for foreign gene expression in yeast. (A) The filled bars represent yeast chromosomal DNA, the hatched bar indicates the yeast 2 μm plasmid DNA, the straight line is DNA from the *E. coli* plasmid pBR322,[14] and the wavy line is the IFN gene. The arrow shows the direction and approximate extent of transcription. The yeast selectable marker is the *TRP1* gene on a 1.4 kb *Eco*RI fragment.[15] It also carries a yeast replication origin. The yeast 2 μm plasmid DNA is the 2.2 kb *Eco*RI fragment from the B form.[16] This fragment also has an origin of replication and is important for plasmid stability in yeast. In addition, it has a transcription termination site. The pBR322 segment is an *Eco*RI/*Pvu*II fragment[14] that has the bacterial origin of replication (ori) and the gene for ampicillin resistance (amp^r). The IFN genes used are those for IFN-αA[12] and IFN-γ.[13] (B and C) All plasmids have the same general configuration shown in (A). The differences are in the promoter used and the junction between the promoter and the IFN coding sequence. pYE7 and pYE8 have an *Eco*RI site engineered just downstream of the transcription start site so that the translation initiation codon is supplied by the IFN gene. This results in the direct expression of the interferon, i.e., with only the NH_2-terminal methionine added to the normal mature interferon protein. Alternatively, pYE71 and pYE72 contain the indicated NH_2-terminal coding regions from *PHO5*[5] in addition to the promoter and result in a fusion protein.

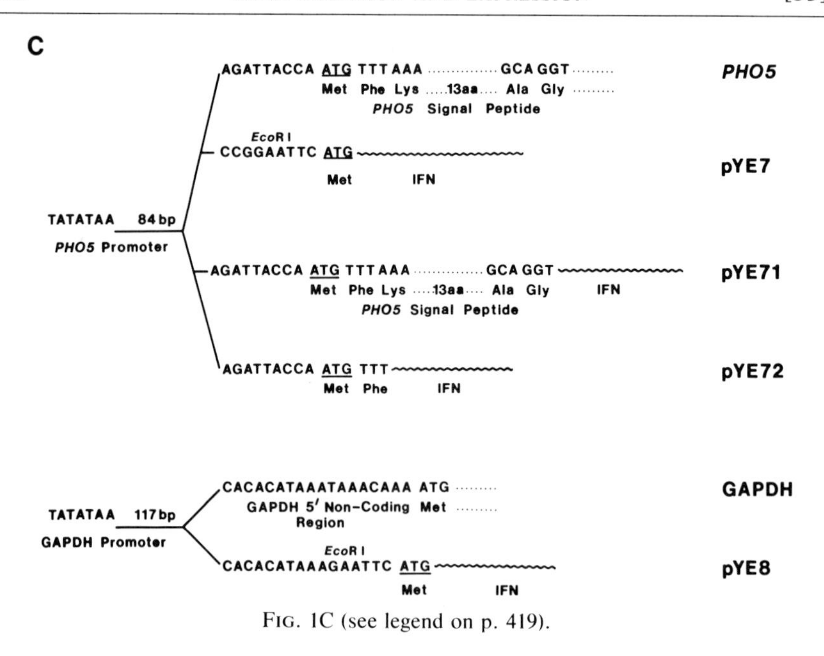

FIG. 1C (see legend on p. 419).

and C was examined. In one case, the coding region for the signal peptide and several NH_2-terminal amino acids of APase was fused to the entire coding region for IFN-αA or IFN-γ. In the other, only the first two codons from the APase signal peptide were fused with the IFN genes as well as several NH_2-terminal acids of mature APase. The vectors with these fusions are referred to as pYE71 and pYE72, respectively.

Synthesis of Interferon in Yeast

Expression of IFN genes was examined by inserting the cloned gene for IFN-αA[12] or IFN-γ[13] into the vectors as depicted in Fig. 1 and per-

[12] D. V. Goeddel, E. Yelverton, A. Ullrich, H. L. Heyneker, G. Miozzari, W. Holmes, P. H. Seeburg, T. Dull, L. May, N. Stebbing, R. Crea, S. Maeda, R. McCandliss, A. Sloma, J. M. Tabor, M. Gross, P. C. Familletti, and S. Pestka, *Nature (London)* **287,** 411 (1980).

[13] R. Crowl, this volume [55].

[14] F. Bolivar, R. L. Rodriguez, P. J. Greene, M. C. Betlach, H. L. Heyneker, H. W. Boyer, J. H. Crosa, and S. Falkow, *Gene* **2,** 95 (1977).

[15] K. Struhl, D. T. Stinchcomb, S. Scherer, and R. W. Davis, *Proc. Natl. Acad. Sci. U.S.A.* **76,** 1035 (1979).

[16] J. R. Broach, *in* "The Molecular Biology of the Yeast Saccharomyces: Life Cycle and Inheritance" (J. N. Strathern, E. W. Jones, and J. R. Broach, eds.), p. 445. Cold Spring Harbor Lab., Cold Spring Harbor, New York, 1981.

forming bioassays for interferon activity on cell extracts (and media). We have previously shown[5] that a vector similar to pYE7 and IFN-αD yielded about 1×10^7 units/liter per OD_{600} unit when induced in phosphate free medium. This represented an induction of about 100- to 200-fold. The pYE7/αA vector described here yielded about $2–5 \times 10^7$ units/liter per OD_{600} unit when induced. Yields for pYE8/αA, pYE71/αA, and pYE72/αA were consistently about 5–10 times higher than for pYE7/αA. The pYE8/αA yields were independent of the phosphate concentration. Similar relative yields were obtained for IFN-γ in these vectors. As illustrated in Fig. 2, this difference in yield corresponds to an actual difference in the amount of IFN protein. The reason for this discrepancy may be in the structure of the DNA (and, therefore, of the mRNA) immediately preceding the initiation codon of the interferon gene). The vector pYE7 contains a cluster of C and G residues just upstream of the translation initiation of the IFN mRNA (Fig. 1C). The same region of pYE71 and pYE72 contains the natural sequence from the highly expressed *PHO5* gene (Fig. 1C). It is possible that the difference in IFN synthesis is due to a decrease in translation initiation efficiency caused by the -C-C-G-G-G- stretch in the 5′-untranslated region of the *PHO5*/IFN mRNA.

Figure 2 illustrates the analysis of yeast cell lysates for the presence of IFN-αA. We observe protein bands which correspond to the expected size for both the direct expression and fusion products of IFN-αA. In addition, we also observe several lower molecular weight proteins which react with IFN-αA antibody. These smaller proteins may correspond to degradation products which result from proteolytic cleavage of the IFN-αA molecule.

Concluding Comments

Since APase is a secreted enzyme in yeast, the pYE71 vectors with IFN-αA and IFN-γ might be expected to direct secretion of interferon, but little or no activity was detected in media from induced cells carrying these plasmids. However, we do have evidence that the presence of the signal peptide does effect the cellular location of the interferon. When spheroplasts of induced cells with pYE7/αA are lysed by either a gentle [resuspension in phosphate-buffered saline (PBS)] or a harsh (7 *M* guanidine · HCl) method, approximately equal yields of activity were obtained. On the other hand, when spheroplasts of cells with pYE71/αA were extracted with PBS, less than 2% of the yield with guanidine · HCl was obtained. Guanidine · HCl extraction of the cell debris pellet from the PBS lysate of pYE71/αA released the expected amount of activity. The presence of the APase signal peptide may be responsible for this disparity. As

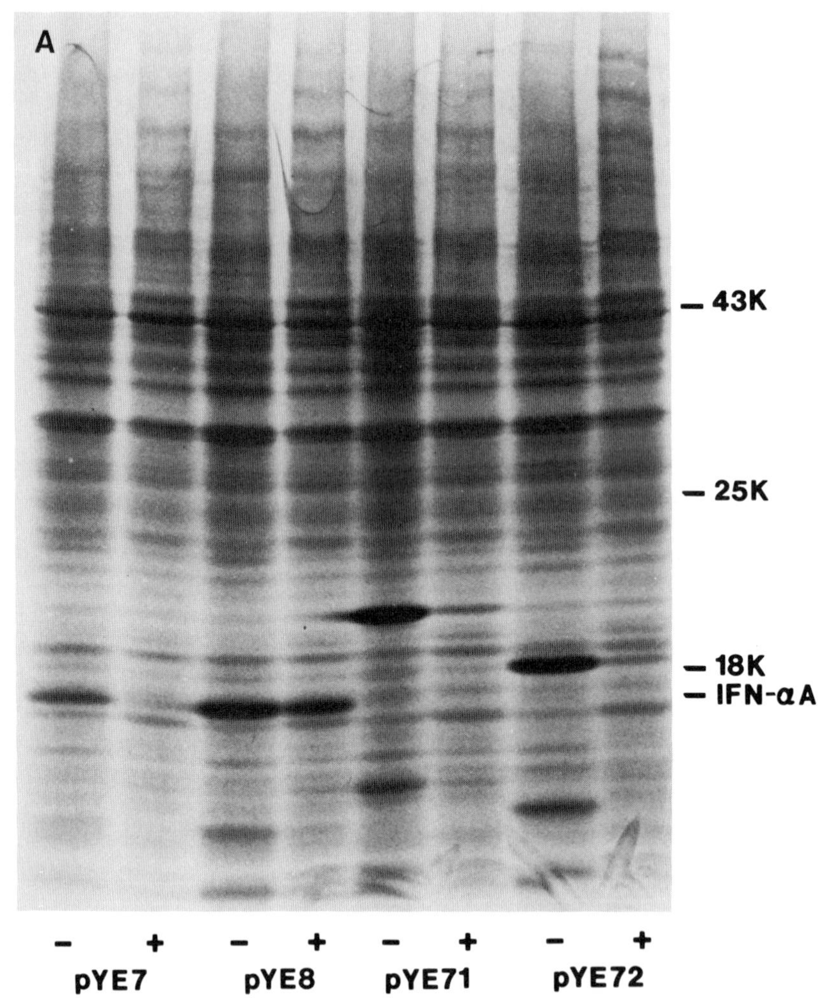

FIG. 2. Analysis of IFN-αA produced in yeast. Yeast cells containing the IFN-αA expression vectors were grown in either phosphate free (−) or high phosphate (+) medium, and lysates were prepared as described in the text. Total [^{35}S]methionine labeled yeast protein from 2 OD_{600} units of cells was applied to each well of duplicate 10–20% gradient gels. (A) An autoradiograph of the dried gel shows the presence of IFN-αA or the expected fusion products in the cell lysates. The position of purified IFN-αA is shown. Molecular weight markers are ovalbumen (43K), α-chymotrypsinogen (25K) and β-lactoglobulin (18K). (B) Immunoblot analysis of yeast cell lysates. The immunoblot was developed as described in the text. The bands we observe in this blot correspond to the same bands identified as IFN-αA related in the autoradiograph from A.

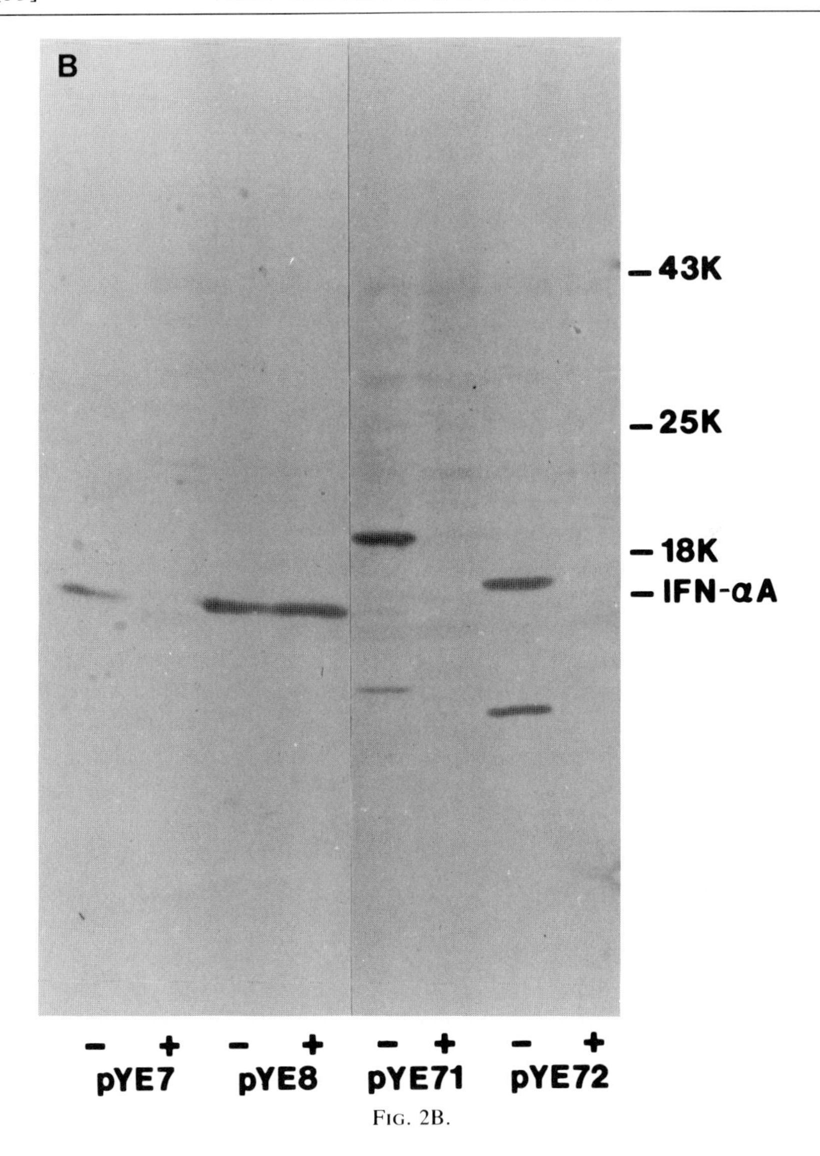

FIG. 2B.

shown by the size of the interferon protein from pYE71/αA in Fig. 2, the signal peptide is apparently not efficiently cleaved. The need for a more disruptive extraction procedure with pYE71 suggests that the protein is either becoming associated with membranes (possibly in the secretory pathway) or is forming globular aggregates inside the cell. In either case, guanidine · HCl is required to solubilize this fusion protein.

[60] Construction of Expression Vectors for Secretion of Human Interferons by Yeast

By RONALD A. HITZEMAN, CHUNG NAN CHANG, MARK MATTEUCCI, L. JEANNE PERRY, WILLIAM J. KOHR, JOHN J. WULF, JAMES R. SWARTZ, CHRISTINA Y. CHEN, and ARJUN SINGH

With the advent of heterologous (non-yeast) gene expression in yeast[1] with yeast plasmids and the 5′- and 3′-flanking DNA control regions of highly expressed yeast genes, it is now possible to use heterologous or homologous amino-terminal signal sequences to produce and secrete heterologous proteins from yeast. Using such expression systems we have previously demonstrated that signal sequences of two human leukocyte interferons (IFN-αA and IFN-αD) and human gamma interferon (IFN-γ) direct the secretion of these protein products into the periplasm and growth medium of yeast.[2] Recently we have used two homologous signals for secretion of the same two leukocyte interferons into the growth medium to obtain the desired processing.[3,4] Such systems may be used for production and secretion of other naturally secreted heterologous protein products with extremely high initial purity and may be useful for the secretion of some homologous and heterologous proteins which are not normally secreted for protein studies.

Materials and Methodology

Materials

Restriction endonucleases, T_4-DNA ligase, AMV reverse transcriptase, and polynucleotide kinase were purchased from BRL or New England Biolabs and used as recommended by the manufacturers. *E. coli* DNA polymerase I (large fragment) was from Boehringer.

[1] R. A. Hitzeman, F. E. Hagie, H. L. Levine, D. V. Goeddel, G. Ammerer, and B. D. Hall, *Nature (London)* **293,** 717 (1981).

[2] R. A. Hitzeman, D. W. Leung, L. J. Perry, W. J. Kohr, H. L. Levine, and D. V. Goeddel, *Science* **219,** 620 (1983).

[3] A. Singh, J. M. Lugovoy, W. J. Kohr, and L. J. Perry, *Proc. Int. Congr. Genet., 15th, 1983* p. 33 (1984).

[4] A. Singh, C. N. Chang, J. M. Lugovoy, M. D. Matteucci, and R. A. Hitzeman, *Proc. Int. Congr. Genet., 15th, 1983* p. 169 (1984).

The synthetic oligodeoxynucleotides were prepared by the phosphotriester method.[5] The 24-mer oligodeoxynucleotide 5′-AGGGAGATCACATCTTTTATCCAA-3′ was used for site-directed mutagenesis to modify the α-factor signal-Met-IFN-αD DNA to obtain properly processed IFN-αD in yeast growth medium. The 15-mer oligonucleotide 5′-TGCCAGGAGCATCAA-3′ was used as a sequencing primer to verify the sequence of this construction. The following two other synthetic oligodeoxynucleotides were used to construct the invertase signal peptide coding sequence:

5′-CCGAATTCATGATGCTTTTGCAAGCTTTCCTTTTCCTTTT-3′
3′-GGAAAAGGAAAACCGACC-
AAAACGTCGGTTTTATAGACGT-5′

Design of Oligodeoxynucleotides Encoding the Invertase Signal Peptide

Two 40-mer oligodeoxynucleotides (see sequence in Materials) were designed based on the known nucleotide sequence to construct the yeast invertase signal.[6,7] A convenient *Eco*RI restriction site was placed in front of the ATGATG with two additional nucleotides (CC) at the 5′-end (see above). These two oligonucleotides have 12 base pairs complementary to each other at their 3′-ends. After repairing these oligonucleotides with the reverse transcriptase, the 3′-end of newly formed double-strand DNA becomes a blunt end which was ligated to a blunt end of a mature methionine–IFN-αA (Met–IFN-αA) or IFN-αA (without the Met) maintaining the proper coding frame. The 5′-end of the signal sequence becomes a sticky end after *Eco*RI digestion, facilitating its ligation to the similar sticky end of an appropriate promoter fragment.

In Vitro Mutagenesis to Make the α-Factor Signal-pro-IFN-αD Modification

A variation of previously described protocols[8] was used for oligonucleotide-directed deletion mutagenesis. The single-stranded DNA template was prepared from the recombinant M13mp8 phage containing the appropriate insert. This template was annealed with the phosphorylated synthetic oligonucleotide of 24 bases in length. This primer-template intermediate was subjected to extension and ligation reaction at 23° for 2 hr

[5] R. Crea and T. Horn, *Nucleic Acids Res.* **8,** 2331 (1980).
[6] D. Perlman and H. O. Halvorson, *Cell* **25,** 525 (1981).
[7] M. Carlson and D. Botstein, *Cell* **28,** 145 (1982).
[8] M. J. Zoller and M. Smith, this series, Vol. 101, p. 468.

in the presence of 500 mM dATP, 100 mM dTTP, 100 mM dGTP, 100 mM dCTP, 20 mM dATP, 3 units DNA polymerase (Klenow), and 400 units T_4-DNA ligase in 10 mM Tris · HCl, pH 7.4, 50 mM NaCl, and 10 mM $MgSO_4$. Then an additional 3 units of DNA polymerase (Klenow) and 400 units of T_4-DNA ligase was added and mixture incubated for 2 hr at 23° followed by incubation at 14° for 15 hr. Aliquots of this mixture were used to transform *E. coli* JM101.

Screening of Phage Plaques

The ^{32}P-labeled[9] 24-mer oligonucleotide described above was used to screen recombinant M13 phages by *in situ* plaque hybridization.[10] Filters were hybridized overnight at 42° in 10 mM Tris · HCl (pH 7.5), 6 mM EDTA, 0.8 M NaCl, 1× Denhardt's solution, 0.5% NP-40, and 0.1 mg/ml *E. coli* tRNA. Filters were washed 3 times for 20 min in 6× SSC, 0.1% SDS at 30°. Dried filters were exposed to Kodak XR-2 X-ray film with DuPont Lightning-Plus intensifying screen at −80°.

DNA Sequence Determination

DNA sequence analysis was carried out by the chain termination method[11] with recombinant phage M13mp8 as a source of single-stranded template DNA as described.[12] A synthetic pentadecanucleotide (see Materials) complementary to the IFN-αD coding strand near the region of interest was used for priming *E. coli* DNA polymerase I (large fragment) in the presence of dideoxynucleotide triphosphates and [α-^{32}P]dCTP for labeling synthesized chains.

Strains and Growth Conditions

Yeast strain 20B-12 (α *trp1 pep4*)[13] was obtained from the Berkeley Yeast Genetics Stock Center. *E. coli* K-12 strain 294 (*endA thi*$^-$ *hsr*$^-$ *hsm*$^+$)[14] was used for bacterial plasmid transformation. *E. coli* strain JM101[12] was used for experiments involving bacteriophage M13.

LB medium for *E. coli* growth was as described by Miller[15] with the addition of 20 μg/ml ampicillin (Sigma) for selection of plasmid transfor-

[9] A. M. Maxam and W. Gilbert, this series, Vol. 65, p. 499.
[10] W. D. Benton and R. W. Davis, *Science* **196,** 180 (1977).
[11] F. Sanger, S. Nicklen, and A. R. Coulson, *Proc. Natl. Acad. Sci. U.S.A.* **74,** 5463 (1977).
[12] J. Messing, this series, Vol. 101, p. 20.
[13] E. Jones, *Genetics* **85,** 23 (1977).
[14] K. Backman, M. Ptashne, and W. Gilbert, *Proc. Natl. Acad. Sci. U.S.A.* **73,** 4174 (1976).
[15] J. H. Miller, "Experiments in Molecular Genetics," p. 433. Cold Spring Harbor Lab., Cold Spring Harbor, New York, 1972.

mants. *E. coli* JM101 was grown on 2YT medium.[15] Yeast were grown on the following media: YEPD medium (nonselective) contained 1% yeast extract, 2% peptone, and 2% glucose with or without 3% Difco agar. YNB + CAA medium (used for Trp^+ selection) contained per liter 6.7 g of Difco yeast nitrogen base (without amino acids) (YNB), 10 mg of adenine, 10 mg of uracil, 5 g Difco casamino acids (CAA), and 20 g glucose. For solid medium 30 g of agar per liter was used. The transformed yeast were always grown in media for selective maintenance of the plasmid. In the fermentor the cells were grown in YNB-like medium with glucose addition to obtain optimal high density growth. NH_4OH was added as required for pH control.

Transformations

E. coli 294 was transformed as described.[16] *E. coli* JM101 cells were transformed as described by Messing.[12] Yeast were transformed essentially as previously described.[17,18]

DNA Preparations

Plasmid DNAs were prepared from *E. coli* by the cleared lysate method[19] and were purified by Bio-Rad Agarose A-50 column chromatography. A quick-screening procedure[20] was used to obtain small amounts of plasmid DNAs from individual *E. coli* transformants. DNA restriction fragments were isolated by electroelution from a 1% agarose gel or a 6% polyacrylamide gel followed by phenol/chloroform extraction and ethanol precipitation.

Interferon Assay

The antiviral activity of yeast extracts was determined by cytopathic effect (CPE) inhibition assays in 96-well microtiter trays with MDBK cells and vesicular stomatitis virus as described elsewhere.[21] Values determined by this method are expressed in units relative to the NIH leukocyte interferon standard G-023-901-527. Yeast cultures were grown to the indicated densities and media were assayed as previously described.[1] The specific activity was 10^8 units/mg of protein for both IFN-αA and IFN-αD.

[16] M. Mandel and A. Higa, *J. Mol. Biol.* **53,** 159 (1970).
[17] A. Hinnen, J. B. Hicks, and G. R. Fink, *Proc. Natl. Acad. Sci. U.S.A.* **75,** 1929 (1978).
[18] J. D. Beggs, *Nature* (*London*) **275,** 104 (1978).
[19] D. B. Clewell and D. R. Helinski, *Biochemistry* **9,** 4428 (1970).
[20] H. C. Birnboim and J. Doly, *Nucleic Acids Res.* **7,** 1513 (1979).
[21] P. Familletti, S. Rubinstein, and S. Pestka, this series, Vol. 78, p. 387.

Interferon Purification

Interferon was purified from culture medium from which cells had been removed by centrifugation. Frozen media were concentrated and dialyzed against 25 m*M* Tris · HCl, 10 m*M* EDTA, pH 8.0, in a 2.5-liter Amicon stirred cell (Amicon 2000) with a YM-5 ultrafiltration membrane. Some of the interferons that were of lower concentration in the media were also subjected to antibody affinity chromatography as previously described.[2,22] One milliliter of the concentrated medium was precipitated with 4 ml acetone, spun in a microfuge, and washed with acetone. The pellet was resuspended in 0.1% (v/v) trifluoroacetic acid (TFA) and further purified by HPLC on a Synchropak RP-P column. The column was eluted with a linear gradient of 0 to 100% acetonitrile in 0.1% TFA in 60 min. A 12 μg sample of purified IFN-αD was chromatographed as a control. The peaks of absorbance at 280 nm were sequenced.

Determination of NH_2-Terminal Amino Acid Sequences

Sequence analysis was based on the Edman degradation.[23] Liquid samples were introduced into the cup of a modified Beckman 8908 spinning cup sequencer. Polybrene was used as a carrier in the cup. Reagents used were Beckman's sequence grade 0.1 molar Quadrol buffer, phenylisothiocyanate, and heptafluorobutyric acid. Norleucine was added during each cycle with the Quadrol buffer to serve as an internal standard. The presence of PTH-norleucine in each chromatogram aided in the identification of PTH amino acids present in each cycle.

Relevance and Usefulness of Expression/Secretion Plasmids

Expression Plasmids Used to Produce Secreted Proteins

YEp1PT (Fig. 1) contains the 2μ origin of replication and yeast *TRP1* gene for its replication and selective maintenance.[2] It has yeast transcription initiation and termination signals flanking the unique *Eco*RI site. Transcription initiation signals are from the 5′-flanking sequence of the yeast 3-phosphoglycerate kinase (PGK) gene[24] and the termination signals are from the 3′-flanking sequence of the yeast 2μ *FLIP* gene.[25]

[22] T. Staehelin, D. S. Hobbs, H.-F. Kung, and S. Pestka, this series, Vol. 78, p. 505.

[23] P. Edman and G. Begg, *Eur. J. Biochem.* **1,** 80 (1967).

[24] R. A. Hitzeman, F. E. Hagie, J. S. Hayflick, C. Y. Chen, P. H. Seeburg, and R. Derynck, *Nucleic Acids Res.* **10,** 7791 (1982).

[25] J. L. Hartley and J. E. Donelson, *Nature (London)* **286,** 860 (1980).

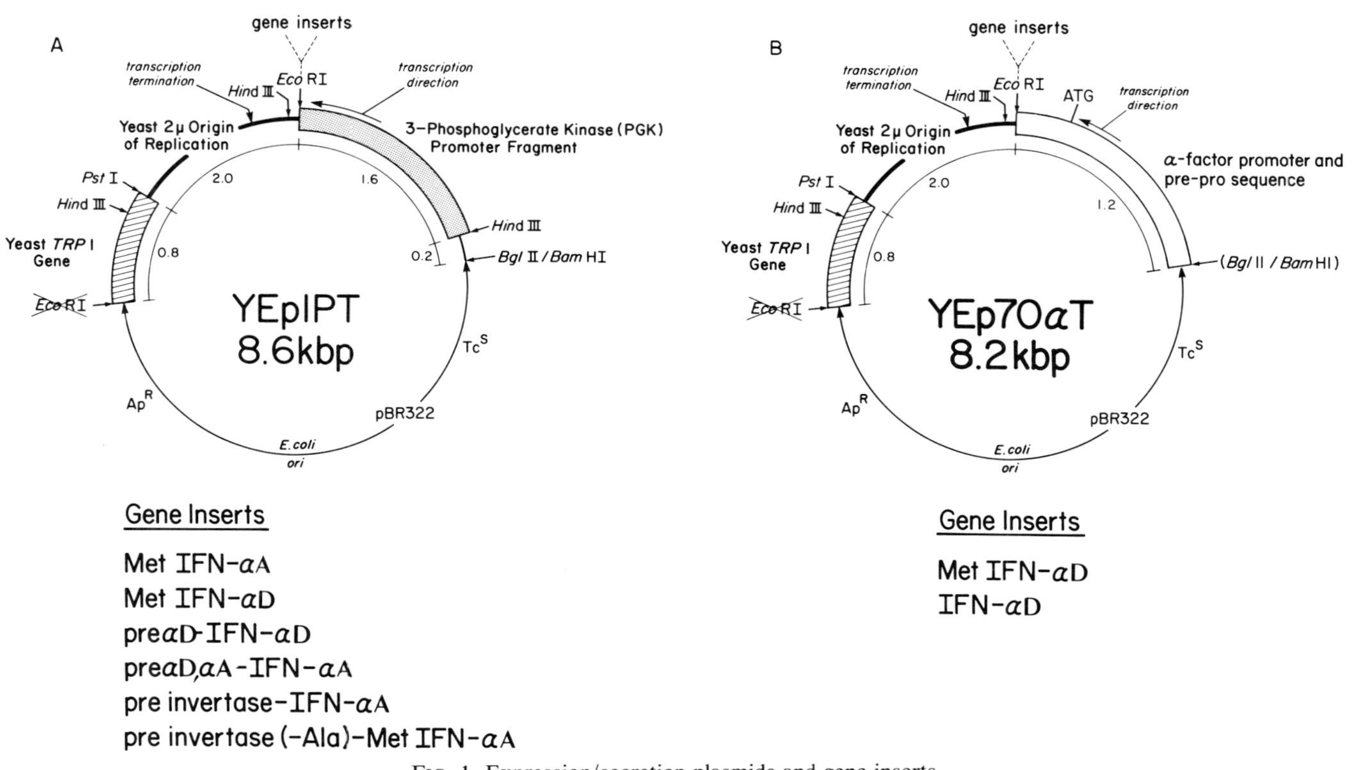

FIG. 1. Expression/secretion plasmids and gene inserts.

Separate expression plasmids were made with each of the gene inserts shown in Fig. 1. About 1% of the total cell protein is produced intracellularly as interferon when expressing Met–IFN-αA or Met–IFN-αD,[2] the mature forms of these leukocyte interferons.[2,26] When these interferons are expressed in the same plasmid with their natural heterologous secretion signals (pre αD-IFN-αD) or with a hybrid signal (pre αD, αA-IFN-αA), only about one-fifth as much interferon is made, but a portion (10–50%) of this is secreted and released into the medium (free of the cell).[2]

When a homologous (yeast) secretion signal from the yeast invertase gene is attached to IFN-αA (the last two gene inserts shown in Fig. 1A), secretion into the yeast medium also occurs.[4] All these constructions are shown in Fig. 2 and their expression levels in fermentors are shown in the table. All produce the desired processed forms in the media as shown in Fig. 2. However, a fraction of the interferon secreted by the transformants containing the interferon signals (constructions 1 and 2) is incorrectly processed.[2]

The plasmid YEp70αT (Fig. 1B) contains the yeast α-factor promoter sequences coding for α-factor prepro (89 amino acids)[27,28] and the sequence for three amino acids (underlined for construction 6 in Fig. 2) due to the introduction of *Xba*I and *Eco*RI sites. The table shows expression and secretion of interferon by yeast transformants that contain this expression system in the plasmid p60.[3,29] YEp70αT has been derived from the plasmid p60 so that any gene can now be inserted into the unique *Eco*RI site to construct appropriate vectors for secretion. We have made shake-flask comparisons of interferon expression and secretion levels from YEp70αT containing Met–IFN-αD gene and from plasmid p60. Both expression plasmids were found to be equivalent. Upon concentration, dialysis, and NH_2-terminal protein sequencing of the secreted form in the medium, processing is observed[3,29] after the dibasic residues (Lys-Arg) as shown in Fig. 2. Protein purification was not required for the protein sequence analysis since about 80 percent of the medium protein is this form of IFN-αD.

In order to get properly processed IFN-αD with the α-factor expression/secretion system, the coding sequence for 8 amino acids (shown with bar in Fig. 2) in construction 6 were removed by the M13 mutagenesis procedure described in Methods. This α-factor-interferon fusion protein

[26] S. Pestka, *Arch. Biochem. Biophys.* **221,** 1 (1983).

[27] J. Kurjan and I. Herskowitz, *Cell* **30,** 933 (1982).

[28] A. Singh, E. Y. Chen, J. M. Lugovoy, C. N. Chang, R. A. Hitzeman, and P. H. Seeburg, *Nucleic Acids Res.* **12,** 4049 (1983).

[29] A. Singh, J. M. Lugovoy, W. J. Kohr, and L. J. Perry, *Nucleic Acids Res.* **12,** 8924 (1984).

	Secretion Signal Sequences																							Mature Protein Sequences					
	-23	-22	-21	-20	-19	-18	-17	-16	-15	-14	-13	-12	-11	-10	-9	-8	-7	-6	-5	-4	-3	-2	-1	1	2	3	4	5	
1. Pre αD, αA - IFN-αA	Met	Ala	Ser	Pro	Phe	Ala	Leu	Leu	Met	Ala	Leu	Val	Val	Leu	Ser	Cys	Lys	Ser	Ser	Cys	↓Ser 36%	Val	Gly	↓Cys 64%	Asp	Leu	Pro	Gln...	IFN-αA
2. Pre αD - IFN-αD	Met	Ala	Ser	Pro	Phe	Ala	Leu	Leu	Met	↓Ala 8%	Leu	Val	Val	Leu	Ser	Cys	Lys	Ser	Ser	Cys	↓Ser 47%	Leu	Gly	↓Cys 45%	Asp	Leu	Pro	Glu...	IFN-αD
3. Preinvertase				Met	Met	Leu	Leu	Glu	Ala	Phe	Leu	Phe	Leu	Leu	Ala	Gly	Phe	Ala	Ala	Lys	Ile	Ser	Ala	↓Ser 100%	Met	Thr	Asn	Glu...	mature invertase
4. Preinvertase - IFN-αA (E3)				Met	Met	Leu	Leu	Glu	Ala	Phe	Leu	Phe	Leu	Leu	Ala	Gly	Phe	Ala Δ	Ala	Lys	Ile	Ser	Ala	↓Cys 100%	Asp	Leu	Pro	Gln...	IFN-αA
5. Preinvertase (-Ala) - Met IFN-αA (F2)					Met	Met	Leu	Leu	Glu	Ala	Phe	Leu	Phe	Leu	Leu	Ala	Gly	Phe	Ala	Lys	Ile	Ser	Ala	↓Met 100%	Cys	Asp	Leu	Pro...	Met IFN-αA
6. α-factor signal - pro-Met IFN-αD	Met	(76 amino acid residues) pre + pro sequences							Glu	Gly	Val	Ser	Leu	Asp	Lys	Arg	↓Glu (pro cleavage site) 100%	Ala	Glu	Ala	Leu	Glu	Phe	Met	Cys	Asp	Leu	Pro...	Met IFN-αD
7. α-factor signal - pro-IFN-αD				Met		(73 amino acid residues)								Ala	Lys	Glu	Glu	Gly	Val	Ser	Leu	Asp	Lys	Arg	↓Cys 100%	Asp	Leu	Pro...		IFN-αD

FIG. 2. Amino-terminal protein sequences encoded by gene inserts of Fig. 1.

YEAST FERMENTATION RESULTS FOR EXPRESSION/SECRETION PLASMIDS[a]

Protein produced	A_{550}	Dry cell weight DCW (g/liter)	Cell associated ($\times 10^{-6}$ units/liter)	Extracellular ($\times 10^{-6}$ units/liter)	Percentage external	Total specific activity ($\times 10^{-6}$ units/gDCW)	Relative expression levels
1. Pre αD, αA-IFN-αA	75	50	430	53	11	10	2.3
2. Pre αD-IFN-αD	70	40	80	90	53	4.3	1.0
4. Preinvertase-IFN-αA (E3)	65	55	1300	80	6	25	5.8
5. Preinvertase (-Ala)-Met–IFN-αA (F2)	80	55	830	150	15	18	4.2
6. α-factor signal pro-Met–IFN-αD	75	40	1000	2500	71	88	20

[a] The data for No. 1 are the average of five fermentations with standard deviations as follows: cell associated, ±230; extracellular, ±35. The data for No. 2 and No. 6 represent the average of two fermentations. The two fermentations for No. 6 were essentially identical. The data for the remainder represent single fermentation experiments.

(construction 7) is expressed at about the same level as construction 6 in shake flasks and is properly processed.[3,29] Recently, another group[30] used a similar yeast α-factor system to obtain secreted and properly processed human epidermal growth factor with the dibasic cleavage site described above.

Variations in Levels of Interferon Produced

The table shows the levels of interferon produced in several fermentations for the same yeast strain containing the various constructions described in Figs. 1 and 2. In all cases interferon is secreted and released from the periplasm into the medium. The fraction found in the medium varies with the construction and with each fermentation. It is interesting to note that constructions 6 and 7 (which contain α-factor prepro sequences) produce proportionately more interferon in the medium than other constructions. It has some advantage even over the yeast invertase signal sequence used in constructions 4 and 5. It is not understood how this signal and pro sequence better facilitate both secretion and release from the periplasm. In addition to more efficient secretion, the α-factor system also produces much greater levels of IFN-α than the other systems (20-fold higher than construction 2 and 3- to 5-fold higher than constructions 4 and 5). This is particularly interesting since it is independent of mRNA levels. All of the constructions shown in the table produce the same steady-state levels of mRNA ($\pm 50\%$) on Northern blot analysis (data not shown). Therefore this dramatic 20-fold difference in IFN-α levels is due either to translational or posttranslational events.

Concluding Comments

We have previously described the use of the natural human IFN-α precursor sequence (heterologous) to signal the secretion and result in partially correct processing of leukocyte interferons when expressed in yeast. Here we also demonstrate that IFN-α secretion and processing can be obtained with two different homologous yeast signal sequences attached to IFN-α, with one signal producing 80% of yeast medium protein as IFN-αD. The two secretion signals used were obtained from the yeast invertase and α-factor genes.

[30] A. J. Brake, J. P. Merryweather, D. G. Coit, U. A. Heberlein, F. R. Masiarz, G. T. Mullenbach, M. S. Urdea, P. Valenzuela, and P. J. Barr, *Proc. Natl. Acad. Sci. U.S.A.* **81,** 4642 (1984).

[61] Procedures for Isolation of Murine Alpha Interferon Genes and Expression of a Murine Leukocyte Interferon in *E. coli*

By BRUCE L. DAUGHERTY and SIDNEY PESTKA

A murine alpha interferon gene (Mu-IFN-αA) was isolated from a genomic DNA library in phage lambda with a human alpha interferon DNA as a probe. The probe consisted of the 642 bp *Ava*II fragment from the plasmid pLeIFA25[1] (also called pIFLrA or pIFN-αA)[2] which represented 90% of the coding region. In this report, the conditions for lambda phage screening and hybridization are described in which this fragment was successfully used to isolate a genomic mouse alpha interferon gene. In addition, the method by which its expression was achieved in *E. coli* is also described.

Materials and General Procedures

Bacterial Strains, Phage, and Plasmids

A murine genomic library in phage lambda Charon 28 was provided by Philip Leder and John Seidman. *E. coli* strain C600 was used as a host to propagate bacteriophage λ. The plasmids pLeIFA25 (coding for human IFN-αA as described[1]), pBR322,[3] and pBR325[4] were used as cloning vectors. *E. coli* strain RRI was used as a recipient in transformation experiments.

Deoxyoligonucleotides

5′AATTCATGTGTGACCTGCCTCAGACTCATAACC3′
3′   GTACACACTGGACGGAGTCTGAGTATTGGAGT3′

These were synthesized by the phosphoramidite method[5,6] used for construction of the Mu-IFN-αA expression vector.[7]

[1] D. V. Goeddel, E. Yelverton, A. Ullrich, H. L. Heyneker, G. Miozzari, W. Holmes, P. H. Seeburg, T. Dull, L. May, N. Stebbing, R. Crea, S. Maeda, R. McCandliss, A. Sloma, J. M. Tabor, M. Gross, P. C. Familletti, and S. Pestka, *Nature* (*London*) **287,** 411 (1980).

[2] E. Rehberg, B. Kelder, E. G. Hoal, and S. Pestka, *J. Biol. Chem.* **257,** 11947 (1982).

[3] F. Bolivar, *Gene* **4,** 121 (1978).

[4] P. Prentki, F. Karch, S. Iida, and J. Meyer, *Gene* **14,** 289 (1981).

[5] S. L. Beaucage and M. H. Caruthers, *Tetrahedron Lett.* **22,** 1859 (1981).

[6] F. Chow, T. Kempe, and G. Palm, *Nucleic Acids Res.* **9,** 2807 (1981).

[7] B. Daugherty, D. Martin-Zanca, B. Kelder, K. Collier, T. C. Seamans, K. Hotta, and S. Pestka, *J. Interferon Res.* **4,** 635 (1984).

Solutions and Media

20× SSPE: 3 *M* NaCl, 20 m*M* EDTA, 0.2 *M* NaH_2PO_4 adjusted to pH 6.8 with NaOH

Denaturation solution: 0.5 *N* NaOH, 1.5 *M* NaCl

Neutralization solution: 0.5 *M* Tris · HCl, pH 7.5, 1.5 *M* NaCl

Hybridization solution: 40% (v/v) formamide, 0.1% (w/v) bovine serum albumin, 0.1% (w/v) Ficoll, 0.1% (w/v) polyvinylpyrrolidone, 0.1% (w/v) SDS, 5× SSPE, and 100 μg/ml of *E. coli* DNA

Washing solutions:

- I: 2× SSPE, 0.2% SDS
- II: 0.2× SSPE, 0.1% SDS
- III: 0.1× SSPE, 0.1% SDS

M9 media: 50 m*M* Na_2HPO_4, 28 m*M* KH_2PO_4, 8.5 m*M* NaCl, 19 m*M* NH_4Cl, 0.1 m*M* $CaCl_2$, 1 m*M* $MgSO_4$, 0.4% (w/v) glucose, 0.4% (w/v) casamino acids, and 2 μg/ml thiamine

NZYM media (per liter): 10 g N-Z-Amine-A (Humko Sheffield), 5 g NaCl, 5 g yeast extract (Difco), 2 g $MgSO_4 \cdot 7H_2O$

Nitrocellulose filters: BA/85 0.45 μm obtained from Schleicher & Schuell

E. coli DNA solution: *E. coli* DNA is prepared by dissolving in water at a concentration of 2 mg/ml for 16 hr at 4° on a tilting platform. The DNA is then sonicated with a Bronwill Biosonic III sonicator with the medium probe at the highest setting for a total of 8 min at 30 sec sequential off and on intervals in ice water. The *E. coli* DNA is put in a boiling water bath for 10 min and then immediately placed at 0° for 10 min.

Transfection

E. coli C600 is grown at 37° to an A_{600} of about 2.0 after which aliquots of 200 μl are pipetted into each of 30 tubes. The lambda phage library is diluted such that 10^4 phage particles are added to each of the tubes. The transfected mixture is then incubated at 37° for 20 min. After incubation, 10 ml of NZYM media containing 0.7% (w/v) agarose prewarmed to 48° is then added to each tube, mixed, and immediately poured onto large (150-mm) petri dishes prepared with NZYM media containing 1.5% agar. The petri dishes are then incubated upside down at 37° for 12–16 hr to permit phage plaques to develop.

Transfer of Lambda Phage DNA onto Nitrocellulose Filters

The agar plates are first allowed to cool to room temperature. To transfer phage DNA from the plaques, nitrocellulose filters are placed on the plates for 10 min. The filters and agar are marked asymmetrically with ink to establish orientation. The nitrocellulose filters are lifted off the agar plates and placed, DNA side up, for 10 min onto Whatman 3MM paper saturated with 0.5 *N* NaOH and 1.5 *M* NaCl to denature the DNA. The DNA is neutralized by transferring the filters onto Whatman 3MM paper saturated with neutralization solution for 10 min. The filters are then rinsed in 1 liter of 2× SSPE for 10 min and allowed to dry on Whatman 3MM paper. The DNA is then fixed to the filters by baking at 80° in a vacuum oven for 2 hr.

Hybridization of ^{32}P-Labeled 642 bp Human IFN-αA AvaII Fragment to the Nitrocellulose Filters

Seal-O-Meal bags are prepared in which 2 filters are placed in each and sealed. One corner is cut off and 10 ml of hybridization solution is added. The *E. coli* DNA is the last addition to the hybridization solution which is added to the bag containing the filters. The Seal-O-Meal bags containing the filters are sealed and then incubated at 42° for 12–16 hr.

The probe to be added to the filters is prepared by nick translation[8] to a specific activity of $1–4 \times 10^8$ cpm/μg. Before use, the probe is placed in a boiling water bath for 10 min and immediately placed at 0° for 10 min. To each bag containing filters, a corner is cut off and 5×10^6 cpm of probe is added and the bag resealed. The filters are then incubated at 42° for 24–48 hr.

Washing the Filters

The hybridization solution is removed from each bag and discarded. The filters are then placed in a tray and washed twice for 15 min with 2 liters of 2× SSPE containing 0.2% SDS at room temperature. The filters are then washed twice for 15 min with 2 liters of 0.2× SSPE containing 0.1% SDS at room temperature, and twice for 15 min with 2 liters of 0.1× SSPE containing 0.1% SDS at 45°. The filters are then dried in air on Whatman 3MM paper and autoradiographed at −70° with Kodak X-OMAT XAR film and an intensifying screen for 24 hr.

Plaque Purification

The films which contain positive signals from the filters are lined up with the agar plates according to the orientation marks. A 10 mm diameter area corresponding to the location of the positive signal is removed from

[8] R. M. Lawn, E. F. Fritsch, R. C. Parker, G. Blake, and T. Maniatis, *Cell* **15,** 1157 (1978).

the plate and resuspended in 1 ml of 0.1 *M* NaCl, 8 m*M* $MgSO_4$, 50 m*M* Tris · HCl, pH 7.5, containing 0.1% (w/v) gelatin. The tube is incubated at room temperature for 5 hr or longer. The phage suspension is diluted in the same buffer and rescreened on NZYM agar plates as described above. The entire procedure is repeated until purification of the positive plaque is achieved. This is accomplished when 100% of the plaques yielded a positive signal in the screening.

Remarks

Several positive signals were found after screening 300,000 plaques. One was purified after 3 rounds of plaque purification. Lambda phage DNA was extracted,[8] digested with several restriction endonucleases for Southern blotting.[9] A single 2.6 kb *Eco*RI–*Pst*I fragment[7] that hybridized with the probe was subcloned into pBR325. The murine IFN-α gene (Mu-IFN-αA) was sequenced by the method of Maxam and Gilbert.[10] The nucleotide sequence as well as the corresponding amino acids of Mu-IFN-αA is shown in Fig. 1.

Construction of Recombinant DNA Molecules and Expression of Mu-IFN-αA in *E. coli*

In order to construct an expression plasmid for Mu-IFN-αA, the 2.6 kb *Eco*RI–*Pst*I fragment was subcloned into pBR322 to provide the plasmid designated pBD1. A unique *Mst*II site was found that corresponds to nucleotide number 25 corresponding to Asn 8 Leu 9 of the mature protein. An oligonucleotide was synthesized which reconstructed the first eight amino acids of the mature protein proximal to the *Mst*II site and contained an ATG methionine initiation codon preceding Cys 1, the first amino acid of the mature protein. The fragment also contained *Eco*RI and *Mst*II restriction sites at the 5′- and 3′-ends, respectively (Fig. 2).

The synthetic oligonucleotides which are complimentary to each other are annealed (20 pmol of each) in a reaction volume of 12.5 μl containing 7 m*M* Tris · HCl, pH 7.5, 7 m*M* $MgCl_2$, and 50 m*M* NaCl in a 1.5-ml Eppendorf tube. The tube is placed in a water bath containing 500 ml of water at 90° for 5 min. The water bath is then allowed to cool to room temperature for 12–16 hr. One picomole of plasmid pBD1 is digested with *Eco*RI and *Mst*II restriction endonucleases and ligated to 5 pmol of the annealed synthetic nucleotide. The ligation is performed in 66 m*M* Tris · HCl, pH 7.5, 5 m*M* $MgCl_2$, 5 m*M* DTT, and 1 m*M* ATP containing 3

[9] E. M. Southern, *J. Mol. Biol.* **98**, 503 (1975).

[10] A. M. Maxam and W. Gilbert, this series, Vol. 65, p. 499.

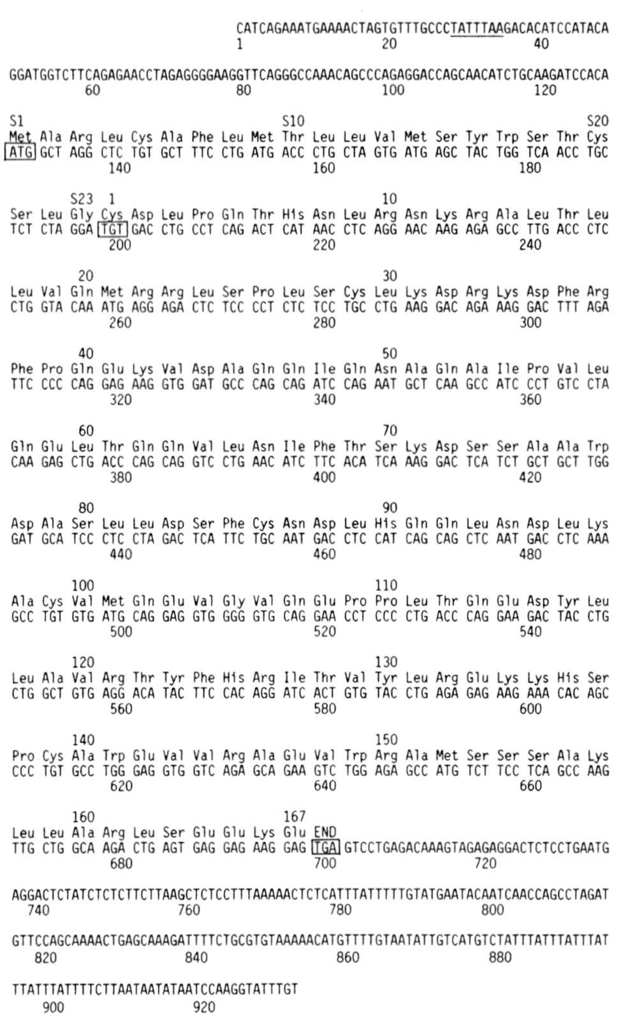

FIG. 1. DNA sequence of murine leukocyte interferon gene IFN-αA. The amino acid sequence predicted from the DNA sequence is shown. The sequence underlined indicates the Goldberg–Hogness box (TATTTAA). The sequence of the signal peptide and the mature protein is designated S1 to S23 and 1 to 167, respectively. The Met (S1) initiation codon, Cys 1 (the first amino acid of the mature protein), and the termination codon TGA are outlined by boxes. Data from Daugherty *et al.*[7]

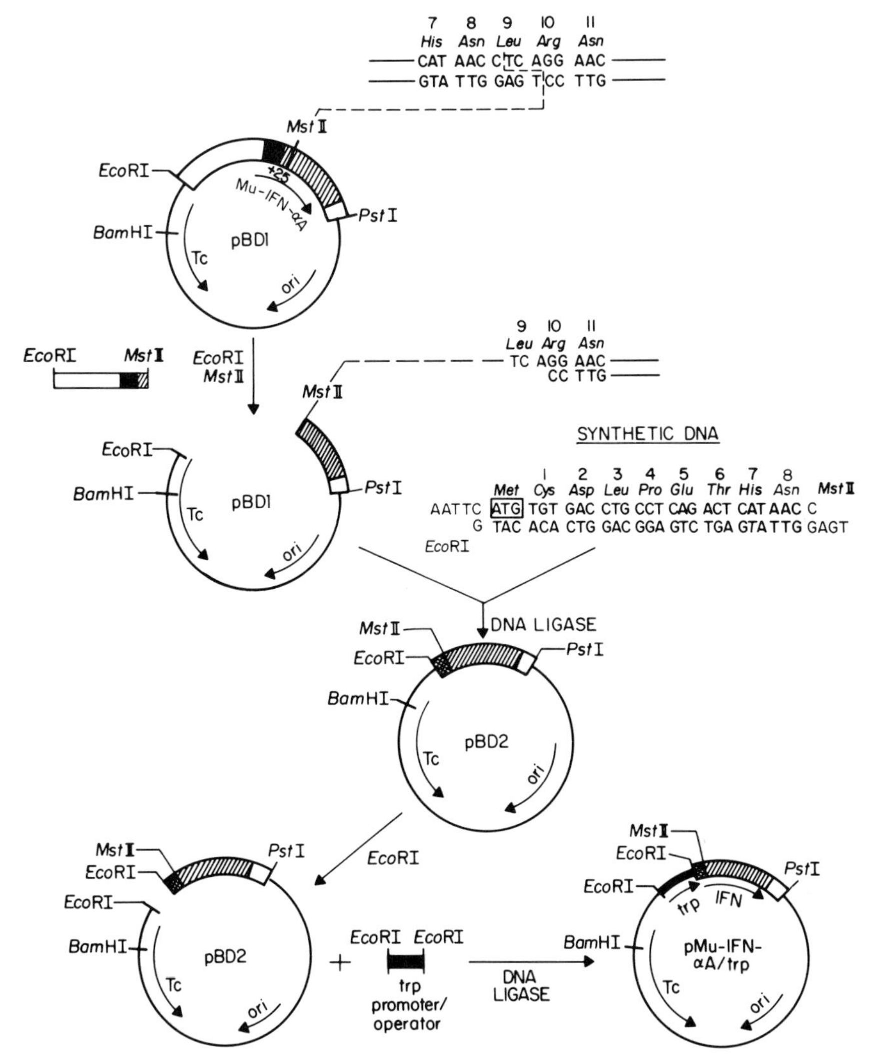

FIG. 2. Construction of an expression vector for murine leukocyte interferon. The plasmid pBD1 which contained the entire sequence of the gene was digested with *Eco*RI and *Mst*II restriction endonucleases. The resulting fragment was then ligated to a synthetic DNA which restored the 8 amino acids proximal to the *Mst*II site, introduced a methionine initiation codon on the 5′-side of Cys 1, and *Eco*RI and *Mst*II sites at the 5′-end and 3′-ends, respectively, to yield plasmid, pBD2. The *trp* promoter/operator fragment[1] flanked by *Eco*RI sites was then inserted into the *Eco*RI site of pBD2. The resulting plasmid designated pMu-IFN-αA/*trp* expressed a protein in *E. coli* which exhibited antiviral activity on mouse L cells.

units of T_4-DNA ligase (Amersham) for 12–18 hr at 16° in a reaction volume of 10 μl.

After transformation,[11] all clones were found to contain the synthetic oligonucleotide by hybridization with ^{32}P-labeled oligonucleotide as described.[12] The resulting plasmid, pBD2, was digested with *Eco*RI, dephosphorylated, and ligated with the 300 bp *trp* promoter/operator *Eco*RI fragment from the expression plasmid containing the human leukocyte interferon gene.[1] Eighty-six tetracycline-resistant colonies were isolated. Colony hybridization was performed with the nick-translated 300 bp *trp* promoter/operator and 38 colonies were found to contain it.

The *trp* fragment has an *Xba*I site adjacent to the *Eco*RI site which must be ligated to the initiator ATG in order for expression to proceed. Thus, the orientation of the *trp* fragment was determined by digesting the plasmids with *Xba*I and *Pst*I which results in distinct size fragments corresponding to each orientation. The correct orientation corresponds to a restriction fragment of 750 bp whereas the opposite orientation yields a fragment of 1050 bp. One clone identified as having the *trp* promoter/operator in the correct orientation was designated pMu-IFN-αA. It was digested with *Xba*I restriction endonuclease, end labeled, and sequenced to confirm the accurate reconstruction of the gene as designated in Fig. 2.

Preparation of *E. coli* Cell Extracts for Antiviral Activity

E. coli RRI containing the expression plasmid was grown at 37° overnight in 10 ml of M9 medium containing 5 μg/ml of tetracycline. The following morning, 25 ml of fresh M9 medium containing 5 μg/ml tetracycline was inoculated with 1 ml of the overnight culture and allowed to grow to an optical density of 1.0 at 550 nm. A volume of 10 ml of each culture was pelleted at 2000 *g* for 10 min and resuspended in 0.5 ml of phosphate-buffered saline in a 1.5-ml Eppendorf tube. The cell suspensions were subsequently broken by sonication with a Bronwill Biosonic III Sonicator twice for 30 sec at the highest setting with the smallest probe at 0°. The extract was assayed for antiviral activity on mouse L cells by the cytopathic effect inhibition assay as previously described.[13] Cell lysates prepared from *E. coli* RRI/pMu-IFN-αA resulted in an antiviral activity of 1300 units/ml.

[11] S. R. Kushner, *in* "Genetic Engineering" (H. W. Boyer and S. Micosia, eds.), p. 17. Elsevier/North-Holland, Amsterdam, 1978.

[12] K. Hotta, K. J. Collier, and S. Pestka, this volume [66].

[13] P. C. Familletti, S. Rubinstein, and S. Pestka, this series, Vol. 78, p. 387.

Concluding Comments

Hybridization of the murine genomic library to the ^{32}P-labeled human alpha interferon gene (642 bp *Ava*II-pLeIFLrA25 fragment) resulted in strong positive signals with virtually no background. The probe chosen represented over 90% of the coding region and contained no poly(A) sequences and no introns. Since we were cross-hybridizing between species and since the extent of the homology between human and mouse interferon genes was unknown, the stringency of the hybridization solution was lowered from a standard concentration of 50% formamide to a lower stringency of 40% formamide. This method proved successful in the isolation of a genomic murine alpha interferon gene.

After the nucleotide sequence of the gene was determined, it was possible to construct an effective expression vector. It should be noted that since the gene isolated was a genomic clone rather than a cDNA clone, it could possibly contain introns. An open reading frame of 190 amino acids was determined from the sequence. By comparison of the nucleotide sequence with that of the nucleotide sequence of the human alpha interferon genes,[1,14] it was concluded that the gene contained no introns, but contained a signal peptide of 23 amino acids followed by a cysteine which corresponded to Cys 1 of the human gene. Therefore, cysteine was concluded to be the first amino acid of the mature protein and the expression vector was appropriately designed and constructed as described above.[7]

[14] S. Pestka, *Arch. Biochem. Biophys.* **221,** 1 (1983).

[62] Cloning, Expression, and Purification of Rat IFN-α1

By PETER H. VAN DER MEIDE, REIN DIJKEMA, MARTIN CASPERS, KITTY VIJVERBERG, and HUUB SCHELLEKENS

This chapter describes the introduction of a structural gene for a rat alpha interferon subtype (rat IFN-α) into *Escherichia coli* to obtain an efficient expression system for the production of large quantities of rat interferon. This would provide a more economic way of production than does the system employing a rat continuous cell line[1] and also make it possible to study the properties of a single gene product. DNA comple-

[1] P. H. van der Meide, J. Wubben, K. Vijverberg, and H. Schellekens, this volume [31].

mentary to rat interferon mRNA isolated from Ratec cells induced by Sendai virus was cloned and amplified in *E. coli*. A purified cDNA fragment identified by hybridization with the mouse $\alpha 2$ gene was then used as a probe for the isolation of the chromosomal genes from a rat gene library. In this way, a 2.5 kb *Eco*RI fragment encoding a complete rat IFN-α gene (rat IFN-$\alpha 1$) was isolated and placed under the control of a hybrid tetracycline–tryptophan promoter into the *Eco*RI site of pMBL604 (a derivative plasmid of PBR322). *E. coli* strains carrying this recombinant plasmid synthesized biologically active interferon.

The procedure for the production and purification of this type of interferon will be described. Also attempts to increase production with other host/vector systems are described and some characteristics of rat IFN-$\alpha 1$ will be discussed.

Cloning of the Chromosomal Rat IFN-α Gene

Preparation, Purification, and Identification of IFN mRNA from Induced Ratec Cells

For the construction of cDNA clones, total cytoplasmic RNA was isolated from Ratec cells actively synthesizing interferon. To establish at what time after virus infection a maximum amount of mRNA coding for rat interferon was present in the cells, poly(A)$^+$ RNA was isolated at different times after Sendai infection and translated in *Xenopus laevis* oocytes (see below). The pattern of IFN activity in this assay revealed an optimum at about 10 hr after infection (see Fig. 1). At that time, total RNA (from 10^{10} cells) was isolated from the cells by extraction with a buffer containing 10 m*M* Tris·HCl (pH 7.9), 1 m*M* EDTA, 150 m*M* NaCl, and 0.65% Nonidet P40 for 5 min at 4°. After centrifugation, the supernatant was mixed with an equal volume of a buffer containing 20 m*M* Tris·HCl (pH 7.9), 20 m*M* EDTA, 300 m*M* NaCl, 7 *M* urea, and 1% SDS. The mixture was then extracted twice with phenol. The poly(A)$^+$ RNA fraction (about 250 μg) was isolated by repeated batchwise absorption onto oligo(dT)-cellulose.[2] Different mRNA size fractions were obtained by use of a SDS–PAGE system as described by Wierenga *et al.*[3] The mRNA fractions were injected into oocytes of *Xenopus laevis* as described by Gordon *et al.* [4] After incubation, the eggs were homogenized

[2] S. Nagata, H. Taira, A. Hall, L. Johnsrud, M. Streuli, J. Ecsödi, W. Boll. K. Cantell, and C. Weissmann, *Nature (London)* **284,** 316 (1980).

[3] B. Wierenga, J. Mulder, A. van der Ende, A. Bruggeman, G. Ab, and M. Gruber, *Eur. J. Biochem.* **89,** 67 (1978).

[4] J. B. Gordon, C. D. Lane, M. Woodlans, and G. Marbaix, *Nature (London)* **233,** 177 (1971).

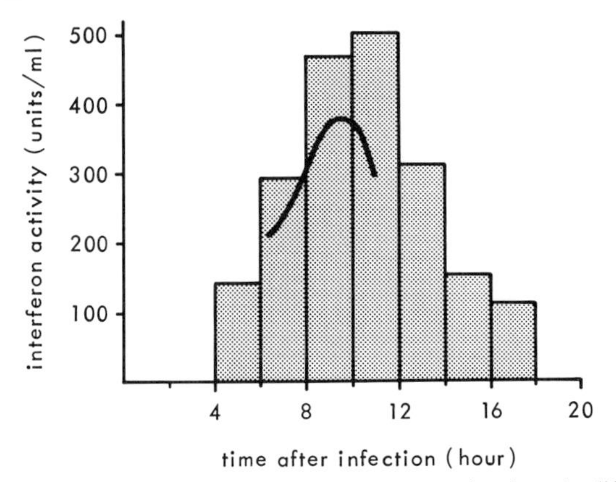

FIG. 1. Time course of IFN activity and IFN mRNA production. At different times, the culture medium was decanted and replaced by fresh medium. The IFN activity was assayed in the withdrawn culture fluid (columns). The relative amount of IFN mRNA in cells was determined by the oocyte assay[4] (——). For experimental details, see text.

and assayed for interferon activity. A single peak of IFN mRNA activity (which contained about 8 μg of RNA) was found at a position of the gel corresponding with a sedimentation coefficient of about 14 S.

Construction and Identification of cDNA Clones Containing Rat Interferon α Sequences

About 14 μg of 14 S mRNA yielded 250 ng of double-stranded cDNA after a reverse transcriptase reaction. The double-stranded cDNAs were size fractionated on a 5% polyacrylamide gel and 125 ng of material ranging from 800 to 1200 bp was obtained from the gel by electroelution. This cDNA was inserted into the *Pst*I site of pBR322 DNA (400 ng) by the G-C tailing procedure of Villa-Komaroff *et al.*[5] After transformation of *E. coli* HB101, 15,000 tetracycline resistant colonies were screened with a 690 bp *Hind*III–*Eco*RI fragment of the chromosomal mouse IFN-α1 DNA as a probe. Three transformants hybridized to the same extent with this probe. To determine whether the cDNA clones contained a complete coding region, DNA sequencing was carried out according to the procedure of Maxam and Gilbert.[6] The nucleotide sequence showed that the three cDNA clones contained rat IFN-α sequences when compared with the chromosomal mouse IFN-α1 gene (Fig. 2). However, they were incomplete with respect to the coding region of this gene. Since the 3′ noncoding

[5] L. Villa-Komaroff, A. Efstratiadis, S. Broome, P. Lomedico, R. Tizard, S. P. Naker, W. L. Chick, and W. Gilbert, *Proc. Natl. Acad. Sci. U.S.A.* **75,** 3727 (1978).

[6] A. M. Maxam and W. Gilbert, *Proc. Natl. Acad. Sci. U.S.A.* **74,** 560 (1977).

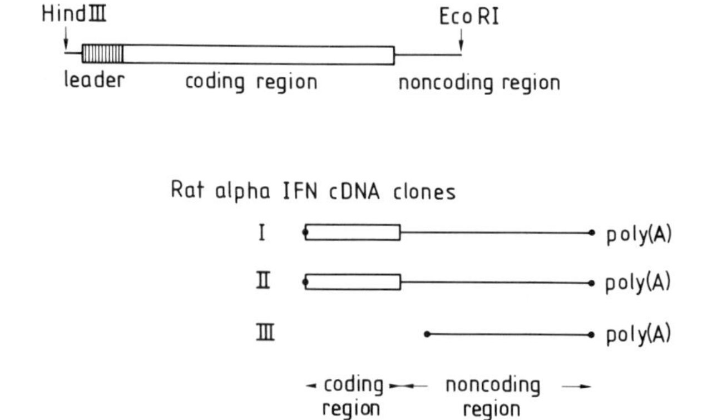

FIG. 2. Size determination of the three different rat IFN-α cDNA clones. On top is shown the structural organization of the chromosomal mouse IFN-α1 gene that was used as a probe. Below are shown the relative positions of the three rat IFN-α cDNA clones (all identical, same subtype).

regions of homologous genes generally diverge more strongly than the coding regions, the 3′ noncoding regions were subjected to sequence determination. The analysis revealed that the three cDNA clones were of the same origin with respect to the 14 S mRNA.

Isolation and Characterization of a Rat IFN-α Chromosomal Gene

A rat gene library cloned in bacteriophage λ harboring DNA fragments with an average size of 15 kb was used for the isolation of the structural gene of rat IFN-α. This library provided by Sargent *et al.*[7] was screened by *in situ* plaque hybridization with nick-translated rat IFN-α1 cDNA clone I. A set of hybrid phages of the library was affixed to nitrocellulose filters and probed with this ^{32}P-labeled cDNA fragment. Eighteen positive phage clones were found among 450,000 and isolated by repeated plaque purification. Hybridization experiments with ^{32}P-labeled cDNA from clone III (containing only the 3′ noncoding region of the respective IFN-α gene) revealed the presence of two phage clones which hybridized specifically to the probe. The DNAs from both phages were mapped by restriction enzyme analysis and found to contain identical parts (15.5 kb) of the rat genome, including the rat IFN-α subtype.

[7] T. D. Sargent, J. R. Wu, J. M. Sala-Trepat, R. B. Wallace, A. A. Reges, and J. Bonner, *Proc. Natl. Acad. Sci. U.S.A.* **76,** 3256 (1979).

Construction of a Plasmid Encoding Rat IFN-α

Digestion of the 15.5 kb insert of the hybrid phage with *Eco* RI revealed fragments of 8.0, 2.9, 2.5, and 2.1 kb. Only the 2.5 kb fragment hybridized to the cDNA probe. This 2.5 kb *Eco*RI fragment was subcloned into the *Eco*RI site of pBR322.

Recombinant plasmid DNA was prepared by the cleared lysate method,[8] digested by *Eco*RI and the 2.5 kb fragment was isolated by polyacrylamide gel electrophoresis. The 2.5 kb segment was cleaved by an appropriate restriction enzyme and the fragments were labeled at the 5′ ends with [γ-^{32}P]ATP and cleaved with a second restriction enzyme to yield single-labeled fragments which served as templates for the determination of the DNA sequence by the method of Maxam and Gilbert.[6]

The locations of the cleavage sites by means of a number of restriction enzymes is given in detail in an earlier publication.[9] The complete nucleotide sequence of rat IFN-α1 DNA is illustrated in Fig. 3. The sequence shows that the coding sequence of the rat IFN-α1 gene contains an open translational reading frame which codes for a protein of 192 amino acids, most likely comprised of a leader or signal peptide and the mature interferon protein.

A sequence 5′ TATTTAA 3′ (Goldberg–Hogness box[10]) was identified at 100 nucleotides upstream from an ATG start codon, suggesting that initiation of transcription occurs about 30 nucleotides downstream from this sequence. More conclusive identification of the coding sequence of the rat IFN-α1 gene was provided by a comparison with the nucleotide sequence of the chromosomal mouse IFN-α1 gene.[11] This gene codes for a mature IFN-α containing 166 amino acids preceded by a 23 amino acid long leader peptide. Because all human[12,13] and mouse[11,14] IFN-α genes code for cysteine at position 1 of the mature protein and since this gene family is highly conserved among vertebrates, the mature rat IFN-α1

[8] H. C. Birnboim and J. Doly, *Nucleic Acids Res.* **7,** 1513 (1979).

[9] R. Dijkema, P. Pouwels, A. de Reus, and H. Schellekens, *Nucleic Acids Res.* **12,** 1227 (1984).

[10] F. Gannon, K. O'Hare, F. Perrin, J. P. LePennec, C. Benoist, M. Cochet, R. Breathnach, A. Royal, A. Garapin, B. Cami, and P. Chambon, *Nature (London)* **278,** 428 (1979).

[11] G. D. Shaw, W. Boll, H. Taira, N. Mantei, P. Lengyel, and C. Weissmann, *Nucleic Acids Res.* **11,** 555 (1983).

[12] D. V. Goeddel, D. W. Leung, T. J. Dull, M. Gross, R. M. Lawn, R. McCandliss, P. H. Seeburg, A. Ullrich, E. Yelverton, and P. W. van Gray, *Nature (London)* **290,** 20 (1981).

[13] S. Pestka, *Arch. Biochem. Biophys.* **221,** 1 (1983).

[14] B. Daugherty, D. Martin-Zanca, B. Kelder, K. Collier, T. C. Seamans, K. Hotta, and S. Pestka, *J. Interferon Res.* **4,** 635 (1984).

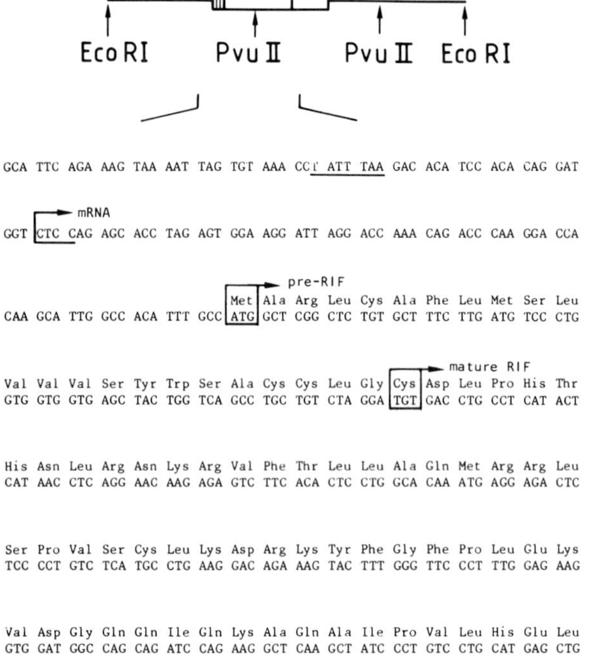

Thr Gln Gln Ile Leu Ser Leu Phe Thr Ser Lys Glu Ser Ser Thr Ala Trp Asp
ACC CAG CAG ATC CTC AGC CTC TTC ACA TCA AAG GAG TCA TCT ACT GCT TGG GAT

Ala Thr Leu Leu Asp Ser Phe Cys Asn Asp Leu Gln Gln Gln Leu Ser Gly Leu
GCA ACC CTC CTA GAC TCA TTC TGT AAT GAC CTC CAG CAG CAG CTG AGT GGT CTG

Gln Ala Cys Leu Met Gln Gln Val Gly Val Gln Glu Ser Pro Leu Thr Gln Glu
CAA GCC TGT CTG ATG CAG CAG GTA GGG GTG CAG GAA TCT CCC CTG ACC CAG GAA

Asp Ser Leu Leu Ala Val Arg Glu Tyr Phe His Arg Ile Thr Val Tyr Leu Arg
GAC TCC CTA CTG GCT GTG AGG GAA TAC TTC CAC AGA ATC ACT GTG TAC CTG AGA
cDNA clone

Glu Asn Lys His Ser Pro Cys Ala Trp Glu Val Val Lys Ala Glu Val Trp Arg
GAG AAT AAA CAC AGC CCC TGT GCC TGG GAG GTG GTC AAA GCA GAA GTC TGG AGA

Ala Leu Ser Ser Ser Ala Asn Leu Met Gly Arg Leu Arg Glu Glu Arg Asn Glu
GCC CTG TCT TCC TCA GCC AAC TTG ATG GGA AGA CTG AGA GAA GAA AGA AAT GAG

Ser ***
TCC TGA GCC ACA TTG GAG AGG ACT CCG

FIG. 3. A partial nucleotide fragment of rat IFN-α1 in a plasmid. The amino acid sequence was deduced from the nucleotide sequence. Indicated are the putative starts of the IFN mRNA, pre-IFN-α1 protein, and mature IFN-α1 protein.

protein most probably starts with the cysteine residue at position 199 (box in Fig. 3) and predicts a protein with 169 amino acids.

The mouse and rat IFN-α1 mature proteins are closely related in their amino acid sequences; 135 of 166 amino acids occupy identical positions (81% homology), whereas the DNA sequences showed 90% homology (for detailed information, see Ref. 9). To determine whether an IFN-like protein was produced, the coding sequence of the mature rat IFN-α1 protein was inserted into the *Eco*RI site of plasmid pMBL604. This vector contains a hybrid tetracycline–tryptophan promoter and a synthetic Shine–Dalgarno (SD) sequence.[9] For that purpose, the 2.5 kb *Eco*RI fragment was digested with *Hph*I to remove the nucleotide sequence which codes for the first 17 amino acids (of 22) of the signal peptide. After electrophoresis on 5% polyacrylamide gels, the *Hph*I fragment was isolated from excised slices and treated with *Bal*31 to remove the remaining 6 amino acid signal codons. An additional ATG start codon (in the form of a synthetic DNA adapter) was ligated to the population of *Bal*31 treated molecules and then cloned into the *Eco*RI site of pMBL604.

Transformants of *E. coli* JA221[9] which became ampicillin resistant were investigated for their capacity to produce substances with interferon activity. Two of 500 transformants were found to exhibit antiviral activity. However, both produced relatively low levels of interferon activity, about 200,000 units per liter of culture (at OD_{600} equal to 1.0). DNA analysis revealed that in the plasmid of the highest producer the initiation codon (AUG) is directly followed by the UGU (Cys) codon of mature IFN. The construction and features of the recombinant plasmid 13.1 are schematically depicted in Fig. 4.

Subcloning of the Chromosomal Rat IFN-α1 Gene

In order to further increase the production level of rat IFN-α in bacteria, other constructions were carried out. The plasmid vectors used in these experiments were pBR322 derived multicopy vectors carrying the powerful major leftward promoter P_L of bacteriophage lambda. This promoter is controlled by a temperature-sensitive repressor protein cI whose gene is located on the bacterial chromosome of a specific host strain (*E. coli* M5219). The P_L promoter gives rise to efficient transcription at 42°, while the promoter is shut off by this thermosensitive repressor at 28°. The vectors carry a 247 base pair fragment containing the leftward operator and promoter and code for the first 114 nucleotides of the P_L transcript.[15] There is no translation initiation signal present on this transcript.

[15] E. Remaut, P. Stanssens, and W. Fiers, *Gene* **15,** 81 (1981).

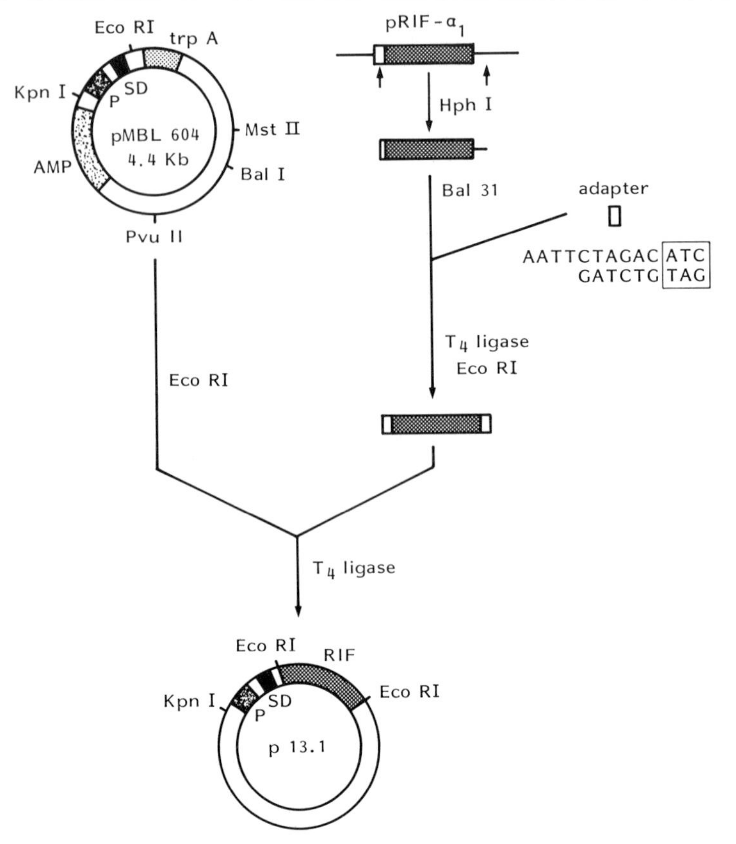

FIG. 4. Schematic illustration of the construction of the rat IFN-α1 expression plasmid p13.1. Amp indicates the region coding for β-lactamase. The operator/promotor region is indicated by a P. SD stands for the Shine–Dalgarno ribosomal binding site.

Three different hybrid plasmids were constructed (as outlined in Fig. 5). At first, an 1400 *Fnu*DII fragment containing the complete sequence coding for the mature rat IFN-α protein plus the synthetic ribosomal binding site of the original vector pMBL604 was isolated by polyacrylamide gel electrophoresis. This fragment was subsequently converted to a *Bam*HI bounded fragment which was ligated to purified *Bam*HI-cut pPLc236 vector DNA (p13.2).

The other two plasmids used (pAT_1 and pAT_{55}) were kindly provided by Dr. W. Fiers. These plasmids are derivatives of pPLc236 and contain a synthetic Shine–Dalgarno (SD) sequence. Plasmid pAT_{55} differs from pAT_1 by a deletion of 26 bp between the promoter/operator region and the

SD site. Lysates from cultures of *E. coli* cells carrying pAT_{55} produce high levels of Hu-IFN-β activity, whereas no activity was found in extracts of *E. coli* carrying pAT_1 (Dr. W. Fiers, personal communication). Plasmid p13.3 was constructed by replacing the *Pst*I–*Sba*I fragment (containing the hybrid tetracycline–tryptophan promoter and the ribosomal binding site) by a 1250 bp *Xba*I–*Pst*I fragment from pAT harboring the P_L promoter/operator region with SD site as shown in Fig. 5. The other subclone p13.4 was constructed in the same way, except that the fragment with the P_L promoter was obtained from pAT_{55}.

To determine whether rat interferon was produced, lysates of transformants were prepared and assayed for antiviral activity. The results clearly showed that all three clones exhibited antiviral activity. However, in no case was the level exceptionally higher than the amount produced by the original construction p13.1 (see Table I).

Preparation of Bacterial Extracts

Overnight cultures were diluted 1:50 with L-broth supplemented with 60 μg/ml ampicillin. Cultures of 500 ml were grown in 1-liter Erlenmeyer flasks at 37° (clone p13.1) or 28° (clones p13.2-4) in a rotary shaking waterbath (100 rpm). When cells of clone p13.1 reached an optical density of about 1.0 at 600 nm they were rapidly cooled on ice, harvested by

TABLE I
RAT IFN LEVELS IN CELLS OF *E. Coli* HARBORING DIFFERENT EXPRESSION PLASMIDS[a]

Strain	Plasmid	IFN activity (units/liter culture) ($\times 10^5$)
JA 221	p13.1	1.7
M5219	p13.2	1.3
M5219	p13.3	1.3
M5219	p13.4	2.4

[a] Cells (containing p13.1) were cultured in rich medium (L broth) and harvested when the culture reached an A_{600} of about 1.0. Cultures with cells containing P_L plasmids (p13.2–4) were heat induced after reaching an A_{600} of about 0.3. The induction time was 150 min. The strains JA 221[9] and M5219[15] have been described.

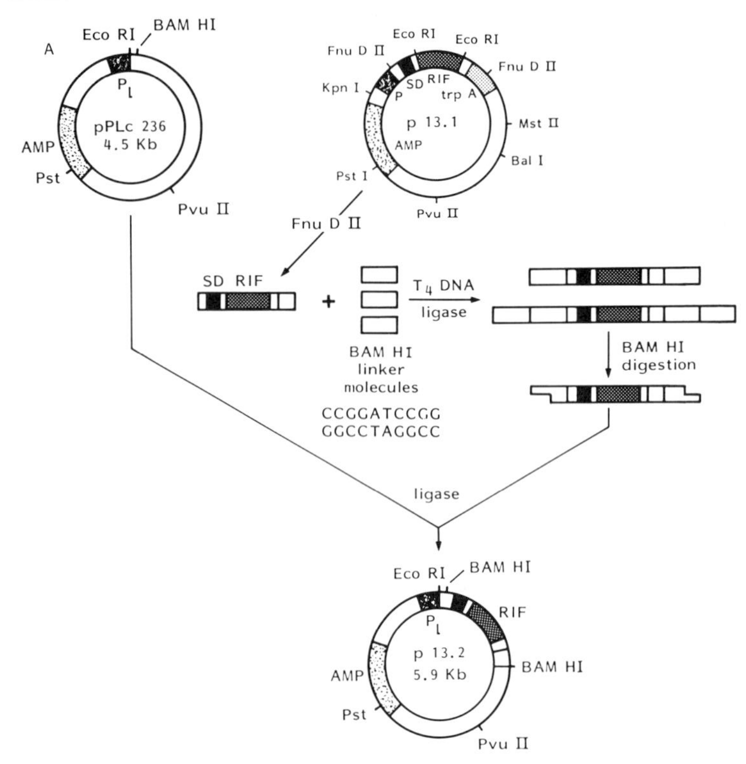

FIG. 5. Schematic illustrations of the construction of two expression plasmids: p13.2 (A) and p13.3 (B). PLc denotes the position of the P_L promoter. Construction and features of plasmid p13.1 is shown in Fig. 4.

centrifugation, and washed once in 200 ml of phosphate-buffered saline (PBS). The other clones were induced after reaching an A_{600} of about 0.3 by a temperature shift from 28 to 42°. The incubation was continued for 150 min at the elevated temperature. They were then chilled, harvested, and washed once with PBS. Cells from the different clones were resuspended in 5 ml of PBS supplemented with 10 m*M* EDTA and 1 mg/ml lysozyme. The mixtures were incubated for 1 hr at 0° and subsequently sonicated by 10 pulses of 30 sec at 30 sec intervals. The resulting suspension, which was designated a crude extract, was clarified by centrifugation at 30,000 *g* for 30 min, yielding a low-speed supernatant. This suspension was centrifuged at 150,000 *g* for 3 hr at 4°, yielding a ribosome-free supernatant (S-100). This S-100 fraction was the starting material for purification.

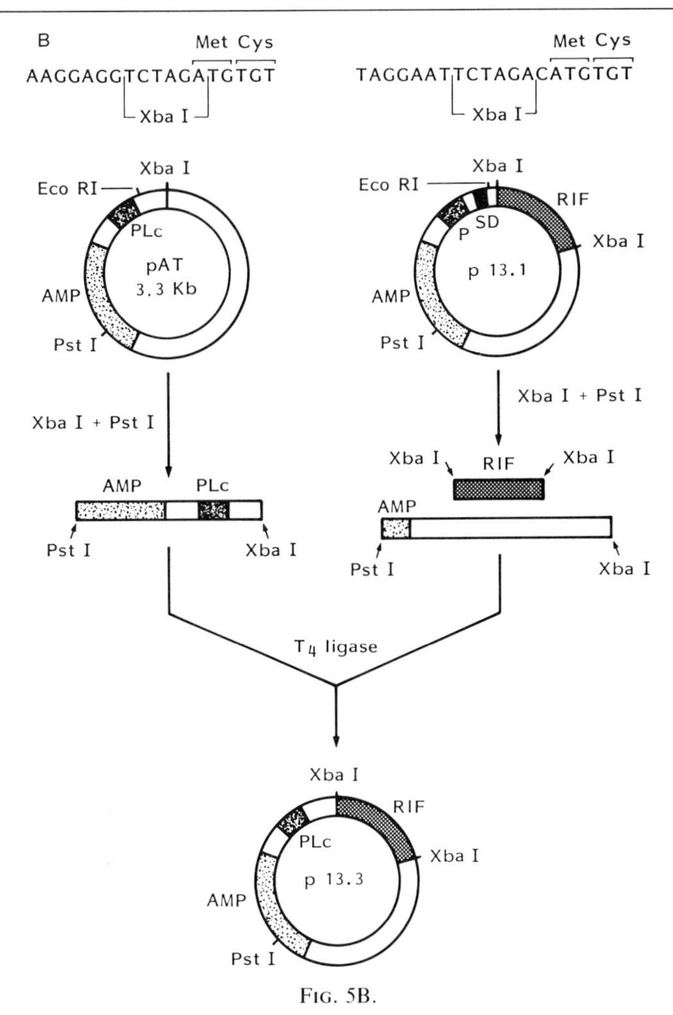

FIG. 5B.

Purification of Recombinant Rat IFN-α with Polyclonal Antibodies Directed against Recombinant Human IFN-α2

The interferon activity of the different clones was neutralized by antiserum from rabbits immunized with recombinant human IFN-α2. As expected from these experiments, the bacterially derived rat IFN-α could be purified by use of immobilized polyclonal antibodies directed against human IFN-α2. S-100 lysates from clone p13.1 were passed through a column to which the purified IgG fraction from rabbit anti Hu-IFN-α2 serum was bound. The procedure for the preparation of the affinity

TABLE II
PURIFICATION OF BACTERIALLY DERIVED RAT IFN-α BY A COLUMN CONTAINING IMMOBILIZED ANTI-Hu-IFN-α2[a]

Purification step	Volume (ml)	IFN (units/ml)	Recovery (%)	Protein (mg/ml)
S-100 lysate	7	1.5×10^4	100	9.8
Antibody column eluate	20	$2\text{–}4 \times 10^3$	38–72	0.005

[a] The column size was 7 ml; the elution buffer was 0.1 *M* glycine · HCl, pH 2.3.

column is described in detail in the chapter which deals with naturally derived rat interferon.[1] The bacterial rat interferon was retained and could be eluted by a 0.1 *M* glycine·HCl buffer of pH 2.3. In this way, recombinant rat IFN-α1 was purified to a specific activity of at least 4×10^5 units/mg protein. However, owing to variation in the recovery (see Table II) on immunochromatography and because there is uncertainty about the protein determination (because of the extremely low protein concentration), this value is a conservative estimate. The eluate (20 ml) from the affinity column was supplemented with 1 ml of a BSA solution containing 500 μg/ml protein. The solution was lyophilized and analyzed by PAGE in the presence of SDS. The gel slices were extracted in a buffer containing 0.5% (w/v) *N*-laurylsarcosine and 1% (v/v) 2-mercaptoethanol and assayed for antiviral activity. These experiments showed that we are dealing with a protein with a mobility corresponding to a M_r of about 16,000.

Comments

Because unambiguous detection of an interferon specific protein band by SDS–polyacrylamide gel electrophoresis was not yet possible, the exact production level of interferon protein in the clones is unknown. However, deduced from several electrophoretic analyses on SDS–PAGE, an amount exceeding 0.5 μg/mg of cellular protein is not likely. We can only speculate about the cause of the low expression in these clones. One of the reasons could be that secondary structure formation of the mRNA may prevent the efficient expression of the gene, but it is also possible that the IFN mRNA is unstable and is rapidly broken down in the cell.

Another aspect of interest in the stability of this protein in the presence of different chemical substances (see Table III). Methods of cell lysis with these detergents revealed that treatment of the bacteria with 6 *M* guanidine·HCl results in a 3-fold higher interferon yield compared to SDS treatment or ultrasonic disruption. Also, the presence of 2-mercapto-

TABLE III
RECOVERY OF RAT IFN ACTIVITY FROM CLONE 13.1 BY TREATMENT WITH DIFFERENT CHEMICALS[a]

Chemical substance	IFN activity (units/liter culture) ($\times 10^5$)
Control (no additions)	1.7
6 *M* guanidine · HCl	4.1
0.5% *N*-laurylsarcosine	3.5
0.1% sodium dodecyl sulfate (SDS)	1.2
9 *M* urea	1.6
1% Nonidet P-40	1.5
0.08% sodium deoxycholate	0.9
1% Triton X-100	1.4

[a] Cell pellets of 1.0 liter cultures were resuspended in 6 ml of 20 m*M* Tris · HCl (pH 7.0) containing 1% (v/v) 2-mercaptoethanol and the different detergents mentioned in the table. After an incubation time of 24 hr at 4°, cells were sonicated and centrifuged at 20,000 *g*. The supernatants were assayed for IFN activity.

ethanol is essential for the recovery of activity from the cells. This illustrates that the level of rat IFN-α is dependent upon the extraction procedure and that the absence of a reducing agent might lead to an inactive aggregated form of interferon, as already suggested by SDS gel electrophoresis of naturally derived rat interferon.[1] The finding that rat IFN-α1 reacts with antibodies against human IFN-α2 indicates that certain interferon subtypes of rat and human origin have strong antigenic similarities.

Acknowledgments

We thank C. Weissmann for generously providing the chromosomal mouse IFN-α1 probe.

[63] Cloning, Expression, and Purification of Rat IFN-γ

By REIN DIJKEMA, PETER H. VAN DER MEIDE, MARTIN DUBBELD, MARTIN CASPERS, JACQUELINE WUBBEN, and HUUB SCHELLEKENS

This chapter deals with the molecular cloning of the chromosomal rat IFN-γ gene from a rat gene library by use of heterologous hybridization with the human IFN-γ cDNA as a probe and the characterization of its primary structure by DNA sequence analysis. The gene was expressed under the control of viral promoters by either transient expression in monkey COS-I/COS-7 cell lines or in Chinese hamster ovary cells (CHO) after cotransformation with the murine *dhfr* (dihydrofolate reductase) gene or *Escherichia coli gpt* gene. The partial purification and some physicochemical characteristics of the rat IFN-γ protein are described.

Identification of the Chromosomal Rat IFN-γ Gene

A recombinant lambda Charon 4A/rat genomic library was screened for IFN-γ sequences by use of a human IFN-γ cDNA clone[1] as a probe. Under low stringency conditions for hybridization and washing (20% formamide; 5× SSC at room temperature), two individual hybridizing signals were observed and the corresponding phages were plaque purified. The number of positive plaques with respect to the total number of hybrid phages is in good agreement with what one would expect for a single copy gene.[2] Restriction enzyme analysis with Southern blotting of both hybrid phages revealed the exact nature of the rat genomic DNA insert; both contained almost identical parts of the rat genome, except that one (lambda RIFN-γ1) contained an additional 2.1 kb *Eco*RI fragment that specifically hybridizes with the 3′ part of the human IFN-γ cDNA clone. The complete chromosomal rat IFN-γ gene present in RIFN-γ1 was subcloned as a 5.5 kb *Pst*I fragment in pBR322.

Primary Structure of the Chromosomal Rat IFN-γ Gene

Because the human IFN-γ cDNA clone is about 1 kb in size and because probes derived from the 5′ and 3′ ends of this clone hybridized to regions of the rat IFN-γ chromosomal gene which is about 5 kb in length, it is highly likely that the rat IFN-γ chromosomal gene has the same structural organization with respect to the exon/intron composition as the human and murine IFN-γ gene. We have sequenced the exons of the rat IFN-γ gene and the exon/intron boundaries in order to compare its primary sequence with IFN-γ sequences of murine and human origin.

Figure 1A shows the chromosomal rat IFN-γ gene structure and Fig. 1B shows the strategy used to determine the nucleotide sequence of the different exons and exon/intron boundaries.

Figure 2 shows the complete nucleotide sequence of the coding parts of the rat IFN-γ gene. Based on what is known about the murine IFN-γ

[1] J. Haynes and C. Weissmann, *Nucleic Acids Res.* **11,** 687 (1983).

[2] R. Dijkema, P. H. Pouwels, A. de Reus, and H. Schellekens, *Nucleic Acids Res.* **12,** 1227 (1984).

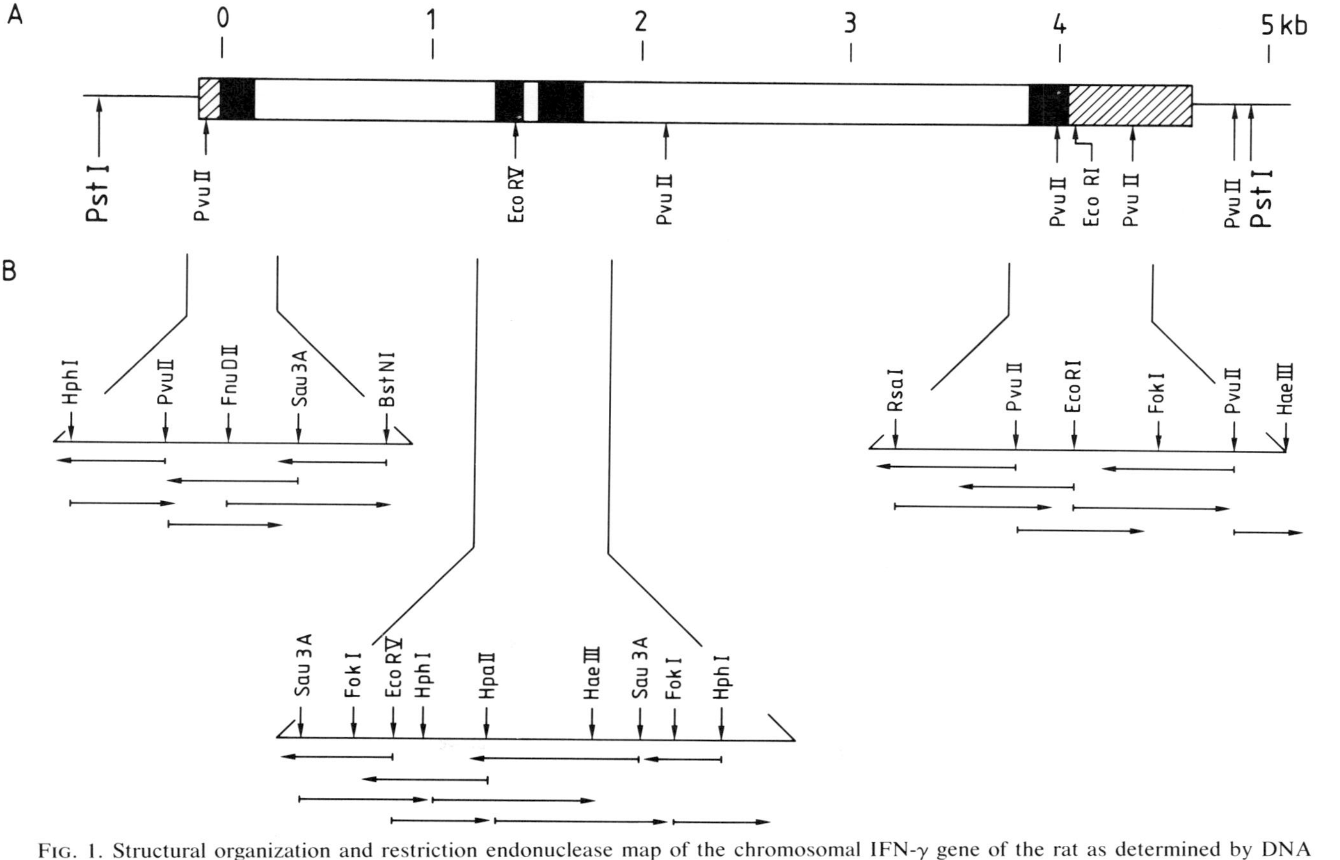

FIG. 1. Structural organization and restriction endonuclease map of the chromosomal IFN-γ gene of the rat as determined by DNA sequence comparison with the murine IFN-γ cDNA (A). The 5′ and 3′ untranslated regions of the rat IFN-γ gene are shown as hatched boxes, coding regions (exons) as solid boxes, and intervening noncoding regions (introns) as open boxes. DNA sequencing was performed as outlined (B). Each sequence was determined from the marked position and sequenced in the direction of the arrow; the length of the sequence determined corresponds to the length of the arrow.

```
GTCACAAACCATAGCTATAATGCAAAGTAACTAGCTCCCGCCACCTATCTTTCACCATCTTAACTTAAAAAAAAACCTGTGAAAATACGTAATCCCAAG

AAGCCTTCGGTCATGTATAAAACTGGAAGCAAGAGAGGTGCAGCGTATAGCTGCCATCGGCTGATCTAGAGAAGACACATCAGCTGATTCTTCGGACTC

                                                        Met Ser Ala Thr Arg Arg Val Leu Val Leu Gln
TCTGACTTAATACAGGAGTTCTGGGCTTTCCCTGCCTTGGCCTAGCTCTGAGACA ATG AGT GCT ACA CGC CGC GTC TTG GTT TTG CAG

Leu Cys Leu Met Ala Leu Ser Gly Cys Tyr Cys Gln Gly Thr Leu Ile Glu Ser Leu Glu Ser Leu Lys Asn Tyr
CTC TGC CTC ATG GCC CTC TCT GGC TGT TAC TGC CAA GGC ACA CTC ATT GAA AGC CTA GAA AGT CTG AAG AAC TAT

Phe
TTT GTAAGTATGATCTTTTCATAGTGCCTGTGGTTGACGGTGGCTGGTGTTGACTCCCTGTAGTGAACGCTAGACTGCCATCTCTGGCCCACAGTCAT

                                                                                           Asn Ser
TTTG-----(INTRON1-1150bp)-----CAGATATTTTCAGGGCAGTTTGGTGAAATAATTACAAATCGATCTTTTCTCTTCTCCTCAG AAC TCA

Ser Ser Met Asp Ala Met Glu Gly Lys Ser Leu Leu Leu Asp Ile Trp Arg Asn Trp Gln Lys
AGT AGC ATG GAT GCT ATG GAA GGA AAG AGC CTC CTC TTG GAT ATC TGG AGG AAC TGG CAA AAG GTGAGCTGAATATCC

                                       -----(INTRON2-103bp)-----                           Asp Gly
CCCCCAACACACTCCCCTGCCTCCCCTGCTTTCCTGTTGTTTCTAATGAACCGGTTCTCACAATACTCTCTTTGTGTTGTTTCCCAAG    GAC GGT

Asn Thr Lys Ile Leu Glu Ser Gln Ile Ile Ser Phe Tyr Leu Arg Leu Phe Glu Val Leu Lys Asp Asn Gln Ala
AAC ACG AAA ATA CTT GAG AGC CAG ATT ATC TCT TTC TAC CTC AGA CTC TTT GAA GTC TTG AAA GAC AAC CAG GCC

Ile Ser Asn Asn Ile Ser Val Ile Glu Ser His Leu Ile Thr Asn Phe Phe Ser Asn Ser Lys Ala Lys Lys Asp
ATC AGC AAC AAC ATA AGT GTC ATC GAA TCG CAC CTG ATC ACT AAC TTC TTC AGC AAC AGT AAA GCA AAA AAG GAT

Ala Phe Met Ser Ile Ala Lys Phe Glu
GCA TTC ATG AGC ATC GCC AAG TTC GAG GTGAGACAGCTTTGCAAACTACCGTATTATTGTTTGTTTCACATTGTCTTTGAATTATCAGAC

AGTAGAAATTAGCTACTCATCAGTTGATAAAGCTGAGAGATGTTTCCCCACCGCAGGCAGATTGGGAGGAATCTGCCTTTTTTTTTTG-----------

-----(intron3-2150bp)-----ATTGCTGTACTACTTTGTTAAGAGGAATATTTTCATTTTCACTGACCATGATGTCAAGAAGAATAGTCCAATG

                                                Val Asn Asn Pro Gln Ile Gln His Lys Ala Val Asn Glu
ACTTATATGCTTGGAATTAATTTCATTTCCCTCCCCACTCCATTAG  GTG AAC AAC CCA CAG ATC CAG CAC AAA GCT GTC AAT GAA

Leu Ile Arg Val Ile His Gln Leu Ser Pro Glu Ser Ser Leu Arg Lys Arg Lys Arg Ser Arg Cys ***
CTC ATC AGA GTG ATT CAC CAG CTG TCA CCA GAA TCT AGC CTA AGG AAG CGG AAA AGG AGT CGG TGC TGA TTCTGGG

GTAGAGAGTGTGCCAATAAGAAGAATTCTGCCAGCACTATTTGAATTTTAAAACTAAACCTATTTATTAATATTTAAATTTATTTATATGGAGAATATA

TTTTAGACTCATCAACCAAAGAAGTATTTATAGTAACAACTTATATGTGATAAGGATGAATTTCTATTAATATATGTGTTATTTATAATCTCTGTGTCC

TTAATTATTCCTCTTTGACCAATCATTCTTTCTGACTAATTAACCCAGACTGTGATTATGAAGTTGTATCTGGGGTGGGGGACAGCCAACCAGCTGAC

TGAACTCACACTGTGGCTTGTGCACTTACTTCACTTGCCAACGGGGAACATTCAGAACTGCAATGACCCCGTGAGGTGCTGCTGACCAGAGGAATGTCT

ATACATCGGCC
```

FIG. 2. Partial nucleotide sequence of the chromosomal IFN-γ gene of the rat. Indicated are the putative mRNA start site (indicated by an asterisk) and the signal peptide cleavage site (indicated by a vertical arrow).

gene, we have extrapolated the presumed translation initiation codon, the signal peptide cleavage site and RNA splicing events to this sequence, resulting in a predicted rat IFN-γ protein containing a 19 amino acid long signal sequence and a 137 amino acid long mature protein.

Expression of the Rat IFN-γ in Eukaryotic Cells

In order to prove the identity of the DNA sequence as being the chromosomal gene for rat IFN-γ, it was necessary to express the genetic information and to show that the resulting protein has biological activities characteristic for rat IFN-γ.

As outlined in Fig. 3, the gene was manipulated in such a way that it could be inserted into both an adenovirus major late promoter and a SV40 early promoter based expression vector (pAdRIFN-γ and pSVRIFN-γ, respectively). In both cases, the genetic information of the rat IFN-γ gene was supplied as two fragments, a proximal part derived from the *Pvu*II site (located in the 5′ untranslated region of the gene) towards *Eco*RV, and a distal part ranging from *Eco*RV toward *Pst*I (in the case of the Ad MLP-construct) or *Eco*RV–*Eco*RI (in the case of the SV40 construct).

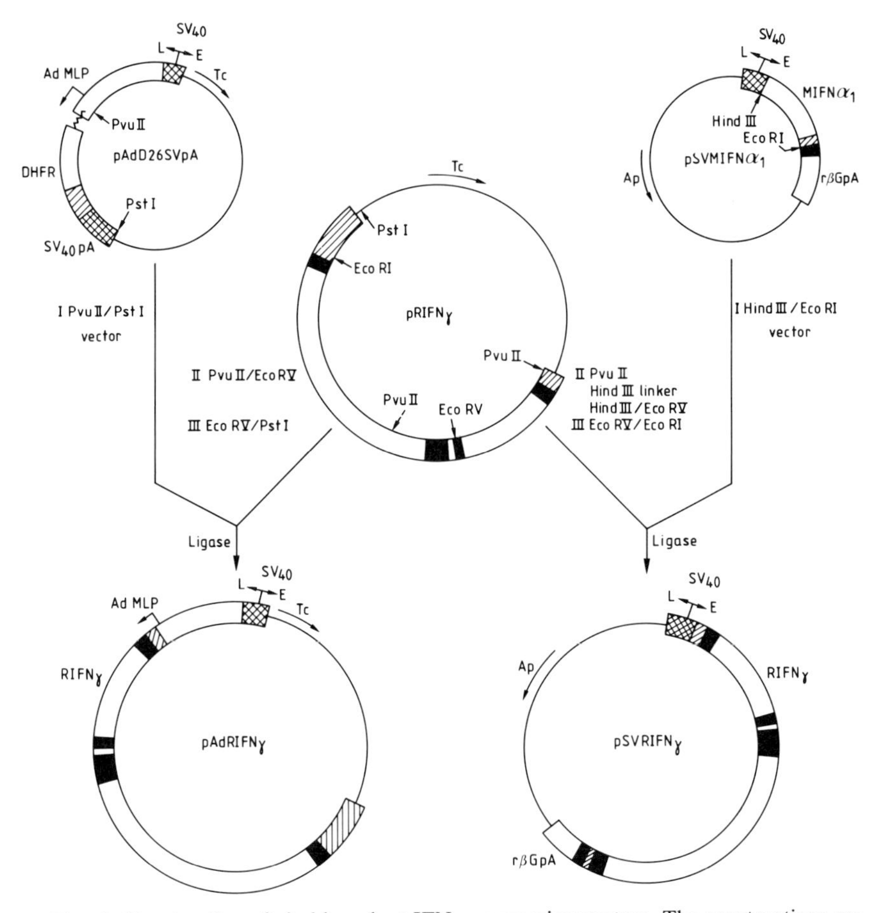

FIG. 3. Construction of viral-based rat IFN-γ expression vectors. The constructions are described in the text. Only the relevant *Pvu*II sites of pRIFN-γ are indicated. Ap and Tc stands for ampicillin and tetracycline resistance, respectively; AdMLP, adenovirus 2 major late promoter; SV40, simian virus 40 origin/promoter region.

In the latter construct, the natural signals for transcription termination and polyadenylation are replaced by those of the rabbit β-globin gene. The construction of these hybrid vectors was checked by restriction enzyme analysis and DNA sequencing. The resulting expression vectors were transfected into the monkey cell lines COS-1 and COS-7, which both endogenously express the SV40 T antigen necessary for replication of such SV40-origin based hybrid vectors.[3]

Transfected cells were grown for 10 days and the culture supernatant was assayed every 24 hr for antiviral activity. Both cell lines exhibited the same transient expression with an optimum at 48 hr after transfection, reaching rat IFN-γ titers of 400 units/ml for pAdRIFN-γ and 1600 units/ml for pSVRIFN-γ. Different IFN-γ levels might well reflect different promoter strengths.

In order to obtain a permanent source of recombinant DNA derived rat IFN-γ, a CHO cell line was cotransformed with pAdRIFN-γ and the *E. coli gpt* gene containing vector pEV_1gptH^+ (kindly provided by Dr. Hoeymakers, Erasmus University, Rotterdam, The Netherlands). In addition, a $CHO\text{-}DHFR^-$ cell line was cotransformed with pSVRIFN-γ and the modular murine *dhfr* gene containing vector pAdD26SVpA-3.[4] Cotransformants of rat IFN-γ with the *E. coli gpt* gene were found to produce about 200 units/ml, whereas cotransformants of rat IFN-γ with the murine *dhfr* gene produced levels ranging from 20 to 10,000 units/ml. Cotransformants with the murine *dhfr* gene were used for further characterization.

Culture of CHO Cell Lines and the Production of Rat IFN-γ

$CHO\text{-}DHFR^-$ cells were maintained in tissue culture flasks (150 cm^2, Costar 3150) at 37° in a humidified atmosphere containing 5% CO_2. The flasks contained 50 ml Dulbecco's modified Eagle's medium (DME) supplemented with fetal calf serum (FCS, 5%), streptomycin (130 μg/ml), penicillin (100 units/ml), glutamine (300 μg/ml), adenosine (10 μg/ml), deoxyadenosine (10 μg/ml), and thymidine (10 μg/ml). $CHO\text{-}DHFR^+$ transformants were cultured under the same conditions, except that the ribonucleosides and deoxyribonucleosides were omitted from the medium. This latter medium is referred as a selective DME. Initially, 15 single colonies were observed in the selective medium after cotransformation of pSVRIFN-γ and pAdD26SVpA-3. These presumed transformants were isolated by using a cloning cylinder and transferred to small plastic

[3] Y. Gluzman, *Cell* **23,** 175 (1981).

[4] S. J. Scahill, R. Devos, J. van der Heyden, and W. Fiers, *Proc. Natl. Acad. Sci. U.S.A.* **80,** 4654 (1983).

flasks (25 cm^2, Falcon 3013F). Cells reached confluency 12 days later. Samples of the culture fluid were screened for antiviral activity on a rat embryonic cell line (Ratec). Two of 15 transformants (defined as γ2 and γ8) contained about 1200 units of interferon per ml culture medium. Confluent monolayers of these two transformed cell lines were treated with trypsin and the cell suspensions were diluted in selective medium to a concentration of 5 cells per ml; 100 μl of these dilutions was transferred to individual wells of a 96-well microtiter plate. Confluency was reached for the individual clones after different incubation times (between 14 and 25 days) at 37° after which the growth medium of the wells was screened for IFN activity.

Detailed screening of 51 wells resulted in the selection of 45 clones which produced IFN at 20–10,000 units per ml (in small culture flasks of 25 cm^2 containing 4 ml selective medium). Six clones produced no detectable amounts of activity and one of them (γ-8N) was maintained to act as a control in experiments described below. One high producing line (constitutively secreting about 10,000 units/ml) was designated as γ-8A and was used for further study. This subline was subjected to a series of passages in selective medium and no decrease in its capacity to synthesize rat interferon was observed up to the eleventh passage. In addition, it was found that IFN activity was secreted at a constant level for at least 7 days, provided that the medium was changed every 24 hr.

For the large-scale production of rat interferon gamma, cells of clone γ8A were cultured in plastic flasks (Costar 3150, 150 cm^2) in selective medium until confluency was reached (1.2×10^7 cells). They were then dispersed by trypsin and suspended in 10 ml of selective medium. The trypsinized suspension was then transferred to a 2-liter roller bottle (Falcon 3027) and replenished with 90 ml of selective medium. Cells were grown to confluency (usually within 4 days) and washed with two successive additions of 50 ml Hanks' balanced salt solution (Flow Lab. Cat. No. 18-104-54), after which 50 ml of serum-free selective medium was added. The medium was changed every 24 hr (for at least 7 days). The decanted culture media were pooled and centrifuged to remove residual cells. The supernatant was called crude IFN-γ and used for further processing.

Concentration and Partial Purification of Rat IFN-γ by Adsorption/Desorption on Controlled-Pore Glass (CPG)

One liter of crude interferon gamma from pooled culture media was aseptically transferred to a 2-liter roller bottle (Falcon 3027). CPG beads (CPG 00120; mesh size, 200/400; Serva, Feinbiochemica) were added at a concentration of 10 mg/ml. The bottle was placed on a roller apparatus

PURIFICATION OF RAT IFN-γ BY CONTROLLED-PORE GLASS (CPG) ABSORPTION

Purification step	Volume	IFN total units	Protein (mg)	Spec. act. (units/mg)	Purification factor
Crude interferon	1000	3×10^6	80	3.8×10^4	1
CPG eluate	22	2.7×10^6	11	2.5×10^5	7

and rotated at 2 rpm at 4° for 20 hr. After the beads had settled for 20 min, the supernatant was removed by aspiration. The beads were washed extensively by four successive additions of 300 ml sterile phosphate-buffered saline (PBS, pH 8.3). They were then collected in a 50-ml plastic tube and 25 ml of a solution containing 50% ethylene glycol plus 1.4 *M* NaCl in PBS (pH 8.3) was added. The mixture was rotated for 24 hr at 4° and the beads were pelleted by low speed centrifugation (5 min, 600 *g*). About 90% of the original activity was recovered in the ethylene glycol fraction. The procedure resulted in a 40-fold concentration and a 7-fold increase in specific activity (see the table). To visualize rat IFN-γ, the crude and partial purified preparations were subjected to SDS–gel electrophoresis.[5] Proteins of known molecular weight were used to determine the molecular weight of rat IFN-γ. The results shown in Fig. 4 strongly suggest that the substance is an IFN species with an M_r of 18,000. However, two additional bands with molecular weights of 35,000 and 55,000 were observed. These bands might well reflect oligomeric forms; however, direct evidence for this interpretation at present is lacking.

Characterization of Rat IFN-γ

It is well known from the literature[6,7] that immune interferon is rapidly inactivated at pH 2. Because it might be useful to work at low pH during purification or characterization, the stability of recombinant rat IFN-γ at different hydrogen ion concentrations was studied.

Figure 5 shows the relationship between the pH and the stability of rat IFN-γ after incubation of a crude preparation for 20 hr at 4°. The results show that a gradual shift to acidity results in an almost linear reduction in the biological activity from 30 to 40% at pH 7 to more than 90% at pH 3. On the other hand, a shift toward alkalinity stabilizes the activity. To

[5] U. K. Laemmli, *Nature (London)* **227,** 680 (1970).

[6] J. Wietzerbin, S. Stefanos, M. Lucero, E. Falcoff, J. O'Malley, and E. Sulkowski, *J. Gen. Virol.* **44,** 773 (1979).

[7] J. Wietzerbin and E. Falcoff, this series, vol. 78, p. 552.

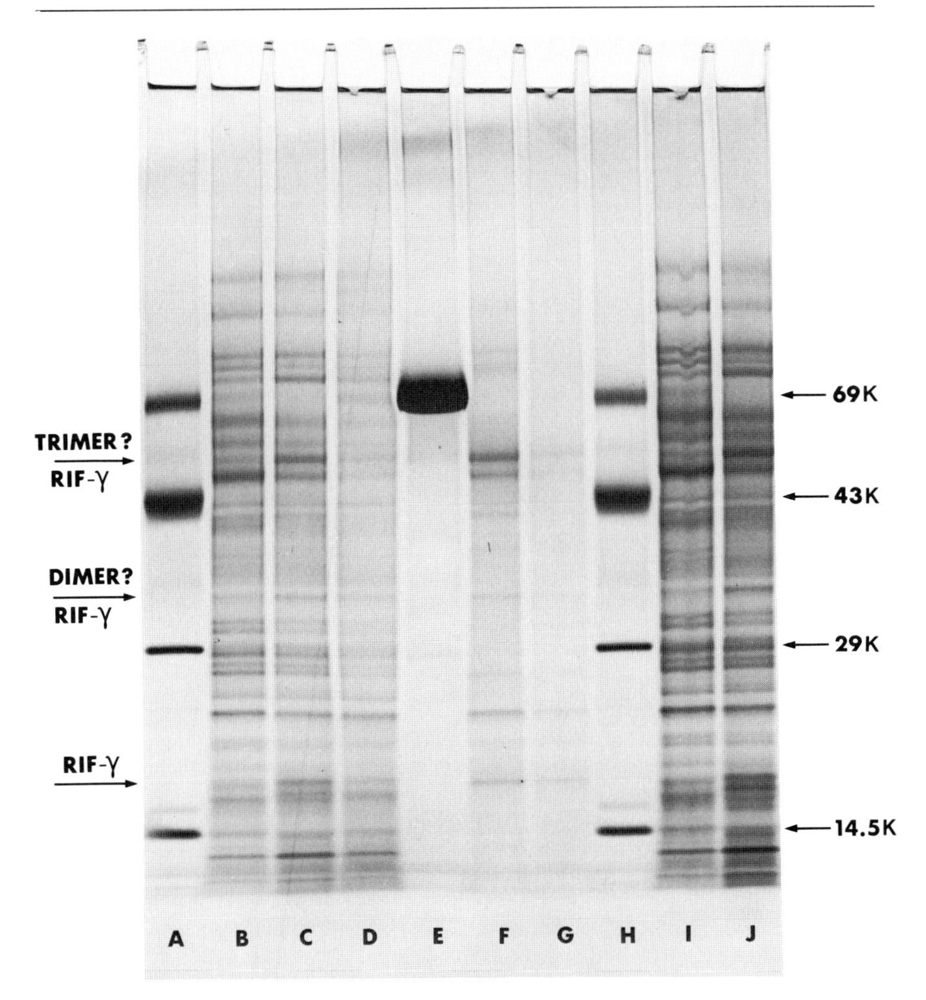

FIG. 4. SDS–polyacrylamide gel electrophoresis of crude and partially purified rat IFN-γ in a 5–30% gradient gel. As molecular weight standards we used (from top to bottom): bovine serum albumin (69K); ovalbumin (43K); carbonic anhydrase (29K) and lysozyme (14.5K). The gel was run as described by Laemmli[5] and stained with Coomassie brilliant blue. The presumed interferon band(s) is (are) indicated by (an) arrow(s). Lane A, marker proteins. Lane B, 40–85% saturated ammonium sulfate (a.s) fraction (100 μg protein) clone γ8N (nonproducing). Lane C, 40–85% saturated a.s. fraction (100 μg protein) clone γ8A (producing). Lane D, 0–85% saturated a.s. fraction (100 μg protein) clone γ8A. Lane E, human serum albumin (100 μg protein). Lane F, CPG fraction (20 μg protein). Lane G, CPG fraction (10 μg protein). Lane H, marker proteins. Lane I, 40–85% saturated a.s. fraction (200 μg protein) from clone γ8N. Lane J, 40–85% saturated a.s. fraction (200 μg protein) from clone γ8A.

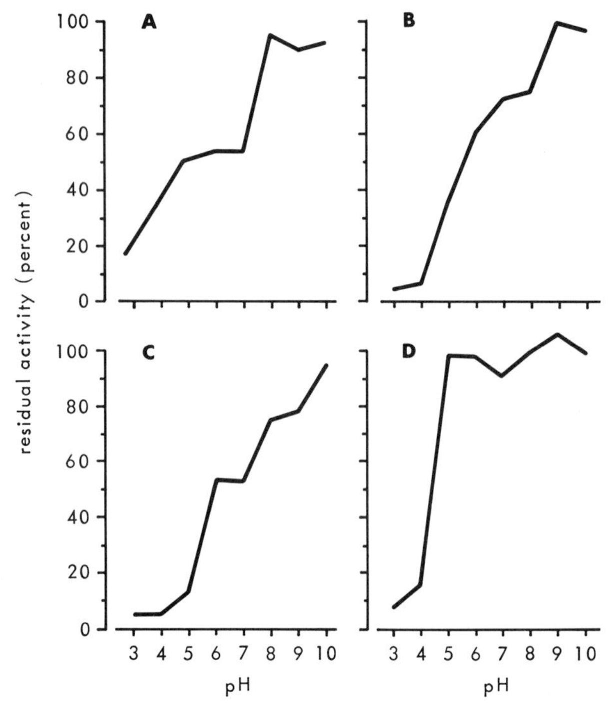

FIG. 5. The effect of pH on the stability of rat IFN-γ in the presence of different chemical substances. Crude IFN-γ samples (culture fluid) supplemented with the different chemicals mentioned below were stored at 4° for 20 hr. After this period, they were directly assayed for interferon activity. (A) No additions; (B) 6 m*M* 2-mercaptoethanol; (C) 6 m*M* 2-mercaptoethanol, 1 m*M* phenylmethylsulfonylfluoride (PMSF); (D) 6 m*M* 2-mercaptoethanol, 1 m*M* PMSF, and 15% glycerol.

determine whether certain additives would increase the stability at neutral and acid pH, the following chemicals were added to the test preparations. Glycerol, phenylmethylsulfonyl fluoride, and 2-mercaptoethanol were added to final concentrations of 15% (v/v), 1 m*M*, and 1% (v/v), respectively. These experiments revealed that the activity is completely maintained after the addition of glycerol at pH values above 5. The other two substances tested were not effective. However, a small but significant destabilizing effect was observed with 2-mercaptoethanol at low pH values (see Fig. 5).

A major characteristic of a glycosylated interferon species is its remarkable stability following a number of different physiochemical treat-

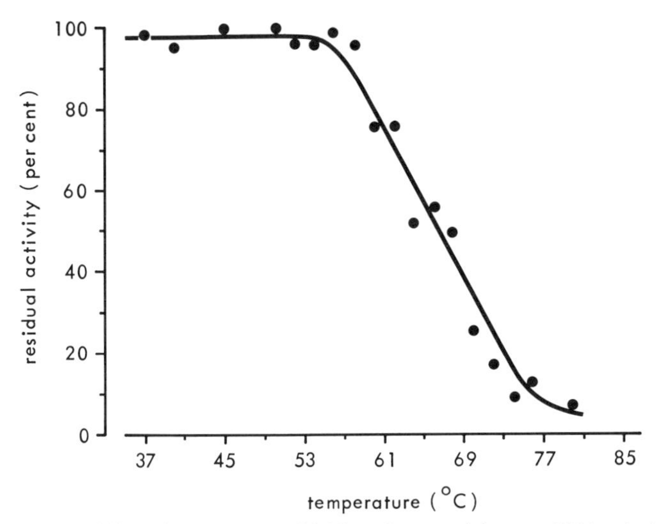

FIG. 6. Heat stability of rat IFN-γ. Initial interferon activity was 3200 units/ml. Protein concentration was 0.05 mg/ml. The interferon activity was determined as described.[9] Details of the experiment are outlined in the text.

ments, including heat.[8] Because there is reason to believe that the rat IFN-γ protein is glycosylated, the thermal stability of crude preparations (at pH 8.0) was investigated; 0.2 ml samples in 10 × 74-mm tightly capped plastic tubes were incubated in a waterbath at temperatures varying from 37 to 85°. After 2 min, the samples were rapidly cooled by the addition of 0.8 ml ice-cold phosphate-buffered saline and stored at 0° before they were assayed for antiviral activity (on the same day). A Ti50 (temperature at which 50% of the biological activity is lost) of 66° indicates that rat IFN-γ is rather stable under these conditions. Thc inactivation curve is shown in Fig. 6.[9]

Another aspect which is important for processing or purification of rat IFN-γ is long-term stability. In the first experiment, 2 ml portions (crude preparations) in plastic tubes were held at 4° for 10 days. On alternate days, samples were placed at −70°. At the end of the 10-day period, all samples were warmed to 4° and assayed for antiviral activity. The experiments revealed a minor reduction in activity, gradually going down to 80%. The same experiment performed at room temperature results in a

[8] S. E. Grossberg, *N. Engl. J. Med.* **287,** 13, 79, 122 (1972).

[9] P. H. van der Meide, J. Wubben, K. Vijverberg, and H. Schellekens, this volume [31].

loss of 90% within 5 days. At 37°, 95% inactivation was observed within 48 hr. When rat IFN-γ was subjected to repeated freezing and thawing, the biological activity remained remarkably stable. Only 30% of the activity was lost after 10 cycles. In the presence of 15% glycerol, this loss was reduced to 10%.

The most effective way of preserving activity is freeze-drying and storage at 4°. No loss of activity has been observed so far (up to 4 months).

[64] Isolation of Bovine IFN-α Genes and Their Expression in Bacteria

By BARUCH VELAN, SARA COHEN, HAIM GROSFELD, and AVIGDOR SHAFFERMAN

Bovine interferons (Bo-IFN) may prove to be ideal broad spectrum antiviral agents in veterinary medicine. Large-scale production of Bo-IFN from bovine cells seems impractical because of the low yields and the complexity of the procedure.[1] The techniques of recombinant DNA technology, leading to the identification and isolation of the human IFN genes,[2–5] permitted the development of industrial processes for production of interferons in heterologous cell systems.

The structural similarity between leucocyte Bo-IFN and human IFN-α[1] indicates that the genes coding for these polypeptides are related. In this chapter we describe the construction of a bovine DNA library, the use of Hu-IFN-α probes for identification of Bo-IFN-α gene sequences in a genomic library, the isolation of these genes, their expression in bacteria, and some biological and physicochemical properties of the bacterial Bo-IFN-α.

[1] S. Cohen, B. Velan, T. Bino, H. Rosenberg, and A. Shafferman, this volume [21].

[2] S. Nagata, H. Taira, A. Hall, L. Johnsrud, M. Streuli, J. Ecsodi, W. Bell, K. Cantell, and C. Weissman, *Nature (London)* **284,** 316 (1980).

[3] T. Taniguchi, M. Sakai, Y. Fuji-Kuriyama, M. Muramatsu, S. Kobayashi, and T. Sudo, *Proc. Jpn. Acad., Ser. B* **55,** 464 (1979).

[4] D. V. Goeddel, E. Yelverton, A. Ullrich, H. L. Heyneker, G. Miozzari, W. Holmes, P. H. Seeburg, T. Dull, L. May, N. Stebbing, R. Crea, S. Maeda, R. McCandliss, A. Sloma, J. M. Tabor, M. Gross, P. C. Familleti, and S. Pestka, *Nature (London)* **287,** 411 (1980).

[5] S. Pestka, *Arch. Biochem. Biophys.* **221,** 1 (1983).

Preparation of Fragmented Bovine DNA for Construction of a Genomic Library

Buffers and Reagents

TSNE buffer: 20 m*M* Tris·HCl, pH 7.6, 0.5% SDS, 1.0 *M* NaCl, 1 m*M* EDTA
TNE buffer: 20 m*M* Tris·HCl, pH 7.6, 1.0 *M* NaCl, 1 m*M* EDTA
TE buffer: 10 m*M* Tris·HCl, pH 7.6, 1 m*M* EDTA
TN buffer: 10 m*M* Tris·HCl, pH 7.6, 0.1 *M* NaCl
UNT buffer: 6.5 *M* urea, 1 *M* NaCl, 10 m*M* Tris·HCl, pH 7.6
RNase, Sigma type III, was dissolved in 10 m*M* Tris·HCl at a concentration of 20 mg/ml and heated to 100° for 15 min
Proteinase K was obtained from Boehringer
T_4-DNA ligase and restriction enzymes were obtained from New England Biolabs.

The commercially available bovine DNA is prepared from bulk treatment of calf thymuses. It is more prudent to construct a genomic library from DNA isolated from an individual animal to minimize allelic variations. A good source for high-molecular-weight DNA is the liver of a freshly slaughtered cow. DNA isolation from the liver is carried out essentially as described by Blin and Stafford.[6] Frozen liver (50 g) is ground to a thin powder in a Waring blendor in the presence of liquid nitrogen. The powder is added in small amounts to a solution of 400 ml phenol and 400 ml of TSNE. The mixture is stirred at room temperature for 30 min and then centrifuged at 4000 rpm for 10 min in a Sorvall GSA rotor. The aqueous phase is reextracted with phenol and then with ether in a separatory funnel (2000 ml of ether are needed per 100 ml of aqueous phase). The DNA preparation is placed in a dialysis bag and RNase is added to make a final concentration of 50 μg/ml. Dialysis is carried on overnight at room temperature versus 10 liters of TNE buffer. EDTA (50 m*M* final concentration), SDS (0.5% final concentration), and proteinase K (100 μg/ml final concentration) are added to the DNA solution which is then incubated for 3 hr at 37°. DNA is then extracted once with phenol and once with chloroform followed by extensive dialysis against TE buffer for 3 days. The dialysis buffer is changed every 12 hours. DNA is precipitated with 2 volumes of ethanol after addition of 1/10 volume of 5 *M* NaCl and resuspended in TE buffer at a concentration of 1 mg DNA per ml.

Bovine DNA fragments (~16 kb) are obtained by partial digestion of high-molecular-weight DNA with the restriction enzyme *Sau*3A as follows: 400 μg of calf liver DNA is incubated for 1 hr at 37° with 100 units of

[6] N. Blin and D. W. Stafford, *Nucleic Acids Res.* **3**, 2303 (1976).

*Sau*3AI. Presence of sufficient amounts of 16 kb fragments in the digest should be verified by agarose gel electrophoresis. The reaction is stopped by adding phenol and the DNA is extracted twice with phenol/chloroform, precipitated with ethanol, and dissolved in 0.5 ml of TE buffer.

To fractionate the DNA, the preparation is heated 10 min at 68°, chilled to 20°, and layered on a 38 ml 10–40% sucrose gradient in 1 *M* NaCl, 20 m*M* Tris·HCl, pH 8.0, and 5 m*M* EDTA. Centrifugation is performed at 26000 rpm in a Beckman SW 27 rotor for 24 hr at 20°. Fractions of 0.5 ml are collected, 10-μl aliquots are analyzed by electrophoresis through an 0.5% agarose gel. Gradient fractions containing DNA in the 12–18 kb size range are pooled. The pooled fractions are dialyzed against 4 liters of TE buffer and then concentrated by chromatography on DEAE cellulose. The DNA is loaded on an 0.3 ml column of DE-52 (Whatman) previously equilibrated with TN buffer. The column is then washed with several volumes of the equilibration buffer. The bound DNA is eluted with UNT buffer: 0.5 ml fractions are collected and those containing DNA are pooled, precipitated, and dried under vacuum. Complete resuspension of the dried pellets at a concentration of 1 mg/ml was achieved by 4 hr incubation at room temperature in TE buffer. Following this treatment 400 μg of chromosomal liver DNA yielded about 60 μg of purified 12–18 kb DNA fragments.

Construction of a Bovine Genomic Library

Bacteriophage λ L47.1 designed by Loenen and Brammar[7] can be used as a cloning vector for the construction of the genomic library. DNA is prepared from CsCl purified phage particles.[8] The mid region of the phage DNA (stuffer fragment) should be removed to accommodate the 12-18 kb *Sau*3A bovine DNA fragments. This is done by digestion of the phage DNA with a restriction enzyme *Bam*HI, followed by sucrose density gradient centrifugation. Three fragments are generated by *Bam*HI digestion. A 23.5 kb fragment representing the left arm, a 10.5 kb fragment representing the right arm, and a 6.6 kb fragment. Quantitative separation of arms from the fragment can be obtained by annealing the right and the left arms of the vector through their cohesive ends to form a 34 kb fragment prior to the sucrose density gradient separation. In addition, the stuffer fragment can be further reduced in size by enzymes which cleave this fragment exclusively. In the case of λ L47.1 *Xho*I and *Sal*I were used to produce fragments smaller than 4.3 kb.

[7] W. A. Loenen and W. J. Brammar, *Gene* **10,** 249 (1980).

[8] K. R. Yamamoto, B. M. Alberts, R. Benzinger, L. Lawhorne, and G. Trieber, *Virology* **40,** 734 (1970).

Digestion with restriction enzymes, annealing of arms, and removal of the stuffer by centrifugation are performed as suggested by Maniatis *et al.*[9] and the purified λ L47.1 arms are ligated to the *Sau*3AI fragments of bovine DNA. First the cohesive ends of the arms are annealed by incubation for 1 hr at 42° in 66 m*M* Tris·HCl, pH 8.0, and 11 m*M* $MgCl_2$; then ATP, DTT, bovine DNA, and ligase are added and the reaction is transferred to 14° for 16 hr. The final reaction in a total volume of 100 μl contains 10 μg of arms, 2.5 μg of bovine DNA 60 m*M* Tris·HCl, pH 8.0, 10 m*M* $MgCl_2$, 1 m*M* ATP, 15 m*M* DTT, and 1000 units of T_4-DNA ligase.

For packaging λ DNA we have used a commercial kit (Amersham) prepared essentially according to Hohn[10] which yielded 7×10^5 PFU per 1 μg of ligated DNA. Packaging of 4 μg of ligated DNA thus results in a primary library of 2.8×10^6 phages which should represent the entire bovine genome. The genomic library was amplified by propagating the bacteriophages in *E. coli* LE 392 on agar plates.

Screening of Bovine Genomic Library for Interferon Genes

Buffers and Reagents

Denaturing solution: 1.5 *M* NaCl, 0.5 *M* NaOH

Neutralizing solution: 1.5 *M* NaCl, 1.0 *M* Tris·HCl, pH 7.6

Denhardt's solution: 0.02% Ficoll, 0.02% polyvinylpyrrolidone, 0.02% BSA

Formamide was deionized by batch treatment with AG501-X8 resin (Bio-Rad Laboratories)

Salmon sperm DNA (Sigma Type III) resuspended in water at a concentration of 10 mg/ml and denatured by boiling for 10 min and quick chilling

Identification of hybrid λ clones containing Bo-IFN-α genes is based on cross hybridization between Bo-IFN-α and Hu-IFN-α sequences. The optimal conditions for the hybridization can be determined by using Hu-IFN-α probes and bovine genomic blots. We have isolated various DNA fragments from a cDNA clone carrying the sequence of Hu-IFN-αJ1[11,12] labeled them with ^{32}P by nick translation,[13] and hybridized them under different stringencies to blots of *Eco*RI digested genomic bovine DNA.

[9] T. Maniatis, R. C. Hardison, E. Lacy, J. Laner, C. O'Connell, G. K. Sim, and A. Efstratiaidis, *Cell* **15,** 687 (1978).

[10] B. Hohn, this series, Vol. 68 [19].

[11] S. Cohen, B. Velan, and A. Shafferman, *in* "Advances in Biotechnological Processes" (A. Mizrahi and A. L. Van Wezel, eds.), Vol. 4, p. 1. Liss, New York, 1985.

[12] A. Shafferman, B. Velan, H. Grosfeld, M. Leitner, Z. Shalita, C. Frist, and S. Cohen, *Dev. Biol. Stand.* **60,** 111 (1985).

[13] P. W. J. Rigby, M. Dieckmann, C. Rhodes, and P. Berg, *J. Mol. Biol.* **98,** 503 (1975).

It was found that the most suitable DNA probe for hybridization spans the sequence between base pairs 70 and 250 at the 5′ end of the coding region of the Hu-IFN-αJ1 gene. These sequences which can be isolated on a 177 bp *Sau*3AI DNA fragment were used to screen the bovine genomic library.

Screening of the bovine genomic library for Bo-IFN sequences is done by *in situ* hybridization essentially according to Maniatis *et al.*[9] The genomic library is plated on thirty 150-mm Petri dishes at a concentration of 30,000 bacteriophage particles per plate. *E. coli* LE392 are used as host bacteria. The plates are incubated at 37° until the plaques reach a diameter of 1 mm (about 7 hr).

The plaques are transferred to nitrocellulose filters by placing the filters on top of the soft agar for 10 min. The filters are then peeled off and floated on top of a denaturing solution for 30 sec. The filters are then dipped into the solution for another 30 sec and transferred to the neutralizing solution for 5 min. The filters are rinsed in 2× SSC, dried at room temperature, and baked at 80° under vacuum for 2 hr.

Hybridization is carried out at 42° for 48 hr in the presence of 50% deionized formamide, 5× Denhardt's solution, 6× SSC, 0.5% SDS, and 100 μg/ml of denatured salmon sperm DNA. The ^{32}P-labeled *Sau*3A 177 bp fragment at a specific activity of 5×10^6 cpm/μg DNA was used as a probe. After hybridization the filters were washed in 2× SSC, 0.1% SDS. The washing procedure consists of three 15 min washes at room temperature followed by two 1.5 hr washes at 42°. The filters are then dried and exposed to an X-ray film. Plaques related to each hybridization spot were identified and phages from each plaque are replated and rehybridized to the 177 bp *Sau*3A probe. When screening about 10^6 clones, 9 hybridization spots were identified.

It was found that three of the clones hybridized to a lesser extent than the other six. DNA was isolated from all the clones, subjected to restriction enzyme digestion and blotted onto nitrocellulose filters.[14] The blots were used to determine the extent of homology between the cloned bovine sequences and the complete Hu-IFN-αJ1 gene. Three probes were used: the 177 bp *Sau*3A probe, an adjacent 270 bp *Sau*3A DNA fragment representing the mid part of the Hu-IFN-αJ1 gene and a 515 bp probe carrying 50 bp of the 3′ end of the coding sequences of the Hu-IFN-αJ1 gene and 3′ nontranslated sequences. Any clone that would hybridize to each of the 3 probes is a good candidate for carrying an IFN-α sequence. Hybridization conditions for Southern blots are identical to those employed for *in situ* hybridization. Out of the 9 clones tested, 6 hybridized

[14] E. Southern, *J. Mol. Biol.* **98,** 503 (1975).

with each of the 3 probes. The clones which did not hybridize with either the 270 or the 515 bp fragments were the same clones which hybridized weakly with the 177 bp probe in our initial *in situ* screening procedure.

Restriction maps of the 6 positive clones were determined by multiple enzyme digestion and the Bo-IFN-α sequences were localized by blot hybridization to the various Hu-IFN-αJ1 probes (Fig. 1). Altogether, at least 5 distinct IFN-α genes or pseudogenes can be isolated from the bovine genome with the 177 bp *Sau*3A probe. (Two of the clones represent overlapping regions on the genome.)

The DNA fragments containing the Bo-IFN-α genes designated A, B, and C were subcloned and their sequences were determined. The sequence of the Bo-IFN-αC gene is given in Fig. 2. It contains an open reading frame of 570 bp encoding for 189 amino acids. The 23 first amino acids in the putative polypeptide probably represent the signal peptide characteristic of IFN molecules. The putative mature protein is related in its amino acid sequence to the Hu-IFN-αs. The average amino acid homology to a consensus Hu-IFN-α sequence[4,5] is 64%, though long regions

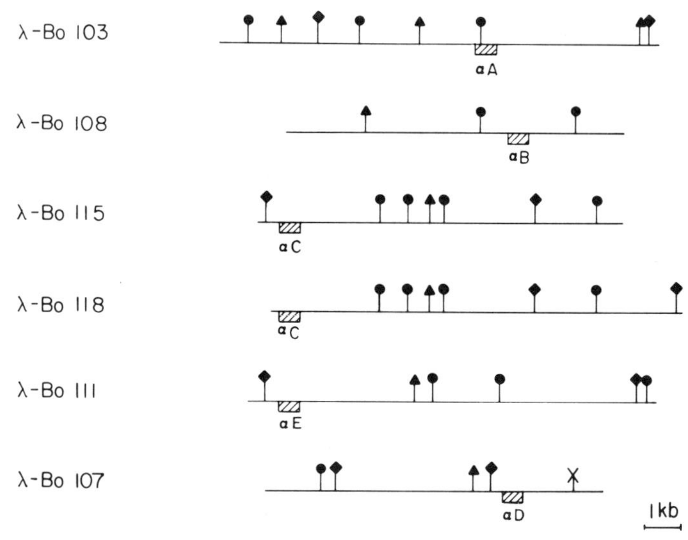

FIG. 1. Restriction maps of bovine genomic λ clones which bear the IFN homologous DNA sequences. The maps do not include phage vehicle sequences. 103, 107, 108, 111, 115, and 118 refer to the clone number. ◆, *Bam*HI; ●, *Eco*RI; ▲, *Hin*dIII; ×, *Sal*I. Shaded bars refer to region hybridizing to Hu-IFN-α sequences. The exact location of IFN related sequences was determined by further mapping, IFN genes or pseudogenes are designated by capital letters.

```
                  S1                                                                         S20
                  met ala pro ala trp ser phe arg leu ala leu leu leu leu ser cys asn ala ile cys
AACCTCCCCAAGGTCCCCA ATG GCC CCA GCC TGG TCC TTC CGC CTG GCC CTG CTG CTG CTC AGC TGC AAT GCC ATC TGC

            1                                                                 20
ser leu gly Cys His Leu Pro His Thr His Ser Leu Ala Asn Arg Arg Val Leu Met Leu Leu Gly Gln Leu Arg
TCT CTG GGC TGC CAC CTG CCT CAC ACC CAC AGC CTG GCC AAC AGG AGG GTC CTG ATG CTC CTG GGA CAA CTG AGG

                                                            40
Arg Val Ser Pro Ser Ser Cys Leu Gln Asp Arg Asn Asp Phe Ala Phe Pro Gln Glu Ala Leu Gly Gly Ser Gln
AGG GTC TCC CCT TCC TCC TGC CTG CAG GAC AGA AAT GAC TTT GCA TTC CCC CAG GAG GCG CTG GGT GGC AGC CAG

                                                60
Leu Gln Lys Ala Gln Ala Ile Ser Val Leu His Glu Val Thr Gln His Thr Phe Gln Leu Phe Ser Thr Glu Gly
TTG CAG AAG GCT CAA GCC ATC TCT GTG CTC CAC GAG GTG ACC CAG CAC ACC TTC CAG CTT TTC AGC ACA GAG GGC

                            80
Ser Ala Thr Met Trp Asp Glu Ser Leu Leu Asp Lys Leu Arg Asp Ala Leu Asp Gln Gln Leu Thr Asp Leu Gln
TCG GCC ACC ATG TGG GAT GAG AGC CTC CTG GAC AAG CTC CGC GAT GCA CTG GAT CAG CAG CTC ACT GAC CTG CAA

        100                                                                      120
Phe Cys Leu Arg Gln Glu Glu Glu Leu Gln Gly Ala Pro Leu Leu Lys Glu Asp Ser Ser Leu Ala Val Arg Lys
TTC TGT CTG AGG CAG GAG GAG GAG CTG CAA GGA GCT CCC CTG CTC AAG GAG GAC TCC AGC CTG GCT GTG AGG AAA

                                                                    140
Tyr Phe His Arg Leu Thr Leu Tyr Leu Gln Glu Lys Arg His Ser Pro Cys Ala Trp Glu Val Val Arg Ala Gln
TAC TTC CAC AGA CTC ACT CTC TAT CTG CAA GAG AAG AGA CAC AGC CCT TGT GCC TGG GAG GTT GTC AGA GCA CAA

                                                160
Val Met Arg Ala Phe Ser Ser Ser Thr Asn Leu Gln Glu Ser Phe Arg Arg Lys Asp END
GTC ATG AGA GCC TTC TCT TCC TCA ACA AAC TTG CAG GAG AGT TTC AGG AGA AAG GAC TGA CCACACACCTGGTTCAACA
```

FIG. 2. Genomic sequence of the Bo-IFN-αC gene and its putative polypeptide molecule. Amino acids are numbered, preceded by the letter S in the signal region. Sequence analysis of both strands was according to Maxam and Gilbert.[20]

of higher homology can be observed close to the COOH terminus and at amino acid positions 48–60 (Fig. 3).

The nucleotide sequences of the Bo-IFN-α genes A and B (data not shown) highly resemble the sequence of Bo-IFN-αC. The nucleotide sequence homology between the three genes is about 95% and the amino acid homology of the putative proteins is 93%.

It should be mentioned that the 3 Bo-IFN-α genes are not spliced thus resembling IFN-αs, from other species.[15,16]

[15] S. Nagata, N. Mantei, and C. Weissman, *Nature (London)* **287,** 401 (1980).

[16] G. D. Shaw, W. Bell, H. Taira, N. Mantei, P. Lengyel, and C. Weissman, *Nucleic Acids Res.* **11,** 555 (1983).

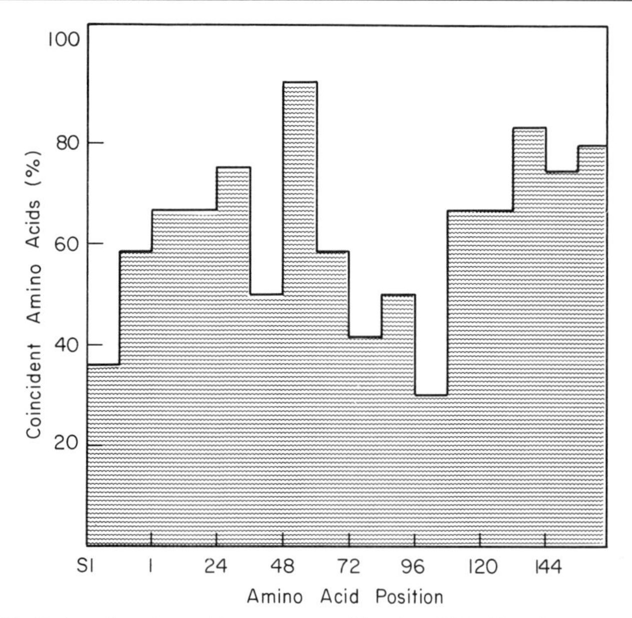

FIG. 3. Similarity of amino acid sequences of bovine IFN-αC and a consensus human IFN-α. The consensus Hu-IFNα was derived from Goeddel *et al.*[4] and Shaw *et al.*[16] Both the consensus and the Bo-IFN-αC sequences were subdivided in segments of 12 amino acids. The percentage of coincident residues was plotted as a function of map position.

Expression of the Bo-IFN-αC gene in *E. coli*

To show that the Bo-IFN-α genes code for a biologically active interferon product(s) the bovine DNA sequences were inserted into expression vehicles. We have constructed several plasmids in which foreign genes can be expressed from the *E. coli trp* promoter as fusion products.[17-19] Such plasmids do not require elaborate construction procedures and are useful for quick evaluation of the biological or biochemical properties of the potential gene product. One such plasmid, pSE2 was used for expressing the Bo-IFN-αC gene product. pSE2 is a derivative of pAS620[17] in which an *Eco*RI site was introduced in the coding sequence (the *Rsa*I site) of the *trp* leader peptide.

The steps involved in the construction of a system for expression of the bovine IFN-αC gene are summarized in Fig. 4 and described in detail

[17] A. Shafferman, R. Kolter, D. Stalker, and D. R. Helinski, *J. Mol. Biol.* **161,** 57 (1982).
[18] J. K. Rose and A. Shafferman, *Proc. Natl. Acad. Sci. U.S.A.* **78,** 6670 (1981).
[19] H. Grosfeld, S. Cohen, B. Velan, Z. Shalita, and A. Shafferman, *Mol. Gen. Genet.* **195,** 358 (1984).

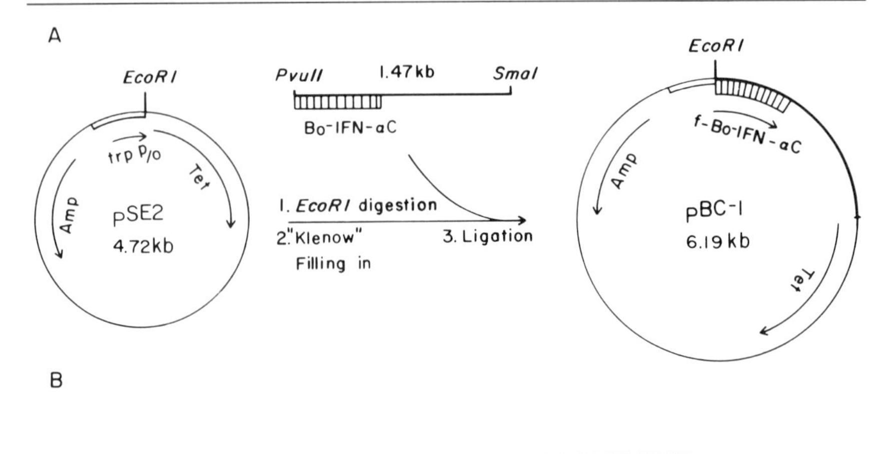

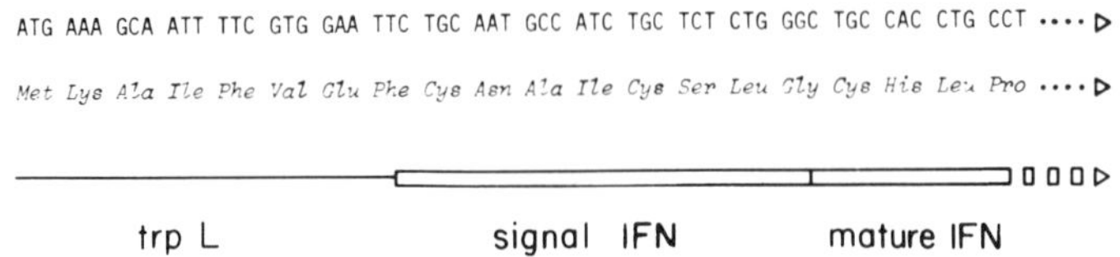

FIG. 4. Construction of a recombinant plasmid expressing a fused Bo-IFN-αC polypeptide. (A) Steps involved in the insertion of the bovine IFN-αC sequences in the *trp* expression vehicle pSE2 (see text). The open bar represents the 360 bp *trp*. *E. coli* promoter fragment containing the first 6 codons of the *trp*L. Shaded bar represents 525 bp of the coding sequence of the Bo-IFN-αC. The insertion of the bovine *Pvu*II–*Sma*I DNA fragment in the filled in *Eco*RI site of pSE2 regenerates the *Eco*RI site. (B) N-terminal sequences of the fused *trp*L-Bo-IFN-αC (f-Bo-IFN-αC) polypeptide.

below. The unique *Eco*RI site of pSE2 is cleaved and the cohesive ends are filled in by DNA polymerase I large fragment. The reaction is carried out at 15° for 1.5 hr in the presence of 50 m*M* Tris·HCl, pH 7.6, 7 m*M* $MgCl_2$, 1 m*M* 2-mercaptoethanol, 50 μg/ml BSA, 50 μ*M* of each of the 4 dNTPs, and 2 units of enzyme (New England Biolabs). The vector is then ligated to a *Pvu*II–*Sma*I fragment isolated from the hybrid λ clone 115. The 1.47 kb *Pvu*II–*Sma*I fragment was eluted from 6% acrylamide gels.[20] This fragment contains the entire coding sequence of the mature Bo-IFN-αC as well as 24 bp coding for part of the signal peptide and about 950 bp downstream from the stop codon of the gene (Fig. 4). For ligation 0.5 μg of filled in vector were incubated with 0.5 μg of the *Pvu*II–*Sma*I fragment for 16 hr at 14° in the presence of 66 m*M* Tris · HCl, pH 7.6, 10 m*M* $MgCl_2$, 10 m*M* DTT, 1 m*M* ATP, 200 μg/ml BSA, and 1000 units of T_4-DNA ligase.

[20] A. M. Maxam and W. Gilbert, this series, Vol. 65 [57].

E. coli HB101 cells are transformed with the recombinant DNA.[21] Ampicillin resistant clones containing Bo-IFN-α sequences are identified by *in situ* colony hybridization[22] to a ^{32}P-labeled IFN probe. Positive clones are analyzed to verify the correct fusion of the Bo-IFN sequences downstream from the *E. coli trp* promoter sequences. Such plasmids designated pBC-1 should express a fused polypeptide (f-Bo-IFN-αC) containing 6 amino acids derived from *trp* L, 2 amino acids from the linking region, 8 amino acids from the Bo-IFN signal peptide, and 166 amino acids of the mature Bo-IFN-αC (Fig. 4B).

To evaluate the biological activity and physicochemical properties of the fused polypeptide bacteria harboring pBC-1 were grown to a density of 2×10^8 cells/ml in 2 liters of M9 synthetic medium[23] supplemented with 0.5% casamino acids. The cells were harvested and resuspended in 40 ml of 10 m*M* Tris · HCl, pH 7.9, 10% sucrose, 5 m*M* EDTA, 0.2 *M* NaCl, and 10 m*M* phenylmethylsulfonyl fluoride,[24] and disrupted mechanically by a "Yeda" press. The debris was removed by centrifugation at 25,000 rpm for 1 hr in a Beckman 30 rotor. The supernatant is tested for antiviral activity in a microassay plate[25] on MDBK cells with vesicular stomatitis virus as challenge virus. NIH standard human leukocyte IFN (G-023-901-527) was used as a reference standard. Extracts from *E. coli* harboring the recombinant pBC-1 plasmid produce about 10^6 IFN units per liter culture. Recently we have constructed a plasmid expressing the Met-mature Bo-IFN-αC taking advantage of a *Fnu*4HI site which is located exactly between the last codon of the signal peptide and the first Cys codon of the mature Bo-IFN polypeptide.

Properties of Bacterial f-Bo-IFN-αC

The activity of the bacterial f-Bo-IFN-αC can be manifested on bovine MDBK cells but not on human fibroblasts. No loss of activity is observed when the interferon is exposed to 1% SDS or to pH 2. When Bo-IFN-αC is subjected to SDS–polyacrylamide gel electrophroesis in nonreducing conditions antiviral activity can be recovered at a molecular weight of 17,000.

The antiviral activity of Bo-IFN-αC is not neutralized by anti-Hu-IFN-α or anti-Hu-IFN-β antibodies. However, Bo-IFN-αC can be at-

[21] M. Dagert and S. D. Ehrlich, *Gene* **6,** 23 (1979).

[22] J. P. Gergen, R. H. Stern, and P. C. Wensink, *Nucleic Acids Res.* **7,** 2115 (1979).

[23] J. H. Miller, "Experiments in Molecular Genetics." Cold Spring Harbor Lab., Cold Spring Harbor, New York, 1972.

[24] T. Staehelin, D. S. Hobbs, H. I. Kung, and S. Pestka, this series, Vol. 78 [72].

[25] J. G. Tilles and M. Finland, *Appl. Microbiol.* **16,** 1706 (1968).

tached to immobilzied anti-Hu-IFN-α. Actually an immunosorbant column made of anti-human lymphoblastoid interferon bound to Sepharose 4B can be used to purify this interferon. It should be noted that the amount of Bo-IFN-αC that saturates such a column is about one-fiftieth of the amount of Hu-IFN-α that is bound under the same conditions. f-Bo-IFN-αC binds strongly to Affi-Gel Blue (Bio-Rad) columns and requires 1 *M* NaCl and 50% ethylene glycol for elution. These physicochemical properties suggest that the genetically engineered Bo-IFN-αC is a member of a subpopulation of bovine interferons produced upon viral induction by fresh leukocytes.[1]

[65] The Detection of Individual Cells Containing Interferon mRNA by *in Situ* Hybridization with Specific Interferon DNA Probes

By RAINER ZAWATZKY, JAQUELINE DE MAEYER-GUIGNARD, and EDWARD DE MAEYER

Introduction

The method of *in situ* hybridization permits localization and identification of nucleic acids of specific sequences in individual cells. This technique was first described by Gall and Pardue[1] for detecting ribosomal gene sequences and has been used by several investigators for detecting mRNAs in cells.[2] With the availability of cDNAs corresponding to the different interferon (IFN) species, it has become possible to set up experimental systems to identify individual cells synthesizing IFN mRNA. To be able to study IFN production at the single cell level is of obvious importance, since it can provide answers to several questions such as the number of IFN producing cells *in vivo* as a function of the inducing agent. Here we describe a method which we have used for detection of murine IFN-α and IFN-β mRNA in butyrate-treated mouse C-243 cells induced with Newcastle disease virus (NDV).

[1] J. G. Gall and M. L. Pardue, *Proc. Natl. Acad. Sci. U.S.A.* **63,** 378 (1969).
[2] P. Szabo and D. C. Ward, *Trends Biochem. Sci.* 425 (1982).

This method identifies individual IFN producing cells, which can at the same time be stained for morphological identification; it is derived from the one described by Brahic and Haase[3] and Haase *et al.*[4] for detection of visna virus RNA in cultures of infected sheep choroid plexus cells.

Preparative Procedures and Methodology

Labeling of cDNA Probes

Since IFN mRNAs are of relatively small size, the theoretical specific activity per molecule of tritium-labeled cDNA probe is rather low, rendering difficult autoradiographic detection because of the long periods of autoradiographic exposure that are required. We have circumvented this problem by using ^{35}S-labeled triphosphate deoxynucleotides for nick translation, enabling us to synthesize probes with specific activities of 2–4 $\times$ 10^8 dpm/μg. Since this represents 8–10 times more than specific activities usually obtained with tritium-labeled nucleotides and since the efficiency of grain development for ^{35}S is 5 times higher than for tritium,[4] the autoradiographic signal theoretically increases by a factor of 50 when ^{35}S-labeled nucleic acids are used, thus reducing autoradiographic exposures to a reasonable period.

We use the recombinant plasmids pMIF1204 and pMβ-3 which carry cDNA inserts that correspond to murine IFN-α or IFN-β, respectively.[5,6] After digestion with *Pst*I the inserts are isolated by electroelution and chromatography on DE52 columns, and labeled by nick translation with [^{35}S]dATP (New England Nuclear, specific activity > 1000 Ci/mmol). 1 25 μl reaction contains 50 m*M* Tris · HCl, pH 7.4, 10 m*M* $MgCl_2$, 1 m*M* dithiothreitol (Boehringer, Mannheim), 50 μg/ml nuclease-free bovine serum albumin (BSA, Bethesda Research Laboratories), 30 μM of each dCTP, dGTP, and dTTP, 250 μCi of [^{35}S]dATP (New England Nuclear), 5 units of DNA polymerase I, 100 pg of DNase I (both from Boehringer, Mannheim), and 400 ng of DNA template. The radioactive precursor is concentrated by lyophilization to approximately 20 m*M* to give a final concentration in the reaction mixture of 10 m*M*.

The mixture is then incubated at 15° and the rate of incorporation of the labeled nucleotide is assessed by TCA precipitation in the presence of

[3] M. Brahic and A. T. Haase, *Proc. Natl. Acad. Sci. U.S.A.* **75,** 6125 (1978).

[4] A. T. Haase, M. Brahic, and L. Stowring, *Methods Virol.* **7,** 189 (1984).

[5] K. A. Kelley, C. A. Kozak, F. Dandoy, F. Sor, D. Skup, J. D. Windass, J. De Maeyer-Guignard, P. M. Pitha, and E. De Maeyer, *Gene* **26,** 181 (1983).

[6] Y. Higashi, Y. Sokawa, Y. Watanabe, Y. Kawade, S. Ohno, L. Takaoka, and T. Taniguchi, *J. Biol. Chem.* **258,** 9522 (1983).

a carrier DNA (salmon sperm DNA, Calbiochem.). We find an incorporation of 30–40% already after 2 hr of incubation, with no further increase. A control DNA of about the same size should be labeled in parallel; for this we have used phage ϕX174 DNA digested by *Hae*III, yielding fragments from 1353 to 72 bp (purchased from BRL).

After chromatography on Sephadex G-75 (Pharmacia), the labeled cDNA is precipitated after addition of 2 μg of yeast tRNA (BRL) and dissolved 10 m*M* in Tris · HCl, pH 7.4, 1 m*M* EDTA at a concentration of 10–15 ng/ml.

The major fragment sizes of the nick translated cDNA probes ranged from 100 to 200 bp as determined by electrophoresis through an 8% polyacrylamide gel under denaturing conditions.

Pretreatment of Glass Slides and Coverslips

To reduce nonspecific binding of the labeled cDNA probe to the slide surface, several procedures for coating of glass slides have been described. We have chosen the procedure described by Haase *et al.*[4] which results in minimal loss of cells during subsequent treatments.

The slides are cleaned in 1 *M* HCl for 1 hr, washed in three changes of distilled water, and dipped in 95% ethanol for 30 min. After wiping dry with gauze, the slides are incubated for 3 hr at 65° in Denhardt's solution,[7] consisting of 0.02% BSA (fraction V), 0.02% polyvinylpyrollidone (both obtained from Sigma), 0.02% Ficoll (Pharmacia Fine Chemicals) in 3× SSC. After a brief wash in distilled water, they are treated with ethanol/acetic acid 3 : 1 for 20 min and air dried. Finally, the slides were acetylated according to the method of Hayashi *et al.*[8]: the slides are soaked in 0.1 *M* triethanolamine · HCl, pH 8.0, and 0.25% (v/v) of acetic anhydride is added under vigorous mixing to quickly dissolve the reagent. After 10 min the slides are washed in distilled water and dehydrated in 95% ethanol. After drying they are stored in a closed box at room temperature until use.

The coverslips used should be siliconized to prevent loss of cDNA probe; we use circular glass coverslips of 16 mm diameter and boil them for 30 min in 1 *M* HCl to remove alkali. After abundant washes in distilled water they are stored overnight in 95% ethanol, dipped in distilled water, then in a 1% solution of Aquasil (Pierce Chemicals) and rinsed briefly in distilled water again. Finally, they are heated to 100° for 2 hr and stored in a Petri dish.

[7] D. T. Denhardt, *Biochem. Biophys. Res. Commun.* **23,** 641 (1966).

[8] S. Hayashi, I. C. Gillam, A. D. Delanoy, and G. M. Tener, *J. Histochem. Cytochem.* **26,** 677 (1978).

Induction of Interferon

Mouse-IFN-α/β was induced in C-243 cells by Newcastle disease virus (NDV). Prior to IFN induction, the cells were incubated for 48 hr in the presence of 1 m*M* sodium butyrate as described.[9] Medium was then removed and the cell monolayer was infected with NDV (Kumarov strain) at an input multiplicity of 15 to 20 PFU per cell.

Preparation of Cells for in Situ Hybridization

We perform hybridization on coated glass slides onto which about 20,000 cells have been deposited by cytocentrifugation (Cytopsin 2, Shandon, UK).

At different times following induction, medium is removed from the culture flasks, and the cell layer is washed with warm PBS. Then, 10 ml of PBS containing 5% Newborn calf serum (NCS) is added, cells are detached from the plastic by vigorous shaking of the flask, and dispersed by repeated pipetting. They are then centrifuged for 10 min at 250 *g*, resuspended in PBS with 2% NCS, and adjusted to 2×10^5 cells per ml. Then 100 μl of the cell suspension is loaded into the vials of the cytocentrifuge and spun for 5 min at 500 rpm. After drying in air and fixing for 20 min with ethanol acetic acid (3:1), the preparations can be stored dry in a closed box at 4° for several weeks.

Prior to the hybridization step, the cells are subjected to treatments that facilitate diffusion of the labeled cDNA probe and enhance the accessibility of the mRNA in the cell. The slides carrying the preparation of fixed cells are treated with 0.2 *N* HCl for 20 min at room temperature, briefly washed in distilled water, and incubated for 30 min at 70° in 2× SSC. After a brief rinse in distilled water, they are transferred for 15 min at 37° to 20 m*M* Tris · HCl, pH 7.4, 2 m*M* $CaCl_2$ containing 1 μg/ml of proteinase K (purchased from Boehringer, Mannheim). We have tested various concentrations of proteinase K from 0.5 to 5 μg/ml and have chosen 1 μg/ml because it provided good hybridization signals with satisfying preservation of cellular morphology. When cells were treated with concentrations of 4 μg/ml or more of proteinase K we often observed complete loss of cytoplasmic morphology. Although the optimal concentration of proteinase K may vary from one type of cell to another, 1 μg/ml of proteinase K seems to provide good results in most of the cell or tissue specimen investigated so far (M. Brahic, personal communication).

After proteinase K digestion, the slides are washed twice in distilled water, and dehydrated through 70 and 95% ethanol for 5 min each time.

[9] A. Cachard and J. De Maeyer-Guignard, *Ann. Virol.* **132E,** 307 (1981).

Finally, for control studies, a slide with NDV-induced and one with non-induced C-243 cells is digested with a mixture containing 100 μg/ml of RNase A and 10 units/ml of RNase T_1 (both obtained from Boehringer, Mannheim) in 2× SSC. Then 20 μl of this solution is placed on the specimen, covered with a coverslip, and incubated at 37° for 20 min. The slides are then washed twice in 2× SSC for 5 min and dehydrated as above.

In Situ Hybridization

The hybridization solution contains 50% deionized formamide (prepared by stirring 50 ml of formamide with 5 g of mixed bed resin AG 501 × 8 (Bio-Rad) for 30 min followed by filtration), 10% dextran sulfate (Pharmacia), 0.6 *M* NaCl, 0.02% (w/v) each of BSA (fraction V), polyvinylpyrrolidone and Ficoll, 10 m*M* Tris · HCl (pH 7.4), 1 m*M* EDTA, 250 μg/ml of sonicated salmon sperm DNA, 1 mg/ml of yeast tRNA (BRL), 100 μg/ml of poly(A) (Boehringer, Mannheim). The solution is mixed vigorously and 0.3 μg/ml of radioactive probe is added, corresponding to 3 ng per slide. To denature the probe, the solution is heated to 95° for 2 min and chilled on ice for 15 min. Then, dithiothreitol is added to 10 m*M* final concentration and the solution is mixed vigorously. To eliminate air bubbles from the rather viscous hybridization mixture, we centrifuged the solution for 10 sec before applying 10 μl to each slide, which is then covered with a siliconized coverslip and sealed with rubber cement to avoid evaporation. For hybridization, sealed slides are then incubated at 22–24° for 72 hr in the dark. Thereafter, coverslips are removed and the slides are rinsed for 10 min in a wash medium consisting of 50% formamide, 0.6 *M* NaCl, 10 m*M* Tris · HCl (pH 7.4), and 1 m*M* EDTA. To achieve low levels of background grains during autoradiography, the slides are washed under stringent conditions in 2× SSC at 55° for 1 hr, dipped in cold 2× SSC, and washed again in the wash medium (see above) for 3 days under stirring. Finally, they are dehydrated through 70 and 95% ethanol containing 0.3 *M* ammonium acetate to stabilize hybrids.

For autoradiography, the slides are dipped into Kodak NTB-2 nuclear track emulsion in complete darkness. The emulsion is melted at 45° in a water bath and diluted 1 : 1 in a beaker with prewarmed 0.6 *M* ammonium acetate. A clean slide is dipped into the beaker and the emulsion is gently mixed. To remove air bubbles, the diluted emulsion is incubated at 45° for another 15–20 min and clean glass slides are dipped into the emulsion from time to time. The experimental slides are then dipped in the emulsion for a few seconds and dried in a rack in a vertical position for 2 hr at 60–70% relative humidity. Several control slides should be included in each experiment. When the surface is dry, the slides are transferred to light-proof slide boxes which are sealed with tape and stored at 4° for 10–14 days in the presence of silica gel as drying agent.

Before developing the slides, the boxes should be removed from the refrigerator and stored at room temperature for 2 hr to prevent condensation on the slide surface. They are then placed in Kodak D-19b developer at 15° for 3 min, rinsed in water for a few seconds, fixed in Kodak Unifix for 5 min at 15°, and rinsed under running water for at least 1 hr. Finally, the slides are stained with Giemsa's stain diluted 1:20 in 10 m*M* sodium phosphate buffer (pH 7.0), and washed for 5 min in water. After drying they are dipped in toluene, and mounted under coverslips with one drop of mounting medium (Eukitt, Serva Biochemicals, Heidelberg, FRG).

Concluding Comments

During autoradiography, specific pairing between labeled cDNA probe and the respective target sequence in the induced cell results in formation of silver grains. There are, however, a variety of artifacts that produce silver grains, too, as for instance nonspecific binding of probe to the specimen or pressure exerted on the emulsion layer during manipulation. Therefore we include a number of controls in our hybridization experiments which are listed in the table. Moreover a blank slide is dipped into the nuclear emulsion before and after dipping of experimental slides and autoradiographed simultaneously.

We have also treated induced and noninduced C-243 cells with RNase (see above) prior to hybridization with the IFN cDNA probes and ob-

SPECIFICITY OF THE HYBRIDIZATION SIGNAL[a]

Treatment of C-243 cells				
Sodium butyrate (1 m*M*)	NDV	RNase	Hybridization with	Number of grains/cell
–	–	–	Control DNA	<10
–	–	–	IFN-α + β cDNA	<10–20
+	–	–	IFN-α + β cDNA	<10–20
+	–	+	IFN-α + β cDNA	<10
+	+	–	Control DNA	<10
+	+	+	IFN-α + β cDNA	<10–50
+	+	–	IFN-α + β cDNA	<10 $\gg$ 200

[a] Cells treated as indicated were hybridized with a mixture of 3 ng of each of the two IFN cDNAs or with 5 ng of control DNA; autoradiographic exposure was for 13 days.

served greatly reduced grain counts per cell. Unfortunately, RNase treatment sometimes leads to complete loss of cellular cytoplasmic morphology probably because we do not fix the preparations with paraformaldehyde following RNase digestion. Paraformaldehyde is a cross-linking fixative which provides good preservation of cellular morphology but under our experimental conditions seems to interfere with hybridization of RNA (M. Brahic, personal communication) since autoradiographic signals are greatly reduced.

For data analysis, grains are counted over at least 200 noninduced cells hybridized with the specific probes and over the same number of induced cells hybridized with the control DNA. These grain counts represent the background level and we consistently observe a slightly higher number of grains in noninduced C-243 cells exposed to the IFN cDNA

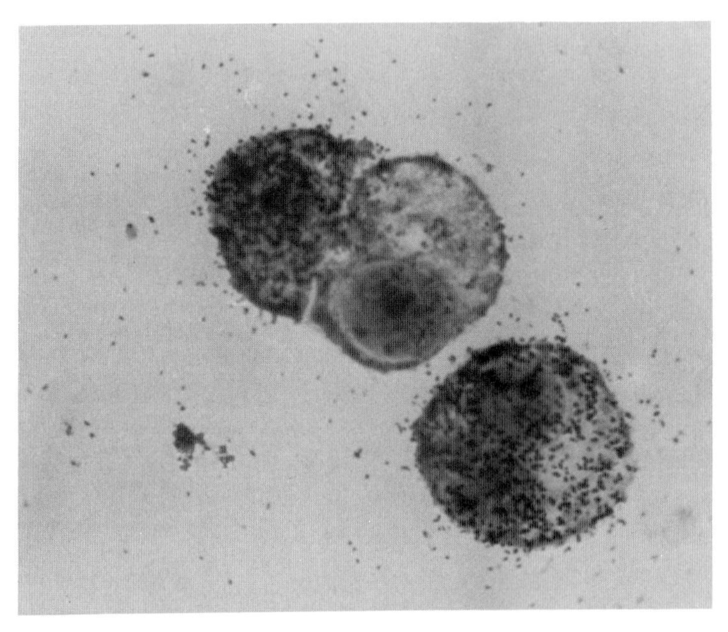

FIG. 1. *In situ* hybridization of NDV-induced mouse C-243 cells with a mixture of ^{35}S-labeled IFN-α and -β probes. Three nanograms of each probe was used. The cells had been harvested 5 hr after induction and prepared as described in the text. Autoradiographic exposure was for 13 days. The picture shows a negative cell, surrounded by two positive cells. Over the course of several experiments we have noticed a marked heterogeneity in the number of grains between the individual cells that can be scored as positive. Whereas the majority of cells show only background levels of grains, like the cell in the center of the picture, from 10 to 15% of the positive cells have grain counts between 30 and 100 and from 15 to 20% of the cells are very strongly labeled, over 100 grains per cell, like the two positive cells in the picture. Microscopic magnification was ×1000.

probes than in induced cells treated with labeled control DNA. This may indicate low levels of IFN mRNA in these cells as we have repeatedly observed that C-243 cells constitutively produce low levels of IFN-α/β. Finally, NDV-induced C-243 cells hybridized with the IFN cDNA probes are analyzed and a cell is scored as positive if the number of grains is at least 1.5 times the background level (Fig. 1).

Acknowledgments

This work was supported by the CNRS (ATP Immunopharmacologie) and a fellowship from the Deutsche Forschungsgemeinschaft (RZ). We are greatly indebted to Michel Brahic for continuous help and guidance.

[66] Detection of a Single Base Substitution between Human Leukocyte IFN-αA and -α2 Genes with Octadecyl Deoxyoligonucleotide Probes

By KUNIMOTO HOTTA, KENNETH J. COLLIER, and SIDNEY PESTKA

Chemically synthesized deoxyoligonucleotides complementary to unique DNA segments have been used to detect cloned DNA sequences.[1-4] Because deoxyoligonucleotide–DNA duplexes with a single base mismatch have a significantly lower thermal stability than those without any mismatch,[5] they can be efficiently applied for detection of site-specific mutations.[6]

There is a single base substitution of A for G between human leukocyte IFN-αA and -α2 genes which results in an amino acid substitution of lysine (αA) for arginine (α2) at the amino acid #23 of the interferon protein. We prepared octadecyl deoxyoligonucleotides complementary to

[1] D. L. Montgomery, B. D. Hall, S. Gillam, and M. Smith, *Cell* **14,** 673 (1978).

[2] R. B. Wallace, P. F. Johnson, S. Tanaka, M. Schöld, K. Itakura, and J. Abelson, *Science* **209,** 1396 (1980).

[3] R. B. Wallace, M. J. Johnson, T. Hirose, T. Miyake, E. H. Kawashima, and K. Itakura, *Nucleic Acids Res.* **9,** 879 (1981).

[4] C. G. Miyada, X. Soberon, K. Itakura, and G. Wilcox, *Gene* **17,** 167 (1982).

[5] R. B. Wallace, J. Shaffer, R. F. Murphy, T. Hirose, and K. Itakura, *Nucleic Acids Res.* **6,** 3543 (1979).

[6] R. B. Wallace, M. Schold, M. J. Johnson, P. Dembek, and K. Itakura, *Nucleic Acids Res.* **9,** 3647 (1981).

the sequences spanning the region of the above single base substitutions to distinguish these. Here, we describe the hybridization conditions to detect these single base changes in plasmids coding for IFN-αA and IFN-α2.

Materials and General Procedures

Octadecyl deoxyoligonucleotides

5′ AGAGAGATTTTCCTCATC 3′ complementary to IFN-αA

5′ AGAGAGATTCTCCTCATC 3′ complementary to IFN-α2

These were prepared by the phosphoramidite method.[7,8]

Plasmids

pLeIFA25 coding for IFN-αA as described[9,10]

pBR325-α2 coding for IFN-α2: the *Eco*RI fragment containing the IFN-α2 coding sequence as reported[11] was subcloned into pBR325. This was supplied by D. V. Goeddel.

Kinase solution: The kinase solution[12] consisted of the following: 50 m*M* Tris · HCl (pH 7.6), 10 m*M* $MgCl_2$, 5 m*M* dithiothreitol, 0.1 m*M* spermidine, 0.1 m*M* EDTA

SSPE solution: 2× SSPE (a stock solution of 20× SSPE consisting of 3 *M* NaCl/0.2 *M* $NaH_2PO_4 \cdot H_2O$/20 m*M* EDTA, pH 7.4, is diluted with water to prepare a solution of 2× SSPE)

Denaturation solution: 0.5 *M* NaOH, 1.5 *M* NaCl

Neutralization buffer: 0.5 *M* Tris · HCl (pH 8.0), 1.5 *M* NaCl

Hybridization buffer: 180 m*M* Tris · HCl (pH 8.0), 0.9 *M* NaCl, 12 m*M* EDTA, 0.1% SDS, 0.2% Ficoll, 0.2% polyvinylpyrrolidone, 0.2% bovine serum albumin (Pentax Fraction V), 100 μg/ml of *E. coli* DNA

Washing solution: 4× SSC (a stock solution of 20× SSC consisting of 3 *M* NaCl/0.3 *M* sodium citrate, pH 7.0, is diluted with water to prepare a solution of 4× SSC)

[7] S. L. Beaucage and M. H. Caruthers, *Tetrahedron Lett.* **22,** 1859 (1981).

[8] F. Chow, T. Kempe, and G. Palm, *Nucleic Acids Res.* **9,** 2807 (1981).

[9] S. Maeda, R. McCandliss, M. Gross, A. Sloma, P. C. Familletti, J. M. Tabor, M. Evinger, W. P. Levy, and S. Pestka, *Proc. Natl. Acad. Sci. U.S.A.* **77,** 7010 (1980); **78,** 4648 (1981).

[10] D. V. Goeddel, E. Yelverton, A. Ullrich, H. L. Heyneker, G. Miozzari, W. Holmes, P. H. Seeburg, T. Dull, L. May, N. Stebbing, R. Crea, S. Maeda, R. McCandliss, A. Sloma, J. M. Tabor, M. Gross, P. C. Familletti, and S. Pestka, *Nature (London)* **287,** 411 (1980).

[11] R. M. Lawn, M. Gross, C. M. Houk, A. E. Franka, P. V. Gray, and D. V. Goeddel, *Proc. Natl. Acad. Sci. U.S.A.* **78,** 5435 (1981).

[12] A. M. Maxam and W. Gilbert, this series, Vol. 65, p. 499.

Preparation of ^{32}P-Labeled Octadecyl Deoxyoligonucleotides

Octadecyl oligonucleotides (25–30 pmol ends) and [γ-^{32}P]ATP (>5000 Ci/mmol, 50–60 pmol) were incubated with T_4 polynucleotide kinase (5 units) in the kinase buffer in a total volume of 30–50 μl at 37° for 45 min in Eppendorf tubes. To precipitate the ^{32}P-labeled octadecyl oligonucleotides, 250 μl of 2.5 *M* ammonium acetate/4 m*M* EDTA/40 μg/ml yeast tRNA solution and 750 μl of 95% ethanol are added, mixed well, kept in dry ice for 30 min, and then centrifuged for 15 min at 4° in an Eppendorf Microfuge. Supernatants are removed and the precipitates are redissolved in 250 μl of 4 m*M* EDTA/0.25 *M* sodium acetate solution. The second ethanol precipitation is performed as above after addition of 750 μl of 95% ethanol. The precipitate is redissolved in 250 μl of 0.25 *M* sodium acetate for the third ethanol precipitation with 750 μl of 95% ethanol. This last precipitate is washed with 70% ethanol. The supernatants are removed and the residual ethanol is evaporated under vacuum in a Speed Vac Concentrator (Savant). The resultant precipitates are dissolved in a small amount of sterile water.

Hybridization of Plasmid DNAs Coding for Human IFN-αA and -α2 to ^{32}P-Labeled Octadecyl Deoxyoligonucleotides

One microliter containing various concentrations of plasmid DNA of pLeIFA25 and pBR325-α2 with or without digestion with restriction endonucleases was spotted on nitrocellulose filters (20 $\times$ 30 mm) presoaked in 2$\times$ SSPE and air dried. The filters are soaked in the denaturation solution for 10 min followed by soaking twice in neutralization solution for 10 min, air dried, and then baked at 80° for 2 hr.

The baked filters are incubated in the hybridization solution at 37° for 2 hr (prehybridization) and then placed into separate plastic bags (Seal-A-Meal) to which 1.0 ml of hybridization solution and ^{32}P-labeled octadecyl deoxyoligonucleotide αA or α2 probe are then added. The oligonucleotides were labeled to a specific activity of $\geq 1 \times 10^8$ cpm/μg and each was present at a concentration of 5 ng/ml. Total volume of the hybridization mixture was 1.0 ml. After sealing, the bags are incubated at the designated temperatures noted in the legend to the figure for 1–2 days. The filters are taken out of the bags, washed with three changes of 100 ml of 4$\times$ SSC at 25° for 1 hr (total time 3 hr), and air dried. X-Ray films (Kodak X-Omat film XAR-2) are exposed to the filters at −80° for 2 hr, overnight or longer.

Concluding Comments

Strong hybridization was observed between the oligonucleotide probe αA and pLeIFA25 or between the probe α2 and pBR325-α2, whereas the hybridization between the probe αA and pBR325-α2 or between the probe α2 and pLeIFA25 where there is a single base pair mismatch was very weak when hybridization was performed at 37° or 42° (Fig. 1). On the other hand, the hybridization of the probe α2 at 25° was stronger with pLeIFA25 than with pBR325-α2. Increasing the washing temperature from 25 to 37 or 42° resulted in an almost complete removal of the radioactivity only from the hybridization between the probe α2 and pLeIFA25. Therefore, a single base substitution between human leukocyte interferon αA and α2 genes can be clearly detected by hybridization with these

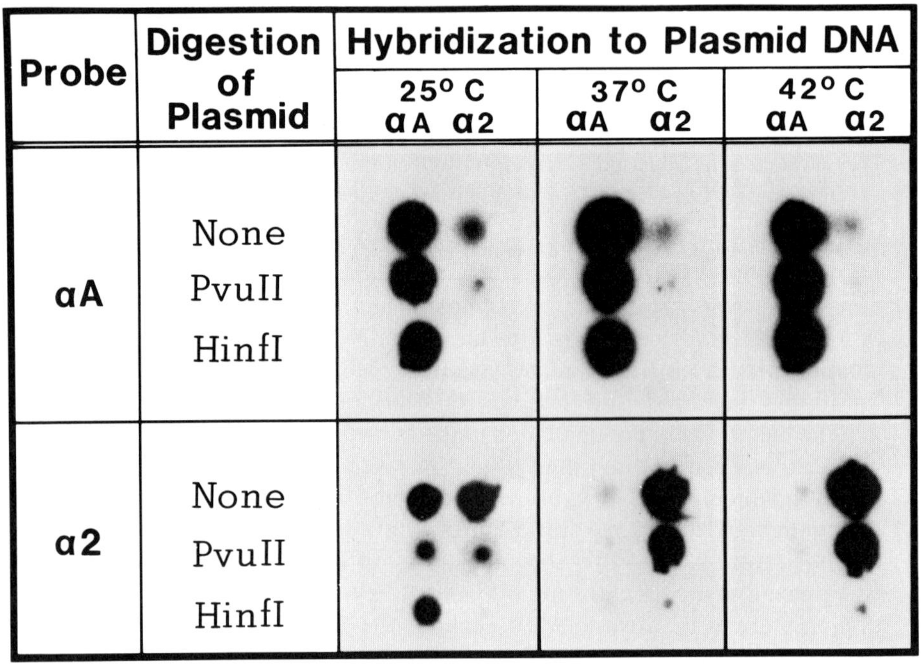

Probe	Digestion of Plasmid	Hybridization to Plasmid DNA		
		25° C αA α2	37° C αA α2	42° C αA α2
αA	None PvuII HinfI			
α2	None PvuII HinfI			

FIG. 1. Hybridization of octadecyl oligonucleotide probes to plasmid DNAs coding for human IFN-αA and -α2. One microliter (50 ng) of plasmid DNAs of pLeIFA25 and pBR325-α2 with or without digestion with *Pvu*II and *Hin*fI was spotted onto nitrocellulose filters and hybridized to the octadecyl oligonucleotide probes for human IFN-αA and -α2 genes at 25, 37, and 42° for 40 hr. After hybridization, the filters were washed with three changes of 4× SSC at 25°. X-Ray film (Kodak X-Omat film XAR-2) was exposed for 12 hr.

deoxyoligonucleotide probes at 37 or 42°. Since only the IFN-α2 sequence contains a recognition site for *Hin*fI at the center position of the sequence hybridizable to the α2 probe, no significant hybridization was detected between *Hin*fI-digested pBR325-α2 and the α2 probe. However, *Pvu*II digestion does not influence the hybridization because there is no *Pvu*II site within the sequences hybridizing to the probes. The probes αA and α2 tightly bind to the IFN-αA and IFN-α2 genes, respectively, so that the probes are not removed even if the nitrocellulose filters are washed with 4× SSC at the calculated dissociation temperature (T_d) as reported[13] (αA probe, 50° and α2 probe, 52°) for 15 min. Therefore, to minimize background, a hybridization temperature of 42° or higher would be better than temperatures lower than 42°. A washing temperature of 42° or higher could also be used although 25° was adequate for the conditions specified. About 2–5 pg of the αA and α2 interferon coding regions in the plasmids pLeIFA25 (10 pg) and pBR325-α2 (50 pg) can be readily detected if the exposure time of autoradiography is increased to about 10 days.

[13] S. U. Suggs, T. Hirose, T. Miyake, E. H. Kawashima, M. J. Johnson, K. Itakura, and R. B. Wallace, *in* "Developmental Biology Using Purified Genes" (D. D. Brown, ed.), p. 683. Academic Press, New York, 1981.

Section VIII

Enzymes in Interferon Action

[67] RNase L, a (2′-5′)-Oligoadenylate-Dependent Endoribonuclease: Assays and Purification of the Enzyme; Cross-Linking to a (2′-5′)-Oligoadenylate Derivative

By Georgia Floyd-Smith and Peter Lengyel

RNase L,[1] an endoribonuclease, is among the mediators of interferon action.[2–4] The level of this enzyme is increased in various cells after exposure to interferons. The increase is 2- to 3-fold in mouse Ehrlich ascites tumor (EAT) cells[3] and about 10- to 20-fold in mouse JLS-V9R cells.[5]

RNase L is latent unless activated by (2′-5′)$(A)_n$. (2′-5′)$(A)_n$ must be a trimer or longer to activate RNase L from EAT, L929, or HeLa cells and a tetramer or longer to activate RNase L from rabbit reticulocyte lysates.[6] Triphosphate or diphosphate moieties at the 5′ termini of (2′-5′)$(A)_n$ are also required for activation of RNase L.[7] If either of the (2′-5′) phosphodiester bonds of (2′-5′)$(A)_3$ are replaced by (3′-5′) phosphodiester bonds the resulting oligoadenylates show greatly reduced ability to activate RNase L.[8] The activation of RNase L by (2′-5′)$(A)_n$ is reversible: The removal of (2′-5′)$(A)_n$ from an activated enzyme by gel filtration results in a reversion of the enzyme to the latent state. The readdition of (2′-5′)$(A)_n$ to the now latent enzyme reactivates it.[9]

[1] Abbreviations: RNase L, (2′-5′)-oligoadenylate-dependent endoribonuclease; EAT, Ehrlich ascites tumor; interferon units, mouse interferon international reference units; (2′-5′)$(A)_n$, (2′-5′)-linked pppA$(pA)_{n-1}$; (2′-5′)$(A)_3$, (2′-5′)pppApApA; (2′-5′)$(A)_3$[^{32}P]pCp, (2′-5′)pppApApA[^{32}P](3′-5′)pCp; SDS, sodium dodecyl sulfate; PMSF, phenylmethylsulfonyl fluoride.

[2] P. Lengyel, *Annu. Rev. Biochem.* **51,** 251 (1982).

[3] G. Floyd-Smith and P. Lengyel, *in* "Interaction of Translational and Transcriptional Controls in the Regulation of Gene Expression" (M. Grunberg-Manago and B. Safer, eds.), p. 417. Elsevier/North-Holland, Amsterdam, 1982.

[4] P. Lengyel, this series, Vol. 79, p. 135.

[5] H. Jacobsen, C. W. Czarniecki, R. M. Friedman, and R. H. Silverman, *Virology* **125,** 496 (1983).

[6] B. R. G. Williams, R. R. Golgher, R. E. Brown, C. S. Gilbert, and I. M. Kerr, *Nature (London)* **282,** 582 (1979).

[7] E. M. Martin, N. J. M. Birdsall, R. E. Brown, and I. M. Kerr, *Eur. J. Biochem.* **95,** 295 (1979).

[8] K. Lesiak, J. Imai, G. Floyd-Smith, and P. F. Torrence, *J. Biol. Chem.* **258,** 13082 (1983).

[9] E. Slattery, N. Ghosh, H. Samanta, and P. Lengyel, *Proc. Natl. Acad. Sci. U.S.A.* **76,** 4778 (1979).

Activation involves the binding of $(2'\text{-}5')(A)_n$ to RNase L. Partially purified preparations of RNase L from EAT cells can retain $(2'\text{-}5')(A)_n$ on nitrocellulose filters whereas free $(2'\text{-}5')(A)_n$ passes through nitrocellulose filters.[9] Binding of a labeled derivative $(2'\text{-}5')(A)_n$ [i.e., $(2'\text{-}5')(A)_3[^{32}P]pCp$ can serve as a convenient assay for RNase L].[10,11] The agent binding $(2'\text{-}5')(A)_3[^{32}P]pCp$ copurifies with RNase L during both ion exchange chromatography and gel filtration.[12] This $(2'\text{-}5')(A)_n$ derivative can also be selectively cross-linked by UV irradiation of a crude cytoplasmic extract from mouse EAT cells to a protein of about 77,000 Da which copurifies with RNase L[12] (see also Ref. 13).

RNase L activated by $(2'\text{-}5')(A)_n$ cleaves single-stranded regions of RNAs to generate products having 3′ phosphate and 5′ hydroxyl termini. The preferred cleavage sites with natural RNAs as substrates are on the 3′ side of UAp, UGp, and UUp.[3,14,15]

Buffers

Buffer A: 25 m*M* Tris · Cl (pH 8.2), 75 m*M* KCl, 5 m*M* $Mg(OAc)_2$, 10 m*M* 2-mercaptoethanol.

Buffer B: 50 m*M* Tris · borate (pH 8.6), 3 m*M* EDTA, 0.5% SDS, 0.05% bromophenol blue, 50% (v/v) glycerol.

Buffer C: 25 m*M* Tris · Cl (pH 7.2), 1 m*M* $Mg(OAc)_2$, 10 m*M* 2-mercaptoethanol.

Buffer D: 25 m*M* potassium phosphate (pH 7.5), 50 m*M* KCl, 5 m*M* $Mg(OAc)_2$, 1 m*M* dithiothreitol.

Buffer E: 35 m*M* Tris · Cl (pH 7.5), 146 m*M* NaCl, 12 m*M* glucose.

Buffer F: 10 m*M* Tris · Cl (pH 7.5), 15 m*M* KCl, 6 m*M* 2-mercaptoethanol, 1.5 m*M* $Mg(OAc)_2$.

Buffer G: 250 m*M* Tris · Cl (pH 7.5), 800 m*M* KCl, 50 m*M* $Mg(OAc)_2$, 60 m*M* 2-mercaptoethanol.

Buffer H: 10 m*M* Tris · Cl (pH 8.2), 10 m*M* 2-mercaptoethanol, 1 m*M* EDTA, 50 μM PMSF, 10% (v/v) glycerol.

[10] M. Knight, P. J. Cayley, R. H. Silverman, D. H. Wreschner, C. S. Gilbert, R. E. Brown, and I. M. Kerr, *Nature (London)* **288,** 189 (1980).

[11] T. W. Nilsen, P. A. Maroney, and C. Baglioni, *J. Biol. Chem.* **256,** 7806 (1981).

[12] G. Floyd-Smith, O. Yoshie, and P. Lengyel, *J. Biol. Chem.* **257,** 8584 (1982).

[13] P. J. Cayley, M. Knight, and I. M. Kerr, *Biochem. Biophys. Res. Commun.* **104,** 376 (1982).

[14] D. A. Wreschner, J. W. McCauley, J. J. Skehel, and I. M. Kerr, *Nature (London)* **289,** 414 (1981).

[15] G. Floyd-Smith, E. Slattery, and P. Lengyel, *Science* **212,** 1030 (1981).

Buffer I: 50 mM Tris · Cl (pH 7.8), 10 mM 2-mercaptoethanol, 1 mM EDTA, 50 μM PMSF, 10% (v/v) glycerol.

Assays

Endonuclease Assay

The cleavage of RNA by RNase L was determined essentially as described by Slattery *et al.*[9] The reaction mixture (30 μl) contained buffer A, 4 μg bacteriophage R17 ^{32}P-labeled RNA (500 to 2000 cpm/μg; prepared according to the procedure of Brownlee[16]), 2 to 20 μl of the protein fraction tested for RNase L activity and either 1 μM (2′-5′)$(A)_n$ or no (2′-5′)$(A)_n$. After incubation at 30° for 1 hr the reaction was stopped by the addition of 10 μl of Buffer B, and applied to the top of a polyacrylamide gel [4% acrylamide, 0.07% bisacrylamide, 90 mM Tris·borate (pH 8.6), 5 mM EDTA, 0.1% SDS, 0.008% ammonium persulfate, 0.01% *N,N,N′,N′*-tetramethylethylenediamine] and electrophoresis performed at 200 V for 1 hr with 45 mM Tris·borate (pH 8.6), 2.5 mM EDTA, 0.1% SDS as running buffer. The polyacrylamide gel was wrapped in Saran wrap and autoradiographed in the presence of an intensifying screen (Dupont Cronex) at −70°. The extent of RNA cleavage during the course of the incubation in the presence or absence of (2′-5′)$(A)_n$ was estimated by excising from each gel track the top 1 cm adjacent to the site of application of the sample (i.e., that portion of the gel which would contain undegraded R17 bacteriophage RNA) and counting the radioactivity in a scintillation counter. The percentage of RNA cleaved by the enzyme was calculated according to the following formula: percentage of RNA cleaved by the enzyme sample = 100 × [cpm in 1 cm gel slice from a reaction mixture without (2′-5′)$(A)_n$] − [cpm in 1 cm gel slice from reaction mixture with (2′-5′)$(A)_n$]/[cpm in 1 cm gel slice from reaction mixture without (2′-5′)$(A)_n$].

To estimate the amount of RNase L activity at each stage of purification serial dilutions were made of each enzyme preparation and the percentage of RNA cleaved was determined. The percentage of RNA cleaved was plotted against the amount of protein present in the assay. The amount of protein required for 50% cleavage of RNA in 1 hr was determined graphically. The results of the determination are shown in the table as "endonuclease specific activity." It can be seen that there is reasonable agreement between the values for the extent of purification which can be calculated from this endonuclease assay and the values

[16] G. G. Brownlee, *in* "Laboratory Techniques in Biochemistry and Molecular Biology" (T. Work and E. Work, eds.), p. 248. North-Holland Publ., Amsterdam, 1972.

PURIFICATION OF RNASE L[a]

Fraction	Total protein (mg)	Specific activity: Endonuclease	Specific activity: (2′-5′)(A)$_3$[^{32}P]pCp binding (fmol/μg protein)	Yield (%)	Purification factor
S-200	5697	—	0.0004	100	1
AS35-47.5	990	—	0.0026	113	6.5
DEAE cellulose I	23	0.3	0.16	161	400
Phosphocellulose I	3.9	1.35	0.72	123	1800
Poly(A)-agarose	0.45	4.1	1.9	38	4750

[a] This table concerns purification according to procedure I. The "endonuclease assay" and "(2′-5′)(A)$_3$[^{32}P]pCp binding assay" are described in the text. The purification factor was determined by measuring the increase in specific (2′-5′)(A)$_3$[^{32}P]pCp binding.

which are calculated from the (2′-5′)(A)$_3$[^{32}P]pCp binding assay described below.

(2′-5′)(A)$_3$[^{32}P]pCp Binding Assay

This was performed essentially according to the procedure of Knight *et al.*[10] and Nilsen *et al.*[11]

Preparation of (2′-5′)(A)$_3$[^{32}P]pCp. (2′-5′)(A)$_n$ trimer [i.e., (2′-5′)(A)$_3$] was prepared as described elsewhere in this volume.[17] This was labeled to high specific activity by ligation with [^{32}P]pCp (2000 to 4000 Ci/mmol; New England Nuclear) in a reaction mixture (0.1 ml) containing 50 m*M* Hepes buffer (pH 7.5), 20 m*M* $MgCl_2$, 3.3 m*M* dithiothreitol, 12 μ*M* ATP, 10 μ*M* [^{32}P]pCp, 10 μ*M* (2′-5′)(A)$_3$, 10% (v/v) dimethyl sulfoxide, 1 μg bovine serum albumin, and 40 units of T$_4$ RNA ligase (PL Biochemicals). After incubation at 4° for 48 to 72 hr the reaction was stopped by the addition of 1 ml 20 m*M* potassium phosphate (pH 7.5), 90 m*M* KCl, and boiling for 5 min. The reaction mixture was centrifuged at 10,000 *g* for 5 min to remove denatured protein. The unreacted [^{32}P]pCp was removed by applying the supernatant fraction to a 0.2 ml DEAE cellulose column which had been equilibrated in 20 m*M* potassium phosphate (pH 7.5), 90 m*M* KCl. The column was washed extensively with the equilibrating solution and the (2′-5′)(A)$_3$[^{32}P]pCp together with the residual (2′-5′)(A)$_3$ were eluted with 1 ml of 20 m*M* potassium phosphate (pH 7.5), 350 m*M* KCl. The (2′-5′)(A)$_3$ was degraded in a reaction mixture (3 ml) containing 1 ml of column eluate, 0.5 ml of EAT S-30 fraction (prepared as described

[17] H. Samanta and P. Lengyel, this volume [69].

in the section on RNase L purification), 10 mM Mg(OAc)$_2$, and 1 mM ATP. [This procedure is based on the finding that on incubation in a crude HeLa cell extract (2′-5′)(A)$_3$ is degraded but (2′-5′)(A)$_3$[^{32}P]pCp is not.[11]] After incubation at 4° for 1 hr the reaction mixture was boiled for 5 min and the denatured protein was removed by centrifugation at 10,000 g for 10 min. The supernatant fraction was made 20 mM in potassium phosphate (pH 7.5) and 90 mM in KCl and was applied to an 0.2 ml DEAE column as described above. The (2′-5′)(A)$_3$[^{32}P]pCp was eluted with 1 ml of 20 mM potassium phosphate (pH 7.5), 350 mM KCl. The eluate typically contained 30,000 to 70,000 cpm/μl and approximately 10 to 20% of the input [^{32}P]pCp was recovered in the form of (2′-5′)(A)$_3$[^{32}P]pCp. The purity and structure of the (2′-5′)(A)$_3$[^{32}P]pCp were checked by thin layer chromatography on polyethyleneimine plates.[11,12]

Binding of (2′-5′)(A)$_3$[^{32}P]pCp. The reaction mixture (0.1 ml) contained Buffer C, 30,000 cpm (2′-5′)(A)$_3$[^{32}P]pCp (in 0.5 to 2 μl of the above preparation) and 2 to 50 μl of the protein fraction assayed. After incubation at 4° for 2–3 hr (for the kinetics of binding see Ref. 11) each reaction mixture was supplemented with 1 ml of Buffer D and the solution was immediately filtered through a nitrocellulose filter fitted to a Millipore filtration apparatus which had been washed and equilibrated with Buffer D containing 100 mM sodium pyrophosphate. The filter was washed with 5 ml Buffer D at room temperature, air dried, and its radioactivity was determined in a scintillation counter.

Purification of RNase L

All the operations were performed at 0° unless otherwise specified.

Preparation of Extracts from Interferon-Treated EAT Cells

Mouse EAT cells were grown in stirred suspension cultures in 6-liter Erlenmeyer flasks containing 4 liters of F-14 medium (Gibco) supplemented with 5% newborn calf serum (Gibco) and 1% (v/v) penicillin (10,000 units/ml), streptomycin (10,000 μg/ml) (Gibco) solution at 37°. For the preparation of cell extracts 40 liters of culture (0.6 × 10^6 cells/ml) was supplemented with 1000 units of mouse interferon[1]/ml of medium, 24 hr before harvesting the cells. (The mouse interferon was produced[18] and purified[19] according to published procedures. Side fractions with a spe-

[18] R. J. Broeze, B. Jayaram, E. Slattery, H. Taira, and P. Lengyel, this series, Vol. 78, p. 143.

[19] B. M. Jayaram, H. Schmidt, O. Yoshie, and P. Lengyel, *J. Interferon Res.* **3**, 177 (1983).

cific activity of 10^8 interferon units/mg protein were used.) The cells were harvested by centrifugation at 630 *g* for 10 min and the cell pellet was resuspended in 500 ml of Buffer E and sedimented by centrifugation at 630 *g* for 10 min. Resuspension of the pellet in Buffer E and sedimentation of the cells from the suspension was repeated twice. Finally the cell pellet was resuspended in Buffer E (100 ml) and sedimented at 2500 *g* for 15 min. The washed, packed cells ("1 volume") were swollen by adding 1.8 volumes of Buffer F, placed on ice for 10 min, and disrupted by homogenization (20 strokes) in a tight fitting Dounce homogenizer. The homogenate was supplemented with 0.16 volume of Buffer G, further homogenized in the Dounce homogenizer (10 strokes), and centrifuged at 10,000 *g* for 5 min. The supernatant fraction was collected and clarified by centrifugation at 30,000 *g* for 30 min. The resulting "postmitochondrial supernatant fraction (S-30 fraction)" was stored at −70°. About 80 ml of S-30 fraction was obtained from 40 liters of suspension culture. The postribosomal supernatant fraction (S-200 fraction) was prepared by centrifuging the S-30 fraction at 200,000 *g* for 3 hr.

Differential Precipitation with $(NH_4)_2SO_4$

S-200 fraction (300 ml) (5700 mg protein) was supplemented with 58.2 g of an $(NH_4)_2SO_4:NH_4HCO_3$ (50:1) mixture gradually with constant stirring [35% $(NH_4)_2SO_4$ saturation]. One hour after all of the $(NH_4)_2SO_4:NH_4HCO_3$ has dissolved the precipitate was collected by centrifugation at 9000 *g* for 30 min. The supernatant fraction (325 ml) was supplemented with 23.4 g of the $(NH_4)_2SO_4:NH_4HCO_3$ (50:1) mixture gradually with constant stirring [47.5% $(NH_4)_2SO_4$ saturation]. One hour after the $(NH_4)_2SO_4:NH_4HCO_3$ has dissolved the precipitate was collected by centrifugation at 9000 *g* for 30 min, dissolved in 20 ml of Buffer H, and dialyzed against 3 changes (each 2000 ml) of Buffer H. The resulting fraction (AS 35-47.5 fraction) was stored at −70°. Two different procedures were developed for the further purification of RNase L from the AS 35-47.5 fraction. Procedure I consists of chromatography on DEAE cellulose, phosphocellulose and poly(A)-agarose. RNase L prepared according to procedure I was used in studies on the cleavage specificity of the enzyme.[3,15] Procedure II consists of gel filtration on Sephacryl S-200 and chromatography on DEAE cellulose and phosphocellulose. RNase L prepared according to procedure II was used in studies on the effect of (2′-5′)$(A)_n$ analogs (containing one or more 3′-5′ linkages) on the enzyme.[8] For these latter studies it was desirable to use RNase L free of (or at least low in) (2′-5′) phosphodiesterase activity. Because (2′-5′) phosphodiesterase has a lower molecular weight (about 40,000)[20] than that of RNase L (about 80,000 as determined by gel filtration and by glycerol gradient centrifugation),[3,12] we used gel filtration to remove (2′-5′) phosphodiesterase from the RNase L preparation.

Procedure I: Purification of RNase L by Chromatography on DEAE Cellulose, Phosphocellulose, and Poly(A)-Agarose

Chromatography on DEAE Cellulose. Protein (990 mg) from the AS 35-47.5 fraction in 20 ml of Buffer H was applied to a 75 ml DEAE cellulose (Whatman DE52) column which had been equilibrated with Buffer H. The column was washed with 150 ml of Buffer H and developed with an 800 ml linear gradient of KCl (0 to 200 m*M*) in Buffer H. The fractions with RNase L activity (eluting at 70 to 100 m*M* KCl) were pooled (DEAE cellulose I fraction) and either purified further as described below, or stored at −70° as such or after concentration involving precipitation with 6 g $(NH_4)_2SO_4$: NH_4HCO_3 (50 : 1) per 10 ml of solution.

Chromatography on Phosphocellulose. Protein (23 mg) from the DEAE cellulose I fraction was dialyzed against Buffer I containing 50 m*M* KCl and applied to a 10 ml phosphocellulose (Whatman P-11) column prepared in a 20-ml plastic syringe packed with baked siliconized glass wool and fitted with an 18-gauge needle and equilibrated with Buffer I containing 50 m*M* KCl. The column was washed with 15 ml of Buffer I and developed with a 100 ml linear gradient of KCl (50 to 500 m*M*) in Buffer I. The fractions with RNase L activity (the peak eluting at 200 m*M* KCl) were pooled (3.9 mg protein in a total volume of 21 ml), dialyzed against Buffer H (phosphocellulose I fraction), and stored at −70° or further purified.

Chromatography on Poly(A)-Agarose. Protein (3.9 mg) from the phosphocellulose I fraction was dialyzed against Buffer H and applied to a 1 ml poly(A)-agarose (PL Biochemicals) column which had been prepared in a 2-ml syringe as described in the previous section and equilibrated in Buffer H. The column was washed with 5 ml of Buffer H and developed with a 40 ml linear gradient of KCl (0 to 1 *M*) in Buffer H. The fractions with RNase L activity (eluting at 200 m*M* KCl) were pooled resulting in 0.45 mg protein in 6 ml [poly(A)-agarose fraction]. An aliquot of this fraction was supplemented with RNase free bovine serum albumin (BRL), dialyzed against Buffer H, and stored at −70°. The enzymatic

[20] A. Schmidt, Y. Chernajovsky, L. Shulman, P. Federman, H. Berissi, and M. Revel, *Proc. Natl. Acad. Sci. U.S.A.* **76,** 4788 (1979).

activity persisted for up to 2 years although some of the activity was lost if aliquots were thawed and refrozen repeatedly. RNase L prepared according to procedure I was purified about 4750-fold (see the table) and was essentially free of contaminating nuclease activity.

In the presence of (2′-5′)$(A)_n$ the purified enzyme cleaved polyuridylic acid but not polyadenylic acid, polycytidylic acid, or polyguanylic acid.[15] When polyuridylic acid was incubated with the enzyme in the presence of (2′-5′)$(A)_n$, small oligouridylates were generated with 3′ phosphate and 5′ hydroxyl termini.[3] In a natural RNA with little secondary structure (a segment from a bactriophage T7 specific RNA) the preferred cleavage sites were at the 3′ sides of UAp, UGp, and to a lesser extent UUp sequences.[15] In a natural RNA with considerable secondary structure (the 3′ terminal region of bacteriophage R17 RNA) the preferred cleavage sites were at the 3′ sides of UUp sequences while a minor cleavage site was at the 3′ side of UCp. All the cleavages took place in single-stranded loop structures, and no cleavages were detected in complementary, base paired (double stranded) regions though these contained several UU and UA sequences.[3]

Procedure II: Gel Filtration on Sephacryl S-200, Chromatography on DEAE Cellulose and Phosphocellulose

Gel Filtration on Sephacryl S-200. Protein (337 mg) from the AS 35-47.5 fraction in 10 ml of Buffer H was dialyzed against Buffer H containing 100 m*M* KCl, and applied to a 510 ml Sephacryl S-200 (Pharmacia) column (K26-100 Pharmacia) equilibrated with Buffer H containing 100 m*M* KCl. The column was developed with Buffer H. The column was calibrated with size markers: blue dextran (about 1,000,000 Da), ferritin (450,000 Da), catalase (240,000 Da), aldolase (158,000 Da), bovine serum albumin (68,000 Da), ovalbumin (45,000 Da), and cytochrome *c* (12,500 Da). RNase L eluted with an apparent molecular weight of 80,000 (Sephacryl S-200 fraction).

Chromatography on DEAE Cellulose. Protein (96.2 mg) from the Sephacryl S-200 fraction in 40 ml of Buffer H containing 100 m*M* KCl was diluted with an equal volume of Buffer H (to lower the KCl concentration to 50 m*M*) and applied to a 10 ml DEAE cellulose column (Whatman DE52) which had been equilibrated with Buffer H containing 50 m*M* KCl. The column was washed with 50 ml of Buffer H containing 50 m*M* KCl and developed with a 100 ml linear gradient of KCl (50 to 250 m*M*) in Buffer H. The fractions with RNase L activity (eluting at 80–100 m*M* KCl) were pooled (DEAE cellulose II fraction).

Chromatography on Phosphocellulose. Protein (8.4 mg) from the DEAE cellulose II fraction in 25 ml was dialyzed against Buffer I containing 50 mM KCl and was applied to a 4 ml phosphocellulose (Whatman P11) column which had been equilibrated with Buffer I. The column was washed with 10 ml of Buffer I containing 50 mM KCl and was developed with a 150 ml linear gradient of KCl (50 to 500 mM) in Buffer I. Fractions with RNase L activity eluting between 150 and 190 mM KCl were pooled (phosphocellulose IIA fraction) separately from those RNase L active fractions eluting between 190 and 200 mM KCl (phosphocellulose IIB fraction). The pooled fractions were dialyzed against Buffer H and stored at −70°. The enzyme activity was stable for up to 2 years but was lost upon repeated thawing and refreezing. The RNase L prepared according to procedure II was purified 6500-fold in the phosphocellulose IIA fraction and 3500-fold in the phosphocellulose IIB fraction. Of the two fractions, the phosphocellulose IIA was lower in $(2'\text{-}5')(A)_n$ and $(3'\text{-}5')(A)_3$ cleaving activity, however the phosphocellulose IIB fraction contained more total RNase L activity.[8]

Cross-Linking of $(2'\text{-}5')(A)_3[^{32}P]pCp$ to RNase L

The selective binding of $(2'\text{-}5')(A)_n$ to RNase L is the basis of the selective crosslinking of $(2'\text{-}5')(A)_3[^{32}P]pCp$ to RNase L in crude cytoplasmic extracts from EAT and HeLa cells[12] (see also Ref. 13).

The reaction mixtures (200 μl) contained Buffer C, 0.05 nM $(2'\text{-}5')(A)_3[^{32}P]pCp$ (60,000 cpm, 1 to 12 μl of the preparation) and 100 μl from the protein fraction assayed. After incubation at 4° for 2–3 hr 100 μl of each reaction mixture was transferred to separate wells of a 96-well tissue culture dish and irradiated with a germicidal UV lamp (with two Sylvania G15T8 tubes and reflectors on the top and sides) at a distance of 9 cm from the light source. The radiation dose (at 254 nm) at this distance was 10 J/m²/sec, as measured by a Latarjet dosimeter connected to an ammeter (Keithley 179 TRMS digital multimeter). After irradiation at 0° for 1 hr (radiation dose 36,000 J/m²) 50 μl of each reaction mixture was added to 30 μl of SDS gel sample buffer [0.25 Tris · Cl (pH 6.8), 1.4 M 2-mercaptoethanol, 0.1% bromophenol blue, 28% glycerol] and electrophoresis was performed through polyacrylamide gels (containing 7.5% acrylamide) in the presence of SDS according to a published procedure.[12,21,22] The gels

[21] G. St. Laurent, O. Yoshie, G. Floyd-Smith, H. Samanta, P. B. Sehgal, and P. Lengyel, *Cell* **33,** 95 (1983).

[22] G. A. Floyd and J. A. Traugh, this series, Vol. 60, p. 911.

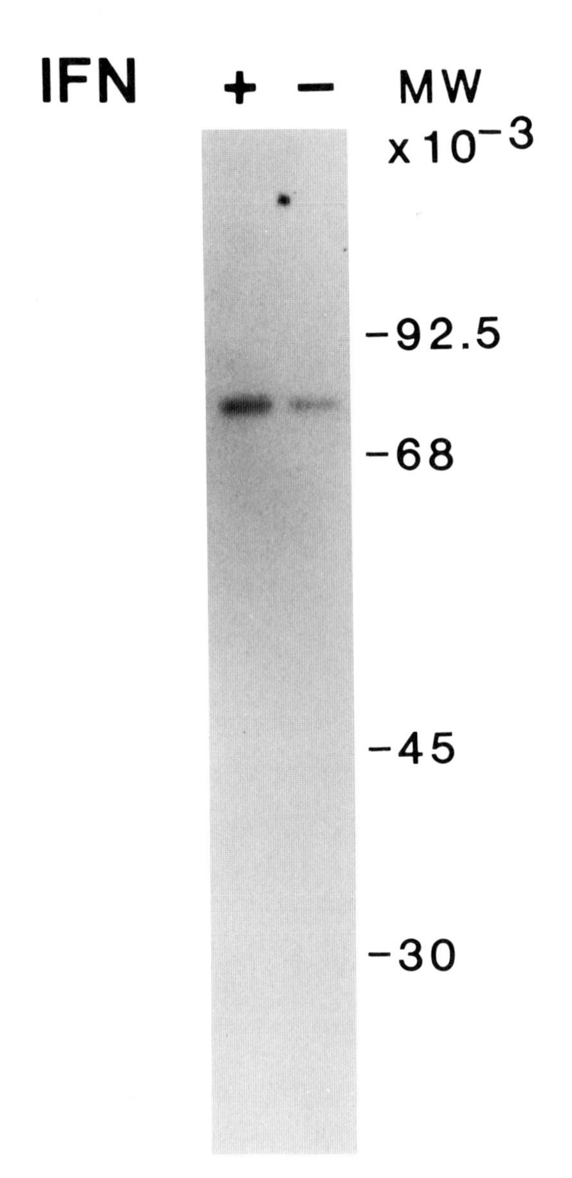

FIG. 1. Ultraviolet light induced cross-linking of (2′-5′)$(A)_3$[^{32}P]pCp to protein in a cytoplasmic extract from interferon-treated and control EAT cells. Cytoplasmic extracts (S-30) from EAT cells treated with 1000 units/ml of mouse interferon (+) and from corresponding control cells (−) were prepared, incubated with (2′-5′)$(A)_3$[^{32}P]pCp, and irradiated with UV light as described in the text. The irradiated protein samples were analyzed by SDS–polyacrylamide gel electrophoresis followed by autoradiography. The position of the size marker proteins [phosphorylase A (92,500 Da), bovine serum albumin (68,000 Da), ovalbumin (45,000 Da), and β-lactoglobulin (30,000 Da)] are indicated.

were stained with Coomassie brilliant blue, dried onto filter paper, and visualized by autoradiography in the presence of an intensifying screen (Dupont Cronex) at −70°. The relative intensities of the labeled bands were determined with a soft laser scanning densitometer (LKB Instruments).

The crosslinking of (2′-5′)$(A)_3$[^{32}P]pCp proved to be a convenient assay for identifying RNase L in crude cytoplasmic extracts from cells as well as in partially purified protein fractions. About 2–5% of the bound (2′-5′)$(A)_3$[^{32}P]pCp was cross-linked to protein. As shown in Fig. 1 only one protein band (about 77,000 Da) was labeled when EAT cell cytoplasmic extracts were incubated with (2′-5′)$(A)_3$[^{32}P]pCp with UV irradiation. In the experiment shown the 77,000 Da protein band from interferon treated cells cross-linked about 2.5 times more (2′-5′)$(A)_3$[^{32}P]pCp than the corresponding protein band from control cells. The (2′-5′)$(A)_3$[^{32}P]pCp could, however, be cross-linked to several proteins from nuclear extracts of EAT cells and some of these proteins were found to be induced by interferon treatment.[21,23]

[23] This work was supported by NIH Research Grants AI-12320 and CA-16038 and a National Research Service Award to G. F. S.

[68] Purification of Double-Stranded RNA-Dependent Protein Kinase from Mouse Fibroblasts[1]

By CHARLES E. SAMUEL, GRACE S. KNUTSON, MARLA J. BERRY, JONATHAN A. ATWATER, and STEPHEN R. LASKY

Protein phosphorylation may play an important role in the antiviral action of interferon.[2,3] Interferon (IFN) treatment of many types of animal cells induces a protein kinase which selectively phosphorylates two pro-

[1] This work was supported in part by research grants from the National Institutes of Health (AI-12520 and AI-20611). C. E. S. was the recipient of Research Career Development Award K04-AI-00340, and J. A. A. was supported by Training Grant 5T32-GM-07621, both from NIH.

[2] C. E. Samuel, *Tex. Rep. Biol. Med.* **41,** 463 (1982).

[3] C. E. Samuel, *in* "Mechanisms of Interferon Actions" (L. Pfeffer, ed.). CRC Press, Cleveland, Ohio, 1985 (in press).

teins, P_1 and α.[4–7] Protein P_1 is a ribosome-associated protein of M_r 67,000 in mouse fibroblast cells[7,8]; α is the M_r 38,000 subunit of protein synthesis initiation factor eIF-2.[9,10] The protein kinase activity that catalyzes the phosphorylation of protein P_1 and eIF-2α in mouse cells is elevated about 5- to 10-fold over a low but frequently detectable basal level of activity by treatment with either natural or cloned IFN, is ribosome associated, and is cyclic AMP independent.[4–9,11,12] The activity *in vitro* of the IFN-induced P_1/eIF-2α protein kinase is dependent upon activation by double-stranded RNA (dsRNA). Both synthetic dsRNA [for example, poly(I)·poly(C)] and natural dsRNA (for example, reovirus genome dsRNA and killer yeast dsRNA) are able to activate the kinase *in vitro*.[4,7,13,14] *In vivo,* the activation step may be fulfilled in intact IFN-treated cells by either virus infection[10,15] or dsRNA treatment[16] of IFN-treated cell cultures. In cell-free extracts prepared from IFN-treated cells, the ability to phosphorylate eIF-2α correlates with the activation of protein P_1 phosphorylation by the addition of dsRNA.[11,17] Under conditions where the IFN-induced P_1/eIF-2α kinase is activated by dsRNA, greater than 75% of the endogenous eIF-2α is phosphorylated *in vitro*; by contrast, the steady-state level of eIF-2α phosphorylation *in vivo* increases to a maximum of only about 30% in IFN-treated virus-infected cell cultures.[10] Two forms of P_1 have been demonstrated in mouse fibroblasts. They differ from each other by the phosphorylation of a major phosphopeptide designated X_{ds}; peptide X_{ds} is phosphorylated in activated P_1 but not in P_1 which has not been activated by dsRNA.[17]

Two procedures are described below for the partial purification of the dsRNA-dependent P_1/eIF-2α protein kinase from mouse fibroblast cells. One procedure yields an enzyme preparation which is dependent upon

[4] B. Lebleu, G. C. Sen, S. Shaila, B. Cabrer, and P. Lengyel, *Proc. Natl. Acad. Sci. U.S.A.* **73,** 3107 (1976).

[5] W. K. Roberts, A. Hovanessian, R. E. Brown, M. J. Clemens, and I. M. Kerr, *Nature (London)* **264,** 477 (1976).

[6] A. Zilberstein, P. Federman, L. Shulman, and M. Revel, *FEBS Lett.* **68,** 119 (1976).

[7] C. E. Samuel, D. A. Farris, and D. A. Eppstein, *Virology* **83,** 56 (1977).

[8] C. E. Samuel, *Virology* **93,** 281 (1979).

[9] C. E. Samuel, *Proc. Natl. Acad. Sci. U.S.A.* **76,** 600 (1979).

[10] C. E. Samuel, R. Duncan, G. S. Knutson, and J. W. B. Hershey, *J. Biol. Chem.* **259,** 13451 (1984).

[11] C. E. Samuel and G. S. Knutson, *J. Biol. Chem.* **257,** 11791 (1982).

[12] J. A. Atwater and C. E. Samuel, *Virology* **123,** 206 (1982).

[13] M. A. Minks, D. K. West, S. Bevin, and C. Baglioni, *J. Biol. Chem.* **254,** 10180 (1979).

[14] P. F. Torrence, M. I. Johnston, D. A. Epstein, H. Jacobsen, and R. M. Friedman, *FEBS Lett.* **130,** 291 (1981).

[15] S. L. Gupta, S. L. Holmes, and L. L. Mehra, *Virology* **120,** 495 (1982).

[16] S. L. Gupta, *J. Virol.* **29,** 301 (1979).

[17] S. R. Lasky, B. L. Jacobs, and C. E. Samuel, *J. Biol. Chem.* **257,** 11087 (1982).

dsRNA for activity; the other procedure involves a dsRNA-Sepharose affinity chromatography step and yields an activated enzyme preparation. The dsRNA-dependent protein kinase activity capable of catalyzing the phosphorylation of protein P_1 and eIF-2α copurifies with protein P_1.

Preparative Procedures and Methodology

Reagents

All chemicals used in the following procedures are of reagent grade. Deionized (Barnstead Nanopure system), glass-distilled water is used in all procedures; buffers are prepared, and the pH is measured at 25° unless otherwise indicated.

- *N*-2-Hydroxyethylpiperazine-*N*-2-ethanesulfonic acid (Hepes), 0.5 *M*, pH 7.5
- KCl, 3 *M*
- NaCl, 4 *M*
- $Mg(OAc)_2$, 0.5 *M*
- Dithiothreitol (DTT), 0.1 *M* (Calbiochem or Sigma)
- ATP, 0.1 *M* (Sigma or P-L Biochemicals)
- Glycerol
- [γ-^{32}P]ATP (prepared as previously described[18])
- Reovirus genome double-stranded RNA, 25 μg/ml (prepared from purified reovirions by the method of Ito and Joklik[19])
- Poly(I) · poly(C) (generously supplied by Dr. G. J. Galasso, Antiviral Substances Program, N.I.H., or purchased from P-L Biochemicals)
- Poly(I) · poly(C)-Sepharose (P-L Biochemicals, or prepared as described by Wagner *et al.*[20])
- Diethylaminoethylcellulose, DEAE-cellulose (Schleicher and Schuell)
- Phosphocellulose (Whatman)
- Hexylamine-agarose (P-L Biochemicals)

Cell Culture, Interferon Treatment, and Cell Harvest

Clone L_{929} mouse fibroblast cells are routinely grown in roller culture (1330-cm^2 growth area/glass bottle) in Eagle's MEM containing 5% fetal calf serum or 5.6% newborn calf serum and 1.4% fetal calf serum. Confluent cells are treated with 300 units of interferon per ml [murine IFN-

[18] C. E. Samuel, this series, Vol. 79 [22].

[19] Y. Ito and W. K. Joklik, *Virology* **50,** 189 (1972).

[20] A. F. Wagner, R. L. Bugianesi, and T. Y. Shen, *Biochem. Biophys. Res. Commun.* **45,** 184 (1971).

(L929/NDV) or Hu-IFN-αA/D (*E. coli*)] at 37° for 18 to 24 hr before harvesting without trypsinization by scraping into cold (0–4°) isotonic buffer (35 m*M* Tris · HCl, pH 6.8 at 25°, 146 m*M* NaCl, 11 m*M* glucose). Cells are then washed three times in isotonic buffer and, after the last wash, are suspended in a small volume of isotonic buffer, transferred to a graduated conical centrifuge tube, and pelleted at 800 *g* for 10 min in an IEC CRU-5000 refrigerated centrifuge. The volume of the cell pellet is recorded, and then as much buffer as possible is carefully removed; the cell pellet can be stored at −80° until the time of extract preparation.

Assay of dsRNA-Dependent Protein Kinase

The activity of the dsRNA-dependent protein kinase is measured by [γ-^{32}P]ATP-mediated phosphorylation of endogenous proteins (e.g., protein P_1) and/or exogenously added proteins (e.g., purified protein synthesis initiation factor eIF-2, histones, etc.). The ^{32}P-labeled polypeptide products are analyzed by sodium dodecyl sulfate–polyacrylamide gel electrophoresis and autoradiography. The procedures have been described in detail.[18] The standard reaction mixture contains 20 m*M* Hepes (pH 7.5), 48 m*M* KCl, 1.5 m*M* DTT, 4 m*M* $Mg(OAc)_2$, 100 μM ATP containing 5–10 μCi of [γ-^{32}P]ATP, enzyme fraction, and, as indicated, dsRNA (1 μg/ml), cyclic AMP (10 μM) and exogenous protein substrate (eIF-2, calf thymus histones, or heat-treated ribosome-associated proteins). Preparations of eIF-2 purified as described,[21] or generously supplied by Drs. W. C. Merrick (rabbit reticulocytes) and J. W. B. Hershey (HeLa cells), are excellent substrates of the dsRNA-dependent protein kinase. Histones purchased from Sigma (calf thymus lysine-rich f_1 subgroup and arginine-rich f_3 subgroup) or generously provided by Drs. R. D. Cole (rabbit thymus H_{2A-B}, H_3, and H_4) and S. Elgin (*Drosophila* H_1) were either not substrates or were very poor substrates. The relative amount of [^{32}P]phosphate transferred to the various protein substrates is determined by SDS–polyacrylamide gel electrophoresis and autoradiography, and is quantitated either by densitometer scanning of a series of autoradiogram films or by directly counting the labeled bands excised from gels with the autoradiogram as a template.

Purification Procedures

General. All operations, unless otherwise indicated, are carried out at 0–4°; preparations should be kept on ice whenever possible, and ice-cold

[21] W. C. Merrick, this series, Vol. 60 [8].

degassed buffers should be utilized in all steps. Cell pellets obtained from about 20 roller bottles of cell culture are typically used for each preparation of the kinase.

Preparation of Cell-Free Extracts by Dounce Homogenization. The cell pellet is suspended in three packed-cell volumes of hypotonic buffer [10 m*M* Hepes, pH 7.5, 15 m*M* KCl, 1.5 m*M* $Mg(OAc)_2$, 1 m*M* DTT], allowed to swell for 10 min, and then disrupted with a tight-fitting Dounce homogenizer (50 strokes). Then 0.1 volume of 10× buffer (100 m*M* Hepes, pH 7.5, 1050 m*M* KCl, 35 m*M* $Mg(OAc)_2$, 10 m*M* DTT) is immediately added, and the homogenate is centrifuged at 5000 *g* for 5 min in a Sorvall RC-5 centrifuge and SS-34 rotor. The pellet is discarded, and the supernatant solution centrifuged at 10,000 *g* for 10 min. The resulting supernatant solution, the Dounce homogenate S10 extract, is used for the preparation of ribosomes and the ribosomal salt-wash fraction.

Preparation of Ribosome and Ribosomal Salt-Wash Fractions. The ribosomes are separated from the cell-sap fraction by centrifugation of nonpreincubated S10 extracts at 150,000 *g* for 2.5 hr in a DuPont Sorvall OTD-50 ultracentrifuge and T865.1 rotor. The crude translucent ribosome P150 pellet is gently suspended in Buffer A.12 [20 m*M* Hepes, pH 7.5, containing 120 m*M* KCl, 5 m*M* $Mg(OAc)_2$, and 1 m*M* DTT], stirred slowly for 1 hr, and centrifuged for 2.5 hr at 150,000 *g*. The resulting 0.12 *M* salt-washed ribosomal pellet is suspended, stirred, and centrifuged as before except that the A.12 buffer is modified to contain 0.8 *M* KCl (Buffer A.8); the 0.8 *M* salt suspension of ribosomes is centrifuged overnight at 120,000 *g*. The resulting supernatant solution, defined as the 0.8 *M* ribosomal salt-wash, is centrifuged for 4 hr at 120,000 *g* and then dialyzed against 250 volumes of Buffer DC.05 [20 m*M* Hepes, pH 7.5, containing 50 m*M* KCl, 5 m*M* $Mg(OAc)_2$, 1 m*M* DTT, and 10% (v/v) glycerol]. The dialysis is routinely carried out for 18 hr with two changes of buffer. After centrifugation of the dialyzed 0.8 *M* ribosomal salt-wash at 5000 *g* for 5 min, the resulting supernatant solution is stored at −80° until further fractionation by either Method A or Method B. The purification from the Dounce homogenate S10 extract to the 0.8 *M* ribosomal salt-wash fraction is estimated to be about 20-fold. However, accurate measurement of the dsRNA-dependent protein kinase activity in S10 extracts is complicated by the presence of phosphoprotein P_1/eIF-2α phosphatase activity which is dsRNA independent; the phosphatase activity is soluble and fractionates with the supernatant S-150 fraction.[7,22]

[22] C. E. Samuel and G. S. Knutson, *J. Interferon Res.* **2,** 441 (1982).

ISOLATION OF dsRNA-DEPENDENT P_1/eIF-2α PROTEIN KINASE FROM MOUSE FIBROBLAST CELLS[a]

Fraction	Protein (mg)	Activity (units)	Specific activity (units/mg)	Purification factor
Dounce S10 homogenate	890			
Ribosomes				
0.8 *M* ribosomal salt-wash	17.7	2222	126	—
DEAE-cellulose pool	1.65	1572	953	7.6
Phosphocellulose pool	0.36	1573	4369	34.7
Hexylamine-agarose pool	0.11	1054	9582	76.0

[a] Summary derived from independent purifications (from Berry *et al.*[24]).

Method A

Fractionation of 0.8 *M* ribosomal salt-wash prepared from IFN-treated mouse fibroblasts by a procedure involving DEAE-cellulose, phosphocellulose, and hexylamine-agarose chromatography yields an enzyme preparation purified about 1500-fold which is dependent upon dsRNA for activity. The isolation of the dsRNA-dependent protein kinase from mouse fibroblasts by Method A is summarized in the table.[24] The column chromatography steps result in about a 75-fold purification; as described above, the purification from Dounce homogenate S10 extract to the 0.8 *M* ribosomal salt-wash fraction is estimated to be about 20-fold. The kinase preparations obtained by Method A are not homogeneous. As detectable by silver staining, they routinely contain from 5–7 polypeptides on SDS–PAGE, one of which corresponds to protein P_1.

DEAE-Cellulose Chromatography. A column (1.2 × 9 cm) of DEAE-cellulose is packed at moderate pressure and equilibrated with Buffer DC.05. The 0.8 *M* ribosomal salt-wash fraction, desalted by dialysis against Buffer DC.05, is applied in a volume of about 2 ml and the column washed with about 10 ml of Buffer DC.05 to remove unbound material. The column may then be developed with a linear gradient of 0.05 to 0.5 *M* KCl in 40 ml of Buffer DC [20 m*M* Hepes, pH 7.5, containing 5 m*M* $Mg(OAc)_2$, 1 m*M* DTT, and 10% (v/v) glycerol]. Fractions of about 1 ml each are collected into plastic tubes. Autoradiograms showing recovered

[23] R. S. Ranu, I. M. London, A. Das, A. DasGupta, A. Majumdar, R. Ralson, R. Roy, and N. K. Gupta, *Proc. Natl. Acad. Sci. U.S.A.* **75,** 745 (1978).

[24] M. J. Berry, G. S. Knutson, S. R. Lasky, S. M. Munemitsu, and C. E. Samuel, *J. Biol. Chem.* **260,** 11240 (1985).

protein kinase activity measured with various substrates, including (A) endogenous activity, (B) purified calf thymus histones, and (C) purified eIF-2, are shown in Fig. 1. As shown in Fig. 1A, the endogenous dsRNA-dependent protein kinase activity does not bind to DEAE-cellulose in Buffer DC.05 and is recovered in the effluent wash fractions 10 to 14.

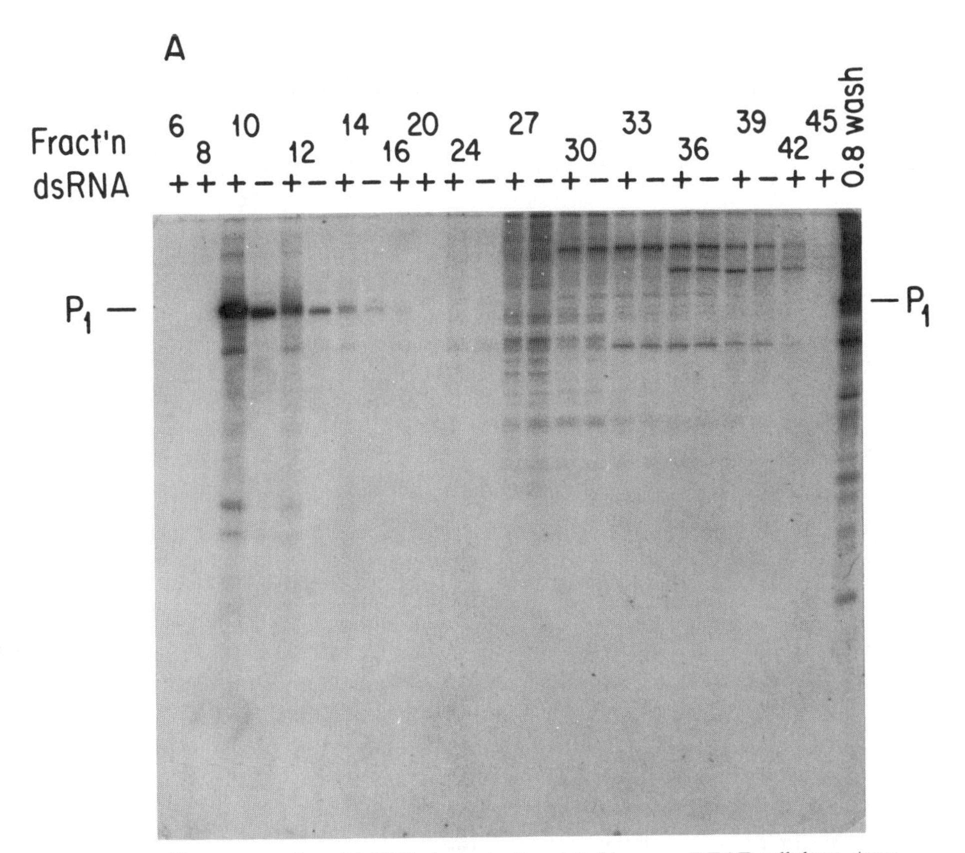

Fig. 1. Chromatography of dsRNA-dependent protein kinase on DEAE-cellulose. Autoradiograms show the protein kinase activities recovered following chromatography on DEAE-cellulose of a 0.8 *M* ribosomal salt-wash fraction prepared from interferon-treated mouse L_{929} cells. The gel lanes show the activity present in the effluent column fractions (Fract'n) and the material applied to the column (0.8 wash; +dsRNA). (A) Protein kinase activity measured in the absence of exogenously added substrate, both in the presence (+) and absence (−) of dsRNA; the position of endogenous protein P_1 is indicated. (B) Protein kinase activity measured with purified histones in the presence of dsRNA, both in the presence (+) and absence (−) of cAMP; the positions of histones H_1, H_2, H_3, and H_4 are indicated. (C) Protein kinase activity measured with purified eIF-2 in the presence of dsRNA; the positions of α subunit of eIF-2 and endogenous protein P_1 are indicated.

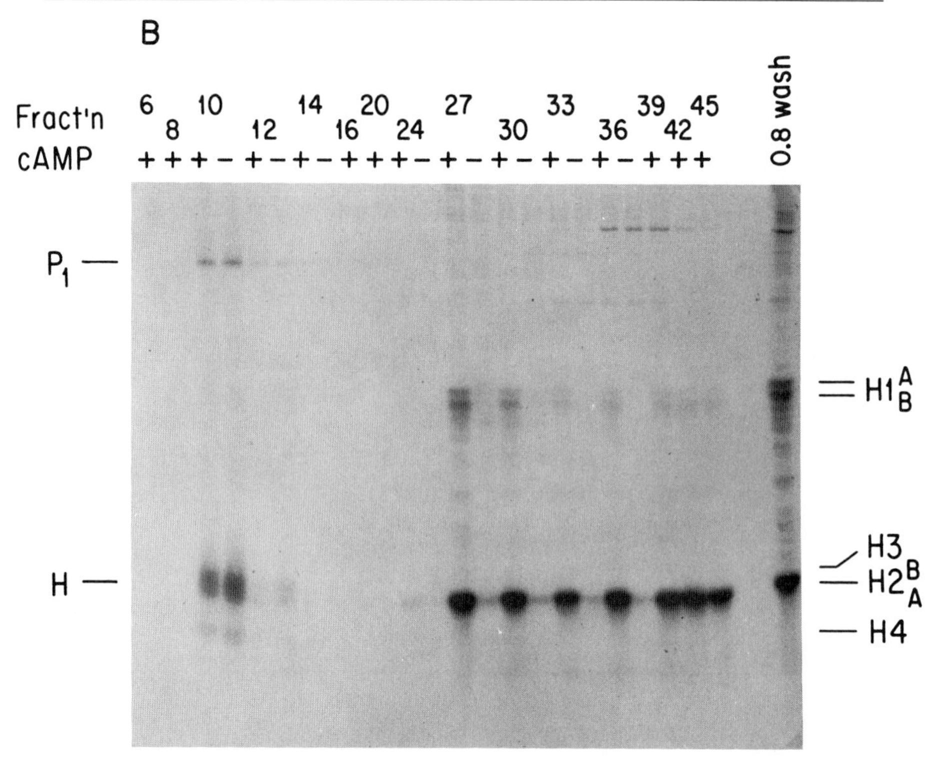

FIG. 1B (see legend p. 505).

Most histone kinase activity present in the 0.8 *M* ribosomal salt-wash fraction binds to DEAE-cellulose and, when measured in the presence of dsRNA, is found to be cAMP dependent (Fractions 24 to 45); column fractions 10 to 14 containing the unbound material do possess detectable histone kinase activity which in the presence of dsRNA is cAMP independent (Fig. 1B). Initiation factor eIF-2α protein kinase activity when measured in the presence of dsRNA is limited to the column fractions containing the unbound effluent material (Fig. 1C); these column fractions also contain the dsRNA-dependent protein kinase activity measured by endogenous protein P_1 phosphorylation (Fig. 1A and C). The elution profile on DEAE-cellulose of protein kinase activity is summarized in Fig. 2. The DEAE-cellulose column chromatography step results in a 5- to 10-fold purification of the dsRNA-dependent P_1/eIF-2α kinase from 0.8 *M* ribosomal salt-wash from IFN-treated cells. For routine preparation of the dsRNA-dependent protein kinase, the linear salt gradient is not required,

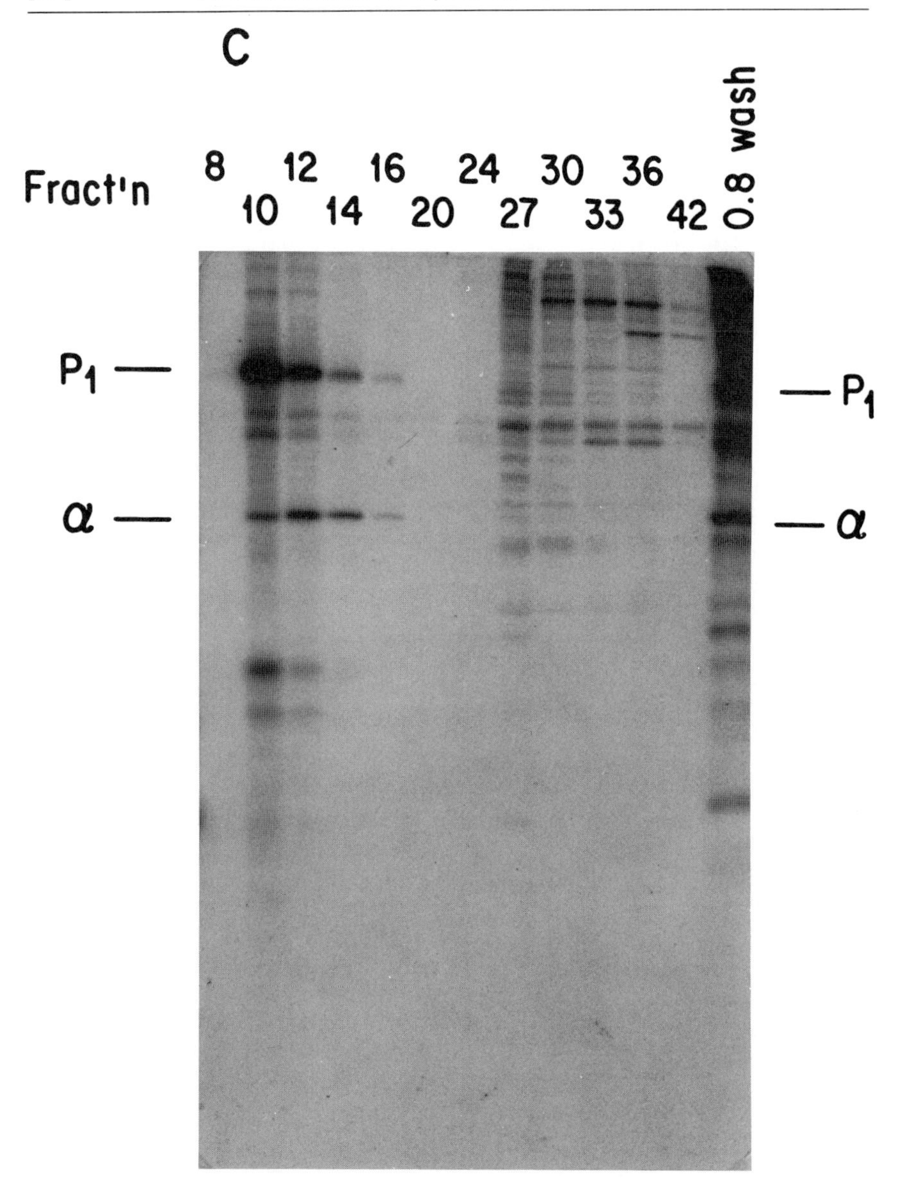

FIG. 1C (see legend p. 505).

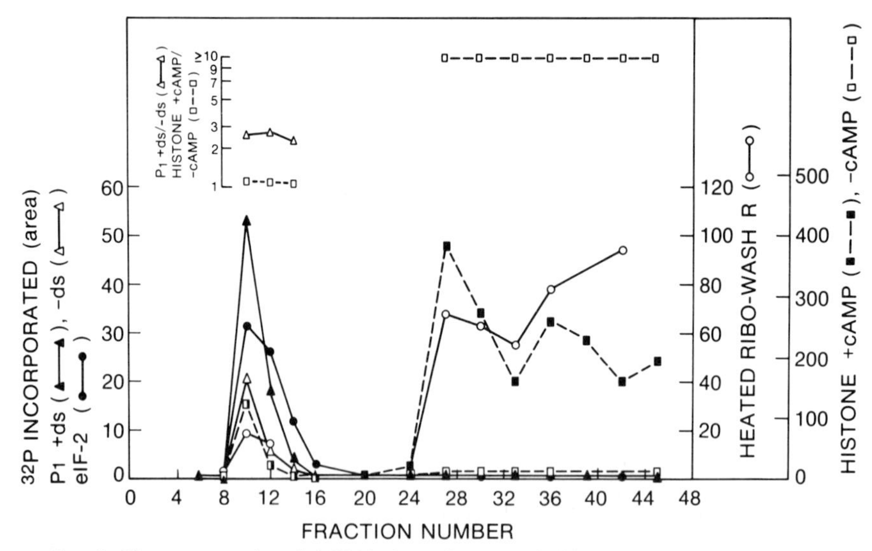

FIG. 2. Chromatography of dsRNA-dependent protein kinase on DEAE-cellulose. Elution profile of protein kinase activity determined with the following substrates: no exogenous substrate, both in the absence (△) and the presence (▲) of dsRNA; purified initiation factor eIF-2 in the presence of dsRNA (●); heat-treated (60°, 20 min) 0.8 *M* ribosomal salt-wash in the presence of dsRNA (○); and unfractionated calf thymus histones in the presence of dsRNA, both in the absence (□) and the presence (■) of cAMP. Quantitation was accomplished by densitometric scanning of autoradiograms.

because all detectable dsRNA-dependent protein kinase activity measured by phosphorylation of endogenous protein P_1 or exogenously added eIF-2 is present in the column wash fractions containing unbound material.

Phosphocellulose Chromatography. A column (Isolab QS, about 1 × 4.5 cm) of phosphocellulose is packed under gravity flow and equilibrated with Buffer PC0.1 [20 m*M* Hepes, pH 7.5, containing 100 m*M* KCl, 1 m*M* DTT, 0.1 m*M* EDTA, and 10% (v/v) glycerol]. The pooled fractions from the DEAE-cellulose column containing the protein P_1/eIF-2α dsRNA-dependent protein kinase activity are dialyzed against two changes of Buffer PC0.1 and applied to the column. The column is washed with about 3 ml of Buffer PC0.1 to remove unbound protein, and then is developed with a linear gradient of 0.1 to 0.75 *M* KCl in 12 ml of Buffer PC [20 m*M* Hepes, pH 7.5, containing 1 m*M* DTT, 0.1 m*M* EDTA, and 10% (v/v) glycerol]. Fractions of about 0.5 ml each are collected into plastic tubes. Autoradiograms showing recovered protein kinase activity measured with various substrates are shown in Fig. 3. The dsRNA-dependent protein kinase activity measured by phosphorylation of endogenous protein P_1 (Fig. 3A) or exogenous purified eIF-2 (Fig. 3C) binds to phosphocellulose in Buffer PC0.1, and routinely is eluted from columns of phosphocellulose with Buffer PC containing about 0.3 *M* KCl. The fractions containing active dsRNA-dependent protein kinase display poor histone kinase activity, and the low level of activity detected with histones is cAMP inde-

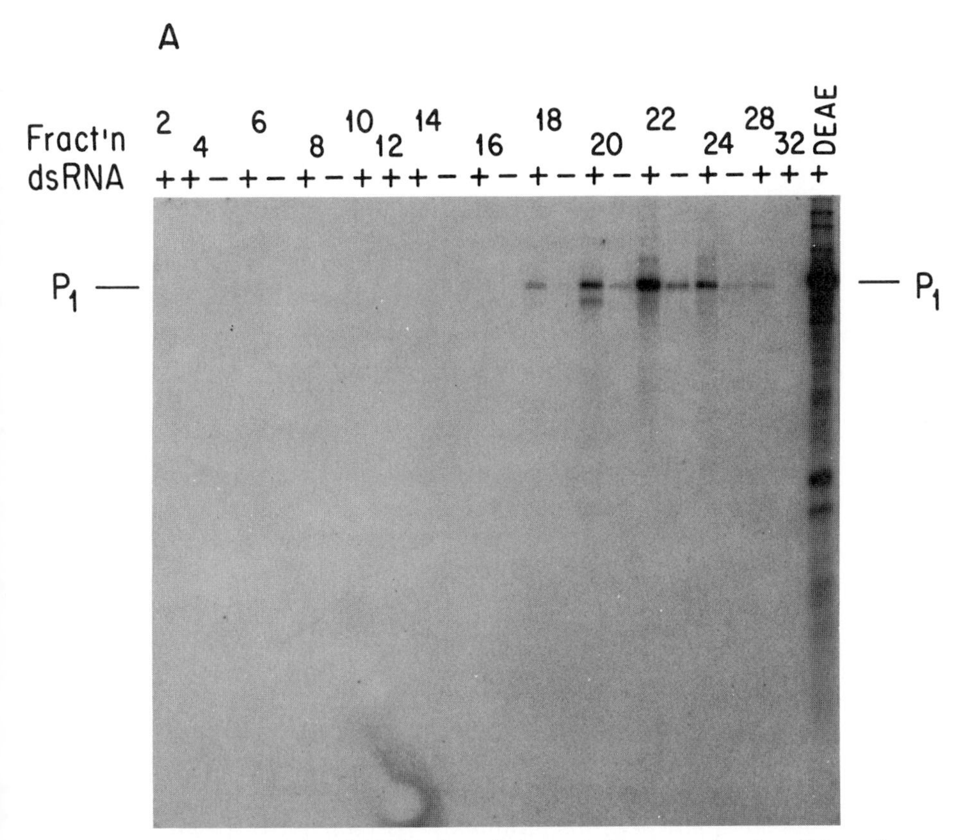

FIG. 3. Chromatography of dsRNA-dependent protein kinase on phosphocellulose. Autoradiograms show the protein kinase activities recovered after chromatography on phosphocellulose of pooled fractions from DEAE-cellulose containing dsRNA-dependent P_1/eIF-2α protein kinase activity. The gel lanes show the activity present in the effluent column fractions (Fract'n) and the material applied to the column (DEAE). (A) Protein kinase activity measured in the absence of exogenously added substrate, both in the presence (+) and absence (−) of dsRNA; the position of endogenous protein P_1 is indicated. (B) Protein kinase activity measured with purified histones in the presence of dsRNA, both in the presence (+) and absence (−) of cAMP; the position of the histone group (H) H_2, H_3, and H_4 is indicated. (C) Protein kinase activity measured with purified eIF-2 in the presence of dsRNA; the positions of the α subunit and endogenous P_1 are indicated.

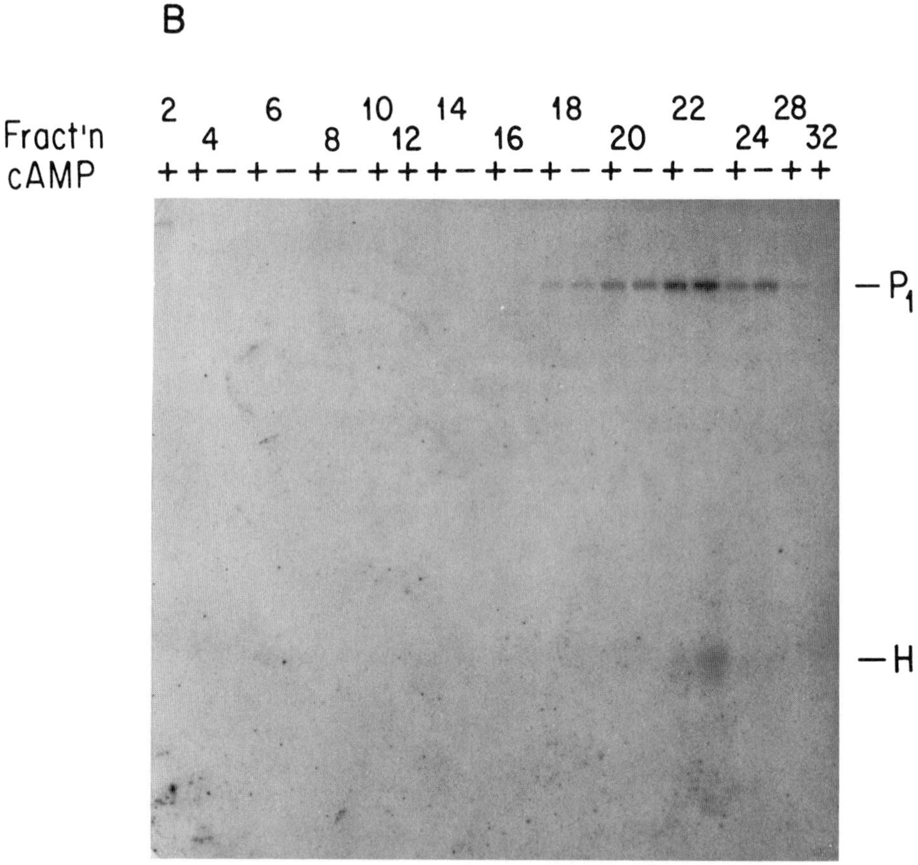

B

Fract'n 2 4 6 8 10 12 14 16 18 20 22 24 28 32
cAMP + + − + − + − + + + − + − + − + − + − + − + +

— P_1

— H

FIG. 3B (see legend p. 509).

pendent (Fig. 3B). The elution profile on phosphocellulose of protein kinase activity and protein is summarized in Fig. 4.

Hexylamine-Agarose Chromatography. A column (Isolab QS, about 1 × 4.0 cm) of hexylamine-agarose is packed under gravity flow and equilibrated with Buffer HA.05 [equivalent to DC.05]. The pooled fractions from the phosphocellulose column containing the protein P_1/eIF-2α dsRNA-dependent protein kinase activity were dialyzed against Buffer HA.05 and applied to the column. The column is washed with about 3 ml of Buffer HA.05 to remove unbound protein and then is developed with a linear gradient of 0.05 to 1.2 M KCl in 12 ml of Buffer HA (equivalent to Buffer DC). The elution profile on hexylamine-agarose of protein kinase activity and protein is shown in Fig. 5. The protein kinase activity mea-

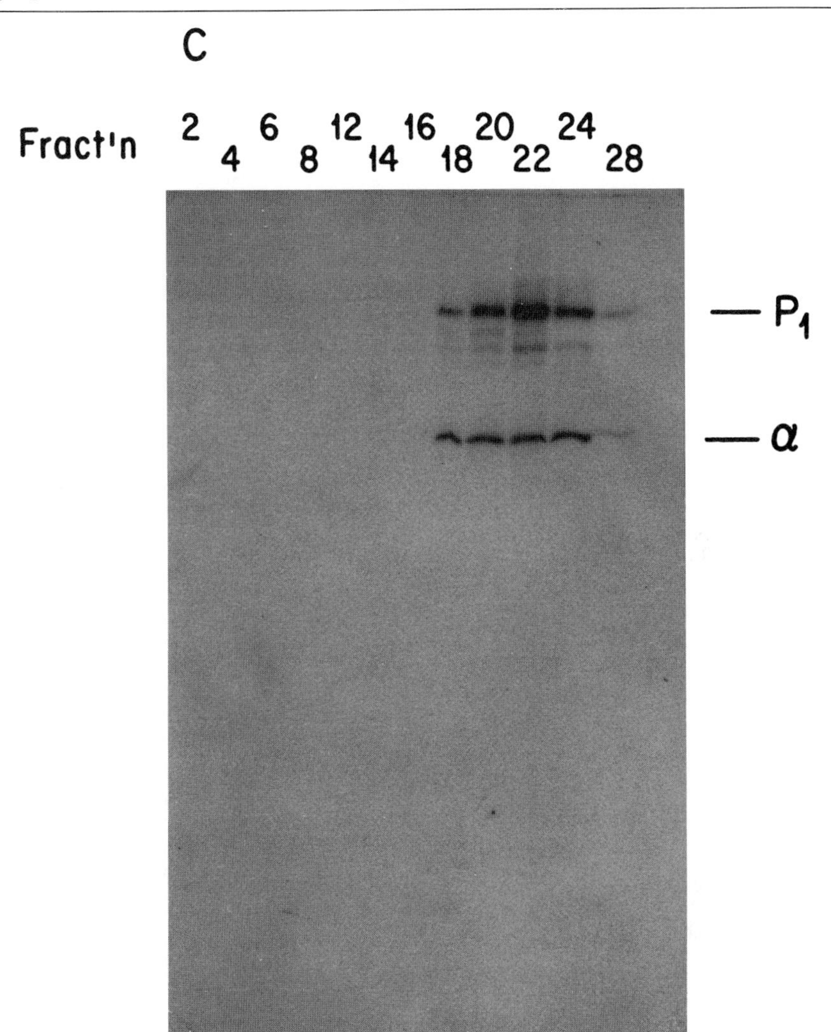

FIG. 3C (see legend p. 509).

sured by phosphorylation of endogenous protein P_1 and exogenously added eIF-2 as substrates elutes from columns of hexylamine-agarose with Buffer HA containing about 0.05 *M* KCl; the recovered P_1/eIF-2α protein kinase activity is dsRNA dependent. The hexylamine-agarose chromatography step gives about a 2-fold purification. Active fractions of the dsRNA-dependent kinase are concentrated by vacuum dialysis against Buffer HA.05.

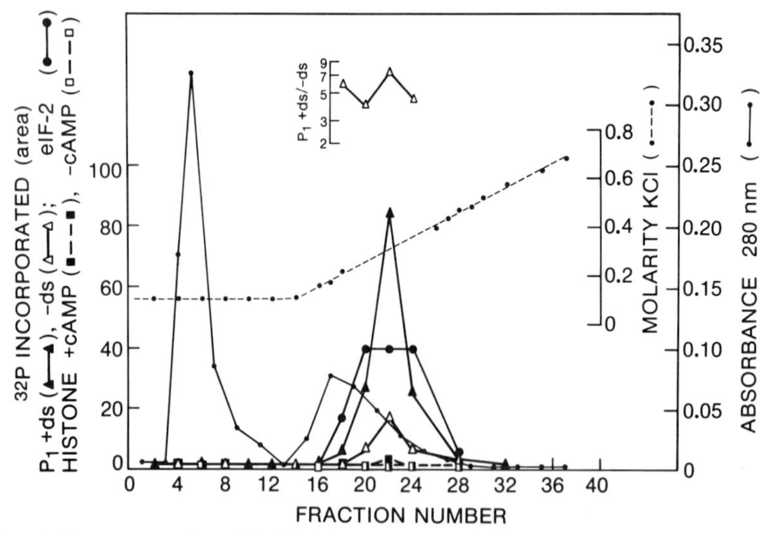

FIG. 4. Chromatography of dsRNA-dependent protein kinase on phosphocellulose. Elution profile of protein kinase activity determined with the following substrates: no exogenous substrate, both in the absence (△) and the presence (▲) of dsRNA; purified initiation factor eIF-2 in the presence of dsRNA (●); and calf thymus histones in the presence of dsRNA, both in the absence (□) and the presence (■) of cAMP. Quantitation was accomplished by densitometric scanning of autoradiograms. (●——●) Protein by A_{280}; (●- - -●) KCl gradient by conductivity.

Method B

Poly(I) · Poly(C)-Sepharose Chromatography. The isolation of ^{32}P-labeled protein P_1 suitable for certain biochemical analyses is efficiently accomplished by poly(I) · poly(C)-Sepharose chromatography and SDS–polyacrylamide gel electrophoresis. Fractionation of 0.8 *M* ribosomal salt-wash prepared from IFN-treated mouse fibroblasts by a procedure involving dsRNA-Sepharose affinity chromatography yields an activated dsRNA-dependent protein kinase preparation tightly complexed with dsRNA. Poly(I) · poly(C)-Sepharose (0.1 ml) is allowed to settle in a 1-ml syringe and then is washed with 2 ml of Buffer IC [30 m*M* Hepes, pH 7.5, containing 3 m*M* $Mg(OAc)_2$, 1 m*M* DTT, and 10% (v/v) glycerol] containing 0.12 *M* KCl. The ribosomal salt-wash fraction or the reaction mixture containing ribosomal salt-wash proteins phosphorylated *in vitro* is applied to the column. The column is incubated for 5 min and then is washed successively with 0.5 ml of Buffer IC containing 0.12, 0.5, 1.0, and 0.12 *M* NaCl. The column is then washed successively with five 0.1-ml volumes of Buffer IC containing 0.12 *M* NaCl and 1 mg/ml poly(I) · poly(C), with a

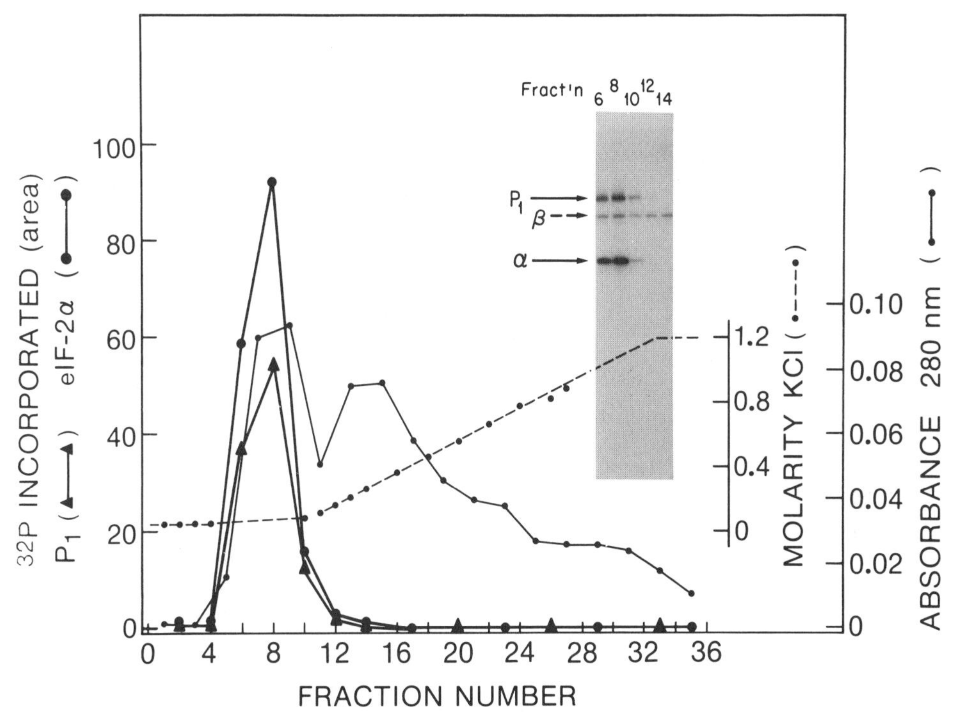

FIG. 5. Chromatography of dsRNA-dependent protein kinase on hexylamine-agarose. Elution profile of protein kinase activity was determined with the following substrates: no exogenous substrate in the presence of dsRNA (▲); purified initiation factor eIF-2 in the presence of dsRNA (●). Quantitation was accomplished by densitometric scanning of autoradiograms. (●——●) Protein by A_{280}; (●- - -●) KCl gradient by conductivity. Autoradiogram inset shows the protein kinase activity measured with added eIF-2 in the presence of dsRNA; the positions of endogenous protein P_1 and the α and β subunits of eIF-2 are indicated. The purified eIF-2 preparation used as a substrate contained a low level of protein kinase activity that phosphorylates the β subunit but not the α subunit of eIF-2.[23] [From Berry *et al.*[24]]

1 hr incubation between the collection of each 0.1 ml effluent fraction. An autoradiogram showing the effluent fractions analyzed by SDS-polyacrylamide gel electrophoresis is shown in Fig. 6. Most of the P_1 phosphorylated in the presence of dsRNA binds to poly(I) · poly(C)-Sepharose (50–80%) and can be selectively eluted with buffers containing free poly(I) · poly(C) ligand. Very little of the bound protein P_1 is eluted from the dsRNA-Sepharose affinity column in the absence of poly(I) · poly(C), and very little of the P_1 remained bound to the column matrix following elution with poly(I) · poly(C). Material which does not initially bind to

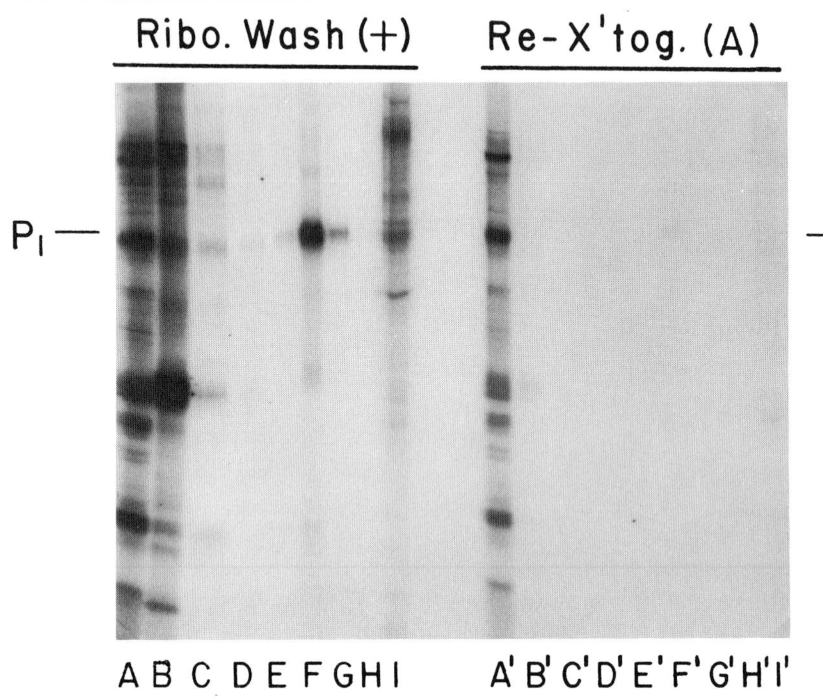

FIG 6. Chromatography of ribosomal salt-wash fraction on poly(I) · poly(C)-Sepharose. The 0.8 *M* ribosomal salt-wash fraction prepared from interferon-treated cells was phosphorylated in the presence of dsRNA, the [γ-^{32}P]ATP remaining in the reaction mixture was depleted by incubation with glucose and hexokinase, and the ^{32}P-labeled products were applied to a column of poly(I) · poly(C)-Sepharose in 0.12 *M* salt. Elution was with buffer containing (A) 0.12 *M* NaCl, (B) 0.5 *M* NaCl, (C) 1.0 *M* NaCl, (D and E) 0.12 *M* NaCl, (F, G, and H) 1 mg/ml poly(I) · poly(C). The fraction represented by gel lane A that contained the material that did not bind to the affinity column in 0.12 *M* salt was applied to a second column (Re-X'tog) of poly(I) · poly(C)-Sepharose; elution was with buffers (A'–I') as described for the first column (A–I). [From Lasky *et al.*[17]]

poly(I) · poly(C)-Sepharose also does not bind to the affinity column matrix upon rechromatography and may represent P_1 tightly complexed with dsRNA.

Concluding Comments

General Comments. The procedures described under Method A for the preparation of dsRNA-dependent protein kinase have been routinely used by our laboratory for kinase purification from mouse and human cell lines grown in culture. Purification procedures involving DEAE-cellulose and phosphocellulose chromatography steps have also been developed by other laboratories.[25,26] Protein kinase preparations obtained by Method A are suitable for biochemical studies which require dsRNA-dependent, catalytically active protein kinase. The poly(I) · poly(C)-Sepharose affinity chromatography procedure described under Method B is suitable for the preparation of activated kinase complexed with dsRNA. Purification of protein kinase by poly(I) · poly(C)-Sepharose chromatography has also been described by other workers,[27] although the conditions required to elute activated protein kinase from dsRNA-affinity columns differ significantly between laboratories.[17,27] The partial proteolytic phosphopeptide map obtained with *Staphylococcus aureus* V8 protease for protein P_1, which copurifies with the dsRNA-dependent protein kinase activity, is indistinguishable for P_1 purified by Method A as compared to Method B.[17,24]

Three lines of evidence are consistent with the possibility that the IFN-induced dsRNA-dependent protein kinase activity resides with protein P_1. First, purification studies have failed to separate the dsRNA-dependent protein kinase activity from protein P_1.[24–27] Second, the apparent native molecular weight of the purified dsRNA-dependent protein kinase as determined by sedimentation analysis (62,000) is comparable to the molecular weight (67,000) determined for denatured phosphoprotein P_1 by SDS–polyacrylamide gel electrophoresis.[8,24] Third, protein P_1 possesses a dsRNA-dependent ATP-binding site as established by the direct photoaffinity labeling of P_1 by ATP.[28]

Specificity. The purified protein kinase is highly selective for the α subunit of protein synthesis initiation factor eIF-2 and protein P_1 which copurifies with the dsRNA-dependent kinase activity.[9,17,24] *O*-Phosphoserine is the major phosphoester linkage both in protein P_1 and in eIF-2α phosphorylated by the activated P_1/eIF-2α kinase.[9,17] Protein synthesis factors other than the α subunit of eIF-2 (EF-1, EF-2, eIF-3, eIF-4A, eIF-4B, and eIF-5), are not substrates.[9,24] Likewise, calf thymus and *Drosophila* histones H1 are not detectable substrates of the purified dsRNA-dependent protein kinase, and calf and rabbit thymus histones H_{2A}, H_{2B}, H_3, and H_4 are very poor substrates.[24] Casein is not a substrate.[24]

Activators and Inhibitors. Phosphorylation of protein P_1 and eIF-2α is dsRNA dependent and cAMP independent.[9,11,24] Phosphorylation is stimulated by Mg^{2+}; Mn^{2+} does not replace Mg^{2+}. High concentrations of salt,

[25] G. C. Sen, H. Taira, and P. Lengyel, *J. Biol. Chem.* **253,** 5915 (1978).

[26] A. Kimchi, A. Zilberstein, A. Schmidt, L. Shulman, and M. Revel, *J. Biol. Chem.* **254,** 9846 (1979).

[27] A. G. Hovanessian and I. M. Kerr, *Eur. J. Biochem.* **93,** 515 (1979).

[28] J. R. Bischoff and C. E. Samuel, *J. Biol. Chem.* **260,** 8237 (1985).

N-ethylmaleimide, EDTA, AMP, pyrophosphate, spermine, and spermidine inhibit both P_1 and eIF-2α phosphorylation, whereas EGTA does not inhibit the phosphorylation of either P_1 or eIF-2α.[24] The phosphorylation of P_1 and eIF-2α depends in a similar manner on the concentration of dsRNA and is optimal at low dsRNA concentrations (0.1–1.0 μg/ml); high concentrations of dsRNA (>100 μg/ml) reduce the phosphorylation of both P_1 and eIF-2α.[24,29]

[29] N. G. Miyamoto, B. L. Jacobs, and C. E. Samuel, *J. Biol. Chem.* **258,** 15232 (1983).

[69] Enzymatic Synthesis of (2′-5′)-Linked Oligoadenylates on a Preparative Scale

By HIMADRI SAMANTA and PETER LENGYEL

Among the enzymes induced by interferons are (2′-5′)$(A)_n$ synthetases[1] (for reviews, see Refs. 2–4). These enzymes are latent unless activated by binding double-stranded RNA. The activated enzymes catalyze the following reaction:

$$n\ \mathrm{ATP} \rightarrow (2'\text{-}5')\mathrm{pppA(pA)}_{n-1} + (n-1)\ \mathrm{PP_i}$$

where PP_i represents pyrophosphate and (2′-5′)pppA$(pA)_{n-1}$ is usually abbreviated as (2′-5′)$(A)_n$, with *n* from 2 to about 15. The equilibrium of this reaction strongly favors synthesis. Over 98% of the ATP can be converted to (2′-5′)$(A)_n$ in spite of the fact that the enzyme does not cleave pyrophosphate.[5]

The only known function of (2′-5′)$(A)_n$ at present is the activation of another interferon induced enzyme the otherwise latent endoribonuclease designated as RNase L.[2,6] In order to activate RNase L the (2′-5′)$(A)_n$ must have either 5′ terminal triphosphate or diphosphate residues.[7]

[1] Abbreviations: (2′-5′)$(A)_n$, (2′-5′)pppA$(pA)_{n-1}$; EAT, Ehrlich ascites tumor; PMSF, phenylmethylsulfonyl fluoride.

[2] P. Lengyel, *Annu. Rev. Biochem.* **51,** 251 (1982).

[3] L. A. Ball, *in* "The Enzymes" (P. D. Boyer, ed.), 3rd ed., Vol. 15, Part B, p. 281. Academic Press, New York, 1982.

[4] P. Lengyel, this series, Vol. 79, p. 135.

[5] H. Samanta, J. P. Dougherty, and P. Lengyel, *J. Biol. Chem.* **255,** 9807 (1980).

[6] G. Floyd-Smith and P. Lengyel, this volume [67].

[7] E. M. Martin, N. J. M. Birdsall, R. E. Brown, and I. M. Kerr, *Eur. J. Biochem.* **95,** 295 (1979).

RNase L from various sources (e.g., L929 cells, EAT cells) can be activated by (2′-5′)$(A)_n$ as short as a trimer [i.e., (2′-5′)pppApApA] whereas RNase L from rabbit reticulocytes requires at least the tetramer for activation [i.e., (2′-5′)pppApApApA].[8]

The activation of RNase L by (2′-5′)$(A)_n$ is reversible. The removal of (2′-5′)$(A)_n$ from the activated enzyme (e.g., by gel filtration) causes the reversion of RNase L to the latent state. The readdition of (2′-5′)$(A)_n$ to the latent enzyme reactivates it.[9]

The binding of (2′-5′)$(A)_n$ to RNase L is tight: labeled (2′-5′)$(A)_n$, if bound to RNase L (but not if free) is retained on nitrocellulose filters.[9] This serves as the basis for a convenient assay for RNase L based on the binding of a labeled derivative of (2′-5′)$(A)_n$ such as (2′-5′)$(A)_3$[^{32}P]pCp to the enzyme.[10,11] The binding is selective enough to allow the crosslinking of (2′-5′)$(A)_3$[^{32}P]pCp essentially uniquely to RNase L by UV irradiation of a crude cytoplasmic extract from EAT or HeLa cells incubated with (2′-5′)$(A)_3$[^{32}P]pCp.[6,12,13]

This binding of (2′-5′)$(A)_n$ might serve as the basis for screening cell extracts for (2′-5′)$(A)_n$ binding proteins other than RNase L. It is conceivable that at least some of these may be proteins whose activities are modulated by (2′-5′)$(A)_n$. This approach has already resulted in the finding of several (2′-5′)$(A)_n$ binding proteins in nuclear extracts from EAT cells. Remarkably, some of these are induced by interferons.[14]

For the purposes of screening for, characterizing, and purifying (2′-5′)$(A)_n$ binding proteins (2′-5′)$(A)_n$ might be required on a preparative (10 to 100 μg) scale. Such amounts can be synthesized chemically.[15–17] However, for biochemical laboratories it may be convenient to generate the (2′-5′)$(A)_n$ from ATP by enzymatic synthesis. This was accomplished on a

[8] B. R. G. Williams, R. R. Golgher, R. E. Brown, C. S. Gilbert, and I. M. Kerr, *Nature (London)* **282,** 582 (1979).

[9] E. Slattery, N. Ghosh, H. Samanta, and P. Lengyel, *Proc. Natl. Acad. Sci. U.S.A.* **76,** 4778 (1979).

[10] M. Knight, P. J. Cayley, R. H. Silverman, D. H. Wreschner, C. S. Gilbert, R. E. Brown, and I. M. Kerr, *Nature (London)* **288,** 189 (1981).

[11] T. W. Nilsen, P. A. Maroney, and C. Baglioni, *J. Biol. Chem.* **256,** 7806 (1981).

[12] G. Floyd-Smith, O. Yoshie, and P. Lengyel, *J. Biol. Chem.* **257,** 8584 (1982).

[13] P. J. Cayley, M. Knight, and I. M. Kerr, *Biochem. Biophys. Res. Commun.* **104,** 376 (1981).

[14] G. St. Laurent, O. Yoshie, G. Floyd-Smith, H. Samanta, P. B. Sehgal, and P. Lengyel, *Cell* **33,** 95 (1983).

[15] S. S. Jones and C. B. Reese, *J. Am. Chem. Soc.* **101,** 7399 (1979).

[16] J. A. J. den Hartog, R. A. Wijnands, J. H. van Boom, and R. Crea, *J. Org. Chem.* **46,** 2242 (1981).

[17] J. Imai and P. F. Torrence, *J. Org. Chem.* **46,** 4015 (1981).

small scale [producing about 35 mg of (2′-5′)$(A)_n$] with crude cell lysates as a source of the enzyme.[18] We report here a procedure for the synthesis of (2′-5′)$(A)_n$ of various chain lengths with a partially purified (2′-5′)$(A)_n$ synthetase from cytoplasmic extracts of EAT cells.

Materials

Buffer A: 17 m*M* *N*-2-hydroxyethyl piperazine-*N*′-2-ethanesulfonic acid (Hepes) (pH 7.5), 30 m*M* 2-mercaptoethanol, 4 m*M* $Mg(OAc)_2$, 1 m*M* EDTA, 10 μ*M* PMSF, 10% (v/v) glycerol.

S-30 fraction (i.e., cytoplasmic extract from EAT cells which had been treated with 1000 units/ml of mouse beta interferon for 24 hr): this was prepared according to the procedure described in this volume by Floyd-Smith and Lengyel.[6]

Partial Purification of (2′-5′)$(A)_n$ Synthetase

About 50 ml of S-30 fraction (from 25 liters of interferon-treated EAT cells) was centrifuged at 200,000 *g* for 2 hr. The resulting pellet was resuspended in 10 ml of buffer A, supplemented with 250 m*M* KOAc and centrifuged at 200,000 *g* for 2 hr. The pellet was dissolved in 10 ml of buffer A, supplemented with 500 m*M* KOAc, and centrifuged at 200,000 *g* for 2 hr. The supernatant fraction (10 ml) was supplemented with 40 ml of buffer A and centrifuged at 10,000 *g* for 10 min. The supernatant fraction was loaded at 4° onto a 10 ml carboxymethylcellulose column (CM52, Whatman) which had been equilibrated with buffer A supplemented with 100 m*M* KOAc. The column was washed with 40 ml of the equilibrating solution. The fractions with (2′-5′)$(A)_n$ synthetase activity were eluted with 30 ml of buffer A supplemented with 250 m*M* KOAc. Analysis of the eluate by polyacrylamide gel electrophoresis in sodium dodecyl sulfate followed by staining with Coomassie blue indicated that the (2′-5′)$(A)_n$ synthetase, a 105,000 Da protein, was among the predominant bands.

Assay of the Enzyme

The concentration of the (2′-5′)$(A)_n$ synthetase preparation required to convert 70 to 80% of the ATP added to (2′-5′)$(A)_n$ in an overnight incubation was determined in a small scale experiment. Each reaction mixture (30 μl) contained 17 m*M* Tris · Cl (pH 8.0), 225 m*M* KOAc, 30 m*M* 2-

[18] A. G. Hovanessian, R. E. Brown, E. M. Martin, W. K. Roberts, M. Knight, and I. M. Kerr, this series, Vol. 79, p. 184.

mercaptoethanol, 12 m*M* $Mg(OAc)_2$, 1 m*M* EDTA, 0.2% Triton X-100, 10 μM PMSF, 5 m*M* ATP (α-^{32}P-labeled, 10^5 cpm/reaction mixture), 5 μg/ml poly(I) · poly(C), and 1 to 20 μl of enzyme. After incubation at 25° for about 16 hr the reaction mixtures were heated at 65° for 3 min, chilled in ice, and clarified by a 15 sec centrifugation in a microfuge. To convert unreacted ATP to ADP aliquots of the clarified solution were supplemented with an equal volume of a second solution to give a final concentration of 7.5 units/ml of hexokinase (EC 2.7.1.1., Sigma type VI, 75 units/mg), 10 m*M* glucose, and 6 m*M* $Mg(OAc)_2$ and were incubated at 30° for 30 min. Aliquots of 2.0 μl were then chromatographed on polyethyleneimine cellulose thin layer plates (Polygram CEL 300 PEI Brinkman) in 0.75 *M* potassium phosphate (pH 3.5) with appropriate standards of ADP and ATP. The spots corresponding to (2'-5')$(A)_n$ were located by autoradiography, eluted with 1 *M* HCl, and counted in a liquid scintillation counter. In this chromatographic system (2'-5')$(A)_n$ migrates close to or together with ATP.

Preparation of (2'-5')$(A)_n$

The composition of the preparative scale reaction mixture (150 ml) was identical to that described for the assay of the enzyme. It included 453 mg of ATP · Na_2 · $3H_2O$ and that volume percentage of the (2'-5')$(A)_n$ synthetase preparation which was found in the assay to convert 70 to 80% of the ATP to (2'-5')$(A)_n$. After incubation at 25° for 16 hr the reaction mixture was heated at 65° for 5 min (to inactivate the enzyme and denature at least some of the proteins), and centrifuged at 10,000 *g* for 10 min. The supernatant fraction was supplemented with 525 ml of buffer A and loaded at a rate of 40 ml/hr onto a 100 ml DEAE cellulose (DE52 Whatman) column which had been equilibrated with buffer A also containing 50 m*M* KOAc. The column was washed with the equilibrating solution until the OD_{260} of the eluate decreased below 0.1 and then with 150 ml of 50 m*M* NH_4HCO_3. The column was eluted with 3 liters of an NH_4HCO_3 gradient (50 to 400 m*M*) which was applied with a peristaltic pump.

A typical elution pattern is shown in Fig. 1. The various peaks were pooled separately, lyophilized, and the residues were dissolved in H_2O and lyophilized again. The dissolution in H_2O and lyophilization were repeated two to four times until all of the NH_4HCO_3 had been removed.

Aliquots from each peak were characterized in the following way: one aliquot from each peak was digested with bacterial alkaline phosphatase (6 units/ml, from Sigma) in 20 m*M* Tris · Cl (pH 8.0) at 37° for 2 hr. The reaction mixture was analyzed by chromatography on polyethyleneimine cellulose sheets with 1 *M* HOAc as the developing solvent (allowing the

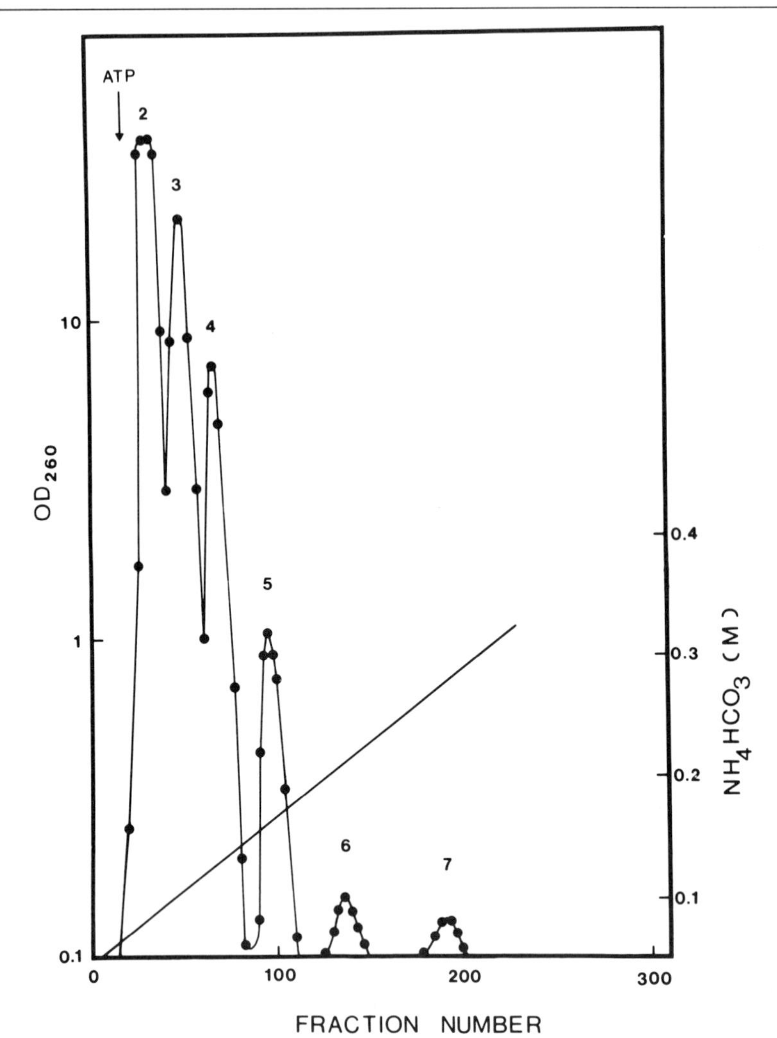

FIG. 1. Fractionation of (2′-5′)$(A)_n$ according to chain length by chromatography on DEAE cellulose. (2′-5′)$(A)_n$ was generated from 453 mg of ATP $Na_2 \cdot 3H_2O$ by enzymatic synthesis. After loading the (2′-5′)$(A)_n$ sample on the column and washing the column, the (2′-5′)$(A)_n$ was fractionated by differential elution with a NH_4HCO_3 gradient (50 to 400 m*M*). Fractions of 10 ml were collected and the A_{260} and the conductivity of each were determined. The fractions comprising the various peaks were separately pooled (omitting the fractions with much cross-contamination). The (2′-5′)$(A)_n$ in the various pools was characterized by enzymatic digestion and chromatographic analysis as described in the text. The OD_{260} of only some of the fractions is shown (●). The straight line indicates the concentration of NH_4HCO_3 as detected by measuring conductivity. The numbers 2, 3, 4, 5, 6, and 7 above the peaks indicate (2′-5′)-linked $(A)_2$, $(A)_3$, $(A)_4$, $(A)_5$, $(A)_6$, and $(A)_7$, respectively. All the oligoadenylates had 5′-terminal triphosphate residues. The position in the gradient at which ATP was eluted is indicated by the arrow.

separation of 2'-5' ApA, ApApA, and ApApApA from each other and from the longer chain oligoadenylates which remain at the origin) or 4 *M* HOAc (which also allows the separation of ApApApApA and ApApApA-pApA from each other and from all of the shorter and longer oligoadenylates).

Other aliquots from each peak were digested with 0.3 *M* KOH at 37° for 15 hr and the products were analyzed by chromatography on polyethyleneimine cellulose sheets with 0.75 *M* potassium phosphate buffer (pH 3.5) as the developing solvent. This system allowed the separation of 5'-terminal pppAp, ppAp, and pAp among other compounds.

The results of the characterization revealed that the products of the preparative scale enzymatic synthesis of (2'-5')$(A)_n$ have 5'-terminal triphosphates. Furthermore the percentages of di-, tri-, tetra-, penta-, hexa-, and heptaadenylates were 38, 29, 6, 1.3, 0.19, and 0.07%, respectively.

Comments

Most of the (2'-5')$(A)_n$- and ATP-degrading enzymes are removed in the course of the purification of the (2'-5')$(A)_n$ synthetase preparation used. This makes it possible to perform the synthesis with poly(I)·poly(C) in solution. The degrading enzymes are usually removed by binding (2'-5')$(A)_n$ synthetase from a crude cell extract to poly(I) · poly(C) linked to a solid support and washing proteins which do not bind tightly off the poly(I) · poly(C).[19] However, the rate of (2'-5')$(A)_n$ synthesis as catalyzed by (2'-5')$(A)_n$ synthetase bound to poly(I) · poly(C) linked to a solid support is much slower than that catalyzed by (2'-5')$(A)_n$ synthetase activated by poly(I) · poly(C) in solution. This might be a consequence of a suboptimal ratio of poly(I) · poly(C) to (2'-5')$(A)_n$ synthetase. Excess poly(I) · poly(C) was found to inhibit (2'-5')$(A)_n$ synthesis.[5]

Further increase in the concentration of the (2'-5')$(A)_n$ synthetase preparation allows the conversion of over 95% of the ATP added to (2'-5')$(A)_n$. Such an approach requires more enzyme, but saves radioactive ATP when generating labeled (2'-5')$(A)_n$. Moreover, as established by homochromatography, the percentage of longer polyadenylates (up to the pentadecamer) is higher in a reaction with 95% conversion of the ATP than in one with only 75% conversion.[5]

Acknowledgments

This work was supported by NIH research Grants AI-12320 and CA-16038.

[19] A. G. Hovanessian and I. M. Kerr, *Eur. J. Biochem.* **93,** 515 (1979).

[70] Methods for the Synthesis of Analogs of (2′-5′)-Oligoadenylic Acid

By Paul F. Torrence, Jiro Imai, Jean-Claude Jamoulle, and Krystyna Lesiak

The discovery of 2-5A,[1] its probable involvement in the mechanism of interferon's antiviral action,[2,3] and its possible role in other aspects of cell regulation[2,3] have generated considerable interest in analogs of 2-5A.[4] Such 2-5A analogs could conceivably find use as antagonists of 2-5A action, permitting the role of 2-5A to be discerned in various situations.[5] In addition, there is an obvious interest in whether or not knowledge of the 2-5A system could be capitalized upon to generate a novel chemotherapeutic approach to viral and/or neoplastic diseases.[4]

A wide variety of approaches have been used in the synthesis of 2-5A and its analogs, ranging from the purely enzymatic to the purely chemical. In a previous volume of this series, we detailed a stepwise chemical approach to the core of 2-5A.[6] In this contribution, we highlight, by means of specific examples, several different approaches, found useful in our laboratory, to the preparation of 2-5A analogs. These methods include (1) modification of a preformed 2-5A molecules; (2) enzymatic phosphorylation of a "core" oligonucleotide; (3) lead ion-catalyzed polymerization of 5′-phosphoroimidazolidates of selected nucleosides.

Chemical Modification of a Preformed (2′-5′)-Oligoadenylate. "Tailing" of p5′A2′p5′A2′p5′A2′p5′A and Subsequent Conversion to the Corresponding 5′-Triphosphate

Reagents

p5′A2′p5′A2′p5′A2′p5′A prepared as described elsewhere[7]
Sodium metaperiodate (0.1 *M*)

[1] I. Kerr and R. E. Brown, *Proc. Natl. Acad. Sci. U.S.A.* **75,** 256 (1978).
[2] P. F. Torrence, *Mol. Aspects Med.* **5,** 129 (1982).
[3] P. Lengyel, *Annu. Rev. Biochem.* **51,** 251 (1982).
[4] P. F. Torrence, K. Lesiak, J. Imai, M. I. Johnston, and H. Sawai, *in* "Nucleosides, Nucleotides, and Their Biological Applications" (J. L. Rideout, D. W. Henry, and L. M. Beacham III, eds.), p. 67. Academic Press, New York, 1983.
[5] P. F. Torrence, J. Imai, and M. I. Johnston, *Proc. Natl. Acad. Sci. U.S.A.* **78,** 5493 (1981).
[6] J. Imai and P. F. Torrence, this series, Vol. 79, p. 233.

n-Hexylamine (Aldrich)
Acetic acid (10% v/v)
Sodium cyanoborohydride (Aldrich)
Triethylammonium bicarbonate buffers as specified
Dimethyl sulfoxide (dried over molecular sieves)
Tri-*n*-butylammonium pyrophosphate
Carbonyl diimidazole (Aldrich)
Sodium iodide

Periodate Oxidation of p5′A2′p5′A2′p5′A2′p5′A, Shiff Base Formation and Cyanoborohydride Reduction.[8] Sodium metaperiodate (0.1 *M*, 120 μl) is added to an ice-cold solution of p5′A2′p5′A2′p5′A2′p5′A [416 A_{258} units, 10 μmol in water (300 μl)]. After the reaction mixture is stirred for 20 min at 0°, hexylamine (8 μl, 60 μmol) is added, and the pH of the resulting solution is immediately adjusted to 8.5 with acetic acid (10%). This reaction mixture is again stirred at 0° for 20 min, then sodium cyanoborohydride (100 μl of 0.5 *M*) is added, and the pH of the solution is adjusted to 6.5 with 10% acetic acid. The resulting mixture is allowed to stir at 0° for 40 min and then is applied to a column of DEAE-Sephadex (HCO_3^-) A25 (1.0 × 20 cm) which subsequently is eluted with a linear gradient (250 ml/250 ml) of 0.35–0.40 *M* triethylammonium bicarbonate (pH 7.6). Proper fractions are pooled, concentrated *in vacuo*, and water is added and removed *in vacuo* several times to insure removal of all triethylammonium bicarbonate buffer. The desired product is isolated as a sodium salt by dissolving the above residue in methanol (300 μl) and pouring this solution into a sodium iodide solution in dry acetone (50 mg/6 ml). The resulting precipitate is centrifuged down, washed twice with dry acetone (each 3 ml), and then dried over P_2O_5 *in vacuo*. The yield should be 85% based on A_{258} determination. This series of reactions produces a tetrameric oligoadenylate 5′-monophosphate in which the last (2′-terminal) adenosine residue has been converted to 2-(9-adenyl)-6-hydroxymethyl-4-hexylmorpholine. Such modified oligonucleotides are referred to as "tailed" derivatives.

Synthesis of the 5′-Triphosphate of 2′-"Tailed" p5′A2′p5′A2′p5′A2′p5′A

The "Tailed tetramer monophosphate (triethylammonium salt, 416 A_{260} units, 10 μmol) is dissolved in dry dimethylformamide (500 μl) and

[7] P. F. Torrence, J. Imai, K. Lesiak, J.-C. Jamoulle, and H. Sawai, *J. Med. Chem.* **27**, 726 (1984).

[8] J. Imai, M. I. Johnston, and P. F. Torrence, *J. Biol. Chem.* **257**, 12739 (1983).

triphenylphosphine (13.1 mg, 50 μmol), imidazole (6.8 mg, 100 μmol), and dipyridinyl disulfide (11.0 mg, 50 μmol) are added to the stirred solution. After 40 min stirring at ambient temperature, the entire reaction mixture is added dropwise to a solution of sodium iodide in acetone (10 ml of 0.1 *M*). The colorless precipitate which forms is centrifuged down, washed with dry acetone (3×) and finally dried over P_2O_5 for 1 hr.

The above sodium salt of the "tailed" tetramer 5′-phosphoroimidazolidate is dissolved in dry dimethylformamide (200 μl) and tri-*n*-butylammonium pyrophosphate in dimethylformamide (200 μl of 0.5 *M*) is added. After 24 hr at room temperature, the dimethylformamide is removed *in vacuo* and the residue is taken up in 0.3 *M* triethylammonium bicarbonate (pH 7.6, 1 ml). This solution is applied to a DEAE-Sephadex A25 column (HCO_3^- form, 1 × 20 cm), and the column is eluted with a linear gradient of 0.3–0.6 *M* triethylammonium bicarbonate buffer (pH 7.6, total volume 500 ml, 137 fractions). Proper fractions (60–85) are pooled and concentrated *in vacuo,* and then water is added to and evaporated from the residue several times to ensure removal of all triethylammonium bicarbonate. The desired "tailed" tetramer 5′-triphosphate is isolated as a sodium salt (335 A_{260}, 8.05 μmol, 80.5% yield) in the same manner as described above for the 5′-phosphoroimidazolidate. The 5′-triphosphate should have the following ^{31}P NMR as determined in D_2O (δ in ppm): −0.65, −0.77, −0.87 (singlets, internucleotide Ps), −5.63 (d, J = 19 Hz, γP), −10.70 (d, J = 18 Hz, αP), and −20.89 (t, J = 18 Hz, βP). When a Beckman HPLC with model 110A pumps is employed, the product should have a retention time of 7.19 min and should be ≥99% pure on a ultrasphere ODS C_{18} column upon elution with an isocratic mixture of 50 m*M* $NH_4H_2PO_3$ (pH 7.2) (30%) and methanol–water (1 : 1) 70% (flow rate, 1 ml/min). The product of this reaction sequence corresponds to a tetrameric oligoadenylate 5′-triphosphate in which the 2′-terminal adenosine unit has been converted to 2-(9-adenyl)-6-hydroxymethyl-4-hexylmorpholine, and its structure is represented in Fig. 1.

Enzymatic Phosphorylation of a "Core" Oligonucleotide. Preparation of 3-5A, a Phosphodiester Linkage Isomer of 2-5A[9]

Reagents

A3′p5′A3′p5′A (Sigma)
Phosphoglucose hexokinase (Sigma)
T_4 polynucleotide kinase (Bethesda Research Laboratories)
KCl
Tris · HCl
Dithiothreitol (DTT)

$MgCl_2$
Bovine serum albumin
ATP
Triethylammonium bicarbonate buffers of specified molarity

Phosphorylation of 3-5A "Core" with T_4 Polynucleotide Kinase. The trinucleotide A3′p5′A3′p5′A is 5′-phosphorylated in a reaction mixture containing the following: 200 A_{260} units (5.56 μmol, corresponding to 8.35 m*M*) of trimer core, 180 μl (180 units) of T_4 polynucleotide kinase (1000 units/ml in 50 m*M* Tris · HCl, pH 7.6, 100 m*M* KCl, 5 m*M* DTT, 200 μg/ml bovine serum albumin), 1333 μl of 0.05 *M* $MgCl_2$ (final conc. 0.01 *M*), 799 μl of 0.5 *M* Tris · HCl, pH 7.6 (final conc. 0.06 *M*), 2556 μl of 38.5 m*M* DTT (final conc. 0.015 *M*), and 16.65 μmol of ATP (final conc. 2.5×10^{-2} *M*). The solution is incubated overnight at 37°. The progress of the reaction can be monitored by HPLC with the system described above with a Beckman ODS column (4.6 mm × 15 cm) and a gradient of 0–70% B in solvent A [B: methanol–water (1 : 1); A: 0.05 *M* $NH_4H_2PO_4$, pH 7] over an elapsed time of 25 min with a 1 ml/min flow rate. In this case the reaction should be finished after overnight incubation.

The reaction mixture then is applied to a DEAE Sephadex A-25 column (1 × 10 cm) which is eluted with a linear gradient of 0.05–0.5 *M* triethylammonium bicarbonate buffer (pH 7.5). Fractions containing p5′A3′p5′A3′p5′A and unreacted ATP are pooled and evaporated to dryness. The residue is dissolved in 600 μl of water and unreacted ATP is converted to ADP by incubation for 1 hr at 30° with an equal amount of phosphoglucose hexokinase solution (0.4 μg/ml hexokinase in 10 m*M* Tris · HCl, pH 7.5). This mixture then is separated on a DEAE-Sephadex column in the same manner as described above. The yield of p5′A3′-

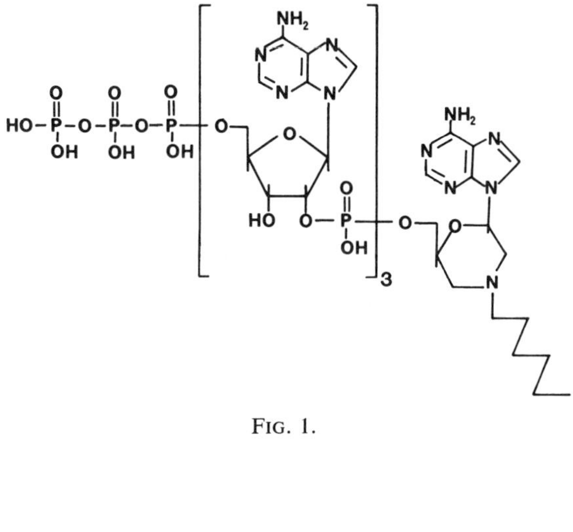

FIG. 1.

p5′A3′p5′A should be about 175 A_{260} units (88% of theoretical). This product gives, upon nuclease P_1 digestion, only 5′ AMP. Alkaline phosphatase digestion regenerates A3′p5′A3′p5′A. Sodium periodate oxidation followed by base elimination with lysine and further alkaline phosphatase digestion gives adenine and A3′p5′A as expected. With the HPLC system described above, p5′A3′p5′A3′p5′A has a retention time of 15.5 min compared with 9.5 min for p5′A2′p5′A2′p5′A.

The trinucleotide 5′-monophosphate can be readily converted to the corresponding 5′-triphosphate (3-5A) through the 5′-phosphoroimidazolidate reaction with pyrophosphate as described above and elsewhere.[9]

Lead Ion Catalyzed Polymerization Synthesis of Tubercidin (c^7A) Analog of 2-5A[10]

Reagents

Tubercidin 5′-monophosphate (Calbiochem)
Imidazole (Aldrich)
Triphenylphosphine (Aldrich)
Dipyridyl disulfide (Aldrich)
Dimethyl sulfoxide (dry)
Sodium iodide
Lead nitrate
Chelex (Bio-Rad)
Ethylenediaminetetraacetic acid
Nuclease P_1 (Worthington)
Triethylammonium bicarbonate buffers as specified

Preparation of Tubercidin 5′-Phosphoroimidazolidate

Tubercidin 5′-monophosphate (255 mg, 7141 A_{270} units, 0.7 mmol) is dissolved in dry dimethyl sulfoxide (5 ml). Thereafter imidazole (255 mg, 3.75 mmol), triphenylphosphine (392 mg, 1.5 mmol), triethylamine (125 μl), and dipyridyl disulfide (330 mg, 1.5 mmol) are added. This yellow solution is kept at room temperature for 40 min, during which time the completion of the reaction can be checked on TLC with silica gel plates and 2-propanol–conc. NH_4OH–H_2O (7 : 1 : 2) as developing solvent. The solution is transferred dropwise to a stirred solution of sodium iodide (565 mg, 3.75 mmol) in dry acetone (25 ml), and the precipitate is collected by centrifugation and washed with acetone (3 × 4 ml). The sodium salt of tubercidin 5′-phosphoroimidazolidate is dried 3 hr *in vacuo* to yield about

[9] K. Lesiak, J. Imai, G. Floyd-Smith, and P. F. Torrence, *J. Biol. Chem.* **258,** 13082 (1983).
[10] J.-C. Jamoulle, J. Imai, K. Lesiak, and P. F. Torrence, *Biochemistry* **23,** 3063 (1984).

6090 A_{270} units (85%) of product and should be homogeneous by TLC on the above silica gel system.

Polymerization of Tubercidin 5'-Phosphoroimidazolidate to Give (2'-5')-Linked Oligodeazaadenylates

The sodium salt of tubercidin 5'-phosphoroimidazolidate (6090 A_{270} units, 261 mg, 0.6 mmol) is dissolved in 0.2 *M* imidazolium buffer (pH 8.0, 14.4 ml) containing lead nitrate (0.25 *M*, 0.75 ml), and the resulting mixture is stirred at 4° for 12 days. The suspension then is treated with Chelex (100–200 mesh, NH_4^+ form, 10 ml) at room temperature for 1 hr. After removal and washing (10 ml of water) of the Chelex by filtration, the pH of the filtrate is adjusted to 7.2 with 10% acetic acid. The clear solution is evaporated to dryness *in vacuo*, the residue is taken up in ethanol (20 ml), and the resulting suspension is filtered. The precipitate is washed with alcohol (1 ml) and then dried for 3 hr *in vacuo*. This residue is dissolved in a buffer (7 ml) of 4-morpholineethanesulfonic acid (50 m*M*, pH 6.0), EDTA (1.0 m*M*) and nuclease P_1 (0.25 mg), and the mixture is incubated overnight at 37°. The incubation mixture then is heated at 100° for 4 min, and the insoluble denatured protein is removed by centrifugation at 10000 *g*. Oligodeazaadenylates are purified from this mixture by HPLC on a reverse-phase column (Zorbax ODS, 9.4 mm × 25 cm). This requires 18 injections of approximately 350 A_{270} units of crude reaction mixture and elution of the column with a linear gradient of 0–35% buffer B in 35 min (buffer A, 50 m*M* ammonium phosphate, pH 7.0; buffer B, methanol–H_2O, 1:1, flow rate 2.7 ml/min). Oligonucleotides up to octamer are eluted from the column in less than 25 min. Appropriate fractions containing each oligomer are pooled and evaporated *in vacuo*, and the residues are applied to a DEAE-Sephadex A-25 (HCO_3^-) column to remove the phosphate salts of the HPLC buffer. Elution is with a linear gradient of 0.2–0.5 *M* (for dimer) or 0.2–0.7 *M* (for trimer or tetramer) triethylammonium bicarbonate (pH 7.5). Fractions containing each oligomer are pooled, evaporated to dryness *in vacuo* at 40°, and then taken up in water and reevaporated. This procedure is repeated twice to remove all triethylammonium bicarbonate buffer. Finally, the residue is dissolved in anhydrous methanol, and the sodium salt is collected as described above for tubercidin 5'-phosphoroimidazolidate. The yields (based on A_{270} readings) are as follows: p5'(c^7A)2'p5'(c^7A), 31.5%; p5'(c^7A)2'p5'(c^7A)2'p5'(c^7A), 7.0%; p5'(c^7A)2'p5'(c^7A)2'p5'(c^7A)2'p5'(c^7A), 2.4%. Also obtained are higher oligomers including pentamer (1.5%), hexamer (1%), and heptamer (<1%).

The oligomers as obtained are completely resistant to degradation by nuclease P_1 but are degraded by snake venom phosphodiesterase to yield tubercidin 5′-monophosphate as the only product. On TLC with PEI-cellulose and 0.25 *M* ammonium bicarbonate, the above trimer has an R_f of 0.32 while on cellulose F plates with isobutyric acid–conc. NH_4OH–EDTA (0.2 *M*) (100:60:0.8), it has an R_f of 0.69. The retention time of the trimer, p5′(c^7A)2′p5′(c^7A)2′p5′(c^7A), with the above HPLC system with a Bondapak C_{18} column (30 × 0.39 cm) and elution with a linear gradient of 0–50% buffer B in buffer A (buffer A, 50 m*M* ammonium phosphate, pH 7.0; buffer B, methanol–H_2O, 1:1; flow rate 1.0 ml/min) in 25 min, is 12.75 min.

The trimeric oligodiazaadenylate 5′-monophosphate can be converted to the corresponding 5′-triphosphate via the 5′-phosphoroimidazolidate as detailed above and elsewhere.[10]

The ^{31}P NMR of ppp5′(c^7A)2′p5′(c^7A)2′p5′(c^7A) as determined in D_2O is as follows (δ in ppm): −0.96, −1.10 (internucleotide), −8.71 (d, γP), −11.08 (d, αP), and −22.10 (t, βP).

Comments

The first method described herein for the preparation of analogs of 2-5A, chemical modification of a preformed 2-5A, represents a facile and efficient route which, in the given example, generates a 2-5A analog which is a significantly more potent inhibitor of protein synthesis[8,11] and viral replication[12] than the parent 2-5A molecule. This increased activity is most probably related to the greatly increased stability of the "tailed" 2-5A molecule in cell extracts and cells due to its resistance to phosphodiesterase action.[8]

The second method detailed above, T_4 polynucleotide kinase[13] catalyzed phosphorylation of "core" oligonucleotides, is a critical means of obtaining the intermediate oligonucleotide 5′-monophosphate used for the preparation of the corresponding 5′-triphosphates. Chemical phosphorylation methods are unsuccessful in these instances unless the 5′-unphosphorylated oligomer can be prepared in a protected form. We have found this enzymatic phosphorylation to be successful in a number of different situations, although the final yield of 5′-phosphorylated oligonucleotide may vary considerably. Using the described method we have been able to prepare the following 5′-monophosphates: p5′A3′p5′A3′p5′A[9] (vide su-

[11] D. A. Epstein, M. A. vander Pas, B. B. Schryver, J. Imai, and P. F. Torrence, unpublished observations.

[12] P. Defilippi, G. Huez, M. Verghaegen-Lawalle, E. De Clercq, J. Imai, P. F. Torrence, and J. Content, unpublished observations.

[13] C. C. Richardson, *in* "Procedures in Nucleic Acid Research" (G. L. Cantoni and D. R. Davies, eds.), Vol. 2, p. 815. Harper & Row, New York, 1971.

pra), p5′A2′p5′A2′p5′A,[14] p5′(3′dA)2′p5′(3′dA)2′p5′(3′dA),[15] p5′(xylo A)2′p5′(xylo A)2′p5′(xylo A),[16] p5′A2′p5′A2′p5′(c^7A),[17] p5′A3′p5′-(br^8A)2′p5′(br^8A),[14] and p5′A2′p5′(br^8A)2′p5′(br^8A).[14] Our experience has been, however, that oligonucleotides with one or more (3′-5′) phosphodiester bonds are superior substrates in this reaction. For instance, phosphorylation of the linkage isomer, A3′p5′(br^8A)2′p5′(br^8A), proceeds nearly quantitatively, but under the same conditions only approximately 5% phosphorylation of A2′p5′(br^8A)2′p5′(br^8A) could be achieved.[14]

The last method outlined here, lead ion-catalyzed polymerization of a nucleoside 5′-phosphoroimidazolidate as developed originally by Sawai and Ohno,[18,19] is an extremely valuable technique to obtain reasonable yields of (2′-5′)-linked oligonucleotide 5′-monophosphate in a very short time. For instance, we have used this method to prepare the following oligonucleotide 5′-monophosphates: (2′-5′)$(pA)_n$ (n = 2–6), (2′-5′)$(pc^7A)_n$[10] (n = 2–7) (vide supra), p5′(m^6A)2′p5′(m^6A)2′p5′(m^6A),[20] and p5′(br^8A)2′p5′(br^8A)2′p5′(br^8A).[14]

[14] K. Lesiak and P. F. Torrence, *J. Med. Chem.*, in press.

[15] H. Sawai, J. Imai, K. Lesiak, M. I. Johnston, and P. F. Torrence, *J. Biol. Chem.* **258,** 1671 (1983).

[16] K. Lesiak, G. Gosselin, J.-L. Imbach, and P. F. Torrence, unpublished observations.

[17] J.-C. Jamoulle, K. Lesiak, and P. F. Torrence, unpublished observations.

[18] H. Sawai and M. Ohno, *Bull. Chem. Soc. Jpn.* **54,** 2752 (1981).

[19] H. Sawai and M. Ohno, *Chem. Pharm. Bull.* **29,** 2237 (1981).

[20] J.-C. Jamoulle, K. Lesiak, and P. F. Torrence, unpublished observations.

Section IX

Assay of Interferons

A. Antiviral Assays

Articles 71 through 74

B. Antiproliferative Assays

Articles 75 and 76

C. Immunoassays

Articles 77 and 78

[71] Objective Antiviral Assay of the Interferons by Computer Assisted Data Collection and Analysis

By ROBERT L. FORTI, SHIRLEY S. SCHUFFMAN, HUGH A. DAVIES, and WILLIAM M. MITCHELL

The problem and purification of human interferon (IFN) as well as the detection of IFNs in various biological samples are contingent upon a reliable, accurate, and rapid means of assay.[1] IFN has historically been assayed by plaque reduction methods, standard to the discipline of virology. This method requires 5–7 days of culture in semimicro wells or Petri dishes under an agar overlay to prevent random dispersion of *de novo* synthesized virus. The method is time consuming, requires large quantities of reagents, and is subject to errors in plaque counting. In 1969, Finter[2] introduced the use of vital dye uptake as a method of estimation of inhibition of viral cytopathic effect (CPE) by IFN. This procedure was cumbersome and time consuming; consequently, it was adopted by only a relatively small number of laboratories. McManus[3] later proposed vital dye uptake in microtiter plates with cells in preformed monolayers as a quantitative measure of IFN titer. This procedure also was relatively time consuming and the methods of sample dilution laborious. Rubinstein *et al.*[4] and Familletti *et al.*[5] more recently have described a microtiter method for the rapid assay of IFN which relies on subjective analysis of viral CPE inhibition. For the last 4 years we have used a method based on the vital dye uptake assay of Finter[2] coupled with this rapid microtiter assay.[4,5] With the use of a microtiter plate colorimeter and a dedicated microcomputer we have devised a format for automated data collection and analysis which provides a rapid and nonsubjective means of data analysis. This method of IFN assay has the following advantages: (1) speed of performance with minimal physical manipulation; (2) objective analysis of quantitative response with statistically generated confidence limits; (3) interface with a dedicated low cost laboratory computer allows a high volume of analyses by a single technician; and (4) timely application to the clinical laboratory on a cost-effective basis.

[1] W. M. Mitchell and R. L. Forti, *Prog. Clin. Pathol.* **9,** 101 (1984).
[2] N. B. Finter, *J. Gen. Virol.* **5,** 419 (1969).
[3] N. H. McManus, *Appl. Environ. Microbiol.* **31,** 35 (1976).
[4] S. Rubinstein, P. C. Familetti, and S. Pestka, *J. Virol.* **37,** 755 (1981).
[5] P. C. Familletti, S. Rubinstein, and S. Pestka, this series, Vol. 78, p. 387.

Solutions and Media

Minimal essential media (MEM) can be obtained from GIBCO as a preformulated dry powder. It is reconstituted with dionized distilled water, filtered through a 0.22-μm Millipore filter, and stored at 4° until use. Sterile fetal bovine serum (FBS) is used to supplement MEM where indicated in the text. Modified Finter[2,6] stock neutral red solution at 1% (w/v) is prepared from dry certified reagent (C.I. No 50040; Matheson, Coleman, and Bell, Inc.) and 100 mM NaH_2PO_4. It is then sequentially processed on Whatman No 1 and 0.22-μm Millipore filters. This sterile stock solution may be stored at room temperature. The daily working solution is prepared from a 1 : 150 dilution of stock neutral red solution in MEM–FBS yielding a final concentration of 67 μg dye per ml.

Acidified ethanol is prepared by a 50 : 50 mixture of absolute ethanol and 1% acetic acid. Dulbecco's phosphate-buffered saline (PBS) is prepared from a preformulated powder supplied by GIBCO.

Assay Cells

Virtually any convenient cell line may be used that is responsive to the antiviral activity of the IFNs being assayed and which yields a uniform cytopathic effect (CPE) to indicator virus by inhibition of cellular uptake of the vital dye, neutral red. In addition one should demonstrate that the uptake of neutral red is a linear function relative to cell concentration up to the maximum number of cells contained in a single microtiter well. Figure 1 demonstrates neutral red adsorbtion as a function of human fibroblast concentration. Between 0.5×10^3 and 4×10^4 cells/well the function is linear with a correlation coefficient of 0.985. In our studies we have used a human foreskin derived fibroblast line (SG-181) between passage 5 and 14 or HEp-2 cells for the assay of human IFN. Mouse L cells have been routinely used for the assay of mouse IFN.

Cytopathic Indicator Virus

Although any IFN sensitive virus capable of distinct CPE should be suitable, we have used vesicular stomatitis virus (VSV), Indiana Strain, and encephalomyocarditis virus (EMCV) for both human and mouse IFN assay. We propagate VSV and EMCV on HeLa cells. Stock virus is adsorbed on HeLa cells in a minimal volume of serum-free MEM at a multiplicity of infection of 0.1 for 1 hr. The culture is washed twice with phosphate-buffered saline (PBS), pH 7.4, MEM with 10% FBS is added

[6] M. D. Johnston, N. B. Finter, and P. A. Young, this series, Vol. 78, p. 394.

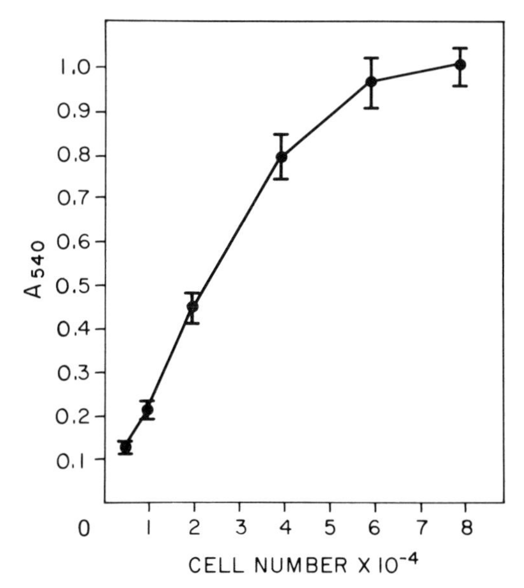

FIG. 1. Neutral red dye uptake by human foreskin fibroblasts (SG-181) cells. Human foreskin fibroblasts were plated in microtiter plates at cell densities ranging from 0.5 to 8 × 10^4 and allowed to adhere to the plastic surface for 4 hr. Cells were stained with neutral red dye as described in the text and dye uptake determined with a Flow Titertek colorimeter. From Forti *et al.*[9]

and incubated for 18 hr for VSV and 20 hr for EMCV or until a distinct viral CPE is produced. The growth medium is harvested, and centrifuged at 500 *g* for 15 min. The virus should be plaque assayed under an agar overlay on any suitable target cell. We have found no differences in plaque-forming units (PFU) between HeLa cells, human fibroblasts, or HEp-2 cells with either VSV or EMCV. Virus is stored at −70° in small aliquots containing approximately 2.0 × 10^7 PFU/ml VSV or EMCV.

Interferon Reference Standards

It is essential that laboratory standards be run daily. International reference standards[7,8] may be obtained from the National Institutes of Allergy and Infectious Diseases and laboratory working standards should be titered against the International Standard and frozen at −70° in small aliquots so that laboratory standards are not refrozen.

[7] N. B. Finter, this series, Vol. 78, p. 14.
[8] S. Pestka, this volume [2].

Interferon Assay Procedure[9]

Assays are performed in standard flat bottom 96-well microtiter plates. The plates are configured in a manner that allow three assays in triplicate as illustrated in Fig 2. Row 1 (8 replicates) serves as a cell control minus virus. Row 2 serves as a virus control minus IFN. Rows 3–5, 6–8, and 9–11 serve as IFN unknown or standard titer test lanes. Row 12 is blank. MEM (100 μl) supplemented with 10% FEBS is added to all wells of rows 1–11; then 100 μl of fluid to be titered is added in triplicate to the first well of the three test lanes and serially diluted 2-fold from an initial 1:2 dilution through a 1:256 dilution. Target cells ($2.5–3.0 \times 10^4$ cells/well) are added from a single cell suspension in 100 μl growth medium and 1 hr later are challenged with VSV or EMCV (50 μl MEM containing 7×10^3 PFU/well). Plates are incubated in an atmosphere of 5% CO_2/air at 100% humidity and 36°. Viral CPE is observed in control wells within 16 to 20 hr after viral infection (about two viral replication cycles). The growth medium is removed by inversion, cell layers washed twice with PBS, pH 7.4, and the medium replaced with 100 μl MEM containing 67 μg/ml of neutral red dye (100 μl MEM containing Finter's stock 1% neutral red dye solution diluted 1 : 150 in MEM).[2,6,9] Following a 2 hr incubation the wells are inverted, washed twice with PBS, pH 7.4, and the dye which was absorbed by the viable cells is extracted into 100 μl acidified ethanol (50% ethanol in 1% acetic acid). The extract dye solution in the original microtiter plate is quantitated colorimetrically at 540 nm with a Flow Titertek colorimeter with row 12 used as a blank.

Data Reduction

Data reduction[10] is carried out on an Apple II computer interfaced to the Titertek type 3100 colorimeter with a type 312B computer interface, both from Flow Laboratories. The Apple II microcomputer must contain 64K bytes of memory, one floppy disk drive, and a printer. Assay data from the Titertek are analyzed in a two step process: first, data from the Titertek are moved into specific computer memory through the instrument interface by a program which mainpulates raw data and transfers assay results to the printer. The data reduction program begins by allowing optical density (OD) readings from specified wells to be deleted from the calculations. Such an operation, although used infrequently, allows

[9] R. L. Forti, R. A. Moldovan, W. M. Mitchell, P. A. Callicoat, S. S. Schuffman, H. A. Davies, and D. M. Smith, *J. Clin. Microbiol.* **21,** 689 (1985).

[10] Data acquisition and reduction software are available on request.

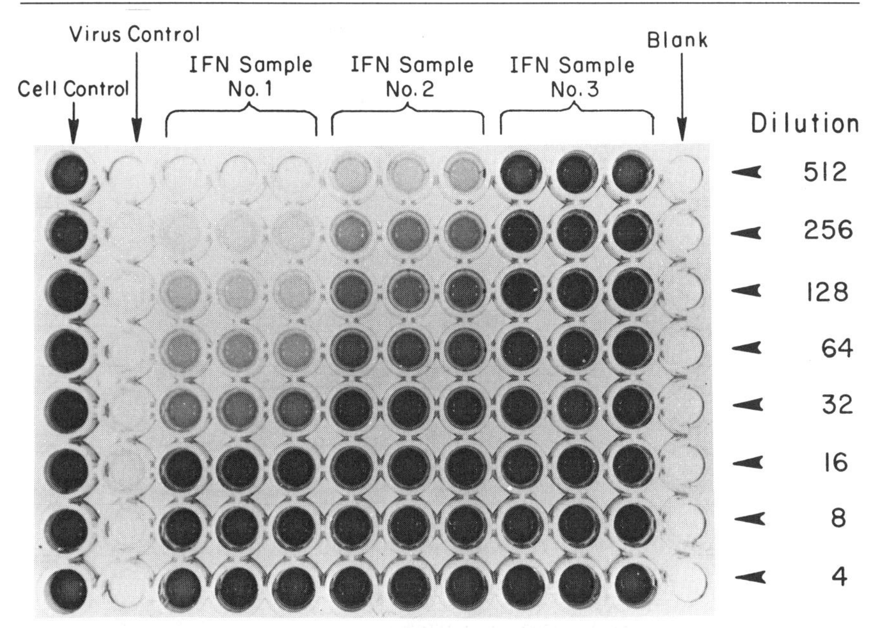

FIG. 2. Microtiter plate configuration used for the computer assisted data acquisition. IFN sample No. 1 demonstrated 50% inhibition at a 1 : 32 dilution to yield 32 laboratory units/ml of IFN antiviral activity; IFN sample No. 2 demonstrated 50% inhibition at a dilution of 1 : 324 to yield 324 laboratory units/ml of IFN antiviral activity; IFN sample No. 3 demonstrated greater than 50% protection at all dilutions or >512 laboratory units/ml.

flexibility in eliminating obvious aberrant values. The eight cell control and eight virus infected control wells are each averaged and an expected range of OD for the assay run is calculated as the difference of the two averages. The range is used to define the OD at which 50% protection by IFN is achieved. Each assay dilution is performed in triplicate yielding three sets of OD values, their mean, and standard deviation. The mean OD value of each dilution is used to calculate the percentage of protection relative to the cell and virus control derived range. The percentage protection and the percentage standard deviation are printed together with the dilution. After each percentage protection is printed, the two values found to bracket the 50% OD level are used by the computer to interpolate linearly the dilution at which 50% protection is estimated to occur. This value is printed. To obtain the actual IFN titer the reciprocal dilution at which 50% protection is achieved must be multiplied by the reciprocal of any prior dilutions of the sample. The observed titer is then adjusted to the titer obtained for the laboratory standard. In each IFN reference

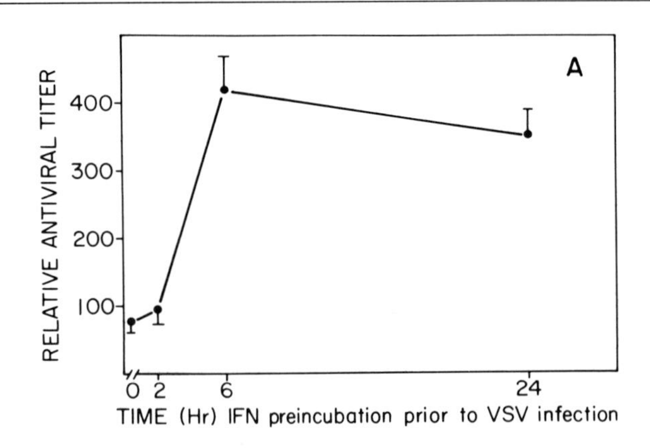

A
RELATIVE ANTIVIRAL TITER
400
300
200
100
0 2 6 24
TIME (Hr) IFN preincubation prior to VSV infection

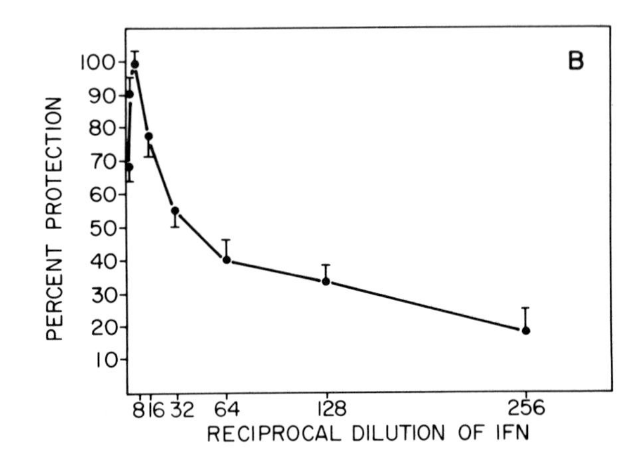

B
PERCENT PROTECTION
100
90
80
70
60
50
40
30
20
10
8 16 32 64 128 256
RECIPROCAL DILUTION OF IFN

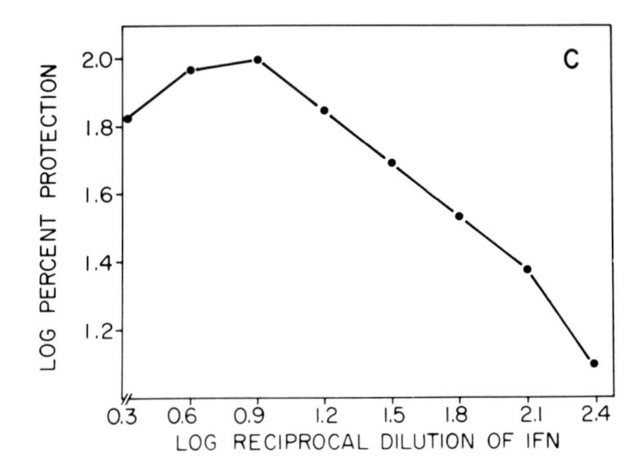

C
LOG PERCENT PROTECTION
2.0
1.8
1.6
1.4
1.2
0.3 0.6 0.9 1.2 1.5 1.8 2.1 2.4
LOG RECIPROCAL DILUTION OF IFN

TABLE I
COMPARATIVE PLASMA HUMAN IFN TITERS[a]

	No. of samples	Mean IFN titer (±SD) (units/ml)	Range (units/ml)
Red Cross blood donor volunteers	104	69 ± 323	≤10–3296
Normal control subjects (nonhospital blood bank)	24	23 ± 29	≤10–111
Clinical virology lab technicians	5	133 ± 12	90–159
Terminal malignancy	15	313 ± 240	63–850
Autoimmune disorders (i.e., 4 lupus and 1 rheumatoid arthritis)	5	513 ± 326	332–873

[a] Data originally reported by Forti *et al.*[9] All values reported as units were normalized to a standard human IFN-α reference preparation.[7,8] Removal of one extraordinarily high IFN value from the Red Cross blood donor samples (3296 units/ml) reduces the mean titer to 38 ± 52 units/ml and the range from <15–2000.

laboratory, the precise conditions of IFN assay must be strictly defined by using NIH or WHO standard reference interferons.[7,8] In our hands, we have found that two parameters are critical to accurate, reproducible IFN titration. First, the length of time of incubation of cells with IFN prior to addition of challenge virus can produce considerable, albeit predictable influence on the levels of IFN activity.[4,5] Second, the choice of target cell likewise is a major determinant of antiviral activity. In conclusion, although virtually any cell line can be used as a target cell for IFN assay, the duration of incubation with IFN and the target cell responsiveness to each IFN subtype must be carefully examined prior to routine assay (Fig. 3).

FIG. 3. Interferon antiviral assay. A WHO standard IFN-α was assayed as described in the text. (A) SG-181 cells were incubated with IFN for 0 to 24 hr prior to infection of the cell culture with VSV. The relative antiviral titer observed is plotted as a function of the time of incubation with IFN prior to viral infection. (B) The percentage protection produced by each IFN dilution at 2 hr incubation with interferon is demonstrated. (C) If the data of B are plotted as the log percentage protection as a function of log reciprocal dilution of IFN, a linear relationship is revealed. The 50% protection point can be linearly extrapolated from such a curve. From Forti *et al.*[9]

TABLE II
COMPARISON OF SERUM AND HEPARINIZED PLASMA LEVELS OF Hu-IFN FROM INDIVIDUALS WITH HIGH ENDOGENOUS LEVELS OF INTERFERON[a]

Serum (units/ml)	Plasma (units/ml)	Ratio plasma/serum
150 ± 10	348 ± 20	2.38
157 ± 5	339 ± 5	2.15
695 ± 24	873 ± 45	1.25
117 ± 11	222 ± 5	1.89
181 ± 7	332 ± 9	1.83
527 ± 28	638 ± 61	1.21
458 ± 71	881 ± 60	1.77
124 ± 31	217 ± 15	1.75

[a] Data originally reported by Forti *et al.*[9] All values reported as units/ml were normalized to a standard human IFN-α reference preparation.[7,8]

Interferon Assay of Clinical Blood Samples

We have found this method of IFN assay to be of considerable utility in detecting IFN activity in venipuncture samples from a variety of human subjects. The results of a recent study are seen in Table I. The details of this work are described elsewhere.[9] Of considerable interest is that plasma IFN levels routinely are higher than serum levels. Table II shows that plasma levels are approximately 2-fold greater than simultaneously collected serum levels.

Concluding Comments

A rapid, quantitative, and nonsubjective method of interferon (IFN) assay is described, which can be readily applied to clinical specimens. Automated data acquisition and data reduction allow a significant increase over existing methodologies in the number of samples that can be processed by an individual in a given unit of time. Plasma always yields higher (usually 2:1) IFN values than serum obtained simultaneously. Clinical virology laboratory technicians and patients with terminal malignancies or autoimmune diseases have significantly higher plasma mean IFN levels compared to blood bank specimens.

[72] Measurement of Interferon in Human Amniotic Fluid and Placental Blood Extract

By PAULETTE DUC-GOIRAN, PIERRE LEBON, and CHARLES CHANY

Introduction

The possible existence of constitutively synthesized interferons stems from observations that alpha interferon can be regularly detected at relatively low titers in human amniotic fluid from the sixteenth week of pregnancy.[1] Furthermore, interferons have also regularly been found in placental blood extracts or released by the human amniotic membrane when suspended in tissue culture medium between the thirty-seventh week and delivery.[2] The exact site of their synthesis is presently unknown. Indirect arguments are in favor of the hypothesis that they are produced in the feto–maternal interface and secreted thereafter into the blood and amniotic fluid. In addition, as shown here, the interferons detected during fetal life are of molecular size and antigenic structure somewhat different from those found in adult tissues or plasma. The biological significance of these constitutively synthesized interferons in fetal development is presently unknown. It can be suggested that they help to maintain maternal tolerance. Indeed, it has been shown that interferons can delay the rejection of allografts,[3,4] exert a protective effect on target cells against cytotoxic lymphocytes,[5,6] and inhibit the transformation of stimulated lymphoblasts by allogeneic cells.[7] It can also be postulated that they could play a role in the development of the fetus at a yet undetermined period and in an unidentified manner. Because of the known contribution of interferons to the decrease of cell replication and the development of the cytoskeleton and extracellular matrix, it is possible that this mechanism could be involved in a manner comparable to the effect on transformed cells.[8]

[1] P. Lebon, S. Girard, F. Thépot, and C. Chany, *J. Gen. Virol.* **59,** 393 (1982).
[2] P. Duc-Goiran, B. Robert-Galliot, J. Lopez, and C. Chany, submitted for publication.
[3] L. E. Mobraaten, E. De Maeyer, and J. De Maeyer-Guignard, York, *Transplantation* **16,** 415 (1973).
[4] M. S. Hirsch, D. A. Ellis, M. H. Black, A. P. Monaco, and M. Wood, *Transplantation* **17,** 234 (1974).
[5] G. Trinchieri and D. J. Santoli, *J. Exp. Med.* **147,** 1314 (1978).
[6] M. Bergeret, A. Grégoire, and C. Chany, *Immunology* **40,** 637 (1980).
[7] I. Heron, K. Berg, and K. Cantell, *J. Immunol.* **117,** 1370 (1976).
[8] C. Chany, *in* "Lymphokines" (E. Pick, ed.), Vol. 4, p. 409. Academic Press, New York, 1981.

Furthermore, we have detected interferon antagonistic substances in placental blood and extracts, which can often mask the antiviral effect of interferons and further complicate the interpretation of the data.

Interferon Assay

Amniotic fluid samples, stored at −20°, are tested without pretreatment. To eliminate cytotoxicity, the sera are heated at 56° for 30 min before interferon assay and are stored at −20°. The assays are performed in plastic microplates for tissue culture using the following cell lines: Madin–Darby bovine kidney (MDBK), human F7000 fibroblasts, human amniotic WISH cells, normal rat kidney (NRK) cells, and rat embryonic fibroblasts (REF). The cells are (1) either grown for 24 hr prior to use, then washed and incubated with serial 2-fold dilutions of the preparations to be tested, or (2) directly mixed with the interferon, incubated together for 24 hr at 37°, then washed and challenged with vesicular stomatitis virus (VSV) at a multiplicity of infection of 0.1. The titer is estimated as the reciprocal of the highest dilution which protects about 50% of the cell population.

Detection of Interferon in Human Amniotic Fluid

Specimen Collection. Seventy-six amniotic fluid samples were collected between the sixteenth and twentieth week of pregnancy. Ten additional samples were collected between the thirtieth and thirty-seventh week for anti-Rhesus agglutinin investigation. Twenty sera from pregnant women were obtained around the sixteenth week.

Level of Antiviral Protection in Different Specimens. The distribution of the number of amniotic fluid samples collected at various stages of pregnancy and the interferon titers observed are given in Table I. Antiviral activity is detected in 74/76 samples collected between the sixteenth and twentieth week. The titers vary from 2 to 16 international reference units/ml. Interferons are not detected in only two cases. It is noteworthy that 10 samples with the highest titers (16 units/ml) on bovine cells are not active when assayed with suspended human F7000 cells. However, if F7000 cells are cultured 24 hr prior to interferon assay, small amounts of interferon (2 to 6 units/ml) can be detected for these amniotic fluid samples. Interferons are also found in 10 amniotic fluid samples collected between the thirtieth and thirty-seventh week, but in somewhat lower amounts, varying from 2 to 8 units/ml. It is important to stress that no interferon was detected in the sera of 20 mothers at 3 months of gestation.

TABLE I
DISTRIBUTION OF THE NUMBER OF AMNIOTIC FLUID SAMPLES IN RELATION TO THE INTERFERON LEVEL[a]

	Interferon titer					
Samples taken	<2	2	4	8	16	Total
Week 16–20	2	4	23	30	17	76
Week 30–37	0	3	4	3	—	10

[a] The interferon titers, the reciprocal of the highest dilution that protects 50% of the cell population against the cytopathic effect of VSV, are expressed in units/ml with respect to the leukocyte interferon reference standard NIH G-025-901-527. The number of specimens found at the indicated titer on MDBK cells is shown in the table.

Identification of the Virus Inhibitor as Human Alpha Interferon. The amniotic fluid samples are (1) either dialyzed for 24 hr at 4 ° against 0.2 *M* glycine·HCl buffer, pH 2, then for 3 hr against Eagle's minimum essential medium (MEM), or (2) incubated for 2 hr at 20° in the presence of alpha interferon antiserum (prepared with alpha interferon, specific activity 10^7 units/mg of protein, in sheep) or beta interferon antiserum prepared in rabbits (kindly provided by Dr. Vilcek), or normal sheep serum. The three sera are diluted 1 : 10 and added at a ratio of 1 vol serum per 5 vol amniotic fluid (final serum dilution 1:60). The anti-beta interferon serum neutralizes 4 units/ml at a 1:1200 dilution. The anti-alpha interferon serum neutralizes 100 units/ml at 1:12,000 dilution. The neutralization assays are performed on MDBK cells for alpha interferon and on human F7000 for beta interferon.

The inhibitory substance in all of the positive specimens tested is nondialyzable, resistant to acid pH, and completely neutralized by antiserum to alpha interferon and not by antiserum to beta interferon. Furthermore, heating at 56° for 30 min does not destroy its activity (data not shown). It can be thus concluded that it represents a species of alpha interferon preferentially protecting bovine cells.

Radioimmunoassay (RIA) and Biological Assay

This identification has been confirmed by testing comparatively 13 amniotic fluid samples with a commercial radioimmunoassay kit (Abbott) and by biological assay. For RIA, 200 μl of reference IFN and samples were incubated for 20 hr with beads coated with antibody to interferon alpha. After washing, 200 μl of anti-interferon alpha antibody, labeled with ^{125}I, was added to each bead. After 3 hr of incubation and washing, the radioactivity of the beads was counted in a gamma counter. The standard curve was constructed with the reference interferon units and net counts per minute (cpm) on log–log graph paper and the interferon titer of the samples was calculated with this curve.

Twelve out of 13 samples (collected at the twenty-second week) have interferon titers between 5 and 22 units/ml. One is less than 5 units/ml with RIA (average 10 units/ml). With the biological assay, all 13 are positive between 2 and 24 units/ml (average of 8 units/ml).

Detection of Constitutively Produced Interferons in the Placenta and Fetal Membranes

Biological Preparations. Fetal membranes, separated from placentas collected shortly after Caesarian section between the thirty-seventh week and delivery, are washed carefully 1 to 5 times in 500 ml of saline, 0.15 *M* NaCl, pH 7, then incubated with MEM, supplemented with 10% heat-inactivated newborn calf serum at 37° for 18 hr or more. Blood from the umbilical cord and from cotyledons is collected. The whole placenta is then compressed under 0.15 kg/cm^2 to squeeze out the blood. Placental extracts (2 ml/g) suspended in MEM supplemented with 10% heat-inactivated newborn calf serum are also prepared. They are homogenized with a Polytron apparatus (type S axis) or ground in a mortar.

As seen in Table II, interferon is frequently detected in placentas (78%), less frequently released from uninduced human amniotic membranes, and even less from umbilical cord blood. This suggests that the site of production is likely to be the placenta. Therefore, most of our effort focused on the interferon content of placental blood.

In some cases, interferon cannot be detected in the crude blood extracts, but is thereafter found at significant and sometimes relatively high titers by different purification procedures. In addition, during some purification steps, an interferon antagonist, reminiscent of the one which has previously been described in sarcomas[9] and normal muscles,[10] has also been detected and characterized as a lectin-like substance.

[9] C. Chany, J. Lemaître, and A. Grégoire, *C.R. Hebd. Seances Acad. Sci.* **269,** 2626 (1969).

[10] P. H. Jiang, F. Chany-Fournier, B. Robert-Galliot, M. Sarragne, and C. Chany, *J. Biol. Chem.* **258,** 12361 (1984).

TABLE II
IFN ACTIVITY IN SAMPLES FROM 37 CAESARIAN SECTIONS[a]

Source of IFN	IFN titer (units/ml) 10	10–60	≥100	Total positive samples
Placenta	9	5	15	29/37 (78%)
Human amniotic membrane	6	5	3	14/29 (48%)
Umbilical cord	1	1	1	3/10 (30%)

[a] The number of positive samples found at the indicated IFN titer on human F7000 or WISH cells is shown in the table.

Purification Procedures

Affinity Chromatography on Concanavalin A (Con A) Sepharose. Interferon preparations are applied to a Sephadex G-25 M PD 10 column, previously equilibrated with 0.02 *M* sodium phosphate (pH 7.2), containing 1 *M* NaCl (E0), and eluted with the same buffer. The breakthrough active fractions (5 ml) are applied to a 5 ml Con A Sepharose column equilibrated with buffer E0 at room temperature at a flow rate of 4.5 ml/hr. After 1 hr contact and washing with E0 buffer, a sequential elution is performed with 0.1 *M* α-methyl-D-mannoside in the E0 buffer and called E1, then with buffer E1 containing 50% (v/v) ethylene glycol and called E2. The bound substances are eluted at a flow rate of 20 ml/hr. Fractions (1 ml) are collected in 0.5 ml of a solution of 1% bovine serum albumin (BSA) in 0.02 *M* sodium phosphate (pH 7.4) in glass tubes. As shown in Fig. 1, the first unbound peak contains alpha interferon and the two latter eluted peaks are of the beta interferon type.

Affinity Chromatography on Immobilized Cibacron Blue F3G-A (Blue Sepharose CL-6B). After buffer exchange with Sephadex G-25 M, as described above, samples are applied to 3.5 ml Blue Sepharose CL-6B gel[11] in a column, equilibrated with starting buffer (E0), and recycled continuously overnight at a flow rate of 4.5 ml/hr. After washing, the bound substances are eluted at a flow rate of 20 ml/hr by increasing the ionic strength of the buffer to 2 *M* NaCl (E1). Finally, residual material is eluted with 50% (v/v) ethylene glycol (E2).

[11] W. J. Jankowski, W. von Muenchhausen, E. Sulkowski, and W. A. Carter, *Biochemistry* **15,** 5182 (1976).

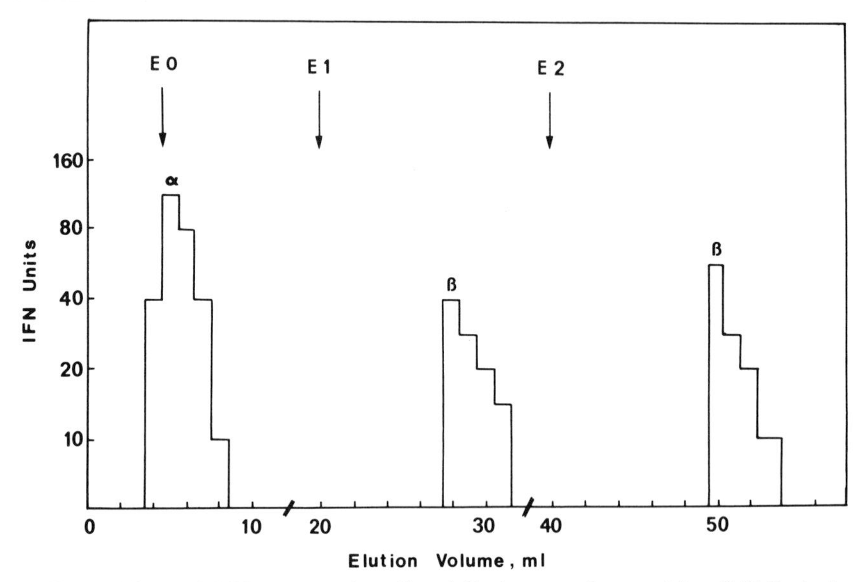

FIG. 1. Placental IFNs separated on Con A Sepharose columns. After G-25 Sephadex filtration, an IFN preparation (40 units/ml and 2.2 mg of protein/ml) was applied to Con A Sepharose in 0.02 *M* sodium phosphate buffer containing 1 *M* NaCl (E0). E1 contains 10 m*M* α-methyl-D-mannoside in starting buffer. E2 contains 10 m*M* α-methyl-D-mannoside and 50% (v/v) ethylene glycol in starting buffer. Biological activity was assayed with WISH cells.

As shown in Fig. 2, four peaks of interferon activity are separated, called respectively A, B, C, and D. Peak A contains breakthrough proteins and alpha interferon. Peaks B, C, and D are further fractioned by sodium dodecyl sulfate–polyacrylamide gel electrophoresis (SDS–PAGE). Peak C also contains an interferon antagonistic substance. Its presence was suspected because the interferon activity appeared only at a high dilution of the fractions. Peak D contains interferon neutralized by both alpha and beta interferon immune sera.

Affinity Chromatography with Immobilized Immune Serum to Alpha Interferon. This procedure is based on methods previously described.[12,13] Antiserum to alpha interferon is produced according to the technique of Pyhälä[14] by immunizing sheep with 10^7 units of Sendai-induced human

[12] J. D. Sipe, J. De Mayer-Guignard, B. Fauconnier, and E. De Maeyer, *Proc. Natl. Acad. Sci. U.S.A.* **70,** 1037 (1973).

[13] K. Berg, C. A. Ogburn, K. Paucker, K. E. Mogensen, and K. Cantell, *J. Immunol.* **114,** 640 (1975).

[14] L. Pyhälä, *Acta Pathol. Microbiol. Scand., Sect. C* **86C,** 291 (1978).

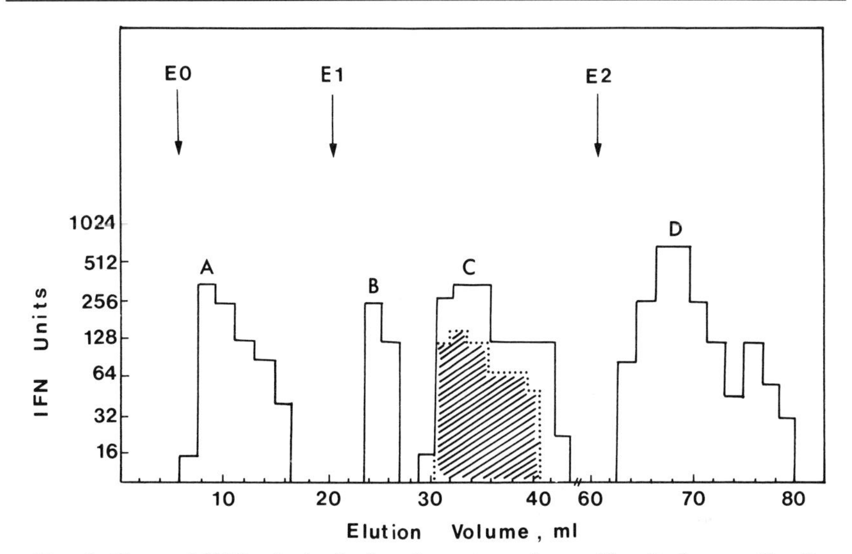

FIG. 2. Placental IFNs obtained after chromatography on Blue-Sepharose. Fractions eluted from Sephadex G-25 M in 0.02 *M* sodium phosphate (pH 7.2), containing 1 *M* NaCl (E0), were applied to a column of Blue-Sepharose equilibrated with this starting buffer. After washing, elution was performed with 0.02 *M* sodium phosphate, pH 7.2, containing 2 *M* NaCl (E1), then with 50% (v/v) ethylene glycol in starting buffer (E2).

leukocyte interferon with a specific activity of 10^7 units/mg of protein. Sheep antisera are decomplemented by heating at 56° for 30 min and γ-globulins precipitated by ammonium sulfate (50%). Antisera are extensively absorbed against impurities (consisting of normal buffy coat, allantoic fluid, human serum and human albumin, Sendai virus) before binding to 4B CNBR-activated Sepharose. One sample of each batch is tested for its capacity to retain alpha interferon.

Placental blood interferons are applied on previously unused anti-alpha interferon Sepharose (2.5 ml) in columns (7.6 cm) equilibrated with phosphate-buffered saline (PBS), pH 7.2. Crude interferons are loaded onto the column at a flow rate of 4 ml/hr at 4° and recycled overnight. After extensive washing with PBS, bound material is eluted at pH 2.2 with 0.05 *M* citrate-phosphate buffer[15] at a flow rate of 20 ml/hr. Elution is achieved by adding 50% (v/v) ethylene glycol in citrate phosphate buffer, pH 2.2. Then, antibody-Sepharose is treated with 100 ml of guanidine (4 *M*) and rinsed with 300 ml of PBS.

[15] K. C. Zoon, M. E. Smith, P. J. Bridgen, D. Zur Nedden, and C. B. Anfinsen, *Proc. Natl. Acad. Sci. U.S.A.* **76,** 5601 (1979).

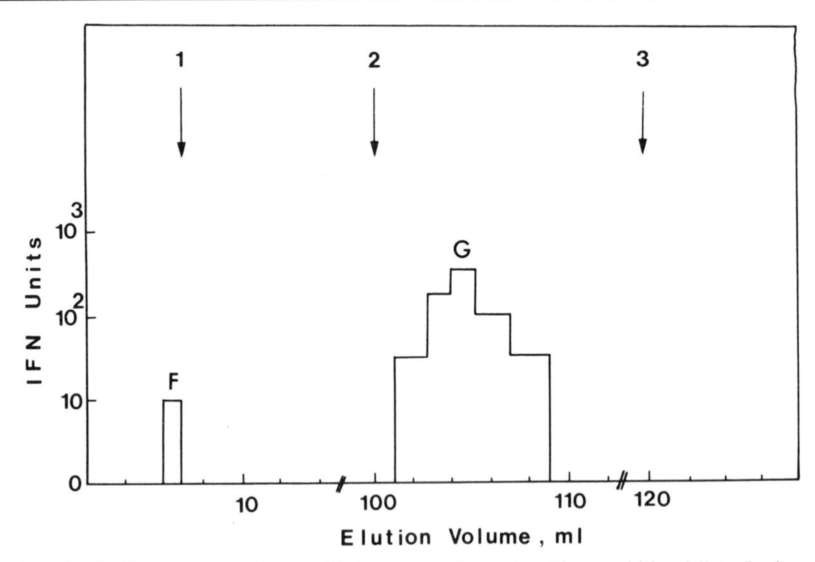

FIG. 3. Purification by antibody affinity chromatography. Placental blood (3 to 5 ml) was loaded onto a 2.5 ml column of anti-IFN-α Sepharose and recycled overnight as described in the text. The Sepharose column was sequentially rinsed with PBS (arrow 1), eluted with 0.05 *M* citrate–phosphate buffer, pH 2.2 (arrow 2), and finally with 50% (v/v) ethylene glycol in citrate–phosphate buffer, pH 2.2 (arrow 3). MDBK and F7000 cells were employed for titration.

The elution profile is seen in Fig. 3. The flow through fraction (F) also contains interferon neutralized by both anti-alpha and beta interferon antibody, while a major peak (G) of relatively high titer consisting of alpha interferon is detached. It is of interest that in some of these samples containing relatively high amounts of interferon, this material cannot be detected in the original crude samples.

SDS–Polyacrylamide Gel Electrophoresis. Discontinuous polyacrylamide gel electrophoresis is performed in slab gels, according to Laemmli[16] and King and Laemmli,[17] with a 5% polyacrylamide stacking gel and a 15% (w/v) polyacrylamide separating gel. Fractions from different chromatographies are concentrated by lyophilization after dialysis against 0.03 *M* ammonium bicarbonate, then heated to 100° for 2 min in presence of 2% SDS and 0.1 *M* 1,4-dithiothreitol in 0.05 *M* Tris·HCl, pH 6.8, 10% glycerol, and 0.002% bromophenol blue as the marker dye. Electrophoresis is carried out at 20 mA/gel for 4 hr at room temperature. One part of

[16] U. K. Laemmli, *Nature (London)* **227,** 680 (1970).

[17] J. King and U. K. Laemmli, *J. Mol. Biol.* **62,** 465 (1971).

TABLE III
ANALYSIS OF THE MOLECULAR FORMS OF PLACENTAL INTERFERONS AFTER SDS–PAGE

Molecular weight component	Cell type	IFN titer (units/ml)		
		After Blue-Sepharose chromatography		After anti-IFN-α chromatography
		Peak B	Peak D	Peak G
80,000	F7000	32	64	32
	MDBK	8	32–64	<8
43,000	F7000	32	32	256
	MDBK	≤8	<8	<8
26,000	F7000	64	<8	<8
	MDBK	16	<8	8–16
21,000–22,000	F7000	128	32	<8
	MDBK	128	16–32	<8
15,000–17,000	F7000	<8	<8	128
	MDBK	32	<8	8

the gel is cut into tracks, then sliced into 2 mm pieces, and eluted for interferon assay. The other part of the gel is stained with 0.25% Coomassie brilliant blue. Gels are calibrated with the following protein standards: phosphorylase *b* (MW 94,000), bovine serum albumin (MW 67,000), ovalbumin (MW 43,000), carbonic anhydrase (MW 30,000), soybean trypsin inhibitor (MW 20,100), and α-lactalbumin (MW 14,400).

Protein concentrations are estimated by absorbance at 280 nm or determined as described by Lowry *et al.*[18] with BSA as standard.

As shown in Table III, the results obtained vary according to the purification procedure and also the cell used for titration, thus showing the heterogeneity of the molecular composition of the interferons. Their molecular weights are comparable to those previously found in the human amniotic membrane after induction.[19]

The physicochemical characteristics of the sustance antagonistic to interferon found in the peak C fraction obtained after Blue-Sepharose chromatography have been studied. The fractions (from 30 to 36 ml effluent volume) are concentrated 8 times by lyophilization, and treated or not

[18] O. H. Lowry, N. J. Rosebrough, A. L. Farr, and R. J. Randall, *J. Biol. Chem.* **193,** 265 (1951).

[19] P. Duc-Goiran, B. Robert-Galliot, T. Chudzio, and C. Chany, *Proc. Natl. Acad. Sci. U.S.A.* **80,** 2628 (1983).

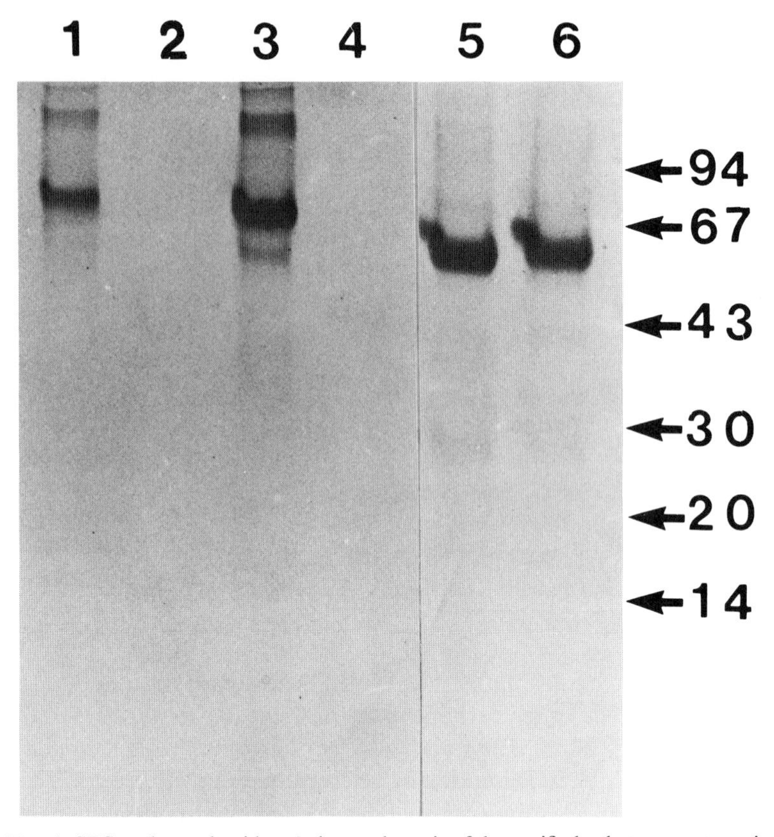

FIG. 4. SDS–polyacrylamide gel electrophoresis of the purified substance antagonistic to IFN. Lanes 1–4 were loaded with bovine serum albumin (BSA) as control: lane 1, BSA 500 μg/ml; lane 2, BSA 500 μg/ml, pepsin treatment; lane 3, BSA 1 mg/ml; lane 4, BSA 1 mg/ml, pepsin treatment. The substance antagonistic to IFN detected in peak C after Blue-Sepharose chromatography is in lanes 5 and 6: lane 5, prior to pepsin treatment; lane 6, after pepsin treatment.

by pepsin (1 mg/ml) at 37°, pH 2, for 20 hr. The enzymatic reaction is stopped by adjusting the pH to 7 with 0.1 *M* Tris. As reference, bovine serum albumin (0.5 and 1 mg/ml) is employed. The protein samples are heated at 100° for $2\frac{1}{2}$ min in the presence of 2% SDS and 0.1 *M* 1,4-dithiothreitol and analyzed by electrophoresis. An interferon antagonist migrates in the 60,000–65,000 Da region (Fig. 4, lane 5). The antagonist is unaltered by a 20 hr pepsin treatment (lane 6), while the control albumin (lanes 1 and 3) is completely destroyed (lanes 2 and 4). For the characterization of this antagonist, see F. Chany-Fournier, P. H. Jiang, and C. Chany, this volume.[20]

Concluding Remarks

Interferons can be detected in the human amniotic membrane and placenta in a sufficiently high proportion of specimens to make the hypothesis of an externally induced origin unlikely. These interferons are highly heterogeneous in their molecular structure. Some resemble usually detected alpha and beta interferons. Others have molecular sizes and antigenic structures reminiscent of both. In many cases, they are only detected after purification of crude placental blood. This might be attributed to the existence of inhibitors or antagonists. One of them, reminiscent of sarcolectins, can be isolated and purified. Although their biological significance is presently unknown, we suggest that these interferons might play a role in fetal development and/or the immune tolerance of the mother.

[20] F. Chany-Fournier, P. H. Jiang, and C. Chany, this volume [91].

[73] Use of Ovine and Caprine Cells to Measure Antiviral Effects of Interferon

By TILAHUN YILMA

Introduction

The biological activity of interferon is often described as being host or species specific. This specificity appears to be mediated in part by specific receptors on cell membranes coded for by specific genes.[1,2] Consequently, the antiviral effects and other biological activities of interferons are assayed primarily on suitable homologous cell lines that are highly sensitive to both the interferon being tested and to the challenge virus.[3] For example, the gene responsible for the sensitivity of human cells to the antiviral effects of interferon α and β has been localized to chromosome

[1] A. Y. Sakguchi, D. Stevenson, and I. Gordon, *Virology* **116,** 441 (1982).
[2] L. B. Epstein and J. C. Epstien, *J. Infect. Dis.* **133,** A56 (1976).
[3] P. C. Familletti, S. Rubinstein, and S. Pestka, this series, Vol. 78, p. 387 (1981).

21. Hence cells derived from patients with Down's syndrome are up to 10-fold more sensitive than normal human cells to human interferons and are often used for the antiviral assay of human interferons.[2] There are exceptions, however, in which heterologous cells are preferred over homologous cells for the assay of the antiviral effects of interferons. For example, Gresser *et al.*[4] demonstrated greater sensitivity of bovine cells than human cells to human leukocyte interferon. Recently we have also shown the greater sensitivity of ovine choroid plexus (OCP)[5] and caprine synovial membrane (CSM)[6] cells when compared with homologous cells to the antiviral effects of various types of human and other animal interferons. Furthermore, variation in the sensitivity of cells to different subtypes of recombinant DNA-derived human interferon alphas[7,8] as well as natural (immune interferon) and recombinant human interferon gammas[5] has been demonstrated. For example, ovine cells are sensitive to the antiviral effects of recombinant human IFN-γ, but not to the natural type. Hence, ovine and caprine cells provide a unique opportunity to increase the sensitivity of the assay in measuring the antiviral effects of various types of interferons. They also provide a convenient system for comparative studies with other cell lines that have limited sensitivity to the antiviral effects of heterologous interferons.

Here the propagation and use of OCP and CSM cells, which are highly sensitive to the antiviral effects of a number of human and other animal interferons, are described.

Preparation of Cell Cultures

Ovine Choroid Plexus Cell Culture. The initiation of a primary culture of OCP cells has been described in detail by Torchio and Trowbridge.[9] A slight modification of the same procedure is herein described. The brain from a fetus (1 to 2 weeks before lambing) or from a recently slaughtered lamb is obtained, and the choroid plexus is removed aseptically from the lateral ventricles. The tissue is then minced with a pair of scalpel blades to a size of 1 mm^3 or smaller in a glass Petri dish containing a small amount of modified Dulbecco's minimum essential medium (DMEM) supplemented with 10% fetal calf serum (FCS) and antibiotics (growth medium).

[4] I. Gresser, T. M. Bandu, D. Brouty-Boye, and M. Tovey, *Nature (London)* **251,** 543 (1974).

[5] T. Yilma, *J. Gen. Virol.* **64,** 2013 (1983).

[6] T. Yilma, R. Breeze, and S. R. Leib, *Am. J. Vet. Res.* **45,** 2094 (1984).

[7] P. K. Weck, S. Apperson, L. May, and N. Stebbing, *J. Gen. Virol.* **57,** 233 (1981).

[8] E. Rehberg, B. Kelder, E. G. Hoal, and S. Pestka, *J. Biol. Chem.* **257,** 11497 (1982).

[9] C. Torchio and R. S. Trowbridge, *In Vitro* **13,** 252 (1977).

The minced tissue is washed with several changes of growth medium until clear of red blood cells, and approximately 40 tissue fragments are seeded in a 25-cm^2 flask containing 1–2 ml of growth medium supplemented with 20% FCS. The flask is incubated at 37° in a 5% CO_2 incubator for 3 to 4 days, without disturbing or jarring, to allow firm anchoring of cells to surface. At the end of the attachment period, cells are then propagated in 5 ml of growth medium until a complete monolayer is formed, which usually takes about a week. It is essential that the growth medium be changed when it turns yellow due to increased acidity. Once a cell monolayer is formed, coarse tissue fragments are removed by vigorous shaking and several rinsings with growth medium. For further propagation, secondary cultures are prepared by trypsinization (Difco, 1 : 250) (Difco Laboratories, Detroit, Michigan). Cells propagated in this manner are useful up to 40 passages for the assay of the antiviral activity of interferons. Long-term supplies of OCP cells are made available by freezing cells in liquid nitrogen. Briefly, 5×10^7 cells in 1 ml of DMEM containing 10% dimethyl sulfoxide (Sigma Chemical Co., St. Louis, Missouri), 10% FCS, and antibiotics are frozen in 2-ml size glass vials or cryotubes (Nunc, Vangard International, Neptune, New Jersey) in liquid nitrogen. Sequential freezing of cells by intervals of a few hours at 4, −20, and −70°, and finally in liquid nitrogen increases the percentage of viable cells when cultures are initiated. For frozen cell activation, rapid thawing of a single vial of cells in a 37° water bath and propagation in a 25-cm^2 plastic flask in growth medium containing 20% FCS are recommended. In our laboratory this protocol has allowed greater than 50% viability of frozen cells after long-term storage.

In addition, we have used an OCP cell line designated as sheep choroid plexus cells (SCP-II) kindly provided by Dr. Richard S. Trowbridge of the New York State Institute for Basic Research in Developmental Disabilities. The SCP-II cell line was once considered an established line because the cells have been successfully subcultured more than 120 times, have undergone 300 generations, and have maintained a fibroblastic appearance and plating efficiency similar to that of their low passage progenitor cells.[9] Although the sensitivity of the SCP-II cell line to the antiviral effects of various interferons is indistinguishable from that of the OCP cell line, we have not been able to propagate the SCP-II cells beyond 136 passages in our laboratory. Similarly, Dr. R. S. Trowbridge was also unable to propagate the SCP-II cells beyond 156 passages (personal communication). Work is presently in progress to transform both OCP and SCP-II cell lines.

Propagation of Caprine Synovial Membrane Cells. The synovial membrane is obtained aseptically from the anterior portion of the radiocarpal joint of a kid or fetal goat. Tissue cultures are prepared by procedures identical to those described for OCP cells. CSM cells also have a useful life span for interferon assay up to 40 passages.

Greater Sensitivity of OCP and CSM Cells to Heterologous Interferons

Antiviral Assay. The greater sensitivity of both OCP and CSM cells to the antiviral effects of the major types of human natural[10–12] and recombinant DNA-derived interferons,[13–15] as well as other natural animal interferons,[11,12] was established with the semimicrotitration method against the cytopathic effects (CPE) of vesicular stomatitis virus (VSV) or encephalomyocarditis virus (EMCV).[3] Because EMCV has a greater sensitivity to natural human immune interferon than VSV, EMCV was also used for the assay of the antiviral effects of natural human immune interferon. Briefly, appropriate dilutions of the interferon sample are made in growth medium in a 96-well microtiter plate (Costar, Cambridge, Massachusetts). Each well is then seeded with 2×10^4 cells, which are allowed to form a monolayer by overnight incubation at 37° in a CO_2 incubator. The minimum dose of challenge virus that gives a 100% CPE in a 24-hr period is added to interferon-treated and virus control wells. The interferon units are expressed as the reciprocal of the dilution of sample which gives 50% protection against the CPE of the appropriate dose of challenge virus for the given cell type. All human interferons, except for immune or gamma interferon, are expressed in reference units in comparison with the National Institutes of Health (NIH) standards for leukocyte interferon (G-023-901-527) and fibroblast interferon (G-023-902-527). The human immune interferon titers obtained in our laboratory are approximately equivalent to those titers provided by the laboratories of origin. All data are analyzed by the Wilcoxon two-sample test for the unpaired case.

The relative sensitivity of various types of cells to the antiviral effects of different types of human interferons is compared to the normalized antiviral titers (100) on human disomic 21 (FS-7) cells in Table I. The relative antiviral activities of different types of animal interferons on OCP

[10] K. Cantell and S. Hirvonen, *Tex. Rep. Biol. Med.* **35,** 138 (1977).

[11] A. Billiau, M. Joniau, and P. DeSomer, *J. Gen. Virol.* **19,** 1 (1973).

[12] M. Wiranowska-Stewart and W. E. Stewart, *J. Interferon Res.* **1,** 233 (1981).

[13] D. V. Goeddel, E. Yelverton, A. Ullrich, H. L. Heyneker, G. Miozzari, W. Holmes, P. H. Seeburg, T. Dull, L. May, N. Stebbing, R. Crea, S. Maeda, R. McCandliss, A. Sloma, J. M. Tabor, M. Cross, P. C. Familleti, and S. Pestka, *Nature (London)* **287,** 411 (1980).

[14] D. V. Goeddel, H. M. Shepard, E. Yelverton, D. Leung, R. Crea, A. Sloma, and S. Pestka, *Nucleic Acids Res.* **8,** 4057 (1980).

[15] P. W. Gray, D. W. Leung, D. Pennica, E. Yelverton, R. Najarian, C. C. Simonsen, R. Derynck, P. J. Sherwood, D. M. Wallace, S. L. Burger, A. D. Levinson, and D. V. Goeddel, *Nature (London)* **295,** 503 (1982).

TABLE I

RELATIVE ANTIVIRAL ACTIVITIES OF HUMAN INTERFERONS ON VARIOUS TYPES OF CELLS[a]

Type of human interferon and source	Ovine choroid plexus cells	Caprine synovial membrane cells	Human trisomic 21 cells (GM-2767)	Human disomic 21 cells (FS-7)	Bovine turbinate cells
Leukocyte interferon[b] (NIH)	6454	ND[c]	152	100	758
Leukocyte interferon[b] (Red Cross)	2352	2822	314	100	460
IFN-αA[d] (Roche/Genentech)	2396	562	485	100	474
Fibroblast interferon[e] (NIH)	994	140	775	100	NA[f]
Fibroblast interferon[e] (WSU)	966	ND	312	100	NA
IFN-β1[d] (Roche/Genentech)	1072	239	451	100	NA
IFN-γ[d] (Genentech)	972	854	331	100	NA
Immune interferon[g]	<3	<3	ND	100	NA

[a] Interferon titers were determined by the semimicrotitration method against the cytopathic effect of vesicular stomatitis virus,[3] and data were normalized relative to the titer of the antiviral activity of interferons on human dermal fibroblasts disomic for chromosome 21 (FS-7).

[b] Leukocyte interferons obtained from the NIH or the American Red Cross was produced in Sendai virus-induced human blood leukocytes.[10]

[c] ND, Not determined.

[d] IFN-αA[13] and IFN-β1[14] were recombinant DNA-derived interferons produced in *Escherichia coli,* whereas IFN-γ was recombinant DNA-derived material produced in monkey cells.[15]

[e] Fibroblast interferon obtained from NIH or Washington State University (WSU) was produced in poly(I) · poly(C)-induced human dermal fibroblast cells (FS-7).[11]

[f] NA, Not applicable.

[g] Immune interferon preparations were produced in phytohemagglutinin-induced blood leukocytes and were gifts of Drs. Vilček and Georgiades.

TABLE II
RELATIVE ANTIVIRAL ACTIVITIES OF VARIOUS TYPES OF ANIMAL INTERFERONS ON OVINE CHOROID PLEXUS AND HOMOLOGOUS CELLS[a]

Source of interferon	Leukocyte interferon[b]		Fibroblast interferon[c]		Immune interferon[d]	
	Ovine	Homologous	Ovine	Homologous	Ovine	Homologous
Bovine cells	740	100	418	100	517	100
Caprine cells	99	100	183	100	114	100
Equine cells	1275	100	99	100	146	100
Porcine cells	163	100	940	100	8900	100

[a] Interferon titers were determined by the semimicrotitration method against the cytopathic effect of vesicular stomatitis virus,[3] and data were normalized relative to the titer of the antiviral activity of interferon on homologous cells.
[b] Leukocyte interferon was produced in Newcastle disease virus-induced blood mononuclear cells.[12]
[c] Fibroblast interferon was produced in poly(I) · poly(C)-induced fibroblast cells.[11]
[d] Immune interferon was produced in phytohemagglutinin-induced blood mononuclear cells.[12]

and CSM cells are compared to the normalized antiviral titers (100) on homologous cells in Table II and Table III, respectively. In summary, OCP cells exhibit greater sensitivity than shown by human cells trisomic or disomic for chromosome 21 to the antiviral effects of a number of human natural and recombinant interferons. They also show greater sensitivity than bovine cells to human leukocyte interferon. Moreover, OCP cells exhibit high sensitivity to the antiviral effects of all 3 natural types of bovine, caprine, equine, and porcine interferon preparations. CSM cells exhibit a similar degree of sensitivity to the antiviral effects of heterologous interferons, with few exceptions. For example, CSM cells exhibit greater sensitivity to the antiviral effects of domestic animal leukocyte interferons than shown by OCP cells (Table III). OCP and CSM cells are insensitive to natural human immune interferon preparations. In our laboratory, human immune interferon preparations (obtained from J. Vilček) with an antiviral activity of 1500 units/ml against VSV in WISH cells showed no detectable activity on OCP and CSM cells. Similarly, another human immune interferon preparation (obtained from J. A. Georgiades) with a titer of 760 and 2170 units/ml against EMCV on WISH and human dermal (HF81-37) cells, respectively, showed no detectable antiviral activity on OCP or CSM cells. A mouse L cell interferon preparation with an antiviral activity of 8170 units/ml on mouse cells (L-929) had 60 units/ml on OCP and no detectable activity on CSM cells. A Newcastle disease

TABLE III
RELATIVE ANTIVIRAL ACTIVITIES OF VARIOUS TYPES OF ANIMAL INTERFERONS ON CAPRINE SYNOVIAL MEMBRANE AND HOMOLOGOUS CELLS[a]

Source of interferon	Leukocyte interferon[b]		Fibroblast interferon[c]		Immune interferon[d]	
	Caprine	Homologous	Caprine	Homologous	Caprine	Homologous
Bovine	38,561	100	172	100	841	100
Ovine	8,480	100	58	100	398	100
Porcine	3,778	100	435[e]	100	25	100

[a] Interferon titers were determined by the semimicrotitration method against the cytopathic effect of vesicular stomatitis,[3] and data were normalized relative to the titer of the antiviral activity of interferon on homologous cells.

[b] Leukocyte interferon was produced in Newcastle disease virus-induced blood mononuclear cells.[12]

[c] Fibroblast interferon was produced in poly(I) · poly(C)-induced fibroblast cells.[11]

[d] Immune interferon was produced in phytohemagglutinin-induced blood mononuclear cells.[12]

[e] Porcine fibroblast interferon was produced in Newcastle disease virus-induced porcine kidney cells (PK-15).[11]

virus-induced canine fibroblast interferon with a titer of 178 units/ml on canine kidney cells had 60 units/ml on OCP and no detectable antiviral activity on CSM cells.

Concluding Remarks

There are several advantages inherent in employing OCP or CSM cell lines to measure the antiviral effects of various types of interferons: (1) their greater sensitivity to the antiviral effects of a number of domestic animal and humans interferons when compared to homologous cells; (2) their ease of propagation; and (3) their pronounced and wide-range sensitivity to the CPE of the most widely used viruses, such as VSV and EMCV, for antiviral assay of interferons. These cell lines, in particular OCP, can provide increased sensitivity for a rapid bioassay to monitor the titer of a number of interferons during production and purification.[3] It may also be convenient and economical to employ a single cell line for measuring antiviral activity where several types of heterologous interferons are used in a study. Furthermore, the mechanisms for host or species specificity of interferons is not fully understood. It is known, however, that efficient binding of interferons to cells alone is not sufficient for the induction of the antiviral state, since some interferons that show significant binding fail to exert their biological activities.[1] Therefore, OCP or CSM cells may provide a unique opportunity to study the role of receptors in the mechanisms of action of interferons by virtue of their sensitivity to interferons of various species.

It is also interesting to note that OCP and CSM cells are sensitive to the antiviral effects of recombinant immune interferon produced in monkey cells, but not to the natural one. Since both types of interferons are glycosylated, it is difficult to attribute this disparity to differences of glycosylation.

Acknowledgments

The author wishes to acknowledge Mr. Stephen Leib and Ms. Barbara Nakata for editorial assistance and Ms. Rose Cheff for secretarial work on the manuscript. The work was supported in part by Public Health Service Grant RR05465-19 and by Grants 11H-2540-8751 and 83-CRSR-2-2189 from the National Institutes of Health and the United States Department of Agriculture, respectively.

[74] Quantitation of Neutralization of Interferon by Antibody

By YOSHIMI KAWADE

Neutralization of the biological activity of interferon (IFN) by antibody (Ab) is one of the most basic techniques in IFN research. Although the IFN–Ab interaction can be studied by immunochemical means[1,2] rather than by bioassays, neutralization by antibody will remain important because it provides not only a simple and reliable criterion for identifying the molecular type of IFN, but also a specific way of analyzing the roles of IFN in various biological systems. Neutralization tests are widely performed and there seems to be little problem as to the qualitative determination of the IFN types (human and mouse α, β, and γ). However, quantitative aspects of the tests have received relatively little attention, and a number of elementary problems need be considered to make the test precise and reproducible.[3,4] In this article, these basic problems in quantitative neutralization tests are discussed, including the experimental method and the expression of the antibody titer. The methodology described is mainly for antibody titration, but it will also serve as a means of

[1] T. Staehelin, C. Stähli, D. S. Hobbs, and S. Pestka, this series, Vol. 79, p. 589.
[2] D. S. Secher, *Nature (London)* **290,** 501 (1981).
[3] Y. Kawade, *J. Interferon Res.* **1,** 61 (1980).
[4] Y. Kawade and Y. Watanabe, *J. Interferon Res.* **4,** 571 (1984).

characterizing IFNs; as an application, a method of analyzing mixed type IFN preparations is presented.

The test can be carried out with any of the IFN bioassay systems, but for convenience, the assay based on reduction of viral cytopathic effects (CPE) is described here. Results with other types of assay can of course be treated similarly.

Experimental Method of Antibody Titration

Perhaps the most commonly used method of titrating a neutralizing antibody is to add a series of mixtures containing a fixed concentration of IFN (e.g., 10 units/ml) to bioassay and various concentrations of antibody to determine the minimum antibody concentration that causes viral CPE to appear[5,6] (the "constant IFN method"). Or, a similar bioassay can be done with a constant quantity of antibody and various IFN concentrations[3,7] (the "constant antibody method"). The two methods may be combined to make a "checkerboard" type assay.[5] In both methods, the titration endpoint must be designated; it is expressed, for instance, as "partial restoration of CPE."[5] Such will be sufficient for qualitative or semiquantitative experiments, but for precise titrations, the same endpoint as in IFN titration without antibody should be used, because it is the most clearly defined point. This endpoint, defined for example by 50% CPE in the CPE reduction assay, represents an IFN concentration of 1 unit/ml. Hereafter, the antiviral unit actually observed is denoted as experimental unit, EU, and is distinguished from the reference unit, RU, which represents an absolute quantity of IFN (the RU refers to the international unit, when international reference IFNs are available, but when they are not, it refers to a certain laboratory standard). Thus, what one determines in quantitative neutralization tests are the Ab dilution $1/f$ and the concentration of IFN, N EU/ml, which is reduced to 1 EU/ml by that antibody. In this way, neutralization assays can be done with essentially the same precision as in IFN titration. Both f and N refer to the final concentrations in the culture medium of assay cells; this practice is recommended to avoid confusion. How the neutralization titer is defined and computed from the observed values of f and N is described in a later section.

Another type of neutralization assay often used is to mix antibody and excess IFN and titrate the "residual IFN" by making a dilution series of

[5] B. Dalton and K. Paucker, this series, Vol. 79, p. 561.

[6] E. A. Havell, this series, Vol. 79, p. 571.

[7] M. de Ley, J. van Damme, H. Claeys, H. Weening, J. W. Heine, A. Billiau, C. Vermylen, and P. DeSomer, *Eur. J. Immunol.* **10,** 877 (1980).

the mixture as in a regular IFN assay. The apparent "IFN titer" obtained is taken to represent IFN not bound to antibody in the original mixture. But this would be true only when the antibody affinity was extremely high. With most actual antibodies, dissociation of the IFN–Ab complex once formed will take place extensively upon dilution,[3,4] and therefore the "residual IFN" does not represent the state in the original mixture. For quantitative purposes, therefore, those assays described above which do not involve dilution of the IFN–Ab mixtures (or only small dilution factors) are recommended since the interpretations are more straightforward.

In both the constant IFN and constant antibody methods, the IFN used must be titrated concomitantly without antibody to determine the exact titer in EU/ml of IFN being reacted with antibody because the IFN sensitivity of the assay system may change from one assay to another. What is important here is the actual antiviral units rather than the absolute quantity of IFN (given in RU).

A typical experiment by the constant antibody method is shown in Fig. 1. The IFN assay is based on the reduction of viral CPE; the CPE is scored with an integer from 0 to 5 and averaged for 2 to 4 replicate wells. In this experiment, Hu-IFN-β was titrated in a 2-fold dilution series with and without fixed concentrations of an antibody to Hu-IFN-β. The CPE scores are plotted on the ordinate as a function of the log of the IFN dilution on the abscissa. From the IFN dose–response curve thus obtained, the apparent IFN titer, I, in the presence of antibody is determined as the point on the abscissa corresponding to 50% CPE. If the IFN

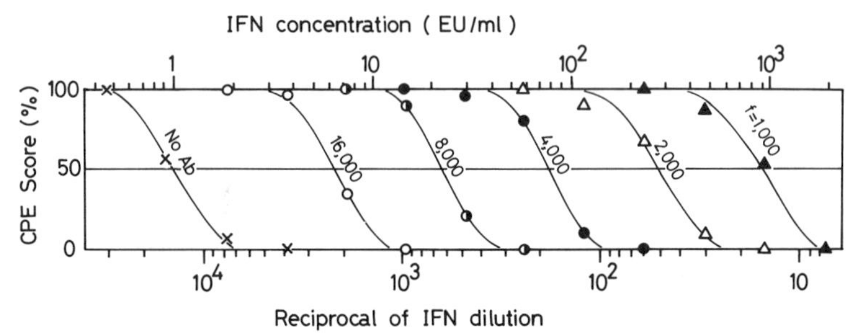

FIG. 1. A neutralization test by the constant antibody method. Hu-IFN-β from diploid fibroblasts induced by poly(I) · poly(C) and the NIH reference antibody to Hu-IFN-β (G-028-501-568) were used, the latter at dilutions ($1/f$) indicated in the graph. Human fibroblasts (trisomic for chromosome 21, GM2504) were seeded in microtiter plates at 2×10^4 cells/well and used 3 days later. They were treated with IFN or IFN–Ab mixture overnight before challenge with encephalomyocarditis virus, and the CPE was scored 2 days later. Four replicate wells were used for control IFN without antibody, and 2 or 3 wells for Ab–IFN mixtures. The average CPE scores are shown as a function of IFN dilution.

TABLE I
NEUTRALIZATION VALUES DETERMINED FROM THE CURVES OF FIG. 1[a]

Antibody dilution ($1/f$)	Apparent IFN titer (I, EU/ml)	N (EU/ml)	Neutralization titer (t)
No antibody	14,000 (= I_0)	—	—
1 : 16,000	2,300	6.1	9,000
1 : 8,000	660	21	18,000
1 : 4,000	180	78	34,000
1 : 2,000	50	280	62,000
1 : 1,000	15	930	103,000

[a] I is the reciprocal of IFN dilution at the point of 50% CPE in the presence of antibody (the lower abscissa of Fig. 1). N is the IFN concentration in EU/ml that is reduced to 1 EU/ml by the antibody ($N = I_0/I$, the upper abscissa of Fig. 1). t is calculated by Eq. (4).

titer without antibody is I_0, the ratio I_0/I gives the IFN concentration, N EU/ml, that is neutralized to 1 EU/ml by that antibody. Table I shows the figures of f, I, and N obtained from the result in Fig. 1.

The constant IFN method is done similarly. The CPE score is plotted against log (antibody dilution), and the antibody dilution $1/f$ that reduces the IFN used (N EU/ml) to 1 EU/ml is read from the graph.

Advantages of Constant Antibody Method

The constant IFN and the constant antibody methods are of course equivalent in principle, but our experience indicates that the constant antibody method is generally more convenient and reliable.[4] The result that is obtained in this method is a family of IFN dose–response curves which are usually parallel to each other (see Fig. 1). This parallelism helps to increase the precision of the assay. It is especially suitable for examining low degrees of neutralization.[3,4,7] If the IFN assay has a high precision, small shifts of IFN dose–response curves may be detected with reasonable certainty, corresponding to N values of 2 to 3. With the constant IFN method, low IFN concentrations must be tested; considering day-to-day fluctuations in IFN sensitivity, several IFN concentrations should be tried, and then, the experimental design will resemble that of the constant antibody method (or the checkerboard assay).[5]

The constant antibody method is also clearly better than the constant IFN method for characterizing monoclonal antibodies and for analyzing IFN preparations containing different antigenic types. These points are discussed in separate sections.

Definition and Computation of Neutralization Titer

The neutralization titer is usually defined as the reciprocal of the antibody dilution that neutralizes a certain IFN concentration, x units/ml, x often being taken to be around 10.[3-6] The choice of the value of 10 is arbitrary, but too low values are inconvenient in view of low precision of IFN bioassays, and too high values will not be suitable for low titer antibodies. Incorporating the endpoint designation, we define the titer t as the reciprocal of antibody dilution, $1/f$, that neutralizes 10 EU/ml of IFN to 1 EU/ml.[3,4]

Note that the IFN unitage in this definition is EU, namely, the actual antiviral units observed, and not RU. It may, however, seem desirable to use RU for the definition, to normalize the results obtained in assays with different IFN sensitivities. There has not been agreement among investigators as to whether EU or RU should be used as the basis of the antibody titer. For instance, suppose that 10 EU/ml was found by experiment to be neutralized to 1 EU/ml by an antibody at a dilution of $1/f$, and in this assay, 1 EU was equal to c RU (c is in general not equal to 1). If only EU is considered,[6,8,9] as in the definition described above, the titer t will be simply equal to f. On the other hand, if RU is adopted instead of EU[5,10-13] [now the titer is defined by neutralization of $(10 - 1) = 9$ RU/ml], the quantity of IFN neutralized must be expressed in RU: $(10 - 1)$ EU/ml = $9c$ RU/ml. Then, the antibody dilution that effects neutralization of 9 RU/ml will be $(1/f) \times (1/c)$; that is, the antibody titer is $f \times c$. Thus, c-fold different values of the titer are derived from the same data, depending on the choice of IFN unitage.

This problem of IFN unitage was examined experimentally as well as theoretically, as detailed elsewhere.[3,4] The antibody titer, being an index of the antibody potency, should ideally be a constant for a given antibody preparation regardless of the experimental conditions for its determination, and so we examined which definition satisfies this condition better in the face of changes in IFN sensitivity in different assays of the same

[8] I. Gresser, M. G. Tovey, M.-T. Bandu, C. Maury, and D. Brouty-Boyé, *J. Exp. Med.* **144,** 1305 (1976).

[9] S. Stefanos, L. Catinot, J. Wietzerbin, and E. Falcoff, *J. Gen. Virol.* **50,** 225 (1980).

[10] K. Paucker, B. J. Dalton, C. A. Ogburn, and E. Törmä, *Proc. Natl. Acad. Sci. U.S.A.* **72,** 4587 (1975).

[11] K. Paucker, B. J. Dalton, and C. A. Ogburn, Research reference reagent notes on the NIH reference antibodies to mouse L cell IFN (No. 19, 1980) and human leukocyte and fibroblast IFNs (Nos. 22 and 24, 1981).

[12] Y. Yamamoto and Y. Kawade, *Virology* **103,** 80 (1980).

[13] H. Okamura, W. Berthold, L. Hood, M. Hunkapiller, M. Inoue, H. Smith-Johannsen, and Y. H. Tan, *Biochemistry* **19,** 3831 (1980).

antibody. The results on various antibodies to human and mouse IFNs indicated that the titer defined in EU generally showed satisfactory constancy. Namely, if a given antibody neutralizes N EU/ml of IFN to 1 EU/ml, then in a second assay where the IFN sensitivity of the cells is different by severalfold, it will again neutralize N EU/ml to 1 EU/ml. On the other hand, if the titer is defined in terms of RU, its value tended to change in proportion to c. In other words, the absolute concentration of IFN neutralized by a given antibody [$c(N - 1)$ RU/ml] will not be constant and will appear greater in assays with lower IFN sensitivities (larger c).

This will appear to contradict the notion that a fixed absolute quantity of IFN (given in RU) is neutralized by a given antibody. However, a theoretical basis for such experimental observations was provided by considering a simplified model of the neutralization reaction, which we call the "simplest model."[4] Here, it is assumed that the only reaction taking place is the bimolecular reaction between antibody and IFN (multiple antibody binding to an IFN molecule is excluded and IFN–receptor binding is neglected), and the IFN–Ab complex is biologically inactive, that is, the antiviral state is determined by the concentration of free IFN. Denoting the molar concentrations of total antibody binding sites, and total and free IFN by $[A]$, $[I_t]$ and $[I_f]$, respectively, the following relationship is readily derived from the law of mass action:[3,4]

$$[A] = ([I_f] + K_d)\left(\frac{[I_t]}{[I_f]} - 1\right) \qquad (1)$$

where K_d is the Ab–IFN dissociation constant. If K_d is much smaller than $[I_f]$ (this means very high antibody affinity), this equation is simplified to

$$[A] = [I_t] - [I_f] \qquad (2)$$

On the other hand, if K_d is much larger than $[I_f]$ (very low antibody affinity),

$$[A] = K_d\left(\frac{[I_t]}{[I_f]} - 1\right) \qquad (3)$$

These equations indicate that, in the former case, a given antibody will neutralize a fixed absolute quantity of IFN, whereas in the latter case, what is neutralized by a given antibody is a function of the ratio of total to free IFN and not an absolute quantity of IFN (total minus free IFN). In neutralization experiments, the total and free IFN are represented by N and 1 EU/ml, respectively, and so the findings with many antibodies mentioned above are consistent with the case of K_d being much larger than $[I_f]$ at the endpoint, 1 EU/ml [Eq. (3)], but not with the other case [Eq. (2)]. Although little is known of the magnitude of K_d of anti-IFN antibodies, this seems reasonable since $[I_f]$ at 1 EU/ml is usually very low, being 10^{-12} to 10^{-14} *M*. The prevalent notion of a fixed quantity of IFN (given in RU) being neutralized by a given antibody is thus equivalent to assuming a very high antibody affinity. In this case, if an antibody is found to neutralize 10 RU of IFN to 1 RU, it will neutralize, for instance, 9 RU to zero, or 100 RU to 91 RU. With most actual antibodies such will not be true; what is predicted from the "simplest model" for low antibody affinity is that when an antibody neutralizes 10 RU to 1 RU, it will neutralize 9 RU to 0.9 RU, or 100 RU to 10 RU.

Thus we adopt EU for the IFN unitage to define the antibody titer; changes in IFN sensitivity will in general not affect the value of the titer much (except perhaps in systems of very low IFN sensitivity, where the condition $K_d \gg [I_f]$ might not hold). The essence here is that the titer is defined by 10-fold reduction of IFN titer, and not by an absolute quantity of IFN neutralized, such as 9 RU. Therefore the question of normalizing the titer based on EU into that based on RU does not occur.

For many purposes not requiring precise data, it will suffice to determine the antibody dilution $1/f$ that neutralizes roughly 10 EU/ml of IFN, and simply take the value of f as the neutralization titer.

More precisely, the following equation may be used to compute t from any experimental values of N and f:[3,4]

$$t = f(N - 1)/9 \qquad (4)$$

Examples of the values of t calculated from the data of Fig. 1 are given in Table I. This equation was derived for the "simplest model" under the condition of $K_d \gg [I_f]$. Although this is clearly an oversimplification of actual neutralization reaction (see below), it has been conveniently used in numerous assays of various antibodies in this laboratory and given consistent values of the titer.

Although the experimental methods used by various investigators for antibody titration are fairly similar to the one described here, and seek the antibody dilution that neutralizes about 10 EU/ml of IFN, some investigators express the antibody titer, not by the reciprocal of antibody dilution, but by the "units of IFN neutralized by 1 ml of the antibody."[13,14] For instance, if a 1 : 1000 dilution of an antibody neutralizes 10 to 1 units/ml (either in EU or RU), 1 ml of the undiluted antibody would be said to neutralize 9000 units of IFN[11] [more generally, for any values of N and f

[14] H. M. Johnson, M. P. Langford, B. Lakhchaura, T.-S. Chan, and G. J. Stanton, *J. Immunol.* **129,** 2357 (1982).

observed, the quantity $f(N - 1)$ is taken]. Such an expression may be misleading, since it appears to imply that a given antibody neutralizes a fixed quantity of IFN, regardless of the quantity of IFN added (i.e., that any quantity of IFN, x units, will be reduced to $x - 9000$ units). In this case, one is tacitly assuming a very high antibody affinity, which will rarely be true; with actual antibodies, the quantity of IFN neutralized will not be constant as was described before, and therefore such statements as given above should be treated with caution. The figure 9000 can actually be regarded to be the antibody titer which is defined by 2-fold (instead of 10-fold) reduction of IFN titer [this can readily be deduced with Eq. (4)]; namely, it would indicate that the antibody at a dilution of 1 : 9000 neutralizes 2 EU/ml to 1 EU/ml. Such a consideration may not altogether be inappropriate, but if one wants to do experiments directly as indicated in the definition, 2-fold reduction in IFN titer is obviously less suitable than 10-fold reduction.

Antibody Concentration Dependence of t Value

The method described above of determining the antibody titer thus resolved the problem of changes in IFN sensitivity in different assays. However, when various antibodies were examined at different concentrations, the t values calculated by Eq. (4) were found to change significantly, except for a few antisera among the many we studied.[3,4] Monoclonal and conventional antibodies differ markedly in this respect, and the former is discussed in the next section. With conventional antibodies, the t value usually increases somewhat with antibody concentration (namely, increased neutralization efficiencies at high antibody concentrations; a possible explanation for this is given in the next section). As an example, our data on the NIH reference antibody to Hu-IFN-β are shown in Fig. 2.

As a quantitative "label" of the antibody, the t value determined from N values not far from 10 (between about 5 and 20) should be adopted.

For fuller characterization of an antibody, N and t values must be determined for a wide antibody concentration range since their concentration dependence may differ for different antibodies. If one wants to neutralize a certain quantity of IFN, the titer determined with N equal to about 10 may serve as a rough guide for the necessary quantity of antibody [through Eq. (4)], but the exact quantity has to be determined empirically (especially in the case of monoclonal antibodies; see next section).

Neutralization Characteristics of Monoclonal Antibodies

Monoclonal antibodies have been found to differ markedly from conventional antibodies, in that neutralization does not increase much with

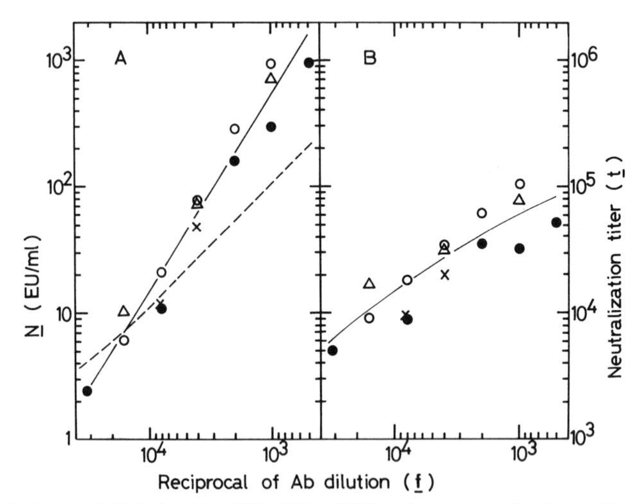

FIG. 2. Values of N (A) and t (B) of the NIH reference antibody to Hu-IFN-β as a function of antibody dilution, determined against Hu-IFN-β from diploid fibroblasts. Different symbols represent different assays, giving an idea about their reproducibility. The broken line in (A) is the theoretical curve for the case of the "simplest model" and $K_d \gg [I_f]$ (its position is arbitrary with respect to the x axis); in this case, $N - 1$ is proportional to $1/f$ [see Eq. (3)] and t should be a constant independent of f [Eq. (4)].

increase in antibody concentration.[15] Monoclonal antibodies to Hu-IFN-α usually react with different subtypes of IFN-α to different degrees,[16] and therefore, even if they have neutralizing capacities to some subtypes, they are generally expected not to neutralize completely natural IFN-α preparations which are mixtures of various subtypes. However, when antigenically homogeneous recombinant IFN-α2 was used, a monoclonal antibody NK-2 was reported to show decreased titers upon increase in antibody concentration.[17] This is contrary to general findings on polyclonal antibodies described in the previous section. We have found the same trend with three monoclonal antibodies to Hu-IFN-α when tested against recombinant IFN-α1, α2, and a hybrid A/D, and also with three monoclonal antibodies to Hu-IFN-β tested against natural IFN-β[15]; these IFNs may be considered antigenically homogeneous.

A typical example of neutralization assay of a monoclonal antibody to Hu-IFN-α (HT-1)[18] against recombinant Hu-IFN-α1 (Fig. 3) was carried

[15] Y. Kawade and Y. Watanabe, *Immunology* **56,** 489 (1985).

[16] T. Staehelin, B. Durrer, J. Schmidt, B. Takacs, J. Stocker, V. Miggiano, C. Stähli, M. Rubinstein, W. P. Levy, R. Hershberg, and S. Pestka, *Proc. Natl. Acad. Sci. U.S.A.* **78,** 1848 (1981).

[17] J. T. D. Whittall, R. M. King, and D. C. Burke, *J. Gen. Virol.* **65,** 629 (1984).

[18] K. Tsukui, S. Uchida, E. Tokunaga, and Y. Kawade, in preparation.

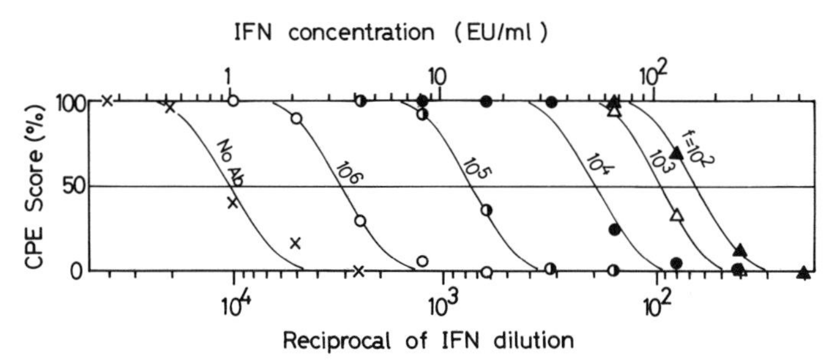

FIG. 3. Neutralization of recombinant Hu-IFN-α1 by a monoclonal antibody to Hu-IFN-α (HT-1) assayed by the constant antibody method. For experimental details, see legend to Fig. 1.

out by the constant antibody method. The neutralization values N determined from these curves are shown in Fig. 4 as a function of antibody dilution. It can be seen that N increases only slowly at high antibody concentrations (compare with Fig. 2) and tends to level off. The $N - 1/f$ relationship is different for different systems; the leveling off is strong with some but moderate for others. In all cases examined, however, the value of t from Eq. (4) decreases with increasing antibody concentration, in marked contrast to conventional antibodies.

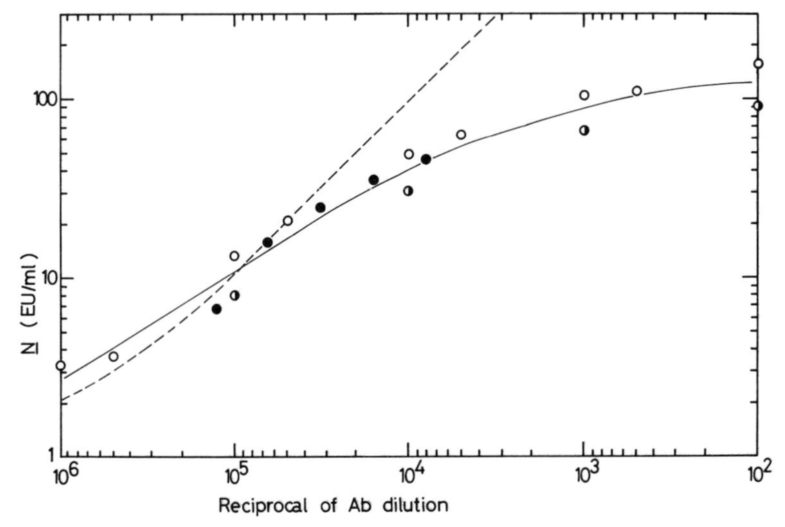

FIG. 4. Extent of neutralization (N) of Hu-IFN-α1 by a monoclonal antibody (HT-1) as a function of antibody dilution. Three independent assays are indicated by different symbols. The broken line is the theoretical curve as in Fig. 1.

For practical purposes, therefore, it seems better to indicate their neutralizing capacities by simply giving several values of N determined at several antibody dilutions, rather than by the values of t.

The constant IFN method is not convenient to characterize monoclonal antibodies, since increases in antibody cause only moderate increases in neutralization, resulting in antibody dose–response curves with gentle slopes, and if too high IFN concentrations are used, some antibodies will fail to show any clear neutralization.[4] The constant antibody method, in contrast, will always give parallel IFN dose–response curves as shown in Fig. 3, and reliable results are readily obtained covering a wide antibody concentration range.

The antibody concentration dependence of t values may be explained as follows.[19] In the case of monoclonal antibodies, an IFN molecule will bind a single antibody molecule, and then its biological activity will in general be diminished but not completely abolished. From consideration of the IFN–Ab–cell receptor equilibrium, it can be shown that, in such cases, the neutralization value N will increase rather slowly with increasing antibody concentration, and approach a limiting value. In the case of conventional antibodies, multiple antibody molecules will bind to a single IFN molecule, causing efficient inactivation of the molecule. This effect will increase with increasing antibody concentration, and hence cause increased efficiencies of neutralization.

Procedure of Antibody Titration

A practical procedure for titrating an antibody with unknown potency is as follows. Dilutions of antibody are made in culture medium with 8-fold steps, starting, for instance, from 1 : 10, then 1 : 80, and so on. A maximum of 6 antibody dilutions can be tested on a single microtiter plate with 96 wells.

IFN is diluted in 2-fold steps. Six samples are made, containing 2–4 to 64–128 EU/ml, to be mixed with each of the antibody dilutions prepared above. For IFN titration without antibody, higher dilutions are also made.

Culture medium of cells in a microtiter plate is removed and 0.05 ml each of antibody and IFN is added with duplicate wells for each sample (a maximum of 6 antibody dilutions $\times$ 6 IFN concentrations $\times$ 2 = 72 wells). The final antibody dilutions are thus 1 : 20, 1 : 160, and so on, and the final IFN concentrations are from 1–2 to 32–64 EU/ml.

For titration of control IFN, quadruplicate wells are used for each of 6 IFN dilutions covering a range of 0.12–0.25 to 4–8 EU/ml (24 wells in total).

[19] Y. Kawade, *Immunology* **56,** 497 (1985).

The rest of the procedure is just the same as in regular IFN assays. In the case of antisera (polyclonal antibodies), one or two of the six antibody dilutions tested will give rise to CPE covering the whole range from 100% to 0 (for the other antibody dilutions, either full CPE in all wells or no significant neutralization). Then the CPE score is plotted against log (IFN dilution), as shown in Figs. 1 and 3, and the value of N is determined as described before. It will not be far from 10; the neutralization titer t is then obtained by Eq. (4).

Notes. 1. Various modifications are of course possible. For instance, 4-fold (instead of 2-fold) IFN dilution steps may be used to obtain a wide range of N values (with less precision). In fact, to obtain precise values of the titer, it is wise to carry out such a preliminary assay first. If the antibody titer is roughly known, the dilution steps of both antibody and IFN in the above procedure can be made narrow to obtain a number of precise values of N.

2. When the IFN dose–response curve is not steep in slope, or when fluctuations in IFN sensitivity in different assays are greater than 2- to 3-fold, more than 6 IFN concentrations will have to be used.

3. High concentrations of antibody preparations (e.g., 1 : 10 dilution of antiserum) might affect cell and virus. To control this, several wells are made containing high antibody concentrations but without added IFN, half of them are challenged with virus, and they are examined as to whether the cells remain healthy and the viral CPE occurs normally.

4. When two or more plates are used, control IFN should be titrated on each plate.

5. The IFN–Ab mixtures are usually preincubated for some time (e.g., for 1 hr at 37°) before being added to assay cells. However, we found that preincubation is usually not necessary for many antibodies to human and mouse IFNs; when antibody and IFN were directly added to cells, the results were exactly the same as those obtained for preincubated mixtures (1 to 5 hr at 37°), except in the case of a single antiserum, which sometimes showed lower neutralization when not preincubated. Thus the Ab–IFN reaction seems a relatively fast one. Omitting preincubation will considerably simplify manipulations in the constant IFN or antibody method, although individual antibody will have to be examined at least once in this regard.

Specificity of Antibody and Antigenic Purity of IFN

As a general precaution, it is preferable to use purified immunoglobulin for neutralization tests, since serum and other body fluids may contain nonspecific inhibitors of IFN. For instance, most rabbit sera we tested had an inhibitory activity to mouse IFN-γ, which was quite potent in some sera; it could be eliminated by purification of the immunoglobulin by protein A-Sepharose (unpublished data).

An important purpose of the neutralization test, besides antibody titration, is the identification of the type of IFN. For this purpose, the antibody prepared against one type of IFN should be checked for any activity against other types of IFN. For instance, some antisera to human leukocyte IFN may contain anti-β activities[20]; the NIH reference antibody to Hu-IFN-α (G-026-502-568) has a neutralization titer of about 2×10^6 against leukocyte IFN-α, but also has a titer of about 500 against Hu-IFN-β. This is because the leukocyte IFN used as immunogen contained a small amount of IFN-β, which was not readily detectable.

To test the antibody specificity, the constant antibody method is recommended, because it is sensitive and convenient to detect weak neutralizing activities.

To test for IFN-α-neutralizing activities, natural IFN-α or certain subtypes of IFN-α may be used as test antigen, but it must be borne in mind that, strictly speaking, apparent lack of activity to the IFN-α used does not necessarily prove complete absence of anti-α activity, because different IFN-α subtypes may differ in fine antigenic specificities. For IFN-β and -γ, there will probably be no such problem, since only single genes have been identified in human and mouse genomes, and these IFNs seem so far to be antigenically uniform.

For the purpose of antibody titration, the IFN used as test antigen must be ascertained for their antigenic purity by use of antibodies with known specificities. For human IFN-α and -β, the NIH reference antibodies are useful for this purpose. The NIH antibody to Hu-IFN-α mentioned above can be used at a dilution of 10^{-4} to 10^{-5} to neutralize leukocyte IFN-α to 1% or less, without affecting IFN-β or -γ; lower dilutions than 10^{-4} must be avoided because some anti-β activity may then appear. The NIH antibody to Hu-IFN-β (G-028-501-568) at a dilution of 10^{-3} will neutralize IFN-β to 1% or less (see Table I and Fig. 2), but has no effect on IFN-α and -γ. If the IFNs tested show lower degrees of neutralization, they must be tested as to whether appropriate mixtures of antibodies to IFN-α, -β, and -γ cause greater neutralization degrees (see next section). (The exact degrees of neutralization may differ for different IFN preparations; even for IFN from fibroblasts, which is considered to be purely of β type, 3- to 4-fold differences were noted in the antibody concentration

[20] E. A. Havell, B. Berman, C. A. Ogburn, K. Berg, K. Paucker, and J. Vilček, *Proc. Natl. Acad. Sci. U.S.A.* **72,** 2185 (1975).

needed to neutralize different preparations to the same degree. The reason for this is not clear.)

For other IFNs, no reference antibodies are available, except for anti-mouse IFN-(α,β) (NIH G-024-501-568), which contains both anti-α and anti-β antibodies[3] but has little anti-mouse γ activity.[21] Therefore, laboratory antibody preparations must be used after checking their specificities.

Analysis of Mixed Type IFN

Throughout the preceding discussions, only the cases of IFN consisting of neutralizable components were considered. Natural IFN preparations often contain antigenically distinct species, and neutralization tests can conveniently be applied to determine the types of constituent IFNs and to estimate their proportions.[21–24] For this purpose, the constant antibody method is superior to the constant IFN method. The principle is simple. In the case of IFN containing only neutralizable species, the IFN dose–response curve will be shifted to the right indefinitely upon increase in antibody, as illustrated in Fig. 1 (with conventional, rather than monoclonal, antibodies). If the IFN contains a nonneutralizable component amounting to 5% of total activity, the IFN dose–response curve cannot be shifted to the right beyond a certain point corresponding to $N = 20$ (100%/5%) however high antibody concentration is used. Thus the limiting value of N indicates the relative proportion of the nonneutralizable component. Its antigenic type can be identified if a second antibody is added together with the first and a further increase in neutralization (an increase in N) is observed.

An example is given in Table II. An IFN preparation from human lymphoblastoid Null cells induced by Sendai virus[25] was titrated in the presence and absence of anti-α and anti-β antibodies (the constant antibody method), and the results are expressed by the apparent IFN titers observed in the presence of antibody. The IFN was largely neutralized by anti-β (Sample 5 compared to 1), but the following points indicated the presence of a small amount of IFN-α: (1) upon 5-fold increase in anti-β, the IFN titer did not decrease much (compare Samples 3 and 5), in contrast to pure IFN-β, which was reduced strongly in titer by this increase in antibody (see Figs. 1 and 2), and (2) with anti-β at 1 : 1600, further addition

[21] M. Tomida, Y. Yamamoto, M. Hozumi, and Y. Kawade, *J. Interferon Res.* **2,** 271 (1982).

[22] M. Saito, T. Ebina, M. Koi, T. Yamaguchi, Y. Kawade, and N. Ishida, *Cell. Immunol.* **68,** 187 (1982).

[23] H. Okamura, K. Kawaguchi, K. Shoji, and Y. Kawade, *Infect. Immun.* **38,** 440 (1982).

[24] E. A. Havell, this series, Vol. 79, p. 575.

[25] M. Matsuyama, Y. Hinuma, Y. Watanabe, and Y. Kawade, *J. Gen. Virol.* **60,** 191 (1982).

TABLE II
ANALYSIS OF MIXED TYPE IFN; NEUTRALIZATION OF IFN FROM SENDAI VIRUS-INDUCED NULL CELLS BY ANTI-α AND ANTI-β ANTIBODIES[a]

Sample number	Antibody added		Apparent IFN titer	
	Anti-α	Anti-β	EU/ml	%
1	None	None	1,200	100
2	1:40,000	None	900	75
3	None	1:8,000	300	25
4	1:40,000	1:8,000	200	17
5	None	1:1,600	170	14
6	1:40,000	1:1,600	16	1.3

[a] The apparent IFN titers were determined in the presence of antibody by the constant antibody method. Anti-α: NIH reference antibody; at the dilution used, it neutralizes leukocyte IFN-α to less than 1%, but has no effect on IFN-β. Anti-β: NIH reference antibody; at dilutions of 1:8,000 and 1:1,600, it neutralizes fibroblast IFN-β to 5–10% and to 1% or less, respectively, but does not affect leukocyte IFN-α.

of anti-α caused a sharp drop (about 10-fold) in IFN titer (compare Samples 5 and 6); the activity of Sample 5 thus represents mostly ($\geqq$90%) IFN-α. From this, the IFN preparation can be estimated to contain about 13% (150 units/1200 units) of IFN-α, the remainder being IFN-β. In this experiment, anti-α alone slightly reduced the IFN titer (compare Samples 2 and 1), but such a small effect of antibody should not be taken at face value (because of possible nonspecific effects of antibody and titration errors) unless supported by other evidence for its reality. For the purpose of detecting a minor IFN component, the second antibody must be used under the condition where the activity of the major component is largely eliminated by the first antibody. It is also obvious that the antibody preparations to be used for these analyses should be characterized well beforehand as to their specificity (effects on heterologous IFNs) and potency against the homologous IFN.

The constant IFN method is not as convenient as the constant antibody method for such analyses. With low IFN concentrations, a single antibody may effectively eliminate the antiviral effects, and mask the presence of a nonneutralizable component. With high enough IFN concentrations, on the other hand, single antibodies will fail to give rise to any CPE since substantial fractions of IFN activity will remain. In this case, the use of antibody mixtures will allow one to identify the constituent IFN types, but for estimation of their relative proportions, several IFN concentrations will have to be examined. Then the experimental design is essentially that of the constant antibody method.

Conclusion

The experimental method and the definition of the antibody titer described above are the outcome of the efforts in this laboratory to make the neutralization tests quantitative and reproducible.[3,4]

The main points for the experimental method are (1) to designate the endpoint of titration clearly, that is, as the same point as that in IFN assay (1 EU/ml), and (2) to use an assay that does not require dilution of the Ab–IFN mixtures (or to express the IFN and antibody concentrations by the final ones in the culture fluid of assay cells). Then the quantities determined in an assay are the antibody dilution $1/f$ and the IFN concentration N EU/ml that is reduced to 1 EU/ml by this antibody. In practice, the constant antibody method is in general more convenient than the constant IFN method.

The definition of antibody titer adopted here (the reciprocal of antibody dilution that reduces 10 EU/ml of IFN to 1 EU/ml) is simply a precise statement of the practice of many investigators. The IFN unitage is a source of confusion among investigators, but what is important here is considered to be the *ratio* of total IFN (N EU/ml) to free IFN (1 EU/ml), and therefore the problem of conversion from EU to RU does not occur; absolute (reference) units of IFN neutralized should not be used as the basis of antibody titer. The titer can be computed from any value of N and f by a simple formula, Eq. (4). The value may change with antibody dilution, and so, as the "label" of the antibody, the value of the titer obtained for N values not far from 10 will be appropriate (this applies to conventional antibodies; monoclonal antibodies with weak neutralizing capacities will require a different approach).

The definition of the antibody titer and the method for computing it should ideally be based on a theoretical formulation of the neutralization reaction, which is difficult at present. Our approach is based on a greatly simplified model and the assumption of K_d much larger than $[I_f]$ at 1 EU/ml; the need for a better method for expressing the neutralization capacity is especially apparent in the data on monoclonal antibodies. Empirically, however, our definition and the formula to compute the antibody titer has proved to be convenient. The above considerations may contribute to making it easier to compare data by different investigators than is possible at present.

[75] A Convenient Microassay for Cytolysis and Cytostasis

By STEPHEN TYRING, W. ROBERT FLEISCHMANN, JR., and SAMUEL BARON

Introduction

In addition to their many other properties, interferons are known to have both cytolytic and cytostatic activities.[1] This is particularly true of gamma interferon (IFN-γ).[1-4] These activities have been measured by inhibition of colony formation, radioisotope uptake or release, trypsinization and cell counting, or other methods generally requiring a relatively large amount of time, material, and preparation. There was a need for an assay that was simple and conserved reagents while preserving the desirable characteristics of previous assays. The method described attempts to meet these needs by employing microhistoplates seeded with 100–200 target cells per well and then incubated with IFNs alone or with effector cells for 1–3 days. Viable cells are counted microscopically to determine cytostasis or cytolysis. The most time-consuming portion of the test is the microscopic cell counting.

Materials and Procedures

Medium. Cells were propagated in RPMI medium containing 10% fetal calf serum (FCS), 50 μM 2-mercaptoethanol, and antibiotics. Medium used in the anticellular assays contained 3% FCS.

Cells. P388 murine lymphoma cells were obtained from Dr. A. E. Boyden, R G and G Mason Research Institute, Worchester, MA. Human K562 cells were obtained from Dr. Gary Klimpel. They were propagated in the above medium at 37° in a humidified atmosphere containing 5% CO_2.

Mouse immune interferon (IFN-γ) was induced in spleen cells of retired breeder C57BL/6 mice (Timco) with staphylococcus enterotoxin A

[1] S. Baron, S. Tyring, W. R. Fleischmann, Jr., G. Klimpel, and G. J. Stanton, *in* "Mechanisms of Lymphocyte Activation" (K. Resch and H. Kirchner, eds.), p. 157. Elsevier/North-Holland Biomedical Press, Amsterdam, 1981.

[2] J. E. Blalock, J. A. Georgiades, M. P. Langford, and H. M. Johnson, *Cell. Immunol.* **49,** 390 (1980).

[3] B. Y. Rubin and S. L. Gupta, *Proc. Natl. Acad. Sci. U.S.A.* **77,** 5928 (1980).

[4] S. Tyring, G. R. Klimpel, W. R. Fleischmann, Jr., and S. Baron, *Int. J. Cancer* **30,** 59 (1982).

(0.04 μg/ml, SEA, Microbial Biochemistry Branch, Division of Microbiology, Food and Drug Administration, Cincinnati, Ohio).[5] Unpurified mouse IFN-γ preparations have a specific activity of about $10^{1.7}$ units/mg of protein. Partially purified mouse IFN-γ (10^4 units/mg of protein) was prepared as follows. Crude IFN-γ was bound to controlled pore glass beads (CPG) (5 g of CPG/1 liter of crude IFN-γ) by overnight stirring at 4°. The CPG was poured into a column. The column was washed with 1 *M* NaCl in phosphate-buffered saline, and the IFN-γ was eluted with 0.3 *M* tetraethylammonium chloride and 1 *M* NaCl in phosphate-buffered saline. The eluate was concentrated by pressure dialysis in an Amicon Stirred Cell (Model 52) and passed through an Ultragel AcA54 column (2.5 × 100 cm) with 1 *M* NaCl in phosphate-buffered saline as the eluent. Peak fractions of IFN-γ activity were pooled, dialyzed against 100 volumes of phosphate-buffered saline, and stored at −70° until used.

Human immune interferon (IFN-γ) obtained courtesy of Dr. V. Papermaster was induced in human peripheral leukocytes by SEA. This interferon was purified 10,000-fold by the method of Papermaster[6] to a product having a specific activity of 10^6 units/mg protein. The preparation was dialyzed against PBS before use.

Assay of Antiviral Activity of Interferon. Antiviral assays were done with mouse L-929 cells in the plaque reduction method according to Campbell *et al.*[7] or with P388 cells in the cytopathic effect (CPE) reduction method. One unit/ml of interferon was defined as the concentration that resulted in a 50% reduction of the number of plaques in the plaque reduction assay. All assay results were expressed in units/ml, corrected by comparison with a laboratory reference preparation of Mu-IFN-α/β that had been calibrated against the W.H.O. International reference preparation G002-904-511 (N.I.H.). In the CPE assay, one unit/ml of interferon equalled the concentration that reduced the virus-induced cytolysis of cells by 50%. Vesicular stomatitis virus (VSV) was used in both assays. The number of plaque-forming units of virus per cell used in the CPE assay was 2.

Microassay for IFN-γ-Mediated Cytolysis

Experiments were conducted to develop an assay system which could be used for detecting IFN-γ-mediated cytostasis as well as cytolysis with

[5] L. C. Osborne, J. A. Georgiades, and H. M. Johnson, *Infect. Immun.* **23,** 80 (1979).

[6] V. Papermaster, *Tex. Rep. Biol. Med.* **41,** 672 (1982).

[7] J. B. Campbell, T. Grunberger, M. A. Kochman, and S. L. White, *Can. J. Microbiol.* **21,** 1247 (1975).

very small quantities of IFN-γ. It was reasoned that good sensitivity and accuracy might be attained as with the virus plaque reduction assay by using very low numbers of cells which could be directly counted.

Such a method was devised by using Histoplates (Dynatech Laboratories, Alexandria, VA) whose individual wells could be completely visualized under one field in a microscope. Two hundred P388 tumor cells in a volume of 10 μl of RPMI 1640 medium were added to each of these wells. Ten microliters of test preparations [either RPMI 1640 medium or IFN-γ (700 units or 2900 units/ml)] was added to the wells containing P388 tumor cells. These Histoplates were incubated at 37° in a humidified 5% CO_2 incubator. At the times specified in Fig. 1, duplicate wells at each point were stained with 5 μl of a 0.4% trypan blue solution (Harleco, Gibbstown, NJ) and the mean number of viable cells determined by their ability to exclude trypan blue.

The results of a representative experiment are illustrated in Fig. 1. Cytolysis of P388 lymphoma cells proceeded rapidly from the fourth to the sixteenth hour of incubation with 2900 units of mouse IFN-γ. By 20 hr all cells had undergone lysis. Incubating cells in only 700 units of IFN-γ resulted in a less rapid reduction in the number of cells during the first 24

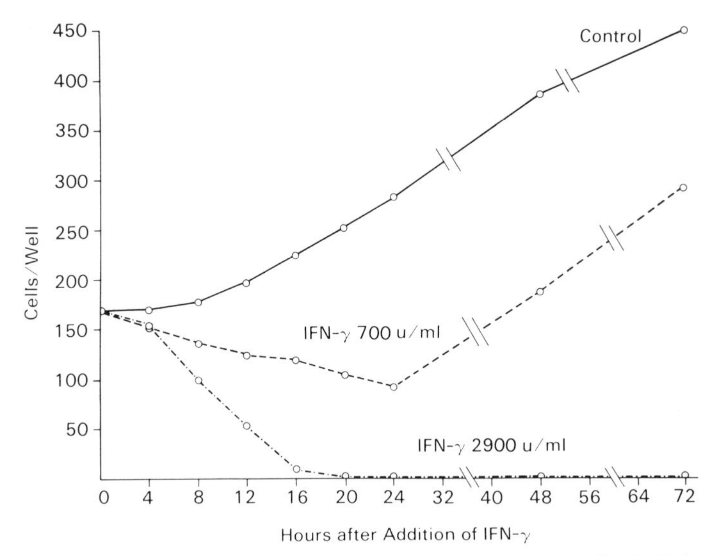

FIG. 1. Kinetics of the anticellular action of mouse IFN-γ on murine P388 lymphoma cells. Cells treated with 700 units/ml of IFN-γ were significantly different from controls at the twelfth through the seventy-second hours ($p < 0.05$; Student's t test). Cells treated with 2900 units/ml of IFN-γ were significantly different from controls at the eighth and twelfth hours ($p < 0.05$) and at the sixteenth through the seventy-second hours ($p < 0.01$).

hr. By 48 hr, however, an increase in cell number was observed and continued through the seventy-second hour. Thus, using this microassay, we have defined conditions in which cytolysis of P388 tumor cells by IFN-γ preparations could be measured within 24 hr of culture. Cytolysis was dependent upon high concentrations of IFN-γ since low concentrations of IFN-γ gave only growth-inhibitory activity. To use this assay for comparing the cytotoxic activity of different IFN-γ preparations and/or the susceptibility of different cell types to IFN-γ, we defined the cytolytic titer as equal to the reciprocal of the dilution of test supernatant fluid which gave a 50% reduction in the number of cells relative to controls at 24 hr of incubation. A 50% reduction in cell number relative to controls in this assay was significant at $p < 0.05$ level as determined by Student's t test. Similar results were obtained with human IFN-γ and K562 cells as well as a number of other human and mouse cell lines with their homologous interferons. This assay can also be employed with lymphoid effector cells when they can be distinguished from the target cells by morphology and or size.

Effects of Target Cell Density

IFN-γ-mediated cytolysis is inversely dependent upon cell density in contrast to antiviral activity which is directly dependent on cell density.[4,8] The experimental evidence was generated by the comparison of the antiviral with the cytotoxic activity of IFN-γ at high and low cell density by the above technique. Two hundred or 2000 cells were added to each of four wells either in culture medium or in culture medium containing different dilutions of a partially purified mouse IFN-γ preparation (IFN-γ plaque reduction titer of 5800 units/ml, specific activity 10^4 units/mg of protein). At 24 hr of culture, the mean number of viable cells was determined in two of the four wells. To the other two wells for each condition, 10 μl of medium containing VSV was added. The dose of VSV employed had previously been determined to result in lysis of 95% of the control cells after 24 hr of incubation. At this time the CPE protection titer was determined. It was seen (data not shown) that an increase in the number of cells from 200 to 2000 per well resulted in an increase in the antiviral titer from 8 to 64. In contrast, the cytolytic titer decreased from 32 for 200 cells per well to 2 for 2000 cells per well. Thus, strong cytotoxic activity of partially purified IFN-γ appeared to be dependent upon a low cell density of P388 tumor cells.

Characterization of the Cytolytic and Antiviral Substance. Since a number of substances other than IFN-γ might have been responsible for

[8] J. E. Blalock and S. Baron, *J. Gen. Virol.* **43,** 363 (1979).

the anticellular actions of IFN-γ preparations seen in this study, the partially purified IFN-γ preparation was subjected to a number of criteria classically used to characterize IFN-γ. The cytolytic activity shared the following properties with IFN-γ: instability at 60° and pH 2, species specificity, inactivation by specific antiserum, production by *E. coli* transfected with a plasmid containing the IFN-γ coding sequence, and copurification with IFN-γ which was purified to virtual homogeneity (10^8 units/mg protein).

Mouse IFN-α/β did not demonstrate cytolytic activity on P388 cells even when antiviral titers exceeding 10^5 units/ml were employed (data not shown).

Comments

The microassay presented here provides several advantages over more conventional methods of assaying cytostatic and cytolytic materials. These include simplicity, convenience, and the use of microquantities of reagents. Accuracy and reproducibility of results is comparable to other assays. The method was developed in our laboratory chiefly due to the need to assay the anticellular activity of a substance available in very limited quantities, IFN-γ. The method should be applicable, however, to assaying the antiproliferative and cytolytic activity of a wide variety of agents.

In the graph presented, it can be seen that the lower concentration of IFN-γ was mainly cytostatic while the higher concentration was clearly cytolytic. This provides a clear advantage over an assay involving inhibition of colony formation in which cytostasis cannot be distinguished from cytolysis.

Although the results presented employ a nonadherent cell line, P388 lymphoma, equally reproducible results have been observed with adherent cell lines such as mouse B16 and human melanomas (results not shown). An advantage of using an adherent cell in this assay is that there is no need to read the results immediately since the adherent cells can be stained with crystal violet (0.1% crystal violet in 20% in ethanol) and examined at a convenient time. The results with the crystal violet staining vary no more than 10% from the results obtained with trypan blue exclusion. This is due to the observation that dead trypan blue stained cells soon become nonadherent.

The finding that the direct anticellular activity of IFN-γ is highly dependent on cell density must be noted when using this assay. Whether the cell density is so critical with other anticellular substances is not known, but it is probably prudent to keep the target cell density constant.

This assay system has also been successfully employed in our laboratory to quantitate the anticellular action of various effector cells from peripheral blood or other sources of lymphocyte cells with or without IFN. The advantages over such methods as the ^{51}Cr release assay are clear: less materials and no need to use radioactive materials. The major disadvantage is that if nonadherent target cells of the same size as effector cells are used, the target cells cannot be accurately distinguished from effector cells. Adherent target cells do not have this problem because of their larger size and because most effector cells are washed away during the staining procedure.

In conclusion, we feel that the major advantages of this microassay for quantitating cytostasis and cytolysis lies in its simplicity and requirement for only minute quantities of test materials, and its convenience while maintaining good sensitivity and reproducibility. Since a great number of assays can be run simultaneously by one person, we feel that one use for this method would be screening biopsy materials (also often available in only minute quantities) in order to determine which patients might benefit from IFN or other therapy. The method would also allow a rapid evaluation of the potential of various immunomodulators to stimulate leukocytes to kill tumor cells from biopsies.

[76] A Simplified Antigrowth Assay Based on Color Change of the Medium

By PAUL AEBERSOLD and SALLY SAMPLE

Assays for antiviral effects of interferon have been made technically straightforward and convenient to score.[1,2] Growth inhibition assays, however, depend on cell counts[3] or [^{3}H]thymidine uptake and are rather inconvenient to score. We therefore developed a simple antigrowth assay which can be scored visually. The assay uses Daudi cells, which are very sensitive to growth inhibition by interferon, and it measures the acidity of the medium due to their metabolism. The acidity of the medium, as detected by the phenol red pH indicator, correlates with the number of cells in the wells, and these parameters are more reliable than [^{3}H]thymidine

[1] J. A. Armstrong, this series, Vol. 78, p. 381.

[2] P. C. Familletti, S. Rubinstein, and S. Pestka, this series, Vol. 78, p. 387.

[3] M. Evinger and S. Pestka, this series, Vol. 79, p. 362.

uptake, which correlates with cell growth only during the thymidine pulse.

Materials and Methods

Daudi cells are grown in suspension culture in RPMI-1640 supplemented with 10% fetal calf serum, 100 units/ml penicillin, and 100 μg/ml streptomycin. They are grown in tissue culture flasks at 37° in a humidified 5% CO_2 incubator. Maintenance of Daudi cells and their use in antigrowth assays have been described in detail by Evinger and Pestka.[3]

Color change assays are set up similarly to cytopathic effect inhibition assays.[1] Dilutions of interferon samples and of a standard reference interferon are made in medium along the rows of microtiter plates; for this work we used 1:3 dilutions, leaving 100 μl of medium per well. The Daudi cells were centrifuged and resuspended in fresh medium, then added to each well at 5×10^4 per well in 100 μl of medium, for a total assay volume of 200 μl. Plates were incubated for 4 days, then removed from the incu-

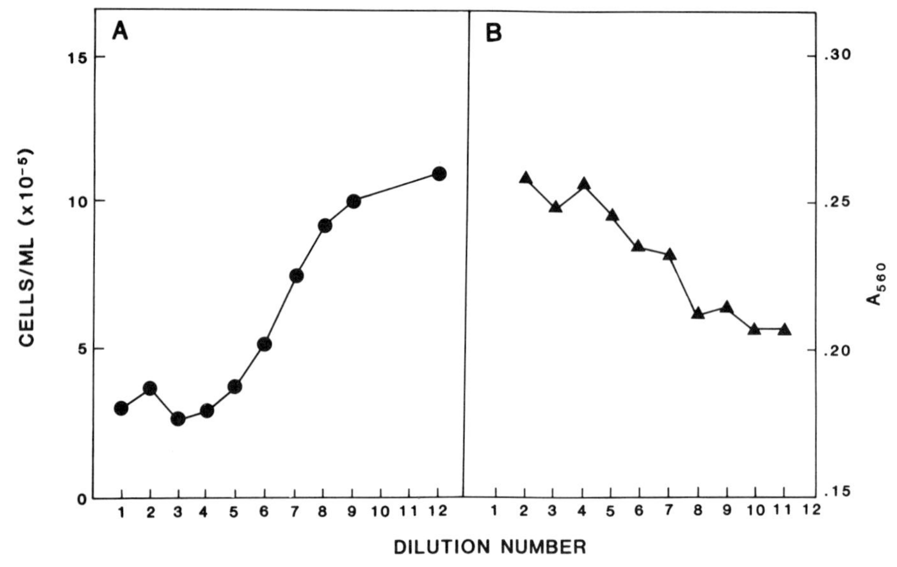

FIG. 1. Effect of interferon on cell number and optical density of the medium. Leukocyte interferon, prepared by the method of Cantell *et al.*,[4] was serially diluted (1:3) in microtiter wells and Daudi cells were added. The leukocyte interferon titer was 500 units/ml in a standard cytopathic effect assay, so the titers for wells No. 6 and 7 were 0.69 and 0.23 units/ml, respectively. After 4 days of incubation, total cells (viable plus nonviable) were counted and optical density of the medium was determined at 560 nm.

Daudi 10-4-82

1

54

61A

6L

76

β1

Cantell's

pBR322

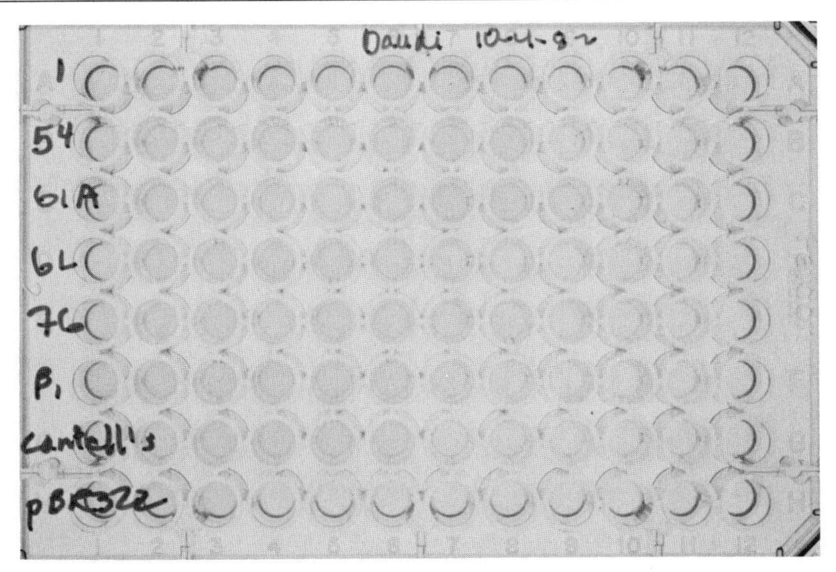

FIG. 2. Photograph of a color change assay plate. Rows A through F contained dilutions of lysates of *E. coli* cultures engineered to express various interferons, row G contained dilutions of 500 units/ml leukocyte interferon,[4] and row H contained dilutions of lysate of an *E. coli* culture that did not express interferon. IFN-α1, IFN-α54, IFN-α61A, IFN-α6L, and IFN-α76 are different alpha interferons with Cetus nomenclature, and IFN-β1 is beta interferon. The end point for leukocyte interferon can be seen between wells No. 6 and 7, for pBR322 between wells No. 3 and 4, and for IFN-α54 at well 9. See page facing p. 580 for color version of this figure.

bator to equilibrate with room air for 2 hr before scoring. The color change from pink (no growth) to yellow (growth) wells can be scored visually.

Comments

The maximum difference in absorbance between acidic and basic medium is at 560 nm (data not shown). Figure 1 shows absorbances and cell counts of a typical color change assay dilution series; the change in absorbance of the medium at 560 nm occurs over the same range of interferon dilutions as the change in number of cells per well. Visual scoring of color change rapidly determines the midpoint for change in absorbance, and in our hands is as accurate as scoring of the 50% cell death point in a viral cytopathic effect inhibition assay. Figure 2 shows a plate from a

[4] K. Cantell, S. Hirvonen, H.-L. Kauppinen, and G. Myllylä, this series, Vol. 78, p. 29.

color change assay. We found that it was difficult to increase the number of cells per well and decrease the incubation time without some loss of pink color in the "no growth" wells due to metabolism of the cells. It is possible to decrease the number of cells per well, but the incubation time must then increase to insure a good yellow color in the growth wells. For titering large numbers of interferon samples for antigrowth activity, this assay substantially decreases the laboratory work load.

[77] A Sandwich Radioimmunoassay for Human IFN-γ

By BRUCE KELDER, ABBAS RASHIDBAIGI, and SIDNEY PESTKA

The use of two different monoclonal antibodies in a solid phase sandwich immunoassay for human leukocyte interferon has been described in an earlier report.[1] Here, we report the development of this technique for the detection and assay of human gamma interferon. The advantage of this solid phase immunoassay for gamma interferon is in reducing the assay time from 36 to 48 hr for the typical biological assay to 3 to 4 hr.

Principle of the Assay

The principle of the assay is identical to that described earlier.[1] Two monoclonal antibodies were selected which recognize a different epitope (antigenic determinant) of human IFN-γ and which bind to IFN-γ simultaneously without mutual interference. One antibody (Ab-1) is immobilized on a solid surface. An interferon solution is incubated with this immobilized Ab-1, then the second antibody (Ab-2), labeled with ^{125}I or with a covalently conjugated enzyme, is added. Binding of Ab-2 is dependent upon and proportional to the interferon bound by the immobilized Ab-1. After incubation and washing, the interferon concentration is determined by measuring the amount of Ab-2 bound. A standard curve is established with known concentrations of IFN-γ.

Selection of Suitable Pair of Monoclonal Antibodies

All hybridomas secreting monoclonal antibodies to Hu-IFN-γ were obtained and subcloned as previously described.[2] Since the radioimmu-

[1] T. Staehelin, C. Stähli, D. Hobbs, and S. Pestka, this series, Vol. 79, p. 589.

[2] T. Staehelin, B. Durrer, J. Schmidt, B. Takacs, J. Stocker, V. Miggiano, C. Stähli, M. Rubinstein, W. P. Levy, R. Hershberg, and S. Pestka, *Proc. Natl. Acad. Sci. U.S.A.* **78,** 1848 (1981).

noassay requires one immobilized monoclonal antibody to bind human IFN-γ, the selection of a suitable pair of monoclonal antibodies was carried out as follows. Each well of a 96-well polyvinylchloride (PVC) microtiter plate (Dynatech Laboratories, Alexandria, VA) was coated with 50 μl (or 100 μl) of affinity purified goat anti-mouse immunoglobulins (Cappel Laboratories) at a concentration of 10 μg/ml overnight at room temperature in a humidified chamber. The following day, the solution was removed and the wells filled with solution A (10% fetal bovine serum, 160 μg/ml human IgG in phosphate-buffered saline, PBS) to block all protein binding sites. After 15 min, this solution was removed and the wells were rinsed three times with phosphate-buffered saline. Tissue culture fluid from various hybridomas producing monoclonal antibodies to human IFN-γ (as determined by antibody binding to IFN-γ bound to wells of a polyvinylchloride microliter plate[2]) was then added to each well (50 μl/well) and incubated for 2 hr at room temperature. Tissue culture fluids were then removed and the wells rinsed three times with PBS. A solution containing 1% BSA plus Hu-IFN-γ which had been radiolabeled with [^{35}S]methionine[3] was then added to each well (50 μl/well) so that each well contained 1000 cpm (approximately 10 units). The plate was incubated for 2 hr at room temperature. The wells were then rinsed three times with PBS and the individual wells were then placed in Bray's scintillation fluid[4] and radioactivity determined in a Beckman LS7800 scintillation spectrometer. Those monoclonal antibodies which bound Hu-IFN-γ were then purified from ascitic fluid as described.[5] A summary of the monoclonal antibodies to Hu-IFN-γ obtained and their neutralizing ability is given in Table I.

Several of the monoclonal antibodies were labeled with ^{125}I by the chloramine-T method.[1,6] Each ^{125}I-labeled monoclonal antibody (as Ab-2) was then screened in a sandwich radioimmunoassay with the other unlabeled antibodies as the immobilized Ab-1. For the assay described below, two monoclonal antibodies that provided a suitable sandwich immunoassay were chosen for use: monoclonal antibody M-Ab-γ-123 as Ab-1 and

[3] Human immune interferon was labeled with [^{35}S]methionine in *E. coli* harboring an expression plasmid with the IFN-γ gene under P_L promoter control similar to the constructions described by Remaut *et al.* and Crowl in chapters [54] and [55], respectively, in this volume. The labeling was performed for 1 min after raising the temperature of the incubation mixture from 30 to 42° for 25 min. The [^{35}S]methionine-labeled interferon was then purified over a monoclonal antibody column as described by Kung *et al.* (this volume [29]).

[4] G. A. Bray, *Anal. Biochem.* **1,** 279.

[5] T. Staehelin, D. Hobbs, H.-F. Kung, C.-Y. Lai, and S. Pestka, *J. Biol. Chem.* **256,** 9750 (1981).

[6] F. C. Greenwood and W. M. Hunter, *Biochem. J.* **89,** 114 (1963).

TABLE I

MONOCLONAL ANTIBODIES TO Hu-IFN-γ[a]

Monoclonal antibody number	cpm [^{35}S]IFN-γ bound	Neutralizing antibody	Monoclonal antibody number	cpm [^{35}S]IFN-γ bound	Neutralizing antibody
6	<0	No	175	174	No
7	<0	No	180	138	No
10	54	No	181	148	No
15	169	No	184	16	No
17	<0	No	185	21	No
19	223	No	186	<0	No
21	193	No	189	36	No
23	<0	No	190	65	No
29	<0	No	191	<0	No
33	<0	No	194	<0	No
36	<0	No	197	112	No
37	74	No	198	<0	No
41	172	No	203	<0	No
45	<0	No	204	<0	No
52	<0	No	206	179	No
54	<0	No	207	97	No
60	150	No	211	<0	No
62	187	No	214	<0	No
65	219	No	218	48	No
69	164	Yes	220	141	Yes
70	1	No	226	<0	No
73	168	Yes	252	179	No
79	<0	No	254	<0	No
80	<0	No	255	210	No
82	<0	No	260	<0	No
87	<0	No	261	88	No
103	<0	No	264	150	No
105	38	No	268	<0	No
106	<0	No	269	137	No
107	<0	No	276	56	No
113	155	Yes	282	180	No
114	154	No	283	9	No
123	208	No	284	<0	Yes
127	186	No	285	163	No
154	<0	No	286	<0	No
168	57	No	287	117	No

[a] Summary of monoclonal antibodies to human gamma interferon. Tissue culture fluids from hybridomas prior to subcloning were screened as described in the text by a solid phase immunoassay dependent on binding monoclonal antibody to Hu-IFN-γ bound to the wells. This assay detects low as well as high affinity antibodies.[11] By testing each monoclonal antibody bound to the plate for its ability to bind [^{35}S]methionine-labeled Hu-IFN-γ, the higher affinity antibodies can be detected.[11] Only 26 of the original 72 monoclonal antibodies bound $\geq$100 cpm. Of the five neutralizing antibodies, four were of the high affinity type. Values <0 represent negative numbers due to subtraction of the blank.

monoclonal antibody M-Ab-γ-127A as Ab-2. For the assay, 10 μg of M-Ab-γ-127A was labeled with ^{125}I by the chloramine-T method[6] to a specific radioactivity of 97 μCi/μg protein (215,000 dpm/ng).

Coating of PVC-Microtiter Plate

Each well of a PVC microtiter plate was coated with 100 μl of M-Ab-γ-123 (10 μg/ml) in PBS) at room temperature overnight ($\geq$0.5 hr could be used) or at 4° for days or weeks in a humidified chamber. Before use, the solution of M-Ab-γ-123 was removed and the wells filled with Solution A for at least 15 min to block all protein-binding sites. The wells were then rinsed three times with PBS.

Interferon and Interferon Titrations

Purified human IFN-γ was prepared as reported.[7] Antiviral interferon assays were performed by a cytopathic effect inhibition assay with human amniotic WISH cells.[8] The laboratory standard of unfractionated human immune interferon was titrated against the reference standard of human IFN-γ (Gg23-901-530) supplied by the Antiviral Substances Program of the National Institute of Allergy and Infectious Diseases, National Institutes of Health, Bethesda, Maryland.

Assay Procedure

A PVC-microtiter plate is coated with 100 μl of monoclonal antibody M-Ab-γ-123 at 10 μg/ml of protein in PBS at room temperature overnight, protein-binding sites are blocked with solution A, and the wells rinsed three times with PBS as described above. Dilutions of Hu-IFN-γ in 100 μl of PBS containing 1 mg/ml BSA are added to each well and incubated for 2 hr at room temperature. The wells are then rinsed three times with PBS and 100 μl of ^{125}I-labeled M-Ab-γ-127A (about 300,000 cpm, 2.5 ng) in PBS containing 1 mg/ml BSA is placed in each well and incubated for an additional 2 hr at room temperature. The wells are then rinsed three times with PBS and the individual wells are counted in a Beckman 300 gamma scintillation spectrometer.

The results of the assay are shown in Fig. 1. The assay is sensitive enough to detect 20 pg (0.3 units) of Hu-IFN-γ in a volume of 100 μl with a linear range of counts bound between 20 pg (0.3 units)/100 μl and 1392 pg

[7] H.-F. Kung, Y.-C. E. Pan, J. Moschera, K. Tsai, E. Bekesi, M. Chang, H. Sugino, and S. Honda, this volume [29].

[8] P. C. Familletti, S. Rubinstein, and S. Pestka, this series, Vol. 78, p. 387.

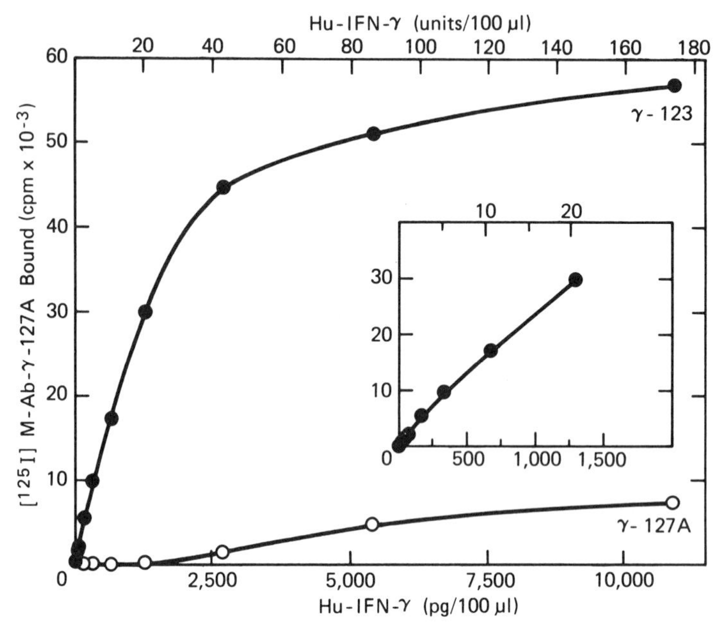

FIG. 1. Radioimmunoassay for human gamma interferon. The assay was performed as described in the text and the data plotted on a linear scale.

(22 units)/100 μl. Interferon concentrations of up to 2714 pg (43 units)/100 μl could be estimated with reasonable accuracy. Higher interferon concentrations bound maximal amounts of ^{125}I-labeled M-Ab-γ-127A. The need for monoclonal antibodies which recognize different epitopes on the interferon molecule is also illustrated in Fig. 1. The use of monoclonal antibody M-Ab-γ-127A as both the immobilized Ab-1 bound to the PVC-plate as well as the ^{125}I-labeled Ab-2 shows a major reduction in the amount of ^{125}I-labeled M-Ab-γ-127A bound. The small amount of binding which occurs when the same monoclonal antibody is used as the collecting Ab-1 and the detecting Ab-2 (Fig. 1, unfilled circles) may be due to the presence of Hu-IFN-γ dimers in the solution.[9,10]

Another aspect of this radioimmunoassay as described previously[11] is that it will only recognize active Hu-IFN-γ molecules as shown in Table II. After incubation for 5 min at 65° or at pH 2.0, the IFN-γ is nearly

[9] S. Pestka, B. Kelder, and S. J. Tarnowski, this volume [78].

[10] S. Pestka, B. Kelder, D. K. Tarnowski, and S. J. Tarnowski, *Anal. Biochem.* **132,** 328 (1983).

[11] S. Pestka, B. Kelder, J. A. Langer, and T. Staehelin, *Arch. Biochem. Biophys.* **224,** 111 (1983).

TABLE II
RECOGNITION OF Hu-IFN-γ AFTER HEAT AND ACID INACTIVATION[a]

Sample	Treatment	cpm ^{125}I bound	Antiviral activity (units/ml)
Hu-IFN-γ (17 ng/ml)	None	16,550	1,482
Hu-IFN-γ (17 ng/ml)	65°, 5 min	390	0.11
Hu-IFN-γ (17 ng/ml)	pH 2.0, 5 min	370	0.11

[a] Human gamma interferon at a concentration of 17 ng/ml was either left untreated, heated at 65° for 5 min, or incubated at pH 2.0 for 5 min. Samples were then assayed in the radioimmunoassay as described in the text. A background of 965, 730, and 775 cpm was subtracted from each value, respectively, to provide the data shown in the table.

completely inactivated as shown by the virtually total loss of ^{125}I-labeled M-Ab-γ-127A counts bound and loss of antiviral activity. The assay is specific for Hu-IFN-γ. Hu-IFN-α or Hu-IFN-β are not detected in this assay (data not shown).

Concluding Comments

The interferon assay shown in Fig. 1 can detect 20 pg (0.3 units) of Hu-IFN-γ in the 0.10 ml reaction mixture. The assay can be made more sensitive by using M-Ab-γ-127A with a higher specific activity or by derivatizing M-Ab-γ-127A with biotin, peroxidase, or β-galactosidase.

Since a pair of monoclonal antibodies are used in this assay, only those IFN-γ molecules that are recognized by both monoclonal antibodies are detected. Thus, Hu-IFN-α, Hu-IFN-β, or Mu-IFN-γ are not detected. A similar sandwich immunoassay for Hu-IFN-γ was described by Chang *et al.*[12]

This radioimmunoassay (or modification thereof, if desired) will prove extremely useful for monitoring IFN-γ during purification or for surveillance of IFN-γ levels in patients for pharmacokinetic studies.

[12] T.-W. Chang, S. McKinney, V. Liu, P. C. Kung, J. Vilček, and J. Le, *Proc. Natl. Acad. Sci. U.S.A.* **81,** 5219 (1984).

[78] Procedures for Measurement of Interferon Dimers and Higher Oligomers by Radioimmunoassay

By SIDNEY PESTKA, BRUCE KELDER, and S. JOSEPH TARNOWSKI

Radioimmunoassays or enzymeimmunoassays have proved to be very useful for the detection and assay of proteins and other antigens and haptens.[1,2] However, these assays do not distinguish between monomer and oligomer forms. With the availability of monoclonal antibodies that recognize a single epitope of a protein (or other antigen), it is possible to devise a simple immunoassay that will measure oligomers (dimer, trimer, etc.) but not monomers. In the usual sandwich-type immunoassay with monoclonal antibodies that recognize different epitopes, one antibody is bound to a solid support to which the antigen binds (Fig. 1). A second monoclonal antibody labeled with ^{125}I or coupled to an enzyme or other group that can be detected readily is used to measure the amount of protein or other antigen present. We have described such a convenient assay for human leukocyte interferon.[3] Appropriate polyclonal antibodies can also be used in such immunoassays. In particular, polyclonal antibodies generated in response to different regions of the antigen would be most useful. These assays detect all the forms of an antigen (monomer, dimer, trimer, and higher oligomers or aggregates). In this report, we describe an assay that detects oligomers of an antigen A (A_n where $n \geq 2$).

Principle of the Assay

In the case of the sandwich immunoassay described in Fig. 1, two separate antibodies recognizing different epitopes are used. If, in fact, the second labeled antibody is identical to the first, then the labeled antibody will not be able to bind to the antigen because its site of binding is already occupied. However, if the antigen were present in oligomeric form (Fig. 2), then there may be a free site available to which the labeled antibody can bind. Dimers and any oligomers will be detected by the assay. Since the larger oligomers have more sites available than the smaller ones, more

[1] S. A. Berson and R. S. Yalow, eds., "Methods in Investigative and Diagnostic Endocrinology," Vol. 2, Part 1. North-Holland Publ., Amsterdam, 1973.

[2] S. A. Berson and R. S. Yalow, *J. Clin. Invest.* **38,** 1996 (1959).

[3] T. Staehelin, C. Stähli, D. S. Hobbs, and S. Pestka, this series, Vol. 79, p. 589.

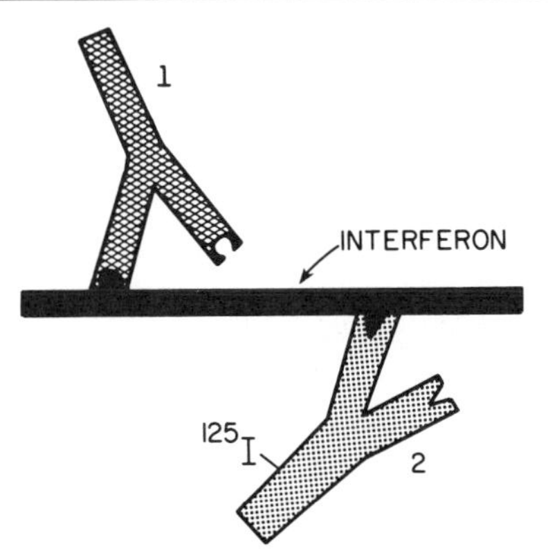

FIG. 1. Schematic diagram of a sandwich radioimmunoassay with two monoclonal antibodies. The monoclonal antibodies bind to different epitopes of the interferon molecule.

label should bind to the higher oligomers. In principle, the number of sites available to the labeled antibody in an oligomer of n subunits is equal to $n - 1$. In practice, however, it can be expected to be less because some of the oligomers may bind to the first antibody through more than one site. Since many monoclonal antibodies against IFN-αA have been isolated,[4] a sandwich radioimmunoassay that would detect the presence of oligomers, but not the monomer, was developed by relying on the ability of a monoclonal antibody to react with one epitope of the IFN-αA molecule.

Monoclonal Antibody and Interferon Preparations

Monoclonal antibodies LI-1 and LI-9 were isolated[4] and prepared[5] as previously described. Recombinant human leukocyte A interferon (IFN-αA) was prepared as reported[5,6] with minor modifications. Monomer, dimer, and oligomer fractions of IFN-αA were separated by gel filtration on Sephadex G-75 as described.[7] Monoclonal antibody to human leukocyte interferon (LI-1) was labeled with ^{125}I as described.[3]

[4] T. Staehelin, B. Durrer, J. Schmidt, B. Takacs, J. Stocker, V. Miggiano, C. Stähli, M. Rubinstein, W. P. Levy, R. Hershberg, and S. Pestka, *Proc. Natl. Acad. Sci. U.S.A.* **78,** 1848 (1981).

[5] T. Staehelin, D. S. Hobbs, H.-F. Kung, C.-Y. Lai, and S. Pestka, *J. Biol. Chem.* **256,** 9750 (1981).

[6] T. Staehelin, D. S. Hobbs, H.-F. Kung, and S. Pestka, this series, Vol. 78, p. 505.

[7] S. Pestka, B. Kelder, D. K. Tarnowski, and S. J. Tarnowski, *Anal. Biochem.* **132,** 328 (1983).

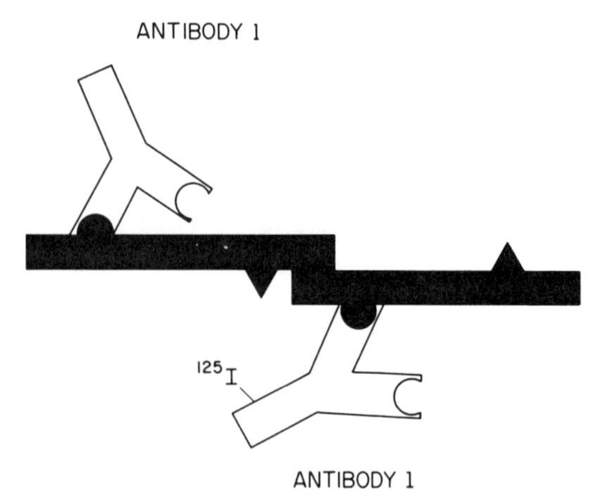

FIG. 2. Schematic diagram of a sandwich radioimmunoassay of a dimer with a monoclonal antibody that recognizes a single epitope. A dimer of the molecules contains two sites of the epitope for recognition. Thus, one monoclonal antibody (unlabeled and attached to a solid support) can be used for collection of the dimers and the same monoclonal antibody labeled with ^{125}I can be used for binding to the other epitopic site. Thus, this assay could detect dimers or higher oligomers, but not monomers that contain a single epitopic site.

Radioimmunoassay of Dimers and Higher Oligomers

Each well of a PVC-microtiter plate is coated with 100 μl of LI-1 and LI-9 (10 μg/ml in PBS) at room temperature overnight or at 4° for several days or weeks in a humidified chamber. Before use, the solution of monoclonal antibody is removed and the wells are filled with Solution I [phosphate-buffered saline (PBS) containing 10% fetal calf serum and 160 μg/ml human IgG] for at least 15 min and then washed four times with PBS. A total of 100 μl per well of interferon test solution in PBS containing 1 mg/ml of bovine serum albumin (BSA) is added and incubated at room temperature for 2 hr. The solution is removed and 100 μl of ^{125}I-labeled LI-1 (approximately 200,000 cpm/50 ng) in Solution I is added to each well. The plates are held at room temperature for 2 hr and then washed four times with PBS. Radioactivity in the individual wells is determined with a gamma scintillation spectrometer. For a convenient and sensitive radioimmunoassay or enzymeimmunoassay of human leukocyte interferon, two different monoclonal antibodies (for example, LI-9 and ^{125}I-labeled LI-1) are used in the sandwich assay.[3] For assay of oligomers, monoclonal antibody LI-1 and ^{125}I-labeled LI-1 are used in succession as described above.

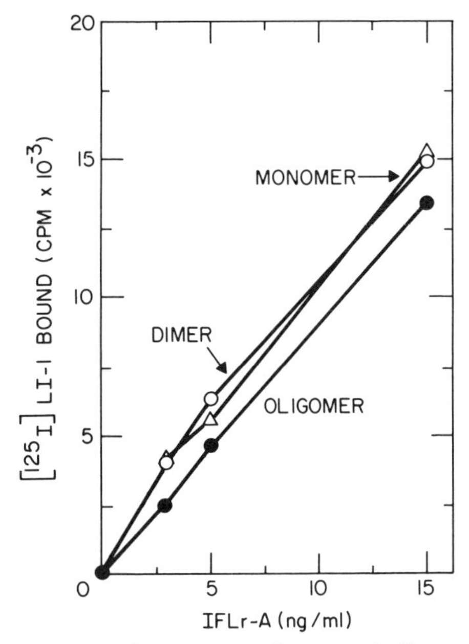

FIG. 3. Radioimmunoassay of monomers, dimers, and oligomers with two different monoclonal antibodies. Monoclonal antibody LI-9, bound to the wells, was incubated with a solution of monomer, dimer, or oligomer fractions of IFN-αA. After the solution containing interferon was removed, ^{125}I-labeled monoclonal antibody LI-1 was added to each well.[3,7] The details of this procedure and determination of ^{125}I-labeled LI-1 bound were described.[3] △, Monomer; ○, dimer; ●, oligomer.

Concluding Comments

The usual two-step radioimmunoassay of IFN-αA with monoclonal antibody LI-9 bound to the polyvinylchloride plate and ^{125}I-labeled LI-1 as the second antibody (Fig. 1) detects monomer, dimer, and oligomer fractions approximately equally (Fig. 3). When monoclonal antibody LI-1 is bound to the plate and ^{125}I-labeled LI-1 used in the second step, only dimers and oligomers are detectable, but not monomer IFN-αA (Fig. 4). The trimer binds more ^{125}I-labeled LI-1 than the dimer. Since the trimer has two and the dimer one available site for ^{125}I-labeled LI-1, the ratio of counts per unit weight of trimer to that of dimer should be 1.3 if binding to LI-1 attached to the plate is through a single site only. The values in the figure show greater binding to oligomers than to the dimer as expected.

IFN-αA produced in bacteria yielded both monomer and oligomers on purification.[5–7] In addition, aggregate forms of human leukocyte interferon have long been observed. For this reason, gel filtration in the pres-

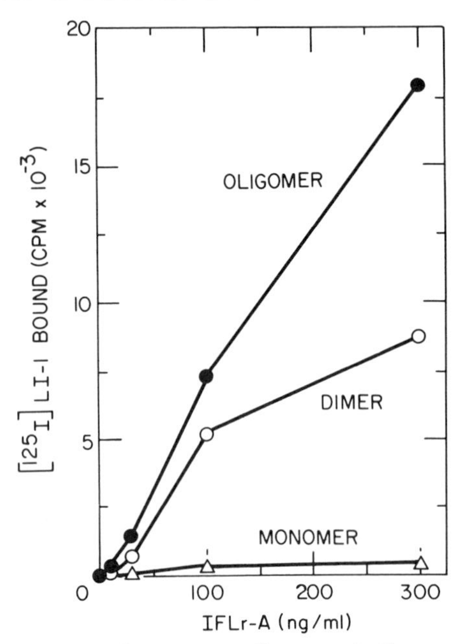

FIG. 4. Radioimmunoassay of monomers, dimers, and oligomers with a single monoclonal antibody. Monoclonal antibody LI-1 (unlabeled), bound to the wells, was incubated with monomer, dimer, or oligomeric fractions of IFN-αA. After the solution containing interferon was removed, ^{125}I-labeled monoclonal antibody LI-1 was added to each well. After incubation, the binding of ^{125}I-labeled LI-1 to the wells was determined. The points plotted represent the averages of duplicate determinations. The average standard deviation for all the values was 7.9% of the counts. The blank subtracted from each value was 133 cpm (an average of six determinations with a standard deviation of ±10). △, Monomer; ○, dimer; ●, oligomer.

ence of urea was used as one of the first steps in purification of the interferon produced by buffy coat cells.[8–10] Oligomeric forms of human fibroblast interferon have been reported as well.[11,12] For this reason, we examined the dimer content of concentrated crude human leukocyte in-

[8] M. Rubinstein, S. Rubinstein, P. C. Familletti, R. S. Miller, A. A. Waldman, and S. Pestka, *Proc. Natl. Acad. Sci. U.S.A.* **76,** 640 (1979).

[9] M. Rubinstein, W. P. Levy, J. A. Moschera, C.-Y. Lai, R. D. Hershberg, R. T. Bartlett, and S. Pestka, *Arch. Biochem. Biophys.* **210,** 307 (1981).

[10] K. C. Zoon, M. E. Smith, P. J. Bridgen, D. Zur Nedden, and C. B. Anfinsen, *Proc. Natl. Acad. Sci. U.S.A.* **76,** 5601 (1979).

[11] H.-J. Friesen, S. Stein, M. Evinger, P. C. Familletti, J. Moschera, J. Meienhofer, J. Shively, and S. Pestka, *Arch. Biochem. Biophys.* **206,** 432 (1981).

[12] E. Knight, Jr. and D. Fahey, *J. Biol. Chem.* **256,** 3609 (1981).

terferon preparations prepared by the procedure of Cantell *et al.*[13] as well as that produced by our procedures.[8,9,14–16] In all these cases, no oligomer forms were detected (data not shown).

The origin of IFN-αA oligomers is uncertain since two disulfide bonds have been assigned to the IFN-αA monomer.[17] Some of these oligomers appear to be crosslinked by disulfide bonds since 2-mercaptoethanol is required for their conversion to the monomer. It is likely that cysteines 1 and 98 are involved in intermolecular disulfide linkages in the oligomeric forms of IFN-αA.[7] This is consistent with the observation that the disulfide bond between residues 1 and 98 is not essential for maximal antiviral activity.[18–20] The use of the oligomer-specific radioimmunoassay could be used to demonstrate whether oligomers are contained in the recombinant *E. coli* cells or if they are generated during the purification procedure. In addition, the oligomer assay could be used to determine the form of interferon circulating in serum of patients during clinical trials.

As described above, the assay can discriminate between monomers and dimers of interferon. In principle, the assay could be used for all sorts of polymeric substances. For the assay to discriminate between monomers and oligomers, it is necessary that two or more identical epitopes be available to the antibody in the oligomeric forms of the molecules.

[13] K. Cantell, S. Hirvonen, H.-L. Kauppinen, and G. Myllylä, this series, Vol. 78, p. 29.

[14] A. A. Waldman, R. S. Miller, P. C. Familletti, S. Rubinstein, and S. Pestka, this series, Vol. 78, p. 39.

[15] R. D. Hershberg, E. G. Gusciora, P. C. Familletti, S. Rubinstein, C. A. Rose, and S. Pestka, this series, Vol. 78, p. 45.

[16] P. C. Familletti, L. Costello, C. A. Rose, and S. Pestka, this series, Vol. 78, p. 83.

[17] R. Wetzel, L. J. Perry, D. A. Estell, N. Lin, H. L. Levine, B. Slinker, F. Fields, M. J. Ross, and J. Shively, *J. Interferon Res.* **1,** 381 (1981).

[18] R. Wetzel, H. L. Levine, D. A. Estell, S. Shire, J. Finer-Moore, R. M. Stroud, and T. A. Bewley, *in* "Chemistry and Biology of Interferons: Relationship to Therapeutics" (T. C. Merigan, R. M. Friedman, and C. F. Fox, eds.), p. 376. Academic Press, New York, 1982.

[19] J. A. Langer and S. Pestka, this volume [34].

[20] S. J. Tarnowski, S. K. Roy, R. A. Liptak, D. K. Lee, and R. Y. Ning, this volume [23].

Section X

Biology of Interferon Action

[79] Selection and Screening of Transformed NIH3T3 Cells for Enhanced Sensitivity to Human Interferons α and β

By VINCENT JUNG and SIDNEY PESTKA

Interferon binding, a prerequisite for its pleiotropic biological effects, is saturable, reversible, and specific for a high affinity receptor.[1] The genetic linkage of interferon sensitivity in human cells to chromosome 21 was originally reported by Tan *et al.*[2] who observed that mouse–human somatic cell hybrids that segregated with human chromosome 21 responded to human interferon as measured by antiviral activity.[3] Similarly, the chromosome governing sensitivity to murine interferon was delineated by somatic cell hybrids into a Chinese hamster[4] or human[5] background; murine interferon sensitivity was enhanced when the segregants carried mouse chromosome 16. The human chromosome 21 linkage was reaffirmed when it was proven that the antiviral,[6–8] the antigrowth,[9] as well as the specific binding of ^{125}I-labeled human interferon αA (IFN-αA)[10] were commensurate with gene dosage in studies with human cells that were monosomic, disomic, and trisomic for chromosome 21. A line trisomic for only the distal half of the long arm of chromosome 21 that was also enhanced for interferon sensitivity further narrowed the locus of the purported receptor.[11,12] Additional data supporting the assignment of interferon sensitivity to chromosome 21 as well as to a receptor came with

[1] For reviews, see M. Aguet and K. E. Mogensen, *Interferon* **5,** 1 (1983); K. C. Zoon and H. Armheiter, *Pharmacol. Ther.* **24,** 259 (1985); and Section VI, this volume.
[2] Y. H. Tan, J. Tischfield, and F. H. Ruddle, *J. Exp. Med.* **137,** 317 (1973).
[3] D. L. Slate, L. Shulman, J. B. Lawrence, M. Revel, and F. H. Ruddle, *J. Virol.* **25,** 319 (1978).
[4] D. R. Cox, L. B. Epstein, and C. J. Epstein, *Proc. Natl. Acad. Sci. U.S.A.* **77,** 2168 (1980).
[5] P. F. Lin, D. L. Slate, F. C. Lawyer, and F. H. Ruddle, *Science* **209,** 285 (1980).
[6] Y. H. Tan, E. L. Schneider, J. Tischfield, C. J. Epstein, and F. H. Ruddle, *Science* **186,** 61 (1974).
[7] J. Weil, G. Tucker, L. B. Epstein, and C. J. Epstein, *Hum. Genet.* **65,** 108 (1983).
[8] C. Chany, M. Vignal, P. Couillin, N. V. Cong, J. Boué, and A. Boué, *Proc. Natl. Acad. Sci. U.S.A.* **72,** 3129 (1975).
[9] Y. H. Tan, *Nature (London)* **260,** 141 (1976).
[10] C. J. Epstein, N. H. McManus, and L. B. Epstein, *Biochem. Biophys. Res. Commun.* **107,** 1060 (1982).
[11] Y. H. Tan and A. E. Green, *J. Gen. Virol.* **32,** 153 (1976).
[12] L. B. Epstein and C. J. Epstein, *J. Infect. Dis.* **133,** Suppl., A56 (1976).

polyclonal[13] and monoclonal[14,15] antisera raised against the mouse–human somatic cell hybrids that blocked the antiviral response to human cells. Furthermore, ^{125}-labeled IFN-αA[16] binds specifically to WA17 cells (from W. T. Brown), a trisomic variant of the mouse–human somatic cell hybrid (J. Langer, unpublished data).

The logical extension of the somatic cell hybrid work is to carry over the gene(s) responsible for cell sensitivity to human interferon into a murine background by DNA-mediated gene transfer. This chapter describes a method to select transformed cells that evince an enhanced sensitivity to human interferon.

Principle of Selection

This positive selection procedure is based on the antiviral properties of interferons. Mouse cells containing human chromosome 21 exhibit less sensitivity (by a factor of 100) to human interferon[3] than human cells, but are generally more sensitive than the parental mouse cells. In comparing the Chinese hamster parent line to the hamster–mouse hybrid containing murine chromosome 16, a 4- to 8-fold enhanced sensitivity of the hybrid cells to murine interferon was observed (see Fig. 1 of CPE inhibition assay in Cox *et al.*[4]).

Theoretically, in a mixed population of cells with various sensitivities to interferon, the subset that is most sensitive to interferon is more likely to survive a viral challenge under low interferon concentrations. That is, a mouse cell containing one or more human genes that increase its sensitivity to human interferon would more likely survive a viral selection in the presence of human interferon than the parental mouse cells. One might expect a bank of mouse NIH3T3 cells transformed with human DNA to represent such a mixed population.

The first step of the procedure entails the generation of a representative bank of murine cells transfected with human DNA. Once the bank is established, the cells are exposed to human interferon. This is followed by selection based on rescue from a viral infection with, for example, vesicular stomatitis virus. The surviving cells are isolated and their sensitivity to human and murine interferons are assessed in comparison to the parental murine cells.

[13] M. Revel, D. Bash, and F. H. Ruddle, *Nature (London)* **260,** 139 (1976).

[14] M. E. Kamarck, D. L. Slate, P. D'Eustachio, B. Barel, and F. H. Ruddle, *Fed. Proc., Fed. Am. Soc. Exp. Biol.* **40,** 1051 (1981).

[15] D. L. Slate and F. H. Ruddle, this series, Vol. 79, 536.

[16] J. Langer and S. Pestka, this volume [44].

Transformation of NIH3T3 Cells

Cells, Plasmids, and Reagents

NIH3T3 cells, a mouse fibroblast line, was maintained in MEM (F11) supplemented with 10% FCS (Gibco), 2× glutamine (Gibco), and 50 μg/ml gentamicin (Schering) unless otherwise specified. GM2504 cells, a human fibroblast line trisomic for chromosome 21, was obtained from NIGMS Human Genetic Cell Repository, Camden, NJ, and was maintained in DMEM, 10% FCS, and 50 μg/ml gentamicin. WaVR4dF94a cells,[3] a somatic cell hybrid containing human chromosome 21 in a mouse background, and the parental A9 cells were obtained from Lester Shulman and Frank Ruddle. These cells were maintained in 10% FCS/MEM with 50 μg/ml gentamicin. Plasmid pSV2-neo,[17] which confers resistance to the aminoglycoside antibiotic G418 was used as a source of the antibiotic-resistance gene and was isolated by the alkaline lysis procedure.[18] The lysate from a 1 liter preparation was precipitated with ethanol, resuspended in 10 ml of TE buffer (10 m*M* Tris · Cl, pH 8.0, 1.0 m*M* EDTA), treated for 0.5 hr with 50 μg of RNAse A followed by 2 hr with 100 μg/ml of proteinase K (Boehringer Mannheim), phenol extracted, reprecipitated with ethanol, and fractionated on a Pharmacia Sephacryl S-1000 column.[19]

Antibiotic G418 was purchased from Gibco and dissolved in 1 *M* Hepes (Sigma), pH 7.3, at 50 mg of active potency per ml. Crystal violet for fixing and staining was prepared as a 0.5% (w/v) solution in 70% methanol. All restriction enzymes were from New England Biolabs. Tissue culture dishes and assay plates were from either Falcon or Corning.

Transformation

High-molecular-weight nuclear DNA from human GM2504 cells was isolated according to Pellicer *et al.*[20] This line was chosen for its trisomy in chromosome 21. Transformation was performed by the calcium phosphate procedure as described by Graham and Van der Eb[21] as modified by Wigler.[22] NIH3T3 cells that have undergone no less than five (1 to 10 split) passages from a freshly resurrected culture are preferred. Cells (5×10^5)

[17] P. J. Southern and P. Berg, *J. Mol. Appl. Genet.* **1,** 327 (1982).
[18] H. C. Birnboim and J. Doly, *Nucleic Acids Res.* **7,** 1513 (1979).
[19] M. Bywater, R. Bywater, and L. Hellman, *Anal Biochem.* **132,** 219 (1983).
[20] A. Pellicer, M. Wigler, R. Axel, and S. Silverstein, *Cell* **14,** 133 (1978).
[21] F. L. Graham and A. J. Van der Eb, *Virology* **52,** 456 (1973).
[22] M. Wigler, A. Pellicer, S. Silverstein, and R. Axel, *Cell* **14,** 725 (1978).

were seeded into each of eight 100-mm (Corning #25020) tissue culture dishes. Transformants were prepared with a mixture of DNA which is composed of 20 μg of DNA from human GM2504 cells and 1 μg of *Eco*RI-digested pSV2-neo DNA per dish.

The medium was changed after a 4 hr exposure to the DNA precipitate. Twenty four hours later the cells were trypsinized, collected, and diluted 3-fold into 24 100-mm dishes and then incubated for 2 hr with intermittent rocking to ensure adherence with a uniform cell distribution. The medium was removed and 5 ml of medium supplemented with 500 μg/ml of antibiotic G418 was added to each plate. The medium with antibiotic was changed after 24 hr and every 48 hr thereafter. This step is needed in order to accelerate the rate of selection because of the indirect density dependent efficacy of antibiotic G418.

To determine the efficiency of selection, 10,000 unselected transformants were plated onto Falcon 60-mm dishes. After allowing 2 hr for the cells to adhere, 2.5 ml of selective medium was added. Colonies were counted after 2 weeks by crystal violet staining. Efficiencies of transformation greater than or equal to 2.5×10^{-3} were routinely obtained by these procedures.

The number of transformants necessary to represent the human genome is dependent on the average size of the integrated fragments.[23–25] Assuming that 1000 to 3000 kb of DNA are integrated per transformant, 2000 to 6000 transformants are necessary to represent the human genome of 6×10^6 kb once. Our library, which consisted of three independent transformations, contained at least 30,000 transformants. The library was maintained in 50 μg/ml of antibiotic G418.

A library represents a heterogeneous population of cells in which population dominance can be gained by cells that exhibit such advantageous phenotypes as a shorter doubling time or enhanced adherence. It is therefore important to culture and freeze stocks of the library as soon as possible to minimize over representation by any individual clones. In our hands, the somatic cell hybrid WaVR4dF94a containing human chromosome 21 exhibits a greater doubling time than does its A9 parent. We therefore never use a library that has gone through more than 5 (1 to 10 split) passages from a freshly resurrected culture given the theoretical possibility that the clone we are seeking might be lost.

[23] M. Perucho, D. Hanahan, and M. Wigler, *Cell* **22,** 309 (1980).

[24] D. M. Robins, S. Ripley, A. S. Henderson, and R. Axel, *Cell* **23,** 29 (1981).

[25] P. Kavathas and L. A. Herzenberg, *Proc. Natl. Acad. Sci. U.S.A.* **80,** 524 (1983).

Endpoint Determination

Viruses and Reagents

Vesicular stomatitis virus (Indiana strain) was isolated as described by Goorha.[26] A titer of 5×10^8 PFU/ml was obtained by the standard plaque assay on mouse L cells.

Recombinant human leukocyte (IFN-αA)[27] and fibroblast (IFN-β) interferon[28] produced in *Escherichia coli* was obtained as described, and was titered on both WISH and MDBK cells as described.[29] IFN-αA and IFN-β interferon concentrations were 1.2×10^8 units/ml (1.8×10^8 units/mg) and 3×10^7 units/ml (2.3×10^7 units/mg), respectively.

Murine interferon was obtained by induction of mouse L cells[30] with poly(I) · poly(C) and had a titer of 10^3 units/ml.

Low melting point agarose was from Seaplaque. A mixture of 0.8% agarose containing 2% FCS/MEM was made by adding equal volumes of a 1.6% low melting point agarose autoclaved in water to 2× MEM supplemented with 4% FCS, 4× glutamine, and 100 μg/ml gentamicin. Phosphate-buffered saline (PBS) is Mg^{2+}/Ca^{2+}-free unless otherwise specified.

Determination of Endpoint Interferon Concentration for Viral Selection

The interferons have characteristic species specificities. Usually they exhibit a high degree of activity on cells from the species from which the interferon was obtained (homologous activity). Nevertheless, the activity of interferons on heterologous cells has long been noted.[31-33] Human IFN-αA[34] and IFN-β are active on murine NIH3T3 cells. It is necessary therefore to establish an endpoint on which selection of the transformants will be based. This endpoint should meet the simple criterion that at this concentration the majority of NIH3T3 cells will not be protected, but that

[26] R. M. Goorha, this series, Vol. 78, p. 309.

[27] T. Staehelin, D. S. Hobbs, H.-F. Kung, C.-Y. Lai, and S. Pestka, *J. Biol. Chem.* **256,** 9750 (1981).

[28] S. Pestka, B. Kelder, P. C. Familletti, J. A. Moschera, R. Crowl, and E. S. Kempner, *J. Biol. Chem.* **258,** 9706 (1983).

[29] P. C. Familletti, S. Rubinstein, and S. Pestka, this series, Vol. 78, p. 387.

[30] H. B. Levy, this series, Vol. 78, p. 242.

[31] J. Desmyter, W. E. Rawls, and J. L. Melnick, *Proc. Natl. Acad. Sci. U.S.A.* **59,** 69 (1968).

[32] R. E. Levy-Koenig, R. R. Golgher, and K. Paucker, *J. Immunol.* **104,** 791 (1970).

[33] I. Gresser, M.-T. Bandu, D. Brouty-Boyé, and M. G. Tovey, *Nature (London)* **251,** 543 (1974).

[34] E. Rehberg, B. Kelder, E. G. Hoal, and S. Pestka, *J. Biol. Chem.* **257,** 11497 (1982).

a 2-fold increase in interferon concentration should lead to some observable sustainance of morphological integrity after virus challenge.

Because competition studies have indicated that leukocyte and fibroblast interferons share a common receptor,[35,36] a mixture of leukocyte and fibroblast interferons was used for the endpoint determination. The procedure is designed to mimic the eventual selection of the library.

In a 24-well plate (Falcon #3047) 20,000 NIH3T3 cells in 0.5 ml of medium are seeded into each well. After 2 hr, with intermittent rocking to ensure an even cell distribution, the medium is removed and fresh medium containing interferon is added at the following concentrations: 36,000 units/ml of Hu-IFN-β for well number 1, diluted 2-fold serially, and 2000 units/ml of Hu-IFN-αA for all wells. The rationale for maintaining a constant IFN-αA concentration is that this interferon only crossreacts with NIH3T3 cells at very high concentrations ($>10^5$ units/ml).[34] A constant level of IFN-αA is maintained for practicality while embracing the possibility that the transformed cell might be sensitive to human IFN-αA at this concentration. Human IFN-β exhibits activity on NIH3T3 cells at lower concentrations (about 10^3 units/ml) than for IFN-αA. The endpoint is therefore more dependent on IFN-β than IFN-αA.

The cells are incubated with interferon for 24 hr, after which the medium is aspirated and 50 μl of 2% FCS/MEM containing 10 PFU/cell of VSV is added and incubated with the cells for 1 hr. It is critical that at the time of infection the cells should be approximately 80% confluent; if not, the original seeding should be adjusted accordingly. While the cells are incubating with VSV, a set of solutions of 0.8% agarose containing 2% FCS/MEM is prepared which match the interferon concentrations of the wells. The mixture can be maintained as a liquid in a 37° bath until use. The inoculant containing VSV was removed from each well and 200 μl of the corresponding agarose/FCS/MEM/IFN mixture was added to the wells. Ten minutes at room temperature permitted the agarose to solidify after which the plate is incubated overnight for 24 hr.

Since the cell confluence is only 80%, cell viability can be readily approximated by scanning under a light microscope. Cells are judged terminally infected by the loss of fibroblastic morphology, by the loss of adhesion, and by lysis. The endpoint has been determined to be approximately 2000 units of IFN-αA and 1000 units of IFN-β. This procedure is repeated for every new preparation of interferon. At twice the endpoint concentration, islets of protected cells can be easily visualized.

[35] A. A. Branca and C. Baglioni, *Nature* (*London*) **294,** 768 (1981).
[36] A. R. Joshi, F. H. Sarkar, and S. L. Gupta, *J. Biol. Chem.* **257,** 13884 (1982).

Cells with enhanced sensitivity to human interferon should exhibit a higher probability of survival at the endpoint concentration. Because the procedure is not absolutely selective, there will be cells that escape VSV infection by chance or by the development of noninterferon-related resistance to VSV. The aim of the selection procedure is to optimize conditions for a transformant with increased sensitivity to human interferon to survive, thus enriching for transformants with enhanced sensitivity to human α and β interferon. The endpoint is the interferon concentration at which selection for all the transformants will take place.

Selection of Transformants

Two methods are outlined in detail below, with Method II being a simple, but effective variation of Method I.

Method I

1. Cells (10,000) were seeded into 100-mm dishes, maintained in 10 ml of medium containing 50 μg/ml of antibiotic G418 and 25% conditioned medium to promote growth of the cells at such low densities. Conditioned medium, obtained from flasks of confluent NIH3T3 cells, was filtered through a 0.2-μm filter (Nalgene #120-0020), resupplemented with an additional 10% FCS, and 2.5 ml added to 7.5 ml of 10% FCS/MEM in each dish. Theoretically, there should be sufficient cells in each dish so that each clone from the library is represented several times. At 80% confluence a 100-mm plate should contain about 3×10^6 cells. Accordingly each clone should grow to an average of 300 cells before being induced for protection by interferon and challenged by VSV.

2. One week later, when the cells have approximated 50% confluence, the medium was replaced with 5 ml of medium supplemented with 2000 units/ml of IFN-αA and 1000 units/ml of IFN-β.

3. After 24 hr, the dishes are infected with 10 PFU/cell in 1 ml of 2% FCS/MEM containing interferon and incubated at 37° for 1 hr with intermittent rocking. It is imperative that at the time of challenge the confluence of the plate be about 80% given the indirect density dependence of cells to VSV infectivity.

4. After 24 hr, as much of the inoculant as possible is aspirated without drying out the cells. The cells are then overlaid with 5 ml of 0.8% low melting point agarose containing 2% FCS/MEM and interferon. The function of the agarose is to prevent the diffusion of VSV from infected cells to neighboring cells. Such a step is critical because the VSV titers in an

infection can reach 10^9 PFU/ml in 24 hr and thus abrogate the interferon-induced protection.

5. After an additional 24 hr, the agarose layer was removed aided by soaking in 2 ml of PBS at 37° for 2 min. The dish was washed three times vigorously with PBS to remove floaters and debris and to drastically reduce the high titer of VSV that had accumulated. The dish was overlaid again with the same agarose/MEM/interferon mixture.

6. Once again after 24 hr, the agarose layer was removed, the dish rinsed vigorously twice with PBS, and each plate trypsinized and transferred to its own 60-mm gridded dish (Falcon #3030) with 2.5 ml IFN-supplemented medium. A gridded dish aids scanning for colonies under a light microscope. After 2 hr, cells that have maintained their integrity should adhere to the plate whereas infected cells will either float or have less adherence. Floaters and weakly adherent cells are removed by multiple vigorous washings with PBS. Again, the cells are overlaid with the 0.8% agarose/FCS/MEM/IFN mixture.

7. After 24 hr, the agarose is removed, the cells are trypsinized, resuspended in 2.5 ml IFN-supplemented media, and allowed to adhere to the same plate for 2 hr. Infected cells are then washed off with PBS.

8. Cells were maintained in 25% conditioned medium with interferon until survivors form colonies and until one is certain that the culture is VSV free. Medium is changed regularly twice each day for the first 3 days with 25% conditioned medium supplemented with interferon, and subsequently every other day. The medium is assayed regularly for VSV contamination by direct transfer of the medium onto normal NIH3T3 cells growing in a 96-well plate and scanning for cytopathic effect. The medium can be changed less frequently if there is consistently no VSV contamination. We have not found it necessary to use VSV neutralizing antibodies in the procedure because of the inherent beneficial selective presence of VSV against cells that might have escaped infection. We have found that frequent washing and the presence of human interferon was sufficient for obtaining the clones we sought. Individual colonies were isolated by cloning cylinders.

Method II

This method represents a variation of Method I. It is designed to ensure that the transformants with enhanced sensitivity to human interferon are selected out faster. It minimizes the eventual screening of these survivors.

1. A number of cells are seeded onto a 100-mm dish such that the probability of obtaining the desired transformant is 0.5 per dish based on

the equation of Clark and Carbon[37]:

$$N = \ln(1 - P)/\ln[1 - (1/n)]$$

where P is the probability, n is the minimal number of transformants necessary to generate a complete library, and N is the number of cells seeded per 100-mm dish. For example, if we are to ascertain by dot blot analysis that there is an average of 3000 kb of integrated human DNA per transformant, then the human genome of 6×10^6 kb can be represented by a library of 2000 clones, that is, $n = 2000$; then for $P = 0.5$, $N = 1385$ cells per dish. When seeding about 1400 cells to a dish, each transformant at 80% confluence will approximate 2000 cells/clone. The larger the clone colony with enhanced interferon protection, the greater will be its chances of survival.

2. The procedure is continued as outlined in Method I from steps 2 to 7, inclusive. At this point all 60-mm dishes with an abundance of survivors (>1000) are maintained as described and the other plates with scant survivors are discarded (<50). This is a major advantage of seeding at probabilities of less than one. The signal-to-noise ratio is magnified for the selection procedure and as a result a great deal of cell cloning and characterization is not necessary. The likelihood that each 60-mm plate with a large number of survivors represents a unique clone is very high. There is therefore no need to isolate an individual colony by cloning cylinders as in Method I, and one can go directly to screening each clone.

3. As before, the cells are maintained in interferon supplemented medium until one is certain that the culture is VSV free as described in step 8.

The next section describes the types and frequencies of clones isolated by each of the above methods.

Screening of Clones

Cytopathic Effect

The initial screening of the cloned transformants was by a standard CPE inhibition assay as outlined by Familletti *et al.*[29] All screening assays were done in antibiotic G418-free media. Reagents are as described under Endpoint Determination.

About 25,000 cells were seeded into each well of a 96-well microtiter plate. Both IFN-αA and IFN-β were introduced into well #1 and serially diluted twofold to well #10; the final concentration of IFN-αA and IFN-β

[37] L. Clarke and J. Carbon, *Cell* **9,** 91 (1976).

in well #1 was 40,000 units/ml for each. All assays were performed in duplicate, with the final two wells as interferon-free virus controls. After 24 hr exposure to interferon, 0.4 PFU/cell of VSV in 10 μl of media was added to each well. The wells were stained with crystal violet when the viral control exhibited complete lysis (approximately 24–36 hr after addition of virus). All cells that are positive by the cytopathic effect inhibition assay are immediately subcloned by serial dilution into 96-well plates in 25% conditioned medium.

Figure 1 is an example of a typical positive response. The parental NIH3T3 cells are included as a control. Further controls included random pSV2-neo transformants from the library (not shown here). The random clones showed no enhancement of sensitivity to the human interferons so that the observed effects cannot be attributed to the presence of the plasmid conferring resistance to antibiotic G418. Clone I was isolated by Method I and Clone II by Method II.

Table I summarizes the frequency of surviving clones as determined by the CPE inhibition assay. Method II is obviously superior to Method I

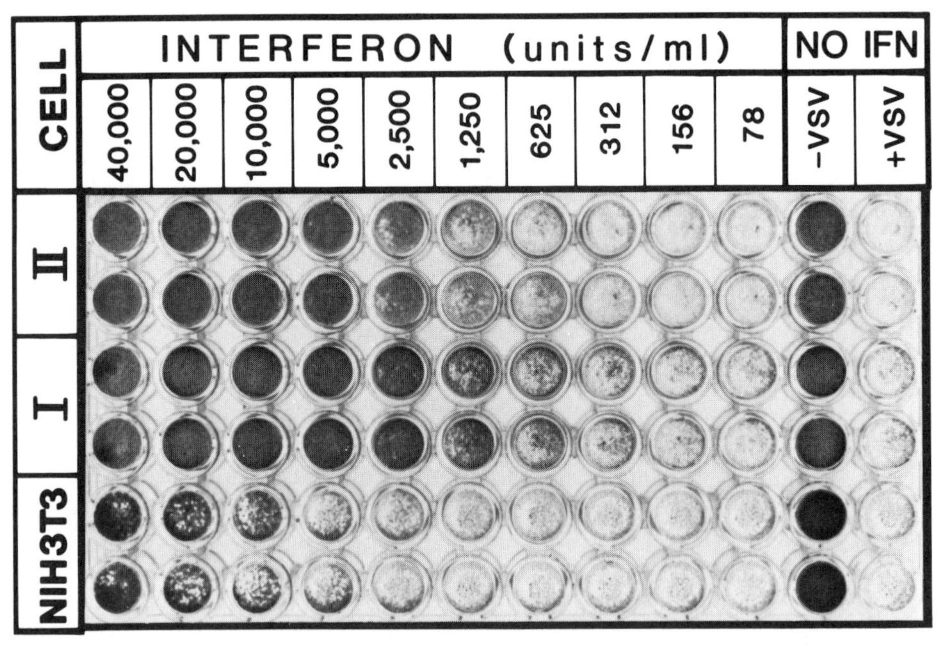

FIG. 1. Cytopathic effect inhibition assay was performed as reported by Familletti *et al.*[29] as described in the text. Clone I was isolated by Method I and clone II by Method II. The concentration refers to units/ml of each of the leukocyte and fibroblast interferons in twofold serial dilutions. Plates were stained after the VSV control exhibited complete lysis.

TABLE I
CHARACTERISTICS OF ISOLATED CLONES[a]

	Method I		Method II	
	Transformed	Control	Transformed	Control
Enhanced sensitivity	6	0	2	0
VSV resistance	11	0	0	0
No change	2	4	0	0
Number screened	1.2×10^5	6×10^4	1.4×10^4	1.4×10^4
Frequency of enhanced sensitivity	5×10^{-5}	0	1.4×10^{-4}	0

[a] Clones isolated directly by selection procedures I and II without subcloning were initially characterized by CPE inhibition assay as described in Fig. 1. Sensitivity enhancement was scored as clones with over 2 well (>4-fold) enhancement. Resistance to VSV is scored as those clones which exhibit no CPE at 0.4 PFU VSV/cell even without the presence of IFN. The frequency of enhancement is the number of clones with enhanced sensitivity divided by the total number of transformants screened. Control cells are untransformed NIH3T3 cells.

in terms of selection frequency and lower background. We cannot explain the preponderance of viral resistance by Method I. However, it is possible that these lines are resistant to VSV by virtue of constitutive production of murine interferon. We cannot explain why these lines are not found by Method II other than the possibility that these lines were discarded with the 60-mm dishes that had scant survivors. It is noteworthy, however, that the control NIH3T3 cells produced neither resistant lines nor cells enhanced for human interferon sensitivity.

A quantitative assay is necessary because the CPE inhibition assay provides only a qualitative result. We therefore used a plaque reduction assay with the cells that were subcloned and reassayed as positive by the CPE inhibition assay.

Plaque Reduction Assay

Two clones, one derived from each of the methods was subcloned, and further characterized by a modification of the microplaque reduction assay as described by Langford *et al.*[38] Clone 1.3.3(I) was isolated by Method I and clone 3.1.4(II) by Method II.

Cells (50,000) were seeded into a 24-well plate. A separate plate was used for each line. The cells were gently rocked every 20 min for 1 hr in

[38] M. P. Langford, D. A. Weigent, G. J. Stanton, and S. Baron, this series, Vol. 78, p. 339.

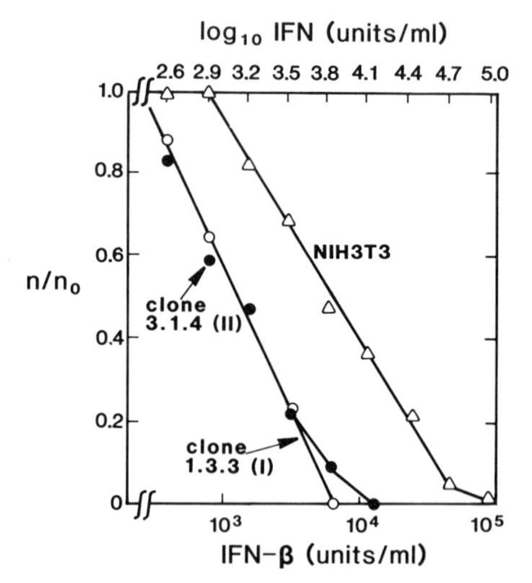

FIG. 2. Relative plaque reduction as a function of IFN-β concentration. A modification of the procedure of Langford *et al.*[38] was used to quantitate the level of enhancement to human interferons as described in the text. Clone 1.3.3(I) was a subclone derived by Method I and clone 3.1.4(II) was a subclone derived by Method II. The relative plaque reduction (n/n_0) for these two transformants compared to the parental NIH3T3 cells as a function of human fibroblast interferon (IFN-β) concentration is presented in the figure. The ratio n/n_0 denotes the relative plaque reduction defined as the number of plaques in the presence of interferon (n) divided by the number of plaques from the interferon-free control (n_0).

order to ensure an even cell distribution. Cell distribution is rather critical in both the CPE inhibition assay as well as the plaque reduction assay to obtain consistent results. No antibiotic G418 is used in these assays.

After 24 hr, the medium is removed and interferon is added at the concentrations described in Figs. 2–4 in 2-fold serial dilutions. All assays were performed in duplicate with the last well free of interferon. The cells should be subconfluent on the day of interferon addition and exactly confluent on the day of viral challenge. NIH3T3 cells do not form well-defined plaques if the cells are overgrown.

After 24 hr in interferon, the wells are washed with PBS supplemented with $MgCl_2 \cdot 6H_2O$ and $CaCl_2$ (0.1 g/liter of each). Then, 200 μl of 2% FCS/MEM with 400 PFU of VSV as titered on mouse L cells is added to each well. The plates are incubated with occasional rocking for 1 to 2 hr. The inoculant is aspirated and 400 μl of a solution consisting of 0.5% methylcellulose containing 2% FCS/MEM is overlaid as described.[38]

After 24 hr, the plate is inverted to drain thoroughly because the viscosity of the methylcellulose can inhibit subsequent staining. The

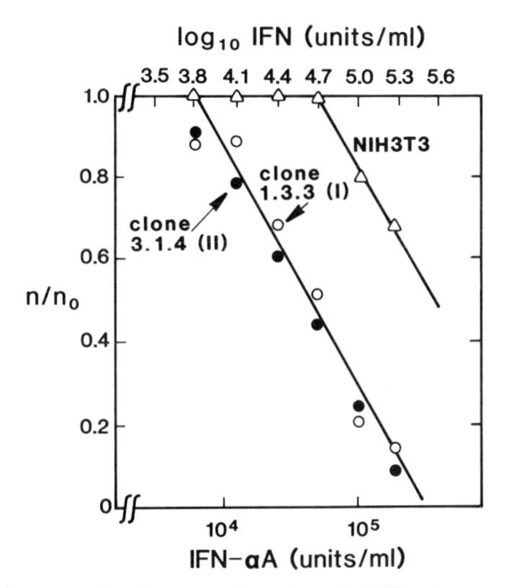

FIG. 3. Relative plaque reduction as a function of IFN-αA concentration. Details are described in the legend to Fig. 2 except that human leukocyte interferon αA (IFN-αA) was substituted for IFN-β.

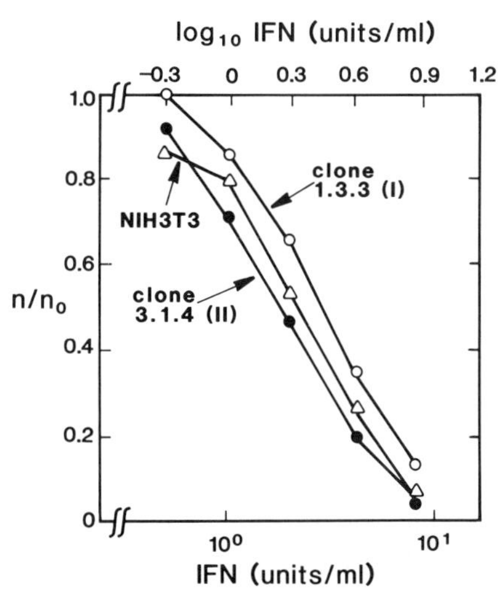

FIG. 4. Relative plaque reduction as a function of murine interferon concentration. Details are described in the legend to Fig. 2 except that murine interferon was substituted for IFN-β.

TABLE II
CONCENTRATION OF INTERFERON FOR 50% VSV PLAQUE REDUCTION (PR_{50})[a]

	Murine IFN		Hu-IFN-αA		Hu-IFN-β	
Cells	Units/ml	Relative enhancement	Units/ml	Relative enhancement	Units/ml	Relative enhancement
NIH3T3	2.2	1.0	400,000	1.0	6,800	1.0
1.3.3(I)	2.9	0.8	41,000	9.8	1,400	4.9
3.1.4(II)	1.8	1.2	41,000	9.8	1,400	4.9

[a] The PR_{50} values are derived from Figs. 2–4. The relative enhancement describes the PR_{50} of the NIH3T3 cells divided by the PR_{50} of the transformants.

plates are fixed and stained with 0.5% crystal violet in 70% methanol. Plaques were scored with the aid of a dissecting microscope. The number of plaques at the specific interferon concentration (n) divided by the number of plaques in the interferon-free control (n_0) was plotted as a function of the interferon concentration in units/ml (Figs. 2–4).

From the curve, the concentration of interferon at which there is a 50% plaque reduction (PR_{50}) can be determined. The ordinate is normalized with reference to the interferon-free control in order to take into account variations in VSV sensitivity for each cell line. Our data indicate that the two clones exhibit enhanced sensitivity to human interferon only. They manifest an apparent 5-fold increase in sensitivity to recombinant human fibroblast interferon (Fig. 2) and a 10-fold enhancement to human leukocyte interferon (Fig. 3). The sensitivity to murine interferon remains relatively the same (Fig. 4) as was found with mouse–human somatic cell hybrids.[3] Table II summarizes some of the data from these plaque reduction assays.

Comments

We have designed a simple selection scheme by which one can isolate transformed NIH3T3 cells that exhibit enhanced sensitivity to the human interferons. Because the sensitivity of these cells to murine interferon remains relatively constant, it would seem that an exegesis for this phenomenon does not include the amplification of the murine receptor. Furthermore, the absence of such clones from the selection of untransformed NIH3T3 cells mitigates against, but does not totally discount, the possibility of a mutant murine receptor that can more readily accommodate the human interferons.

Slate *et al.*[3] had shown that human cells are approximately 100-fold more sensitive to human interferon than somatic cell hybrids of mouse A9 cells containing human chromosome 21. On the other hand, human cells are approximately 1000- and 10,000-fold more sensitive to IFN-β and IFN-αA, respectively, than our transformed NIH3T3 cells. This variation may be due partly to the different murine lines used in these respective experiments and to differences in the regulation or expression of human DNA in the mouse cell host. Chany *et al.*[8] had described the helper effects of extra chromosomes in boosting the level of interferon sensitivity in somatic cell hybrids. If our cells do contain receptor-related sequences, the presence of only a fragment of chromosome 21 might explain the low level of enhancement to human interferon. The cells, however, do fall into the 4- to 8-fold sensitivity enhancement observed for hamster–mouse somatic hybrids in comparison to the parent Chinese hamster line.[4]

We do not know if these clones contain the human receptor or even genes related to the interferon system. A great deal remains to be done before these clones are definitively characterized. Nevertheless, the methods should prove useful for dissecting the mechanisms of sensitivity to the interferons.

[80] Measurement of the Effect of Interferons on Cellular Differentiation in Murine and Human Melanoma Cells

By PAUL B. FISHER, ARMAND F. MIRANDA, and LEE E. BABISS

Although originally identified on the basis of its antiviral properties, more recent investigations indicate that both crude and recombinant interferons are potent modulators of numerous cellular processes (for review, see Refs. 1–4). Of potential clinical relevance is the observation that various interferon preparations can inhibit the proliferation of tumor

[1] I. Gresser and M. G. Tovey, *Biochim. Biophys. Acta* **516,** 231 (1978).
[2] J. Taylor-Papadimitriou, *in* "Interferon" (I. Gresser, ed.), Vol. 2, p. 13. Academic Press, New York, 1980.
[3] P. B. Fisher and S. Grant, *Pharmacol. Ther.* **27,** 143 (1985).
[4] J. Greiner, J. Schlom, P. Giacomini, M. Kusama, S. Ferrone, and P. B. Fisher, *Pharmacol. Ther.* (in press).

cells *in vitro* and *in vivo*.[1,5–10] The mechanism by which interferon induces its antitumor activity is not known. However, important insights into this process have come from studies with cell culture systems capable of differentiating *in vitro*. Interferon has been shown to inhibit several programs of differentiation, including granulopoiesis (human myeloid progenitor cells),[11] melanogenesis (murine B-16 melanoma cells),[3,12,13] erythrogenesis (murine erythroleukemia cells),[14,15] adipogenesis (murine 3T3 cells),[16,17] and myogenesis (chicken embryo muscle cells).[18] In contrast, interferon, used alone or in combination with specific inducers of differentiation, enhances differentiation in several human cell systems, including adult skeletal muscle,[19] histiocytic lymphoma,[20] promyelocytic leukemia,[21–23] melanoma,[24,25] and murine cell systems, including macrophage,[26]

[5] S. Baron, F. Dianzani, and G. J. Stanton, eds., "The Interferon System: A Review to 1982-Part I and II," Tex. Rep. Biol. Med., Vol. 41, p. 1. University of Texas Medical Branch, Galveston, 1982.

[6] E. N. Fish, K. Banerjee, and N. Stebbing, *Biochem. Biophys. Res. Commun.* **112,** 537 (1983).

[7] S. Grant, K. Bhalla, I. B. Weinstein, S. Pestka, and P. B. Fisher, *Biochem. Biophys. Res. Commun.* **108,** 1048 (1982).

[8] J. Gutterman and J. Quesada, *Tex. Rep. Biol. Med.* **41,** 626 (1982).

[9] J. R. Quesada, D. A. Swanson, A. Trindade, and J. U. Gutterman, *Cancer Res.* **43,** 940 (1983).

[10] J. R. Quesada, J. Reuben, J. T. Manning, E. M. Hersh, and J. U. Gutterman, *N. Engl. J. Med.* **310,** 15 (1984).

[11] D. S. Verma, G. Spitzer, A. Zander, J. U. Gutterman, K. B. McCredie, K. A. Dicke, and D. A. Johnston, *Exp. Hematol.* **9,** 63 (1981).

[12] P. B. Fisher, R. A. Mufson, and I. B. Weinstein, *Biochem. Biophys. Res. Commun.* **100,** 823 (1981).

[13] P. B. Fisher, H. Hermo, Jr., D. R. Prignoli, I. B. Weinstein, and S. Pestka, *Biochem. Biophys. Res. Commun.* **119,** 108 (1984).

[14] G. B. Rossi, A. Dolei, L. Cioe, A. Benedetto, G. P. Matarese, and F. Belardelli, *Proc. Natl. Acad. Sci. U.S.A.* **74,** 2036 (1977).

[15] G. B. Rossi, G. P. Matarese, C. Grappelli, and F. Belardelli, *Nature (London)* **267,** 50 (1977).

[16] L. Cioe, T. G. O'Brien, and L. Diamond, *Cell Biol. Int. Rep.* **4,** 255 (1980).

[17] S. Keay and S. E. Grossberg, *Proc. Natl. Acad. Sci. U.S.A.* **77,** 4099 (1980).

[18] J. Lough, S. Keay, J. L. Sabran, and S. E. Grossberg, *Biochem. Biophys. Res. Commun.* **109,** 92 (1982).

[19] P. B. Fisher, A. F. Miranda, L. E. Babiss, S. Pestka, and I. B. Weinstein, *Proc. Natl. Acad. Sci. U.S.A.* **80,** 2961 (1983).

[20] T. Hattori, M. Pack, P. Bougnoux, Z. Chang, and T. Hoffman, *J. Clin. Invest.* **72,** 237 (1983).

[21] M. Tomida, Y. Yamamoto, and M. Hozumi, *Biochem. Biophys. Res. Commun.* **104,** 30 (1982).

[22] S. Grant, M. D. Mileno, I. B. Weinstein, S. Pestka, and P. B. Fisher, *Blood* **62,** 503 (1983).

[23] E. D. Ball, P. M. Guyre, L. Shen, J. M. Glynn, C. R. Maliszewski, P. E. Baker, and M. W. Fanger, *J. Clin. Invest.* **73,** 1072 (1984).

erythroleukemia,[27] and myeloid leukemia.[28] These findings provide further support for the hypothesis that in certain cases inhibition of tumor growth *in vivo* by interferon may involve a direct modulation of differentiation of the tumor cells.[3]

Melanocytes and melanotic melanoma cells express the required genetic information needed to synthesize melanin from tyrosine. This process is regulated by a single enzyme, tyrosinase, which catalyzes three separate reactions in the conversion process.[29] During active growth the C3 clone of mouse B-16 melanoma cells synthesizes low levels of melanin, whereas at confluence there is a cessation of growth and large quantities of melanin are synthesized and secreted into the culture medium (Fig. 1).[12,30–32] Growth of B-16 cells in 5×10^{-7} *M* melanocyte stimulating hormone (α-MSH) results in a suppression of growth and a more rapid onset of melanogenesis in B-16 cells.[12,30,31] Exposure of B-16 cells to 0.03 to 30 units/ml of crude mouse L cell interferon or 100 units/ml of a hybrid recombinant human leukocyte interferon, IFN-αA/D(*Bgl*),[33] results in a concentration-dependent inhibition of both spontaneous and α-MSH-induced melanogenesis without a significant alteration in the growth of B-16 cells.[12,13] The antidifferentiation effect of either type of interferon on melanin synthesis in B-16 cells is enhanced when these compounds are used in combination with the tumor-promoting agent TPA, 12-*O*-tetradecanoyl phorbol-13-acetate, or the antileukemic compound mezerein (MEZ).[3,12] In contrast to murine B-16 melanoma cells, growth of several human melanoma cell lines is suppressed to various degrees by TPA, MEZ, or interferon and the combination of these agents results in a dramatic inhibition in cellular proliferation (Fig. 2).[24,25] Interferon, by itself, enhances melanin synthesis in several human melanoma cell lines, and the combination of MEZ (or TPA) plus interferon results in a synergistic induction of

[24] P. B. Fisher, H. Hermo, Jr., S. Pestka, and I. B. Weinstein, *Pigm. Cell* **7,** 325 (1985).

[25] P. B. Fisher, D. R. Prignoli, H. Hermo, Jr., I. B. Weinstein, and S. Pestka, *J. Interferon Res.* **5,** 11 (1985).

[26] S. N. Vogel, L. L. Weedon, R. N. Moore, and D. L. Rosenstreich, *J. Immunol.* **128,** 380 (1982).

[27] A. Dolei, G. Colletta, M. R. Gapobianchi, G. B. Rossi, and G. Vecchio, *J. Gen. Virol.* **46,** 227 (1980).

[28] M. Tomida, Y. Yamamoto, and M. Hozumi, *Cancer Res.* **40,** 2919 (1980).

[29] A. Korner and J. Pawelek, *Science* **217,** 1163 (1982).

[30] P. B. Fisher, A. F. Miranda, R. A. Mufson, L. S. Weinstein, H. Fujiki, T. Sugimura, and I. B. Weinstein, *Cancer Res.* **42,** 2829 (1982).

[31] R. A. Mufson, P. B. Fisher, and I. B. Weinstein, *Cancer Res.* **39,** 3915 (1979).

[32] J. Kreider and M. Schmoyer, *J. Natl. Cancer Inst.* (*U.S.*) **55,** 641 (1975).

[33] E. Rehberg, B. Kelder, E. G. Hoal, and S. Pestka, *J. Biol. Chem.* **257,** 11497 (1982).

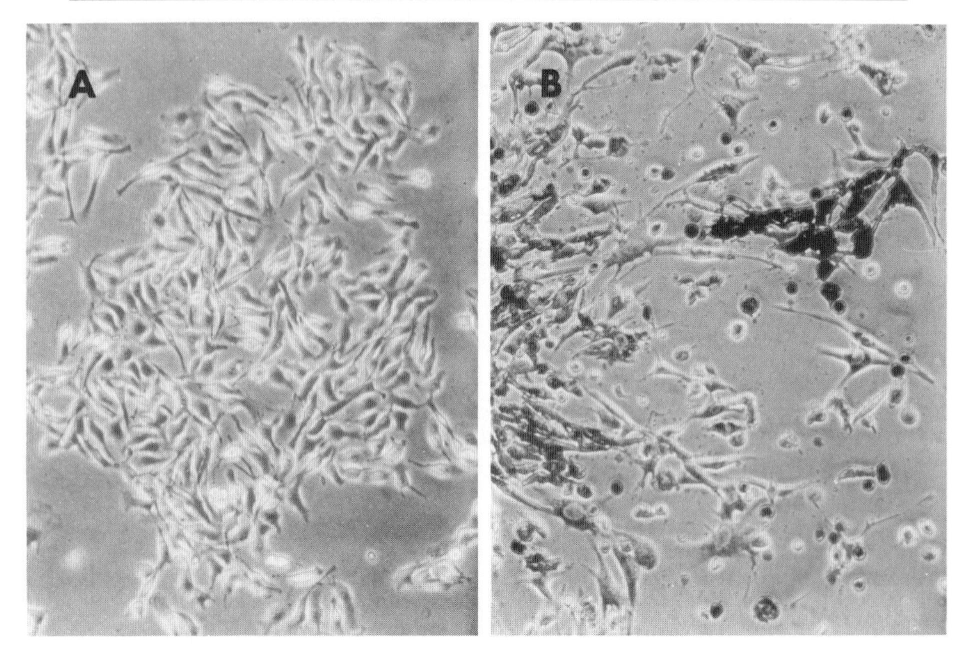

FIG. 1. Morphology of B-16 mouse melanoma cells during active proliferation (A) and following terminal differentiation (B). Phase contrast photomicrograph.

melanogenesis.[3,24,25] These mouse and human systems should prove useful in investigating the effect of modulators of differentiation, such as phorbol esters, MEZ, and interferon, on growth, gene expression, and differentiation as well as serve as models for studying potential interrelationships between these processes. In this chapter we describe some techniques for evaluating the effect of interferon or other compounds on melanogenesis in murine and human melanoma cell cultures.

Cell Cultures and Melanin Assays

B-16 Mouse Melanoma Culture. The C_3 clone of B-16 mouse melanoma cells at passage 35 was originally supplied by Dr. John Kreider, Hershey Medical Center, Hershey, PA.[32] A specific subclone of B-16, B-16 S3, was generated from a single cell clone of C_3 and displays a higher basal level of melanin production, and in contrast to the C_3 clone, detaches from the substrate when it undergoes melanogenesis.[30] Cells are maintained in Dulbecco's modified Eagle's medium (DMEM) supplemented with 10% bovine serum (GIBCO, Grand Island, N.Y.). Serum lots

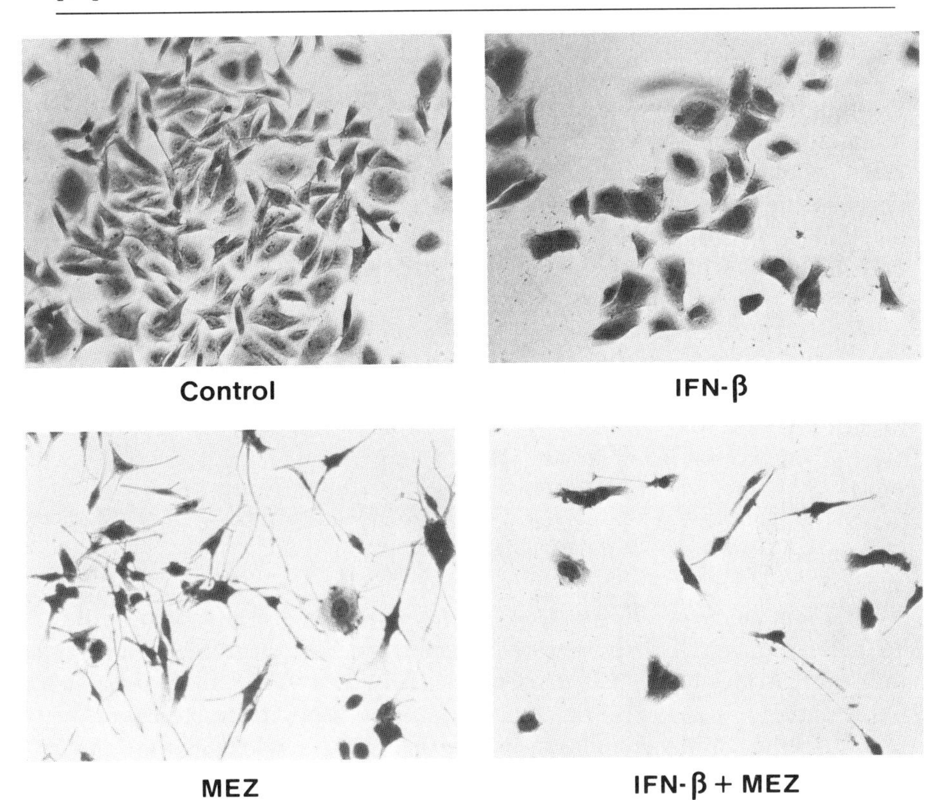

FIG. 2. Effects of recombinant human fibroblast interferon (IFN-β) and mezerein (MEZ), used alone and in combination, on the morphology of BO-2 human melanoma cells (Giemsa, $\times 150$). IFN-β (2000 units/ml), MEZ (10 ng/ml), or the combination was applied to cells for 24 hr.

must be tested since secretion as well as intracellular accumulation of melanin is inhibited when certain batches of fetal bovine serum or bovine serum are used. The final level of melanin produced by B-16 cells can be enhanced by supplementing the medium with tyrosine (final concentration of 2 mM). Cells should be passaged prior to confluency with 0.2% versene (EDTA) prepared in PBS. If left at confluency for prolonged periods, the cells will terminally differentiate and lose viability. It should be noted that passage of the C_3 or S3 clones of B-16 cells with trypsin results in an immediate reduction in cell viability. Thus, cells should not be subcultured with trypsin.

Melanin Determination in B-16 Mouse Melanoma Cells. Total melanin content in both cells and culture medium is determined spectrophotometrically by the method of Oikawa and Nakayasu,[34] however, Protosol (New England Nuclear, Boston, MA) is used instead of Soluene to solubilize the pigment. A simple modification for determining the effect of differentiation modifying compounds on B-16 cells is to quantitate directly melanin content in the culture medium spectrophotometrically (A_{400} units) because these murine melanoma cells secrete melanin into the culture medium. Levels of melanin can be expressed relative to cell number (A_{400} units/10^6 cells), protein (A_{400} units/μg protein), or DNA (A_{400} units/μg DNA).

Human Melanoma Cultures. The four human melanoma cell lines we have employed include BO-2, DU-2, FO-1, and HO-1.[35] These cell lines were kindly provided by Dr. Beppino C. Giovanella, Stehlin Foundation for Cancer Research, Houston, Texas. Cells are grown at 37° in a 95% air 5% CO_2-humidified incubator in DMEM supplemented with 5% (DU-1, FO-1, and HO-1) or 15% (BO-2) fetal bovine serum. Cultures are maintained in the logarithmic phase of growth by subculturing (1 : 5) with trypsin/versene (0.125%/0.02%, w/w) prior to confluency approximately every 5 days.

Melanin Determination in Human Melanoma Cells. Unlike mouse B-16 melanoma cells, which secrete melanin into the culture medium, the four human melanoma cell lines we have studied accumulate melanin intracellularly with appropriate stimulation, but do not secrete appreciable quantities of melanin into the culture medium. Total cellular melanin content is determined by the method of Whittaker.[36] Cells (5 to 30 $\times$ 10^6) are resuspended by trypsinization, pelleted by centrifugation at 1000 rpm for 10 min, transferred to Eppendorf tubes, and washed once with PBS. Cell preparations are then precipitated twice with 5% (w/v) trichloroacetic acid, extracted twice with 3 : 1 ethanol : diethylether, resuspended in diethyl ether, and dried in air overnight. The samples are then dissolved in hot 0.85 *N* KOH and absorbance at 400 nm is determined with a spectrophotometer. The quantity of melanin is determined by comparing experimental values to a standard curve prepared with a melanin preparation (Sigma Chemical CO., MO). Protein is determined by a dye-binding assay[37] (Bio-Rad, Rockefeller Center, New York, NY) protein determination kit.

[34] A. Oikawa and M. Nakayasu, *Yale J. Biol. Med.* **46,** 500 (1973).

[35] B. C. Giovanella, J. S. Stehlin, C. Santamaria, S. O. Yim, A. C. Morgan, L. J. Williams, A. Leibovitz, P. Y. Fialkow, and D. M. Mumford, *J. Natl. Cancer Inst.* (*U.S.*) **56,** 1131 (1976).

[36] J. R. Whittaker, *Dev. Biol.* **8,** 99 (1963).

[37] M. M. Bradford, *Anal. Biochem.* **72,** 248 (1976).

Measuring Effects of Interferon

Compounds modifying differentiation can either be added directly to cells during the initial cell plating, in fresh medium 4 hr after plating or 24 hr after plating. In the case of B-16 cells, the degree of inhibition of differentiation induced by phorbol esters or interferon is directly related to the time compounds are added to cultures. If added 48 or 72 hr after plating, inhibition of differentiation may be marginal or absent. B-16 cells are most sensitive to perturbations in differentiation during the phase of growth prior to the commitment stage of differentiation which may occur 48 or 72 hr following cell plating. In general, when performing assays with B-16 cells 2.5×10^5 cells are seeded per 35-mm plate and 4 hr after plating the medium is replaced with new medium containing the test substance. Depending on the study performed, triplicate plates are either assayed daily (intracellular melanin) or the melanin secreted into the medium is collected and quantified spectrophotometrically (A_{400} units/mg DNA, μg protein, or 10^6 cells). Results are expressed as a percentage of control melanin levels (which generally increase intracellularly between days 4 to 7 and extracellularly between days 5 to 9).[12,13,30,31]

In the case of human melanomas, which often do not secrete melanin into the culture medium, total cell melanin is extracted from intact cells.[25] Since many of the compounds which induce melanogenesis, such as the combination of MEZ and interferon, also suppress growth of human melanomas, the number of cultures required to generate adequate numbers of cells must be adjusted accordingly. In general, 1 to 2×10^6 cells are seeded per 10-cm tissue culture plate and 4 or 24 hr after plating, the medium is replaced with fresh medium containing the desired compounds. After 4 or 7 days (depending on the growth rate of the melanoma, i.e., 7 days for slower growing cells; with a medium change at 4 days for the 7-day assay), three plates from each series are counted with a hemocytometer (with trypan blue dye exclusion to test for viability) and a model Z_f Coulter Counter. Cells from additional replicate plates (4 to 10) are resuspended by trypsinization and melanin is extracted as described above. To compensate for growth suppression, a minimum of two times as many plates are used for test compounds (usually ten 10-cm plates) as for control cultures (usually five 10-cm plates).

Concluding Comments

With the B-16 and human melanoma cell culture systems it is possible to study the effect of virtually any compound or combination of compounds which is soluble in tissue culture medium on the process of melanogenesis. These model systems are valuable in defining the role of extrinsic signals in modulating differentiation, in examining at the genetic level (i.e., transcription and processing of mRNA) the role of specific genes in the differentiation process, and in examining the role of transmembrane signaling in regulating differentiation. The fact that the same compound (interferon or TPA) induces a different response in murine cells than in human cells should also facilitate an evaluation of the biochemical events occurring distal to receptor occupancy which result in either an inhibition (B-16) or an induction (various human melanomas) of differentiation.

Cell culture systems are currently available for analyzing the molecular and biochemical events associated with specific programs of cell differentiation. With these systems it should now be possible to determine the role of endogenous compounds, such as hormones and interferon, as well as the effect of various exogenously applied agents, such as chemotherapeutic drugs, growth factors and viruses, on differentiation. With an increased understanding of the molecular events controlling normal cellular differentiation, new insights should be forthcoming on the role of aberrant patterns of differentiation in the generation of developmental abnormalities and in the initiation, promotion and progression stages of cancer.

Acknowledgments

This research was supported by an Award from Hoffmann-La Roche Inc. (PBF) and NIH Grant NS 1844602 (AFM). We thank Dr. Sidney Pestka (Roche Institute of Molecular Biology, Nutley, NJ) for supplying natural and recombinant interferons for our studies and Bruce Kelder for assays of interferon preparations. We are also indebted to Drs. Sidney Pestka and I. Bernard Weinstein (Columbia University, College of Physicians and Surgeons) for their constant support and encouragement of our research and Ms. Mary Tortorelis for assistance in the preparation of this manuscript. The excellent technical assistance of Diane R. Prignoli and Robert J. McDonald is also appreciated.

[81] Measurement of the Effect of Interferons on Cellular Differentiation of Human Skeletal Muscle Cells

By ARMAND F. MIRANDA, LEE E. BABISS, and PAUL B. FISHER

Improved cell culture techniques now permit the routine cultivation of human skeletal muscle cells *in vitro*.[1-3] During ontogeny and regeneration of skeletal muscle *in vivo,* myogenic mononuclear cells fuse spontaneously to form multinucleated syncytia (myotubes) that no longer undergo conventional DNA synthesis (Fig. 1). As myoblasts fuse into myotubes, several muscle-specific proteins are produced, while the synthesis of non-muscle proteins is reduced. A similar process also occurs in cultures prepared from both fetal or neonatal tissue. It is even possible to establish differentiating muscle cultures from mature skeletal muscle, because dormant mononuclear myogenic cells (called "muscle-satellite cells"), which are present with the muscle fibers, remain present throughout life. When adult muscle is injured *in vivo* or cells from adult skeletal muscle are seeded *in vitro,* muscle-satellite cells are stimulated to proliferate and differentiate into myotubes in a similar manner as myogenic stem cells derived from fetal tissue.[1] These "regenerating" muscle cultures can be used to evaluate the effects of various agents, such as interferon, on cellular differentiation of human muscle derived from individuals of all ages. Utilizing this system we have recently demonstrated that two modifiers of differentiation, interferon and phorbol ester tumor promoters, exert opposing effects on myogenesis.[3] In addition to morphologic criteria for evaluating muscle differentiation, a very useful biochemical marker of myogenesis is creatine kinase (CK; ATP : N-creatine phosphotransferase; EC 2.7.3.2) which occurs in three molecular forms made up of two subunits (Fig. 2). In differentiating muscle transition from a non-muscle isozyme to muscle-specific forms occur.

Preparative Procedures and Methodology

Muscle Culture Method. One or more pieces of skeletal muscle (the size of a pencil eraser) are collected aseptically in complete nutrient me-

[1] A. F. Miranda, S. DiMauro, and H. Somer, *in* "Muscle Regeneration" (A. Mauro, ed.), p. 453. Raven Press, New York, 1979.

[2] A. F. Miranda, T. Mongini, and S. DiMauro, *Adv. Cell Cult.* **4,** 1 (1985).

[3] P. B. Fisher, A. F. Miranda, L. E. Babiss, S. Pestka, and I. B. Weinstein, *Proc. Natl. Acad. Sci. U.S.A.* **80,** 2961 (1983).

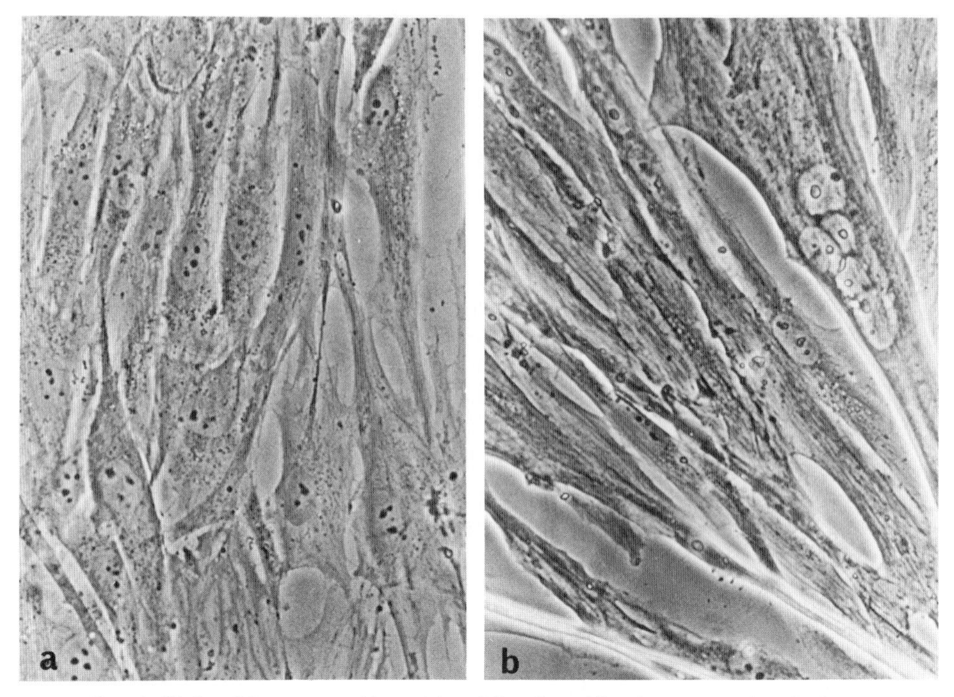

FIG. 1. Unfused human myoblasts (a) and fused, multinucleate myotubes (b) in clonal cultures.

dium (see below) and stored at 4°, preferably for no longer than 72 hr. The muscle is transferred to a 35-mm tissue culture dish in approximately 0.6 ml of "proliferation medium" and carefully cleaned from adherent connective tissue under a dissecting microscope with fine pointed forceps. The teased tissue is then cut with a curved scalpel, (#15) into 0.5-mm pieces, with a rotating motion, and transferred to 35-mm cluster dishes (Costar, Cambridge, MA) containing no more than 0.75 ml of "proliferation medium" per dish. A wide bore Pasteur pipette is used to position 5–6 explants in each dish (floating explants will not grow). The dishes are grown at 36.5° in a humidified incubator in 5% CO_2 and 95% air. The cultures are fed twice weekly (taking precaution that the explants continue to remain in contact with the plastic). After sufficient cellular outgrowth has occurred, but prior to myoblast fusion (12–14 days), the explants are removed with a wide bore Pasteur pipette and the outgrowth dislodged with trypsin–EDTA. The same explants can be transferred to a new set of dishes to obtain a second set of cultures. The pooled trypsinized cells are transferred to a 100-mm dish in 8–10 ml medium and are

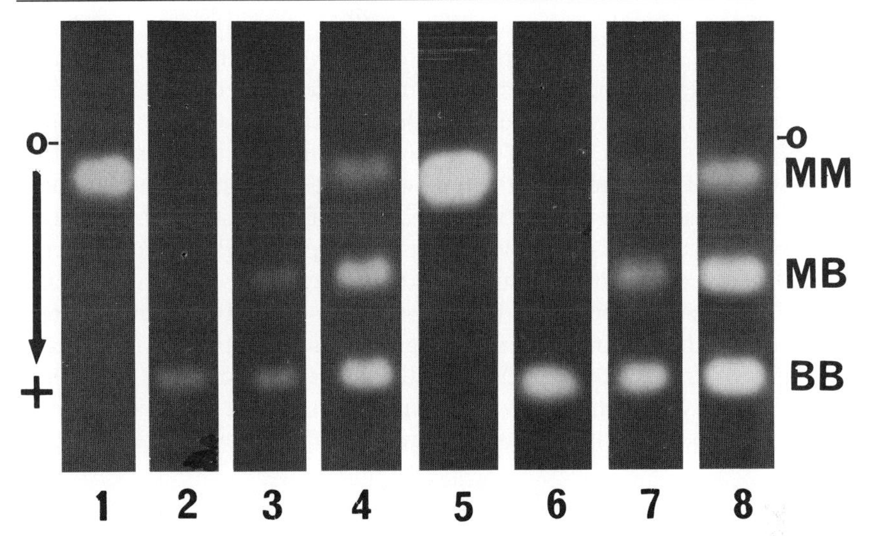

FIG. 2. Cellulose acetate electrophoresis to demonstrate CK isozyme activities. Lanes 1 and 5, control, fully mature intact muscle. Lanes 2 and 6, muscle culture before myoblast fusion. Lanes 3 and 7, muscle culture during the period of myoblast fusion. Lanes 4 and 8, muscle culture 2 days after optimal myoblast fusion. Densitometry was carried out in a Gelman ACD-18 densitometer. Selected lanes from two membranes were cut and remounted for clarity.

allowed to settle and adhere ("preplating") for 10–12 min in a CO_2 incubator. Contaminating fibroblasts that may be present adhere to the plastic before the myoblasts and the myoblast-enriched supernate is siphoned off and placed in another dish for further growth and differentiation.[1] The cultures should not be fed more than twice a week in order to allow "conditioning" of the medium by the cells to promote optimal proliferation. As the cells approach confluency they can be trypsinized again, counted, and placed in appropriate dishes for further experimentation. The cells should always be transferred before there is evidence of myotube formation since myotubes no longer proliferate. At confluency cultures are fed with maintenance medium containing 5% serum, and no enrichments, thereby reducing "growth factors" that might stimulate continued proliferation.

Some laboratories prefer to dissociate the muscle tissue with collagenase and/or trypsin, thus initiating primary cultures form a suspension of dissociated satellite cells. Moreover, it is feasible to obtain pure muscle cultures by growing the dissociated cells at clonal density and then pooling the myogenic colonies for further growth and differentiation (see Refs.

2, 4, 5 for additional citations). This method eliminates all remaining fibroblasts that escape the preplating step: 400–600 cells are seeded into 100-mm dishes in "proliferation medium" and allowed to grow undisturbed for 2–2.5 weeks without a medium change. Since the myogenic cells of human muscle are difficult to distinguish from fibroblasts, the colonies should be selected at the time when the myogenic clones are just beginning to fuse, generally in the central area of the colony. The selected myogenic cells can be isolated with glass cloning rings (Brandel, Gaithersburg, MD) glued over the colonies with nontoxic sterilized silicone grease (Dow Corning, Midland, MI). The cells are dissociated by adding a drop of trypsin solution into the cloning rings and they are transferred to culture flasks or dishes. It is worth noting that since these cultures are rather slow to establish, the humidity in the incubator should approach saturation because media evaporation may severely affect growth due to hypertonicity.

As the cells approach confluency the cultures are fed with complete Eagle's minimal essential medium (MEM) and 5% fetal bovine serum (FBS) or horse serum (HS). In this "nutrient-poor" medium the myogenic cells will differentiate more precociously. It is usually necessary to pretest the batches of sera and select those that are best capable of promoting good myotube formation.

"Fusion efficiency" can be estimated by counting the percentage of total nuclei that are incorporated into myotubes ("fusion index"). For this purpose some cultures are fixed with absolute ethanol (10 min), rinsed with water, stained with Giemsa solution (Fisher Scientific Co. Fairlawn, NJ), diluted with tap water 1 : 10 for 15 min, rinsed with water several times to remove excess dye, and examined with an inverted microscope equipped with 20× or 40× objectives, and an ocular grid insert. The fusion index is determined in several random fields.

The components for formulating 100 ml of media and trypsin solution are given below. Other formulations appear satisfactory as well (see Ref. 2 for review).

Proliferation Medium

MEM (Earle's base), 81 ml
Nonessential amino acids (100×), 1 ml
Vitamins (100×), 1 ml

[4] A. F. Miranda and T. Mongini, *in* "Myology" (A. G. Engel and B. Q. Banker, eds.). McGraw-Hill, New York (in press).

[5] H. M. Blau and C. Webster, *Proc. Natl. Acad. Sci. U.S.A.* **78,** 5623 (1981).

Sodium pyruvate (100×), 1 ml
FBS, 15 ml
Gentamicin sulfate, 50 mg/ml (Whittaker M. A. Bioproducts, Walkersville, MD), 0.05 ml
Amphotericin B (25 μg/ml), 1 ml

Maintenance Medium

MEM (Earle's base), 94 ml
FBS or HS, human serum, (pretested), 5 ml
Gentamicin sulfate, 0.05 ml
Amphotericin B (25 μg/ml), 1 ml

Trypsin Solution

Trypsin solution (2.5% stock), 5 ml
EDTA (2% stock), 7.5 ml
Earle's balanced salt solution, 87.5 ml

Media components were obtained from GIBCO (Grand Island, NY). Trypsinization is done by applying trypsin–EDTA solution at 37° for 10–20 sec and removing most of the trypsin so the cells are barely moist. After 3–5 min, when cells are dislodged as confirmed by microscopic examination, the cells are resuspended in proliferation medium.

Creatine Kinase (CK). CK consists of a dimer with two subunits, M and B. In most tissues CK activity is due to the CK-BB isozyme. In developing skeletal muscle, CK-BB (CK_1) is gradually replaced by CK-MM (CK_3) and the hybrid isozyme CK-MB (CK_2) as myoblasts fuse and myotubes mature. In mature skeletal muscle total CK activity is almost entirely due to the CK-MM (CK_3) isozyme (Fig. 2) (see Ref. 1 for review). In conventional aneurally grown muscle cultures complete transition from CK-BB to CK-MM does not occur, since neuronal influences yet to be identified appear to be required for complete CK-isozyme maturation.[6]

Electrophoresis chambers, sample applicators, and premixed substrates produced specifically for CK isozyme electrophoresis are available commercially. These kits are used in the clinical laboratory to test for presence of CK isozymes in serum. Both agarose gels and cellulose acetate membranes are produced for this purpose. The cellulose acetate membranes have the advantage over agarose in that they can be stored at room temperature for long periods without any apparent deterioration.

[6] A. F. Miranda, E. R. Peterson, and E. B. Masurovsky, *J. Neuropathol. Exp.* **42,** 350 (1983).

Our procedure closely follows the NADPH fluorescence method for serum samples recommended by Helena Inc. with only slight modifications.[7,8]

Demonstration of CK activity depends on the following reactions:

$$\text{Phosphocreatine} + \text{ADP} \xrightleftharpoons{\text{creatine kinase}} \text{creatine} + \text{ATP}$$

$$(2\ \text{ADP} \xrightleftharpoons{\text{adenylate kinase}} \text{ATP} + \text{AMP})$$

$$\text{ATP} + \text{glucose} \xrightleftharpoons{\text{hexokinase}} \text{ADP} + \text{glucose 6-phosphate}$$

$$\text{Glucose 6-P} + \text{NADP}^+ \xrightleftharpoons[\text{dehydrogenase}]{\text{Glucose-6-phosphate}} \text{6-phosphogluconate} + \underset{\text{(fluorescent)}}{\text{NADPH}} + \text{H}^+$$

Newly formed NADPH fluoresces when excited with an ultraviolet (360 nm) light source.[1,7,8] Excess AMP is present in the substrate mix to inhibit adenylate kinase (AK) activity which may show up as a fluorescent band, slightly overlapping with CK-MM isozyme.[9,10] In tissue and tissue culture homogenates excess AMP is insufficient to block AK activity completely. It is necessary, therefore, to add a specific AK inhibitor, such as diadenosine pentaphosphate (AP_5A; 25–50 μM).[1,9,10] Several CK kit manufacturers now add AP_5A to the premixed CK reagents. If AP_5A is not present, freshly prepared stock solution is added to the substrate mix.

Materials

Complete CK reagents and buffers are commercially available (e.g., Abbott Labs; Gelman Sciences; Helena Labs). All these formulations appeared satisfactory when reconstituted, as specified for serum samples. (Corning Diagnostics manufactures a CK kit that utilizes thin agarose gels. The results are also good, but gels must be used prior to the expiration date indicated.)

CK standard
Two 9.5 × 7.5-cm glass plates (or slightly larger)
Waterproof marker

[7] H. Somer and A. Konttinen, *Clin. Chim. Acta* **40,** 133 (1972).

[8] T. Golias, "Electrophoresis Manual." Copyright Helena Labs Corp., Beaumont, Texas, Procedure 20, 1984.

[9] G. Szasz, W. Gerhardt, and E. Bernt, *Clin. Chem.* (*Winston-Salem, N.C.*) **22,** 650 (1976).

[10] G. Szasz, W. Gerhardt, and W. Gruber, *Clin. Chem.* (*Winston-Salem, N.C.*) **23,** 1888 (1977).

2 weights (10 × 7.5-cm slabs of lead; weight 10–15 lb) (Helena Labs)
2-ml pipette
10-μl pipettor
Two 9.5 × 7.5-cm cellulose acetate membranes (e.g., Helena Labs; Gelman Sciences)
2 large (15-cm) Petri dishes or other container (for soaking membranes)
1 heavy glass test tube (approximately 1.5 cm diameter)

Equipment

Electrophoresis chamber with 350 V power source (Helena Labs)
37° incubator (Helena Labs; Corning)
Hair dryer or airflow incubator (45°) (Helena Labs; Corning)
Densitometer with UV light source (Gelman Sciences; Helena Labs)
Spectrophotometer
Sonicator (Sonic 300 Dismembrator: Artek Systems Inc.)

Equipment can be obtained from several manufacturers (e.g., Corning Diagnostics; Gelman Sciences; Helena Labs). Electrophoresis is performed at 4° (coldroom or insulated icebath).

Procedure

The origin of application is indicated with a water-proof marker pen on the mylar side of one cellulose acetate (CA) membrane. One marked and one unmarked CA membrane are soaked in buffer for 30 min or longer. They are immersed *slowly* to avoid trapping of air and used only after the membranes are completely wet.

The cell samples are prepared from 35-mm tissue culture dishes by brief trypsinization and transferred to a centrifuge tube. Serum or serum-containing medium is added to stop trypsin activity and centrifuged for 5 min (1000 *g*). The pellet is rinsed three times with phosphate-buffered saline (PBS; pH 7.3) and centrifuged at 5 min (1000 *g*) each time. The pellet is resuspended in a small centrifuge tube with 50 μl of PBS and the cells are disrupted with an ultrasonicator operated at 40–50% for 15 sec with an intermediate size tip (e.g., Sonic 300 Dismembrator; Artek Systems Corp. Farmingdale, NY). The samples should be kept in ice during the procedure to avoid overheating.

The disrupted cells are pelleted in a refrigerated centrifuge for 10 min at 10,000 *g*. Total CK activity is measured in the supernatant of identical samples at 30° with CK reagent mix in a spectrophotometer at 340 nm. Readings are recorded when the reaction becomes linear (about 4–5 min).

The samples should be diluted to no more than 800–1000 IU CK/liter activity in PBS and 10 μl transferred to each well of the sample tray. The marked CA membrane is removed from the buffer, blotted thoroughly between filter paper, and placed in the holder wet with one or two drops of water to prevent sliding. The samples are applied immediately with the applicator, primed by releasing the first delivery of samples on filter paper. The applicator is depressed again in the sample wells and released onto the membrane. If CK activity is low a second application is needed. The membrane is placed in the electrophoresis chamber with cellulose acetate layer downward and the side of application toward the cathode. Some coins or a small glass plate are placed on top of the membrane to assure good contact with the wicks. Electrophoresis is carried out at 350 V for 10–12 min.

The second membrane, soaked in buffer, is blotted between filter paper, placed on a glass slide, and wet with one or two drops of water to prevent sliding. Then 2 ml of substrate–sucrose mix is pipetted onto the membrane making certain that the entire membrane is coated. The membrane is covered with a large Petri dish to avoid evaporation.

After electrophoresis, the membrane is removed from the chamber, blotted on filter paper, and sandwiched on the membrane covered with the CK reagent. In order to avoid sliding the membranes over one another, the CA side is lowered slowly against the substrate, starting at one end and taking care to avoid air bubbles. Two or three layers of filter paper are then placed on top of the sandwich and rolled firmly over the membranes with a test tube to remove excess substrate. The glass plate and the two membranes, kept in firm contact, are placed between the two weights and incubated for 25–30 min at 37°. The membranes are separated and dried in air for 5 min in a 45° oven or with a hair dryer held at a distance to avoid overheating and examined under a 360 nm UV light source. The electropherograms can be stored in a desiccator or plastic bag for several months in the refrigerator, but it is advisable to scan the membranes within a day or two, since some fading of the fluorescent bands may occur during storage.

Although densitometry of the electropherograms appears to be adequate for estimating the isozyme ratios (see the table), more accurate quantitation of the isozyme activities is feasible by cutting out each isozyme band, eluting the enzymes, and measuring activities spectrophotometrically. This method is described in detail by Roberts *et al.*[11] for serum

[11] R. Roberts, P. D. Henry, S. A. G. J. Witteveen, and B. E. Sobel, *Am. J. Cardiol.* **33,** 650 (1974).

Relative Percentages of CK Isozymes in Differentiating Human Muscle Cultures[a]

	Lanes							
	1	2	3	4	5	6	7	8
MM	100	0	14.1	25.4	98.6	0	18.7	26.5
MB	0	0	42.1	37.4	0.4	0	34.5	37.6
BB	0	100	43.8	37.2	0	100	46.8	35.9

[a] A multiunit applicator (8 lanes, Helena Labs) was used. Densitometer tracings from the electropherograms shown in Fig. 2 yielded the values given in the table. Lanes 1–4 represented a single application of approximately 0.4 μl whereas lanes 5–8 represented a double application of about 0.8 μl.

and skeletal muscle homogenates, but is equally satisfactory if muscle culture samples are analyzed.

Permanent records of the electropherograms can be obtained by photography with type 107 Polaroid film (3000 ASA) in ultraviolet light with a yellow Kodak 2c Wratten filter.

Concluding Comments

Although several muscle-specific markers can be analyzed for evaluating myogenesis in culture, CK isozyme analysis is among the most convenient since isozyme transitions begin to appear rather early in the differentiation process and the fluorescent isozyme assay is sensitive, rapid, and rather inexpensive. Moreover, CK isozyme test kits, produced for serum assays in the clinical laboratory by several companies, can be used satisfactorily for tissue culture homogenates, thereby, eliminating the time-consuming process of preparing reagents. Several investigators have reported the presence of nonspecific fluorescent bands in CK-electropherograms of clinical samples.[12] In our tissue culture studies we have not encountered this difficulty: fluorescence caused by AK activity, slightly overlapping the CK-MM, represented a problem in earlier studies, but was eliminated by addition of the AK inhibitor, AP_5A. Since gradual loss of CK activity may occur, it is advisable to analyze the homogenized samples on the same day. Undisrupted cell pellets may be stored at $-70°$ for about 1 month without significant loss of activity (our

[12] T. C. Kwong and D. A. Arvan, *Clin. Chim. Acta* **115,** 3 (1981).

observation), but long-term storage or repeated freezing and thawing of the cell *homogenates* should be avoided.

Both morphologic (e.g., myoblast fusion and the appearance of cross-triations) and biochemical criteria (e.g., appearance and maturation of CK isozyme patterns) are useful for evaluating the effects of agents on cellular differentiation.[1] It is also possible to analyze the effects of drugs on clonal (low density) growth of myogenic cells. For these studies it is essential, however, that the drugs be applied in medium previously exposed to high density cultures for 24 hr (conditioned medium)[3,13] since repeated feeding with fresh medium will severely affect the ability of the cells to proliferate at low density.

It is now possible to analyze critically the effect of diverse environmental agents, including viruses, on specific programs of cell differentiation.[14-16] In previous studies we have demonstrated that IFN-αA (100 to 5000 units/ml), but not heat-inactivated or trypsin-treated interferon, induces a dose-dependent acceleration in myotube formation and CK isoenzyme transition from CK-BB to CK-MM in human skeletal muscle cultures.[3] In this chapter, we described techniques for growing and analyzing the effect of exogenously applied compounds on the process of myogenesis in human skeletal muscle cultures. The techniques are applicable to investigations employing both normal and genetically diseased human muscle. These cell culture systems should prove valuable in defining the molecular basis of both normal and abnormal myogenesis.

Acknowledgments

This research was supported by an Award from Hoffmann-La Roche, Inc. (PBF) and NIH Grant NS 18446-02 (AFM). We thank Dr. Sidney Pestka (Roche Institute of Molecular Biology, Nutley, NJ) for supplying natural and recombinant interferons for our studies and Bruce Kelder for assays interferon preparations. We are also indebted to Drs. Sidney Pestka and I. Bernard Weinstein (Columbia University, College of Physicians and Surgeons) for their constant support and encouragement of our research and Ms. Mary Tortorelis for assistance in the preparation of this manuscript. The excellent technical assistance of Diane R. Prignoli and Robert J. McDonald is also appreciated.

[13] P. B. Fisher, A. F. Miranda, R. A. Mufson, L. S. Weinstein, H. Fujiki, T. Sugimura, and I. B. Weinstein, *Cancer Res.* **42,** 2829 (1982).

[14] P. B. Fisher and S. Grant, *Therapeutics* **27,** 143 (1985).

[15] P. B. Fisher, H. Hermo, Jr., S. Pestka, and I. B. Weinstein, *Pigm. Cell* **7,** 325 (1985).

[16] A. F. Miranda, L. E. Babiss, and P. B. Fisher, *Proc. Natl. Acad. Sci. U.S.A.* **80,** 6581 (1983).

[82] Measurement of the Effect of Interferons on the Proliferative Capacity and Cloning Efficiency of Normal and Leukemic Human Myeloid Progenitor Cells in Culture

By PAUL B. FISHER, STEVEN GRANT, JOHN W. GREINER, and JEFFREY SCHLOM

In addition to its antiviral activity interferon is also capable of inhibiting the *in vitro* proliferation of a wide variety of normal and tumor cells.[1-3] Partially purified preparations of interferon have been shown to decrease the *in vitro* growth of normal human bone marrow granulocyte–macrophage progenitor cells (CFU-GM).[4] Although the mechanism underlying this inhibitory effect is not known, experimental evidence has been presented by Verma *et al.*[5] suggesting that interferon may exert this growth suppressive effect by blocking early granulopoietic differentiation. Various interferon preparations have also been shown to modify other programs of differentiation, including melanogenesis, myogenesis, and adipogenesis, in normal and tumor-derived cell cultures (for review, see Refs. 6, 7). In the case of human leukemic myeloblasts, interferon has been shown to decrease colony formation in leukemic myeloblasts from patients with acute myeloblastic leukemia as well as clonigenicity in soft agar of various leukemic cell lines, such as HL-60, KG-1, and K-562 cells.[2,8-10] Interferon also induces a decrease in the self-renewal capacity of leukemic myeloblasts, i.e., the ability of cloned cells to form secondary colonies when recultured.[8] Growth inhibition induced by crude and recombinant IFN-α and IFN-β in the established human promyelocytic leukemic cell line HL-60 does not involve an induction of terminal differentiation, whereas recent studies indicate that IFN-γ can induce antigenic

[1] I. Gresser and M. G. Tovey, *Biochim. Biophys. Acta* **516,** 231 (1978).
[2] S. Grant, K. Bhalla, I. B. Weinstein, S. Pestka, and P. B. Fisher, *Biochem. Biophys. Res. Commun.* **108,** 1048 (1982).
[3] E. N. Fish, K. Banerjee, and N. Stebbing, *Biochem. Biophys. Res. Commun.* **112,** 537 (1983).
[4] P. L. Greenberg and S. A. Mosny, *Cancer Res.* **37,** 1974 (1977).
[5] D. S. Verma, G. Spitzer, A. R. Zander, J. U. Gutterman, K. B. McCredie, K. A. Dicke, and D. A. Johnston, *Exp. Hematol.* **9,** 63 (1981).
[6] P. B. Fisher, H. Hermo, Jr., S. Pestka, and I. B. Weinstein, *Pigm. Cell* **7,** 325 (1985).
[7] P. B. Fisher and S. Grant, *Pharmacol. Ther.* **27,** 143 (1985).
[8] R. Taetle, R. N. Buick, and E. McCulloch, *Blood* **56,** 549 (1980).
[9] S. Grant, K. Bhalla, I. B. Weinstein, S. Pestka, and P. B. Fisher, *Blood* **62,** 503 (1983).
[10] E. C. Bradley and F. W. Ruscetti, *Cancer Res.* **41,** 244 (1981).

and biochemical changes normally associated with the differentiation of HL-60 cells into monocytes and granulocytes.[9,11–13]

The methods employed for evaluating the effect of interferon on colony formation of normal and leukemic myeloid progenitor cells, as well as for continuously cultured human leukemic cell lines, share several common features. Cells are suspended in supplemented medium prepared in a semisoft agar (0.3%) or methylcellulose (0.8%) matrix. This suspension system permits cells to absorb essential nutrients but renders them immobile, thereby permitting progenitor cells to give rise to easily identifiable colonies consisting of progeny cells. In the case of normal and leukemic bone marrow cells, as well as some continuously cultured cell lines, a source of colony stimulating activity (CSA) is mandatory for colony formation. CSA may be supplied by growing cells over a feeder layer containing normal immobilized human granulocytes or mononuclear cells[14] or by incorporating commercial preparations of colony stimulating activity such as GCT medium into the agar preparations.[15] In the case of human leukemic myeloblasts, phytohemagglutinin (PHA) or PHA-stimulated medium is often required for optimal growth.[16] The use of this mitogen requires that T lymphocytes (T cells) be removed from the initial cell inoculum to prevent T cell colonies from interfering with accurate enumeration of blast colonies. In all of the methods, the effect of various interferon preparations on colony formation is determined by comparing the growth of interferon-treated cells with untreated controls after incubation at 37° in a 5% CO_2–95% air humidified incubator for 7 to 14 days.

Materials and Reagents

12 well 18 mm tissue culture plates (Costar, Cambridge Mass)
RPMI tissue culture medium (GIBCO, Grand Island, NY)
α-MEM tissue culture medium (GIBCO)
McCoy's 5A culture medium (GIBCO)
Phytohemagglutinin (PHA, lyophilized) (GIBCO)
L-Glutamine (GIBCO)

[11] M. Tomida, Y. Yamamoto, and M. Hozumi, *Biochem. Biophys. Res. Commun.* **104,** 30 (1982).

[12] T. Hattori, M. Pack, P. Bougnoux, Z. Chang, and T. Hoffman, *J. Clin. Invest.* **72,** 237 (1983).

[13] E. D. Ball, P. M. Guyre, L. Shen, J. M. Glynn, C. R. Maliszewski, P. E. Baker, and M. W. Fanger, *J. Clin. Invest.* **73,** 1072 (1984).

[14] B. L. Pike and W. A. Robinson, *J. Cell. Comp. Physiol.* **76,** 77 (1970).

[15] T. F. Di Persio, J. K. Brennan, M. A. Lichtman, C. Y. Affoud, and F. H. Kirkpatrick, *Blood* **56,** 717 (1980).

[16] M. T. Aye, Y. Niko, J. E. Pill, and E. A. McCulloch, *Blood* **44,** 205 (1974).

Asparagine (GIBCO)
Serine (GIBCO)
Sodium pyruvate (GIBCO)
Essential amino acids (GIBCO)
Nonessential amino acids (GIBCO)
MEM vitamins (GIBCO)
Bacto-agar (Difco, Detroit, MI)
Methylcellulose (Sigma Chemicals, St. Louis, MO)
Lymphocyte separation medium
(sp. grav. 1.077–1.081) (Bionetics, Kensington, MD)
GCT conditioned medium (GIBCO)
Fetal calf serum (GIBCO)
Preservative-free heparin
(500 IU/ml) (GIBCO)

Procedure for Normal Human Bone Marrow Progenitor Cells

The following technique represents a minor modification of the method of Broxmeyer *et al.* for the culture of human bone marrow progenitor cells.[17] Bone marrow samples are aspirated into a sterile syringe containing 500 IU of preservative-free heparin. The sample is passed three times through a 26-gauge needle to disperse clumps and diluted 1 : 3 in McCoy's 5a medium. The cell suspension is gently layered over a 3 ml cushion of lymphocyte separation medium (sp. grav. 1.077–1.081) in sterile 50-ml centrifuge tubes and centrifuged at 400 *g* for 38 min at room temperature to remove granulocytes and red blood cells. The interface layer, consisting of mononuclear cells, is extracted with a sterile Pasteur pipette, and the cells are washed twice with McCoy's 5a medium. The mononuclear cells, which contain the granulopoietic progenitors, are cultured by a bilayer agar technique. The bottom agar layer consists of 0.5 ml of McCoy's 5a medium supplemented with 20% fetal calf serum, sodium pyruvate, essential amino acids, nonessential amino acids, MEM vitamins, L-glutamine, serine, asparagine, and 0.5% Bacto-agar. To produce the bottom layer, a 1.25% solution of agar in sterile distilled water is prepared by boiling and is then allowed to cool to 45–50° in a fixed temperature waterbath. The agar is then mixed with 2× strength McCoy's 5a and fetal calf serum in a ratio of 4 : 4 : 2. Aliquots of 0.5 ml of the agar-media mixture are pipetted into the wells of a 12-well 18-mm tissue culture plate, gently swirled, and allowed to gel at room temperature. The top layer, containing the cells, consists of 0.5 ml of supplemented McCoy's

[17] H. E. Broxmeyer, M. de Sousa, A. Smithyman, P. Ralph, J. Hamilton, J. I. Kurland, and J. Bognacki, *Blood* **55,** 324 (1980).

medium, 0.3% Bacto-agar, 20% fetal calf serum, 2×10^5 mononuclear cells, and the appropriate concentration of interferon. To produce the top layer, a 1% agar solution, 2× McCoy's 5a medium, fetal calf serum, and single strength McCoy's 5a containing cells (10^6 cells/ml) are combined in a ratio of 3.7 : 3.7 : 2.5 : 2.5. Aliquots of 0.5 ml of this mixture are pipetted over the bottom layer and allowed to gel. After the agar hardens, 0.1 ml of GCT medium is gently added to each plate as a source of colony stimulating activity. The plates are then placed in a 37° 5% CO_2–95% air fully humidified incubator (Napco, Portland, Oregon) for 10–14 days. At the end of the incubation period, the plates are removed and the number of colonies, consisting of groups of 50 or more cells, are scored with the aid of an inverted microscope. The effect of interferon on granulocyte–macrophage progenitor cells (CFU-GM) is determined by comparing colony formation in control and interferon treated cultures.

Procedure with Continuously Cultured Human Leukemic Cell Lines

Cultured human leukemic cell lines such as HL-60, KG-1, and K-562 also maintain the capacity to form colonies when grown in soft agar or methylcellulose.[18] In some cell lines (e.g., KG-1), an exogenous source of colony stimulating activity is required whereas for other cell types (e.g., HL-60) spontaneous colony formation can occur in the absence of such factors.[19] The following bilayer agar technique is based upon a previously described method by Grant *et al.*[2] Cells are grown in RPMI medium supplemented with sodium pyruvate (100 m*M*), nonessential amino acids (100 m*M*), and 20% fetal calf serum. The base layer, consisting of 0.5 ml of supplemented RPMI medium containing 20% fetal calf serum and 0.5% Bacto-agar, is pipetted into 12-well 18-mm plates and allowed to harden at room temperature. The top layer consists of 0.5 ml of RPMI medium, 20% fetal calf serum, 0.3% agar, the appropriate concentration of interferon, and 2000–5000 cells. For cells requiring a source of colony stimulating activity, 0.1 ml of GCT medium can be added to each plate after the agar solidifies. The plates are then placed in a 37°, 5% CO_2–95% air fully humidified incubator for 7–10 days after which colonies, consisting of groups of 50 or more cells are scored with the aid of an inverted microscope. Cloning efficiency of control cells may vary from 3 to 20% depending upon the cell type and the presence or absence of colony stimulating activity. As with normal bone marrow progenitor cells, the effect of interferon is assessed by comparing colony formation for control and interferon treated cells.

[18] H. P. Loeffler and D. W. Golde, *Blood* **56,** 344 (1980).

[19] R. Gallagher, S. Collins, T. Prigillo, K. McCredie, M. Alearn, S. Tsai, R. Metzgar, C. Aulakh, R. Ting, F. Ruscetti, and R. Gallo, *Blood* **54,** 713 (1979).

Procedure with Human Leukemic Myeloblasts

The technique used for determining the effect of interferon on the growth of human leukemic blast progenitor cells is based on a modification of the method of Minden *et al.*[20] Leukemic myeloblasts are obtained from the bone marrow or peripheral blood of leukemic patients and aspirated into sterile syringes containing 500 IU of preservative free heparin. The samples are diluted 1 : 4 with α-MEM medium and layered over a 3 ml cushion of Ficoll–Hypaque (sp. grav. 1.077–1.081) in a 50-ml centrifuge tube. The cells are spun at 400 *g* for 38 min at room temperature and the interface layer, containing myeloblasts, is removed with a Pasteur pipette. After an additional two washes with α-MEM medium, the cells are suspended in medium and the cell density adjusted to 5×10^6 cell/ml. The cell suspension is then mixed with an equal volume of 0.5% neuraminidase treated sheep red blood cells[21] and allowed to stand for 10 min at room temperature. After a 5 min centrifugation at 200 *g* to pellet the cells, the tubes are placed at 4° for an additional hour. The cells are then resuspended and layered over lymphocyte separation medium as before. After centrifugation at 400 *g* for 30 min, the interface layer, containing the T-rosette-depleted cell fraction, is extracted and washed with α-MEM medium. Cells are then plated in methylcellulose in 18 mm 12-well plates as described above. To each well is added 1 ml of α-MEM medium containing 0.8% methylcellulose, 20% fetal calf serum, 10^5 cells and the appropriate concentration of interferon. This medium is prepared by combining 2.6% methylcellulose, 2× MEM medium, 1× MEM containing 10^6 cells/ml, and fetal calf serum in a ratio of 1 : 1 : 0.6 : 0.6. PHA may be added directly to the culture medium to yield a final concentration of 0.5%[16] and 0.1 ml of GCT conditioned medium added to each plate as a source of colony stimulating activity. The plates are then placed in a 37°, 5% CO_2–95% air fully humidified incubator for 10–14 days, after which colonies, consisting of groups of 20 or more cells, are scored with an inverted microscope. Colonies should consist primarily of immature cells (e.g., myeloblasts and occasional promyelocytes). The effect of interferon on blast progenitor cell growth is assessed in the same manner as for the other cell types described in this chapter.

Remarks

Recent advances in recombinant DNA technology have led to the cloning of several human interferon genes, including those for IFN-α,

[20] M. D. Minden, R. N. Buick, and E. A. McCulloch, *Blood* **54,** 186 (1979).
[21] M. Madsen and H. E. Johnsen, *J. Immunol. Methods* **27,** 61 (1979).

IFN-β, and IFN-γ, and their efficient expression in *Escherichia coli*.[22-24] The demonstration that IFN-αA exhibits antitumor activity *in vitro*[2,6,7] and *in vivo*[25,26] suggests that interferon may provide physicians with a relatively nontoxic compound for use against specific human malignancies. The potential antitumor activity of the various interferon preparations, when used alone, in combination with other interferons or in combination with other chemotherapeutic agents, can only be evaluated if well defined systems can be developed for comparing toxicity toward tumor cells and normal host target tissues, such as normal bone marrow progenitor cells versus myeloid leukemic cells. When potentially useful combinations have been identified, these combinations can then be evaluated for potential *in vivo* efficacy in nude mice. Short term tests employing cultured human leukemic cells (or other primary tumor cells) from patients may prove useful in designing a specific drug regimen which is directly tailored to the needs of the particular patient to achieve efficient eradication of tumor cell populations and to minimize reappearance of the tumor resulting from the development of resistant leukemic populations. The techniques described in this article should, therefore, prove useful to investigators interested in evaluating the potential clinical utility of various interferon preparations, as well as other potential antitumor agents, toward hematopoietic malignancies. In addition, cell culture systems from normal and leukemic myeloblasts will facilitate studies designed to determine the molecular mechanism(s) involved in the conversion of normal bone marrow progenitor cells into leukemic cell populations.

Acknowledgments

We thank Barbara Hamilton for assistance in the preparation of this manuscript. Research support was supplied by an award to Drs. Paul B. Fisher and I. Bernard Weinstein from Hoffmann La Roche, Inc.

[22] D. V. Goeddel, E. Yelverton, A. Ullrich, H. L. Heyneker, G. Miozzari, W. Holmes, P. H. Seeburg, T. Dull, L. May, N. Stebbing, R. Crea, S. Maeda, R. McCandliss, A. Sloma, J. M. Tabor, M. Gross, P. C. Familletti, and S. Pestka, *Nature* (*London*) **287,** 411 (1980).

[23] C. Weissman, S. Nagata, W. Boll, M. Fountoulakis, A. Fujisawa, J.-I. Fujisawa, J. Haynes, K. Henco, N. Mantei, H. Ragg, C. Schein, J. Schmid, G. Shaw, M. Streuli, H. Toira, K. Todokoro, and U. Weidlle, *in* "Interferons" (T. C. Merrigan and R. M. Friedman, eds.), Vol. 4, p. 295. Academic Press, New York, 1982.

[24] S. Pestka, *Arch. Biochem. Biophys.* **221,** 1 (1983).

[25] J. R. Quesada, D. A. Swanson, A. Trindade, and J. U. Gutterman, *Cancer Res.* **43,** 940 (1983).

[26] J. R. Quesada, J. Reuben, J. T. Manning, E. M. Hersh, and J. U. Gutterman, *N. Engl. J. Med.* **310,** 15 (1984).

[83] Measurement of Effect of Interferons on Cloning Efficiency of Primary Tumor Cells in Culture

By SYDNEY E. SALMON

The biological properties initially associated with interferons were their antiviral activities,[1,2] but they were subsequently shown to have antiproliferative effects against a variety of cell types both *in vitro* and *in vivo*.[3,4] Attempts to purify interferons and prepare them in large quantities were begun in the early 1970s.[5] Initial clinical trials in cancer patients carried out in the 1970s suggested that the interferons might have antitumor activity.[6,7] Recently, recombinant DNA methods have been successfully applied to interferon production.[8–13]

In order to evaluate the antiproliferative activity of natural and recombinant human interferons in a model that might prove predictive of clinical antitumor response, we have utilized the human tumor clonogenic assay (HTCA) on fresh biopsies obtained from cancer patients. HTCA has previously been applied to chemosensitivity testing with standard cytotoxic

[1] A. Isaacs and J. Lindenmann, *Proc. R. Soc. London, Ser. B* **147,** 258 (1957).

[2] I. Gresser, M. G. Tovey, M. T. Bandu, C. Maury, and D. F. Brouty-Boyé, *J. Exp. Med.* **144,** 1305 (1976).

[3] I. Gresser, *Cell. Immunol.* **34,** 406 (1977).

[4] S. Einhorn and H. Strander, *J. Gen. Virol.* **35,** 573 (1977).

[5] K. Cantell, S. Hirvonen, K. E. Mogensen, and L. Pyhälä, *In Vitro Monogr.* **3,** 35 (1974).

[6] H. Mellstedt, A. Ahre, M. Bjorkholm, G. Holm, B. Johansson, and H. Strander, *Lancet* **1,** 245 (1979).

[7] J. U. Gutterman, G. R. Blumenschein, R. Alexanian, H. Y. Yap, A. U. Buzdar, F. Cabanillas, G. N. Hortobagyi, E. M. Hersh, S. L. Rasmussen, M. Harmon, M. Kramer, and S. Pestka, *Ann. Intern. Med.* **93,** 399 (1980).

[8] D. V. Goeddel, H. M. Shepard, E. Yelverton, D. Leung, R. Crea, A. Sloma, and S. Pestka, *Nucleic Acids Res.* **8,** 4057 (1980).

[9] T. Taniguchi, S. Ohno, F. Kuriyama, and M. Muramatsu, *Gene* **10,** 11 (1980).

[10] D. V. Goeddel, E. Yelverton, A. Ullrich, H. L. Heyneker, G. Miozzari, W. Holmes, P. H. Seeburg, T. Dull, L. May, N. Stebbing, R. Crea, S. Maeda, R. McCandliss, A. Sloma, J. M. Tabor, M. Gross, P. C. Familletti, and S. Pestka, *Nature (London)* **287,** 411 (1980).

[11] S. Nagata, H. Taira, A. Hall, L. Johnsrud, M. Streuli, J. Escodi, W. Boll, K. Cantell, and C. Weissmann, *Nature (London)* **284,** 316 (1980).

[12] E. Yelverton, D. Leung, P. Weck, P. W. Gray, and D. V. Goeddel, *Nucleic Acids Res.* **9,** 731 (1981).

[13] P. W. Gray, D. W. Leung, D. Pennica, E. Yelverton, R. Najarian, C. C. Simonsen, R. Derynck, P. J. Sherwood, D. M. Wallace, S. L. Berger, A. D. Levinson, and D. V. Goeddel, *Nature (London)* **295,** 503 (1982).

anticancer drugs and found to be predictive of clinical response in cancer patients.[14–19] In addition, the assay has been applied to "*in vitro* phase II testing" of a variety of new anticancer drugs.[20–23] The results of *in vitro* studies suggest that both natural and recombinant interferons tested have antitumor activity against a variety of human cancers but may differ somewhat in their potency and specificity for tumors of differing histologic origin.[24,25]

Reagents and Media

Interferons

With the advent of recombinant interferons, we have focused primarily on these pure interferons and standardized their use on a weight basis. The recombinant human leukocyte interferons IFN-αA and IFN-αD were produced in *Escherichia coli* transformed with the LeIFA25 plasmids[10] and LeIFD3 plasmids,[26] respectively. The bacteria were grown in mass culture and the interferons were purified to homogeneity by affinity chromatography with immobilized monoclonal antibodies against leukocyte

[14] S. E. Salmon, "Cloning of Human Tumor Stem Cells." Alan R. Liss, Inc., New York, 1980.

[15] S. E. Salmon, A. W. Hamburger, B. Soehnlen, B. G. M. Durie, D. S. Alberts, and T. E. Moon, *N. Engl. J. Med.* **298,** 1321 (1978).

[16] D. S. Alberts, S. E. Salmon, H. S. G. Chen, E. A. Surwit, B. Soehnlen, L. Young, and T. E. Moon, *Lancet* **2,** 340 (1980).

[17] D. S. Alberts, H. S. G. Chen, S. E. Salmon, E. A. Surwit, L. Young, T. E. Moon, and F. L. Meyskens, *Cancer Chemother. Pharmacol.* **6,** 279 (1981).

[18] F. L. Meyskens, T. E. Moon, B. Dana, E. Gilmartin, W. J. Casey, H. S. G. Chen, D. H. Franks, L. Young, and S. E. Salmon, *Br. J. Cancer* **44,** 787 (1981).

[19] D. D. von Hoff, J. Casper, E. Bradley, J. Sandbach, D. Jones, and R. Makuch, *Am. J. Med.* **70,** 1027 (1981).

[20] S. E. Salmon, *in* "Cloning of Human Tumor Stem Cells" (S. E. Salmon, ed.), Chapter 22. Alan R. Liss, Inc., New York, 1980.

[21] S. E. Salmon, F. L. Meyskens, D. S. Alberts, B. Soehnlen, and L. Young, *Cancer Treat. Rep.* **65,** 1 (1981).

[22] D. D. von Hoff, C. A. Coltman, and B. Forseth, *Cancer Chemother. Pharmacol.* **6,** 141 (1981).

[23] S. E. Salmon, *in* "Cancer: Achievements, Challenges and Prospects for the 1980's" (J. H. Burchenal and H. F. Oettgen, eds.), p. 33. Grune & Stratton, New York, 1980.

[24] L. B. Epstein, J. S. Shen, J. S. Abele, and C. C. Reese, *in* "Cloning of Human Tumor Stem Cells" (S. E. Salmon, ed.), Chapter 21. Alan R. Liss, Inc., New York, 1980.

[25] S. E. Salmon, B. G. M. Durie, L. Young, R. Liu, P. W. Trown, and N. Stebbing, *J. Clin. Oncol.* **1,** 217 (1983).

[26] P. W. Weck, S. Apperson, N. Stebbing, P. W. Gray, D. Leung, H. M. Shepard, and D. V. Goeddel, *Nucleic Acids Res.* **9,** 6153 (1981).

interferon as described previously.[27] IFN-αA and IFN-αD had specific activities of 2.0×10^8 and 5.0×10^6 units/mg protein, respectively, as determined in the microtiter assay on human amnion cells (WISH) challenged with vesicular stomatitis virus with reference to the NIAID human leukocyte interferon standard (G023–901–527).

Materials for Preparation of Overlayer (Enriched CMRL Medium)

Medium: CMRL 1066 (Gibco, Grand Island, New York). Enrichments: horse serum (15%) (KC Biologicals, Lenexa, Kansas); insulin, 2 units/ml (Lilly); penicillin–streptomycin solution, 10,000 units and 10 mg/ml, respectively (Gibco); glutamine (200 m*M*), 100× stock solution used; L-ascorbic acid, 42 μg/ml; Petri dishes, plastic 35 mm; Bacto-agar (Difco, Detroit, Michigan); Trypan blue, 0.4% solution (Gibco).

Materials for Preparation of Underlayers

McCoy's 5A medium (Gibco). Sodium pyruvate (0.22 mg/ml, tissue culture tested; L-asparagine (0.08 mg/ml, anhydrous); and L-serine (42 μg/ml, Gibco) (final concentrations).

Materials for Preparation of Tumor Cells

Hanks' balanced salt solution (10×) (Gibco). Heparin, sodium salt, preservative-free (O'Neal, Jones and Feldman, Maryland Heights, Missouri). Collagenase, Type IA (Sigma). DNase, Type I (Sigma). Nitex nylon bolting cloth, various meshes (Tekto, Elmsford, New York). Ficoll–Paque (Pharmacia, Piscataway, New Jersey).

Preparation of Cell-Free Underlayers

1. Prepare Bacto-Agar in double glass distilled water as 3% stock.
   a. Place 0.75 g agar in 25 ml distilled water in 100-ml bottles.
   b. Boil for 30 min prior to use to dissolve and sterilize agar.
2. Prepare enriched McCoy's 5A medium.
   a. Make up all supplements: L-serine (21 mg/ml), penicillin–streptomycin, and sodium pyruvate (2.2%) in glass distilled water.
   b. Sterilize by filtration in Nalgene units with 0.22-μm grids.
   c. Add to 500-ml bottle of McCoy's 5A medium the following: 50 ml of FBS, 25 ml horse serum, 1 ml L-serine, 5 ml penicillin–streptomycin, 5 ml sodium pyruvate.

[27] T. Staehelin, D. S. Hobbs, H.-F. Kung, C. Y. Lai, and S. Pestka, *J. Biol. Chem.* **256,** 9750 (1981).

3. Dilute agar to 0.5% in enriched McCoy's 5A medium.
   a. Keep agar in water bath at 42° and medium at 37° before mixing.
   b. Vortex the mixture or pipette vigorously.
   c. Distribute 1 ml final mixture to each 35-mm plastic Petri dish.
   d. Allow agar to gel for about 10–20 min before applying the second layer containing tumor cells.

Collection and Preparation of Tumor Cells

Tumor specimens obtained during the course of routine diagnostic surgical procedures were transferred promptly and under aseptic conditions to the laboratory in McCoy's 5A tissue culture medium.

1. Cells from pleural effusions:
   a. Collect pleural effusions in bottles containing heparin (10 U/ml).
   b. Remove cells from pleural effusions by centrifuging at 400 *g* for 20 min.
   c. Lyse excess red blood cells if necessary as follows. Add 45 ml of sterile double-distilled water to 5 ml packed cell pellet in a 50 ml centrifuge tube. Rotate tube gently and monitor for hemolysis (1–2 min). Add 5 ml of a 10× Hanks' balanced salt solution and centrifuge at 400 *g* for 7 min. Wash cells twice in McCoy's 5A medium and 10% FBS.
   d. Count cells in a hemocytometer and determine viability.
   e. Resuspend cells at 3×10^6 cells/ml in enriched CMRL medium.
   f. Prepare slides for morphologic studies.
2. Cells from solid tumor biopsies:
   a. Mince solid tumor biopsies with scissors or scalpel into 1–3 mm pieces and place in a 25-ml trypsinizing flask containing enzyme solution. Incubate for 1–2 hr at 37° in a final concentration of 0.15% collagenase IA and 0.015% DNase I. Filter cells released into the enzyme solution through 30μ–80μ nylon mesh to remove cell clumps. Wash cells twice in McCoy's 5A medium and 10% FBS.
   b. Draw filtrate up and down pipette to break up any remaining cell clumps.
   c. Count in a hemocytometer for viability and cell number.
   d. Resuspend to 3×10^6 cells/ml in enriched CMRL 1066 medium.
   e. Remove a small aliquot for morphological studies.
   f. We have obtained successful cultures from effusions or bone marrows kept at 4° for as long as 48 hr after collection. Solid tumors should be plated within 24 hr. Although trypan blue viability cannot predict cloning efficiency, we have found it use-

ful to enrich for viable cells. When cell suspensions contain excess debris or viability is less than 10%, cells (2×10^6/ml in McCoy's 5A medium without serum) are placed on an equal volume of Ficoll and spun at 800 g for 20 min at 15°. Viable cells pellet and debris remain in the supernatant. Cells are then washed twice with McCoy's 5A medium containing 10% FBS.

Preparation of Overlayers

1. Prepare enriched CMRL.
2. For each 3 ml final mixture, add 0.3 ml 3% agar stock in 2.2 ml enriched CMRL medium.
   a. Vortex or pipette vigorously.
   b. Add 0.5 ml tumor cell suspension when the agar medium mixture has cooled to about 39°.
   c. Pipette 1 ml of resultant mixture onto 1 ml feeder layers in 35-mm Petri dishes. Each plate should contain 5×10^5 cells.

Evaluation of Colony Growth

1. Incubate cultures at 37° in 5% CO_2 in a humidified atmosphere with no additional feeding. Incubators should be kept very humid to prevent cultures from drying out.
2. Examine cultures twice weekly with an inverted phase microscope at 100× and 200×.
3. Make final colony counts when colonies reach maximum size (10–21 days after plating). Both colonies (aggregates of 30 or more cells) and clusters (3–30 cells) should be scored. Colonies may be counted with an inverted phase microscope or an automated image-analyzer such as the Bausch and Lomb FAS II. The automated instrument provides a histogram of colony size versus frequency and calculates percentage growth inhibition as compared to controls.
4. Dried slides may be prepared from agar cultures to maintain permanent records of colony growth.
   a. Carefully fill each Petri dish with HBSS and incubate for 15 min at room temperature to elute most of the extracellular proteins and reduce background staining. Aspirate supernatants.
   b. Fill each dish with fixative solution consisting of 3% glutaraldehyde in HBSS and allow to stand at room temperature for 5–10 min. Remove fixative.
   c. Fill each dish with distilled water, agitate, and submerge in a tray containing at least 50 ml distilled water. Incubate for 10

min. Agitate plates to displace the plating layer from Petri dish so that it floats in the water. Introduce a clean microscope slide into an undrained tray and allow the plating layer to spread on the slide.

d. Carefully place a prewetted cellulose acetate electrophoresis membrane on the slide. This strip provides for uniform evaporation of water from the agar. Dry the slide for 4–12 hr and remove the strip carefully.

e. Stain the slide. Papanicolau staining is used for routine purposes with standard staining time. Colonies can be stained for peroxidase activity prior to fixation.

Interferon Testing

For routine studies, three plates are used per point. Natural interferons are used in the most highly purified form available, however, whenever possible, recombinant interferons are used because they are of extremely high purity and have a known specific activity. Natural interferons are titrated on a unit basis, whereas recombinant interferon concentrations are expressed primarily by weight and secondarily as interferon units. Interferons are diluted in McCoy's 5A medium containing 10% heat-inactivated fetal calf serum. For most studies, clinically relevant concentrations are used. For example, with recombinant IFN-αA (Hoffmann-La Roche Inc., Nutley, New Jersey), useful final testing concentrations range from 0.4 to 40 ng/ml (these dosages are the equivalent of 80 to 8000 units/ml of interferon activity). The interferons are tested by incorporation into the semisolid agar culture medium. Test results with the recombinant interferons are reported on a weight basis since they are pure proteins of known related molecular structure that have somewhat differing biologic activities in various antiviral and cellular test systems.[26,28–30]

Each experiment included six control plates plus three replicate plates containing the respective interferon at each concentration tested. Stability experiments (data not shown) established that the antiviral activity of the recombinant interferons was stable at 37° in the agar culture plates for at least 14 days and could be quantitatively recovered by saline extrac-

[28] P. K. Weck, S. Apperson, L. May, and N. Stebbing, *J. Gen. Virology* **57,** 233 (1981).

[29] M. J. Kramer, R. Denoin, C. Kramer, G. Jones, R. Kale, A. Gruarin, M. Timmes, H. F. Kung, and P. W. Trown (submitted for publication) (1982).

[30] M. Streuli, A. Hall, W. Boll, W. E. Stewart, S. Nagata, and C. Weissmann, *Proc. Natl. Acad. Sci. U.S.A.* **78,** 2848 (1981).

tion. Separate aliquots from a number of the tumor cell suspensions were also tested simultaneously against various standard cytotoxic drugs with standard dosing, reported previously in relation to prediction of clinical response with the HTCA.[14,31,32] All plates were monitored for aggregation on the morning after plating and discarded if over 15 aggregates per plate were present. Three additional control plates were tested with 10 μg/ml of abrin as a positive control.[33] Cultures were subsequently reviewed by inverted microscopy every 3 days. Leukemic colonies were counted on days 5–7 by inverted microscopy, while solid tumor colonies were counted between days 14 and 17 with a Bausch and Lomb FAS II image analyzer specially equipped and programmed for tumor colony counting.[34]

Results reported are for all tumors tested which gave rise to at least 30 colonies per control plate (500,000 cells plated). All colonies counted were at least 60 μm in diameter. Results were calculated as percentage change from the number of control tumor colonies. As an operational definition, tumor colony-forming units (TCFU) were considered to show some sensitivity to interferon if the number of tumor colonies was reduced to 50% of the control or less, and to show marked sensitivity if the percentage was reduced to 30% of control or less at the 4.0 ng/ml dosage. These definitions are analogous to those used for cytotoxic drugs *in vitro*.[15,19,32] Based on preliminary pharmacokinetic studies with IFN-αA,[35] peak serum levels in the range of 4.0 ng/ml (800 units) are pharmacologically achievable in patients receiving single intramuscular injections of this agent. The dosage range tested in vitro was selected to encompass such serum levels, but would yield a significantly higher total concentration-time product as a result of incorporation of these stable interferons into the agar.

[31] S. E. Salmon, D. S. Alberts, F. L. Meyskens, B. G. M. Durie, S. E. Jones, B. Soehnlen, L. Young, H. S. G. Chen, and T. E. Moon, *in* "Cloning of Human Tumor Stem Cells" (S. E. Salmon, ed.), Chapter 18. Alan R. Liss, Inc., New York, 1980.

[32] T. E. Moon, S. E. Salmon, C. S. White, H. S. G. Chen, F. L. Meyskens, B. G. M. Durie, and D. S. Alberts, *Cancer Chemother. Pharmacol.* **6,** 211 (1981).

[33] S. E. Salmon, R. Liu, C. Hayes, J. Persaud, and R. Roberts, *IND* **1,** 277 (1983).

[34] B. E. Kressner, R. R. A. Morton, A. E. Martens, S. E. Salmon, D. D. von Hoff, and B. Soehnlen, *in* "Cloning of Human Tumor Stem Cells" (S. E. Salmon, ed.), Chapter 15. Alan R. Liss, Inc., New York, 1980.

[35] J. U. Gutterman, S. Fein, J. Quesada, S. J. Horning, J. L. Levine, R. Alexanian, L. Bernhardt, M. J. Kramer, H. Spiegel, W. Colburn, P. W. Trown, T. Merigan, and Z. Dziewanowska, *Ann. Intern. Med.* **96,** 549 (1982).

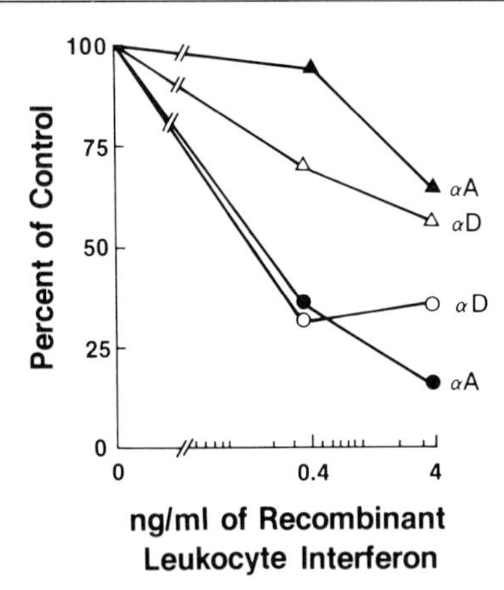

FIG. 1. *In vitro* dose–response curves exhibiting inhibition of proliferation of human ovarian cancer TCFU from two patients (△, ▲, one patient; ○, ●, second patient) tested simultaneously with IFN-αA and IFN-αD. (Data from Salmon *et al.*[25] with permission of the publisher.)

Representative Results

Results of such testing typically show increasing inhibition of tumor colony growth with increasing dosage of IFN. A series of representative growth inhibition curves with fresh human tumor biopsies is depicted in Fig. 1. We have observed significant differences in inhibition curves in selected instances whereas simultaneous testing has been carried out with multiple IFNs (e.g., IFN-α, IFN-β, and IFN-γ). This type of testing has also been carried out with human tumor cell lines in HTCA for either IFNs alone or in combination with anticancer drugs. In the latter circumstance, additive or synergistic effects have sometimes been observed.[36]

[36] M. S. Aapro, D. S. Alberts, and S. E. Salmon, *Cancer Chemother. Pharmacol.* **10,** 161 (1983).

[84] Measurement of the Antiproliferative Effect of Interferon: Influence of Growth Factors

By NICOLETTE EBSWORTH, ENRIQUE ROZENGURT, and JOYCE TAYLOR-PAPADIMITRIOU

In the study of interferon action, cell types are often referred to as being "sensitive" or "resistant" to its antiproliferative activity. However, the responsiveness of a cell to IFN is determined by a complex set of experimental conditions as well as by fundamental properties of the cell, and it is thus very important to be aware of such problems when undertaking studies of this nature.

The "sensitivity" of a cell to any ligand will be the result of the combination of a number of variables. First, is the ability of the cell to recognize the ligand by virtue of the presence of specific cell surface receptors. The ligand–receptor complex generates the corresponding intracellular signals which then elicit biological responses in the cell. In the case of many hormones, neurotransmitters, and other pharmacological agents this sequence of events determines the overall "sensitivity" of the cell to the ligand. When studying the antiproliferative activity of interferon, however, a further problem has to be taken into account, as the degree of mitogenic stimulation to which the cell is exposed has a profound effect on its response (see Ref. 1 and below). Thus, the "sensitivity" of a cell to the antiproliferative action of IFN is determined not only by the presence of biologically active receptors for this ligand, but also by the number of and "sensitivity" to the growth factors used. Here, we describe some of the problems associated with the measurement of IFN-mediated inhibition of cell growth and demonstrate how these may be overcome.

General Methods

Cell Culture. Stock cultures of Swiss 3T3 cells[2] were maintained in Dulbecco's modified Eagle's (DME) medium plus 10% fetal bovine serum (FBS), penicillin (100 units/ml) and streptomycin (100 μg/ml) in a humidified atmosphere of 10% CO_2 : 90% air at 37°. Stocks were seeded at 5×10^4 cells per 90 mm Nunc petri dish and were subcultured every 3 days. In

[1] J. Taylor-Papadimitriou, M. Shearer, and E. Rozengurt, *J. Interferon Res.* **1,** 401 (1981).
[2] G. J. Todaro and H. Green, *J. Cell Biol.* **17,** 299 (1963).

order to obtain quiescent cultures, cells were plated at 2×10^4 cells per 30 mm Nunc petri dish in DME supplemented with 10% FBS. The medium was changed 2 days later and the cells used after a further 5 days. These cells were arrested in G_1 as judged by cytofluorometric analysis and by the fact that less than 1% of the cells were autoradiographically labeled after a 40 hr exposure to [^{3}H]thymidine.

Measurement of [^{3}H]Thymidine Incorporation into Acid-Insoluble Material. Confluent and quiescent cultures of Swiss 3T3 cells were washed twice with DME medium at 37° to remove residual serum. The cultures were then incubated in 2 ml DME/Waymouth medium (1 : 1)[3] containing [^{3}H]thymidine and various additions as indicated. After 40 hr at 37°, cultures were washed twice with ice-cold phosphate-buffered saline (0.15 *M* NaCl in 0.1 *M* potassium phosphate buffer, pH 7.4) and acid-soluble radioactivity was removed by a 20 min treatment with 5% trichloroacetic acid (TCA) at 4°. Cultures were then washed twice in ethanol, cells were solubilized by a 30 min incubation in 0.1 *M* NaOH, 2% Na_2CO_3, and the radioactivity incorporated into acid-insoluble material was determined.

Autoradiography of Labeled Nuclei. Confluent and quiescent cultures of 3T3 cells were washed and incubated as described above except that [^{3}H]thymidine was added to a concentration of 0.2 μM and 5 μCi/ml. After a 40 hr incubation, cultures were washed twice with isotonic saline, extracted with 5% TCA twice for 5 min, washed three times with ethanol, and dried. The dishes were coated with chrome alum and left to dry. Then Kodak AR10 stripping film was laid on the dishes which were then stored in the dark for 1–3 weeks. The film was developed with Kodak D19 developer (4 min) and fixed for 5 min with Hypam fixer (Ilford, Basildon, Essex, UK) diluted 1 : 4. Cells were then stained with Giemsa stain.

Chrome alum solution: 5 g gelatin was heated in 400 ml glass distilled water to dissolve; 0.5 g chrome alum was dissolved separately in 400 ml of water. Solutions were mixed when cool, and made up to 1 liter.

Cytofluorimetry. Cultures of quiescent cells were treated with various agents as indicated and incubated for 40 hr. In order to ensure that cells which were stimulated to synthesize DNA were arrested in G_2, the microtubule disrupting agent, colchicine, was added at 18 hr to those cultures which had not already been treated with it. Colchicine has no mitogenic effect when added so late after the commencement of the experiment. After the 40 hr incubation, cells were trypsinized from the monolayer, washed, passed through a 27-gauge needle to ensure a single-cell suspension, and incubated in lysis buffer (0.5% Triton X-100, 4 m*M* $MgCl_2$, 0.6

[3] K. Mierzejewski and E. Rozengurt, *Exp. Cell Res.* **106,** 394 (1977).

M sucrose, 10 m*M* Tris · HCl, pH 7.5) for 3 min at room temperature. The nuclei were incubated at 37° for 30 min in buffer (0.25 *M* sucrose, 5 m*M* $MgCl_2$, and 20 m*M* Tris · HCl, pH 7.4) containing 0.5 mg/ml ribonuclease A, and then stained with propidium iodide (0.05 mg/ml in 0.1% trisodium citrate). The stained nuclei were spun down at 4°, resuspended in cold phosphate-buffered saline, and the DNA content measured immediately on a fluorescence activated cell sorter (Becton Dickinson FACS-1) at 488 nm excitation. The results shown in the table were obtained by analysis of histograms of DNA content versus cell number.

Materials. Bovine insulin (26 units/mg), Arg-vasopressin, colchicine, ribonuclease A, and propidium iodide were obtained from Sigma Chemical Co. Ltd, Poole, Dorset, UK. Chrome alum (chromic potassium sulfate) was purchased from BDH Chemicals Ltd, Poole, Dorset, UK and Giemsa stain from Gurr Ltd., London SW6, UK. Epidermal growth factor (EGF), receptor grade, was supplied by Collaborative Research. Mouse interferon (IFN) was obtained from Lee Biomolecular Res. Inc., San Diego, CA, and had a specific activity of 5.3×10^7 units/mg. Fibro-

INHIBITION BY IFN OF CYCLING OF 3T3 CELLS STIMULATED BY GROWTH FACTORS[a]

	Percentage of cells in ($S + G_2 + M$)		
	Treatment		
Additions	None	A	B
None	11	51	83
IFN	13	16	63

[a] Cultures of quiescent cells were washed and incubated in DME/Waymouth medium containing (A) insulin and EGF or (B) insulin, EGF, vasopressin, FDGF, and colchicine, in the absence or presence of 1000 units/ml interferon. In order to arrest the stimulated cells in mitosis, colchicine was added at 20 hr after the start of the experiment to dishes exposed to medium alone or to EGF and insulin. Colchicine does not have mitogenic activity when added so late in the incubation period.[7] After a 40 hr incubation the cells were harvested and stained for cytofluorimetric analysis as described. Concentrations of mitogenic agents are given in the legend to Fig. 1.

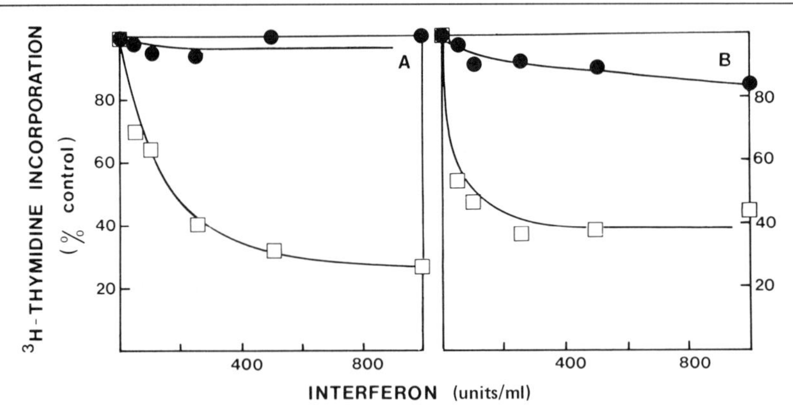

FIG. 1. Dose–response curve for the effect of interferon on the growth of 3T3 cells. Cultures of quiescent cells were stimulated with EGF and insulin (□) or EGF, insulin, vasopressin, FDGF, and colchicine (●) with IFN at the concentrations indicated, and the incorporation of [^{3}H]thymidine into acid-insoluble material was measured as described in the text. Incubation medium containing [^{3}H]thymidine at 1 μM, 0.25 μCi/ml (A) or 10 μM, 2.5 μCi/ml (B). Stimulating agents were added at the following concentrations: EGF, 5 ng/ml; insulin, 10 μg/ml; vasopressin, 20 ng/ml; colchicine, 1 μM; FDGF, 3 μg/ml.

blast-derived growth factor (FDGF) was isolated and purified from the medium conditioned by simian virus 40-infected baby hamster kidney cells as described.[4]

Measurement of Antiproliferative Activity

We have found it convenient to use the mouse Swiss 3T3 cell system to investigate the antiproliferative activity of IFN. These cells can become arrested in G_1 as described above, and subsequent stimulation of the cells with serum or growth factors results in a fairly synchronous population allowing the analysis of events at the molecular level.[5] The inhibition by interferon of this stimulation has been well documented. Figure 1A shows two dose–response curves for the effect of interferon on [^{3}H]thymidine incorporation into acid-precipitable material by Swiss 3T3 cells. The cells were allowed to become quiescent, then stimulated with either insulin and EGF (open squares) or insulin, EGF, vasopressin, FDGF, and colchicine (closed circles), all of which are mitogenic to 3T3 cells.[6] It is clear from this diagram that the categorization of this cell type as interferon "sensitive" or "resistant" depends very largely on how the cells

[4] H. R. Bourne and E. Rozengurt, *Proc. Natl. Acad. Sci. U.S.A.* **73,** 4555 (1976).

[5] T. Sreevalsan, E. Rozengurt, J. Taylor-Papadimitriou, and J. Burchell, *J. Cell Physiol.* **104,** 1 (1980).

[6] E. Rozengurt, *Curr. Top. Cell. Regul.* **17,** 59 (1980).

have been stimulated. When only two growth factors are used, [^{3}H]thymidine incorporation is reduced to 50% of control by 100 units/ml of IFN. In sharp contrast, there is very little growth inhibition even at 1000 units/ml of IFN when five factors are used. It is thus very important to define the mitogenic stimulus when making investigations of this nature. The use of unfractionated serum in experiments of this type makes interpretation of the results very difficult, as the number and type of growth factors present in the serum are not only unknown, but probably vary among batches.

The incorporation of [^{3}H]thymidine into acid-insoluble material has been widely used as a convenient method of assessing the inhibitory effects of IFN on the growth of synchronized cells. However, this technique not only assesses effects on DNA synthesis, but can also reflect alterations in the transport of thymidine across the plasma membrane, and in the phosphorylation of this nucleoside by thymidine kinase which occurs before it can be incorporated into DNA. In some systems, agents previously thought to have antiproliferative activity have subsequently been shown merely to inhibit thymidine transport; however, this technique provides a convenient measurement of changes in DNA synthesis, provided that appropriate controls are included.

An easy way to circumvent any effects on thymidine transport is to vary the concentration of thymidine used in the experiment. If the apparent inhibition of DNA synthesis is entirely due to a decrease in thymidine transport, then the effect should be overcome by increasing the concentration of this nucleoside, because at high concentrations, the nucleoside permeates the cell membrane by simple diffusion rather by carrier-mediated transport. Figure 1 shows dose–response curves for the effect of interferon on the growth of Swiss 3T3 cells at two different concentrations of thymidine. Clearly, the dose–response relationships are the same and the difference between the curves obtained with two and five growth factors are seen at two concentrations of thymidine (1 and 10 μM), thus showing that in this system the effect of IFN is not at the level of thymidine transport.

To substantiate further that neither the transport nor the phosphorylation of thymidine is affected by a putative inhibitory agent, the proportion of cells actually synthesizing DNA can be measured by autoradiographic techniques. This method is much less sensitive to changes in the specific radioactivity of the precursor pool. Figure 2 shows that the results presented in Fig. 1A are confirmed by this technique. Figure 2A shows the number of labeled nuclei in cultures treated with EGF and insulin in the absence (open bars) or in the presence (closed bars) of 1000 units/ml of IFN while in Fig. 2B the cells were stimulated with five growth factors. Clearly, interferon inhibits entry into DNA synthesis much more

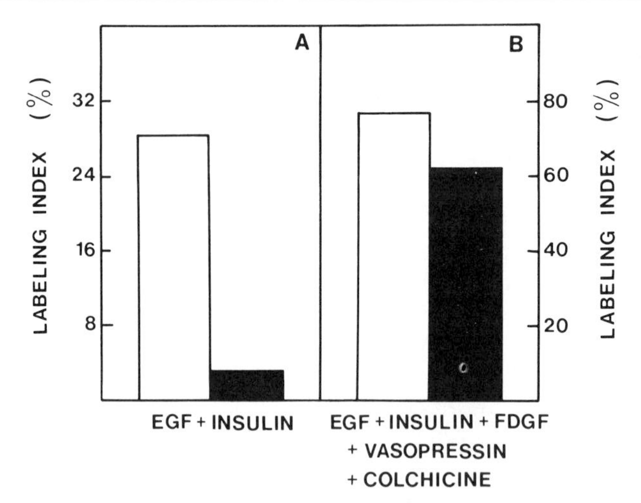

FIG. 2. Effect of IFN on the labeling indices of 3T3 cultures exposed to two (A) or five (B) growth factors. Cultures of quiescent cells were treated with growth factors as described at the concentrations shown for Fig. 1 in the absence (open bars) or presence (shaded bars) of 1000 units/ml of IFN. After a 40 hr incubation, cultures were fixed and prepared for autoradiography as described in the text.

efficiently in cells exposed to the two factors than in those treated with the five mitogenic agents.

An important approach to ascertain whether the results obtained with the [^{3}H]thymidine incorporation method are an accurate reflection of the state of DNA synthesis is to use a completely independent technique, such as cytofluorimetry. By adding colchicine, which blocks the cells in mitosis but does not potentiate the stimulation of DNA synthesis by EGF and insulin if added 20 hr after these factors,[7] the effect of IFN on the movement of the cells through the cycle from G_1 through S phase and into G_2 can be readily assessed. The table shows, once again, that the proliferation of the cells effected by the addition of the two growth factors is markedly inhibited by 1000 units/ml IFN. If the cells are stimulated with the five factors, on the other hand, then the inhibitory effect of interferon is substantially reduced.

The data presented here illustrate some of the pitfalls which may be encountered when attempting to measure the antiproliferative activity of

[7] Although colchicine does stimulate the incorporation of [^{3}H]thymidine into DNA in 3T3 cells it has been shown by M. Friedkin, A. Legg, and E. Rozengurt [*Exp. Cell Res.* **129,** 23 (1980)] that its mitogenic activity is lost if it is added to the cells once S phase has been initiated, i.e., 20 hr after stimulation with other factors.

interferon. Clearly, useful and relevant measurements can be made, providing the appropriate control experiments indicate that the results are not due to one of the artifacts described. However, even if these problems have been overcome, it is also important to work with materials and media which are chemically defined, as the degree to which cell growth is inhibited by interferon, and the ability of growth factors to reverse this antiproliferative activity both vary according to the presence or absence of additional stimulating agents.

[85] Animal Models for Investigating Antitumor Effects of Interferon

By FRANCES R. BALKWILL

Animal tumor models have been important in the development of many cancer therapies and catalytic in the current upsurge of clinical interest in interferons (IFNs). Murine IFNs of varying degrees of purity increased survival time or reduced tumor size of mice implanted with many murine tumor cell lines, showing an activity broadly comparable to single agent chemotherapy.[1] Investigations into the mechanisms of this antitumor activity are complex because IFNs are regulatory agents capable of directly influencing cellular proliferation and differentiation, directly protecting cells from viral infection and transformation, and indirectly inhibiting tumor growth by enhancing host reaction to the tumor. In animal model systems all three mechanisms may be important.[1]

However, there are several problems in relating the data obtained with these animal tumors to clinical experience in man. Many experiments used highly selected cell lines maintained *in vitro* and therapy commenced at the time of cell injection. Also, because IFNs are generally species specific, such models cannot test the potential of the large number of natural and recombinant human IFNs which are available for preclinical and clinical studies. Therefore human tumors growing as xenografts in nude mice provide another useful model, and this chapter will describe how such a system can be used to study (1) the direct antitumor effects of a range of human IFNs alone, or in combination with chemotherapy, and (2) the indirect effects of murine IFNs on human tumor growth, hence

[1] F. R. Balkwill, *in* "The Interferons" (N. Finter and R. Oldham, eds.), Vol. 4, p. 23. Elsevier/North-Holland Biomedical Press, Amsterdam, 1985.

providing an insight into the mechanism of IFN action. The methodology described here would also be suitable for studying other animal tumor models.

Animals, Tumors, Interferons, and Reagents

Mice

Six- to eight-week-old female *nu/nu* mice of mixed genetic background[2] were housed in negative pressure isolators (Vickers Medical, Basingstoke, UK) and all routine supplies were sterilized by heat or irradiation. All items entering or leaving the isolator were sprayed with 2% Tegodor (T H Goldschmidt Ltd, Eastcote, Middlesex, UK) in the entry lock. The life span of mice kept under these conditions was comparable to normal SPF mice.

Tumors

Human tumor xenografts in the mice were established from subcutaneous implantation of 1 mm cubes of fresh pretreatment human tumor with a Bashford needle. Xenografts that became established were passaged in the mice by subcutaneous injection of minced tumor suspension with 19-gauge needles.

Tumor 1068 is a mucoid carcinoma of the breast derived from a postmenopausal woman who had received no prior therapy. The establishment of this tumor has been previously described by Balkwill *et al.*[3]

Tumor EF3 was derived from an adenocarcinoma of the cecum from a pretreatment male. Both tumors were karyotyped at frequent intervals to confirm their human origin as described previously.[3]

Tumors were excised after growing subcutaneously (sc) in the mice for 6 weeks, minced finely, diluted with 1 ml PBS per tumor, and 0.05 ml tumor suspension inoculated sc ventrally on each mouse with a 19-gauge needle. Mice were then given terramycin powder (2%) (Pfizer Ltd, Kent, UK) in their drinking water for 1 week to prevent any infection at the site of implant. After 1–3 weeks growth, two tumor diameters at right angles were measured with calipers, the product of which gave the tumor size index. Mice bearing tumors of approximately equal size index (±20% SD) were selected and divided randomly into treatment groups of 4–5 mice.

[2] A. Sebesteny, J. Taylor-Papadimitriou, R. Ceriani, R. Millis, C. Schmidtt, and D. Trevan, *JNCI, J. Natl. Cancer Inst.* **63,** 1331 (1979).

[3] F. R. Balkwill, E. M. Moodie, V. Freedman, and K. H. Fantes, *Int. J. Cancer* **30,** 231 (1982).

Interferons

Purified recombinant Hu-IFN-αA and Hu-IFN-αD were a gift from Dr. S. Pestka, Roche Institute of Molecular Biology, Nutley New Jersey.[4] Mouse IFN-β was made from Ehrlich ascites cells as described previously[5] and was a gift from Dr. S. Mowshowitz (Mount Sinai Hospital, New York, New York). Hu-IFN-α (from Namalwa cells) was obtained from Dr. K. H. Fantes (Wellcome Research Laboratories, Beckenham, UK) and had a specific activity of 2×10^8 units/mg.

The IFN titers were confirmed in a biological assay which measured inhibition of viral (Semiliki forest virus) RNA synthesis on WISH cells (Hu-IFN-αA and Hu-IFN-α, Namalwa, MDBK cells (Hu-IFN-αD), or L cells (Mu-IFN-β). The human interferons were calibrated against the British human leukocyte interferon reference standard 69/19 (National Institute of Biological Standards & Controls, London, UK) and Mu-IFN-β, against mouse interferon reference standard. G002-904-511 (from the Research Resources Branch, National Institute of Allergy and Infectious Diseases, National Institutes of Health, Bethesda, Maryland). As Hu-IFN-αD has low antiviral activity on human cells,[4] the human IFNs were diluted on a weight basis, in PBS containing 3 mg/ml bovine serum albumin (BSA) (Sigma Chemical Co. Poole, UK) and stored in appropriate aliquots at $-70°$. A fresh vial was used for each injection. IFNs were administered daily sc at a site distant from the tumor.

Drugs

Cyclophosphamide (Endoxana, W. B. Pharmaceuticals Ltd, Bracknell, UK) was freshly diluted for each injection and was administered intraperitoneally, once weekly.

Assay of the Effects of IFNs on the Growth of Human Tumor Xenografts

All three IFNs inhibited the human breast tumor xenograft in a dose-dependent fashion as shown by the examples in Fig. 1. A comparison of the effects of the two recombinant human IFNs showed that in agreement with its low antiviral activity on human cells, Hu-IFN-αD had less activity on the human xenograft than Hu-IFN-αA. The latter IFN caused

[4] S. Pestka, *Arch. Biochem. Biophys.* **221,** 1 (1983).

[5] F. R. Balkwill, E. M. Moodie, S. Mowshowitz, and K. H. Fantes, *in* "The Biology of the Interferon System 1983" (D. DeMaeyer and H. Schellekens, eds.), p. 443. Elsevier, Amsterdam, 1983.

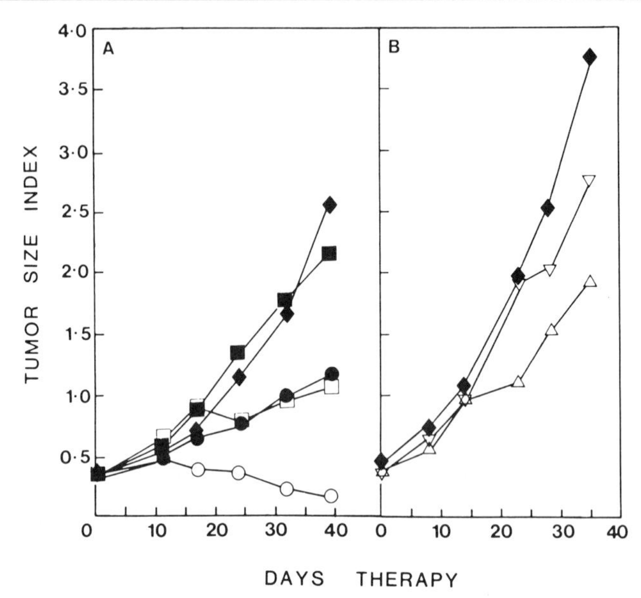

FIG. 1. The effect of human and mouse IFNs on the growth of human breast cancer xenograft 1068. Daily therapy with IFN given sc commenced 14 days after the tumors were transplanted. (A) ♦, Control; ■, 250 ng/day Hu-IFN-αD; ●, 1 μg/day Hu-IFN-αD; ○, 1 μg/day Hu-IFN-αA; □, 250 ng/day Hu-IFN-αA. (B) ♦, control; ▽, 5×10^4 units/day of mouse IFN-β; △, 2×10^5 units/day mouse IFN-β.

complete regression and disappearance of tumor (Fig. 1A) and these tumors did not reappear for at least 365 days after treatment. The effects of the Mu-IFN-β were less dramatic but reproducible and statistically significant (Fig. 1B, control versus 2×10^5 units Mu-IFN-β, $p < 0.01$).

All tumor inhibition occurred after a lag period of 7–14 days following initiation of therapy. Previous experiments with Hu-IFN-α (Namalwa) have shown that daily therapy gives best results.[3]

Mechanisms of Tumor Inhibitory Effects of IFNs

As described in the introduction, there is evidence from animal model systems that IFNs may exert an antitumor effect by direct effects on tumor cells or indirectly, by enhancing host responses. In this model system we can measure the direct effect of the IFNs binding to tumor and host tissues by assay of the IFN induced enzyme, 2-5A synthetase, and the immunostimulatory effects of IFNs by assay of murine spleen natural killer (NK) cell activity. For the 2-5A synthetase assay, tissues were stored at −70° immediately after removal from the mouse, then mashed

2-5A SYNTHETASE LEVELS IN HUMAN TUMOR AND MOUSE SPLEEN[a]

Therapy	pmol ATP/OD260/hr	
	Human tumor	Mouse spleen
Control	11	81
Hu-IFN-αD	89	85
Hu-IFN-αA	152	102

[a] Mice were treated subcutaneously with a daily dose of 250 ng of Hu-IFN-αA or αD for 35 days. Each value represents the mean of levels obtained from two mice.

with scalpels, homogenized, and centrifuged at 10,000 *g* for 1 hr to obtain S10 extracts. The enzyme (2′-5′)-oligoadenylate synthetase was assayed in these extracts by its ability to generate 2-5A which was measured in a radiobinding assay. The methodology for this assay has been described in detail in a previous volume[6] except that poly(I) · poly(C) cellulose was used to bind the enzyme instead of poly(I) · poly(C) paper. The table shows the effect of human IFNs on the levels of 2-5A synthetase in the human tumor xenograft and the spleen from the tumor bearing mouse in a typical experiment. Both the human IFNs have a strong stimulatory effect on human tumor 2-5A synthetase levels, but have no effect on mouse spleen, whereas in previous experiments we have shown that mouse IFN has no consistent effect on the human tumor but increases the level of 2-5A synthetase in mouse spleen cells.[3]

To measure murine NK cell activity, mice were killed by cervical dislocation and the spleens removed immediately, placed into tissue culture medium, and teased apart with needles. Cells were then layered onto Ficoll–Hypaque density 1.14, and centrifuged at 2500 rpm for 15 min at room temperature. Nucleated cells were then counted, diluted appropriately to give effector target cell ratios of 200 : 1–25 : 1 in RPMI 1640 medium (Gibco Biocult Ltd) with 5% fetal bovine serum, FBS (Sera Labs, Crawley, UK), and added in 100 μl volumes to 96-well V-bottomed microculture plates (Linbro, Flow Laboratories, Irvine, UK). The human breast cancer cell line MDA-157 was labeled with ^{51}Cr (100 Ci/ml; sp. act. 350–600 mCi/mg; Amersham Radiochemical Centre, UK) for 1 hr at 37°. Cells were washed, resuspended in the RPMI medium, and added in 100-

[6] G. R. Stark, R. E. Brown, and I. M. Kerr, this series, Vol. 79, p. 194.

μl aliquots to give 10^4 cells/well. The plates were centrifuged gently to pack the cells and incubated at 37° for 16 hr. The plates were then centrifuged again at 1000 rpm for 1 min and 100 μl of supernatant removed from each well for counting in a γ counter. The specific cytolysis was calculated from the following formula:

$$\frac{(\text{cpm in test wells}) - (\text{spontaneous released cpm})}{(\text{total releasable counts cpm}) - (\text{spontaneous released cpm})}$$

Figure 2 shows typical results obtained after 5 weeks therapy with human or mouse IFN. In this assay the two human IFNs had contrasting effects. Hu-IFN-αA had no effect on murine spleen NK cell activity whereas Hu-IFN-αD therapy of 1 μg/day caused a stimulation of mouse NK cell activity that was comparable to that obtained with 2 × 10^5 units/day of Mu-IFN-β.

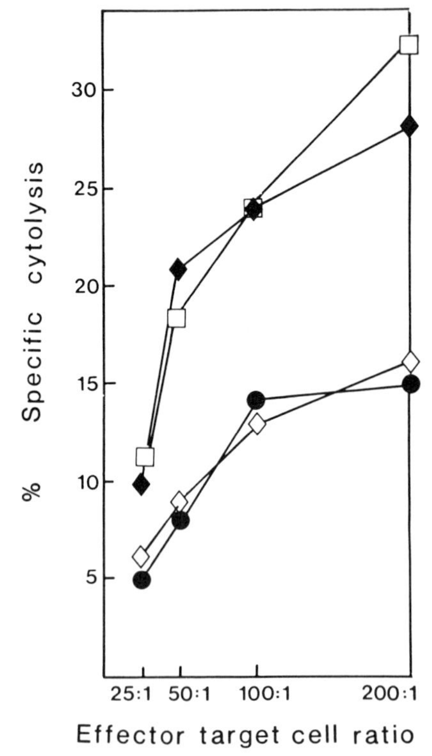

FIG. 2. The effect of daily IFN therapy on nude mouse spleen NK cell activity against a human breast cancer cell line, MDA 157. Daily IFN therapy was given for 5 weeks before removal of spleens. ●, Control; ◇, Hu-IFN-αA, 1 μg; ◆, Hu-IFN-αD, 1 μg; □, Mu-IFN-β, 2 × 10^5 units.

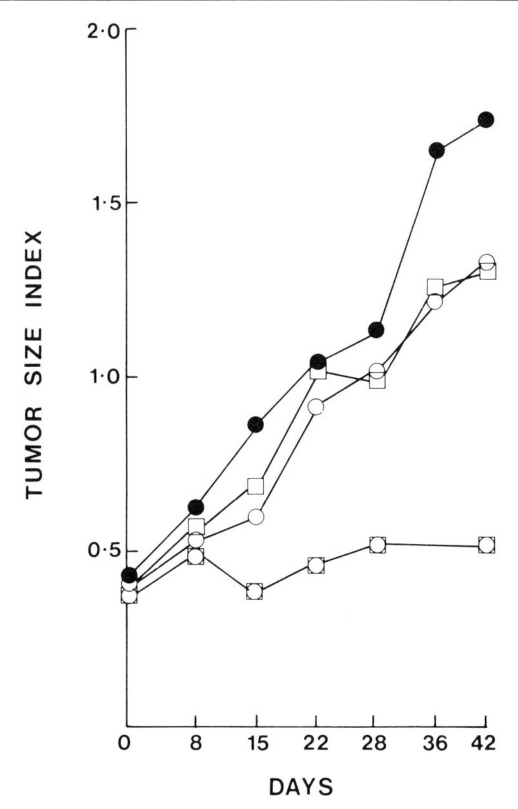

FIG. 3. The effect of a combination of IFN and cyclophosphamide on the growth of bowel cancer xenograft EF3. IFN was given daily sc, cyclophosphamide once weekly. ●, Control; ○, 2×10^5 units Hu-IFN-α (Namalwa); □, 1 mg cyclophosphamide; ⊡ 2×10^5 units Hu-IFN-α (Namalwa) and 1 mg cyclophosphamide.

Effect of IFN Chemotherapy Combinations

We have found positive interactions between IFN and commonly used chemotherapeutic agents against both the breast and bowel xenografts. In the breast xenograft subline that is less sensitive to IFNs, we have shown that a combination of cytostatic doses of Hu-IFN-α (Namalwa) and cyclophosphamide or adriamycin could cause complete regression and disappearance of tumors.[7] Maximal therapeutic effect was obtained when the IFN was administered daily and the cytotoxic drug once weekly.[7] An example of positive interactions between Hu-IFN-α (Namalwa) and cyclophosphamide on the bowel cancer xenograft is shown in Fig. 3. The

[7] F. R. Balkwill and E. M. Moodie, *Cancer Res.* **44,** 904 (1984).

cyclophosphamide and Hu-IFN-α (Namalwa) given as single agents had little effect, but in combination cause cytostasis.

Concluding Comments

This assay system permits examination of direct effects of human IFNs on a range of human tumor cells, many of which will not grow *in vitro* and permits screening a range of human IFN types and subtypes for antitumor activity.[8] The two IFN subtypes, Hu-IFN-αA and Hu-IFN-αD, have markedly different effects on both the human tumor and the host. In keeping with its low specific activity on human cells, Hu-IFN-αD had lower antitumor activity, but at the high (1 μg/day/mouse) dose stimulated mouse spleen NK cell activity almost as efficiently as mouse IFN when given for a 5-week period. However, the lower dose (250 ng/day/mouse) had no effect (F. Balkwill, unpublished results) and this dose, which strongly stimulated the human tumor 2-5A synthetase levels, had no effect on enzyme levels in mouse spleen. Further experiments are in progress to look at the effect of the high dose of Hu-IFN-αD on 2-5A synthetase levels in nude mouse tissues. Thus, although it appears that the antitumor effects of Hu-IFN-αA, like those of Hu-IFN-α (Namalwa) as previously published,[3] are due to direct growth inhibition of the human tumor, it is not certain whether high dose Hu-IFN-αD has a direct effect on the tumor or stimulates the host NK cells to reject the tumor.

It may be possible to use this model system to investigate host mediated antitumor effects with homologous mouse IFN. We have shown that *in vivo* mouse IFN therapy inhibits human tumor cell growth, stimulates mouse NK cell activity and IFN-induced enzymes in mouse tissues, but has no effect on enzyme levels in human tumor tissues.[3,5] The mechanism of tumor growth inhibition seen with mouse IFN is uncertain but is presumably T cell independent in these athymic mice.

The dissociation of IFN effects on the tumor from effects on the host that can be achieved by using mouse and human IFNs has also proved useful in investigating the mechanism of the synergy seen between IFN and chemotherapy. Chemotherapeutic drugs are metabolized, by the host, particularly in the liver, and the balance between activation and detoxification is important in determining activity. It has been reported that IFNs can inhibit the activity of drug-metabolizing enzymes[9] and we have found that mouse, but not human IFNs, alter the activity of two

[8] F. R. Balkwill, L. Goldstein, and N. Slebbing, Int. J. Cancer **35,** 613 (1985).

[9] G. Singh, K. W. Renton, and N. Stebbing, *Biochem. Biophys. Res. Commun.* **106,** 1256 (1982).

enzyme systems important in the *in vivo* activation and detoxification of drugs, cytochrome *P*-450 and glutathione transferases (Balkwill *et al.*, submitted for publication). However, addition of mouse IFN to the human IFN/chemotherapy combination did not alter the synergy seen so we conclude that a direct interaction is occurring in the tumor tissues.

Although this model system does not allow us to study immunoregulatory properties of human IFNs, we are able to look at direct antitumor effects of these IFNs on human tumors *in vivo* both alone and in combination with chemotherapy, and to look at effects of mouse IFN on the tumor-bearing host which may influence the progress of tumor growth.

[86] Measurement of Antagonistic Effects of Growth Factors and Interferons

By ANNA D. INGLOT and ELŻBIETA PAJTASZ

The proliferation of cells and tissues probably cannot occur without the participation of growth factors (GFs) or substances that either mimic or substitute for GFs. GFs have been defined as endogenous substances produced by various tissues which have multiplication stimulating activity and/or other trophic effects but are not nutrients, antibodies, enzymes, or attachment factors. The majority of GFs are polypeptides of various molecular weight and structure which have a hormone-like mode of action.[1] These include the interleukins which stimulate proliferation of cells of the immune system.[2]

Interferons (IFNs), as well as GFs, fulfill the main biological and biochemical criteria of hormones. However, the response of various cells and tissues to IFNs is inhibition of the proliferation and other complex reactions not necessarily connected with the cell cycle but clearly antagonistic to GFs.[3] Thus, it has been suggested that IFNs and GFs can be grouped together as nonclassical hormones with opposite actions.[3,4]

In this chapter we describe techniques for measurement of interactions between IFNs and GFs *in vitro,* in tissue culture of either human

[1] R. A. Bradshaw and J. S. Rubin, *J. Supramol. Struct.* **14,** 183 (1980).

[2] K. A. Smith, *in* "Lymphokines" (E. Pick and M. Landy, eds.), Vol. 7, p. 203. Academic Press, New York, 1982.

[3] A. D. Inglot, *Arch. Virol.* **76,** 1 (1983).

[4] A. D. Inglot, *in* "The Physiology and Pathology of Interferon System Contributions to Oncology (L. Borecký and V. Lackovič, eds.), Vol. 20, p. 72. S. Karger, Basel, 1984.

diploid fibroblasts or blasts of the cortisone resistant mouse thymocytes, and also *in vivo,* in mice infected with Moloney sarcoma virus.

Assay of Interactions between IFN and GF in Fibroblasts

Fibroblasts. Human embryonic diploid fibroblasts (HEF) strains are used between fifth and twentieth passage levels in culture. The cells are found to be sensitive to both human IFN and several GFs.[5,6]

HEF are grown as monolayer cultures in Eagle's minimal essential medium (MEM) supplemented with 10% calf serum, penicillin 100 units/ml, and streptomycin 100 μg/ml at 37°. Monolayers are routinely dispersed for passage or use in experiments, with 0.05% trypsin and 0.02% ethylenediaminetetraacetic acid (EDTA) in solution A (8 g NaCl, 0.4 g KCl, 0.35 g $NaHCO_3$, 1 g glucose, and 0.005 g phenol red in 1 liter of water).

BALB/c 3T3 cells, clone A1 selected in our laboratory for titration of GFs,[5,6] are cultured and passaged in the same way as HEF.

Growth Factors. Platelet-derived growth factor (PDGF) is prepared by the method developed in our laboratory by Narczewska *et al.*[7,8] Briefly, concentrated platelets from human or bovine blood are disrupted by freezing and thawing and extracted with acid/ethanol according to Davoren's procedure useful for preparation of many protein hormones and transforming growth factors.[9] PDGF is further purified to electrophoretic homogeneity in sodium dodecyl sulfate–polyacrylamide gels by consecutive ion-exchange chromatography on DEAE-Sephadex, CM-Sephadex, and molecular sieving on Sephadex G-100.[7,8]

A transforming growth factor (TGF) is isolated and purified from mouse C_{243} cells propagated as subcutaneous tumors in the irradiated BALB/c mice (W. Popik, quoted in Ref. 6 and unpublished results). It resembles TGFs isolated from tumors by Todaro *et al.*[9] The mitogenic activity of TGF is measured in the same way as PDGF.

Interferons. Human leukocyte IFN was supplied by Dr. M. Krim and prepared from leukocytes induced with Sendai virus; titer 1×10^6 units/ml. Human lymphoblastoid, Namalwa IFN, Hu-IFN-αLy, was supplied

[5] E. Oleszak and A. D. Inglot, *J. Interferon Res.* **1,** 37 (1980).

[6] A. D. Inglot and M. Albin, *J. Interferon Res.* **3,** 75 (1983).

[7] B. Narczewska, J. Czyrski, and A. D. Inglot, *Can. J. Biochem. Cell Biol.* **63,** 187 (1985).

[8] J. Czyrski, B. Narczewska, and A. D. Inglot, *Arch. Immunol. Ther. Exp.* **32,** 589 (1984).

[9] G. J. Todaro, J. E. De Larco, C. Fryling, P. A. Johnson, and M. D. Sporn, *J. Supramol. Struct.* **15,** 287 (1981).

by Dr. K. H. Fantes and prepared from the cells induced with Sendai virus; titer 3.8×10^6 units/ml. Recombinant human leukocyte A IFN (IFN-αA) was supplied by Dr. S. Pestka; titer 3×10^6 units/ml. Mouse fibroblast IFN-α/β was prepared in our laboratory from mouse C_{243} cells induced with Newcastle disease virus; titer 3×10^6 units/ml.

Microassay for the Mitogenic Activity of Growth Factors

The rationale of the assay is as follows. A strain of cells is selected that cannot proliferate or multiplies very slowly if maintained in a medium deficient for certain GFs. The medium either for BALB/c 3T3 or HEF cells is MEM supplemented with 2% platelet poor plasma serum (PPPS) instead of conventional serum. The cells can multiply vigorously if the medium is supplemented with an appropriate concentration of PDGF or TGF.

We have found that for the short-term experiments lasting 2–3 days 3T3 cells are suitable. For long-term experiments lasting 1–3 weeks human fibroblasts are more convenient because the cells attach firmly to the surface of culture vessels and the cultures do not require frequent exchange of media. The assay is carried out in Micro Test II Tissue culture plates with 96 flat bottom wells (3040 Falcon, Oxnard, California).

BALB/c 3T3 cells, clone A1, are dispersed by trypsinization and suspended at a concentration of 1×10^5 cells/ml in 2% PPPS medium (MEM).

PPPS is prepared according to the standard method from fresh human donor plasma from which the platelets are removed by centrifugation.[5,6,10] The plasma was heated at 56° for 30 min, centrifuged to remove fibrin, dialyzed against Dulbecco's phosphate-buffered saline (PBS) for 2 days at 4°, and sterilized by UV irradiation. Only batches of PPPS which do not support the proliferation of 3T3 cells without PDGF added to the medium are used for the experiments.

The serial dilutions of PDGF are prepared in the plates in 50 μl of 2% PPPS in MEM. The plates are then exposed to UV irradiation (Phillips UV lamp, 30 W) at a distance of about 20 cm from the plate for 5 min at room temperature. The UV irradiation allows the rapid and efficient sterilization of the samples.

The 3T3 cells (100 μl at a concentration of 1×10^5 cells/ml) are placed into each well of the microtiter plate and incubated at 37° in a humidified atmosphere of 5% CO_2 in air for 2–3 days. The cultures are examined with an inverted Zeiss-Jena microscope. Our PDGF mitogenic unit is taken as

[10] R. B. Rutherford and R. Ross, *J. Cell Biol.* **69,** 196 (1976).

RESULTS OF TITRATION OF THE MITOGENIC ACTIVITY OF GROWTH FACTORS BY THE RAPID MICROASSAY[a]

GF	Formation of 3T3 cell monolayer at the indicated dilution of the preparation of GF							
	10^{-1}	$10^{-1.5}$	10^{-2}	$10^{-2.5}$	10^{-3}	$10^{-3.5}$	10^{-4}	$10^{-4.5}$
PDGF, crude	Tox	4	3	2	2	2	±	0
PDGF, purified	4	4	4	3	3	2	2	±
TGF, crude	Tox	Tox	Tox	3	3	2	0	0
TGF, purified	3	3	2	2	2	±	0	0
None	0	0	0	0	0	0	0	0

[a] Procedure of the assay is described in the text. Incubation time was 2 days. Arbitrary scale: 4, confluent monolayer; 3, partially confluent monolayer; 2, approximately 50% of the surface of the well covered with cells; + or ±, negligible cell growth; 0, no cell growth; Tox, toxic for the cells.

the reciprocal of the highest dilution of a preparation of GF that stimulates cell multiplication and spreading to an extent that allows the formation of a monolayer covering approximately 50% of the bottom of the well (see the table).

Measurement of Antagonistic Effects of PDGF and Hu-IFN

Cell proliferation studies in the presence of PDGF and IFN are done in 35-mm tissue culture dishes (Corning, New York) by serial cell counts. PDGF and/or Hu-IFN at the indicated dilution in a volume of 20 μl are introduced as two separate drops into each dish. At least two dishes are taken for each value. The dishes are UV irradiated for 5 min at room temperature for sterilization.

HEF cells are dispersed from the monolayer, diluted with 2% PPPS in MEM to a concentration of 6×10^4 cells/ml, and 2 ml of the cell suspension is added to each. The cultures are incubated at 37° in a humidified atmosphere of 5% CO_2 in air for 6–8 days without any exchange of medium or reagents. The cultures are inspected daily with the aid of a microscope. The cells are dispersed with the 0.05% trypsin/EDTA solution and counted in a hemocytometer in the presence of trypan blue to determine cell viability.

Comments

The proliferation of human fibroblasts maintained in 2% PPPS MEM is almost completely dependent on the presence of a growth factor, in our case on PDGF used at a concentration of 10 units/ml. Hu-IFN clearly inhibits the mitogenic action of PDGF and the degree of inhibition is dose related (Fig. 1). The assay is very sensitive because it detects as little as 1–10 units of Hu-IFN-αLy. Under standard conditions in the presence of 5–10% serum human fibroblasts are much less sensitive to the antiproliferative action of Hu-IFN-α or -β.[5]

The assay may be applied to studies of various preparations of human or animal IFN interacting with the mitogenic actions of several GFs. We have investigated the antagonistic effects of Hu-IFN-αLe, Hu-IFN-αLy, and recombinant IFN-αA versus PDGF or TGF. The reproducibility of the results is satisfactory.

In contrast to PDGF, our preparation of TGF caused profound alteration in the morphology of HEF cells which rapidly detached from the surface of the dish after 3–4 days. Therefore, it may be necessary to count not only the attached cells but also floating cells.

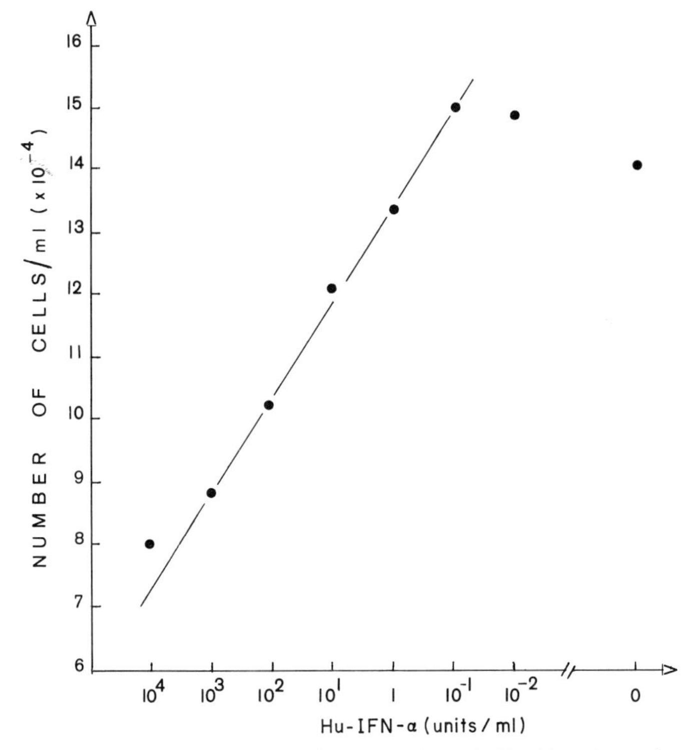

FIG. 1. Inhibition of the proliferation of human embryonic fibroblasts by various concentrations of human lymphoblastoid interferon (Hu-IFN-αLy) were cultured in 2% PPPS medium supplemented with 10 units of PDGF. Details as described in the text. Incubation time was 8 days at 37°.

We have consistently observed that human IFN-α as well as mouse IFN-α/β at concentration of 0.1–0.01 units/ml may enhance the multiplication of HEF cells (unpublished results). It has been suggested that IFN at very low concentration may positively cooperate with PDGF and/or other GFs in stimulating cell division.[3,6]

On the other hand, we have also observed that even highly purified preparations of human or animal IFNs may contain a mitogenic activity that can be demonstrated in HEF or BALB/c 3T3 cells maintained in 2% PPPS MEM supplemented with the submitogenic concentrations of PDGF.[6] We have suggested that the preparations of IFN may be contaminated with various GFs originating either from cells and media used for the production of IFN or even from additives employed for the stabilization of IFN activity such as human or bovine albumin.[6]

Inhibition of the Proliferative Response of Thymocytes to Interleukin-2 by Mouse Interferon

Preparation of Blasts from Thymus. Female BALB/c mice, 8–10 weeks old, were obtained from the Breeding Center of Inbred Animals at our Institute. Mice are treated with hydrocortisone acetate (Polfa, Poland) at a dose of 150 mg/kg body weight. Two days later thymuses are removed aseptically into RPMI 1640 medium supplemented with 2.5% fetal calf serum (FCS; Flow Labs; heated at 56° for 30 min), penicillin (100 units/ml), and streptomycin (100 μg/ml). Cortisone-resistant thymocytes (CRT) are prepared by pressing the organs through a mesh nylon screen and washed twice in the same medium. Then, the cells are suspended at a concentration of 1×10^6 cells/ml and cultured on Linbro tissue culture plates (Flow Labs) in 2 ml volumes of RPMI 1640 medium containing 100 units/ml penicillin, 100 μg/ml streptomycin, 2 m*M* glutamine, 1 m*M* sodium pyruvate, 5 m*M* Hepes, 5×10^{-5} *M* 2-mercaptoethanol, 10% FCS, and 4 μg/ml concanavalin A (Con A; Pharmacia, Uppsala), and cultured in humidified atmosphere of 5% CO_2 in air at 37° for 48 or 72 hr.[11,12]

Source of Interleukin-2 (IL-2) and Other Reagents. Standard conditioned medium (CM) is prepared from the culture of BALB/c mouse spleen cells (10^7 cells/ml) which are incubated with 4 μg/ml of Con A for 24 hr in the glass bottles. The CM is harvested and centrifuged (10 min, 1500 rpm) and supernatant is titrated for the IL-2 activity on CRT blasts[11,12] and stored at −20°. α-Methyl-D-mannoside (α-MM; Carl Roth D

[11] E. Pajtasz, P. Kuśnierczyk, and P. Kisielow, *Arch. Immunol. Ther. Exp.* **31,** 437 (1983).
[12] E. Pajtasz, P. Kuśnierczyk, and P. Kisielow, *Arch. Immunol. Ther. Exp.* **31,** 443 (1983).

7500 Karlsruhe 21, W. Germany) is added into the culture of CRT blasts at concentration of 50 m*M* as a competitive inhibitor of Con A. The [*methyl*-^{3}H]thymidine was obtained from the Radiochemical Center (Amersham, England) with a specific activity of 5 Ci/mmol.

Measurement of Inhibition of CRT Blasts Proliferation by Mu-IFN-α/β

The CRT-Con A stimulated blasts are harvested, washed, resuspended at concentration of $0.5–1 \times 10^6$ cells/ml, and incubated further for 24 hr in fresh culture medium supplemented with 50 m*M* α-MM and 25% (v/v) of CM. To assay the inhibitory effect of IFN, the cultures of CRT blasts are set up as described above with addition of different concentration of Mu-IFN ranging from 0.001 to 10,000 units/ml. Growth of the cells is assayed by incorporation of [^{3}H]thymidine added at concentration of 0.7 μCi/well for the last 6–7 hr of culture. Then, the cells are collected on a glass fiber filter (Whatman) with a multiple cell harvester and radioactivity determined in a liquid scintillation counter. The results are expressed as mean cpm of triplicate cultures.

Value of Δcpm is calculated as follows: Δcpm = cpm value of culture with CM − cpm value of culture without CM. Percentage inhibition of blasts proliferation in the presence of IFN is calculated according to formula:

$$\% \text{ inhibition} = 100 - \frac{\text{cpm value of culture with Mu-IFN}}{\text{cpm value of culture without Mu-IFN}} \times 100$$

Analysis of Data. Mu-IFN is found to inhibit the proliferation of CRT blasts stimulated by IL-2 present in CM (Fig. 2A). It is advisable to use 72-hr-old blasts rather than 48-hr-old blasts for the assay. Although 48 hr blasts respond to IL-2 with higher cpm values, the 72 hr blasts give lower background values in the absence of CM. Therefore, stimulation indices with CM are higher. The eventual effect of Con A remaining in the CM is blocked by α-MM. Thus, the system seems to be useful for studies of the inhibitory effect of IFN on T cell proliferative response to a growth factor. Its sensitivity permits detection of as little as one antiviral unit of Mu-IFN (Fig. 2A and B). In the conventional assays, not entirely dependent on the exogenous IL-2, the inhibitory effects of IFNs on lymphocyte proliferation are usually much less.[13]

[13] T. Leanderson, V. Hillörn, D. Holmberg, E. L. Larsson, and E. Lundgren, *J. Immunol.* **129,** 490 (1982).

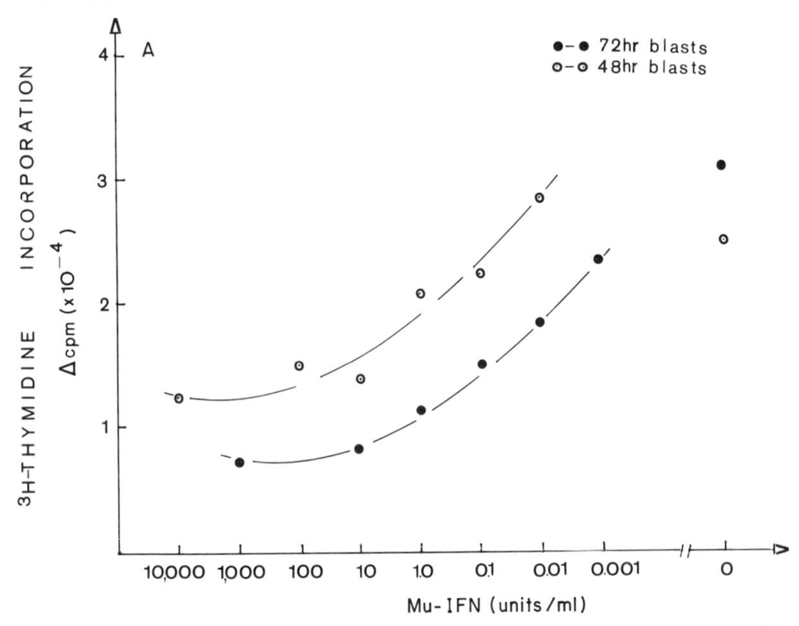

FIG. 2. Inhibition of the IL-2 response of CRT blasts by Mu-IFN. (A) CRT Con A blasts (48 or 72 hr) are cultured for 24 hr in the presence of CM and different doses of Mu-IFN-α/β. The proliferative response of blasts is presented as net cpm value (△cpm). Control cpm values in the absence of CM for 48 and 72 hr blasts are 20,641 and 1,815, respectively. (B) Percentage inhibition of the IL-2 response of CRT blasts by Mu-IFN-α/β is shown for two independent experiments with 72 hr blasts which had lower background proliferation in the absence of CM than 48 hr blasts (see text). The differences in the magnitude of inhibition observed from experiment to experiment are dependent on the level of the proliferative response, nevertheless, all data showed dose-dependent inhibition by Mu-IFN.

Measurement of the Antagonistic Effect of PDGF and Mu-IFN *in Vivo*

The Moloney sarcoma virus (MSV)-induced disease in mice appears to be an appropriate model to study interactions between Mu-IFN and PDGF. The MSV-induced tumors are composed of cells that were found to be sensitive to both PDGF and Mu-IFN.[14]

MSV is passaged in 1-week-old BALB/c mice. The stock virus in 20% suspension of tumor tissue in MEM is stored in liquid nitrogen.

[14] A. D. Inglot and O. Inglot, *Arch. Immunol. Ther. Exp.* **31,** 243 (1983).

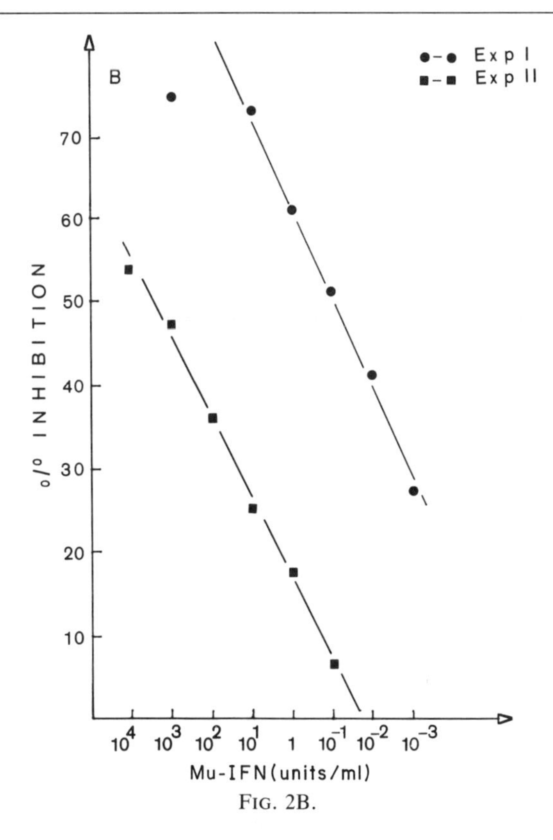

FIG. 2B.

For experiments 5- to 12-week-old male BALB/c mice are used. Groups of 6–10 animals are injected intramuscularly (im) into the right thigh with 0.1 ml of the preparation of MSV diluted 1 : 4 with MEM. Two days after the infection the mice are injected im at the site of MSV inoculation with 0.1 ml of Mu-IFN and/or PDGF. The injections are repeated daily for 2 weeks. Control mice receive 0.1 ml of PBS daily, also im at the site of MSV inoculation.

The growth of the MSV-induced early tumors is estimated by daily measurements of the size of each mouse thigh at the site of virus inoculation. Results are analyzed statistically by the Student's t test.

Results of one typical experiment are shown in Fig. 3. While Mu-IFN or PDGF given alone inhibits or enhances, respectively, the tumor proliferation, the mixture of both factors yields an intermediate effect. Thus, the administration of PDGF may diminish the antitumor effects of IFN.[14]

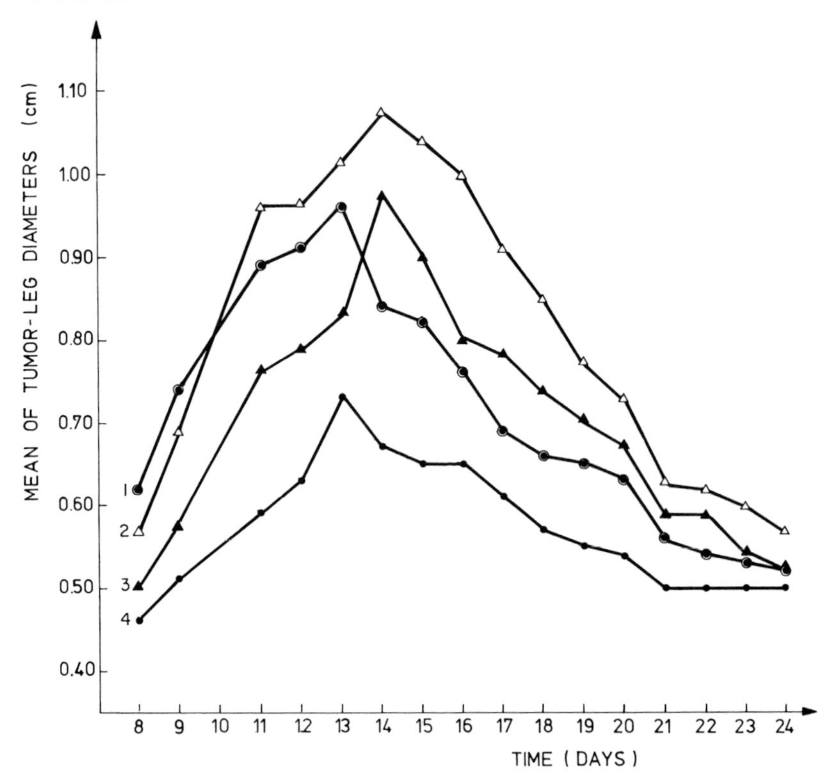

FIG. 3. Inhibition of the antitumor effect of Mu-IFN by simultaneous administration of PDGF to MSV-infected mice. (1) PBS-treated control mice infected with MSV; (2) PDGF, 500 units per mouse; (3) Mu-IFN, 10^5 units per mouse plus PDGF, 500 units per mouse; (4) Mu-IFN, 10^5 units per mouse. All of the preparations are administered im daily for 14 days. From Inglot and Inglot.[14]

Acknowledgments

Anna D. Inglot would like to thank M. Albin and O. Inglot for expert technical assistance, Drs. K. H. Fantes, M. Krim, and S. Pestka for samples of human interferons, and B. Narczewska, J. Czyrski, and W. Popik for preparation of growth factors and for helpful suggestions. Elżbieta Pajtasz thanks Dr. P. Kuśnierczyk for critical comments. This work was supported by 10.5 grant of the Polish Academy of Sciences.

[87] Measurement of Effect of (2′-5′)-Oligoadenylates and Analogs on Protein Synthesis and Growth of Cells

By ROBERT J. SUHADOLNIK, CHOONGEUN LEE, and DAVID H. WILLIS, JR.

Since the discovery of the (2′-5′)-oligoadenylate system in interferon-treated cells, the potential role of these unique 2′-5′-linked molecules in the regulation of cell growth and metabolism has been strongly implicated.[1–5] Several laboratories have also investigated the role of the dephosphorylated (2′-5′)-oligoadenylate cores[5a] on cellular processes.[6] The enzymatic and chemical syntheses of structurally modified (2′-5′)-oligoadenylates have been reported as a means to enhance the biological activity of these interesting molecules.[6–15]

This laboratory has been involved in the study of 2,5-A analogs with

[1] P. Lengyel, *Annu. Rev. Biochem.* **51,** 251 (1982).

[2] M. Etienne-Smekens, P. Vandenbussche, J. Content, and J. E. Dumont, *Proc. Natl. Acad. Sci. U.S.A.* **80,** 4609 (1983).

[3] A. Kimchi, H. Shure, and M. Revel, *Eur. J. Biochem.* **114,** 5 (1981).

[4] A. A. Creasey, D. A. Eppstein, Y. V. Marsh, Z. Khan, and T. C. Merigan, *Mol. Cell. Biol.* **3,** 780 (1983).

[5] P. B. Fisher, H. Hermo, Jr., D. R. Prignoli, I. B. Weinstein, and S. Pestka, *Biochem. Biophys. Res. Commun.* **119,** 108 (1984).

[5a] The abbreviations used are 2,5-p_3A_n, oligomer of adenylic acid with 2,5-phosphodiester linkages and a triphosphate at the 5′-end; 2,5-$p_3 3'dA_n$, oligomer of 3′-deoxyadenylic acid with (2′-5′)-phosphodiester linkages and a triphosphate at the 5′-end; core, 5′-dephosphorylated oligonucleotide.

[6] Y. Devash, A. Gera, D. H. Willis, M. Reichman, W. Pfleiderer, R. Charubala, I. Sela, and R. J. Suhadolnik, *J. Biol. Chem.* **259,** 3483 (1984), and references 15–31 therein.

[7] C. Baglioni, S. B. D'Allesandro, T. W. Nilsen, J. A. J. den Hartog, R. Crea, and J. H. van Boom, *J. Biol. Chem.* **256,** 3253 (1981).

[8] R. H. Silverman, R. H., D. H. Wreschner, C. S. Gilbert, and I. M. Kerr, *Eur. J. Biochem.* **115,** 79 (1981).

[9] P. W. Doetsch, J. M. Wu, Y. Sawada, and R. J. Suhadolnik, *Nature (London)* **291,** 355 (1981).

[10] J. Imai, M. I. Johnston, and P. F. Torrence, *J. Biol. Chem.* **257,** 12739 (1982).

[11] B. G. Hughes, P. C. Srivastava, D. D. Muse, and R. K. Robins, *Biochemistry* **22,** 2116 (1983).

[12] J. Justesen, D. Ferbus, and M. N. Thang, *Proc. Natl. Acad. Sci. U.S.A.* **77,** 4618 (1980).

[13] H. Samanta, J. P. Dougherty, and P. Lengyel, *J. Biol. Chem.* **255,** 9807 (1980).

[14] M. C. Haugh, P. J. Cayley, H. T. Serafinowska, D. G. Norman, C. B. Reese, and I. M. Kerr, *Eur. J. Biochem.* **132,** 77 (1983).

[15] J.-L. Drocourt, C. W. Dieffenbach, P. O. P. Ts'o, J. Justesen, and M. N. Thang, *Nucleic Acids Res.* **10,** 2163 (1982).

respect to the correlation between structure and biological activity. We have reported that cordycepin 5′-triphosphate (3′dATP) is a substrate for the 2,5-A synthetase in reticulocyte lysates, HeLa cells, and L cells.[9,16,17] The (2′-5′)-cordycepin analogs are metabolically more stable than the naturally occurring (2′-5′)-oligoadenylates. At 6.7×10^{-10} M, the 2,5-$p_3 3'dA_4$ inhibits protein synthesis in reticulocyte lysates by 61%.[16] The (2′-5′)-cordycepin trimer and tetramer triphosphates degrade globin ^{32}P-labeled mRNA as well as virus ^{32}P-labeled mRNA.[16,18] However, when assay conditions are modified (i.e., one amino acid < 50 μM), little or no inhibition of protein synthesis is observed with 2,5-p_3A_n or 2,5-$p_3 3'dA_n$.[16,18–21] When the amino acid concentration is increased to ≥ 50 μM, the (2′-5′)-oligoadenylates and analogs are most efficient inhibitors of protein synthesis.[16,18,19] Thus, caution must be exercised in interpreting results when protein synthesis is studied in the presence of the 2,5-A system.

In this chapter two aspects of the effect of (2′-5′)-oligoadenylates and analogs on cell growth are described: (1) inhibition of protein synthesis and cell growth in intact mammalian cells and (2) inhibition of tumor growth *in vivo*.

Inhibition of Protein Synthesis and Cell Growth by the Cordycepin Analog of 2,5-p_3A_n

Inhibition of protein synthesis through degradation of mRNA or rRNA by the 2,5-A dependent endonuclease is a well-established mechanism of the antiviral and anticellular action of the 2,5-adenylates.[1,22] Accordingly, the ability of 2,5-p_3A_n or its analogs to inhibit protein synthesis in cell-free extracts or in intact cells has been used as an index of biological activity

[16] R. J. Suhadolnik, Y. Devash, N. L. Reichenbach, M. B. Flick, and J. M. Wu, *Biochem. Biophys. Res. Commun.* **111,** 205 (1983).

[17] R. J. Suhadolnik, P. Doetsch, J. M. Wu, Y. Sawada, J. D. Mosca, and N. L. Reichenbach, this series, Vol. 79, p. 257.

[18] Y. Devash, R. J. Suhadolnik, B. Eslami, and J. M. Wu, *J. Biol. Chem.* (submitted for publication).

[19] R. J. Suhadolnik, Y. Devash, P. Doetsch, E. E. Henderson, J. M. Wu, W. Pfleiderer, and R. Charubala, *in* "Nucleosides, Nucleotides, and Their Biological Applications" (J. L. Rideout, D. W. Henry, and L. M. Beacham, III, eds.), p. 147. Academic Press, New York, 1983.

[20] P. F. Torrence, K. Lesiak, J. Imai, M. I. Johnston, and H. Sawai, *in* "Nucleosides, Nucleotides and Their Biological Applications" (J. L. Rideout, D. W. Henry, and L. M. Beacham, III, eds.), p. 67. Academic Press, New York, 1983.

[21] H. Sawai, J. Imai, K. Lesiak, M. I. Johnston, and P. F. Torrence, *J. Biol. Chem.* **258,** 1671 (1983).

[22] G. C. Sen, *Prog. Nucleic Acids Res. Mol. Biol.* **27,** 105 (1982).

of these nucleotides. However, the inhibitory effects of 2,5-p_3A_n on *in vitro* or *in vivo* protein synthesis via 2,5-A-dependent endonuclease activation are transient, probably due to the short half-life of 2,5-p_3A_n. Therefore, we have used the cordycepin analog of 2,5-p_3A_n,[17] in view of its increased resistance to degradation and its ability to inhibit *in vitro* translation by the cleavage of mRNA through endonuclease activation, to study the anticellular action of the (2′-5′)-oligoadenylates.

Reagents

2,5-$p_3 3'dA_3$ and 2,5-$p_3 3'dA_4$ (enzymatically[9,16,17] or chemically[23] synthesized as described)
2,5-p_3A_3 and 2,5-p_3A_4 (P-L Biochemicals Inc.)
MEM/F-10, Eagle's minimal essential medium supplemented with 10% fetal calf serum, penicillin (50 units/ml), streptomycin (50 μg/ml), and fungizone (2 μg/ml) (pH 7.0) (Grand Island Biological Co.)
Hepes buffered saline (8 g/liter NaCl, 0.37 g/liter KCl, 0.125 g/liter $Na_2HPO_4 \cdot 2H_2O$, 1 g/liter glucose, and 5 g/liter Hepes, pH 7.05)
Cell lines: mouse L929 and normal human fibroblasts (GM 731) maintained in monolayer culture in humidified 5% CO_2-in-air incubator
Calcium chloride ($CaCl_2$), 600 mM
Trichloroacetic acid, 5% (w/v)
Trypan blue, 0.1% in glass distilled H_2O
[^{35}S]Methionine, 1057 Ci/mmol (Amersham)

The inhibitory effect of 2,5-$p_3 3'dA_n$ on cellular protein synthesis has been studied by the calcium phosphate coprecipitation technique to introduce these highly negatively charged molecules into intact mammalian cells as has been described for 2,5-p_3A_n.[24] A preliminary report of these procedures has appeared.[25] 2,5-p_3A_n or 2,5-$p_3 3'dA_n$ (20 μl) is first mixed with $CaCl_2$ (20 μl of a 600 mM solution). Hepes-buffered saline (160 μl of a 1.25× solution, prepared from recrystallized Hepes) is then added dropwise. The resulting solutions (final $CaCl_2$ conc. 60 mM) are allowed to form fine coprecipitates before being applied to cell monolayers, maintained in MEM/F-10, which had been seeded in 24-well microtiter plates at 1.5×10^5 cells/well in a volume of 1 ml 24 hr earlier. The concentration of 2,5-p_3A_n or 2,5-$p_3 3'dA_n$ (in 200 μl) is 10^{-7} to 10^{-10} M for the first 45 min incubation at 25°. An equal volume (200 μl) of serum-free medium is then added to each well and the cells are incubated for an additional 90 min at 37°. The liquid in each well is then aspirated, [^{35}S]methionine (2 μCi in 1 ml of complete MEM/F-10 per well) is then added, and the cells are

[23] R. Charubala and W. Pfleiderer, *Tetrahedron Lett.* **21,** 4077 (1980).
[24] A. G. Hovanessian and J. N. Wood, *Virology* **101,** 81 (1980).
[25] C. Lee and R. J. Suhadolnik, *FEBS Lett.* **157,** 205 (1983).

labeled for 60 min at 37°. The [^{35}S]methionine incorporated into the TCA-precipitable material is determined.[14]

When introduced into intact mouse L cells and human fibroblasts by the calcium phosphate coprecipitation technique described above, 2,5-$p_3 3'dA_n$ (either enzymatically or chemically synthesized) results in inhibition of cellular protein synthesis in a dose-dependent manner. The inhibition caused by 2,5-$p_3 3'dA_n$ was comparable to that caused by the naturally occurring 2,5-p_3A_n (70% inhibition at 10^{-7} *M*) (Fig. 1). Adenosine, cordycepin, ATP, or 3′dATP (at 10^{-7} *M*) had no effect on protein

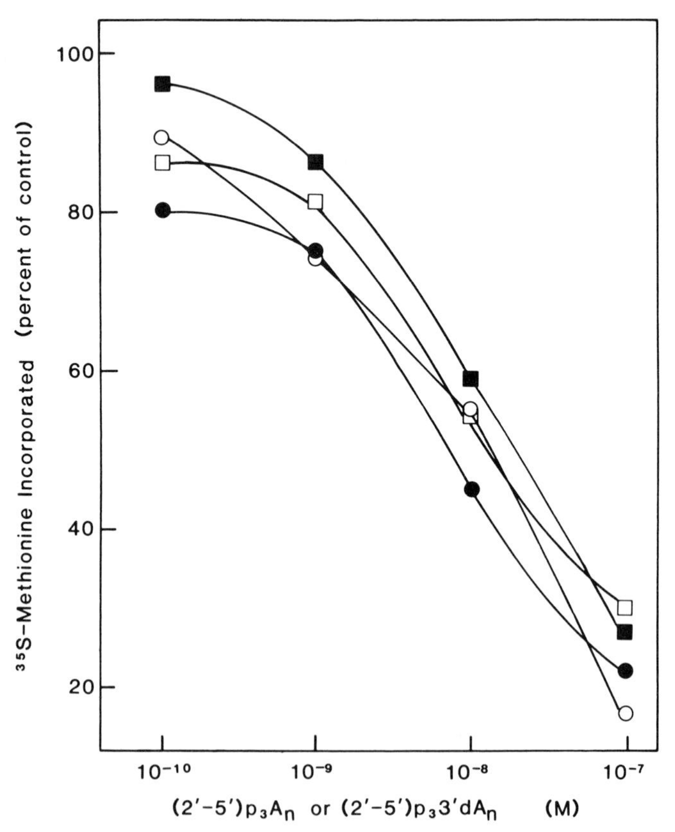

FIG. 1. Inhibition of protein synthesis by (2′-5′)p_3A_n and (2′-5′)$p_3 3'dA_n$ in mouse L929 cells. Inhibition of protein synthesis by (2′-5′)p_3A_3 (●), (2′-5′)p_3A_4 (○), (2′-5′)$p_3 3'dA_3$ (■), and (2′-5′)$p_3 3'dA_4$ (□) was determined in intact L929 cells by the calcium phosphate coprecipitation technique as described in the text. Control cultures were treated with calcium phosphate alone and the incorporation of [^{35}S]methionine into control cultures was taken as 100%. Reprinted with permission from Lee and Suhadolnik.[25]

synthesis. Furthermore, upon incubation of cells up to 96 hr after treatment followed by extensive washing, a more persistent inhibition is observed in cultures treated with 2,5-$p_3 3'dA_n$ than with 2,5-p_3A_n. This inhibition is manifested by a slower rate of recovery of protein synthesis and a longer duration of inhibition of protein synthesis (Fig. 2). Despite the persistent inhibitory action on cellular protein synthesis, no cytotoxicity of 2,5-$p_3 3'dA_n$ is observed, whereas 2,5-p_3A_n results in about 50% cell death after 48–96 hr. Cytotoxicity is determined by counting viable cell number after trypan-blue dye exclusion.

Inhibition of Murine Swarm Chondrosarcoma Growth *in Vivo* by 2,5-3′dA_3 Core

In view of our observations on the inhibition of mammalian cellular processes by 2,5-$p_3 3'dA_n$[25] and transformation of Epstein–Barr virus

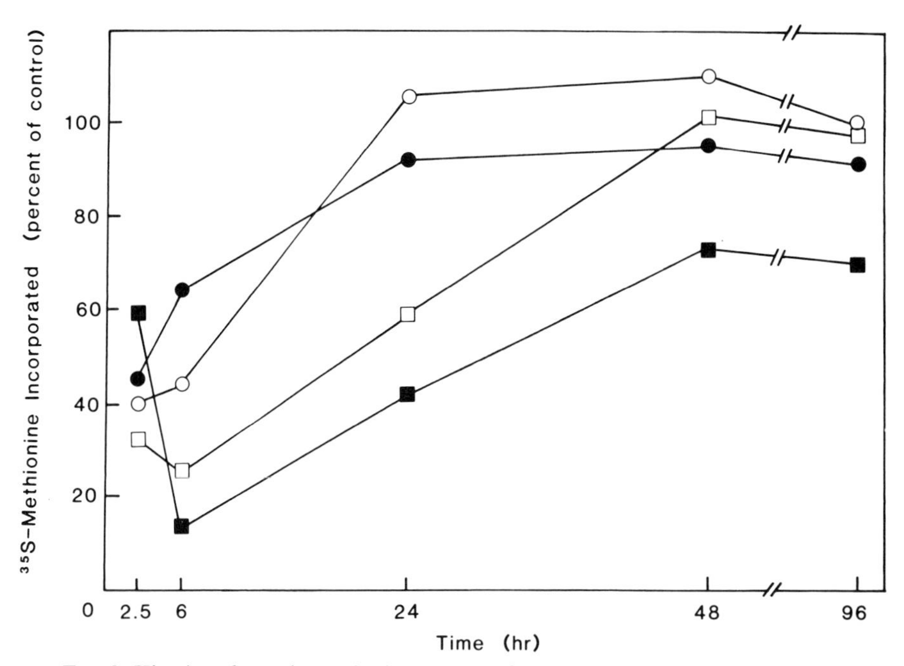

FIG. 2. Kinetics of protein synthesis recovery after (2′-5′)p_3A_n and (2′-5′)$p_3 3'dA_n$ treatment. Mouse L929 cell cultures were treated with oligonucleotides (1×10^{-8} *M*) as described in the text. Cellular protein synthesis after various times of recovery (abscissa) was calculated considering the total radioactivity incorporated per viable cell. Control cultures treated with calcium phosphate alone were assayed in parallel at each time point. (2′-5′)p_3A_3 (●), (2′-5′)p_3A_4 (○), (2′-5′)$p_3 3'dA_3$ (■), (2′-5′)$p_3 3'dA_4$ (□). Reprinted with permission from Lee and Suhadolnik.[25]

(EBV) infected lymphocytes in culture by the 2,5-3′dA_3 core,[26,27] the direct injection of this core analog might be an effective treatment of tumor growth *in vivo*. We selected chondrosarcoma as a tumor which occurs in humans and is quite resistant to chemotherapy and radiation therapy.[28] There is an equivalent animal tumor model, the murine Swarm chondrosarcoma.[29] Therefore, we investigated the ability of the cordycepin analog of 2,5-A_3 core to inhibit growth of this murine chondrosarcoma *in vivo*.[30,31]

Reagents

100–125 g male, Sprague–Dawley rats (≥3 per treatment group)
2,5-A_3 core and 3,5-A_3 core (P-L Biochemicals Inc.)
2,5-3′dA_3 core (chemically synthesized as described[23])
2′-Deoxycoformycin (dCF) (obtained from Warner-Lambert/Parke-Davis)
Cordycepin and 3′dAMP (prepared according to Suhadolnik *et al.*[32])
Phosphate-buffered saline (PBS)
RPMI-1640 medium (Grand Island Biological Co.)

Preparation of Tumor-Bearing Animals

The murine Swarm chondrosarcoma is transferred to test animals by the following procedure. Tumors (30–100 g) are excised from a tumor-bearing animal and finely minced asceptically with a Stately–Riggs cutting blade while keeping tissue moist with PBS. The minced tissue (free of fibrous tissue) is collected in a sterile flask and rinsed 3–4 times with PBS (PBS : tumor volume, 10 : 1) to remove blood cells. Tumor pieces in a minimal volume of RPMI-1640 medium (5 ml medium/3 g tumor pieces) are passed sequentially through a 50-ml syringe bearing a 16-gauge butterfly needle. This procedure ensures smooth delivery of tumor pieces to the

[26] P. W. Doetsch, R. J. Suhadolnik, Y. Sawada, J. D. Mosca, M. B. Flick, N. L. Reichenbach, A. Q. Dang, J. M. Wu, R. Charubala, W. Pfleiderer, and E. E. Henderson, *Proc. Natl. Acad. Sci. U.S.A.* **78,** 6699 (1981).

[27] E. E. Henderson, P. W. Doetsch, R. Charubala, W. Pfleiderer, and R. J. Suhadolnik, *Virology* **122,** 198 (1982).

[28] W. D. McCumbee, J. M. Harrelson, and H. E. Lebowitz, *Cancer Res.* **43,** 513 (1983).

[29] B. D. Smith, G. R. Martin, E. J. Miller, A. Dorfman, and R. Swarm, *Arch. Biochem. Biophys.* **166,** 181 (1975).

[30] D. H. Willis, W. Pfleiderer, R. Charubala, and R. J. Suhadolnik, *Fed. Proc., Fed. Am. Soc. Exp. Biol.* **42,** 1833 (Abstr. 443) (1983).

[31] D. H. Willis, Jr., R. Charubala, W. Pfleiderer, and R. J. Suhadolnik, *Cancer Res.* (in press).

[32] R. J. Suhadolnik, M. B. Lennon, T. Uematsu, J. E. Monahan, and R. Baur, *J. Biol. Chem.* **252,** 4125 (1977).

animal. The resulting slurry is injected in 2-ml aliquots at two subcutaneous sites in ether-anesthetized animals. Best tumor growth is obtained in ventral abdominal sites. Following tumor transfer, animals are maintained on sawdust bedding to minimize irritation and are fed and watered *ad libitum*.

Treatment of Tumor-Bearing Animals with 2,5-A_3 Core and Analog

Five to six days after tumor implantation, animals are treated with 2,5-A_3 core or 2,5-3′dA_3 core. Since each animal bears two tumors, three animals will provide six tumors for statistical analysis. 2,5-A_3 core or core analog is diluted in PBS. Anesthetized animals are injected intratumorally with the test compound or PBS (control) with a 1-ml tuberculin syringe (26.5-gauge needle). Tumor size is monitored prior to treatment and subsequently at 4 day intervals by measuring tumor diameter with a ruler. The logarithm of tumor diameter increases proportionately with time after a 4–5 day lag.

Animals treated with 0.5 or 5 μmol of 2,5-3′dA_3 core/100 g body weight/tumor show significant inhibition of tumor growth for 7 or 13 days, respectively. There is no inhibition of tumor growth by 2,5-A_3 core, 3,5-A_3 core, adenosine, or AMP. Five micromoles of cordycepin and 3′dAMP have slight effects on tumor growth, with 3′dAMP having a greater effect than cordycepin. The antitumor effects of the 2,5-3′dA_3 core are locally mediated as demonstrated by the inhibition of only one tumor following injection of 2,5-3′dA_3 core (10 μmol/100 g body weight) into one of two tumors per animal. Because the 2,5-3′dA_3 core may be acting as a prodrug of cordycepin, we sought to inhibit the deamination of cordycepin by adenosine deaminase. This was accomplished by the simultaneous addition of 2,5-3′dA_3 core with 2′-deoxycoformycin (dCF), a known potent inhibitor of adenosine deaminase.[33] When animals were treated simultaneously with 0.25 μmol 2,5-3′dA_3 core plus 0.1 μmol dCF/100 g body weight/tumor (Fig. 3, ▲), the inhibition of tumor growth by the 2,5-3′dA_3 core was strongly potentiated. Of the 6 tumors treated in this manner, 4 tumors were not visible by day 23 after treatment and the remaining 2 tumors were barely discernable. Neither 2,5-3′dA_3 core (0.25 μmol/100 g body weight, △) nor dCF (0.1 μmol/100 g body weight, ●) showed any effect on tumor growth when added alone. When tumors were treated with 0.75 μmol cordycepin plus 0.1 μmol dCF/tumor/100 g body weight (equivalent to 0.25 μmol 2,5-3′dA_3 core), no additional inhibition of tumor growth was observed (◆). Thus, the regression of tumors observed with

[33] R. J. Suhadolnik, "Nucleosides As Biological Probes." Wiley, New York, 1979.

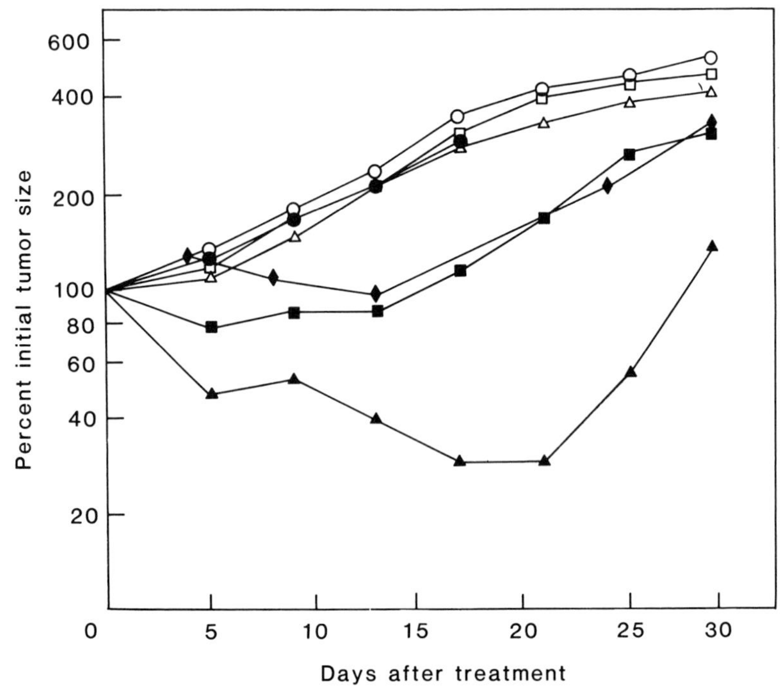

FIG. 3. Synergistic effect of (2′-5′)3′dA$_3$ core and dCF on chondrosarcoma growth. Animals (100–125 g) were treated as described in the text. Control, ○; 0.1 μmol dCF/tumor/100 g body weight, ●; 0.25 μmol 2,5-3′dA$_3$ core/tumor/100 g body weight, △; 0.25 μmol cordycepin/tumor/100 g body weight, □; 0.25 μmol 2,5-3′dA$_3$ core plus 0.1 μmol dCF/tumor/100 g body weight, ▲; 0.25 μmol cordycepin plus 0.1 μmol dCF/tumor/100 g body weight, ■; 0.75 μmol cordycepin plus 0.1 μmol dCF/tumor/100 g body weight, ◆. Results are expressed as a percentage of the initial tumor size at the time of treatment (day zero).

2,5-3′dA$_3$ core plus dCF is apparently due to the activity of the oligomer as a prodrug.

Concluding Comments

There appear to be several possible explanations for the anticellular effects of (2′-5′)-oligoadenylate cores in different cell systems. Initially it was thought that the prolonged inhibitory effect of 2,5-p$_3$3′dA$_n$ on protein synthesis in L cells was due to the increased stability of the analog and, therefore, activation of the 2,5-A-dependent endonuclease for an extended time. However, the differential cytotoxicity of 2,5-p$_3$A$_n$ and 2,5-

$p_3 3'dA_n$ in L cells suggested to us that the cordycepin analog might cause a preferential hydrolysis of mRNA in addition to the persistent activation of the endonuclease. Such selectivity of RNA hydrolysis might result in a pronounced inhibition of protein synthesis by 2,5-$p_3 3'dA_n$, but a relatively rapid turnover of mRNA would prevent cells from irreversible damage. Indeed, we have observed preferential hydrolysis of mRNA over rRNA by 2,5-$p_3 3'dA_n$ in reticulocyte lysates and L cell extracts.[34] The possibility that the effect of 2,5-$p_3 3'dA_n$ on protein synthesis in intact cells is due to the same preferential hydrolysis is under investigation. In similar studies, Hughes and Robins reported that the 2-5-linked analogs of tubercidin, sangivamycin, toyocamycin, and 8-bromoadenosine also show preferential degradation of mRNA over rRNA.[35] As demonstrated with the murine Swarm chondrosarcoma, the 2,5-$3'dA_3$ core is a potent, nontoxic inhibitor of tumor growth in animals, acting as a prodrug of cordycepin. Glazer and co-workers reported that 2,5-$3'dA_3$ core also acts via a prodrug mechanism in HT-29 human colon carcinoma cells in culture.[36] Eppstein *et al.* reported a similar mechanism for the xyloadenosine core.[37] The mechanism by which 2,5-$3'dA_3$ core inhibits transformation of human lymphocytes differs in that the 2,5-$3'dA_3$ core has been isolated intact from TCA-soluble cytoplasmic extracts of human lymphocytes, with *no* evidence of 5'-rephosphorylation.[38] These data suggest that a mechanism independent of the 2',5'-A/interferon system is operating in EBV-infected lymphocytes. It remains to be determined if the 2,5-$3'dA_3$ core isolated is due to binding and/or uptake.

Acknowledgments

This work was supported in part by U.S. Public Health Service Grant GM26134, National Science Foundation Research Grant PCM-8111752, and NRSA Training Grant 5 T32 AM07162.

[34] Y. Devash and R. J. Suhadolnik, *Fed. Proc., Fed. Am. Soc. Exp. Biol.* **43,** 2020 (Abstr. 3501) (1984).

[35] B. G. Hughes and R. K. Robins, *Biochemistry* **22,** 2127 (1983).

[36] M. S. Chapekar and R. I. Glazer, *Biochem. Biophys. Res. Commun.* **115,** 137 (1983).

[37] D. A. Eppstein, Y. V. Marsh, and B. B. Schryver, *Virology* **131,** 341 (1983).

[38] R. J. Suhadolnik, P. W. Doetsch, Y. Devash, E. E. Henderson, R. Charubala, and W. Pfleiderer, *Nucleosides Nucleotides* **2,** 351 (1983).

[88] Assay of Effect of (2′-5′)-Oligoadenylate on Macrophages

By XIN-YUAN LIU, HONG DA ZHENG, NING WANG, WEN HUA REN, and T. P. WANG

The main component of the (2′-5′)-oligo(A) family is $pppA_2'p_5'A_2'p_5'A$ abbreviated $(2'\text{-}5')P_3A_3$. It is produced by the (2′-5′)-oligo(A) synthetase from interferon-induced cells or from reticulocytes. William *et al.*[1] and Hovanessian and Wood.[2] found that (2′-5′)-oligo(A) can inhibit virus multiplication. We found that (2′-5′)-oligo(A) can protect cells from virus infection to some extent[3,4] and can inhibit virus multiplication *in vitro* in vertebrates,[3–7] in plant tissue[8] and *in vivo* in insects (silkworm).[4,9,10] These antiviral effects on animal and plant cells suggests that (2′-5′)-oligo(A) and related compounds may be natural antiviral substances.[4,11–14]

The effect of interferon on immune cells is an important aspect of its action. (2′-5′)-Oligo(A) can regulate the action of natural killer cells[4,14–16]

[1] B. R. G. Williams, R. R. Golgher, and I. M. Kerr, *FEBS Lett.* **105,** 47 (1979).

[2] A. G. Hovanessian and J. N. Wood, *Virology* **101,** 81 (1980).

[3] X.-Y. Liu, Y.-M. Wen, Y.-C. Lou, H.-J. Lou, and T. P. Wang, *Kexue Tongbao* **26,** 850 (1981) (in Chinese).

[4] X.-Y. Liu, Y.-M. Wen, Y.-D. Hou, K. Yao, Y.-C. Lou, Z.-Q. Chen, H.-D. Zheng, W.-H. Ren, T.-Z. Lin, Z.-R. Huang, and D.-B. Wong, *in* "The Biology of Interferon System" (E. De Maeyer *et al.,* eds.), p. 115. Elsevier/North-Holland Biomedical Press, Amsterdam, 1981.

[5] K. Yao, Y.-Z. Shou, J.-R. Jin, X.-Y. Liu, Y.-C. Lou, B.-L. Li, and T. P. Wang, *Acta Biochim. Biophys. Sin.* **15,** 72 (1983).

[6] X.-Y. Liu, K. Yao, Y.-Z. Shou, J.-R. Jin, Y.-M. Wen, Y.-C. Lou, B.-L. Li, and T. P. Wang, *Sci. Sin.* **26,** 809 (1983).

[7] X.-X. Zhao, Y. D. Hou, Y.-L. Su, and X.-Y. Liu, *Acta Acad. Med. Sin.* **4,** 367 (1982).

[8] X.-H. Zhang, B. Tian, Z.-Q. Chen, and X.-Y. Liu, *Acta Biochim. Biophys. Sin.* **15,** 586 (1983) (in Chinese).

[9] Y.-C. Lou, Z.-R. Huang, and X.-Y. Liu, *Acta Biochim. Biophys. Sin.* **17,** 206 (1985) (in Chinese).

[10] W.-B. Zhong, Z.-R. Huang, Y.-L. Lu, Y.-C. Lou, and X.-Y. Liu, *Kexue Tongbao* **27,** 761 (1982) (in Chinese).

[11] Y. Devash, I. Sela, and R. J. Suhadolnik, this volume [99].

[12] Y. Devash, R. J. Suhadolnik, and I. Sela, this volume [100].

[13] E. M. Martin, D. M. Reisinger, A. G. Hovanessian, and B. R. G. Williams, this series, Vol. 79, p. 273.

[14] X.-Y. Liu, *J. Exp. Pathol.* (in press).

[15] P.-K. Tian, T.-Z. Lin, J.-X. Hong, Q.-J. Zhang, H.-Y. Wu, X.-S. Yu, X.-Y. Liu, H.-D. Zheng, Y.-C. Lou, and M.-H. Yao, *Acta Cell. Biol. Sin.* **1,** 36 (1984).

and macrophages.[4,17,18] $(2'\text{-}5')P_3A_3$ enhances the activity of macrophages from different species (mice, rats, guinea pigs, rabbits, and human) and antagonizes the inhibitory effect of α-fetoprotein on macrophages.[17] An assay for the direct effect of (2′-5′)-oligoadenylate on macrophages is described in this chapter. This action of $(2'\text{-}5')P_3A_3$ on the macrophages appears to be mediated by a receptor.[19,20] Measurement of this receptor is described in this volume.[21]

Materials and Reagents

(2′-5′)-Oligo(A) was synthesized and separated by a method which can yield about 1 g of total (2′-5′)-oligo(A) and 400–500 mg of $(2'\text{-}5')P_3A_3$ from reticulocytes from four rabbits. The yeast *Candida albicans* (Robin) Berk was obtained from the Department of Microbiology, Shanghai Second Medical College, stored at −20°, and inactivated by boiling at 100° for 5 min before use. The yeast suspension consisted of 9×10^6 cells/ml in cell culture medium. ICR mice were supplied by the Animal Center, Shanghai Branch, Academia Sinica. Wild-type mice, rats, and guinea pigs were from the Animal Department of the Shanghai Institute of Biochemistry. Human macrophages were induced under the skin of the forearm by cantharidin.[22] RPMI-1640 medium was purchased from Serva. Sodium mercaptoacetate, calf serum, and heparin were from the same sources described by Liu *et al.*[21] Culture medium for macrophages contained RPMI-1640 medium, 20% calf serum (inactivated at 56°, 30 min), and 10 units heparin/ml, pH 7.0. Giemsa stain was prepared routinely[23,24] and diluted 10- to 20-fold with Sörensen solution (0.067 *M* sodium phosphate, pH 6.8) before use.

---

[16] R. B. Herberman, J. R. Ortaldo, C. Riccardi, T. Timonen, A. O. Schmidt, A. Maluish, and J. Djev, *in* "Interferons" (T. C. Merigan and R. M. Friedman, eds.), p. 287. Academic Press, New York, 1982.

[17] N. Wang, H.-D. Zheng, B.-L. Li, X.-Y. Liu, and T. P. Wang, *Acta Biochim. Biophys. Sin.* **14,** 623 (1982).

[18] X.-Y. Liu, H.-D. Zheng, N. Wang, B.-L. Li, W.-H. Ren, R.-L. Kong, and T. P. Wang, *Sci. Sin.* **26,** 1057 (1983).

[19] B.-L. Li and X.-Y. Liu, *Sci. Sin.* **28,** 697 (1985).

[20] B.-L. Li and X.-Y. Liu, *Sci. Sin.* **28,** 844 (1985).

[21] X.-Y. Liu, B.-L. Li, and S. W. Li, this volume [52].

[22] Tumor Hospital, Chinese Academy of Medical Science, *Nat. Med. J. China (Peking)* **56,** 229 (1976).

[23] G. Giemsa, *Zentralbl. Bakteriol. Parasitenkd., Infektionskr. Hyg., Abt. 1: Orig.* **31,** 429 (1902).

[24] G. Giemsa, *Zentralbl. Bakteriol., Parasitenkd., Infektionskr. Hyg., Abt. 1: Orig.* **32,** 307 (1902).

Preparation of Macrophages

Preparation of Peritoneal Macrophages

Wild type mice (20 g) can be used directly. ICR mice need to be induced first by mercaptoacetate but their macrophages yield more reproducible results than macrophages from wild-type mice. Male ICR mice (about 20 g) were injected intraperitoneally with 0.3–0.4 ml of 3% sodium mercaptoacetate dissolved in normal saline containing 0.2% agar and killed 3–4 days later. Peritoneal macrophages were washed out by injecting 2 ml of culture medium into the abdomen and massaging it. Immediately thereafter, peritoneal fluid was removed and the number of macrophages was counted and adjusted to 3×10^6/ml with culture medium. Precisely 0.1 ml of the macrophage suspension was seeded onto cleaned cover glasses (1×2 cm) which were put on a stack of three microscope slides in a Petri dish with cover. After a 30 min incubation at 37° under 3% CO_2, the cover glasses were taken out with forceps and washed 4–5 times with 2 ml of Hanks' solution to remove the nonadherent cells; the adherent macrophages on the cover glasses can be used for assay of phagocytosis. Peritoneal macrophages prepared without mercaptoacetate from guinea pigs were obtained as described above.

Preparation of Macrophages from Lung

Rats were injected intramuscularly with 0.8 ml of 2.5% sodium pentobarbital (for rabbits, the pentobarbital was administered to the exposed neck muscle) and the trachea exposed by surgical operation. Lung macrophages were washed out twice by perfusion with Hanks' solution (100 ml in total for the rabbit, 8 ml for the rat), collected by low speed centrifugation, and resuspended in culture medium to a concentration of 3×10^5/ml. Because the macrophages from lung are rather pure, they do not need further purification by adhesion as is necessary with murine peritoneal macrophages.

Assay of Phagocytosis

Assay of Effect of (2′-5′)P_3A_3 on Peritoneal Macrophages

To the adherent macrophages, 0.1 ml of freshly diluted (2′-5′)P_3A_3 in culture medium was added. The effect of (2′-5′)P_3A_3 on phagocytosis by macrophages is dependent on the concentration of (2′-5′)P_3A_3 with an optimum at about 10^{-7} *M* (Fig. 1). Therefore, it is usually diluted with

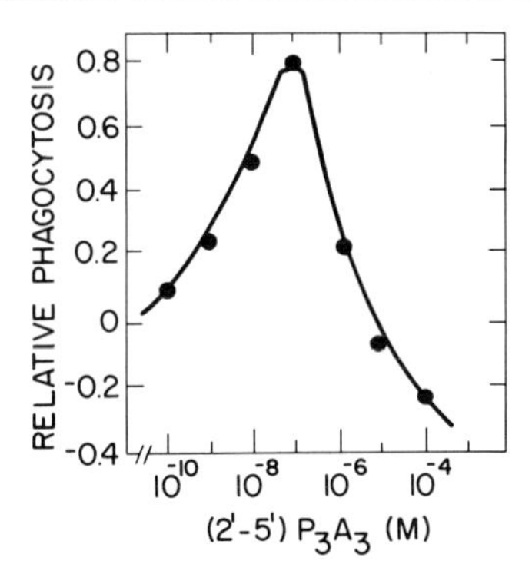

FIG. 1. Effect of (2′-5′)P_3A_3 concentration on the phagocytosis of yeast by macrophages. The concentration of (2′-5′)P_3A_3 is calculated on the basis of 45 A_{260} units/μmol of trimer.

culture medium to this concentration just before use. For a control, 0.1 ml of culture medium without (2′-5′)P_3A_3 was added. Cover glasses were returned to the Petri dish as before and incubated at 37° for 30 min. Phagocytosis by macrophages was activated during this incubation.

To measure phagocytosis, the culture medium on the cover glasses was drained, replaced with 0.1 ml of yeast suspension, and incubated at 37° for 30 min. Higher concentrations of yeast will stimulate nonspecific activation and are not recommended. Cells were stained with Giemsa as follows. The cover glasses were washed 4–5 times with Sörensen solution to remove free yeast, dried completely, fixed for 10 min in methanol, dried again, soaked in Sörensen solution for 10 min, transferred to Giemsa solution for 8–10 min, washed with water, dried, and sealed by neutral glue on microscope slides. The phagocytosis was examined with a microscope. A typical ingestion of yeast by macrophages is shown in Fig. 2. The arrow shows the ingested yeast particles. The percentage of cells exhibiting phagocytosis is much higher for the (2′-5′)P_3A_3 group than for the control cells (see the table).

Assay of Effect of (2′-5′)P_3A_3 on Macrophages from Lung

Lung macrophages obtained above were treated with 10^{-7} *M* (2′-5′)P_3A_3 in 1 ml cell suspension at 37° for 20 min. Yeast were added at ratio of 3 : 1 (yeast : macrophage) and the incubation was continued for another 30 min. After centrifugation, the macrophages were spread onto a microscope slide, then fixed and stained as described above.

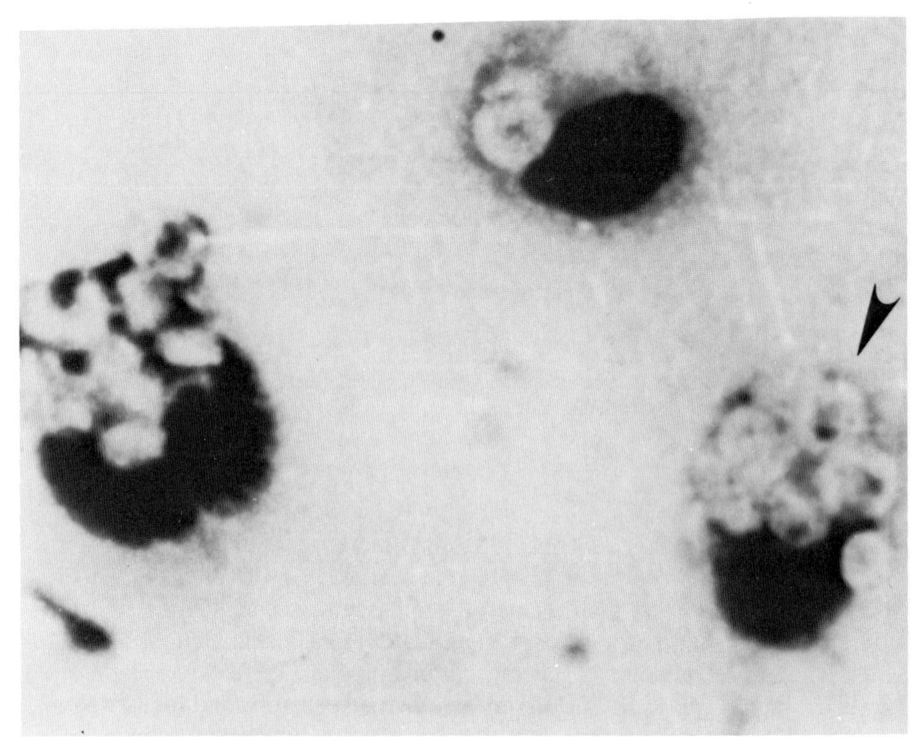

FIG. 2. Yeast ingested by macrophages. The arrow points to ingested yeast particles.

Measurement of Phagocytic Activity

Phagocytic activity was expressed in terms of two parameters, the fraction of cells exhibiting phagocytosis (F) and relative phagocytic activity (RP), defined as

$$F = \frac{\text{Number of macrophages which have ingested yeast}}{\text{Total number of macrophages counted}}$$

$$RP = \frac{F\ (\text{sample}) - F\ (\text{control})}{F\ (\text{control})}$$

Some typical results are shown in the table. A positive value for relative phagocytosis (RP) indicates an increase of phagocytosis, whereas a negative value indicates a decrease relative to control cells. To score a macrophage positive for ingestion of yeast, at least one yeast particle

ACTIVATION OF MACROPHAGES FROM DIFFERENT ORIGINS BY (2′-5′)P_3A_3[a]

Source of macrophages	Control (F)	Concentration of (2′-5′)P_3A_3 (M)				
		10^{-10}	10^{-9}	10^{-8}	10^{-7}	10^{-6}
		Relative phagocytosis (RP)				
Rabbit	0.18	0.44	0.38	0.50	1.33	0.88
Mouse	0.04	0.22	0.26	1.04	2.22	0.31
Guinea pig	0.10	0.13	0.33	0.28	1.13	0.45
Human	0.20	0.35	0.80	1.40	0.60	0.45

[a] Peritoneal macrophages from wild type rabbits and guinea pigs were used. Macrophages from ICR mice and from humans were prepared as described in the text. F is defined as the fraction of macrophages ingesting yeast after examining at least 100 macrophages. Relative phagocytic activity (RP) is defined in the text. Data from Liu *et al.*[18]

must be seen within the macrophage. Sometime more than 10 yeast particles can be seen within the macrophages.

Remarks

Previous studies on the biological effect of (2′-5′)-oligo(A) were always done with the help of adding $CaCl_2$ or other agents to introduce (2′-5′)-oligo(A) into the cells.[1,2] Such procedures do not provide any *in vivo* information. Measuring the direct effect of (2′-5′)P_3A_3 on macrophages as described here without the use $CaCl_2$ or other agents can provide some information for the possible effect of (2′-5′)P_3A_3 and its derivatives on the immune system which may be useful clinically.

Acknowledgments

This work was supported by the Chinese Academy of Science. The authors wish to thank Dr. Sidney Pestka for his support and critical review of the manuscript, Dr. Jerome Langer for reviewing the manuscript, and Ms. Sophie Cuber and Ms. Wendy Ewald for typing the manuscript.

[89] Radioimmunoassay for Detection of Changes in Cell Surface Tumor Antigen Expression Induced by Interferon

By J. W. GREINER, P. HORAN HAND, D. WUNDERLICH, and D. COLCHER

Introduction

The interferons are multifaceted biological modifiers that exert antiviral, antiproliferative, and several different immunomodulatory actions.[1,2] Their ability to increase the amount of tumor antigens on the surface of human carcinoma cells would have potential utility in enhancing the *in vivo* detection and/or therapy of tumors with monoclonal antibodies in clinical trials. These antigens are recognized by a panel of monoclonal antibodies that were generated in our laboratory with membrane-enriched fractions from liver metastases from patients with primary breast carcinoma. Partially purified preparations of human interferon have been shown to increase the amount of HLA-A,B,C and β_2-microglobulin on the surface of several different human cell types.[3,4] A radioimmunoassay with live cells was used to determine whether human recombinant leukocyte (alpha) A interferon (IFN-αA) could increase the presentation of tumor antigens on the surface of human breast and colon carcinomas as detected by our monoclonal antibodies.

Materials for Routine Cell Culture

Equipment

Humidified 37° Lunaire (model B106136) incubator (Lunaire Environmental, Inc., Williamsport, PA)
Laminar flow biological safety cabinet (Contamination Control, Inc., Lansdale, PA)
Liquid nitrogen storage system (for long-term −170° storage of cells and other biologics) (Union Carbide, Inc., Gatlinburg, TN)

[1] S. Baron, F. Dianzani, and G. J. Stanton, eds., *Tex. Rep. Biol. Med.* **41,** 1 (1982).
[2] I. Gresser and M. G. Tovey, *Biochim. Biophys. Acta* **516,** 231 (1978).
[3] A. Dolei, M. R. Capobianchi, and F. Ameglio, *Infect. Immun.* **40,** 172 (1983).
[4] T. Y. Basham, M. F. Bourgeade, A. A. Creasey, and T. C. Merigan, *Proc. Natl. Acad. Sci. U.S.A.* **79,** 3265 (1982).

Cell Culture Media and Supplements

Dulbecco's modified Eagle's medium (DMEM), with 4.5 g glucose/liter (GIBCO, Grand Island, NY, Catalog No. 320-1965)
Bovine pancreas insulin (Sigma Chemical, Inc., St. Louis, MO, catalog No. I-5500). A stock solution of 2.5 mg/ml in 0.1 *M* HCl is stored at −20° after sterile filtration
Sodium pyruvate (GIBCO, Catalog No. 890-1840). A stock solution of 100 m*M* in triple-distilled water is stored at −20° after sterile filtration
MEM nonessential amino acids (100×) (GIBCO, Catalog No. 320-1140).
Gentamicin sulfate (50 mg/ml) (M. A. Bioproducts, Walkersville, MD, Catalog No. 17-5182)

Cell Culture

Several human cells were used as representative target cells during the present study. The human breast carcinoma cell line, MCF-7, was grown in DMEM supplemented with 10% heat-inactivated fetal bovine serum (FBS), 5.0 μg/ml bovine insulin, 1× nonessential amino acids, 1 m*M* sodium pyruvate, and 50 μg/ml gentamicin. The human colon carcinoma cell line, WiDr, was grown in EMEM with 10% FBS, while the A375 cells (human melanoma) grew in DMEM with 10% FBS. All three cell types were subcultured once a week with trypsinization in 0.1% trypsin/0.5 m*M* EDTA (Meloy Laboratories, Inc., Springfield, VA, Catalog No. SR 1017).

Monoclonal Antibodies

The generation and subsequent purification of the monoclonal antibodies have been described elsewhere.[5,6] Briefly, somatic cell hybrids were prepared by fusing splenic lymphocytes from mice immunized with membrane enriched fractions of two human breast tumor metastases to the liver with the non-immunoglobulin-secreting myeloma cell line P3-NS1-AG4. All hybridomas were cloned twice by limiting dilution. The monoclonal antibodies were purified from ascites fluid obtained from pristane-primed BALB/c mice inoculated with 10^7 hybridoma cells. The immunoglobulins were precipitated with saturated ammonium sulfate and

[5] D. Colcher, P. Horan Hand, M. Nuti, and J. Schlom, *Proc. Natl. Acad. Sci. U.S.A.* **78,** 3199 (1981).
[6] P. Horan Hand, M. Nuti, D. Colcher, and J. Schlom, *Cancer Res.* **43,** 728 (1983).

the appropriate antibodies eluted from an ion-exchange column. The fractions containing immunoglobulin were pooled and dialyzed against Dulbecco's phosphate-buffered saline (PBS), pH 7.2 (GIBCO, Catalog No. 310-4040). The protein concentration was determined and the purified antibodies stored in aliquots of 0.1 to 1.0 ml in Dulbecco's PBS at −20°.

Production, Characterization, and Storage of Human Recombinant Leukocyte Interferon

The expression and purification of IFN-αA (specific activity 2–4 $\times 10^8$ units/mg protein) were previously described.[7–9] The interferon was provided in ampules of 3×10^6 units/ml and upon receipt was diluted with RPMI 1640 containing 25 m*M* Hepes, 1% BSA, and stored in 1×10^6 units/ml aliquots at −70°. Immediately prior to use an aliquot was thawed, diluted, and added to the growth medium at the indicated amount. Periodically, an aliquot of the leukocyte interferon was rechecked for antiviral activity which remained essentially unchanged when stored at −70°.

Cell Preparation (MCF-7 Cells) for Live Cell Radioimmunoassay

Cells were maintained in complete growth medium as monolayer cultures in T-75 flasks (Costar, Cambridge, MA, Catalog No. 3275) and harvested in log phase growth at 70–80% confluency by the procedure outlined below.

1. The growth medium is removed by aspiration.
2. The T-75 flask(s) are washed with 8.0 ml 0.1% trypsin/0.5 m*M* EDTA.
3. Eight milliliters of the trypsin–EDTA solution is then added to the cells and the T-75 flask returned to the humidified 37° incubator.
4. After 4–5 min the flasks are removed from the incubator, returned to the biological hood, and released from the surface of the flask by gentle tapping
5. Ten–15 ml of the complete growth medium containing 10% FBS is added to the flask and the cells transferred to 50-ml sterile centrifuge tubes (Falcon, Oxnard, CA, Catalog No. 2070).

[7] S. Pestka, *Arch. Biochem. Biophys.* **211,** 1 (1983).

[8] T. Staehelin, D. S. Hobbs, H.-F. Kung, and S. Pestka, this series, Vol. 78, p. 505.

[9] T. Staehelin, D. S. Hobbs, H.-F. Kung, C.-Y. Lai, and S. Pestka, *J. Biol. Chem.* **256,** 9750 (1981).

6. The cells are centrifuged at 600 *g* for 5 min in a table-top clinical centrifuge (IEC, Needham Heights, MA).

7. The cell pellets are resuspended in 10 ml of complete medium and a cell count done with a hemocytometer.

8. The cells are diluted to a final concentration of 5×10^5 cells/ml in complete growth medium and 5×10^4 in 100 μl seeded in each well of a 96-well microtiter plate (Costar, Catalog No. 3596). Following seeding, the wells are checked visually to ascertain approximately equal numbers of cells per well.

As described above, cells to be evaluated for surface tumor antigen expression are seeded in hard plastic, tissue culture grade, Costar 96-well plates at a concentration of 5×10^4 cells/well. The cells are seeded 18–24 hr in advance of the assay and should be approximately 80% confluent at the time of the live cell radioimmunoassay. Recombinant human leukocyte interferon is added in various amounts for specified periods of time.

Iodination of Goat Anti-Mouse IgG or Protein A

Materials

Affinity purified goat anti-mouse IgG (heavy chain, gamma) (Kirkegaard and Perry Labs, Gaithersburg, MD, Catalog No. 011802)
Staphylococcus aureus Protein A (Pharmacia Fine Chemicals, Piscataway, NJ, Catalog No. 17-0770-01)
Iodo-gen (1,3,4,6-tetrachloro-3α,6α-diphenylglycouril) (Pierce Chemical Co., Rockford, IL, Catalog No. 28600)
Sephadex G-25 (Pharmacia Fine Chemicals, Catalog No. 17-003302)
Sodium phosphate, dibasic
Na ^{125}I, specific activity 17.4 Ci/mg (New England Nuclear, Boston, MA, Catalog No. NEZ 033A)
12 × 75 siliconized glass tubes (Corning, Corning, NY)

Procedure

1. Ten milliliter Sephadex G-25 columns can be prepared in 10-ml disposable pipettes. The columns are equilibrated with 0.01 *M* Na_2HPO_4 (adjusted to pH 7.2 with HCl) and several can be prepared and stored in 0.02% sodium azide (Sigma Chemical Co., St. Louis, MO, Catalog No. S-2002).

2. Prior to iodination the G-25 column is washed with 3 column volumes of 0.01 *M* sodium phosphate, pH 7.2.

3. When iodinating the goat-anti-mouse IgG antibody, 50 μg of the antibody is dissolved in 0.1 *M* sodium phosphate (pH 7.2) and brought up

to a final volume of 100 μl with distilled H_2O. When Protein A is used, 20 μg is usually iodinated.

4. Iodo-gen tubes are prepared by adding the Iodo-gen to chloroform at a final concentration of 10 mg/ml. Aliquot 20 μl (200 μg) into 12 × 75 siliconized glass tubes and allow to evaporate to dryness. The Iodo-gen tubes can be stored sealed (cork stoppers) at −20°. The mixture in step #3 is added to the Iodo-gen just prior to iodination.

5. To the Iodo-gen tube containing either the antibody or Protein A, 1 mCi of $Na^{125}I$ is added and the reaction allowed to continue for 2 min at room temperature.

6. In order to separate the free from bound ^{125}I the reaction mixture is taken up into a Pasteur pipette and layered onto a G-25 column. The reaction tube is rinsed with a small amount (1.0 ml or less) of 0.01 *M* sodium phosphate (pH 7.2) which is added to the column.

7. The column is eluted with the same buffer and 0.5 ml fractions are collected with the labeled antibody generally in fractions 5–7.

8. The two or three fractions of highest activity are combined, diluted with two volumes of 0.01 *M* sodium phosphate containing 1% BSA and dialyzed overnight against 4 liters of 0.01 *M* sodium phosphate containing 5 m*M* NaN_3 (pH 7.2).

9. The iodinated protein is then stored at 4°.

Radioimmunoassay of Surface Antigen

1. With an 8- or 12-part manifold [Drummond Scientific Co., Broomall, PA, Catalog No. 30-00-093 (8-part), 30-00-096 (12-part)], the medium is aspirated from each well of the 96-well plate. During all aspirations care should be taken to place the manifold at approximately the same position in the well and avoid contact with the cells.

2. With a multi-channel Titertek multipipetter (Flow Laboratories, McLean, VA, Catalog No. 77-888-0) 100 μl of RPMI-1640 containing 10% bovine serum albumin (BSA) and 0.08% (w/v) NaN_3 is added to each well and the 96-well plate covered and incubated with this "blocking" medium at 37° for 1 hr.

3. The 10% BSA-containing medium is aspirated and the plate washed once with RPMI-1640 containing 25 m*M* Hepes and 1% BSA. The washing consists of submerging the plate in wash buffer and inverting the plate over a sink. The excess wash buffer is drained onto absorbant paper.

4. Each monoclonal antibody at appropriate dilutions is added in 50 μl RPMI-1640 containing 25 m*M* Hepes, 1% BSA, and 0.08% NaN_3 to duplicate wells and incubated at 37° for 1 hr.

5. After 1 hr, the solution is removed by aspiration and the cells washed twice as described.

6a. When ^{125}I-labeled goat-anti-mouse IgG is used for the second antibody, the stock solution is diluted to approximately 75,000 cpm/50 μl which is added to each well and incubated at 37° for 1 hr.

6b. If ^{125}I-labeled protein A is to be used, the monoclonal antibody must be first reacted with a F(ab')$_2$ fragment of a rabbit anti-mouse IgG antibody. A stock solution of the rabbit antibody (1 mg/ml) is diluted 1 : 1000 in RPMI-1640 containing 1% BSA and 50 μl is added to each well and incubated at 37° for 1 hr. The rabbit anti-mouse F(ab')$_2$ is removed, the cells washed, 50 μl of the ^{125}I-labeled protein A diluted to 50,000 cpm/50 μl is added and the plates incubated at 37° for 1 hr.

7. After incubation with either radiolabeled protein the cells are washed three times and examined for inappropriate cell loss (i.e., no more than 5% loss should occur during the assay).

8. In order to quantitate the radioactivity the cells are lysed by adding 100 μl of 2 *N* NaOH to each well. The lysed cells and resulting solution are then absorbed with a cotton Q-tip which is added to 12 × 75 tubes and counted in a gamma scintillation spectrometer (LKB Instruments, Gaithersburg, MD, Model 1274).

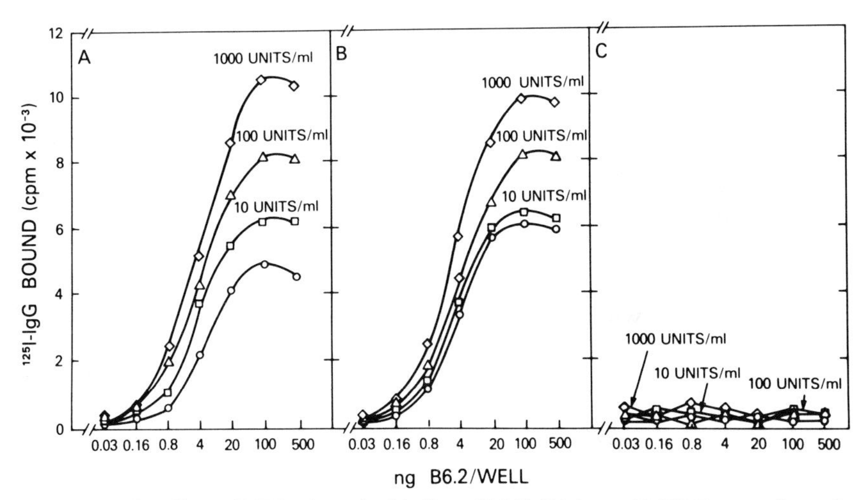

FIG. 1. The effect of IFN-αA on the binding of MAb B6.2 to a 90,000 Da protein on the surface of human cells. Human breast (MCF-7, A), colon (WiDr, B) carcinoma cells, and human melanoma (A375, C) cells were assayed for the surface binding of MAb B6.2 after a 24 hr incubation in the presence of 10 (□), 100 (△), or 1000 (◇) units/ml of IFN-αA. The binding of B6.2 to the untreated cells is denoted by the circles. All data represent the mean of at least four experiments with a standard error less than 10%.

Concluding Comments

Monoclonal antibody B6.2 recognizes a 90,000 Da determinant which is primarily expressed on the surface of human breast and colon carcinoma cells. Figure 1 summarizes the effect of human IFN-αA on the binding of B6.2 to various human cell types. In Fig. 1A and B, the 90K surface antigen which binds B6.2 is constitutively expressed on the human breast (MCF-7) and colon (WiDr) cells, but was undetectable on the surface of human melanoma (Fig. 1C) and normal fibroblasts (WI-38 or Flow 4000 cells; data not presented). The treatment of MCF-7 or WiDr cells with 10, 100, or 1000 units/ml of IFN-αA resulted in a dose-dependent increase in B6.2 binding to the surface of both the MCF-7 and WiDr cells. In contrast, the addition of high amounts of interferon (up to 10,000 units/ml) was unable to initiate binding (i.e., induce surface tumor expression) to the A375, WI-38, or Flow 4000 cells. The results suggest that recombinant human leukocyte interferon can increase the expression of tumor antigens on the surface of human breast and colon carcinoma cells. This enhanced presentation of surface antigens thus aids in overcoming the tumor cell heterogeneity of expression of many tumor antigens and may increase the effectiveness of labeled monoclonal antibodies for tumor localization and therapy in clinical trials.

[90] Measurements of Changes in Histocompatibility Antigens Induced by Interferons

By MARIANNE HOKLAND, IVER HERON, PETER HOKLAND, PER BASSE, and KURT BERG

It now generally is agreed that a major nonantiviral effect, among others, of interferon is the modulation of histocompatibility antigens on the surface of a variety of cells.[1-3] Concerning MHC class I antigens (HLA-A, -B, and -C), all three types of Hu-IFNs (alpha, beta, and gamma) induce an increase in expression and synthesis of these antigens.[4,5] However, when looking at MHC class II antigens, Hu-IFN-γ is

[1] P. Lindahl-Magnusson, P. Leary, and I. Gresser, *Nature (London), New Biol.* **237,** 120 (1972).

[2] I. Heron, M. Hokland, and K. Berg, *Proc. Natl. Acad. Sci. U.S.A.* **75,** 6215 (1978).

[3] M. Hokland, I. Heron, and K. Berg, *J. Interferon Res.* **1,** 483 (1981).

[4] T. Y. Basham and T. C. Merigan, *J. Immunol.* **130,** 1492 (1983).

[5] V. E. Kelley, W. Fiers, and T. B. Strom, *J. Immunol.* **132,** 240 (1984).

the only type of interferon able to induce a substantial increase in the expression of these antigens and, furthermore, cells from the monocyte/macrophage lineage are especially sensitive to this effect of immune IFN.

Several methods have been reported for the demonstration of MHC enhancement, but we have found an indirect immunofluorescence method with subsequent measurement in a fluorescence-activated cell sorter (FACS-II)[6] to be easy, convenient, and sensitive to very small changes. Furthermore, by storing the data in the FACS, it is possible to recall them in a computer system for further calculations.

Treatment of Cell Suspensions with Interferons

Leukocytes from healthy donors were separated from heparinized venous blood by centrifugation on Ficoll/Isopaque. Mononuclear cells were washed twice, counted, and cultured in RPMI-1640 containing 5% normal pooled, heat-inactivated human AB-serum at a concentration of 3×10^6 cells/ml. Use of human serum is necessary to block Fc-IgG receptors and avoid nonspecific antibody binding. IFN was added to these cultures in microliter quantities at final concentrations ranging from 100 to 1000 units of IFN/ml.

In addition to PBLs, two human cell lines were treated with IFNs: the K562 and HL-60 myeloid leukemia cell lines. These were kept in continuous culture in RPMI-1640 containing 10% heat-inactivated FCS, penicillin (100 units/ml), streptomycin (100 μg/ml), and Hepes. At the time of addition of IFN (which was done as for PBL), they were in an exponential growth phase. The IFN preparations were either recombinant Hu-IFN-α2 (kindly donated by Schering Inc., New Jersey), or native immune IFN (kindly donated by Dr. Jerzy Georgiades, IML, Stafford, Texas).

Immunofluorescence Staining

Before conjugation, the cells were washed twice (10 min at 150 *g*) in Medium 199 containing 5% heat-inactivated FCS and 15 m*M* NaN_3 in order to avoid capping. After the last wash cells were resuspended in 1 ml of medium and labeled with either a monoclonal antibody directed against a framework of MHC class I antigen (W6/32 purchased from Seralab, Sussex Downs, UK) or a monoclonal antibody directed against a framework MHC class II antigen (I2, kindly provided by Dr. L. M. Nadler, Harvard Medical School, Boston, Massachusetts). As negative controls,

[6] M. R. Loken and L. A. Herzenberg, *Ann. N.Y. Acad. Sci.* **254,** 163 (1975).

cells were incubated with ascites from a nonproducing hybridoma at the same concentrations as used for the specific antibodies. After a 45 min incubation period at 0° and two subsequent washing cycles in the cold, the cells were incubated with a FITC-conjugated goat anti-mouse antibody (Dakopatts, Copenhagen, Denmark) at 0° for another 45 min before they were washed and processed for the fluorescence-activated cell sorter (FACS-II, Becton-Dickinson, Mountain View, California).

Measurement of Cell Fluorescence Intensity and Computation of the Results

Cells to be analyzed in a FACS-II were washed three times and resuspended in Medium 199 containing 5% heat-inactivated FCS and 19 mM NaN_3 at a concentration of approximately 1×10^6 cells/ml. Cells (90,000–100,000 of each sample) were analyzed at 525 nm after excitation at 588 nm with an argon laser (500 mW). The mean fluorescence intensity per cell was computed by using the formula:

$$\sum_{255}^{x=1} C_x \cdot N_x \Big/ \sum_{255}^{x=1} N_x$$

in which C_x is the relative fluorescence intensity per cell (expressed as the channel number) and N_x is the number of cells counted in the appropriate channel. The photograph of a histogram from the screen of the FACS-II showing the number of stained cells against the fluorescence intensity (Fig. 1) was projected on the digitizer plateau of a Hewlett-Packard 9815–9864 calculator system and data collected for computation of the number of cells at different fluorescence intensities. Scatter signals given by the analyzed cells were likewise recorded. Controls analyzed included non-stained cell suspensions to correct for background fluorescence and cells stained with ascites from nonproducing hybridomas as described above. When the same suspension of labeled cells was analyzed twice, less than 2% difference in the computed mean fluorescence intensity was found.

Alternatively, we also defined the median fluorescence intensity as that channel, above and below which lay 50% of the cell population, in order to calculate the median fluorescence per cell. As for the mean fluorescence intensity, this method was used to compare the expression of MHC class I and II antigens of cells incubated with and without IFN.

Finally, the number of cells positive for a given antigen was determined by subtracting the reactivity of the nonproducing ascites mentioned above.

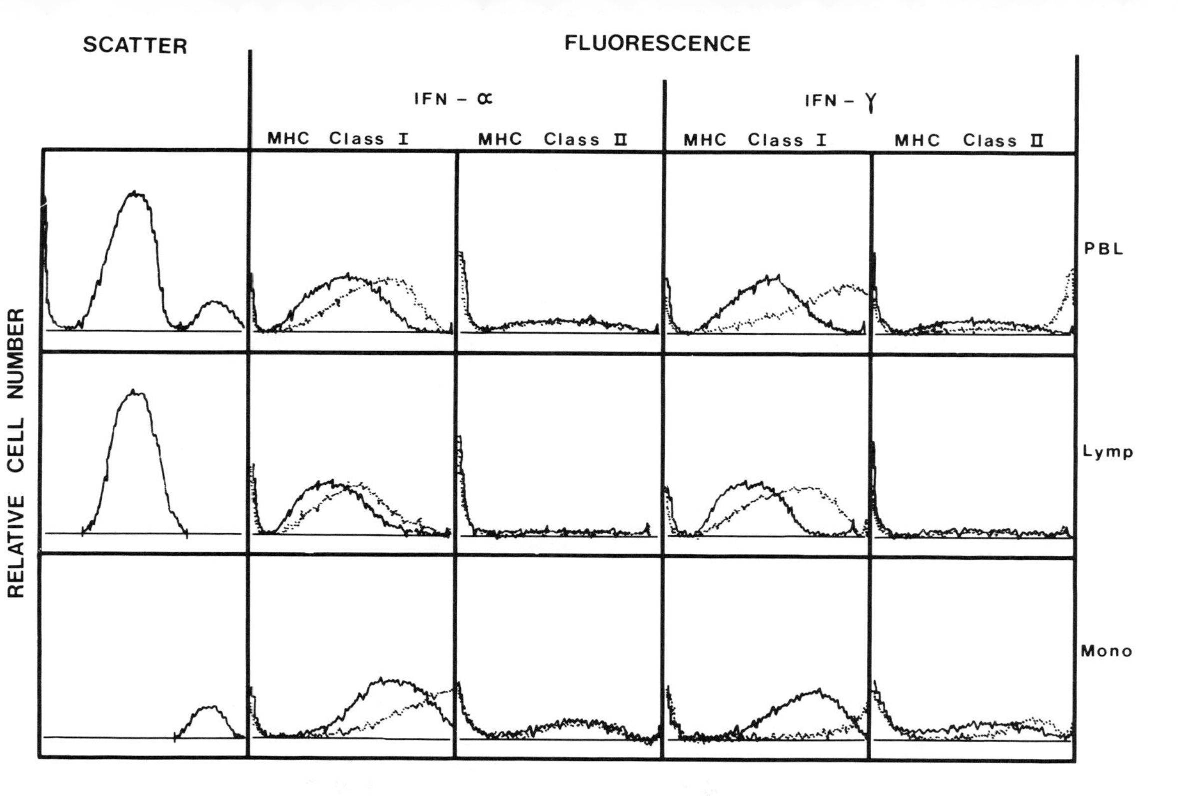

FIG. 1. Peripheral blood mononuclear cells were gated according to forward light scatter into lymphocytes after interferon treatment and labeling with anti-MHC monoclonal antibodies as described. Ungated cells were analyzed as controls (upper row). Each box shows the expression of the MHC antigen in question (x axis) and the cell number (y axis) for IFN-treated (···) and untreated (——) cells. Approximately 100,000 cells were analyzed.

Concluding Comments

One advantage of the method described here is that it makes it possible to distinguish between the behavior of different subpopulations present in the cell suspension in question after IFN treatment. This is illustrated in Fig. 1, showing the different effects of IFN-α and IFN-γ on the human lymphocyte and monocyte fractions separated according to forward light scatter (cell size) with respect to change in the expression of MHC class I and II antigens. It appears that while both the Hu-IFN-α and Hu-IFN-γ increase the MHC class I expression, only IFN-γ strongly increased the expression of MHC class II antigens, and only on the monocyte-rich cell fraction. On the other hand, no effect was seen on the lymphocyte-rich fraction, indicating that immune IFN-γ does not affect the MHC class II expression on resting lymphocytes. Thus, from the data generated by this method, the selective effect of different types of IFNs on lymphocyte subpopulations can be demonstrated easily.

In addition to the direct "visual" information provided immediately by the FACS-II, quantitative information about the mean or median fluorescence intensity also is obtainable easily. As seen in Table I, which shows the median fluorescence intensities (calculated as described above), this means of expressing the IFN enhancement of MHC expression gives additional information when seen together with the histogram (Fig. 1).

Finally, analysis by cell fluorescence can yield the percentage of cells positive for a given surface antigen. Such information is important when

TABLE I

EFFECT OF INTERFERONS ON MHC ANTIGEN EXPRESSION MEASURED BY CYTOFLUOROMETRY[a]

		Median fluorescence intensity					
		Unseparated cells		Lymphocytes		Monocytes	
Antigen	IFN (units/ml)	Hu-IFN-α2	Hu-IFN-γ	Hu-IFN-α2	Hu-IFN-γ	Hu-IFN-α2	Hu-IFN-γ
Class I	0	151	150	132	135	172	169
	100	210	242	170	215	239	252
Class II	0	112	113	76	78	165	163
	100	110	135	75	80	168	229

[a] The data in the table indicate the median fluorescence intensity of the cells determined by the median channel number as described in the text. The amount of Hu-IFN-α is expressed in units/ml with respect to the international leukocyte interferon reference standard (69/19B, MRC, UK). Since no international reference standard existed for IFN-γ at the time the experiments were executed, the 69/19B standard was used. Fibroblasts and VSV were employed for interferon titrations.

TABLE II
EFFECT OF INTERFERONS ON MHC ANTIGEN EXPRESSION BY MYELOID CELL LINES MEASURED BY CYTOFLUOROMETRY[a]

		Percentage of positive cells			
		K562 cells		HL-60 cells	
Antigen	IFN (units/ml)	Hu-IFN-α2	Hu-IFN-γ	Hu-IFN-α2	Hu-IFN-γ
Class I	0	2 ± 2	3 ± 2	91 ± 5	95 ± 4
	100	10 ± 1	24 ± 3	98 ± 3	97 ± 3
Class II	0	3 ± 3	1 ± 1	4 ± 2	4 ± 2
	100	5 ± 2	3 ± 3	3 ± 1	15 ± 2

[a] The data in the table indicate the percentage of cells positive for the antigen in question. The means ±SD of triplicate analyses are shown. See footnote to Table I for interferon units and assays.

testing the ability of different types of IFNs to induce the expression of antigens poorly expressed by the cell types in question. This is exemplified in Table II which shows the percentage of cells positive for class I and class II antigens upon IFN treatment of two myeloid cell lines. It appears that on K562 cells both IFN-α and IFN-γ are able to induce the expression of class I antigens which virtually are absent on these cells before IFN treatment. Furthermore, it can be seen that IFN-γ, but not IFN-α, induces an increase in MHC class II antigens on the human myelocytic leukemia cell line, HL-60. Interestingly, IFN-γ did not seem to enhance class II antigen expression on K562 cells (this has been found in several other experiments).

In conclusion, the effects of interferons on cell membrane antigen expression are easily and reliably evaluated by cell fluorescence. Even though we have provided data only from experiments evaluating the effect of interferon on MHC antigen expression, the method is generally applicable to other antigens as well.[7,8]

Acknowledgment

We thank Ms. Anni Skovbo, Mr. Søren Petersen, and Ms. Monia Pilgaard for excellent technical assistance, and Prof. Lars Bolund for access to the FACS-II machine. This work was supported by the Danish Cancer Society and the Danish MRC.

[7] M. Hokland, P. Hokland, I. Heron, and S. F. Schlossman, *Scand. J. Immunol.* **17,** 559 (1983).

[8] M. Hokland, J. Ritz, and P. Hokland, *J. Interferon Res.* **3,** 199 (1983).

[91] Purification, Assay, and Characterization of the Interferon Antagonist: Sarcolectin

By FRANÇOISE CHANY-FOURNIER, PAN HONG JIANG, and CHARLES CHANY

Introduction

In an adult organism, interferon is produced either when a virus penetrates into the cells or when some immune competent cells are triggered by foreign antigen(s). The interferon secreted leads to an antiviral state in the cells. However, under natural conditions, in the organism, the cells recover virus sensitivity relatively rapidly. It can be postulated, therefore, that a regulatory system may account at least in part for the recovery of normal status. This intracellular regulation could also be affected by mediators acting from the outside.

It has been possible to isolate, purify, and identify a group of substances present in a great variety of tissues which, although lacking completely any structural or biological resemblance with interferon, contribute to the decay of the antiviral state it establishes.[1-6] It is presently unknown to what extent these tissue antagonists modify other biological functions of interferon. It is clear, however, that these substances, presently named "sarcolectins," are not all identified and that other inhibitors of the different interferon functions might also exist.[6]

Preparative Procedures and Methodology

Preparation of Normal and Tumor Extracts

The normal and tumor tissues of human or animal origin are dissected, cut into pieces of about 2–3 g, washed 3 times with PBS, dried by lyophili-

[1] F. Fournier, S. Rousset, and C. Chany, *Proc. Soc. Exp. Biol. Med.* **132,** 943 (1969).

[2] C. Chany, A. Grégoire, and J. Lemaítre, *C. R. Hebd. Seances Acad. Sci.* **269,** 1236 (1969).

[3] C. Chany, J. Lemaítre, and A. Grégoire, *C. R. Hebd. Seances Acad. Sci.* **269,** 2626 (1969).

[4] P. Brown, M. Sarragne, C. Chany, C. J. Gibbs, and D. C. Gajdusek, *in* "Effects of Interferon on Cells, Viruses and the Immune System" (A. Geraldes, ed.), p. 337. Academic Press, New York, 1975.

[5] F. Chany-Fournier, A. Pauloin, and C. Chany, *Proc. Natl. Acad. Sci. U.S.A.* **75,** 2333 (1978).

[6] P. H. Jiang, F. Chany-Fournier, B. Robert-Galliot, M. Sarragne, and C. Chany, *J. Biol. Chem.* **258,** 12361 (1984).

TABLE I
CELL AGGLUTINATING ACTIVITY OF CRUDE TISSUE EXTRACTS (NORMAL AND TUMOR) ON DIFFERENT CELL LINES AND ON T AND B LYMPHOCYTES

Crude tissue extracts	Cell lines					Lymphocytes	
	MEF	L-929	Fr-3T3	XC	NRK	T	B
	(cytoagglutinating units)[a]						
Normal extracts	256	256–512	512	256	256	256	32
Tumor extracts	512	512–1024	1024	512	512	256–512	32

[a] The cytoagglutinating units (CA units/0.05 ml) are expressed as the inverse of the dilution of tissue extracts that agglutinate about 50% of the cell population. Mouse T and B lymphocytes are assayed in the same test.

zation, and kept at −20° before extraction. Under these conditions the material is stable up to 10 years. For extraction, the dried material is ground to obtain a fine powder. The fine powder is rehydrated in Eagle's medium at a final concentration of 4 g/100 ml. The suspension is agitated at 4° overnight and clarified by low-speed centrifugation (6000 *g* for 30 min at 4°), followed by high-speed centrifugation (90,000 *g* for 2 hr at 4°) in the 30 rotor of a preparative Spinco ultracentrifuge. The supernatant is stored at −20°. Prior to assay, the extracts are filtered through a 1.2-μm Millipore membrane for clarification.

Estimation of the Biological Activity of Tissue Extracts

Cell Agglutinating Activity. The cell agglutinating activity of normal or tumor extracts can be tested by using different cell lines: mouse embryonic fibroblasts (MEF), mouse L cells, clone 929 (L-929), Fisher 3T3 rat cells (Fr-3T3), infected by a thermo-sensitive mutant of Rous virus, rat embryonic fibroblasts transformed by the Prague strain of Rous virus (XC), and normal rat kidney (NRK) cells (Table I). Normal splenic lymphocytes from Swiss mice were separated by filtration on scrubbed nylon fiber (type 200, Fenwall Laboratories) to obtain selected B and T lymphocytes, according to a previously described method,[7] and were finally resuspended in RPMI serum-free medium at the concentration of 10^6 cells/

[7] G. Janossy and M. F. Greaves, *Clin. Exp. Immunol.* **9,** 483 (1971).

TABLE II
INHIBITION OF CELL AGGLUTINATION OF HAMSTER OR HUMAN SARCOLECTIN BY SUGARS

	Sarcolectin (cytoagglutinating units)[a]	
	Hamster	Human
Control	256	64
D-Galacturonic	2	16
D-Glucosamine · HCl	256	16
D-Galactosamine · HCl	256	8
N-Acetylneuraminic acid	256	2
Fucose	16–32	64

[a] The cytoagglutination titer is expressed as the 50% agglutination end point per 0.05 ml with Fr-3T3 cells.

ml (Table I). It is of importance to use V-shaped plastic microtiter plates usually employed for hemagglutination and untreated for tissue culture.

The cytoagglutinating activity is estimated with an inverted light microscope by 2-fold dilutions of tissue extract in the presence of either 5×10^5 murine or rat cells or 1×10^6 murine lymphocytes. The mixture is incubated at 4° for 2 hr. The results are expressed as the reverse of the dilution that agglutinates about 50% of the cell population. The average agglutinating titers vary from 256 to 512 cell units (CA units).

Cytoagglutination Inhibition Assay and Sugar Solutions. The sugars are dissolved at 0.05 *M* in MEM adjusted to pH 7.3, filtered for sterility through a 0.22-μm Millipore membrane, and stored at −4°. Tissue extracts (0.05 ml) are diluted 2-fold with MEM in microtiter plates. Sugars (0.05 ml) are mixed with the extracts 1:2 at 4° for 20 min. The cells are then added and the microtiter plates replaced at 4° for 2 hr. The inhibitory effect of the sugars is determined by the dilution which inhibits about 50% of cell agglutination (Table II).

Antagonistic Effect of Normal or Tumor Tissue Extracts on the Interferon-Induced Antiviral State. In order to induce antiviral protection, mouse L 929 cells (10^5 cells/ml) are treated with murine interferon[8] (200 units/ml) at 37° for 5 hr. Interferon is then discarded and the cells incubated with crude normal or tumor extracts (4 g/100 ml) for 20 hr at 37°.

[8] E. Dussaix, A. Grégoire, C. Chany, D. C. Thang, and M. N. Thang, *J. Gen. Virol.* **64,** 285 (1983).

TABLE III
ANTAGONISTIC EFFECT OF CRUDE TISSUE EXTRACTS (NORMAL AND TUMOR) ON THE INTERFERON-INDUCED ANTIVIRAL STATE

Time of additions		EMC yield (challenge virus yield)[a]
5 hr	20 hr	
Medium	Medium	4096
Interferon	Medium	2
Interferon	Normal extracts	256
Interferon	Tumor extracts	2048–4096
Medium	Normal extracts	4096
Medium	Tumor extracts	4096

[a] EMC yield expressed as hemagglutinating units (HA units/0.05 ml) is titered with human erythrocytes. The experimental values in the table represent the mean of three assays.

The tissue extracts are removed, and the cells are infected with EMC virus (at a multiplicity of infection of 0.2) diluted in medium containing 2% calf serum. The yield of infectious virus is measured after 18 hr at 37°. As shown in Table III, both tissue extracts restore almost (normal extracts) or completely (tumor extracts) the yield of the challenge virus in the resistant cells. As shown in the controls, in the absence of interferon, tissue extracts do not affect virus multiplication.

Enzymes and Inhibitors

Proteases. A stock solution of Trypsin (Choay, twice purified, 180 UAE/mg) is prepared at a concentration of 2.5 mg/ml in Eagle's MEM at pH 7.8. Pepsin solution (Worthington, twice crystallized, 2.675 units/mg) is prepared at a concentration of 2 mg/ml diluted in Eagle's MEM at pH 2 (Eagle's MEM: 0.01 *N* HCl at a ratio of 1 : 50). The reaction mixture is stopped by adjusting the pH to 7.3. Pronase (Kaken Chemical Co. Ltd., 45,000 p.u.k./g) is prepared at a concentration of 1 mg/ml in Eagle's MEM at pH 7.8.

In these three cases, the enzyme–substrate reaction (v/v) is assayed for 1 hr at 37° and stopped by the addition of protease inhibitor overnight at 4°. However, pepsin can be added to crude extracts (v/v) at pH 2 for 20 hr at 37° without altering the molecule in a detectable manner.

Protease Inhibitors. Iniprol (Choay, 10^6 inhibitory units) is used at the concentration of 10^5 units/ml of proteases overnight at 4°. Phenylmethylsulfonyl fluoride (Calbiochem-Behring) is prepared at a concentration of 0.1% (w/v) in 95% ethanol; this stock solution of phenylmethylsulfonyl fluoride used at 0.1 ml inhibits 1 ml of the protease reaction mixtures.

Nucleases. DNase I (Worthington, 2000 units/mg) is assayed at a concentration of 1 unit/μl with 500 μg of DNA in a buffer containing 0.1 *M* $MgSO_4$ in a total volume of 1 ml. After an incubation time of 1 hr at 37°, the enzymatic activity is stopped by the addition of 0.2 *M* EDTA at pH 8.

RNase (Sigma, type II/A from bovine pancreas) with an enzymatic activity of 50–75 units/mg of protein is used in a medium containing Ca^{2+} and Mg^{2+} for 1 hr at 37° and pH 7.3.

Purification Procedures

Gel Filtration. A column (30 × 1.5 cm, Pharmacia) is packed with Sephacryl S-200 (Pharmacia) and washed with PBS. The absorbance of the proteins present in the effluent volume is monitored at 280 nm with an ISCO type 6 optical unit and UA-5 absorbance monitor. For the determination of the molecular weight of components in crude extracts (normal or tumor extracts), a standardization curve is previously established. A 1-ml sample of pepsin-treated extract is added to the top of the column and eluted with PBS (flow rate 20 ml/hr). The collection of fractions is started immediately after the loading of the column. Each fraction is tested for biological activity.

Ion-Exchange Chromatography. Pepsin-treated and Sephacryl S-200 filtered active fractions are dialyzed against Tris acetate buffer (20 m*M* NaCl, 0.1 m*M* EDTA, 15 m*M* 2-mercaptoethanol, 20 m*M* Tris acetate, pH 7.6). The dialyzed fractions are loaded onto a DEAE-cellulose column (30 × 1.5 cm, Pharmacia) previously rinsed with the same buffer and eluted until the absorption at 280 nm of the eluate is less than 0.2. Proteins are then eluted for 4 hr by the buffer supplemented with a 0–0.5 *M* NaCl solution representing a linear pH gradient of 7.6 to 4.0. The effluent rate is 40 ml/hr and fraction volume is 3.0 ml. Each fraction is separately dialyzed against PBS before biological assay.

Hydrophobic Chromatography. The fractions purified by DEAE-cellulose-chromatography are applied to octyl-agarose column (1 × 1 cm) (Miles Laboratories, Inc.). The column is washed sequentially with 0.02 *M* sodium phosphate, pH 7.4, containing 0.15 *M* NaCl (indicated as eluant E0); 0.02 *M* sodium phosphate, pH 7.4, containing 1 *M* NaCl (E1); and 0.02 *M* sodium phosphate, pH 7.4, containing 1 *M* NaCl and 8.5 *M* ethylene glycol (E2). Fractions (1 ml) are collected at a flow rate of 30 ml/cm^2/hr. The material from each fraction is used for biological assay.

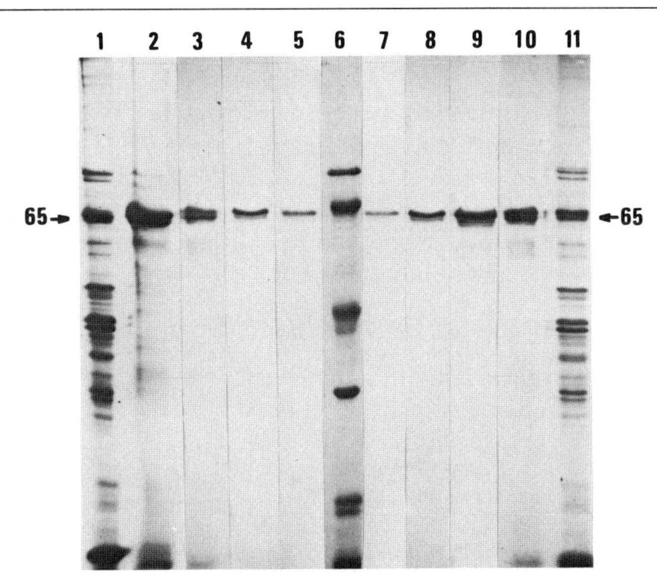

FIG. 1. Polyacrylamide gel electrophoresis of tissue extracts. The migration of tumor extract constituents is illustrated in lanes 1 to 5: lane 1, crude extract; lane 2, pepsin treatment; lane 3, Sephacryl gel filtration (10–20 μg); lane 4, DEAE-cellulose (5 μg); lane 5, octyl-agarose (3 μg). Lane 6 shows the migration of protein markers: phosphorylase *b* (MW 94,000), bovine serum albumin (MW 67,000), ovalbumin (MW 43,000), carbonic anhydrase (MW 30,000), soybean trypsin inhibitor (MW 20,100), α-lactalbumin (MW 14,400). The migration of normal muscle extract constituents is illustrated in lanes 7 to 11: lane 7, octyl-agarose (3 μg); lane 8, DEAE-cellulose (5 μg); lane 9, Sephacryl gel filtration; lane 10, pepsin treatment; lane 11, crude extract (10–20 μg). After the last purification step, one band is no longer detectable in the 65,000-Da region.

Monitoring the Purification Steps by Sodium Dodecyl Sulfate (SDS)–Polyacrylamide Gel Electrophoresis of Tissue Extracts

Electrophoresis of the proteins is carried out in 15% polyacrylamide slab gels.[9] The proteins are initially denatured in the presence of 0.05 *M* Tris · HCl, pH 6.8, 1% SDS, 2% dithiothreitol, 0.001% bromophenol blue, and 20% glycerol for 2 min at 100° and then loaded onto slab gels (10-cm running gel and 1-cm stacking gel). Electrophoresis is carried out at 20 mA/gel at 20° for about 4 hr. At the end of the running periods, the gels are fixed with 45% ethanol and 7% acetic acid for 30 min.

Representative gels are stained by Coomassie blue for 2 hr at room temperature (Fig. 1).

[9] U. K. Laemmli, *Nature* (*London*) **227,** 680 (1970).

Purification Steps of Tumor Extracts, Estimation of the Specific Activity, and Calculation of Yields

The purification procedures are summarized in Table IV and are planned to separate the lectins from IgG and albumin, the major contaminants. The prolonged pepsin treatment clarifies the preparation. Gel filtration separates albumin and the DEAE cellulose step retains the lectin. A further hydrophobic chromatographic step with octyl agarose can also be employed but is not always necessary.

Physicochemical Properties of Tissue Extracts

The two biological activities, cytoagglutinating (expressed by CA units) and interferon antagonistic (expressed by EMC HA units), are not affected by ultracentrifugation (90,000 *g*) and dialysis (Table V). Treatment by trypsin and pronase destroy completely both biological activities

TABLE IV
PURIFICATION OF CRUDE NORMAL AND TUMOR EXTRACTS; ESTIMATION OF SPECIFIC ACTIVITY

Purification steps	Proteins (mg)	Interferon antagonistic activity[a]	Agglutinating activity[b]	Specific activity	Yield (%)	Purification factor
Normal extracts	30	256	3,840	128	100	1
Normal extracts + pepsin	25	128	3,840	154	100	1.2
Normal extracts + Sephacryl S-200	1.5	16	1,600	1,053	42	8.2
Normal extracts + DEAE cellulose	0.1	4	320	3,200	8	25
Tumor extracts	33	1024	15,360	465	100	1
Tumor extracts + pepsin	27	512	15,360	568	100	1.2
Tumor extracts + Sephacryl S-200	2.4	128	7,680	3,160	50	6.8
Tumor extracts + DEAE cellulose	0.14	32	2,560	18,300	17	39.3

[a] Antagonistic effect of tissue extracts is expressed as the ratio of virus yield in cells exposed to the extract and interferon to the yield in cells treated with interferon only. EMC yield on control cells was 4096 HA units/ml. EMC yield on interferon-treated cells was ≤ 2.

[b] One cytoagglutinating unit (CA) is defined as the amount of tissue extract required to agglutinate 50% of the cell population. Specific activity is calculated in terms of CA activity/mg of protein. Crude extracts are partially purified followed by sequential chromatography on Sephacryl S-200 and on DEAE-cellulose.

TABLE V
PHYSICOCHEMICAL PROPERTIES OF TISSUE EXTRACTS

	Normal extract		Tumor extract	
	CA units[a]	EMC HA units[b]	CA units[a]	EMC HA units[b]
Control tissue extracts	512	256	1024–2048	4096
Ultracentrifugation, 90,000 *g*	512	256	1024–2048	4096
Dialysis	512	256	1024–2048	4096
Proteases				
Trypsin	<2	<2	<2	<2
Pronase	<2	<2	<2	<2
Pepsin	256–512	256	1024	2048
Nucleases				
DNase	256	ND[c]	1024	ND
RNase	256	ND	1024	ND
Temperature				
−80°, 30 days	256–512	256	1024–2048	4096
4°, 48 hr	256–512	256	1024–2048	4096
37°, 1–2 hr	256–512	256	1024–2048	4096
56°, 1 hr	256–512	ND	1024–2048	ND
100°, 2 min	256–512	ND	512–1024	ND
SDS + heat, 100°, 2 min	256	ND	512	1024
Dithiothreitol + heat, 100°	ND	ND	64–128	256–512
2-mercaptoethanol	256	ND	512	1024
pH				
2	512	256	1024–2048	4096
8	512	256	1024–2048	4096

[a] The CA units/0.05 ml are expressed as the inverse of the dilution of tissue extracts which agglutinate 50% of Fr-3T3 cells.
[b] EMC yield is shown as HA titer/0.05 ml with human erythrocytes.
[c] ND, not determined.

of tissue extracts, while pepsin does not modify them. In parallel, the nuclease treatment, DNase and RNase, does not affect cytoagglutinating activity. Incubation at different temperatures between −80 and +100° in the presence of 1% SDS under nonreducing or reducing condition (2% dithiothreitol or 15 m*M* 2-mercaptoethanol) does not alter the molecule. Resistance to pH variation between 2 and 8 is complete.

Concluding Comments

Definition of Sarcolectins. The lectin-like interferon antagonists presently characterized agglutinate a variety of normal or transformed cells including T and B lymphocytes. This cell agglutination is inhibited by competitive sugars (Table II). These tissue lectins are probably glycoproteins with a molecular weight of 60,000–65,000 as estimated by SDS–gel electrophoresis under reducing conditions. They are destroyed by proteases except pepsin, resist pH variations from 2 to 8, and are stable to heating at 100° in presence or absence of SDS. They have been isolated from hamster,[2,6] and human[3] sarcomas and normal muscles,[2,3] but comparable substances have been already detected in human amnion and chorion,[1] brain,[4] murine costal cartilage,[5] human placenta, and blood.[10]

The characterization of the interferon antagonistic substances which might be found in different tissues is only beginning. We show here a relatively well characterized group of substances which seem to exist in a number of tissues and animal species. Their biological importance is unknown. Besides their effect on blocking antiviral protection in the interferon-treated cells,[11] their impact on other interferon functions has not been determined. Furthermore, their presence in placental blood suggests they might have other biological roles during fetal development.[10]

[10] P. Duc-Goiran, P. Lebon, and C. Chany, this volume [72].
[11] F. Fournier, C. Chany, and M. Sarragne, *Nature (London), New Biol.* **235,** 47 (1972).

[92] Assay of Effect of Interferon on Virus-Induced Cell Fusion

By YOSHIMI TOMITA and TSUGUO KUWATA

Interferon is known to exert multiple effects on cells.[1] Suppression by interferon of virus-induced cell fusion is one of such pleiotropic actions of interferon.[2–8] These anti-cell fusion actions were observed with human alpha, beta, and gamma interferons, cloned human IFN-αA and mouse interferon, but all molecular species of interferons from various animals

[1] J. Taylor-Papadimitriou, *in* "Interferons" (D. C. Burke and A. G. Morris, eds.), p. 109. Cambridge Univ. Press, London and New York, 1983.
[2] Y. Tomita and T. Kuwata, *J. Gen. Virol.* **43,** 111 (1979).
[3] Y. Tomita and T. Kuwata, *Ann. N.Y. Acad. Sci.* **350,** 625 (1980).
[4] S. Chatterjee and E. Hunter, *Virology* **104,** 487 (1980).
[5] Y. Tomita and T. Kuwata, *J. Gen. Virol.* **55,** 289 (1981).
[6] Y. Tomita, J. Nishimaki, F. Takahashi, and T. Kuwata, *Virology* **120,** 258 (1982).
[7] S. Chatterjee, H. C. Cheung, and E. Hunter, *Proc. Natl. Acad. Sci. U.S.A.* **79,** 835 (1982).
[8] Y. Tomita, A. Fuse, T. Miki, and T. Kuwata, *J. Biochem. (Tokyo)* **95,** 495 (1984).

probably exhibit this action. Paramyxoviruses and other enveloped viruses are known to fuse cultured cells *in vitro*. Cell-to-cell fusion of tissue culture cells is mediated by virus adsorbed to cells (fusion from without) or by viral glycoproteins expressed on the surface of infected cells (fusion from within). Interferon seems to inhibit both types of cell fusion.[4,5] Sendai virus is known as a most potent cell-fusing virus and such cell fusion is known as a typical fusion from without.[9]

Here we describe methods for the assay of effect of interferon on cell fusion induced by Sendai virus and several retroviruses.

Preparation of Sendai Virus

Sendai virus is grown in the allantoic sac of 10-day-old chick embryos. After incubation at 36° for 3 days, the eggs are chilled at 4° overnight and the allantoic fluid is harvested. For the partial purification of the virus, the allantoic fluid is centrifuged at 3000 rpm for 30 min and the supernatant is centrifuged at 20,000 *g* for 30 min to sediment the virus. The virus is suspended in BSS (140 m*M* NaCl, 54 m*M* KCl, 0.34 m*M* Na_2HPO_4, 0.44 m*M* KH_2PO_4, 1 m*M* Tris · HCl, pH 7.6) and homogenized by pipeting. After removal of insoluble aggregates by low-speed centrifugation at 3000 rpm for 10 min, the virus is resedimented by high speed centrifugation at 20,000 *g* for 30 min. Finally, the virus is suspended in BSS containing 2 m*M* $CaCl_2$ and hemagglutination units (HAU) is determined. In our experiments,[5] we used Sendai virus for cell fusion after inactivation by UV irradiation, because Sendai virus rapidly loses infectivity after UV irradiation, but the cell–fusion ability is relatively stable. To inactivate Sendai virus, about 2 ml of virus in a 6 cm diameter dish was exposed to UV light at a distance of 150 mm with a germicidal lamp (about 60 $ergs/mm^2/sec$) for 4 min.

Preparation of Retroviruses

Retroviruses from various species of animals including humans[10,11] induce formation of syncytia in certain indicator cells. Cocultivation of retrovirus releasing cells with indicator cells also produces syncytia,[10,11] but for quantitative experiments use of cell-free virus preparations is desirable. To test the anti-cell fusion action of human interferon we have

[9] Y. Hosaka and K. Shimizu, *in* "Virus Infection and Cell Surface" (G. Poste and G. L. Nicolson, eds.), p. 129. Elsevier/North-Holland Biomedical Press, Amsterdam, 1977.

[10] K. Nagy, P. Clapham, R. Cheingsong-Popov, and R. A. Weiss, *Int. J. Cancer* **32,** 321 (1983).

[11] H. Hoshino, M. Shimoyama, M. Miwa, and T. Sugimura, *Proc. Natl. Acad. Sci. U.S.A.* **80,** 7337 (1983).

used chiefly RD114 virus.[2,6] For a preparation of this virus, culture fluids of RD114 cells are collected at 24 hr intervals and clarified by low-speed centrifugation at 8000 *g* for 10 min. The virus particles are sedimented by centrifugation at 80,000 *g* for 90 min and the pellet is resuspended in culture media to make a concentrated suspension of one-twentieth of the original volume by shaking at room temperature. If it is hard to homogenize the precipitate, a Teflon-glass homogenizer can be used in ice. Finally, insoluble materials are removed by centrifugation at 3000 rpm for 10 min. Fusion activity of the RD114 virus stored at $-70°$ is fairly stable. When the RD114 virus is sedimented through a cushion of 30% (w/v) sucrose in TEN buffer (10 m*M* Tris · HCl, pH 7.4, 1 m*M* EDTA, 100 m*M* NaCl) or concentrated by banding in a sucrose gradient in TEN buffer, the cell-fusing ability is occasionally destroyed. For the inactivation of RD114 virus, β-propiolactone (BPL) at a 1 : 1200 dilution is mixed with RD114 virus in serum-free medium and incubated at 5° for 16 hr. The virus is sedimented by centrifugation to remove BPL and resuspended in one-twentieth of the original volume.

Likewise, baboon endogenous virus (strain M7) can be harvested from human A204 cells[6,12] and Mason–Pfizer monkey virus from monkey foreskin B7 cells.[4,12]

Selection of Indicator Cells

Since interferon actions are species specific, indicator cells and interferon should be from homologous species. However, Sendai virus or retroviruses are able to form syncytia in cells of various species. Therefore, if human interferon is to be tested, human cells should be adopted as indicator cells. Thus, we have used two clones of RS cells,[13] RSa and RSb, or IFN^r cells[14] as the indicator cells. These cell lines are sensitive to cell fusion actions of various retroviruses of different origins[10,12,15,16]

Assay for Anti-Cell Fusion Activity of Interferon

General procedures for the assay of the anti-cell fusion activity of human interferon are as follows. Monolayer cultures of human cells in 3.5

[12] J. Cogniaux, R. Olislager, S. Sprecher-Goldberger, and L. Thiry, *J. Virol.* **43,** 664 (1982).
[13] T. Kuwata, T. Oda, S. Sekiya, and N. Morinaga, *J. Natl. Cancer Inst. (U.S.)* **56,** 919 (1976).
[14] T. Kuwata, A. Fuse, and N. Morinaga, *J. Gen. Virol.* **33,** 7 (1976).
[15] H. Ogura, T. Tanaka, M. Ocho, T. Kuwata, and T. Oda, *Arch. Virol.* **57,** 195 (1978).
[16] T. Tanaka, H. Ogura, M. Ocho, M. Namba, S. Omura, and T. Oda, *Virology* **108,** 230 (1981).

or 6-cm dishes are treated with various concentrations of interferon for 20 hr. The cells are washed with fresh medium and 0.7 ml/3.5 cm dish or 1.5 ml/6 cm dish of virus sample is added. After chilling on ice for 10 min (only in the case of cell fusion by Sendai virus), the cells are incubated at 37° for 3 to 20 hr in a humidified incubator. Then the cells are washed with phosphate-buffered saline, fixed with methanol, and stained with Giemsa solution. The number of syncytia which contain four or more nuclei is counted with a microscope at 100× magnification.

In our experiments, RSa or IFN^r cell cultures were used at about 90% confluence, because confluent cells did not fuse well, especially in the case of cell fusion by RD114 virus.[2] Prior treatment of cells with 20 μg/ml of polybrene often facilitated the induction of syncytia with various retroviruses and cell systems,[11,12] but it was not always essential for cell fusion. The suppressive effect of human IFN-α at 250 units/ml was observed in IFN^r cells about 6 hr after the treatment, but more than 12 hr was required for a full induction of the action.[5] The total number of syncytia per dish was calculated from the average number in 20 randomly chosen square fields (about 38 mm^2) in two dishes. The number of multinucleated giant cells which appeared spontaneously without addition of virus in control cultures was subtracted from the total number in each test culture. Typical experimental results are depicted in Fig. 1.

When HeLa-S3 cells were used as the indicator of cell fusion, monolayer cells were treated with interferon. Then the cells were removed from the dish with trypsin, washed with EMEM, and suspended in EMEM (2×10^7 cells/ml), chilled on ice with Sendai virus for 10 min and incubated at 37° for 45 min. Decrease in cell number as a result of cell fusion was calculated and expressed as the fusion index (FI).[17] Prior treatment of the cells with human IFN-α suppressed their fusion significantly.[3]

Comments

The assay of the anti-cell fusion action of interferon is simple and sensitive. With a suitable assay system, 10 units/ml of interferon may be sufficient to detect the suppressive effect of interferon on cell fusion. Recently, we found that this activity of interferon is induced in a clonal line in which (2′-5′)-oligo(A) synthetase,[6] dsRNA-dependent protein kinase,[6] and 56K protein[8] were not induced. Thus, tests for the sensitivity of cells to the anti-cell fusion action of IFN are useful for the characteriza-

[17] Y. Okada and J. Tadokoro, *Exp. Cell Res.* **32,** 417 (1963).

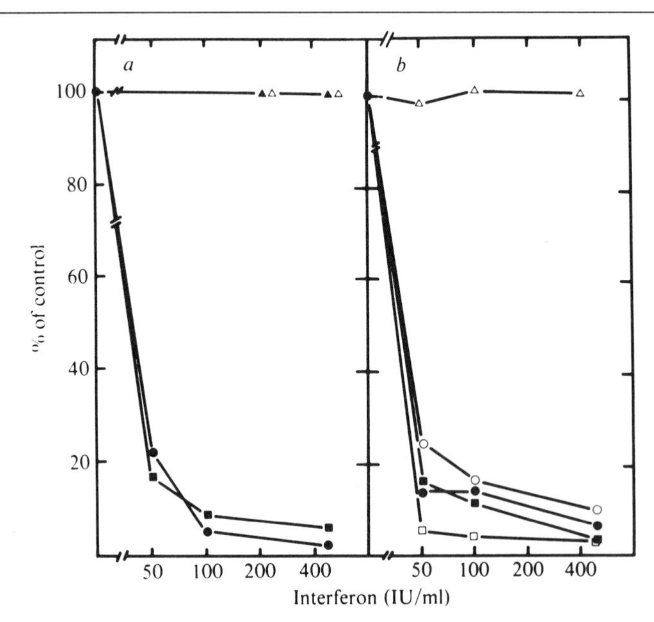

FIG. 1. Inhibition of syncytium formation as a function of IFN-α concentration. The ordinate represents the number of syncytia in the presence of interferon as a percentage of the number in its absence.[5] (a) RSa or IFN^r cells were treated with various amounts of IFN-α for 16 hr and incubated with 1600 HAU/ml of UV-inactivated Sendai virus for 4 hr: (●) IFN-α-treated RSa cells; (■) IFN-α-treated IFN^r cells; (▲) mock IFN-α-treated IFN^r cells; (△) mock IFN-α-treated RSa cells. (b) Cells were treated with various amounts of IFN-α for 20 hr, and incubated with UV-inactivated Sendai virus for 16 hr. The HAU/ml of UV-inactivated Sendai virus used for RSb, RD114, VA13, IFN^r, and L cells were 60, 20, 20, 60, and 60, respectively. (●) RSb; (○) RD114; (■) VA13; (□) IFN^r; (△) L. Crude IFN-α from peripheral blood leukocytes was used in these experiments. IU/ml represents the units of IFN-α in units/ml with reference to the international human leukocyte interferon reference standard.

tion of cells. Furthermore, the suppressive effect of chicken interferon has been also observed on nonviral cell fusions which occur during differentiation of chicken myoblasts in vitro,[18,19] though different results were reported with human interferon and a human myoblast system.[20] On the other hand, syncytium formation by polyethylene glycol is not inhibited by interferon treatment of cells (unpublished data).

[18] J. Lough, S. Keay, J. L. Sabran, and S. E. Grossberg, *Biochem. Biophys. Res. Commun.* **109,** 92 (1982).

[19] Y. Tomita and S. Hasegawa, *Biochim. Biophys. Acta* **20,** 370 (1984).

[20] P. B. Fisher, A. F. Miranda, L. E. Babiss, S. Pestka, and I. B. Weinstein, *Proc. Natl. Acad. Sci. U.S.A.* **80,** 2961 (1983).

[93] Measurement of Hyporesponsiveness to Interferon and Interferon Induction with Prostaglandins

By DALE A. STRINGFELLOW

Data generated over the past 10 years suggest that a relationship exists between interferons and prostaglandins. The first clues to this relationship were contained in publications by Yaron *et al.*[1] and Stringfellow.[2] The first publication[1] indicated that when human synovial cells were stimulated to produce interferon or were exposed to exogenously administered interferon, they responded in several ways including an increased level of intracellular prostaglandin E. The second publication[2] suggested that prostaglandins could enhance the amount of interferon produced by cells which were otherwise hyporeactive in their ability to respond to interferon induction. In these studies exposure of cells to prostaglandin E_1 or A_1 restored the amount of interferon produced by hyporeactive cells to levels made by comparable normal cells suggesting a regulatory role of prostaglandins in the interferon induction process. These data stimulated additional research into this area and the possible interrelationship between interferons and prostaglandins were explored to consider the relationship with regard to the biologic activity of either class of agents. This research has included studies of the effect of interferons and prostaglandins on metastasis of malignant cells, cellular division, and virus replication. In each case studies were done to delineate if this relationship (increased prostaglandin levels in cells exposed to interferon) was coincidental or if prostaglandins actually mediated any of the biologic activities of interferons.[3] The information covered in this chapter will be restricted to consideration of hyporesponsiveness to induction of interferon and measurement of this hyporesponsiveness by prostaglandins.

Hyporesponsiveness

Reduced ability of cells or animals to respond to interferon induction can be easily monitored. It is now clear that hyporesponsiveness to interferon induction develops as a consequence of virus infection, neoplastic

[1] M. Yaron, I. Yaron, D. Gurari-Rotman, M. Revel, H. R. Lindner, and U. Zor, *Nature (London)* **267,** 457 (1977).

[2] D. A. Stringfellow, *Science* **201,** 376 (1978).

[3] D. A. Stringfellow and R. Brideau, *in* "Interferon" (J. Vilček and E. De Maeyer, eds.), p. 147. Elsevier, Amsterdam, 1984.

TABLE I
In Vitro INTERFERON RESPONSE OF NORMAL AND HYPOREACTIVE MURINE CELLS[a]

	Without PGE_1		With PGE_1	
	Inducer		Inducer	
Cell source	NDV	None	NDV	None
EMC PECs	<50	<50	600	<50
L_{1210} PECs	<50	<50	350	<50
Normal PECs	780	<50	850	<50
L_{929} (NDV)	<50	<50	1400	<50
L_{929} (1×10^5 units/ml IFN)	<50	<50	2700	<50
L_{929} (controls)	3500	<50	3500	<50

[a] The values in the table represent the comparative interferon response (units/ml) in peritoneal exudate cells (PECs) from EMC virus infected (72 hr after infection) or L_{1210} leukemic (72 hr after injection) mice and in L_{929} cells previously induced to produce interferon by UV inactivated NDV (1×10^6 PFU) or exposed to 1×10^5 units of murine interferon. Units of interferon are given in international units/ml. PGE_1 (0.5 μg/ml) was added to cells at time of NDV induction. In all cases media was collected for interferon assay 24 hr after NDV induction.

disease, repeated exposures to interferon inducing agents, or exposure to large doses of interferons themselves (Table I). In each case cells which had been treated to modulate interferon responsiveness (for example, when exposed to high levels of interferons) were induced to produce interferon in a side by side comparison with control cells which had not been exposed to the modulating agent. In these studies normal and hyporeactive cells were induced in the same assay because of the variability of that assay. Likewise in whole animal studies the ability of potentially compromised animals to respond to inducers was monitored in a side by side manner with matched normal controls. For example, in Table II the ability of EMC virus infected mice to respond to several inducers including poly(I) · poly(C), tilorone hydrochloride, and a pyrimidinone (2-amino-5-bromo-6-phenyl-pyrimidinone, ABPP) are summarized. The onset of hyporeactivity can vary with inducer, virus infection, etc. and each system must be monitored closely with a matched set of controls in each study.[4,5]

[4] D. A. Stringfellow, *in* "Augmenting Agents in Cancer Therapy" (E. M. Hersh *et al.*, eds.), p. 215. Raven Press, New York, 1981.

[5] E. F. Wheelock, *Science* **149,** 310 (1965).

TABLE II
SERUM INTERFERON RESPONSE OF NORMAL AND EMC VIRUS INFECTED MICE[a]

	Interferon response (units/ml)			
	EMC-infected mice		Normal mice	
Inducer	With PGE_1	Without PGE_1	With PGE_1	Without PGE_1
Poly(I) · poly(C)	5500	<50	4000	5000
Tilorone HCl	7500	<50	8000	12000
ABPP	5600	<50	9000	7500
NDV	5000	<50	4500	3300
PBS control	<50	<50	<50	<50

[a] Poly(I) · poly(C) (100 μg/mouse ip), Tilorone HCl (250 mg/kg, po), ABPP (500 mg/kg, po), and NDV (10^7 PFU ip) were injected and mice were bled 6 hr after poly(I) · poly(C) and NDV and 12 hr after Tilorone and ABPP. Mice were injected with inducers 72 hr after infection with 50 LD_{50} of encephalomyocarditis (EMC) virus. PGE_1 was injected at 2 mg/kg ip at time of administration of inducer.

Interferon Assays

All interferon assays described in this chapter used plaque reduction of vesicular stomatitis virus on L929 murine fibroblasts.[4] In each case the international murine interferon reference standard was included and all titers are expressed as international units.

Prostaglandin Assays

The arachidonic acid (AA) cascade is illustrated in Fig. 1. Membrane bound phospholipids are enzymatically converted to arachidonic acid by the enzyme phospholipase. Arachidonic acid can then be enzymatically shunted into the lipooxygenase or cyclooxygenase pathways which likewise are enzymatically regulated by these two enzymes. The cyclooxygenase pathway is discussed in this chapter although it is acknowledged that lipooxygenase by-products may be as interesting. At the time of this writing, not enough was known about the products of the lipooxygenase systems with regard to an interrelationship with interferons to make any meaningful statements.

Prostaglandins and thromboxane were monitored by a radioimmunoassay.[6,7] To monitor the arachidonic acid cascade of hyporeactive and

[6] F. A. Fitzpatrick, D. A. Stringfellow, J. Maclouf, and M. Rigaud, *J. Chromotgr.* **177,** 51 (1979).

[7] F. A. Fitzpatrick and D. A. Stringfellow, *J. Immunol.* **125,** 431 (1980).

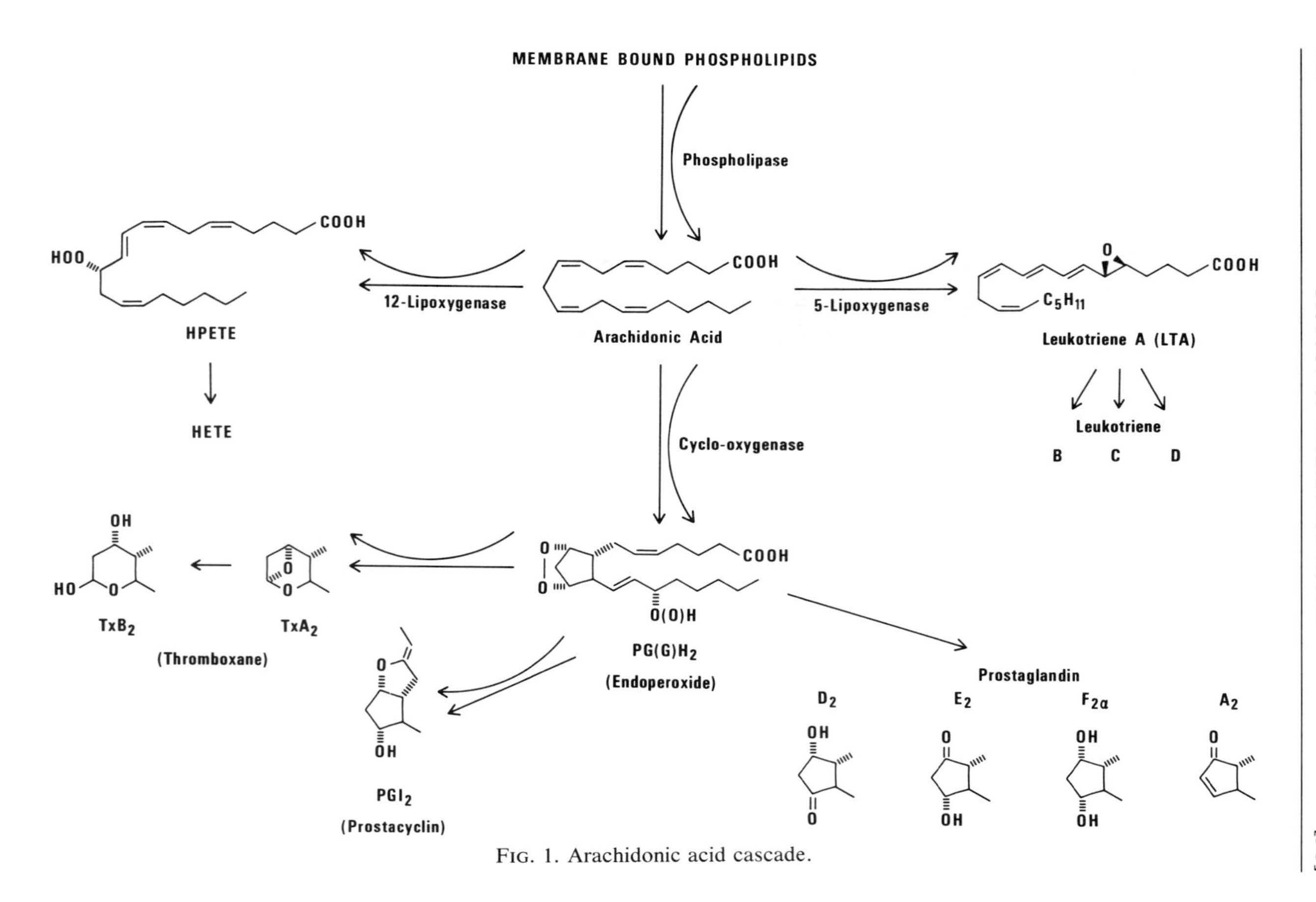

FIG. 1. Arachidonic acid cascade.

normal cells, it is possible to follow cellular metabolism with ^{14}C-labeled arachidonic acid as previously described by Hammarström[8] or follow metabolism of the prostaglandin endoperoxide H2 intermediate with radiolabeled endoperoxide H2.[7] In each case care must be taken to monitor the hyporesponsiveness of cells in question during the labeling and harvesting studies.

Overcoming Hyporesponsiveness by Prostaglandins

One method to measure hyporesponsiveness to interferon induction or the biologic activity of interferons is by either blocking cellular cyclooxygenase activity by treating cells with nonsteroidal antiinflammatory agents such as aspirin (100 μg/ml) or indomethacin (20 μg/ml) or by exposing cells to high concentration of prostaglandins ($\geq$1 μg/ml depending upon the prostaglandin and length of exposure). Neither aspirin or indomethacin at those concentrations was toxic to cells and 100 μg/ml of aspirin or 20 μg/ml of indomethacin will for short periods of time block cyclooxygenase activity. Aspirin is a reversible and indomethacin an irreversible inhibitor of the enzyme so the length of exposure needed to maintain the inhibitor activity in the cells requires careful monitoring. Once cells have been treated for 1–2 hr the inhibitor can be removed if it is not desired to have the drug present during collection of samples or measurement of biologic activity.

Treatment of cells with exogenous prostaglandins can be done but is less rigorous than the selective use of the specific enzyme inhibitors. However, it is often necessary to do direct addition studies and prostaglandins can be added at concentrations of approximately 1 μg/ml without cytotoxicity. It should, however, be remembered that each of the prostaglandins have different profiles of activity and should be titrated on the specific cells used to obtain the desired biologic effect. Prostaglandins in an aqueous environment are not totally stable and should be stored in absolute ethanol at $-20°$ until used. Just prior to use they should be diluted in an aqueous culture media without serum and should be added directly to the cells. We routinely prepare prostaglandins at a 10$\times$ concentration and then add 0.1 ml of prostaglandins per 0.9 ml of tissue culture media to get the desired final concentration.

Concluding Comments

Cells exposed to interferons or interferon inducers respond with increased intracellular levels of prostaglandins. The type of prostaglandin

[8] S. Hammarström, *Eur. J. Biochem.* **74,** 7 (1977).

produced by cells differs with each cell type and exposure of cells to interferons does not change the prostaglandin profile of those cells only the relative levels made. Interferons appear to selectively enhance the arachidonic acid cascade toward prostaglandin production. This appears to be a fairly ubiquitous response of a variety of mammalian cells to interferons[7] and is more than a coincidental relationship since prostaglandins can modulate the ability of hyporeactive cells to respond to interferon induction and can modulate the biologic activity of interferons.[3] The most thoroughly studied system to date is the ability of prostaglandins to modulate the interferon responsiveness of otherwise hyporeactive cells.[2] Cells from hyporeactive animals or cells made hyporeactive *in vitro* can have their hyporeactivity circumvented by coadministration of prostaglandins with the interferon inducer in question. Interestingly, the interferon response of normal cells is not affected by administration of prostaglandins at similar concentrations. Thus one means of measuring hyporeactivity is by inducing cells in the presence and absence of exogenously added prostaglandins. If cells are hyporeactive, they will respond by producing higher concentrations of interferon when the inducer is given with the prostaglandin than without.[2] This may be particularly important in studies with human cells where a normal control cell may not be available. By inducing those cells in the presence and absence of prostaglandins, it is possible to measure the level of hyporesponsiveness.

Section XI

Measurement of Effect of Interferons on Drug Metabolism

[94] Measurement of Effect of Interferon on Metabolism of Diphenylhydantoin

By GERALD SONNENFELD and DONALD E. NERLAND

The induction and passive transfer of interferons have been shown to inhibit cytochrome *P*-450 activity in the liver of rodents.[1–5] Cytochrome *P*-450 is one of the major enzyme systems for the metabolism of lipophilic compounds.[6] The inhibition of cytochrome *P*-450 by interferon has great implications in two major areas. One area is the activation of carcinogens by cytochrome *P*-450, as many polycyclic aromatic hydrocarbon carcinogens require activation to the final carcinogenic form by the cytochrome *P*-450 system.[7] Inhibition of cytochrome *P*-450 activity by interferon could inhibit the activation of carcinogens.[7,8]

The second area of interest is the effects of interferon on concomitantly administered drugs that are metabolized and detoxified by cytochrome *P*-450.[9,10] Inhibition of cytochrome *P*-450 activity by interferon could result in prolonged retention of drugs that could yield either detrimental or beneficial effects for the host.[11] Here we describe techniques for measuring the effects of interferon induction and passive transfer on the metabolism of a drug, diphenylhydantoin, that is detoxified by cytochrome *P*-450.[12,13]

[1] K. W. Renton and G. J. Mannering, *Drug Metab. Dispos.* **4,** 223 (1976).
[2] K. W. Renton and G. J. Mannering, *Biochem. Biophys. Res. Commun.* **73,** 343 (1976).
[3] G. Sonnenfeld, C. L. Harned, S. Thaniyavarn, T. Huff, A. D. Mandel, and D. E. Nerland, *Antimicrob. Agents Chemother.* **17,** 969 (1980).
[4] G. Singh, K. W. Renton, and N. Stebbing, *Biochem. Biophys. Res. Commun.* **106,** 1256 (1982).
[5] A. Parkinson, J. L. Lagker, M. J. Kramer, M.-T. Huang, G. E. Thomas, D. E. Ryan, L. M. Reik, R. L. Norman, W. Levin, and A. H. Conney, *Drug Metab. Dispos.* **10,** 579 (1982).
[6] G. J. Mannering, *in* "Drug Metabolism" (P. Jenner and B. Testa, eds.), p. 53. Dekker, New York, 1981.
[7] J. J. Reiners, Jr., D. Crowe, C. McKeown, D. E. Nerland, and G. Sonnenfeld, *Carcinogenesis (N.Y.)* **5,** 125 (1984).
[8] K. W. Renton, D. E. Keyler, and G. J. Mannering, *Biochem. Biophys. Res. Commun.* **88,** 1017 (1979).
[9] K. W. Renton, *in* "Advances in Immunopharmacology" (J. W. Hadden, L. Chédid, P. W. Mullen, and F. Spreafico, eds.), p. 17. Pergamon, Oxford, 1981.
[10] G. Sonnenfeld, P. K. Smith, and D. E. Nerland, *in* "Interferons" (T. C. Merigan and R. M. Friedman, eds.), p. 329. Academic Press, New York, 1982.
[11] G. Sonnenfeld, C. L. Harned, and D. E. Nerland, *Tex. Rep. Biol. Med.* **41,** 363 (1982).

The clearance of diphenylhydantoin (DPH, also known as phenytoin) from the blood of rats has been shown to be inhibited by treatment of the rats with polyriboinosinic–polyribocytidylic acid or *Bordatella pertussis* vaccine, both inducers of IFN-α/β.[12] In addition, metabolism of diphenylhydantoin to its metabolites by the cytochrome P-450 system was inhibited in these animals. Mice sensitized with *Mycobacterium bovis* strain BCG and challenged with tuberculin, which induces IFN-γ, or mice treated with preparations containing IFN-γ activity also showed inhibition of metabolism of diphenylhydantoin to its derivatives by the cytochrome P-450 system.[13] Described below are techniques for the measurement of the effects of interferon induction or passive transfer on the *in vitro* metabolism of diphenylhydantoin to two derivatives, p-hydroxydiphenylhydantoin and 5-(3,4-dihydroxy-1,5-cyclohexadien-1-yl)-5-phenylhydantoin, that requires action of the cytochrome P-450 system.[13]

Treatment of Mice with Interferon or Interferon Inducers

The protocol used to treat mice with IFN inducers to measure effects on metabolism of diphenylhydantoin is the same as that used for induction of IFN in these animals. For IFN-α/β, the mice are intravenously injected with 10 μg of polyriboinosinic–polyribocytidylic acid.[14] For IFN-γ induction, mice are sensitized by intravenous injection of *Mycobacterium bovis* strain BCG, and then intravenously challenged with 50 mg of tuberculin (Jensen-Salsbery Laboratories, Kansas City, MO) 3 weeks after initial sensitization.[13] For treatment of mice with exogenous preparations of IFN-γ, mice should be intravenously injected with 2000 units of the IFN-γ preparation three times in a 1-day period.[13]

Removal and Preparation of Liver Microsomes

Livers are surgically excised from mice 18–24 hr after the induction process for IFN was begun, or after the final injection of exogenous IFN. This time period has been shown to yield the maximum effects of IFN induction or passive transfer on metabolism of diphenylhydantoin.[12,13] The livers are weighed, and then placed in isotonic (1.15%) KCl. All procedures should be carried out in the cold (4°). The livers are then homogenized with a Dounce homogenizer, then centrifuged at 9000 g for

[12] K. W. Renton, *J. Pharmacol. Exp. Ther.* **208,** 267 (1979).

[13] C. L. Harned, D. E. Nerland, and G. Sonnenfeld, *J. Interferon Res.* **2,** 5 (1982).

[14] G. Sonnenfeld, D. Meruelo, H. O. McDevitt, and T. C. Merigan, *Cell. Immunol.* **57,** 427 (1981).

20 min.[15] The 9000 g supernatant is next centrifuged at 105,000 g for 1 hr. The microsomal pellet of the 105,000 g centrifugation is resuspended in KCl solution, and microsomal protein content determined by the standard method of Lowry with bovine serum albumin as the standard.[16]

Measurement of the Metabolism of Diphenylhydantoin

Measurement of the metabolism of diphenylhydantoin is carried out according to a modification of the method of Gabler and Hubbard.[17] One-milliliter aliquots (1 mg protein/ml) of the microsomal fraction are placed in disposable culture tubes. To this microsomal suspension is added 1.7 ml of 0.1 M phosphate buffer, 0.1 ml of 150 μM $MgCl_2$, and 0.2 ml of a NADPH generating system containing 1.0 μmol NADP, 25 μmol glucose 6-phosphate, and 0.5 units of glucose-6-phosphate dehydrogenase. The reaction mixture is preincubated for 2 min in a reciprocating water bath at 37°. At this point, the reaction is initiated by adding [^{14}C]diphenylhydantoin (New England Nuclear, Boston, MA) (0.04 μCi in 0.1 N NaOH in a volume of 20 μl) to each sample to yield a final concentration of 150 μM.[12,13,17] The reaction is allowed to progress for 15 min at 37°, and is then terminated by addition of 8 ml of cold ethyl acetate. The reaction mixture is extracted for 10 min by shaking the tubes, the tubes are centrifuged to separate the organic and aqueous layers, and the organic layer is removed. This procedure is repeated twice. The combined organic extracts are evaporated to dryness in a Concentratube (Laboratory Research Co., Los Angeles, CA) under a stream of nitrogen and the residue redissolved in 50 μl of methanol. The methanol solution is applied to a silica gel GF thin-layer chromatography plate (Analtech, Newark, DE) and developed in a benzene–methanol–acetic acid (45 : 8 : 4) solvent system. Standards containing diphenylhydantoin and p-hydroxydiphenylhydantoin (Sigma Chemical Co., St. Louis, MO) are simultaneously developed. The spots can be visualized under ultraviolet light,[13] or alternatively, by autoradiography.[12] The spots corresponding to the metabolites are scraped from the plates and quantitated with a liquid scintillation spectrometer. A decrease in radioactivity in spots corresponding to the metabolites is indicative of inhibition of metabolism of diphenylhydantoin by IFN induction or passive transfer.[12,13]

[15] S. El Defrawry El Masry, G. M. Cohen, and G. J. Mannering, *Drug Metab. Dispos.* **2,** 267 (1974).

[16] O. H. Lowry, N. J. Rosebrough, A. L. Farr, and R. J. Randall, *J. Biol. Chem.* **193,** 265 (1951).

[17] W. L. Gabler and G. L. Hubbard, *Biochem. Pharmacol.* **21,** 3071 (1972).

In addition, a determination of cytochrome *P*-450 levels by the spectrophotometric technique of Omura and Sato[18] should be carried out on the microsomes. This is to ensure that the depression of cytochrome *P*-450 activity correlates with the depression of metabolism of diphenylhydantoin in IFN-treated mice.

Concluding Comments

The inhibition of diphenylhydantoin metabolism by IFN induction and passive transfer was one of the readily available techniques for demonstrating that inhibition of cytochrome *P*-450 activity by interferon actually affected drug levels.[12,13] This observation should be taken into consideration in the planning of any clinical trials involving dual treatment with interferon and chemotherapeutic agents, as the presence of interferon may result in alteration of the effects of the drug. This has already been observed when individuals who were treated with theophylline were vaccinated with influenza virus, which induces interferon.[19] Several of these individuals developed theophylline toxicity, and the use of the drug had to be suspended.

Acknowledgment

Studies with techniques described here performed in the authors' laboratories were funded in part by grants from the Kentucky Tobacco and Health Research Institute.

[18] T. Omura and R. Sato, *J. Biol. Chem.* **238,** 2370 (1964).
[19] K. W. Renton, J. D. Gray, and R. I. Hall, *Can. Med. Assoc. J.* **123,** 288 (1980).

[95] Measurement of Effect of Interferon on Drug Metabolism

By GILBERT J. MANNERING

The cytochrome *P* 450 monooxygenase systems of the hepatic endoplasmic reticulum are comprised of a single reductase (NADPH–cytochrome *P* 450 reductase) and several cytochrome *P* 450 isozymes.[1] Under certain conditions, electrons can be donated by cytochrome(s) *P* 450 via NADH–cytochrome b_5 reductase and cytochrome b_5,[1] but since the con-

[1] G. J. Mannering, *in* "Concepts of Drug Metabolism" (P. Jenner and B. Testa, eds.), Part B, p. 53. Dekker, New York, 1981.

tribution of electrons by this system is not always predictable, monooxygenase reactions are usually carried out with NADPH as their only source of electrons. The substrate specificities of hepatic cytochrome *P* 450 systems are highly unselective; almost all lipophilic drugs and other xenobiotics that are capable of being oxidized are biotransformed by these systems. The oxidative reactions catalyzed by cytochrome *P* 450 systems not only render drugs more polar, and therefore more excretable, but in doing so, they add functional groups that can participate in conjugation reactions (e.g., glucuronidation) that further detoxify and favor excretion. In short, the activities of cytochrome *P* 450 systems largely determine the duration and intensity of drug action and thereby play an important role in drug therapy and toxicity.

The administration of a wide variety of unrelated interferon inducing agents has been shown to depress hepatic *P* 450 systems of rats and mice.[2-4] This strongly suggests that endogenous interferon per se depresses these systems although the possibility cannot be ignored that interferon inducing agents may cause this effect by mechanisms other than, or in addition to, the induction of interferon. In any event, it has been shown that cytochrome *P* 450 systems are depressed in mice by partially purified murine alpha and beta interferon[5,6] and by IFN-αA/D,[5,6] a recombinant hybrid human interferon that possesses antiviral activity in mice.

The effect that interferon may have on drug metabolism in human patients becomes an important issue when large doses of human interferon are administered with drugs that are metabolized by the hepatic cytochrome *P* 450 systems. At this time, studies that would determine the effect of interferon on drug metabolism in humans have not been published. The methods to be described here have been used in rat and mouse studies.

Materials

Ethylmorphine · HCl (Merck and Co., Rahway, NJ)
Glucose 6-phosphate (Sigma Chemical Co., St. Louis, MO)

[2] K. W. Renton and G. J. Mannering, *Biochem. Biophys. Res. Commun.* **73,** 343 (1976).
[3] G. J. Mannering, K. W. Renton, R. el Azhary, and L. B. Deloria, *Ann. N.Y. Acad. Sci.* **350,** 314 (1980).
[4] G. Sonnenfield, C. L. Harned, S. Thaniyavarn, T. Hatt, A. D. Mandel, and D. E. Nerland, *Antimicrob. Agents Chemother.* **17,** 969 (1980).
[5] G. Singh, K. W. Renton, and N. Stebbing, *Biochem. Biophys. Res. Commun.* **106,** 1256 (1982).
[6] G. J. Mannering and L. B. Deloria, unpublished results (1983).

Glucose-6-phosphate dehydrogenase (Boehringer Mannheim, Indianapolis, IN)
NADP (Sigma Chemical Co., St. Louis, MO)
Nash reagent (150 g ammonium acetate, 1 ml of acetyl acetone, 1.5 ml glacial acetic acid q.s. 500 ml with distilled water). Store overnight in an amber bottle before using
Aminopyrine (ICN Pharmaceuticals, Inc., Plainview, NY)
Aniline · HCl (Eastman Kodak Co., Rochester, NY)
Benzo[*a*]pyrene (Aldrich Chemical Co., Milwaukee, WI)
Quinine sulfate (Sigma Chemical Co., St. Louis, MO)
7-Ethoxycoumarin (Aldrich Chemical Co., Milwaukee, WI)
7-Hydroxycoumarin (Aldrich Chemical Co., Milwaukee, WI)

Preparative Procedures and Methodology

Preparation of Microsomes

Fed mice are killed and bled by cutting through the jugular vessels and spinal cord with scissors. The liver is excised and the gall bladder is removed. The liver is blotted, weighed, and homogenized with 5 ml of ice cold 1.15% KCl in a Dounce homogenizer with a loose fitting pestle (15 strokes). The homogenate is centrifuged at 9000 *g* for 20 min. The supernatant fraction is diluted to 10 ml with 1.15% KCl solution and centrifuged at 105,000 *g* for 60 min. The entire pellet is homogenized with 3 ml of 1.15% KCl solution in a Dounce homogenizer with a tight fitting pestle (10 strokes). Both centrifugations are conducted at 4° and the preparation is kept on ice until used. Unless it has been established that monooxygenase activity is not lost when the microsomes are stored at −80°, the microsomes should be used the day they are prepared. The method is applicable to livers of laboratory animals other than mice. The major advantage of differential centrifugation is that it is the most commonly used procedure for preparing microsomes and therefore provides a large repository of data that can be used for comparing results. Its disadvantages are that it is time consuming, it requires an ultracentrifuge, and the recovery of microsomes from the liver is only about 40–50%.[7]

The procedure is less satisfactory for harvesting microsomes from fetal or early neonatal livers, in which cases, recovery of microsomes is only 10–15%.[7] This is due largely to the failure of homogenization to rupture a large percentage of the hepatocytes. This can be remedied by homogenizing the liver in distilled water and then slowly adjusting to

[7] M. S. Robbins and G. J. Mannering, *Biochem. Pharmacol.* **33,** 1213 (1984).

isotonicity with hypertonic KCl solution. When this homogenate is passed through a Sepharose 2B column, about 90% of the microsomes in the liver is recovered in the eluted solvent front.[7] The method is particularly useful when only small amounts of liver are available. It has the advantage of not requiring an ultracentrifuge. About 50% of the hepatic mitochondria are recovered with the microsomal fraction; this may be a disadvantage in some studies.

Microsomes from rat livers have been recovered from the 17,000 *g* supernatant fraction by aggregation with $CaCl_2$.[8,9] The method is rapid (about 1 hr), an ultracentrifuge is not required, and the recovery and biochemical properties of the microsomes appear to be very similar to those obtained by ultracentrifugation. The disadvantage is that the method has not as yet been applied extensively; this limits the basis for comparison of results.

Assay of Cytochrome P 450

The assay of cytochrome *P* 450[10] is based on the observation that carbon monoxide shifts the maximum absorption of reduced cytochrome *P* 450 from about 415 to about 450 nm. Difference spectroscopy is used to visualize this shift. Carbon monoxide is bubbled for 15 sec into 6 ml of microsomal preparation (1 mg of protein/ml of 0.2 *M* potassium phosphate in 1.15% KCl, pH 7.4). The preparation is divided into 3 ml sample and reference cuvettes which are then scanned between 400 and 500 nm in a dual beam spectrophotometer (e.g., an Aminco DW-2). A pinch of sodium dithionite is added to the sample cuvette and the absorbance scanned between 400 and 500 nm after the cytochrome *P* 450 is fully reduced, i.e., when the absorbance at 450 nm is no longer increasing. The concentration of cytochrome *P* 450 is calculated by using $\epsilon_{450-490\,\mathrm{nm}} = 104\ \mathrm{m}M^{-1}\ \mathrm{cm}^{-1}$. The original method of assaying cytochrome *P* 450[11] used dithionite in both cuvettes and carbon monoxide in the sample cuvette only and an $\epsilon_{450-490}$ of 91 $\mathrm{m}M^{-1}\ \mathrm{cm}^{-1}$. The advantage of the method described here is that contaminating hemoglobin does not interfere with the spectrum; in fact, this method can be used to measure the cytochrome *P* 450 content of whole homogenate and the 9000 *g* supernatant fraction.

The following procedures for the assay of monooxygenase activities provide optimal conditions with respect to enzyme content and linearity

[8] S. A. Kamath, F. A. Kummerow, and K. A. Narayan, *FEBS Lett.* **17,** 90 (1971).

[9] J. B. Schenkman and D. L. Cinti, *Life Sci.* **11,** 247 (1972).

[10] T. Matsubara, M. Koike, A. Touchi, Y. Tochino, and K. Sugeno, *Anal. Biochem.* **75,** 596 (1976).

[11] T. Omura and R. Sato, *J. Biol. Chem.* **239,** 2379 (1964).

of reaction velocity throughout the period of incubation. All incubations are performed in a Dubnoff metabolic shaker (120 oscillations/min) at 37°.

Ethylmorphine N-Demethylase Activity

This assay is based on measurement of HCHO formed by the oxidative N-demethylation of ethylmorphine. O-Deethylation also occurs but the CH_3CHO formed does not interfere with the measurement of HCHO by the Nash method.[12] The reaction mixture (5.0 ml) in a 25-ml Erlenmeyer flask contains 2–3 mg of microsomal protein and the following components dissolved in 1.15% KCl (μmol): semicarbazide (37.5), sodium potassium phosphate buffer, pH 7.4 (200), glucose 6-phosphate (20), 2 units of glucose-6-phosphate dehydrogenase, NADP (2), magnesium chloride (10), and ethylmorphine HCl (10). The reaction is started by adding microsomes, continued for 10 min, and terminated by pouring the mixture into a 40-ml conical centrifuge tube containing 3 ml of 5% zinc sulfate. Three milliliters of 4.5% barium hydroxide heptahydrate is added and the mixture is centrifuged for 10 min (International 2K centrifuge; about 1500 rpm). Five milliliters of the supernatant fraction is transferred to a 20-ml test tube, 3.0 ml of Nash reagent is added, and the contents are agitated on a vortex mixer. The tube is placed in a 60° bath for 20 min, cooled to room temperature, and read at 412 nm in a spectrophotometer ($\epsilon_{412\,nm} = 8.64\ mM^{-1}\ cm^{-1}$). A blank value obtained by incubating the same microsome-containing mixture without ethylmorphine is subtracted from this reading.

Aminopyrine N-Demethylase Activity

The HCHO formed from the oxidative demethylation of the two N-methyl groups of aminopyrine is measured by the same procedure described for the N-demethylation of ethylmorphine. The reaction mixture (5 ml) contains 15 μmol of aminopyrine. A more sensitive method which uses [^{14}C]aminopyrine has been published.[13]

Aniline Hydroxylase Activity

The method measures the 4-hydroxylation of aniline.[14] The reaction mixture (3.0 ml) in a 25-ml Erlenmeyer flask contains 3 mg of microsomal protein and the following components (μmol): potassium phosphate buffer, pH 7.4 (50), magnesium chloride (6), glucose 6-phosphate (12), 1.2

[12] T. Nash, *Biochem. J.* **55,** 416 (1953).
[13] A. P. Poland and D. W. Nebert, *J. Pharmacol. Exp. Ther.* **184,** 269 (1973).
[14] Y. Imai, A. Ito, and R. Sato, *J. Biochem. (Tokyo)* **60,** 417 (1966).

units of glucose-6-phosphate dehydrogenase, NADP (1.2), and aniline HCl (0.75). The reaction is started by adding microsomes, continued for 20 min, and terminated by adding 1.5 ml of 20% TCA. The mixture is decanted into a 15 ml conical centrifuge tube and centrifuged for 10 min (International 2K centrifuge; about 1500 rpm). Two milliliters of the supernatant fraction is transferred to a 20-ml centrifuge tube and 2 ml of 0.5 *M* NaOH in 1% phenol and 2 ml of 1 *M* Na_2CO_3 are added. The contents are agitated on a vortex mixer. The tube is allowed to stand at room temperature for 40 min and then read at 630 nm in a spectrophotometer ($\epsilon_{630\,nm}$ = 31.28 μM^{-1} cm^{-1}). A blank value obtained by incubating the same microsome-containing mixture without aniline is subtracted from this reading.

Benzo[a]pyrene Hydroxylase Activity

The method measures the 3-hydroxylation of benzo[*a*]pyrene.[15] The reaction mixture (1.0 ml in a 20-ml test tube) contains 0.167 mg of microsomal protein and the following components (μmol): potassium phosphate buffer, pH 7.4 (67), magnesium chloride (2), glucose 6-phosphate (4), 0.4 units of glucose-6-phosphate dehydrogenase, NADP (0.4), and benzo[*a*]pyrene (0.167) in 33 μl of acetone. The reaction is started by adding NADP, continued for 10 min in the dark, and terminated by adding 1 ml of ice cold acetone. The test tube containing the reaction mixtures is transferred to an ice bath, 3 ml of cold hexane is added, the tube covered with a cap plug, and the contents agitated on a vortex mixer. The tube is allowed to stand in the ice bath for 30 min. Two milliliters of the upper layer is transferred to a 13 × 100 mm test tube, 1 ml of 1 *M* NaOH is added, and the tube is agitated on a vortex mixer. The entire contents of the tube is transferred to a cuvette which exposes only the lower aqueous phase to the excitation light. In exactly 10 min, the aqueous layer is read in a Farrand photoelectric fluorometer equipped with a primary filter that transmits light maximally at 400 nm and a secondary interference filter that absorbs light maximally at 522 nm. A quinine sulfate solution (0.3 μg/ml in 0.05 *M* H_2SO_4) is used as a fluorescence standard. The fluorometer is set such that a reading of 0.01 μA is equivalent to 0.166 μmol of 3-hydroxybenzo[*a*]pyrene in 1 *M* NaOH. The fluorescence observed with 1 μg/ml of quinine sulfate is the same as that observed with 4100 pmol of 3-hydroxybenzo[*a*]pyrene.[16] The value obtained at 0 min incubation serves as the blank.

[15] L. W. Wattenberg, J. L. Leong, and P. J. Strand, *Cancer Res.* **22,** 1120 (1962).
[16] D. W. Nebert, *Pharmacol. Ther.* **6,** 395 (1979).

7-Ethoxycoumarin O-Deethylase Activity

The method measures the formation of 7-hydroxycoumarin.[17] The reaction mixture (2.0 ml in a 20-ml test tube) contains 0.2 mg of microsomal protein and the following components (μmol): potassium phosphate buffer, pH 7.4 (130), glucose 6-phosphate (8), 0.8 units of glucose-6-phosphate dehydrogenase, magnesium chloride (4), NADP (0.8), and 7-ethoxycoumarin in 2.5 μl of acetone (1). The reaction is started with NADP, continued for 10 min, and terminated by adding 0.25 ml of 15% TCA. Four milliliters of chloroform is added, the contents of the test tube are agitated on a vortex mixer and then centrifuged for 10 min (International 2K centrifuge; about 1500 rpm), and 0.125 ml of the chloroform layer is transferred into a 20-ml test tube containing 3.0 ml of 0.01 *M* NaOH–1 *M* NaCl. The mixture is agitated on a vortex mixer and centrifuged for 10 min as described previously. The aqueous phase is transferred to a cuvette and read in an Aminco-Bowman spectrophotofluorimeter (excitation maximum: 368 nm; emission maximum, 456 nm). 7-Hydroxycoumarin is used as the standard.

Concluding Comments

A much larger number of monooxygenase activities could have been included than the four that have been described. There is considerable individuality in the way the preparation of microsomes and the evaluation of cytochrome *P* 450-linked monooxygenase activities are performed in different laboratories. The methods described here have been used successfully in the study of the effects of interferon and interferon-inducing agents on drug metabolism in this laboratory. The inclusion of four rather than a single monooxygenase activity offers a choice, but more importantly, it serves to emphasize the need to consider more than one substrate when the effects of interferon on drug metabolism are being evaluated. Although individual cytochrome *P* 450 isozymes show little qualitative substrate selectivity, they vary quantitatively in their patterns of activity for a wide variety of substrates. The four substrates were selected because each is known to serve as a major substrate for one or more of the major cytochrome *P* 450 isozymes.

Temporal aspects of the evaluation of the effects of interferon and interferon inducing agents are vital. When interferon inducing agents are used, it is essential to know the extent of the induction period. For example, poly(I) · poly(C) induces two peak serum levels of interferon in mice,

[17] W. Greenlee and A. Poland, *J. Pharmacol. Exp. Ther.* **205,** 596 (1978).

one at about 2 hr and another at about 12 hr whereas the induction of interferon by *C. parvum* requires several days. Whether induced or administered per se, time must be allowed for interferon to depress the synthesis of cytochrome *P* 450. Because the turnover rates of cytochrome *P* 450 isozymes may differ considerably, the collection of microsomes must be timed to anticipate a measurable decline of cytochrome *P* 450. This requires that collections be made serially, e.g., 12, 24, and 48 hr after the injection of interferon or the appearance of endogenous interferon. The possibility is also to be considered that not all forms of interferon will affect all forms of cytochrome *P* 450; accordingly, their effects on some monooxygenase activities may be greater than on others. A very small effect on total cytochrome *P* 450 does not necessarily mean that one can conclude that drug metabolism was not affected; a quantitatively minor cytochrome *P* 450 isozyme may be responsible for a quantitatively major monooxygenase activity.

Section XII

Interferon and Plant Cells

[96] Production, Preparation, and Assay of an Antiviral Substance from Plant Cells

By ABDULLAH GERA, SARA SPIEGEL, and GAD LOEBENSTEIN

Various resistance and interference phenomena of virus multiplication are known to occur in plants, whereby virus multiplication and spread are inhibited. Thus in the localized infection, the virus after inoculation invades several hundred cells but does not spread to neighboring tissues. Apparently virus multiplication in these cells is retarded by an active process requiring translation of the host genome.[1,2] The association of resistance with antiviral substances produced by the plant cell has so far been hampered by the lack of suitable test methods. Thus "antiviral" agents have been extracted in the past from both infected and uninfected-resistant tissue. However, their connection with localization or induced resistance is still an open question, especially as they have been tested as *inhibitors of infection* given during inoculation (by mixing with the test virus) and not as *inhibitors of virus multiplication,* which should be applied to the infected cell 5–12 hr (or later) after inoculation.

For reliable evaluation of substances as inhibitors of virus multiplication, it is necessary to have a test system where cells that have been synchronously infected can be treated uniformly after different times of inoculation with the test substance. The protoplast system seems to be most suitable, as it is possible to inoculate simultaneously all cells and to obtain "one-step growth curves," in contrast to the intact plants where only a small percentage of the cells becomes infected during the primary inoculation. It is also possible to apply the test material uniformly to every protoplast, in contrast to a parallel application in the intact plant. Furthermore, the protoplast system can be used to evaluate antiviral compounds associated with resistance, which are released from the protoplast into the medium, thereby obviating the necessity of obtaining and testing such substances from crude plant homogenates which contain many other compounds and their oxidation products.

Recently we reported that a substance(s) inhibiting virus replication (IVR) is released into the medium from tobacco mosaic virus (TMV)-infected protoplasts of a cultivar in which infection in the intact plant is localized. IVR inhibited virus replication in protoplasts from both local-

[1] G. Loebenstein, *Annu. Rev. Phytopathol.* **10,** 177 (1972).

[2] I. Sela, this volume [97].

lesion-responding resistant Samsun NN and systemic-responding susceptible Samsun plants, when applied up to 18 hr after inoculation. IVR was not produced in protoplasts from susceptible plants or from noninoculated protoplasts of the resistant cultivar. IVR was partially purified by $ZnAc_2$ precipitation, and yielded two biologically active principles with molecular weights of about 26,000 and 57,000, respectively.[3]

IVR was found to be sensitive to trypsin and to chymotrypsin but not to RNase. It is of a proteinaceous nature. IVR was found also to inhibit virus replication in tobacco and tomato leaf tissue disks infected with TMV. IVR also inhibited replication of cucumber mosaic virus (CMV) and potato virus X (PVX) in different host tissues, indicating that it is neither virus nor host specific.[4]

Reagents

VIM: 0.2 *M* KH_2PO_4, 10 m*M* $CaCl_2$, 1 m*M* KNO_3, 1 m*M* KI, 0.1 m*M* $MgSO_4$, 13.5% mannitol
Macerozyme R-10 (Japan Biochemical, Nishinomiya, Japan)
Cellulase R-10 (Japan Biochemical, Nishinomiya, Japan)
Sucrose, crystalline (Sigma)
Poly-L-ornithine, approximate MW 120,000 (Sigma)
Carbenicillin (Beecham Res. Lab., England)
Mycostatin (Squibb & Sons LTD, England)
Millipore filters (0.2 μm)
"Nytex" filter (250 μm) (ZBF, 8803 Ruschlikon, Switzerland)
Tobacco mosaic virus purified by density gradient centrifugation
Sephadex G-75 (Superfine, Pharmacia)
Omnimixer with a microattachment (Sorvall)

Growing of Plants

Nicotiana tabacum L. varieties Samsun and Samsun NN were grown in 10-cm pots in a soil–sand mixture in a glasshouse. Three feedings with a complete nutrient solution were given at weekly intervals, starting 1 week after planting. Leaves 9–12 from the top, which were fully expanded, were selected 10–14 days after the last feeding. Selected leaves had a "light green" color and care was taken not to overfertilize the plants and to prevent all diseases and pests with minimum pesticide application.

[3] G. Loebenstein and A. Gera, *Virology* **114,** 132 (1981).
[4] A. Gera and G. Loebenstein, *Phytopathology* **73,** 114 (1983).

Preparation of Protoplasts and IVR

Protoplasts were obtained by the mixed enzyme procedure with 0.02–0.06% macerozyme R-10 and 0.2–0.6% cellulase R-10 in incubation medium (VIM) overnight at 25°.[5] The mixture was filtered through a filter (Nytex, 250 μm) and the protoplasts were washed twice with 13.5% mannitol, floated on 23% sucrose, and again washed with 13.5% mannitol.

For preparation of IVR, Samsun NN tobacco protoplasts were used. Each preparation of protoplasts was divided into two batches and each was suspended in 10 ml 13.5% mannitol to a concentration of 5×10^5/ml. Concurrently, a solution of TMV (2 μg/ml) was adjusted to 10 ml of 0.02 *M* potassium citrate buffer, pH 5.2, containing 13.5% mannitol and poly-L-ornithine (2 μg/ml) and incubated at 25° for 10 min with gentle rocking (30 strokes/min). This solution was then mixed with an equal volume of the protoplast suspension and incubated for 10 min at 25° with gentle rocking. The second batch was sham inoculated with the same infection medium without TMV. The two mixtures were incubated at 25° for 10 min while being rocked gently. Protoplasts were then washed twice with 13.5% mannitol containing 0.1 m*M* $CaCl_2$, and twice with VIM containing carbenicillin (200 μg/ml) and mycostatin (25 units/ml); they were then divided into 10-ml portions in 100-ml Erlenmeyer flasks at a concentration of 1×10^5 protoplasts/ml, and incubated at 25° under continuous illumination of about 2500 lux.

Partial Purification of IVR

The incubation media were collected after 72 hr incubation, sterilized through a Millipore filter (0.2 μm), and dialyzed against 0.1 *M* phosphate buffer, pH 7.0 for 48 hr with two changes of the buffer. The solution was then centrifuged at 3500 *g* for 20 min, the precipitate was discarded, and $ZnAc_2$ was added to the solution to a final concentration of 0.02 *M*. After 3 hr the preparation was centrifuged for 45 min at 2000 *g*, the supernatant discarded, and the precipitate dissolved with 0.02 *N* HCl by bringing the solution to pH 2.5–2.6. After centrifugation for 15 min at 2500 *g*, the supernatant was dialyzed overnight against 0.01 *M* phosphate buffer, pH 6.0, then centrifuged for 15 min at 2500 *g*. The precipitate was discarded and the supernatant lyophilized and kept at −20° until use. The IVR preparation originating from 1×10^6 protoplasts was defined as 1 unit IVR.

[5] G. Loebenstein, J. Cohen, S. Shabtai, R. H. A. Coutts, and R. K. Wood, *Virology* **81,** 117 (1977).

Assay of Inhibitory Potency of IVR

The inhibitory effect of IVR was determined either on protoplasts or on leaf disks. The lyophilized preparation was dissolved in 1 ml of 0.01 *M* phosphate buffer, pH 6.0, dialyzed overnight against 0.1 *M* phosphate buffer, pH 6.0, centrifuged for 15 min at 2500 *g*, and then dialyzed overnight against VIM. All steps were conducted in the cold (4°). After dialysis, IVR was diluted appropriately with VIM.

Assay on Protoplasts. IVR was added to protoplasts preparations (10^5/ml, 10 ml) which had been inoculated 5 hr earlier with TMV. A preparation from similarly treated noninoculated protoplasts served as a control. The protoplasts were incubated at 25° for 72 hr at 2000 lux. They were then collected by centrifugation and homogenized for 2.5 min in 4 ml of 0.05 *M* phosphate buffer, pH 7.5, in the cold, with the aid of a microattachment to an Omnimixer. The homogenate was inoculated onto 12 half-leaves of *N. glutinosa* plants and compared with a standard solution of purified TMV on the opposite half-leaf. Lesions counted were adjusted to 1×10^6 live protoplasts and calibrated with the following equation:

$$N_c = \frac{N_m M_T}{N_T}$$

where N_c is the calibrated number of lesions per half-leaf for homogenate, N_m is the mean number of lesions per half-leaf produced by homogenate, N_T is the mean number of lesions produced by standard TMV on opposite half leaf, and M_T is the mean number of lesions produced by standard TMV from all experiments. The survival rate of the protoplasts after 72 hr was generally around 90%, and those experiments with survival rates lower than 80% were discarded.

Assay on Leaf Disks. Samsun tobacco plants were inoculated with a purified TMV solution (2.5 μg/ml) and kept in the greenhouse. After 5 hr or later, disks (11 mm in diameter) were punched out of the inoculated leaves and floated on IVR and control preparations dissolved in VIM (without mannitol) in Petri dishes (5 cm in diameter). Disks were incubated at 25° under continuous illumination. After incubation, disks from IVR and control samples were washed with distilled water and homogenized in 2 ml of 0.05 *M* phosphate buffer, pH 7.5. The homogenate was used to inoculate 12 half-leaves of *N. glutinosa* L. plants and compared with a standard solution of TMV on the opposite half-leaf.

Virus titers in leaf disks and in protoplasts were also determined by the enzyme-linked immunosorbent assay (ELISA)[6] calibrated to purified TMV.

[6] M. F. Clark and A. N. Adams, *J. Gen. Virol.* **34,** 475 (1977).

Dose–Response of IVR

The effect of increasing concentrations of IVR on inhibition of TMV replication was tested both with the leaf disk and protoplast assays. Figure 1 shows the level of inhibition obtained with increasing concentrations of IVR in both assays as determined by the local lesion assay 72 hr after inoculation.

Effect of IVR on CMV and PVX

Samsun NN tobacco plants, *Cucumis sativus* L. 'Bet Alpha' and *Capsicum fructescens* L. 'Vindale' plants were inoculated with CMV and *N. glutinosa* plants with PVX. Plants were kept in the greenhouse, and disks were removed from inoculated leaves and floated on IVR and control preparations as described above. Infectivity of CMV was assaysed on *Vigna sinensis* Endl. 'Blackeye' and that of PVX on *Gomphrena globosa* L. plants by comparing the number of local lesions on opposite half-leaves in IVR and control treatments. Inhibition with 1 unit of IVR ranged between 60 and 80% of CMV and 60 and 70% of PVX.

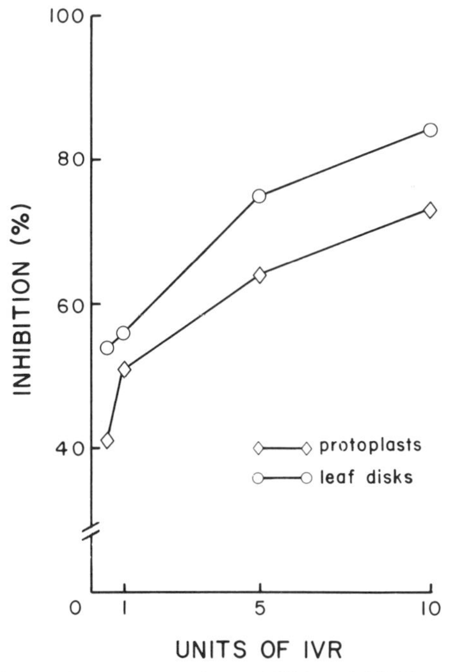

FIG. 1. Effect of increasing concentrations of IVR on inhibition of TMV replication in protoplasts and tobacco leaf disks determined by local lesion assay on *N. glutinosa*.

Further Purification and Molecular Weight Estimation of IVR

Further purification and estimation of the molecular weight were done by gel filtration through a Sephadex G-75 column.[7] A lyophilized IVR preparation obtained from 3×10^7 inoculated protoplasts was dissolved in 1 ml of 0.1 *M* phosphate buffer, pH 6.0 ($OD_{280} = 2$), placed on a precalibrated Sephadex G-75 column (33×2.3 cm), and eluted with the same buffer at a flow rate of 100 ml/hr. A control preparation obtained from a similar number of noninoculated protoplasts with the OD_{280} adjusted to 2 was passed through the same column in a similar way. Inhibitory potency of each third fraction, in comparison with the respective control fraction, was tested as before on inoculated Samsunn NN protoplasts. Peaks of activity were eluted at two positions, equivalent to molecular weights of 26,000 and 57,000.

Characteristics of IVR

IVR was stable in pH 2.5 and the antiviral activity was conserved after dialysis against 0.05 *M* phosphate buffer, pH 7.5, or against VIM. IVR was inactivated completely or its activity markedly reduced by treatment with trypsin or chymotrypsin, but not with ribonuclease.[4] IVR was found to be thermolabile and was inactivated by heating at 60° for 10 min. These data suggest that IVR is of a proteineous nature.

[7] J. R. Whitaker, *Anal. Chem.* **35,** 1950 (1963).

[97] Preparation and Measurement of an Antiviral Protein Found in Tobacco Cells after Infection with Tobacco Mosaic Virus

By ILAN SELA

AVF (antiviral factor) is a cumulative name given by Sela and Applebaum[1] to what turned out to be a group of closely related proteinaceous plant substances with antiviral activity. AVF, or its precursors, also present in noninfected tobacco leaves, is easily stimulated by various

[1] I. Sela and S. W. Applebaum, *Virology* **17,** 543 (1962).

agents,[2] notably poly(I) · poly(C)[3] and especially TMV.[4] AVF stimulation is most pronounced following TMV infection of plants carrying the N-gene, which is responsible for localizing TMV infection in these plants[5,6] (for discussion of the N-gene see Sela[2]). A substance similar to AVF was later described in N-gene-carrying tobacco mesophyll protoplasts.[7,8]

AVF, or some of it, is a phosphorylated glycoprotein.[9,10] Its activity in tissues, estimated at 0.1–10 molecules per cell, triggers some primary reactions whose products communicated to neighboring cells impose an "antiviral state."[2] A major activity of AVF (and human leukocyte interferon) in tobacco cells is the stimulation of a synthetase activity which polymerizes ATP to oligoadenylates in the presence of dsRNA.[11] The plant oligonucleotides though antivirally active[12] are not identical with 2,5-A_n.[13,14] AVF resembles interferon[15] which was indeed found to protect plant tissues from TMV,[16,17] as well as 2,5-A_n[18] and its derivatives.[19]

Assaying Antiviral Activity in Plants

General Comments on Quantitation

Measuring infectivity is a prerequisite for measuring antiviral activity. The lack of clonal cell lines in plants makes it impossible to develop a truly quantitative assay. Although a newly developed assay with protoplasts from standard tobacco cell suspensions,[20] is, so far, the best assay

[2] I. Sela, *Adv. Virus Res.* **26,** 201 (1981).
[3] O. Gat-Edelbaum, A. Altman, and I. Sela, *J. Gen. Virol.* **64,** 211 (1983).
[4] TMV: tobacco mosaic virus.
[5] Y. Antignus. I. Sela, and I. Harpaz, *J. Gen. Virol.* **35,** 107 (1977).
[6] I. Sela, A. Hauschner, and R. Mozes, *Virology* **89,** 1 (1978).
[7] G. Loebenstein and A. Gera, *Virology* **114,** 132 (1981).
[8] A. Gera, S. Spiegel, and G. Loebenstein, this volume [96].
[9] Y. Antignus, I. Sela, A. Hauschner, and I. Harpaz, *Physiol. Plant Pathol.* **6,** 159 (1975).
[10] R. Mozes, Y. Antignus, I. Sela, and I. Harpaz, *J. Gen. Virol.* **38,** 241 (1978).
[11] Y. Devash, A. Hauschner, I. Sela, and K. Chakraburtty, *Virology* **111,** 103 (1981).
[12] M. Reichman, Y. Devash, R. J. Suhadolnik, and I. Sela, *Virology* **128,** 240 (1983).
[13] Y. Devash, M. Reichman, I. Sela, and R. J. Suhadolnik, *Biochemistry* **24,** 593 (1985).
[14] Y. Devash, I. Sela, and R. J. Suhadolnik, this volume [99].
[15] I. Sela, *Perspect. Virol.* **11,** 129 (1981).
[16] P. Orchansky, M. Rubinstein, and I. Sela, *Proc. Natl. Acad. Sci. U.S.A.* **79,** 2278 (1982).
[17] N. Rosenberg, M. Reichman, A. Gera, and I. Sela, *Virology* **140,** 173 (1985).
[18] Y. Devash, S. Biggs, and I. Sela, *Science* **216,** 1415 (1982).
[19] Y. Devash, A. Gera, M. Reichman, W. Pfleiderer, R. Charubala, I. Sela, and R. J. Suhadolnik, *J. Biol. Chem.* **259,** 3482 (1984).
[20] I. Sela, M. Reichman, and A. Weissbach, *Phytopathology* **74,** 385 (1984).

for reliable quantitation of infection, it should still be considered as no more than an estimate.

Measuring the antiviral activity of AVF and interferon is further complicated by the odd dose–response of tobacco tissues and cells. Interferon protects tobacco cells at a very low dosage, but its antiviral activity diminishes gradually with increase in interferon concentration.[17] With AVF at least two waves of antiviral activity are usually observed upon dilution.

Infectivity Assay: The Local Lesion Test

Some viruses react with their host plants by producing local lesions (mostly necrotic) on the infected leaves. The number of local lesions is proportional to virus concentration, but, since leaves vary considerably in their response to viral infection, it is impossible to draw an accurate standard curve. In order to overcome this inherent shortcoming, at least partially, we usually combine the half-leaf test with one involving a dilution end-point. Purified TMV (prepared according to Devash *et al.*[11]) is added to AVF or mock AVF[21] solutions to a final concentration of 5 μg TMV per ml 0.01 *M* sodium phosphate, pH 7.6, and carborundum (400–600 mesh) is added to a concentration of 0.1 g/ml. The content of the tube is stirred with a sterile painter's brush which is then squeezed against the inner tube wall to remove excess liquid. Half-leaves of *Datura stramenium* L. are then inoculated by gently moving the brush from tip to petiol, creating 10-mm-wide inoculated strips. Control and AVF solutions are applied to opposite halves of the same leaves (Fig. 1). A series of AVF dilutions to about 0.01 pg protein per ml are thus tested (the TMV concentration remains constant), each dilution being applied to 12 half-leaves. The plants are kept at 24° under constant illumination and local lesions are counted 3–4 days later. Activity at each dilution is expressed as a percentage of protection of the AVF-treated half-leaves compared to the control half-leaves.

The Leaf Disk Test

In this test the viral content of leaf disks is determined either by the above described infectivity tests of disk homogenates, or, more reliably, by the enzyme-linked immunosorbent assay (ELISA).[22]

[21] Mock AVF: A fraction prepared similarly to the particular AVF fraction but from noninfected plants.

[22] M. F. Clark and A. H. Adams, *J. Gen. Virol.* **34,** 475 (1977).

FIG. 1. Half-leaf test. Leaves of *D. stramonium* were inoculated with 5 μl/ml TMV containing 100 ng protein of mock DEAE–AVF preparation (left half) and 100 ng protein of DEAE–AVF (right half).

ELISA Reagents

PBS-Tween: 0.5 ml Tween 20 (Sigma), 0.2 g NaN_3 in 1 liter PBS

Coating buffer: 1.59 g Na_2CO_3, 2.93 g $NaHCO_3$, 0.2 g NaN_3 in 1 liter H_2O

PBS-Tween–PVP: 20 g PVP-40 (polyvinylpyrrolidone, Sigma) in 1 liter PBS-Tween

PBS-Tween–PVP–BSA: 0.2% bovine serum albumin (BSA) in PSB-Tween–PVP

Substrate buffer: 97 ml diethanolamine, 800 ml H_2O, 0.2 g NaN_3. Titrate with HCl to pH 9.8 and bring to 1 liter with H_2O

Substrate solution: *p*-nitrophenyl phosphate. Keep solid at −20°. Prepare a fresh solution of 0.1 mg/ml in substrate buffer before each use

Microplates: Dynatech 129B or Nunc Immunoplates

ELISA Procedure

The γ-globulin fraction of chicken and rabbit sera immunized against TMV is prepared according to Clark and Adams[22] and stored at 1 mg/ml at −20°. To each well of a microtiter plate 200 μl of chicken γ-globulin, diluted to 1 μg/ml in coating buffer, is added and incubated for 3 hr at 30° or overnight at 4°. The wells are thoroughly washed (3 times) with PBS-Tween and 200 μl samples of TMV or disk homogenates (both prepared in PBS-Tween–PVP) are added to every well and incubated as above. Following washing, rabbit γ-globulin (1 μg/ml in PBS-Tween–PVP) is added to the wells and incubated as above. Commercially available goat anti-rabbit IgG (conjugated to alkaline phosphatase and diluted 1:1000 in PBS-Tween–PVP–BSA) is added next. Following incubation and washing as above, 250 μl of substrate solution is added and incubated at room temperature. Absorption at 410 nm is determined in a Microelissa Autoreader (Dynatech MR-580). Since a 96-well plate is read in less than a minute, the reaction is not stopped and readings are taken at 10 min intervals until 1 μg/ml TMV yields an absorption of about 0.8. TMV standards for calibration curves are included in every plate, and every sample is applied in triplicate. Absorbance is linear with TMV concentration bewteen approximately 10 and 1000 ng/ml. TMV content is expressed as ng/disk. Since the TMV concentration must fall in the linear range, serial dilutions are made from each disk homogenate.

Antiviral Assay by ELISA

Leaves of tobacco var. "Samsun" are dusted with carborundum and are inoculated with TMV (1 μg/ml in 0.01 *M* sodium phosphate, pH 7.6). After extensive washing with water, leaf disks (6 or 8 mm in diameter) are punched out with a paper-punch into a beaker containing the standard buffer. Twenty disks are collected at random from the common pool and placed in Petri dishes floating on 20 ml of the above buffer. Some plates contain disks from noninfected leaves, some contain untreated inoculated disks, and the rest contain the various materials to be tested for antiviral activity (added to the disks within 1 hr after inoculation). Plates are incubated at 24° under constant illumination. The disks are then frozen at −20°, homogenated with a Teflon-coated, motor-driven homogenizer, and their TMV content is determined by ELISA. TMV is first detected (in homogenate of 20 disks) 36 hr after inoculation, and increases rapidly between 48 and 96 hr after inoculation. It is better to follow the TMV growth curve in control and tested solutions, but since, in many cases, this is too elaborate a task, a single time-point between 60 and 80 hr is

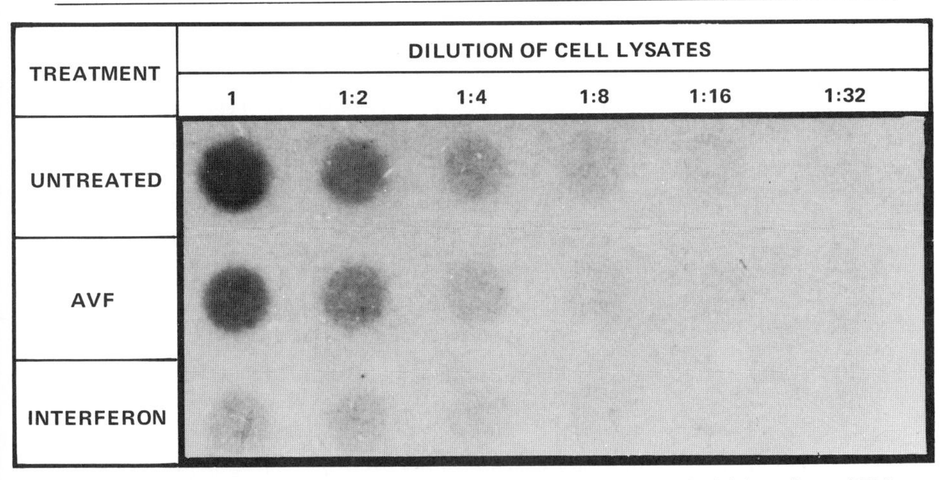

FIG. 2. Dot-blot assay. TMV-inoculated protoplasts were treated with interferon (IFN-αA; 0.1 units/ml) and with DEAE–AVF (60 ng protein/ml). Cell lysates were serially diluted (from left to right), spotted onto nitrocellulose, and hybridized as described in the text.

chosen. The homogenate of 20 disks, 72 hr after inoculation, should be diluted 1:10–1:100 for ELISA.

Antiviral Assay by Dot-Blot Test

This test, developed in order to detect TMV–RNA in crude leaf and protoplast extracts, is about 100 times more sensitive than the ELISA.[20] This assay is based on a specific hybridization between radioactive cDNA (reverse-transcribed from TMV–RNA) and the RNA in the extracts and measures TMV–RNA accumulation. The detailed procedure is described elsewhere.[23]

AVF and human interferon suppress TMV–RNA replication. The availability of highly purified interferon enables the expression of its activity in the protoplast system in terms of molecules per cell. By comparing cruder AVF preparations to interferon in the same experiment, interferon units could be assigned to AVF (Fig. 2).

Purification

AVF elutes as a broad peak of antiviral activity upon DEAE-cellulose chromatography and HPLC. After electrophoresis it appears as a broad

[23] I. Sela, this volume [98].

zone rather than a sharp band. The better resolving HPLC procedure exhibits, within the area of AVF activity, a number of protein peaks, some of which appear homogeneous on electrophoresis. Hence, it seems that AVF is a family of closely related molecules. The AVF preparations described herein cannot, therefore, be regarded as homogeneous. One HPLC fraction, which was studied a number of times was found to be homogeneous; however, it is premature to draw any definitive conclusions in this respect.

Reagents

Standard buffer: 0.01 *M* sodium phosphate, pH 7.6

HCP (hydrated calcium phosphate[24]): a volume of 0.1 *M* Na_2HPO_4 is added to the same volume of 0.1 *M* $CaCl_2$. A precipitate is formed and washed 10 times with H_2O by decantation. HCP is centrifuged prior to its addition to plant material to remove excess liquid

Con A buffer: 0.1 *M* sodium acetate, pH 6.0; 1 m*M* $MnCl_2$; 1 m*M* $MgCl_2$; 1 m*M* $CaCl_2$

HPLC buffer: 1 *M* pyridine, 2 *M* formic acid, pH 4.0

Protein Determinations

A 50 μl sample of an AVF preparation is mixed with 50 μl of a Coomassie blue reagent[25,26] in an ELISA microtiter plate and the developed color is read at 570 nm in the automated ELISA reader. A calibration curve of known BSA concentrations is read in every plate. Green plant homogenate is precipitated with 7% perchloric acid and then redissolved in 0.1 *N* NaOH for protein determination. Dilutions (at least 1:10 in H_2O) of this basic solution can be added to the reagent and read quantitatively.

Plant Material

Nicotiana glutinosa plants (carrying the N-gene) are grown in an insect-proof green house at 24 ± 2° under constant illumination. At the stage of 4–6 fully expanded leaves and following dusting with carborundum, the leaves are rubbed gently with TMV (5 μg/ml in the standard buffer). Leaves are harvested 2 days later when distinct local lesions appear but are not yet fully expanded.

[24] R. W. Fulton, *Virology* **9,** 522 (1959).

[25] J. J. Sedmak and S. E. Grossberg, *Anal. Biochem.* **79,** 544 (1977).

[26] Commercial kit with protocol from Bio-Rad Laboratories.

Procedure

Step 1: Preparation of Crude AVF. The leaves are homogenized in a blender in the standard buffer at a ratio of 1:1 (w/v). The homogenate is squeezed through 3 layers of cheesecloth and centrifuged at 5000 *g* for 10 min. A half-volume of HCP is added to the supernatant fluid, shaken vigorously to obtain a homogeneous emulsion, and recentrifuged as above. Another half-volume of HCP is added to the supernatant and the procedure is repeated. HCP absorbs TMV and removes it from the preparations. Total removal of TMV is ascertained by always setting aside samples for infectivity tests. HCP also absorbs 80–95% of the homogenate protein, which results in 5- to 20-fold purification. However, since this is the first fraction devoid of TMV, it is also the first one for which antiviral activity can be estimated and is designated "crude AVF."

Crude AVF is dialyzed against H_2O and lyophilized. It is stable in this form in a desiccator at room temperature for many months.

Step 2: Chromatography on DEAE-Cellulose. A column of DEAE-cellulose (Whatman DE-23 or DE-52, 0.6 × 15 cm) is equilibrated with the standard buffer. Crude AVF (200 mg protein) is dissolved in the buffer and applied to the column. A gradient of 0–0.75 *M* NaCl in the standard buffer is applied. Fractions, diluted with the standard buffer (1:1000), are tested for antiviral activity. Mock AVF preparations are similarly prepared and assayed (Fig. 3). AVF is eluted from DEAE-cellulose in a very peculiar position for a protein (0.48–0.55 *M* NaCl) probably because it is phosphorylated. Indeed, only 2% of the protein is eluted at the AVF position. At this stage, however, the purification factor is far greater than expected from the amount of protein removed (see the table), indicating that inhibitors to AVF activity are apparently removed.

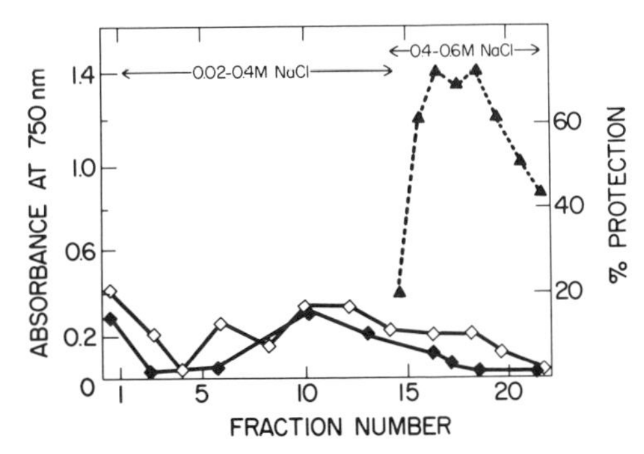

FIG. 3. Elution pattern of AVF from DEAE cellulose. Absorbancy of proteins from a mock preparation (◇) and from AVF preparation (◆). Antiviral activity of AVF (▲). Data from Antignus *et al.*[9]

SUMMARY OF PURIFICATION OF AVF

Fraction	Total protein (μg)	Dilution end-point of AVF activity (pg protein/ml)	Estimated purification factor
Leaf homogenate	200,000	—	—
Crude AVF	30,000	150,000	1
DEAE–AVF	600	6	25,000
Con A–AVF	80	0.8	187,500
HPLC–AVF (Fraction No. 39)	2.4	0.24	625,000

In routine preparations, AVF is eluted from DEAE-cellulose in a stepwise manner between 0.45 and 0.60 *M* NaCl. It is dialyzed against H_2O, lyophilized, and stored as crude AVF, designated DEAE–AVF.

Step 3: Affinity Chromatography on Concanavalin A-Sepharose. A column (0.6 × 10 cm) of concanavalin A-Sepharose (Pharmacia) is equilibrated with the Con A buffer. DEAE–AVF, 10–50 mg protein in the same buffer, is applied to the column and washed with 30 ml of the Con A buffer. AVF is eluted with 5% α-methyl-D-mannoside (Sigma) in the above buffer. A minute quantity of concanavalin A is constantly leaking from the column, and, since there is very little AVF at this stage, con-

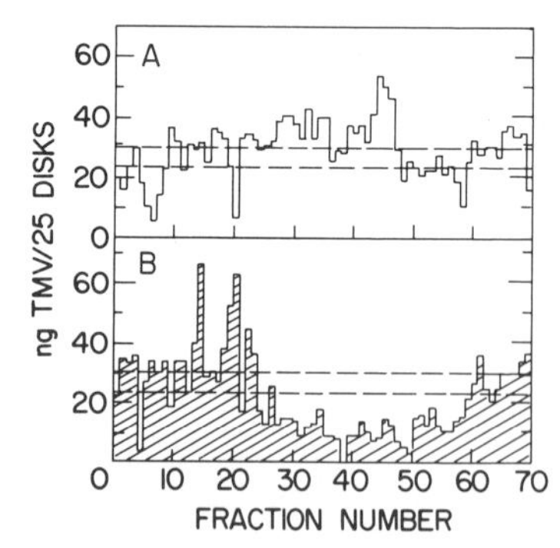

FIG. 4. Elution patterns of antiviral activity from reverse-phase HPLC (see text for details). Disks of TMV-infected tobacco leaves were immersed in the standard buffer containing the various column fractions for 72 hr at which time the virus content was determined by ELISA. (A) Fractions of mock AVF preparation. (B) Fractions of AVF preparation. The TMV content of buffer-treated leaf disks fell between the dashed lines. The antiviral activity of the AVF fractions accumulated in a wide zone, and was especially strong in fraction No. 39.

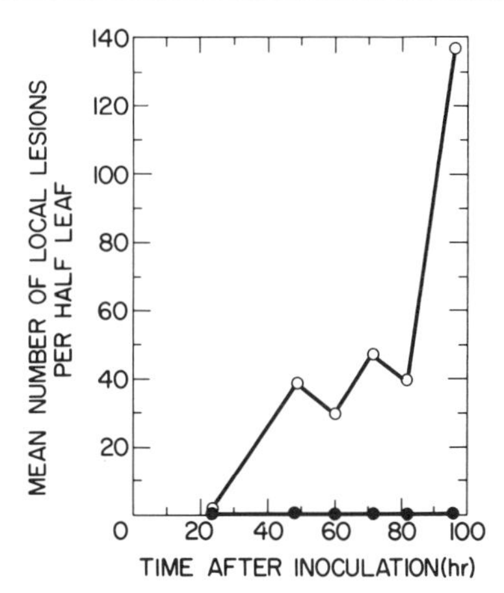

FIG. 5. TMV growth curve of control leaf disks and of disks treated with fraction No. 39 of Fig. 4. TMV in leaf disks was estimated by the infectivity of the disk homogenates. (●) Fraction No. 39; (○) control.

canavalin A contamination is considerable. The preparation is therefore dialyzed against the standard buffer and reapplied to DEAE-cellulose. Concanavalin A is eluted from the column with 0.45 *M* NaCl and AVF is then eluted with 0.60 *M* NaCl. The AVF fraction (designated Con A–AVF) is dialyzed against water, lyophilized, and stored in a desiccator at room temperature.

Purification by HPLC

This is still at the experimental stage and should be so regarded. Crude AVF is dissolved in the HPLC buffer, loaded on a reverse-phase column (octyl silica, Merck RP-8, 10 μm), and chromatographed with a linear gradient (0–40%) of *n*-propanol as described for human leukocyte interferon by Rubinstein and Pestka.[27] A great number of fractions eluted in the area exhibiting antiviral activity (Fig. 4). The most active fraction (No. 39) was further tested for antiviral activity (Fig. 5). Since about 10^7 cells are in a leaf disk and the molecular weight of AVF is estimated to be 22,000 (see below), it seems that less than 1 AVF molecule per cell is sufficient to confer protection from TMV.

[27] M. Rubinstein and S. Pestka, this series, Vol. 78, p. 464.

Properties

Stability

AVF, at every stage, can be stored, dry, at room temperature for months without loss of activity. Con A–AVF remains active after being dialyzed at pH 2 for 48 hr and then back to pH 7.6. DEAE–AVF and Con A–AVF remain active in 0.1% SDS but must be diluted for the antiviral assay. AVF can be eluted in an active form from SDS–polyacrylamide gels.

Purity

DEAE–AVF appears as a broad zone on polyacrylamide gel electrophoresis indicating heterogeneity of the active molecules. At least 20 protein bands, on a background of a smear, are observed upon staining. The expansion of the stained area beyond the active zone indicates that AVF is far from pure at this stage.

Con A-AVF upon electrophoresis also appears as a zone of activity, but in this case the stained area is superimposed on that of antiviral activity. The center of the AVF zone migrates as a protein with a MW of 22,000. Fraction No. 39 of Fig. 4 migrates as a highly active homogeneous protein of MW 21–22,000 on SDS–polyacrylamide gel electrophoresis.

[98] Assay of Effect of Human Interferons on Tobacco Protoplasts

By ILAN SELA

Human interferons were found to protect tobacco leaf disks from TMV[1] infection[2] and induce the polymerization of ATP to antivirally active oligonucleotides.[3,4] The interferon (and AVF[5])-induced plant nucleotides are not identical to the animal 2,5-A_n,[4,6] but the latter and many of its derivatives inhibit TMV multiplication.[7,8]

[1] TMV; tobacco mosaic virus.

[2] P. Orchansky, M. Rubinstein, and I. Sela, *Proc. Natl. Acad. Sci. U.S.A.* **79,** 2279 (1982).

[3] M. Reichman, Y. Devash, R. J. Suhadolnik, and I. Sela, *Virology* **128,** 240 (1983).

[4] Y. Devash, I. Sela, and R. J. Suhadolnik, this volume [99].

[5] AVF: TMV-induced antiviral factor from *Nicotiana glutinosa*.

The lack of homogeneity in the various plant systems makes accurate quantitation of plant virus infectivity impossible. However, cell suspension-derived protoplasts seem to be quite homogeneous. Since protoplasts are always prepared from the same cell population grown under controlled conditions, genetic and physiological variations are minimal. It is easy to lyse protoplasts and to determine their TMV content without extensive manipulations. In this system, interferon activity can be expressed in terms of molecules/cell rather than units/ml.

Protoplast Preparation and Inoculation

All solution must be prepared in glass-distilled water. All stock solutions are sterilized by filtration and stored at $-20°$.

Stock Solutions

Myoinositol: 10% (w/v) in H_2O
Naphthalene acetic acid: 60 mg in 4 ml ethanol brought to 10 ml with H_2O
Kinetin: 0.1% (w/v) in 0.1 *N* NaOH
Casein hydrolysate: 8% (w/v) in H_2O
N^6-Benzyladenine (6-benzylaminopurine): 20 mg in 20 ml dimethyl sulfoxide brought to 100 ml with H_2O
Vitamin solution 1: 10 mg thiamine · HCl, 50 mg pyridoxine, 50 mg nicotinic acid, and 200 mg glycine in 100 ml H_2O
Vitamin solution 2: 200 mg thiamine · HCl and 10 mg biotin in 80 ml H_2O. Add 10 ml of folic acid (1 mg/ml in 0.1 *N* NaOH) and bring to 100 ml with H_2O

MSS Medium for Tobacco Cell Suspension[9]

1 pouch of plant salt mixture[10]
10 ml of casein hydrolyzate stock solution

---

[6] Y. Devash, M. Reichman, I. Sela, and R. J. Suhadolnik, *Biochemistry* (in press).
[7] Y. Devash, S. Biggs, and I. Sela, *Science* **216,** 1415 (1982).
[8] Y. Devash, S. Gera, M. Reichman, W. Pfleiderer, R. Charubala, I. Sela, and R. J. Suhadolnik, *J. Biol. Chem.* **259,** 3482 (1984).
[9] T. Murashige and F. Skoog, *Physiol. Plant* **15,** 485 (1962).
[10] The plant salt mixture used in this laboratory is from Flow Laboratories, Irvine KA12 8NB, Scotland, or Gibco Laboratories, Grand Island, New York. The composition of the salt mixture in mgs is as follows: NH_4NO_3, 1650; KNO_3, 1900; $CaCl_2 \cdot 2H_2O$, 440; $MgSO_4 \cdot 7H_2O$, 370; KH_2PO_4, 170; Na_2-EDTA, 37.3; $FeSO_4 \cdot 7H_2O$, 27.8; H_3BO_3, 6.2; $MnSO_4 \cdot H_2O$, 16.9; $ZnSO_4 \cdot 7H_2O$, 8.6; KI, 0.83; $Na_2MoO_4 \cdot 2H_2O$, 0.25; $CuSO_4 \cdot 5H_2O$, 0.025; $CoCl_2 \cdot 6H_2O$, 0.025.

1 ml of myoinositol stock solution
1 ml of N^6-benzyladenine stock solution
0.1 ml of naphthalene acetic acid stock solution
1 ml of vitamin solution 1
5 ml of vitamin solution 2
30 g sucrose
Add H_2O to make 1 liter. If necessary, the pH is adjusted with KOH to 5.6–5.8. Autoclave and store at 4°
MSS-M: 12% (w/v) mannitol in MSS. Autoclave

CPW for Protoplasts

KH_2PO_4, 27.2 mg/liter
KNO_3, 101 mg/liter
$CaCl_2 \cdot 2H_2O$, 1480 mg/liter
$MgSO_4 \cdot 7H_2O$, 246 mg/liter
KI, 0.16 mg/liter
$CuSO_4 \cdot 5H_2O$, 0.025 mg/liter
Mannitol, 135 g/liter

The protoplast medium is prepared from stock solutions ($CaCl_2$ 200×; all the rest 1000×). Mannitol is added as a solid. The pH is adjusted to 5.6–5.8 with KOH and the solution sterilized by filtration.

Citrate buffered mannitol: 13.5% mannitol in 0.01 *M* potassium citrate, pH 5.8. Autoclave.

$CaCl_2$/mannitol solution: 0.1 m*M* $CaCl_2$ in 13.5% (w/v) mannitol. Sterilize by filtration.

PEG-$CaCl_2$: 40% (w/v) polyethylene glycol 6000, 20 m*M* $CaCl_2$ in 13.5% (w/v) mannitol. Sterilized by filtration.

Enzymes. Cellulase "Onozuka" R-10 and Macerozyme R-10 are from Yakult Pharmaceutical Industry, Nishinomiya, Japan. Pectolyase Y-23 is from Seishin Pharmaceutical Co. Tokyo, Japan.

Enzyme solution consists of 1.5% (w/v) cellulase, 0.1% (w/v) macrozyme, and 0.01% (w/v) pectolyase in CPW. The pH is adjusted with KOH to 5.4–5.8 and the solution sterilized by filtration.

Interferons. The following human alpha interferons were tried successfully in this system: (1) subspecies γ_3 (2.4×10^8 units/mg) of Rubinstein *et al.*[11]; (2) recombinant human alpha interferon D (IFN-αD, 2.2×10^8 units/mg),[12] similar to the above γ_3 natural interferon; (3) recombinant

[11] M. Rubinstein, S. Rubinstein, P. C. Familleti, R. S. Miller, A. A. Waldman, and S. Pestka, *Proc. Natl. Acad. Sci. U.S.A.* **76,** 640 (1979).

[12] E. Rehberg, B. Kelder, E. G. Hoal, and S. Pestka, *J. Biol. Chem.* **257,** 11497 (1982).

human alpha interferon A (IFN-αA, 2.9 $\times$ 10^8 units/mg),[13] similar to the natural leukocyte IFN-α2 and -β1.[14,15] The recombinant interferon preparations were obtained from Roche Research Center, Nutley, New Jersey.

Procedure

All steps are performed aseptically. All centrifugations are carried out in a swing-out rotor at room temperature.

Cell Suspension. Tobacco var. Wisconsin 38 is used. Cells are suspended in MSS at 25–28° and shaken orbitally at 100 rpm in the dark. Even so, cells tend to clump together in small agaregates. Cell-doubling time is about 36 hr. Transfers are made every 3–4 days by diluting the cells 1:4 into fresh MSS.

Preparation of Protoplasts for TMV Inoculation. Cells, packed by centrifugation at 300 *g* for 5 min, are resuspended in 10 volumes of CPW, centrifuged again, and resuspended in CPW as before. It is essential to remove all traces of MSS for successful protoplast preparation.

The cells are suspended in freshly prepared enzyme solution (2–3 volumes of that of the packed cells), incubated for 3 hr at 37° with shaking (100 rpm), then centrifuged at 300 *g* for 5 min. The protoplasts are resuspended in 13.5% mannitol, recentrifuged, and resuspended again in a minimal volume of 13.5% mannitol. The protoplast suspension is placed on top of a column of 5–10 volumes of 23% sucrose and centrifuged 10 min at 600 *g*. The top layer is aspirated with a wide-mouth pipette, 5 volumes of 13.5% mannitol is added, and the protoplasts are centrifuged 5 min at 300 *g* and finally resuspended in a small volume of 13.5% mannitol. The protoplasts are counted with a hemocytometer.

Inoculation of Protoplasts with TMV. Solution A: 10 μl TMV (2 mg/ml, sterilized by filtration) and 10 μl poly-L-ornithine (2 mg/ml, sterilized by filtration) are added to 1 ml citrate-buffered mannitol.

Solution B: 5 $\times$ 10^6 protoplasts in 10 ml 13.5% mannitol.

One volume of solution A and 1 volume of solution B are mixed and shaken at 25° (100 rpm) for 10 min. Protoplasts are pelleted at 300 *g* for 5 min and resuspended in $CaCl_2$/mannitol solution (4 ml per 10^6 protoplasts). The protoplasts are centrifuged and resuspended once again in a sufficient volume of MSS-M to maintain 10^6 protoplasts/ml.

Protoplast droplets, 25 μl containing 25,000 protoplasts, are distributed in Petri dishes with a sterile micropipette (with tips which were cut

[13] T. Staehelin, D. S. Hobbs, H.-F. Kung, and S. Pestka, this series, Vol. 78, p. 505.

[14] W. P. Levy, M. Rubinstein, J. Shively, U. Del Valle, C.-Y. Lai, J. Moschera, L. Brink, L. Gerber, S. Stein, and S. Pestka, *Proc. Natl. Acad. Sci. U.S.A.* **78,** 6186 (1981).

[15] S. Pestka and S. Baron, this series, Vol. 78, p. 3.

about 1/4 distance from the narrow end). The plates are floated in a Pyrex baking pan, containing about 0.5 cm height of water, and sealed tightly with Saran wrap to make a humid chamber. They are incubated at 25°.

The supply of energy (MSS-M) and access to oxygen (small droplets) support TMV multiplication for up to 120 hr. Maintaining protoplasts in CPW with shaking[16] allows only about 48 hr of metabolism.

Inoculating Protoplasts with TMV–RNA

TMV–RNA (100 μg) is dissolved in 1 ml PEG–$CaCl_2$ and sterilized by filtration. The protoplasts are pelleted, resuspended for 10 sec in the above RNA solution, diluted immediately into 100 volumes of 13.5% cold mannitol, and kept on ice for 20 min. It is important that the mannitol solution is cold. The protoplasts are pelleted again, washed twice with 13.5% mannitol brought to 10^6 protoplasts/ml, and distributed in Petri dishes as above.

ELISA Assays for TMV Multiplication

At the specified times, 20 droplets are collected (5×10^5 protoplasts) and frozen. After thawing, 1 ml of PBS-Tween–PVP[17] is added for 10 min at room temperature. A few strokes with a Teflon-coated homogenizer aids lysis. A series of 10-fold dilutions are made and applied as a virus source to the microplate. ELISA is performed as previously[17] described.

Dot-Blot Hybridization Assay from TMV Multiplication

Solutions

Reverse transcription mixture: 3.8 μl 1 *M* NaCl, 9 μl 1 *M* Tris · HCl, pH 8.3, 4.5 μl 0.2 *M* $MgCl_2$, 6 μl 0.1 *M* DTT, 2 μl of 5 m*M* each of dATP, dGTP, dTTP; 0.7 μl H_2O

Primer DNA: 5 mg of calf-thymus DNA is incubated with 70 μg of DNAse I in 1 ml of 10 m*M* Tris · HCl, pH 7.6, 10 m*M* $MgCl_2$ at 37° for 1 hr. The reaction mixture is autoclaved at the end of the reaction

Column buffer: 0.01 *M* Tris · HCl, pH 7.5, 0.1 *M* NaCl, 0.1% SDS

Prehybridization solution: 50% formamide, 5× SSC, 0.05 *M* sodium phosphate, pH 6.5, 0.02% each of BSA, Ficoll, 400,000, and PVP-360, 100 μg/ml salmon-sperm denatured DNA, 0.1% SDS

[16] I. Sela, M. Reichman, and A. Weissbach, *Phytopathology* **74,** 385 (1984).

[17] I. Sela, this volume [97].

Hybridization solution: identical to the prehybridization solution with sodium phosphate reduced to 0.02 *M* and dextran sulfate added to 10%

TMV–RNA: 3–5 mg/ml, prepared according to Devash *et al.*[18]

Reverse transcriptase: from Avian myoblastosis virus (Life Sciences)

Preparation of Random Primed cDNA to TMV–RNA

[α^{32}P]dCTP (200 μCi; 3000 Ci/mmol) is dried in a vacuum. To the dry tube 10 μl of primer DNA, 10 μl reverse transcription mixture, 6 μl TMV–RNA, 2 μl of 30 m*M* sodium pyrophosphate (to prevent second strand synthesis), and 2 μl (20–30 units) of reverse-transcriptase are added and incubated at 37° for 15 min. The reaction is stopped by adding 6 μl of 0.2 *M* EDTA. Salmon sperm DNA is added to a concentration of 100-200 μg/ml and then 200 μl of column buffer. The mixture is passed through a Sephadex G-50 column (0.6 × 20 cm) equilibrated and washed with the column buffer and 0.5 ml fractions are collected. Each fraction is monitored with a Geiger counter. The fractions comprising the first radioactive peak are pooled, 0.5 volume of 1 *N* NaOH is added, incubated at 37° for 30 min to hydrolyze RNA, then 0.1 volume of 3 *M* sodium acetate is added followed by 1 *N* HCl (same volume as the previously added NaOH; sodium acetate also buffers the HCl so that pH cannot drop below 4.5). The cDNA is precipitated with two volumes of ethanol at −20°, centrifuged, dried in a vacuum, and dissolved in a minimal volume of sterile H_2O. In this reaction we usually obtain 0.7–1.2 × 10^8 dpm as cDNA, with fragments length of 70–200 bases, which is totally sensitive to nuclease S1 and of which 50–70% hybridizes to TMV–RNA in the presence of excess RNA.

Procedure

TMV-inoculated protoplasts (8 droplets, 200 μl) are lysed by adding 2 μl of 10% SDS and incubating for 10 min at 42°. This also strips TMV of its coat protein exposing the TMV–RNA. To this lysate, 120 μl of 20× SSC and 80 μl of 37% formaldehyde are added (according to White and Bancroft[19]) and incubated at 65° for 15 min. A sheet of nitrocellulose is wet with 2× SSC and placed on 3MM paper in a vacuum suction manifold. Samples of 100 μl are applied with suction and the filter is then baked at 80° under vacuum. The nitrocellulose sheet is placed in a plastic bag, prehybridization solution is added, the bag sealed, and incubated at 42°

[18] Y. Devash, A. Hauschner, I. Sela, and K. Chakraburtty, *Virology* **111,** 103 (1981).

[19] B. A. White and F. C. Bancroft, *J. Biol. Chem.* **257,** 8569 (1982).

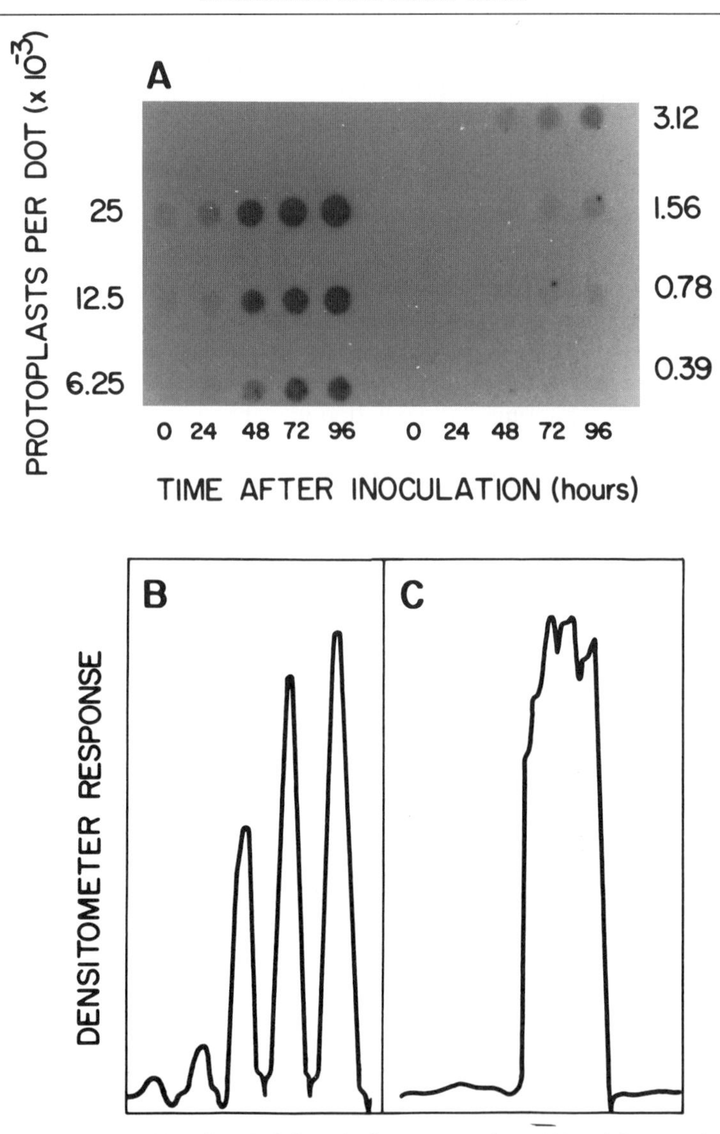

FIG. 1. Dot-blots of lysed TMV-infected tobacco protoplasts (A) and the scanning of the row of the highest protoplast density (B) and the lowest one (C). Densitometer sensitivity in C is 10 times higher than in B. From Sela *et al.*[16]

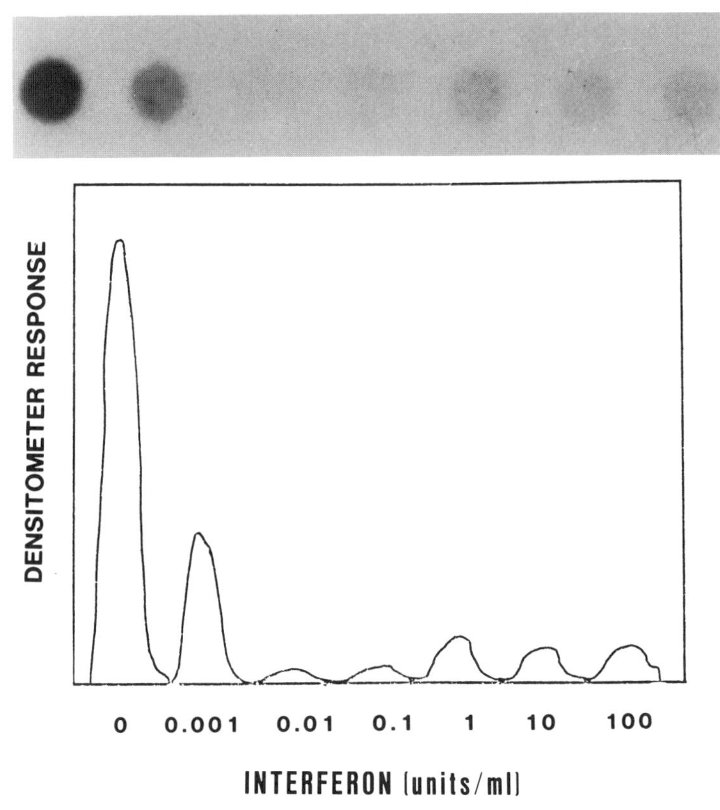

FIG. 2. Dot-blots of TMV-infected protoplasts and their corresponding densitometer tracings. Protoplasts were treated with human leukocyte interferon (IFN-αA) for 3 hr prior to TMV inoculation and were dotted 72 hr after inoculation. Each dot corresponds to 25,000 protoplasts. From Rozenberg *et al.*[20]

for 1–2 hr. The solution in the bag is replaced with hybridization solution and incubated at 42° for 15–20 hr. The nitrocellulose sheet is placed into 1 liter of 0.1× SSC containing 0.1% SDS and shaken at 37° for 1 hr. The wash buffer is changed once, the filter washed for an additional hour, dried, and autoradiographed. The TMV–RNA content of 400 protoplasts can thus be visualized 96 hr after infection (Fig. 1). This procedure is 80–100 times more sensitive than ELISA.

Assay of Effect of Interferon

Interferon is diluted to the desired concentration with MSS-M (pH 5.8) containing 1% fetal calf serum (FCS); 0.1 volume of the interferon dilution is added to the protoplasts so that the FCS concentration in the protoplast suspension is 0.1%. Higher FCS concentrations cause the appearance of filamentous structures and protoplast aggregation. MSS-M containing FCS without interferon serves as a control. Protoplasts, incubated with interferon for 3 hr at 25° (80 rpm), washed with 13.5% mannitol and inoculated with TMV or TMV–RNA, are lysed 48–72 hr after inoculation, dot-hybridized, and autoradiographed. The autoradiogram can also be scanned (Fig. 2).[20] If TMV coat protein is to be measured, a larger sample of protoplasts is taken (20 droplets) and the protein determined by ELISA.

Comments

The effect of interferon on protoplasts is transient, gradually disappearing 60–90 hr after the addition of interferon. The dose–response of tobacco protoplasts is peculiar, with maximum activity observed at 0.01–0.1 units/ml of interferon. Further increase in interferon concentration reduces its effect.

[20] N. Rozenberg, M. Reichman, A. Gera, and I. Sela, *Virology* **140,** 173 (1985).

[99] Enzymatic Synthesis of Plant Oligoadenylates *in Vitro*

By YAIR DEVASH, ILAN SELA, and ROBERT J. SUHADOLNIK

Tobacco plants carrying the N-gene for virus localization (*Nicotiana glutinosa*) respond to tobacco mosaic virus (TMV) infection by producing an antiviral factor (AVF) that resembles interferon.[1-10] A factor (DF)

[1] I. Sela, *Adv. Virus Res.* **26,** 201 (1981).
[2] I. Sela, *Trends Biochem. Sci.* **6,** 31 (1981).
[3] Y. Devash, A. Hauschner, I. Sela, and K. Chakraburtty, *Virology* **111,** 103 (1981).
[4] P. Orchansky, M. Rubinstein, and I. Sela, *Proc. Natl. Acad. Sci. U.S.A.* **79,** 2278 (1982).
[5] Y. Devash, S. Biggs, and I. Sela, *Science* **216,** 1415 (1982).
[6] M. Reichman, Y. Devash, R. J. Suhadolnik, and I. Sela, *Virology* **128,** 240 (1983).
[7] Y. Devash, A. Gera, D. H. Willis, M. Reichman, W. Pfleiderer, R. Charubala, I. Sela, and R. J. Suhadolnik, *J. Biol. Chem.* **249,** 3482 (1984).
[8] Y. Devash, M. Reichman, I. Sela, N. L. Reichenbach, and R. J. Suhadolnik, *Biochemistry* **24,** 593 (1985).
[9] A. Gera, S. Spiegel, and G. Loebenstein, this volume [96].
[10] I. Sela, this volume [98].

which discharges histidine from histidinyl TMV–RNA is induced by AVF in tobacco leaves or tobacco callus tissue cultures.[3] Another enzyme which is stimulated by AVF in these plants is the dsRNA-dependent plant oligoadenylate synthetase.[3,6,8] This enzyme converts ATP in the presence of dsRNA to a family of plant oligoadenylates which (1) regulate the transient discharging factor,[3] (2) inhibit TMV replication in tobacco tissues,[6] and (3) inhibit *in vitro* protein synthesis in lysates from rabbit reticulocytes and the wheat germ cell-free system.[8] This chapter describes several systems from which the plant oligoadenylate synthetase can be partially purified and the synthesis of plant oligoadenylates from ATP *in vitro*.

Reagents

Plant material: Several sources can be used as the source of the plant oligoadenylate synthetase: (1) *Nicotiana glutinosa* or *N. tabacum* (5–10 leaves), (2) tobacco callus tissue culture maintained on Murashige and Skoog agar medium,[11] and (3) tobacco cell suspension maintained on Murashige and Skoog suspension medium.[11]

Poly(rI) · poly(rC)-agarose (P-L Biochemicals Inc.), 0.5 × 1.5 cm columns (0.3 ml) in buffer A

DEAE-cellulose (DE-52) (Whatman)

ATP (Sigma Chemical Co.)

[α-^{32}P]ATP, 410 Ci/mmol (Amersham)

[8-^{3}H]ATP, 38 Ci/mmol (Amersham)

Buffer S: 0.01 *M* sodium phosphate, pH 7.6

IFN-αA, recombinant human leukocyte interferon (from Dr. S. Pestka, Roche Institute of Molecular Biology)

Buffer A: 0.02 *M* Hepes (pH 7.5), 0.1 *M* KCl, 0.002 *M* $Mg(OAc)_2$, 0.002 *M* dithiothreitol, 10% glycerol

Buffer B: 0.02 *M* Hepes (pH 7.5), 0.09 *M* KCl

Buffer C: 0.02 *M* Hepes (pH 7.5), 0.35 *M* KCl

Hydrated calcium phosphate (HCP), prepared from $CaCl_2$ and Na_2HPO_4 as described[3]

Dialysis tubing, Spectrapor 6, MW cut-off 1000 (Spectrum Medical Industries)

Carborundum, 400 mesh

Triethylammonium bicarbonate (TEAB), pH 7.6, 2 *M*

Buffer D: 50 m*M* ammonium phosphate, pH 7.0

Buffer E: methanol : H_2O, 1 : 1, v/v

[11] I. Sela, M. Reichman, and A. Weissbach, *Phytopathology* (in press).

Induction and Preparation of Plant Oligoadenylate Synthetase

The dsRNA-dependent plant oligoadenylate synthetase is prepared from TMV-infected leaves of *N. glutinosa,* 48 hr after inoculation with TMV (5 μg/ml purified TMV plus 0.1 g/ml carborundum in buffer S).[3,12] In the case of *N. tabacum,* tobacco callus tissue cultures or tobacco cell suspension, 10 g of plant tissue is immersed in buffer S or buffer S containing the antiviral agents [i.e., AVF (DE-0.65 fraction, 10 ng protein/ml),[3] mock AVF (control from untreated leaves), or IFN-αA (1 unit/ml)[4,6]]. The plant tissue is then homogenized in a Waring blender in 10 ml of buffer S, 1 min, 4°. Plant cell suspensions are washed three times with 50 ml of buffer S at 4° and broken by passage through a French pressure cell (5000 psi) at 4°. All homogenates are then filtered through cheese cloth and centrifuged (10,000 *g,* 10 min, 4°). The supernatants are treated with hydrated calcium phosphate (HCP).[3] The fractions that do not adsorb to HCP are used as the source of plant proteins for further purification.[3] The plant proteins (20–50 μg protein/ml) are chromatographed on DEAE-cellulose columns (1 × 20 cm) in buffer S at room temperature. The proteins that do not adsorb to the DEAE-cellulose column (fraction DE-0)[3] are adjusted to 500 ng protein/ml with buffer S following protein determination by the fluorescence method of Udenfriend *et al.*[13] This fraction serves as the source of the plant oligoadenylate synthetase.

Enzymatic Synthesis of Plant Oligoadenylates *in Vitro*

Poly(rI) · poly(rC)-agarose columns bound with plant oligoadenylate synthetase are prepared by adding 5 ml of fraction DE-0 (500 ng protein/ml) to the column (0.5 × 1.5 cm) and washing with 25 ml of buffer A. The poly(rI) · poly(rC)-agarose bound with the enzyme is transferred into sterilized 1.5-ml Eppendorf tubes. The reaction mixture [poly(rI) · poly(rC)-agarose bound with enzyme, [8-^{3}H]ATP (10 μCi, 38 Ci/mmol), and [α-^{32}P]ATP (10 μCi, 410 Ci/mmol), 500 nmol of ATP] is incubated at 25° for 5 hr with constant rotation. After incubation, the reaction tubes are centrifuged (10,000 *g,* 4°, 10 min). An alternative method for the *in vitro* synthesis on poly(rI) · poly(rC)-agarose columns is exactly as described for the synthesis of the cordycepin analog of 2,5-p_3A_n.[14] After centrifugation, the supernatant is adjusted to 0.09 *M* KCl applied to DEAE-cellulose

[12] A. Gera, G. Loebenstein, and S. Shabtai, *Virology* **127,** 475 (1983).

[13] S. Udenfriend, S. Stein, P. Bohlen, W. Dairman, W. Leimgruber, and M. Weigele, *Science* **178,** 871 (1972).

[14] R. J. Suhadolnik, P. Doetsch, J. M. Wu, Y. Sawada, J. D. Mosca, and N. L. Reichenbach, this series, Vol. 79, p. 257.

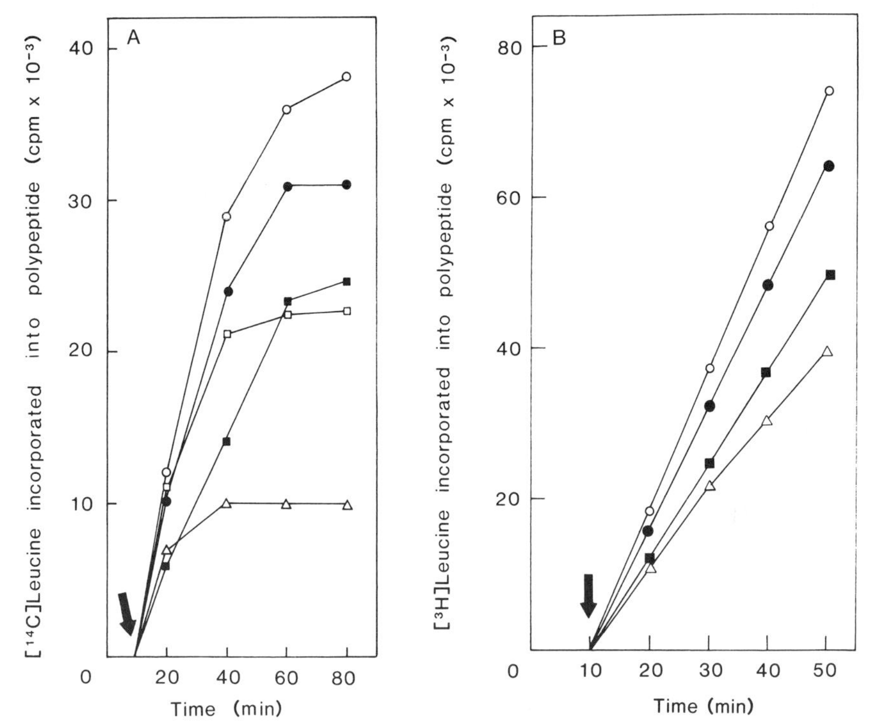

FIG. 1. Inhibition of protein synthesis in lysates from rabbit reticulocytes (A) and in the wheat germ cell-free system (B) by plant oligoadenylates synthesized *in vitro*. Tobacco cell suspensions were incubated with 1 unit/ml of IFN-αA for 24 hr prior to extracting the plant oligoadenylate synthetase and *in vitro* synthesis of plant oligoadenylates. *In vitro* translation assays were as described.[8] The inhibition of protein synthesis by the plant oligoadenylates at 2×10^{-7} *M* (●), 4×10^{-7} *M* (■), and 8×10^{-7} *M* (△) is indicated. The arrow indicates the addition of master mix to the translation assay. The ability of the plant oligoadenylates to inhibit protein synthesis is compared to that of 2,5-p_3A_4 (1×10^{-8} *M*, □) and control (○) assays. Reprinted with permission from Devash *et al.*[8] Copyright 1985 American Chemical Society.

columns (0.5 × 17 cm) and washed with buffer B (100 ml, 1 ml fractions). The plant oligoadenylates are displaced from the column with buffer C (5 ml, 1 ml fractions). The plant oligoadenylates displaced with buffer C are extensively dialyzed (in 4 liters H_2O replaced each hour for 4 hr, 0°). The yield of plant oligoadenylates is determined as described.[14] These *in vitro* synthesized plant oligoadenylates are potent inhibitors of TMV replication, inhibitors of *in vitro* protein synthesis in reticulocyte lysates and in the wheat germ cell-free system (Fig. 1) and regulators of discharging factor activity (Fig. 2).[3,6,8]

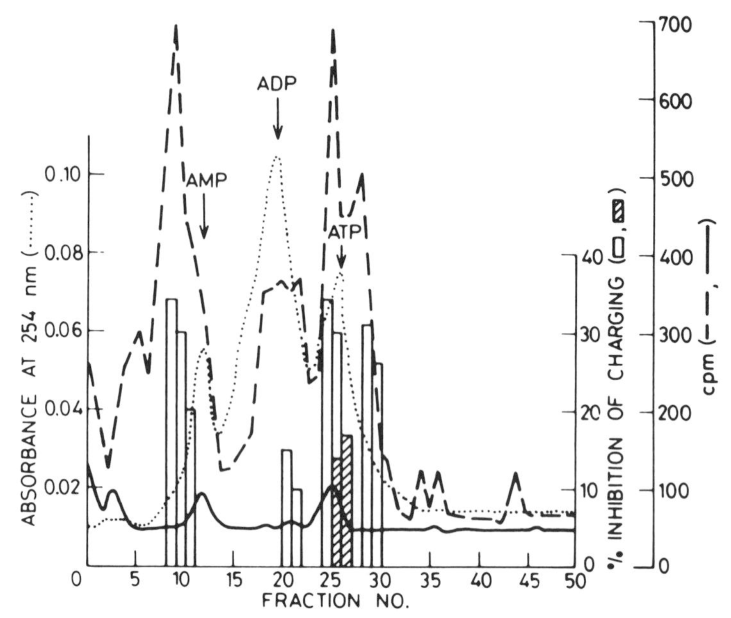

FIG. 2. Elution profile of plant oligodenylates synthesized *in vitro*. Plant oligoadenylate synthetase was induced in AVF-treated tobacco leaves and plant oligoadenylates were synthesized as described in the presence of [^{3}H]ATP. Following elution from DEAE-cellulose columns, the plant [^{3}H]oligoadenylates were mixed with AMP, ADP, and ATP as UV markers. DEAE-cellulose column chromatography in a gradient of 0.01–1 *M* TEAB was performed and absorption at 254 nm was monitored. An aliquot (0.1 ml) was removed from each 1 ml fraction to test for discharging factor (DF) activity.[3] The fractions were then lyophilized for radioactive determination. Plant [^{3}H]oligoadenylates synthesized from mock AVF-treated tobacco leaves (control, solid line), from AVF-treated leaves (treatment, broken line), DF activity in fractions derived from control (dashed columns), and treatment (open columns). Reprinted with permission from Devash *et al.*[3]

Further Purification of Plant Oligoadenylates

DEAE-Cellulose Chromatography

The dialyzed plant oligoadenylates are separated by chromatography on DEAE cellulose (0.5 × 17 cm) with a 0.05 to 0.15 *M* linear gradient of NaCl (40 ml/40 ml) 0.05 *M* Tris · HCl (pH 8.0) in 7 *M* urea as has been described.[14] Alternatively, a linear gradient of 0.1 to 1 *M* triethylammonium bicarbonate (pH 7.6) can be used.[3] The fractions are lyophilized and the residue is resuspended in glass distilled H_2O and stored at −20°.

High-Performance Liquid Chromatography (HPLC)

The HPLC system is composed of a μBondapak C_{18} column, two 6000A pumps, a 660 solvent programmer, a U6K injector, and a fixed wavelength (254 nm) spectrometric detector (Waters Associates) linked to a double pen recorder. A linear gradient of buffer D and buffer E is employed. Elution (1 ml/min) is performed by a linear gradient of 0 to 45% buffer E in 20 min. When plant [^3H,^{32}P]oligoadenylates were applied to HPLC, five radioactive peaks were resolved, three of which inhibited protein synthesis in reticulocyte lysates (Fig. 3).[8] The molar $^3H:^{32}P$ ratio remained 1:1 in all purification steps.

Concluding Comments

The *in vitro* conversion of ATP into oligoadenylates by the plant oligoadenylate synthetase is approximately 0.2–0.4%, compared to the con-

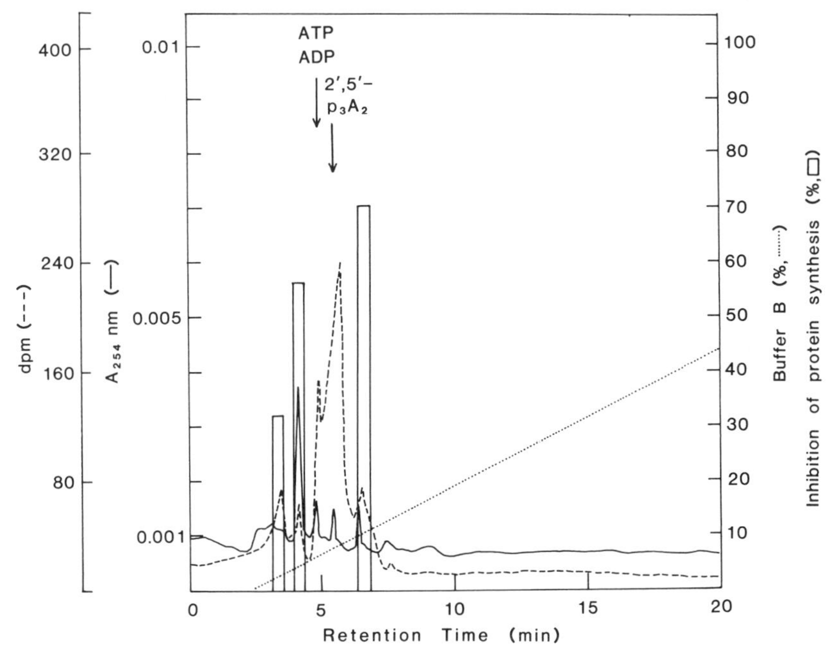

FIG. 3. HPLC profile of plant oligoadenylates synthesized *in vitro* by plant oligoadenylate synthetase purified from an IFN-αA-treated tobacco cell suspension. The ability of the various radioactive peaks to inhibit protein synthesis in lysates from rabbit reticulocytes is indicated (open bars). Reprinted with permission from Devash et al.[8] Copyright 1985 American Chemical Society.

version of ATP to the mammalian 2,5-oligoadenylates which is 15–40%. Based on enzymatic and chemical degradations, the plant oligoadenylates synthesized *in vitro* from ATP differ in structure from the mammalian 2′,5′-oligoadenylates.[8] In addition, the inhibition of protein synthesis by the plant oligoadenylates does *not* proceed via the activation of the 2,5-A-dependent endonuclease as occurs with the 2,5-oligoadenylates. The plant oligoadenylates do *not* compete with 2,5-p_3A_n for binding sites on the 2,5-A-dependent endonuclease, nor do plant extracts contain binding sites for 2,5-p_3A_n.[8] This explains why other investigators were not able to detect 2,5-p_3A_n in plants by radiobinding assay.[15]

In addition to the *in vitro* synthesis of oligoadenylates from ATP by the plant oligoadenylate synthetase, antiviral activity has been demonstrated following TCA extraction of TMV-infected leaves of *N. glutinosa*.[3] The natural substrate for the *in vivo* synthesis remains to be determined. It has also not been determined if the plant oligoadenylates with antiviral activity are the same oligoadenylates that inhibit protein synthesis.

We have demonstrated that the mammalian, 2,5-oligoadenylate cores can inhibit TMV replication in plants[16] and that plant oligoadenylates synthesized *in vitro* from ATP inhibit mammalian protein synthesis (Figs. 1 and 3). Future studies will be needed to determine if there is an evolutionary link in the development of the antiviral state in plants and animals.

Acknowledgments

This work was supported in part by U.S. Public Health Service Grant GM26134 and National Science Foundation Research Grant DMB 84-15002.

[15] P. J. Cayley, R. F. White, J. F. Antoniw, N. J. Walesby, and I. M. Kerr, *Biochem. Biophys. Res. Commun.* **108,** 1243 (1982).

[16] Y. Devash, R. J. Suhadolnik, and I. Sela, this volume [100].

[100] Measurement of Effect of (2′-5′)-Oligoadenylates and Analogs on Tobacco Mosaic Virus Replication

By YAIR DEVASH, ROBERT J. SUHADOLNIK, and ILAN SELA

In addition to their numerous effects on cellular processes in mammalian systems,[1] the (2′-5′)-oligoadenylate cores have been demonstrated to be very potent inhibitors of tobacco mosaic virus (TMV)[1a] replication in tobacco leaf disks.[2] Sela and co-workers reported that both AVF and interferon induce a dsRNA-dependent plant oligoadenylate synthetase which converts ATP to plant oligoadenylates with antiviral activity.[3,4] More recently, we reported that selected 2,5-A_3 core analogs inhibit TMV replication in tobacco protoplasts and in intact plants.[1,5] This chapter describes the inhibition of TMV replication in leaf disks, protoplasts, and intact plants by 2,5-A_3 core and selected analogs.

Reagents

2,5-Trimer cores of adenosine, inosine and cordycepin (chemically synthesized as described[6,7])
Buffer A: sodium phosphate buffer, 0.01 *M*, pH 7.6
Buffer B: Hepes-KOH, 0.01 *M*, pH 7.05; 0.125 g/liter $Na_2HPO_4 \cdot H_2O$
Buffer C: $CaCl_2$, 0.6 M

Preparation of Plant Material, Treatment and Measurement of Infectivity

Growth of *Nicotiana glutinosa* and *N. tabacum* var. "Samsun" and preparation and inoculation of protoplasts are as described.[8–11] 2,5-A_3

[1] R. J. Suhadolnik, C. Lee, and D. H. Willis, this volume [87], and references therein.
[1a] Abbreviations: 2,5-p_3A_n, oligomer of adenylic acid with 2,5-phosphodiester linkages and a triphosphate at the 5′-end; 2,5-$p_3$3′dA_n, oligomer of 3′-deoxyadenylic acid with 2′,5′-phosphodiester linkages and a triphosphate at the 5′-end; core, 5′-dephosphorylated oligonucleotide.
[2] Y. Devash, S. Biggs, and I. Sela, *Science* **216**, 1415 (1982).
[3] M. Reichman, Y. Devash, R. J. Suhadolnik, and I. Sela, *Virology* **128**, 240 (1983).
[4] Y. Devash, A. Hauschner, I. Sela, and K. Chakraburty, *Virology* **111**, 103 (1981).
[5] Y. Devash, A. Gera, D. H. Willis, M. Reichman, W. Pfleiderer, R. Charubala, I. Sela, and R. J. Suhadolnik, *J. Biol. Chem.* **259**, 3483 (1984), and references 15–31 therein.
[6] R. Charubala and W. Pfleiderer, *Tetrahedron Lett.* **21**, 4077 (1980).
[7] R. Charubala and W. Pfleiderer, *Tetrahedron Lett.* **23**, 4789 (1982).
[8] G. Loebenstein and A. Gera, *Virology* **114**, 122 (1981).
[9] G. Loebenstein, A. Gera, A. Barnett, S. Shabtai, and J. Cohen, *Virology* **100**, 110 (1980).

core and core analogs are added to protoplast suspensions (10 ml, 1×10^5 cells/ml) after inoculation with TMV.[5] Protoplasts are then incubated at 25° under continuous illumination of about 1500 lux. At various times, aliquots are collected by centrifugation and homogenized.[8] The TMV content of the extracts is determined by an enzyme-linked immunosorbent asay (ELISA) as described.[2,11,12]

The assays for antiviral activity in TMV-infected tobacco leaf disks are done with leaves of the *N. tabacum* var. "Samsun" in which TMV spreads systemically. Leaves are mechanically inoculated with purified virus (5 μg/ml in 0.01 *M* sodium phosphate buffer, pH 7.6, containing 0.1 g/ml of carborundum). Leaf disks (6.5 mm diameter) are punched out of the inoculated leaves and placed in a beaker containing Buffer B. Groups of 20 disks are selected at random from this common pool and placed into separate Petri dishes, each containing 18 ml of Buffer B. Disks are treated with 2,5-A_3 core or analogs for 1 hr by adding the antiviral compounds and 2 ml of Buffer C as described.[2] The Petri dishes are kept for 72 hr at room temperature under constant illumination. The disks are homogenized and TMV content determined by ELISA.[2,12]

The ability of the 2,5-A_3 core and analogs to inhibit TMV replication in intact *N. glutinosa* is determined by infectivity tests. Solutions containing 5 μg/ml TMV, 0.1 g/ml carborundum (400 mesh), and 2,5-A_3 core or analog are applied to half-leaves of *N. glutinosa*.[5] Cheese cloth (5×5 cm) is dipped into the TMV/carborundum/2,5-A_3 core solution then rubbed lightly with the finger tips two times on the top of the leaf surface. The untreated half-leaves serve as controls (inoculated with a solution containing no cores). The infection is allowed to proceed 48 hr under continuous illumination, at which time local virus lesions appear.[5] Inhibition of TMV replication is calculated as the percent of local lesions produced in 2,5-A_3 core or analog treated half-leaves compared to control half-leaves.

The 5′-dephosphorylated 2,5-A_3 and its analogs inhibit TMV replication in TMV-infected tobacco leaf disks (Fig. 1), TMV-infected protoplasts, and whole plants.[5] 2,5-A_3 core analogs modified at the 6-amino position (2,5-inosine trimer core), modified at the 2′-terminus [2,5-adenylyl(2,5)-adenylyl(2,5)-9β-D-arabinofuranosyladenine or 2,5-adenylyl-(2,5)adenylyl(2,5)-tubercidin] or the 3′-position (2,5-3′dA_3 core) were potent inhibitors of TMV replication as determined by ELISA or lesion counting.[5] No toxicity (chlorosis or necrosis) was observed when noninfected plants were treated with the 2,5-A_3 core or analogs. Nucleosides or

[10] I. Sela, this volume [97].

[11] I. Sela, this volume [98].

[12] P. Orchansky, M. Rubinstein, and I. Sela, *Proc. Natl. Acad. Sci. U.S.A.* **79,** 2278 (1982).

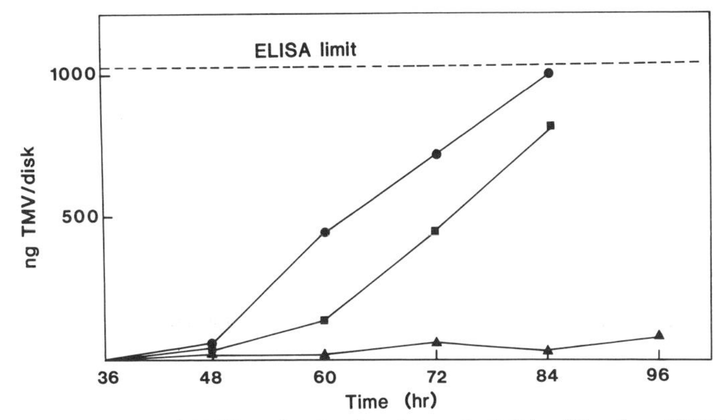

FIG. 1. Time course of TMV replication in tobacco leaf disks. Kinetics of TMV replication in the presence of 2,5-A_3 core (2×10^{-7} *M*) (■) or 2,5-3′dA_3 core (2×10^{-7} *M*) (▲) is compared to control (no oligonucleotide) (●). TMV content was determined by ELISA.[2,11,12] Reprinted with permission from Devash *et al.*[5]

nucleotides that are putative degradation products of the 2,5-A_3 core and analogs were not antivirally active at 1×10^{-6} *M*.

Concluding Comments

The potency of 2,5-A_3 core and its analogs as inhibitors of TMV replication in infected protoplasts, TMV-infected leaf disks, and intact plants has been demonstrated.[2,3,5] The effectiveness of the 2,5-A_3 core and analogs is markedly decreased when applied to leaves following TMV infection.[2] Although the mechanism by which the mammalian (2′-5′)-oligoadenylate cores inhibit TMV replication is not as yet known, it must be emphasized that the concentrations of 2,5-A_3 core and analogs required to inhibit TMV replication in plants are 1/1000th of that required for inhibition of mammlaian viruses or mammalian cell growth.[3] It is possible that 2,5-A_3 cores activate the plant discharging factor[4] that discharges aminoacylated TMV RNA and thereby inhibits virus replication. Abrasion of plant leaf epidermis with a carborundum solution containing dilute solutions of 2,5-A_3 core or its analogs provides a novel approach to the inhibition of virus replication in the plant kingdom.

Acknowlegments

This work was supported in part by U.S. Public Health Service Grant GM-26134 and National Science Foundation Research Grant PCM-8111752.

Author Index

Numbers in parentheses are footnote reference numbers and indicate that an author's work is referred to although the name is not cited in the text.

A

Aapro, M. S., 642
Ab, G., 442
Abele, J. S., 636
Abelson, J., 481
Ackerman, G. A., 333, 334(12)
Ackerman, S. A., 415
Ackerman, S. K., 255
Adams, A. H., 736, 738
Adams, A. N., 732
Adelman, J., 3(12, 13, 14), 4, 5(12, 13, 14), 7(12, 13, 14)
Adolf, G. R., 37, 48, 234
Affoud, C. Y., 630
Aggarwal, B. B., 52
Aguet, M., 263, 269, 271(4), 272(4), 273(4), 274(4), 275(4), 276, 281, 305, 310(1), 312, 322, 323(2), 324(2), 342(1, 3), 597
Ahearn, M., 306, 309(15), 632
Ahl, R., 137, 211, 212(6), 213, 214, 215, 217, 219(14, 15), 220
Ahmad, M., 244
Ahre, A., 635
Akashi, K., 93
Alberts, B. M., 466
Alberts, D. S., 636, 641, 642
Albin, M., 658, 659(6), 662(6)
Alexanian, R., 242, 635, 641
Allen, G., 30, 38, 234, 235(16), 237(16), 268, 275(3)
Allen, P. T., 238
Allet, B., 368, 369(10), 370, 371(17), 372(10), 375(10)
Alm, G. V., 58
Alperin, I., 69
Altman, A., 735
Alton, K., 30, 240, 241
Altrock, B., 240, 241(77)
Ameglio, F., 682
Ammerer, G., 183, 424, 427(1)
Anderson, N., 417
Anderson, P., 235, 276, 281, 282(5), 283, 312, 315, 317(1), 332, 336(5), 340, 341(6), 342(6), 346(6)
Anderson, R. G. W., 332, 336
Anfinsen, C. B., 198, 236, 277, 312, 547, 592
Ankel, H., 239
Antignus, Y., 735, 741
Antonelli, G., 63
Antoniw, J. F., 758
Apontoweil, P., 176
Apperson, S., 27, 28(10), 31(10), 33(10), 241, 552, 636, 640
Applebaum, S. W., 734
Armstrong, J. A., 14, 184, 185(9), 193, 221, 284, 400, 579, 580(1)
Arnheiter, H., 276, 281, 286, 290(2), 307, 309, 312, 314, 315, 324, 326, 329, 330, 331, 332, 333(4), 335, 336(4, 14, 15), 337(14), 338(4, 14, 15), 339, 340, 341(4), 342(4), 346(4)
Arnott, S., 378
Aronson, D. L., 256
Arvan, D. A., 627
Ashwell, G., 257
Atkins, G. J., 98
Atwater, J. A., 500
Aulakh, G., 306, 309(15), 632
Axel, R., 599, 600
Aye, M. T., 630, 633(16)

B

Babiss, L. E., 612, 619, 628, 706
Babiuk, L. A., 137, 211
Backman, K., 379, 426
Baer, G., 103, 104(1)
Baglioni, C., 267, 276, 281, 286, 305, 307(4), 310(4, 5, 9), 312, 332, 336(6), 340, 341(2, 7), 342(2), 346(2), 490, 492(11), 493(11), 500, 517, 602, 667

Baker, J. B., 263
Baker, P. E., 612, 630
Balazs, I., 265, 266(11), 343(20), 345
Baldwin, R. W., 239
Balkwill, F. R., 649, 650, 651, 652(3), 653(3), 655, 656, 657
Ball, E. D., 612, 630
Ball, G. D., 166
Ball, L. A., 516
Ball, W., 240
Bancroft, F. C., 749
Bandu, M.-T., 211, 263, 269, 271, 272(4, 7), 273(4), 274(4, 7), 275(4, 7), 276, 281, 305, 307(13), 310(1, 13), 311(13), 312, 325, 340, 341(3), 342(3), 347, 348(1), 562, 601, 635
Banerjee, K., 612, 629
Barakat, F., 199, 236
Barel, B., 598
Barnett, A., 759
Baron, S., 3, 14, 76, 83, 84, 103, 104(1), 129, 200, 574, 577, 607, 608(38), 612, 682, 747
Barr, P. J., 433
Barraon, E. A., 359, 363(4)
Barrie, P. A., 9
Barrowclough, B. S., 84, 90, 193, 235, 282, 283, 297
Bartlett, R. T., 27, 28(8), 161, 234, 235(17), 237(17), 592, 593(9)
Bash, D., 598
Basham, T. Y., 682, 688
Basnal, R. P., 137
Basu, S. K., 336
Bate, J., 28
Baur, R., 672
Bayse, G. S., 264
Baze, W. B., 103
Beaucage, S. L., 434, 482
Beavo, J. A., 296
Begg, G., 428
Beggs, J. D., 427
Begon-Lours, J., 211
Bekesi, E., 5, 297, 316, 585
Belardelli, F., 612
Belfrage, G., 199, 200(2), 201(2), 203(2)
Bell, W., 464, 470, 471(16)
Bellett, A. J. D., 113
Bello, J., 103, 105(7)
Benedetto, A., 612
Benoist, C., 445
Benton, W. D., 362, 426
Benzinger, R., 466
Berends, W., 176
Berent, S. L., 359, 363(4)
Berg, K., 166, 238, 541, 546, 570, 688
Berg, P., 384, 398, 467, 599
Berger, S. L., 3(36), 5, 10(36), 143, 197, 204, 243, 282, 301, 376, 635
Bergeret, M., 541
Bergstróm, L., 28
Berissi, H., 495
Berman, B., 233, 399, 570
Bernard, H.-U., 370, 371(20), 373(20), 374(20), 377, 378(8)
Bernhard, H., 205
Bernhardt, L., 242, 641
Bernstein, A., 384
Bernt, E., 624
Berry, M. J., 504, 513, 515(24), 516(24)
Bersch, N., 52
Berson, S. A., 588
Berthold, W., 199, 214, 236, 562, 564(13)
Besançon, F., 239
Betlach, M. C., 419(14), 420
Bevin, S., 500
Bewley, T. A., 235, 238(28), 239(28), 249, 250(17), 252(17), 253(17), 254(17), 269, 593
Bhalla, K., 612, 629, 630(9), 632(2)
Biggs, S., 735, 744(7), 745, 752, 759, 760(2), 761(2)
Billiau, A., 38, 63, 66(5), 76, 86, 90, 92(5), 96, 145, 193, 199, 218, 233, 259, 389, 395(23), 554, 555(11), 556(11), 557(11), 559, 561(7)
Bino, T., 464, 474(1)
Birch, J. R., 3(31), 4, 5(31), 9(31)
Birdsall, N. J. M., 489, 516
Birkenmeier, C. S., 143
Birnboim, H. C., 367, 409, 427, 445, 599
Birrows, J. W., 287
Bischoff, J. R., 515
Bjorkholm, M., 635
Black, M. H., 541
Blake, G., 362, 436, 437(8)
Blalock, J. E., 574, 577
Blanchard, B., 276, 312, 322, 323(2), 324(2), 325, 340, 341(1), 342(1)
Blangy, D., 325
Blau, H. M., 622
Blin, N., 465

Blumenschein, G. R., 635
Bocci, V., 275
Bodo, G., 35, 163, 166, 235
Bognacki, J., 631
Bohlen, P., 754
Bolivar, F., 379, 434
Boll, W., 3, 5(7), 7(7), 9, 13(48), 166, 240, 241(67), 249, 268, 269(2), 273(2), 275(2), 359, 442, 445, 634, 635, 640
Bollin, E., 218
Bollon, A. P., 3(20, 21), 4, 5, 7(20, 21, 38), 359, 363, 364, 365(11)
Bolton, A. E., 282, 307
Bonner, J., 444
Borst, P., 368
Bose, S., 235
Boseley, P., 3(26, 34), 4, 5(26, 34), 364
Bostian, K. A., 416, 418
Botstein, D., 425
Boué, A., 597, 611(8)
Boué, J., 597, 611(8)
Bougnoux, P., 612, 630
Bourgeade, M. F., 682
Bourne, H. R., 646
Bowden, D. W., 3(19), 5(19)
Bowman, B. H., 199
Boyer, H. W., 419(14), 420
Brack, C., 135, 240
Bradford, M. M., 157, 407, 409(12)
Bradley, E., 636, 641(19)
Bradley, E. C., 629
Bradshaw, R. A., 657
Brady, R. O., 341
Braestrup, C., 258
Bragg, P. W., 359, 363(4)
Brahic, M., 475, 476(4), 477
Brake, A. J., 433
Brammar, W. J., 466
Branca, A. A., 267, 276, 281, 286, 305, 307(4), 310(4, 5, 9), 312, 332, 336(6), 340, 341(2, 7), 342(2), 346(2), 602
Braude, I. A., 54, 57, 62(6), 63, 72, 73, 75, 76, 84, 193, 197(7), 198(4), 234, 237, 238
Braun, R., 54
Bray, G. A., 583
Breathnach, R., 445
Breeze, R. G., 134, 552
Brennan, J. K., 630
Brideau, R., 707, 712(3)
Bridgen, P. J., 36, 198, 236, 277, 312, 547, 592
Brink, L., 27, 206, 235, 415, 747
Broach, J. R., 419(16), 420
Brodeur, B. R., 130
Broeze, R., 111(21), 114, 493
Broome, S., 443
Brouty-Boyé, D. F., 211, 552, 562, 601, 635
Brown, M. S., 332, 336
Brown, P., 694, 702(4)
Brown, R. E., 351, 489, 490, 492(10), 500, 516, 517, 518, 522, 653
Brownlee, G. G., 491
Broxmeyer, H. E., 631
Bruggeman, A., 442
Bruns, W., 3(33), 4, 5(33), 398
Buckler, C. E., 103, 104(1), 277
Bugianesi, R. L., 501
Buick, R. N., 629, 633
Builder, S. E., 5, 63
Bunn, P. A., 49, 54(11)
Burchell, J., 646
Burger, S. L., 554, 555(15)
Burke, D. C., 3(26), 4, 5(26), 9, 28, 30, 98, 114, 166, 234, 236, 364, 566
Burman, C. J., 166
Burnett, J. P., 168
Burnette, W. N., 209, 210(15), 418
Buzdar, A. U., 635
Bywater, M., 599
Bywater, R., 599

C

Cabanillas, F., 635
Cabrer, B., 500
Cachard, A., 477
Callicoat, P. A., 535(8), 536, 539(8), 540(8)
Campbell, F. E., 133
Campbell, J. B., 575
Canaani, D., 384, 398
Canni, B., 445
Cantell, K., 27, 28, 29(6), 31, 33(7), 39, 41, 54, 55(1), 62, 63, 73, 74, 88, 137, 141, 166, 233, 235, 237, 240, 248, 249(1), 269, 359, 442, 464, 541, 546, 554, 555(10), 580, 581, 593, 635
Capobianchi, M. R., 63, 69, 682
Capon, D. J., 5, 9, 210, 220
Carbon, J., 605
Cardamone, J. J., Jr., 114
Carey, N. H., 3(31), 4, 5(31), 9(31)

Carlson, K., 187
Carlson, M., 425
Carlsson, J., 199, 200(2), 201(2), 203(2)
Carlström, G., 62
Carmichael, L. E., 211
Carter, W. A., 91, 214, 218, 219, 234, 248, 389, 395(24)
Cartwright, T., 3(31), 4, 5(31), 9(31)
Caruthers, M. H., 434, 482
Carver, D. H., 107, 109(6), 112(6), 116, 117, 127
Casadaban, M. J., 414
Casey, W. J., 636
Casper, J., 636, 641(19)
Caspers, M., 9, 10(57, 58), 13(56), 228, 230(8)
Catinot, L., 562
Cayley, P. J., 490, 492(10), 497(13), 517, 667, 670(14), 758
Ceriani, R., 650
Cesareni, G., 400
Chada, K. C., 171, 199, 234
Chakraburtty, K., 735, 736(11), 749, 752, 753(3), 754(3), 755(3), 756(3), 758(3), 759, 761(4)
Chambon, P., 445
Chamers, D. A., 292
Chan, T.-S., 564
Chang, C. N., 424, 430
Chang, M., 297, 585
Chang, N., 295, 415
Chang, N. T., 3(10, 15), 4, 5(10, 15), 7(10, 15), 9(10), 241, 385
Chang, R. L. S., 103
Chang, T.-W., 587
Chang, Z., 265, 266(11), 343(20), 345, 612, 630
Chany, C., 5, 239, 301, 316, 541, 544, 549, 551, 597, 611, 694, 696, 702
Chany, E. H., 143
Chany-Fournier, F., 544, 551, 694, 702(5, 6)
Chapekar, M. S., 675
Charrey, J., 211
Charubala, R., 667, 668, 669, 672, 675, 735, 744(8), 745, 752, 759, 760(5), 761(5)
Chatterjee, S., 702, 703(4), 704(4)
Cheingsong-Popov, R., 703, 704(10)
Chen, C. Y., 428
Chen, E. Y., 430
Chen, H. S. G., 636, 641
Chen, Y., 93
Chen, Z.-Q., 351, 676, 677(4)
Cheng, Y.-S. E., 377
Cheong, L. C., 292
Chernajovsky, Y., 3(27), 4, 5, 495
Cheroutre, H., 3(35), 5, 10(35), 197, 204, 241, 243(15), 244, 301, 370, 384
Cheung, H. C., 702
Chevalier, M. J., 239
Chiang, T.-R., 3(10), 5(10), 7(10), 9(10), 241, 385
Chick, W. L., 443
Chow, F., 434, 482
Chroboczek Kelker, H., 235
Chudzio, T., 233, 237, 549
Cinti, D. L., 721
Cioe, L., 612
Ciufo, D. M., 398
Claeys, H., 76, 88, 90, 92(5), 96, 193, 259, 559, 561(7)
Clapham, P., 703, 704(10)
Clark, M. F., 732, 736, 738
Clarke, L., 605
Clemens, M. J., 500
Clereq, E., 376
Clewell, D. B., 293, 427
Clifford, P., 305, 306(3)
Cochet, M., 445
Cogniaux, J., 704, 705(12)
Cohen, G. M., 717
Cohen, J., 731, 759
Cohen, L. W., 400, 409
Cohen, S., 9, 144, 464, 467, 471, 474(1)
Cohen, S. H., 114
Cohen, S. N., 367, 414
Cohn, Z. A., 264, 332
Coit, D. G., 433
Colburn, W., 242, 641
Colcher, D., 683
Colletta, G., 613
Collier, K., 3(37), 5, 9, 13(49), 240, 249, 434, 437(7), 438(7), 441(7), 445
Collins, J., 3(33), 4, 5(33), 398
Collins, S., 306, 309(15), 632
Collins, S. J., 306, 309(32), 311
Coltman, C. A., 636
Comens, P. G., 263
Commoy-Chevalier, M. J., 5, 301, 316
Confalone, D., 244
Cong, N. V., 597, 611(8)

Connell, E. V., 177
Conney, A. H., 715
Content, J., 3(30), 4, 5(30), 9(30), 240, 368, 369, 376, 397, 528, 667
Cooper, J. A., 296
Cooper, P. D., 113
Coppenhaver, D. H., 199, 200
Corbin, J. D., 296
Corley, L., 235
Cortese, R., 400
Costello, L., 3(10), 4, 5(10), 7(10), 9(10), 166, 241, 385, 593
Couillin, P., 597, 611(8)
Coulson, A. R., 363, 426
Coupin, G., 220
Coutts, R. H. A., 731
Cox, D. R., 597, 598, 611(4)
Crawford, M. P., 103
Crea, R., 3(9, 32), 4, 5(9, 32), 7(9), 9(32), 11(32), 153, 166(17), 167, 178, 183(6), 184, 192(6), 240, 242, 288, 289(12), 293, 376, 397, 405(7), 406, 408(7), 419(12), 420, 425, 434, 440(1), 441(1), 464, 469(4), 471(4), 482, 517, 554, 555(13, 14), 634, 635, 636(10), 667
Creasey, A., 183, 184(1), 192(1), 397
Creasey, A. A., 343(22), 345, 667, 682
Crosa, J. H., 419(14), 420
Cross, M., 554, 555(13)
Crowe, D., 715
Crowl, R., 84, 204, 205(3), 243(16), 244, 257, 259(10), 260(10), 319, 378, 379(17), 404, 407(2), 408, 414(2), 419(13), 420, 583, 585(3), 601
Cuatrecasas, P., 321, 324(1), 325(1)
Cullen, B. R., 385
Cuncliffe, B., 233, 234(3)
Czarniecki, C. W., 238, 239(56), 276, 291, 489
Czech, M. P., 341, 346(15)
Czyrski, J., 658

D

Dagert, M., 473
Dahl, H.-H. M., 3(33), 4, 5(33)
Dairman, W., 754
Dalbadie-McFarland, G., 400, 403, 409
D'Alessandro, S. B., 281, 305, 332, 336(6), 340, 559, 561(5), 562(5), 667
Dalton, B. J., 227, 562, 564(11)
Dana, B., 636
Dandoy, F., 475
Dang, A. Q., 672
Das, A., 504, 513(23)
Das, H. K., 370, 371(20), 373(20), 374(20)
DasGupta, A., 504, 513(23)
Daubas, P., 15, 274
Daugherty, B., 9, 13, 240, 249, 434, 437(7), 438, 441(7), 445
David, G. S., 264, 322
Davies, H. A., 535(8), 536, 539(8), 540(8)
Davis, L. R., Jr., 199
Davis, R. W., 362, 419(15), 420, 426
Dawidzik, J., 103, 105(7), 106
Debarbouille, M., 378
De Brabander, M., 333
DeChiara, T. M., 153, 159(4), 239, 242, 245(8), 255, 382, 417, 418(5), 419(5), 421(5)
De Clercq, E., 3(30), 4, 5(30), 9(30), 237, 240, 249, 368, 369, 375, 397, 528
Defilippi, P., 528
DeGrado, W., 286
Degrave, W., 3(35), 5, 10(35), 197, 204, 241, 243(15), 244, 301, 370, 384
DeLange, R. J., 253
Delanoy, A. D., 476
De Larco, J. E., 658
De Ley, M., 63, 66(5), 76, 88, 90, 92(5), 96, 145, 193, 197, 218, 259, 559, 561(7)
Delgarno, L., 377
Deloria, L. B., 719
Del Valle, U., 27, 179, 190, 235, 389, 415, 747
De Maeyer, E., 475, 541, 546
De Maeyer-Guignard, J., 475, 477, 541, 546
Dembek, P., 413, 481
DeMey, J., 333
Denhardt, D. T., 476
den Hartog, J. A. J., 517, 667
Dennin, R., 66, 67(12), 177
de Reus, A., 9, 13(55), 63, 445, 447(9), 449(9), 454
Derom, C., 372
Derynck, R., 3(30), 4, 5(30), 9, 10(61), 197, 204, 210, 240, 243, 282, 301, 368, 369, 376, 397, 428, 554, 555(15), 635

Desmyter, J., 237, 238, 389, 395(23), 601
DeSomer, P., 76, 86, 90, 92(5), 96, 145, 193, 199, 218, 233, 237, 238, 249, 259, 375, 389, 395(23), 554, 555(11), 556(11), 557(11), 559, 561(7)
de Sousa, M., 631
D'Eustachio, P., 598
Devash, Y., 667, 668, 669(16), 675, 676, 735, 736, 744, 745, 749, 752, 753(3, 6, 8), 754(3, 6), 755, 756, 757, 758, 759, 760(2, 5), 761
Devos, R., 3(30, 35), 4, 5, 9(30), 10(35), 197, 204, 240, 241, 243(15), 244, 301, 368, 369, 370, 372(10), 375, 384, 458
Diamond, L., 612
Dianzani, F., 14, 63, 69, 200, 259, 612, 682
DiBerardino, L., 389, 395(24)
Dicke, K. A., 612, 629
Dickson, R. B., 339
Dieckmann, B., 397, 398, 399(5, 13), 403(5)
Dieckmann, M., 467
Dieffenbach, C. W., 667
Dijkema, R., 9, 10(57, 58), 13, 228, 230(8), 445, 447(9), 449(9), 454
DiMauro, S., 619, 621(1), 622(2), 623(1), 624(1), 628(1)
Di Persio, T. F., 630
Dixon, C. B., 248
Dixon, D., 359, 363(4)
Doel, S. M., 3(31), 4, 5(31), 9(31)
Doetsch, P., 668, 669(17), 754, 755(14), 756(14)
Doetsch, P. W., 667, 668(9), 669(9), 672, 675
Dolei, A., 612, 613, 682
Doly, J., 367, 409, 427, 445, 599
Donelson, J. E., 428
Dorfman, A., 672
Dougherty, J. P., 516, 521(5), 667
Douhan, J., 240, 376
Dreyer, W. I., 206
Drocourt, J.-L., 667
Drummond, R. J., 343(22), 345
Dubbeld, M., 9, 10(57, 58)
Duc-Goiran, P., 541, 549, 702
Duerinck, F., 241, 398
Dull, T., 3(9), 4, 5(9), 7(9), 153, 166(17), 167, 183(6), 184, 192(6), 242, 288, 289(12), 293, 376, 405(7), 406, 408(7), 419(12), 420, 434, 440(1), 441(1), 464, 469(4), 471(4), 482, 554, 555(13), 634, 635, 636(10)
Dumont, J. E., 667
Duncan, R., 500
Dungworth, D. L., 310
Durie, B. G. M., 636, 641, 642(25)
Durrer, B., 15, 566, 582, 583(2), 589
Dussaix, E., 696
Dworkin, M. B., 3(25), 4, 5(25)
Dworkin-Rastl, E., 3(25), 4, 5(25)
Dziewanowska, Z., 242, 641

E

Easton, A., 3(26, 34), 4, 5(26, 34), 9, 364
Eaton, M. A. W., 3(31), 4, 5(31), 9(31)
Ebina, T., 93, 571
Ecsödi, J., 359, 442, 464
Edelman, G. M., 49, 50(9)
Edman, P., 428
Edy, V. G., 86, 199, 218, 238, 389, 395(23)
Efstratiadis, A., 443, 467, 468(9)
Ehrlich, S. D., 473
Eid, P., 305, 348, 349(2), 350, 351(3)
Einhorn, S., 635
el Azhary, R., 719
El Defrawry Masry, S., 717
Ellis, D. A., 541
Emtage, J. S., 3(31), 4, 5(31), 9(31)
Enders, J. F., 40
Engel, K., 103
Eppstein, D. A., 500, 503(7), 667, 675
Epstein, C. J., 551, 552(2), 597, 598(4), 611(4)
Epstein, D. A., 500, 528
Epstein, L. B., 551, 552(2), 597, 598(4), 611(4), 636
Erlichman, J., 301
Erlitz, F., 153, 159(4), 239, 242, 245(8), 255, 382
Ersson, B., 58
Escodi, J., 166, 240, 635
Eshhar, Z., 84
Eslami, B., 668
Estell, D. A., 192, 235, 238(28), 239(28), 240, 241(70), 242, 249, 250(17), 252(17), 253(17), 254(17), 276, 291, 404, 593
Etienne-Smekens, M., 667
Evans, S., 66, 67(12)
Evinger, M., 3, 5(8), 7(8), 166, 179, 190, 199, 259, 375, 389, 405(6), 406, 482, 579, 580, 592

F

Facchini, J., 69
Fahey, D., 214, 219(10), 235, 259, 286, 375, 592
Falcoff, E., 146, 233, 399, 460, 562
Falkow, S., 419(14), 420
Faltynek, C. R., 281, 305, 310(9), 332, 336(6), 340, 341(7)
Familletti, P. C., 3, 4, 5(8, 9), 7(8, 9), 14, 18(9), 54, 64, 65, 66, 67(11), 124, 132, 153, 162, 166, 167, 178, 179, 180(4), 183(6), 184, 190, 192(6), 198, 199, 201, 206, 210(12), 242, 243, 250, 256, 259, 260(10), 287, 288, 289(12), 293, 299, 307, 319, 375, 376, 380, 385, 389, 405(6, 7), 406, 407, 408(7), 417, 419(12), 420, 434, 440, 441(1), 464, 469(4), 471(4), 482, 533, 539(4, 5), 551, 554, 555(3, 13), 556, 557(3), 579, 585, 592, 593, 601, 605, 606, 634, 635, 746
Fanger, M. W., 612, 630
Fantes, K. H., 30, 38, 233, 234, 235(16), 236(22), 237(16, 22), 248, 650, 651, 652(3), 653(3), 656(3), 659
Fárkkilá, M., 28
Farr, A. L., 40, 178, 180(3), 277, 549, 717
Farris, D. A., 500, 503(7)
Fauconnièr, B., 220, 546
Faulk, W. P., 333
Federman, P., 495, 500
Feeney, R. E., 253
Fein, S., 641
Feinstein, S. I., 3(27), 4, 5(27)
Feldman, H. A., 310
Felix, A. M., 175, 244
Fennie, C. W., 276, 291
Ferbus, D., 667
Ferguson, B., 30, 240, 241(77)
Ferguson-Miller, S., 256
Ferrone, S., 611
Fersht, A. K., 409
Fialkow, P. Y., 616
Fields, F., 192, 249, 593
Fiers, W., 3(30, 35), 4, 5, 9(30), 10(35), 63, 197, 204, 240, 241, 243(15), 244, 301, 368, 369, 370, 371(5, 8), 372, 374(8, 19), 375, 376, 384, 397, 398, 447, 449, 458, 688
Figurski, D., 367
Fine, S., 242
Finer-Moore, J., 235, 238(28), 239(28), 250(17), 252(17), 253(17), 254(17), 593
Fink, G. R., 417, 427
Finland, M., 138, 473
Finter, N. B., 14, 38, 104, 125, 166, 533, 534, 535, 536, 539(6), 540(6)
Fischer, D. G., 315, 317(2)
Fish, E. N., 305, 612, 629
Fisher, P. B., 611, 612, 613, 614(3, 24, 25, 30), 617(12, 13, 25, 30, 31), 619, 628, 629, 630(9), 632(2), 667, 706
Fishman, P. H., 341
FitzGerald, D. J. P., 315, 330, 331, 332, 333(4), 336(4), 338(4), 339(4)
Fitzpatrick, F. A., 709, 711(7), 712(7)
Flannery, G. R., 239
Fleischmann, W. R., Jr., 83, 84(1), 114, 574, 575(4)
Flick, M. B., 668, 669(16), 672
Flowers, D., 28
Floyd, G. A., 497
Floyd-Smith, G., 489, 490, 493(12), 494(3, 8, 15), 495(3, 12), 496(3, 15), 497, 499(21), 516, 517, 518, 524(9), 526, 528(9)
Fluke, D. J., 256
Fogy, I., 163
Fong, C. K. Y., 114, 126, 128, 129(5, 7)
Fontaine, A., 168
Forseth, B., 636
Forti, R. L., 533, 535, 536, 539, 540
Fountoulakis, M., 3, 5(7), 7(7), 249, 268, 269(2), 273(2), 275(2), 326, 634
Fournier, F., 694, 702
Franceschini, T., 415
Francoeur, A. M., 128
Franke, A. E., 3(12, 13, 14), 4, 5(12, 13, 14), 7(12, 13, 14), 415, 482
Franke, J., 79, 81(5)
Franklin, H. A., 35
Franklin, N., 370, 371(20), 373(20), 374(20)
Franks, D. H., 636
Fraser, M., 183
Frey, T. K., 114
Friedlander, J., 83, 84
Friedman, L. I., 77, 79, 81(2, 5), 82(2)
Friedman, R. M., 143, 489, 500
Friesen, H.-J., 179, 190, 199, 218, 259, 375, 389, 592
Frist, C., 467
Fritsch, E. F., 287, 362, 400, 436, 437(8)
Fryling, C., 658

Fujii-Kuriyama, Y., 3(28, 29), 4, 5(28, 29), 9(29), 190, 464
Fujiki, H., 97, 613, 614(30), 617(30), 628
Fujisawa, A., 3, 5(7), 7(7), 249, 268, 269(2), 273(2), 275(2), 634
Fujisawa, J.-I., 3, 5(7), 7(7), 227, 249, 268, 269(2), 273(2), 275(2), 634
Fuke, M., 3(20, 21, 38), 4, 5, 7(20, 21, 38), 359, 363(4), 364, 365(11)
Fuller, F. J., 114, 119
Fulton, R. W., 211, 740
Funa, K., 58
Fuse, A., 702, 704, 705(8)

G

Gabay, J., 378
Gabler, W. L., 717
Gajdusek, D. C., 694, 702(4)
Gall, J. G., 474
Gallagher, R. E., 306, 309(32), 311
Gallati, H., 15, 157
Galliot, B., 239
Gallo, R. C., 49, 54(11), 306, 309(32), 311
Gannon, F., 445
Gapobianchi, M. R., 613
Garapin, A., 445
Gat-Edelbaum, O., 735
Gazdar, A. F., 49, 54(11)
Geoghegan, W. D., 333, 334(12)
Georges, P., 5, 301, 316
Georgiades, J. A., 54, 62(5), 63, 69, 83, 84, 85, 200, 259, 389, 395(23), 574, 575, 689
Gera, A., 12, 667, 730, 734(4), 735, 736(17), 744(8), 745, 751(20), 752, 754, 759, 760(5, 8)
Gerber, L., 27, 235, 415
Gergen, J. P., 473
Gerhardt, W., 624
Gerner, R. E., 35
Geuns, G., 333
Gewert, D. R., 305
Gheysen, D., 241, 372, 398
Ghosh, N., 489, 490(9), 491(9), 517
Giacomini, P., 611
Gibbs, C. J., 103, 104(1), 694, 702(4)
Gibson, K. D., 235
Gibson, M., 384
Giemsa, G., 677, 679
Gifford, G. E., 123
Gilbert, C. S., 489, 490, 492(10), 517, 667
Gilbert, W., 363, 386, 390, 426, 437, 443, 445, 470, 472, 482
Gill, T., 3(19), 4, 5(19)
Gillam, I. C., 476
Gillam, S., 481
Gilmartin, E., 636
Giovanella, B. C., 616
Giovannella Moscovici, M., 385
Girard, S., 541
Giron, D. J., 14, 101
Glazer, A. N., 253
Glazer, R. I., 675
Gluzman, Y., 458
Glynn, J. M., 612, 630
Goeddel, D. V., 3(9, 11, 12, 13, 14, 32, 36), 4, 5, 7, 9, 10(36, 54), 11(32), 27, 153, 166, 167, 178, 183, 184, 189, 192(6), 197, 204, 210, 220, 240, 241, 242, 243, 282, 288, 289(12), 293, 301, 359, 376, 377, 378(13), 397, 405(7), 406, 408(7), 415, 419(12), 420, 424, 427(1), 428(2), 430(2), 434, 440(1), 441(1), 445, 464, 468(4), 471, 482, 554, 555(13, 14, 15), 634, 635, 636, 640(26)
Goff, S. P., 384
Gold, P., 9, 13, 14
Goldberg, A. L., 168, 210
Golde, D. W., 52, 306, 632
Goldstein, J. L., 332, 336
Goldstein, L., 30, 240, 241(77)
Golgher, R. R., 489, 517, 601, 676, 681(1)
Golias, T., 624
Gonzalez-Noriega, A., 336
Goodman, H. M., 225
Goorha, R. M., 601
Gordon, I., 551, 557(1)
Gordon, J. B., 418, 442, 443(4)
Goren, T., 27
Gorovsky, M., 187
Gosselin, G., 529
Gottschalk, M., 214, 215, 217, 219(14, 15), 220
Graham, F. L., 599
Granados, E. N., 103, 105
Grant, S., 611, 612, 613(3), 614(3), 628, 629, 630(9), 632
Grappelli, C., 612
Gray, J. D., 239, 718

Gray, P. W., 3(11, 14, 36), 4, 5, 7(11, 14), 9, 10(36, 54), 27, 197, 204, 210, 240, 241, 243, 282, 301, 359, 376, 482, 554, 555(15), 635, 636, 640(26)
Greaves, M. F., 695
Green, A. E., 597
Green, H., 643
Greenberg, P. L., 629
Greene, P. J., 419(14), 420
Greenlee, W., 724
Greenwood, F. C., 583, 585(6)
Grégoire, A., 541, 544, 694, 696, 702(2, 3)
Greiner, J., 611
Gresser, I., 134, 211, 234, 263, 269, 271(4), 272(4), 273(4), 274(4), 275(4), 276, 281, 305, 310(1), 312, 325, 340, 341(3), 342(3), 552, 562, 601, 611, 612(1), 629, 635, 682, 688
Grob, P. M., 199, 234
Groopman, J. E., 52
Grosfeld, H., 9, 144, 467, 471
Gross, G., 3(33), 4, 5(33), 398
Gross, M., 3, 4, 5(8, 9, 12, 13), 7(8, 9, 12, 13), 27, 153, 166, 167, 183(6), 184, 192(6), 240, 242, 288, 289(12), 293, 359, 376, 405(6, 7), 406, 408(7), 419(12), 420, 434, 440(1), 441(1), 445, 464, 469(4), 471(4), 482, 634, 635
Grossberg, S. E., 14, 234, 249, 463, 612, 706, 740
Grosveld, F., 3(33), 4, 5(33)
Gruarin, A., 640
Grubb, J. H., 336
Gruber, M., 442
Gruber, W., 624
Grunberger, T., 575
Guarente, L., 240, 376
Guidon, P. T., Jr., 107, 111(5), 114(5), 118, 119, 120(15), 122(15)
Gupta, N. K., 504, 513(23)
Gupta, S. L., 264, 267, 281, 305, 307(7), 312, 315, 317(3), 332, 340, 341(5, 8), 342(5), 343, 345, 346, 347(10), 500, 574, 602
Gurari-Rotman, D., 235, 277, 707
Gusciora, E. G., 593
Gutte, B., 312, 314(8), 324, 326, 329(3)
Gutterman, J. U., 177, 242, 612, 629, 634, 635, 641
Guyre, P. M., 612, 630

H

Haase, A. T., 36, 475, 476
Hack, A. M., 286, 288(4), 289(4)
Hagie, F. E., 424, 427(1), 428
Hahon, N., 14
Haigler, H. T., 257
Hall, A., 166, 240, 241(67), 359, 442, 464, 635, 640
Hall, B., 183
Hall, B. D., 424, 427(1), 481
Hall, M. N., 378
Hall, R. I., 718
Halvorson, H. O., 416, 425
Hamburger, A. W., 636, 641
Hamill, R. L., 234
Hamilton, J., 631
Hammarström, S., 711
Hanahan, D., 600
Hannigan, G. E., 305
Hao, Y. L., 59
Hardison, R. C., 467, 468(9)
Hardman, J. G., 296
Hardy, R. W. F., 179, 190, 295, 300(12), 301
Harmon, J. T., 258
Harmon, M., 635
Harned, C. L., 715, 716, 717(13), 718(13), 719
Harpaz, I., 735, 741(9)
Harrelson, J. M., 672
Harrington, D., 103
Hartley, J. L., 428
Hasegawa, S., 706
Hathaway, G. M., 296
Hatt, T., 719
Hattori, T., 612, 630
Haugh, M. C., 667, 670(14)
Hauschner, A., 735, 736(11), 741(9), 749, 752, 753(3), 754(3), 755(3), 756(3), 758(3), 759, 761(4)
Hauser, H., 398
Havell, E. A., 38, 135, 141, 227, 233, 234, 284, 389, 395(22), 399, 559, 562(6), 570, 571
Hawke, D., 206
Hawkins, R. E., 30, 33
Hayashi, S., 476
Hayes, C., 641
Hayes, T. G., 136
Hayflick, J. S., 428

Haynes, J., 3, 5(7), 7(7), 249, 268, 269(2), 273(2), 275(2), 454, 634
Hayward, G. S., 398
Hearon, J. Z., 258
Heberlein, U. A., 433
Heimer, E. P., 244
Heine, J. W., 76, 90, 92(5), 193, 214, 218, 219, 259, 559, 561(7)
Heine, K. J. W., 86
Heintzelman, M., 255, 415
Helinski, D., 205
Helinski, D. R., 293, 367, 370, 371(20), 373(20), 374(20), 377, 378(8), 427, 471
Hellman, J. M., 211
Hellman, L., 599
Henco, K., 3, 5(7), 7(7), 135, 249, 268, 269(2), 273(2), 275(2), 634
Henderson, A. S., 600
Henderson, E. E., 668, 672, 675
Hendrix, C., 359, 363(4)
Hendrix, L. C., 3(38), 5, 7(38)
Henle, G., 36
Henle, W., 36
Henriksen, D., 48, 49(3, 4), 50(3), 51(3), 52(3), 53(4)
Henriksen-DeStefano, D., 52
Henry, P. D., 626
Herberman, R. B., 19, 27, 241, 305, 676
Heremans, H., 145, 389, 395(23)
Hermann, T. E., 70
Hermo, H., Jr., 612, 613, 614(24, 25), 617(13, 25), 628, 629, 667
Heron, I., 541, 688, 693
Herschberg, R., 566
Hersh, E. M., 612, 634, 635
Hershberg, R. D., 27, 28(8), 234, 235(17), 237(17), 592, 593
Hershey, J. W. B., 500
Hershfield, M. V., 370, 371(20), 373(20), 374(20)
Herskowitz, I., 430
Herzenberg, L. A., 600, 689
Hewick, R. M., 206
Heyneker, H., 183(6), 184, 192(6)
Heyneker, H. L., 3(9), 4, 5(9), 7(9), 153, 166(17), 167, 242, 288, 289(12), 293, 376, 405(7), 406, 408(7), 419(12, 13), 420, 434, 440(1), 441(1), 482, 554, 555(13), 634, 635, 636(10)
Hicks, J. B., 417, 427
Hiervonen, S., 88
Higa, A., 427
Higashi, N., 63, 93
Higashi, Y., 9, 240, 475
Hilliker, S., 417, 418(5), 419(5), 421(5)
Hillörn, V., 663
Hilmas, D. E., 103
Hinnen, A., 417, 427
Hinuma, Y., 571
Hirose, T., 481, 485
Hirsch, M. S., 541
Hirvonen, S., 27, 28(6), 29(6), 39, 41(2, 3, 4), 54, 55(1), 63, 73, 74(2), 137, 141, 554(10), 580(4), 581, 592, 635
Hiscott, J., 27, 33(7)
Hiscott, J. B., 3(15), 4, 5(15), 7(15)
Hitchcock, M. J. M., 143
Hitzeman, R., 183
Hitzeman, R. A., 210, 424, 427(1), 428, 430
Ho, M., 40, 248, 400
Hoal, E. G., 18, 240, 241(71), 243, 434, 552, 601, 602(34), 613, 746
Hobbs, D., 305, 582, 583
Hobbs, D. S., 3(15), 4, 5(15), 7(15), 15, 153, 157, 166, 175, 205, 242, 245(2), 250, 258, 287, 288(10), 291(10), 295, 300(11), 301, 306, 307(18), 385, 407, 409(9), 428, 473, 558, 588, 589, 590(3), 591(3, 5, 6), 601, 637, 684, 747
Hochkeppel, H., 398
Hochkeppel, H. K., 197
Hoffman, T., 265, 266(11), 343(20), 345, 612, 630
Hoffmann, J. W., 384
Hohn, B., 467
Hoken, M., 693
Hoken, P., 693
Hokland, M., 688, 693
Hokland, P., 693
Hollenberg, M. D., 321, 324(1), 325(1)
Holm, G., 635
Holmberg, D., 663
Holmes, S. L., 500
Holmes, W., 3(9), 4, 5(9), 7(9), 153, 166(17), 167, 183(6), 184, 192(6), 242, 288, 289(12), 293, 376, 405(7), 406, 408(7), 419(12), 420, 434, 440(1), 441(1), 464, 469(4), 471(4), 482, 554, 555(13), 634, 635, 636(10)
Holsti, L. R., 28
Honda, S., 297, 585
Hong, J.-X., 676

Honore, T., 258
Hood, L., 562, 564(13)
Hood, L. E., 179, 190, 206, 295, 300(12), 301
Horan Hand, P., 683
Horn, T., 425
Horning, S. J., 242, 641
Horoszewicz, J. S., 389, 395(24)
Horowitz, B., 47
Horowitz, M., 47
Hortobagyi, G. N., 635
Hosaka, Y., 703
Hoshino, H., 703, 705(11)
Hotta, K., 9, 13(49), 249, 440, 445
Hou, Y.-D., 676, 677(4)
Hou, Y. T., 351
Houchuli, E., 242
Houck, C. M., 3(12, 13, 14), 4, 5(12, 13, 14), 7(12, 13, 14), 415
Houghton, M., 3(31), 4, 5(31), 9(31)
Houk, C. M., 482
Hovanessian, A., 500
Hovanessian, A. G., 325, 351, 515, 518, 521, 669, 676, 681(2)
Hovi, T., 54
Howard, L., 183
Howley, P. M., 398
Hozumi, M., 571, 612, 613, 630
Hsiao, K., 3(19), 4, 5(19)
Hsiung, G. D., 114, 126, 128, 129(5, 7)
Huang, M.-T., 715
Huang, Z.-R., 351, 676, 677(4)
Hubbard, A. L., 264
Hubbard, G. L., 717
Huez, G., 528
Huff, T., 715
Hughes, B. G., 667, 675
Hui, D. Y., 257
Hunkapiller, M., 255, 415, 562, 564(13)
Hunkapiller, M. W., 179, 190, 206, 285, 295, 300(12), 301
Hunter, E., 702, 703(4), 704(4)
Hunter, T., 296
Hunter, W. M., 267, 270(1), 282, 307, 583, 585(6)
Hutchinson, D. W., 235

I

Iadarola, M. J., 103, 104(1)
Iida, S., 434
Iisuka, M., 14
Iivanainen, M., 28
Iizuka, M., 101
Illinger, D., 220
Imai, J., 489, 494(8), 497(8), 517, 522, 523, 524(9), 526, 528, 529, 667, 668
Imai, Y., 722
Imbach, J.-L., 529
Ingham, K. C., 59
Ingimarsson, S., 62
Inglot, A. D., 657, 658, 659(5, 6), 661(5), 662(3, 6), 664, 665(14), 666(14)
Inglot, O., 664, 665(14), 666(14)
Innerarity, T. L., 257
Innis, M., 397, 398(5), 399(5), 403(5)
Innis, M. A., 3(23), 4, 5(23), 7(23), 241, 345
Inoue, M., 15, 562, 564(13)
Inouye, M., 415
Inouye, S., 415
Isaacs, A., 234, 635
Iscove, N., 55
Iserentant, D., 372
Ishida, N., 93, 571
Ishii, A., 276, 305, 310(11), 332, 336(7)
Itakura, K., 400, 409, 413, 481, 485
Ito, A., 722
Ito, M., 389, 395(24)
Ito, Y., 501
Iwakura, Y., 236

J

Jacobs, B. L., 500, 514(17), 515(17), 516
Jacobsen, H., 489, 500
Jameson, P., 14
Jamieson, J. D., 341
Jamoulle, J.-C., 522(7), 523, 526, 528(10), 529
Jankowski, W. J., 214, 218, 545
Janossy, G., 695
Jariwalla, R. J., 249
Jayaram, B., 493
Jayaram, B. M., 493
Jeffrey, A., 377
Jeffreys, A. J., 9
Jiang, P. H., 544, 551, 694, 702(6)
Jimenez, H., 385
Jin, J.-R., 676
Johansson, B., 635
Johnsen, H. E., 633
Johnson, A. D., 377

Johnson, H., 84
Johnson, H. M., 76, 193, 200, 259, 564, 574, 575
Johnson, J. S., 36
Johnson, M. J., 413, 481, 485
Johnson, P. A., 658
Johnson, P. F., 481
Johnson, S., 252
Johnsrud, L., 166, 359, 442, 464, 635
Johnston, D. A., 612, 629
Johnston, M. D., 14, 36, 37, 98, 234, 536
Johnston, M. I., 500, 522, 523, 528(8), 529, 667, 668
Johnston, P. D., 238, 239(56), 249, 250(18), 252(18), 254(18)
Johnston, P. N., 176
Joklik, W. K., 123, 501
Jones, D., 636, 641(19)
Jones, E., 426
Jones, E. V., 114
Jones, G., 640
Jones, S. E., 641
Jones, S. S., 517
Joniau, M., 554, 555(11), 556(11), 557(11)
Jordan, G. W., 114
Joshi, A. R., 264, 281, 305, 307(7), 312, 340, 341(5), 342(5), 345(5), 346(5), 602
Joshi, R. C., 137
Joyner, A., 384
Ju, G., 385
Jung, V., 3(18), 4, 5(18), 7(18)
Justesen, J., 667

K

Kabayo, J. P., 235
Kagg, H., 268, 269(2), 273(2), 275(2)
Kahn, C. R., 258
Kahn, M., 367
Kale, R., 640
Kamarck, M. E., 598
Kamath, S. A., 721
Karch, F., 434
Kasal, J. A., 36
Kashima, H., 103
Katsoyannis, P. G., 252
Kauppinen, H.-L., 27, 28, 33(13), 54, 55(1), 63, 73, 74(2), 88, 137, 580(4), 581, 593
Kaupponen, H.-L., 39, 41(4)
Kavathas, P., 600
Kawade, Y., 9, 18, 93, 227, 234, 236, 240, 322, 475, 558, 559(3), 560(3, 4), 561(3, 4), 562, 563(3, 4), 564(3, 4), 565(3, 4), 566, 568, 571, 573(3, 4)
Kawaguchi, K., 571
Kawashima, E. H., 481, 485
Ke, Y. H., 248
Keay, S., 612, 706
Kelder, B., 9, 13(49), 15, 18, 19, 21(20), 27, 84, 153, 162(3), 171, 177(23), 239, 240, 241, 242, 243, 245(5), 249, 251, 252(24), 257, 259, 260(10), 306, 309(20), 319, 434, 437(7), 438(7), 441(7), 445, 552, 584(11), 586, 589, 591(7), 593(7), 601, 602(34), 613, 746
Kelker, H. C., 52, 283
Kelley, K. A., 9, 13, 475
Kelley, V. E., 688
Kellum, M., 398
Kempe, T., 400, 434, 482
Kempner, E., 84, 257, 258
Kempner, E. S., 19, 27, 241, 256, 257, 258, 259(10), 260(10), 319, 601
Kenny, C., 179, 190, 214, 389
Kern, J., 79, 81(5)
Kerr, I., 522
Kerr, I. M., 351, 489, 490, 492(10), 497(10), 500, 515, 516, 517, 518, 521, 653, 667, 670(14), 676, 681(1), 758
Ketoh, J., 160
Keyler, D. E., 715
Khan, Z., 667
Khosrovi, B., 184
Kiesel, G. K., 211
Kimchi, A., 515, 667
Kimelman, D., 240, 376
Kincaid, R. L., 258
King, J., 548
King, R. M., 566
Kirchner, H., 54
Kirkpatrick, F. H., 630
Kisielow, P., 662
Klein, E., 305, 306(3)
Klein, F., 166
Klein, G., 48, 305, 306(3)
Klimpel, G., 574
Klimpel, G. R., 574, 577(4)
Klotz, I. M., 310, 325
Knight, E., 190, 214, 219(9, 10), 235, 236, 300(12), 301

Knight, E., Jr., 179, 198, 259, 278, 285, 286, 307, 375, 592
Knight, M., 351, 490, 492, 497(13), 517, 518
Knutson, G. S., 500, 503, 504, 513(24), 515(24), 516(24)
Kobayashi, S., 3(28), 4, 5(28), 14, 101, 464
Kochman, M. A., 575
Koeffler, H. P., 306
Kohase, M., 114
Kohr, W. J., 210, 424, 428(2), 430, 433(3, 29)
Koi, M., 93, 571
Koike, M., 721
Koistinen, V., 27, 28(6), 29(6), 141, 233
Kolter, R., 367, 471
Kong, R.-L., 351, 352(5), 353(5), 677, 681(18)
Konrad, M., 184
Konttinen, A., 624
Kopchick, J. J., 384, 385, 388(10)
Korant, B. D., 179, 190, 295, 300(12), 301
Korner, A., 613
Koyama, S., 235
Kozak, C. A., 475
Kramer, C., 66, 67(12), 640
Kramer, M., 242, 635
Kramer, M. J., 66, 67(12), 177, 640, 641, 715
Kramer, R. A., 417, 418, 419(5), 421(5)
Krebs, E. G., 296
Kreider, J., 613, 614
Kressner, B. E., 641
Krisch, H. M., 370
Kronenberg, L. H., 227
Kummerow, F. A., 721
Kung, H.-F., 5, 15, 153, 166, 175, 205, 242, 245(2), 250, 287, 288(10), 291(10), 292, 295, 297, 300(11), 301, 306, 307(18), 316, 317(5), 321(5), 414, 415, 428, 583, 585, 589, 591(5, 6), 601, 637, 640, 684, 747
Kung, P. C., 587
Kurimoto, M., 235
Kuriyama, F., 635
Kurjan, J., 430
Kurland, J. I., 631
Kushner, S. R., 440
Kuśnierczyk, P., 662
Kuwata, T., 702, 703(5), 704, 705(2, 3, 5, 6, 8), 706(5)
Kwoh, D. Y., 377
Kwong, T. C., 627

L

Labdon, J. E., 235
Lacy, E., 467, 468(9)
Lad, P. M., 257
Laemmli, U. K., 157, 159, 173, 178, 186, 206, 208, 225, 250, 265, 266, 278, 285, 287, 296, 344, 380, 407, 409(10), 418, 548, 699
Lagker, J. L., 715
Lai, C.-Y., 27, 28(8), 153, 175, 205, 234, 235, 237(17), 242, 245(2), 250, 259, 287, 288(10), 291(10), 300(11), 301, 306, 307(18), 407, 409(9), 415, 583, 589, 591(5), 592, 593(9), 601, 637, 684, 747
Lakhchaura, B., 564
Lam, T., 128
Lambros, T. J., 244
Lane, C. D., 442, 443(4)
Lane, E., 467
Laner, J., 467, 468(9)
Langer, J., 165, 414, 598
Langer, J. A., 3, 5(6), 7, 15, 21(20), 239, 251, 252(24), 306, 307, 309(20), 310, 311, 321, 584(11), 586
Langford, M., 83, 84
Langford, M. P., 14, 200, 564, 574, 607, 608
Langford, P. M., 259
Larsson, E. L., 663
Lasky, S. R., 500, 504, 513(24), 514, 515(17, 24), 516(24)
Lavie, V., 5
Lawhorne, L., 466
Lawn, R. M., 3(11, 12, 13, 14), 4, 5(11, 12, 13, 14), 7(11, 12, 13, 14), 240, 359, 362, 415, 436, 437(8), 445, 482
Lawrence, J. B., 597, 598(3), 599(3), 610(3), 611(3)
Lawyer, F. C., 597
Le, H. V., 214, 219(13)
Le, J., 48, 49, 53(4, 5), 63, 235, 265, 266(11), 283, 343(20), 345, 587
Leanderson, T., 663
Leary, P., 688
Lebleu, B., 500
Lebon, P., 541, 702
Lebowitz, H. E., 672
Lederberg, E. M., 367
Lee, C., 669, 670, 671, 759
Lee, D. K., 245

Lee, F., 384, 398, 399(13)
Lefkowitz, S. S., 211
Le Goff, S., 238
Lei, M. T., 123
Leib, S. R., 134, 552
Leibovitz, A., 616
Leimgruber, W., 754
Leist, T., 326
Leitner, M., 467
Lemaítre, J., 694, 702(2, 3)
Lemire, J. M., 416
Lengyel, P., 9, 13(48), 111(21), 114, 249, 340, 445, 470, 471(16), 489, 490, 491(9), 492, 493, 494(3, 15), 495(3, 12), 496(3, 15), 497, 499(21), 500, 515, 516, 517, 518, 521(5), 522, 667, 668(1)
Lennon, M. B., 672
Leong, J. L., 723
Leong, S. S., 389, 395(24)
Le Pennec, J. P., 445
Leroux, M., 54
Lesiak, K., 489, 494(8), 497(8), 522, 523, 524(9), 526, 528(9, 10), 529, 668
Letchworth, G. J., 211
Leung, D., 3(32), 4, 5(32), 9(32), 11(32), 178, 240, 241, 293, 376, 397, 554, 555(14), 635, 636, 640(26)
Leung, D. W., 3(11, 36), 4, 5, 7(11), 9, 10(36), 27, 197, 204, 210, 220, 240, 243, 282, 301, 359, 376, 415, 424, 428(2), 430(2), 445, 554, 555(15), 635
Leventhal, B. G., 103
Levin, W., 715
Levine, A. S., 103
Levine, H. L., 235, 238(28), 239(28), 242, 249, 250(17), 252(17), 253(17), 254(17), 269, 404, 424, 427(1), 428(2), 430(2), 593
Levine, J. F., 242
Levine, J. L., 641
Levinson, A. D., 3(36), 5, 10(36), 197, 204, 243, 282, 301, 376, 399, 554, 555(15), 635
Levy, H. B., 36, 103, 104, 601
Levy, W. P., 3, 4, 5(8, 10, 15), 7(8, 10, 15), 9(10), 27, 28(8, 10), 166, 234, 235, 237(17), 241, 385, 405(6), 406, 415, 482, 566, 582, 583(2), 589, 592, 593(9), 747
Levy-Koenig, R. E., 601
Lewis, H. M., 3(31), 4, 5(31), 9(31)
Li, B.-L., 351, 352, 353(5), 355(7, 8), 356(7, 8), 676, 677, 681(18)
Li, S. W., 677
Lichtman, M. A., 630
Lieb, M., 377
Lilquist, J. S., 3(19), 4, 5(19)
Lin, L., 84, 183, 184, 192, 397
Lin, L. S., 233, 234, 237, 238, 345
Lin, N., 192, 249, 593
Lin, P. F., 597
Lin, T.-Z., 351, 676, 677(4)
Lindahl-Magnusson, P., 688
Lindenmann, J., 234
Lindner, H. R., 707
Liptak, R. A., 242, 243(3), 245, 407, 408(8)
Little, J. W., 377
Liu, C. M., 70
Liu, R., 636, 641, 642(25)
Liu, V., 587
Liu, X.-Y., 3(18, 37), 4, 5, 7(18), 351, 352, 353, 355(7, 8), 356(7, 8), 676, 677, 681
Lloyd, R. E., 14
Loan, R. W., 211
Loebenstein, G., 12, 729, 730, 731, 734(4), 735, 752, 754, 759, 760(8)
Loeffler, H. P., 632
Loenen, W. A., 466
Loken, M. R., 689
Lomedico, P., 378, 379(17), 443
Lomedico, P. T., 396
Lomniczi, B., 114
London, I. M., 504, 513(23)
London, W. T., 103, 104(1)
Long, W. F., 114
Lopez, J., 541
Lou, H. J., 353
Lou, Y.-C., 351, 353, 676, 677(4)
Lough, J., 612, 706
Lowry, O. H., 40, 178, 180, 277, 549, 717
Lu, S., 183, 184(1), 192(1)
Lu, S. D., 345, 397
Lu, Y.-L., 676
Lubiniecki, A. S., 5, 63, 77, 81(2), 82(2)
Lucerno, M., 460
Lugovoy, J. M., 424, 430, 433(3, 29)
Luk, A. D. H., 211
Lundak, T. S., 296
Lundberg, L., 199, 201(1), 203(1)
Lundgren, E., 663
Luria, S. E., 287
Lvovsky, E., 103

M

McAndrew, S., 378, 379(17)
McAndrew, S. J., 396
McCandliss, R., 3, 4, 5(8, 9, 10, 11, 15), 7(8, 9, 10, 11, 15), 9(10), 27, 143, 153, 166, 167, 183(6), 184, 192(6), 241, 242, 288, 289(12), 293, 359, 376, 385, 405(6, 7), 406, 408(7), 419(12), 420, 434, 440(1), 441(1), 445, 464, 469(4), 471(4), 482, 554, 555(13), 634, 635, 636(10)
McCandliss, R. M., 240
McCandliss, S., 281, 305, 310(9), 340, 341(7)
McCauley, J. W., 490
McClain, D. A., 49, 50(9)
McCormick, F., 241, 345, 397, 398(5), 399(5), 403(5)
McCray, J. W., 236
McCredie, K., 306, 309(15), 632
McCredie, K. B., 612, 629
McCulloch, E., 629
McCulloch, E. A., 630, 633
McCumbee, W. D., 672
McDevitt, H. O., 716
McGarry, M., 103, 105(7)
McGregor, W. C., 160
McGuire, T. C., 130, 131(4), 132(4)
McInnes, J., 397
McKeown, C., 715
McKinney, S., 587
Maclouf, J., 709
McManus, N. H., 533, 597
McPherson, I. A., 97
Madsen, J. A., 77, 81(2), 82(2)
Madsen, M., 633
Maeda, S., 3, 4, 5, 7(8, 9), 9(10), 153, 166(17), 167, 183(6), 184, 192(6), 241, 242, 288, 289(12), 293, 376, 385, 405(6, 7), 406, 408(7), 419(12), 420, 434, 440(1), 441(1), 464, 469(4), 471(4), 482, 554, 555(13), 634, 635, 636(10)
Mahley, R. W., 257
Mahmoudi, M., 359, 363(4)
Majarian, R., 301
Majumdar, A., 504, 513(23)
Makino, S., 174
Makuch, R., 636, 641(19)
Maliszewski, C. R., 612, 630
Maluish, A., 676
Mandal, N. C., 293
Mandel, A. D., 715, 719
Mandel, M., 427
Manganiello, V. C., 258
Maniatis, T., 287, 362, 384, 398, 400, 436, 437(8), 467, 468
Mannering, G. J., 715, 717, 718, 719, 720, 721(7)
Manning, J. T., 612, 634
Mantei, N., 3, 4, 5(7, 16), 7(7, 16), 9, 13(48), 27, 183, 192, 240, 249, 268, 269(2), 273(2), 275(2), 445, 470, 471(16), 634
Mantovani, A., 305
Mao, J., 3(19), 4, 5(19)
Marbaix, G., 442, 443(4)
Marchalonis, J. J., 264
Marcus, P. I., 107, 108, 109, 111, 112, 113, 114, 116, 117, 118, 119, 120, 122(15), 123(7, 13, 16), 124(9), 125(6, 16), 127
Margolis, S., 36
Mark, D., 183, 184(1), 192(1), 397
Mark, D. F., 345
Maroney, P. A., 490, 492(11), 493(11), 517
Maroteaux, L., 5, 398
Marsh, Y. V., 667, 675
Marshall, L. W., 248
Martens, A. E., 641
Martin, E. M., 351, 489, 516, 518, 676
Martin, G. E. M., 351
Martin, G. R., 672
Martin-Zanca, D., 9, 13(49), 240, 249, 434, 437(7), 438(7), 441(7), 445
Masiarz, F. R., 433
Mastro, A. M., 49
Masuda, K., 235
Masurovsky, E. B., 623
Matarese, G. P., 612
Matsubara, S., 93
Matsubara, T., 721
Matsuyama, M., 571
Matteucci, M. D., 424, 430(4)
Mattson, K., 28
Mauer, R., 377
Maury, C., 562, 635
Maxam, A. M., 363, 390, 426, 437, 443, 445, 470, 472, 482
May, L., 3(9), 4, 5(9), 7(9), 27, 28(10), 31(10), 33(10), 153, 166(17), 167, 183(6), 184, 192(6), 242, 288, 289(12), 293, 376, 405(7), 406, 408(7), 419(12), 420, 434, 440(1), 441(1), 464, 469(4), 471(4), 482, 552, 554, 555(13), 634, 635, 636(10), 640
May, L. T., 241

Mayr, U., 3(33), 4, 5(33), 398
Meager, A., 3(34), 4, 5(34), 23
Means, G. E., 253
Mehra, L. L., 500
Meienhofer, J., 179, 190, 199, 259, 375, 389, 592
Meindl, P., 235
Melchers, F., 55
Mellman, I. S., 332
Mellon, P., 384, 398
Mellstedt, H., 635
Melnick, J. L., 601
Merigan, T., 242, 641
Merigan, T. C., 130, 248, 667, 682, 688, 716
Merrick, W. C., 502
Merrifield, R. B., 244
Merril, C. R., 178
Merryweather, J. P., 433
Meruelo, D., 716
Messing, J., 363, 400, 426, 427
Metzgar, R., 306, 309(15), 632
Metzger, J. F., 89, 145
Meyer, B. J., 377
Meyer, F., 3(24), 4, 5(24), 7
Meyer, J., 434
Meyer, R., 367
Meyskens, F. L., 636, 641
Mierzejewski, K., 644
Miggiano, V., 566, 582, 583(2), 589
Miki, T., 702, 705(8)
Mikulski, A. J., 199, 219
Mileno, M. D., 612
Miller, E. J., 672
Miller, J. H., 256, 258, 426, 427(15), 473
Miller, L., 30, 240, 241(77)
Miller, R. S., 54, 64, 166, 198, 243, 592, 593, 746
Millis, R., 650
Minai, M., 36
Minden, M. D., 633
Minks, M. A., 500
Minna, J. D., 49, 54(11)
Miozzari, G., 3(9), 4, 5(9), 7(9), 153, 166(7), 167, 183(6), 184, 192(6), 242, 288, 289(12), 293, 376, 405(7), 406, 408(7), 419(12), 420, 434, 440(1), 441(1), 464, 469(4), 471(4), 482, 554, 555(13), 634, 635, 636(10)
Miranda, A. F., 612, 613, 614(30), 617(30), 619, 621(1), 622, 623, 624(1), 628, 706
Mitchell, W. M., 533, 535(8), 536, 539(8), 540(8)
Mitrani-Rosenbaum, S., 398
Miwa, M., 703, 705(11)
Miyada, C. G., 481
Miyake, T., 481, 485
Miyamoto, N. G., 516
Mizrahi, A., 35, 36, 141, 166, 234, 239
Mobraaten, L. E., 541
Moeremans, M., 333
Mogensen, K. E., 31, 88, 166, 235, 237, 248, 249(1), 263, 269, 271, 272(4, 7), 273(4), 274, 275, 276, 281, 286, 305, 307(13), 310(1, 13), 311(13), 312, 325, 340, 341(3), 342(3), 347, 348, 349(2), 350, 351(3), 546, 597, 635
Moldave, K., 296
Moldovan, R. A., 535(8), 536, 539(8), 540(8)
Monaco, A. P., 541
Monahan, J. E., 672
Monahan, T. M., 69
Mongini, T., 619, 622
Montgomery, D. L., 481
Moodie, E. M., 650, 651, 652(3), 653(3), 655, 656(3, 5)
Moon, T. E., 636, 641
Moore, G. E., 35
Moore, R., 97
Moore, R. N., 612(26), 613
Mooren, A. T. A., 9, 13(51)
More, D. J., 378
Morehead, H., 249, 250(18), 252(18), 254(18)
Morgan, A. C., 616
Mori, M., 97
Morin, C., 400, 409
Morinaga, N., 704
Morinaga, Y., 415
Morris, A., 96, 101(4, 5), 102
Morrison, M., 264
Morrissey, J. H., 207
Morser, J., 30, 233, 234, 235
Morton, R. R. A., 641
Mory, Y., 3(27), 4, 5, 398
Mosca, J. D., 668, 669(17), 672, 754, 755(14), 756(14)
Moschera, J., 27, 84, 179, 190, 199, 235, 259, 297, 375, 389, 415, 585, 592, 747
Moschera, J. A., 3(15), 4, 5(15), 7(15), 19, 27, 28(8), 179, 214, 234, 235(17), 237(17), 240, 257, 260(10), 319, 385, 592, 593(9), 601

Moscovici, C., 385
Mosny, S. A., 629
Mount, D. W., 377
Mowshowitz, S., 651, 656(5)
Mozes, R., 735
Mueller, G. C., 49
Mufson, R. A., 612, 613, 614(30), 617(12, 30, 31), 628
Mukhopadhyay, M., 293
Mulder, J., 442
Mullenbach, G. T., 433
Muller, W. A., 332
Mulligan, R., 384
Mumford, D. M., 616
Munemitsu, S. M., 504, 513(24), 515(24), 516(24)
Munk, K., 54
Munn, R. J., 310
Munson, P. J., 315
Muramatsu, M., 3(28, 29), 4, 5(28, 29), 9(29), 190, 192, 464, 635
Murashige, T., 745
Murphy, R. F., 481
Murray, H. W., 96
Muscettola, M., 275
Muse, D. D., 667
Musti, A. M., 418
Muth, W. L., 168
Myllylä, G., 27, 39, 41(4), 54, 55(1), 73, 74(2), 88, 137, 580(4), 581, 593

N

Nachbar, M. S., 48, 49(3), 50(3), 51(3), 52(3)
Nadkarni, J. J., 305, 306(3)
Nadkarni, J. S., 305, 306(3)
Nagata, S., 3, 4, 5(7, 16, 17), 7(7, 16, 17), 27, 135, 136, 166, 192, 240, 241(67), 249, 268, 269(2), 273(2), 275(2), 305, 310(11), 359, 442, 464, 470, 634, 635, 640
Nagler, C., 281, 283
Nagy, K., 703, 704(10)
Najarian, R., 3(12, 13, 36), 4, 5, 7(12, 13), 10(36), 197, 204, 243, 282, 376, 554, 555(15), 635
Nakashima, H., 174
Nakayasu, M., 97, 616
Naker, S. P., 443
Namba, M., 704
Narayan, K. A., 721
Narczewska, B., 658
Nash, T., 722
Nathan, G., 96
Nathan, I., 52
Nebert, D. W., 722, 723
Nedden, D. Z., 415
Nelson-Rees, W. A., 310
Nerland, D. E., 715, 716, 717(13), 718(13), 719
Neurath, A. R., 15
Ng, M. H., 240
Nichols, B. P., 289, 377
Nicklen, S., 363, 426
Nicolas, J. C., 339
Nielsen, M., 258
Nielsen, T. B., 257, 258
Niiranen, A., 28
Niko, Y., 630, 633(16)
Nilsen, T. W., 490, 492, 493(11), 517, 667
Nilsson, K., 306, 316
Ning, R. Y., 245
Nir, U., 5
Nishimaki, J., 702, 704(6), 705(6)
Nishizuka, Y., 48
Norman, D. G., 667, 670(14)
Norman, R. L., 715
Norrander, J., 400
Novick, D., 84, 315, 317(2)
Nuti, M., 683
Nuydens, R., 333

O

O'Brien, T. G., 612
Ocho, M., 704
O'Connell, C., 467, 468(9)
O'Connor, B. H., 7, 10(46)
Oda, T., 704
O'Farrell, P. H., 225
O'Farrell, P. Z., 225
Offord, R. E., 235
Ogawa, K., 93
Ogburn, C. A., 166, 236, 238, 546, 562, 564, 570
Ogura, H., 704
O'Hare, K., 445
Ohno, M., 312, 314(8), 324, 326, 329(3), 529
Ohno, S., 3(29), 4, 5(29), 9, 183, 190, 240, 384, 398, 475, 635
Oikawa, A., 616

Oka, A., 409
Okada, Y., 705
Okamura, H., 562, 564(13), 571
Okubo, Y., 93
Oleszak, E., 658, 659(5), 661(5)
Oleszek, D., 199
Olin, B., 203
Olislager, R., 704, 705(12)
Olson, R. A., 77, 81(2), 82(2)
Olsson, I., 199, 200(2), 201(2), 203(2)
O'Malley, J., 103, 105(7), 460
O'Malley, J. A., 91, 234
Omedeo-Sale, F., 341
Omura, S., 704
Omura, T., 718, 721
O'Neill, C. F., 234, 236(22), 237(22), 248
Oppenheim, J. D., 48, 49(3), 50(3), 51(3), 52(3)
Opsomer, C., 375
Orchansky, P., 27, 315, 317(2), 752, 754(4), 760, 761(12)
O'Rourke, E. C., 343(22), 345
Ortaldo, J. R., 19, 27, 241, 305, 676
Osborn, M., 141, 142
Osborne, J. C., Jr., 258
Osborne, L. C., 575
Otto, B., 235
Otto, M. J., 234, 235(23)

P

Pabo, C. O., 377
Pacini, A., 275
Pack, M., 612, 630
Paglin, S., 341
Pajtasz, E., 662
Palleroni, A. V., 177
Palm, G., 434, 482
Palter, K., 278
Pan, Y. C., 297
Pan, Y.-C. E., 585
Panayotatos, N., 167(19), 168
Panem, S., 3(16), 4, 7(16)
Pang, R. H. L., 48, 49(3), 50(3), 51(3), 52, 77, 259, 282
Papermaster, V., 84, 575
Papermaster, V. M., 200
Pardue, M. L., 474
Parker, R. C., 362, 436, 437(8)
Parkinson, A., 715
Pastan, I., 339
Pastan, I. H., 332
Patel, T. P., 3(31), 4, 5(31), 9(31)
Paterson, B. M., 418
Paucker, K., 62, 166, 227, 238, 546, 559, 561(5), 562, 564(11), 570, 601
Paucker, M., 236
Paulesu, L., 275
Pauloin, A., 694, 702(5)
Pawelek, J., 613
Payess, B., 239
Pearlstein, K., 52
Pearson, N. J., 211
Pelham, J. M., 239
Pellicer, A., 599
Pennica, D., 3(36), 5, 10(36), 197, 204, 243, 282, 301, 376, 554, 555(15), 635
Perlman, D., 425
Perrin, F., 445
Perry, L., 192
Perry, L. J., 210, 249, 424, 428(2), 430, 433(3, 29), 593
Perryman, L. E., 130, 131(4), 132(4)
Persaud, J., 641
Perucho, M., 600
Pessina, G. P., 275
Pestka, S., 3, 4, 5, 7, 9, 11(32), 13(49, 50), 14, 15, 18, 19, 21(20), 23, 27, 28(3, 8), 39, 54, 64, 65, 84, 124, 125, 129, 132, 135, 143, 153, 157, 162, 165, 166, 167, 171, 175, 177, 178, 179, 180(4), 183(6), 184, 190, 192(6), 198, 199, 201, 205, 206, 210(12), 218, 234, 235, 237(17), 239, 240, 241, 242, 243, 245, 249, 250, 251, 252(24), 257, 259, 260, 265, 269(8), 287, 288, 289(12), 292, 293, 295, 299, 300(11), 301, 305, 306, 307, 309(20), 311, 316, 317(5), 319, 321(5), 359, 375, 376, 380, 385, 386(20), 389, 397, 404, 405, 406, 407, 408(7), 409(9), 414, 415, 417, 419(12), 420, 427, 428, 430, 434, 437(7), 440, 441, 445, 464, 469(4, 5), 471(4), 473, 482, 533, 535, 539(4, 5, 7), 540(7), 551, 552, 554, 555(3, 13), 556(3), 557(3), 558, 566, 579, 580, 582, 583, 584(11), 585, 586, 588, 589, 590(3), 591(3, 5, 6, 7), 592, 593, 598, 601, 602(34), 605(29), 606(29), 612, 613, 614(24, 25), 617(13, 25), 619, 628, 629, 630(9), 632(2), 634,

635, 637, 651, 659, 667, 683, 684, 706, 743, 746, 747
Peterhans, E., 211
Peterson, E. R., 623
Pfleiderer, W., 667, 668, 669, 672, 675, 735, 744(8), 745, 752, 759, 760(5), 761(5)
Phillips, A. W., 166
Pickering, L. A., 227
Pike, B. L., 630
Pilch, P. F., 341, 346(15)
Pill, J. E., 630, 633(16)
Pimm, M. V., 239
Pitha, P. M., 9, 13, 15, 234, 248, 398, 475
Plaetinck, G., 241
Poiesz, B. J., 49, 54(11)
Poindron, P., 220
Poland, A., 724
Poland, A. P., 722
Poppie, M. J., 130
Poráth, J., 199, 200, 201, 203
Porter, A. G., 3(31), 4, 5(31), 9(31)
Pouwels, P., 9, 13(55), 445, 447(9), 449(9)
Pouwels, P. H., 9, 10(58), 454
Powers, D. B., 276, 291
Powledge, T. M., 166
Preiss, J. W., 256
Prensky, W., 48, 49(4, 5), 53(4, 5), 343(20), 345
Prentki, P., 434
Preston, M. S., 257
Prigillo, T., 632
Prignoli, D. R., 612, 613, 614(25), 617(13, 25), 667
Prior, C. P., 235
Ptashne, M., 240, 376, 377, 384, 398, 426
Pyhälä, L., 31, 63, 88, 235, 546, 635

Q

Quan, S. G., 52
Quesada, J. R., 177, 242, 612, 634

R

Ragg, H., 3, 5(7), 7(7), 249, 634
Raj, N. B. K., 15, 234, 398
Ralph, P., 631
Ralson, R., 504, 513(23)
Randall, R. J., 40, 178, 180(3), 277, 549, 717
Ranu, R. S., 504, 513(23)
Rashidbaigi, A., 3(18), 4, 5, 7(18), 15, 316, 317(5), 321(5)
Rashidbargi, A., 301
Rasmussen, S. L., 635
Rawls, W. E., 601
Read, S. E., 305
Reese, C. B., 517, 667, 670(14)
Reese, C. C., 636
Reges, A. A., 444
Rehberg, E., 3(15), 4, 5(15), 7(15), 18, 19, 27, 240, 241, 243, 434, 552, 601, 602(34), 613, 746
Reichenbach, N. L., 668, 669(16, 17), 672, 752, 753(8), 754, 755(8, 14), 756(14), 757(8), 758(8)
Reichman, M., 667, 735, 736(17), 739(20), 744, 745, 748, 750(16), 751(20), 752, 753, 754(6), 755(6), 757(8), 758, 759, 760(5), 761(3, 5)
Reik, L. M., 715
Reiners, J. J., Jr., 715
Reiser, H., 54
Reisfeld, R. A., 264, 322
Reisinger, D. M., 351, 676
Reiss, B., 370
Remaut, E., 241, 367, 368, 369(8, 10), 370, 371(5, 8, 20), 372(10), 373(20), 374(8, 19, 20), 375(10), 376, 397, 447, 449(15), 583, 585(3)
Ren, W.-H., 351, 352(5), 353(4), 676, 677, 681(18)
Reno, D. L., 287
Renton, K. W., 656, 715, 716, 717(12), 718, 719
Reuben, J., 612, 634
Reuveny, S., 239
Revel, M., 3(27), 4, 5, 27, 141, 234, 495, 500, 515, 597, 598, 599(3), 610(3), 611(3), 667, 707
Reyes, G. R., 398
Reyes Luna, V. E., 211
Rhodes, C., 467
Riccardi, C., 676
Rice, J. A., 103, 104(1)
Richards, J. H., 400, 403, 409
Richards, R., 30, 240, 241(77)
Richardson, C. C., 528
Richoz, Y., 211

Ricketts, R. T., 166
Rigaud, M., 709
Rigby, P. W. J., 467
Riggs, A. D., 400, 409
Riley, F., 103
Rinderknecht, E., 7, 10, 52, 240, 241(70)
Ringold, G., 384, 397, 398, 399, 403(5)
Ripley, S., 600
Ritz, J., 693
Robbins, M. S., 720, 721(7)
Robert, M., 239
Robert-Galliot, B., 5, 301, 316, 541, 544, 549, 694, 702(6)
Roberts, R., 626, 641
Roberts, T. M., 240, 376, 377
Roberts, W. K., 351, 500, 518
Robins, D. M., 600
Robins, R. K., 667, 675
Robinson, W. A., 630
Rodbell, M., 257, 258(9)
Rodriguez, H., 7, 10(46)
Rodriguez, R. L., 419(14), 420
Rönnholm, L., 58
Rose, C. A., 166, 593
Rose, J. K., 471
Rose, K., 235
Rosebrough, N. J., 40, 178, 180(3), 277, 549, 717
Rosen, O. M., 301
Rosenbaum, J., 187
Rosenberg, H., 464, 474(1)
Rosenberg, N., 735, 736(17)
Rosengurt, E., 643, 644, 646
Rosenquist, B. D., 211
Rosenstreich, D. L., 612(26), 613
Ross, M., 192
Ross, M. J., 249, 593
Ross, R., 659
Rossi, C. R., 211
Rossi, G. B., 612, 613
Roszkowski, M., 244
Rothstein, R., 416
Rouse, B. T., 137, 211
Rousset, S., 694, 702(1)
Roy, R., 504, 513(23)
Roy, S. K., 159, 160, 245
Royal, A., 445
Rozenberg, N., 751, 752
Rubin, B., 52
Rubin, B. Y., 96, 574
Rubin, C. S., 300(12), 301
Rubin, J. S., 657
Rubin, Y., 283
Rubinstein, M., 27, 83, 84, 166, 198, 234, 235, 237(17), 243, 301, 315, 317(2), 415, 566, 582, 583(2), 589, 592, 593(8, 9), 735, 743, 744, 746, 747, 752, 754(4), 760, 761(12)
Rubinstein, S., 14, 18(9), 28(8), 54, 64, 65, 124, 132, 162, 166, 178, 180(4), 198, 201, 206, 210(12), 243, 250, 287, 299, 307, 380, 385, 407, 417, 427, 440, 533, 539(4, 5), 551, 554(3), 555(3), 556(3), 557(3), 579, 592, 593, 601, 605(29), 606(29), 746
Ruddle, F. H., 597, 598, 599(3), 610(3), 611(3)
Ruegg, U. T., 235
Rupp, W. D., 286, 287, 288(4), 289(4)
Ruscetti, F., 306, 309(15), 632
Ruscetti, F. W., 49, 54(11), 309(32), 311, 629
Rusckowski, M., 236
Rush, J. D., 310
Rutherford, R. B., 659
Ryan, D. E., 715

S

Sabran, J. L., 612, 706
Sada, E., 160
Sadlik, J. R., 265, 266(11), 343(20), 345
Sagar, A. D., 241
St. John, A. C., 168, 210
St. Laurent, G., 497, 499(21), 517
Saito, M., 93, 571
Sakai, M., 3(28), 4, 5(28), 464
Sakguchi, A. Y., 551, 557(1)
Saksela, E., 54
Sala-Trepat, J. M., 444
Salmon, S. E., 636, 641, 642
Saman, E., 368, 376, 397
Samanta, H., 489, 490(9), 491(9), 492, 497, 499(21), 516, 517, 521(5), 667
Sambrook, J., 287, 400
Samuel, C. E., 499, 500, 501, 503, 504, 513(24), 514, 515, 516
Sancar, A., 286, 288(4), 289(4)
Sandbach, J., 636, 641(19)
Sanders, J. P. H., 368
Sanger, F., 363, 426

Santamaria, C., 616
Santiano, M., 69
Santoli, D. J., 541
Sargent, T. D., 444
Sarkar, F. H., 264, 267, 281, 305, 307(7), 312, 315, 317(3), 332, 340, 341(5, 8), 342(5), 343, 345, 346, 347(10), 602
Sarragne, M., 544, 694, 702
Sasaki, H., 409
Sato, R., 718, 721, 722
Sato, Y., 235
Sauer, R. T., 377
Sawada, Y., 667, 668, 669(9, 17), 672, 754, 755(14), 756(14)
Sawai, H., 522, 523, 529, 668
Saxinger, C., 48, 49(5), 53(5)
Scahill, S., 241
Scahill, S. J., 240, 375, 458
Scatchard, G., 310, 314, 318, 325, 341
Schaber, M. D., 417, 418(5), 419(5), 421(5)
Schafferman, A., 9
Schambrock, A., 135
Schein, C., 3, 5(7), 7(7), 249, 268, 269(2), 273(2), 275(2), 634
Schellekens, H., 9, 10(57, 58), 13(55, 56), 63, 228, 230(8), 441, 445, 447(9), 449(9), 452(1), 453(1), 454, 463
Schenkman, J. B., 721
Scherer, G. F. E., 378
Scherer, S., 419(15), 420
Scherwood, P. J., 197
Schlegel, W., 257, 258
Schlom, J., 611, 683
Schlossman, S. F., 693
Schmid, J., 3, 5(7), 7(7), 249, 268, 269(2), 273(2), 275(2), 634
Schmidt, A., 495, 515
Schmidt, A. O., 676
Schmidt, H., 493
Schmidt, J., 15, 566, 582, 583(2), 589
Schmidtt, C., 650
Schmoyer, M., 613, 614(32)
Schneider, E. L., 597
Schober, I., 54
Schold, M., 413, 481
Schonne, E., 233
Schryver, B. B., 528, 675
Schuffman, S. S., 535(8), 536, 539(8), 540(8)
Schwartz, D. E., 386
Schwartz, M., 378
Schwartzstein, M., 3(16), 4, 5(16), 7(16), 192
Scolnick, E. M., 384
Scott, G. M., 28
Seamans, T. C., 9, 13(49), 240, 249, 434, 437(7), 438(7), 441(7), 445
Sebesteny, A., 650
Secher, D., 28, 166
Secher, D. S., 15, 28, 30, 33(20), 54, 558
Sedmak, J. J., 14, 234, 235(23), 249, 740
Seeburg, P. H., 3(9, 11), 4, 5(9, 11), 7(9, 11), 27, 153, 166(17), 167, 240, 242, 288, 289(12), 293, 359, 376, 405(7), 406, 408(7), 419(12), 420, 428, 430, 434, 440(1), 441(1), 464, 469(4), 471(4), 482, 554, 555(13), 634, 635, 636(10)
Segev, D., 3(27), 4, 5(27)
Sehgal, P. B., 241, 497, 499(21), 517
Sekellick, M. J., 107, 108, 109(7), 111(5), 112, 113, 114, 118(11), 119, 120(11, 15), 122(15), 123(7), 124(9), 127
Sekiya, S., 704
Sela, E., 735
Sela, I., 12, 667, 676, 729, 734, 735, 736(11, 17), 739, 741(9), 744, 745, 748, 749, 750, 751(20), 752, 753, 754(3, 4, 6), 755(3, 6, 8), 756(3), 757(8), 758, 759, 760, 761(2, 3, 4, 5, 11, 12)
Sen, G. C., 500, 515, 668
Serafinowska, H. T., 667, 670(14)
Shabtai, S., 731, 754, 759
Shaffer, J., 481
Shafferman, A., 144, 464, 467, 471, 474(1)
Shaila, S., 500
Shalita, Z., 467, 471
Shaw, G., 3, 5(7), 7(7), 249, 268, 269(2), 273(2), 275(2), 634
Shaw, G. D., 9, 13, 249, 445, 470, 471
Shearer, M., 643
Shen, J. S., 636
Shen, L., 612, 630
Shen, T. Y., 501
Shepard, H. M., 3(32), 4, 5, 9(32), 11(32), 178, 240, 241, 293, 376, 377, 378(13), 397, 415, 554, 555(14), 635, 636, 640(26)
Sherwood, P. J., 3(36), 5, 10(36), 204, 243, 282, 301, 376, 554, 555(15), 635
Shimizu, K., 703
Shimotohno, K., 384
Shimoyama, M., 703, 705(11)
Shine, J., 377

Shire, S., 174, 242, 249, 250(17), 252(17), 253(17), 254(17), 404, 593
Shire, S. J., 162
Shively, J., 27, 179, 190, 192, 199, 235, 249, 259, 375, 389, 415, 592, 593, 747
Shively, J. E., 206
Shizawa, M., 160
Shoji, K., 571
Shou, Y.-Z., 676
Shulman, L., 495, 500, 515, 597, 598(3), 599(3), 610(3), 611(3)
Shure, H., 667
Shuttleworth, J., 234
Sidhu, R. S., 359, 363(4)
Siggens, K., 3(34), 4, 5(34)
Sigman, D. S., 253
Silverman, R. H., 489, 490, 492(10), 517, 667
Silverstein, S., 599
Sim, G. K., 467, 468(9)
Simmer, R. L., 263
Simon, E. H., 114
Simona, M. G., 235
Simons, G., 241, 368, 369(10), 370(10), 372(10), 375(10)
Simonsen, C., 282
Simonsen, C. C., 3(36), 5, 10(36), 197, 204, 243, 301, 376, 399, 554, 555(15), 635
Sims, J., 83
Singh, A., 204, 424, 430, 433(3, 29)
Singh, G., 656, 715, 719
Sipe, J. D., 546
Skalka, A. M., 385
Skehel, J. J., 490
Skiftas, S., 275
Sklarz, B., 237
Skoog, F., 745
Skup, D., 475
Slate, D. L., 597, 598, 599(3), 610(3), 611
Slattery, E., 111(21), 114, 489, 490, 491, 493, 494(15), 496(15), 517
Slinker, B., 192, 249, 593
Slocombe, P., 3(26, 34), 4, 5(26, 34), 364
Slocombe, P. M., 9
Sloma, A., 3, 4, 5(8, 9, 22, 32), 7(8, 9, 22), 9(32), 11(32), 143, 153, 166, 167, 178, 183(6), 184, 192(6), 240, 242, 288, 289(12), 293, 376, 397, 405(6, 7), 406, 408(7), 419(12), 420, 434, 440(1), 441(1), 464, 469(4), 471(4), 482, 554, 555(13), 634, 635, 636(10)
Sly, W. S., 336
Smith, B. D., 672
Smith, D. M., 535(8), 536, 539(8), 540(8)
Smith, G., 183
Smith, J. C., 3(31), 4, 5(31), 9(31)
Smith, K. A., 657
Smith, M., 312, 314(8), 324, 326, 329(3), 409, 425, 481
Smith, M. E., 36, 198, 236, 277, 312, 547, 592
Smith, P. K., 715
Smith-Johannsen, H., 199, 236, 562, 564(13)
Smithyman, A., 631
Sobel, B. E., 626
Soberon, X., 481
Soehnlen, B., 636, 641
Sokawa, Y., 9, 240, 475
Sollenne, N. P., 199
Soltvedt, B. C., 377
Somer, H., 619, 621(1), 623(1), 624, 628
Sonnenfeld, G., 126, 715, 716, 717(13), 718(13)
Sor, F., 475
Southern, E., 468
Southern, E. M., 363, 437
Southern, P. J., 599
Spear, P. G., 384
Spears, C., 292, 414
Spero, L., 89, 145
Spiegel, H., 242, 641
Spiegel, S., 12, 735, 752
Spitzer, G., 612, 629
Sporn, M. D., 658
Spragg, J. S., 30, 33(20)
Sprecher-Goldberger, S., 704, 705(12)
Sreevalsan, T., 646
Srivastava, P. C., 667
Stabinsky, Y., 30, 240, 241(77)
Stacey, D. W., 384, 385, 388(10), 389(16)
Staehelin, T., 3(15), 4, 5(15), 7(15), 15, 21(20), 153, 157, 166, 175, 205, 242, 245(2), 250, 251, 252(24), 259, 287, 288(10), 291(10), 295, 300(11), 301, 306, 307(18), 309(20), 385, 407, 409(9), 418, 428, 473, 558, 566, 582, 583, 584(11), 586, 588, 589, 590(3), 591(3, 5, 6), 601, 637, 684, 747
Stafford, D. W., 465
Stähli, C., 15, 250, 287, 558, 566, 582, 583(2), 588, 589, 590(3), 591(3)

Stalker, D., 471
Stanners, C. P., 128
Stanssens, P., 241, 368, 369, 370, 371(8), 374(8, 19), 376, 397, 447, 449(15)
Stanton, G. J., 14, 193, 200, 259, 564, 574, 607, 608(38), 612, 682
Stanton, G. Y., 76
Stark, G. R., 653
Stebbing, N., 3(9), 4, 5(9), 7(9), 27, 28(10), 30, 31(10), 33(10), 153, 166(17), 167, 183(6), 184, 192(6), 240, 241, 242, 288, 289(12), 293, 376, 405(7), 406, 408(7), 419(12), 420, 434, 440(1), 441(1), 464, 469(4), 471(4), 482, 552, 554, 555(13), 612, 629, 634, 635, 636, 640, 642(25), 656, 715, 719
Steer, C. J., 257
Stefanos, S., 146, 460, 562
Stehlin, J. S., 616
Stein, S., 3(15), 4, 5(15), 7(15), 27, 179, 190, 199, 205, 206, 214, 218, 235, 259, 375, 385, 389, 415, 592, 747, 754
Steinman, R. M., 332
Stephen, E. S., 103
Stern, R. H., 473
Stevenson, D., 551, 557(1)
Stevenson, H. C., 265, 266(11), 343(20), 345
Stewart, A. G., 3(31), 4, 5(31), 9(31)
Stewart, W., II, 84
Stewart, W. E., 227, 233, 234, 237, 238, 240, 241(67), 364, 554, 556(12), 557(12), 640
Stinchcomb, D. T., 419(15), 420
Stocker, J., 566, 582, 583(2), 589
Stoker, M. G. P., 97
Stone-Wolff, D. S., 52
Stowring, L., 475, 476(4)
Strand, P. J., 723
Strander, H., 62, 166, 635
Streuli, M., 3, 4, 5(7, 16, 17), 7(7, 16, 17), 136, 166, 240, 241, 249, 268, 269(2), 273(2), 275(2), 359, 442, 464, 634, 635, 640
Strick, N., 15
Stringfellow, D. A., 707, 708, 709, 711(7), 712(2, 3, 7)
Strom, T. B., 688
Stromberg, R. R., 77, 79, 81(2, 5), 82(2)
Stroud, R. M., 235, 238(28), 239(28), 249, 250(17), 252(17), 253(17), 254(17), 593
Struhl, K., 419(15), 420
Studier, F. W., 418
Su, Y.-L., 676
Sudo, T., 3(28), 4, 5(28), 464
Sugeno, K., 721
Suggs, S. U., 485
Sugimoto, K., 409
Sugimura, T., 97, 613, 614(30), 617(30), 628, 703, 705(11)
Sugino, H., 297, 585
Suhadolnik, R. J., 667, 668, 669, 670, 671, 672, 673, 675, 735, 744, 745, 752, 753(6, 8), 754, 755(6, 8, 14), 756(14), 757(8), 758(8), 759, 760(5), 761(3, 5)
Sulea, I. T., 48, 49(2), 50(2), 76
Sulkowski, E., 91, 189, 199, 214, 218, 219, 234, 460, 545
Sumar, H. K. L., 137
Summers, M., 183
Sun, S., 235
Sundström, C., 306, 316
Surwit, E. A., 636
Suzuki, F., 93
Suzuki, J., 14, 101
Svitlik, C., 108, 112, 113, 118(11), 119, 120(11)
Swanson, D. A., 612, 634
Swarm, R., 672
Swetly, P., 3(25), 4, 5(25), 48, 234
Swistok, J., 244
Switzer, J. W., 310
Switzer, R. C., III, 178
Szabo, P., 474
Szasz, G., 624

T

Tabin, C. J., 384
Tabor, J. M., 3, 4, 5(8, 9), 7(8, 9), 153, 166, 167, 288, 289(12), 293, 376, 405(6, 7), 406, 408(7), 419(12), 420, 434, 440(1), 441(1), 464, 469(4), 471(4), 482, 554, 555(13), 634, 635, 636(10)
Tadokoro, J., 705
Taetle, R., 629
Tahara, S. M., 296
Taira, H., 3, 5(7), 7(7), 9, 13(48), 111(21), 114, 166, 240, 249, 268, 269(2), 273(2), 275(2), 359, 442, 445, 464, 470, 471(16), 493, 515, 635

Takacs, B., 566, 582, 583(2), 589
Takahashi, F., 702, 704(6), 705(6)
Takanami, M., 409
Takaoka, C., 9, 240
Takaoka, L., 475
Takatsuki, A., 91, 234
Talkad, V., 336
Tamura, G., 234
Tan, C., 199, 214, 236
Tan, Y. H., 15, 199, 214, 218, 236, 400, 562, 564(13), 597
Tanaguchi, T., 384
Tanaka, S., 481
Tanaka, T., 704
Taniguchi, T., 3(28, 29), 4, 5(28, 29), 9, 183, 192, 240, 376, 464, 475, 635
Tanimoto, T., 235
Tarnowski, D. K., 15, 153, 162(3), 171, 177(23), 238, 242, 243(5), 245(5), 259, 586, 589, 591(7), 593(7)
Tarnowski, S. J., 15, 19, 27, 153, 154, 155, 159(4), 161, 162(3), 165(5), 171, 175, 177(23), 238, 242, 243(3, 5), 245(5, 8), 255, 259, 382, 407, 408(8), 586, 589, 591(7), 593(7)
Tarr, C., 63, 193
Tarutani, T., 93
Tavernier, J., 3(30), 4, 5(30), 9(30), 240, 241, 369, 398
Taya, Y., 3(35), 5, 10(35), 204, 243(15), 244, 301, 370, 384
Taylor, G. M., 333
Taylor-Papadimitriou, J., 611, 643, 646, 650, 702
Temin, H., 384
Tener, G. M., 476
Tenser, R. B., 126
Teruda, M., 97
Testa, D., 3(19), 4, 5(19)
Thang, D. C., 696
Thang, M. N., 667, 696
Thaniyavarn, S., 715, 719
Thatcher, D. R., 235
Theilen, G. H., 310
Thépot, F., 541
Thill, G. P., 417
Thiry, L., 704, 705(12)
Thomas, C., 367
Thomas, G. E., 715
Thomas, R. M., 326
Thuring, R. W. J., 368
Tian, B., 676
Tian, P.-K., 676
Tilles, J. G., 138
Timmes, M., 640
Timonen, T., 676
Ting, R., 306, 309(15), 632
Tinsley, J., 3(34), 4, 5(34)
Tischfield, J., 597
Tizard, R., 386, 443
Tochino, Y., 721
Todaro, G. J., 643, 658
Todokoro, K., 3, 5(7), 7(7), 249, 268, 269(2), 273(2), 275(2), 634
Toira, H., 634
Tokunaga, E., 566
Tometsko, A., 252
Tomida, M., 571, 612, 613, 630
Tomita, Y., 702, 703(5), 704(2, 6), 705(2, 3, 5, 6, 8), 706
Torchio, C., 552, 553(9)
Torczynski, R., 364, 365(11)
Torczynski, R. M., 3(20, 21), 4, 5(20, 21), 7, 359, 363(4)
Torma, E., 235
Törmä, E., 562
Torrence, P. F., 489, 494(8), 497(8), 500, 517, 522, 523, 524(9), 526, 528, 529, 667, 668
Touchi, A., 721
Tovey, M., 134, 552, 601, 611, 612(1), 629
Tovey, M. G., 211, 234, 322, 562, 635
Towbin, H., 418
Trahey, M., 397, 398(5), 399(5), 403(5)
Trakatellis, A. C., 252
Trapman, J., 9, 13(51)
Traub, A., 36, 143, 239
Traugh, J. A., 296, 497
Trevan, D., 650
Trieber, G., 466
Trinchieri, G., 541
Trindade, A., 612
Trowbridge, R. S., 552, 553
Trown, P., 242
Trown, P. W., 177, 636, 640, 641, 642(25)
Trujillo, J., 306, 309(15)
Truong, K., 167(19), 168
Tsai, K., 297, 585
Tsai, S., 306, 309(15), 632
Tsao, H., 368, 371(5)

Ts'o, P. O. P., 667
Tsujimoto, M., 63, 93
Tsukui, K., 566
Tucker, G., 597
Tuppy, H., 235
Tyring, S., 574, 577(4)
Tyring, S. K., 211
Tyrrell, D., 28

U

Uchida, S., 566
Udenfriend, S., 754
Uematsu, T., 672
Ullrich, A., 3(9, 11), 4, 5(9, 11), 7(9, 11), 27, 153, 166(17), 167, 183(6), 184, 192(6), 240, 242, 288, 289(12), 293, 359, 376, 405(7), 406, 408(7), 419(12), 420, 434, 440(1), 441(1), 445, 464, 469(4), 471(4), 482, 554, 555(13), 634, 635, 636(10)
Urban, C., 52, 77, 84, 90, 193, 210, 235, 259, 282, 283, 297
Urdea, M. S., 433
Uzé, G., 263, 286

V

Vaks, B., 3(27), 4, 5(27)
Valenzuela, P., 433
van Boom, J. H., 517, 667
van Damme, J., 63, 66(5), 76, 96, 193, 259, 389, 395(23), 559, 561(7)
Vandenbussche, P., 667
Van der Eb, A. J., 599
van der Ende, A., 442
van der Heyden, J., 240, 241, 375, 458
van der Meide, P. H., 9, 10(57, 58), 13, 228, 230(8), 441, 452(1), 453(1), 463
vander Pas, M. A., 528
van Frank, R. M., 168
van Gray, P. W., 445
Van Heuverswyn, H., 3(35), 5, 10(35), 197, 204, 243(15), 244, 301, 370, 384
van Lente, F., 278
van Reis, R., 5, 77, 79, 81, 82
Vastola, K., 199
Vaughn, M., 258
Vecchio, G., 613
Velan, B., 9, 144, 464, 467, 471, 474(1)
Verghaegen-Lawalle, M., 528
Verma, D. S., 612, 629
Vermylen, C., 76, 90, 92(5), 96, 193, 259, 559, 561(7)
Victor Rebois, R., 341
Vieira, J., 363
Vignal, M., 597, 611(8)
Vignaux, F., 263, 269, 271(4), 272(4), 273(4), 274(4), 275(4), 276, 281, 305, 310(1), 312, 340, 341(3), 342(3)
Vijverberg, K., 9, 13(56), 228, 230(8), 441, 452(1), 453(1), 463
Vilček, J., 23, 48, 49(2, 3, 4), 50(2, 3), 51(3), 52, 53(4, 5), 76, 77, 84, 90, 114, 136, 141, 193, 210, 233, 234, 235, 259, 265, 276, 281, 282, 283, 284, 297, 312, 315, 317(1), 332, 336(5), 340, 341(6), 342(6), 343(20), 345, 346(6), 389, 395(22), 397, 399, 570, 587
Villa-Komaroff, L., 443
Vincent, C., 234
Virtanen, I., 54
Vogel, S. N., 612(26), 613
Volckaert, G., 3(30), 4, 5(30), 9(30), 240, 369
Volvovitz, F., 48, 49(2), 50(2), 76
von Hoff, D. D., 636, 641
von Muenchhausen, W., 199, 214, 545
von Wussow, P., 93
Vovis, G. F., 3(19), 4, 5(19)

W

Wagner, A. F., 501
Wainwright, N. R., 3(15), 4, 5(15), 7(15)
Waldman, A. A., 54, 64, 166, 198, 243, 592, 593, 746
Walesby, N. J., 758
Walkinshaw, M. D., 378
Wallace, D. M., 3(36), 5, 10(36), 197, 243, 282, 301, 376, 554, 555(15), 635
Wallace, R. B., 413, 444, 481, 485
Wang, C.-T., 244
Wang, D. B., 351
Wang, J. L., 49, 50(9)
Wang, L.-H., 385
Wang, N., 351, 352(5), 353(4, 5), 677, 681(18)

Wang, T. P., 351, 352(5), 353, 676, 677, 681(18)
Ward, D. C., 474
Watanabe, Y., 9, 227, 234, 240, 475, 558, 560(4), 561(4), 562(4), 564(4), 565(4), 566, 568(4), 571, 573(4)
Wattenberg, L. W., 723
Weber, D. V., 160
Weber, K., 141, 142
Webster, C., 622
Weck, P., 635
Weck, P. K., 27, 28(10), 31(10), 33(10), 240, 241, 552, 640
Weedon, L. L., 612(26), 613
Weening, H., 76, 90, 92(5), 193, 259, 559, 561(7)
Wei, C.-M., 384
Weidele, U., 249
Weidle, U., 3, 5(7), 7(7), 268, 269(2), 273(2), 275(2)
Weidlle, U., 634
Weigele, M., 754
Weigent, D., 83, 84(1)
Weigent, D. A., 14, 607, 608(38)
Weil, J., 597
Weil, R., 236
Weinberg, R. A., 384
Weinstein, I. B., 612, 613, 614(24, 25, 30), 617(12, 13, 25, 30, 31), 619, 628, 629, 630(9), 632(2), 667, 706
Weinstein, L. S., 613, 614(30), 617(30), 628
Weintraub, H., 278
Weismann, C., 192
Weiss, R. A., 703, 704(10)
Weissbach, A., 735, 739(20), 748, 750(16), 753
Weissbach, H., 292, 414
Weissman, C., 27, 33(7), 135, 136, 464, 470, 471(16), 634
Weissmann, C., 3, 4, 5(7, 16, 17), 7, 9, 13(48), 166, 183, 240, 241(67), 249, 268, 269(2), 273(2), 275(2), 327, 359, 442, 445, 454, 635, 640
Wen, Y.-M., 351, 353, 676, 677(4)
Wensink, P. C., 473
West, D. K., 500
West, M. D., 103
Westmacott, L. M., 98
Wetzel, R., 176, 192, 235, 238, 239(28, 56), 242, 249, 250(17, 18), 252, 253, 254(17, 18), 269, 404, 593
Wheelock, E. F., 48, 708
Whitaker, J. R., 734
White, B. A., 749
White, C. S., 641
White, R. F., 758
White, S. L., 575
Whittaker, J. R., 616
Whittall, J. T. D., 566
Wickerhauser, M., 59
Wieke, M. E., 96
Wierenga, B., 442
Wietzerbin, J., 146, 460, 562
Wigler, M., 599, 600
Wigzell, H., 305, 306(3)
Wijnands, R. A., 517
Wilcox, G., 481
Wilde, C. E., III, 287
Wilkinson, A. J., 409
Wilkinson, M., 96, 101(4, 5), 102
Williams, B. R. G., 305, 351, 489, 517, 676
Williams, D. C., 168
Williams, L. J., 616
Willingham, M. C., 315, 332, 333(4), 336(4), 338(4), 339
Willis, D. H., 667, 672, 752, 759, 760(5), 761(5)
Wilson, V., 9
Windass, J. D., 475
Winget, C. A., 248
Winship, T. R., 108, 109, 112(9), 114, 118, 119, 123(13), 128, 129(5, 7)
Winter, G., 409
Wiranowska-Stewart, M., 84, 93, 96, 233, 237, 238, 554, 556(12), 557(12)
Witteveen, S. A. G. J., 626
Wold, F., 296
Wolfe, R. A., 205
Wong, D.-B., 676, 677(4)
Wood, J. N., 669, 676, 681(2)
Wood, M., 541
Wood, R. K., 731
Woodlans, M., 442, 443(4)
Wreschner, D. A., 490
Wreschner, D. H., 490, 492(10), 517, 667
Wu, H.-Y., 676
Wu, J. M., 667, 668, 669(9, 16, 17), 672, 754, 755(14), 756(14)
Wu, J. R., 444
Wubben, J., 9, 10(57), 441, 452(1), 453(1), 463
Wyler, R., 211

Y

Yalow, R. S., 588
Yamaguchi, T., 93, 571
Yamamoto, K. R., 466
Yamamoto, R., 183, 184(1), 192(1), 345, 397
Yamamoto, Y., 227, 384, 562, 612, 613, 630
Yamazaki, S., 141, 233
Yanofsky, C., 289, 370, 371(20), 373(20), 374(20), 377
Yao, K., 351, 676, 677(4)
Yao, M.-H., 676
Yap, H. Y., 635
Yaron, I., 707
Yaron, M., 707
Yelverton, E., 3(9, 11, 32, 36), 4, 5, 7(9, 11), 9(32), 10(36), 11(32), 27, 153, 166(17), 167, 178, 183(6), 184, 192(6), 197, 204, 240, 242, 243, 282, 293, 301, 359, 376, 377, 378(13), 397, 405(7), 406, 408(7), 419(12), 420, 434, 440(1), 441(1), 445, 464, 469(4), 471(4), 482, 554, 555(13, 14, 15), 634, 635, 636(10)
Yilma, T., 130, 131, 132(4), 134, 552
Yim, S. O., 616
Yip, Y. K., 38, 48, 49(2, 3), 50(2, 3), 51(3), 52, 63, 76, 77, 90, 193, 210, 234, 235, 259, 265, 266(11), 276, 281, 282, 283, 297, 312, 315, 317(1), 332, 336(5), 340, 341(6), 342(6), 343(20), 345, 346(6), 399
Yoichi, T., 197
Yokobayashi, K., 235
Yonehara, S., 236, 276, 305, 310(11), 332, 336(7)
Yonehara-Takahashi, M., 276, 305, 310(11), 332, 336(7)
Yoshie, O., 490, 493, 495(12), 497, 499(21), 517
Yoshikawa, S., 235
Young, L., 636, 641, 642(25)
Young, P. A., 14, 536
Youngner, J. S., 114
Yu, X.-S., 676
Yuan, P. M., 206

Z

Zajac, B. A., 36
Zalut, C., 252
Zander, A., 612
Zander, A. R., 629
Zehner, Z., 418
Zerebeckyj-Eckhardt, I., 48, 49(3), 50(3), 51(3), 52(3)
Zhang, Q.-J., 676
Zhang, X.-H., 676
Zhao, X.-X., 676
Zheng, H.-D., 351, 352(5), 353(4, 5), 676, 677, 681(18)
Zhong, W.-B., 676
Zilberstein, A., 500, 515
Zinn, K., 384
Zinn, K. P., 398
Zipser, D., 377
Zoller, M., 409
Zoller, M. Z., 425
Zoon, K. C., 36, 133, 143, 198, 276, 277, 309, 312, 313, 314, 315, 324, 326, 327, 329, 330, 331, 332, 333, 335, 336(4, 14, 15), 337(14), 338(4, 14, 15), 339, 340, 341(4), 342(4), 346(4), 547, 592
Zor, U., 707
Zubay, G., 292
Zur Nedden, D., 198, 255, 276, 277, 281, 307, 309(24), 312, 313, 314, 315, 327, 329, 330(5), 331, 332, 333, 335, 336(4), 338(4), 339(4), 340, 341(4), 342(4), 346(4), 547, 592
Zwarthoff, E. C., 9, 13

Subject Index

A

A-23187, *see* Calcium ionophores, A-23187
A_{260} unit, definition, 22
Abbreviations, 14–23
amino acids, 20–21
general, 22–23
interferons, 19–20
Acid phosphatase gene, promoter, 418
Acidification, effect on alpha helix of Hu-IFN-αA, 239
Actimomycin D
in equine IFN-β induction, 133
in superinduction of Hu-IFN-β, 284, 400
Adenocarcinoma of the cecum, xenograft in nude mice, 650–653, 655
Adenosine triphosphate, polymerization induced by TMV, 744
Adenovirus major late promoter, insertion of rat IFN-γ gene, 457
Adenylate kinase
diadenosine pentaphosphate inhibition, 624, 627
in measurement of creatine kinase activity, 624, 627
2-(9-Adenyl)-6-hydroxymethyl-4-hexylmorpholine, derivative of oligoadenylic acid, 523
Affi-Gel Blue chromatography
in Bo-IFN-α purification, 138–140, 474
removal of Sendai virus from, 138
Affigel-10-lactoperoxidase, in radiolabeling of Mu-IFN, 323
Affinity chromatography, *see also* Immunoaffinity chromatography
blue Sepharose in purification of,
Bo-IFN-β, 214–216
Hu-IFN-γ, 91, 92
concanavalin A-Sepharose, 99–102, 545–548
ethylene glycol in, 545–547
in purification of
Hu-IFN-γ, 84, 90–91, 92
Mu-IFN-γ, 146–147, 149
α-methyl-D-mannoside in, 545–546
goat anti-mouse immunoglobulins, 583
Hu-IFN-α antiserum in, 546–548
immobilized Cibacron Blue F3G-A (Blue Sepharose CL-6B) in, 545–546
poly(U)-Sepharose
in Hu-IFN-γ purification, 91, 92
in purification of AVF, 742–743
purification factor, 742
Affinity purification, *see* Affinity chromatography
Agglutination, lectin-like interferon antagonists in, 701–702
Aggregation, of rat interferon, 225
Alkaline phosphatase, digestion of (2′-5′)-oligoadenylic acid, 526
Allografts, interferon in delaying rejection of, 541
Alpha factor
in vitro mutagenesis of, 425
prepro sequence, 430
promoter sequences, 430, 433
Alpha interferon A, *see* Interferon alpha A
Alpha interferon D, *see* Interferon alpha D
Alpha interferon, *see* Interferon alpha
Amino acid composition
IFN-α, 11
IFN-β, 11
IFN-γ, 7, 11, 209
Amino acids
abbreviations, 20–21
sequencing by Edman degradation, 206
Amino terminal protein sequences, encoded by gene inserts, 431–432
2-Amino-5-bromo-6-phenyl-pyrimidinone, inducer of interferon, 708–709
Aminoglycoside antibiotic G418, resistance to by plasmid pSV2-neo, 599–600

Aminopeptidase, effect on interferon, 234
Aminopyrine, in measurement of drug metabolism, 720, 722
Aminopyrine N-demethylase, in assay for cytochrome P-450, 722
Ammonium chloride, in lysis of erythrocytes, 40, 56, 65, 68, 78, 88–89, 137
Ammonium sulfate
  in Hu-IFN-α2 purification, 171
    yield, 172
  in IFN-γ concentration, 58–59
  in IFN-δ concentration, 99
  in precipitation of RNase L, 494–495
  in purification of rat IFN, 222
2,5-A_n, *see* (2′-5′)-oligoadenylate
Aniline HCl, in measurement of drug metabolism, 720, 722
Aniline hydroxylase, in assay for cytochrome P-450, 722–723
Animal models, for antitumor effects of interferon, 649–657
Antagonism of interferons and growth factors, 657–666
Anti-Hu-IFN-α2 antibody
  binding to
    mouse interferon, 227–230
    rat interferon, 227–230
  neutralization with
    mouse interferon, 227–230
    rat interferon, 227–230
Anti-IFN-γ-Sepharose, in Hu-IFN-δ purification, 97, 99–101
Antibody, *see also* Antiserum; Monoclonal and Polyclonal antibodies
  binding to receptor-bound interferons, 326–332
  effect of dilution on neutralization of Hu-IFN-α1, 567
  goat-anti-mouse, 685–687
    flourescein conjugated, 690
    F(ab′)$_2$ fragment, 687
  Hu-IFN-γ specific
    neutralization of interferon gamma, 53
  titration, 559–571
    constant antibody method, 560–561, 573
      advantages, 561, 573
    constant IFN method, 560–561, 573
    procedure, 568–569
Antibody affinity chromatography, see Immunoaffinity chromatography
Anticoagulant, in Hu-IFN-γ production, 50
Antigens
  cell surface tumor, 682–688
    radioimmunoassay for, 682–688
  HLA-A,B,C, 688–689, 692
  IFN-induced, 682–688
  MHC class I, 688–689, 692–693
    W6/32, 689
  MHC class II, 688–689, 692–693
    I2, 689
  radiolabeling, 686–687
Antigrowth assay
  Daudi cells in, 579–582
  medium color change in, 579–582
  phenol red pH indicator in, 579
Antiproliferative assay, 574–582
Antiserum, *see also* Monoclonal antibody; Polyclonal antibody
  bovine anti-Hu-IFN-α, 17
  goat-anti mouse immunoglobulins, 583
    affinity purified, 583
  to Hu-IFN-α, 546–548
    in affinity chromatography, 546–548
  to Hu-IFN-α2, 227–230, 451–452
    neutralization of
      mouse IFN-α, 227–230
      rat IFN-α, 227–230, 451–452
    in purification of rat IFN-α, 451
  to IFN-γ, 97
  inhibitory activity, 570
  to interferons, 15, 17–18
  to Mu-IFN, 17
  neutralization of interferon, 558–573
    quantitation of, 558–573
  neutralization titer
    assay, 18
    definition, 18, 562–571
  polyclonal sheep anti-leukocyte interferon
    in interferon assay, 31
  to poly(I) · poly(C)-induced human diploid cell interferon, 17
  rabbit anti-Hu-IFN-γ, 17
  rabbit anti-Mu-IFN-γ, 17
  to Sendai virus, 138
  to Sendai virus-induced buffy coat interferon, 17

to Sendai virus-induced Namalva cell interferon, 17
sheep anti-Hu-IFN-α, 17, 451
sheep anti-Hu-IFN-β, 17, 566
sheep anti-mouse L-cell interferon, 17
standards, 15, 17–18
to staphylococcal enterotoxin A-induced Hu-IFN-γ, 17
Antitumor effects
animal models in investigating, 649–657
of interferon, 649–657
Antiviral activity
Hu-IFN-α fractions, 28–29
of IFN-αA, 18–19, 251, 278–279, 404–408, 414
effect of radiolabeling on, 278–279
of IFN-gold colloid conjugate, 334
relationship to enzyme immunoassays, 14–15
relationship to radioimmunoassays, 14–15
Antiviral assay, 533–573, *see also* Cytopathic effect inhibition
caprine cells in, 551–558
clinical blood samples, 540
comparative plasma Hu-IFN titers, 539
computer assisted analysis of, 533–540
data reduction, 536–540
EMCV in, 534–540
HEp-2 cells in, 534
human fibroblast SG-181 cell line in, 534
interferon preincubation, 538
ovine cells in, 551–558
Semliki Forest Virus, 98–99
VSV in, 534–540
Antiviral factor (AVF), *see also* IVR substance
ELISA for, 736–739
infectivity assay, 736
inhibitors, 741
molecular weight, 744
preparation from tobacco cells, 734–744
production in *N. glutinosa,* 752–754
properties, 744
purification, 739–744
reverse phase HPLC in, 742–743
purification factor, 742
purity, 744
stability, 744
stimulation by poly(I) · poly(C), 734–735
stimulation by TMV, 734–735
Antiviral state, IFN-induced, antagonistic effects, 696–697
Arachidonic acid cascade
arachidonic acid in, 710
cyclooxygenase in, 710
endoperoxide in, 710
leukotriene in, 710
5-lipoxygenase in, 710
12-lipoxygenase in, 710
phospholipase in, 710
prostaglandin in, 710
thromboxane in, 710
Aspirin
in measuring hyporesponsiveness to IFN, 711
reversible inhibition of cyclooxygenase, 711
ATP-mediated phosphorylation
in assay for dsRNA dependent protein kinase, 502
in phosphorylation of IFN-γ, 296–301, 316
Autoradiography, of labeled nuclei, 644
AVF, *see* Antiviral factor
Avian reovirus strain S1133(P40)
in chicken IFN induction, 119
in dose-response curve generation, 109, 112
5-Azacytidine, in enhancement of Hu-IFN-α synthesis, 234

B

Baboon endogenous virus, harvesting from human A204 cells, 704
Baby hamster kidney cells
for FDGF preparation, 646
simian virus 40-infected, 646
Bacterial fermentation
production of IFN-γ by, 204–205
temperature sensitive λ cI repressor in, 205
Bacterial synthesis of interferon, 364, 370–371, 376–383, 407, 437–441, 445–450, 471–473
Bacteriophage Charon 28, murine genomic library in, 434

Bacteriophage Charon 4A
human genomic library in, 359–362
screening, 360–363
Bacteriophage lambda
in isolation of
Bo-IFN-γ genes, 466–467
Hu-IFN-α genes, 360
Mu-IFN-α genes, 434
rat IFN-α genes, 444
rat IFN-γ genes, 454
Bacteriophage M13
for *in vitro* mutagenesis, 400
M13mp8, 363
M13mp11, 363
Bacteriophage R17 radiolabeled RNA, in endonuclease assay for RNase L, 491
Bacteriophage T7 DNA
in IFN-α2 vector, 167, 168
ribosome binding site in, 167, 168
BALB/c 3T3 cells, for titration of growth factors, 658–659, 662, 664
Benzo[*a*]pyrene hydroxylase, in assay for cytochrome *P*-450, 723
Benzo[*a*]pyrine, in measurement of drug metabolism, 720, 723
N^6-Benzyladenine (6-benzylaminopurine), in preparation of protoplasts, 745–746
Beta interferon, *see* Human interferon beta; Interferon beta
BIIK cells, in Bo-IFN-α assay, 141
BHK-21 cell cultures, 212
for growth of
bluetongue virus, 212
foot-and-mouth disease virus, 212
Binding
of antibodies to receptor-bound interferons, 326–332
of Hu-IFN-α2 to Daudi cells, 348–349
of Hu-IFN-α2 to receptor, 342–343, 345, 347
competition, 342
of Hu-IFN-γ
to U937 cells, 315–321
time course of, 319–320
of interferon, 321–326
mouse cells in, 321–325
Iodine-125 radiolabeled interferon, 323–325
in monolayer cultures, 325
receptor mediated, 332–339
Binding affinity, of radiolabeled IFN-α2, 274
Binding equation
derivation, 318–319
Scatchard, 318–319
Blood samples, interferon assay of, 539–540
Blue Sepharose
affinity chromatography with, 214–216
purification of Bo-IFN-β by, 214–216
in purification of
Hu-IFN-β, 179
Hu-IFN-γ, 91, 92
radiolabeled Hu-IFN-β, 285
in tandem chromatography with Phenyl-Sepharose, 218
Blue Tongue Virus, in induction of Bo-IFN-β, 212–213, 220
Bo-IFN-β, *see* Bovine interferon beta
Bo-IFN-γ, *see* Bovine interferon gamma
Bolton-Hunter reagent
effect on Hu-IFN-β activity, 236
in radiolabeling, 263, 276–281, 281–283, 286, 290
effect on antiviral activity, 278–279, 283
Hu-IFN-αA, 286, 290, 306, 312–313, 327, 333
Hu-IFN-αD, 312–313
Hu-IFN-α1, 327
Hu-IFN-α2, 312–313, 333
Hu-IFN-γ, 315–316
monoclonal antibody, 327
Bovine, antiserum, *see* Antiserum
Bovine cells
antiviral activity of IFN-αA on, 18–19
BL-3 cells, 310–311
EBTr cells,
in Bo-IFN-α assay, 141
in Hu-IFN-δ assay, 97, 101, 102
in interferon assays, 102, 556–557
embryo kidney cells, interferon activity on, 87
monolayer culture, binding of interferon in, 312–315
NBL cells in assay for antiviral activity, 28–29

sensitivity to
Hu-IFN-γ, 552
human interferons, 134
turbinate CCL44 cells, *see* Bovine cells, EBTr
Bovine DNA, for construction of genomic library, 465–466
Bovine fibroblast interferon, *see* Bovine interferon beta
Bovine genomic lambda clones, interferon alpha restriction maps, 469
Bovine genomic library, construction of, 466–470
Bovine IFN-α, *see* Bovine interferon alpha
Bovine IFN-β, *see* Bovine interferon beta
Bovine IFN-γ, *see* Bovine interferon gamma
Bovine immune interferon, *see* Bovine interferon gamma
Bovine interferon alpha
acid stability, 138
fraction I, 140, 143
fraction II, 140, 143
amino acid sequence, 13
homology, 470–471
with Hu-IFN-α's, 144
antiviral activity
on bovine EBTr cells, 141
on MDBK cells, 140–141
specificity to bovine cells, 144
binding, 143
characterization, 136–144
cross hybridization with Hu-IFN-α, 467–468
cross-reactivity with human interferons, 143
cross-species antiviral activity, 140–141
on BHK cells, 141
on HeLa cells, 141
on human FSII cells, 141
on murine L 929 cells, 141
on simian Vero cells, 141
on WISH cells, 141
cytopathic effect inhibition assay, 138
with MDBK cells, 138
vesicular stomatitis virus in, 138
elution with ethylene glycol, 139
expression in bacteria, 464–474
genes, 469
induction, 136–144
by Sendai virus, 137
mature protein, 13
molecular weight determination, 141–142
by SDS–polyacrylamide gel electrophoresis, 141–142
neutralization, 143, 219
with anti-Hu-IFN-α, 143
production
effect of priming with bovine or human IFN-α, 137
kinetics, 138, 139
large-scale, 137
rate, 139
yields, 138
purification
by Affi-Gel Blue chromatography, 138–140
by immunoaffinity chromatography, 143
with anti-Hu-IFN-α, 143
signal peptides, 13
specific activity, 140, 141
storage, 137
subclasses, 138, 139–140
translated from 13 S mRNA, 143
Bovine interferon alpha C
antiviral activity, 473–474
neutralization by
anti-Hu-IFN-α, 473
anti-Hu-IFN-β, 473
characterization, 473–474
SDS–PAGE in, 473
expression in *E. coli,* 471–473
trp leader peptide in, 471
trp promoter in, 473
expression plasmid for, 471–473
construction, 471–473
fused trpL-Bo-IFN-αC polypeptide in, 472
genomic DNA sequence for, 470
homology to Hu-IFN-α, 469
nucleotide sequence homology, 470
restriction map, 469
signal peptide of DNA sequence for, 469
Bovine interferon alpha gene
isolation, 464–474
restriction maps, 469

Bovine interferon beta
amino acid sequence, 8–9
mature protein, 8
signal peptides, 8
analysis of
SDS–polyacrylamide gel electrophoresis in, 218
chelate Sepharose
nickel, lack of binding, 218
zinc, lack of binding, 218
hydrophobicity, 219
induction of, 212–213, 220
neutralization of, 219
rabbit antiserum in, 219
production, 211–220
in Bovine kidney cell cultures, 211–213
time course of production, 212–213
purification, 211–220
by affinity chromatography, 213–217
by hydrophobic chromatography, 213, 217
recovery, 215–217
various induced species, 220
Bovine interferon gamma
amino acid sequence, 10–11
mature protein, 10
signal peptide, 10
Bovine kidney CCL22 cells, *see* MDBK cells
Bovine leukocyte interferon, *see* Bovine interferon alpha
Bovine leukocytes
purification, 137
yield, 137
Bowel cancer xenograft EF3
effect of
cyclophosphamide, 655–656
interferon, 655–656
8-Bromoadenosine, 2-5-linked analogs, preferential degradation of mRNA over tRNA, 675
5-Bromodeoxyuridine, in enhancement of Hu-IFN-α synthesis, 234
BTV, *see* Blue Tongue Virus
Buffy coats, *see* Leukocytes
1,4-Butanediol diglycidylether, in nickel chelate column preparation, 200–201
Butyrate
in enhancement of
Hu-IFN-α synthesis, 234
Mu-IFN synthesis, 477, 479
n-Butyric acid, stimulation of IFN production in Namalva cells, 37

C

Calcium channel blockers
Verapamil
in interferon gamma induction, 70, 71
Calcium depletion, in inhibition of Hu-IFN-γ induction, 70, 71
Calcium ionophores
A-23187
in Hu-IFN-γ induction, 69–72, 72–76
calcium channel blockers
Verapamil, 70, 71
ionomycin
in Hu-IFN-γ induction, 70
Calcium phosphate coprecipitation technique, with intact mammalian cells, 669–670
Calf serum
in chicken IFN production, 121
in Hu-IFN-γ production, 75–76
Caprine synovial membrane cells
in antiviral assay, 551–558
propagation of cell line, 553–554
sensitivity to
heterologous interferons, 552, 554–558
human interferons, 134
Capsicium fructescens, inoculation with cucumber mosaic virus, 733
Carbohydrate
residues
on Hu-IFN-α, 235
synthesis
inhibitors, 233–234
Carboxymethyl BioGel agarose, in IFN-γ purification, 194, 195, 197
Carboxymethylcellulose
in Hu-IFN-αA purification, 160
protein loss in, 157, 161
in partial purification of (2′-5′)-oligoadenylate synthetase, 518

in poly ICL-CM Dextran preparation, 103, 104
Carboxypeptidase, effect on interferon, 234
Cation-exchange chromatography, *see* Ion-exchange chromatography
Cavia cobaya, see Guinea pig
CCL25 cells, *see* WISH cells
Cell agglutinating activity, *see* Cytoagglutinating activity
Cell cultures
preparation of, 212
lactalbumin hydrolysate in, 212
kidneys in, 212
Cell fusion
virus induced
effect of interferon on, 702–706
Cell growth
effect of
(2′-5′)-oligoadenylates, 667
analogs, 667–675
inhibition by cordycepin 5′-triphosphate, 668–671
Cell number, effect on optical density, 580–582
Cell proliferation, cytoflourimetry in measurement of, 644–645, 648
Cell surface
histocompatibility antigens HLA-A,B,C, 682
effect of interferon, 682
Cell-free protein synthesis, in radiolabeling of interferons, 292–296
Cellulase
for protoplasts, 730–731
medium, 746
preparation of, 730–731
Cellulose acetate electrophoresis, in measurement of creatine kinase activities, 621
Chaotropic agents, effect on interferon activity, 248–249
Charon 28 bacteriophage, murine genomic library in, 434
Charon 4A bacteriophage
human genomic library in, 359–362
screening, 360–363
Chemical cross-linking, of interferon receptors, 340–347
Chemotherapy, synergy between interferon, 656–657
Chick embryo cells
in interferon production, 115–117, 120–121
effect of calf serum, 121
effect of fetal bovine serum, 121–122
preparation, 115–116
yields, 116
standard growth medium, 116
Chick interferon, standards, 16
Chick proteins, as Hu-IFN-γ impurity, 62–63
Chick embryo muscle cells, inhibition of differentiation, 612
Chicken interferon
acid-stability, 124
cytopathic effect inhibition assay of, 124–125
vesicular stomatitis virus in, 124–125
induction, 115–125
with avian reovirus strain S1133(P40), 119
with defective virus particles, 119
double-stranded RNA in, 118–119
high titer, 115–125
with human reovirus, 119
ts mutants, 119
multiplicity of virus, 119–120
with Newcastle disease virus
AV strain, 119
California strain, 119
N.J.-LaSato strain, 119
at nonpermissive temperatures, 119
with poly(rI) · poly(rC), 119
ratio of interferon-inducing particles, 119–120
serum-free NCI medium in, 120
with Sinbis wildtype virus, 119
temperature, 120
with *ts* mutants of vesicular stomatitis virus, 119
with vesicular stomatitis virus [±]DI-011, 119, 120–125
with viral inducer, 118–119
production
kinetics, 120, 122

removal of viral inducer, 121–124
fetal bovine serum in, 122
perchloric acid in, 122
temperature, 123
standards, 16
MRC Research Standard A 62/4, 109, 112
storage, 124, 125
Chimeric clones, construction of, 386–395
Chinese hamster ovary cells
amplification of genes in, 398–400
cotransformation, 454, 458
E. coli gpt gene in, 454, 458
culture of, 458
DHFR⁻ cell line
cotransformation with *E. coli* gpt gene
in metabolic labeling of Hu-IFN-β, 400–403
in production of Hu-IFN-β, 397–403
transfection in, 398–400
Chloramine T
effect on Hu-IFN-β activity, 236
in radiolabeling, 263, 267–276, 286
of Hu-IFN-α, 267–276
limitations, 275
of Hu-IFN-αA, 286, 307
effect on activity, 307
of monoclonal antibodies, 583
p-Chloromercuribenzoate, effect on interferon activity, 237
Chloroquine, inhibition of IFN internalization, 336–337
Chromosome 21, human
cells trisomic for, 96, 551
GM2504, 97, 101, 599–600
GM2767, 97–102
sensitivity to interferon, 551, 597, 611
long arm in, 597
Chymotrypsin
effect on interferon, 234–235
sensitivity of IVR substance to, 730, 734
cI repressor, regulation of P_L promoter, 378
Cibacron Blue F3G-A (Blue Sepharose CL-6B)
in affinity chromatography, 545–546
Clathrin-coated pits, IFN-colloidal gold complexes in, 338
Clearance curves, of radiolabeled Hu-IFN-α2, 274
Cloned human cells
screening for sensitivity to interferons, 605–610
Cloning, effect of interferon on efficiency, 629–642
CM-Sephadex C-50, in IFN-γ purification, 59–61
CM-Sepharose CL-6B, in Hu-IFN-γ purification, 86
Colchicine
in cytoflourimetry, 644, 648
in measuring antiproliferative effects of interferon, 645–646, 648
Collagenase, in preparation of tumor cells, 638
Colloidal gold, *see* Gold colloid
Colony stimulating activity, 630, 632–633
Con A, *see* Concanavalin A
Con A-Sepharose, *see* Concanavalin A-Sepharose
Concanavalin A
induction of
equine IFN-γ, 134
Hu-IFN-γ, 52, 53, 58, 83, 88–92
preparation, 89
as mitogen, 58
as phytohemagglutinin substitute, 52, 53
in preparation of thymic blasts, 662
in purification of AVF, 742–743
purification factor, 742
sterilization, 89
Concanavalin A-agarose, in Hu-IFN-γ purification, 87
Concanavalin A-Sepharose
in affinity chromatography, 545–546
Mu-IFN-γ, 146–147, 149
purification of
Hu-IFN-β, 401–403
elution with ethylene glycol, 403
elution with α-methylmannoside, 401–403
Hu-IFN-γ, 90–91, 92
Hu-IFN-δ, 97, 99–102

Consensus interferons
beta interferons, 8–9
Hu-IFN-α, 6–7
Controlled pore glass
purification of
Hu-IFN-α, 43
Hu-IFN-γ, 62, 193–195, 459–460, 575
rat IFN-γ, 459–460
ethylene glycol in, 460
COOH-terminal peptides, HPLC reverse phase, 209
Copper chelate chromatography
purification of
interferon, 199
Hu-IFN-α2, 171, 172
yield, 172
preparation, 171
Cordycepin 5′-triphosphate
in degradation of mRNA, 668–669
inhibition of protein synthesis, 668
substrate for 2,5-A synthetase, 668
CPE, *see* Cytopathic effect
CPG, *see* Controlled pore glass
Creatine kinase, *see also* Creatine phosphotransferase isoenzyme, 621, 623, 627
Creatine phosphotransferase, *see also* Creatine kinase
as marker of myogenesis, 619–628
cellulose acetate electrophoresis in, 621
Cross hybridization, Bo-IFN-α with Hu-IFN-α DNA sequences, 467–468
Cross-linking reagents
disuccinimidyl suberate, 341
dithiobis(succinimidyl propionate), 341
Cross-reactivity of antiserum
to human interferon, 227
to mouse interferon, 227
to rat interferon, 227
CSM cells, *see* Caprine synovial membrane cells
Cucumber mosaic virus, replication inhibition by IVR substance, 730
Cucumis sativus L.
inoculation with cucumber mosaic virus, 733
Cyanoborohydride reduction, of oligoadenylic acid, 523
Cyanogen bromide
peptide fragmentation with, 205
in IFN-γ characterization, 197
Cyclic-AMP-dependent protein kinase, *see* Protein kinase
Cyclohexamide
in equine IFN-β induction, 132
in superinduction of Hu-IFN-β, 284, 400
Cyclooxygenase activity
in arachidonic acid cascade, 710
blocking by
aspirin, 711
indomethacin, 711
Cyclophosphamide, in xenograft tumor growth, 651, 655
D-Cycloserine, *in vivo* radiolabeling of interferon, 288
Cysteine residues
in human interferons, 6, 8, 10–11, 13, 249, 252, 404–407, 414
in murine interferons, 8, 10, 13, 249
substitution in Hu-IFN-αA, 404–407
Cytoagglutination
characterization of sarcolectin, 695–696, 700–701
inhibition assay, 696, 700–701
Cytochrome *P*-450
assay of, 721–724
aminopyrine N-demethylase in, 722
aniline hydroxylase in, 722–723
benzo[*a*]pyrene hydroxylase in, 723
7-ethoxy *O*-deethylase in, 724
ethylmorphine *N*-demethylase in, 722
effects of interferon on, 657
inhibition of, 715–718
Cytoflourimetry, 690–693
measurement of cell proliferation, 644–645, 648
colchicine in, 644, 648
Cytolysis
microassay for, 574–579
human K562 cells in, 574
murine lymphoma P388 cells in, 574–578

Cytopathic effect inhibition
in antiviral assay for interferons, 533–540
assay of
Bo-IFN-α, 138, 473
of chicken IFN, 124–125
heterologous interferons, 554–557
EMCV in, 554
VSV in, 554
Hu-IFN-α, 265, 278–279, 417, 580
human fibroblasts in, 265
MDBK cells in, 279, 291, 417
Hu-IFN-αA, 162, 250, 287, 291, 306–307, 554–558
MDBK cells in, 250, 287, 291, 307, 417, 427
VSV in, 291, 427
WISH cells in, 250, 307
Hu-IFN-αD, 427
MDBK cells in, 427
Hu-IFN-αWA, 364
VSV in, 364
WISH cells in, 364
Hu-IFN-α2, 175
Hu-IFN-β, 265, 372, 385, 554
EMC virus in, 372, 554
FS4 cells in, 372
human fibroblasts in, 265
WISH cells in, 385
Hu-IFN-γ, 65, 81, 94, 201, 265, 283, 299, 372, 417, 427, 554–558, 585
effect of radiophosphorylation, 299
FL amnion cells in, 94
sinbis virus, 94
T21 cells in, 372
WISH cells in, 265, 299, 585
IFN-γ, 380, 417, 427, 585
WISH cells in, 380, 585
radiolabeled interferon, 278–279, 283, 287
MDBK cells in, 279, 287
recombinant Hu-IFN-β, 178, 184
with human fibroblast AG-1732 cells, 178
with human fibroblast GM2504 cells, 184
with vesicular stomatitis virus, 178, 184
caprine synovial membrane cells in, 554–558
ovine choroid plexus cells in, 554–558
rat embryonic cells in, 221–222
vesicular stomatitis virus in, 206, 417, 427, 473, 534, 554
WISH cells in, 201, 206
Cytoskeleton, effect of interferon on, 541
Cytostasis
microassay for, 574–579
human K562 cells in, 574
murine lymphoma P388 cells in, 574–578
Cytotoxic lymphocytes, interferon protection against, 541
C-18 reverse phase column
in separating COOH-terminal peptides, 209

D

Dansylchloride, modification of Hu-IFN-α with, 236
Datura stramenium L., in assay for AVF, 736–737
Daudi cells, 306, 309, 341
in antigrowth assay, 579–581
binding of
Hu-IFN-αA to, 309
time course of, 310
Hu-IFN-α2 in, 348
extraction of IFN-α-receptor complex from, 347–349
in identification of IFN receptor, 341
Delta interferon, *see* Interferon delta
Denaturants, effect on interferon activity, 248–249
2′-Deoxycoformycin
effect on cell growth, 672–673
inhibitor of adenosine deaminase, 673
2-Deoxyglucose, in modification of interferons, 233
Dexamethasone, in enhancement of Hu-IFN-α synthesis, 234
Dextrans
carboxymethylation, 103
in poly ICL-CM dextran preparation, 103–104
molecular weights, 103

Diadenosine pentaphosphate, inhibitor of adenylate kinase, 624, 627
Diethylaminoethyl cellulose
in fractionation of (2′-5′)-oligoadenylate, 520
in purification of
AVF, 741–743
purification factor, 742
plant oligoadenylates, 753–756
P_1/eIF-2α protein kinase, 501–504, 508
RNase L, 495–496
sarcolectin, 698
Diethylaminoethyl-Sephadex
in oligoadenylate analogs separation, 523–525, 527
Diethylaminoethyl-Dextran
in equine IFN-α induction, 133
in equine IFN-β production, 132
Diethylaminoethyl-Sephacel, in Hu-IFN-γ purification, 90, 92
Diethylaminoethyl-Sephadex A20
in Hu-IFN-α2 purification, 171
yield, 172
Differentiation
effect of interferon, 611–618
of human skeletal muscle cells
effect of interferon on, 619–628
Digitonin, in extraction of IFN-α-receptor complexes, 347–351
Dihydrofolate reductase, 398
in amplification, 397
in cotransformation of CHO cells, 454
mutant gene for, 399
Dimers
of Hu-IFN-αA, 242, 245–246, 407
HPLC analysis of, 245–246, 248
of IFN-γ, 207, 586
of interferon
measurement by radioimmunoassay, 588–593
Dimethyl sulfoxide
enhancement
in Hu-IFN-α synthesis, 234
in Namalva cells, 37
Diphenylhydantoin
alteration of effects by interferon, 718
effect of interferon on metabolism of, 715–718
Discharging factor, activated by 2,5A_3, 761
Dissociation constant
binding of Hu-IFN-αA to suspension cells, 309
interferon receptor, 312, 314–315
of radiolabeled Hu-IFN-γ, 317–320
Disuccinimidyl suberate, as cross-linking reagent, 341
Disulfide bonds
in activity of interferons, 237–239, 404–408, 414
in cross-linking of oligomers, 593
in Hu-IFN-αA, 238–239, 242, 249, 254–255, 404–408, 414, 593
Disulfide reducing agents, effect on interferon activity, 248
5,5′-Dithiobis(2-nitrobenzoic acid), reaction with –SH groups, 236
Dithiobis(succinimidyl propionate), as cross-linking reagent, 341
DNA polymerase I, 366, 413
Klenow fragment, 413
DNA-dependent protein synthesis, in radiolabeling of interferons, 292–296
DNA-mediated gene transfer, 598
DNase
in preparation of tumor cells, 638
in sarcolectin characterization, 701
Dose-response curves, 109, 111, 113
determination of
interferon-inducing particles, 107, 113
quantum yield, 107, 112–113
generation, 107–112
addition of virus antiserum, 108
avian reovirus-S1133(P40) in, 109, 112
defective virus particles in, 108
heat-inactivated virus in, 108
interferon production time-course in, 108
mengovirus is-1 (*ifp*$^+$) in, 111, 112
Newcastle disease virus in, 108, 112–113
in nonpermissive cells, 108
primary chick embryo cells in, 109, 112
vesicular stomatitis virus mutant tsG41 in, 108
vesicular stomatitis virus [±]DI-011 in, 108

induction by viral dsRNA, 107
in interferon-inducing capacity determination, 114
for interferon induction, 106–114
Dot-blot assay, for TMV RNA, 739, 748–749
Double-stranded RNA, activation of oligoadenylate synthetase, 516
Drug metabolism, effect of interferon, 713–725

E

EBTr cells, *see also* Bovine EBTr cells
in Hu-IFN-γ assay, 101
Edman degradation, 206
Egg proteins, as Hu-IFN-γ impurity, 62–63
EGTA, inhibition of Hu-IFN-γ induction, 70, 71
Ehrlich ascites tumor (EAT) cells
RNase L in, 489–490
purification of RNase L, 493–494
EIA, *see* Enzyme immunoassay; Enzyme-linked immunosorbent assay
Electron microscopic visualization, of internalized Hu-IFN-αA complex, 337–339
ELISA, *see* Enzyme-linked immunosorbent assay
EMCV, *see* Encephalomyocarditis virus
Encephalomyocarditis virus
in antiviral assays, 534–540
heterologous interferon assay, 554
Hu-IFN-α2 assay, 175
IFN-β, 372
in characterization of sarcolectin, 697, 700–701
infection of mice, 708–709
storage, 535
Endoperoxide, in arachidonic acid cascade, 710
Endoproteases, inactivation of interferon, 234–235
Enzyme immunoassays (EIA)
of Hu-IFN-αA, 157
for interferon measurement, 14–15
relationship to antiviral activity, 14–15
Enzyme-linked immunosorbent assay (ELISA)
for AVF, 736–739
for TMV content, 760
in determining virus titers, 732
for TMV multiplication, 748
Enzymobeads, in radiolabeling of Hu-IFN-αA, 286
Epidermal growth factor, in measuring cell proliferation, 645–648
Epitope, for Hu-IFN-γ, 582
Epstein-Barr virus, infection of lymphocytes, 671–672
Equilibrium binding, 313, 318
of radiolabeled Hu-IFN-γ, 318–319
Equine alpha interferon, *see* Equine interferon alpha
Equine beta interferon, *see* Equine interferon beta
Equine dermal cells
in cytopathic effect inhibition assay, 132, 133
preparation of cultures, 130–131
Equine fibroblast interferon, *see* Equine interferon beta
Equine gamma interferon, *see* Equine interferon gamma
Equine immune interferon, *see* Equine interferon gamma
Equine interferon
antiviral activity
on bovine cells, 134–135
on feline cells, 135
on murine cells, 135
on ovine cells, 134–135
cross-species antiviral activities, 134–135
cytopathic effect inhibition assay, 132, 133
equine dermal cells in, 132, 133
induction, 130–136
production
isolation of mononuclear cells from equine blood, 131
yield, 131
Equine interferon alpha
cross-species antiviral activity, 134–135
induction, 133
DEAE-Dextran in, 133
with Newcastle disease virus, 133
acid stability, 133
with polyinosinic acid · polycytidylic acid, 133
acid stability, 133

stability, 133
storage, 133
Equine interferon beta
acid resistance, 133
cross-species antiviral activity, 134–135
induction, 132–133
with Newcastle disease virus, 133
with polyinosinic acid · polycytidylic acid, 132
priming, 132–133
production
with DEAE-Dextran, 132
superinduction, 132–133
actinomycin D in, 133
cyclohexamide in, 132
yields, 133
Equine interferon gamma
cross-species antiviral activity, 134–135
induction, 134
with concanavalin A, 134
mononuclear cells in, 134
with phytohemagglutinin, 134
loss of activity, 136
stability, 134
storage, 134, 136
Equine leukocyte interferon, *see* Equine interferon alpha
Equine rhinopneumonitis, 130
Erythrocytes
removal from leukocytes
with hydroxyethyl starch, 64, 88–89
lysis with ammonium chloride, 40, 56, 64, 65, 68, 78, 88, 137
Erythrogenesis, inhibition by interferon, 612
Escherichia coli
acid-treated
in Hu-IFN-αA production, 154
cell breakage, 170
with lysozyme, 170
cell extracts, 440
preparation of, 440, 449–450
endotoxin
as recombinant Hu-IFN-β impurity, 177
in expression of
Bo-IFN-α, 471–473
trp leader peptide in, 471
Hu-IFN-αA, 154, 410
Hu-IFN-α2, 168
fermentation, 168
yield, 168
human interferons in, 366–375, 636
phage lambda P_L promoter in, 366–375
IFN-γ in, 204, 380–383, 636
pRC23 in, 380–382
Mu-IFN-αA in, 437–441
recombinant IFN-β in, 178, 183, 184
frozen cell paste, 178
in production of
human interferon, 376–383, 636
lambda P_L promoter in, 376
IFN-γ, 204–205
Mu-IFN-αA, 434–441
rat IFN-α, 449
in radiolabeling of Hu-IFN-αA, 288
use of D-cycloserine in, 288
storage of, 205
strain C600, 368, 434–435
as host to bacteriophage lambda, 434–435
strain HB101
in expression of Bo-IFN-αC, 473
trp promoter in, 473
strain JA 221
in production of rat IFN-α, 449
strain JM103, 364
host for M13 bacteriophage, 364
strain K12ΔH1Δtrp, 374
strain LE392, 468
in screening of genomic library, 468
strain M5219
in production of rat IFN-α, 449
strain RR1, 380, 383, 434
expression of IFN-γ in, 380, 383
pRK248cIts in, 380, 383
recipient in transformation, 434, 440
Escherichia coli gpt gene, in cotransformation of CHO cells, 454, 458
7-Ethoxycoumarin *O*-deethylase, in assay for cytochrome *P*-450, 724
7-Ethoxycoumarin, 720
Ethylene glycol
in affinity chromatography, 545–547
for elution from concanavalin A-Sepharose, 194–195, 401–403
elution of
Bo-IFN-α, 139, 474
Bo-IFN-β, 214, 215

Hu-IFN-γ, 90, 207
Hu-IFN-δ, 99
Mu-IFN-γ, 146
in purification of
Hu-IFN-α, 43
IFN-γ, 194, 460
radiolabeled Hu-IFN-β, 285, 401–403
rat IFN-γ, 460
recombinant Hu-IFN-β, 178
removal from Hu-IFN-β, 180
sarcolectin, 698
Ethylmorphine, in measurement of drug metabolism, 719, 722
Ethylmorphine *N*-demethylase, in assay for cytochrome *P*-450, 722
Expression of interferons, 364, 370–375, 383–403, 414, 416–441, 449, 458–459, 471–473
Bo-IFN-α, 471–473
effect of medium on, 380, 382
Hu-IFN-α, 364, 403–415, 420–421, 428–433
Hu-IFN-β, 370–375, 383–396, 397–403
Hu-IFN-γ, 370–375, 380–383, 420–421
Mu-IFN-α, 434–441
Ra-IFN-α1, 441–453
Ra-IFN-γ, 453–464
Expression plasmids
construction of, 368–370, 376–379, 418–420, 424–441, 445–449, 453–458, 471–473
for Hu-IFN-αA, 410
for Hu-IFN-β, 369
for Hu-IFN-γ, 369, 371, 374, 379–381
for Mu-IFN-α, 437–439
Extinction coefficients, for Hu-IFN-α, 268–269
Extracellular matrix, effect of interferon on, 541

F

Fast migrating monomer (FMM), of Hu-IFN-αA, 242, 244
Fast protein liquid chromatography
on mono Q column, 147–149
in Mu-IFN-γ purification, 147–149
rat IFN purification by, 223–225
Fatty acids
short-chain
stimulation of interferon production, 37
Feline cells, antiviral activity of IFN-αA on, 18–19
Fetal membranes, constitutively produced interferon, 544–545
α-Fetoprotein, effect of oligoadenylate on action of, 677
Fibroblast cells
induction of, 284
in radiolabeling of Hu-IFN-β, 284
Fibroblast interferon, *see* Interferon beta
Fibroblast-derived growth factor, in measuring cell proliferation, 645–646, 648
Fibroblasts, for IFN titrations, 28–29, 56, 94, 97, 101, 175, 178, 184, 372, 534, 542, 658–662, 692
Fisher 3T3 rat cells, in characterization of sarcolectin, 695–696
FL cells, in cytopathic effect inhibition assay, 94
Fluorescamine
effect on
Mu-IFN-α, 236
Mu-IFN-β, 236
protein analysis by, 206
Fluorescein isothiocyanate, conjugated with goat anti-mouse antibody, 690
Fluorescence activated cell sorter, in measuring DNA content, 645
Foot-and-mouth disease virus (FMDV), in induction of Bo-IFN-β, 212–213, 220
FPLC, *see* Fast protein liquid chromatography
Freeze-drying, effect on rat IFN-γ, 463
FS4 cells, in assay of IFN-β, 372
Fused bovine interferon alpha C
properties, 473–474
binding to Affi-Gel Blue, 474
ethylene glycol in elution of, 474
Fusion index, in muscle cell cultures, 622

G

Gamma interferon, *see* Human interferon gamma; Interferon gamma
Gamma interferon, *see* Interferon gamma

Gel filtration
in purification of
IFN-γ, 197
Ultrogel AcA 54 in, 194, 195, 197
rat IFN, 222–224
Ultrogel AcA 44 in, 222–224
Gene transfer, DNA-mediated, 598
D-Glucosamine
in modification of interferons, 233
in radiolabeling of Hu-IFN-β, 401
Glucose-6-phosphate dehydrogenase, in measurement of creatine kinase activity, 624
Glutaraldehyde, in preparation of Hu-IFN-β/immunoglobulin conjugate, 240
Glutathione transferases, effect of interferons on, 657
Glyceraldehyde-3-phosphate dehydrogenase gene, promoter for, 418
Glycerol, effect on stability of rat IFN-γ, 462
Glycoproteins, 235
effect of glycosidases on, 235
tissue lectins, 702
molecular weight, 702
Glycosidases
effect on
Hu-IFN-β, 235
Hu-IFN-γ, 235
neuraminidase, 235
in modification of interferons, 235
GM2504 cells, in Hu-IFN-δ assay, 97, 101
GM2767 cells
in assay for
Hu-IFN-γ, 101
Hu-IFN-δ, 101
in interferon assays, 102
Gold colloid
preparation, 333–334
storage, 334
Gold colloid-interferon conjugates, 333–335, 337–339
antiviral activity of, 334
intracellular localization of, 339
preparation, 333–335
coupling to IFN-αA, 334
Goldberg-Hogness box, 445
Gomphrena globosa, in Potato virus X assay, 733
Granulopoiesis, inhibition of by interferon, 612
Growth factors
antagonistic effect of interferons, 657–666
effect of interferon on, 643–649
Guanidine hydrochloride, 179
in Bo-IFN-β production, 212
effect on activity of rat IFN-α, 453
in purification of
Hu-IFN-αA, 153, 154
Hu-IFN-α2, 169
IFN-β, 179, 189
in IFN-γ extraction, 206, 207, 209
in restoration of Hu-IFN-αA activity, 251
Guinea pig cells, antiviral activity of Hu-IFN-αA on, 18–19
Guinea pig embryo cells
aging, 127
culture preparation, 126–127
interferon induction, 127–128
interferon-producing capacity, 126
Guinea pig fibroblasts, in interferon induction, 125–129
Guinea pig interferon
assay, 128–129
harvesting, 128
induction
Newcastle disease virus in, 128
in serum-free medium, 128
Sindbis virus in, 128
in Vero cells, 128
vesicular stomatitis virus in, 128
plaque reduction assay, 128–129
with passaged guinea pig embryo cells, 128
production, 125–129
virus inactivation, 129
irradiation, 129
stability at pH 2, 129
storage, 128

H

Hairy cell leukemia, T cell line Mo derived from, 52
Half-leaf test, for AVF, 737
Halomethyl ketone derivatives, effects on interferon activities, 236

HeLa cells
  in anti-cell fusion assay of interferon, 705
  in assay for
    Bo-IFN-α, 141
    Hu-IFN-γ, 101
    Hu-IFN-δ, 97, 101
  inhibition of protein synthesis in, 668
Hemagglutinating units, for EMCV yield, 697, 700–701
Hemoglobin
  contaminating Hu-IFN-α, 43
  removal from Hu-IFN-α, 43
Hemopexin, as Hu-IFN-γ impurity, 62
HEp-2 cells
  in assay of
    human interferon, 28–29, 56, 97, 101, 175, 534
    Hu-IFN-α2 assay, 175
    Hu-IFN-γ assay, 101
    Hu-IFN-δ assay, 97, 101
Heparin
  as anticoagulant in Hu-IFN-γ production, 50
Heparin-Sepharose
  in IFN-γ purification, 194, 195
Hepatic cytochrome *P*-450 system
  depression by IFN-inducers, 719
Hetastarch, *see* Hydroxyethyl starch
Heteroduplex DNA, 404, 409–411
  formation of, 409–411
Heterologous cells, in interferon assay, 552, 554–558
Hexokinase, in measurement of creatine kinase activity, 624
Hexylamine-agarose
  in purification of P_1/eIF-2α protein kinase, 501, 510–513
HFF cells
  in assay of
    Hu-IFN-γ, 101
    Hu-IFN-δ, 97, 101
High performance liquid chromatography
  in analysis of
    Hu-IFN-αA, 242–248, 252
    Hu-IFN-αA/D, 243, 245–247
    Hu-IFN-αD, 243, 245–246
    Hu-IFN-γ, 243, 245–247
    synthetic peptides, 244–246
  in purification of
    AVF, 742–743
      purification factor, 742
    oligodeazaadenylates, 527–528
    plant oligoadenylates, 757
High pressure liquid chromatography, *see* high performance liquid chromatography
Histiocytic lymphoma, differentiation by interferon, 612
Histocompatibility antigens
  increase of on cell surface
    HLA-A,B,C, 682
  interferon induced changes, 688–693
HL-60 cells, 306, 309, 311
  binding of Hu-IFN-αA to, 309
  differentiation of interferon receptors, 311
  in Hu-IFN-γ production, 65, 66–68
Horse cells, antiviral activity of Hu-IFN-αA on, 18–19
HPLC, *see* High performance liquid chromatography
HT1080 tumor cells, in cytopathic effect inhibition assay, 81
HTLV-I, production in Hut 102-B2 cells, 54
Hu-IFN-α, *see* Human interferon alpha
Hu-IFN-β, *see* Human interferon beta
Hu-IFN-γ, *see* Human interferon gamma
Hu-IFN-δ, *see* Human interferon delta
Human A204 cells, in harvesting baboon endogenous virus
Human amnion CCL25 cells, *see* WISH cells
Human amniotic fluid
  interferon in, 541–551
    detection of, 542–543
Human bone marrow progenitor cells, effect of interferon on, 631–632
Human breast carcinoma
  cell line MCF-7
    effect of IFN on cell surface antigens, 682, 684, 687–688
      preparation of monoclonal antibodies, 683
  cell line MDA
    in measurement of murine natural killer cell activity, 653

xenograft
effect of interferons on, 652–654
Human cells, *see also* specific cells
antiviral activity of interferon, 18–19, 102, 141, 178, 184, 265, 372, 534, 542, 554
sensitivity to interferon
chromosome 21 in, 551
Human colon carcinoma
cell line WiDr
effect of IFN on cell surface antigens, 682, 684, 687–688
Human delta interferon, *see* Human interferon delta
Human embryo lung fibroblasts CCL23, *see* Hep2c cells
Human embryonic diploid fibroblasts
in assay of growth factors and IFN interactions, 658–662
enhancement of replication, 662
Human fibroblasts
AG-1732 cells
in recombinant Hu-IFN-β assay, 178
foreskin
in interferon assays, 102
FS-7 cells
sensitivity to interferons
comparison to caprine cells, 554–557
comparison to ovine cells, 554–557
FSII cells
in Bo-IFN-α assay, 141
F700 cells
interferon assays with, 542
GM731 cells
inhibition of protein synthesis in, 669–671
GM2504 cells
in recombinant Hu-IFN-β assay, 184
in transformation of NIH3T3 cells, 599–600
GM 2767 cells
in antiviral assay, 555
SG-181 cell line
in antiviral assay for interferon, 534
trisomic chromosome 21, 372, 599
in antiviral assay for IFN-γ, 372
Human fibroblast interferon, *see* Human interferon beta
Human genomic library
in Charon 4A bacteriophage, 359–362
screening, 360–363
Human hybrid interferons, 241
Human IFN-α, *see* Human interferon alpha
Human IFN-β, *see* Human interferon beta
Human IFN-γ, *see* Human interferon gamma
Human IFN-δ, *see* Human interferon delta
Human immune interferon, *see* Human interferon gamma
Human interferon
antiviral activity
on caprine synovial membrane cells, 134
on ovine choroid plexus cells, 134
degradation of, 315
effect on tobacco protoplasts, 744–752
induction, 25–106
in plasma, 539
Human interferon alpha, *see also* specific species
activity
effect of
^{125}I-labeling, 236
periodate, 237–238
pH, 268
radiolabeling, 278–279
SDS, 237–238
urea, 237–238
amino acid composition, 11
amino acid sequence, 6–7, 13
mature protein, 6
signal peptides, 6
antitumor effects in animal model, 651–656
mechanisms, 652–654
antiviral activity, 29, 543
on bovine cells, 134
plants, tobacco, 12
antiviral assay for, 66–67, 250, 265
binding to MDBK cells, 314
optimum pH, 314
binding to WISH cells, 314

carbohydrate residues on, 235
changes in antigens induced by, 688–693
composition, 27
concentration, 28, 41
with acid ethanol, 28, 29, 42
with potassium thiocyanate, 28, 29, 41, 42, 45
conjugation to albumin, 239
with *N*-succinimidyl-3-(2-pyridylthio)propionate, 239
consensus sequence, 6–7, 13
cysteine residues in, 6–7, 13, 249, 252
effect on
tobacco protoplasts, 744–752
virus induced cell fusion, 702–706
expression in *E. coli*, 364, 376, 403–415, 420–421, 428–433
extinction coefficient of, 268–269
gene for
cross hybridization with Bo-IFN-α genes, 467–468
hybridization with, 436
harvesting, 37
hemoglobin contaminant, 43
removal, 43
homology
to Bo-IFN-αC, 469
to Mo-IFN-α, 13
to Mu-IFN-α, 13
to Ra-IFN-α, 13
induction
Newcastle disease virus in, 27
priming with interferon alpha, 41
by Sendai virus, 27, 41, 101
isoelectric point, 47
labile species, 44
large-scale production, 35–38, 39–47
modification
[^{3}H]dansylchloride in, 236
2-deoxyglucose in, 233
2-hydroxy-5-nitrobenzyl bromide in, 236
D-glucosamine in, 233
tunicamycin in, 233
molecular weight, 3, 11, 43, 258–259
from Namalwa cells, 651
oligomers of, 259
periodate treatment, 237
p*I*, 47
production
for clinical use, 35
enhancement by
5-azacytidine, 234
5-bromodeoxyuridine, 234
butyrate, 234
dexamethasone, 234
dimethyl sulfoxide, 37, 234
tetradecanylphorbol acetate, 234
in Hut 102-B2 cells, 53
from leukocytes, 39–47
from lymphoblastoid cells, 35–38
stimulators, 37
yields, 38, 44, 46
production scale in Namalva cell system, 38
pseudogene, 6–7
purification, 27–35, 38
controlled pore glass, 43–44
by ethanol fractionation, 28
ethylene glycol in, 43
of Hu-IFN-αA, 43
by immunoaffinity chromatography, 28–33, 43–44, 45–47
yields, 46
with monoclonal antibodies, 28–33, 43
LI-8, 155–160
NK2, 28, 30–33
YOK, 30–33
preparation of P-IF, 41–43, 45
yields, 45, 47
zinc chelate chromatography in, 199
purification fractions
antiviral activity, 28–29
comparisons from different sources, 33, 34
P-IF, 28, 41
P-IF I, 33, 34
P-IFA, 28, 41, 43
P-IFB, 28
radiolabeling, 266, 269–273, 276–281
Bolton-Hunter reagent in, 276–281
effect on antiviral activity, 278–279
chloramine T in, 267–276
characterization of, 269
limitations of, 275

with ^{125}I, 263–267, 276–281
effect on antiviral activity of, 278–279
lactoperoxidase in, 263–267
specific radioactivity of, 266, 275, 281
receptor mediated binding, 332–339
recombinant
purification, 153–165
large-scale, 153–165
reduction of, 248–255
relative antiviral activity, 18–19
reoxidation of, 248–255
signal peptides, 6–7
specific activity, 45, 46, 651
stability, 44
standards, 16
G-023-901-527, 94, 138, 543
human interferon alpha (leukocyte/Sendai), 16
human interferon alpha (Namalva/Sendai), 16
recombinant Hu-IFN-αA, 16
recombinant Hu-IFN-αD, 16
recombinant Hu-IFN-α2, 16
sterilization, 43, 44
subspecies γ3 in assay with tobacco protoplasts, 746
target molecular weight of, 260
truncated, 235
ultrafiltration, 41
Human interferon alpha A (Hu-IFN-αA)
acidification of
effect on α-helix, 239
active fragment, 255
activity
restoration by SDS, 251
amino acid sequence, 6
analysis
high-performance liquid chromatography in, 242–248, 252
immunoblot in, 422–423
SDS–polyacrylamide gel electrophoresis in, 287, 290, 422
antitumor effects in animal model, 651–656
mechanisms, 652–654
antiviral activity
effect of disulfide bonds on, 405, 407
attachment to secretion signal, 430
base substitution between Hu-IFN-α2, 481–485
octadecyl deoxyoligonucleotide in detection of, 481–485
binding to
cells, 597
time course, 310
MDBK cells, 314
optimum pH, 314
suspension cells, 307–309
dissociation constant, 309
WISH cells, 314
characterization, 251
by SDS–PAGE, 153, 157–159, 164–165
coupling to gold colloid, 334
cysteine residues in, 159, 404–405, 407
cytopathic effect inhibition assay, 162, 407, 554–557
caprine synovial membrane cells in, 554–557
EMCV in, 554–557
ovine choroid plexus cells in, 554–557
with MDBK cells, 162, 407
with vesicular stomatitis virus, 162, 407, 554
deletions with synthetic DNA, 411–414
dimers, 157, 158, 242, 245–246, 407
disulfide bonds in, 238–239, 404
effect on
activity, 242, 249, 252, 254–255, 404
cyanogen bromide cleavage on, 238–239
monomer, 238–239
oligomers, 238–239
effect on
binding of monoclonal antibody B6.2, 687
human breast cancer xenograft, 652, 654, 656
human breast carcinoma, 682, 684, 687–688
human colon carcinoma, 682, 684, 687–688
tumor cells, 640–641
virus induced cell fusion, 702–706

fast migrating monomer of, 242, 244
gene fragments
replacement with synthetic DNA, 406
gene isolation, 166
gold colloid, coupling to, 334
higher oligomers, 157, 158
intermolecular disulfide bond scrambling, 153, 159, 404–407
internalization by MDBK cells, 336–339
intracellular reduced molecule, 159
irradiation of, 259
modification of carboxy terminus, 409
through site-specific mutagenesis, 409
molecular weight, 157, 158, 289, 307
monomers of, 153, 157–158, 242, 407
oligomers of, 153, 159, 242, 245
HPLC analysis of, 245, 248
oxidized, antiviral activity, 251
in plaque reduction assay, 609
preparation of expression vectors
from human leukocyte mRNA, 167
removal of signal peptide coding region, 167
production
in *Escherichia coli,* 153, 407, 636
in yeast cells, 416–423, 426–433
incorrect processing
purification, 153–165, 288–289, 290, 306–307, 427
acid-treated *Escherichia coli* in, 154
guanidine hydrochloride in, 153, 154
by immunoaffinity chromatography, 155–160, 288, 306, 407
coupling of antibody to polyhydroxyphase silica, 160
immobilized anti-IFN-α monoclonal antibody, 407
longevity of immobilized antibodies, 160
with monoclonal antibody LI-8, 155–160
purification factor, 157
recovery, 157
Sepharose 4B in, 159
by ion-exchange chromatography, 157, 160–161
with carboxymethyl (CM)-52 cellulose, 160
precipitation of polymeric forms of rIFN-αA, 161
protein loss, 157, 161
removal of cleavage enzyme, 161
large-scale, 153–165
monoclonal antibodies in, 153, 155–160, 288–289, 306–307, 407
purification factor, 157
recovery, 157, 163
removal of Triton X-100, 161
by SDS–PAGE, 157–159, 407–408, 163–165
under nonreducing conditions, 157
by Sephadex G-50 chromatography, 157, 161–163
monomers, 157, 161, 162–165
oligomers, 161, 162–165
summary of steps, 157
from unfractioned natural Hu-IFN-α, 43
radiolabeled
analysis by SDS–PAGE, 287
purification, 288–291
yield, 291
with monoclonal antibody affinity column, 288–289, 306
specific binding to WA17 cells, 598
specific radioactivity, 289, 291, 307
storage, 281
radiolabeling, 276–281, 286–291, 306–307, 327
Bolton-Hunter reagent in, 276–281, 286, 306, 312–313, 327, 333
effect on antiviral activity, 278–279
chloramine T in, 286, 307
effect on activity, 307
enzymobeads in, 286
lactoperoxidase in, 307
in maxicells, 288
Escherichia coli in, 288
reaction with iodoacetate
antiviral activity of, 254
SDS–PAGE analysis of, 254
receptor bound, 335–339
degradation, 335
kinetics, 335

internalization, 335–339
kinetics, 335
receptor mediated endocytosis, 332–333
reduced
antiviral activity, 251
regenerated, 252
antiviral activity of, 253
guanidine hydrochloride in, 251
HPLC analysis of, 253
SDS–PAGE analysis of, 253
relative antiviral activity, 18–19
schematic representation of, 405
in screening of IFN-sensitive cells, 601–603, 605, 609–611
secretion from yeast cells, 424–433
signal sequence, 424
slow and fast moving monomers, 157, 158, 242, 244, 407
specific activity, 157, 427, 637, 684
storage, 156, 162
prevention of proteolytic degradation, 156
structure
relation to activity, 404
substrate for protein kinase, 299–301
sulfitolysis of, 252–253
tetramers of, 242
tetrasulfonate form, 252, 253
characterization of, 253
tobacco protoplast assay for, 746, 751
in treatment of tobacco cells, 753–755, 757
trimers of, 157–158, 242, 245–246
truncated, 411–414
antiviral effects, 414
by thermolysin, 235
Human interferon alpha A-colloidal gold complex
in clathrin coated pits, 338–339
internalization
electron micrographs, 337–339
in receptosomes, 338–339
Human interferon alpha A/D (Hu-IFN-αA/D)
in depression of hepatic *P*-450 system, 719
HPLC analysis, 243, 245
in inhibition of melanogenesis, 613
12-*O*-tetradecanoylphorbol-13-acetate in, 613
purification, 165
in treatment of mouse L929 cells, 502
Human interferon alpha C (Hu-IFN-αC), purification, 165
amino acid sequence, 6
Human interferon alpha D (Hu-IFN-αD)
amino acid sequence, 6
antitumor effects in animal model, 651–656
mechanisms, 652–654
binding to cells, 314
optimum pH, 314
effect on human breast cancer xenograft, 652, 654, 656
HPLC analysis, 243, 245–246
lack of binding to monoclonal antibody NK2, 30
pH 2 stability, 33
production
in *E. coli,* 636
in yeast cells, 421, 424–433
incorrect processing, 430
purification, 165, 427
radiolabeling, 281, 312–313
Bolton-Hunter reagent in, 312–313
secretion from yeast cells, 424–433
signal sequence, 424
specific activity, 427, 637, 656
tobacco protoplast assay with, 746
Human interferon alpha F (Hu-IFN-αF)
amino acid sequence, 6
lack of binding to monoclonal antibody NK2, 30
Human interferon alpha I (Hu-IFN-αI)
amino acid sequence, 6
purification, 165
Human interferon alpha J (Hu-IFN-αJ)
amino acid sequence, 6
cross hybridization of gene with Bo-IFN-α gene, 467–468
purification, 165
Human interferon alpha K (Hu-IFN-αK)
amino acid sequence, 6
purification, 165
Human interferon alpha L (Hu-IFN-αL)
sequence homology, 6, 364
Human interferon alpha WA (Hu-IFN-αWA)
amino acid sequence, 365

expression in *E. coli,* 364
M13mp11 lac-fusion expression, 364
gene sequence homology, 364
Human interferon alpha 1 (Hu-IFN-α1)
in antigrowth assay, 581
monoclonal antibody binding to, 326
neutralization of, 567
antibody dilution in, 567
monoclonal antibody in, 567
in preparation of monoclonal antibodies, 326
Human interferon alpha 2 (Hu-IFN-α2), 327, 329–331
base substitution between Hu-IFN-αA gene, 481–485
octadecyl deoxyoligonucleotide in detection of, 481–485
binding to MDBK cells, 314–315
equilibrium binding, 314
optimum pH, 314
time course of, 314
binding to WISH cells, 314
charge heterogeneity, 273
correctly refolded, 177
separation of, 177
cross reactivity with rat IFN, 227
cytopathic effect inhibition assay with, 175
with encephalomyocarditis virus, 175
with HEp2 cells, 175
degradation
inhibition by monensin, 336–337
expression vector for, 167–168
in *Escherichia coli,* 168
yield, 168
fermentation, 168
from human leukocyte mRNA, 167
removal of signal peptide coding region, 167
p*l* promoter in, 167, 168
ribosome binding site in, 167, 168
extraction from cells, 170
formation in *E. coli,* 168–169
purification, 168–169
gene isolation, 166
identification
chemical cross-linking in, 342
internalization, 330
inhibition by chloroquine, 336–337
intracellular insoluble aggregate, 168–169, 176
formation in *E. coli,* 168–169
use in purification, 168–169
isoelectric focussing, 176
isoelectric point, 175, 176, 273–274
microcrystal formation, 171, 172
yield, 172
molecular weight, 171, 173, 174
monoclonal antibody binding to, 326, 329
partitioning into inclusion bodies, 170, 176
purification, 165–177
ammonium sulfate in, 171
yield, 172
by column chromatography, 170–173
with SP Sephadex A-50, 171, 172
copper chelate chromatography in, 171, 172
yield, 172
DEAE-Sephadex A20 in, 171
yield, 172
guanidine HCl in, 169
by native polyacrylamide gel electrophoresis, 173–174
SDS–polyacrylamide gel electrophoresis in, 170–171, 173
Sephacryl S-200 in, 171, 172
yield, 172
purity, determined by SDS–PAGE, 173
radiolabeled, 265, 267, 273, 275–281, 327, 330
binding affinity of, 274
binding to Daudi cells, 348–349
binding to receptor, 342–343, 345–347
competition, 342
displacement, 343, 345
internalization, 330
molecular weight, 279
separation of, 271–272
yield, 272
storage of, 265, 267, 275, 281
radiolabeling of, 264–266, 276–281, 327
Bolton-Hunter reagent in, 276–281, 312–313, 327, 333
yield, 278–279
lactoperoxidase in, 264

receptor bound, 326–331
reverse phase HPLC of, 174–175
specific activity, 172, 175
Human interferon alpha 6L (Hu-IFN-α6L), in antigrowth assay, 581
Human interferon alpha 54 (Hu-IFN-α54), in antigrowth assay, 581
Human interferon alpha 61A (Hu-IFN-α61A), in antigrowth assay, 581
Human interferon alpha 76 (Hu-IFN-α76), in antigrowth assay, 581
Human interferon alpha gene, *in vitro* mutagenesis of, 403–415
Human interferon alpha-receptor complex
chemical cross-linking in identification, 340
digitonin in extraction, 347–351
molecular weight, 349
separation, 349–350
stability, 349
Human interferon beta
activity
effect of
Bolton-Hunter reagent on, 236
chloramine T on, 236
^{125}I-labeling on, 236
periodate on, 237–238
SDS on, 238
urea on, 237–238
amino acid composition, 11
amino acid sequence, 8–9, 181, 191
mature protein, 8
signal peptide, 8
antiviral assay for, 66–67, 265
human fibroblasts in, 265
plants, tobacco, 12
binding to receptor, 343
changes in antigens induced by, 688–693
concanavalin A-Sepharose chromatography of, 401–403
conjugation with immunoglobulin, 240
glutaraldehyde in preparation of, 240
effect of glycosidases on, 235
effect on virus induced cell fusion, 702–706
expression plasmid for, 369–370, 374
glycosylated, 397
expression in CHO cells, 397
modification, 397–403
secretion from CHO cells, 403
induction, 400
CHO-cells in, 400
modification
2-deoxyglucose in, 233
D-glucosamine in, 233
in heterologous cells, 397–403
tunicamycin in, 233–234
molecular weight of, 258–259, 403
oligomers of, 259
in plaque reduction assay, 608
production
CHO cells in, 400
in heterologous cells, 307–403
QT35 cells in, 385, 388, 392, 395
Rous sarcoma virus in, 383–396
protein kinase
poor substrate for, 299–301
purification, 177–183, 183–192, 204–210, 285
radiolabeled, 284–286
characterization, 285
DL-[^{14}C]glucosamine in, 401
in vivo, 284–286
human diploid fibroblast cells in, 284
DL-[^{3}H]mannose in, 401
[^{35}S]methionine in, 284, 400–401
purification, 285
Blue Sepharose in, 285
effect on activity, 285
ethylene glycol in, 285
specific radioactivity of, 285
recombinant
amino acid composition, 11, 190
characterization, 180–182, 188–192
HPLC, 187–189
concentration, 180
cytopathic effect inhibition assay, 178
with human fibroblast AG-1732 cells, 178
with human fibroblast GM2504 cells, 184
with vesicular stomatitis virus, 178
disulfide bonds in, 181, 192
fragments, 186
hydrophobicity, 181, 189

methionine, amino terminal, 181, 190
molecular weight, 180, 188
oligomers, 192
production
in *E. coli,* 177–183, 183–192
purification, 177–183, 183–192, 204–210
acid precipitation in, 185
aggregates in, 184, 188, 189
Blue Sepharose chromatography in, 179
cell disruption, 178–179
ethylene glycol in, 178
from *E. coli* endotoxin, 177
from frozen live *E. coli* cell paste, 178
guanidine hydrochloride in, 179, 183
Manton-Gaulin homogenizer in, 179, 184
organic extraction in, 185
with 2-butanol, 185
precipitation of interferon, 178–179
SDS–polyacrylamide gel electrophoresis in, 178, 188
silver staining in, 178, 182
Sephacryl S-200 chromatography, 185, 186
Sephadex G-75 chromatography, 185–187
sonication in, 184
specific activity, 179, 181, 184, 185
ultrafiltration in, 179, 180
yields, 181, 185
SDS–polyacrylamide gel electrophoresis, 182, 188
sensitivity of caprine synovial membrane cells to, 552, 555–558
comparison to human FS-7 cells, 554–557
sensitivity of ovine choroid plexus cells to, 552, 555–558
comparison to human FS-7 cells, 554–557
storage, 179, 180, 184
in screening of IFN-sensitive cells, 601–603, 605, 608–611
[Ser^{17}]IFN-β
characterization, 183–192
HPLC of, 187–189
properties, 183–192
purification, 183–192
selective extraction, 185, 189
standards, 16
superinduction of, 284
actinomycin D in, 284, 400
cyclohexamide in, 284, 400
poly(I) · poly(C) in, 284, 400
target molecular weight of, 260
truncated, 235
unglycosylated
secretion from CHO cells, 403
specific biological activity of, 403
Human interferon beta gene, 385–388
amplification
DHFR gene in, 403
methotrexate in, 403
positioning into viral genome for expression, 386–395
near splice acceptor site, 393–396
near viral ATG, 386–389
near viral LTR, 391–393
Human interferon beta receptor, chemical cross-linking
in identification, 340
Human interferon beta Ser, *see* Human interferon beta, [Ser^{17}]IFN-β
Human interferon delta
antigenic specificity, 102
antiviral activity, 101
on bovine cells, 97, 101, 102
in cells trisomic for chromosome, 21, 96
GM2504, 97, 101
on human cells, 97–102
on murine L 929 cells, 97, 101, 102
biological properties, 102
comparison to interferon gamma, 101
concentration
ammonium sulfate precipitation in, 99
induction
phytohemagglutin in, 97, 98
teleocidin in, 97–98
molecular weight, 102
pH 2 stability, 98–99

physicochemical properties, 102
preparation, 96–102
production
with interferon gamma, 98
properties of, 102
purification, 96–102
affinity chromatography
with concanavalin A-Sepharose, 97, 99, 100, 101, 102
with matrex blue, 97, 99–102
ammonium sulfate precipitation, 99
concanavalin A-Sepharose, 99–102
elution with ethylene glycol, 99
immunoaffinity chromatography
anti-IFN-γ-Sepharose, 97, 99–101
yields, 100, 101
separation from interferon gamma, 99–102
by pH 2 stability, 99–102
serological properties, 102
stability
heat, 102
pH, 102
SDS, 102
Human interferon gamma
amino acid composition, 7, 11, 209
amino acid sequence, 5–7, 10–11
from cDNA clone, 10
mature protein, 10
modified COOH terminus, 10
from protein, 10
signal peptide, 10
analysis
of amino acid sequence, 206
HPLC in, 243, 245
antiviral activity, 101, 299
on bovine cells, 101
comparison to interferon delta, 101
effect of radiophosphorylation, 299
on human cells, 101
on murine L 929 cells, 101
assay, 65–66, 67, 81, 94, 201, 265, 283, 299, 380
WISH cells in, 265, 380
binding to receptor, 343
displacement, 343
time course for, 319
U937 cells, 315–321
antigens induced by, 688–693
characterization, 197, 198
immunoabsorbent chromatography in, 197
western blot in, 197
concentration
adsorption to silicic acid, 90, 92, 200, 202
large-scale, 77–83
filtration, 78–79
polyethylene glycol-ammonium sulfate in, 55, 58–59
recovery, 59, 61
small-scale, 78
by ultrafiltration, 68, 78–79
yields, 60
crude
specific activity, 82
storage, 76
after concentration, 58–59, 78, 81
after induction, 57
cyanogen bromide digest of, 197
cytopathic effect inhibition assay, 65–66, 67, 81, 94, 201, 265, 283, 299, 380
FL amnion cells, 94
Sindbis virus in, 94
WISH cells in, 265, 299, 380
degradation of, 83, 209
during production, 83
dimers, 207, 586
molecular weight, 207
effect of glycosidases on, 235
effect on virus induced cell fusion, 702–706
epitope, 582
expression in *E. coli,* 376–383
lambda P_L promoter in, 376–383
in pRC23, 380–382
expression plasmid for, 374, 381
construction of, 381
extraction from *E. coli,* 206, 207
guanidine hydrochloride in, 206, 207, 209
filtration, 61, 63
fragmentation of, 205
cyanogen bromide in, 205
harvesting, 90
impurities in, 62, 87
chick proteins, 62–63

egg proteins, 62–63
hemopexin, 62
inducers, 87
interleukin-1, 87
interleukin-2, 87
interleukin-3, 87
lens culinaris lectin, 62
lymphotoxins, 87
Sendai virus, 62–63
T cell growth factor (IL-2), 87
transferrin, 62
tumor necrosis factor, 63
induction, 57–58, 65, 78, 93–95
calcium flux, 69
calcium ionophore A-23187 in, 69–72, 72–76, 193
calcium ionophore ionomycin in, 70
concanavalin A in, 52, 53, 83, 88–92
human serum albumin in, 43
as serum substitute, 95
inhibition
by calcium channel blocker Verapamil, 70, 71
by calcium depletion EGTA, 70, 71
by EGTA, 70, 71
mezerein in, 72–76, 77, 193
monoclonal antibody OKT-3 in, 52
oxidation of galactose, 69
phorbol esters in, 48–54, 64, 66, 68
phytohemagglutinin in, 48–54, 64, 66, 68, 78, 83, 85, 98
plant lectins in, 52
pokeweed mitogen in, 52
in serum-free medium, 93
serum substitute, 95
staphylococcal enterotoxin A in, 83, 88–92, 575
staphylococcal enterotoxin B in, 70, 71–72, 83, 85, 93, 95
with streptococcal preparation OK-432, 93, 95
T cell mitogens in, 93–95
12-*O*-tetradecanoylphorbol-13-acetate in, 64, 66, 68
isoelectric point, 282
modification
tunicamycin in, 233–234
modified molecules, 298
molecular weight, 197, 258–259
monoclonal antibodies to, 582–587
assay with, 585–587
monomer
molecular weight, 207
native form, 83
neutralization, 53, 67, 95
by monoclonal antibodies, 584
by polyclonal antibody, 67
oligomers of, 259
phosphorylation of, 296–301, 316, 321
cAMP-dependent protein kinase in, 316, 321
polyclonal antibody to, 67
production, 54–63, 88–92
λ cI repressor in, 205
bacterial fermentation in, 204–205
calf serum in, 75–76
characterization of product, 66–67
E. coli in, 204–205, 376
from human leukocytes, 50–52, 63–69, 72–76
yield, 51
in Hut 102-B2 cells, 48–54
with interferon delta, 98
kinetics, 82, 94
large-scale, 77–83, 83–87
leukocyte culture conditions, 74
mitogen activation, 83
mixed lymphocyte reaction, 83
in Mo cells, 52
multiple harvest/reinduction scheme, 81
natural, 77–83
yields, 81
with OK-432 and staphylococcal enterotoxin B
kinetics, 94
synergistic effect, 95
from peripheral blood mononuclear cells, 93–95, 51
phytohemagglutinin in, 282
from plateletpheresis residues, 49, 50–52
preparation of
leukocytes for, 77–78
lymphocytes for, 77–78
nylon wool column, 78
mononuclear cells in, 51
staphylococcal enterotoxin in, 200
by T cell hybridomas, 49, 50–52

12-*O*-tetradecanoyl-phorbol-13-acetate in, 282
in yeast cells, 416–423, 424–433
yields, 66–67
purification, 59–63, 85–86, 88–92, 193–210, 427, 575
affinity chromatography, 84, 86
on blue Sepharose, 91, 92
on concanavalin A-Sepharose, 90–91, 92
monoclonal antibody in, 63, 84, 207
on poly(U)-Sepharose, 91, 92
automated, 198
carboxymethyl BioGel agarose in, 194, 195, 197
cation exchange chromatography, 196–197
CM-Sephadex C-50 in, 59–61
CM-Sepharose CL-6B in, 86
concanavalin A-agarose column in, 87
concanavalin A-Sepharose in, 194 195
α-methyl-D-mannopyranoside for elution from, 195, 196
ethylene glycol for elution, 194, 195
controlled pore glass in, 62, 84, 193–195
regeneration of, 194
copper chelate chromatography, 199
DEAE-Sephacel in, 90, 92
ethylene glycol in, 90
gel-filtration chromatography, 197
heparin-Sepharose in, 194, 195
immunoaffinity chromatography, 63, 84
ion exchange, 59–61, 86
metal chelate chromatography, 199
monoclonal antibody in, 63, 84, 207
nickel chelate chromatography, 199–204
effect of temperature, 202
partition chromatography, 83
recovery, 60, 61, 62
sequential chromatography, 193–199
silica in, 207
sonication in, 206–207
specific activity, 198
tetraethylammonium chloride in, 90
ultrafiltration, 85–86
Ultrogel AcA 54 in, 194, 200, 202
yields, 60, 61, 198
zinc chelate chromatography in, 86–87, 199
radiolabeled, 316–321
specific radioactivity of, 266, 283, 316
storage of, 265, 316
radiolabeling, 264–266, 281–283, 296–301, 315
Bolton-Hunter reagent in, 315–316
^{125}I in, 263–267, 281–283
effect on antiviral activity, 283
yield, 283
lactoperoxidase in, 263–267
^{32}P in, 316
protein kinase in, 296–301, 316
receptor for, 283, 301, 317
recovery, large-scale, 77–83
removal of impurities, 87
sandwich radioimmunoassay, 582–587, 588–589
principle, 582, 588–589
SDS–PAGE of, 197
signal sequence, 424
specific activity, 85–87, 210, 427
stability
heat, 87
low pH, 87
standards, 16
Gg 23-901-530, 16, 66, 299, 585
recombinant Hu-IFN-γ, 16
sterilization, 61, 63
storage, 62
target molecular weight of, 260
tetrameric functional unit, 84, 87, 260
yield from mononuclear cells, 50
Human interferon gamma receptor, chemical cross linking in identification, 340
Human K562 cells, in microassay for cytolysis, 574
Human leukemic cell lines, effect of interferon on, 632
Human leukemic myeloblasts, effect of interferon on, 633
Human leukocyte interferon, *see also* Human interferon alpha

composition, 27
definition, 19–20
large-scale production, 39–47
production
kinetics, 139
purification, 27–35
Human leukocytes, *see also* Leukocytes
in Hu-IFN-γ production, 54–58, 62–63, 72–76, 83, 85, 575
Human lymphoblastoid interferon, *see* Human interferon alpha
Human lymphoblastoid Namalwa interferon, 658–659, 661
see also Human interferon alpha
Human melanoma cells
cell line A375
effect of IFN on cell surface antigens, 683, 687–688
cell line BO-2, 616
cell line DU-2, 616
cell line FO-1, 616
cell line HO-1, 616
cultures, 616
differentiation
effect of interferon on, 611–618
secretion of melanin, 616
Human myeloid progenitor cells
effect of interferon on
cloning efficiency, 629–634
differentiation, 612
proliferative capacity, 629–634
Human serum albumin
in Hu-IFN-γ production, 93
serum substitute, 95
Human skeletal muscle cells
cell culture, 619–624
fusion efficiency, 622
differentiation
effect of interferon on, 619–628
Human T cell leukemia virus, see HTLV-I
Hut 102-B2, *see* T cell lines, Hut 102-B2
Hybrid interferons, 241
Hybridization
oligonucleotides, 363
plasmid DNA to octadecyl deoxyoligonucleotides, 483
Hydrated calcium phosphate
adsorption of TMV, 740–741
in preparation of plant oligoadenylate, 753–754
Hydrophobic chromatography
phenyl Sepharose, 214
Bo-IFN-β purification by, 213, 217–218
in purification of sarcolectin, 698
2-Hydroxy-5-nitrobenzyl bromide, modification of Hu-IFN-α with, 236
7-Hydroxycoumarin, 720
Hydroxyethyl starch, in human leukocyte isolation, 64–65, 88–89
Hydroxyphenylpropionic acids, N-hydroxysuccinimide ester
see Bolton-Hunter reagent
Hyporesponsiveness to interferon, 707–712

I

IFN-α, *see* Interferon alpha; *also* specific species
IFN-αA, *see* Human interferon alpha A
IFN-αD, *see* Human interferon alpha D
IFN-αF, *see* Human interferon alpha F
IFN-β, *see* Interferon beta; Human, Murine, etc. interferon beta
IFN-γ, *see* Interferon gamma; Human, Murine, etc. interferon gamma
IFN-δ, *see* Interferon delta
IF^r cells, 704
IL-1, *see* Interleukin 1
IL-2, *see* T cell growth factor (Interleukin 2)
IL-3, *see* Interleukin 3
Image-analyzer, for estimating colony growth, 639
Iminodiacetic acid
in nickel chelate column preparation, 200, 201, 203
activation of Sepharose, 201
Immune interferon, *see* Interferon gamma
Immune serum, *see* antiserum
Immunoadsorbent chromatography, *see* Immunoaffinity chromatography
Immunoaffinity chromatography
anti-IFN-γ-Sepharose, 99–101
in Hu-IFN-δ purification, 99–101
column preparation, 205
Affi-gel 10 in, 205
column reuse, 165
coupling of antibody to polyhydroxyphase silica, 160

in Hu-IFN-γ characterization, 197
longevity of immobilized antibodies, 160
monoclonal antibodies in, 63, 84, 155–160, 207, 288, 306, 407
of mouse IFN-α, 227–230
physical supports, 159
in purification of
Bo-IFN-α, 143
with anti-Hu-IFN-α, 143
Hu-IFN-α, 28–33, 43–44, 45–47
automation, 155, 156
efficiency, 47
interferon yields, 46
Hu-IFN-αA, 155–160
Hu-IFN-γ, 63, 84
Hu-IFN-δ
anti-IFN-γ-Sepharose in, 97, 99–101
of rat IFN-α, 227–230, 452
Immunoassay, 582–593, *see also* EIA; ELISA
Immunoblot analysis, 418–423
of IFN-αA, 422–423
Immunoglobulin
goat anti-mouse
in radioimmunoassay, 686–687
radiolabeling of, 685–687
storage, 686
In situ hybridization, 478–481
effect of paraformaldehyde, 480
preparation of cells, 477–478
specificity of, 479
Indolyl-3-acrylic acid, *in vivo* radiolabeling of interferon, 288
Indomethacin
irreversible inhibition of cyclooxygenase, 711
in measuring hyporesponsiveness to IFN, 711
Induction, 212–213
Bo-IFN-α, 136–144
Bo-IFN-β
with bluetongue virus, 212–213, 220
with Newcastle disease virus, 212–213, 220
with foot-and-mouth disease virus, 212–213, 220
chicken interferon, 115–125
equine IFN-α, 133
equine IFN-β, 132–133
equine IFN-γ, 134
guinea pig IFN, 128
Hu-IFN-α, 27, 35, 41, 101, 234
Hu-IFN-β, 284, 400
Hu-IFN-γ, 50, 63, 72, 77, 83, 94, 200, 282
Mu-IFN-α/β, 322, 477, 480
NDV in, 322
Mu-IFN-γ, 574
rat IFN-α, 442–443
Iniprol, protease inhibitor, 697
Insulin, in measuring antiproliferative effects of interferon, 645–646, 648
Interferon, *see also* specific interferons
abbreviations
amino acid additions, 20
amino acid deletions, 20
amino acid substitutions, 20
animal species of orign, 20
crude leukocyte interferon, 19–20
modified interferons, 20–21
prefixes, 20
suffixes, 20
action, biology of, 595–712
activity
effect of
p-chloromercuribenzoate, 237
α-chymotrypsin, 234–235
endoproteases, 234–235
performic acid, 237
essentiality of disulfide bonds in, 237–239
allografts delaying rejection of, 541
alpha, *see* Interferon alpha
in amniotic fluid, human, 541–551
antagonistic effect of growth factors, 657–666
anti-cell fusion activity, 704–705
antibody to, *see* Antiserum, to interferons
antiproliferative assays, 574–582, 643–649
antiserum to, *see* Antiserum, to interferons
antitumor effects
animal models in, 649–657
mechanisms, 652–654
antiviral assays, 533–573

assay, 66–67, 543–544
comparison of MDBK and WISH cells, 67
on F700 human fibroblasts, 542, 549
on HEp2 cells, 56
on L-929 cells, 66–67
on MDBK cells, 66–67, 417, 542, 549
on normal rat kidney cells, 542
plaque-reduction assay, 607–610
on rat embryonic fibroblasts, 542
vesicular stomatitis virus in, 56, 221–222, 417, 542–543
on WISH cells, 66–67, 542
bacterial synthesis of, 370–371
beta, *see* Interferon beta
binding, 597–598
to cells in monolayer culture, 312–315
receptor mediated, 332–339
to suspension cells, 305–311
biosynthesis, cell-free, *see* protein biosynthesis
Bo-IFN-β, *see* Bovine interferon beta
Bo-IFN-γ, *see* Bovine interferon gamma
cell-surface bound
detection by antibody, 330
changes in histocompatibility antigens induced by, 688–693
chicken, *see* Chicken interferon
concentration
endpoint determination, 601–603
constitutive in fetal membranes, 544–545
molecular weight, 549
in placenta, 544–545
delta, *see* Interferon delta
designation, 19–23
effect of
aminopeptidases on, 234
carboxypeptidase on, 234
effect on
adipogenesis, 629
cell replication, 541
cloning efficiency
of myeloid progenitor cells, 629–634
of primary tumor cells, 635–642
cytoskeleton, 541
differentiation, 612, 629–634
of human melanoma cells, 611–618
of murine melanoma cells, 611–618
of skeletal muscle cells, 619–628
drug metabolism, 713–725
extracellular matrix, 541
HL-60 cell line, 629, 632
hyporesponsiveness, 708
K-562 cell line, 629, 632
KG-1 cell line, 629, 632
melanogenesis, 629
metabolism of diphenylhydantoin, 715–718
myogenesis, 629
(2′-5′)-oligoadenylate synthetase levels, 653
proliferative capacity
of leukemic myeloid progenitor cells, 629–634
of normal myeloid progenitor cells, 629–634
equine, *see* Equine interferon
extraction, 371–372
fibroblast, *see* Interferon beta
gamma, *see* Interferon gamma
gene expression, 364, 366–375, 376–383, 383–396, 397–403, 414, 416–423, 424–433, 434–441, 445–449, 453–464, 471
histocompatibility antigens on cell surface increase due to, 682
HLA-A,B,C, 662
Hu-IFN-α, *see* Human interferon alpha
Hu-IFN-β, *see* Human interferon beta
Hu-IFN-γ, *see* Human interferon gamma
Hu-IFN-δ, *see* Human interferon delta
IFN-α, *see* Interferon alpha
IFN-β, *see* Interferon beta
IFN-γ, *see* Interferon gamma
IFN-δ, *see* Human interferon delta
immune, *see* Interferon gamma
immunoassays, 582–593, *see also* EIA, ELISA
inactivation by radiation, 255–260
molecular weight determination by, 258–260

induction, 25–149, 200, 212, 220, 234, 284, 322, 400
dose-response curves, 106–114
human interferons, 25–106, 200, 234, 282, 284, 400
non-human interferons, 106–149, 212–213, 220, 322, 442–443, 477, 480, 574
poly ICL-CM dextran, 103–106
prostaglandins, 707–712
influence of growth factors, 643–649
murine Swiss 3T3 cells in, 643–648
inhibition of cytochrome *P*-450, 715–718
inhibition of tumor cell colony growth, 642
large-scale production, 35–38
leukocyte, *see* Leukocyte interferon; Interferon alpha
lymphoblastoid, *see* Interferon alpha
methionine labeling, 400–401
β-microglobulin increase by, 682
Mo-IFN-α, *see* Monkey interferon alpha
Mo-IFN-β, *see* Monkey interferon beta
modification
glycosidases in, 235
neuraminidase, 235
of murine interferon, 233–241
Mu-IFN-α, *see* Murine interferon alpha
Mu-IFN-β, *see* Murine interferon beta
Mu-IFN-γ, *see* Murine interferon gamma
from Namalva cell, *see* Human interferon alpha
neutralization by antibody, 558–573
quantitation of, 558–573
in placental blood extract, 541–551
production, 25–149
kinetics, 70, 71
lymphoblastoid cells in, 35–38
large scale, 38
yeast cells in, 416–423, 424–433
in protection against cytotoxic lymphocytes, 541
purification, 151–230, 427
purity, 569–571
Ra-IFN-α, *see* Rat interferon alpha
Ra-IFN-γ, *see* Rat interferon gamma
rabbit, *see* Rabbit interferon
radioimmunoassay, 543–544
receptor mediated binding, 332–339
receptor mediated endocytosis, 332
recombinant, *see* Recombinant interferon; specific interferons
relation with plant cells, 727–761
sensitivity to
screening of cells for, 597–611
standards, *see also* reference reagents, interferon standards
international reference preparation, 15–18
laboratory reference preparation, 15
storage, 15
synergy with chemotherapy, 656–657
synthesis in yeast, 420–421
in transformation inhibition, 541
tumor inhibition
mechanisms, 652–654
update, 3–14
Interferon-albumin conjugate, 239
Interferon alpha, *see also* Leukocyte interferon; specific interferon alphas
amino acid sequences
Bo-IFN-α, 47
Hu-IFN-α, 6, 13
Mo-IFN-α, 13
Mu-IFN-α, 13
rat IFN-α, 13
antigenic specificity, 17, 31, 102, 227–230, 451–452, 546–548, 567
antiviral activity, 29, 66, 138, 162, 175, 250, 265, 278, 287, 306, 364, 417, 427, 473, 543, 554, 580
on bovine EBTr cells, 102
on human foreskin fibroblasts, 102
on human trisomic 21 GM 2767 cells, 102
on murine L 929 cells, 102
on WISH cells, 102
biological properties, 102
chromatographic behavior
concanavalin A-Sepharose, 102
matrex blue, 102
cloning of genes for, 5–7, 364
expression in *E. coli*, 364, 382–383, 403–415, 420–421, 428–441, 441–453
human, *see* Human interferon alpha

induction
by Newcastle disease virus, 54
by Sendai virus, 54
messenger RNA probes, 33
molecular weight, 7, 102
mouse, *see* Murine interferon alpha
physicochemical properties, 44, 47, 102, 153, 157, 238, 242, 252, 259, 268, 273, 407
production
human leukocytes in, 54–58, 62–63
in monocytes, 54
by Sendai virus induction of Namalva cells
interferon beta contaminant, 38
yeast cells in, 416–423, 424–433
stability
heat, 102
pH, 102
SDS, 102
standards, see Reference reagents, interferon standards
tumor cell colony inhibition, 642
Interferon alpha genes
characterization, 363–364
isolation, 359–365
short oligonucleotide probes in, 359–364
sequence homologies, 364
Interferon antagonist
lectin-like, 701–702
in agglutination of cells, 701–702
molecular weight, 550
Interferon beta, *see also* specific beta interferons
amino acid sequences
bovine interferon beta, 8–9
human interferon beta, 8–9
monkey interferon beta, 8–9
murine interferon beta, 8–9
antigenic specificity, 17, 102, 566
antigrowth assay, 581
antiviral activity, 66–67, 265
on bovine EBTr cells, 102
on human foreskin fibroblasts, 102
on human trisomic 21 cells GM2767, 102, 555
on murine L 929 cells, 102
on WISH cells, 102
biological properties, 102
bovine, *see* Bovine interferon beta
chromatographic behavior
concanavalin A-Sepharose, 102
matrex blue, 102
cloning of genes for, 5–7
consensus sequence, 8–9
as contaminant in Namalva/Sendai interferon alpha, 38
expression in *E. coli,* 370–375, 382–383, 397–403
expression plasmid for, 369–370, 374
human, *see* Human interferon beta
molecular weight, 102
monkey, *see* Monkey interferon beta
murine, *see* Murine interferon beta
physicochemical properties, 102, 181, 189, 192, 259
signal peptides, 8–9
stability
heat, 102
pH, 102
SDS, 102
standards, *see* Reference reagents, interferon standards
tumor cell colony inhibition, 642
Interferon delta, *see* Human interferon delta
Interferon gamma, *see also* specific gamma interferons
amino acid sequences
Bo-IFN-γ, 10–11
Hu-IFN-γ, 10–11
Mu-IFN-γ, 10–11
rat IFN-γ, 10–11
antigenic specificity, 17, 53, 67, 97, 102, 582
antisera to, *see* Antiserum, to interferon gamma
antiviral activity, 67, 81, 94, 101, 201, 265, 283, 299, 380
on bovine EBTr cells, 102
on human foreskin fibroblasts, 102
on human trisomic 21 cells GM2767, 102, 555
on murine L 929 cells, 102
on WISH cells, 102
biological properties, 102
bovine, *see* Bovine interferon gamma
chromatographic behavior
concanavalin A-Sepharose, 102
matrex blue, 102
cloning of genes for, 5–7

E. coli expressed, 370–375, 380–383, 420–421, 453
activity of, 383
human, see Human interferon gamma
induction
by ionophore, 69, 72, 193
by mitogens, 54, 55, 57–58, 62, 65, 68, 78, 83, 98
mitogen and tumor promotor in, 64, 68, 72, 77, 96, 193, 282
molecular weight, 102
murine, *see* Murine interferon gamma
physicochemical properties, 84, 87, 102, 259, 282
production, 54–63, 88–92
nonpurified leukocytes, 57
purified leukocytes, 57
by yeast cells, 416–423, 424–433
yields, 57
rat, *see* Rat interferon gamma
signal peptides, 10–11
stability
heat, 102
pH, 102
SDS, 102
standards, *see also* Reference reagents; Interferon standards
Gg 23-901-530, 51, 56, 60
tumor cell colony inhibition, 642
Interferon inducing agents
depression of cytochrome *P*-450, 719
effect on hyporesponsiveness to IFN, 708
Interferon-inducing capacity, determination from dose-response curves, 114
Interferon-inducing particles
definition, 107
determination from dose-response curves, 107, 113
in interferon induction, 107, 113
Interferon receptor, 283, 301, 305, 317, 332
characterization of, 346
chemical cross-linking of, 340–347
differentiation in HL-60 cells, 311
dissociation constants, 312, 314–315
identification
by chemical cross-linking, 340–347
Interferon-receptor complexes, 326, 347
molecular weight, 345
Interferon resistant cells, 704
Interferon·RNA
detection *in situ* of cells containing, 474–481
specific DNA probes in, 474
Interleukin 1 (IL-1)
as Hu-IFN-γ impurity, 87
IFN-γ induction product, 83
Interleukin 2 (IL-2), *see* T cell growth factor
Interleukin 3 (IL-3)
as Hu-IFN-γ impurity, 87
IFN-γ induction product, 83
Internalization
chloroquine inhibition, 336–337
of Hu-IFN-αA into suspension cells, 309
of human interferons, 315
of radiolabeled Hu-IFN-α2, 330
of receptor-bound IFN-α, 335–339
kinetics, 335
of receptor-IFN-γ complex, 321
International reference preparation, *see* Reference reagents
International standards, 14–23
International unit, distinction from experimental unit, 559
Iodination
effect on activity of, 278, 283
Hu-IFN-α, 236, 266, 271, 307
Hu-IFN-β, 236
Hu-IFN-γ, 266
Mu-IFN, 236, 322
in radiolabeling
interferons, 263–283, 290, 322–324
monoclonal antibody, 327
Iodine-125 labeled Bolton-Hunter reagent, *see* Bolton-Hunter reagent
Iodo-gen, in radiolabeling, 685–686
Iodoacetamide, modification of Hu-IFN-αA with, 253–254
Ion exchange chromatography
by fast protein liquid chromatography, 147–149
on mono Q column, 147–149
in Hu-IFN-αA purification, 157, 160–161
carboxymethyl (CM)-52 cellulose in, 160
precipitation of polymeric forms of Hu-IFN-αA, 161
protein loss in, 157, 161

removal of cleavage enzyme, 161
removal of Triton X-100, 161
of Hu-IFN-γ, 59–61, 84, 194–197
carboxymethyl BioGel agarose in, 194–197
in Mu-IFN-γ purification, 147–149
in purification of sarcolectin, 698
Ionomycin, *see* Calcium ionophores, ionomycin
Ionophores, *see* Calcium ionophores
Isoelectric point
Hu-IFN-α, 47, 175, 176
Hu-IFN-γ, 282
rat interferon, 227
IVR substance, *see also* antiviral factor
chymotrypsin sensitivity, 730
definition of unit, 731
dose-response curve, 733
effect on
cucumber mosaic virus, 733
potato virus X, 733
molecular weight, 730, 734
pH stability, 734
purification, 731–732
Sephadex G-75 in, 734
zinc acetate in, 731
RNase sensitivity, 730
from TMV-infected protoplasts, 729–734
trypsin sensitivity, 730

K

KG-1 cells, 306, 309
binding to Hu-IFN-αA to, 309
Kidneys, in preparation of cell culture, 212
Killer yeast dsRNA, in activation of P_1/eIF-2α protein kinase, 500
Kinase activity, of protein P_1, 515
Kinetin, in preparation of protoplasts, 745

L

L 265-K cells, *see* T Cell hybridoma lines, L 265-K
L 265-O cells, *see* T Cell hybridoma lines, L 265-O
L 929 cells, *see* Murine L 929 cells
L cells, *see* Murine L 929 cells
Labeling
effect on activity of
Hu-IFN-α, 236, 278, 307
Hu-IFN-β, 236
Hu-IFN-γ, 266
Mu-IFN, 236, 322
Lactalbumin hydrolysate, in preparation of cell cultures, 211, 212
Lactoperoxidase
immobilized, 264
in radiolabeling of
Hu-IFN-α, 263–267
effect on activity, 307
Hu-IFN-γ, 263–267
Mu-IFN, 322–323
Lambda Charon 4A/rat genomic library, identification of rat IFN-γ gene in, 454
Lambda cI repressor, in production of IFN-γ, 205
Lambda P_L promoter, 376, 378–379
in *E. coli,* 376
regulation by cI repressor, 378
LCL, *see Lens culinaris* lectin
Lead ion, in synthesis of tubercidin (c^7A) analog of 2-5A, 526
Leaf disks, in assay of IVR, 732, 736–737
Lens culinaris lectin
as Hu-IFN-γ impurity, 62
in IFN-γ induction, 55, 57–58, 62–63
as mitogen, 55, 57–58, 62–63
Lentil lectin, *see Lens culinaris* lectin
Leukapheresis, 68
Leukocyte buffy coats, *see* Leukocytes
Leukocyte interferon, *see also* Interferon alpha
crude, 19–20
definition, 19–20
production, 376
Leukocytes
bovine, *see* Bovine leukocytes
collection, 40–41, 56–57
from human blood, 40–41
culture conditions, 74
erythrocyte removal, 40, 56, 64, 65, 68, 78, 88, 137
human buffy coat, 210
induced with phytohemagglutinin, 210
in Hu-IFN-α production, 39–47

in Hu-IFN-γ production, 50–52, 63–69, 88–90, 92
yields, 51
interferon induction by Sendai virus, 41
isolation, 64–65, 68
obtained by leukapheresis, 68
preparation
for Hu-IFN-γ, 77–78
purification, 56–57
recovery, 56
Leukotriene, in arachidonic acid cascade, 710
LI-8, see Monoclonal antibody, interferon alpha specific, LI-8
12-Lipoxygenase, in arachidonic acid cascade, 710
Long terminal repeat sequence, 383–385, 396
for expression of Hu-IFN-β, 384–385, 396
Lymphoblastoid cells, *see also* Namalva cells
induction
with Newcastle disease virus, 36
with Sendai virus, 36
defective interfering particles in, 36
in large scale Hu-IFN-α production, 35–38
spontaneous formation of interferon, 36
Lymphoblastoid Interferon, *see* Human interferon alpha; Interferon alpha
Lymphocyte
EBV infected, 671–672
transformation of, 671
preparation, 77–78
nylon wool column, 78
for Hu-IFN-γ production, 77–78
subpopulations
effect of IFN on, 692
Lymphokines, *see also* specific lymphokines
Hu-IFN-γ induction product, 52, 82, 83
interleukin 2, *see* T cell growth factor
lymphotoxin, *see* Lymphotoxin
monocyte cytotoxin, 52
purification, 87
separation from Hu-IFN-γ, 52
T cell growth factor, 52
Lymphotoxin, Hu-IFN-γ induction product, 52, 82
Lysine
halomethyl ketone derivatives
effect on interferons, 236

M

Macerozyme
in preparation of protoplasts, 730–731
for protoplast medium, 746
Macrophages
activation by oligoadenylate, 681
binding of (2′-5′)-oligo-A to, 354–356
measurement of, 355
properties of, 355–356
time course of, 355
effect of (2′-5′)-oligoadenylate on, 676–681
guinea pig, 681
human, 681
ingestion of yeast, 680
inhibitory effect of α-fetoprotein, 677
lung
preparation of, 678
mouse, 681
(2′-5′)-oligoadenylate receptor for, 351–356
peritoneal, 678
preparation of, 353, 678
rabbit, 681
Madin-Darby bovine kidney cells, *see* MDBK cells
Magnesium ion, stimulation of phosphorylation, 515
Mannitol, in preparation of IVR substance, 731
DL-[^{3}H]Mannose, in radiolabeling of Hu-IFN-β, 401
Manton-Gaulin homogenizer, 407
in recombinant Hu-IFN-β purification, 179, 184
Mason-Pfizer monkey virus, harvesting from monkey foreskin B7 cells, 704
Matrex blue, in Hu-IFN-δ purification, 97, 99–102
Maxicells
in radiolabeling of Hu-IFN-αA, 288
D-cycloserine in, 288

MDBK cells, 313, 333, 336–339
in assay of
Bo-IFN-α, 138, 140–141, 473
Hu-IFN-αA, 162, 407, 414, 417, 427, 601
Hu-IFN-αD, 427
Hu-IFN-γ, 101, 417, 427
Hu-IFN-δ, 101
binding of Hu-IFN-α2, 329
in interferon assays, 65, 66–67, 250, 278, 287, 407, 414, 417, 542, 601
comparison to WISH cells, 67
internalization of Hu-IFN-αA in, 336–339
Melanin
assay for with mouse B-16 cells, 614–617
enhancement by interferon, 613
secretion from mouse B-16 cells, 613, 616
Melanoma cells
differentiation, 612
tyrosinase in, 613
Mengovirus is-1 (*ifp*$^+$), in dose-response curve generation, 111–112
2-Mercaptoethanol
effect on stability of rat IFN-γ, 462
in sarcolectin characterization, 701
Metal chelate chromatography
in IFN-α purification, 171–172, 199
in IFN-γ purification, 86, 199
[^{35}S]Methionine
in cell culture labeling, 669–670
in radiolabeling of
Hu-IFN-αA, 286–291, 417–418
Hu-IFN-β, 400–401
in vivo labeling with, 417–418
Methotrexate
in amplification, 398–399
Hu-IFN-β gene, 403
resistance to, 399
2-Methoxy-5-nitrobenzyl bromide, tryptophan reagent, 236
α-Methyl-D-mannoside
in affinity chromatography
elution with, 401–403, 545–546
of AVF, 742
of Hu-IFN-γ, 195, 196
of Mu-IFN-γ, 146
Mezerein
in differentiation, 613
in Hu-IFN-γ induction, 49, 51, 52, 72–76, 193
large-scale, 77
Mice
in poly ICL-CM dextran assay, 104–106
survival, 105
Microassay
for cytolysis, 574–579
for cytostasis, 574–579
for mitogenic activity of growth factors, 659–660
rationale of, 659
β_2-Microglobulin, increase on cell surface, 682
Microsomes, preparation, 720–721
Mitogenic activity, 83
microassay for, 659–660
Mitogens
concanavalin A, *see* Concanavalin A
in induction of
equine IFN-γ, 134
Hu-IFN-γ, 48–55, 57–58, 62–64, 66, 68, 78, 83, 85, 209, 282
Hu-IFN-δ, 97, 98
LCL, *see Lens culinaris* lectin
phytohemagglutinin, *see* Phytohemagglutinin
staphylococcal enterotoxin A, *see* Staphylococcal enterotoxin A
staphylococcal enterotoxin B, *see* Staphylococcal enterotoxin B
Mixed lymphocyte reaction, 83
Mo-IFN-α, *see* Monkey interferon alpha
Mo-IFN-β, *see* Monkey interferon beta
Molecular size determination
by radiation inactivation
applications to interferons, 258–260
theory, 256–258
Molecular weight
Hu-IFN-α, 11, 260, 279
Hu-IFN-αA, 11, 289
Hu-IFN-α2-receptor complex, 349
Hu-IFN-β, 11, 260, 285, 403
Hu-IFN-γ, 11, 260
of interferon-receptor complexes, 345
Mu-IFN-α, 322

Mu-IFN-β, 322
by target size analysis, 257–258
equation for, 257
Moloney sarcoma virus-induced tumors, antagonism of
PDGF and Mu-IFN, 664–666
Monensin, inhibition of ligand degradation, 336–337
Monkey
in poly ICL-CM dextran assay, 104–106
survival, 105
Monkey cells
antiviral activity of Hu-IFN-αA on, 18–19
COS-1
in expression of rat IFN-γ, 454, 458
transient expression, 458
Monkey fibroblast interferon, *see* Monkey interferon beta
Monkey foreskin B7 cells, 704
in harvesting of Mason-Pfizer monkey virus, 704
Monkey IFN-β, *see* Monkey interferon beta
Monkey interferon alpha (Mo-IFN-α)
amino acid sequence, 13
mature protein, 13
signal peptide, 13
Monkey interferon beta (Mo-IFN-β)
amino acid sequence, 8–9
mature protein, 8
signal peptide, 8
Mono-Q column chromatography, 147, 223–224
Monoclonal antibodies, *see also* Antibody; Antiserum
anti-Hu-IFN-αA, 155–160
anti-Hu-IFN-α1, 567
anti-Hu-IFN-β, 566
characterization of, 561, 565–573
antibody concentration dependence, 565
in detection of cell-surface bound interferon, 330–331
dissociation constant, 564
in purification of
Hu-IFN-α, 28–33, 43–44, 45–47, 153, 155–160, 288, 306, 407
Hu-IFN-γ, 63, 84, 207
to human interferon receptor
Hu-IFN-α, 598
Hu-IFN-β, 598
IFN-α specific
LI-8, 155
in Hu-IFN-αA purification, 155–160
NK2, 30–33, 43–44
YOK, 30–33
binding of IFN-α1, 30–31
binding of IFN-αD, 30–31
neutralization of
Hu-IFN-α1, 326
Hu-IFN-γ, 584
titer, 326
computation, 558–573
definition, 558–573
OKT-3
in Hu-IFN-γ induction, 52
physical supports for, 159
preparation
against human breast tumor metastases, 683–684, 686
against COOH-terminal peptide of Hu-IFN-γ, 205
purification, 205, 326–327
in radioimmunoassay, 582–587, 588–593, 686–687
radiolabeling, 327, 589
Bolton-Hunter reagent in, 327
chloramine T in, 583
specificity, 569–571, 587
storage, 327
Monoclonal antibody affinity column
in purification of
Hu-IFN-αA, 155–160, 288–289, 306, 407
Hu-IFN-γ, 63, 84, 207
radiolabeled interferon, 288–289
Monoclonal antibody B6.2
effect of Hu-IFN-αA on binding to cells, 687–688
recognition of 90,000 Da determinant, 688
Monocyte cytotoxin, Hu-IFN-γ induction product, 52
Monocytes
effect of IFN-γ on MHC class II antigens, 692
in IFN-α production, 54

Monomers
of IFN-γ, 207
of Hu-IFN-αA, 242, 246
Mononuclear cells, *see also* Lymphocytes; Monocytes
Hu-IFN-γ yields in, 50
in interferon induction
with phytohemagglutinin, 48
isolation from equine blood, 131
yield, 131
preparation of, 51
for production of Hu-IFN-γ, 51
Monooxygenase activities, 724
Mouse alpha interferon, *see* Murine interferon alpha
Mouse beta interferon, *see* Murine interferon beta
Mouse cells
antiviral activity of Hu-IFN-αA on, 18–19
B-16 melanoma
cell culture, 614–615
effect of interferon on differentiation, 611–618
inhibition of differentiation by interferon, 612–618
in melanin assays, 614–616
secretion of melanin, 613, 616
C-243 cells
interferon mRNA in, 474–481
in situ hybridization, 477–478
induction of interferon, 477, 480
NDV in, 477, 480
erythroleukemia cells
inhibition of differentiation by IFN, 612
fibroblast A9 cells
in NIH3T3 transformation, 599–600, 611
JLS-V9R cells, RNase in, 489
L cells
in antiviral assay for murine interferon, 534
in characterization of sarcolectin, 695–696
poly(I) · poly(C) in induction of, 601
L929 cells
in assay of
Bo-IFN-α, 141
Hu-IFN-γ, 101
Hu-IFN-δ, 101
inhibition of protein synthesis in, 668–671, 675
in interferon assays, 65, 66–67, 102
L(Y) strain
aging, 118
interferon yield after, 118
in plaque reduction assay for IFN-γ, 575
in purification of P_1/eIF-2α protein kinase, 501–503
L1210 cells
interferon receptor number, 324
Lymphoma P388 cells
in microassay for cytolysis, 574–578
Mu-IFN-γ in, 576
in microassay for cytostasis, 574–578
Swiss 3T3 cells
antiproliferative effects of interferon on, 643–649
[^{3}H]thymidine incorporation into, 644, 647
Mouse chromosome 16, effect on IFN sensitivity, 597
Mouse IFN-α, *see* Murine interferon alpha
Mouse IFN-β, *see* Murine interferon beta
Mouse IFN-γ, *see* Murine interferon gamma
Mouse immune interferon, *see* Murine interferon gamma
Mouse interferon alpha, *see* Murine interferon alpha
Mouse interferon beta, *see* Murine interferon beta
Mouse interferon gamma, *see* Murine interferon gamma
Mouse leukocyte interferon, *see* Murine interferon alpha
Mouse macrophages
binding of (2′-5′)-oligo-A to, 354–356
measurement of, 355
properties of, 355–356
Mouse-human somatic WA17 cells, 598
binding of radiolabeled IFN-αA, 598
Mu-IFN-α, *see* Murine interferon alpha
Mu-IFN-β, *see* Murine interferon beta
Mu-IFN-γ, *see* Murine interferon gamma
Mucoid carcinoma of the breast, xenograft in nude mice, 650–653, 655
Murine fibroblast interferon, *see* Murine interferon beta

Murine IFN-α, *see* Murine interferon alpha
Murine IFN-β, *see* Murine interferon beta
Murine IFN-γ, *see* Murine interferon gamma
Murine immune interferon, *see* Murine interferon gamma
Murine interferon
activity
effect of
^{125}I-labeling, 236
periodate, 237–238
SDS, 238
urea, 237–238
antitumor effects, 652, 654, 664–666
binding
to cells in monolayer culture, 325
to receptor, 321–326
to suspension cells, 323–324
cysteine residues in, 8, 10, 13, 249
effect on virus induced cell fusion, 702–706
induction, 322
by NDV, 322
temperature in, 121
inhibition of proliferative response to IL-2, 662–663
in measuring cell proliferation, 645
modification
2-deoxyglucose in, 233
D-glucosamine in, 233
tunicamycin in, 233–234
from mouse L cells, 601
periodate treatment of, 237
in plaque-reduction assay, 609
radiolabeled
specific radioactivity of, 323
radiolabeling, 323
Affigel-10-lactoperoxidase in, 323
with ^{125}I, 322–323
yield in, 322–323
in screening of IFN-sensitive cells, 601, 609–610
sensitivity to
enhancement by chromosome 16, 597
specific activity, 645
specific binding sites, 324
standards
G-002-904-511, 111
Murine interferon alpha, 475, 481
amino acid sequence, 13, 438
antagonistic effects of growth factors, 659, 662–663
consensus sequence, 13
constitutive production in C-243 cells, 481
cytochrome *P*-450 depression by, 719
effect of fluorescamine on, 236
effect on 2-5A synthetase levels, 653
expression in *E. coli,* 434–441
expression vector, 437, 439
construction, 439
homology to rat IFN-α, 13, 447
immunoaffinity chromatography of, 227–230
induction with NDV, 477, 480
mature protein, 13, 438
molecular weight, 322
signal peptides, 13, 438
standards, 16
Murine interferon alpha gene
isolation in *E. coli,* 434–441
DNA sequence, 438
Murine interferon beta, 475, 481
amino acid sequence, 8–9
mature protein, 8
signal peptide, 8
antagonistic effects of growth factors, 659, 662–663, 666
antitumor effects in mice, 651–656
mechanisms, 652–654
constitutive production in C-243 cells, 481
cytochrome *P*-450 depression by, 719
effect of fluorescamine on, 236
effect on human breast cancer xenograft, 652, 654
induction with NDV, 477, 480
molecular weight, 322
standards, 16
Murine interferon beta mRNA
detection in butyrate-treated mouse C-243 cells, 474
Murine interferon gamma
amino acid sequence, 10–11
mature, 146
signal peptide, 10
induction by staphylococcus enterotoxin A, 574
in microassay for cytolysis, 574–578

purification, 145–149
by affinity chromatography, 146–147, 149
on concanavalin A-Sepharose, 146–147, 149
by ion-exchange chromatography, 147–149
fast protein liquid chromatography on mono Q column, 147–149
purification factor, 149
polyethylene glycol MW 20,000 in, 146
recovery, 149
on silicic acid, 146, 149
elution with ethylene glycol, 146
purification factor, 149
specific activity, 149
standards, 16
storage, 146
Murine natural killer cells, activity, 653–656
Murine spleen cell cultures, preparation, 145
Murine swarm chondrosarcoma, *in vivo* inhibition by IFN, 671–674
Mutagenesis
of Hu-IFN-α gene *in vitro,* 403–415
in vitro, 400
of α-factor, 425, 430
M13 in, 400
site-directed
plasmid DNA heteroduplex in, 410
Myeloid cell line HL-60, effect of IFN on MHC antigen, 693
Myeloid cell line K562, effect of IFN on MHC antigen, 693
Myogenesis, effects of interferon, 612
Myoinositol, in preparation of protoplasts, 745–746

N

NADPH-cytochrome *P*-450 reductase, 718
Namalva cell interferon, *see* Human interferon alpha
Namalva cells
induction
with Sendai virus, 36
defective interfering particles, 36
interferon beta contaminant, 38
in interferon production, 38, 651, 655
maintenance, 35–36
for production of Hu-IFN-α, 651
stimulation by
dimethyl sulfoxide, 37
n-butyric acid, 37
short-chain fatty acids, 37
Namalwa cells, *see* Namalva cells
Naphthalene acetic acid, in preparation of protoplasts, 745–746
NCI medium, in chick embryo cell preparation, 116
NDV, *see* Newcastle disease virus
Neoplastic disease, effect on hyporesponsiveness to IFN, 707–708
Neuraminidase, in modification of interferon, 235
Neutral red dye, in antiviral assay for interferons, 534–535
Neutralization, 559–571
of Bo-IFN-β, 219
of Hu-IFN-α, 17, 567
of Hu-IFN-β, 17
of Hu-IFN-γ, 17, 53, 67, 95, 584
of interferon, 326
quantitation of, 558–573
of mouse interferon, 227–230
rabbit antiserum in, 219
of rat interferon, 227–230
standards, 17
Neutralization assays, types of, 559–561
Newcastle disease virus
AV strain
in chicken interferon induction, 119
in dose-response curve, 112
B_1-Hitchner strain
in dose-response curve, 112
California strain
in chicken IFN induction, 119
in dose-response curves, 108, 112–113
heat inactivation before IFN induction, 133
in induction of
Bo-IFN-β, 212–213, 220
chicken interferon, 119
equine IFN-α, 133
equine IFN-β, 133
guinea pig IFN, 128
IFN-α, 54
mouse C-243 cells, 474, 480

murine L 929 cells, 17
Mu-IFN, 322
irradiation before IFN induction, 133
N.J.-LaSota strain
in chick embryo cell aging, 118
in chicken interferon induction, 119
in dose-response curve, 112
in production of Hu-IFN-α, 27
Nickel chelate chromatography
column preparation
1,4-butanediol diglycidyl ether in, 200, 201
iminodiacetic acid in, 200, 201, 203
Sepharose in, 200, 201, 203
sodium borohydride in, 200, 201
in IFN-γ purification, 199–204
effect of temperature on, 202
Nickel ion scavenging, by iminodiacetic acid-activated Sepharose, 201
Nicotiana glutinosa
in assay of IVR, 732–733
carrying the N-gene, 740
preparation of protoplasts from, 759–760
production of AVF, 752–754
Nicotiana tabacum
preparation of
IVR from, 731
protoplasts from, 759–760
NIH3T3 cells
isolation of transformed cells, 599–604
screening for sensitivity to Hu-IFN-α/β, 597–611
principle, 598
transformation of, 599–600
plasmid pSV2-neo in, 599–600
NK2, *see* Monoclonal antibody, interferon alpha specific, NK2
Nu/nu mice, as host for human tumor xenografts, 649–653, 655
Nuclease P_1, in characterization of (2'-5')-oligoadenylate, 526–527
Nucleases, in sarcolectin characterization, 701
Nuclei
radiolabeled
autoradiography of, 644
staining with propidium iodide, 645
Nylon wool column, in lymphocyte preparation, 78

O

Octyl-agarose column, in purification of sarcolectin, 698
OK-432, *see* Streptococcal preparation OK-432
OKT-3, *see* Monoclonal Antibody, OKT-3
Oligo(dT)-cellulose, in isolation of poly(A)$^+$ RNA, 442
(2'-5')-Oligoadenylate
3'-5'A phosphodiester linkage isomer, 524–526
2-(9-adenyl)-6-hydroxymethyl-4-hexylmorpholine derivative of, 523
alkaline phosphatase digestion, 526
analogs, 522–529
effect on
cell growth, 667–675
protein synthesis, 667–675
TMV replication, 759–761
synthesis, 522–529
5'-phosphoroimidazolidates in, 522
chemical modification, 522–528
cyanoborohydride reduction, 523
periodate oxidation, 523
Schiff base formation, 523
cores
inhibition of TMV replication, 758
effect on
cell growth, 667–675
macrophages, 676–681
protein synthesis, 667–675
TMV replication, 759–761
enzymatic synthesis of, 516–521
preparative scale, 516–521
fractionation by chain length, 520
DEAE cellulose in, 520
nuclease P_1 digestion in, 526
phosphorylation, 524–526
T_4 polynucleotide kinase in, 525–526, 528
in plant cells, 735
preparation of, 519–521
receptor, 351–356
synthesis in human tumor, 652–655
synthesis of 5'-triphosphate, 523–524
tritium labeled, preparation of, 353
(2'-5')-Oligoadenylate synthetase
activation by dsRNA, 516

assay, 518–519
poly(I) · poly(C) in, 519
effect of interferon on
in xenograft, 653
partial purification of, 518
carboxymethylcellulose in, 518
SDS–PAGE in, 518
Oligoadenylic acid, *see* (2′-5′)-Oligoadenylate
(2′-5′)-Oligodeazaadenylates
from tubercidin 5′-phosphoroimidazolidate, 527–528
Oligonucleotide probes
labeling, 360–362
storage, 362
Optical density, effect of cell number on, 580–582
Ovine choroid plexus cells
in antiviral assay, 551–558
cell cultures, 552–553
sensitivity to
heterologous interferons, 552, 554–558
human interferons, 134
Oxidants, generation from chloramine T, 267
Oxidation, of Hu-IFN-α, 248–255
Oxidative sulfitolysis, of Hu-IFN-αA disulfide bonds, 238–239

P

P-IF, *see* Human interferon alpha
Paraformaldehyde, interference with hybridization, 480
Paramyxoviruses, induction of cell fusion, 703
Partition chromatography, in Hu-IFN-γ purification, 83
PBMC, *see* Peripheral blood mononuclear cells
pBR322
in IFN-α2 vector, 167, 168, 204, 379
PDGF, *see* platelet-derived growth factor
Pectolyase, in tobacco protoplast assay for interferon, 746
PEG, *see* Polyethylene glycol
Pepsin, in sarcolectin characterization, 701
Peptides
analysis of HPLC, 244–246
separation, 205
C-18 reverse phase column in, 205
HPLC in, 205
Perchloric acid
in chicken interferon production
removal of viral inducer, 122
effect on interferon activity, 237
Periodate
effect on interferon activity, 237–238
oxidation of oligoadenylic acid, 523
treatment of
Hu-IFN-α, 237
Mu-IFN, 237
Peripheral blood leukocytes, *see* Leukocytes
Peripheral blood mononuclear cells
in Hu-IFN-γ induction, 70, 72
in Hu-IFN-γ production, 93–95
induction with mitogen and tumor promotor, 96
preparation, 97–98
Peripheral white blood cells, *see* Leukocytes
Peritoneal macrophages, 353–356
binding of (2′-5′)-oligoadenylate to, 354–356
measurement of, 355
properties of, 355–356
pH
effect on
alpha helix of Hu-IFN-αA, 239
sarcolectin, 701–702
stability of
Hu-IFN-δ, 98, 102
IFN-α, 102
IFN-β, 102
IFN-γ, 102
IVR substance, 734
rat IFN-γ, 462
PHA, *see* Phytohemagglutinin
Phagocytosis
assay of, 678–681
effect of oligoadenylate on, 678–681
measurement of, 680–681
Phenol red pH indicator, in antigrowth assay, 579
Phenyl-Sepharose
hydrophobic chromatography with, 214, 217
purification of bovine IFN-β, 214, 217

in tandem chromatography with Blue-Sepharose, 218
Phenylalanine, halomethyl ketone derivatives
effect on interferons, 236
Phenylmethylsulfonyl fluoride
effect on stability of rat IFN-γ, 462
protease inhibitor, 697–698
Phenytoin, *see* diphenylhydantoin
Phorbol esters, *see also* 12-*O*-Tetradecanoylphorbol-13-acetate; Mezerein
effect on differentiation, 619
in Hu-IFN-γ induction, 48–54
Phorbol myristate acetate, *see* 12-*O*-Tetradecanoylphorbol-13-acetate
Phosphocellulose
in purification of
dsRNA dependent protein kinase, 508–510
RNase L, 495–497
Phosphodiesterase
snake venom
degradation of (2′-5′)-oligoadenylate, 527
Phospholipase, in arachidonic acid cascade, 710
Phosphorylation
of (2′-5′)A analogs, 528–529
cAMP dependence, 515–516
of Hu-IFN-γ, 296–301, 316, 321
O-Phosphoserine, in protein P_1 and eIF-2α, 515
Phytohemagglutinin
in colony stimulating assays, 630, 633
in induction
of equine IFN-γ, 134
of Hu-IFN-γ, 48–54, 58, 64, 66, 68, 78, 83, 85, 209, 282
optimal concentration, 50
of Hu-IFN-δ, 97, 98
of interferon in human mononuclear cells, 48
as mitogen, 58, 83
p*I*, *see* Isoelectric point
Placenta
constitutively produced interferon in, 544–547
molecular weight, 549
Placental blood extract, interferon in, 541–551
Plant cells
antiviral factor (IVR, AVF) from, 729–734
purification of, 729–734
effect of interferon on, 727–761
Plant lectins, *see also* specific lectins
as phytohemagglutinin substitute, 52
in Hu-IFN-γ induction, 52
Plant oligoadenylate synthetase
dsRNA dependent
stimulation by AVF, 753
induction, 754
preparation, 754
Plant oligoadenylates
enzymatic synthesis of, 752–758
inhibition of *in vitro* protein synthesis, 753
inhibition of TMV replication, 753
purification of, 754–758
DEAE cellulose in, 753–756
HPLC in, 757
Plants, in assays for antiviral activity, 735–736
Plaque reduction assay, 607–610
for IFN-γ
mouse L929 cells in, 575
Plasmid
amplification in CHO cells, 398–400
bacterial transformation with, 367
construction, 368–370
encoding rat IFN-α, 444–448
construction of, 444–448
for expression of interferons, 167, 369, 379, 410, 419, 429, 439, 448–451, 472
pBR322, 167, 204, 379
pcI857, with temperature sensitive cI repressor, 371
Plasmid DNA
coding for Hu-IFN-αA, 483
coding for Hu-IFN-α2, 483
hybridization to ^{32}P-labeled octadecyl deoxyoligonucleotides, 483
ligation of, 367–370
preparation, 427
for cell-free DNA-dependent biosynthesis, 293
transfection in CHO cells, 398–400
Plasmid pSV2-neo, controlling resistance to aminoglycoside antibiotic, 599–600

Platelet derived growth factor (PDGF)
antagonism by
Hu-IFN, 660–662
Mu-IFN, 664–666
in vivo, 664
in microassay for mitogenic activity, 659–660
purification, 658
Platelet poor plasma, in microassay for mitogenic activity, 659
Plateletpheresis residues
in Hu-IFN-γ production, 49, 50–52
prevention of clotting, 50
PMA, see 12-*O*-Tetradecanoylphorbol-13-acetate
Pokeweed mitogen
in Hu-IFN-γ induction, 52
as phytohemagglutinin substitute, 52
Poly ICL-CM dextran
assay, 103–106
in interferon induction, 103–106
in mice, 104–106
survival, 105
in rhesus monkeys, 104–106
survival, 105
preparation, 103–106
Poly ICLC, 103–106
Poly(A)$^+$ RNA
isolation
oligo(dT)-cellulose in, 442
Poly(A)-agarose, in purification of RNase L, 495–496
Poly(I) · poly(C)
in assay for (2′-5′)-oligoadenylate synthetase, 519
in chicken interferon induction, 119
induction of mouse L cells with, 601
interferon inducer, 708, 716
of human diploid cells, 17
in poly ICL-CM dextran preparation, 103, 104
P_1/eIF-2α protein kinase, activation of, 500–501
stimulation of AVF, 734–735
in superinduction of Hu-IFN-β, 284
in synthesis of oligoadenylate, 753–754
Poly(I) · poly(C)-Sepharose
in purification of P_1/eIF-2α protein kinase, 501–503, 512–514
Poly(U)-Sepharose, in Hu-IFN-γ purification, 91, 92
Poly-L-ornithine, in preparation of IVR substance, 731
Polyacrylamide gel electrophoresis
sodium dodecyl sulfate, *see* SDS–polyacrylamide gel electrophoresis
Polyadenylation signals
in Hu-IFN-β expression vectors, 386–387
in rat IFN-γ expression vectors, 458
Polybrene, in cell fusion, 705
Polyclonal antibodies, *see also* Antibody; Antiserum
anti Hu-IFN-α2
in purification of rat IFN-α, 451
to Hu-IFN-α receptor, 598
to Hu-IFN-β receptor, 598
Hu-IFN-γ specific, 67
Polyethylene glycol
in IFN-γ concentration, 55, 58–59
in Mu-IFN-γ purification, 146
for protoplast medium, 746
Polylysine
in poly ICLC preparation, 103, 104
in poly ICL-CM dextran preparation, 103, 104
Polyriboinosinic · polyribocytidylic acid, *see* poly(I) · poly(C)
Porcine cells, in antiviral assay, 556–557
Potato virus X, inhibition by IVR substance, 730
Primary chick embryo cells
aging, 112, 117
interferon yield after, 117
in dose-response curve, 109, 112
effect of refrigeration
on interferon yields, 117
on monolayers, 117
preparation, 116–117
secondary chick embryo cells
low interferon yields, 117
Primary tumor cells, effect of interferon on cloning efficiency of, 635–642
Promoter
for acid phosphatase, 418
for expression
hybrid tetracycline-tryptophan promoter, 447
P_L promoter, 366–375, 376–383, 410, 447, 449–450

regulation by temperature sensitive repressor cI, 447
for glyceraldehyde-3-phosphate dehydrogenase, 418
long terminal repeat sequence, 383–396
for expression of Hu-IFN-β, 384–385, 396
M13mp11 lac-fusion, 378
phage lambda P_L
in expression of human interferons, 366–375
trp promoter
in expression of Bo-IFN-αC, 473
trp promoter/operator, 440
Promyelocytic cell lines
HL-60, *see* HL-60 cells
in Hu-IFN-γ production, 65, 66–68
Promyelocytic leukemia, enhancement of differentiation, 612
Propidium iodide, in nuclei staining, 645
β-Propiolactone, in inactivation of RD114 virus, 704
Prostaglandin
in arachidonic acid cascade, 710
assay, 709–711
induction of interferon, 707–712
intracellular, 707, 711
in measuring hyporesponsiveness to IFN, 711
modulation of IFN biological activity, 712
Prostaglandin E, 707
Protease inhibitors
iniprol, 697
phenylmethylsulfonyl flouride, 697–698
Protein A, radioiodinated, 685–687
Protein eIF-2α
molecular weight, 500
phosphorylation
activators, 515–516
inhibitors, 515–516
Protein kinase
bovine heart, 296–297, 299, 301, 316
catalytic subunit, 299, 301
in radiolabeling of Hu-IFN-γ, 316
cyclic AMP-dependent, 296–297, 301, 316
double-stranded RNA-dependent
assay of, 502
ATP-mediated phosphorylation in, 502
purification, 499–516
mouse fibroblasts in, 499, 501–503
ribosomal salt-wash fractions in, 503–508
poor substrate for, 299–301
in radiolabeling of interferon, 296–301, 316
Protein P_1
molecular weight, 500
phosphorylation
activators of, 515–516
inhibitors of, 515–516
Protein P_1 kinase, activity, 515
Protein P_1/eIF-2α kinase
activation
by killer yeast dsRNA, 500
by poly(I) · poly(C) dsRNA, 500–501
by reovirus genome dsRNA, 500–501
specificity, 515
Protein synthesis
cell-free DNA dependent
procedure for, 292–295
effect of (2′-5′)-oligoadenylates and analogs, 667–675
inhibition by cordycepin 5′-triphosphate, 668–671
Protoplasts, in assay for inhibitory potency of IVR, 732
PWBC, *see* Leukocytes

Q

Quantum yield
definition, 107
determination from dose-response curves, 107, 112–113
Quinine sulfate, in measurement of drug metabolism, 720

R

Ra-IFN-γ, *see* Rat interferon gamma
Rabbit antiserum, *see also* Antibody; Antiserum; Polyclonal antibodies
in neutralization of bovine IFN-β, 219
Rabbit cells, antiviral activity of Hu-IFN-αA on, 18–19
Rabbit interferon, standards, 16

Radiation
in inactivation of interferon, 255–260
in molecular size determination, 256
Radioimmunoassay
in assay for prostaglandins, 709–711
of Hu-IFN-γ, 582–587
of IFN-induced cell surface tumor antigen, 682–688
preparation of cells for, 684
of interferon, 543–544, 582, 590
of interferon monomers, 14–15
of interferon dimers, 590–593
of oligomers, 590–593
relationship to antiviral activity, 14–15, 587
of surface antigen, 686–687
Radioiodination, *see* Radiolabeling, Iodination
Radiolabeling
Affigel-10-lactoperoxidase in, 323
cell-free system in, 292–296
DNA-dependent system in, 292–296
of Hu-IFN-α, 267–276, 276–281
Bolton-Hunter reagent in, 276–281
effect on antiviral activity, 278–279
with ^{125}I, 263–267, 276–281
effect on antiviral activity, 278–279
lactoperoxidase in, 263–267
protein kinase in, 296–301, 316
Hu-IFN-αA, 286–291, 306–307, 333, 327, 417
Bolton-Hunter reagent in, 286, 290, 312–313, 333, 327
chloramine T in, 286
in *E. coli* cells
indolyl-3-acrylic acid in, 288
enzymobeads in, 286
maxicells in, 288
Hu-IFN-αD
Bolton-Hunter reagent in, 312–313
Hu-IFN-α1, 327
Bolton-Hunter reagent in, 327
Hu-IFN-α2, 264–266, 276–281, 333
Bolton-Hunter reagent in, 276–281, 312–313, 333
lactoperoxidase in, 264
Hu-IFN-β, 284–286
glucosamine in, 401
DL-[^{3}H]mannose in, 401
[^{35}S]methionine in, 284, 400
in vivo, 284–286
human diploid fibroblast cells in, 284
Hu-IFN-γ, 264–266, 281–283, 316
with [γ-^{32}P]ATP, 316
cyclic-AMP-dependent protein kinase in, 316
Bolton-Hunter reagent in, 281–283
effect on antiviral activity, 283
yield, 283
with ^{125}I, 263–267, 281–283
effect on antiviral activity, 283
lactoperoxidase in, 263–267
in vivo, 275, 284–291, 417
of yeast proteins, 417–418
limitations, 275
of monoclonal antibody, 327, 583
Bolton-Hunter reagent in, 327
Mu-IFN, 322–323
lactoperoxidase in, 322–323
of yeast proteins, 417–418
Radiospecific activity
of Hu-IFN-αA, 289, 291, 307
of Hu-IFN-β, 285
of Hu-IFN-γ, 316
Rat cells, antiviral activity of Hu-IFN-αA on, 18–19
Rat embryonic cell line (Ratec), 220–221, 422
cytopathic effect assay with vesicular stomatitis virus, 221
for detection of antiviral activity, 459
induction of IFN, 220–221, 422
kinetics, 221
with Sendai virus, 220, 422
mRNA from, 442–443
sedimentation coefficient, 443
Rat embryonic fibroblasts
in characterization of sarcolectin, 695–696
interferon assays with, 542
Rat gene library
in bacteriophage λ, 444
in cloning of rat IFN-γ, 453–454
Rat IFN-γ, *see* Rat interferon gamma
Rat immune interferon, *see* Rat interferon gamma

Rat interferon
  aggregation of, 225
  characterization of, 220–230
    SDS–PAGE in, 225–226
  classes of, 227
  cross reactivity with Hu-IFN-α2, 227
  elution from gel, 225
    sarcosine in, 225
  isoelectric focussing, 225, 227
    recovery from, 227
  isoelectric points, 227
  production, 220–221
    serum free medium in, 221
    kinetics, 221
  purification, 222–225
    ammonium sulfate in, 222
    with fast protein liquid chromatography, 223–225
    loss of activity in, 225
    by mono-Q column chromatography, 223–224
    recovery, 223
    Ultrogel AcA 44 in, 222–224
  stability of, 225
Rat interferon alpha (Ra-IFN-α)
  clone containing DNA for, 443
  homology to Hu-IFN-α, 13, 227–230
  induction
    mRNA for, 443
    time course of, 443
  molecular weight, 452
  plasmid coding for, 445–451
    construction of, 445–448
  purification, 451–452
  recombinant
    purification with monoclonal antibodies, 451
  specific activity, 449, 452–453
    recovery by treatment with chemicals, 453
Rat interferon alpha 1 (Ra-IFN-α1)
  amino acid sequence, 13, 446
  expression of, 230, 440–453
    plasmid coding for, 441–453
      construction, 443–449
  homology
    to Hu-IFN-α, 13, 227–230
    to Mu-IFN-α1, 13, 447
  mature protein, 13, 445–447
  purification of, 441–453
    anti-Hu-IFN-α2 in, 451–452
    immunochromatography in, 452
    with polyclonal antibodies, 451
  signal peptide, 13, 445
Rat interferon alpha 1 gene
  expression
    P_L promoter in, 447, 449–450
  subcloning, 447–449
Rat interferon alpha chromosomal gene
  characterization, 444
  isolation, 444
    bacteriophage λ in, 444
Rat interferon alpha mRNA, 230, 442
  from Ratec cells induced with Sendai virus, 442–443
    sedimentation coefficient, 443
    time course of production, 443
Rat interferon gamma
  activity
    effect of
      freeze drying, 463
      freezing and thawing, 463
  amino acid sequence, 10–11
    mature protein, 10
    signal peptide, 10
  characterization, 460–464
    SDS–PAGE in, 460–461
  cloning of DNA for, 453–459
    rat gene library in, 453–454
  expression, 456–459
    in CHO cells, 454, 458
      dihydrofolate reductase in, 454
      E. coli gpt gene in, 454
    in eukaryotic cells, 456–459
      yields, 458
    in monkey COS-1/COS-7 cell line, 454, 458
      transient expression, 458
  expression vector, 457–458
    construction of, 457–458
    polyadenylation signal in, 458
    transcription termination signal in, 458
  mature protein, 10, 456
  molecular weight, 460
  oligomeric forms, 460
  production, 458–459

purification, 459–464
controlled-pore glass beads in, 459–460
signal sequence, 10, 456
stability, 460, 462
effect of
glycerol, 462
2-mercaptoethanol, 462
pH, 462
phenylmethylsulfonyl flouride, 462
temperature, 462–463
storage, 463
Rat interferon gamma gene
expression vector for
with adenovirus major late promoter, 457
with SV40 early promoter, 457–458
identification, 454
Hu-IFN-γ cDNA clone in, 454
initiation codon, 456
nucleotide sequence, 454–456
restriction endonuclease map of, 454
RNA splicing events, 456
signal peptide cleavage site, 456
RD114 virus
cell fusion by
effect of interferon on, 704–705
inactivation with β-propiolactone, 704
Receptor
for Hu-IFN-α, 274, 307–310, 314, 326–331, 335–339, 342–343, 345–348, 597
IFN-receptor complex, 340, 347–351
antibody bonding to, 326–339
for Hu-IFN-β, 343
for Hu-IFN-γ, 283, 301, 315–321, 343
for murine interferon, 324
number per L1210 cell, 324
for (2′-5′)-oligoadenylate, 351–356
on macrophages, 351–356
Receptor mediated endocytosis, 332–333
of Hu-IFN-αA, 332
of interferon, 332
Receptosomes, IFN-colloidal gold complex in, 338–339
Recombinant human interferon alpha, *see* Human interferon alpha, specific species
Recombinant human interferon beta, *see* Human interferon beta
Recombinant human interferon gamma, *see* Human interferon gamma
Recombinant interferon, *see* specific interferons
standards
human interferon alpha A (Hu-IFN-αA), 16
human interferon alpha D (Hu-IFN-αD), 16
human interferon alpha 2 (Hu-IFN-α2), 16
human interferon beta (Hu-IFN-β), 16
human interferon gamma (Hu-IFN-γ), 16
[Ser^{17}]Hu-IFN-β, 16
Red blood cells, *see* Erythrocytes
Reduction, of Hu-IFN-α, 248–255
Reference reagents
antiserum to interferon
anti-Hu-IFN-α, 571–572
anti-Hu-IFN-β, 571–572
bovine-anti-Hu-IFN-α, 17
rabbit anti-Hu-IFN-γ, 17
rabbit anti-Mu-IFN-γ, 17
sheep anti-Hu-IFN-α, 17
sheep anti-Hu-IFN-β, 17
sheep anti-mouse L-cell interferon, 17
interferon standards, 14–19, 535, 539
chick interferon, 16
chicken interferon MRC Research Standard A 62/4, 109, 112
Hu-IFN-α (leukocyte/Sendai), 16, 94, 138, 427, 473, 543, 637, 651, 692
Hu-IFN-α (Namalva/Sendai), 16
Hu-IFN-β, 16, 184, 372, 385
Hu-IFN-γ, 16, 51, 56, 60, 66, 193, 585
Mu-IFN-α, 16, 575
Mu-IFN-α/β, 16, 111, 651
Mu-IFN-β, 16, 575, 651
Mu-IFN-γ, 16
murine interferon, 709
rabbit interferon, 16
recombinant Hu-IFN-αA, 16

recombinant Hu-IFN-αD, 16
recombinant Hu-IFN-α2, 16
recombinant Hu-IFN-β, 16
recombinant [Ser[17]]Hu-IFN-β, 16
recombinant Hu-IFN-γ, 16
storage, 15, 535
virus standards
vesicular stomatitis virus
MRC Research Standard A 62/4, 121, 125
Reference unit, distinction from experimental unit, 559
Remazol brilliant blue R
Sepharose coupled, 218
binding of bovine IFN-β to, 218
Reovirus
genome dsRNA
in activation of P_1/eIF-2α protein kinase, 500–501
strain *ts*C(447)
in chicken interferon induction, 119
strain *ts*G(107)
in chicken interferon induction, 119
ts mutants
in chicken interferon induction, 119
Repressor, temperature sensitive, 377
Reticulocyte lysate
hydrolysis of mRNA in, 675
protein synthesis in, 668
inhibition, 668
by plant oligoadenylates, 755
Retroviruses
cell fusion by
effect of IFN on, 703–706
preparation, 703–704
Reverse phase high performance liquid chromatography, *see also* HPLC
C-18 column in, 209
of Hu-IFN-α2, 174–175
in purification of AVF, 742–743
Rhesus monkey
in poly ICL-CM dextran assay, 104–106
survival, 105
Ribosomal salt-wash, in purification of dsRNA-dependent protein kinase, 503–508
Ribosome binding site, on expression plasmid, 369, 371–372, 377–389
Ribosomes
preparation, 293
in cell-free protein biosynthesis, 295
RNase L
activation by (2′-5′)-oligoadenylate, 489, 516
assays, 489, 491–493
(2′-5′)$(A)_3$[^{32}P]pCp binding assay, 492–493
endonuclease assay, 491–492
bacteriophage R17 radiolabeled RNA in, 491
cross-linking to (2′-5′)-oligoadenylate derivative, 489, 497–499
molecular weight, 495, 496
purification, 489, 493–497
ammonium sulfate precipitation in, 494–495
DEAE cellulose in, 495–496
phosphocellulose in, 495–497
poly(A)-agarose in, 495–496
Sephacryl S-200 gel filtration in, 496
source of, Ehrlich ascites tumor cells as, 493–494
storage, 495
Rous sarcoma virus, expression of Hu-IFN-β with, 383–396
RPMI-1788 cells, in IFN-γ production, 65, 68

S

S1 nuclease, 366, 368
Saccharomyces cerevisae, host for interferon production, 416–423
Sandwich radioimmunoassay, 588–589
for Hu-IFN-γ, 582–587
principle, 582
Sangivamycin
2-5A analogs
preferential degradation of mRNA, 675
Sarcolectin
assay for, 694–702
blocking of antiviral activity, 697, 701–702
characterization, 694–702
definition of, 701–702
interferon antagonist, 700

molecular weight, 702
physiochemical properties, 700–701
in placental blood, 702
purification, 694, 698–700
gel filtration in, 698
hydrophobic chromatography in, 698
ion-exchange chromatography in, 698
SDS–PAGE in, 699
specific activity, 700–701
stability, pH, 701–702
Scatchard plot, equation for, 318–319
SDS–PAGE, *see* SDS–polyacrylamide gel electrophoresis
SDS–polyacrylamide gel electrophoresis
analysis of
alpha interferons, 31–32, 287
AVF, 744
bacterial cells, 372–373, 380
Bo-IFN-α, 141–142, 473
Bo-IFN-β, 218
cell-free DNA-dependent products, 294
Hu-IFN-α, 47, 173–174, 278, 280–281, 418
Hu-IFN-αA, 153, 157–159, 163–165, 418
Hu-IFN-α2, 170–171, 173
Hu-IFN-β, 178, 188, 401
(2′-5′)-oligoadenylate synthetase, 518
P-IFA, 47
placental interferon, 548–550
radiolabeled Hu-IFN-αA, 287, 290, 407–408, 418
rat IFN-α, 452
rat IFN-γ, 460–461
rat interferon, 225, 452, 460
RNase L, 498
antibodies, 328
in assay for dsRNA dependent protein kinase, 502
in identification of IFN receptors, 341
in purification of sarcolectin, 699
recovering interferon activity from, 249
SEA, *see* Staphylococcal enterotoxin A
SEB, *see* Staphylococcal enterotoxin B
Secretion signals, 6, 8, 10, 13, 433
yeast
attachment to
Hu-IFN-αA gene, 430–433
Hu-IFN-αD gene, 430–433
Semliki Forest Virus
in Hu-IFN-δ assay, 98–99
in interferon assays, 651
Sendai virus
antiserum to, 138
effect of IFN on cell fusion by, 703–706
preparation, 703
inactivation
by acidification, 138
with antiserum, 138
by UV light, 703, 706
induction
defective interfering particles in, 36
of leukocytes, 17, 137
of lymphoblastoid null cells, 571
of Namalva cells, 17, 101
of rat embryonic cells, 220
in production of
Bo-IFN-α, 137
Hu-IFN-α, 27, 40, 41, 54, 101
rat interferon, 221–222
removal from Bo-IFN-α, 138
by chromatography on Affi-Gel Blue, 138–140
Sephacryl S-200 gel filtration
in Hu-IFN-α2 purification, 171, 172
yield, 172
in purification of
RNase L, 496
sarcolectin, 698
Sephadex G-50 chromatography
in Hu-IFN-αA purification, 157, 161–163
resolution of oligomers and monomers, 162–165
Sephadex G-75, in purification of IVR substance, 734
Sepharose
iminodiacetic acid-activated
in nickel chelate column preparation, 200, 201
for nickel ion scavenging, 201

Remazol brilliant blue R coupled, 218
binding of Bo-IFN-β to, 218
Sepharose 4B, in Hu-IFN-αA purification, 159
Sequential chromatography, *see* Tandem chromatography
Serine, substitution for cysteine in
Hu-IFN-αA, 404, 406
Hu-IFN-β, 183–192
Serum free medium
in Hu-IFN-γ induction, 93
in production of rat interferon, 221
Serum substitute, in Hu-IFN-γ induction, 95
Severe combined immunodeficiency disease
in horses, 130
interferon alpha in, 130
interferon beta in, 130
interferon gamma in, 130
Sheep, antiserum from, *see* Antibody; Antiserum; Polyclonal antibodies
Shine-Delgarno sequence, 377, 381, 447–449
in expression plasmid
Hu-IFN-γ, 379
Rat IFN-α1, 447–449
Signal sequence, of interferons, 6, 8, 10, 13, 424, 445
Silica, in purification of IFN-γ, 207
Silicic acid
in Hu-IFN-γ concentration, 90, 92, 200, 202
in Mu-IFN-γ purification, 146, 149
Silver stain, of SDS–polyacrylamide gels, 208
Simian virus 40 (SV40)
fibroblast-derived growth factor
preparation of, 645–646
infection of baby hamster kidney cells, 646
Sindbis virus
in cytopathic effect inhibition assay, 94
in induction of
guinea pig interferon, 128
chicken interferon, 119
Size exclusion chromatography
see also, Gel filtration; HPLC
Skeletal muscle, differentiation by interferon, 612
Slow migrating monomer, of Hu-IFN-αA, 242, 244, 407
Sodium borohydride
in nickel chelate column preparation, 200, 201
Sodium butyrate, *see* Butyrate; Butyric acid
Sodium dodecyl sulfate–polyacrylamide gel electrophoresis; *see* SDS–polyacrylamide gel electrophoresis
Sodium dodecyl sulfate
effect on
interferon, 102, 237–238
sarcolectin, 701–702
regeneration of controlled pore glass, 194
in restoration of Hu-IFN-αA activity, 251
Solid-phase enzyme immunoassay, *see also* EIA, ELISA
for Hu-IFN-αA, 157
Sonication, in IFN-γ purification, 206–207
SP Sephadex A-50, in Hu-IFN-α2 purification, 171, 172
Specific activity, *see also* specific interferons
Hu-IFN-α, 45, 46, 651
Hu-IFN-αA, 157, 427, 637, 684
Hu-IFN-α2, 172, 175
Hu-IFN-β, 179, 181, 184, 185
unglycosylated, 403
Hu-IFN-γ, 198
radiolabeled Hu-IFN-αA, 307
Specific binding
definition, 308, 313, 317, 340
sites, *see also* receptors
Specific cytolysis
definition, 654
measurement, 654
Standards
antiserum, *see* Reference reagents, antiserum to interferons
interferon, *see* Reference reagents, interferon standards
Staphylococcal enterotoxin A
induction
Hu-IFN-γ, 17, 58, 83, 85, 88–92, 200, 574–575
Mu-IFN-γ, 17, 145–146, 575

as mitogen, 58, 83, 85
production, 89
from *Staphylococcus aureus*, 89, 145
sterilization, 89
storage, 89
Staphylococcal enterotoxin B, in Hu-IFN-γ induction, 70, 71–72, 83, 93, 95
Staphylococcus aureus
protein A, 685
source of staphylococcal enterotoxin A, 89, 145
Streptococcal preparation *OK*-432
in Hu-IFN-γ induction, 88–92, 93, 95
from *Streptococcus pyogenes* A3, 89
from *Streptococcus pyogenes* Su, 93
Streptococcus pyogenes
strain A3
streptococcal preparation *OK*-432, 89
strain Su
streptococcal preparation *OK*-432, 93
N-Succinimidyl-3-(4-hydroxy,5-[^{125}I]iodophenyl)propionate
see Bolton-Hunter reagent
N-Succinimidyl-3-(2-pyridylthio)propionate, in conjugation of Hu-IFN-α to albumin, 239
Sulfitolysis, of Hu-IFN-αA, 252–253
Superinduction
of Hu-IFN-β, 284, 400
actinomycin D in, 284, 400
cycloheximide in, 284, 400
poly(I) · poly(C) in, 284, 400
SV40 early promoter, expression of rat IFN-γ, 457
Syncytium formation, inhibition by Hu-IFN-α, 706
Synergy, between chemotherapy and interferon, 656–657

T

T cell growth factor (interleukin 2, IL-2)
coinduction with Hu-IFN-γ, 52, 82, 83
impurity in Hu-IFN-γ, 87
separation from Hu-IFN-γ, 52
T cell hybridoma lines
culture conditions, 53
establishment of, 53
in Hu-IFN-γ production, 49, 52–54
L 265-K cells, 49, 52–54
L 265-O cells, 49, 52–54
yields, 53
T cell lines
Hut 102-B2
in Hu-IFN-γ production with TPA, 48–54
production of HTLV-I, 54
Mo
in Hu-IFN-γ production, 52
from patient with hairy cell leukemia, 52
SH9, 49
Hut 102-B2 derivative, 49
6-thioguanine resistant mutant, 49
T cell mitogens, in Hu-IFN-γ induction, 93–95
T21 cells, *see* human fibroblasts, trisomic
Tandem chromatography, 193–199
purification of Hu-IFN-γ, 193–199
Blue-Sepharose and Phenyl-Sepharose in, 218
concanavalin A-Sepharose in, 194, 195
controlled pore glass in, 193–195
heparin-Sepharose in, 194, 195
Target size analysis, 255–260
of interferons, 258–260
theory, 256–258
Teleocidin, in Hu-IFN-δ induction, 97, 98
Temperature
effect on stability of
rat IFN-γ, 463
sarcolectin, 701–702
1,3,4,6-Tetrachloro-3α,6α-diphenylglycouril (Iodo-gen)
in radiolabeling, 685–686
12-*O*-Tetradecanoylphorbol-13-acetate, 218
in differentiation
synergy with interferon, 613
in Hu-IFN-γ induction, 48–54, 64, 66, 68, 77, 78, 234
optimal concentration, 50
tumor-promoter, 613
Tetraethylammonium chloride, elution of Hu-IFN-γ, 90
Tetramers, of Hu-IFN-αA, 242
Tetrasulfonate of IFN-αA
antiviral activity of, 253

on MDBK cells, 253
on WISH cells, 253
HPLC analysis of, 253
SDS–PAGE analysis of, 253
Thermolysin, in preparation of truncated Hu-IFN-αA, 235
Thiocyanate, inhibition of iodination, 269
6-Thioguanine resistance, SH9 cell line, 49
[^{3}H]Thymidine
in assay for cell growth, 663
with Swiss 3T3 cells, 644, 647
transport across plasma membrane, 647
Thymocytes
proliferation with interleukin-2, 662–663
inhibition by interferon, 662
Tilorone hydrochloride, interferon inducer, 708–709
Tobacco cells
preparation of AVF (antiviral factor) with, 734–744
source of plant oligoadenylate synthetase, 753
Tobacco mosaic virus (TMV)
in assay for IVR substance, 732–733
ELISA assay of, 748, 760
growth curve, 743
induction of
AVF with, 740, 752–754
oligoadenylates with, 744
inoculation of protoplasts with, 747–748
in preparation of antiviral factor (AVF; IVR), 730–731, 734–744
replication, inhibition
by AVF, 739, 755
by human interferon, 739, 744
by (2′-5′)-oligoadenylate analogs, 759–761
in protoplasts, 733
RNA
dot-blot assay, 749–751
sensitivity, 751
inoculation of protoplasts with, 748
in preparation of cDNA, 749
time course of replication, 761
Tobacco protoplasts
effect of human interferons on, 744–752
inoculation with
TMV, 746–747
TMV-RNA, 748
preparation of, 745–748
Toyocamycin
2,5-A analogs
preferential degradation of mRNA, 675
TPA, *see* 12-*O*-Tetradecanoylphorbol-13-acetate
Transcription terminator
on constructed plasmids, 370, 372
in rat IFN-γ expression vectors, 458
Transferrin, as Hu-IFN-γ impurity, 62
Transformation
inhibition by interferon, 541
of NIH3T3 cells, 599–600
Transforming growth factor
antagonism by interferon, 658, 661
purification, 658
Trimers
Hu-IFN-αA, 242, 245–246
HPLC analysis of, 245–246, 248
Triton X-100
in purification of Hu-IFN-αA, 153
removal from Hu-IFN-αA, 161
Trypan blue, measurement of cell viability, 576
Tryptophan modifying reagent, 2-methoxy-5-nitrobenzyl bromide, 236
Tubercidin
2,5-A analogs
5′-monophosphate, 526
preferential degradation of mRNA, 675
synthesis, 526–528
Tumor cells
from biopsies, 638
collection of, 638–639
colony growth, 639–640
image-analyzer for, 639
inhibition by interferon, 642
from pleural effusions, 638
preparation of, 638–639
collagenase in, 638
DNase I in, 638
Tumor growth
inhibition *in vivo*
by (2′-5′)-oligoadenylates, 668–675
potentiation, 673
xenografts in nude mice, 649–653, 655
Tumor necrosis factor, as Hu-IFN-γ impurity, 63

Tumor promotors
induction of mononuclear cells, 96
in preparation of
Hu-IFN-γ, 48–54, 64, 66, 68, 77, 78, 97
Hu-IFN-δ, 96–97
Tunicamycin, in modification of interferons, 233
Tyrosinase, in regulation of melanin synthesis, 613

U

U937 histiocytic lymphoma cells, 306, 309, 316
binding of
Hu-IFN-αA to, 309
Hu-IFN-γ to, 316–320
preparation for binding assay, 316
Ultrafiltration
concentration of
Hu-IFN-αA, 154
Hu-IFN-γ, 69
Ultraviolet light, inactivation of viruses, 703, 706
Ultrogel AcA 44, in rat IFN purification, 222–224
Ultrogel AcA 54, in IFN-γ purification, 194, 200, 202, 575
Urea, effect on interferon activity, 237–238

V

Vasopressin, in cell growth medium, 645–646, 648
Verapamil
calcium channel blocker, 70, 71
in Hu-IFN-γ induction
blocking of, 70, 71
Vero cells
in Bo-IFN-α assay, 141
in guinea pig interferon induction, 128
for VSV production, 128
Vesicular stomatitis virus
in CPE assay, 56, 65, 81, 124, 138, 178, 184, 206, 221–222, 299, 364, 407, 417, 427, 473, 533–540, 542–543, 554, 639
for interferon titrations, 692
mutant tsG41 in dose-response curves, 108
in plaque reduction assay, 31
in poly ICL-CM dextran assay, 104
resistance to, 607
selection of IFN sensitive cells with, 601–608, 610
storage, 535
strain [±]DI-011
in chicken interferon induction, 119, 120–125
in dose-response curves, 108
strain T1026R1
in guinea pig interferon induction, 128
wild type
in guinea pig interferon induction, 128
Vigna sinensis, in cucumber mosaic virus assay, 733
Virus antiserum, in dose-response curves, 108
Virus inactivation
in guinea pig interferon preparation, 129
by irradiation, 129
Virus infection, effect on hyporesponsiveness to IFN, 707
VSV, *see* Vesicular stomatitis virus

W

Western blot, in IFN-γ characterization, 197
Wheat germ cell-free system
protein synthesis
inhibition by plant oligoadenylates, 755
WISH cells
in assay of
amniotic fluid, 542
Bo-IFN-α, 141
Hu-IFN-α, 102, 250, 601, 637, 651
Hu-IFN-αWA, 364
Hu-IFN-β, 102, 385, 601
Hu-IFN-γ, 65–67, 101–102, 201, 206, 585
Hu-IFN-δ, 97, 101, 102
in identification of IFN receptor, 341
viral RNA synthesis in, 651

X

Xenografts, in nude mice, 649–653, 655
Xenopus laevis oocytes
assay of mRNA from
bovine leukocytes, 143
rat cells, 442

Y

Yeast
expression of
[Met]Hu-IFN-αA
incorrect processing, 430
yield, 430
[Met]Hu-IFN-αD
incorrect processing, 430
yield, 430
expression vectors
construction of, 418–420, 424–433
for production of interferon, 416–423, 424–433
Host strain W301-18A, 416
ingestion by macrophages, 680
secretion signal, 433
in strain 20B-12, 426–427
YOK, *see* Monoclonal antibody, interferon alpha specific, YOK

Z

Zinc acetate, in purification of IVR substance, 731
Zinc chelate chromatography
in Hu-IFN-γ purification, 86–87, 199
lack of Hu-IFN-α adsorption, 199